Tasman's Psyc

MW01553934

Allan Tasman · Michelle B. Riba ·
Renato D. Alarcón · César A. Alfonso ·
Shigenobu Kanba ·
Dusica Lecic-Tosevski ·
David M. Ndetei · Chee H. Ng ·
Thomas G. Schulze

Editors

Tasman's Psychiatry

Fifth Edition

Volume 5

With 320 Figures and 628 Tables

 Springer

Editors
Allan Tasman
Department of Psychiatry and
Behavioral Sciences
University of Louisville
Louisville, KY, USA

Renato D. Alarcón
Mayo Clinic College of Medicine
Rochester, MN, USA

Shigenobu Kanba
Kyushu University
Fukuoka, Japan

David M. Ndetei
Department of Psychiatry
University of Nairobi
Nairobi, Kenya

Thomas G. Schulze
Department of Psychiatry
University of Munich
München, Bayern, Germany

Michelle B. Riba
Department of Psychiatry
University of Michigan
Ann Arbor, MI, USA

César A. Alfonso
Columbia University Medical Center
New York, NY, USA

Dusica Lecic-Tosevski
Serbian Academy of Sciences and Arts
Belgrade, Serbia

Chee H. Ng
Department of Psychiatry
University of Melbourne
Melbourne, VIC, Australia

ISBN 978-3-030-51365-8   ISBN 978-3-030-51366-5 (eBook)
https://doi.org/10.1007/978-3-030-51366-5

1st edition: © W.B. Saunders Company, USA 1997
2nd to 4th edition: © John Wiley & Sons, Ltd, England 2003, 2008, 2015
© Springer Nature Switzerland AG 2024, corrected publication 2024

This work is subject to copyright. All rights are solely and exclusively licensed by the Publisher, whether the whole or part of the material is concerned, specifically the rights of translation, reprinting, reuse of illustrations, recitation, broadcasting, reproduction on microfilms or in any other physical way, and transmission or information storage and retrieval, electronic adaptation, computer software, or by similar or dissimilar methodology now known or hereafter developed. The use of general descriptive names, registered names, trademarks, service marks, etc. in this publication does not imply, even in the absence of a specific statement, that such names are exempt from the relevant protective laws and regulations and therefore free for general use.
The publisher, the authors and the editors are safe to assume that the advice and information in this book are believed to be true and accurate at the date of publication. Neither the publisher nor the authors or the editors give a warranty, expressed or implied, with respect to the material contained herein or for any errors or omissions that may have been made. The publisher remains neutral with regard to jurisdictional claims in published maps and institutional affiliations.

This Springer imprint is published by the registered company Springer Nature Switzerland AG.
The registered company address is: Gewerbestrasse 11, 6330 Cham, Switzerland

If disposing of this product, please recycle the paper.

Allan Tasman
To my wife Cathy, talented artist and teacher, and nurturing mom and grandma, with my love and thanks for being my partner in all things, for having faith in me when I didn't, for being the most thorough proofreader and editor of all my lectures and publications, and for being a wonderful role model for our kids and hundreds of others through your advocacy for women's rights and mental health for all. Here's looking at you, kid.

Michelle B. Riba
With deep appreciation to my family, friends, colleagues, teachers, and patients who all have been sources of inspiration, with gratitude.

Renato D. Alarcon
To my wife and children for their permanent love, to my teachers and mentors for their unparalleled wisdom, to my patients as source of constant inspiration, and to our readers as searchers of truth and carriers of hope.

Cesar A. Alfonso
With gratitude to my students and trainees, the future leaders of Psychiatry, and to the Alfonso family children, Alejandro, Cala, Camilo, and Ciro, as they continue their quest of making this a safer world.

Shigenobu Kanba
To my wife, and psychiatrist, Kiyoko, my two daughters, Rumiko and Mikako, who are also psychiatrists, and my youngest daughter, Mamiko, who is a resident physician.

Dusica Lecic-Tosevsky
To the patients who gave purpose to my life devoted to psychiatry.
Giving is receiving.

David M. Ndetei
To all mental health workers of different categories and cadres
who see a clear link between public health and clinical practice.

Chee H. Ng
To my family, especially my wife, both daughters, and parents.

Thomas G. Schulze
To my wife Cordula, my daughter Mathilda, and my son Joey for
generously cutting me a lot of slack for always being out and
about. . .

Preface

All psychiatrists know that excellent patient care starts with a thorough and multifocal evaluation. This evaluation is informed by the clinician's synthesis of an amalgam of a broad base of knowledge ranging from molecular genetics as it pertains to our field, physiology and other pertinent aspects of the science of bodily functions, normal development of cognitive, emotional, and self-regulation capabilities, the impact of life experiences based on individual characteristics of perception and processing, familial and peer group interactions, and the sociocultural and economic environment in which they historically and currently live.

The information gathered is then combined with an understanding of the nature of psychiatric disorders and the findings from specific exploration for their presence. Finally, the clinician develops a preliminary formulation to create the treatment plan, foster and maintain a collaborative therapeutic alliance, and assess treatment effectiveness and modify it when necessary. We also know, however, that the psychosocial disruptions of the COVID-19 pandemic have adversely affected the preexisting limitations in available time, personnel, treatment resources, and facilities. Thus the ability to provide optimal care has been diminished in essentially every setting. Because of the increased volume of those seeking care often has exceeded our already maximized capacity, staying up to date on the latest advances is even more difficult both in required time and cost.

Thus, the need exists for the broadest-based reference that a psychiatrist at any career stage can utilize for any level of depth of review. *Tasman's Psychiatry*, the fifth edition of this textbook, published previously with the title *Psychiatry*, has been created to meet this need. When work on the first edition of this book started in the early 1990's, the structure of the book was developed to assist the psychiatrist in staying up to date in the increasingly diverse areas needed to practice most effectively using the framework of patient care described above. In addition, we aimed to make the book easily readable through a clear communication style and structure for each chapter, an attribute that has been enhanced in this edition with more graphic elements to break up the text and highlight key information. The use of full color in graphic elements helps make the information even more accessible.

The first section of the book, "Approaches to the Patient," always has emphasized the central importance of the therapeutic alliance in treatment outcomes. Thus, the first chapter continues to be "Listening to the Patient"

(▸ Chap. 1). This chapter highlights the variety of ways we are able to listen, reinforcing the understanding that it remains our greatest source of essential information about both who is the person with the illness and what is the illness, both of which are essential to maximize positive treatment outcomes. This part of the chapter in this edition has been enhanced by more thorough attention to the role of cultural aspects in patient presentation and communication styles.

Much has changed in psychiatry and the wider range of mental health services since the first edition was published in 1997. These advances in our knowledge and skills have resulted in ongoing changes in both the scope and depth of practice, which are addressed in this edition in a variety of ways. One aspect that has not changed, except through increased depth of discussion, is the original organizing principle for this book regarding an integrative biopsychosocial model of assessment and treatment. We believe this approach is of central importance for optimal understanding of the patient's needs and to best foster the therapeutic relationship between the psychiatrist and patient. The centrality of this approach is reaffirmed by the developing focus in our field on the impact of social determinants of mental health and illness. One needs to look no further than the epidemic of increased mortality due to suicide, addiction, and overdose, the so-called "deaths of despair," brought about by the social isolation and social disruptions of the COVID-19 pandemic, to see the importance of these factors. The pandemic years occurred during a time of the largest human migrations in recorded history, itself a significant aspect of social determinants. The mental health impacts of these migrations were amplified not only by the pandemic but also by the attendant societal disruptions of other recent natural and man-made disasters. The increasing coverage of social determinants in this edition benefits from the burgeoning scholarship in this area.

International collaboration in psychiatry has expanded greatly with the advent of changes in communication technology over the last three decades. For example, textbooks in many fields, not only in medicine, were historically the products of experts based primarily in a few countries yet marketed around the world. Cultural and other societal influences on mental health were often, therefore, not optimally represented. To amplify the increased knowledge and skills such international collaboration can produce, the entire group of authors, section editors, and editors in this edition was constructed to address this historical challenge.

The nine editors, responsible for developing the conceptual framework for the content of this edition, represent every continent (except Antarctica). The specific chapter titles and subject matter topics in each section were developed with input not only from the editors but also from the 27 section editors as they reviewed draft manuscripts prior to sending them to production. Members of that group also represent every continent. The majority of chapters, both updated and new, are authored by at least two individuals from at least two different countries or continents and involve well over 500 authors writing in their areas of expertise. These editorial decisions result in a book that embodies state-of-the-art knowledge and skills from around the world. The availability of the online updating capacity also means this aspect of the book will be augmented regularly.

Another essential benefit of this global collaboration is the extensive attention in this edition to the role of culture and other societal aspects regarding the way people with psychiatric illnesses present themselves for care, are assessed, and are treated. We also address, from a more global perspective, how service systems operate and how psychiatric educational structure and content might optimally be implemented by highlighting the situation in low- and middle-income countries. Another example of the shift of emphasis in this edition is the fact that ICD 11, the international standard for diagnostic classification from the World Health Organization, is used as the primary frame of reference for diagnostic categorization, which is also compared and contrasted with the other major global standard, the DSM 5TR from the American Psychiatric Association.

The expanding base of knowledge and skills also is reflected in the fact that all chapters from the last edition have been updated and revised, and that there are 37 brand new chapters. There are now 22 chapters devoted to neuroscientific and social/psychological foundational knowledge, compared to 8 chapters in the first edition. The current number of treatment chapters in the broad topics of psychotherapy, psychopharmacotherapy, and other somatic interventions continues to increase and is 50% more than in the first edition. In addition, these chapters are supplemented by over 20 other chapters addressing specialized areas of care. In the "Disorders" section, the cultural aspects of the patient's presentation and the impact on the therapeutic alliance are addressed. Where possible, authors have also included information about modifications to treatment for those practicing in low- and middle-resource areas. This aspect of the book will certainly grow in the future as well. Other sections that are of particular importance and new in this edition are Prevention, Systems of Care, and Psychosocial Aspects of Treatment, Treatment Issues in Specific Populations and Settings, and Digital Mental Health Services and Technologies.

As mentioned above, a major technical advance in this edition is the Springer online version of this major reference work. This system allows for ongoing revisions where needed since the manuscripts were first submitted, meaning this edition includes the most up-to-date content of any major psychiatric textbook. The online revision system will continue to foster the inclusion of state-of-the-art information without needing to wait for the publication of the next print edition. The readership access made possible by the online version resulted in thousands of chapter views, even before this print edition was published.

A work of this magnitude could not have been produced without the participation of an exemplary group of psychiatric scholars. The fact that this work was done during a global pandemic which placed increasing demands on an already highly productive group is a testament to our contributors' commitment to excellence. A great vote of thanks goes to all those who provided their expert scholarship. Particular recognition goes to authors whose untimely deaths, acknowledged in their chapters, prevented them from seeing the results of their outstanding scholarship published here.

Finally, this work could not have been brought to fruition without the consistent support of many of our collaborators at Springer. There are too many to list individually, but two people in particular deserve special recognition. Sunali Mull has been consistently outstanding in serving as the primary person coordinating the flow of manuscripts from the authors before they went to production, including liaison with me and the other editors and section editors. She has been readily available, always helpful, and consistently able to answer and help resolve the multitude of questions and concerns that have arisen. Andrew Spencer, who was in charge of production for this book, has been an expert in providing support and answers to what must have at times seemed like a never-ending stream of questions and concerns from me. He has performed his multiple duties in interacting with our chapter authors and editors with diplomacy, skill, and a deep and broad understanding of the production constraints and processes. The quality of this book owes much to the efforts of both of these terrific collaborators.

As with many projects of this magnitude, the most important thanks go to those closest to us, whose support, encouragement, and tolerance of time spent on this book, which otherwise would have gone to them, have helped make our work possible.

Louisville, USA   Allan Tasman, MD
August 2024   Editor in Chief

Contents

Volume 3

Volume 4

Volume 6

About the Editors

Allan Tasman, M.D. is Emeritus Professor and Chair of the Department of Psychiatry and Emeritus Schwab Endowed Chair in Social and Community Psychiatry at the University of Louisville. He completed residency at the University of Cincinnati and psychoanalytic training at the Western New England Institute for Psychoanalysis. He became known nationally and internationally while a faculty member at the University of Connecticut (1976–1991) for his expertise in teaching, mentorship, and educational program development. Through his subsequent national and international work, he has been involved in a broad range of educational, clinical, strategic planning, and mental health policy issues, particularly addressing disparities in health and mental health.

His commitment to serving the disadvantaged started in medical school where he founded and was CEO of the Lexington Free Clinic. While psychiatry chair in Louisville, to address health and mental health disparities in the low-income Medicaid population, he conceptualized and spearheaded implementation of Passport Health Plan in the early 1990s. This innovative non-profit managed care organization, which prioritized the quality of health outcomes and attention to health inequities, was consistently ranked nationally in the highest levels. Passport became the second largest Medicaid system in Kentucky, with a two billion dollar annual budget. He later was a 2018–2023 appointee to the National Advisory Council of the US Substance Abuse and Mental Health Services Administration (SAMHSA).

His funded research over four decades has focused on the neurophysiology of cognitive processes. His laboratory at the University of Connecticut was one of the first to describe functional neural abnormalities in offspring of alcohol addict fathers. More recent research in Louisville focused on investigation of innovative neuromodulation treatments for autism and substance abuse. He has authored or edited 38 books, over 260 peer-reviewed articles, chapters, and abstracts, nearly 80 editorials in psychiatric publications, and over 400 national and international presentations. He is founding Editor in Chief of all editions of this textbook, now named Tasman's Psychiatry, with earlier editions called "the best current textbook of psychiatry" by the *New England Journal of Medicine* and the "gold standard" by the *American Journal of Psychiatry*. He also was Editor in Chief from 2014 to 2019 of *Psychiatric Times*, the psychiatric publication most widely read by psychiatrists in the United States.

He was president of the American Association of Directors of Psychiatric Residency Training, the Association for Academic Psychiatry, and two terms for the American Association of Chairs of Departments of Psychiatry. As American Psychiatric Association (APA) president elect and president, he conceptualized and established the American Psychiatric Institute for Research and Education, implemented the initial planning process for DSM 5, and initiated and oversaw a complete corporate reorganization. In 1991, he founded and was Deputy Editor of the *American Psychiatric Press Journal of Psychotherapy Practice and Research*.

In 2005, he was elected to a 6-year term as Secretary for Education of the World Psychiatric Association, where he produced WPA global guidelines for medical student and resident psychiatric education, with special attention to programs in resource poor countries. As President of the Pacific Rim College of Psychiatrists (2006–2008), he conceptualized and established the journal *Asia Pacific Psychiatry* (Editor in Chief 2014–2023), which is now also the official journal of the Asian Federation of Psychiatric Associations. It is the first transnational English language psychiatric journal focused on the entire Pacific Rim region.

He currently serves as Treasurer and Vice President for North America and the Caribbean of the World Federation for Mental Health, collaborating with the WHO and UN.

He has received multiple national and international awards for leadership, educational excellence, and distinguished professional service, including an APA special presidential commendation for career leadership and service. He received distinguished alumnus awards from the University of Kentucky Medical School and Franklin and Marshall College. He is a Fellow of the Royal College of Psychiatrists of the United Kingdom and a number of other organizations including Distinguished Life Fellow of the APA, and an Honorary Fellow of the World Psychiatric Association. He received the American College of Psychiatrists Distinguished Service to Psychiatry Award, their highest honor. He is the 2024 recipient of the prestigious C. Charles Burlingame Award for excellence in administration, research, and education.

Michelle B. Riba, M.D., M.S., DFAPA, FAPM is Professor of Psychiatry, University of Michigan Medical School; Director of the PsychOncology Program at the University of Michigan Rogel Cancer Center; and Co-director of the Workplace Mental Health Program at the University of Michigan Eisenberg Family Depression Center. Dr. Riba is a board-certified consultation-liaison psychiatrist. She formerly served as Associate Chair for Education and Academic Affairs, Associate Chair for Integrated Medical and Psychiatric Services, Director of Residency Training and Director of the Consultation Liaison Fellowship in the University of Michigan Department of Psychiatry, and Associate Chair in the University of Michigan Comprehensive Depression Center. She has served as President of the American Psychiatric Association, Association for Academic Psychiatry, and American Association of Directors of Psychiatric Residency Training, as well as Secretary for Scientific Publication of the World Psychiatric Association and on the Board of the Directors of the Pacific Rim College of Psychiatrists. Dr. Riba is Chair of the National Comprehensive Cancer Network (NCCN) Distress Guidelines, as well as a member of the NCCN

Fatigue Guidelines. She has served as the American Medical Association's representative to the American College of Surgeons Commission on Cancer, the AAMC Council of Academic Societies-Leadership Development Committee, the World Federation for Mental Health Scientific Committee and Program Chair, and the Association of Women Psychiatrists Executive Council; Trustee of the American Association for Emergency Psychiatry, the Society for Clinical and Translation Science, American College of Psychiatrists PRITE Commission, and Institute of Medicine Committee on Incorporating Research into Psychiatry Residency Training.

Dr. Riba received the Nancy C.A. Roeske, MD APA Award for Excellence in Medical Student Teaching from the University of Connecticut and the University of Michigan; the Hedwig van Ameringen Executive Leadership in Academic Medicine (ELAM) Program for Women; the APA Irma Bland Award for Excellence in Teaching Residents; the Outstanding Clinician Award from the University of Michigan; the University of Michigan Making a Difference Award; Special Recognition Award from the Academy of Psychosomatic Medicine; Inaugural member, Distinguished Life Fellows, Association for Academic Psychiatry; American College of Psychiatrists Distinguished Service Award; American Psychiatric Association Distinguished Service Award; Association of Women Psychiatrists/APA Alexandra Symonds Award; and Special Recognition Award, Indo-American Psychiatric Association for Leadership Exceptional Achievements and Dedication to Minority Psychiatrists. She is the 2024 recipient of the Academy of Consultation-Liaison Psychiatry Distinguished Service Award.

Dr. Riba is the author or editor of over 300 scientific articles, books, chapters, and scientific abstracts. She is Editor-in-Chief, *Current Psychiatry Reports*, Springer. Dr. Riba has served on the editorial board of *Psychiatric Services and Cancer News on the Net*, *Current Psychiatry*, *Academic Psychiatry*, and *Hospital and Community Psychiatry*, and has served on the editorial advisory board of the *American Psychiatric Press*, Inc. She is a reviewer for many international journals, including *Psycho-Oncology, Academic Psychiatry, Psychiatric Services, Journal of Psychiatric Practice, Psychosomatic Medicine*, and Breast Cancer Research and Treatment. She has co-edited fifteen editions of the American Psychiatric Press Review of Psychiatry series. She has co-edited *Psychopharmacology and Psychotherapy: A Collaborative Approach*, APPI, Inc.; *Primary Care Psychiatry*, Saunders; *The Doctor-Patient Relationship in Pharmacotherapy: Improving Treatment Effectiveness*, Guilford; and *Psychopharmacology in Oncology and Palliative Care: A Practice Manual*, Springer. Dr. Riba has edited or co-written over 40 books. She is the senior author of *Competency in Combining Pharmacotherapy and Psychotherapy*, APPI, Inc.; *Clinical Manual of Emergency Psychiatry*, APPI, Inc. (First and Second Edition); *Psychiatry and Heart Disease*, John Wiley and Sons, Inc.; *Physician Mental Health and Well-Being*, Springer; *Workplace Mental Health*, Springer; and *College Psychiatry*, Springer. She has served on the international advisory board of the journal *Academic Psychiatry* and was Deputy Editor of *Psychiatric Times*.

Dr. Riba has a wonderful family—married to Arthur L. Riba, M.D. for over 52 years. They have two daughters, Alissa B. Roger, J.D., and Erica B. Riba, MSW, son-in-law, James M. Roger, DDS, and the most marvelous grandchildren, Oscar Thomas, and Jean Badian Roger. Michelle loves to play tennis, kayak, and piano; and be with family, friends, and colleagues. Dr. Riba is indebted to her patients and their families for the honor and privilege to serve as their physician.

Renato D. Alarcón, MD, MPH is Distinguished Emeritus Professor in the Department of Psychiatry and Psychology at Mayo Clinic School of Medicine, Rochester, Minnesota. He is also Emeritus Professor and Honorio Delgado Chair at the Universidad Peruana Cayetano Heredia (UPCH) in Lima, Perú, where he obtained his MD degree in 1965. He completed his Psychiatric Residency and Fellowships in Psychosomatic Medicine and Clinical Psychopharmacology at Johns Hopkins Hospital and earned his MPH degree at the Hopkins School of Public Health (1967–1972).

Between 1972 and 1980, he organized and conducted the first academically based psychiatric Residency program in Peru at the UPCH and was Associate Dean and Director of Academic Programs before returning to the U.S. He occupied leadership positions in the Departments of Psychiatry at the University of Alabama in Birmingham (1980–1992) and Emory University School of Medicine in Atlanta, GA (1992–2002), before joining Mayo Clinic where he was Chief of the Adult Psychiatry Division and the Mood Disorders Program (2002–2015).

He was Secretary General of the World Association of Cultural Psychiatry (WACP) between 2015 and 2018 and has also served in numerous Committees of the American Psychiatric Association, including the DSM Steering Committee and its Personality Disorders Work Group and Equity and Inclusion Subcommittee.

Author or co-author of over 270 articles and author/editor of 26 books and 130 book chapters, Dr. Alarcón is Senior Editor of the most widely read psychiatric textbook in Latin America (four editions) and board member of several publications in the continent. He has received, among other distinctions, the APA Simon Bolivar and George Tarjan Awards, and the WACP Weng Shing Tseng and Jean Garrabé Awards. He has received honorary membership of several international academic institutions in the U.S., Europe, and Latin America. His academic and clinical interests include personality and mood disorders, PTSD, psychiatric diagnosis, global mental health, and cultural psychiatry.

César A. Alfonso, M.D. is Clinical Professor of Psychiatry at Columbia University, Adjunct Professor at the University of Indonesia, and Visiting Professor at Prince of Songkla University in Thailand, and at the National University of Malaysia. He serves as Editor of *Psychodynamic Psychiatry* (2020–present). Born in Cuba, he spent formative years in Spain and Puerto Rico before pursuing undergraduate studies at Yale University, and graduate and postgraduate studies in medicine, psychiatry, psychosomatic medicine, and psychoanalysis at New York Medical College. During the first half of his career, he was

Director of the Division of Consultation-Liaison Psychiatry at Metropolitan Hospital-New York Medical College, with a focus on HIV Psychiatry and Psycho-oncology. He also chaired the hospital's Ethics Committee. Dr. Alfonso served as President of the American Academy of Psychodynamic Psychiatry and Psychoanalysis in 2010–2012 and remains Chair of the Psychoanalytic Education Committee. He served as Chair of the Psychotherapy Section of the World Psychiatric Association in 2017–2023 and is the current President of the World Federation for Psychotherapy (2023–2026). His recent work includes biopsychosocial aspects of suicide, the care of visually impaired persons, bidirectionality and comorbidity, psychotherapy as a biological treatment, psychoanalysis and creativity, cultural anthropology and psychotherapy, and the design and implementation of psychotherapy training programs worldwide. He is a Distinguished Fellow of the American Psychiatric Association and is a Fellow of the New York Academy of Medicine, the Academy of Consultation-Liaison Psychiatry, and the American Academy of Psychodynamic Psychiatry and Psychoanalysis. He served as a Teichner Scholar with Visiting Professorships at Northwell Staten Island and the University of Iowa. He has over 100 publications, including books: *Advances in Psychodynamic Psychiatry* (Guilford Press, 2018), *Suicide by Self-Immolation: Biopsychosocial and Transcultural Aspects* (Springer Nature, 2021), and the 5th Edition of *Tasman's Psychiatry* (Springer Nature, 2024). He is Editor of the 3rd Edition of *Cohen's Comprehensive Textbook of AIDS Psychiatry* (Springer Nature, est. 2027). For the past 15 years, he has been the Chief Psychiatrist at the Lighthouse Guild Behavioral Health Clinic, a mental health clinic dedicated to the care of persons with vision loss or who are blind.

Shigenobu Kanba, M.D., Ph.D. Current appointments are Professor Emeritus, Kyushu University, Japan; Japan Depression Center, President; Iida Hospital (Nagano), Advisor; and Fukuoka Institute of behavioral Medicine, all since 2019.

Education and Employment

Keio University School of Medicine, Tokyo: 1980

Keio University Hospital, Intern: 1980–1981

Mayo Clinic, Research Fellow (Pharmacology): 1982–1983

Mayo Clinic, Resident (Psychiatry) and Assistant professor (Psychiatry): 1984–1987

Keio University School of Medicine, Assistant and Assistant Professor (Psychiatry): 1987–1996

Yamanashi University School of Medicine, Professor (Psychiatry): 1996–2003

Kyushu University Graduate School of Medicine, Professor and Chairman (Psychiatry): 2003–2019

Research and Editorial Activity

Research areas have focused on mood disorders, psychoneuroimmunology, and psychopharmacology. He has published more than 400 scientific papers (h-index = 68) and 14 book chapters in English (as of January 2024). He is emeritus editor-in-chief of *Psychiatry and Clinical Neurosciences*, editor of *Current Opinion of Psychiatry*, *International Journal of Bipolar Disorders*, *Irish Journal of Psychological Medicine*, *Asia-Pacific Psychiatry*, *Pharmacopsychiatry*, *Neurology, Psychiatry and Brain Research*, *Asian Journal of Psychiatry*, and *Indian Journal of Psychiatry*.

Organization Leadership

He served as President, Japanese Society of Psychiatry and Neurology; Executive Secretary, Japanese Association of Medical Sciences and Japanese Medical Science Federation; Vice President, International Society of Bipolar Disorders; President, Asian Federation of Psychiatric Societies; and Associate Secretary-Treasurer, World

Federation of Societies of Biological Psychiatry, WFSBP.

Government Leadership

Currently, he is program supervisor and program officer of the Japan Agency for Medical Research and Development and an advisory member of the Japan Science and Technology Agency. He serves the Tokyo Metropolitan Institute of Medical Science as a director and serves/served on the Science Council of Japan, Strategy Committee of Health and Medicine of Japanese Cabinet Office, Social Security Committee of Japanese Ministry of Health, Labor and Welfare, Drug and Food Administration, and the Brain Science Committee of the Japanese Ministry of Education, Culture, Sports, Science and Technology. He was an advisory member of Japan Society for the Promotion of Science.

Awards and Honors

American Psychiatric Association, Pennwalt Award, 1985

Mayo Clinic, Balfour Award, 1987

Mayo Clinic, Rome, H. Award, 1987

CINP, Rafaelsen Award, 1988

Kyushu University Research Award, 2012–2016

Minister of Health, Labor and Welfare Commendation, 2020

Honorary Member of the World Psychiatric Association

Dusica Lecic-Tosevski, MD, PhD is a full member of the Serbian Academy of Sciences and Arts, psychiatrist, psychotherapist and professor of psychiatry. She was the director of the Institute of Mental Health, the university psychiatric hospital, from 2004 to 2019, and was internationally awarded as the best leader. She has been the founder and head of the WHO Collaborating Centre for Mental Healthcare Force Development and the WHO national counterpart for mental health. Chairing the National Committee for Mental Health, she has made a significant contribution to the reform of mental health care in the country and has been an editor of the National Strategy for Mental Health Care. In the past, she was a coordinator of several programs for mental health care of refugees and torture victims.

Prof. Lecic-Tosevski has been actively involved in WPA, as past zonal representative for Central Europe, past chair of the Section on Preventive Psychiatry, and co-chair of the Committee for Scientific Publications, and is a member of the board of the Section Psychoanalysis and Psychiatry, as well as of other sections.

She is APA distinguished fellow, International Associate of the Royal College of Psychiatry, EPA Fellow, WPA honorary member, president of the Psychiatric Association of Eastern Europe and the Balkans, of the Serbian Association of Social Psychiatry, and past president of the Serbian Psychiatric Association (now its honorary president). She has been a member of the WHO Working Group on ICD-11 Classification of Personality Disorders.

Prof. Dusica Lecic-Tosevski has been a coordinator and investigator of many international, multicentric, and national research projects. She authored numerous scientific articles and chapters in international handbooks and books; is editor of two international books, the national textbook of psychiatry, three books at the Academy of Sciences and Arts, as well as guidelines for evidence-based treatment of depression and schizophrenia and many practical manuals. She is a member of editorial boards and reviewer of numerous international journals and has given invited talks at many international congresses and universities, some of which she has organized. She has translated seven professional and six literature works.

Professor Lecic-Tosevski's professional interests are personality disorders, affective disorders, traumatic stress, and comorbidity of mental and somatic disorders. She has devoted her professional life to the noble discipline of psychiatry, improvement of person-centered treatment of mental disorders, as well as their prevention and mental health promotion.

David M. Ndetei is a Professor in Department of Psychiatry, University of Nairobi, Kenya, the Founding Director of Africa Mental Health Research and Training Foundation (AMHRTF) now rebranded as "Africa Institute of Mental and Brain Health" and the Director of World Psychiatric Association Collaborating Centre for Research and Training, Kenya. Over the last 20 years, AMHRTF has become a leading center in the region in mental health and brain health research. It has partnerships and collaborators across Africa and many countries, both low and middle-income countries and high-income countries. Our emphasis is on the community approach for innovative interventions that are social-culturally context appropriate, affordable, available, accessible, and evidence-based with potential for critical reach. AMHRTF takes a family oriented, multi-disciplinary, multi-stakeholder, and policymaker approach to research from design, implementation, and validation of results for shared ownership and therefore seamless transition from research/programs to policy and practice uptake. He has trained nearly all senior psychiatrists in Kenya and some from the region. Recently, he was ranked as one of the two top scientific researchers in Kenya https://research. com/scientists-rankings/social-sciences-and-humanities/ke. He has 417 scholarly publications including authoring 6 books and 21 monographs. He is also editing a book titled "Global Mental Health in Africa: Towards Inclusivity, Innovations and Opportunities." It is to be published in the United Kingdom (UK). It has 150 contributors drawn globally but mainly from Africa.

He has pioneered local training for psychiatrists in Kenya; training of clinical psychologists at the University of Nairobi, postgraduate diplomas in psychiatric social work, psychotrauma, substance use and abuse, and clinical psychiatry (just below the level of a psychiatrist) at the University of Nairobi. In his academic career, he has put emphasis on mentoring students for their Masters, Ph.D., and Doctor of Science (D.Sc.), which is a higher doctorate, as well as postdoc fellowships from across the globe.

He is a member of various professional bodies and honorary official of various academic institutions and associations across the globe.

Link to CV: https://africamentalhealthresearch andtrainingfoundation.org/wp-content/uploads/ 2023/02/Prof-David-M-Ndetei-CV-2023.pdf

Prof. Chee Ng, MBBS (Melb), MMed (Psych), FRANZCP, MD currently holds the position of Healthscope Chair of Psychiatry at the Melbourne Clinic and Professor of Psychiatry in the Department of Psychiatry, the University of Melbourne. The Melbourne Clinic is Australia's largest private psychiatric hospital, which provides a comprehensive inpatient program, day program center, and an outreach service. The Professorial Unit of the Melbourne Clinic delivers specialized treatment for mood and anxiety disorders with complex problems through a multidisciplinary team, which include biological, psychosocial, and cognitive approaches. It conducts academic teaching for university MD students and psychiatric trainees, as well as psychiatric research innovations providing synergy between clinical practice and research for best evidence treatments.

He has considerable research experience in psychopharmacology, pharmacogenetics, mood and anxiety disorders, schizophrenia, old age, and cross-cultural psychiatry. He has published over 400 original articles in peer-reviewed journals and more than 25 book chapters and is the Editor-in-Chief of the *Asia-Pacific Psychiatry Journal*. He is currently an Executive Committee Member of the APEC Mental Health Digital Hub and Past President of the Pacific Rim College of Psychiatrists, and Fellow of the Royal Australian and New Zealand College of Psychiatrists.

He is also appointed as Director of World Health Organisation Collaborating Center in Mental Health Research and Training at St. Vincent's Hospital Melbourne, where he is also senior Consultant Psychiatrist and Director of the International Psychiatry Unit which was awarded the St. Vincent's Australia National Award 2010 for Community Service. He had led the Postgraduate Overseas Specialists Training (POST) Program which has trained over 500 international psychiatrists, nurses, and mental health professionals. In this capacity, he co-founded Asia-Australia Mental Health which supported mental health development in the Asia-Pacific region. He was an Advisory Committee member of the China-Australia Community Mental Health Program for China's national community mental health program and a consultant of Community Mental Health at Shenzhen Mental Health Center. He previously led the Asia-Pacific Community Mental Health Development Project, involving 18 Asia-Pacific countries. He also served as an international mental health consultant for WHO, the Commonwealth of Nations, and Asia-Pacific Economic Cooperation. He has collaborated in major national projects in numerous countries, including China, India, Japan, Vietnam, Laos, Cambodia, Malaysia, Myanmar, Qatar, and Solomon Islands.

Prof. Thomas G. Schulze born in 1969, studied medicine in Erlangen (Bavaria), Manama (Bahrain), Barcelona (Catalonia), and Chapel Hill and Winston-Salem (both North Carolina). He trained as a psychiatrist and held various positions in Germany (Bonn, Mannheim, Göttingen, Munich) and the USA (Chicago, IL; Bethesda, MD; Baltimore, MD; Syracuse, NY).

Since 2014, he has held the position of Chair and Director of the Institute of Psychiatric Phenomics and Genomics (www.ippg.eu) at the Ludwig-Maximilians-University of Munich (IPPG). He is a research affiliate with the National Institute of Mental Health (NIMH) in Bethesda, MD, and on Faculty with the Department of Psychiatry and Behavioral Sciences of Johns Hopkins University, Baltimore, MD. In 2019, he also joined the Faculty of the Department of Psychiatry and Behavioral Sciences of SUNY Upstate Medical University, Syracuse, NY, USA, where he holds an appointment as clinical professor. He is licensed to practice medicine in the European Union and the State of New York.

Dr. Schulze's research focuses on genotype-phenotype relationships and personalized medicine approaches in psychiatric disorders. He coordinates a German-wide center grant on longitudinal psychosis research (www.PsyCourse.de) and spearheads an international study on the genetic basis of response to lithium treatment in bipolar disorder (www.ConLiGen.org), comprising several research groups from Europe, the Americas, Asia, and Australia. He has authored over 300 papers, and his h-index being 70 (Web of Science) and 95 (Google Scholar), respectively. In addition to national German awards, he is the 2006 recipient of the Robins-Guze Award of the American Psychopathological Association (APPA), the 2006 recipient of the Theodore-Reich-Award of the International Society of Psychiatric Genetics (ISPG), and the winner of the Colvin Prize 2016 of the Brain and Behavioral Research Foundation (BBRF). He is a Fellow of the American College of Neuropsychopharmacology (ACNP) and the APPA and served as President of the APPA from 2015 through 2016. Between 2016 and 2020, he also held the office of President of the ISPG. From 2011 through 2017, he served as the Chair of the Section on Psychiatric Genetic of the World Psychiatric Association (WPA), in which he is an honorary member. In 2017, he was elected to the Executive Committee of the WPA, starting a 6-year term as Secretary of Scientific Sections. In 2021, he was elected to the German Academy of Sciences (Leopoldina.org). In 2023, he was elected President-Elect of the WPA, with the 3-year term as President commencing in 2026.

Thomas G. Schulze speaks German, English, French, Catalan, Spanish, and Latin and has a basic knowledge of Arabic.

Section Editors

Section I: Approaches to the Patient
Renato D. Alarcón Mayo Clinic, Rochester, MN, USA

Maria I. Lapid Mayo Clinic, Rochester, MN, USA

Robert J. Ursano Uniformed Services University of the Health Sciences, Bethesda, MD, USA

Section II: Human Development Through the Life Cycle
César A. Alfonso Columbia University Medical Center, New York, NY, USA

Alma L. Jimenez University of the Philippines, Manila, Philippines

Section III: Neuroscientific Foundations of Psychiatry
Thomas G. Schulze University of Munich, Munich, Germany

Gonzalo Laje Washington Behavioral Medicine Associates, LLC, Chevy Chase, MD, USA

Section IV: The Psychological and Social Scientific Foundations of Psychiatry
Richard F. Summers University of Pennsylvania, Philadelphia, PA, USA

Renato D. Alarcón Mayo Clinic, Rochester, Rochester, MN, USA

Section V: Evaluation and Assessment in Psychiatry
Andrea Fiorillo University of Campania "L. Vanvitelli", Naples, Italy

Patrick D. McGorry Orygen, Parkville, VIC, Australia

Umberto Volpe Polytechnic University of Marche, Ancona, Italy

Section VI: Disorders
Chee H. Ng University of Melbourne, Melbourne, VIC, Australia

Dusica Lecic-Tosevski Serbian Academy of Sciences and Arts, Belgrade, Serbia

César A. Alfonso Columbia University Medical Center, New York, NY, USA

Ihsan M. Salloum University of Texas Rio Grande Valley, Harlingen, TX, USA

Editorial Assistant:
Sanya Virani Department of Psychiatry, University of Massachusetts Medical School, Worcester, MA, USA

Section VII: Prevention, Systems of Care, and Psychosocial Aspects of Treatment
David Ndetei Africa Mental Health Research and Training Foundation, Nairobi, Kenya

Jacqueline Maus Feldman University of Alabama at Birmingham, Birmingham, AL, USA

Section VIII: Psychotherapeutic Treatments
César A. Alfonso Columbia University Medical Center, New York, NY, USA

Reham Aly Ministry of Health Cairo, Cairo, Egypt

David Choon Liang Teo Changi General Hospital, Singapore, Singapore

Section IX: Pharmacological and Other Somatic Treatments
Shigenobu Kanba Kyushu University, Fukuoka, Japan

Rif S. El-Mallakh University of Louisville, Louisville, KY, USA

Joseph Zohar Chaim Sheba Medical Center, Tel Hashomer, Israel

Section X: Consultation Liaison and Collaborative Care in Medical Illness and Comorbid Psychiatric and Medical Illnesses
Michelle Riba University of Michigan, Ann Arbor, MI, USA

Luigi Grassi Institute of Psychiatry, University of Ferrara, Ferrara, Italy

Section XI: Treatment Issues in Specific Populations and Settings
Andrea Fiorillo University of Campania "L. Vanvitelli", Naples, Italy

Tarek A. Okasha Ain Shams University, Cairo, Egypt

Marianne Kastrup Copenhagen University Hospital, Copenhagen, Denmark

Jack Drescher Columbia University, New York, NY, USA

Section XII: Emergency Psychiatry
Michelle Riba University of Michigan, Ann Arbor, MI, USA

Shigenobu Kanba Kyushu University, Fukuoka, Japan

Section XIII: Digital Mental Health Services and Technologies
Umberto Volpe Polytechnic University of Marche, Ancona, Italy

Donald Hilty Northern California Veterans Administration Health Care System, Mather, CA, USA

Section XIV: Appendices
Allan Tasman Department of Psychiatry and Behavioral Sciences, University of Louisville School of Medicine, Louisville, KY, USA

Contributors

Sabra M. Abbott Department of Neurology, Northwestern University Feinberg School of Medicine, Chicago, IL, USA

Center for Circadian and Sleep Medicine, Northwestern University Feinberg School of Medicine, Chicago, IL, USA

Mohammad T. Abou-Saleh Division of Mental Health, St George's, University of London, London, UK

Somya Abubucker Department of Psychiatry and Behavioral Sciences, Johns Hopkins School of Medicine, Baltimore, MD, USA

Kristina Adorjan Department of Psychiatry and Psychotherapy, University Hospital, LMU Munich, Munich, Germany

Institute of Psychiatric Phenomics and Genomics (IPPG), University Hospital, LMU Munich, Munich, Germany

Adekunle G. Ahmed Department of Psychiatry, University of Ottawa, Ottawa, ON, Canada

Nicole Akramoff Department of Psychological and Brain Sciences, Boston University, Boston, MA, USA

Renato D. Alarcón Department of Psychiatry and Psychology, Mayo Clinic Alix School of Medicine, Rochester, MN, USA

Universidad Peruana Cayetano Heredia, Lima, Peru

Sheila M. Alessi Departments of Medicine and Psychiatry, University of Connecticut School of Medicine, Farmington, CT, USA

César A. Alfonso Department of Psychiatry, Columbia University Medical Center, New York, NY, USA

Department of Psychiatry, Faculty of Medicine, National University of Malaysia, Cheras, Kuala Lumpur, Malaysia

Department of Psychiatry, Universitas Indonesia, Jakarta, Indonesia

Department of Psychiatry, Prince of Songkla University, Songhkla, Thailand

Amina Z. Ali University of Toronto, Centre for Addiction and Mental Health, Toronto, ON, Canada

Yesne Alici Department of Psychiatry and Behavioral Sciences, Memorial Sloan Kettering Cancer Center, New York, NY, USA

Michael H. Allen University of Colorado School of Medicine, Denver, CO, USA

Kelly Allott Orygen, Melbourne, VIC, Australia

Centre for Youth Mental Health, The University of Melbourne, Melbourne, VIC, Australia

Michaela Amering Department of Psychiatry and Psychotherapy, Medical University of Vienna, Vienna, Austria

Nina J. Anderson Department of Psychology, University of Denver, Denver, CO, USA

Martin M. Antony Department of Psychology, Toronto Metropolitan University, Toronto, ON, Canada

Kevin M. Antshel Department of Psychology, Syracuse University, Syracuse, NY, USA

Yumi Aoki Department of Psychiatric & Mental Health Nursing, Graduate School of Nursing, St. Luke's International University, Tokyo, Japan

Elie G. Aoun Division of Law, Ethics, and Psychiatry, Columbia University Vagelos College of Physicians and Surgeons, New York, NY, USA

Paul S. Appelbaum Center for Law, Ethics & Psychiatry, New York State Psychiatric Institute and Columbia University Irving Medical Center, New York, NY, USA

Angeline Monica A. Arcenas Department of Psychiatry and Behavioral Medicine, University of the Philippines, Manila, Philippines

Paul Arnold Department of Psychiatry and Medical Genetics, University of Calgary, Calgary, AB, Canada

Martijn Arns Brainclinics Foundation, Nijmegen, The Netherlands

Department of Cognitive Neuroscience, Faculty of Psychology and Neuroscience, Maastricht University, Maastricht, The Netherlands

Yoni K. Ashar Weill-Cornell Medical Center, New York, NY, USA

Gordon J. G. Asmundson Department of Psychology, University of Regina, Regina, SK, Canada

Pratyusha Attaluri Tanenbaum Centre for Pharmacogenetics, Campbell Family Mental Health Research Institute, Centre for Addiction and Mental Health (CAMH), Toronto, ON, Canada

Evelyn Attia Department of Psychiatry, Columbia University College of Physicians and Surgeons, New York, NY, USA

Division of Anxiety, Mood, Eating & Related Disorders, New York State Psychiatric Institute, New York, NY, USA

Department of Psychiatry, Weill Cornell Medical College, New York, NY, USA

Jura Augustinavicius School of Population and Global Health, McGill University, Montreal, QC, Canada

Marc Auriacombe Addiction Treatment and Research Center at Hospital Charles Perrens CHU de Bordeaux SANPSY CNRS UMR 6033, University of Bordeaux, Bordeaux, France

Jimmy N. Avari Weill Cornell Institute of Geriatric Psychiatry, Weill Cornell Medical College, White Plains, NY, USA

Michael Avissar Columbia University College of Physicians and Surgeons, New York, NY, USA

Thomas J. Babor Department of Public Health Sciences, University of Connecticut School of Medicine, Farmington, CT, USA

Amy Bachyrycz College of Pharmacy, University of New Mexico, Albuquerque, NM, USA

Ruben D. Baler National Institute on Drug Abuse, Bethesda, MD, USA

Kirrie J. Ballard Sydney School of Health Sciences in the Faculty of Medicine and Health and the Brain and Mind Centre, The University of Sydney, Camperdown, NSW, Australia

Cornelio G. Banaag Jr Department of Psychiatry and Behavioral Medicine, University of the Philippines, Manila, Philippines

Lucy C. Barker Women's College Research Institute and Department of Psychiatry, University of Toronto, Toronto, ON, USA

Debasish Basu Department of Psychiatry, Postgraduate Institute of Medical Education and Research, Chandigarh, India

William R. Beardslee Department of Psychiatry, Boston Children's Hospital, Boston, MA, USA
Department of Psychiatry, Harvard Medical School, Boston, MA, USA

Ina Becker Columbia University Medical Center, New York, NY, USA
New York - Presbyterian Hospital, New York, NY, USA

Jessica E. Becker Division of Child and Adolescent Psychiatry, Department of Psychiatry, Massachusetts General Hospital, Boston, MA, USA
Harvard Medical School, Boston, MA, USA

Jean C. Beckham Durham VA Health Care System, Department of Psychiatry and Behavioral Sciences, Duke University Medical Center, Durham, NC, USA

Gillinder Bedi Orygen, Parkville, Australia
University of Melbourne, Centre for Youth Mental Health, Parkville, Australia

Antonello Bellomo Department of Clinical and Experimental Medicine, University of Foggia, Foggia, Italy

Robert H. Belmaker School of Medicine, Ben Gurion University of the Negev, Beersheva, Israel

Martino Belvederi Murri Department of Neuroscience and Rehabilitation, Institute of Psychiatry, University of Ferrara, Ferrara, Italy

David M. Benedek Department of Psychiatry, Uniformed Services University of the Health Sciences, Bethesda, MD, USA

Giuseppe Berardino Department of Clinical and Experimental Medicine, University of Foggia, Foggia, Italy

Michael Berk IMPACT The Institute for Mental and Physical Health and Clinical Translation, Deakin University, Geelong, VIC, Australia

Centre for Youth Mental Health, University of Melbourne, Parkville, VIC, Australia

Department of Psychiatry, University of Melbourne, Parkville, VIC, Australia

Orygen, the National Centre for Excellence in Youth Mental Health, Parkville, VIC, Australia

Florey Institute for Neuroscience and Mental Health, Parkville, VIC, Australia

David Bienenfeld Department of Psychiatry, Wright State University Boonshoft School of Medicine, Dayton, OH, USA

Shahaf Bitan Chaim Sheba Medical Center, Tel Hashomer, Israel

Shannon M. Blakey Durham VA Health Care System, VISN 6 Mental Illness Research, Education and Clinical Center, Durham, NC, USA

Carlos Blanco National Institute on Drug Abuse, Bethesda, MD, USA

Elisabeth H. Bos Department of Developmental Psychology, University of Groningen, Groningen, The Netherlands

Interdisciplinary Center Psychopathology and Emotion Regulation, University Medical Center Groningen, University of Groningen, Groningen, The Netherlands

Mario Braakman Pro Persona Mental Health & Tilburg University, Wolfheze, Gelderland, The Netherlands

Paolo Brambilla Department of Neurosciences and Mental Health, Fondazione IRCCS Ca' Granda Ospedale Maggiore Policlinico, Milan, Italy

Department of Pathophysiology and Transplantation, University of Milan, Milan, Italy

Peer Briken Institute for Sex Research, Sexual Medicine, and Forensic Psychiatry, University Medical Centre, Hamburg-Eppendorf, Hamburg, Germany

Eileen Britt University of Cantebury, Te Whare Wānanga o Waitaha, Christchurch, New Zealand

Beth S. Brodsky Department of Psychiatry, Columbia University Irving Medical Center, New York, NY, USA

Ellie Brown Orygen, Parkville, Australia

University of Melbourne, Centre for Youth Mental Health, Parkville, Australia

Gregory K. Brown Department of Psychiatry, Perelman School of Medicine of the University of Pennsylvania, Philadelphia, PA, USA

Richard P. Brown Department of Psychiatry, Columbia University Vagelos College of Physicians and Surgeons, New York, NY, USA

Timothy A. Brown Center for Anxiety and Related Disorders, Department of Psychological and Brain Sciences, Boston University, Boston, MA, USA

Barbara Burton Department of Psychiatry, Beth Israel Deaconess Medical Center, Harvard Medical School, Boston, MA, USA

Deborah L. Cabaniss Department of Psychiatry, Columbia University College of Physicians and Surgeons, New York, NY, USA

Kristin L. Callahan Department of Psychiatry, Louisiana State University, School of Medicine, New Orleans, LA, USA

Alessandra Canuto University Hospitals of Geneva, Geneva, Switzerland

Lora Capobianco Research and Innovation, Greater Manchester Mental Health NHS Foundation Trust, Manchester, UK

School of Psychological Sciences, University of Manchester, Manchester, UK

Sara Carucci Department of Biomedical Sciences, Section Neuroscience and Clinical Pharmacology, University of Cagliari, Cagliari, Italy

Child and Adolescent Neuropsychiatry Unit, "A. Cao" Paediatric Hospital, Cagliari, Italy

Manuel F. Casanova Department of Biomedical Sciences, University of South Carolina School of Medicine, Greenville, SC, USA

João Mauricio Castaldelli-Maia Department of Neuroscience, Medical School, Fundação do ABC, Santo André, SP, Brazil

Department of Psychiatry, Medical School, University of São Paulo, São Paulo, SP, Brazil

Robert L. Caudill Department of Psychiatry and Behavioral Sciences, University of Louisville, School of Medicine, Louisville, KY, USA

Prabha S. Chandra Department of Psychiatry, National Institute of Mental Health and Neurosciences, Bangalore, India

International Association for Women's Mental Health, Bangalore, India

Andrew Chanen Orygen, Parkville, VIC, Australia

Centre for Youth Mental Health, The University of Melbourne, Parkville, VIC, Australia

Lisa N. Chen Department of Psychiatry and Behavioral Science, Johns Hopkins University School of Medicine, Baltimore, MD, USA

Samuel Eng Teck Cheng Department of Psychological Medicine, Changi General Hospital, Singapore, Singapore

Duke National University of Singapore Graduate Medical School, Singapore, Singapore

Yong Loo Lin School of Medicine, National University of Singapore, Singapore, Singapore

Sabrina Cherry Department of Psychiatry, Columbia University College of Physicians and Surgeons, New York, NY, USA

Cezar Cimpeanu Department of Psychiatry, University of Massachusetts Medical School, Worcester, MA, USA

David Coghill Department of Paediatrics and Psychiatry, The University of Melbourne, Parkville, VIC, Australia

Wilson M. Compton National Institute on Drug Abuse, National Institutes of Health Department of Health and Human Services, Bethesda, MD, USA

Raul Condemarín Harvard University, Boston, MA, USA

John N. Constantino Division of Child Psychiatry, Washington University School of Medicine, St. Louis, MO, USA

Alessandra Costanza Faculty of Medicine, University of Geneva, Geneva, Switzerland

Janna Cousijn Department of Psychology, Education & Child Studies, Erasmus University Rotterdam, Rotterdam, The Netherlands

Elizabeth Cox Department of Psychiatry, The University of North Carolina at Chapel Hill, Chapel Hill, NC, USA

Glenn W. Currier Morsani College of Medicine, University of South Florida, Tampa, FL, USA

Vivek Datta Department of Psychiatry, Weill Institute for Neurosciences, University of California, San Francisco, CA, USA

Chris Davey University of Melbourne, Department of Psychiatry, Parkville, Australia

E. David Leonardo Department of Psychiatry, Columbia University College of Physicians and Surgeons, New York, NY, USA

New York State Psychiatric Institute, New York, NY, USA

Rachel E. Davis-Martin Departments of Emergency Medicine and Family Medicine and Community Health, University of Massachusetts Medical School, Worcester, MA, USA

Danilo De Gregorio Neurobiological Psychiatry Unit, Department of Psychiatry, McGill University, Montreal, QC, Canada

Division of Neuroscience, Vita-Salute San Raffaele University, Milan, Italy

Peter de Jonge Department of Developmental Psychology, University of Groningen, Groningen, The Netherlands

Interdisciplinary Center Psychopathology and Emotion Regulation, University Medical Center Groningen, University of Groningen, Groningen, The Netherlands

Claire De Souza Department of Psychiatry, The Hospital for Sick Children, Toronto, ON, Canada

Department of Psychiatry, University of Toronto, Toronto, ON, Canada

Ymkje Anna de Vries Department of Child and Adolescent Psychiatry, University Medical Center Groningen, University of Groningen, Groningen, The Netherlands

Sebastian Del Corral Winder Department of Psychiatry, Louisiana State University, School of Medicine, New Orleans, LA, USA

Kristina M. Deligiannidis Psychiatry and Obstetrics & Gynecology, Zucker School of Medicine at Hofstra/Northwell, Hempstead, NY, USA

Feinstein Institutes for Medical Research, Manhasset, NY, USA

Katz Institute for Women's Health, Queens, NY, USA

Psychiatry, University of Massachusetts Medical School, Worcester, MA, USA

Women's Behavioral Health, Zucker Hillside Hospital, Northwell Health, Glen Oaks, NY, USA

Constantine Della College of Medicine, Department of Psychiatry and Behavioral Medicine, University of the Philippines Manila, Manila, Philippines

Giuseppe Delvecchio Department of Pathophysiology and Transplantation, University of Milan, Milan, Italy

Koen Demyttenaere Psychiatry Research Group, Department of Neurosciences, Faculty of Medicine, KU Leuven, Leuven, Belgium

University Psychiatric Center KU Leuven, Leuven, Belgium

Nilifa Desilva Applied Health Research Centre, Unity Health Toronto in Toronto, ON, Canada

D. P. Devanand Division of Geriatric Psychiatry, New York State Psychiatric Institute, College of Physicians and Surgeons of Columbia University, New York, NY, USA

Carolyn S. Dewa Department of Psychiatry and Behavioral Science, University of California, Davis, Davis, CA, USA

Mental Health Services and Evaluation Program, University of California, Davis, Davis, CA, USA

Department of Public Health Sciences, University of California, Davis, Davis, CA, USA

Mantosh J. Dewan Department of Psychiatry, State University of New York Upstate Medical University, Syracuse, NY, USA

Aryeh Dienstag Department of Psychiatry, Hadassah Hebrew University Medical Center, Jerusalem, Israel

Kirsten H. Dillon Durham VA Health Care System, Department of Psychiatry and Behavioral Sciences, Duke University Medical Center, Durham, NC, USA

Vaibhav A. Diwadkar Department of Psychiatry & Behavioral Neurosciences, Wayne State University, Detroit, MI, USA

Seetal Dodd IMPACT The Institute for Mental and Physical Health and Clinical Translation, Deakin University, Geelong, VIC, Australia
Centre for Youth Mental Health, University of Melbourne, Parkville, VIC, Australia
Department of Psychiatry, University of Melbourne, Parkville, VIC, Australia

Coreen B. Domingo Department of Psychiatry and Behavioral Sciences, Baylor College of Medicine, Houston, TX, USA

Jack Drescher Department of Psychiatry, Columbia University, College of Physicians and Surgeons, New York, NY, USA

Dante M. Durand Department of Psychiatry and Behavioral Sciences, University of Miami Miller School of Medicine, Miami, FL, USA

Joy Duxbury Manchester Metropolitan University, Manchester, UK

Christine Ecker Department of Child and Adolescent Psychiatry, Goethe-University, Frankfurt am Main, Germany

Jane L. Eisen McLean Hospital, Belmont, MA, USA

Stuart J. Eisendrath Department of Psychiatry, Weill Institute for Neurosciences, University of California, San Francisco, CA, USA

Hussien Elkholy Okasha Institute of Psychiatry, Department of Neurology and Psychiatry, Faculty of Medicine, Ain Shams University, Cairo, Egypt

Rif S. El-Mallakh School of Medicine, University of Louisville, Louisville, KY, USA

Susan Evans Weill-Cornell Medical Center, New York, NY, USA

Anita Everett Center for Mental Health Services, US HHS Substance Abuse and Mental Health Services, Rockville, MD, USA

Peter Falkai Department of Psychiatry and Psychotherapy, University Hospital, LMU Munich, Munich, Germany
Max-Planck-Institute of Psychiatry, Munich, Germany

Brian A. Fallon Columbia University College of Physicians and Surgeons, New York, NY, USA

Stephen V. Faraone Departments of Psychiatry and of Neuroscience and Physiology, SUNY Upstate Medical University, Syracuse, NY, USA

Ali A. Farooqui Department of Psychiatry and Behavioral Sciences, Integrative Psychiatry, PLLC, University of Louisville, School of Medicine, Louisville, KY, USA

Michael Farrell National Drug and Alcohol Research Centre (NDARC), UNSW Sydney, Sydney, NSW, Australia

Seena Fazel Department of Psychiatry, University of Oxford, Oxford, UK

Jacqueline Maus Feldman Department of Psychiatry and Behavioral Neurobiology, University of Alabama at Birmingham, Birmingham, AL, USA

Marc D. Feldman Department of Psychiatry and Behavioral Medicine, University of Alabama, Tuscaloosa, AL, USA

Justin Fiala Center for Circadian and Sleep Medicine, Northwestern University Feinberg School of Medicine, Chicago, IL, USA

Andrea Fiorillo Department of Psychiatry, University of Campania "L. Vanvitelli", Naples, Italy

Michael B. First Department of Psychiatry, Columbia University Irving Medical Center, New York, NY, USA

Paul B. Fitzgerald School of Medicine and Psychology, Australian National University, Canberra, ACT, Australia

W. Wolfgang Fleischhacker Department of Psychiatry, Psychotherapy and Psychosomatics, Medical University Innsbruck, Innsbruck, Austria

Malcolm Forbes IMPACT The Institute for Mental and Physical Health and Clinical Translation, Deakin University, Geelong, VIC, Australia

Department of Psychiatry, University of Melbourne, Parkville, VIC, Australia

Thomas Fovet University of Lille, Inserm, CHU Lille, U1172 – Lille Neuroscience & Cognition, Lille, France

R. Fremont Department of Psychiatry, Icahn Mount Sinai School of Medicine, New York, NY, USA

Nora D. B. Friedman Massachusetts General Hospital for Children, Lurie Center for Autism, Lexington, MA, USA

Department of Psychiatry, Harvard Medical School, Boston, MA, USA

Richard E. Frye Department of Neurology, Phoenix Children's Hospital, Phoenix, AZ, USA

Andrés E. Fuenmayor Department of Psychiatry, Columbia Duke University, Durham, NC, USA

Michiko Fujimoto Department of Psychiatry, Osaka University Graduate School of Medicine, Osaka, Japan

Thomas Gargot Centre Universitaire de Pédopsychiatrie, Excellence Center in Autism and Neurodevelopmental Disorders – Tours ExAC-T, CHRU de Tours, France

iBrain U1253 Imagerie et Cerveau, Tours, France

Christina Kitt Garza Department of Psychiatry, Columbia University Irving Medical Center, New York, NY, USA

Patricia L. Gerbarg Department of Psychiatry, New York Medical College, Valhalla, NY, USA

Luk Gijs Institute for Family and Sexuality Studies, Department of Neurosciences, KU Leuven, Leuven, Belgium

Hartej Gill Mood Disorders Psychopharmacology Unit, University Health Network, Toronto, ON, Canada

Institute of Medical Science, University of Toronto, Toronto, ON, Canada

Gabriella Gobbi Neurobiological Psychiatry Unit, Department of Psychiatry, McGill University, Montreal, QC, Canada

Kris Goethals Collaborative Antwerp Psychiatric Research Institute (CAPRI), University of Antwerp, Antwerp, Belgium

University Forensic Centre (UFC), Antwerp University Hospital, Edegem, Belgium

Guy M. Goodwin Department of Psychiatry, University of Oxford, Warneford Hospital, Oxford, UK

Kimberly Gordon-Achebe Department of Psychiatry, Division of Child and Adolescent Psychiatry, University of Maryland School of Medicine, Baltimore, MD, USA

Department of Psychiatry, Tulane University School of Medicine, New Orleans, LA, USA

Joseph S. Goveas Department of Psychiatry and Behavioral Medicine, Medical College of Wisconsin, Milwaukee, WI, USA

Iris Tatjana Graef-Calliess Research Group Social and Transcultural Psychiatry, Department of Psychiatry, Social Psychiatry and Psychotherapy, Hannover Medical School, Hannover, Germany

Jon E. Grant Department of Psychiatry & Behavioral Neuroscience, Pritzker School of Medicine, University of Chicago, Chicago, IL, USA

Larrilyn Grant Department of Psychiatry, Boonshoft School of Medicine, Wright State University, Fairborn, OH, USA

Luigi Grassi Department of Neuroscience and Rehabilitation, Institute of Psychiatry, University of Ferrara, Ferrara, Italy

David Gratzer Centre for Addiction and Mental Health, Toronto, ON, Canada

Department of Psychiatry, University of Toronto, Toronto, ON, Canada

Ruth L. Graver Department of Psychiatry, Columbia University College of Physicians and Surgeons, New York, NY, USA

Roger P. Greenberg Department of Psychiatry, State University of New York Upstate Medical University, Syracuse, NY, USA

James L. Griffith Department of Psychiatry and Behavioral Sciences, The George Washington University School of Medicine and Health Sciences, Washington, DC, USA

Roland R. Griffiths Department of Psychiatry and Behavioral Sciences, Johns Hopkins University School of Medicine, Baltimore, MD, USA

Department of Neuroscience, Johns Hopkins University School of Medicine, Baltimore, MD, USA

Sophie Grigoriadis University of Toronto, Toronto, ON, Canada

Women's Mood and Anxiety Clinic: Reproductive Transitions, Toronto, ON, Canada

Department of Psychiatry, Sunnybrook Health Sciences Centre, Toronto, ON, Canada

Sunnybrook Research Institute, Toronto, ON, Canada

Women's College Research Institute, Toronto, ON, Canada

Matthew W. Grover Department of Psychiatry, University of Michigan, Ann Arbor, MI, USA

Frank H. Guenther Departments of Speech, Language, & Hearing Sciences and Biomedical Engineering, Boston University, Boston, MA, USA

Jeffrey Guina Beaumont Health, Graduate Medical Education, Dearborn, MI, USA

Easterseals Michigan, Auburn Hills, MI, USA

William Beaumont School of Medicine, Oakland University, Rochester, MI, USA

Boonshoft School of Medicine, Wright State University, Dayton, OH, USA

Oye Gureje University of Ibadan, Ibadan, Nigeria

Jaswant Guzder Division of Social & Transcultural Psychiatry, McGill University, Montreal, QC, Canada

Elizabeth Haase University of Nevada School of Medicine, Reno, NV, USA

Outpatient Behavioral Health, Carson Tahoe Regional Medical Center, Carson City, NV, USA

Ute Habel Department of Psychiatry, Psychotherapy and Psychosomatics, RWTH Aachen University, Aachen, Germany

Phoebe Brosnan Hall Department of Psychological and Brain Sciences, Boston University, Boston, MA, USA

Jeffrey M. Halperin Icahn School of Medicine at Mount Sinai, New York, NY, USA

Queens College of the City University of New York, Flushing, NY, USA

Bridget Hamilton University of Melbourne, Melbourne, VIC, Australia

Hannah R. Hamilton Master of Arts Program in the Social Sciences, University of Chicago, Chicago, IL, USA

Jibril I. M. Handuleh Psychiatry Resident, Department of Psychiatry, Saint Paul's Hospital Millennium Medical College, Addis Ababa, Ethiopia

Department of Medicine, School of Medicine, Amoud University, Borama, Somalia

Charlotte Hanlon King's College London, London, UK

Kelli J. Harding Columbia University College of Physicians and Surgeons, New York, NY, USA

John Hardy Institute of Neurology, University College London, London, UK

Martin Härter Department of Medical Psychology, University Medical Center Hamburg-Eppendorf, Hamburg, Germany

Georgina Hartzell Weill Cornell Medical College, Payne Whitney Clinic, New York, NY, USA

Carol Harvey Psychosocial Research Centre (Department of Psychiatry), The University of Melbourne, Melbourne, VIC, Australia

North West Area Mental Health Service, Melbourne, VIC, Australia

Katie Hayes Trinity College, University of Toronto, Toronto, ON, Canada

Eric P. Hazen Division of Child and Adolescent Psychiatry, Department of Psychiatry, Massachusetts General Hospital, Boston, MA, USA

Harvard Medical School, Boston, MA, USA

Daniela Heddaeus Department of Medical Psychology, University Medical Center Hamburg-Eppendorf, Hamburg, Germany

Elke Heirman University Psychiatric Center KU Leuven, Leuven, Belgium

Helen Herrman Orygen, Melbourne, VIC, Australia

Centre for Youth Mental Health, The University of Melbourne, Melbourne, VIC, Australia

Donald Hilty Northern California Veterans Administration Health Care System, Mather, CA, USA

Department of Psychiatry and Behavioral Sciences, UC Davis, Davis, CA, USA

Eric Hollander Department of Psychiatry and Behavioral Sciences, Albert Einstein College of Medicine, Bronx, NY, USA

Kevin Ann Huckshorn Kevin Huckshorn & Associates, Inc., Chapel Hill, NC, USA

Edward D. Huey Department of Psychiatry and Human Behavior, Brown University, Providence, RI, USA

Larimer V. Hugo Department of Psychiatry and Behavioral Medicine, University of the Philippines, Manila, Philippines

Leslie Hulvershorn Department of Psychiatry, Indiana University School of Medicine, Indianapolis, IN, USA

Kathryn L. Humphreys Department of Psychology and Human Development, Vanderbilt University, Nashville, TN, USA

Department of Psychiatry and Behavioral Sciences, Tulane University School of Medicine, New Orleans, LA, USA

Ken Inada Department of Psychiatry, Kitasato University, Kanagawa, Japan

H. Yavuz Ince Department of Psychiatry, University of Michigan Medical School, Ann Arbor, MI, USA

Antonio Inserra Neurobiological Psychiatry Unit, Department of Psychiatry, McGill University, Montreal, QC, Canada

Salvatore Iuso Department of Clinical and Experimental Medicine, University of Foggia, Foggia, Italy

Ana Ivkovic Massachusetts General Hospital, Boston, MA, USA

Allison Jackson Department of Psychiatry, Beth Israel Deaconess Medical Center, Harvard Medical School, Boston, MA, USA

Amir Hossein Jalali Nadoushan Mental Health Research Center, Psychosocial Health Research Institute (PHRI), Department of Psychiatry, Iran University of Medical Sciences, Tehran, Iran

L. Fredrik Jarskog Department of Psychiatry, University of North Carolina School of Medicine, Chapel Hill, NC, USA

G. Eric Jarvis Division of Social & Transcultural Psychiatry, McGill University, Montreal, QC, Canada

Department of Psychiatry, Jewish General Hospital, Montreal, QC, Canada

Daniel C. Javitt Columbia University College of Physicians and Surgeons/Nathan Kline Institute for Psychiatric Research, New York, NY, USA

Rachel Jenkins Kings College London, London, UK

Alma L. Jimenez Department of Psychiatry and Behavioral Medicine, University of the Philippines, Manila, Philippines

Matthew W. Johnson Department of Psychiatry and Behavioral Sciences, Johns Hopkins University School of Medicine, Baltimore, MD, USA

Lamia Jouini Mental Health, Center of Competencies in Psychiatry and Psychotherapy, Wallis Hospital, Sion, Switzerland

Kyle K. Kampman Addiction Treatment and Research Center at the Perelman School of Medicine, University of Pennsylvania, Philadelphia, PA, USA

Shigenobu Kanba Kyushu University, Fukuoka, Japan

Japan Depression Center, Tokyo, Japan

Denise B. Kandel Mailman School of Public Health, Columbia University Irving Medical Center, New York, NY, USA

Department of Psychiatry, Columbia University College of Physicians and Surgeons, New York, NY, USA

New York State Psychiatric Institute, New York, NY, USA

Meg S. Kaplan Department of Psychiatry, Columbia University Vagelos College of Physicians & Surgeons, New York, NY, USA

Siegfried Kasper Center for Brain Research, Medical University of Vienna, Vienna, Austria

Marianne C. Kastrup Anti Torture Support Foundation, Copenhagen, Denmark

Tadafumi Kato Department of Psychiatry and Behavioral Science, Juntendo University Graduate School of Medicine, Tokyo, Japan

Jerald Kay Department of Psychiatry, Wright State University Boonshoft School of Medicine, Dayton, OH, USA

Martin B. Keller Department of Psychiatry and Human Behavior, The Warren Alpert Medical School of Brown University, Providence, RI, USA

James L. Kennedy Tanenbaum Centre for Pharmacogenetics, Campbell Family Mental Health Research Institute, Centre for Addiction and Mental Health (CAMH), Toronto, ON, Canada

Institute of Medical Science, Toronto, ON, Canada

Department of Psychiatry, University of Toronto, Toronto, ON, Canada

Philipp Khaitovich V. Zelman Center for Neurobiology and Brain Restoration, Skolkovo Institute of Science and Technology, Moscow, Russia

Gabriela Khazanov Corporal Michael J Crescenz Veterans Affairs Medical Center, Philadelphia, PA, USA

Eóin Killackey Orygen, Melbourne, VIC, Australia

Centre for Youth Mental Health, The University of Melbourne, Melbourne, VIC, Australia

Euitae Kim Department of Psychiatry, Seoul National University College of Medicine, Seoul, Korea

Hyun-Hee Kim Department of Psychiatry, Massachusetts General Hospital, Boston, MA, USA

Youl-Ri Kim Department of Psychiatry, Seoul Paik Hospital, Inje University, Seoul, South Korea

Robert A. King Yale Child Study Center, Yale University School of Medicine, New Haven, CT, USA

Laurence J. Kirmayer Division of Social & Transcultural Psychiatry, McGill University, Montreal, QC, Canada

Department of Psychiatry, Jewish General Hospital, Montreal, QC, Canada

Keith P. Klein McLean Hospital, Belmont, MA, USA

William M. Klykylo Department of Psychiatry, Wright State University Boonshoft School of Medicine, Dayton, OH, USA

Chih-Hung Ko Department of Psychiatry, Kaohsiung Medical University Hospital, Kaohsiung Medical University, Kaohsiung City, Taiwan

Department of Psychiatry, Kaohsiung Municipal Siaogang Hospital, Kaohsiung Medical University, Kaohsiung City, Taiwan

Department of Psychiatry, Faculty of Medicine, College of Medicine, Kaohsiung Medical University, Kaohsiung City, Taiwan

Je D. Ko Child and Adolescent Psychiatry, Massachusetts General Hospital, Boston, MA, USA

Robert Kohn Department of Psychiatry and Human Behavior, The Warren Alpert Medical School of Brown University, Providence, RI, USA

Department of Psychiatry and Human Behavior, The Miriam Hospital, Brown University, Providence, RI, USA

Brandon Kohrt Global Mental Health, George Washington University, Washington, DC, USA

George F. Koob National Institute on Alcohol Abuse and Alcoholism, National Institutes of Health, Bethesda, MD, USA

Susan G. Kornstein Department of Psychiatry and Institute for Women's Health, Virginia Commonwealth University School of Medicine, Richmond, VA, USA

Thomas R. Kosten Departments of Psychiatry and Behavioral Sciences, Neuroscience, Pharmacology, and Immunology, Baylor College of Medicine, Houston, TX, USA

Rachel L. Krakauer Department of Psychology, University of Regina, Regina, SK, Canada

Beth Krone Icahn School of Medicine at Mount Sinai, New York, NY, USA

Rachel Kronick Division of Social & Transcultural Psychiatry, McGill University, Montreal, QC, Canada

Division of Child and Adolescent Psychiatry, McGill University, Montreal, QC, Canada

Lady Davis Research Institute, Jewish General Hospital, Montreal, QC, Canada

Richard B. Krueger Department of Psychiatry, Columbia University Vagelos College of Physicians & Surgeons, New York, NY, USA

Kathleen Kruse University of Michigan Medical School, Ann Arbor, MI, USA

Eric Kuhn National Center for PTSD, Dissemination and Training Division, VA Palo Alto Healthcare System, Menlo Park, CA, USA

Department of Psychiatry and Behavioral Sciences, Stanford University School of Medicine, Stanford, CA, USA

Simon Kung Department of Psychiatry and Psychology, Mayo Clinic Alix School of Medicine, Rochester, MN, USA

Gonzalo Laje Department of Psychiatry, Texas Tech University, Lubbock, TX, USA

Washington Behavioral Medicine Associates, LLC, Chevy Chase, MD, USA

Maryland Institute for Neuroscience and Development, Inc (MIND), Chevy Chase, MD, USA

Maria I. Lapid Department of Psychiatry & Psychology, Mayo Clinic, Rochester, MN, USA

Mark E. Larsen Black Dog Institute, University of New South Wales, Sydney, NSW, Australia

Ryan E. Lawrence Columbia University Medical Center, New York, NY, USA

New York - Presbyterian Hospital, New York, NY, USA

Janice LeBel Department of Mental Health, Commonwealth of Massachusetts, Boston, MA, USA

James F. Leckman Yale Child Study Center, Yale University School of Medicine, New Haven, CT, USA

Frances R. Levin Department of Psychiatry, Columbia University College of Physicians and Surgeons/New York State Psychiatric Institute, New York, NY, USA

Daniel Minkin Levy The Jerusalem Mental Health Center, Jerusalem, Israel

Roberto Lewis-Fernández Department of Psychiatry, Columbia University and New York State Psychiatric Institute, New York, NY, USA

Wen Li Academy of Forensic Science, Shanghai, China

Jeffrey A. Lieberman Department of Psychiatry, New York State Psychiatric Institute, Columbia University College of Physicians and Surgeons, New York, NY, USA

Michael R. Liebowitz Department of Psychiatry, Columbia University College of Physicians and Surgeons, New York, NY, USA

New York State Psychiatric Institute, New York, NY, USA

Keh-Ming Lin Department of Psychiatry and Biobehavioral Sciences, David Geffen School of Medicine at UCLA, Los Angeles, CA, USA

Shih-Ku Lin Department of Psychiatry, Taipei Chang Gung Memorial Hospital, Taipei, Taiwan

Mari Lloyd-Williams Department of Primary Care and Mental Health/Liverpool Marie Curie Hospice and Liverpool Health Partners, University of Liverpool, Liverpool, UK

Samantha M. Loi Department of Psychiatry, Melbourne Neuropsychiatry Centre, The University of Melbourne, Parkville, VIC, Australia

The Royal Melbourne Hospital, Neuropsychiatry, Parkville, VIC, Australia

Fernando Lolas Interdisciplinary Center for Studies in Bioethics, University of Chile, Santiago, Chile

Central University of Chile, Santiago, Chile

Jiann Lin Loo Betsi Cadwaladr University Health Board, Wrexham, UK

Francis Lu University of California, Davis, Sacramento, CA, USA

Mario Luciano Department of Psychiatry, University of Campania "L. Vanvitelli", Naples, Italy

Rogelio Luna-Zamora Departamento de Sociología, Universidad de Guadalajara, Guadalajara, Mexico

David D. Luxton Department of Psychiatry and Behavioral Sciences, University of Washington School of Medicine, Seattle, WA, USA

Dean F. MacKinnon Department of Psychiatry and Behavioral Science, Johns Hopkins University School of Medicine, Baltimore, MD, USA

Nasuh Malas Department of Psychiatry, University of Michigan Medical School, Ann Arbor, MI, USA

Department of Pediatrics, University of Michigan Medical School, Ann Arbor, MI, USA

Dolores Malaspina Departments of Psychiatry, Genetics and Genomics, and Neuroscience, Icahn School of Medicine at Mount Sinai, New York, NY, USA

Jan Malat Department of Psychiatry, University of Toronto, Toronto, ON, Canada

Centre for Addiction and Mental Health, Toronto, ON, Canada

José R. Maldonado Psychiatry and Behavioral Sciences, Division of Medical Psychiatry, Consultation-Liaison Psychiatry, Critical Care Psychiatry Service, Stanford University School of Medicine, Stanford, CA, USA

Mirko Manchia Unit of Psychiatry, Department of Medical Sciences and Public Health, University of Cagliari, Cagliari, Italy

Department of Pharmacology, Dalhousie University, Halifax, NS, Canada

Abby Adler Mandel Department of Psychology, The Catholic University of America, Washington, DC, USA

John C. Markowitz Columbia University Vagelos College of Physicians & Surgeons, New York State Psychiatric Institute, New York, NY, USA

Veronica Martinez-Cerdeño Department of Pathology and Laboratory Medicine, UC Davis, Davis, CA, USA

Steve Martino Yale University School of Medicine, New Haven, CT, USA
VA Connecticut Healthcare System, West Haven, CT, USA

Daniel Martins-de-Souza Laboratory of Neuroproteomics, Department of Biochemistry and Tissue Biology, University of Campinas (UNICAMP), Campinas, SP, Brazil
D'Or Institute for Research and Education (IDOR), São Paulo, Brazil
Experimental Medicine Research Cluster (EMRC), University of Campinas, Campinas, SP, Brazil
Instituto Nacional de Biomarcadores em Neuropsiquiatria (INBION), Conselho Nacional de Desenvolvimento Científico e Tecnológico, São Paulo, Brazil

Manuel Mattheisen Institute of Psychiatric Phenomics and Genomics (IPPG), LMU University Hospital, LMU Munich, Munich, Germany
Department of Community Health and Epidemiology, Dalhousie University, Halifax, NS, Canada
Faculty of Computer Science, Dalhousie University, Halifax, NS, Canada

Oscar Rodriguez Mayoral Palliative Care Unit, Cancer Institute of Mexico, Mexico City, Mexico

Gillian A. McCabe Department of Psychology, University of Kentucky, Lexington, KY, USA

Randi E. McCabe Department of Psychiatry and Behavioural Neurosciences, McMaster University, Hamilton, ON, Canada
Anxiety Treatment and Research Clinic, St. Joseph's Healthcare, Hamilton, ON, Canada

Christopher J. McDougle Massachusetts General Hospital for Children, Lurie Center for Autism, Lexington, MA, USA
Department of Psychiatry, Harvard Medical School, Boston, MA, USA

Daniel C. McFarland Department of Medicine, Northwell Health Cancer Institute, Lenox Hill Hospital, Manhattan Eye Ear Throat Hospital, New York, NY, USA
Department of Psychiatry, Division of Collaborative Care and Wellness, Department of Medicine, Division of Hematology and Oncology, University of Rochester Medical Center, Rochester, NY, USA

Alexander McFarlane Discipline of Psychiatry, Adelaide Medical School, The University of Adelaide, Adelaide, SA, Australia

Patrick D. McGorry Orygen, Parkville, VIC, Australia

Centre for Youth Mental Health, The University of Melbourne, Parkville, VIC, Australia

Roger S. McIntyre Mood Disorders Psychopharmacology Unit, University Health Network, Toronto, ON, Canada

Institute of Medical Science, University of Toronto, Toronto, ON, Canada

Department of Pharmacology, University of Toronto, Toronto, ON, Canada

Department of Psychiatry, University of Toronto, Toronto, ON, Canada

Brain and Cognition Discovery Foundation, Toronto, ON, Canada

Mia A. McLean Department of Pediatrics, University of British Columbia, Vancouver, BC, Canada

BC Children's Hospital Research Institute, Vancouver, BC, Canada

Francis J. McMahon Human Genetics Branch, Intramural Research Program, National Institute of Mental Health, National Institutes of Health, Bethesda, MD, USA

Marta Meana Department of Psychology, University of Nevada, Las Vegas, Las Vegas, NV, USA

Michael J. Meaney Douglas Hospital Research Center, Montreal, QC, Canada

Sandra Melanie Meier Department of Psychiatry, Dalhousie University, Halifax, NS, Canada

Child and Adolescent Mental Health Centre, Copenhagen University Hospital – Mental Health Services CPH, Copenhagen, Denmark

IWK Health Centre Department of Psychiatry & Specific Care Clinics, Halifax, NS, Canada

Tando Abner Sivile Melapi Department of Psychiatry, University of the Witwatersrand, Johannesburg, South Africa

Samantha Meltzer-Brody Department of Psychiatry, The University of North Carolina at Chapel Hill, Chapel Hill, NC, USA

Glenn A. Melvin Centre for Social and Early Emotional Development, School of Psychology, Deakin University, Geelong, VIC, Australia

Kathleen R. Merikangas Genetic Epidemiology Research Branch, National Institute of Mental Health, Bethesda, MD, USA

Andreas Meyer-Lindenberg Department of Psychiatry and Psychotherapy, Central Institute of Mental Health, Medical Faculty Mannheim, Heidelberg University, Mannheim, Germany

Frank A. Middleton SUNY Upstate Medical University, Syracuse, NY, USA

Dana Mihaila SUNY Upstate Medical University, Syracuse, NY, USA

Pamela Mirsky Psychiatry, University of Arizona College of Medicine, Tucson, AZ, USA

Kazuo Mishima Department of Neuropsychiatry, Akita University Graduate School of Medicine, Akita-city, Akita, Japan

Philip B. Mitchell Discipline of Psychiatry and Mental Health, University of New South Wales, Sydney, NSW, Australia

Maria Mody Department of Radiology, MGH/MIT/HMS Athinoula A. Martinos Center for Biomedical Imaging, Charlestown, MA, USA

Diana E. Moga Department of Psychiatry, Columbia University College of Physicians and Surgeons, New York, NY, USA

Ayeshah G. Mohiuddin Tanenbaum Centre for Pharmacogenetics, Campbell Family Mental Health Research Institute, Centre for Addiction and Mental Health (CAMH), Toronto, ON, Canada

Paul C. Mohl Department of Psychiatry, University of Texas Southwestern Medical Center, Dallas, TX, USA

Emily Morelli University of New Mexico, Albuquerque, NM, USA

Claudia Moreno Department of Child and Adolescent Psychiatry, Yale University School of Medicine Child Study Center, New Haven, CT, USA

Joshua C. Morganstein Department of Psychiatry, Uniformed Services University of the Health Sciences, Bethesda, MD, USA

Adrian P. Mundt Medical Faculty, Universidad Diego Portales, Santiago, Chile

Department of Psychiatry and Mental Health, Medical Faculty, Universidad de Chile, Santiago, Chile

Jennifer M. Mundt Department of Neurology, Northwestern University Feinberg School of Medicine, Chicago, IL, USA

Center for Circadian and Sleep Medicine, Northwestern University Feinberg School of Medicine, Chicago, IL, USA

Jade Murray Turner Institute for Brain and Mental Health, Monash University, Notting Hill, VIC, Australia

Manisha Murugesan Women's Mental Health, Bangalore, India

Department of Psychiatry, NIMHANS, Bangalore, India

Philip R. Muskin Department of Psychiatry, Columbia University Irving Medical Center, New York, NY, USA

Maria Muzik Departments of Psychiatry, Obstetrics & Gynecology, University of Michigan- Michigan Medicine, Ann Arbor, MI, USA

Zero To Thrive & Women and Infant Mental Health Program, Ann Arbor, MI, USA

Perinatal Psychiatry Service, Ann Arbor, MI, USA

MC3 Perinatal Psychiatry Assess Program Michigan, Ann Arbor, MI, USA

Abhijit Nadkarni London School of Hygiene & Tropical Medicine, London, UK

Sangath, Goa, India

Sanjiv Nair Department of Psychological Medicine, Changi General Hospital, Singapore, Singapore

Zui Narita Department of Behavioral Medicine, National Institute of Mental Health, National Center of Neurology and Psychiatry, Tokyo, Japan

Ramzi Nasir Massachusetts General Hospital for Children, Lurie Center for Autism, Lexington, MA, USA

Department of Psychiatry, Harvard Medical School, Boston, MA, USA

Sandeep M. Nayak Department of Psychiatry and Behavioral Sciences, Johns Hopkins University School of Medicine, Baltimore, MD, USA

Randolph M. Nesse University of Michigan, Ann Arbor, MI, USA

Paul Nestadt Departments of Psychiatry, Johns Hopkins University School of Medicine, Baltimore, MD, USA

Departments of Mental Health, Johns Hopkins Bloomberg School of Public Health, Baltimore, MD, USA

Jeffrey H. Newcorn Icahn School of Medicine at Mount Sinai, New York, NY, USA

Roger Ng Department of Psychiatry, Kowloon Hospital, Kowloon, Hong Kong, China

Thomas Nickl-Jockschat Departments of Psychiatry, Neuroscience and Pharmacology, Carver College of Medicine, University of Iowa, Iowa City, IA, USA

Pedro Nobre Center for Psychology (CPUP), Faculty of Psychology and Educational Sciences, University of Porto, Porto, Portugal

Kimberly Nordstrom University of Colorado School of Medicine, Denver, CO, USA

Guido Nosari Department of Neurosciences and Mental Health, Fondazione IRCCS Ca' Granda Ospedale Maggiore Policlinico, Milan, Italy

Lisa A. Nowinski Massachusetts General Hospital for Children, Lurie Center for Autism, Lexington, MA, USA

Department of Psychiatry, Harvard Medical School, Boston, MA, USA

Edward V. Nunes Department of Psychiatry, Columbia University College of Physicians and Surgeons/New York State Psychiatric Institute, New York, NY, USA

John M. Oldham Department of Psychiatry and Behavioral Sciences, Baylor College of Medicine, Houston, TX, USA

Maria A. Oquendo Department of Psychiatry, Perelman School of Medicine, University of Pennsylvania, Philadelphia, PA, USA

Mojtaba Oraki Kohshour Institute of Psychiatric Phenomics and Genomics (IPPG), LMU University Hospital, LMU Munich, Munich, Germany

Department of Immunology, Faculty of Medicine, Ahvaz Jundishapur University of Medical Sciences, Ahvaz, Iran

Sharone Ornstein Department of Psychiatry, Columbia University College of Physicians and Surgeons, New York, NY, USA

Laura Orsolini Unit of Clinical Psychiatry, Department of Neurosciences/DIMSC, Polytechnic University of Marche, Ancona, Italy

Sami Ouanes Department of Psychiatry, Hamad Medical Corporation, Doha, Qatar

Bruno Pacciardi Università di Pisa, Pisa, Italy

Alex Palffy Department of Psychiatry, University of Michigan Medical School, Ann Arbor, MI, USA

Michelle L. Palumbo Departments of Pediatrics and Psychiatry, The Lurie Center for Autism, Massachusetts General Hospital for Children, Lexington, MA, USA

Sergi Papiol Department of Psychiatry and Psychotherapy, University Hospital, LMU Munich, Munich, Germany

Institute of Psychiatric Phenomics and Genomics, University Hospital, LMU Munich, Munich, Germany

Max-Planck-Institute of Psychiatry, Munich, Germany

Sagar V. Parikh John F. Greden Chair in Depression and Clinical Neuroscience, University of Michigan, Ann Arbor, Ann Arbor, MI, USA

Department of Psychiatry, Michigan Medicine, University of Michigan, Ann Arbor, Ann Arbor, MI, USA

Health Management and Policy, School of Public Health, University of Michigan, Ann Arbor, Ann Arbor, MI, USA

Workplace Mental Health Solutions, Eisenberg Family Depression Center, University of Michigan, Ann Arbor, Ann Arbor, MI, USA

Seon-Cheol Park Department of Psychiatry, Hanyang University College of Medicine, Seoul, Republic of Korea

Department of Psychiatry, Hanyang University Guri Hospital, Guri, Republic of Korea

Vikram Patel Harvard Medical School, Boston, MA, USA

Sangath, Goa, India

Michele T. Pato Department of Psychiatry, Rutgers Biomedical and Health Sciences, Robert Wood Johnson Medical School, Piscataway, NJ, USA

Jennifer L. Payne Department of Psychiatry and Neurobehavioral Sciences, University of Virginia, Charlottesville, VA, USA

Abraham Peled Psychiatry, Mental Health Center, Davis, CA, USA

"Technion" Israel Institute of Technology, Haifa, Israel

Dorian Peters Leverhulme Centre for the Future of Intelligence, University of Cambridge, Cambridge, UK

Dyson School of Design Engineering, Imperial College London, London, UK

Robin L. Peterson Department of Physical Medicine and Rehabilitation, University of Colorado School of Medicine, Aurora, CO, USA

Annamaria Petito Department of Clinical and Experimental Medicine, University of Foggia, Foggia, Italy

Katharine A. Phillips New York-Presbyterian Hospital, Weill Cornell Medical College, New York, NY, USA

David Pilkey Yale University School of Medicine, New Haven, CT, USA

VA Connecticut Healthcare System, West Haven, CT, USA

Debra A. Pinals Department of Psychiatry, University of Michigan, Ann Arbor, MI, USA

Mariana Pinto da Costa South London and Maudsley NHS Foundation Trust, London, UK

Institute of Biomedical Sciences Abel Salazar (ICBAS), University of Porto, Porto, Portugal

Institute of Psychiatry, Psychology & Neuroscience, King's College London, London, UK

Cara Anne Poland Department of Obstetrics, Gynecology and Reproductive Biology, College of Human Medicine, Michigan State University, East Lansing, MI, USA

Michele Poletti Department of Mental Health and Pathological Addiction, Child and Adolescent Neuropsychiatry Service, Azienda USL-IRCCS di Reggio Emilia, Reggio Emilia, Italy

Laura C. Politte Massachusetts General Hospital for Children, Lurie Center for Autism, Lexington, MA, USA

Department of Psychiatry, Harvard Medical School, Boston, MA, USA

David Popovic Department of Psychiatry and Psychotherapy, University Hospital, LMU Munich, Munich, Germany

International Max Planck Research School for Translational Psychiatry (IMPRS-TP), Munich, Germany

Max-Planck-Institute of Psychiatry, Munich, Germany

Kelly Posner Department of Psychiatry, Columbia University Medical Center, New York City, NY, USA

James B. Potash Department of Psychiatry and Behavioral Sciences, Johns Hopkins School of Medicine, Baltimore, MD, USA

Seth Powsner School of Medicine, Yale University, New Haven, CT, USA

Abhishek Pratap Krembil Center for Neuroinformatics, Center for Addiction and Mental Health, Toronto, ON, Canada

Vector Institute, Toronto, ON, Canada

Department of Biomedical Informatics and Medical Education, University of Washington, Seattle, WA, USA

King's College London, London, UK

Florian Raabe Department of Psychiatry and Psychotherapy, University Hospital, LMU Munich, Munich, Germany

International Max Planck Research School for Translational Psychiatry (IMPRS-TP), Munich, Germany

Max-Planck-Institute of Psychiatry, Munich, Germany

Andrea Raballo Chair of Psychiatry and Psychotherapy, Faculty of Biomedical Sciences, Università della Svizzera Italiana (USI), Lugano, Switzerland

Cantonal Socio-psychiatric Organization (OSC), Public Health Division, Department of Health and Social Care, Repubblica e Cantone Ticino, Mendrisio, Switzerland

Geoffrey S. Rachor Department of Psychology, University of Regina, Regina, SK, Canada

Nevena V. Radonjić Department of Psychiatry, SUNY Upstate Medical University, Syracuse, NY, USA

Mark Ragins Los Angeles County Department of Health Services, Los Angeles, CA, USA

Elmars Rancans Department of Psychiatry and Narcology, Riga Stradins University, Riga, Latvia

Tayyab Rashid Centre for Wellbeing Science, Melbourne Graduate School of Education, University of Melbourne, Melbourne, Victoria, Australia

Aswin Ratheesh Orygen, Parkville, VIC, Australia

Centre for Youth Mental Health, The University of Melbourne, Parkville, VIC, Australia

Vibhay Raykar Child and Adolescent Mental Health Service, GV Health, Shepparton, VIC, Australia

Greg Reger VA Puget Sound Healthcare System, Seattle, WA, USA

Department of Psychiatry and Behavioral Sciences, University of Washington School of Medicine, Seattle, WA, USA

Gary Remington Department of Psychiatry, University of Toronto, Toronto, ON, Canada

Nancy Renn-Bugai Corewell Health, Grand Rapids, MI, USA

Mary Roary Substance Abuse Mental Health Service Administration (SAMHSA), Center for Behavioral Health Quality and Statistics (CBHSQ), Rockville, MD, USA

The Catholic University of America, Washington, DC, USA

Carolyn I. Rodriguez Veterans Affairs Palo Alto Health Care System, Palo Alto, CA, USA

Stanford University School of Medicine, Palo Alto, CA, USA

Annelieke M. Roest Department of Developmental Psychology, University of Groningen, Groningen, The Netherlands

Interdisciplinary Center Psychopathology and Emotion Regulation, University Medical Center Groningen, University of Groningen, Groningen, The Netherlands

Stephanie A. Rolin Center for Law, Ethics & Psychiatry, New York State Psychiatric Institute and Columbia University Irving Medical Center, New York, NY, USA

Anthony J. Rosellini Center for Anxiety and Related Disorders, Department of Psychological and Brain Sciences, Boston University, Boston, MA, USA

Paul Rosenfield Departments of Psychiatry and Medical Education, Icahn School of Medicine at Mount Sinai, New York, NY, USA

Daniel A. Rossignol Rossignol Medical Center, Melbourne, FL, USA

Cécile Rousseau Division of Social & Transcultural Psychiatry, McGill University, Montreal, QC, Canada

Fabian M. Saleh Department of Psychiatry, Beth Israel Deaconess Medical Center, Harvard Medical School, Boston, MA, USA

Ihsan M. Salloum Department of Neuroscience, University of Texas Rio Grande Valley, Harlingen, TX, USA

Paul Salmon Department of Psychological and Brain Sciences, University of Louisville, Louisville, KY, USA

Gaia Sampogna Department of Psychiatry, University of Campania "L. Vanvitelli", Naples, Italy

Jerome Sarris NICM Health Research Institute, Western Sydney University, Westmead, NSW, Australia

Akira Sawa Departments of Psychiatry, Johns Hopkins University School of Medicine, Baltimore, MD, USA

Departments of Mental Health, Johns Hopkins Bloomberg School of Public Health, Baltimore, MD, USA

Departments of Neuroscience, Johns Hopkins University School of Medicine, Baltimore, MD, USA

Departments of Biomedical Engineering, Johns Hopkins University School of Medicine, Baltimore, MD, USA

Departments of Genetic Medicine, Johns Hopkins University School of Medicine, Baltimore, MD, USA

Departments of Pharmacology, Johns Hopkins University School of Medicine, Baltimore, MD, USA

Adelle Schaefer Department of Psychiatry and Behavioral Sciences, Mount Sinai Beth Israel, New York, NY, USA

Alan Schatzberg Department of Psychiatry, Stanford University, Stanford, CA, USA

Stephen W. Scherer The Center for Applied Genomics, The Hospital for Sick Children, Toronto, ON, Canada

Andrea Schmitt Department of Psychiatry and Psychotherapy, University Hospital, LMU Munich, Munich, Germany

Laboratory of Neuroscience (LIM27), Institute of Psychiatry, University of Sao Paulo, São Paulo, Brazil

Sonja W. Scholz National Institute of Neurological Disorder and Stroke, Bethesda, MD, USA

Department of Neurology, Johns Hopkins University, Baltimore, MD, USA

Meryam Schouler-Ocak Research Division Intercultural Migration and Care Research, Social Psychiatry at Charité-Universitätsmedizin Berlin, Psychiatric University Clinic of Charité at St. Hedwig Hospital, Berlin, Germany

Eva C. Schulte Department of Psychiatry, University Medical Center, University of Munich, Munich, Germany

Institute of Psychiatric Phenomics and Genomics (IPPG), University Medical Center, University of Munich, Munich, Germany

Institute of Human Genetics, University Hospital, Faculty of Medicine, University of Bonn, Bonn, Germany

Department of Psychiatry, University Hospital, Faculty of Medicine, University of Bonn, Bonn, Germany

Thomas G. Schulze Institute of Psychiatric Phenomics and Genomics (IPPG), LMU University Hospital, LMU Munich, Munich, Germany

Department of Psychiatry and Behavioral Sciences, Norton College of Medicine, SUNY Upstate Medical University, Syracuse, NY, USA

Department of Psychiatry and Behavioral Sciences, Johns Hopkins University School of Medicine, Baltimore, MD, USA

Emanuel Schwarz Department of Psychiatry and Psychotherapy, Central Institute of Mental Health, Medical Faculty Mannheim, Heidelberg University, Mannheim, Germany

Eric L. Scott Department of Pediatrics, Department of Anesthesiology, University of Michigan Medical School, Ann Arbor, MI, USA

Soraya Seedat Department of Psychiatry, Faculty of Medicine and Health Sciences, Tygerberg Campus, Stellenbosch University, Cape Town, South Africa

South African Research Chair in PTSD, Stellenbosch University, Cape Town, South Africa

SAMRC Genomics of Brain Disorders Research Unit, Stellenbosch University, Cape Town, South Africa

Martin E. P. Seligman Department of Psychology, University of Pennsylvania, Philadelphia, PA, USA

Kevin A. Sevarino Department of Psychiatry, Yale University School of Medicine, New Haven, CT, USA

Melania Severo Department of Clinical and Experimental Medicine, University of Foggia, Foggia, Italy

Theodore Shapiro Weill Cornell Medical College, Payne Whitney Clinic, New York, NY, USA

Samantha Shaw Department of Psychiatry, Rachel Upjohn Building, Ann Arbor, MI, USA

Department of Psychiatry, University of Michigan- Michigan Medicine, Ann Arbor, MI, USA

M. Katherine Shear Columbia University School of Social Work, Columbia University College of Physicians and Surgeons, New York, NY, USA

Tomer Shechner School of Psychological Sciences, University of Haifa, Haifa, Israel

Winston Wu-Dien Shen Department of Psychiatry, Wan Fang Medical Center, Taipei, Taiwan

College of Medicine, Taipei Medical University, Taipei, Taiwan

Yelizaveta Sher Psychiatry and Behavioral Sciences, Transplant Psychiatry Service, Stanford University School of Medicine, Stanford, CA, USA

Kazutaka Shimoda Department of Psychiatry, Dokkyo Medical University, Shimotsuga-gun, Tochigi, Japan

Gen Shinozaki Department of Psychiatry and Behavioral Sciences, Stanford University, Stanford, CA, USA

Jay H. Shore Department of Psychiatry, University of Colorado Aschutz Medical Campus, Denver, CO, USA

Joseph Silvio American Academy of Psychodynamic Psychiatry and Psychoanalysis, Bethessda, MD, USA

Naomi M. Simon Department of Psychiatry, New York University Grossman School of Medicine, New York, NY, USA

Scott A. Small Department of Neurology, Columbia University, New York, NY, USA

David E. Smith University of California San Francisco, San Francisco, CA, USA

Joshua R. Smith Department of Psychiatry and Behavioral Sciences, Division of Child and Adolescent Psychiatry, Vanderbilt University Medical Center, Nashville, TN, USA

Estate M. Sokhadze Department of Neurology, Duke University School of Medicine, Durham, NC, USA

Suzan J. Song Department of Psychiatry and Behavioral Sciences, The George Washington University School of Medicine and Health Sciences, Washington, DC, USA

Stephen M. Sonnenberg Department of Psychiatry, Uniformed Services University of the Health Sciences, Bethesda, MD, USA

Department of Psychiatry and Behavioral Sciences, Dell Medical School, Austin, TX, USA

Jennifer Sotsky Department of Psychiatry, NYU Grossman School of Medicine, New York, NY, USA

David Spiegel Center on Stress and Health & Center for Integrative Medicine, Department of Psychiatry & Behavioral Sciences, Stanford University School of Medicine, Stanford, CA, USA

A. Benjamin Srivastava Department of Psychiatry, Columbia University College of Physicians and Surgeons/New York State Psychiatric Institute, New York, NY, USA

Alexandra Stainton Orygen, Parkville, Australia

University of Melbourne, Centre for Youth Mental Health, Parkville, Australia

Jordan Standlee Department of Neurology, Northwestern University Feinberg School of Medicine, Chicago, IL, USA

Barbara Stanley Division of Molecular Imaging and Neuropathology, New York State Psychiatric Institute, New York, NY, USA

Department of Psychiatry, Columbia University Medical Center, New York, NY, USA

Brett N. Steenbarger Department of Psychiatry, State University of New York Upstate Medical University, Syracuse, NY, USA

Dan J. Stein University of Cape Town, Cape Town, South Africa

Christopher M. Stewart Department of Psychiatry, University of Louisville, Louisville, KY, USA

Donna E. Stewart University Health Network, University of Toronto, Toronto, ON, Canada

Julia Stoll Institute of Biomedical Ethics and History of Medicine, University of Zurich, Zurich, Switzerland

Michael H. Stone Department of Psychiatry, University of Columbia College of Physicians and Surgeons, New York, NY, USA

Jeffrey R. Strawn Department of Psychiatry & Behavioral Neuroscience, University of Cincinnati, College of Medicine, Cincinnati, OH, USA

Nora I. Strom Institute of Psychiatric Phenomics and Genomics (IPPG), LMU University Hospital, LMU Munich, Munich, Germany

Department of Psychology, Humboldt-Universität zu Berlin, Berlin, Germany

Centre for Psychiatry Research, Department of Clinical Neuroscience, Karolinska Institutet & Stockholm Health Care Services, Region Stockholm, Sweden

Department of Biomedicine, Aarhus University, Aarhus, Denmark

Dorothy Stubbe Department of Child and Adolescent Psychiatry, Yale University School of Medicine Child Study Center, New Haven, CT, USA

Richard F. Summers Department of Psychiatry, Perelman School of Medicine, University of Pennsylvania, Philadelphia, PA, USA

Charlene Sunkel Global Office, Global Mental Health Peer Network, Paarl, South Africa

Mary M. Sweeney Department of Psychiatry and Behavioral Sciences, Johns Hopkins University School of Medicine, Baltimore, MD, USA

Anna Szombathy SUNY Upstate Medical University, Syracuse, NY, USA

Hiroyoshi Takeuchi Department of Neuropsychiatry, Keio University School of Medicine, Tokyo, Japan

Allan Tasman Department of Psychiatry and Behavioral Sciences, University of Louisville School of Medicine, Louisville, KY, USA

Danielle S. Taubman Eisenberg Family Depression Center, University of Michigan, Ann Arbor, MI, USA

Inês Tavares Center for Psychology (CPUP), Faculty of Psychology and Educational Sciences, University of Porto, Porto, Portugal

Andre Teck Sng Tay Department of Psychological Medicine, Changi General Hospital, Singapore, Singapore

Duke National University of Singapore Graduate Medical School, Singapore, Singapore

Yong Loo Lin School of Medicine, National University of Singapore, Singapore, Singapore

Steven Taylor Department of Psychiatry, University of British Columbia, Vancouver, BC, Canada

Howard Tennen Department of Public Health Sciences, University of Connecticut School of Medicine, Farmington, CT, USA

David Choon Liang Teo Department of Psychological Medicine, Changi General Hospital, Singapore, Singapore

Kowsar Teymouri Tanenbaum Centre for Pharmacogenetics, Campbell Family Mental Health Research Institute, Centre for Addiction and Mental Health (CAMH), Toronto, ON, Canada

Institute of Medical Science, Toronto, ON, Canada

Michael E. Thase Corporal Michael J Crescenz Veterans Affairs Medical Center, Philadelphia, PA, USA

Department of Psychiatry, Perelman School of Medicine, University of Pennsylvania, Philadelphia, PA, USA

Robyn P. Thom Massachusetts General Hospital for Children, Lurie Center for Autism, Lexington, MA, USA

Department of Psychiatry, Harvard Medical School, Boston, MA, USA

Kenneth Thompson University of Pittsburgh, Pittsburgh, PA, USA

Anna Tkachev V. Zelman Center for Neurobiology and Brain Restoration, Skolkovo Institute of Science and Technology, Moscow, Russia

Martha C. Tompson Department of Psychological and Brain Sciences, Boston University, Boston, MA, USA

Julio Torales Department of Psychiatry, School of Medical Sciences, National University of Asunción, Asunción, Paraguay

Merete S. Torvanger Division of Psychiatry, Haukeland University Hospital, Bergen, Norway

Heike Tost Department of Psychiatry and Psychotherapy, Central Institute of Mental Health, Medical Faculty Mannheim, Heidelberg University, Mannheim, Germany

Manuel Trachsel Clinical Ethics Unit, University Hospital Basel (USB), University Psychiatric Clinics (UPK) Basel, Geriatric University Hospital (UAFP) Basel, University Children's Hospital Basel (UKBB), Basel, Switzerland

Medical Faculty, University of Basel, Basel, Switzerland

Bryan J. Traynor Laboratory of Neurogenetics, National Institute on Aging, Bethesda, MD, USA

Department of Neurology, Johns Hopkins University, Baltimore, MD, USA

Peter Tyrer Community Psychiatry, Imperial College, London, UK

Hiroyuki Uchida Department of Neuropsychiatry, Keio University School of Medicine, Tokyo, Japan

Blair Uniacke Department of Psychiatry, Columbia University College of Physicians and Surgeons, New York, NY, USA

Division of Anxiety, Mood, Eating & Related Disorders, New York State Psychiatric Institute, New York, NY, USA

Jürgen Unützer Department of Psychiatry and Behavioral Sciences, University of Washington, Seattle, WA, USA

Amy M. Ursano Department of Psychiatry, University of North Carolina, Chapel Hill, NC, USA

Robert J. Ursano Department of Psychiatry, Uniformed Services University of the Health Sciences, Bethesda, MD, USA

Christina van der Feltz-Cornelis Department of Health Sciences, Hull York Medical School, University of York, York, UK

Susan C. Vaughan Department of Psychiatry, Columbia Vagelos College of Physicians and Surgeons, Columbia University, New York, NY, USA

Columbia University Center for Psychoanalytic Training and Research, Columbia University, New York, NY, USA

Antonio Ventriglio Department of Clinical and Experimental Medicine, University of Foggia, Foggia, Italy

Simone N. Vigod Women's College Hospital and Women's College Research Institute, University of Toronto, Toronto, ON, Canada

Sergio Villaseñor Bayardo Departamento de Clínicas Médicas, Universidad de Guadalajara, Guadalajara, Mexico

Lawrence A. Vitulano Yale Child Study Center, Yale University School of Medicine, New Haven, CT, USA

Michael L. Vitulano Yale Child Study Center, Yale University School of Medicine, New Haven, CT, USA

University of California, Los Angeles, CA, USA

Lan Chi Vo Department of Psychiatry, Perelman School of Medicine, University of Pennsylvania, Philadelphia, PA, USA

Department of Child and Adolescent Psychiatry and Behavioral Sciences, Children's Hospital of Philadelphia, Philadelphia, PA, USA

Nora D. Volkow National Institute on Drug Abuse, Bethesda, MD, USA

Umberto Volpe Unit of Clinical Psychiatry, Department of Neurosciences/DIMSC, Polytechnic University of Marche, Ancona, Italy

Lisa Wagels Department of Psychiatry, Psychotherapy and Psychosomatics, RWTH Aachen University, Aachen, Germany

Tanner Waldman SUNY Upstate Medical University, Syracuse, NY, USA

John T. Walkup Pritzker Department of Psychiatry and Behavioral Health, Ann and Robert H. Lurie Children's Hospital of Chicago, Chicago, IL, USA

B. Timothy Walsh Department of Psychiatry, Columbia University College of Physicians and Surgeons, New York, NY, USA

Division of Anxiety, Mood, Eating & Related Disorders, New York State Psychiatric Institute, New York, NY, USA

Koichiro Watanabe Department of Neuropsychiatry, Kyorin University School of Medicine, Tokyo, Japan

Kerstin Weber Faculty of Medicine, University of Geneva, Geneva, Switzerland

Donald R. Wesson Medications Development, Oakland, CA, USA

James C. West Department of Psychiatry, Uniformed Services University of the Health Sciences, Bethesda, MD, USA

Jessie Whitfield Department of Psychiatry and Behavioral Sciences, University of Washington, Seattle, WA, USA

Natalia Widiasih Raharjanti Department of Psychiatry, Faculty of Medicine, Dr. Cipto Mangunkusumo General Hospital, Universitas Indonesia, Jakarta, Indonesia

Thomas A. Widiger Department of Psychology, University of Kentucky, Lexington, KY, USA

Angelika Wieck Greater Manchester Mental Health NHS Foundation Trust & University of Manchester, Manchester, UK

Laureate House, Manchester, UK

Wythenshawe Hospital, Manchester, UK

Michael P. Wilson University of Arkansas for Medical Sciences, Little Rock, AR, USA

Arnold Winston Department of Psychiatry and Behavioral Sciences, Mount Sinai Beth Israel, New York, NY, USA

Department of Psychiatry, Icahn School of Medicine at Mount Sinai, New York, NY, USA

Department of Psychiatry, St. George's Medical School, St. George, Grenada

Elizabeth Winter Department of Veterans Affairs, Office of Inspector General, Baltimore, MD, USA

George E. Woody Addiction Treatment and Research Center at the Perelman School of Medicine, University of Pennsylvania, Philadelphia, PA, USA

Jesse H. Wright Department of Psychiatry and Behavioral Sciences, University of Louisville School of Medicine, Louisville, KY, USA

Yu-Tao Xiang Unit of Psychiatry, Department of Public Health and Medicinal Administration, Faculty of Health Sciences, University of Macau, Macao, SAR, China

Shigeto Yamawaki Center for Brain, Mind and KANSEI Sciences Research, Hiroshima University, Hiroshima, Japan

Eric Yarbrough Private Practice or AGLP: The Association of LGBTQ Psychiatrists, New York, NY, USA

Norio Yasui-Furukori Department of Psychiatry, Dokkyo Medical University, Shimotsuga-gun, Tochigi, Japan

Firas Yatim Beaumont Health, Graduate Medical Education, Dearborn, MI, USA

M. Yanki Yazgan Yale Child Study Center, Yale University School of Medicine, New Haven, CT, USA

GuzelGunler Clinic, Istanbul, Turkey

Ju-Yu Yen Department of Psychiatry, Kaohsiung Medical University Hospital, Kaohsiung Medical University, Kaohsiung City, Taiwan

Department of Psychiatry, Faculty of Medicine, College of Medicine, Kaohsiung Medical University, Kaohsiung City, Taiwan

Department of Psychiatry, Kaohsiung Municipal Siaogang Hospital, Kaohsiung Medical University, Kaohsiung City, Taiwan

Charles H. Zeanah Institute of Infant and Early Childhood Mental Health, Tulane University School of Medicine, New Orleans, LA, USA

Department of Psychiatry and Behavioral Sciences, Tulane University School of Medicine, New Orleans, LA, USA

Scott L. Zeller Department of Psychiatry, University of California-Riverside School of Medicine, Riverside, CA, USA

Sasson Zemach The Jerusalem Mental Health Center, Jerusalem, Israel

Melvyn Weibin Zhang National Addictions Management Service, Institute of Mental Health, Singapore, Singapore

Lee Kong Chian School of Medicine, Nanyang Technological University Singapore, Singapore, Singapore

Xudong Zhao Shanghai Pudong New Area Mental Health Center, School of Medicine, Tongji University, Shanghai, China

Douglas Ziedonis University of New Mexico, Albuquerque, NM, USA

Mark A. Zinn NeuroCognitive Research Institute, Chicago, IL, USA

Joseph Zohar Department of Psychiatry, Sheba Medical Center, Tel-Hashomer, and Sackler School of Medicine, Tel Aviv University, Tel Aviv-Yafo, Israel

Alessandro Zuddas Department of Biomedical Sciences, Section Neuroscience and Clinical Pharmacology, University of Cagliari, Cagliari, Italy

Child and Adolescent Neuropsychiatry Unit, "A. Cao" Paediatric Hospital, Cagliari, Italy

Section IX

Pharmacological and Other Somatic Treatments

General Principles of Pharmacologic Therapy

120

Ken Inada, Shigeto Yamawaki, Shigenobu Kanba, Gen Shinozaki, and Siegfried Kasper

Contents

K. Inada
Department of Psychiatry, Kitasato University, Kanagawa, Japan
e-mail: inadaken@kitasato-u.ac.jp

S. Yamawaki (✉)
Center for Brain, Mind and KANSEI Sciences Research, Hiroshima University, Hiroshima, Japan
e-mail: yamawaki@hiroshima-u.ac.jp

S. Kanba
Kyushu University, Fukuoka, Japan

Japan Depression Center, Tokyo, Japan
e-mail: kanba.shigenobu.921@m.kyushu-u.ac.jp

G. Shinozaki
Department of Psychiatry and Behavioral Sciences, Stanford University, Stanford, CA, USA
e-mail: gens@stanford.edu

S. Kasper
Center for Brain Research, Medical University of Vienna, Vienna, Austria
e-mail: siegfried.kasper@meduniwien.ac.at

© Springer Nature Switzerland AG 2024
A. Tasman et al. (eds.), *Tasman's Psychiatry*,
https://doi.org/10.1007/978-3-030-51366-5_128

Abstract

The use of psychotropic drugs in clinical psychiatry began in the late nineteenth century, changing the goal of treatment of mental illness from symptom improvement to social functioning. Psychotropic drugs are still widely used today. Clinicians prescribe the right drug, in the right dose, and in the right manner to the right patient population to lead them to recovery. This goal requires a proper patient-provider relationship, accurate diagnosis, and appropriate drug selection. Understanding the effects and side effects of drug administration requires knowledge of adherence, pharmacokinetics, pharmacodynamics, and pharmacogenomics. Pharmacokinetics is affected by processes such as drug administration, absorption, distribution, metabolism, and exertion, and is colored by environmental and genetic factors such as aging, comorbidities, and concomitant use of other drugs. Knowledge of these factors is necessary. In order to establish an appropriate patient-provider relationship, it is necessary to understand the concept of evidence-based treatment planning and shared decision making (SDM) when making treatment decisions with patients. Finally, after drug administration, evaluation and monitoring of efficacy, adverse drug reaction, and side effects are necessary. This chapter aims to summarize the basic knowledge required for effective pharmacotherapy, which will ultimately contribute to effective and safe pharmacotherapy.

Keywords

Clinical pharmacology · Pharmacological therapy · Drug selection · Shared decision making · Adherence · Pharmacokinetics · Pharmacodynamics · Pharmacogenomics · Adverse drug reaction

Introduction

The use of psychotropic drugs in clinical psychiatry began in the late nineteenth century, with the introduction of morphine and other drugs that shifted treatment from physical restraint to behavioral regulation, and the clinical introduction of chlorpromazine in the 1950s, which made shock and surgical therapy a thing of the past and subsequently shifted the goal of treatment from symptom improvement to restoration of social functioning (Ban, 2001).

Today, psychotropic drugs are widely used. 7.2% of the general population takes at least one psychotropic medication (Beck et al., 2005);

nearly one in five community-dwelling elderly (Aparasu et al., 2003), and 47% of institutionalized elderly (Snowdon et al., 2006) regularly take psychotropic medications. In addition, 87% of psychiatric patients are prescribed some form of psychotropic medication, and 58% are taking two or more psychotropic medications, a rate that is increasing each year (Mojtabai & Olfson, 2010).

The clinician prescribes the right drug, in the right dose, to the right person, in the right way, and leads the patient to recovery. This requires knowledge of pharmacokinetics(PK) and pharmacodynamics(PD), starting with the appropriateness of diagnosis, drug selection, and adherence to treatment.

The factors that alter PK include processes such as drug administration, absorption, distribution, metabolism, and exertion. These factors may include environmental factors such as aging, coexisting diseases, and concomitant use of other drugs, as well as the involvement of genetic background.

Pharmacological actions are explained by the concept of neurotransmitter-receptor-concept, which describes how various drugs act on receptors to modulate neurotransmitters. The behavior of neurotransmitters in the receptor is visualized by positron emission tomography (PET), which enables scientific investigation of PK and clinical effects and side effects of drugs.

It is well known that there are large individual differences in the effects and adverse reactions of psychotropic drugs. This difference can be explained by both PK and PD factors, and genetics is involved in both. Advances in pharmacogenomics in this area are beginning to be implicated in predicting drug response and warning of adverse drug reactions.

In order to change from "drug administration" to "pharmaco-therapy," it is important to have a patient-provider relationship in which the patient accepts the drug. The concept of shared decision making (SDM), which is evidence-based decision making and shared decision making between provider and patient, is necessary to make treatment decisions with patients. Post-medication evaluation and monitoring of efficacy and side effects are necessary. Psychiatric medication may be administered over a long period of time and should be monitored for the possibility of metabolic and other side effects in addition to the direct effects of pharmacological action.

This chapter summarizes the basic knowledge required for effective pharmacotherapy and ultimately aims to help in effective and safe pharmacotherapy (Fig. 1).

Drug Selection

In clinical psychopharmacologic therapy, the goal is to choose the right drug for the right patient, also called precision medicine, and to achieve a good balance between high benefit and low risk.

Evidence-based medicine (EBM) is a means of searching for the best solution to reach this goal. The characteristics of EBM are that it introduces the concept of hierarchy into the evidence for clinical decisions, it proposes methods that allow clinicians to judge the strongest possible evidence for themselves, and it proposes ways to utilize evidence in clinical practice. It also proposes ways to make use of the evidence in clinical practice. Furthermore, it also suggests ways to make use of evidence in clinical practice. In addition, it proposes a method for making clinical decisions about medical care based on four factors: the values of each patient, the individual clinical situation, evidence, and the professional skills of the clinician (expertise) (Guyatt et al., 2014).

In drug selection, decisions must be based not only on scientific rationality, but also on the patient's values and wishes. In order to make an appropriate choice, it is necessary to carefully examine the information. The information needs to include the patient's information such as diagnosis, condition, and wishes, and the drug's information such as its characteristics. Drugs are those whose efficacy and safety have been proven through clinical trials. If a drug is prescribed to a patient who meets the inclusion criteria in a clinical trial, and the dosage and administration are followed, the same level of efficacy as in the trial can certainly be obtained. However, in actual clinical situations, patients and medical conditions

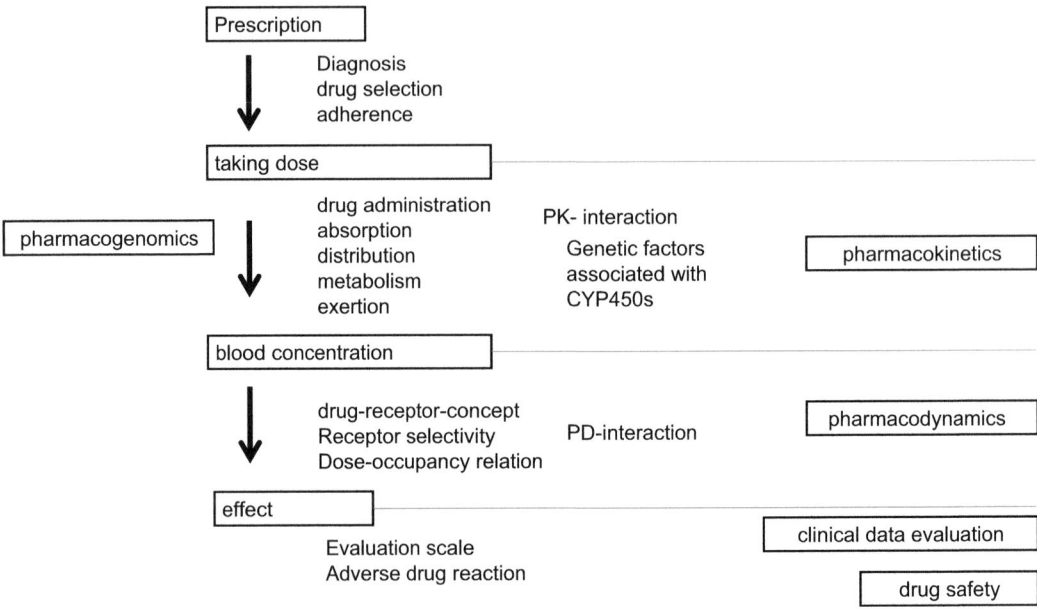

Fig. 1 Factors associated with clinical psychopharmacologic therapy

that cannot be included in clinical trials are treated, so we need to understand the discrepancies between the clinical trials and the drug in everyday life of the patient, with all his comorbidities and personal circumstances.

Pharmacokinetics

Even if a drug is administered as described in the package insert, it may not produce the expected effect or may cause unexpected side effects. The differences in drug efficacy among individuals are thought to be related to the extent to which the drug reaches the target organ, the site of action of the drug, and the blood concentration of the drug in the peripheral blood. For this reason, pharmacokinetics (PK), which describes how the blood concentration of an administered drug changes over time, is important.

Drug Administration

Drug forms and route of administration are important factors in PK. The fastest route to the

brain is generally achieved by inhalation. Our patients who smoke know this from experience. For psychiatric clinicians, there is only one medication that is routinely administered by this route, that is inhaled loxapine for agitation. More often, intravenous (IV) administration is the fastest way to get drugs into the bloodstream and is the route of administration with the most control over drug exposure, but it is not a common route in psychiatric practice. Another route of injectable administration is intramuscular (IM) injection, which includes both short-acting injections used for sedation of agitated states and long-acting injections used in maintenance treatment.

A long-acting injection is one in which, after injection, blood levels rise gradually and are maintained for weeks or months. This has the advantages of ensuring adherence, avoiding first-pass effects, and reducing peak-trough variability. Therefore, it can be a highly effective treatment modality in long-term psychiatric treatment, such as antipsychotic treatment for schizophrenia. In fact, when patients with schizophrenia are introduced to long-acting injectable drugs, readmissions decrease (Kishimoto et al., 2018).

The most common route of administration is oral (PO) administration. The dosage forms of orally administered drugs include tablets, orally disintegrating tablets, aqueous solutions, suspensions, and extended-release formulations. Oral administration is characterized by the active administration of medication by the patient, which makes it easier to respect the patient's free will, but adherence is more likely to vary.

Sustained release formulations take a longer time to reach maximum blood concentration (Tmax) compared to immediate release formulations. Therefore, the number of doses taken per day can be reduced and the range of peak and trough blood concentrations in the steady state after repeated dosing is reduced. Finally, the side effects associated with the peak values and the lack of efficacy associated with the troughs can be reduced.

Transdermal patches have been developed as a new route of administration. These products are not subject to first-pass effects and have stable plasma concentrations, which is an intermediate property between long-acting injectable and oral sustained-release products.

Recently, an intranasal route of administration has been introduced for esketamine as antidepressant augmentation additionally to conventional medications for depression which enables rapid absorption (Kasper et al., 2023). This safe and effective route of administration is not associated with pain for patients and also does not exhibit a first pass effect.

Absorption

Absorption from the gastrointestinal tract into the portal vein is mediated by diffusional transport along a concentration gradient. After the first-pass effect in the liver from the portal vein bloodstream, the drug enters the body circulation and is distributed throughout the body. Oral mucosal absorbents, patches, and injections, which are not subject to the first-pass effect, have high bioavailability, but may be limited by ability to cross barriers like the skin.

Distribution

Administered drugs are distributed to tissues through the body circulation. Factors that affect distribution include the amount of blood return to each tissue, lipid solubility of the drug, permeability of the blood-brain barrier (BBB), binding to plasma proteins, and affinity to P-glycoprotein.

Drugs with high lipid solubility (high lipophilicity) are more widely distributed than those with low solubility; since the BBB is a lipid membrane and tight junctions do not allow chemicals to diffuse between cells, increasing lipid solubility increases permeability through the membrane, which leads to more rapid onset of action. P-glycoprotein is also present in the BBB and influences drug distribution. The P-glycoprotein is intended to protect the body (gut) and brain (BBB) from xenochemicals but its effects are limited to actually having adequate affinity to that chemical(Benet et al.,1999).

Drugs can be divided into those bound to plasma proteins such as plasma albumin and α1-acid glycoprotein in the bloodstream and free drugs that are not bound. Only free drugs are distributed outside the bloodstream to tissues, exert pharmacological effects, and undergo hepatic metabolism or renal filtration (Benet & Hoener, 2002). Because many psychotropic drugs are highly bound to plasma proteins, fluctuations in plasma proteins affect blood levels and the onset of drug effects. Competition for plasma protein binding is a major source of potential drug-drug interactions, and one that is very poorly studied since affinity to binding sites are generally not studied.

Metabolism

Metabolism of drugs in the liver occurs via two phases: phase I and phase II. Phase I metabolism is mainly oxidative reactions mediated by cytochrome P450 (CYP), including dealkylation, hydroxylation, oxidation, deamination, desulfurization, and sulfoxide formation. Phase II metabolism involves conjugation or synthesis reactions that make the molecule water soluble and available for excretion by the kidneys.

CYP is a general term for oxidative enzymes that are important in plant, animal, and human physiology. These enzymes exist predominantly to metabolize xenobiotics, hence their preferred presence in the liver and the design of the portal circulation, where all drainage from the gut must first pass through the liver. The genetic variation in these enzymes is, in part, due to the xenobiotics in local foods of the ancestors of the patient. CYP enzymes that are needed to break down endogenous compounds tend to have much less genetic variation (e.g., CYP 3A4). (Nemeroff et al., 1996).

CYPs include more than 30 related enzymes, of which five human CYPs (CYP-1A2, -2C9, -2C19, -2D6, and -3A4) metabolize more than 90% of the psychiatric drugs in use today. A variety of substances can serve as substrates for reactions by CYPs and promote or inhibit these reactions. Detailed information on these and a drug interaction table are available on the website (Cytochrome P450 Drug Interaction table; http://medicine.iupui.edu/flockhart/table.htm).

Excretion

Drugs are eliminated from the body by renal excretion through the processes of glomerular filtration, active tubular secretion, and passive tubular reabsorption, after being converted to hydrophilic metabolites through metabolism in the liver. To evaluate the excretory capacity, clearance defined as "the amount of blood from which a drug is removed per unit time" is used. From a clinical perspective, the clearance of creatinine (CrCl) is used to estimate the clearance of drugs. Renal dysfunction or age-related decline in renal function results in a decrease in clearance.

Steady State

After a period of repeated drug administration and metabolic excretion, a steady state is reached in which the plasma concentration of the drug is within a certain range of peaks and troughs. Adverse effects occur when the peak concentration is greater than or equal to the blood concentration at which dose-dependent adverse events occur.

The peak and trough are determined not only by the type of drug, but also by the dosage form, dose, and dosing interval. In general, the difference between the peak and trough increases with decreasing dosing frequency, and the difference between the peak and trough decreases for extended-release formulations, patch formulations, and long-acting injectable formulations with longer dosing intervals.

Clinically, if a short-acting drug was causing side effects due to high blood levels when it needed to be administered three times a day to achieve efficacy, switching to a long-acting injection that is administered once every 4 weeks may provide stable blood levels and reduce side effects (Fig. 2).

Pharmacodynamics (PD) describes what a drug does to the body over time and with intensity. Here, the drug-receptor concept, receptor selectivity, and dose-effect relationships are considered.

Drug-Receptor Concept

Most of the current psychotropic drugs exert their effects and side effects by modulating synaptic transmission in the brain. When a drug acts on a receptor at a presynapse or postsynapse to mimic or enhance the function of an endogenous ligand, it is called an agonist. Conversely, when a drug binds to a receptor and inhibits the function of that receptor, it is called an antagonist. Partial agonists are active at the receptor to a lesser degree than the endogenous ligand. Inverse agonists will inactivate the receptor, so that even spontaneous activity is eliminated.

Antagonists or partial agonists of the dopamine D2 receptor as a whole inhibit neurotransmission in the dopaminergic nervous system. These drugs are called antipsychotics and are used to treat schizophrenia and related disorders. Partial agonists may be as effective as antagonists with fewer side effects by modulating the action or antagonism of their target receptors.

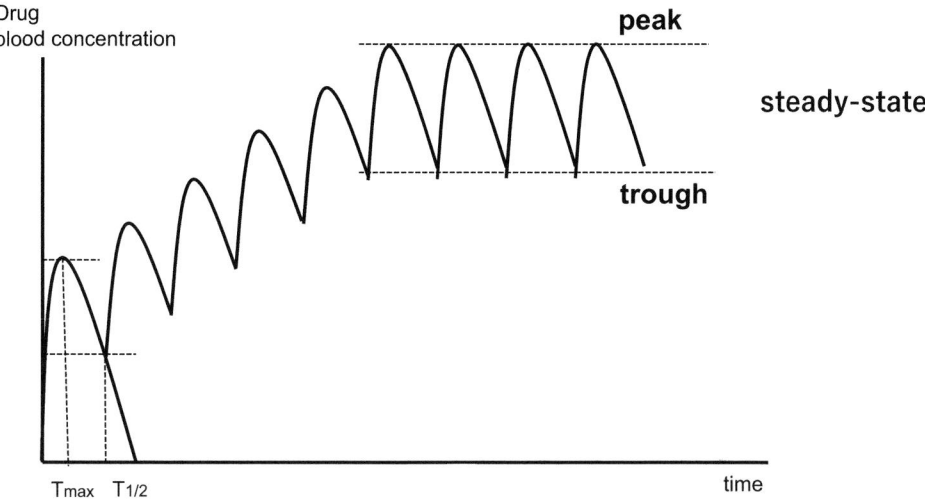

Fig. 2 Pharmacodynamics (PD)

Aripiprazole is a partial agonist of the dopamine system, causing antagonism when it occupies the majority of the available receptors (95% receptor occupancy), or agonism when it binds a smaller number of available receptors (e.g., 50% receptor occupancy). Clinically, dose is a surrogate of receptor occupancy, so that a higher, full antipsychotic dose (say 15 mg) will occupy nearly all the receptors and reduce the dopamine signal thereby acting as an antipsychotic, while a lower dose (say 5 mg) will leave sufficient receptors unoccupied to preserve the endogenous signal while adding to it with the partial activation, thereby acting as an antidepressant. The intrinsic activity of the partial agonist replaces the normal tonic dopamine signal resulting in fewer side effects (Kasper et al., 2003).

Drugs that facilitate the release of neurotransmitters from the presynaptic endoplasmic reticulum are called neurostimulants or monoamine releasers. Monoamine releasers are used in the treatment of narcolepsy and attention-deficit hyperactivity disorder (ADHD).

After synaptic transmission, some neurotransmitters are broken down by degrading enzymes, while others are reuptaken by transporters and reused. Drugs that inhibit the transporter are called reuptake inhibitors. Originally used to treat depression, they are also used to treat anxiety or ADHD.

Some neurotransmitters are degraded by monoamine oxidase. Monoamine oxidase inhibitors inhibit degradation and thus facilitate synaptic transmission. Such drugs are used as medications for depression and in the treatment of Parkinson's and Alzheimer's disease.

Drugs can also modify the function of ion channels (e.g., sodium channels or calcium channels), second messenger proteins (e.g., lithium, valproate), or gene expression (e.g., cortisol) (Fig. 3).

Receptor Selectivity

The affinity and selectivity of a drug for its receptor define the characteristics of the drug. The beneficial effects of a drug derive from its activity at its target receptors, while its side effects partly derive from its activity at other receptors or receptors in other brain regions. Depending on which receptors the drug binds to and whether it acts as an agonist or antagonist, the effects and side effects produced will differ.

Antipsychotics are thought to alleviate the positive symptoms of schizophrenia and produce extrapyramidal symptoms by acting on the

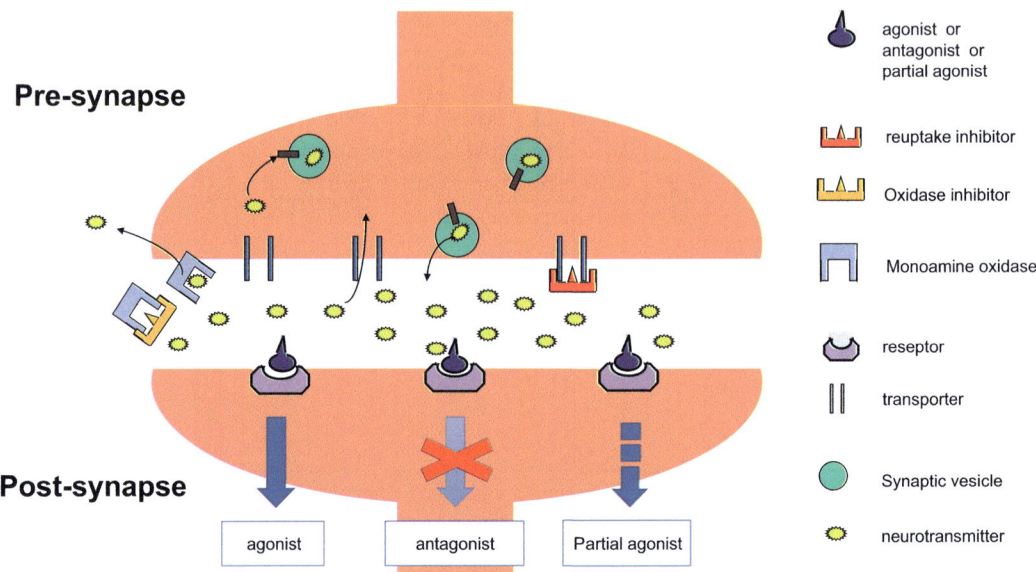

Fig. 3 Schematic illustration of receptor concept

dopamine D2 receptor. When combined with antagonism of serotonin 2A receptors, extrapyramidal symptoms are reduced.

Based on this idea, drugs with antagonism of both dopamine D2 receptor and serotonin 2A receptor have been termed atypical antipsychotics and it is evident that they produce extrapyramidal symptoms to a lower degree than so-called typical antipsychotics like haloperidol. Furthermore, the antagonism of the serotonin2A receptor exhibits a very potent antidepressant and anxiolytic mechanism of action which gives these compounds an additional value.

On the other hand, excessive D2 blockade can lead to parkinsonism, while antagonism of non-target receptors such as histamine 1 (H1) or muscarinic receptors leads to side effects such as sedation, weight gain, constipation, urinary retention, dry mouth, and cognitive impairment. Not all side effects are undesirable, anticholinergic and antihistaminic activity can also reduce parkinsonism that may result from excessive dopamine blockade. In addition, antagonism at alpha1-adrenergic receptors can cause orthostatic hypotension. Drugs with high receptor selectivity are thought to have limited side effects.

Similarly, multidrug combinations act on multiple receptors and can produce a wide variety of side effects.

However, some drugs, such as clozapine, act on such a wide variety of receptors that the target receptors are unknown and are effective in the treatment of treatment-resistant schizophrenia.

Dose or Concentration–Effect Relationships

The development of PK-PD studies has increased our understanding of the relationship between drug concentration and efficacy. Several correlations between blood concentration and efficacy have been found. From this, therapeutic drug monitoring (TDM) was born to optimize blood concentrations, aiming for therapeutic efficacy and avoidance of toxicity (Hiemke et al., 2018). The correlation between blood concentration and efficacy has been discussed for so-called typical antipsychotics such as haloperidol, antiepileptic drugs, and lithium carbonate, and blood concentration is measured. TDM aims to plan the administration of a drug so that its blood concentration is within the range of the minimum effective blood concentration and the blood concentration at which adverse events occur. In other words, TDM can be specifically useful in avoiding side effects and to detect low or ultrarapid metabolizers.

Pharmacogenetic studies, which will be discussed later, have shown that genetic polymorphisms of metabolic enzymes have different metabolic abilities, which in turn change the likelihood of side effects. Pharmacogenetic testing aims to avoid problems with toxicity or inefficacy that are discovered only after the fact when one relies exclusively on TDM.

Dose-Occupancy Relationships

In order to better define the relationship between receptors/transporters and drugs, receptor/transporter occupancy is being measured.

Positron emission tomography (PET) uses radioisotopes to measure the occupancy of specific receptors in the brain by drugs and to estimate the required dose.

Studies on D2 receptor occupancy, therapeutic efficacy, and side effects revealed that doses of 60–95% dopamine D2 receptor occupancy are therapeutically effective, while doses of 80–90% or more cause side effects of extrapyramidal symptoms (Kapur et al., 2000). Consequently, D2 receptor occupancy is a robust biomarker for predicting the therapeutic effects and adverse effects of antipsychotics (de Greef et al., 2011). However, clozapine and quetiapine with their low affinity "loose binding" capacity suggest that other mechanisms play a substantial role in efficacy.

Using this evidence, attempts have been made to design dose-determined clinical trials targeting doses that result in 65–90% receptor occupancy, or to convert clinical doses to equivalent (Lako et al., 2013).

In medications for depression, the medication's dose and its ability to bind to transporters have been studied (Arakawa et al., 2019). In venlafaxine users, dose and noradrenaline transporter occupancy were examined, and occupancy increased in a dose-dependent manner. At low doses, venlafaxine had an effect on the serotonin system comparable to that of SSRIs, and at higher doses, venlafaxine was found to have properties that affect the noradrenaline system as well as the serotonin system. From these results, it is clear that if one expects the drug to act on both the serotonin and noradrenalin systems, it is necessary to administer a higher dose.

It is important to note that when a medication is blocking a neurotransmitter receptor or a transporter, that its efficacy occurs in a threshold fashion. Once the receptor occupancy reaches 60–70% the clinical effect is observed and increasing the dose does not result in any additional benefit through that receptor. However, when a drug acts as an agonist (e.g., opiates or benzodiazepines), the effect occurs in a continuous dose-response fashion, so that modification of the dose will frequently result in modification of receptor function. Interference with other proteins (e.g., ion channels) also seems to produce a continuous dose-response effect.

Drug-Drug Interaction

Pharmacokinetic Drug Interaction

As previously stated, one of the most common sites of unforeseen drug-drug interactions is protein binding. The best example of this is the difficulty in using warfarin. The affinity of warfarin to circulating proteins is so low, many other pharmaceuticals will increase free concentrations, and lead to excessively reduced clotting times. Unfortunately, since routine investigation of affinity to plasma protein sites is not currently performed, this particular type of drug-drug interaction will continue to be unavoidable.

Since many drugs are metabolized by CYPs in the liver, when drugs that induce or inhibit CYPs are used in combination, the metabolism of the drug is affected, resulting in pharmacokinetic interactions. For example, suppose that drugA is metabolized by CYPs and converted to metaboliteA. When drugB, which inhibits CYPs, is used in combination with drugA, the metabolism of drugA is inhibited and its blood concentration is higher than when used alone. When used with drugC, which induces CYPs, the metabolism of drugA is accelerated and the blood concentration of drugA is lower than when used alone.

Various drugs or exogenous substances can be substrates, inducers (increasing enzyme activity), or inhibitors (decreasing enzyme activity) of CYPs. Variations can be expected when two or more drugs are used in combination, but when multiple drugs are used, variations in blood concentrations due to interactions are difficult to predict. Predictability becomes more difficult when one considers that some foods also have similar effects on CYP enzymes (Fig. 4).

The increase in the plasma concentration of the affected drug is immediate, and this second drug reaches a new steady state. For example, if fluvoxamine is given to a patient who is already taking clomipramine, the clomipramine concentration will increase. This occurs because fluvoxamine inhibits the demethylation of clomipramine via CYP2C19. If the plasma concentration of the affected clomipramine is too high, toxicity (side effects) may result.

Induction requires the synthesis of CYP isozymes, and this effect may not be immediate; when CYP450 isozymes are produced, the plasma concentration of the affected drug decreases accordingly. This enzyme induction can occur not only by drugs but also by xenobiotics. An example is the known relationship between olanzapine and clozapine and smoking. Olanzapine and clozapine are primarily metabolized by CYP1A2, but smoking induces CYP1A2, resulting in plasma concentrations that are approximately 50% lower in smokers than in nonsmokers (Haslemo et al., 2006). This type of reaction is made even more complicated since some individuals will have a hyperinducible gene variant of CYP1A2. The clinical picture of this condition will be discussed in the clinical vignette. Similar findings have also been demonstrated for the medication for depression agomelatonin.

Clinical Vignette 1
[Case]

A 28-year-old man with schizophrenia was admitted with a relapse of his psychosis. He was last well approximately 6 weeks ago when he started hearing voices and feeling like people were

(continued)

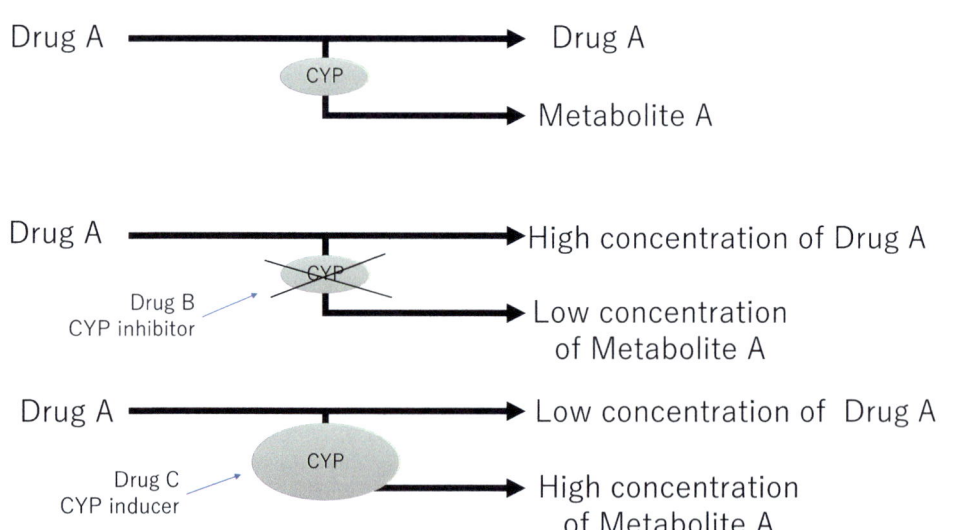

Fig. 4 Concept of drug metabolism by CYP

watching him. This was similar to his first episode of illness. He deteriorated to the point that he was agitated and started walking around with a knife. He became acutely paranoid and pulled the knife on a bystander. The police were called and he was brought to the hospital. He was stabilized in the inpatient unit with a combination of olanzapine 15 mg PO QHS and over the next 2 weeks his auditory hallucinations resolved and his paranoia diminished. He was discharged home at that time with follow-up in clinic. Four weeks later, there was an urgent telephone call from his parents stating that he was once again psychotic. They insisted that he had been taking his medications. Of note is the fact that he had been smoking a lot of cigarettes since his discharge.

[Discussion]

The patient was stabilized in hospital in a nonsmoking unit, although he did smoke a little toward the end of his hospitalization when he went on passes. Upon discharge, he resumed his previous smoking regime, which induced CYP1A2 metabolism. With increased activity of this isoenzyme, there was increased metabolism of olanzapine, which then decreased to subtherapeutic concentrations, resulting in relapse. This is an example of a pharmacokinetic drug interaction often seen in clinical practice.

Pharmacodynamic Drug Interactions

Pharmacodynamic drug interactions occur as a result of the clinical effects of multiple drugs at the same receptor site. For example, the concurrent use of alcohol and benzodiazepines may exacerbate psychomotor disorders due to their superimposed agonist effects on GABA receptors. Similarly, the concomitant use of monoamine oxidase inhibitors (MAOIs) with SSRIs can cause a serotonin syndrome due to their superimposed effects on serotonin receptors. Concomitant use of multiple drugs with anticholinergic effects, which are more frequently used for the elderly, may result in delirium due to cumulative anticholinergic effects(Tune, 2000).

Clinical Vignette 2
[Case]

A 76-year-old woman with dementia is stable in long-term care on a regime of quetiapine 50 mg BID, clonazepam 0.25 mg BID, and donepezil 10 mg QD. She is moved to a different floor in the nursing home and after a few days is noted to be listless, less interested in socializing, and more sedated. She is noticed to be drinking quite a lot of grapefruit juice, which was freely available in this unit. After several days, she is diagnosed with depression and treated with fluoxetine 10 mg PO QD. Two days later, she becomes excessively sedated such that her drive to breathe is decreased and she is difficult to rouse. She is sent to hospital.

[Discussion]

In this case, the addition of grapefruit juice results in inhibition of CYP3A4, which metabolizes both clonazepam and quetiapine. Increasing blood levels of those drugs result in sedation and listlessness, which are diagnosed as depression. The addition of fluoxetine complicates matters as it further inhibits CYP3A4 through its active metabolite norfluoxetine, resulting in a complex pharmacokinetic drug interaction and hospitalization of this patient.

Pharmacogenomics

Pharmacogenomics is the study of the effects of genetic variation on drug metabolism, mechanism of action, and site of action during the course of drug administration and the development of clinical effects and adverse reactions.

This may have been triggered by a twin study of nortriptyline metabolism by Alexanderson (Alexanderson et al., 1969). Since there was little difference in blood concentration between monozygotic twins and a large difference between dizygotic twins, the involvement of genetic factors in drug metabolism was considered. Since then, pharmacokinetic pharmacogenetic studies have been conducted to examine the relationship between genetic polymorphisms of various metabolic enzymes and blood concentrations of drugs, and the importance of genetic polymorphisms of CYPs, which are involved in the metabolism of many psychotropic drugs, has been recognized (Hiemke et al., 2018). The genotypes of CYPs differ among individuals as a function of their ancestorship and selective pressure on different CYP enzymes as a function of toxins in available foods, and these individual differences can lead to differences in drug metabolism and contribute to differences in blood drug concentrations.

Genetic polymorphisms of neurotransmitter receptors are also known to exist. For example, mutations in the dopamine D2 receptor have been shown to be associated with therapeutic effects in ADHD (Myer et al., 2018).

As discussed above, genetic differences can affect all aspects of pharmacokinetic and pharmacodynamic processes. Ultimately, they may be involved in both drug efficacy and toxicity. Therefore, the aim of precision medicine (personalized medicine/tailor-made medicine) is to understand genetic differences in advance and to consider drug selection and dosage. For example, if the presence or absence of genetic mutations is confirmed prior to drug administration, fluctuations in blood concentration can be generally predicted, which may help prevent the occurrence of addiction symptoms that are difficult to predict. Clinical guidelines recommend that prescription doses should be reduced or increased according to metabolic capacity.

Evaluating Effectiveness and Side Effects

After the administration of a drug, it needs to be observed for efficacy and side effects. If the expected efficacy has not been achieved, the diagnosis and treatment plan, including medication, needs to be reconfirmed. Even if the expected efficacy is obtained, monitoring should be done to see if there are any side effects. At that time, several factors should be considered in EBM, including the current evaluation of drug efficacy, the nature of clinical trials, and the interpretation of data.

Diagnosis and Rating Scales

Appropriate case selection is necessary in order to prescribe the right drug at the right dose. In current efficacy trials of psychotropic drugs, subjects are selected by diagnosis based on operational diagnostic criteria, and the change in the symptom rating scale is used as the primary endpoint. Operative diagnostic criteria and symptom rating scales are intended to obtain high inter-examiner agreement when performed by well-trained raters using a set procedure. In order to select cases in clinical situations, it is important to understand the diagnostic criteria and to consider the diagnosis before making drug choices.

In recent years, attempts to establish more objective assessment items in addition to symptom rating scales have been made. For example, the MATRICS Consensus Cognitive Battery (MCCB), developed by the Measurement and Treatment Research to Improve Cognition in Schizophrenia (MATRICS), has been used as a primary endpoint in the development of drugs targeted at improving cognitive dysfunction. It is recommended by the United States Food and Drug Administration (FDA) to be used as the primary endpoint (Nuechterlein et al., 2008) although insurance coverage for such evaluations is usually difficult to obtain.

In addition, evaluation of receptor binding capacity by PET, genome, cytokines, etc. are being investigated as biomarkers of therapeutic efficacy, but they have not been adopted as primary efficacy endpoints.

Inclusion

In clinical trials for the development of novel drugs, only a very limited number of patients are included. However, in real world clinical settings, there are many patients who would not be included in the clinical trials. It is therefore necessary to examine in detail whether the results of the trials can be extrapolated to the patients in front of us. For example, the elderly are often excluded from clinical trials, and it may be inappropriate to extrapolate adult data to this population (Lotrich & Pollock, 2005).

Placebo-Controlled Trials

In efficacy trials of psychotropic drugs, placebo controls are required as a means of assessing true treatment effects. However, placebo treatments involve many unmeasured psychosocial factors that can produce both efficacy and adverse events. Consequently, this may reduce the sensitivity and power of trials to detect significant effects of drugs and confound the results. Placebo response rates in clinical trials for depression are generally around 30%, and the view that placebo response rates are increasing (Walsh et al., 2002) has been concluded to be a problem of trial design (Furukawa et al., 2018).

Meta-Analysis

Meta-analysis is a method of analyzing results by integrating the results of multiple clinical studies and increasing the number of subjects, thereby increasing statistical power. Meta-analyses of randomized controlled trials (RCTs) are sometimes falsely considered to be the highest quality evidence in evidence-based medicine (EBM) (Hasan et al., 2019), however cannot replace RCTs with direct group-to-group comparisons. For instance there might be 3 trials with no statistical group differences, but if the 3 studies are grouped together, suddenly a positive effect might emerge, due to increase of sample size. On the other hand, meta-analyses may also compensate for problems in individual RCT design.

While placebo trials have the potential to confound results as described above, network meta-analysis, which integrates and compares placebo trials, can compare and characterize drugs that have not been directly compared, again raising further limitations. A further problematic issue is that meta-analysis integrates the results of clinical trials with different inclusion criteria.

Observational Study Using National Database Analysis

Large observational studies have been conducted to obtain data on populations that are closer to actual clinical practice rather than populations that are too special to be included in RCTs. Systematic Treatment Enhancement Program for Bipolar Disorder (STEP-BD) (Sachs et al., 2003), Clinical Trials of Intervention Effectiveness of medications for psychosis (CATIE) (Lieberman et al., 2005), and Sequenced Treatment Alternatives to Relieve Depression (STAR*D) (Rush et al., 2006) are well known.

While these studies have the advantage of being more clinically relevant, they are more ambiguous in terms of population characteristics, and while they can examine more robust outcomes such as rehospitalization, they are less reliable in terms of symptom change outcome data as assessed by rating scales due to the failure of randomization, placebo control as well as blind rating .

Medical database analysis is an attempt to verify the effects of pharmacotherapy from observational studies in actual clinical practice. This is a research method that can be established by referring to the medical database of the entire population, mainly in Northern Europe. A study examining the relationship between medications for psychosis and mortality (Tiihonen et al., 2009) showed that the mortality rate was lower when medications for psychosis were taken continuously. This result dispelled the fears of patients and their families that the side effects of medications for psychosis might worsen their prognosis.

Clinical Practice Guidelines

EBM is the selection of optimal treatment based on evidence, but it is difficult for clinicians to be familiar with all the evidence. According to the IOM (Institute of Medicine, now the National Academy of Medicine) report, clinical practice guidelines are defined as "statements that include recommendations, intended to optimize patient care, that are informed by a systematic review of evidence and an assessment of the benefits and harms of alternative care options" (Medicine, 2011).

A variety of organizations have created guidelines, like the World Federation of Societies of Biological Psychiatry (WFSBP; Hasan et al., 2019) or the Collegium Internationale Neuropychopbarmacolorum (CINP, Fountoulakis et al.2017). When using the guidelines, it is necessary to confirm the entity that created the guidelines and the target audience. This is because healthcare systems differ from country to country and region to region. Nonetheless, it is important to understand that guidelines are not absolute, and clinicians may deviate from guidelines to address unique aspects of their patients.

Adherence

Compliance/adherence is an important factor when a medication is prescribed but does not have the full expected effect. Patients, whether intentionally or inadvertently, often forget to take their medications or take them inappropriately, leading to non-adherence. Delusions due to psychotic disorders, cognitive dysfunction, denial of illness, and refusal to accept treatment due to stigma can also cause noncompliance. Even when adherence is an issue, healthcare providers may attribute lack of treatment response to treatment resistance. It is important to always consider and confirm adherence when reassessing the efficacy and safety of medications. Caregivers can be asked to provide information, and checking the amount of remaining medication may also be helpful.

A study estimating adherence to prescription refills in patients with schizophrenia found that readmission rates correlated with increased medication gaps, and that both nonadherence and partial adherence were significantly associated with the clinical event of readmission (Weiden et al., 2004).

In order to improve adherence, it is desirable that patients themselves independently choose their own treatment. Shared decision making (SDM) has been proposed as a concept to support therapeutic decision making, and is gradually gaining popularity. The SDM approach is expected to be more widely adopted in the selection of drug therapy (Stovell et al., 2016) (Beitinger et al., 2014).

Drug Safety

Adverse Drug Reactions (ADR)

Adverse drug reactions (ADRs) are harmful and unintended reactions caused by drugs used in humans. Side effect refers to an effect other than the primary purpose of the drug, but is generally undesirable and is used as a synonym for ADR. Side effects are generally undesirable and are used synonymously with ADRs.

ADRs are classified according to their mechanism of occurrence as shown in Table 1.

Table 1 Adverse drug reactions

1. ADR due to drug mechanisms or inappropriate doses
2. ADR due to idiosyncrasies
3. ADR due to allergy

Dose-related ADRs are those that occur when the effect increases as a function of increasing dose. When ADRs occur at typical therapeutic doses clinicians should consider impaired hepatic or renal function, or genetically reduced or deficient activity of drug-metabolizing enzymes. One example of this is when a person who is genetically deficient in alcohol-degrading enzymes falls into intoxication because alcohol is not metabolized after ingestion. ADRs due to idiosyncrasies are unpredictable and may occur regardless of the dose. Immune-mediated adverse reactions have been shown to correspond to about four different forms of allergy. The occurrence of severe drug rash and anaphylactic shock are included here.

While the majority of ADRs are considered minor, some result in clinically important outcomes such as hospitalization, changes in medication regimes, morbidity, and mortality. In terms of frequency, European studies have shown that 3.5–6.5% of hospitalized patients are admitted because of ADRs, and nearly 15% of hospitalized patients will experience one or more ADRs (Davies et al., 2009) (Bouvy et al., 2015).

A 2015 National Institute for Health and Care Excellence (NICE) optimization statement (https://www.nice.org.uk/guidance/ng5/evidence/full- guideline-6,775,454) points out that 72% of ADRs are potentially avoidable and that the estimated cost of these ADRs in the UK alone is over £500 million, concluding that ADRs are creating an inherently avoidable burden on the health system and need to be right-sizing.

One cause of ADRs is medication errors; a review of 5366 medication errors from 1993 to 1998 reported that 469 were fatal. The most common types of errors resulting in patient death were administering the wrong dose (40.9%), giving the wrong drug (16%), and using the wrong route of administration (9.5%)(Phillips et al., 2001).

In 2000, the National Academy of Medicine (NAM) published a report on medical errors, "To Err Is Human: Building a Safer Health System," which outlined a table of specific methods to reduce medical errors (Institute of Medicine Committee on Quality of Health Care in, 2000).

It is known that the risk of ADRs increases with the number of drugs prescribed (Rodrigues & Oliveira, 2016) and that multidrug use is associated with an increased risk of ADR-related hospitalizations (Marcum et al., 2012).

Especially in the elderly, the risk of developing ADRs is likely to be higher, and the risk is likely to be higher than the benefit gained, which is called Potentially Inappropriate Medication (PIM) and calls for caution in its use ("American Geriatrics Society, 2019 Updated AGS Beers Criteria® for Potentially Inappropriate Medication Use in Older Adults," 2019) (Guillot et al., 2020).

The use of multiple medications increases the risk of ADRs in terms of both PK and PD, and causes many problems such as decreased adherence due to increased medication complexity, medication errors, difficulty in confirming drug efficacy, and difficulty in identifying the causative agent when ADRs occur. The simplest and most effective way for clinicians to ensure drug safety is to prescribe a single drug (Table 2).

Table 2 Selected strategies to improve medication safety

Adopt a system-oriented approach to medication error reduction.
Implement standard processes for medication doses, dose timing, and dose scales in a given patient care unit.
Standardize prescription writing and prescribing rules.
Limit the number of different kinds of common equipment.
Implement physician order entry.
Use pharmaceutical software.
Implement unit dosing.
Have the central pharmacy supply high-risk intravenous medications.
Use special procedures and written protocols for the use of high-risk medications.
Do not store concentrated solutions of hazardous medications on patient care units.
Ensure the availability of pharmaceutical decision support.
Include a pharmacist during rounds of patient care units.
Make relevant patient information available at the point of patient care.
Improve patients' knowledge about their treatment.

Table 3 Points which should be considered when administering medication

When deciding whether or not to use medication, consider the potential benefits and harms.
When making a decision, present options and aim for consensus through discussion.
Avoid multidrug therapy whenever possible.
Avoid concomitant use of multiple medications if possible, and pay attention to interactions when concomitant use is unavoidable.
Use the lowest effective dose.
Pay attention to side effects and monitor.

Summary

General principles for psychopharmacotherapy are discussed. The goal of psychopharmacotherapy is to administer the most appropriate drug in the most appropriate dose to achieve safe and highly effective results and to bring the patient to a better state. As discussed in this chapter, good psychopharmacotherapy requires a wide range of knowledge that is not limited to pharmacodynamics. For detailed discussion of each topic, please refer to the separate chapters. Throughout, Table 3 notes the points to be considered in daily clinical practice.

References

Alexanderson, B., Evans, D. A., & Sjöqvist, F. (1969). Steady-state plasma levels of nortriptyline in twins: Influence of genetic factors and drug therapy. *British Medical Journal, 4*(5686), 764–768. https://doi.org/10.1136/bmj.4.5686.764

American Geriatrics Society. (2019). Updated AGS beers criteria® for potentially inappropriate medication use in older adults. (2019). *Journal of the American Geriatrics Society, 67*(4), 674–694. https://doi.org/10.1111/jgs.15767

Aparasu, R. R., Mort, J. R., & Brandt, H. (2003). Psychotropic prescription use by community-dwelling elderly in the United States. *Journal of the American Geriatrics Society, 51*(5), 671–677. https://doi.org/10.1034/j.1600-0579.2003.00212.x

Arakawa, R., Stenkrona, P., Takano, A., Svensson, J., Andersson, M., Nag, S., ... Lundberg, J. (2019). Venlafaxine ER blocks the norepinephrine transporter in the brain of patients with major depressive disorder: A PET study using [18F]FMeNER-D2. The

International Journal of Neuropsychopharmacology, 22(4), 278–285. https://doi.org/10.1093/ijnp/pyz003.

Ban, T. A. (2001). Pharmacotherapy of mental illness – A historical analysis. *Progress in Neuro-Psychopharmacology & Biological Psychiatry, 25*(4), 709–727. https://doi.org/10.1016/s0278-5846(01)00160-9

Beck, C. A., Williams, J. V., Wang, J. L., Kassam, A., El-Guebaly, N., Currie, S. R., ... Patten, S. B. (2005). Psychotropic medication use in Canada. Canadian Journal of Psychiatry, 50(10), 605–613. https://doi.org/10.1177/070674370505001006.

Beitinger, R., Kissling, W., & Hamann, J. (2014). Trends and perspectives of shared decision-making in schizophrenia and related disorders. *Current Opinion in Psychiatry, 27*(3), 222–229. https://doi.org/10.1097/yco.0000000000000057

Benet, L. Z., & Hoener, B. A. (2002). Changes in plasma protein binding have little clinical relevance. Clin Pharmacol Ther, 71(3), 115–21. https://doi.org/10.1067/mcp.2002.121829. PMID: 11907485.

Bouvy, J. C., De Bruin, M. L., & Koopmanschap, M. A. (2015). Epidemiology of adverse drug reactions in Europe: A review of recent observational studies. *Drug Safety, 38*(5), 437–453. https://doi.org/10.1007/s40264-015-0281-0

Davies, E. C., Green, C. F., Taylor, S., Williamson, P. R., Mottram, D. R., & Pirmohamed, M. (2009). Adverse drug reactions in hospital in-patients: A prospective analysis of 3695 patient-episodes. *PLoS One, 4*(2), e4439. https://doi.org/10.1371/journal.pone.0004439

de Greef, R., Maloney, A., Olsson-Gisleskog, P., Schoemaker, J., & Panagides, J. (2011). Dopamine D2 occupancy as a biomarker for antipsychotics: Quantifying the relationship with efficacy and extrapyramidal symptoms. *The AAPS Journal, 13*(1), 121–130. https://doi.org/10.1208/s12248-010-9247-4

Furukawa, T. A., Cipriani, A., Leucht, S., Atkinson, L. Z., Ogawa, Y., Takeshima, N., . . . Salanti, G. (2018). Is placebo response in antidepressant trials rising or not? A reanalysis of datasets to conclude this long-lasting controversy. Evidence-Based Mental Health, 21(1), 1–3. https://doi.org/10.1136/eb-2017-102827.

Guillot, J., Maumus-Robert, S., Marceron, A., Noize, P., Pariente, A., & Bezin, J. (2020). The burden of potentially inappropriate medications in chronic polypharmacy. *Journal of Clinical Medicine, 9*(11). https://doi.org/10.3390/jcm9113728

Guyatt, G., Rennie, D., Meade, M. O., & Cook, D. J. (2014). *Users' guides to the medical literature: A manual for evidence-based clinical practice* (3rd ed.). McGraw-Hill Education.

Hasan, A., Bandelow, B., Yatham, L. N., Berk, M., Falkai, P., Möller, H. J., & Kasper, S. (2019). WFSBP guideline task force chairs. WFSBP guidelines on how to grade treatment evidence for clinical guideline development. World J Biol Psychiatry, 20(1), 2–16. https://doi.org/10.1080/15622975.2018.1557346. Epub 2019 Feb 4. PMID: 30526182.

Haslemo, T., Eikeseth, P. H., Tanum, L., Molden, E., & Refsum, H. (2006). The effect of variable cigarette consumption on the interaction with clozapine and olanzapine. *European Journal of Clinical Pharmacology, 62*(12), 1049–1053. https://doi.org/10.1007/s00228-006-0209-9

Hiemke, C., Bergemann, N., Clement, H. W., Conca, A., Deckert. J., Domschke, K., et al. (2018). Consensus guidelines for therapeutic drug monitoring in neuropsychopharmacology: Update 2017. *Pharmacopsychiatry, 51*(1–02), 9–62. https://doi.org/10.1055/s-0043-116492. Epub 2017 Sep 14. Erratum in: Pharmacopsychiatry. 2018 Jan;51(1–02):e1. PMID: 28910830.

Institute of Medicine Committee on Quality of Health Care in, A. (2000). In L. T. Kohn, J. M. Corrigan, & M. S. Donaldson (Eds.), *To err is human: Building a safer health system*. National Academies Press (US). Copyright 2000 by the National Academy of Sciences. All rights reserved.

Kapur, S., Zipursky, R., Jones, C., Remington, G., & Houle, S. (2000). Relationship between dopamine D (2) occupancy, clinical response, and side effects: A double-blind PET study of first-episode schizophrenia. *The American Journal of Psychiatry, 157*, 514–520.

Kasper, S., Lerman, M. N., McQuade, R. D., Saha, A., Carson, W. H., Ali, M., et al. (2003). Efficacy and safety of aripiprazole vs. haloperidol for long-term maintenance treatment following acute relapse of schizophrenia. *The International Journal of Neuropsychopharmacology, 6*(4), 325–337. https://doi.org/10.1017/s1461145703003651

Kishimoto, T., Hagi, K., Nitta, M., Leucht, S., Olfson, M., Kane, J. M., & Correll, C. U. (2018). Effectiveness of long-acting injectable vs Oral antipsychotics in patients with schizophrenia: A meta-analysis of prospective and retrospective cohort studies. *Schizophrenia Bulletin, 44*(3), 603–619. https://doi.org/10.1093/schbul/sbx090

Lako, I. M., van den Heuvel, E. R., Knegtering, H., Bruggeman, R., & Taxis, K. (2013). Estimating dopamine D2 receptor occupancy for doses of 8 antipsychotics: A meta-analysis. *Journal of Clinical Psychopharmacology, 33*(5), 675–681. https://doi.org/10.1097/JCP.0b013e3182983ffa

Lieberman, J. A., Stroup, T. S., McEvoy, J. P., Swartz, M. S., Rosenheck, R. A., Perkins, D. O., et al. (2005). Effectiveness of antipsychotic drugs in patients with chronic schizophrenia. *The New England Journal of Medicine, 353*(12), 1209–1223. https://doi.org/10.1056/NEJMoa051688

Lotrich, F. E., & Pollock, B. G. (2005). Aging and clinical pharmacology: Implications for antidepressants. *Journal of Clinical Pharmacology, 45*(10), 1106–1122. https://doi.org/10.1177/0091270005280297

Marcum, Z. A., Amuan, M. E., Hanlon, J. T., Aspinall, S. L., Handler, S. M., Ruby, C. M., & Pugh, M. J. (2012). Prevalence of unplanned hospitalizations caused by adverse drug reactions in older veterans. *Journal of the American Geriatrics Society, 60*(1), 34–41. https://doi.org/10.1111/j.1532-5415.2011.03772.x

Medicine, I. O. (2011). *Clinical practice guidelines we can trust*. The National Academies Press.

Mojtabai, R., & Olfson, M. (2010). National trends in psychotropic medication polypharmacy in office-based psychiatry. *Archives of General Psychiatry, 67*(1), 26–36. https://doi.org/10.1001/archgenpsychiatry.2009.175

Myer, N. M., Boland, J. R., & Faraone, S. V. (2018). Pharmacogenetics predictors of methylphenidate efficacy in childhood ADHD. *Molecular Psychiatry, 23*(9), 1929–1936. https://doi.org/10.1038/mp.2017.234

Nemeroff, C. B., DeVane, C. L., & Pollock, B. G. (1996). Newer antidepressants and the cytochrome P450 system. *The American Journal of Psychiatry, 153*(3), 311–320. https://doi.org/10.1176/ajp.153.3.311

Nuechterlein, K. H., Green, M. F., Kern, R. S., Baade, L. E., Barch, D. M., Cohen, J. D., et al. (2008). The MATRICS consensus cognitive battery, part 1: Test selection, reliability, and validity. *The American Journal of Psychiatry, 165*(2), 203–213. https://doi.org/10.1176/appi.ajp.2007.07010042

Phillips, J., Beam, S., Brinker, A., Holquist, C., Honig, P., Lee, L. Y., & Pamer, C. (2001). Retrospective analysis of mortalities associated with medication errors. *American Journal of Health-System Pharmacy, 58*(19), 1835–1841. https://doi.org/10.1093/ajhp/58.19.1835

Rodrigues, M. C., & Oliveira, C. (2016). Drug-drug interactions and adverse drug reactions in polypharmacy among older adults: An integrative review. *Revista Latino-Americana de Enfermagem, 24*, e2800. https://doi.org/10.1590/1518-8345.1316.2800

Rush, A. J., Trivedi, M. H., Wisniewski, S. R., Nierenberg, A. A., Stewart, J. W., Warden, D., . . . Fava, M. (2006). Acute and longer-term outcomes in depressed outpatients requiring one or several treatment steps: A STAR*D report. The American Journal of Psychiatry, 163(11), 1905–1917. https://doi.org/10.1176/appi.ajp.163.11.1905.

Sachs, G. S., Thase, M. E., Otto, M. W., Bauer, M., Miklowitz, D., Wisniewski, S. R., . . . Rosenbaum, J. F. (2003). Rationale, design, and methods of the systematic treatment enhancement program for bipolar disorder (STEP-BD). Biological Psychiatry, 53(11), 1028–1042. https://doi.org/10.1016/s0006-3223(03)00165-3.

Snowdon, J., Day, S., & Baker, W. (2006). Current use of psychotropic medication in nursing homes. *International Psychogeriatrics, 18*(2), 241–250. https://doi.org/10.1017/s1041610205002449

Stovell, D., Morrison, A. P., Panayiotou, M., & Hutton, P. (2016). Shared treatment decision-making and empowerment-related outcomes in psychosis: Systematic review and meta-analysis. *The British Journal of Psychiatry, 209*(1), 23–28. https://doi.org/10.1192/bjp.bp.114.158931

Tiihonen, J., Lonnqvist, J., Wahlbeck, K., Klaukka, T., Niskanen, L., Tanskanen, A., & Haukka, J. (2009). 11-year follow-up of mortality in patients with schizophrenia: A population-based cohort study (FIN11 study). *Lancet, 374*, 620–627.

Tune, L. E. (2000). Serum anticholinergic activity levels and delirium in the elderly. *Seminars in Clinical Neuropsychiatry, 5*(2), 149–153. https://doi.org/10.153/SCNP00500149.

Walsh, B. T., Seidman, S. N., Sysko, R., & Gould, M. (2002). Placebo response in studies of major depression: Variable, substantial, and growing. *JAMA, 287*(14), 1840–1847. https://doi.org/10.1001/jama.287.14.1840

Weiden, P. J., Kozma, C., Grogg, A., & Locklear, J. (2004). Partial compliance and risk of rehospitalization among California Medicaid patients with schizophrenia. *Psychiatric Services, 55*(8), 886–891. https://doi.org/10.1176/appi.ps.55.8.886

Neuroscience-Based Nomenclature (NbN): A New Pharmacological Driven Classification of Psychotropics

Daniel Minkin Levy, Sasson Zemach, Guy M. Goodwin, Michele T. Pato, and Joseph Zohar

Contents

D. M. Levy · S. Zemach
The Jerusalem Mental Health Center, Jerusalem, Israel
e-mail: daniel@minkinlevy.com;
sasson.zemach@mail.huji.ac.il

G. M. Goodwin
Department of Psychiatry, University of Oxford,
Warneford Hospital, Oxford, UK

M. T. Pato
Department of Psychiatry, Rutgers Biomedical and Health
Sciences, Robert Wood Johnson Medical School,
Piscataway, NJ, USA

J. Zohar (✉)
Department of Psychiatry, Sheba Medical Center,
Tel-Hashomer, and Sackler School of Medicine, Tel Aviv
University, Tel Aviv-Yafo, Israel
e-mail: jozohar@gmail.com

© Springer Nature Switzerland AG 2024
A. Tasman et al. (eds.), *Tasman's Psychiatry*,
https://doi.org/10.1007/978-3-030-51366-5_129

Abstract

The contemporary application of psychotropics is based on the growing body of knowledge of their neuroscientific properties. The terminology used to describe these agents is based on the conditions they were first discovered to be useful for. This terminology, which was patched together unofficially over the years, created classes of drugs, e.g., "antipsychotics," "antidepressants," and "mood stabilizers," suggesting a single indication for each group of drugs. The "indication concept" misrepresents each drug's wide range of applications and often confuses students, doctors, and patients. For example, clinicians often prescribe "antidepressants" for patients with anxiety disorders or "second-generation anti-psychotics" for depressed patients who show no evidence of psychosis.

Neuroscience-based nomenclature (NbN) is a pharmacologically driven nomenclature that focuses on each drug's pharmacology and mode of action. NbN aims to reform the current terminology by bridging contemporary neuro-science to psychotropic classification and clinical practice. By creating practical tools (i.e., the NbN app), NbN increases the "tool-box" and may increase prescription precision as it brings about more professional pharmaco-logic driven concepts and improves communication with patients (i.e., no more: "Why I'm getting antidepressants for my anxiety?").

Keywords

Nomenclature · Terminology · Classification · Psychopharmacology · Psychotropics

Introduction

As in many scientific breakthroughs, the discovery of the first psychiatric medications was by serendipity. Such is the case of chlorpromazine, which was first synthesized in 1951 as an aid for initiating general anesthesia (Charpentier et al., 1952). Its efficacy in treating psychosis was revealed when it was used experimentally to treat agitation in a manic patient who exhibited complete remission after a few weeks of treatment (Laborit et al., 1952). This serendipitous discovery could not have happened intentionally since, at the time, the knowledge of the molecular processes occurring in psychosis was scant. Dopamine itself was identified as a neurotransmitter only 7 years later in 1958 by Arvid Carlsson, and the link between the D2 receptor binding and clinical antipsychotic efficacy was described two decades later (Creese et al., 1975; Snyder, 1976).

The discovery of the first monoamine oxidase inhibitor was also related to serendipity (López-Muñoz et al., 2007). In 1952, clinicians noted that patients given iproniazid, an isoniazid analogue,

displayed euphoria and overactive behavior during a tuberculosis trial (Krieser et al., 1953). These observation-initiated studies on the efficacy of isoniazid derivates in psychiatric disorders helped establish the connection between monoamines and depression. In 1956, aiming to discover another substance with an antipsychotic action, Roland Kuhn studied imipramine, only to find it lacked efficacy in treating psychosis. Disappointed, Kuhn attempted to treat depressed patients with imipramine and found its effectiveness for depression (Kuhn, 1957). These newly found substances were divided into two groups: the "antipsychotics" and the "antidepressants." At the time, this division made sense. The groups reflected the indications for which the drugs were being used – and not much was known on *how* they worked.

The Current Terminology of Psychotropics

Currently, disease-based terminology is in use by several official classification systems. Internationally, the Anatomical Therapeutic Chemical (ATC), first published by the World Health Organization in 1976 as a monitoring tool, is the most prominent classification system (WHO Collaborating Centre for Drug Statistics Methodology, 2021). The ATC was built as a strict hierarchy and consisted of five levels, each level defined by a different property: by anatomical site, therapeutic effect, indication, and chemical structure (Table 1). The ATC uses terms as "psycholeptics" and "psychoanaleptics," which are no longer in clinical use, to describe psychiatric medications, alongside other CNS medication (e.g., "anesthetics," "analgesics," and "anti-Parkinson drugs").

Although the ATC was not meant for practice, it affects clinical terminology to this day as it has been widely adopted in some form by different organizations. The Physicians' Desk Reference (PDR) attributes substances to classes primarily by indication but occasionally by mentioning its mechanism of action or even by date of invention; while haloperidol is classified solely as a "first-

Table 1 The classification of haloperidol and fluvoxamine in the anatomical therapeutic chemical (ATC) system

	Haloperidol	Fluvoxamine
1st level: anatomical site of action	Nervous system (N)	Nervous system (N)
2nd level: general therapeutic effect	Psycholeptics (N05)	Psychoanaleptics (N06)
3rd level: indication	Antipsychotics (N05A)	Antidepressants (N05A)
4th level: chemical structure/ mechanism	Butyrophenone derivatives (N05AD)	Selective serotonin reuptake inhibitors (N05AB)
5th level: agent	Haloperidol (N05AD01)	Fluvoxamine (N06AB08)

generation antipsychotic" (FGA), chlorpromazine is classed both as an FGA and under "phenothiazine antiemetics." Similarly, risperidone is classified under "serotonin-dopamine antagonist (SDA) antipsychotics." The Food and Drug Administration (FDA) approach on the classification of psychotropics is inconsistent as well, sometimes referring to a drug by indication and at other times to its chemical structure. The use of some version of disease-based terminology by the ATC, PDR, FDA, and other organizations and publications such as Medscape contributed to turning it into a field standard.

For close to seven decades after the discoveries that defined psychiatry, the field of neuroscience has seen an impressive expansion of knowledge. In the light of the introduction of newer agents and the accumulation of evidence on their efficacy, it seems disease-based definitions misrepresent the field's contemporary knowledge – and the diverse range of applications psychotropics have in clinical practice.

As a result of the gap between clinical practice and terminology, medications termed as "antidepressants" are often prescribed to patients who are not depressed – and medications termed "antipsychotics" are given to patients not suffering from psychosis. Other than the terminological inaccuracy, there are possible practical implications of calling a drug by an indication. Patients are often reluctant to take medications labelled for a disorder they do not have, which may facilitate poor treatment adherence – especially in a social climate where the stigma on mental health and psychiatry is present. From the clinician's point of view, grouping these drugs by indication has little use when prescribing. Whether they are called "antidepressants," "antipsychotics," or "mood stabilizers," this vocabulary does not represent

the way these agents differ from each other, either by neuroscientific attributes or by clinical application.

What Defines a Drug as an "Antidepressant"?

The treatment of depression relies on several factors, including the symptomatology exhibited by a patient and the neuroscientific attributes of each medication. This approach was achieved by evolution and is secondary to the growing understanding of the biology of depression – from the early, revolutionary yet unrefined, monoamine hypothesis of depression (Schildkraut, 1965) to multifactorial approaches considering complicated neuronal interactions.

The term "antidepressant" was coined to describe medications that raise the concentration of different monoamines. Slowly, the concept that only these agents are beneficial for the depressed patient was being put into doubt, as some of the drugs for psychosis and some of the drugs for bipolar disorder have proven to possess independent antidepressive properties.

For instance, aripiprazole, a partial dopamine D2 agonist and a partial serotonin 5-HT$_{1A}$ receptor agonist, has been approved by the FDA as adjunctive therapy for major depressive disorder. This is one example of an "antipsychotic" medication (as per FDA/PDR language) used in patients not suffering from psychosis but from depression, based on its unique dopaminergic and serotonergic effects while administered in small dose.

Quetiapine, first approved for the treatment of schizophrenia in 1996, can also be used in the treatment of depression and is considered a first-line treatment for bipolar depression in part for its activity as a serotonin 5-HT$_{1A}$ receptor partial agonist (Lieberman et al., 2005). Other examples of drugs that are widely used either as monotherapy or as augmenting agents for depression despite not being considered "antidepressants" (according to current FDA/PDR nomenclature) are olanzapine, lithium, and lamotrigine (Fig. 1).

Indeed, some medications have an antidepressive activity while not being considered "antidepressants." Other medications termed "antidepressants" are often prescribed for conditions different from depression (e.g., trazodone and doxepin in low doses are prescribed for insomnia). Moreover, not all "antidepressants" are the same. Some of them are effective in one disorder but not in another one. In the treatment of the obsessive-compulsive disorder (OCD), noradrenergic medications, which are very effective in

treating depression, are entirely ineffective. Only part of the "antidepressants" effectively treats OCD, namely, the serotonin reuptake inhibitors. Accordingly, OCD is not treated with "antidepressant" but with serotonin reuptake inhibitors (SRI) (Fineberg et al., 2020).

In addition to being used as "antidepressants" (e.g., the first-line therapy for depression), SRIs are also indicated for treatment of anxiety (e.g., panic disorder, social anxiety disorder) and stress-related disorders (e.g., adjustment disorder and post-traumatic stress disorder). Similarly, the class of serotonin-norepinephrine reuptake inhibitors (SNRI) (e.g., duloxetine, venlafaxine) is also named "antidepressants." However, it also has an indication for chronic pain (Welsch et al., 2018), and the norepinephrine-dopamine reuptake inhibitor (NDRI) bupropion is indicated both for depression and as an aid to smoking cessation (Wilkes, 2008).

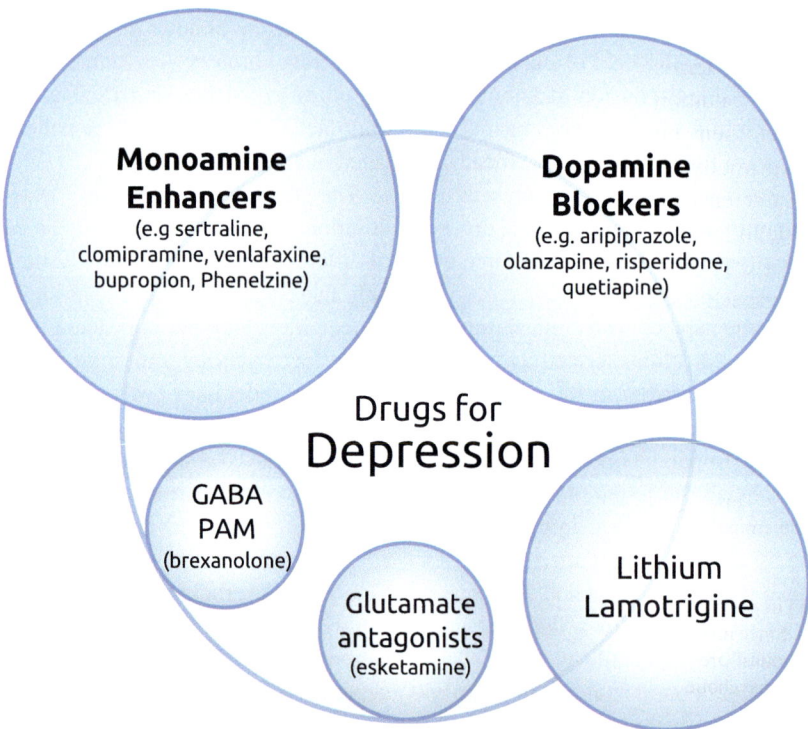

Fig. 1 Drugs for depression. According to the current nomenclature, only monoamines enhancers would be under the category of "antidepressants" while (**a**) many of them are used in other disorders (e.g., anxiety, OCD, PTSD) and (**b**) many other drugs that are not called "antidepressants" are actually effective and used for the treatment of depression

As the neuroscience knowledge increases, it is realized that depression is a heterogeneous disorder, often needing individualized treatment. Since many patients do not respond to first-line treatment for depression (Sinyor et al., 2010; Souery et al., 2006), there is often a need to select a specific treatment for their specific depression (considering the symptoms, stage of the disorder, genetics, medical history) with a single or a combination of medications. At this point in therapy, overinformed or misinformed patients are likely deterred by the fact they are being prescribed an "antipsychotic" or "mood stabilizer," wondering if the treatment is appropriate since they are not psychotic. Another example is patients suffering from concurrent depression and psychosis, which may be best treated with an "antipsychotic," despite not being diagnosed with schizophrenia.

As a result, during the initiation and management of psychiatric disorder, there is often a need for clarification, explaining to a confused patient that it is indeed the appropriate treatment despite the drug "class." This confusion may add further distress to an already tense situation, as the patient may feel their condition is worse than they thought or that their problem was not well-understood, potentially harming the trust needed for treatment adherence and success.

Another side to the problem is that the current terms may "narrow" patients', physicians', and researchers' perspectives. For example, a patient who is not responding to an "antidepressant" may be reluctant to stop a particular medication because it is an "antidepressant." Similarly, a researcher may be unable to make a conceptual leap that an "antidepressant" or an "antipsychotic" may be pro-depressant or pro-psychosis, ideas that have been proposed in the literature. This phenomenon may have delayed physicians' ability to understand and identify that an antibiotic may increase the likelihood of some infections.

The Use of "Antipsychotics"

Modulation of the dopamine receptor (e.g., the use of dopamine receptor antagonists, partial agonists) is the cornerstone on which modern psychiatric treatment of schizophrenia is built. The "antipsychotics" class of medications comprises dozens of drugs, all with the common denominator of dopamine modulation. As in the class of the "antidepressants," current evidence shows that the therapeutic application of the "antipsychotic" extends beyond treating schizophrenia, as dopamine was found to be involved not only in psychosis but also in nonpsychotic disorders such as depression, addiction, bipolar disorder, tics, and more (Fig. 2).

Dopamine blockers are an essential component in the treatment of affective disorders. In addition to their importance in treating unipolar depression, described earlier, dopamine blockers are widely

Fig. 2 Indications for using dopamine blockers: dopamine blockers is an umbrella term for drugs acting mainly on the dopamine receptor. This term includes dopamine receptor antagonists, dopamine-serotonin antagonists, and dopamine partial agonists. Since many times "antipsychotics" are prescribed for nonpsychotic conditions, the term "antipsychotics" is confusing both to the clinicians, patients, and their families

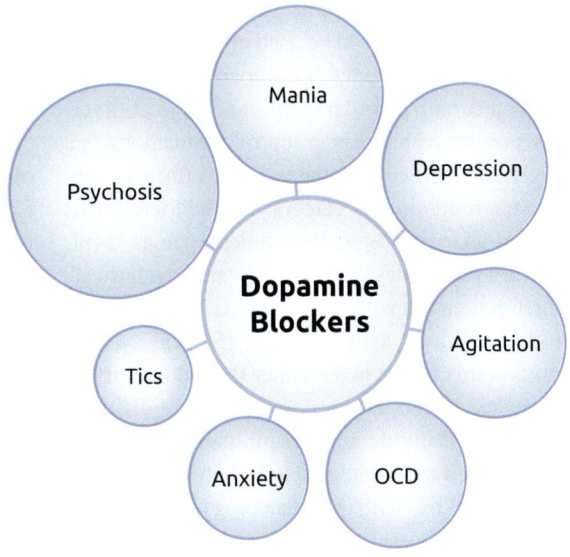

used to treat different stages of bipolar disorder alongside lithium, valproic acid, and carbamazepine (Vieta & Sanchez-Moreno, 2008) (National Collaborating Centre for Mental Health (UK), 2014), regardless of the presence of concordant psychotic symptoms. While it is not conclusive whether dopamine blockers are better for the control of manic episodes with psychotic features, clinical practice and guidelines suggest haloperidol, olanzapine, quetiapine, and risperidone can be especially useful for agitation in severely ill, but not psychotic, patients (Goodwin et al., 2016).

The application of dopamine blockers is also indicated in disorders outside of the affective and psychotic spectrum. Tic disorders are often treated with dopamine blockers as pimozide, haloperidol, risperidone, and aripiprazole and are considered as first-line treatment for those conditions (Pringsheim et al., 2019). Some dopamine blockers are used for behavioral modification, as risperidone is frequently used to reduce aggression and conduct problems in children and adolescents (Loy et al., 2017). Interestingly, treating the obsessive-compulsive disorder with a dopamine blocker can be beneficial or hazardous – depending on the specific drug; while low doses of risperidone and aripiprazole are proven helpful augmentation strategies to serotonergic enhancers, especially for OCD patients with a comorbid tic disorder (Gerasch et al., 2016), clozapine and to some extend olanzapine, were reported to worsen OCD symptoms (Lykouras et al., 2000) (Schirmbeck & Zink, 2012).

If psychosis is just one indication for which dopamine blockers can be used and the efficacy of some dopamine blockers is superior to others in treating certain disorders, pharmacological diversity present within this group of medications is a likely concept. Unfortunately, past efforts to create subgroups in this category birthed terms that are not pharmacologically minded such as "typical antipsychotics," "atypical antipsychotics," "first-generation antipsychotic (FGA)," and "second-generation antipsychotic (SGA)." These terms were created in a manner that added further confusion to the terminology, as vastly different drugs (amisulpride and aripiprazole are some examples) were considered to be similar, although they differ by both pharmacology and overall clinical utility, while others that were actually very similar like sulpiride and amisulpride were classed separately as first- and second-generation, respectively. While these denominations appear to be effective marketing terms, they carry neither scientific merit nor help make for informed prescribing.

Thus, the current taxonomy for psychotropics consists of official and unofficial terms and has a loose connection between the names of drug groups, their mechanisms, and the indications for which they are being used. Naming a category of psychotropics according to a single indication fails to properly represent the neuroscientific basis of psychiatric disorders and the immense versatility of the drugs used in treating them.

The continued use of the current indication-based terminology does not encourage psychiatrists to utilize the most contemporary evidence-based knowledge for prescribing medications and may encourage habitual prescribing of a limited number of drugs. Moreover, the continual use of the disease-based nomenclature extends our patients' confusion, contributing to the stigma on mental illness and mistrust toward psychiatry, resulting in poor adherence. To address these problems, a new approach toward the terminology of psychotropics is needed.

Neuroscience-Based Nomenclature (NbN): History and Mission

In 2008, the "Nomenclature Taskforce," composed of representatives from five major international organizations (the European College of Neuropsychopharmacology [ECNP], the American College of Neuropsychopharmacology [ACNP], the Asian College of Neuropsychopharmacology [AsCNP], the International College of Neuropsychopharmacology [CINP], and the International Union of Basic and Clinical Pharmacology [IUPHAR]), was launched (Zohar et al., 2015). The mission was "to examine ways of improving the current nomenclature in Psychopharmacology." This nonprofit endeavor was funded by the ECNP, without any commercial support from any pharmaceutical company (Zohar & Kasper, 2016).

When considering which principles should guide the building of the new terminology, relying on the growing neuroscience knowledge of recent decades appeared to be logical. Compared to the arbitrary nature of diagnostic criteria, pharmacology provides better and sound foundations for psychotropics' nomenclature. Moreover being pharmacology-driven, the nomenclature does not conflict with the clinical use of medications – rather enhances its rationale use, providing a potential bridge between neuroscience and clinical psychiatry (Möller et al., 2016).

Importantly, since the terminology roots are based on scientific principles, it allows the flexibility needed to accommodate future scientific advancements, including newly discovered evidence on existing drugs and discovering novel ones, without a need for another complete overhaul, thus contributing to the system's longevity.

Structure of the NbN

The Two Main Domains

To best represent the complexity of psychotropics, a dual-axial system is implemented, as every drug is defined by two main domains: (1) "pharmacology" and (2) "mode of action" (MOA) (Fig. 3). These domains are attributed to all the drugs widely used in psychiatry.

Pharmacology includes ten different domains referring to the neurotransmitter, molecule, or system being modified by the drug. This enables grouping together medications that modulate one (or more) pharmacological domain.

Mode of Action includes nine different modes which specify the nature of the action exerted by the drug, such as receptor antagonist, reuptake inhibitor, enzyme modulator, and so forth

The Nomenclature

 Pharmacology

 Mode of Action

5 Additional Dimentions

 Approved Indication

 Efficacy & Side Effect

 Practical Notes

 Neurobiology

 Pregnancy

Fig. 3 Structure of the Neuroscience-Based Nomenclature. The proposed nomenclature allows medications to be attributed to more than one domain, and the richness of psychopharmacology is better represented

10 Pharmacological Domains

1. Acetylcholine
2. Dopamine
3. GABA
4. Glutamate
5. Histamine
6. Melatonin
7. Norepinephrine
8. Opioid
9. Orexin
10. Serotonin

9 Modes of Action

1. Enzyme inhibitor
2. Enzyme modulator
3. Ion channel blocker
4. Neurotransmitters releaser
5. Positive allosteric modulator (PAM)
6. Receptor agonist
7. Receptor antagonist
8. Receptor partial agonist
9. Reuptake inhibitor

Fig. 4 The NbN two main domains, the pharmacological domain and the mode of action. These two domains reflect the state of art knowledge as of 2022

(Fig. 4). Naturally, some medications have more than one pharmacological domain and/or more than one mode of action. To better reflect on their versatility, compounds can be assigned, based on clinical relevance, to more than one domain. Those ten pharmacological domains and nine modes of action listed in the current version of the nomenclature, NbN third edition revised (NbN3R), can be used in different combinations to describe close to 150 compounds currently (2022) listed (Nutt & Blier, 2016; Frazer & Blier, 2016; Uchida et al., 2016).

For example, olanzapine under the previous nomenclature was named as an "atypical/second-generation antipsychotic." Under the NbN, it is now defined under the pharmacological domain of "dopamine (2), serotonin (10)," and as a "receptor antagonist (7)" under the mode of action. In comparison, brexpiprazole, another "atypical" in the current system, is attributed under the pharmacology domain "dopamine (2), serotonin (10)" as well – but with two modes of action: "partial agonist (8) and antagonist (7)" (Fig. 4). This approach better illustrates the commonalities and differences between the two drugs and lays out the rationale for selecting either of them for a given indication.

To turn NbN from an "academic exercise" to a practical, convenient clinical tool, the taskforce focused on the clinically relevant pharmacology

and mode of action details. For example, while different serotonin reuptake inhibitors differ from each other in receptor binding profile, all are assigned as serotonin (pharmacology) and reuptake inhibitors (mechanism of action), while the finer differences are spelt out in the neurobiology section. When there are more than three pharmacological domains/modes of action (as in the case of amitriptyline, mirtazapine, trazodone), the term "multimodal" is used, while the full detailed relevant pharmacology and mode of actions can be found in the neurobiology domain. In cases where the mode of action is not fully understood (e.g., valproate, topiramate), it is noted as "unclear," stating only the pharmacology domain. Lithium is the only exception as it stands as its own pharmacological domain (which is still unknown). By acknowledging the current scientific gaps, NbN presents not only the significant advances but also the limitations and leaves the door open for ongoing updates, updates that take place on a biannual basis.

Four Additional Dimensions

To enhance the practicality of NbN in daily use, the taskforce added five clinical dimensions to each drug.

The **approved indications** layer states which indications are officially approved by the major regulatory bodies (e.g., FDA, EMA (European Medicines Agency)) the **efficacy and side effects** layer specifies whether there is evidence to support the compound's use in additional indications to the ones approved by regulators (e.g., narcolepsy for methylphenidate, mirtazapine as a sleep promoter). Under "side effects," only frequent or life-threatening side effects are listed; the **practical note** layer contains clinical knowledge deemed useful by the taskforce, and it includes issues such as drug-drug interaction, metabolism, and dosing, to name a few. The **pregnancy** layer displays current information on safety of use during pregnancy and lactation. Finally, the layer termed **neurobiology** summarizes the empirical, clinically relevant data on pharmacology and modes of action. Highlights from preclinical and clinical studies, physiological neurotransmitter effects, and brain circuits are also included.

Clinical Aids: The NbN3 Apps and Website

In parallel to the nomenclature launch, two clinical aids were made public – the NbN application and the NbN booklet. The app is available through Android and iOS devices free of charge. It is constantly updated based on new scientific insights and clinicians' feedback. It is meant for daily practice. It has an entirely searchable database, taking advantage of the multiple axes the classification is built on. An additional app is also available: NbN C&A contains information relevant to medications used for children and adolescents.

The search process, in a nutshell, is illustrated in the following link: https://youtu.be/rE-VBnIj0TA. The use of the NbN app is done with the search bar, in a free search, or by first selecting an axis (Fig. 5a). Since the different domains and dimensions of the classification are searchable, the user can search for drugs by any parameter – including but not limited to pharmacology, mode

of action, approved indication, efficacy, side effects, and former terminology (e.g., antidepressants, antipsychotics). To better refine the results, it is possible to add more search axes. For example, if the term "depression" is searched, the app would display all medications used for treating this disorder. If the patient experienced sexual dysfunction as a side effect of previous medication and is having trouble sleeping, it is possible to refine the search by adding the axis "efficacy and side effects" twice – once for "sexual dysfunction low" and again for "sleep disturbance." As shown in Fig. 5b, the results list all the medications that fit these criteria. Once a compound is selected, all information is presented to the user clearly (Fig. 5c).

Since one of the goals of the NbN is to improve patient adherence to treatment and is driven by the belief that a well-informed patient is more likely to be motivated and compliant, a similar app called NbN-p&f)patient and family) was developed. The NbN-p&f is linked to NbN. After prescribing the medications to the patients, the clinician can either give the patient a PDF with the relevant information or refer him to the app/site. All the information is written in an approachable language, free from medical jargon yet following the spirit of NbN. The website NbNscience.com incorporates both NbN3, NbN C&A, and NbN-p&f, in addition to an educational module designed to introduce the classification and its tools to new users.

Recognition

In addition to the founding organizations mentioned (ECNP, ACNP, AsCNP, CINP, and IUPHAR), position statements supporting the use of NbN in clinical and academic practice were issued by the American Psychiatric Association (APA), the European Psychiatric Association (EPA), the Japanese Society of Psychiatry and Neurology (JSPN), the German Association for Psychiatry, Psychotherapy and Psychosomatics (DGPPN), and the Spanish Psychiatry Society (SEP), to name a few.

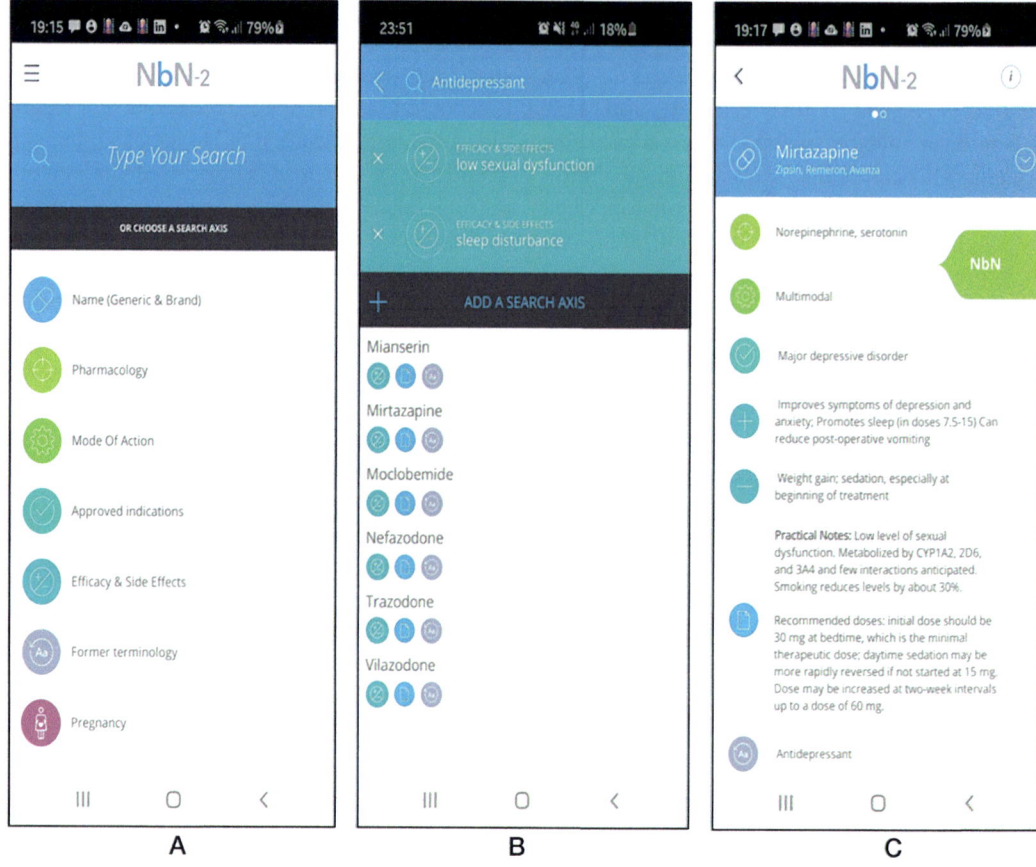

Fig. 5 The use of the NbN app. (**a**) shows the app home screen; (**b**) shows the search results after filtration by two axes; (**c**) shows the drug information screen

Conclusion

NbN presents a shift from disease-based to pharmacology-driven nomenclature. It supports the application of neuroscientific rationale to the prescription process for psychiatric medications, accommodates contemporary neuroscience insights, avoids patient confusion (no more "antipsychotic" for your depression or "antidepressants" for your anxiety), decreases stigma by helping to bridge the gap between psychiatry and other fields of medicine, and presents to students the rich fabric of the neuroscience basis of psychotropics.

References

Charpentier, P., Gailliot, P., Jacob, R., et al. (1952). Recherches sur les diméthylaminopropyl-N phénothiazines substituées. *Comptes rendus de l'Académie des sciences (Paris), 235,* 59–60.

Creese, I., Burt, D. R., & Snyder, S. H. (1975). Dopamine receptor binding: Differentiation of agonist and antagonist states with 3H-dopamine and 3H-haloperidol. *Life Sciences, 17*(6), 933–1001. https://doi.org/10.1016/0024-3205(75)90454-3

Fineberg, N. A., Hollander, E., Pallanti, S., Walitza, S., Grünblatt, E., Dell'Osso, B. M., Albert, U., Geller, D. A., Brakoulias, V., Janardhan Reddy, Y. C., Arumugham, S. S., Shavitt, R. G., Drummond, L., Grancini, B., De Carlo, V., Cinosi, E., Chamberlain, S. R., Ioannidis, K., Rodriguez, C. I., et al. (2020). Clinical advances in obsessive-compulsive disorder:

A position statement by the International College of Obsessive-Compulsive Spectrum Disorders. *International Clinical Psychopharmacology*. https://doi.org/10.1097/YIC.0000000000000314. Publish Ahead of Print.

Frazer, A., & Blier, P. (2016). A neuroscience-based nomenclature (NbN) for psychotropic agents. *The International Journal of Neuropsychopharmacology, 19*(8), pyw066. https://doi.org/10.1093/ijnp/pyw066

Gerasch, S., Kanaan, A. S., Jakubovski, E., & Müller-Vahl, K. R. (2016). Aripiprazole improves associated comorbid conditions in addition to tics in adult patients with Gilles de la Tourette Syndrome. *Frontiers in Neuroscience, 10*. https://doi.org/10.3389/fnins.2016.00416

Goodwin, G., Haddad, P., Ferrier, I., Aronson, J., Barnes, T., Cipriani, A., Coghill, D., Fazel, S., Geddes, J., Grunze, H., Holmes, E., Howes, O., Hudson, S., Hunt, N., Jones, I., Macmillan, I., McAllister-Williams, H., Miklowitz, D., Morriss, R., et al. (2016). Evidence-based guidelines for treating bipolar disorder: Revised third edition recommendations from the British Association for Psychopharmacology. *Journal of Psychopharmacology, 30*(6), 495–553. https://doi.org/10.1177/0269881116636545

Krieser, A. E., Sanderson, A. G., Vik, M., & Myers, J. A. (1953). Effects of isonicotinic acid hydrazide in mentally ill patients. *Diseases of the Chest, 23*(1), 28–35. https://doi.org/10.1378/chest.23.1.28

Kuhn, R. (1957). [Treatment of depressive states with an iminodibenzyl derivative (G 22355)]. *Schweizerische Medizinische Wochenschrift, 87*(35–36), 1135–1140.

Laborit, H., Huguenard, P., & Alluaume, R. (1952). [A new vegetative stabilizer; 4560 R.P.]. *La Presse Médicale, 60*(10), 206–208.

Lieberman, J. A., Stroup, T. S., McEvoy, J. P., Swartz, M. S., Rosenheck, R. A., Perkins, D. O., Keefe, R. S. E., Davis, S. M., Davis, C. E., Lebowitz, B. D., Severe, J., & Hsiao, J. K. (2005). Effectiveness of antipsychotic drugs in patients with chronic schizophrenia. *New England Journal of Medicine, 353*(12), 1209–1223. https://doi.org/10.1056/NEJMoa051688

López-Muñoz, F., Álamo, C., Juckel, G., & Assion, H.-J. (2007). Half a century of antidepressant drugs: On the clinical introduction of monoamine oxidase inhibitors, tricyclics, and tetracyclics. Part I: monoamine oxidase inhibitors. *Journal of Clinical Psychopharmacology, 27*(6), 555–559. https://doi.org/10.1097/jcp.0b013e3181bb617

Loy, J. H., Merry, S. N., Hetrick, S. E., & Stasiak, K. (2017). Atypical antipsychotics for disruptive behaviour disorders in children and youths. *Cochrane Database of Systematic Reviews, 2017*. https://doi.org/10.1002/14651858.CD008559.pub3

Lykouras, L., Zervas, I. M., Gournellis, R., Malliori, M., & Rabavilas, A. (2000). Olanzapine and obsessive-compulsive symptoms. *European Neuropsychopharmacology: The Journal of the European College of Neuropsychopharmacology, 10*(5), 385–387. https://doi.org/10.1016/s0924-977x(00)00096-1

Möller, H. J., Schmitt, A., & Falkai, P. (2016). Neuroscience-based nomenclature (jNbN) to replace traditional terminology of psychotropic medications. *European Archives of Psychiatry and Clinical Neuroscience, 266*, 385–386. https://doi.org/10.1007/s00406-016-0699-0

National Collaborating Centre for Mental Health (UK). (2014). *Bipolar disorder: The NICE guideline on the assessment and management of bipolar disorder in adults, children and young people in primary and secondary.* Care. http://www.ncbi.nlm.nih.gov/books/NBK498655/

Nutt, D. J., & Blier, P. (2016). Neuroscience-based Nomenclature (NbN) for Journal of Psychopharmacology. *Journal of Psychopharmacology (Oxford, England), 30*(5), 413–415. https://doi.org/10.1177/0269881116642903

Pringsheim, T., Okun, M. S., Müller-Vahl, K., Martino, D., Jankovic, J., Cavanna, A. E., Woods, D. W., Robinson, M., Jarvie, E., Roessner, V., Oskoui, M., Holler-Managan, Y., & Piacentini, J. (2019). Practice guideline recommendations summary: Treatment of tics in people with Tourette syndrome and chronic tic disorders. *Neurology, 92*(19), 896–906. https://doi.org/10.1212/WNL.0000000000007466

Schildkraut, J. J. (1965). The catecholamine hypothesis of affective disorders: A review of supporting evidence. *The American Journal of Psychiatry, 122*(5), 509–522. https://doi.org/10.1176/ajp.122.5.509

Schirmbeck, F., & Zink, M. (2012). Clozapine-induced obsessive-compulsive symptoms in schizophrenia: A critical review. *Current Neuropharmacology, 10*(1), 88–95. https://doi.org/10.2174/157015912799362724

Sinyor, M., Schaffer, A., & Levitt, A. (2010). The sequenced treatment alternatives to relieve depression (STAR*D) trial: A review. *Canadian Journal of Psychiatry. Revue Canadienne De Psychiatrie, 55*(3), 126–135. https://doi.org/10.1177/070674371005500303

Snyder, S. H. (1976). The dopamine hypothesis of schizophrenia: Focus on the dopamine receptor. *The American Journal of Psychiatry, 133*(2), 197–202. https://doi.org/10.1176/ajp.133.2.197

Souery, D., Papakostas, G. I., & Trivedi, M. H. (2006). Treatment-resistant depression. *The Journal of Clinical Psychiatry, 67 Suppl 6*, 16–22.

Uchida, H., Yamawaki, S., Bahk, W. M., & Jon, D. I. (2016). Neuroscience-based Nomenclature (NbN) for clinical psychopharmacology and neuroscience. *Clinical Psychopharmacology and Neuroscience, 14*(2), 115–116. https://doi.org/10.9758/cpn.2016.14.2.115

Vieta, E., & Sanchez-Moreno, J. (2008). Acute and long-term treatment of mania. *Dialogues in Clinical Neuroscience, 10*(2), 165–179.

Welsch, P., Üçeyler, N., Klose, P., Walitt, B., & Häuser, W. (2018). Serotonin and noradrenaline reuptake inhibitors (SNRIs) for fibromyalgia. *Cochrane Database of Systematic Reviews, 2020*. https://doi.org/10.1002/14651858.CD010292.pub2

WHO Collaborating Centre for Drug Statistics Methodology. (2021). *Guidelines for ATC classification and DDD assignment* (p. 2020).

Wilkes, S. (2008). The use of bupropion SR in cigarette smoking cessation. *International Journal of Chronic Obstructive Pulmonary Disease, 3*, 45–53. https://doi.org/10.2147/COPD.S1121

Zohar, J., Stahl, S., Moller, H. J., Blier, P., Kupfer, D., Yamawaki, S., Uchida, H., Spedding, M., Goodwin, G. M., & Nutt, D. (2015). A review of the current nomenclature for psychotropic agents and an introduction to the Neuroscience-based Nomenclature. *European Neuropsychopharmacology: The Journal of the European College of Neuropsychopharmacology, 25*(12), 2318–2325. https://doi.org/10.1016/j.euroneuro.2015.08.019

Zohar, J., & Kasper, S. (2016). Neuroscience-based Nomenclature (NbN): A call for action. *The World Journal of Biological Psychiatry, 17*(5), 318–320. https://doi.org/10.1080/15622975.2016.1193626

Pharmacogenomics and Precision Psychiatry

122

Pratyusha Attaluri, Ayeshah G. Mohiuddin, Kowsar Teymouri, and James L. Kennedy

Contents

P. Attaluri · A. G. Mohiuddin
Tanenbaum Centre for Pharmacogenetics, Campbell Family Mental Health Research Institute, Centre for Addiction and Mental Health (CAMH), Toronto, ON, Canada

K. Teymouri
Tanenbaum Centre for Pharmacogenetics, Campbell Family Mental Health Research Institute, Centre for Addiction and Mental Health (CAMH), Toronto, ON, Canada

Institute of Medical Science, Toronto, ON, Canada

J. L. Kennedy (✉)
Tanenbaum Centre for Pharmacogenetics, Campbell Family Mental Health Research Institute, Centre for Addiction and Mental Health (CAMH), Toronto, ON, Canada

Institute of Medical Science, Toronto, ON, Canada

Department of Psychiatry, University of Toronto, Toronto, ON, Canada
e-mail: jim.kennedy@camh.ca

© Crown 2024
A. Tasman et al. (eds.), *Tasman's Psychiatry*,
https://doi.org/10.1007/978-3-030-51366-5_130

Abstract

Globally psychiatric disorders present a major health concern. Currently, innovations in treatment are limited by (i) lack of biomarkers for symptom change and diagnosis, (ii) suboptimal pharmacotherapeutic options that are limited to a trial-and-error approach, and (iii) insufficient leveraging of predictive technologies. This chapter documents developments in the field of pharmacogenetics, which is based on examining the genetic variants of a given individual in order to predict the medications that they will have the best probability of responding to, with a minimum of side effects. Newer approaches that integrate predictive computational technologies (i.e., artificial intelligence and machine learning) are also described. When pharmacogenetic (PGx) testing information is used to guide drug selection, treatment adherence increases, adverse reactions are reduced, and significantly increased rates of remission are achieved. However, the uptake of PGx testing faces several challenges. The chapter discusses how challenges are being met with increasingly larger datasets, including randomized controlled trials, clinical utility analyses, and cost-effectiveness.

Keywords

Personalized medicine · Pharmacogenetic testing · Psychiatry · Clinical implementation · AI · Machine learning

Introduction

Knowledge of human genetics has been advancing over the last several decades facilitating the identification of genetic risk factors underlying diseases. In 2003, the completion of the Human Genome Project further advanced our understanding of genetic variability, susceptibility to a disease, and probability of drug response (Aworunse et al., 2018). Genomics can now be thought of as an integral part of modern medicine where novel drug targets can be identified and genetically guided therapeutic approaches can be used (Dugger et al., 2018). Pharmacogenetics (PGx) or pharmacogenomics (single gene vs all genes) is a field of precision medicine exploring genetic causes for an individual's variable response to a drug. Precision medicine aims to tailor treatment based on empirical evidence, and PGx uses genetic information predict side effects and treatment efficacy such as response and remission rates. The approach is based on peer-reviewed evidence from research on genetic factors that influence blood levels of a given drug via liver enzyme activity (pharmacokinetics) as well as the pharmacodynamic action of the drug in the brain (Corponi et al., 2018). This chapter presents a background and summary of advances in PGx research and testing in psychiatry. The section on PGx and psychiatry reviews metabolism of chemicals entering the body from the environment (xenobiotics) and the influence various enzymes have on drug metabolism and the effects of genetic polymorphisms on the level of expression and functioning of these enzymes. The section will also describe enzymes of particular clinical relevance in psychiatry and provide background and updates on the relatively new application of epigenetic methods, as well as the integration of environmental factors. Future approaches combining PGx with predictive technologies such as artificial intelligence and machine learning will be covered in the subsequent section. Finally, a description on the implementation challenges of PGx testing will be explored from a wider lens along with the many significant opportunities that will drive the field forward.

PGx and Psychiatry

Xenobiotic Metabolism

Foreign compounds to the body, such as plant compounds, food additives, and synthetic compounds including drugs, are all classified as xenobiotics. If the body recognizes them as potentially dangerous, it will attempt to chemically modify them to an inactive molecule for excretion via the kidney or intestines. Biotransformation of xenobiotics occurs in two phases: phase I produces polar, water-soluble compounds by reduction and oxidation and hydrolysis reactions. Phase I, in a limited number of cases, may create an active form from the parent drug molecule (e.g., conversion of codeine to morphine) or most commonly a substrate for phase II biotransformation. The phase II reaction produces larger hydrophilic compounds by methylation, acetylation, glucuronidation, sulfation, or conjugation with amino acids and glutathione, virtually always inactivating a drug molecule. Additionally, drug transporters alter pharmacokinetics of and drug level differences at the cellular or tissue levels. Phase I reactions are catalyzed by a family of heme-containing proteins called the cytochrome enzymes. In the past two decades, 57 CYP genes encoding 18 CYP families have been identified (Nebert et al., 2013). The nomenclature of CYP enzymes consists of the spectrometry absorption wavelength (450 nm) followed by number-letter-number format, which is based on the amino acid sequence homology. In humans, ~75% of enzymes involved in drug metabolism are CYP families 1–4 (Furge & Guengerich, 2006). It is understandable that variations in the protein-coding or regulatory regions of the genes encoding CYP enzymes could have a significant effect on drug metabolism.

Clinical Impact of CYP450 Metabolism Genes

CYP2D6, CYP1A2, CYP3A4, CYP2C9, and CYP2C19 enzymes are most commonly involved in metabolism of medications for psychosis and Major Depressive Disorder (MDD). Decades of pharmacological research on interindividual variations in medication levels have shown CYP450 enzymes to have a major role in phase I metabolism of the majority of drugs. Phase II processing further alters drug molecules by attaching compounds (e.g., glucuronidation) that makes the original drug molecule more hydrophilic and thus easier to process for excretion from the body. The fact that genetic variation of these liver enzymes can affect the blood level of medications represents the core rationale for pharmacogenetic testing.

The clinical implications of the genetic variation in the drug-metabolizing enzyme genes depend on the genotype of the polymorphism. For example, a loss of function variant can reduce the amount or the activity of the enzyme, and a gain of function variant does the opposite. To explain that in terms of metabolizer status, based on the genotype results for a specific pharmacokinetic gene, individuals are classified as ultrarapid metabolizer (UM) if they carry two increased functioning alleles or more than two copies of normal-functioning alleles. Extensive metabolizers (EM) carry two normal-functioning alleles. Intermediate metabolizers (IM) could carry one normal allele and a decreased functioning allele or two decreased function alleles, and a poor metabolizer (PM) could carry one decreased functioning allele and a nonfunctioning allele or two nonfunctioning alleles. Ums tend to have low blood levels of their metabolized drug, IM and EM have relatively normal blood levels, and PM usually have blood levels that are too high leading to side effects. Evidence from a real-life setting suggests that the cost of healthcare per year for an extreme metabolizer (i.e., PM or UM) is $4000–6000 higher than an IM or EM (Chou et al., 2000). Therefore, using PGx testing to identify which patients are extreme metabolizers before prescribing should considerably reduce costs among the extreme metabolizers, by up to 28% (Herbild et al., 2013). Distribution of metabolizer status can vary based on ancestry of the population. For instance, reduced *CYP2D6* activity occurs in 7–10% of European Caucasians and 1–4% of Asians and 2–3% with African ancestry, which can help explain the drug response variation in these populations (Kitada, 2003).

PGx in psychiatry is particularly beneficial in view of the substantial clinical heterogeneity and individual variation in treatment outcomes, as well as the fact that there are no biomarkers for measuring response. For example, choosing a pharmacological treatment for depression is largely trial-and-error, with less than 40% of patients achieving remission during the first medication trial. In addition, many patients on dopamine and serotonin blockers develop drug-induced weight gain, increased risk of diabetes mellitus type 2, metabolic syndrome, and cardiovascular disease that leads to a 20% shorter life span for an individual with SCZ than the general population (Lett et al., 2012). Combinatorial pharmacogenomic-guided treatment, which consists of a multigene testing panel, integrates pharmacodynamic and pharmacokinetic genetic factors to predict therapeutic response, side effects, medication exposure, and appropriate dosing (see Table 1). This can result in more clinically favorable treatment decisions and increase the number of patients that are medication adherent (Tiwari et al., 2022).

Relevance of the ABCB1 Transporter Gene

ABCB1 (ATP-binding cassette subfamily B member 1) is *a* pharmacokinetic transporter gene, encoding the P-glycoprotein 1 (permeability glycoprotein, P-gp) efflux pump involved in drug elimination. P-gp is an important protein located in the cell membrane that pumps many xenobiotic substances out of cells. P-gp is expressed in the intestinal epithelium where it can move drugs or toxins back into the intestinal lumen. It performs similar functions in liver cells performing efflux into the bile ducts and in the capillary endothelial cells in the blood-brain barrier where it pumps compounds back into the capillary lumen.

Some medications for psychosis that are substrates of the P-gp are rapidly removed from the brain by cells in the blood-brain barrier (e.g., risperidone, paliperidone, amisulpride); consequently they are dosed to override the effect of the P-gp. However, since the anterior pituitary does not have a blood-brain barrier, it experiences a relatively higher antipsychotic concentration than the rest of the brain, leading to extensive inhibition of the dopamine D2 receptor and elevations of prolactin release (El-Mallakh & Watkins, 2019).

Polymorphisms in *ABCB1* were originally discovered in their role in treatment efficacy in cancer (Juliano & Ling, 1976), followed by demonstration of their role in antipsychotic medication efflux (Mas et al., 2012). *ABCB1* is also involved in efflux of certain antidepressants from the brain. It is hypothesized that higher expression of P-glycoprotein will result in a lower brain concentration of the substrate, due to increased drug efflux. Thus, a person who is genetically predisposed to faster efflux by P-gp will tend to have poorer response to drugs that are substrates of P-gp.

Future of Pharmacogenomics

Even though pharmacogenomic-guided treatment is still in its infancy, awareness and education are spreading widely, and many commercial testing services are currently available. Along with pharmacogenomics and data from multi-omics approaches (including transcriptomics, metabolomics, epigenomics, and proteomics) are developing at an exponential rate. Application of these multi-omics approaches to pharmacology has given birth to pharmaco-omics which looks at the effects of pharmacokinetics and pharmacodynamics from multiple dimensions. Combination of physicians' clinical assessment and the patient's individual multi-omics data can help in finding a suitable medication, individualized approach to prescribing drugs, and drug target discovery. To get such answers from vast multidimensional data, artificial intelligence/machine learning may be very useful.

It is not possible to cover the full range of pharmaco-omics in this chapter, and there is limited literature across the various omic data types. However, for the subtopic of pharmaco-epigenetics, there is a somewhat larger set of

Table 1 Common genes from current commercial PGx tests

Genes	Physiological role	Impact of variant	Treatment impact	Test provider
CYP1A2, CYP2B6, CYP2C9, CYP2C19, CYP2D6, CYP3A4/5	Hepatic enzymes that metabolize drugs	Majority of psychiatric medications are metabolized by these cytochrome enzymes	Patient's dose may be adjusted based on their metabolizer status (Winner et al., 2013; Furge & Guengerich, 2006)	GST GCT NPG NID ITE QTD
ABCB1	Encodes for the multidrug efflux pump P-glycoprotein. Responsible for pumping drug out of target tissue	Slow efflux variants might lead to intracerebral drug toxicity due to inadequate clearance of the drug out of the brain	Drug dosage may need to be reduced for slow clearance patients (Mas et al., 2012)	NPG QTD
UGT2B15	Phase II xenobiotic enzyme	Polymorphisms in this gene can lead to variance in systemic clearance of certain psychiatric drugs, e.g., lorazepam (Ching et al., 2018)	Clinicians can decide drug and dosage according to the carrier status of individuals	NPG QTD
SLC6A4 (5-HTTLPR)[a]	Encodes for serotonin transporter, for moving serotonin from the synaptic cleft back into the presynaptic neuron	Polymorphisms in this gene can lead to either an insertion or a deletion resulting in a "long" L allele or a "short" S allele, respectively. Individuals with "S" allele have reduced expression of the serotonin transporter protein	Some evidence suggests that the S allele predicts reduced SSRI efficacy and increased likelihood of adverse effects (van Schaik et al., 2020)	GST GCT NPG NID QTD
HTR2A[a]	Encodes for a postsynaptic serotonin receptor and is a target for many antidepressants and some antipsychotics	Variants may lead to adverse effects and poor medication response particularly with paroxetine (Kato et al., 2006; Wilkie et al., 2009)	Evidence suggests HTR2A polymorphisms can be a predictor of antidepressant response (Garriock & Hamilton, 2009; Tiwari et al., 2013)	GST NID ITE QTD
COMT[a]	Encodes for catechol-O-methyltransferase, enzyme required to breakdown dopamine, epinephrine, and norepinephrine	Imbalances in dopamine levels are known to affect mood/behavior. Polymorphisms in this gene are reported to impact response to dopaminergic agents. Genotypes of the Val158/Met polymorphism have been identified to affect gene expression Val/Val = high Val/Met = intermediate Met/Met = low	Use caution using dopaminergic agents in Met/Met individuals, e.g., bupropion (Fawver et al., 2020)	NID NPD GCT ITE QTD
DRD2[a]	Encodes for Dopamine receptor D2 type. It is one of the five types of dopamine receptors, associated with inhibitory functions	141C Ins/Del polymorphism is well studied in antipsychotic medication response Taq 1A polymorphisms have been associated with	Inc C allele is associated with favorable treatment outcome, and Del C allele carrier has significantly poorer antipsychotic response (Zhang & Malhotra, 2011)	GCT

(continued)

Table 1 (continued)

Genes	Physiological role	Impact of variant	Treatment impact	Test provider
		antipsychotic medication response	Higher Taq A2 allele frequency is suggestive for potential tardive dyskinesia (Zhang & Malhotra, 2011)	
OPRM1[a]	Encodes for the mu-opioid receptor (MOR) that is reported to regulate pain response to opioids and influence reward functions thus may predispose to addiction	The most studied genetic variant, A118G, appears to have functional effects on OPRM1 gene expression. It has shown association with the efficacy of opioid medications and risk for addiction (Crist & Berrettini, 2014)	Patients with higher-risk variants may be prescribed alternative pain medication	NPG GCT ITE QTD

[a]These pharmacodynamic genes should be viewed with caution as they do not have support from meta-analyses
GST – GeneSight PGx panel offered by Myriad Genetics, Inc.
NID – NeuroIDgenetix PGx panel offered by AltheaDx, Inc.
NPG – Neuropharmagen PGx panel offered by Neuropharmagen
GCT – Genecept PGx panel offered by Genomind
ITE – Invitae PGx panel (not psychiatry-specific)
QTD – Quest Diagnostics PGx panel (not psychiatry-specific)
Note: Adapted from the company websites

publications. Therefore, a brief overview and a few examples of epigenetic applications in personalizing medication prescriptions will be provided.

Epigenetics: Fine-Tuning Precision Medicine

Epigenetics is the study of changes in gene expression brought about by processes including DNA methylation, posttranslational histone modifications, and noncoding RNAs (ncRNAs) as opposed to standard genetic polymorphisms that alter the DNA nucleotide sequence.

Approximately 60 pharmacokinetic genes are currently considered to be affected by DNA methylation, histone modification, and miRNAs (Kacevska et al., 2011). Epigenetic landscapes change over years for multiple reasons, including physiological, pathological, and environmental influences leading to alterations in gene expression. One of the potential categories of environmental influences is the psychiatric medications themselves. Several medications for depression and for bipolar disorder have been described to modulate epigenetic patterns by inhibiting histone deacetylases (HDAC) and DNA methyltransferases (DNMT), leading to DNA methylation changes at many locations across the genome, which are directly responsible for gene expression and function. Evidence shows that cytochrome genes such as *CYP2C9*, *CYP2C19*, and *CYP3A4* are subject to epigenetic modification that can increase or decrease gene expression. CYP2D6 is not currently recognized as having significant influence by epigenetic mechanisms (Lauschke et al., 2018).

DNA methylation occurs on the cytosines of a DNA sequence (CpG islands). This results in gene silencing or reduced gene expression by altering the transcription factor binding capacity or by changing the chromatin organization (Liu et al., 2018). This mechanism has been identified in disease pathology and as a marker for treatment response. For example, rats administrated olanzapine showed altered DNA methylation on dopamine system genes (Melka et al., 2013). Additionally, methylation pattern changes were observed in pre- and post-pharmacotherapy in patients with SCZ, providing evidence that epigenetic alterations can be used to measure disease and response to therapy (Castellani et al., 2015). Another gene of interest is *BDNF*, which is important for nerve growth and maturation. This gene encodes the protein brain-derived neurotropic factor which is active in creating, remodeling, and maintaining synapses. Methylation status of *BDNF* promoters has also been studied as a predictor for antidepressant treatment. Patients showing hypomethylation of the *BDNF* promoter region are unlikely to benefit from antidepressant pharmacotherapy (Tadic et al., 2014). Thus, understanding the epigenetic landscape in nonresponder groups of patients can open pathways for identifying new biomarkers and refine the personalized medicine approach.

Long noncoding RNAs, a type of RNA which does not code for a protein, creates epigenetic marks in gene promoters by recognizing and recruiting enzymes to the targeted site on selected genes, resulting in histone protein modifications. These modifications (either acetylation or methylation) alter gene expression by a mechanism that alters the 3D folded structure of the DNA. This organized network of epigenetic enzymes and noncoding RNAs, known as a "histone code," can affect the expression of pharmacokinetic genes by altering the levels of their mRNA expression. This alteration leads to changes in CYP450 enzyme levels and interindividual differences in drug response (Ivanov et al., 2012). MicroRNAs (miRNAs), which are smaller noncoding RNAs, play a role in posttranscriptional gene silencing by labeling specific mRNA gene transcripts for destruction and thus can modulate expression of drug-metabolizing enzymes (Yu & Pan, 2012). Thus, epigenetic markers at specific locations on the genome that are transcribed into a coding or noncoding RNA can influence the interindividual difference in the enzyme activity levels. In addition, chronic use of the antidepressant imipramine has been shown to reverse certain histone modifications caused by chronic stress that altered the expression of the *BDNF* gene in mouse hippocampus (Tsankova et al., 2006). The implication of this line of research is that

epigenetics may reveal a new mechanism of action in relation to medications for depression and also a potential epigenetic mechanism of depression pathology. In terms of pharmacogenomics, it can be seen that epigenetic mechanisms may influence a given pharmacokinetic or pharmacodynamic gene's expression and/or translation into its protein product in the form of a metabolizing enzyme or drug receptor. A research challenge for the future is to determine where and when to measure, in a given patient, the dynamically changing epigenetic marks on genes involved in drug action and metabolism.

Emerging Role of Artificial Intelligence and Machine Learning in Pharmacogenomics

In recent years, advances in technology have provided the opportunity to use multiple data types and consolidate the information to drive a better decision on diagnosis, prognosis, and treatment choices. As the future progresses, pharmacogenomic-guided medication selection will likely leverage multiple data types such as whole genome sequencing, epigenetic modifications, RNA sequencing to measure the transcriptome, and proteomics. On top of these, molecular "omic" data may be added neuroimaging and electrophysiology measures to separate populations of patients to into subgroups based on specific biomarkers, to further deliver a tailored pharmacotherapy (Lin et al., 2020). Functional neuroimaging is already being used to target stimulation sites for transcranial magnetic stimulation (Cole et al., 2022). However, in order to deliver a more personalized treatment, conventional statistical analysis is unable to capture the heterogeneous and complex nature of underlying psychiatric phenotypes, pointing to modern machine learning (ML) and artificial intelligence (AI) approaches as promising methods (Koppe et al., 2021). AI has supported clinicians in the provision of personalized therapy for individual patients by integrating ML with electronic health records and analyzing other biometrics (Reddy et al., 2019).

Despite the surge for AI use in the physical health domains of medicine, implementation of these methods has been slow in mental health (Jiang et al., 2017; Miller & Brown, 2018), mainly because of the qualitative and subjective nature of the clinical data. According to a review by Shatte et al. (2019), ML has mainly been applied to four domains of mental health: (i) detection and diagnosis; (ii) prognosis, treatment, and support; (iii) public health applications; and (iv) research and clinical administration. Most of the recent research has been on the application of ML in detection and diagnosis of depression, suicide risk, and cognitive decline.

There has been an increasing interest in the employment of AI and ML techniques in psychiatric pharmacotherapy response prediction, albeit it is in its early stages due to insufficient human studies in investigating the predictive models of drug response (Lin et al., 2020). Psychiatric disorders have heterogeneous symptoms and treatment response, which makes the selection of an appropriate treatment, tailored to a patient challenging (Squarcina et al., 2021). Squarcina et al. found promising results in their review for predicting treatment response with deep learning methods. Similarly, Maciukiewicz et al. investigated the treatment response of patients with MDD to duloxetine, using ML techniques applied to GWAS data (Maciukiewicz et al., 2018). The results showed that while response models were unaffected, remission models were promising with accuracy of 0.66. Recently, deep learning has been a popular method to predict treatment response as it is able to effectively analyze large neuroimaging datasets and integrate them with other biomarkers including clinical, molecular, and multi–omics (Squarcina et al., 2021). Another study deployed conventional ML methods to clinical and demographic data to predict antidepressant treatment response with 89% accuracy (Patel et al., 2015). Additionally, AI/ML techniques were employed to analyze clinical data (Nunes et al., 2020) and gene expression data (Eugene et al., 2018) to predict lithium response in patients with bipolar disorder.

Overall, as clinicians and the scientific community are progressing toward precision psychiatry and personalized treatment, efforts are being put forward to improve medication response and discover new targets based on population and toxicology data. PGx testing has emerged as a mitigation to the increasing problems of adverse drug reactions and side effects. The complex combination of data leveraged alongside pharmacogenomics, including environmental factors, demographics, and nongenetic characteristics of drug response, presents the rationale for leveraging AI/ML as a promising tool to embrace and integrate these large datasets.

Implementing PGx Testing in Psychiatry

Prior literature has demonstrated the benefits of PGx testing in the medical field, including psychiatry. However, the adoption of PGx testing in routine clinical practice remains slow. This is largely attributable to resistant behavior toward innovation as it can disrupt established routines. For example, PGx-guided treatment aims to personalize medications and change the trial-and-error approach of prescribing medications. In this section, common barriers as well as enablers of PGx testing in psychiatric care settings will be explored.

Mental health illnesses cost over US$1 trillion each year worldwide (The Lancet Global Health, 2020). In many patients, pharmacological interventions do not meet the desired therapeutic response. A large prospective trial, Sequenced Treatment Alternatives to Relieve Depression (STAR*D), found that initial treatment response to medications for depression was only 49.6%, reflecting the need for an improved treatment approach (Rush et al., 2006). In another study, only six in ten patients diagnosed with SCZ, bipolar disorder, and MDD reached symptom remission (Corponi et al., 2018). Thus, at present, psychotropic medications are only effective in a subset of a patient population. Recent studies in psychiatry indicate the benefits and clinical relevance of PGx testing over treatment as usual (TAU) for MDD (Table 2). Further, a meta-analysis found that PGx-guided MDD treatment versus TAU improves response rate (50% vs 36%) and remission rate (40% vs 25%)

Table 2 Clinical trials assessing impact of PGx-guided treatment for MDD

Author, year	Number of participants	Depression scale	Outcome summary
Hall-Flavin et al., 2013	PGx-guided arm: 114 TAU arm: 113	HDRS-17, QIDS-C16, PHQ-9	– PGx-guided arm had greater rates of remission at 8 weeks – Scores for all three scales declined in PGx-guided arm more than the TAU by week 4
Winner et al., 2013	PGx-guided arm: 26 TAU arm: 25	HAMD-17, PHQ-9, QIDS-SR, QIDS-CR	– PGx-guided arm had greater remission, response rate, and symptom improvement at week 10 – PGx can help identify patients with treatment-resistant depression
Singh, 2015	PGx-guided arm: 74 TAU arm: 74	HDRS	– PGx-guided arm had a higher remission rate than the TAU arm at 12 weeks – Antidepressant efficacy is improved in guided vs non-guided treatment groups
Perez et al., 2017	PGx-guided arm: 155 TAU arm: 161	PGI-I (primary outcome), FIBSER, HDRS-17, SDI, SATMED-Q, CGI-S	– PGx-guided arm had a higher response rate compared to the TAU arm at 12 weeks – Tolerability of side effects higher in PGx-guided group than in TAU at 6 and 12 weeks – PGx tools are more useful in those with prior unsuccessful medication trial
Bradley et al., 2018	PGx-guided arm: 352 TAU arm: 333	HAM-D17 and HAM-A	– PGx-guided arm had a trending higher response rate at 4, 8, and 12 weeks – PGx-guided treatment improves depression and anxiety symptoms in various healthcare settings
Greden et al., 2019	PGx-guided arm: 681 TAU arm: 717	HDRS-17 (primary outcome), QIDS-C16, PHQ-9	– PGx-guided arm had improved response and remission rates; however, symptom improvement was not statistically significant at 8 weeks – Reduced side effects when medication was changed to PGx-guided recommendations – At 24 weeks, there was a continued response and remission improvement in PGx-guided arm
Dunlop et al., 2019	PGx-guided arm: 621 TAU arm: 677	HAM-D6	– PGx-guided arm had greater remission, response rate, and symptom improvement at week 8
Tiwari et al., 2022	PGx-guided arm: 65 TAU arm: 69	HDRS-17	– PGx-guided arm had greater symptom improvement (27.6% versus 22.7%), response (30.3% versus 22.7%), and remission rates (15.7% versus 8.3%) at week 8

Abbreviations: *PGI-I* patient global impression of improvement, *FIBSER* frequency, intensity, burden of side effects rating, *HDRS-17* 17-item Hamilton depression rating scale, *CGI-S* clinical global impression-severity scale, *SDI* Sheehan disability inventory, *SATMED-Q* treatment satisfaction with medicines questionnaire, *QIDS-C16/CR/SR* 16-item quick inventory of depression symptomology, clinician-rated, subject-rated, *PHQ-9* 9-item patient health questionnaire, *SIGH-D-17*, *C-SSRS* columbia-suicide severity rating scale, *HAM-D6* Hamilton depression rating scale, *HAM-A* Hamilton rating scale for anxiety

(Rosenblat et al., 2018). In terms of action toward implementation of PGx testing, the Dutch Pharmacogenetics Working Group (DPWG) and the Clinical Pharmacogenetics Implementation Consortium (CPIC) utilize a combination of literature to provide evidence-based treatment recommendations guided by genetic information including *CYP2D6* and *CYP2C19* genes. CPIC has published recommendations based on hepatic enzyme variant information on over 160 medications, and DPWG provides dosing advice for over 94 medications. Other developing guidelines with recommendations include the Canadian Pharmacogenomics Network for Drug Safety (CPNDS) and the French National Network (Réseau) of Pharmacogenetics (RNPGx). All these guidelines can be found on the website of PharmGKB: www.pharmgkb.org.

Despite such advances in research, translating PGx testing into clinical decisions remains a challenge. To move from discovery of an innovation to implementation in healthcare has many layers of complexity requiring clinical validation and acceptance by key stakeholders involved (Fig. 1). The main barriers to adoption include (1) lack of knowledge among users, (2) clinical utility and validity misconceptions, and (3) cost. On the other hand, an enabling opportunity is provided by feedback from participatory action research. This feedback from patients and clinicians generally expresses optimism toward PGx, increased awareness and availability of PGx testing, and motivation resulting from reduction in adverse reactions and financial savings.

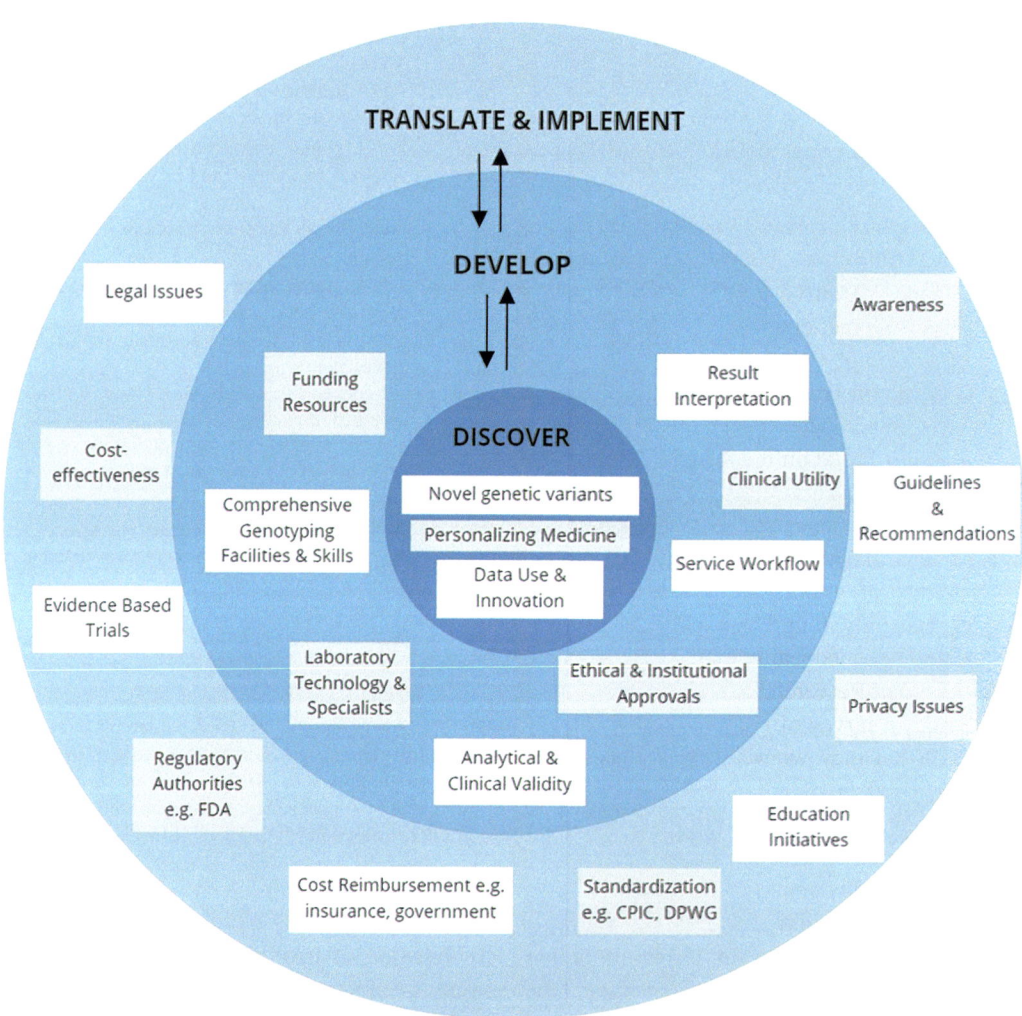

Fig. 1 Complexity of translating PGx testing from bench to beside (*and beyond*)

Challenges and Opportunities to Implementing PGx Testing in Psychiatry

Lack of Knowledge Among Users

The circle of care in psychiatry is often multidisciplinary requiring several healthcare providers to agree with a certain treatment plan. Healthcare professionals include physicians, nurses, pharmacists, and trainees, and the translation of their knowledge of PGx testing to patients is crucial. An informed patient is empowered to not only understand the benefits and limitations of such a test but also voice any concerns. Lack of PGx testing knowledge has presented as a common barrier to its clinical adoption. Several studies show that physicians prescribing psychotropic medications (mainly primary care providers and psychiatrists) felt they had limited knowledge of PGx use, thus affecting their attitudes toward its utility (Dunbar et al., 2012; Ilona et al., 2015; Thompson et al., 2015). Chan et al. surveyed 167 physicians and 27 pharmacists practicing in psychiatric settings and reported 53.6% were concerned regarding not feeling competent to order/suggest the test while 59% felt competent to make treatment recommendations based on the results (Chan et al., 2017). In another survey with 107 psychiatric pharmacists, 47% believed they were not knowledgeable regarding the test (Shishko et al., 2015). The findings of a poll of 1000 US citizens revealed that only 1 in 5 patients believe their physician is knowledgeable with genomic information (Cogent Research, 2010). This finding is supported by a survey of 113 psychiatrists where 72.6% felt that they would benefit if a genetic counselor were involved in their patient care routine (Thompson et al., 2015). These trends appear to continue given that a recent study assessing pharmacogenetic knowledge among pediatricians and child psychiatrists reported that most felt they did not have adequate training to interpret results (Jessel et al., 2022). Interestingly, in a survey with 168 general practitioners and psychiatrists, 76% were satisfied with their comprehension of the results (Walden et al., 2019). Thus, ongoing education and counseling programs are needed to prevent further impediments to the adoption of PGx-guided care in psychiatry. Of note, only 21 out of 74 medical schools across the USA and Canada offer over 4 h of PGx teachings (Green et al., 2010). Furthermore, the need for standardization and harmonization across guidelines such as CPIC and DPWG with emerging literature and ongoing trials has been noted (Laplace et al., 2021).

Clinical Utility and Validity Misconceptions

Clinical utility is the ability of a medical test to reduce morbidity and mortality, whereas clinical validity refers to the accuracy and reproducibility of the test. Thus, clinical utility explores the risk vs benefit for a patient resulting from the test, and clinical validity reflects the delivery and accuracy of the technology. Both these criteria can be studied through randomized controlled trials (RCTs) that provide high quality of evidence-based findings. Psychiatric RCTs are few, thus limiting the estimation of clinical utility and validation of PGx testing. To date, there have been a few published meta-analyses of RCTs. A review of five RCTs totaling 1727 patients found that PGx-guided treatment was 1.7 times more likely to have symptom remission compared to TAU (Bousman et al., 2018). Correspondingly, another analysis found treatment response and symptom remission to be 1.36 and 1.74 times higher in the PGx-guided group compared to TAU (Rosenblat et al., 2018). Both these studies evaluated MDD. A Canadian RCT of a combinatorial PGx test vs TAU showed a similar effect size of 1.7 favoring PGx-guided treatment in increasing the remission rate in a sample of 270 treatment-resistant MDD patients (Tiwari et al., 2022). While such positive supports are promising, they are still in their infancy in terms of shifting clinical practice away from trial-and-error approaches. Among the reasons RCTs are limited in size and number are their large scale and high budget demands. Another reason may be that PGx results are not always generalizable, which will be discussed in the following section.

Further refinement of existing genotyping technology is required for the wider adoption of PGx testing. Currently, all the genes and gene variants affecting treatment response are not known. One relatively new tool for combining genetic variants associated with the risk for a psychiatric disorder (or other phenotypes) is polygenic risk score (PRS). This score provides a quantitative index of genomic burden of a large number of risk variants in a particular individual to help predict whether that person will develop a specified disorder in their lifetime (Fullerton & Nurnberger, 2019). PRS has strong potential to impact PGx testing in the case where susceptibility to a disorder is associated with treatment outcomes. These advances have been implicated in other medical fields, for example, the use of statins in high-risk cardiovascular patients (Natarajan et al., 2017). The clinical utility of PRS in psychiatry is not yet clear, in large part due to the small number of studies thus far as well as the large sample size required to create an effective PRS.

Pharmacoeconomics: Cost-Effectiveness and Reimbursement

The cost of PGx testing continues to present a concern among both patients and clinicians. This can, in part, be due to costs seeming higher than the benefits. Pharmacoeconomics is an area of health economics that links efficacy, safety, and quality of pharmacotherapeutics with economics. This field serves as a valuable tool for implementing novel treatment approaches in a financially equitable manner. Types of pharmacoeconomic evaluations include cost minimization analysis (CMA), cost benefit analysis (CBA), cost-effectiveness analysis (CEA), and cost utility analysis (CUA) (Areda et al., 2011). There have been several studies evaluating cost-effectiveness of psychiatric PGx; however, they are limited in sample sizes and consistency (Karamperis et al., 2021). Here the focus is on CEA and the influence of insurance reimbursement (complete or partial) on PGx testing in psychiatry.

CEA compares the relative cost with outcomes in terms of years of life saved, days of disability avoided, and units of lives saved. Pharmacogenetic testing in treatment-resistant depression patients was found to increase quality-adjusted life years (QALYs) by 0.316 years, which can be translated to a saving of $3711 in medical costs per patient (Hornberger et al., 2015). Additionally, patients with an improved quality of life can return to work, further contributing to the economy. PGx-directed treatment for clinically impairing depression and anxiety can save around $3500 and $7000, respectively, on medication costs (Winner et al., 2015). Improved adherence to medications due to enhanced treatment accuracy reduces adverse effects, hospital visits and admissions, and healthcare costs. Clearly, pharmacogenomics has great economic potential.

Cost-effectiveness of PGx testing is also influenced by differences in genetic variant frequencies across populations. For example, HLA-B genotyping is commonly used to prevent adverse reactions (e.g., Stevens Johnson syndrome) in Pacific Oceanic and Asian populations given that a variant allele (HLA-B*15:02) is relatively common (1–12%) in this subpopulation, but it is rare in European and African populations (0–0.02%). Another example is *CYP2C19* poor metabolizer status patients prescribed clopidogrel following acute coronary syndrome; clopidogrel has a reduced effect in *CYP2C19* poor metabolizers, and the more expensive ticagrelor is recommended (Sorich et al., 2013). Consequently, this is one example of how CEA of PGx testing can differ between subpopulations. Another consideration is that healthcare funding can be public and/or private, depending on the jurisdiction, impacting cost-saving assessments. Overall, more pharmacoeconomic large-scale studies with geographic diversity and measurement of economic variables are required to improve estimates of cost-effectiveness of PGx testing.

Cost reimbursement, by government or private insurance companies, is an additional deciding factor for PGx testing implementation in routine psychiatric care. Several psychiatrists expressed cost as a deterrent in ordering PGx testing for their patients (Dunbar et al., 2012; Laplace et al., 2021). Additionally, patients would be more inclined to have the test if they were financially supported (Liko et al., 2020). This is especially relevant in psychiatry, where illnesses present with a wide range of behavioral and other disabilities preventing patients from attaining sustainable income sources. One of the reasons coverage is not provided is because PGx testing is often deemed experimental and investigational (Phillips et al., 2017). With additional pharmacoeconomic research, genetic testing will likely be more highly incorporated into routine clinical care including psychiatry, where the burden of trial-and-error approaches remains high. The cost of running PGx testing is expected to decrease with competition among the steadily growing number of test providers entering the market and technological advances including but not limited to constantly improving efficiency of DNA measurement instrumentation, as well as efficiencies from application of machine learning. It is worth noting that a PGx test only needs to be performed once for a given patient since their DNA sequence does not change and thus the information provided can be used to personalize appropriate treatment lifelong. However, it is likely that new markers and new technologies will arise that would require additional testing.

To learn more regarding perspectives on PGx testing, qualitative research strategies such as surveys and focus groups are now being conducted. Early reports show persuasive evidence supporting PGx testing from the physician and patient end users. For example, physicians reported feeling comfortable discussing and explaining treatment options with their patients using PGx data (Dunbar et al., 2012). This in turn, improved engagement between patients and healthcare personnel, helping overcome treatment resistance and increase adherence (Vest et al., 2020). As mentioned in the previous section, psychotropics present with a range of potentially overwhelming side effects, e.g., weight gain, tremors, and insomnia. Patients in psychiatry are more eager to advocate for PGx testing to avoid poorly tolerated medications (McCarthy et al., 2020). In an analysis of 598 patients in psychiatry, interest in the promise of PGx testing was high,

despite familiarity being low (Laplace et al., 2021). This is strengthened by physicians having expectations that PGx testing will become a standard part of clinical care in the future (Chan et al., 2017; Ilona et al., 2015; Vest et al., 2020; Walden et al., 2015). Additional qualitative research aiming to document the pros and cons of PGx testing as experienced by patients is warranted.

Ethical Considerations

Ethical dilemmas pertaining to PGx testing are important topics for discussion. First, given the price point of the testing and limited reimbursement options for lower-income individuals, there is a disparity in the equitable distribution of the test. Second, the PGx genetic status of the patient could be interpreted by insurance companies or other third parties as a healthcare liability (although this would generally imply unfair discrimination if applied), and patients may be concerned about disclosure of this information. Third, pharmacogenetic testing was largely developed using participants of European ancestry, limiting generalizability to other ancestral backgrounds. As such, information pertaining to other ancestral backgrounds is less definitive, potentially resulting in mistrust of genetic research. Lastly, misinterpretation of the test results may occur either by the physician or patient, and as is the case with most other medical interventions, treatment success is not guaranteed. In summary, such issues need to be continuously discussed and evaluated with regulatory bodies to ensure patient safety and satisfaction.

Concluding Remarks and Recommendations

Current pharmacogenetic research shows promising evidence that measurable interindividual differences play a role in predicting medication response for patients in psychiatry. Research continues to identify important pathways composed of genetic and epigenetic variants that may also provide important targets for discovery of novel treatments in the future. In addition, the recent application of polygenic risk scores for categorizing patients into separate risk groups and predicting outcomes in mental health disorders appears very promising to be applied more specifically to medication response and side effects. Based on the material presented in this chapter, actionable suggestions to progress the field of personalized medicine in psychiatry are provided:

1. Updating pharmacogenomic testing panels as further genetic and epigenetic variants are identified and validated, as well as their interactions.
2. Strategically engaging key stakeholders, e.g., patients and government policymakers, in the equitable distribution of PGx tests.
3. Increase effort and funding for AI/ML approaches to encompass the size and complexity of genomic, other omic, clinical, and environmental datasets in PGx research.
4. Careful evaluation of clinical utility and validity of PGx tests by using high-quality randomized controlled trials.

Through such actions, patients with psychiatric disorders can hope to be on more comfortable and precise pathways to recovery, with reduced healthcare costs and improved quality of life.

Declaration of Interest/Funding JLK is a member of the Scientific Advisory Board of Myriad Neuroscience (unpaid) and holds several patents relating to pharmacogenetic tests for psychiatric medications. The remaining authors have no conflicts of interest to disclose.

References

Areda, C. A., Bonizio, R. C., & Freitas, O. D. (2011). Pharmacoeconomy: an indispensable tool for the rationalization of health costs. *Brazilian Journal of Pharmaceutical Sciences, 47*, 231–240.

Aworunse, O. S., Adeniji, O., Oyesola, O. L., Isewon, I., Oyelade, J., & Obembe, O. O. (2018). Genomic interventions in medicine. *Bioinformatics and Biology Insights, 12*, 1177932218816100. https://doi.org/10.1177/1177932218816100

Bousman, C. A., Arandjelovic, K., Mancuso, S. G., Eyre, H. A., & Dunlop, B. W. (2018). Pharmacogenetic tests and depressive symptom remission: a meta-analysis of randomized controlled trials. *Pharmacogenomics, 20*(1), 37–47. https://doi.org/10.2217/pgs-2018-0142

Bradley, P., Shiekh, M., Mehra, V., Vrbicky, K., Layle, S., Olson, M. C., ... Lukowiak, A. A. (2018). Improved efficacy with targeted pharmacogenetic-guided treatment of patients with depression and anxiety: A randomized clinical trial demonstrating clinical utility. *Journal of Psychiatric Research, 96*, 100–107. https://doi.org/10.1016/j.jpsychires.2017.09.024

Castellani, C. A., Melka, M. G., Diehl, E. J., Laufer, B. I., O'Reilly, R. L., & Singh, S. M. (2015). DNA methylation in psychosis: Insights into etiology and treatment. *Epigenomics, 7*(1), 67–74. https://doi.org/10.2217/epi.14.66

Chan, C. Y. W., Chua, B. Y., Subramaniam, M., Suen, E. L. K., & Lee, J. (2017). Clinicians' perceptions of pharmacogenomics use in psychiatry. *Pharmacogenomics, 18*(6), 531–538. https://doi.org/10.2217/pgs-2016-0164

Ching, N. R., Alzghari, A. K., & Alzghari, S. K. (2018). The Relationship of UGT2B15 Pharmacogenetics and Lorazepam for Anxiety. *Cureus, 10*(8), e3133. https://doi.org/10.7759/cureus.3133

Chou, W. H., Yan, F. X., de Leon, J., Barnhill, J., Rogers, T., Cronin, M., et al. (2000). Extension of a pilot study: Impact from the cytochrome P450 2D6 polymorphism on outcome and costs associated with severe mental illness. *Journal of Clinical Psychopharmacology, 20*(2), 246–251. https://doi.org/10.1097/00004714-200004000-00019

Cogent Research. (2010). *Cogent Genomics, Attitudes & Trends study (CGAT)*. Retrieved from http://www.councilforresponsiblegenetics.org/GeneWatch/GeneWatchPage.aspx?

Cole, E. J., Phillips, A. L., Bentzley, B. S., Stimpson, K. H., Nejad, R., Barmak, F., Veerapal, C., Khan, N., Cherian, K., Felber, E., Brown, R., Choi, E., King, S., Pankow, H., Bishop, J. H., Azeez, A., Coetzee, J., Rapier, R., Odenwald, N., Carreon, D., et al. (2022). Stanford neuromodulation therapy (SNT): A double-blind randomized controlled trial. *The American Journal of Psychiatry, 179*(2), 132–141. https://doi.org/10.1176/appi.ajp.2021.20101429

Corponi, F., Fabbri, C., & Serretti, A. (2018). Pharmacogenetics in psychiatry. *Advances in Pharmacology, 83*, 297–331. https://doi.org/10.1016/bs.apha.2018.03.003

Crist, R. C., & Berrettini, W. H. (2014). Pharmacogenetics of OPRM1. *Pharmacology, Biochemistry, and Behavior, 123*, 25–33. https://doi.org/10.1016/j.pbb.2013.10.018

Dugger, S. A., Platt, A., & Goldstein, D. B. (2018). Drug development in the era of precision medicine. *Nature Reviews. Drug Discovery, 17*(3), 183–196. https://doi.org/10.1038/nrd.2017.226

Dunbar, L., Butler, R., Wheeler, A., Pulford, J., Miles, W., & Sheridan, J. (2012). Clinician experiences of employing the AmpliChip® CYP450 test in routine psychiatric practice. *Journal of Psychopharmacology,*

26(3), 390–397. https://doi.org/10.1177/0269881109106957

Dunlop, B. W., Parikh, S. V., Rothschild, A. J., Thase, M. E., DeBattista, C., Conway, C. R., ... Greden, J. F. (2019). Comparing sensitivity to change using the 6-item versus the 17-item Hamilton depression rating scale in the GUIDED randomized controlled trial. *BMC Psychiatry, 19*(1), 420. https://doi.org/10.1186/s12888-019-2410-2

El-Mallakh, R. S., & Watkins, J. (2019). Prolactin elevations and the permeability glycoprotein. *Primary Care Companion for CNS Disorders, 21*(3), pii: 18nr02412.

Eugene, A. R., Masiak, J., & Eugene, B. (2018). Predicting lithium treatment response in bipolar patients using gender-specific gene expression biomarkers and machine learning. *F1000Res, 7*, 474. https://doi.org/10.12688/f1000research.14451.3

Fawver, J., Flanagan, M., Smith, T., Drouin, M., & Mirro, M. (2020). The association of COMT genotype with buproprion treatment response in the treatment of major depressive disorder. *Brain and Behavior: A Cognitive Neuroscience Perspective, 10*(7), e01692. https://doi.org/10.1002/brb3.1692

Fullerton, J. M., & Nurnberger, J. I. (2019). Polygenic risk scores in psychiatry: Will they be useful for clinicians? *F1000Research, 8*, 1293. https://doi.org/10.12688/f1000research.18491.1

Furge, L. L., & Guengerich, F. P. (2006). Cytochrome P450 enzymes in drug metabolism and chemical toxicology: An introduction. *Biochemistry and Molecular Biology Education, 34*(2), 66–74. https://doi.org/10.1002/bmb.2006.49403402066

Garriock, H. A., & Hamilton, S. P. (2009). Genetic studies of drug response and side effects in the STAR*D study, part 2. *The Journal of Clinical Psychiatry, 70*(9), 1323–1325. https://doi.org/10.4088/JCP.09ac05522

Greden, J. F., Parikh, S. V., Rothschild, A. J., Thase, M. E., Dunlop, B. W., DeBattista, C., ... Dechairo, B. (2019). Impact of pharmacogenomics on clinical outcomes in major depressive disorder in the GUIDED trial: A large, patient- and rater-blinded, randomized, controlled study. *Journal of Psychiatry Research, 111*, 59–67. https://doi.org/10.1016/j.jpsychires.2019.01.003

Green, J. S., O'Brien, T. J., Chiappinelli, V. A., & Harralson, A. F. (2010). Pharmacogenomics instruction in US and Canadian medical schools: Implications for personalized medicine. *Pharmacogenomics, 11*(9), 1331–1340. https://doi.org/10.2217/pgs.10.122

Hall-Flavin, D. K., Winner, J. G., Allen, J. D., Carhart, J. M., Proctor, B., Snyder, K. A., ... Mrazek, D. A. (2013). Utility of integrated pharmacogenomic testing to support the treatment of major depressive disorder in a psychiatric outpatient setting. *Pharmacogenetics and Genomics, 23*(10), 535–548. https://doi.org/10.1097/FPC.0b013e3283649b9a

Health, T. L. G. (2020). Mental health matters. The Lancet. *Global Health, 8*(11), e1352.

Herbild, L., Andersen, S. E., Werge, T., Rasmussen, H. B., & Jurgens, G. (2013). Does pharmacogenetic testing for CYP450 2D6 and 2C19 among patients with diagnoses within the schizophrenic spectrum reduce treatment costs? *Basic & Clinical Pharmacology & Toxicology, 113*(4), 266–272. https://doi.org/10.1111/bcpt.12093

Hornberger, J., Li, Q., & Quinn, B. (2015). Cost-effectiveness of combinatorial pharmacogenomic testing for treatment-resistant major depressive disorder patients. *The American Journal of Managed Care, 21*(6), 12.

Ilona, S., Almeida, K., Silvia, R., & Tataronis, G. (2015). Psychiatric pharmacists' perception on the use of pharmacogenomic testing in the mental health population. *Pharmacogenomics, 16*(9), 949–958. https://doi.org/10.2217/pgs.15.22

Ivanov, M., Kacevska, M., & Ingelman-Sundberg, M. (2012). Epigenomics and interindividual differences in drug response. *Clinical Pharmacology & Therapeutics, 92*(6), 727–736. https://doi.org/10.1038/CLPT.2012.152

Jessel, C. D., Al Maruf, A., Oomen, A., Arnold, P. D., & Bousman, C. A. (2022). Pharmacogenetic Testing Knowledge and Attitudes among Pediatric Psychiatrists and Pediatricians in Alberta, Canada. *Journal of the Canadian Academy of Child and Adolescent Psychiatry = Journal de l'Academie canadienne de psychiatrie de l'enfant et de l'adolescent, 31*(1), 18–27.

Jiang, F., Jiang, Y., Zhi, H., Dong, Y., Li, H., Ma, S., et al. (2017). Artificial intelligence in healthcare: Past, present and future. *Stroke and Vascular Neurology, 2*(4), 230–243. https://doi.org/10.1136/svn-2017-000101

Juliano, R. L., & Ling, V. (1976). A surface glycoprotein modulating drug permeability in Chinese hamster ovary cell mutants. *Biochimica et Biophysica Acta, 455*(1), 152–162. https://doi.org/10.1016/0005-2736(76)90160-7

Kacevska, M., Ivanov, M., & Ingelman-Sundberg, M. (2011). Perspectives on epigenetics and its relevance to adverse drug reactions. *Clinical Pharmacology and Therapeutics, 89*(6), 902–907. https://doi.org/10.1038/clpt.2011.21

Karamperis, K., Koromina, M., Papantoniou, P., Skokou, M., Kanellakis, F., Mitropoulos, K., et al. (2021). Economic evaluation in psychiatric pharmacogenomics: A systematic review. *The Pharmacogenomics Journal, 21*(4), 533–541. https://doi.org/10.1038/s41397-021-00249-1

Kato, M., Fukuda, T., Wakeno, M., Fukuda, K., Okugawa, G., Ikenaga, Y., et al. (2006). Effects of the serotonin type 2A, 3A and 3B receptor and the serotonin transporter genes on paroxetine and fluvoxamine efficacy and adverse drug reactions in depressed Japanese patients. *Neuropsychobiology, 53*(4), 186–195. https://doi.org/10.1159/000094727

Kitada, M. (2003). Genetic polymorphism of cytochrome P450 enzymes in Asian populations: Focus on CYP2D6. *International Journal of Clinical*

Pharmacology Research, 23(1), 31–35. Retrieved from https://www.ncbi.nlm.nih.gov/pubmed/14621071

Koppe, G., Meyer-Lindenberg, A., & Durstewitz, D. (2021). Deep learning for small and big data in psychiatry. *Neuropsychopharmacology: official publication of the American College of Neuropsychopharmacology, 46*(1), 176–190. https://doi.org/10.1038/s41386-020-0767-z

Laplace, B., Calvet, B., Lacroix, A., Mouchabac, S., Picard, N., Girard, M., & Charles, E. (2021). Acceptability of pharmacogenetic testing among French psychiatrists, a National Survey. *Journal of Personalized Medicine, 11*(6), 446. https://doi.org/10.3390/jpm11060446

Lauschke, V. M., Barragan, I., & Ingelman-Sundberg, M. (2018). Pharmacoepigenetics and toxicoepigenetics: Novel mechanistic insights and therapeutic opportunities. *Annual Review of Pharmacology and Toxicology, 58*, 161–185. https://doi.org/10.1146/annurev-pharmtox-010617-053021

Lett, T. A., Wallace, T. J., Chowdhury, N. I., Tiwari, A. K., Kennedy, J. L., & Muller, D. J. (2012). Pharmacogenetics of antipsychotic-induced weight gain: Review and clinical implications. *Molecular Psychiatry, 17*(3), 242–266. https://doi.org/10.1038/mp.2011.109

Liko, I., Lai, E., Griffin, R. J., Aquilante, C. L., & Lee, Y. M. (2020). Patients' perspectives on psychiatric pharmacogenetic testing. *Pharmacopsychiatry, 53*(06), 256–261. https://doi.org/10.1055/a-1183-5029

Lin, E., Lin, C.-H., & Lane, H.-Y. (2020). Precision psychiatry applications with pharmacogenomics: Artificial intelligence and machine learning approaches. *International Journal of Molecular Sciences, 21*(3), 969. Retrieved from https://www.mdpi.com/1422-0067/21/3/969

Liu, C., Jiao, C., Wang, K., & Yuan, N. (2018). DNA methylation and psychiatric disorders. *Progress in Molecular Biology and Translational Science, 157*, 175–232. https://doi.org/10.1016/bs.pmbts.2018.01.006

Maciukiewicz, M., Marshe, V. S., Hauschild, A. C., Foster, J. A., Rotzinger, S., Kennedy, J. L., et al. (2018). GWAS-based machine learning approach to predict duloxetine response in major depressive disorder. *Journal of Psychiatric Research, 99*, 62–68. https://doi.org/10.1016/j.jpsychires.2017.12.009

Mas, S., Gassò, P., Álvarez, S., Parellada, E., Bernardo, M., & Lafuente, A. (2012). Intuitive pharmacogenetics: Spontaneous risperidone dosage is related to CYP2D6, CYP3A5 and ABCB1 genotypes. *The Pharmacogenomics Journal, 12*(3), 255–259. https://doi.org/10.1038/tpj.2010.91

McCarthy, M. J., Chen, Y., Demodena, A., Fisher, E., Golshan, S., Suppes, T., & Kelsoe, J. R. (2020). Attitudes on pharmacogenetic testing in psychiatric patients with treatment-resistant depression. *Depression and Anxiety, 37*(9), 842–850. https://doi.org/10.1002/da.23074

Melka, M. G., Castellani, C. A., Laufer, B. I., Rajakumar, R. N., O'Reilly, R., & Singh, S. M. (2013). Olanzapine induced DNA methylation changes support the dopamine hypothesis of psychosis. *Journal of Molecular Psychiatry, 1*(1), 19. https://doi.org/10.1186/2049-9256-1-19

Miller, D. D., & Brown, E. W. (2018). Artificial intelligence in medical practice: The question to the answer? *The American Journal of Medicine, 131*(2), 129–133. https://doi.org/10.1016/j.amjmed.2017.10.035

Natarajan, P., Young, R., Stitziel, N. O., Padmanabhan, S., Baber, U., Mehran, R., et al. (2017). Polygenic risk score identifies subgroup with higher burden of atherosclerosis and greater relative benefit from statin therapy in the primary prevention setting. *Circulation, 135*(22), 2091–2101. https://doi.org/10.1161/CIRCULATIONAHA.116.024436

Nebert, D. W., Wikvall, K., & Miller, W. L. (2013). Human cytochromes P450 in health and disease. *Philosophical Transactions of the Royal Society of London. Series B, Biological Sciences, 368*(1612), 20120431. https://doi.org/10.1098/rstb.2012.0431

Nunes, A., Ardau, R., Berghöfer, A., Bocchetta, A., Chillotti, C., Deiana, V., et al. (2020). Prediction of lithium response using clinical data. *Acta Psychiatrica Scandinavica, 141*(2), 131–141. https://doi.org/10.1111/acps.13122

Patel, M. J., Andreescu, C., Price, J. C., Edelman, K. L., Reynolds, C. F., 3rd, & Aizenstein, H. J. (2015). Machine learning approaches for integrating clinical and imaging features in late-life depression classification and response prediction. *International Journal of Geriatric Psychiatry, 30*(10), 1056–1067. https://doi.org/10.1002/gps.4262

Perez, V., Salavert, A., Espadaler, J., Tuson, M., Saiz-Ruiz, J., Saez-Navarro, C., … Menchon, J. M. (2017). Efficacy of prospective pharmacogenetic testing in the treatment of major depressive disorder: results of a randomized, double-blind clinical trial. *BMC Psychiatry, 17*(1), 250. https://doi.org/10.1186/s12888-017-1412-1

Phillips, K. A., Deverka, P. A., Trosman, J. R., Douglas, M. P., Chambers, J. D., Weldon, C. B., & Dervan, A. P. (2017). Payer coverage policies for multigene tests. *Nature Biotechnology, 35*(7), 614–617. https://doi.org/10.1038/nbt.3912

Reddy, S., Fox, J., & Purohit, M. P. (2019). Artificial intelligence-enabled healthcare delivery. *Journal of the Royal Society of Medicine, 112*(1), 22–28. https://doi.org/10.1177/0141076818815510

Rosenblat, J. D., Lee, Y., & McIntyre, R. S. (2018). The effect of pharmacogenomic testing on response and remission rates in the acute treatment of major depressive disorder: A meta-analysis. *Journal of Affective Disorders, 241*, 484–491. https://doi.org/10.1016/j.jad.2018.08.056

Rush, A. J., Trivedi, M. H., Wisniewski, S. R., Nierenberg, A. A., Stewart, J. W., Warden, D., et al. (2006). Acute and longer-term outcomes in depressed outpatients requiring one or several treatment steps: A STAR*D report. *The American Journal of Psychiatry, 13*, 1905.

Shatte, A. B. R., Hutchinson, D. M., & Teague, S. J. (2019). Machine learning in mental health: A scoping review of methods and applications. *Psychological Medicine, 49*(9), 1426–1448. https://doi.org/10.1017/S0033291719000151

Shishko, I., Almeida, K., Silvia, R. J., & Tataronis, G. R. (2015). Psychiatric pharmacists' perception on the use of pharmacogenomic testing in the mental health population. *Pharmacogenomics, 16*(9), 949–958. https://doi.org/10.2217/pgs.15.22

Singh, A. B. (2015). Improved Antidepressant Remission in Major Depression via a Pharmacokinetic Pathway Polygene Pharmacogenetic Report. *Clinical Psychopharmacology and Neuroscience, 13*(2), 150–156. https://doi.org/10.9758/cpn.2015.13.2.150

Sorich, M. J., Horowitz, J. D., Sorich, W., Wiese, M. D., Pekarsky, B., & Karnon, J. D. (2013). Cost–effectiveness of using *CYP2C19* genotype to guide selection of clopidogrel or ticagrelor in Australia. *Pharmacogenomics, 14*(16), 2013–2021. https://doi.org/10.2217/pgs.13.164

Squarcina, L., Villa, F. M., Nobile, M., Grisan, E., & Brambilla, P. (2021). Deep learning for the prediction of treatment response in depression. *Journal of Affective Disorders, 281*, 618–622. https://doi.org/10.1016/j.jad.2020.11.104

Tadic, A., Muller-Engling, L., Schlicht, K. F., Kotsiari, A., Dreimuller, N., Kleimann, A., et al. (2014). Methylation of the promoter of brain-derived neurotrophic factor exon IV and antidepressant response in major depression. *Molecular Psychiatry, 19*(3), 281–283. https://doi.org/10.1038/mp.2013.58

Thompson, C., Steven, P. H., & Catriona, H. (2015). Psychiatrist attitudes towards pharmacogenetic testing, direct-to-consumer genetic testing, and integrating genetic counseling into psychiatric patient care. *Psychiatry Research, 226*(1), 68–72. https://doi.org/10.1016/j.psychres.2014.11.044

Tiwari, A. K., Zai, C. C., Sajeev, G., Arenovich, T., Müller, D. J., & Kennedy, J. L. (2013). Analysis of 34 candidate genes in bupropion and placebo remission. *The International Journal of Neuropsychopharmacology, 16*(4), 771–781. https://doi.org/10.1017/s1461145712000843

Tiwari, A. K., Zai, C. C., Altar, C. A., Tanner, J. A., Davies, P. E., Traxler, P., et al. (2022). Clinical utility of combinatorial pharmacogenomic testing in depression: A Canadian patient- and rater-blinded, randomized, controlled trial. *Translational Psychiatry, 12*(1), 101. https://doi.org/10.1038/s41398-022-01847-8

Tsankova, N. M., Berton, O., Renthal, W., Kumar, A., Neve, R. L., & Nestler, E. J. (2006). Sustained hippocampal chromatin regulation in a mouse model of depression and antidepressant action. *Nature Neuroscience, 9*(4), 519–525. https://doi.org/10.1038/nn1659

van Schaik, R. H. N., Muller, D. J., Serretti, A., & Ingelman-Sundberg, M. (2020). Pharmacogenetics in psychiatry: An update on clinical usability. *Frontiers in Pharmacology, 11*, 575540. https://doi.org/10.3389/fphar.2020.575540

Vest, B. M., Wray, L. O., Brady, L. A., Thase, M. E., Beehler, G. P., Chapman, S. R., et al. (2020). Primary care and mental health providers' perceptions of implementation of pharmacogenetics testing for depression prescribing. *BMC Psychiatry, 20*(1), 518. https://doi.org/10.1186/s12888-020-02919-z

Walden, L. M., Brandl, E. J., Changasi, A., Sturgess, J. E., Soibel, A., Notario, J. F., et al. (2015). Physicians' opinions following pharmacogenetic testing for psychotropic medication. *Psychiatry Research, 229*(3), 913–918. https://doi.org/10.1016/j.psychres.2015.07.032

Walden, L. M., Brandl, E. J., Tiwari, A. K., Cheema, S., Freeman, N., Braganza, N., et al. (2019). Genetic testing for CYP2D6 and CYP2C19 suggests improved outcome for antidepressant and antipsychotic medication. *Psychiatry Research, 279*, 111–115. https://doi.org/10.1016/j.psychres.2018.02.055

Wilkie, M. J., Smith, G., Day, R. K., Matthews, K., Smith, D., Blackwood, D., et al. (2009). Polymorphisms in the SLC6A4 and HTR2A genes influence treatment outcome following antidepressant therapy. *The Pharmacogenomics Journal, 9*(1), 61–70. https://doi.org/10.1038/sj.tpj.6500491

Winner, J., Allen, J. D., Altar, C. A., & Spahic-Mihajlovic, A. (2013). Psychiatric pharmacogenomics predicts health resource utilization of outpatients with anxiety and depression. *Translational Psychiatry, 3*(3), e242. https://doi.org/10.1038/tp.2013.2

Winner, J. G., Carhart, J. M., Altar, C. A., Goldfarb, S., Allen, J. D., Lavezzari, G., et al. (2015). Combinatorial pharmacogenomic guidance for psychiatric medications reduces overall pharmacy costs in a 1 year prospective evaluation. *Current Medical Research and Opinion, 31*(9), 1633–1643. https://doi.org/10.1185/03007995.2015.1063483

Yu, A. M., & Pan, Y. Z. (2012). Noncoding microRNAs: Small RNAs play a big role in regulation of ADME? *Acta Pharmaceutica Sinica B, 2*(2), 93–101. https://doi.org/10.1016/j.apsb.2012.02.011

Zhang, J. P., & Malhotra, A. K. (2011). Pharmacogenetics and antipsychotics: therapeutic efficacy and side effects prediction. *Expert Opinion Drug Metabolism and Toxicology, 7*(1), 9–37. https://doi.org/10.1517/17425255.2011.532787

Cultural and Ethnic Perspectives in Psychopharmacology

123

Keh-Ming Lin, Mario Braakman, Kazutaka Shimoda, and Norio Yasui-Furukori

Contents

This chapter is an update from the 4th edition. Previous edition authors were Keh-Ming Lin and Margaret T. Lin

K.-M. Lin (✉)
Department of Psychiatry and Biobehavioral Sciences,
David Geffen School of Medicine at UCLA, Los Angeles,
CA, USA
e-mail: linkeh@g.ucla.edu

M. Braakman
Pro Persona Mental Health & Tilburg University,
Wolfheze, Gelderland, The Netherlands

K. Shimoda · N. Yasui-Furukori
Department of Psychiatry, Dokkyo Medical University,
Shimotsuga-gun, Tochigi, Japan

© Springer Nature Switzerland AG 2024
A. Tasman et al. (eds.), *Tasman's Psychiatry*,
https://doi.org/10.1007/978-3-030-51366-5_131

Abstract

Culture and ethnicity represent indispensable dimensions in patient care. Reflecting their cultural and personal backgrounds, patients' concepts on the cause and course of illnesses and treatment expectations often diverge substantially from their physicians deeply immersed in biomedical traditions directing their attention, assessment, and treatment goals. Such discrepancies, pervasive in most clinical encounters, further aggravated where patients and physicians come from distinct cultural systems, contribute to clinically significant nonadherence and placebo response. Efforts in exploring and bridging such gaps are crucial in enhancing therapeutic alliance, treatment engagement, optimization of placebo response, and thus better outcome. At the same time, cross-ethnic variations in pharmacological responses, caused by both genetic and epigenetic factors, are also ubiquitous and clinically significant. Genes controlling both pharmacokinetics and pharmacodynamics are highly polymorphic, whose patterns vary across ethnicity. Superimposed on such genetic variations, exposure to various xenobiotics including herbs, as well as psychosocial stresses, also affect drug responses through pharmacokinetic and pharmacodynamic mechanisms. The rapidly developing fields of pharmacogenetics and pharmacogenomics hold promise for clinicians to take both genetic and epigenetic factors into consideration at the same time, in order to predict the dosing, side-effect profiles, and efficacy of psychiatric medications. Ethnicity is an important dimension in the construction of such pharmacogenomic panels and in interpreting testing results. Finally, while striving to be cognizant of the importance of culture and ethnicity in clinical care, clinicians should also be mindful of the fact that cross-ethnic variations are embedded in interindividual variations and guard against the pitfalls of stereotyping.

Keywords

Culture · Ethnicity · Psychopharmacology · Adherence · Placebo · Nocebo · Pharmacogenetics · Pharmacogenomics · Pharmacokinetics · Pharmacodynamics · Herb · Drug-herb interaction · Cytochrome P-450 enzyme (CYP) · Neurotransmitter · Serotonin · Dopamine · Catechol-O-methyltransferase (COMT) · Serotonin transporter gene (5-HTT; SLC6A4) · Human leukocyte antigen (HLA) · Stevens-Johnson syndrome · Agranulocytosis

Introduction

In the 1950s, a number of serendipitous discoveries heralded the dawning of modern psychopharmacology, revolutionizing the care of hundreds of millions of the mentally ill worldwide (Lin et al., 1993; Lin & Lin, 2013). In the ensuing decades, a large body of literature has emerged to demonstrate the following: (1) used appropriately, all classes of psychotropics are safe and effective across ethnicity and culture; (2) significant cross-cultural/cross-ethnic variations exist regarding specific aspects of drug effects, including dose ranges and side-effect profiles; (3) these differences are embedded in and overlap with substantial interindividual variations in treatment responses; and (4) both cross-ethnic and interindividual variations are determined by specific biological (genetic and epigenetic) and sociocultural mechanisms.

Although traditionally the term "race" has often been used in discussing interpopulation differences, it is avoided in this chapter as much as possible because of the ambiguity of the concept and its enormous potential for misunderstanding and misuse and its association with racism and even genocides (Tishkoff & Kidd, 2004). Instead, "ethnicity," a more accurate, more specific, and more neutral term, denoting people who share common ancestry and identity, is used in association with genetic and biological variations in drug responses.

Dispelling the Myth of "Color-Blindness"

Progress notwithstanding, variations in drug responses at both cross-ethnic and interindividual levels have been largely neglected in medical education and practices. Rooted in modern scientific traditions, biomedicine strives at distilling what may be regarded as "universal," often discounting variations as "noises." While this may be necessary in research, in practice it encourages a "Procrustean," or "one size fits all," approach. As a result, patients not responding to standard therapeutic regimens or manifesting

unusual adverse effects risk being branded as "uncooperative" or "difficult." Neglected is the possibility that their treatment responses may be "atypical" due to specific reasons, ranging from differential rates of drug metabolism to clashes in beliefs and expectations.

As shown in Fig. 1, pharmacological responses are importantly influenced by many factors, both intrinsic (e.g., patients' genetic endowment, age, gender, exposure to xenobiotics, physiological and pathological conditions) and extrinsic (environmental exposures to herbs and toxins, lifestyles, physicians' prescription patterns, medication adherence, and placebo responses). For heuristic purposes, these factors are grouped into "biological" and "non-biological" categories in this chapter, respectively, although it should be noted at the outset that they are not meant to be mutually exclusive, as "non-biological"

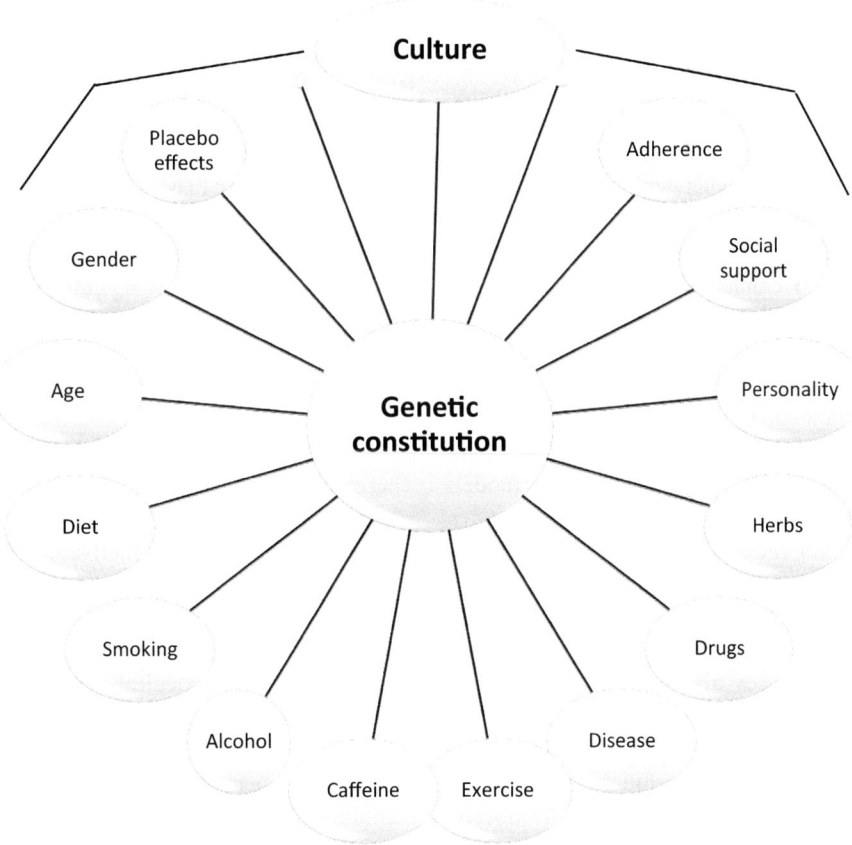

Fig. 1 Factors that affect drug response

phenomena, such as placebo effects, often are associated with prominent physiological changes.

The Sociocultural Context of Psychopharmacotherapy

In conceptualizing how social and cultural forces impact clinical practices, it is important to keep in mind that these considerations are not only crucial when working with "special populations," such as patients from "non-Western" societies and ethnic minority populations, but should be regarded as an indispensable dimension of successful clinical care with all patients. This is so because, going through the intense process of professionalization, modern medical practitioners become deeply immersed in the increasingly technology-driven biomedical traditions and are far removed from their patients in terms of their focus of attention, priorities, and the conceptualization of problems. In this sense, biomedicine is itself a distinct culture, and all clinical encounters are thus "cross-cultural." To highlight such contrasts, George Engel (1977) coined the term "folk models" to represent patients' views on health and illnesses, which are often at odds with the "biomedical model" guiding physicians' thinking and actions. In a similar vein, Arthur Kleinman (Kleinman & Benson, 2006) emphasized that, while patients primarily focus on their "illnesses" (subjective experiences of suffering) and seek "care" (removal of or coping with suffering), physicians are "acculturated" to search for underlying "diseases" and to strive for "cure" (elimination of pathology). Divergence in the explanatory models (EMs) between patients and physicians are often substantial yet hidden, causing miscommunication, dissatisfaction, and poor clinical outcome (Weiss et al., 1992). Such difficulties are further exaggerated where patients' cultural backgrounds and life experiences diverge significantly from their treating physicians (Lin et al., 1993).

In the following, these issues are further elaborated in the context of prescription patterns, medication nonadherence, placebo and nocebo responses, and the use of herbal and other "natural products."

Sociocultural Influences on Clinicians' Assessment and the Prescription of Medications

Patients' cultural/ethnic backgrounds significantly influence how they are diagnosed and treated. Using case vignettes that are identical except ethnic group identification, studies have shown that those identified as non-whites tend to be given more severe diagnoses. Reflecting such a bias, many clinical studies demonstrated that African American psychiatric patients are more likely diagnosed with schizophrenia. However, when structured interviews were used to reassess these patients, such differences disappeared, suggesting that the "overdiagnosis" was determined by clinicians' biases rather than patients' clinical manifestations (Adebimpe, 1981).

For similar reasons, African American patients also often receive higher doses of neuroleptics, irrespective of diagnosis. Further, they are more likely to be treated with older (first-generation) and less expensive antipsychotics (Schwartz & Blankenship, 2014), and with depot antipsychotics, presumably owing to a general preconception of their being less compliant to treatment. These diagnostic and treatment biases are far from innocuous. Several large-scale studies have documented significantly higher rates of tardive dyskinesia in African Americans (Wonodi et al., 2004), possibly due to their excessive exposure to antipsychotics.

Adherence

The magnitude of nonadherence to psychotropic medications ranges from 20 to 90%. Although factors such as insight and motivation may render treatment adherence particularly challenging for psychiatric patients, the phenomenon is far from unique to psychiatry, as similarly high rates also exist in other chronic medical conditions requiring long-term pharmacotherapy (Kleinsinger, 2018).

Factors that contribute to difficulties in treatment engagement and adherence are myriad, including financial and logistic limitations (e.g., lack of transportation or childcare), health literacy and miscommunication, structural barriers such as long wait time and difficulties negotiating the bureaucratic and confusing healthcare systems, and lack of family and social support (Acosta et al., 2012). Superimposed on these already daunting structural and logistic challenges, an even more crucial determinant of adherence and treatment outcome is the discrepancy between patients' and their physicians' understanding of what may be at the root of their sufferings, what might be most helpful in relieving their distress, and how they prioritize their treatment needs (Kleinman & Benson, 2006; Weiss et al., 1992). In clinical encounters where patients' sociocultural backgrounds differ from those of the physicians, such gaps are further magnified, leading to delays in help-seeking, early and frequent dropouts, and poor medication adherence (Diaz et al., 2005; Sue, 1977). Supporting such findings, a large body of literature based on the health belief model (HBM) further demonstrates the importance of patients' beliefs in determining their health-related behaviors, including treatment engagement and response (Budd et al., 1996).

The Explanatory Model EM represents an effective approach to improve adherence and treatment outcomes (Kleinman & Benson, 2006). By systematically eliciting the patient's perspectives on their symptoms, attributions, help-seeking experiences, and preferences, the model facilitates the bridging of the "biomedical" and the "folk" models, minimizing nonadherence. Such an approach serves as the basis for the cultural formulation that is part of the *Diagnostic and Statistical Manual of Mental Disorders* (DSM-5; American Psychiatric Association, 2013).

To further highlight the importance of sociocultural factors in determining medication adherence, several examples are presented below.

At one of the country's most well-established psychiatric clinics specializing in the care of refugee patients, Kinzie and his associates (1987) observed that 85% of their depressed patients receiving adequate doses of tricyclic antidepressants (TCAs) showed either no or very low TCA serum levels. When confronted, these Southeast Asian patients admitted to nonadherence for a variety of reasons. Education emphasizing the importance of long-term medication and the maintenance of appropriate blood levels resulted in significant

improvements in adherence in some, but not all, of these refugee patients.

Similarly, a study in South Africa (Gillis et al., 1987) followed 406 patients belonging to 3 ethnic groups (Whites, Blacks, and Asian Indians) for 2 years after discharge from a psychiatric hospital. They found that approximately two-thirds of black patients, half of the Asian-Indian patients, and one-quarter of white patients were non-adherent for oral phenothiazines. Black and Asian-Indian patients, and their families, had a poorer understanding of the treatment protocols because of cultural gaps and difficulties in communication. Similar observations have been reported in African Americans (Price et al., 1985).

Depending on patients' beliefs and expectations, drug effects could be interpreted as either negative or positive, resulting in different levels of adherence. For example, in a study of Hong Kong Chinese bipolar patients treated with lithium, Lee et al. (1992) found that, unlike their Western counterparts, Chinese patients rarely complained of "missing the highs" and "loss of creativity" and regarded polydipsia, polyuria, and weight gain as signs suggesting that the medicine was working. In contrast, lethargy, drowsiness, and poor memory represented serious concerns for many of these patients, even though such symptoms occurred at a similar rate among their matched controls. Such findings highlight the importance of culturally based beliefs and expectations in determining how drug effects are interpreted, irrespective of their chemical and physiological properties.

"Placebo" and "Nocebo" Effects

The "placebo" phenomenon represents a paradox that has long been shrouded in mysteries (Louhiala, 2020). In most randomized, blinded, placebo-controlled drug trials and intervention research, placebo accounts for 30–50% of the response, leaving not much room for the experimental drugs to show their efficacy. The power of placebo is such that more than half of industry-sponsored phase III drug studies end up as "failed trials" (Jadad & Enkin, 2007).

Despite its potency, placebo effects are often seen in clinical settings as annoyance or threats to physicians' authority and credibility and discounted as "tricks of the mind." However, by ignoring or minimizing the role of placebo, clinicians are likely to misattributing efficacy to interventions that may be unnecessary or even harmful.

Further, by trivializing the placebo phenomenon, clinicians may also unknowingly inflict "nocebo" effects, the opposite of "placebo," upon their patients. Although rarely discussed, "nocebo" is neither rare nor innocuous. In clinical trials, subjects on the placebo arms commonly experience substantial "side effects" identical to those that are attributed to active drugs (Petersen et al., 2014; Wood et al., 2020). "Voodoo death" serves as an extreme example of the power of nocebo, where "curses" alone sufficed to induce disastrous outcomes in susceptible individuals (Sternberg, 2002).

Although psychological mechanisms including suggestion, expectation, and conditioning have been shown to facilitate "placebo" response, it is not *only* a *purely* psychological phenomenon, as it is typically accompanied by concurrent biological changes. Such responses, including alterations in endocrine and brain functions, are often indistinguishable from those induced by active drugs (Louhiala, 2020).

Even more importantly, "placebo" does not depend on "deception." This is most clearly demonstrated in well-designed open-labeled placebo studies where patients were given sugar pills and informed as such (Kaptchuk & Miller, 2018), which nevertheless induced outcomes comparable to the placebo administered in blinded drug studies, where subjects were not sure if they were receiving placebo.

Given the power of placebo, a strong case could be made for physicians to embrace it, regard it as a reflection of patient's trust in and therapeutic alliance with them, instead of leaving it in the dark or seeing it as a threat to their authority and expertise. By using the EM or similar approaches to elucidate patients' beliefs and expectations, clinicians could more easily explain their proposed treatment approaches in ways congruent with the patient's belief system. This should foster trust, hope, and positive expectations conducive to placebo responses, healing, and recovery.

As an example, the concept of Qi ("vital energy") remains salient in how many modern Chinese view their health and illnesses, even for those who may be regarded as highly "Westernized" (Lin, 1981). When suffering from depression, they may interpret their symptoms as consequences of deficiency, or disturbances in the circulation, of Qi. While working with such patients, physicians might do well incorporating the Qi theories to explain how antidepressants work, in lieu of, or in addition to, neurotransmitter hypotheses, which may also be regarded as equally speculative.

Similarly, most traditional medical systems emphasize the importance of maintaining a dynamic balance between "coldness" and "hotness" (Castro et al., 1984) or between "yin" and "yang" in the case of the East Asian belief systems (Lin, 1981). For patients who subscribe to such beliefs, a perceived mismatch between the therapeutic agents and the traditional beliefs may diminish the placebo response or even induce nocebo effects and nonadherence. For example, red-colored pills might be regarded as capable of enhancing the "hot" element and thus less effective in treating conditions perceived as the result of excessive "hotness" (e.g., fever, anxiety state, or mania).

Buckalew and Coffield (1982) reported significant ethnic differences in response to placebo pills with different colors. In this well-controlled study, white capsules were seen as analgesics by Caucasian subjects but as stimulants by their African American counterparts. In contrast, black capsules were seen as stimulants by Caucasians and as analgesics by African Americans.

The Concomitant Use of Complementary/Alternative/Indigenous Treatment and Healing Methods

Contrary to common assumptions, the worldwide ascendancy of modern Western medicine has not led to the eclipse of traditional medical and healing systems. Instead, "alternative" traditions (e.g., Chinese and Ayurvedic medicines) have continued to thrive, coexisting and competing with modern bio-medicine. Such practices are not only limited to "non-Western" societies and ethnic minority populations but have also become increasingly popular in North American and Western European mainstream communities (Fjær et al., 2020).

Herbs and other "natural" substances, used widely in traditional practices, are bioactive and often possess potent physiological effects that are not necessarily "benign" or nontoxic. This should not come as a surprise since most of the widely used contemporary pharmacotherapeutic agents, such as digitalis, opium, ephedrine, and penicillin, have their roots in nature.

In addition to their inherent biological properties, herbs also alter the effects of prescribed drugs (Niv et al., 2010). For example, herbal preparations possessing potent anticholinergic properties may cause atropine psychosis, especially when used concomitantly with psychotropics with similar side-effect profiles. Similarly, herbs with distinctive stimulant or sedative properties may either potentiate or attenuate the effects of psychiatric medications prescribed for the same purposes.

Herbs also often interact with prescribed drugs through pharmacokinetic mechanisms. This is expected since xenobiotics (natural substances), abundant in our ancestors' environments and current non-processed foods, were the original targets for the "drug-metabolizing enzymes," whose expression and functioning are often modulated by herbal products, leading to widespread "herb-drug interactions."

Since patients from different cultural backgrounds may vary in their exposure to divergent herbs and other natural substances, the inquiry into such possibilities should be regarded as an important aspect of patient care.

Genetic Diversity, Ethnicity, and Drug Responses

"Symbolic" effects aside, psychotropics do exert powerful "instrumental" effects that are biology-based and are regulated by pharmacokinetic and pharmacodynamic processes. While pharmacokinetics dictates organisms' processing and elimination of drugs and other xenobiotics (what does the body do to the drug), pharmacodynamics governs medications' effect on organisms' physiological functions (what does the drug do to the body). Together, they determine medications' dosing consideration, efficacy, toxicity, and other adverse effects.

Due to remarkable progress in pharmaco-genetics in recent decades, much has been clarified regarding the genetic basis of both pharmacokinetics and pharmacodynamics. Importantly, most, if not all, of the genes responsible for these processes are highly polymorphic, usually caused by either the substitution of a single nuclear acid-base pair (SNP) or other similarly minor alterations. Although structurally the changes distinguishing variant allele types and genotypes are minuscule (the length of most human genes ranges from 10 to 15,000 base pairs), they often alter the structure and functions of the genes, leading to up to 40 folds of variations in drug responses among individuals within any given population. Evolutionarily, such diversity was advantageous for our ancestors who were constantly exposed to divergent xenobiotics and other environmental challenges, and heterogeneity favors populations' survival; and conversely, some specific variants were selected for in different geographic populations due to the available food sources.

Reflecting the history of human migration in the past hundred thousand years, similar evolutionary forces also led to remarkable cross-ethnic variations in the frequency of these genetic polymorphisms. Such patterns have long been utilized by researchers to trace the origin and migration of populations (e.g., the populating of Polynesians across the whole Pacific Oceans in the past 10,000 years) (Cavalli-Sforza & Cavalli-Sforza, 1995) and have more recently been commercialized for those curious about their ancestral origins (Horton, et al., 2019). Medically, knowledge of cross-group variations is important because they in part determine disease susceptibility and treatment responses (Cooper, 2004).

Ethnic and Individual Variations in Pharmacokinetics

Dramatic cross-country and cross-ethnic differences in the optimal dosing and side-effect profiles were reported by different pioneers in the field soon after modern psychotropics were introduced in clinical care (Lin et al., 1993). However, such claims were based on clinical impressions and observations, whose magnitude and mechanisms had remained unexplored.

By measuring serum levels of psychotropics, several studies conducted in the 1980s showed that such differences were at least in part biologically determined. One of these studies, Lin et al. (1988), showed marked differences in serum haloperidol concentrations between Asian and Caucasian volunteers after the administration of a small dose of haloperidol (see Fig. 2), with the Asian group achieving up to 2.6 times higher haloperidol concentrations. These results indicate pharmacokinetic factors, including absorption and hepatic first-pass metabolism, may contribute to the difference in haloperidol responses between Caucasians and Asians.

Similar ethnic differences have been shown with other psychotropics including desipramine, clomipramine, adinazolam, and clozapine and medications used in other medical fields (Ng et al., 2008). For example, a review of clozapine concludes that East and South Asians (together represent 50% of the world's population) need only half the clozapine dosage used in Caucasian patients in Western countries to achieve similar levels of blood concentrations (de Leon et al., 2020). In a study using a population pharmacokinetic approach, Shimoda et al. (1999) also reported that Japanese depressed patients were three to six times slower in their clearance of clomipramine and desmethyl-clomipramine, its major metabolite, as compared to Swedes (both clozapine and clomipramine share CYP1A2 in their metabolism; see discussion below).

As shown in Fig. 2, although cross-ethnic contrasts in the pharmacokinetics of these medications are substantial in aggregate, they are embedded in remarkable interindividual variations, such that some Asians showed blood concentrations lower than Caucasians, and vice versa, highlighting the peril of stereotypical interpretation of ethnic variation in drug responses.

Genetic Polymorphism of Genes Encoding "Drug-Metabolizing Enzymes"

As shown in Fig. 3, of the four factors (absorption, distribution, metabolism, and excretion) that

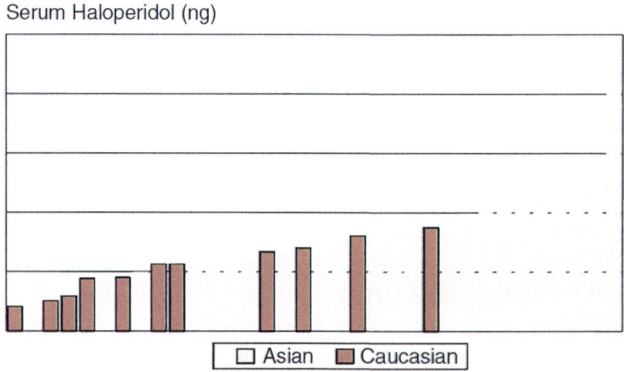

Fig. 2 Variability of haloperidol concentrations. Variability of haloperidol concentrations in normal volunteers after the administration of haloperidol (0.5 mg, IM). The graph shows (1) substantial interindividual variability within each of the ethnic groups, (2) dramatic differences in the pharmacokinetics of haloperidol between the two ethnic groups, and (3) overlap of the pharmacokinetics between the two groups. (From Lin et al. (1988). Reproduced with permission of Lippincott Williams & Wilkins)

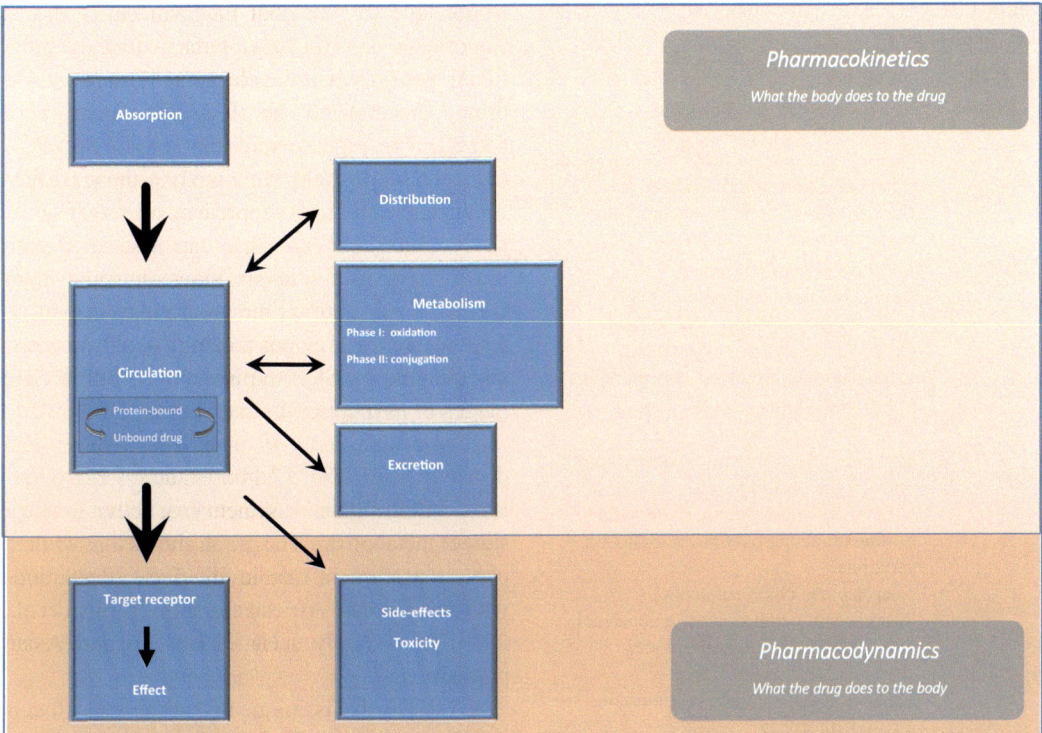

Fig. 3 Pharmacokinetics and pharmacodynamics

together determine the fate and disposition of most drugs, variability in the process of metabolism is most substantial and is usually the reason for interindividual and cross-ethnic variation in drug responses (Lin & Lin, 2013).

Most drugs are metabolized via two phases: phase I, commonly mediated by one or more of the cytochrome P-450 enzymes (CYPs), leads to the oxidation of the substrates; phase II involves conjugation and is usually mediated by one of the transferases. In general, both phases render the substrates more water-soluble and hence more ready for excretion. There is clear evidence of interindividual and cross-ethnic variations in the activities of enzymes in both phases, the genetic basis of which has been increasingly elucidated in recent years. Since far more information is currently available regarding the CYPs than the phase II enzymes and since the CYPs appear to control the rate-limiting steps in the metabolism of most psychotropics, the following discussions will focus mainly on these enzymes.

Table 1 includes a list of major CYPs that handle the phase I metabolism of commonly used psychotropics and selected substances that are psychoactive and are commonly used by psychiatric patients. With few exceptions (e.g., lithium does not require biotransformation; gabapentin is eliminated unchanged by the kidney; lorazepam is directly conjugated without first going through oxidation), the pharmacokinetics of practically all psychotropics depend on one or more of the CYPs, whose activities significantly influence the tissue concentrations, dose requirement, and side-effect profiles of their substrates.

Functionally significant genetic polymorphisms exist in most of the CYPs, leading to extremely large variations in the activity of these enzymes in any given population. CYP2D6 represents the most dramatic example. One hundred and forty-five variants have been discovered so far that significantly affect the functioning of this enzyme (Consortium, The Pharmacogene Variation, 2021). Significantly, many of these mutant

Table 1 Major human cytochrome P450 enzymes and their substrates

Enzyme	Substrates
CYP1A2	*Antidepressants*: amitriptyline, clomipramine, imipramine, fluvoxamine *Neuroleptics*: haloperidol, phenothiazines, thiothixene, clozapine, olanzapine *Others*: tacrine, caffeine, theophylline, acetaminophen, phenacetin
CYP2C19	*Benzodiazepines*: diazepam *Antidepressants*: imipramine, amitriptyline, clomipramine, citalopram *Others*: propranolol, hexobarbital, mephobarbital, proguanil, omeprazole, (*S*)-mephenytoin
CYP2D6	*Antidepressants*: amitriptyline, clomipramine, imipramine, desipramine, nortriptyline, trimipramine, *N*-desmethylclomipramine, fluoxetine, norfluoxetine, paroxetine, venlafaxine, sertraline *Neuroleptics*: chlorpromazine, thioridazine, perphenazine, haloperidol, reduced haloperidol, risperidone, clozapine, sertindole *Others*: codeine, opiate, propranolol, dextromethorphan
CYP2E1	*Ethanol and acetaldehyde* *Neuropsychiatric medications:* zopiclone, eszopiclone, verapamil *Others*: acetaminophen, theophylline
CYP3A4	*Antidepressants*: mirtazapine, nefazodone, sertraline *Neuroleptics*: thioridazine, haloperidol, clozapine, quetiapine, risperidone, sertindole, ziprasidone *Mood stabilizers*: carbamazepine, gabapentin, lamotrigine *Benzodiazepines*: alprazolam, clonazepam, diazepam, midazolam, triazolam, zolpidem *Calcium channel blockers*: diltiazem, nifedipine, nimodipine, verapamil *Steroids*: androgens, estrogens, cortisol *Others*: erythromycin, terfenadine, cyclosporine, dapsone, lovastatin, lidocaine, alfentanil, amiodarone, codeine, sildenafil

Note: mainly psychopharmacological substrates and some additional substrates relevant for psychiatry
For a full overview, see https://drug-interactions.medicine.iu.edu/MainTable.aspx

alleles are to a large extent ethnically specific. For example, *CYP2D6*4*, which leads to the production of defective proteins, is found in approximately 25% of Caucasians but is rarely identified in other ethnic groups. This mutation is mainly responsible for the poor metabolizers (PMs) in Caucasians (5–9%), who possess two dysfunctional genes and are exceptionally sensitive to drugs metabolized by CYP2D6. Instead of *CYP2D6*4,* high frequencies of *CYP2D6*10* and *CYP2D6*17* are found among those of East Asian and sub-Saharan African origins, respectively. Both of these alleles are associated with lower enzyme activities and slower metabolism of CYP2D6 substrates (Fig. 3) and may in part be responsible for previous findings of slower pharmacokinetic profiles and lower therapeutic dose ranges of psychotropics in Asians.

Figure 4 presents an overview of the global distribution of *2D6*17*, one of the CYP2D6 variants conferring decreased enzyme activities and a slower metabolism. The graph shows that while it plays a prominent role in all of the populations with sub-Saharan African ancestry (Rajman et al., 2017), it is rarely seen in Western and Asian populations.

CYP2D6 also is unique in that the gene often is duplicated or multiplied (up to 13 copies). Those possessing these duplicated or multiple genes, termed "ultrarapid" metabolizers (UMs), have proportionally more enzymes and faster enzyme activities. They are found in 1% of Swedes, 5% of Spaniards (white Americans are in between these two figures), 8.8% of Ashkenazi Jews (Scott et al., 2007), 13–18% of non-European Jews (Luo et al., 2004), 19% of Arabs, and 29% of Ethiopians (Aklillu et al. 1996). UM patients are likely to fail to respond to usual doses of medications biotransformed by CYP2D6, since they typically fail to achieve therapeutic drug levels unless treated with unusually high doses of the same drugs. Consequently, they may be mistakenly blamed as nonadherent (Aklillu et al., 1996). On the other hand, "regular" doses of prodrugs such as codeine which depend on CYP2D6 for their conversion to their active forms (morphine in the case of codeine) could lead to unexpected toxicities in those with the UM status (Gasche et al., 2004).

CYP2C19 represents another dramatic example of the existence of both cross-ethnic and interindividual variations in drug metabolism. This enzyme is involved in the metabolism of

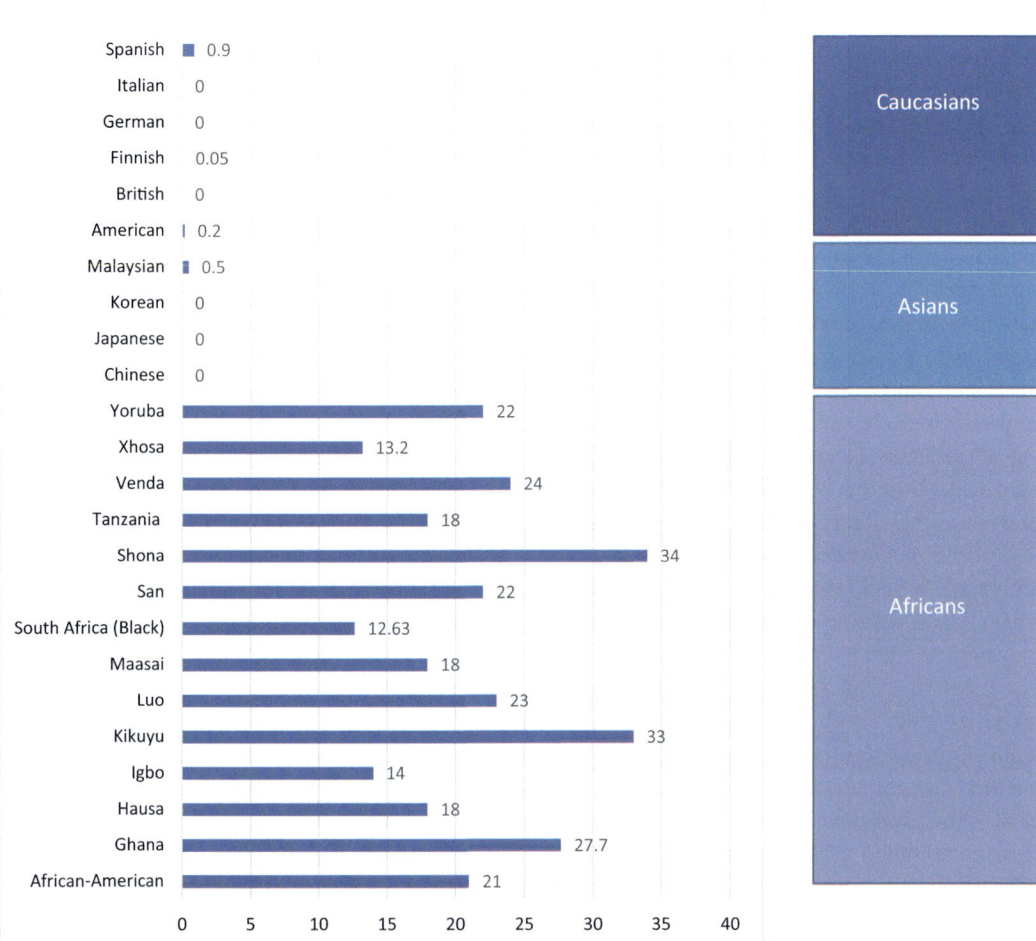

Data extracted from supplementary files: Rajman I, Knapp L, Morgan T, Masimirembwa C. African Genetic Diversity: Implications for Cytochrome P450-mediated Drug Metabolism and Drug Development. EBioMedicine. 2017;17:67-74

Fig. 4 *CYP2D6*17* variations in different populations. (Data extracted from supplementary files: Rajman et al. (2017))

commonly used psychotropics such as diazepam and tertiary TCAs, as well as citalopram. Using S-mephenytoin as probe, earlier studies demonstrated that up to 20% of East Asians (Chinese, Japanese, and Koreans), but only in 3–5% of Caucasians, are PMs (Qin et al., 1999). The enzyme deficiency is mainly caused by two unique mutations (*CYP2C19*2* and *CYP2C19*3*). While *2 can be found in all ethnic groups, *3 is specific to those with Eastern Asian origins. The presence of *3, together with a higher frequency of *2, is responsible for the higher rate of PMs among Asians, rendering them more sensitive to drugs that are substrates of CYP2C19.

In contrast, *CYP2C19*17*, encoding a variant enzyme with higher activities, is found in about one-quarter of European and sub-Saharan African populations but is negligible in East Asians (Biswas, 2020).

Genetic polymorphism also exists in many other drug-metabolizing enzymes. It is interesting to note that, almost without exception, wherever genetic polymorphism is identified, the allele frequency of the mutations typically shows substantial ethnic variations.

Factors Affecting the Expression of Drug-Metabolizing Enzymes

Besides genetic endowments, a large number of nongenetic factors, both external and internal, also significantly influence the expression of drug-metabolizing genes. External factors include nutrients, various plant products, pharmaceutical agents, and other chemicals. Internal factors include steroid hormones and other endogenous substances. They could either inhibit or induce the activities of the enzymes and thus affect the metabolism of drugs that are their substrates. Such shifts in the enzyme status could at times lead to serious clinical consequences. The following are selected examples that have caught the field's attention in recent years: (1) Many antidepressants, including fluoxetine and paroxetine, two of the most widely used selective serotonin reuptake inhibitors (SSRIs), are potent inhibitors of CYP2D6, capable of converting an extensive metabolizer into a PM. When these drugs are prescribed for a patient who has already been taking a CYP2D6 substrate (e.g., TCAs or neuroleptics), the concentration of the latter could be pushed unexpectedly into the toxic range (Lam et al., 2002). Of note is that both fluoxetine and paroxetine inhibit CYP2D6 *and* CYP3A4 which collectively metabolize over 80% of psychotropic drugs, suggesting that a drug-drug interaction is nearly guaranteed if either of these antidepressants is prescribed. (2) Smoking has long been known to significantly reduce the serum concentration of psychotropics and may be responsible for the relapse of many patients soon after discharge from the hospital because they resumed smoking (Guengerich et al., 1994). This effect is now known to be due to the induction of CYP1A2 by constituents of tobacco or marijuana (essentially any burned hydrocarbon). (3) Many drugs and natural substances significantly inhibit the activity of CYP3A4, altering its ability to metabolize drugs that depend on this enzyme for their biotransformation. A widely known example is the grapefruit juice (Bailey et al., 2013), which can increase the blood level of antiviral drugs and psychotropics such as nefazodone and alprazolam

by several folds if taken concurrently. (4) A large body of literature indicates that macronutrients (i.e., high-protein versus high-carbohydrate diet) also significantly influence the activity of CYPs. A high-protein diet accelerates the metabolism of drugs such as antipyrine and theophylline and appears to have the opposite effect (Kirkpatrick et al., 2019).

These examples demonstrate the importance of environmental factors in substantially modifying the activity of drug-metabolizing enzymes. Patients from different ethnic/cultural backgrounds live divergent lifestyles and are likely to be exposed to unique substances that may have powerful effects on the expression and activity of drug-metabolizing enzymes. For example, studies have shown that Asian Indians and Africans were significantly slower in metabolizing substrates of CYP1A2, such as theophylline, antipyrine, and clomipramine. However, after they immigrated to Europe and adapted to the new dietary habits, their metabolic profiles for these drugs became indistinguishable from the "native" Westerners (Allen et al., 1977).

As discussed earlier, herbal medicines are routinely and extensively used by people worldwide. Since patients typically are not aware of the potential of herb-drug interactions, they often combine the use of herbs and "Western medicines." When severe toxic effects subsequently emerge, they are likely to attribute such effects on drugs prescribed by clinicians, rather than on herbal preparations obtained over the counter, or from traditional/ alternative practitioners. Since thousands of herbs are widely in use and the indication and popularity of these herbs vary greatly across cultural traditions, the potential for interactions between herbs and modern pharmaceutical agents is endless and largely unexplored (Li & Weng, 2017).

As a prominent example, St. John's wort, a popular over-the-counter herb with possible antidepressant effects, markedly decreases the effectiveness of medications such as oral contraceptives and warfarin (Nicolussi et al., 2020), through the induction of CYP3A4 enzyme and the P-glycoprotein drug transport system. In contrast, by inhibiting CYP3A4, a small glass of grapefruit juice could cause severe toxicity of medications depending on this enzyme for their metabolism. The ingredient responsible for such interactions, such as furanocoumarins and flavonoids, is also found in other citrus fruits and commonly used herbs.

Ethnic and Individual Variations due to Absorption, Distribution, and Excretion

Although less well studied, evidence also points to ethnic variations in other dimensions of pharmacokinetics, including absorption, distribution, and excretion.

Absorption: Contrary to common perceptions, drug is not only absorbed by passive diffusion but often also involves active transporter processes. For example, MDR1/ABCB1, the gene encoding P-glycoprotein responsible for transmembrane drug transport, is highly polymorphic, with the C3435T allele leading to overexpression of the gene. This variant allele is twice as frequent (83–84%) in populations of African descent compared to British Caucasians (48%), Portuguese (43%), and Javanese patients from Indonesia (47%) (Ameyaw et al., 2001, Istikharah et al., 2020).

Distribution: Regarding distribution, drugs binding only to albumin do not show ethnic differences in plasma protein binding. In contrast, drugs binding to alpha-1-acid glycoprotein (AGP) consistently show higher binding (and lower plasma-free fractions) in Caucasian populations than other ethnic groups (Johnson, 2000).

Excretion: Although not extensively studied, there is also evidence suggesting ethnic differences affecting excretion. As an example, expression levels of some drug transporter genes (ATP7B and KCNJ8) in human kidneys were found to be significantly higher in African American females compared to European American females (Joseph et al., 2015), confirming a greater clearance and lower blood levels of certain drugs.

Genetic Polymorphism of Genes Affecting Pharmacodynamics

Monoamines, especially dopamine (DA) and serotonin (5-HT), have been the primary focus of biological research of various psychiatric disorders, including schizophrenia and mood disorders, and also are targets for the development of novel psychotropic agents since the 1950s. The functioning of these systems depends on genes encoding the receptors and transporters of these monoamines, as well as those involved in their biosynthesis and catabolism. These genes also are highly polymorphic, and the patterns of these polymorphisms vary substantively across ethnicity. The following sections summarize relevant information on genes associated with the function of these two monoamine neurotransmitter systems.

Dopaminergic System

Catechol-O-Methyltransferase (COMT): A single nucleotide change from guanine to adenine (G > A) at codon position 158, resulting in the replacement of valine by methionine (Val158Met) in the encoded protein responsible for the catabolism of dopamine and other catecholamines, confers a fourfold reduction of the activity of the enzyme. This polymorphism has been shown to significantly influence responses to levodopa and psychostimulants and may also modulate the effect of neuroleptics and antidepressants on the dopaminergic system. Numerous studies have suggested an association between the gene and many psychiatric disorders, including schizophrenia (González-Castro et al., 2016), mood and anxiety disorders, attention-deficit hyperactive disorder, and addiction disorders (Hall et al., 2019). The allele frequency of this genotype is lower in African Americans (26%) and East Asians (18%) in comparison with Caucasians (50%). Reflecting these differences, patients with African American and Asian heritages achieve lower levodopa plasma levels than their Caucasian counterparts and are more likely to suffer from dyskinesia. Such discrepancies may also be responsible for the inconsistent results of genetic association studies conducted in different populations. For example, Taylor (2018) reported that the Val allele was associated with substance-use disorders only in East Asian populations and with attention-deficit hyperactivity disorder and panic disorder in Caucasians.

Dopamine D2 Receptor (DRD2): In addition to being the primary target of most neuroleptics, this receptor also plays an important role in mediating many brain functions, especially attention, learning, motivation, and response to rewards, and has been a major focus of schizophrenia and addiction research.

Two polymorphisms, in particular, are highly associated with the expression of the gene in the brain. The distribution of their allele frequencies varies significantly across ethnic groups: (1) The *TaqIA1* allele ranges from 9–22% in Caucasians to approximately 36% in African Americans and 42–47% in East Asians. As a promising candidate allele for the risk of alcoholism, *TaqIA1* has been a focus of interest in many such studies. However, by neglecting ethnicity as an important variable, some of the earlier studies yielded inconsistent results, highlighting the importance of population stratification in genetic studies (Munafò et al., 2007). (2) A cytosine (C) insertion/deletion (Ins/Del) at nucleotide position -141C (-141C Ins/Del) polymorphism is found in 9% in Caucasians, 22% in East Asians, and 50% in sub-Saharan Africans. It has been identified as a risk marker for schizophrenia in Han Chinese, but not in Caucasian, Japanese, and Indian populations (Zhao et al., 2016).

Dopamine D3 Receptor (DRD3): DRD3 has a strong affinity not only to first-generation neuroleptics but also to atypical antipsychotics (second-generation neuroleptics), such as sulpiride, risperidone, and clozapine. It is particularly abundant in the limbic system of the brain, an area important for regulating emotion and the modulation of stress. A *Bal* I RFLP polymorphism at the first exon of the D3 receptor gene, with A-to-G transition and the substitution of serine by glycine at codon 9, is associated with responses to risperidone and clozapine and is a potential risk factor for schizophrenia and tardive dyskinesia. The frequency of the allele *Bal I A1* is 66% in East Asians and 20% in sub-Saharan Africans (Dubertret et al., 1998). Among Caucasian populations, it ranges from 72% in Swedes (Jönsson et al., 1999) and French (Lannfelt et al., 1992) to 57% in Italians (Catalano et al., 1994), showing sizable variations even within the European continent.

Dopamine Transporter (DAT; SLC6A3): The polymorphism of a 40 bp variable number tandem repeat (VNTR) in the 3′ untranslated region of the dopamine transporter gene (DAT1-VNTR) significantly affects the basal level of expression of the transporter and is associated with many psychiatric and addictive conditions, including bipolar disorder, depression, ADHD, and alcohol withdrawal symptoms. The frequency of the 9-copy allele, conferring increased production of the transporter, ranges from 35% in Caucasian and 24% in Africans to 5% in East Asians and is absent in American Indians and Australo-Melanesians (Mitchell et al., 2000).

Serotonergic (5-HT) System

Serotonin Transporter Gene (5-HTT; SLC6A4): As the target of selective serotonin reuptake inhibitors (SSRIs) and other newer anti-depressants, the serotonin transporter has been a focus of intense research efforts for the last three decades. In particular, the serotonin transporter gene-linked polymorphic region (5-HTTLPR) polymorphism, a 44 base-pair insertion/deletion in the promoter region of the gene, includes two major variants: the short (*s*) variant with14 repeats shows less transcriptional activity and lower serotonin uptake, as compared to the long (*l*) variant with 16 or more repeats. The frequency of the *s* allele ranges from 26% African Americans to 40% in Caucasians and 80% in East Asians (Poland et al., 2013; Porcelli et al., 2012). A study on Eastern European populations showed *s* allele frequency of 38% in Croats, 44% in Russians, and 50% in Tatars and Bashkirs (Turkish) (Noskova et al., 2008).

5-HTTLPR polymorphism has been shown to correlate with SSRI responses in many studies. In

Caucasian populations, the *l* allele is consistently associated with better treatment outcomes. However, this appears to be less clear in other ethnic groups, with most East Asian studies showing better responses in those with the *s* allele, again highlighting the importance of considering ethnic backgrounds when interpreting genetic findings (El-Mallakh & Ali, 2019).

In addition to its relationship with drug response, the polymorphism is also intriguing for its effects on stress response. Those possessing the *ss* allele may be more sensitive to social cues and may respond with greater fear and anxiety when confronted with adverse stimuli (Beevers et al., 2010). For example, *s* allele has been reported to interact with childhood trauma and stressful life events to increase the general risk for PTSD (Walsh et al., 2014). Paralleling this, in rhesus monkey -variant also confers increased vulnerability to adverse effects of psychosocial stress associated with subordinate status, while those with the l/ll/l genotype benefit the most from the absence of stress conferred by dominant social status (Lindell et al., 2012).

Genetic Polymorphism of Human Leukocyte Antigens (HLA)

Human leukocyte antigen (HLA) proteins, encoded within the major histocompatibility complex (MHC) locus, play an essential role in the adaptive immune systems. Immune cells compare these self-antigens with foreign antigens (Itoh et al., 2005), triggering immune responses when there is a mismatch. Such responses are crucial for the organism's defense against pathogens. Besides, these HLA markers also determine the suitability of tissues for transplantation. Further, specific regions in the HLA site on chromosome 6 have been identified to be responsible for histocompatibility issues of specific autoimmune diseases (such as insulin-dependent diabetes mellitus, myasthenia gravis, rheumatoid arthritis, and ankylosing spondylitis). The over 240 HLA genes are classified into distinct classes (I, II, or III) based on their structure and function. Both class I (e.g., HLA-A, HLA-B, and HLA-C) and class II (e.g., HLA-DR, HLA-DQ, and HLA-DP) genes

Table 2 Global population carrier frequencies for HLA class I risk alleles (Mullan et al., 2019)

Ethnicity (Country/Continent)	B*15: 02	A*24: 02	A*31: 01
Han Chinese (China)	0.0607	0.1488	0.0299
Han Chinese (Taiwan)	0.0450	0.1580	0.0280
Taiwanese (Taiwan)	0.0404	0.2939	0.0135
Caucasians (Europe)	0.0002	0.0998	0.0262
Japanese (Japan)	0.0004	0.3642	0.0843
Thai (Thailand)	0.0846	0.0390	0.0181
Indian (India)	0.0182	0.1418	0.0411
Korean (South Korea)	0.0146	0.2205	0.0559

exhibit a high degree of allelic polymorphisms (Prugnolle et al., 2005). These polymorphisms are believed to be caused by differentiation among human populations over time to adapt to a vast diversity of pathogens and allergens (Table 2).

Antiepileptic Drugs Induce Cutaneous Adverse Reactions

HLA-B*15:02: The first report showing a strong association between *HLA-B*15:02* and carbamazepine-induced Stevens-Johnson syndrome and toxic epidermal necrolysis (SJS/TEN) was from a study conducted in a Han Chinese population. Over the last decade, many other studies have shown similar associations in Roma in Spain, Javanese/Sundanese, and other populations in Thailand, India, Vietnam, and Malaysia, but not in Europeans and Japanese (Mullan et al., 2019), which is not surprising, given the relatively low frequency of this HLA allele in these two populations.

HLA-B*15:02 showed no association with carbamazepine-induced eosinophilia and systemic symptoms (DRESS) in studies involving European, Han Chinese, Vietnamese, Singaporean, or Malaysian populations (Mullan et al., 2019). It also was not associated with carbamazepine-induced maculopapular exanthema (MPE) in a comprehensive genome-wide association study (GWAS) in either European or Asian populations. Although the validity of these findings may be questioned because of the small sample size of these studies, in aggregate they suggest that *HLA-B*15:02* is a risk factor for the more serious SJS/TEN, but it is not a universal marker for the full range of carbamazepine-induced skin side effects.

In a study with Han Chinese patients, *HLA-B*15:02* did not appear to be a significant risk factor for lamotrigine-induced SJS/TEN (Wang et al., 2014).

HLA-A*24:02: This is another HLA gene that may also be associated with antiepileptic drugs (AED)-induced cutaneous adverse drug reactions (cADRs). Studies examining the relationship between *HLA-A*24:02* and lamotrigine-induced pathologies showed that it is a statistically significant risk for SJS/TEN in Han Chinese, DRESS in Spanish Romani, and MPE in Koreans. However, *HLA-A*24:02* seems to play no role in lamotrigine-induced MPE in Han Chinese or European populations (Shear & Dodiuk-Gad, 2019).

HLA-A*24:02 appears to be a risk for the entire range of carbamazepine-induced cADRs across diverse ethnic groups. Such an association is further substantiated by a GWAS in Japanese, perhaps reflecting the higher carrier frequency of *HLA-A*24:02* in this population.

In summary, although *HLA-A*24:02* allele appears to be a risk factor for various types of cADRs in diverse ethnic groups, such an assertion has yet to be confirmed with better-designed studies involving larger sample sizes.

HLA-A*31:01: While *HLAB*15:02* may be associated only with the more severe form of AED-induced cADR pathology, several studies suggested that *HLA-A*31:01* is associated with both mild and severe multiorgan/multisystem cADR: a pooled cohort of carbamazepine-induced MPE/HSS patients showed *HLA-A*31:01* to be a significant risk factor; two independent GWAS demonstrated that the allele is significantly associated with carbamazepine-induced skin side effects. Such an association may be limited to carbamazepine and is not generalizable to other antiepileptics, since there are no positive GWAS reports of phenytoin-induced MPE in either European or Asian populations.

Mushiroda et al. (2018) recently demonstrated the utility of genotyping *HLA-A*31:01* in patients treated with carbamazepine. Compared with historical cohorts as controls, groups prescreened for this allele showed a significantly decreased incidence of carbamazepine-induced cADRs, suggesting that such preemptive testing may be warranted in routine clinical practice among Japanese patients.

Antipsychotics Induce Granulocytosis and/or Neutropenia

HLA-DQB1: Clozapine is the only effective antipsychotic drug for refractory schizophrenia, but it is not widely prescribed because of the risks of agranulocytosis (CIA) and neutropenia (CIN). The genetic mechanism of clozapine-related agranulocytosis and neutropenia is complex and likely involves mutations in several genes, including HLA-DQB1, HLA-B, and SLCO1B3/SLCO1B7 (Legge & Walters, 2019).

There is currently considerable support for the role of mutations within HLA-DQB1 in clozapine-induced CIA/CIN. This association was first identified in candidate gene studies and was replicated in studies with independent samples. Large-scale genomic studies provided further evidence for the role of HLA-B mutants in CIN (Legge &Walters, 2019).

HLA-B: A large-scale genomic study conducted by the CIA Consortium (CIAC) demonstrated a relationship between HLA-B (158 T) polymorphism and CIN, independent of HLA-DQB1 (126Q), driven primarily by those of Ashkenazi Jewish ancestry (Yunis et al., 1995). Further evidence of the role of HLA-B comes from Saito et al. (2016), who also found a significant association between HLA-B*59:01 and CIN in the Japanese population. As with HLA-DQB1, it is currently unknown how risk variants of HLA-B lead to CIN, but it is presumed to be due to immune-related mechanisms.

Liver Transporter Genes: Apart from HLA variants, the intron SNPs for the liver transporter genes, SLCO1B3 and SLCO1B7, are present in 7.37% of CIN cases vs. 1.52% of controls in the CLOZUK sample and 4.20% of CIN cases vs. 1.67% of controls in the CIAC sample

(Legge & Walters, 2019). These genes encode a liver-specific organic anion transporter polypeptide that facilitates the uptake of drugs such as clozapine from the portal vein into hepatocytes and is subsequently metabolized or excreted; thus they may be associated with CIN by modulating the pharmacokinetics of clozapine.

Conclusions

This brief survey serves to highlight the significance as well as the complexity of issues surrounding cultural and ethnic influences on psychotropic responses. Taken together, the literature reviewed demonstrates the importance of these factors in determining the process and outcome of psychopharmacotherapy. These influences are mediated through both "non-biological" and biological mechanisms. On the culture side, patients' views on their illness experiences and treatment expectations often diverge significantly from their clinicians. Such discrepancies are further aggravated in cross-cultural situations, leading to treatment dropouts, medication nonadherence, diminished placebo responses, and nocebo effect. Systematic elicitations of patients' "folk" explanatory models represent an effective strategy in bridging the gaps, thereby facilitating clinical communication and collaboration.

On the biological side, progress in the field of pharmacogenetics has led to the discovery of numerous genetic polymorphisms influencing the pharmacokinetic and the pharmacodynamic processes, thereby regulating how medications are metabolized and eliminated and how they interact with therapeutic targets (major neurotransmitter systems). Together, these genes determine the dosing, side-effect profiles, and treatment responses of psychotropics. Of note, the patterns of these genetic polymorphisms vary significantly across ethnic groups, highlighting the importance of taking ethnicity (i.e., one's ancestral origin) into consideration in the practice of pharmacotherapy.

Since most psychotropics act on multiple targets (e.g., neurotransmitter receptors and transporters) and are metabolized through more than one pathway, the effect of a large number of genes needs to be considered together for comprehensive, and clinically useful, approaches for the prediction of psychotropic treatment response. Many pharmacogenomic panels have been developed to pursuing such a goal (Owusu Obeng et al., 2019). Ethnicity represents an important factor both for the development of these tools (e.g., decisions on which genes and alleles to be included in these panels) and the interpretation of the testing results (Lin et al., 2008).

Ethnicity also represents an important consideration in the epigenetic regulation of both pharmacokinetic and pharmacodynamic genes. On the pharmacokinetic side, patients with different ethnic/cultural backgrounds are often exposed to different kinds of herbs, natural substances, and environmental chemicals, which could either induce or inhibit the expression of these drug-metabolizing enzymes. On the pharmacodynamic side, stresses and other life experiences could interact with genes controlling the function of neurotransmitter systems, suggesting that sociocultural processes and other environmental factors might also influence these therapeutic targets' responses to prescribed medications.

While mindful of the importance of culture and ethnicity in the practice of psychopharmacology, it is equally crucial for clinicians to guard against stereotyping. In this regard, it is useful to keep in mind that almost all ethnic and cultural contrasts are superimposed on substantial interindividual variations within any population group. Stereotypic representations of cultural and ethnic differences could lead to biases and prejudices that may be as detrimental to patient care as the neglect of such influences. To avoid these pitfalls, it is important not to sequester culture and ethnicity to "special populations," but to incorporate such considerations into the care of all patients. By seeing a patient not just as an isolated entity, but as an individual with a unique genetic heritage, embedded in his or her particular cultural traditions, belief systems, and expectations, the clinician would be in a much better position to offer care that is customized, individualized, effective, and cost-efficient and thus gratifying for patients and those taking care of them.

Bibliography

Acosta, F. J., Hernández, J. L., Pereira, J., Herrera, J., & Rodríguez, C. J. (2012). Medication adherence in schizophrenia. *World Journal of Psychiatry, 2*(5), 74–82. PubMed. https://doi.org/10.5498/wjp.v2.i5.74

Adebimpe, V. R. (1981). Overview: White norms and psychiatric diagnosis of black patients. *American Journal of Psychiatry, 138*(3), 279–285. https://doi.org/10.1176/ajp.138.3.279

Aklillu, E., Persson, I., Bertilsson, L., Johansson, I., Rodrigues, F., & Ingelman-Sundberg, M. (1996). Frequent distribution of ultrarapid metabolizers of debrisoquine in an Ethiopian population carrying duplicated and multiduplicated functional CYP2D6 alleles. *Journal of Pharmacology and Experimental Therapeutics, 278*(1), 441. http://jpet.aspetjournals.org/content/278/1/441.abstract

Allen, J. G., Rack, P., & Vaddadi, K. (1977). Differences in the effects of clomipramine on English and Asian volunteers: Preliminary report on a pilot study. *Postgraduate Medical Journal, 53*, 79–85.

American Psychiatric Association. (2013). *Diagnostic and statistical manual of mental disorder* (5th ed.). American Psychiatric Publishing.

Ameyaw, M.-M., Regateiro, F., Li, T., Liu, X., Tariq, M., Mobarek, A., Thornton, N., Folayan, G. O., Githang'a, J., Indalo, A., Ofori-Adjei, D., Price-Evans, D. A., & McLeod, H. L. (2001). MDR1 pharmacogenetics: Frequency of the C3435T mutation in exon 26 is significantly influenced by ethnicity. *Pharmacogenetics and Genomics, 11*(3). https://journals.lww.com/jpharmacogenetics/Fulltext/2001/04000/MDR1_pharmacogenetics__frequency_of_the_C3435T.5.aspx

Bailey, D. G., Malcolm, J., Arnold, O., & David Spence, J. (1998). Grapefruit juice–drug interactions. *British Journal of Clinical Pharmacology, 46*(2), 101–110. https://doi.org/10.1046/j.1365-2125.1998.00764.x

Bailey, D. G., Dresser, G., & Arnold, J. M. (2013). Grapefruit-medication interactions: Forbidden fruit or avoidable consequences? *CMAJ, 185*(4), 309–316. https://doi.org/10.1503/cmaj.120951

Beevers, C. G., Ellis, A. J., Wells, T. T., & McGeary, J. E. (2010). Serotonin transporter gene promoter region polymorphism and selective processing of emotional images. *Biological Psychology, 83*(3), 260–265. https://doi.org/10.1016/j.biopsycho.2009.08.007

Biswas, M. (2020). Global distribution of CYP2C19 risk phenotypes affecting safety and effectiveness of medications. *The Pharmacogenomics Journal, 21*, 190. https://doi.org/10.1038/s41397-020-00196-3

Buckalew, L., & Coffield, K. (1982). An investigation of drug expectancy as a function of capsule color and size and preparation form. *Journal of Clinical Psychopharmacology, 2*(4), 245–248. PubMed. http://europepmc.org/abstract/MED/7119132

Budd, R. J., Hughes, I. C. T., & Smith, J. A. (1996). Health beliefs and compliance with antipsychotic medication.

British Journal of Clinical Psychology, 35(3), 393–397. https://doi.org/10.1111/j.2044-8260.1996.tb01193.x

Castro, F. G., Furth, P., & Karlow, H. (1984). The Health Beliefs of Mexican, Mexican American and Anglo American Women. *Hispanic Journal of Behavioral Sciences* [Internet]. Dec 1 [cited 2020 Dec 12], *6*(4), 365–383. https://doi.org/10.1177/07399863840064003

Catalano, M., Sciuto, G., Di Bella, D., Novelli, E., Nobile, M., & Bellodi, L. (1994). Lack of association between obsessive-compulsive disorder and the dopamine D3 receptor gene: Some preliminary considerations. *American Journal of Medical Genetics, 54*(3), 253–255. https://doi.org/10.1002/ajmg.1320540312

Cavalli-Sforza, L. L., & Cavalli-Sforza, F. (1995). *The great human diasporas: The history of diversity and evolution*. Addison-Wesley.

Consortium, The Pharmacogene Variation (PharmVar). (2021). *The Human Cytochrome P450 (CYP) Allele Nomenclature Database*. https://www.pharmvar.org/gene/CYP2D6

Cooper, R. S. (2004). Genetic factors in ethnic disparities in health. In *Critical perspectives on racial and ethnic differences in health in late life* (pp. 269–309). National Academy of Sciences.

de Leon, J., Rajkumar, A. P., Kaithi, A. R., Schoretsanitis, G., Kane, J. M., Wang, C.-Y., Tang, Y.-L., Lin, S.-K., Hong, K. S., Farooq, S., Ng, C. H., Ruan, C.-J., & Andrade, C. (2020). Do Asian patients require only half of the clozapine dose prescribed for Caucasians? A critical overview. *Indian Journal of Psychological Medicine, 42*(1), 4. https://doi.org/10.4103/IJPSYM.IJPSYM_495_19

Diaz, E., Woods, S. W., & Rosenheck, R. A. (2005). Effects of ethnicity on psychotropic medications adherence. *Community Mental Health Journal, 41*(5), 521–537. https://doi.org/10.1007/s10597-005-6359-x

Dubertret, C., Gorwood, P., Ades, J., Feingold, J., Schwartz, J.-C., & Sokoloff, P. (1998). Meta-analysis of DRD3 gene and schizophrenia: Ethnic heterogeneity and significant association in Caucasians. *American Journal of Medical Genetics, 81*(4), 318–322. https://doi.org/10.1002/(SICI)1096-8628(19980710)81:4<318::AID-AJMG8>3.0.CO;2-P

El-Mallakh, R., & Ali, Z. (2019). Therapeutic implications of the serotonin transporter gene in depression. *Biomarkers in Neuropsychiatry, 1*, 100004. https://doi.org/10.1016/j.bionps.2019.100004

Engel, G. (1977). The need for a new medical model: A challenge for biomedicine. *Science, 196*(4286), 129. https://doi.org/10.1126/science.847460

Fjær, E. L., Landet, E. R., McNamara, C. L., & Eikemo, T. A. (2020). The use of complementary and alternative medicine (CAM) in Europe. *BMC Complementary Medicine and Therapies, 20*(1), 108–108. PubMed. https://doi.org/10.1186/s12906-020-02903-w

Gasche, Y., Daali, Y., Fathi, M., Chiappe, A., Cottini, S., Dayer, P., & Desmeules, J. (2004). Codeine intoxication associated with Ultrarapid CYP2D6 metabolism.

New England Journal of Medicine, 351(27), 2827–2831. https://doi.org/10.1056/NEJMoa041888

Gillis, L. S., Trollip, D., Jakoet, A., & Holden, T. (1987). Non-compliance with psychotropic medication. *South African Medical Journal, 72*(9), 602–606.

González-Castro, T. B., Hernández-Díaz, Y., Juárez-Rojop, I. E., López-Narváez, M. L., Tovilla-Zárate, C. A., & Fresan, A. (2016). The role of a Catechol-O-methyltransferase (COMT) Val158Met genetic polymorphism in schizophrenia: A systematic review and updated meta-analysis on 32,816 subjects. *Neuromolecular Medicine, 18*(2), 216–231. https://doi.org/10.1007/s12017-016-8392-z

Guengerich, F. P., Shimada, T., Yun, C. H., Yamazaki, H., Raney, K. D., Thier, R., Coles, B., & Harris, T. M. (1994). Interactions of ingested food, beverage, and tobacco components involving human cytochrome P4501A2, 2A6, 2E1, and 3A4 enzymes. *Environmental Health Perspectives, 102 Suppl, 9*(Suppl 9), 49–53. https://doi.org/10.1289/ehp.94102s949

Hall, K. T., Loscalzo, J., & Kaptchuk, T. J. (2019). Systems pharmacogenomics – Gene, disease, drug and placebo interactions: A case study in COMT. *Pharmacogenomics, 20*(7), 529–551. https://doi.org/10.2217/pgs-2019-0001

Horton, R., Crawford, G., Freeman, L., Fenwick, A., Wright, C. F., & Lucassen, A. (2019). Direct-to-consumer genetic testing. *BMJ, 367*, l5688. https://doi.org/10.1136/bmj.l5688

Istikharah, R., Hartienah, S. D., Vitriyani, S., & Ningrum, V. D. A. (2020). Allele frequency of carbamazepine major efflux transporter encoding gene ABCB1 C3435T among Javanese-Indonesian population. *Open Access Macedonian Journal of Medical Sciences, 8*(A), 406–413. https://doi.org/10.3889/oamjms.2020.4184

Itoh, Y., Mizuki, N., Shimada, T., Azuma, F., Itakura, M., Kashiwase, K., Kikkawa, E., Kulski, J. K., Satake, M., & Inoko, H. (2005). High-throughput DNA typing of HLA-A, -B, -C, and -DRB1 loci by a PCR–SSOP–Luminex method in the Japanese population. *Immunogenetics (New York), 57*(10), 717–729. https://doi.org/10.1007/s00251-005-0048-3

Jadad, A. R., & Enkin, M. W. (2007). *Randomized controlled trials: Questions, answers, and musings (second)*. BMJ Books/Blackwell Publishing.

Johnson, J. (2000). Predictability of the effects of race or ethnicity on pharmacokinetics of drugs. *International Journal of Clinical Pharmacology and Therapeutics, 38*(2), 53–60. https://doi.org/10.5414/cpp38053

Jönsson, E. G., Nöthen, M. M., Grünhage, F., Farde, L., Nakashima, Y., Propping, P., & Sedvall, G. C. (1999). Polymorphisms in the dopamine D2 receptor gene and their relationships to striatal dopamine receptor density of healthy volunteers. *Molecular Psychiatry, 4*(3), 290–296. https://doi.org/10.1038/sj.mp.4000532

Joseph, S., Nicolson, T. J., Hammons, G., Word, B., Green-Knox, B., & Lyn-Cook, B. (2015). Expression of drug transporters in human kidney: Impact of sex, age, and ethnicity. *Biology of Sex Differences, 6*(1), 4–4. https://doi.org/10.1186/s13293-015-0020-3

Kaptchuk, T. J., & Miller, F. G. (2018, October 2). Open label placebo: can honestly prescribed placebos evoke meaningful therapeutic benefits? BMJ [Internet], *363*, k3889. http://www.bmj.com/content/363/bmj.k3889.abstract

Kinzie, J. D., Leung, P., Boehnlein, J. K., & Flick, J. (1987). Antidepressant Blood Levels in Southeast Asians Clinical and Cultural Implications. *The Journal of Nervous and Mental Disease* [Internet], *175*(8). https://journals.lww.com/jonmd/Fulltext/1987/08000/Antidepressant_Blood_Levels_in_Southeast_Asians.6.aspx

Kirkpatrick, C. F., Bolick, J. P., Kris-Etherton, P. M., Sikand, G., Aspry, K. E., Soffer, D. E., Willard, K.-E., & Maki, K. C. (2019). Review of current evidence and clinical recommendations on the effects of low-carbohydrate and very-low-carbohydrate (including ketogenic) diets for the management of body weight and other cardiometabolic risk factors: A scientific statement from the National Lipid Association Nutrition and Lifestyle Task Force. *Journal of Clinical Lipidology, 13*(5), 689–711.e1. https://doi.org/10.1016/j.jacl.2019.08.003

Kleinman, A., & Benson, P. (2006). Anthropology in the clinic: The problem of cultural competency and how to fix it. *PLoS Medicine, 3*(10), e294–e294. PubMed. https://doi.org/10.1371/journal.pmed.0030294

Kleinsinger, F. (2018). The unmet challenge of medication nonadherence. *The Permanente Journal, 22*, 18–033. https://doi.org/10.7812/TPP/18-033

Lam, Y. W. F., Gaedigk, A., Ereshefsky, L., Alfaro, C. L., & Simpson, J. (2002). CYP2D6 inhibition by selective serotonin reuptake inhibitors: Analysis of achievable steady-state plasma concentrations and the effect of ultrarapid metabolism at CYP2D6. *Pharmacotherapy: The Journal of Human Pharmacology and Drug Therapy, 22*(8), 1001–1006. https://doi.org/10.1592/phco.22.12.1001.33603

Lannfelt, L., Sokoloff, P., Martres, M.-P., Pilon, C., Giros, B., Jönsson, E., Sedvall, G., & Schwartz, J.-C. (1992). Amino acid substitution in the dopamine D3 receptor as a useful polymorphism for investigating psychiatric disorders. *Psychiatric Genetics, 2*(4) https://journals.lww.com/psychgenetics/Fulltext/1992/10000/Amino_acid_substitution_in_the_dopamine_D3.3.aspx, 249.

Lee, S., Wing, Y. K., & Wong, K. C. (1992). Knowledge and compliance towards lithium therapy among Chinese psychiatric patients in Hong Kong. *Australian & New Zealand Journal of Psychiatry, 26*(3), 444–449. https://doi.org/10.3109/00048679209072068

Legge, S. E., & Walters, J. T. (2019). Genetics of clozapine-associated neutropenia: Recent advances, challenges and future perspective. *Pharmacogenomics, 20*(4), 279–290. https://doi.org/10.2217/pgs-2018-0188

Li, F. S., & Weng, J.-K. (2017). Demystifying traditional herbal medicine with modern approach. *Nature Plants, 3*(8), 17109. https://doi.org/10.1038/nplants.2017.109

Lin, K. M. (1981). Traditional Chinese medical beliefs and their relevance for mental illness and psychiatry. In *Normal and abnormal behavior in Chinese culture* (pp. 95–114). Reidel.

Lin, K. M., & Lin, W. L. (2013). Culture, ethnicity and psychopharmacology. In E. Sorel (Ed.), *21st century global mental health* (pp. 95–118). Jones & Bartlett Learning.

Lin, K. M., Poland, R. E., Lau, J. K., & Rubin, R. T. (1988). Haloperidol and prolactin concentrations in Asians and Caucasians. *Journal of Clinical Psychopharmacology, 8*(3), 195–201. https://journals.lww.com/psychopharmacology/Fulltext/1988/06000/Haloperidol_and_Prolactin_Concentrations_in_Asians.8.aspx

Lin, K. M., Poland, R. E., & Nakasaki, G. (1993). *Psychopharmacology and psychobiology of ethnicity.* American Psychiatric Press.

Lin, K. M., Perlis, R. H., & Wan, Y. J. Y. (2008). Pharmacogenomic strategy for individualizing antidepressant therapy. *Dialogues in Clinical Neuroscience, 10*(4), 375–382. https://doi.org/10.31887/DCNS.2008.10.4/kmlin

Lindell, S. G., Yuan, Q., Zhou, Z., Goldman, D., Thompson, R. C., Lopez, J. F., Suomi, S. J., Higley, J. D., & Barr, C. S. (2012). The serotonin transporter gene is a substrate for age and stress dependent epigenetic regulation in rhesus macaque brain: Potential roles in genetic selection and Gene × Environment interactions. *Development and Psychopathology, 24*(4), 1391–1400. Cambridge Core. https://doi.org/10.1017/S0954579412000788

Louhiala, P. (2020). *Placebo effects: The meaning of Care in Medicine.* Springer International Publishing. https://doi.org/10.1007/978-3-030-27329-3

Luo, H.-R., Aloumanis, V., Lin, K.-M., Gurwitz, D., & Wan, Y.-J. Y. (2004). Polymorphisms of CYP2C19 and CYP2D6 in Israeli ethnic groups. *American Journal of Pharmacogenomics: Genomics-Related Research in Drug Development and Clinical Practice, 4*(6), 395–401. PubMed. https://doi.org/10.2165/00129785-200404060-00006

Mitchell, R. J., Howlett, S., Earl, L., White, N. G., McComb, J., Schanfield, M. S., et al. (2000). Distribution of the 3' VNTR polymorphism in the human dopamine transporter gene in world populations. *Human Biology* [Internet], *72*(2), 295–304. Available from: http://www.jstor.org/stable/41465826

Mullan, K. A., Anderson, A., Illing, P. T., Kwan, P., Purcell, A. W., & Mifsud, N. A. (2019). HLA-associated antiepileptic drug-induced cutaneous adverse reactions. *HLA, 93*(6), 417–435. https://doi.org/10.1111/tan.13530

Munafò, M. R., Matheson, I. J., & Flint, J. (2007). Association of the DRD2 gene Taq1A polymorphism and alcoholism: A meta-analysis of case–control studies and evidence of publication bias. *Molecular Psychiatry, 12*(5), 454–461. https://doi.org/10.1038/sj.mp.4001938

Mushiroda, T., Takahashi, Y., Onuma, T., Yamamoto, Y., Kamei, T., Hoshida, T., et al. (2018, July 1). Association of HLA-A*31:01 screening with the incidence of carbamazepine-induced cutaneous adverse reactions in a Japanese population. *JAMA Neurology* [Internet], *75*(7), 842–849. Available from: https://pubmed.ncbi.nlm.nih.gov/29610831

Ng, C. H., Lin, K.-M., Singh, B. S., & Chiu, E. Y. K. (Eds.). (2008). *Ethno-psychopharmacology: Advances in current practice.* Cambridge University Press; Cambridge Core. https://doi.org/10.1017/CBO9780511544149

Nicolussi, S., Drewe, J., Butterweck, V., & Meyer zu Schwabedissen, H. E. (2020). Clinical relevance of St. John's wort drug interactions revisited. *British Journal of Pharmacology, 177*(6), 1212–1226. https://doi.org/10.1111/bph.14936

Niv, N., Shatkin, J. P., Hamilton, A. B., Unützer, J., Klap, R., & Young, A. S. (2010). The use of herbal medications and dietary supplements by people with mental illness. *Community Mental Health Journal, 46*(6), 563–569. https://doi.org/10.1007/s10597-009-9235-2

Noskova, T., Pivac, N., Nedic, G., Kazantseva, A., Gaysina, D., Faskhutdinova, G., Gareeva, A., Khalilova, Z., Khusnutdinova, E., Kovacic, D. K., Kovacic, Z., Jokic, M., & Seler, D. M. (2008). Ethnic differences in the serotonin transporter polymorphism (5-HTTLPR) in several European populations. *Progress in Neuro-Psychopharmacology and Biological Psychiatry, 32*(7), 1735–1739. https://doi.org/10.1016/j.pnpbp.2008.07.012

Owusu Obeng, A., Samwald, M., & Scott, S. A. (2019). Chapter 13—Reactive, point-of-care, preemptive, and direct-to-consumer pharmacogenomics testing. In Y. W. F. Lam & S. A. Scott (Eds.), *Pharmacogenomics* (2nd ed., pp. 369–384). Academic Press. https://doi.org/10.1016/B978-0-12-812626-4.00013-9

Petersen, G. L., Finnerup, N. B., Colloca, L., Amanzio, M., Price, D. D., Jensen, T. S., & Vase, L. (2014). The magnitude of placebo effects in pain: A meta-analysis. *Pain, 155*(8), 1426–1434. https://doi.org/10.1016/j.pain.2014.04.016

Poland, R. E., Lesser, I. M., Wan, Y.-J. Y., Gertsik, L., Yao, J., Raffel, L. J., Lin, K.-M., & Myers, H. F. (2013). Response to citalopram is not associated with SLC6A4 genotype in African-Americans and Caucasians with major depression. *Life Sciences, 92*(20–21), 967–970. https://doi.org/10.1016/j.lfs.2013.03.009

Porcelli, S., Fabbri, C., & Serretti, A. (2012). Meta-analysis of serotonin transporter gene promoter polymorphism (5-HTTLPR) association with antidepressant efficacy. *European Neuropsychopharmacology, 22*(4), 239–258. https://doi.org/10.1016/j.euroneuro.2011.10.003

Price, N., Glazer, W., & Morgenstern, H. (1985). Demographic predictors of the use of injectable versus oral antipsychotic medications in outpatients. *American Journal of Psychiatry, 142*(12), 1491–1492. https://doi.org/10.1176/ajp.142.12.1491

Prugnolle, F., Manica, A., Charpentier, M., Guégan, J. F., Guernier, V., & Balloux, F. (2005). Pathogen-driven selection and worldwide HLA class I diversity. *Current Biology, 15*(11), 1022–1027. https://doi.org/10.1016/j.cub.2005.04.050

Qin, X.-P., Xie, H.-G., Wang, W., He, N., Huang, S.-L., Xu, Z.-H., et al. (1999, December 1). Effect of the gene dosage of CYP2C19 on diazepam metabolism in Chinese subjects. *Clinical Pharmacology & Therapeutics* [Internet]. [cited 2021 Feb 15]*, 66*(6):642–646. https://doi.org/10.1053/cp.1999.v66.103379001

Rajman, I., Knapp, L., Morgan, T., & Masimirembwa, C. (2017). African genetic diversity: Implications for cytochrome P450-mediated drug metabolism and drug development. *eBioMedicine, 17*, 67–74. https://doi.org/10.1016/j.ebiom.2017.02.017

Saito, T., Ikeda, M., Mushiroda, T., Ozeki, T., Kondo, K., Shimasaki, A., Kawase, K., Hashimoto, S., Yamamori, H., Yasuda, Y., Fujimoto, M., Ohi, K., Takeda, M., Kamatani, Y., Numata, S., Ohmori, T., Ueno, S., Makinodan, M., Nishihata, Y., et al. (2016). Pharmacogenomic study of clozapine-induced Agranulocytosis/Granulocytopenia in a Japanese population. *Schizophrenia Biomarkers, 80*(8), 636–642. https://doi.org/10.1016/j.biopsych.2015.12.006

Schwartz, R. C., & Blankenship, D. M. (2014). Racial disparities in psychotic disorder diagnosis: A review of empirical literature. *World Journal of Psychiatry, 4*(4), 133–140. PubMed. https://doi.org/10.5498/wjp.v4.i4.133. https://pubmed.ncbi.nlm.nih.gov/25540728

Scott, S. A., Edelmann, L., Kornreich, R., Erazo, M., & Desnick, R. J. (2007). CYP2C9, CYP2C19 and CYP2D6 allele frequencies in the Ashkenazi Jewish population. *Pharmacogenomics, 8*(7), 721–730. https://doi.org/10.2217/14622416.8.7.721

Shear, N. H., & Dodiuk-Gad, R. P. (2019). *Advances in diagnosis and management of cutaneous adverse drug reactions: Current and future trends.* Springer Nature.

Shimoda, K., Jerling, M., Böttiger, Y., Yasuda, S., Morita, S., & Bertilsson, L. (1999). Pronounced differences in the disposition of clomipramine between Japanese and Swedish patients. *Journal of Clinical Psychopharmacology, 19*(5), 393. https://journals.lww.com/psychopharmacology/Fulltext/1999/10000/Pronounced_Differences_in_the_Disposition_of.2.aspx

Sternberg, E. M. (2002). Walter B. Cannon and "'Voodoo' Death": A perspective from 60 years on. *American Journal of Public Health, 92*(10), 1564–1566. https://doi.org/10.2105/AJPH.92.10.1564

Sue, S. (1977). Community mental health services to minority groups: Some optimism, some pessimism. *American Psychologist, 32*(8), 616–624. https://doi.org/10.1037/0003-066X.32.8.616

Taylor, S. (2018). Association between COMT Val158Met and psychiatric disorders: A comprehensive meta-analysis. *American Journal of Medical Genetics Part B: Neuropsychiatric Genetics, 177*(2), 199–210. https://doi.org/10.1002/ajmg.b.32556

Tishkoff, S. A., & Kidd, K. K. (2004). Implications of biogeography of human populations for "race" and medicine. *Nature Genetics, 36*(11), S21–S27. https://doi.org/10.1038/ng1438

Walsh, K., Uddin, M., Soliven, R., Wildman, D. E., & Bradley, B. (2014). Associations between the SS variant of 5-HTTLPR and PTSD among adults with histories of childhood emotional abuse: Results from two African American independent samples. *Journal of Affective Disorders, 161*, 91–96. https://doi.org/10.1016/j.jad.2014.02.043

Wang, W., Hu, F.-Y., Wu, X.-T., An, D.-M., Yan, B., & Zhou, D. (2014). Genetic predictors of Stevens–Johnson syndrome and toxic epidermal necrolysis induced by aromatic antiepileptic drugs among the Chinese Han population. *Epilepsy & Behavior, 37*, 16–19. https://doi.org/10.1016/j.yebeh.2014.05.025

Weiss, M. G., Doongaji, D. R., Siddhartha, S., Wypij, D., Pathare, S., Bhatawdekar, M., et al. (1992). The Explanatory Model Interview Catalogue (EMIC): Contribution to cross-cultural research methods from a study of leprosy and mental health. *British Journal of Psychiatry* [Internet]. 2018/01/02 ed.*, 160*(6), 819–830. https://www.cambridge.org/core/article/explanatory-model-interview-catalogue-emic/396594FC69FE714B5C960C5FA9B5CA0C

Wonodi, I., Adami, H. M., Cassady, S. L., Sherr, J. D., Avila, M. T., & Thaker, G. K. (2004). Ethnicity and the course of tardive dyskinesia in outpatients presenting to the motor disorders clinic at the Maryland psychiatric research center. *Journal of Clinical Psychopharmacology, 24*(6), 592–598. https://doi.org/10.1097/01.jcp.0000144888.43449.54

Wood, F. A., Howard, J. P., Finegold, J. A., Nowbar, A. N., Thompson, D. M., Arnold, A. D., Rajkumar, C. A., Connolly, S., Cegla, J., Stride, C., Sever, P., Norton, C., Thom, S. A. M., Shun-Shin, M. J., & Francis, D. P. (2020). N-of-1 trial of a statin, placebo, or no treatment to assess side effects. *New England Journal of Medicine, 383*(22), 2182–2184. https://doi.org/10.1056/NEJMc2031173

Yunis, J., Corzo, D., Salazar, M., Lieberman, J., Howard, A., & Yunis, E. (1995). HLA association in clozapine induced agranulocytosis. *Blood, 86*, 1177–1183. https://doi.org/10.1182/blood.V86.3.1177.bloodjournal8631177

Zhao, X., Huang, Y., Chen, K., Li, D., Han, C., & Kan, Q. (2016). −141C insertion/deletion polymorphism of the dopamine D2 receptor gene is associated with schizophrenia in Chinese Han population: Evidence from an ethnic group-specific meta-analysis. *Asia-Pacific Psychiatry, 8*(3), 189–198. https://doi.org/10.1111/appy.12206

Adherence with Medication Treatment in Psychiatric Disorders

124

Yumi Aoki, Hiroyoshi Takeuchi, Koichiro Watanabe, and Allan Tasman

Contents

This chapter is an update from the 4th edition. Previous edition authors were Dawn I. Velligan, Stephanie A. Riolo and Christina G. Weston

Y. Aoki
Department of Psychiatric & Mental Health Nursing, Graduate School of Nursing, St. Luke's International University, Tokyo, Japan
e-mail: yumiaoki@slcn.ac.jp

H. Takeuchi
Department of Neuropsychiatry, Keio University School of Medicine, Tokyo, Japan
e-mail: hirotak@dk9.so-net.ne.jp

K. Watanabe (✉)
Department of Neuropsychiatry, Kyorin University School of Medicine, Tokyo, Japan
e-mail: koichiro@tke.att.ne.jp

A. Tasman
Department of Psychiatry and Behavioral Sciences, University of Louisville School of Medicine, Louisville, KY, USA
e-mail: allan.tasman@louisville.edu

© Springer Nature Switzerland AG 2024
A. Tasman et al. (eds.), *Tasman's Psychiatry*,
https://doi.org/10.1007/978-3-030-51366-5_132

Abstract

While pharmacological treatment is effective for a number of psychiatric disorders, medication adherence has been a clinical concern for a long time since nonadherence is a major cause of ineffectiveness. Ensuring medication adherence is particularly difficult in cases in which medication is taken to prevent the recurrence of symptoms. Given the preponderance of data supporting the high prevalence of medication nonadherence across all age, racial, and disease groups, health-care providers need to accept the reality that perfect medication adherence is unattainable and conduct regular assessments, including asking about the patients' real-life barriers, feelings, beliefs, and attitudes regarding medication, all of which contribute to both unintentional and intentional non-adherence. These assessments are best performed in the context of a favorable therapeutic alliance. Maintaining a collaborative relationship is crucial, and the concept of concordance is gaining attention as a patient-centered advanced approach, which reflects patients' preferences and values regarding medication regimens. This concept overlaps shared decision-making, which involves discussion and collaboration between patients and health-care providers. Evidenced-based interventions such as environment support, new technologies, and smartphone applications appear to be promising for the promotion of medication adherence and should be further investigated and implemented in treatment settings. This chapter describes measurement of medication adherence, the risk factors for non-adherence, and evidenced-based intervention strategies to improve adherence, along with descriptions of the related key concepts such as concordance, shared decision-making, and therapeutic alliance.

Keywords

Psychiatric disorders · Medication adherence · Therapeutic alliance · Concordance · Shared decision-making · Risk factors · Interventions · Strategies

Introduction

Adherence refers to the extent to which a person follows the advice given to them by a medical professional. Adherence is important because, as pointed out by former surgeon general of the USA, C. Everett Coop, "drugs don't work in patients who don't take them." Adherence is poor across a wide range of physical and psychiatric disorders and is particularly poor in cases in which medication is taken to prevent the onset or recurrence of symptoms (Velligan et al., 2009).

Research suggests that half to two-thirds of persons with psychiatric illnesses who may benefit from treatment either fail to seek help or, when they do, are nonadherent to treatment (Regier et al., 1993; Kessler et al., 2001; Velligan et al., 2009). The consequences of poor adherence can be both devastating for the individual and costly for society. Patients can become suicidal, risk hospitalization, jeopardize relationships with family and friends, or compromise their school performance or employment. Moreover, poor adherence can complicate prescribing decisions. Physicians, unaware of adherence problems, may incorrectly assume that poor therapeutic outcomes are due to failure of the prescribed treatment and increase the dose, add medication, or prescribe medication with potentially more serious adverse reactions (Velligan et al., 2006; Katzung, 2007).

Reasons for problem adherence are numerous and varied. Factors contributing to problem adherence include patient-related factors such as poor insight, problems with remembering and organizing, and concerns over the stigma of having a mental illness. Medication-related factors include side effects and complicated dosing strategies. Finally, issues related to the provider and service delivery system can lead to problems with adherence. The therapeutic alliance between the patient and prescriber can impact the patient's willingness to take medication. Such an alliance is often more difficult to forge in a health care system in which prescriber visits last no longer than 15 min and can be spaced months apart. Not having insurance coverage or not being able to scrape up the money for insurance co-pays can significantly interfere with adherence to medication (Zeber et al., 2007; Velligan et al., 2009).

Novel interventions to improve adherence are being developed and tested. Different interventions target different factors contributing to problem adherence.

This chapter describes the scope of problems with medication adherence and the impact of poor adherence on treatment and outcomes for psychiatric disorders. Methods for quantifying adherence and for identifying risk factors for poor adherence have been reviewed. This chapter also reviews emerging strategies that providers can use to maximize patient adherence to psychiatric medication. Practical and specific suggestions are given with clinical examples.

Compliance, Adherence, and Concordance

The term "compliance" was often used to describe a patient's continuation to medications. However, it has largely been replaced by the term "adherence" in recent years. The word "compliance" is derived from the Latin word "complire," meaning to fill up, i.e., to complete an action, transaction, or process and to fulfill a promise (Aronson,

2007). Accordingly, compliance refers to the degree of consistency between a patient's behavior and the clinician's instructions. The patient should follow the medication regimen developed by the clinician; poor compliance is interpreted as the patient's inability to follow the instructions or having problems violating the instructions. Gradually, the idea of "compliance," which is contrary to patient-centered care, has been replaced by the concept of "adherence." The word "adherence" is derived from the Latin word "adhaerere," meaning to cling to, keep close, or remain constant (Aronson, 2007). Adherence refers to the degree of consistency between a patient's behavior and the clinician's recommendations. This concept treats a patient as an individual, who decides whether to accept the clinician's recommendations. With an increasing value of the independence and autonomy of individuals, the term adherence is generally used in discussions about the continuation of medications. In this chapter, the term adherence is used instead of compliance.

It is also important to mention "concordance" as a more recent, patient-centered care advanced concept. The relationship between patients and clinicians becomes more equal in concordance, and the prescribing behavior is based on a partnership between both parties. This concept is characterized by the patient and the clinician discussing and deciding the medication regimen, which completely reflects the patient's preferences and values (Horne, 2006b). "Concordance" includes the active involvement of the patient in the decision-making process, and this concept is similar to shared decision-making, which is discussed in the next section. The reflection of the patient's opinion in the medication regimen indicates that the patient is willing to consume it, which improves the medication adherence (Fig. 1).

Shared Decision-Making

No discussion of adherence can take place in the absence of an understanding about how patients and providers arrive at a treatment decision. Traditional models of patient–provider interaction that focus on the "expert" provider informing the

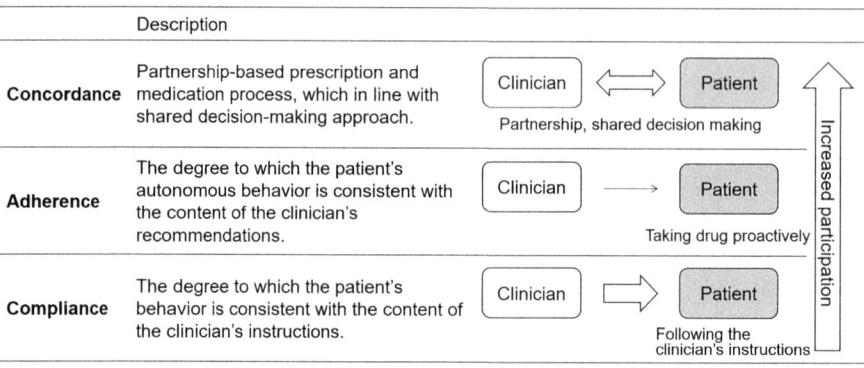

Fig. 1 A description of compliance, adherence, and concordance

patient about the best course of action for a specific condition have given way to a model of shared decision-making (SDM). Shared decision-making is a process involving both the clinician and patient exchanging information and treatment preferences, with both parties working toward consensus and ending with an agreed upon treatment approach (Charles et al., 1997). This approach is based upon a collaborative approach to establishing and maintaining the therapeutic alliance. For more in-depth information about establishing and maintaining this type of therapeutic alliance, please see also Part I, ▶ Chap. 1, "Listening to the Patient," Part VII, ▶ Chap. 103, "The Recovery Model and Other Rehabilitative Approaches," and Part VIII, ▶ Chap. 119, "Combined Psychotherapy and Somatic Treatments."

Shared decision-making is particularly important in mental health prescribing. (Table 7) A focus on understanding the preferences and values of the patient is central to treatment outcomes and adherence to medications in situations in which there is no clear "best" treatment and there are many choices with variable side effect profiles (Mahar, 2007; Joosten et al., 2008; Roe et al., 2009). This is exactly the case in mental health treatment. Research has demonstrated that consumers with mental illness want more input into decisions regarding their treatment and want treatment decisions to be derived in a shared manner (Adams et al., 2007; Klein et al., 2007; Curtis et al., 2010). A Cochrane systematic review which included 15 SDM studies in mental health

field suggests that people exposed to SDM interventions may perceive greater levels of involvement immediately compared with those in control groups without extending the length of consultations (Aoki et al., 2022). Following a shared decision-making approach is a key first step in improving adherence. A more patient-centered definition describes adherence as the extent to which a patient follows through on an agreed upon treatment plan sufficiently to achieve symptom remission and prevent relapse (Velligan et al., 2009).

Theoretical Models of Health Behavior

Various conceptual models have been developed to describe patients' health behavior during treatment of chronic conditions (Conrad, 1985; Leventhal & Cameron, 1987). Major models of health behavior that have been developed from social cognitive theory include: (1) health belief model, (2) theory of reasoned action and theory of planned behavior, (3) stages of change theory, and (4) protection motivation theory. Although these models differ in important ways, they all assume that medication adherence can be predicted by patients' perceptions of threat from the medical condition and their expectancy regarding the consequences of taking medication for that condition.

The health belief model (HBM) assumes that medication-taking is a rational decision made by the patient based on a cost–benefit analysis (Janz

& Becker, 1984; Rosenstock, 1988). Regardless of the "objective" view of the prescribing physician, it is the patient's perception of the benefits and risks of medication that determines their adherence behavior.

The HBM has been demonstrated to have predictive value with regard to medication adherence in a variety of chronic medical conditions, including tuberculosis, diabetes, and hypertension (Becker & Maiman, 1975; Becker, 1985). More recently, the HBM has been used in studies of medication adherence in psychiatric disorders (Budd et al., 1996). Lingam and Scott (2002) conducted a review of research (1976–2001) on treatment nonadherence in adults with affective disorders. They concluded that patients' attitudes and beliefs are at least as important as side effects in predicting adherence to medication for affective disorders and that the HBM is a promising tool for the identification of individuals at risk for nonadherence.

Bush and Iannotti (1988, 1990) adapted the HBM for use in children and adolescents. They added key developmental and contextual components to the model, including caretaker attitudes (perceived threat to child from illness and perceived benefit of medication), parent–child interactions, level of child/adolescent autonomy, and child readiness factors (illness concern and expectancy for efficacy of medication treatment). Their model, called the children's health belief model, was based on data from 420 elementary-age children and was tested with a sample of 270 adolescents. The model accounted for 64% of the variance in adolescents' behavioral response to illness. Based on these findings, the best predictors of adolescent adherence to treatment were their perception of the severity of the illness and the predicted/perceived benefit of the medication. They found that the primary caretaker's attitudes influenced their adolescents' readiness and expectancies regarding medications, but that the adolescents' own cognitions and attitudes were most important in determining adherence to medication. Bush and Iannotti (1988, 1990) concluded that although parents do influence their children's health beliefs and

health behavior, these influences are small compared with other developmental influences on children's attitudes and behaviors.

The theory of reasoned action (TRA) by Ajzen and Fishbein (1980) and its more recent extension, the theory of planned behavior (TPB) by Ajzen (1988), resemble the HBM in the assumption that people make rational decisions (i.e., cost–benefit analyses) about their behavior based on information and beliefs about the behavior, and that health behaviors are under volitional control. The TRA and TPB differ from the HBM in two major ways. TRA and TPB assert that patient intentions are the most important determinant of their behavior. Furthermore, they state that intentions are a function of a person's attitude toward the behavior and the person's perceptions of social norms regarding the behavior. When applied to patient adherence to psychiatric medication, a patient's attitude toward medication adherence is based on their predicted consequences of taking medication and the value (costs and benefits) that they attribute to these consequences. For example, if a person believes that a medication is effective in the treatment of a particular symptom/illness, then they may believe that the consequence of *not* taking the medication will be persistence or return of the symptom/illness. If so, depending on the degree of distress from the symptom/illness, they may see a benefit in taking the medication; however, the "cost" of taking the medication may outweigh this benefit. According to TRA/TPB, however, an important factor influencing their decision is a person's expectation of whether significant others in their life (i.e., family and friends) will support their taking psychiatric medication. Of course, the person's motivation to adhere to the preferences of their significant others will determine the impact of perceived social norms on predicting their medication adherence. To some degree, this model assumes a rational response to an important decision, which in reality is often complicated by other factors.

Stages of change theory is based on the premise that behavior change occurs in stages or phases and the factors that influence behavior may change from one stage to another. Implicit in stage models is the assumption that, initially, behavior change is largely under the control of deliberate cognitive processes. As the new behavior is repeated over time, situational cues to decisions begin to elicit the cognitive processes that result in decisions and intentions. Over time, actions may come to be performed automatically in response to situational cues such that they are no longer under volitional control (Prochaska et al., 1994).

Protection motivation theory (PMT) was originally proposed to provide conceptual clarity to the understanding of fear appeals. It was later extended to a theory of persuasive communication, with an emphasis on the cognitive processes mediating behavioral change. According to the theory, appraisal of a health threat and appraisal of possible coping responses result in the intention to perform adaptive responses (protection motivation). The intention to protect oneself depends upon four factors: (1) perceived severity of the threat, (2) perceived vulnerability, (3) perceived response efficacy – expectancy that carrying out recommendations can remove the threat, and (4) perceived self-efficacy – the belief in one's ability to execute the recommended courses of action successfully (Rogers, 1983).

In general, these models provide a framework with which to understand health care decision-making. However, it should be kept in mind that the intention to take medication is not the same thing as actually taking medication. Many of our health care beliefs are not based on rational ideation, and are influenced by many factors, including cultural, familial, religious, and other factors. Because these factors often may not be based on "rational" considerations, they may adversely influence positive adherence behaviors. The factors influencing daily life, including the simplicity of the regimen prescribed, the nature of side effects, and the rapidity of the medication response also play a role. A patient may think that it is a good idea to take all the antibiotics prescribed for a particular condition because it is understood that failure to do so can lead to antibiotic-resistant strains of bacteria that can be bad for everyone. However, he/she may still forget to take these pills once the condition improves.

Prevalence of Poor Adherence to Medication

Adherence to psychiatric medication depends on many factors, for example, whether the prescribed regimen is short term or long term, and whether the medication is for prophylaxis or for the treatment of disease. For long-term psychiatric illnesses, individuals take only about 50% of prescribed medication (Velligan et al., 2009). A meta-analysis of studies investigating the adherence to oral antipsychotics using electronic adherence monitoring (EAM) in patients with schizophrenia demonstrated that the oral antipsychotic adherence rates were in the range of 70% (Yaegashi et al., 2020) though clinical experience suggests this may vary widely among individual patients.

Patterns of Medication Nonadherence

There are many patterns of medication nonadherence that range from complete refusal of medication to simply omitting the occasional dose. Individuals who refuse medication are often easier to spot. They may tell the prescriber that they are not going to accept a particular medication treatment. However, about 13% of prescriptions are never filled, indicating that some patients may not let the doctor know that they are refusing treatment. Despite its apparent simplicity, taking medication as prescribed is a complex and dynamic behavior. Empirical evidence demonstrates that adherence is best understood as a variable behavior, rather than a trait characteristic. Rudd et al. (1989) reported marked inter- and intra-subject variability in medication adherence on a week-to-week basis. Patients may continue to keep appointments while not taking prescribed medications, so attendance at appointments should not be overly reassuring. Patients prescribed more than one medication may take one but not others. Patients also may initially adhere to medication recommendations, but later may discontinue their medication without informing their physician. Based on a comprehensive review of reported rates, patterns, and correlates of medication nonadherence, the following have been observed (Acosta et al., 2012; Tham et al., 2016):

1. Medication adherence tends to decline over time.
2. Baseline adherence is the strongest predictor of long-term adherence to medication.
3. Patients who have been poorly adherent in the past are more likely to be poorly adherent in the future.

Many studies treat adherence as dichotomous and univariate, rather than as a continuous phenomenon with multiple dimensions. A patient may adhere to one part of treatment recommendations but not to others (Orme & Binik, 1989; Fotheringham & Sawyer, 1995). Failure to adhere to medication taking may include omissions of doses, taking more than the prescribed dose, errors in dosage or prescribed frequency of doses, taking medication for the wrong reason, taking medication as needed rather than on a regular schedule, and discontinuing medication early.

Both the pattern and rate of missed medication doses are important. Converging evidence suggests that medication adherence is viewed on a continuum with many different patterns of partial adherence, but 100% adherence is abnormal. Bachmann et al. (1999) studied methods to achieve high therapeutic success despite less-than-perfect adherence. They suggested that this may be achieved by selection of an agent that, based on its pharmacological properties, is "more forgiving" of delayed or omitted doses, such as fluoxetine. Forgiveness is defined as the drug's post-dose duration of action minus the prescribed interval between doses. Moreover, because prescribing occurs in an atmosphere of poor adherence, dosage may creep up over time because symptoms are not well managed. In this case, an individual may not need all the medication prescribed to treat their symptoms adequately once they are more adherent (Velligan et al., 2013).

Delays and omissions in dose administration are the predominant medication errors. Electronic monitoring shows a skew toward dose omission, with widely variable intervals between doses, although dosing in the day or two prior to scheduled visits is usually correct (Feinstein, 1990). Appropriate administration of the medication for the few days preceding an appointment, the so-called "white coat" adherence, does not reflect long-term adherence. One in six patients is very punctual in the timing of prescribed doses. Three in six make minor errors in timing that are too small to reduce the full-dose action of any but the least forgiving medications (Urquhart, 1995, 1997).

Medication errors can be further classified according to whether the error was unintentional (forgetting or misunderstanding the dose or frequency) or intentional (conscious decision to modify medication dose/frequency for any reason). There is a substantial difference between forgetting to take a pill despite knowing the proper regimen and believing, incorrectly, that only two rather than three pills are needed daily and adjusting the regimen without consulting the doctor (Rudd, 1993). Patients may alter their dose of medication(s) without consulting their physician, or may take medication only "as needed." Lapses between prescribed doses of three or more days are called "drug holidays." Drug holidays occur monthly to quarterly in two-thirds of patients (Urquhart, 1995, 1997). If patients have prolonged or more frequent drug holidays, not only will the medication likely be ineffective, but also patients may experience side effects due to the recurrent rebound and/or first-dose effects. These side effects may, in turn, contribute to persistent or worsened adherence to the medication regimen.

Measurement of Medication Adherence

Measurement of medication adherence is difficult. Physician estimate is a poor measure of adherence and more experienced clinicians do no better than trainees. Faithfully attending appointments does not ensure adherence to prescribed treatment, although adherence is much worse in those who fail to keep appointments. Velligan et al. (2007) demonstrated that physicians' estimated rate of their patient's adherence was weakly correlated with rates obtained by objective measurement or electronic monitoring. In the same study, physicians correctly identified less than half of nonadherent patients.

Although there have been hundreds of studies debating the merits of different measurement methods, no existing measure is entirely fool-proof. In general, measures of adherence to medication treatment can be classified as direct or indirect, as outlined in the following.

achieve target levels after a few doses so that appropriate administration for a few days before a blood draw ("white coat" adherence) could result in normal levels. Although a concentration of zero shows no recent medication ingestion, "spot concentrations" are of limited use.

Direct Measures

Supervised Doses

Watching a person ingest medication is one way to ensure that medication is making it into the patient. However, this method tends to be impractical. With longer acting medications that would be taken weekly rather than daily or several times per day, this method becomes more feasible. Supervised doses can be unreliable. Even under direct observation by trained staff on an inpatient psychiatric unit, patients can feign medication ingestion. For adolescent patients, direct observation, particularly by parents, may provoke negative emotions in the adolescent struggling for autonomy.

Blood Levels

Blood levels of drugs and/or their metabolites are only useful when accurate measurements of all significant active metabolites of the drug are available and there is a known relationship between the clinical effects of the drug and blood/serum concentration (Morselli & Bianchetti, 1983; Gualtieri & Golden, 1984). In general, blood levels are clinically useful for lithium carbonate, tricyclic antidepressants, and certain mood stabilizer medications. None of the selective serotonin reuptake inhibitor (SSRI) medications have serum levels that are sufficiently reliable to warrant the cost, inconvenience, and pain of blood draws. Similarly, the use of blood levels to examine adherence for antipsychotic medications is not warranted. When obtained, typically the only useful information derived from an SSRI level or levels of antipsychotic medications is the gross presence or absence of the medication at the time of the blood draw (Green, 1995; Dulcan, 1998; Velligan et al., 2006). Even for medications with reliable pharmacokinetic information, many drugs

Indirect Measures

Patient/Other Self-Report

Patient self-report is the most widely used method of measuring adherence to prescribed treatment in both clinical and research settings (Velligan et al., 2006). In usual practice, clinicians do not use standardized methods; instead, questions regarding adherence, if asked, are often closed and relatively leading, for example, "are you taking your medication?" One difficulty of self-report data is the risk of over-reporting. Cramer (1995) concluded that direct questioning during a patient interview has been proven inadequate to evaluate medication adherence because patients tend to tell doctors what they think the doctor wants to hear. A second limitation is lack of reliability of patient recall. Daily diaries have been used to circumvent this problem; however, few data are available on their utility or acceptability to patients. Parents or other caregivers can be asked whether the adolescent is taking their medication as prescribed; however, unless these individuals are present at the time of administration of the medication, their report is presumptive. To increase the reliability and validity of self-report data, a number of researchers have attempted to develop and use structured tools for gathering patient self-report information (Morisky et al., 1986; Svarstad et al., 1999; Byerly et al., 2007). The Brief Adherence Rating Scale is one such example. This scale sensitizes the patient to missed doses prior to asking about adherence. Asking about ingestion of each dose of medication at the time it is supposed to be taken is an approach gaining wider acceptability and aided by technology. "Did you take your medication today?" can be asked via cell phone or smart pill container. Technological methods are described below. These methods asking about each dose limit the problems with poor recall that plague self-report measures, and have been found to be better correlated with objective measures than other forms of self-report.

Table 1 shows validated patient-reported measurement tools, which are widely used to measure medication adherence.

Pill Count

This refers to counting tablets that remain in the patient's bottles and comparing the number of pills with the number that should be missing based on the fill date. Pill counts can be done in the office, but this requires patients to bring in their bottles. This method has been shown by many studies to be inadequate in measuring actual medication adherence (Cramer, 1989; Rudd et al., 1989; Pullar & Feely, 1990; Kruse et al., 1993; Urquhart, 1995). Patients often forget or are reluctant to bring bottles to appointments. Some patients put their medication in other containers. Patients, perhaps in an attempt to please their doctor, may discard or hoard remaining tablets and bring the empty bottle. In several recent studies, in-home pill counts done unannounced at random intervals 3–4 weeks apart have been recommended (Velligan et al., 2007, 2008, 2009, 2013). In this way, patients are unlikely to discard pills prior to a visit and pills both in prescription bottles and other containers can be counted.

Pharmacy Records

Prescription renewals can be used to monitor the frequency of refills. This method tends to underestimate actual ingestion of medication. Patients sometimes obtain medications from more than one pharmacy, more than one doctor, or from friends or family. Health maintenance organizations (HMOs) and insurance plans may require patients to refill their medication quarterly. Patients may accept unneeded prescriptions because they are embarrassed to admit that they are not taking the medication as prescribed (Rudd et al., 1989). However, in large samples, prescription refill data have proven to be a valid indicator of risk for poor outcome. For patients with schizophrenia, gaps as short as 11 days in which medication is not available for the patient to take have

been found to increase the risk of hospitalization (Valenstein et al., 2006). The longer the gap, the higher is the risk for rehospitalization.

Electronic Monitoring (EM)

Available since 1986, EM is sometimes considered the best "gold standard" for measuring adherence to prescribed medication (Cramer, 1989; Riekert & Rand, 2002). Medication is dispensed from a container that has a microprocessor (usually in the cap) that records the date, time, and duration of each container opening. Connecting to a computer using commercially available software retrieves data. Using EM, the variability of dosing in ambulatory trials practice has been shown to be far greater than previously indicated by other methods (Feinstein 1990; Pullar & Feely, 1990; Stichele, 1990; Bond & Hussar, 1991). EM provides more complete longitudinal information than any other currently available method and it does not rely on the memory of patients or doctors. EM has also been demonstrated to be superior to other methods in demonstrating important findings. For example, in their study comparing SSRIs with tricyclic antidepressants (TCAs), Thompson et al. (2000) found that only electronic monitoring allowed the collection of data sufficiently complete to measure prolonged periods of nonadherence. Using pill count and patient questionnaires, they found no significant differences in adherence between SSRI and TCA medication, despite marked differences in side effect profiles and dose regimens. However, using survival analysis of data from EM, they found an association between adherence and efficacy and showed superior adherence and efficacy of SSRIs versus TCAs.

There are also potential disadvantages of EM. First, it is an expensive system which is unavailable even in high-resource regions, so it is unlikely to be widely available. Further, it is not completely accurate – opening the EM cap does not correlate one-to-one medication ingestion. Data are lost if a patient forgets to put the cap back on the bottle. Many patients do not return caps for download of information. This limitation has been addressed by monitors that download the data directly using cellular technology. EM is expensive – the substantial minimum investment needed to purchase sufficient devices for simultaneous

Table 1 Medication adherence measurement tools

Developer	Measurement	Description	Item number	Response
Hogan et al. (1983)[a]	Drug Attitude Inventory, DAI-30	The patient's perceptions and experiences of treatment	30	True/false
Morisky (1986)	Morisky, Green, and Levine Self-Reported Medication Taking Scale, MGLS	The patient's perceptions and experiences of treatment	4	Yes/no
Van Putten (1978)	Neuroleptic Dysphoria Scale; NDS	The response to antipsychotics graded on a euphoric-dysphoric continuum	4	Yes/no
Awad (1993)	Drug Attitude Inventory, DAI-10	A simplified version of DAI-30	10	True/false
Weiden (1994)	Rating of Medication Influences in schizophrenia; ROMI	The reasons for medication adherence and nonadherence of patients with schizophrenia	20	None/mild/strong/not assessable
Naber (1995, 2001)	Subjective Well-Being Under Neuroleptic Treatment Self-Applied Scale; SWN	The well-being of patients receiving antipsychotic medications, independent of the improvement in their psychotic symptoms	Original 38/Short ver. 20	Yes/no
Horne (1999)	Beliefs about Medicines Questionnaire, BMQ	The cognitive representation of medication	Specific 10/General 8	Strongly/agree/uncertain/disagree/strongly disagree
Svarstad (1999)	The Brief Medication Questionnaire, BMQ	Screening adherence and associated barriers	14	Yes/no
Thompson (2000)	Medication Adherence Rating Scale; MARS	The patient's behaviors and attitudes toward psychoactive medications	10	Yes/no
Kampman (2000)	Attitudes toward Neuroleptic Treatment; ANT	The patient's attitudes and insights toward neuroleptic treatment	10	Visual analogue scale
Voruganti (2002)	Personal Evaluations of Transitions in Treatment; PETiT	Changes perceived by the patient under antipsychotic drugs and the effects of second-generation antipsychotic drugs on outcomes, such as subjective well-being	6 (about medication) 24 (general questions)	Often/sometimes/never
Dolder (2004)	Brief Evaluation of Medication Influences; BEMIB	Identifying patients who are more likely to be nonadherent to antipsychotic medications	8	Completely disagree/generally disagree/undecided/generally agree/completely agree
Morisky (2008)	Medication Adherence Scale-8, MMAS-8	To address adherence concerns, such as forgetting medication consumption or discontinuing medications without guidance	8	Yes/no

(continued)

Table 1 (continued)

Developer	Measurement	Description	Item number	Response
Gabriel (2010)[b]	Antidepressant Adherence Scale, AAS	To assess the degree to which forgetting, carelessness, and stopping owing to feeling worse or better interfere with antidepressant adherence	4	Times

[a]High average correlation with items of the Kuder-Richardson Formula 20 (Nunnally, 1976)
[b]Developed based on MGLS (Morisky, 1986)

patients is compounded because patients may lose or damage the devices (Riekert & Rand, 2002; Kreyenbuhl et al., 2016).

In recent years, EM has gained more "bells and whistles." Containers that monitor up to five medications and download all data to secure Web servers are available. These systems can also e-mail caregivers and treatment teams when adherence drops below a particular threshold. These devices are particularly helpful because they do not just monitor adherence but cue the patient to take medication by using alarms and LCD displays (Velligan et al., 2013). These devices also ask the patient when a drawer is opened whether that medication was taken or not. This type of self-report is highly correlated with more objective measures such as the electronic calculation from container openings and in-home pill counts (Velligan et al., 2013).

A recent addition to the use of electronic monitoring is the development of a smart pill that includes a sensor transmits information to a skin patch which then transmits that information to a smart phone app (Cohen et al. 2022). When exposed to stomach acids, a 1 mm $\times$ 1 mm sensor embedded within the pill is activated and sends a signal to a patch worn by the patient on the upper body (Papola et al., 2018). The information about medication ingestion is then available for review by both the patient and the treatment team (Cohen et al. 2022). The system is available for only aripiprazole and is marketed as Abilify MyCite. In a phase 3 study involving 277 patients with schizophrenia, maintenance in the study for the entire 3 months period was less than 40%, but patients who were more engaged in their treatment actually had greater symptom improvement

than those who were less involved (Cochran et al., 2022). Clinical acceptance of this approach has been limited by a variety of reasons.

Risk Factors for Nonadherence

Commonly reported reasons for medication discontinuation include distress associated with side effects, low perceived need for medication or poor insight, feeling better, and perceived medication ineffectiveness. Table 2 summarizes more factors commonly associated with poor medication adherence (Velligan et al., 2009).

Just as individual patients' medication adherence pattern may vary over the course of treatment, the relative significance of specific barriers to adherence varies greatly between patients. Based on their review of the pediatric literature, Logan et al. (2003) identified the most important correlates of nonadherence: (1) disease and regimen factors (duration and course of illness, symptom severity, type and complexity of regimen, efficacy, and side effects); (2) patient factors (developmental characteristics such as cognitive developmental level and level of autonomy); (3) interpersonal or attributional tendencies (depression, psychological coping strategies, and self-efficacy); (4) peer/family influences (stigma of disease/regimen, perceived need for secrecy, family cohesion/conflict, parental support, and shared responsibility for regimen); and (5) relationship with medical team. Using these a priori factors, they developed the Illness Management Survey, which is a self-report measure designed to identify the relative

Table 2 Factors Associated with Poor Medication Adherence

Level of distress/low perceived necessity for medication/poor insight; e.g., "Only really sick people take medication," "I feel better so I don't need medication"
Lack of or partial efficacy with continued symptoms
Medication is not perceived as beneficial
Ongoing substance use problems
Negative influence of social media, family, and friends
Learned helplessness
Poor treatment alliance
Complex medication regimen
Lack of daily routines
Practical problems (e.g., finances and transportation)
Distress associated with persistent side effects; e.g., "I don't feel like myself," weight gain, and sexual dysfunction
Fear of potential side effects; e.g., "I won't be able to feel my emotions," "Medication will make me a zombie," and "The medication will change my personality"
Avoidance of dependency
Concern about stigma/discrimination; e.g., "Only crazy people take psychiatric medications"

Source: Adapted from Velligan et al. (2009)

importance of common adherence barriers for individual patients. Their results were striking; they discovered that four of the five most important barriers to adherence cited by adolescents could be classified as due to internal processes (cognitive ability, perceived support from their family and doctor, and perceived stigma of disease/treatment). Adolescents' perceived burden of the disease/treatment (complexity of regimen, interference with normal life, and unpleasantness of side effects) was also an important factor. Figure 2 presents an integrated model of factors that influence patients' adherence to medication. Many of these same risk factors apply to adults. Additionally for serious mental illnesses, comorbid substance abuse, comorbid medical conditions, resource and logistical problems such as transportation and insurance, chaotic environments and a lack of daily routine, and problems with the therapeutic alliance have all been found to predict adherence problems (Velligan et al., 2009).

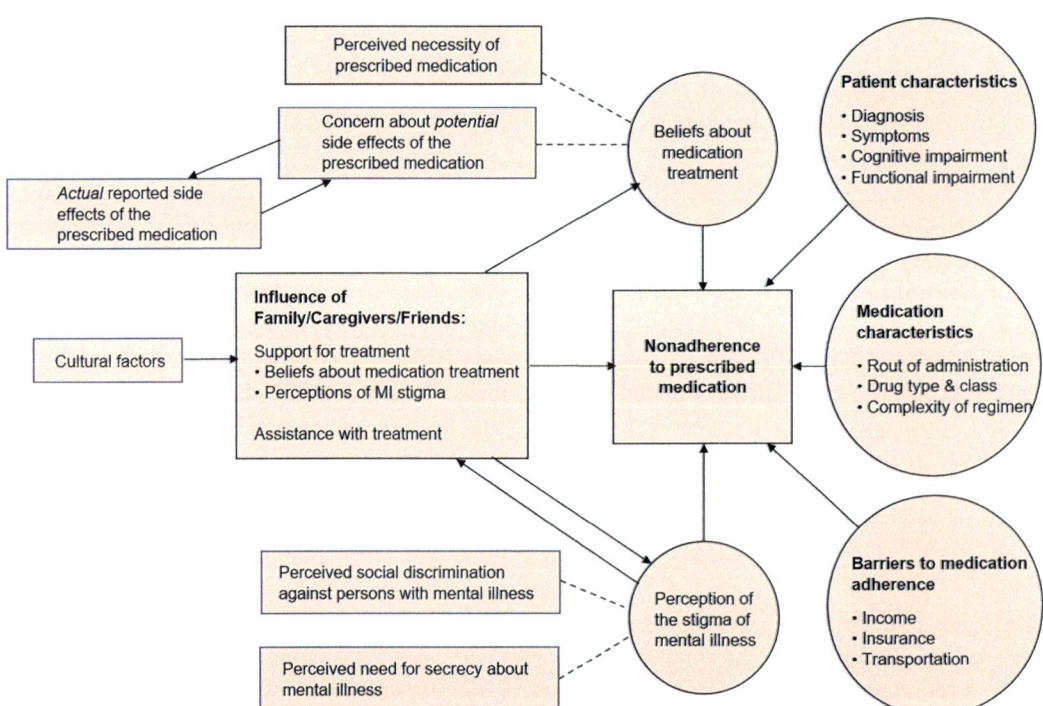

Fig. 2 Factors related to patients' medication adherence based upon research

Demographic Factors

Demographic factors are associated with adherence problems in adults with serious mental illness (Velligan et al., 2009; El Abdellati et al., 2020). In fact, although there is some variability in the literature, possibly due to sampling differences, studies have generally found that younger age, male gender, lower socioeconomic status, minority status, and poorer social functioning are all associated with adherence problems (for a review, see Velligan et al., 2009). However, clinicians need to be wary of jumping to conclusions based upon demographic variables.

Medication adherence also appears to be influenced by age/developmental level in children and adolescents. Past research has shown that adherence to medication in children is related to the mother's perception of the severity of her child's illness, but not to the physician's estimate of severity of the child's illness. Data also suggest that adolescents are less adherent to medical recommendations than younger children or adults, which may be due to developmental factors (Cromer & Tarnowski, 1989; Riekert, 2002). Developmental factors that may influence adherence (see Table 3) in adolescents (and developmentally delayed adults) include: (1) brain maturation and development of cognitive processes and (2) psychological development (see Table 3).

Table 3 Developmental issues in medication adherence

1. Brain maturation and development of cognitive processes

  (a) Age related; e.g., Piagetian stage, Eriksonian stage, psychological autonomy, legal competency, executive functioning

  (b) Biologically related; e.g., head injury, IQ, mental illness, substance abuse, encephalopathy, attention-deficit/hyperactivity disorder, impulse control disorders

2. Psychological development

  (a) Age related; e.g., development of ego strength and psychological defense mechanisms

  (b) Biologically related; e.g., innate temperament, family history of psychiatric illness, genetic variation of drug metabolism, organ maturity, physical growth and development, disorders of kidney and/or liver functioning, drug–drug interaction

Adolescence is characterized by experimentation and risk taking, limited insight, a sense of invulnerability, increased influence of peers, and lack of future orientation. Although a sense of autonomy is a positive developmental milestone, a child who feels autonomous may view health-promoting influences as a restriction to autonomy.

Only a small proportion of mental health treatment in adolescents is self-initiated. Adherence to medication among children and adolescents requires the consent and cooperation of a parent and also the assent and cooperation of the youth. The involvement of adults may create conflict for teens because help-seeking and utilization of adult-provided services is inconsistent with their desire for autonomy, privacy, and independence.

Patient-Related Factors

Pragmatics
Some examples of pragmatic barriers to improved medication adherence include income, stability/stress in home life, homelessness, transportation, telephone service, medical and prescription coverage, travel time to the doctor's office, waiting times to see physicians, access to prescription refills, access to nurse/physician to answer questions, practices of third-party payers, and national health care policies.

Ability
A patient's ability to take medications as prescribed is influenced by the patient's cognitive functioning, motor functioning, and knowledge about the medication (Tham et al., 2016; Velligan et al., 2017). Their cognitive functioning is determined by level of alertness, orientation to person/place/time, sustained attention, memory, ability to use a calendar/reminders/organizers, thought processes, processing speed, problem-solving skills, capacity for logical operations (understanding of cause and effect), intelligence, literacy, math skills, executive functioning, problems with initiation (amotivation and negative symptoms of schizophrenia), and impulse control. A patient's motor functioning consists of fine motor control, physical mobility, ability to drive or obtain

transportation, and ability to swallow medication, and can affect the physical ability to obtain and take medication. Finally, a patient's knowledge about medication influences their adherence to medication. A number of studies have examined the impact of patient knowledge about their illness and treatment on medication adherence. Most studies have concluded that although patient instruction is an important component of treatment and may improve adherence to short-term medication treatment, patient education alone does not assure adherence to long-term medication treatment (Morris & Halperin, 1979). Sackett and Snow (1979) showed that even "mastery learning" about medication treatment does not necessarily improve medication-taking behavior. Psychoeducation involving family members in addition to the person with the illness has been shown to improve medication adherence for individuals with schizophrenia and bipolar disorders (Velligan et al., 2009). However, psychoeducation involving only the person with the illness has not proven beneficial in serious mental illness.

Motivation

A patient's motivation to take prescribed medication is influenced by many complex and interrelated factors, as listed in Table 4.

Patient Beliefs/Attitudes About Medication

In a landmark publication, Frank (1973) described the close association between beliefs and attitudes

Table 4 Factors which influence motivation to take medication

Severity of psychiatric symptoms – impacts the patient's cognitive functioning, insight, future orientation, and confidence

Past experience with prescribed medication, especially psychiatric

Personal history of mental illness

Personal beliefs – health/illness beliefs, acceptance of illness, understanding of illness, concerns about medication, perceived necessity for medication, perceived medication efficacy, and concerns about mental illness stigma

Treatment goals

Temperament/Personality

toward treatment seeking, expectancies of change, and actual illness outcome. Subsequent research supports the importance of medication beliefs in determining patient adherence to medication recommendations. For example, Lingam and Scott (2002) conducted a review of research from 1976–2001 on treatment nonadherence in affective disorders and reported that attitudes and beliefs are at least as important as side effects in predicting nonadherence to antidepressant treatment. Research in patients with chronic medical illness suggests that how patients understand their illness has a significant impact on treatment adherence (Leventhal et al., 1992). Horne and Weinman (1999) proposed a theoretical model for the relationship between patient beliefs about medication and adherence to medication treatment in adults with chronic illness. In their study of 324 patients with chronic illness (asthma, renal, cardiac, and cancer), they found that patient beliefs about medication were more powerful predictors of adherence than clinical or demographic factors. Based on their findings, the authors concluded that patients' general beliefs about medication influenced their initial orientation toward prescribed medication, but it was patients' specific beliefs about their prescribed medication that is most strongly related to their medication adherence. These data point out the importance of assessing a person's beliefs about medication in addition to their medication-taking behavior when trying to predict whether someone will follow through with taking prescribed medication on a regular basis.

Kemp et al. (1996, 1998) developed a treatment designed to take into account the person's beliefs about medication and treatment in addition to their history and experience with medication. Adherence therapy is a cognitive behavioral talk therapy that includes components of motivational interviewing and psychoeducation and focuses on helping the person alter their beliefs and choices regarding adherence. Adherence therapy showed early promise for promoting adherence in serious mental illnesses such as schizophrenia, but was not found to be effective in later studies (O'Donnell et al., 2003; El-Mallakh & Findlay, 2015).

Table 5 Attitudinal factors influencing medication adherence

Testing – the patient stops medication in order to "test" whether it is needed and/or effective
Control of dependency – the fear that medication is a crutch; fear of dependency on medication; medication becomes a symbol of the dependence created by a chronic illness and is a constant reminder of the illness
Destigmatization – the fear of societal stigma of illness and "flight to health"; avoiding medication is a way to avoid the stigma of the illness
Practical considerations – for example, perceived expense and value and/or burden of the regimen; wish to avoid problems that might occur by mixing alcohol or illicit drugs with the medication

Studies have documented the resistance of adolescents to psychiatric medication. Scott et al. (1992) found that 58% of mentally ill adolescents (with diagnosed schizophrenia, unipolar depression, and bipolar depression) disagreed with the need to take prescribed medication. Williams et al. (1998) reported that 61% of adolescents with various psychiatric diagnoses had a bias against taking medication and a lack of positive expectancy for effect of medication.

Problem adherence can be classified into subgroups based on a person's particular attitudinal risk factors or barriers to adherence (Conrad, 1985), as given in Table 5:

Patient Concerns About Medication

Many patients have substantial suspicions about prescribed medication and these beliefs may influence how information about medication is interpreted and acted upon. Certain beliefs about medication are highly prevalent across diverse populations; for example, medications are addictive and accumulate in the body to produce adverse long-term effects.

Williams et al. (1998) studied attitudes about psychiatric medications among incarcerated female adolescents. The most common concerns reported by this sample were as follows:

1. "Medicine might be hard to swallow or taste bad," 44.4%
2. "I feel strongly that people should solve problems on their own – medicines are a crutch," 33.2%

3. "Medicine might change my personality or not let me be myself," 30.8%. Other concerns included: "I wouldn't want others to know I was taking medication to help with my feelings or behavior – I might get teased by my peers," and "Medicine might cost too much for my family to afford."

Patient Concerns About Mental Illness Stigma

The word "stigma" literally means "mark of shame" (El-Mallakh and Potter 2019). Link and Phelan (2001) defined stigma as any trait of an individual or group that evokes a negative or punitive social response. In common parlance, people refer to the stigma associated with individual traits (e.g., skin color and obesity), events (e.g., personal history of psychiatric hospitalization, incarceration, or substance abuse), or groups (e.g., persons with mental illness and persons who are HIV positive). Mental illness stigma specifically refers to status loss and discrimination due to negative stereotypes about people labeled with mental illness.

The mentally ill are one of the most stigmatized social groups (Sayre, 2000; Link et al., 2004). Derogatory messages about mental illness are ubiquitous in society and seen in colloquial expressions, cartoon characters, and mass advertisement (Wahl, 1995; Philo, 1996; Wilson et al., 2000). Link (1987) found that people learn negative stereotypes long before they themselves become patients and that the experience of illness triggers internalized negative beliefs.

The *1999 Surgeon General's Report on Mental Health* cited stigma as a powerful obstacle to seeking care (Satcher, 1999). Jorm and Korten (1997) found that the public more often perceives psychiatric medication as more harmful than helpful. Studies of stigma and adherence to depression treatment have found that perceived stigma predicts treatment discontinuation (Lai et al., 1997; Sirey et al., 2001). If a person believes that the mentally ill are less competent than other people, it is not surprising that they may wish to distance themselves from such a label. If taking psychiatric medication implies mental illness, nonadherence to prescribed medication can reflect denial or ambivalence about a mental illness diagnosis (Chakrabarti, 2019).

Perceived Necessity for Medication Treatment

Studies across medical conditions show that people prescribed the same medication for the same condition differ in their perceptions of personal need for medication (Horne & Weinman, 1999). The factors determining perceived necessity for medication in depression are not fully understood. According to the common sense model of self-regulation, patients do not blindly follow medical advice, but assess the threat posed by their illness to determine whether treatment is needed and, if so, what treatment is appropriate (Leventhal et al., 1992). A number of factors may contribute to perceived need for medication, including perceived severity of illness, level of distress/impact of illness, understanding of illness (e.g., beliefs about causation, expected duration, and beliefs about amenability to treatment), perceived efficacy of medication, and perceptions of personal resilience (Horne, 2006a).

Adams and Scott (2000) found that perceived benefit of treatment and perceived severity of illness together accounted for 43% of variance in adherence. Using the HBM, Budd et al. (1996) found that perceived severity of symptoms was associated with greater medication adherence.

Perceived Efficacy of Medication Treatment

Patients' perceptions of the efficacy of medication may not be the same as those of their doctor. An individual's perception of the effectiveness of a treatment depends on whether the outcome of treatment matches their expectations. For example, a patient who believes that effective treatment of depression means that they will be "happy" after taking medication for 1 week is likely to be disappointed.

Perceived efficacy of prescribed medication has a complex relationship to perceived necessity for medication and adherence to prescribed medication (Kikkert & Dekker, 2017). In their study of depressed adults in primary care, Brown et al. (2001) found that, although most patients acknowledged the negative impact of depression on their lives, less than 40% viewed antidepressants as effective in treating their symptoms.

Moreover, clinical experience and published data suggest that regardless of their clinical response to medication treatment, patients commonly have attitudes and beliefs about psychiatric medication that are inconsistent with medication adherence (Williams et al., 1998).

Physicians often assume that if a medication produces a significant reduction in a patient's symptoms, then the medication is "necessary"; however, it is not uncommon for patients who have experienced some benefit from medication to discontinue it prematurely, stating, "I feel better so I don't need to take the medication any more."

Provider-Related Factors

Pragmatics

Some examples of pragmatic barriers to improved medication adherence include the quality/quantity of office staff, location of practice, practices of third-party payers, national health care policies, continuity of care, and communication between various medication providers. Goldfinger et al. (1984) examine the role of service systems breakdown in patients whose treatment fails.

Ability

A physician's ability to facilitate patient adherence to medication is influenced by innate talents, interpersonal skills, empathy, psychological mindedness, confidence, openness, quality of training, and clinical experience.

Motivation

A physician's motivation to improve their patients' medication adherence is influenced by personal history of mental illness, personal beliefs (especially about the stigma of mental illness and beliefs about medication efficacy and safety), family and/or friends with mental illness, ethical principles, perceived professional mission, and, on the darker side, perhaps avoidance of legal liability and high malpractice rates.

Therapeutic Alliance

Patients' medication adherence is substantially influenced by perceived support from their physician. Doctors and patients develop a relationship that is based on communication that includes information, affect, and social learning. Therapeutic alliance is as important for success in psychopharmacology as it is for psychotherapy. Patients' high trust in clinicians is indirectly associated with better adherence by mediating improved medication attitudes (McCabe et al., 2012; De Las Cuevas et al., 2014, 2017). With increasing pressure from all sides for doctors to perform only brief "medication checks," the interpersonal aspects of the doctor–patient relationship are likely dismissed. Optimal therapeutic alliance depends on mutual trust, respect, honesty, openness, and comfort between doctor and patient, all of which take time.

Velligan et al. (2009) conducted a survey of 41 experts in adherence to medication in serious mental illness. Problems with the therapeutic alliance were rated among the most important risk factors in predicting problem adherence. Medication adherence is affected by continuity of care. Charney et al. (1967) demonstrated that adherence was worse when patients saw the partner of their regular physician than when they saw their regular physician. Other factors that contribute to the quality of the doctor–patient relationship include: (1) availability of the physician (or appropriate medical staff) between appointments, (2) length, frequency, and quality of appointments, and (3) waiting times to see the doctor. Finally, the quality of office support staff can have a significant impact on patient care.

Patients who are poorly adherent to treatment frequently elicit strong countertransference responses in their providers. Doctors may experience feelings of disappointment, anger, and/or defensiveness. They may blame the patient, as exemplified by the statement; the patient is "resistant" or "doesn't want to get better." These responses are counterproductive and serve only to erode the therapeutic alliance further. In this light, medication nonadherence can be seen as a failure of the relationship, rather than as a failure of the patient (Weiden & Rao, 2005).

Contextual Factors

Family Factors

Several systematic reviews reported an association between relatives' beliefs about illness and their beliefs about treatment, and found evidence that family beliefs influence patient adherence (El Abdellati et al., 2020; Edgcomb & Kima, 2018; Marrero et al., 2020). They observed that family members who consider mental illness to be caused by a chemical imbalance strongly reinforced the use of medication in the patient. In a cross-cultural study sponsored by the World Health Organization, Bush and Hardon (1990) found that medication use, family beliefs, and health-related practices are instilled in children by school age and influence their readiness to take prescribed medication. In the survey conducted by Velligan et al. (2009), experts in serious mental illness cited lack of family support for medication as one of the main risk factors for patients intentionally not taking medication as prescribed.

Stine (1994) reported several reasons for nonadherence to psychiatric medication in children and adolescents despite clinical improvement and absence of side effects, including parental discomfort with the idea of psychiatric medication, patient fear of stigma, and child refusal and resistance. Public awareness of the controversy regarding the efficacy and safety of SSRI medications in children and adolescents may impact parent/family concerns about antidepressant medications. Prescribing providers need to be sensitive to concerns about medication safety, and concerns about medicating children. Similarly, stimulant medications are often overprescribed, leading parents to wonder whether their child with attention-deficit disorder is really in need of these medications. Sharing research results with family members regarding the safety and efficacy of medications with a focus on long-term outcomes can be of some benefit in improving their children's adherence to medication.

Financial Factors

For patients without health insurance and/or prescription coverage, the cost of medication and medical care may be a significant barrier to medication adherence. Patients with health insurance may still sustain significant medical costs that affect medication adherence. Many health insurance plans do not include prescription coverage or have restricted formularies. Managed care formularies increasingly restrict medication choices. Many patients are required to pay deductibles and co-pays, which vary in amount, but can be as high as 20–80% of the actual cost of the prescription or service. A study conducted in the Veterans Administration system by Zeber et al. (2007) found that when the co-pay was raised from $2 to $7, there was a 25% decrease in refills of psychiatric medications. Some insurance plans require patients to obtain prescriptions for a three-month quantity of medication and receive their medications by mail. Patients with Medicaid and Medicare insurance are now required to enroll in a managed care plan that regulates prescription benefits. These plans have restricted formularies and require prior authorization for all nonformulary medications (Kihlstrom 1998).

Clinical Factors

Medication Regimen

Studies of the number, type, dose, frequency, and time of medication prescribed in relationship to medication adherence have revealed variable results. In a review of studies using electronic monitoring to measure adherence, Claxton et al. (2001) found an inverse relationship between number of daily doses and adherence. Kruse et al. (1993) reported higher patient adherence to twice daily (85%) versus four times daily regimens (65%), and that evening doses are omitted twice as often as morning doses. Horne and Weinman (1999) found no relationship between the total number of prescribed medications and adherence. Yaegashi et al. (2020) also reported that more frequent regimen was one of the reasons for poor adherence.

Severity of Illness/Level of Distress

Multiple studies of patient medication adherence, across a range of chronic illnesses, have shown that physicians' assessments of medication necessity or illness severity often have little correlation with patients' level of medication adherence (Lemanek, 1990; Phipps & Decuir-Whalley, 1990; Kiley et al., 1993). Even when the patient's reported symptoms underlie the physician's assessment (e.g., using the Beck's Depression Inventory (Beck et al., 1961) to assess severity of depression), this may not predict patient adherence to prescribed medication.

One predictor of medication nonadherence is the absence of patient distress (Ellison, 2000). Without the motivation provided by distress, taking medication may seem pointless and not worth the side effects. Even patients who are sufficiently motivated by distress to schedule an evaluation with a psychiatrist may, in a "flight to health," experience relief prior to starting medication. Many patients, after taking medications and experiencing relief from symptoms, may discontinue medication prematurely – with or without the physician's knowledge. Other patients, perhaps pressured into seeing a psychiatrist, may not acknowledge the distress observed by others for a number of reasons (e.g., denial of illness and lack of insight).

Finally, depending on the type of disorder and severity of symptoms, psychiatric illness itself may interfere with patients' willingness and/or ability to adhere to medication (Chen, 1991; Velligan et al., 2017; El Abdellati et al., 2020). These factors are discussed at length later in the section Adherence in Specific Patient Populations, for example, schizophrenia, bipolar disorder, and attention-deficit disorder.

Side Effects

It is a mistake to ignore the possibility that a patient's nonadherence to medication is based on concerns about current or potential side effects of treatment. Side effects to prescribed medication, even if considered "mild" by the medical provider, may have a significant impact on medication adherence. Patients may discontinue medication that has been effective if the side effect is sufficiently noxious, such as nausea or sexual dysfunction. Patients' concerns about the possibility of side effects or adverse events (even in the absence of past or current problems) may also be a significant factor in adherence to prescribed medication (Kikkert & Dekker, 2017; Velligan et al., 2017). For example, fear of severe adverse events (potentially fatal or causing disability, for example, hepatotoxicity, cardiac arrhythmia, and tardive dyskinesia), even if rare, may be a significant deterrent to patient adherence.

Even when a patient offers no reason for not taking the medication, or points to some other cause, the physician should make specific inquiries about embarrassing (e.g., sexual dysfunction) or difficult-to-describe side effects (e.g., akathisia). If no current side effects can be elicited, past experiences with medications should be reviewed, and also whether patients have witnessed or heard reports of bad experiences with psychiatric medications from family, friends, the media, or Internet reports (Appelbaum & Gutheil, 1982).

Side effects are associated with nonadherence to medication, but to a lesser extent than initially believed. Prevalence of antidepressant nonadherence has not changed substantially with the advent of newer medications (Lingam & Scott, 2002). In their study of medication nonadherence in severe affective disorders, Adams and Scott (2000) reported that side effect fears were more predictive of nonadherence than actual side effects. It is important to remember that it is subjective distress about the side effect, rather than the side effect itself, that is most linked to medication nonadherence.

Adherence in Specific Patient Populations

Psychiatric diagnoses are associated with poorer medication adherence than are nonpsychiatric illnesses (Haynes et al., 1979; Litt & Cuskey, 1980). Psychiatric illness, by virtue of impact on patient ability and motivation, presents particular challenges for adherence to medication treatment. Although it is beyond the scope of this chapter to address adherence issues related to every psychiatric disorder, factors unique to a few specific diagnostic groups are outlined in the following.

Depression

Approximately 8–10% of patients never fill their antidepressant prescription (Rashid, 1982).

Subsequent antidepressant adherence decays over time and medication discontinuation is greatest within the first month of treatment. Johnson (1981) reported antidepressant discontinuation rates of 16% within the first week, 41% within 2 weeks, 59% within 3 weeks, and 68% within 4 weeks. In their meta-analysis of discontinuation of SSRIs and TCAs, Hotopf et al. (1997) found that 30% of patients stopped taking their antidepressant within 1 month and 45–60% stop by 3 months.

Rapid cessation of SSRIs can produce a discontinuation syndrome consisting of somatic and psychological symptoms. In some situations, symptoms related to medication discontinuation may reinforce the need for adherence to medication; however, if these symptoms are attributed to the medication itself, they may serve as a barrier to future adherence.

Depression has been shown to predict nonadherence independently. Depression-specific factors in nonadherence include reduced motivation, reduced task initiation, and cognitive impairment (slowed processing speed, poor concentration, and impaired executive functioning). The severely depressed patient may lack the energy and motivation to participate in treatment of any kind. Feelings of hopelessness or frank suicidal ideation related to depression may prevent patients from adhering to medication or other treatments prescribed for comorbid medical illnesses. This phenomenon can be particularly devastating to the treatment of patients with chronic or potentially fatal illness (diabetes, end-stage renal disease, cancer, and transplant patients).

Bipolar Disorder

The problem of medication nonadherence in the treatment of bipolar disorder has likely been experienced by many readers. Barriers to adherence in bipolar disorder can be classified as follows: (1) pragmatic/ability to take medication – in both the manic and depressed phases of the illness, there are a number of core symptoms that can decrease the patient's ability to take the medication as prescribed, such as impaired concentration/memory, disorganization, anergia, and altered sleep cycle; (2) willingness – patients with bipolar disorder may be hesitant to take medication that reduces hypomanic periods that they associate with exuberance, peaks of creativity, periods of increased goal-directed behavior, and decreased need for sleep; the manic phase of bipolar disorder is characterized by grandiosity, and occasionally frank psychosis, both of which impact the patient's insight and willingness to adhere to treatment; (3) aversive nature of treatment – medications typically used to treat bipolar disorder (i.e., mood stabilizers and antipsychotics) commonly have unpleasant side effects (e.g., weight gain, sedation and movement disorders) or require frequent blood level tests, which can be a barrier for some patients.

Rates of medication nonadherence among patients with bipolar disorder range from 18–52% (Keck et al., 1998; Scott & Pope, 2002; Rosa et al., 2007). In a study of patients prescribed mood stabilizers, 50% of subjects admitted to some degree of medication nonadherence (Scott & Pope, 2002). Scott and Pope (2002) reported that 32% of patients admitted to only partial adherence to medication, which was defined as missing 30% or more of their medication. Factors associated with nonadherence were greater length of time on mood stabilizers, past history of nonadherence, denial of the severity of the illness, fear of side effects (not actual side effects), and negative attitudes toward medication (Scott & Pope, 2002; Rosa et al., 2007). Gitlin et al. (1989) reported increased medication nonadherence in patients reporting coordination problems and/or cognitive side effects in response to lithium. In their study of medication adherence in patients with bipolar disorder, Keck et al. (1998) reported increased medication nonadherence in patients with comorbid substance abuse disorders.

Schizophrenia

Based on numerous studies, it is estimated that patients with schizophrenia take approximately half of their prescribed medication (Novak-Grubic & Tavcar, 2002; Rettenbacher et al.,

2004; Valenstein et al., 2006), although this may be an underestimate. Valenstein et al. (2006) found that poor adherence was associated with young age, non-White race, substance abuse diagnosis, psychiatric hospitalization, and predominant treatment with first-generation antipsychotics. Rettenbacher et al. (2004) found adherence to be positively associated with patients' feelings of a positive effect of the drug on the illness, and greater negative symptoms.

In addition to deficits in judgment and insight present during active psychosis, persons suffering from schizophrenia have a number of cognitive deficits that are present not only in acute phases of illness, but also during periods of remission (Gold & Harvey 1993). Specific areas of impairment include attention, memory, and executive function. Similarly to patients with other mental disorders (e.g., major depression and bipolar disorder), patients with schizophrenia often have irregular sleep and meal patterns and/or chaotic lives, making taking medications on a regular schedule very challenging. An intervention developed by Velligan et al. (2008) demonstrated that the use of environmental supports such as calendars, signs, checklists, voice alarms, and specialized pill containers can improve adherence to medication and reduce rates of relapse for individuals with schizophrenia. This suggests that a fair amount of problem adherence in this population is not intentional.

Olfson et al. (2000) found that schizophrenic patients with substance use, difficulty recognizing their own symptoms, a weak alliance with inpatient staff, or poor family support were at increased risk for nonadherence following discharge from inpatient treatment.

Attention-Deficit/Hyperactivity Disorder (ADHD)

The characteristic features of ADHD, such as inattention, disorganization, and distractibility, can all contribute to nonadherence to medication. Since ADHD is a chronic illness with significant morbidity, it requires adherence to a medication regimen for many years. Owing to the strong genetic etiology, parents of children with ADHD frequently have ADHD themselves and suffer from cognitive and temperamental characteristics similar to those of their children, which can contribute to nonadherence (Stine, 1994).

One study examined the length of time children with ADHD were adherent to stimulant treatment over a three-year period (Thiruchelvam et al., 2001). The authors reported an 81% adherence rate at 1 year, but only 52% at 3 years. Children who were adherent were younger and tended to have more severe symptoms of ADHD, but no oppositional/defiant symptoms. Children who were adherent to medication showed more improvement in teacher-rated symptoms at five-year follow-up (Charach et al., 2004). In an older study of adherence to ADHD medications, Firestone (1982) found that 26% of participants refused medication at the start of the study. Of those who accepted medication, 20% stopped it after 4 months and 55% after 10 months. Even more striking, fewer than 10% of families talked to their physician prior to discontinuing medication.

The reasons for nonadherence among patients with ADHD are multiple. Thiruchelvam et al. (2001) reported greater rates of nonadherence in older children and children with oppositional defiant disorder (ODD) symptoms. This suggests that older and more defiant children may resist and ultimately stop taking medication. Reported side effects were not related to adherence in this study.

Stine (1994) published an eloquent discussion of psychodynamic and psychosocial factors related to psychiatric medication adherence in the treatment of ADHD. The most commonly reported reasons for medication discontinuation were: (1) the child's impression that the medication makes them feel strange or uneasy; (2) the belief that the medication makes the child less popular or entertaining when they take it; (3) the difficulty of having to take the medication at school; (4) the belief that taking medication makes the child different from friends or that friends will treat them differently because of their medication; (5) the desire to disavow symptoms or the need for treatment; and (6) beliefs about the dangers of drugs (these fears are sometimes extreme, for example, "medications are poisonous").

Substance Abuse Disorders

It is not within the scope of this chapter to discuss specific treatment regimens for the various substance abuse disorders. The key point is that comorbid substance abuse (in addition to any other medical or psychiatric disorder) increases the risk for medication nonadherence to prescribed medications, either psychiatric or nonpsychiatric. A number of factors likely contribute. Patients who are actively using alcohol and/or illicit drugs are often fearful that the prescribed medication(s), particularly psychiatric medications, will interact dangerously with alcohol or other substances. Physicians, family, and friends frequently reinforce many of these beliefs the advice to avoid substances that can backfire and a number of patients have reported to their doctor, "I didn't take it because I wanted to drink (or use) those days." Many patients, especially when feeling slightly better, will simply stop their prescribed medication(s), or take them irregularly or as needed, in order to use alcohol or other substances "safely." Data support an increased risk of secondary depression/anxiety/insomnia in patients with significant alcohol or cannabis abuse; however, alcohol does not directly interact with many of the newer generation antidepressants. As physicians, it is important to assess the degree of the substance abuse openly with the patient, monitor for abuse of prescription medications and also signs of CNS toxicity/oversedation, and collaborate with the patient – initially to provide nonjudgmental guidance and education, and later to explore the patient's motivation to engage in treatment for substance abuse, generally while considering *continuing* medication for the patient's psychiatric disorder. Unfortunately, use of substances greatly muddies the clinical picture and makes prescribing more difficult. Doctors will often ask patients to avoid alcohol and substances at least initially in treatment so that they can work with the patient to find the medication and dose that will work best for their symptoms.

Developmentally Disabled/Cognitively Impaired

Cognitive developmental level affects the capacity and motivation for treatment adherence in a number of ways, including through memory, perception of time, and understanding of causality and consequences. Patients may not adhere to physician's recommendations because of the latter's use of abstract rather than concrete reasoning in education about the illness and treatment required. Complex cognitive processes are required to understand the varied presentation of psychiatric symptoms, the multiple causes of mental illness, and the factors that contribute to severity of symptoms. Patients must have an adequate understanding of causality to understand how adherence (or nonadherence) affects the course of illness. They must have an adequate perception of time to understand the short- and long-term consequences of depression and to understand the need to take medication regularly despite the delayed time-to-effect of medications. They must possess the ability for formal operations to understand fully the multiple abstract causes and consequences involved in illness and treatment of depression. Pediatric patients differ considerably in these cognitive skills.

In early adolescence, youth are able to think about illness in increasingly complex ways. By later adolescence, most youth have the capacity for a somewhat scientific understanding of illness and treatment. Even adult patients may not have the health knowledge or logical operations necessary to understand the necessity to understand fully various components of treatment. Bush and Iannotti (1990), using their children's health belief model, found that adolescent' perceptions of the severity of illness and benefit of medication were the most important factors contributing to their adherence to medication. Both of these factors require the understanding of causality and consequences. It is clear that efforts to enhance treatment adherence should match the intervention with the cognitive level of the patient.

Intervention Strategies

Haynes (1976) identified eight major strategies to improve medication almost five decades ago and these recommended approaches are still pertinent today to enhance adherence (see Table 6).

Much later, in a systematic review of interventions for enhancing medication adherence, Haynes et al. (2005) concluded that improving short-term adherence is relatively successful with a variety of simple interventions; however, interventions for enhancing long-term adherence in chronic health problems have been more complex and not very effective. Unfortunately, many medication regimens in psychiatry are long term. It may be that having a positive and collaborative therapeutic alliance with the primary clinician prescribing the medications plays a more important role not only in the initiation of treatment but also during long-term treatment. The alliance provides the context in which all these strategies are best implemented.

Given the number of past intervention trials that have failed or produced results of marginal clinical utility for chronic medication regimens, it seems probable that past paradigms must be examined. The limited success of past interventions may be due to attempts to address unintentional but not intentional nonadherence. The simple strategies listed above fail to consider the power of psychosocial factors, such as patient beliefs, therapeutic alliance, and the influence of family, friends, and the community. In fact, the role of the therapeutic alliance may be of the greatest importance.

Table 6 Eight major strategies to enhance adherence

1. Informational and instructional
2. Reduction of barriers to adherence
3. Behavior modification approaches – problem-solving, skills training
4. Increased patient supervision
5. Medication reminders and cues
6. Pharmacy interventions
7. Increasing resources/supports
8. Collaborative treatment teams

Selected Promising Intervention Strategies

Patient Self-Regulation of Medication

For some patients, self-regulation of dosage can be beneficial (Epstein & Cluss, 1982). Nessman et al. (1980) utilized a self-control package that included self-monitoring of blood pressure and self-selection of dose regimen. Patients with insulin-dependent diabetes mellitus have long used the "sliding scale" method for fine-tuning blood sugar control to avoid dangerous highs and lows (Epstein et al., 1981). In pediatrics and child psychiatry, parents are commonly trained to administer "as needed" doses of medication (e.g., the use of nebulized medication treatments as needed in asthma and the use of anxiolytic or antipsychotic medication as needed for acute agitation in severely behaviorally disturbed patients).

Motivational Interviewing

Motivational interviewing uses a collaborative style of interaction between patient and provider to assess patient motivation to change (Chakrabarti, 2019; Abdellati et al., 2020; Loots et al., 2021). Over 70 clinical trials have established motivational interviewing as an effective method to promote behavioral change, including increasing treatment adherence. Miller and Rollnick (2002) demonstrated that adding even one motivational interviewing session in early treatment can improve patient retention, increase adherence, and improve outcome. The three principles of motivational interviewing are the use of collaboration between the physician and the patient, evocation of patient responses, and autonomy for the patient. Tailoring Medication Regimens: Using Pharmacokinetics to Optimize Medication Adherence.

What is good enough adherence? Unfortunately, most published studies set an arbitrary cutoff point for what constitutes adherence. Many studies even treat adherence as a dichotomous construct (e.g., adherent versus nonadherent). Even those studies that begin with a continuous measure of adherence (e.g., percentage of doses taken as prescribed) generally set an arbitrary cutoff point for satisfactory adherence (e.g., 80%). In reality, the minimum level of adherence necessary to achieve the desired clinical outcome (e.g., remission of depression symptoms) depends on many factors, including the pharmacokinetic properties of the medication, individual differences in drug absorption and metabolism, and individual variation in response to medication, with respect to efficacy and side effects (Epstein & Cluss, 1982). For some conditions/medications, 50% adherence may be sufficient, whereas for others 80% or more may be necessary.

The pharmacokinetic (absorption, distribution, and elimination) and pharmacodynamic (pharmacological effect, clinical response, toxicity, and efficacy) profile of any medication varies across patients and can vary within the same patient due to many factors, for example, age, race, gender, organ functioning due to maturation/disease, disease severity and comorbidity, and pregnancy (Katzung, 2007).

Individual patterns of drug taking ("medication adherence") influence the pharmacokinetics and pharmacodynamics of a medication; known pharmacokinetic properties of medications can be used to select more feasible, "forgiving" medication regimens that can be tailored to individual patients (Urquhart, 1992; Levy, 1993; Rudd & Lenert, 1995).

Weiden and Rao (2005) proposed the use of a mnemonic of the "four Fs" to guide psychopharmacological treatment to increase medication adherence: (1) effectiveness, (2) flexibility, (3) forgiveness, and (4) user-friendliness (Table 7).

This method of prescribing medication utilizes evidence-based practices for selecting a range of medication regimens, all of which have sufficient empirical and/or clinical data support, and then integrates known pharmacokinetic properties of medications, awareness of risk factors for nonadherence, and substantive understanding of the individual patient's goals and priorities to compare the ability of the selected medication regimens to minimize barriers to medication adherence and compensate for imperfect patient behaviors. This process is tailored to individual patients and must be dynamic and interactive, just as medication adherence is a dynamic process, influenced by many factors.

Table 7 The model of shared decision-making

	"Traditional Medical Model" (paternalistic) (paternalistic model)	Shared decision-making	Informed choice
Role of the doctor	Active: gives selected information to patient and chooses treatment *for* the patient	Active: gives all information to patient and chooses treatment *with* the patient	Passive: gives all information to patient. Patient chooses treatment
Role of the patient	Passive: expected to accept and adhere to prescribed treatment	Active: receives all information. Discusses concerns and preferences. Chooses treatment *with* the doctor	Active: receives all information. Chooses treatment alone
Information flow	One-way patient → physician	Bidirectional patient ↔ physician	One-way patient ← physician
Responsibility	Lies with the physician alone; high risk of litigation	Lies with physician and patient together	Lies with patient alone; low risk of litigation

Source: Hamann J, Leucht S & Kissling W (2003) Shared decision-making in psychiatry. *Acta Psychiatrica Scandinavica*, **107**, 403–409. Reproduced with permission of John Wiley & Sons

Some examples are as follows. If a depressed patient is bothered by insomnia and weight loss, mirtazapine or paroxetine might be reasonable choices. For patients who are obese and/or at risk for diabetes or elevated cholesterol, the physician should avoid medications that are apt to increase weight, glucose, cholesterol, and/or glucose (e.g., second-generation antipsychotic agents). Optimally, through the use of patient education regarding symptom targets and potential side effects, several potential medication regimens can be identified and the physician can collaborate with the patient to select the regimen that best suits that individual.

Environmental Supports to Improve Adherence

In a study of serious mental illness, Velligan and colleagues found in two different studies that the use of cues (alarms, signs, checklists, and the organization of belongings) can improve adherence to medication (Velligan et al., 2008, 2013). These cues are established on weekly home visits or use electronic devices such as smart pill containers that download information on adherence to a secure server and alert the case worker when adherence is poor. Adherence in groups treated with these supports remains between 80 and 90% for 9 months compared with adherence in individuals not receiving these interventions of around 60–65%. Moreover, when home visits stopped, adherence continued to remain above 80% in treated groups. This suggests that using environmental supports can help to form habit behaviors that are performed more automatically. Although home visits are expensive, in illnesses where hospitalization costs due to poor adherence are high, environmental supports established on weekly home visits can be cost-effective.

Long-Acting Medications

Long-acting medications available by injection or tablet may assist in the quest for better adherence (El Abdellati et al., 2020; Curto et al., 2021). With injectable medications, the treatment team knows whether or not medication has made it into the patient (Weiden, 2011). This information can lead the treatment team to intervene quickly once problem adherence has been identified.

Conclusions

Given the preponderance of data supporting the high prevalence of medication nonadherence across all age, racial, and disease groups, physicians must accept the reality that perfect medication adherence is unattainable. Identifying adherence is very difficult unless an individual is on an injection. Regular assessment of adherence is important, including prescribers asking about the patient's real-life barriers, feelings, beliefs, and attitudes regarding medication, all of which contribute to both unintentional and intentional nonadherence. These assessments are best carried in the context of a positive and collaborative therapeutic alliance. It is incumbent on the psychiatrist to assess the state of the alliance as regularly as assessing all these other factors. Some of the more recent efforts in treatment for adherence problems appear promising, for example, environmental supports, new technologies, and smart phone applications, and should be further investigated and implemented in treatment settings.

References

Acosta, F. J., Hernández, J. L., Pereira, J., Herrera, J., & Rodríguez, C. J. (2012). Medication adherence in schizophrenia. *World Journal of Psychiatry, 2*(5), 74–82.

Adams, J., & Scott, J. (2000). Predicting medication adherence in severe mental disorders. *Acta Psychiatrica Scandinavica, 101*, 119–124.

Adams, J. R., Drake, R. E., & Wolford, G. L. (2007). Shared decision-making preferences of people with severe mental illness. *Psychiatric Services, 58*(9), 1219–1221.

Ajzen, I. (1988). *Attitudes, personality, and behavior.* Dorsey Press.

Ajzen, I., & Fishbein, M. (1980). *Understanding attitudes and predicting social behavior.* Prentice Hall.

Aoki, Y., Yaju, Y., Utsumi, T., et al. (2022). Shared decision making interventions for people with mental health conditions. *Cochrane Database of Systematic Reviews, 11*(11), CD007297.pub3.

Appelbaum, P. S., & Gutheil, T. G. (1982). Clinical aspects of treatment refusal. *Comprehensive Psychiatry, 23*, 560–566.

Aronson, J. K. (2007). Compliance, concordance, adherence. *British Journal of Clinical Pharmacology, 63*, 383–384.

Awad, A. G. (1993) Subjective response to neuroleptics in schizophrenia. *Schizophrenia Bulletin, 19*, 609–618.

Bachmann, L. H., Stephens, J., Richey, C. M., et al. (1999). Measured versus self-reported compliance with doxycycline therapy for chlamydia-associated syndromes: High therapeutic success rates despite poor compliance. *Sexually Transmitted Diseases, 26*, 272–278.

Beck, A. T., Erbaugh, J., Ward, C. H., et al. (1961). An inventory for measuring depression. *Archives of General Psychiatry, 4*, 561–571.

Becker, M. H. (1985). Patient adherence to prescribed therapies. *Medical Care, 23*, 539–555.

Becker, M. H., & Maiman, L. A. (1975). Sociobehavioral determinants of compliance with health and medical care recommendations. *Medical Care, 13*, 10–24.

Bond, W. S., & Hussar, D. A. (1991). Detection methods and strategies for improving medication compliance.

American Journal of Hospital Pharmacy, 48, 1978–1988.

Brown, C., Dunbar-Jacob, J., Palenchar, D. R., et al. (2001). Primarycare patients' personal illness models for depression: A preliminary investigation. *Family Practice, 18*, 314–320.

Budd, R. J., Hughes, I. C., & Smith, J. A. (1996). Health beliefs and compliance with antipsychotic medication. *British Journal of Clinical Psychology, 35*, 393–397.

Bush, P. J., & Hardon, A. P. (1990). Toward rational medicine use: Is there a role for children? *Social Science and Medicine, 31*, 1043–1050.

Bush, P. J., & Iannotti, R. J. (1988). Origins and stability of children's health beliefs relative to medicine use. *Social Science and Medicine, 27*(4), 345–352.

Bush, P. J., & Iannotti, R. J. (1990). A children's health belief model. *Medical Care, 28*(1), 69–86.

Byerly, M. J., Nakonezny, P. A., & Rush, A. J. (2007). The Brief Adherence Rating Scale (BARS) validated against electronic monitoring in assessing the antipsychotic medication adherence of outpatients with schizophrenia and schizoaffective disorder. *Schizophrenia Research, 100*, 60–69.

Chakrabarti, S. (2019). Treatment attitudes and adherence among patients with bipolar disorder: A systematic review of quantitative and qualitative studies. *Harvard Review of Psychiatry, 27*, 290–302.

Charach, A., Ickowicz, A., & Schachar, R. J. (2004). Stimulant treatment over five years: Adherence, effectiveness, and adverse effects. *Journal of the American Academy of Child and Adolescent Psychiatry, 43*, 559–567.

Charles, C., Gafni, A., & Whelan, T. (1997). Shared decision-making in the medical encounter: What does it mean? (or it takes at least two to tango). *Social Science and Medicine, 44*, 681–692.

Charney, E., Bynum, R., & Eldredge, D. (1967). How well do patients take oral penicillin? A collaborative study in private practice. *Pediatrics, 40*, 188–195.

Chen, A. (1991). Noncompliance in community psychiatry: A review of clinical interventions. *Hospital and Community Psychiatry, 42*, 282–287.

Claxton, A. J., Cramer, J. A., & Pierce, C. (2001). A systematic review of the associations between dose regimens and medication compliance. *Clinical Therapeutics, 23*, 1296–1301.

Cochran, J. M., Fang, H., Le Gallo, C., et al. (2022). Participant engagement and symptom improvement: Aripiprazole tablets with sensor for the treatment of schizophrenia. *Patient Preference and Adherence, 16*, 1805–1817.

Cohen, E. A., Skubiak, T., Hadzi Boskovic, D., et al. (2022). Phase 3b multicenter, prospective, open-label trial to evaluate the effects of a digital medicine system on inpatient psychiatric hospitalization rates for adults with schizophrenia. *Journal of Clinical Psychiatry, 83*(3), 21m14132. d.

Conrad, P. (1985). The meaning of medications: Another look at compliance. *Social Science and Medicine, 20*, 29–37.

Cramer, J. (1989). How often is medication taken as prescribed? *JAMA, 261*, 3273–3277.

Cramer, J. (1995). Microelectronic systems for monitoring and enhancing patient compliance with medication regimens. *Drugs, 49*, 321–327.

Cromer, B. A., & Tarnowski, K. J. (1989). Noncompliance in adolescents – A review. *Journal of Developmental and Behavioral Pediatrics, 10*, 207–215.

Curtis, L. C., Well, S. M., Penney, D. J., et al. (2010). Pushing the envelope: Shared decision-making in mental health. *Psychiatric Rehabilitation Journal, 31*(1), 14–22.

Curto, M., Fazio, F., Ulivieri, M., Navari, S., Lionetto, L., & Baldessarini, R. J. (2021). Improving adherence to pharmacological treatment for schizophrenia: A systematic assessment. *Expert Opinion on Pharmacotherapy, 22*, 1143–1155.

De Las Cuevas, C., Peñate, W., & Sanz, E. J. (2014). The relationship of psychological reactance, health locus of control and sense of self-efficacy with adherence to treatment in psychiatric outpatients with depression. *BMC Psychiatry, 14*, 324.

De Las Cuevas, C., de Leon, J., Peñate, W., & Betancort, M. (2017). Factors influencing adherence to psychopharmacological medications in psychiatric patients: A structural equation modeling approach. *Patient Preference and Adherence, 11*, 681–69090.

Dolder, C. R., Lacro, J. P., Warren, K. A., Golshan, S., Perkins, D. O., & Jeste, D. V. (2004). Brief evaluation of medication influences and beliefs: development and testing of a brief scale for medication adherence. *Journal of Clinical Psychopharmacology, 24*, 404–409.

Dulcan, M. K. (1998). Treatment of childhood and adolescent disorders. In A. F. Schatzberg & C. B. Nemeroff (Eds.), *The American Psychiatric Press textbook of psychopharmacology* (2nd ed.). American Psychiatric Press.

Edgcomb, J. B., & Zima, B. (2018). Medication adherence among children and adolescents with severe mental illness: A systematic review and meta-analysis. *Journal of Child and Adolescent Psychopharmacology, 28*, 508–520.

El Abdellati, K., De Picker, L., & Morrens, M. (2020). Antipsychotic Treatment Failure: A Systematic Review on Risk Factors and Interventions for Treatment Adherence in Psychosis. *Frontiers in Neuroscience, 14*, 531763.

Ellison, J. M. (2000). Enhancing adherence in the pharmacotherapy treatment relationship. In A. Tasman, M. B. Riba, & K. R. Silk (Eds.), *The doctor-patient relationship in pharmacotherapy: Improving treatment effectiveness* (pp. 71–94). Guilford Press.

El-Mallakh, P., & Findlay, J. (2015). Strategies to improve medication adherence in patients with schizophrenia: The role of support services. *Neuropsychiatric Disease Treatment, 16*, 1077–1090.

El-Mallakh, R. S., & Potter, N. N. (2019). Stigma in neuropsychiatric illness, cultural norms, and social involvement. *Current Trends in Neurology, 13*, 101–108.

Epstein, L. H., & Cluss, P. A. (1982). A behavioral medicine perspective on adherence to long-term medical regimens. *Journal of Consulting and Clinical Psychology, 50*, 950–971.

Epstein, L. H., Beck, S., Figueroa, J., et al. (1981). The effects of targeting improvements in urine glucose on metabolic control in children with insulin dependent diabetes. *Journal of Applied Behavioral Analysis, 14*, 365–375.

Feinstein, A. R. (1990). On white-coat effects and the electronic monitoring of compliance. *Archives of Internal Medicine, 150*, 1377–1378.

Firestone, P. (1982). Factors associated with children's adherence to stimulant medication. *American Journal of Orthopsychiatry, 52*, 447–457.

Fotheringham, M. J., & Sawyer, M. G. (1995). Adherence to recommended medical regimens in childhood and adolescence. *Journal of Paediatrics and Child Health, 31*, 72–78.

Frank, J. D. (1973). *Persuasion and healing: A comparative study of psychotherapy* (2nd ed.). Johns Hopkins University Press.

Gabriel, A., & Violato, C. (2010) Knowledge of and attitudes towards depression and adherence to treatment: the Antidepressant Adherence Scale (AAS). *Journal of Affective Disorders, 126*, 388–394.

Gitlin, M. J., Cochran, S. D., & Jamison, K. R. (1989). Maintenance lithium treatment: Side effects and compliance. *Journal of Clinical Psychiatry, 50*, 127–131.

Gold, J. M., & Harvey, P. D. (1993). Cognitive deficits in schizophrenia. *Psychiatric Clinics of North America, 16*, 295–312.

Goldfinger, S. M., Hopkin, J. T., & Surber, R. W. (1984). Treatment resisters or system resisters? Toward a better service system for acute care recidivists. *New Directions in Mental Health Services, 21*, 17–27.

Green, W. H. (1995). *Child and adolescent clinical psychopharmacology*. Williams & Wilkins.

Gualtieri, C. T., & Golden, R. N. (1984). Blood level measurement of psychoactive drugs in pediatric psychiatry. *Therapeutic Drug Monitoring, 6*, 127–141.

Haynes, R. B. (1976). Strategies for improving compliances: A methodologic analysis and review. In D. L. Sackett & R. B. Haynes (Eds.), *Compliance with therapeutic regimens* (pp. 69–82). Johns Hopkins University Press.

Haynes, R. B., Taylor, D. W., & Sackett, D. L. (Eds.). (1979). *Compliance in health care*. Johns Hopkins University Press.

Haynes, R. B., Yao, X., Degani, A., et al. (2005). Interventions for enhancing medication adherence. *Cochrane Database of Systematic Reviews, 4*, CD000011.

Hogan, T. P., Awad, A. G., & Eastwood, R. (1983). A self-report scale predictive of drug compliance in schizophrenics: reliability and discriminative validity. *Psychological Medicine, 13*, 177–83.

Horne, R., Weinman, J., Hankins, M. (1999). The Beliefs about Medicines Questionnaire: The development and evaluation of a new method for assessing the cognitive representation of medication. *Psychology & Health, 14*, 1–24.

Horne, R. (2006a). Beliefs and adherence to treatment: The challenge for research and clinical practice. In P. W. Halligan & M. Aylward (Eds.), *The power of belief: Psychosocial influence on illness, disability, and medicine* (pp. 115–136). Oxford University Press.

Horne, R. (2006b). Compliance, adherence, and concordance: Implications for asthma treatment. *Chest, 130*(Suppl), 65S–72S.

Horne, R., & Weinman, J. (1999). Patients' beliefs about prescribed medication and their role in adherence to treatment in chronic physical illness. *Journal of Psychosomatic Research, 47*, 555–567.

Hotopf, M., Lewis, G., & Normand, C. (1997). Putting trials on trial – The costs and consequences of small trials in depression: A systematic review of methodology. *Journal of Epidemiology and Community Health, 51*, 354–358.

Janz, N. K., & Becker, M. H. (1984). The health belief model: A decade later. *Health Education Quarterly, 11*, 1–47.

Johnson, D. A. (1981). Depression: Treatment compliance in general practice. *Acta Psychiatrica Scandinavica, 6*, 447–453.

Joosten, E. A. G., Defuentes-Merillas, L., De Weert, G. H., et al. (2008). Systematic review of the effects of shared decision-making on patient satisfaction, treatment adherence and health status. *Psychotherapy and Psychosomatics, 77*(4), 219–226.

Jorm, A. F., & Korten, A. E. (1997). Mental health literacy: A survey of the public's ability to recognise mental disorders and their beliefs about the effectiveness of treatment. *Medical Journal of Australia, 166*, 182–186.

Kampman, O., Lehtinen, K., Lassila, V., Leinonen, E., Poutanen, O., & Koivisto, A. (2000) Attitudes towards neuroleptic treatment: reliability and validity of the attitudes towards neuroleptic treatment (ANT) questionnaire. *Schizophrenia Research. 45*, 223–234.

Katzung, B. G. (2007). *Basic and clinical pharmacology.* McGraw-Hill Lange Medical.

Keck, P. E., Jr., Mcelroy, S. L., Strakowski, S. M., et al. (1998). 12-month outcome of patients with bipolar disorder following hospitalization for a manic or mixed episode. *American Journal of Psychiatry, 155*, 646–652.

Kemp, R., Hayward, P., Applewhaite, G., et al. (1996). Compliance therapy in psychotic patients: Randomised controlled trial. *BMJ, 312*(7027), 345–349.

Kemp, R., Kirov, G., Everitt, B., et al. (1998). Randomised controlled trial of compliance therapy: 18-month follow-up. *British Journal of Psychiatry, 172*, 413–419.

Kessler, R. C., Berglund, P. A., Bruce, M. L., et al. (2001). The prevalence and correlates of untreated serious mental illness. *Health Services Research, 36*, 987–1007.

Kihlstrom, L. C. (1998). Managed care and medication compliance: Implications for chronic depression. *Journal of Behavioral Health Services and Research, 25*, 367–376.

Kikkert, M. J., & Dekker, J. (2017). Medication adherence decisions in patients with schizophrenia. *The Primary Care Companion for CNS Disorders, 19*, 17n02182.

Kiley, D. J., Lam, C. S., & Pollack, R. (1993). A study of treatment compliance following kidney transplantation. *Transplantation, 55*, 51–56.

Klein, E., Rosenberg, I., & Rosenberg, S. (2007). Whose treatment is it anyway? The role of consumer preferences in mental health care. *American Journal of Psychiatric Rehabilitation, 10*, 65–80.

Kreyenbuhl, J., Record, E. J., & Palmer-Bacon, J. (2016). A review of behavioral tailoring strategies for improving medication adherence in serious mental illness. *Dialogues in Clinical Neuroscience, 18*, 191–201.

Kruse, W., Eggert-Kruse, W., Rampmaier, J., et al. (1993). Compliance and adverse drug reactions: A prospective study with ethinylestradiol using continuous compliance monitoring. *Clinical Investigator, 71*, 483–487.

Lai, K. Y., Chan, T. S., Pang, A. H., et al. (1997). Dropping out from child psychiatric treatment: Reasons and outcome. *International Journal of Social Psychiatry, 43*, 223–229.

Lemanek, K. (1990). Adherence issues in the medical management of asthma. *Journal of Pediatric Psychology, 15*, 437–458.

Leventhal, H., & Cameron, L. (1987). Behavioral theories and the problem of compliance. *Patient Education and Counselling, 10*, 117–138.

Leventhal, H., Diefenbach, M., & Levensky, E. A. (1992). Illness cognition using common sense to understand treatment adherence and affect cognition interaction. *Cognitive Therapy and Research, 16*, 143–163.

Levy, G. (1993). A pharmacokinetic perspective on medication noncompliance. *Clinical Pharmacology and Therapeutics, 54*, 242–244.

Lingam, R., & Scott, J. (2002). Treatment non-adherence in affective disorders. *Acta Psychiatrica Scandinavica, 105*, 164–172.

Link, B. G. (1987). Understanding labeling effects in the area of mental disorders: An assessment of the effects of expectations of rejection. *American Sociological Review, 52*, 96–112.

Link, B. G., & Phelan, J. C. (2001). Conceptualizing stigma. *Annual Review of Sociology, 27*, 363–385.

Link, B. G., Yang, L., Phelan, J. C., et al. (2004). Measuring mental illness stigma. *Schizophrenia Bulletin, 30*, 511–541.

Litt, I. F., & Cuskey, W. R. (1980). Compliance with medical regimens during adolescence. *Pediatric Clinics of North America, 27*, 3–15.

Logan, D., Zelikovsky, N., Labay, L., et al. (2003). The illness management survey: Identifying adolescents' perceptions of barriers to adherence. *Journal of Pediatric Psychology, 28*, 383–392.

Loots, E., Goossens, E., Vanwesemael, T., Morrens, M., Van Rompaey, B., & Dilles, T. (2021). Interventions to improve medication adherence in patients with schizophrenia or bipolar disorders: A systematic review and meta-analysis. *International Journal of Environmental Research and Public Health, 18*, 10213.

Mahar, M. (2007). Braveheart: Jack Wennberg. *Dartmouth Medicine, 32*(2), 31–68.

Marrero, R. J., Fumero, A., de Miguel, A., & Peñate, W. (2020). Psychological factors involved in psychopharmacological medication adherence in mental health patients: A systematic review. *Patient Education and Counseling, 103*, 2116–2131.

McCabe, R., Bullenkamp, J., Hansson, L., et al. (2012). The therapeutic relationship and adherence to antipsychotic medication in schizophrenia. *PLoS One, 7*(4), e36080.

Miller, W. R., & Rollnick, S. (2002). *Motivational interviewing: Preparing people for change* (2nd ed.). Guilford Press.

Morisky, D. E., Ang, A., Krousel-Wood, M., & Ward, H.J. (2008) Predictive validity of a medication adherence measure in an outpatient setting. *Journal of Clinical Hypertension, 10*, 348–354.

Morisky, D. E., Green, L. W., & Levine, D. M. (1986). Concurrent and predictive validity of a self-reported measure of medication adherence. *Medical Care, 24*, 67–74.

Morris, L. A., & Halperin, J. A. (1979). Effects of written drug information on patient knowledge and compliance: A literature review. *American Journal of Public Health, 69*, 47–52.

Morselli, P. L., & Bianchetti, G. (1983). Therapeutic drug monitoring of psychotropic drugs in children. *Pediatric Pharmacology, 3*, 149–156.

Naber, D. (1995) A self-rating to measure subjective effects of neuroleptic drugs, relationships to objective psychopathology, quality of life, compliance and other clinical variables. *International Clinical Psychopharmacology, Suppl 3*, 133–138.

Naber, D., Moritz, S., Lambert, M., Pajonk, F. G., Holzbach, R., Mass, R., & Andresen, B. (2001) Improvement of schizophrenic patients' subjective well-being under atypical antipsychotic drugs. *Schizophrenia Research, 50*, 79–88.

Nessman, D. G., Carnahan, J. E., & Nugent, C. A. (1980). Increasing compliance: Patient-operated hypertension groups. *Archives of Internal Medicine, 140*, 1427–1433.

Novak-Grubic, V., & Tavcar, R. (2002). Predictors of noncompliance in males with first-episode schizophrenia, schizophreniform and schizoaffective disorder. *European Psychiatry, 17*, 148–154.

O'Donnell, C., Donohoe, G., Sharkey, L., et al. (2003). Compliance therapy: A randomised controlled trial in schizophrenia. *BMJ, 327*(7419), 834.

Olfson, M., Mechanic, D., Hansell, S., et al. (2000). Predicting medication noncompliance after hospital discharge among patients with schizophrenia. *Psychiatric Services, 51*, 216–222.

Orme, C. M., & Binik, Y. M. (1989). Consistency of adherence across regimen demands. *Health Psychology, 8*, 27–43.

Papola, D., Gastaldon, C., & Ostuzzi, G. (2018). Can a digital medicine system improve adherence to antipsychotic treatment? *Epidemiology and Psychiatric Sciences, 27*(3), 227–229.

Philo, G. (1996). *Media and mental distress*. Longman.

Phipps, S., & Decuir-Whalley, S. (1990). Adherence issues in pediatric bone marrow transplantation. *Journal of Pediatric Psychology, 15*, 459–475.

Prochaska, J. O., Velicer, W. F., Rossi, J. S., et al. (1994). Stages of change and decisional balance for 12 problem behaviors. *Health Psychology, 13*, 39–46.

Pullar, T., & Feely, M. P. (1990). Reporting compliance in clinical trials. *The Lancet, 336*, 1252–1253.

Rashid, A. (1982). Do patients cash prescriptions? *BMJ, 284*, 24–26.

Regier, D. A., Narrow, W. E., Rae, D. S., et al. (1993). The de facto U.S. mental and addictive disorders service system: Epidemiologic catchment area prospective 1-year prevalence rates of disorders and services. *Archives of General Psychiatry, 50*, 85–94.

Rettenbacher, M., Hofer, A., Eder, U., et al. (2004). Compliance in schizophrenia: Psychopathology, side effects, and patients' attitudes toward the illness and medication. *Journal of Clinical Psychiatry, 65*, 1211–1218.

Riekert, K. (2002). The beliefs about medication scale: Development, reliability, and validity. *Journal of Clinical Psychology in Medical Settings, 9*, 177–184.

Riekert, K., & Rand, C. S. (2002). Electronic monitoring of medication adherence: When is high-tech best? *Journal of Clinical Psychology in Medical Settings, 9*, 25–34.

Roe, D., Goldblatt, H., Baloush-Klienman, V., et al. (2009). Why and how people decide to stop taking prescribed psychiatric medication. Exploring the subjective process of choice. *Psychiatric Rehabilitation Journal, 33*, 366–374.

Rogers, R. W. (1983). Cognitive and physiological processes in fear appeals and attitude change: A revised theory of protection motivation. In J. Cacioppo & R. Petty (Eds.), *Social psychophysiology* (pp. 153–176). Guilford Press.

Rosa, A. R., Marco, M., Fachel, J., et al. (2007). Correlation between drug treatment adherence and lithium treatment attitudes and knowledge by bipolar patients. *Progress in Neuro-Psychopharmacology and Biological Psychiatry, 31*, 217–224.

Rosenstock, I. M. (1988). Enhancing patient compliance with health recommendations. *Journal of Pediatric Health Care, 2*, 67–72.

Rudd, P. (1993). The measurement of compliance: Medication taking. In N. A. Krasnegor, L. Epstein, S. B. Johnson, et al. (Eds.), *Developmental aspects of health compliance behavior* (pp. 185–213). Lawrence Erlbaum Associates.

Rudd, P., & Lenert, L. (1995). Pharmacokinetics as an aid to optimising compliance with medications. *Clinical Pharmacokinetics, 28,* 1–6.

Rudd, P., Byyny, R. L., Zachary, V., et al. (1989). The natural history of medication compliance in a drug trial: Limitations of pill counts. *Clinical Pharmacology and Therapeutics, 46,* 169–176.

Sackett, D. L., & Snow, J. C. (1979). The magnitude of compliance and noncompliance. In R. B. Haynes, D. W. Taylor, & D. L. Sackett (Eds.), *Compliance in health care* (pp. 11–22). Johns Hopkins University Press.

Satcher, D. (1999). *Mental health: A report of the surgeon general.* Department of Health and Human Services, National Institutes of Health.

Sayre, J. (2000). The patient's diagnosis: Explanatory models of mental illness. *Qualitative Health Research, 10,* 71–83.

Scott, J., & Pope, M. (2002). Nonadherence with mood stabilizers: Prevalence and predictors. *Journal of Clinical Psychiatry, 63,* 384–390.

Scott, C. S., Lore, C., & Owen, R. G. (1992). Increasing medication compliance and peer support among psychiatrically diagnosed students. *Journal of School Health, 62,* 478–480.

Sirey, J. A., Bruce, M. L., Alexopoulos, G. S., et al. (2001). Perceived stigma as a predictor of treatment discontinuation in young and older outpatients with depression. *American Journal of Psychiatry, 158,* 479–481.

Stichele, R. V. (1990). Measurement of patient compliance and the interpretation of randomized clinical trials. *European Journal of Clinical Pharmacology, 41,* 27–35.

Stine, J. J. (1994). Psychosocial and psychodynamic issues affecting noncompliance with psychostimulant treatment. *Journal of Child and Adolescent Psychopharmacology, 4,* 75–86.

Svarstad, B. L., Chewning, B. A., Sleath, B. L., et al. (1999). The brief medication questionnaire: A tool for screening patient adherence and barriers to adherence. *Patient Education and Counselling, 37,* 113–124.

Tham, X. C., Xie, H., Chng, C. M., Seah, X. Y., Lopez, V., & Klainin-Yobas, P. (2016). Factors affecting medication adherence among adults with schizophrenia: A literature review. *Archives of Psychiatric Nursing, 30,* 797–809.

Thiruchelvam, D., Charach, A., & Schachar, R. J. (2001). Moderators and mediators of long-term adherence to stimulant treatment in children with ADHD. *Journal of the American Academy of Child and Adolescent Psychiatry, 40,* 922–928.

Thompson, K., Kulkarni, J., & Sergejew, A. A. (2000). Reliability and validity of a new medication adherence rating scale (MARS) for the psychoses. *Schizophrenia Research, 42,* 241–247.

Urquhart, J. (1992). Ascertaining how much compliance is enough with outpatient antibiotic regimens. *Postgraduate Medical Journal, 68,* S49–S59.

Urquhart, J. (1995). Correlates of variable patient compliance in drug trials: Relevance in the new health care environment. *Advances in Drug Research, 26,* 238–257.

Urquhart, J. (1997). The electronic medication event monitor: Lessons for pharmacotherapy. *Clinical Pharmacokinetics, 32,* 345–356.

Valenstein, M., Ganoczy, D., McCarthy, J., et al. (2006). Antipsychotic adherence over time among patients receiving treatment for schizophrenia: A retrospective review. *Journal of Clinical Psychiatry, 67,* 1542–1550.

Van, Putten T., May, P. R. (1978) Subjective response as a predictor of outcome in pharmacotherapy: the consumer has a point. *Archives of general psychiaty, 35,* 477–480.

Velligan, D. I., Lam, Y. W., Glahn, D. C., et al. (2006). Defining and assessing adherence to oral antipsychotics: A review of the literature. *Schizophrenia Bulletin, 32*(4), 724–742.

Velligan, D. I., Wang, M., Diamond, P., et al. (2007). Relationships among subjective and objective measures of adherence to oral antipsychotic medications. *Psychiatric Services, 58*(9), 1187–1192.

Velligan, D. I., Diamond, P. M., Mintz, J., et al. (2008). The use of individually tailored environmental supports to improve medication adherence and outcomes in schizophrenia. *Schizophrenia Bulletin, 34*(3), 483–493.

Velligan, D. I., Weiden, P. J., Sajatovic, M., et al. (2009). Expert consensus panel on adherence problems in serious and persistent mental illness. The expert consensus guideline series: Adherence problems in patients with serious and persistent mental illness. *Journal of Clinical Psychiatry, 70*(Suppl 4), 1–46.

Velligan, D. I., Mintz, J., Maples, N., et al. (2013). A randomized trial comparing in person and electronic interventions for improving adherence to oral medication in schizophrenia. *Schizophrenia Bulletin, 39*(5), 999–1007.

Velligan, D. I., Sajatovic, M., Hatch, A., Kramata, P., & Docherty, J. P. (2017). Why do psychiatric patients stop antipsychotic medication? A systematic review of reasons for nonadherence to medication in patients with serious mental illness. *Patient Preference and Adherence, 11,* 449–468.

Voruganti, L. N., & Awad, A. G. (2002). Personal evaluation of transitions in treatment (PETiT): a scale to measure subjective aspects of antipsychotic drug therapy in schizophrenia. *Schizophrenia Research, 56,* 37–46.

Wahl, O. (1995). *Media madness: Public images of mental illness.* Rutgers University Press.

Weiden, P., Rapkin, B., Mott, T., Zygmunt, A., Goldman, D., Horvitz-Lennon, M., & Frances, A. (1994) Rating of medication influences (ROMI) scale in schizophrenia. *Schizophrenia Bulletin, 20,* 297–310.

Weiden, P. J. (2011). Long-acting injectable antipsychotics and the management of nonadherence. *Psychiatric Annals, 41*(5), 271–278.

Weiden, P. J., & Rao, N. (2005). Teaching medication compliance to psychiatric residents: Placing an orphan topic into a training curriculum. *Academic Psychiatry, 29*, 203–210.

Williams, R. A., Hollis, H. M., & Benoit, K. (1998). Attitudes toward psychiatric medications among incarcerated female adolescents. *Journal of the American Academy of Child and Adolescent Psychiatry, 37*, 1301–1307.

Wilson, C., Nairn, R., Coverdale, J., et al. (2000). How mental illness is portrayed in children's television. *British Journal of Psychiatry, 176*, 440–443.

Yaegashi, H., Kirino, S., Remington, G., et al. (2020). Adherence to oral antipsychotics measured by electronic adherence monitoring in schizophrenia: A systematic review and meta-analysis. *CNS Drugs, 34*, 579–598.

Zeber, J. E., Grazier, K. L., Valenstein, M., et al. (2007). Effect of a medication copayment increase in veterans with schizophrenia. *American Journal of Managed Care, 13*(6 Pt 2), 335–346.

Medications for Depression: Monoamine Enhancers and Esketamine (Antidepressants)

Seon-Cheol Park and Winston Wu-Dien Shen

Contents

S.-C. Park (✉)
Department of Psychiatry, Hanyang University College of
Medicine, Seoul, Republic of Korea

Department of Psychiatry, Hanyang University Guri
Hospital, Guri, Republic of Korea
e-mail: psc76@hanyang.ac.kr

W. W.-D. Shen
Department of Psychiatry, Wan Fang Medical Center,
Taipei, Taiwan

College of Medicine, Taipei Medical University, Taipei,
Taiwan

© Springer Nature Switzerland AG 2024
A. Tasman et al. (eds.), *Tasman's Psychiatry*,
https://doi.org/10.1007/978-3-030-51366-5_133

Abstract

In this chapter, "drugs for depression," the authors are covering the description of neuroscience-based nomenclature (NbN), pharmacological action, therapeutic effect, side effect, and drug-drug interaction, based on the conventional subgrouping of "antidepressant." They include selective serotonin reuptake inhibitors (SSRIs), serotonin partial agonist/reuptake inhibitors (SPARIs), serotonin antagonist/reuptake inhibitor (SARI), serotonin-norepinephrine reuptake inhibitors (SNRIs), norepinephrine-dopamine reuptake inhibitor (NDRI), selective norepinephrine reuptake inhibitor (NRI), melatonin receptor agonist, noradrenergic and specific serotonergic antidepressant (NaSSA), serotonin reuptake enhancer (SRE), monoamine oxidase inhibitors (MAOIs), and tricyclic antidepressant/tetracyclic antidepressant (TCA/TeCA). Moreover, the novel drugs for depression (brexanolone, zuranolone, and mifepristone) or adjunctive treatments (olanzapine, quetiapine, aripiprazole, brexpiprazole, ketamine/esketamine) on top of existing SNRIs or SSRIs are also included. In addition, for the monoamine enhancers, the relationship between neurogenesis and antidepressant action has been described. Along with the conventional subgrouping of "antidepressant," the potential influences of sex, race, and ethnicity on the antidepressant psychopharmacokinetics and pharmacodynamics are also comprehensively covered. Then, the symptom-based selection algorithm of drugs for depression has been introduced, based on neuroanatomical and neurochemical reductions for the symptoms of major depression. Finally, a pharmacological strategy algorithm for treatment-resistant major depression has been introduced. The algorithm includes optimization, switching, combination, and augmentation.

Keywords

Neuroscience-based nomenclature (NbN) ·
Antidepressant · Monoamines · Esketamine ·
Symptom-based selection

Introduction

In this chapter, the traditional term "antidepressants" is replaced with a new term "drugs for depression," reflecting the adoption of a pharmacologically driven nomenclature rather than an indication-based nomenclature (Moller et al., 2016; Frazer & Blier, 2016; Uchida, 2018; Zohar & Kasper, 2016). A newly defined background on the pharmacologically driven nomenclature is described.

Based on the Anatomical Therapeutic Chemical (ATC) classification system, psychotropic medications have been traditionally classified into antipsychotics (N05A), mood stabilizers (N03A), antidepressants (N06A), antianxiety drugs (N05B), hypnotics (N05C), antidementia drugs (N06C), psychostimulants (N06B), antiparkinson drugs (N04), and other drugs. The ATC code, originally published in 1976, was developed by the World Health Organization Collaborating Centre for Drug Statistics Methodology (WHOCC) (WHO 2021). The ATC code aids assortment of psychotropic medications according to the affected neuroanatomical structures and chemical characteristics. As shown in Fig. 1, in the ATC code, antipsychotics (N05A), antianxiety drugs (N05B), and hypnotics (N05C) are included in the section on psycholeptics (N05), whereas antidepressants (N06A), antidementia drugs (N06C), and psychostimulants (N06B) are included in the section on psychoanaleptics (N06). But several limitations in the ATC code have been proposed (Moller et al., 2016; Frazer & Blier, 2016; Uchida, 2018; Zohar & Kasper, 2016):

Limitations in Anatomical Therapeutic Chemical Classification System

• The ATC code is partly inconsistent with the psychotropic drug use patterns in psychiatric practices and psychopharmacological studies.

Hence, the boundaries among various categories of psychotropic medications are increasingly more ambiguous when the current nomenclature of the ATC code is used. Second-generation (atypical) antipsychotics have been usually recommended as adjunctive agents for treatment-resistant depression and psychotic depression in most evidence-based pharmacological treatment guidelines for depressive disorders. Furthermore, drugs for depression can be used not only for treating depressive disorder but also for improving anxiety disorder, obsessive-compulsive disorder (OCD), attention deficit/hyperactivity disorder (ADHD), and other mental disorders in the evidence-based pharmacological treatment guidelines.

• Both psycholeptics (N05A) and psychoanaleptics (N06) are considered outdated in the current nomenclature for psychotropic medications. Further, in the ATC code, lithium (N05AN01) and clonazepam (N03AE01) are included in the sections on antipsychotics (N05A) and antiepileptics (N03), respectively.

• The current nomenclature prescribed by the ATC code cannot be used to provide the relevant pharmacological anchors to help psychiatrists formulate an informative decision in clinical practice, although it is partly based on

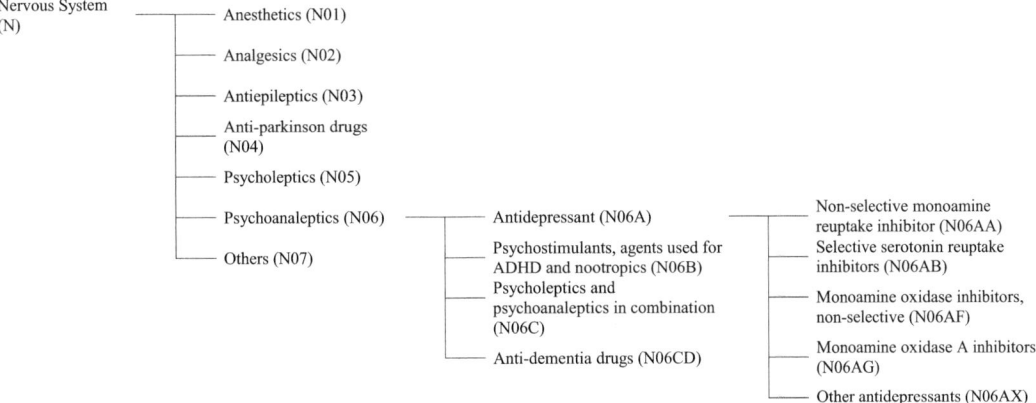

Fig. 1 Classification and nomenclature of antidepressants according to the ATC classification index. (adopted from the World Health Organization [WHO]. ATC classification system [internet]. Oslo, Norway: WHO; 2018 [cited: Feb 17, 2021]. Available from: https://www.whocc.no/atc/structure_and_principles, according to the Creative Commons license WHO). ADHD attention deficient/hyperactivity disorder; ATC Anatomical Therapeutic Chemical

an indication-based nomenclature (Muscholl, 1995; Nutt, 2009; Stahl, 2013a).

Proposed Neuroscience-Based Nomenclature System

Neuroscience-based nomenclature (NbN) has shown to avoid the limitations of the ATC classification system, for an alternative classification system for psychotropic medications in 2008 by a task force team consisting of a collaboration among five major international neuropsychopharmacological scientific organizations, namely, the European College of Neuropsychopharmacology (ECNP), the American College of Neuropsychopharmacology (ACNP), the Asian College of Neuropsychopharmacology (AsCNP), the International College of Neuropsychopharmacology (CINP), and the International Union of Basic and Clinical Pharmacology (IUPHAR). An outline of NbN was originally published in 2014. NbN was developed to help achieve a pharmacologically driven nomenclature rather than an indication-based nomenclature, focusing on both the pharmacological domain and mode of action (https://nbn2r.com/) (ECNP, 2017; Moller et al., 2016; Frazer & Blier, 2016; Uchida, 2018; Zohar & Kasper, 2016). NbN consists of ten pharmacological domains (i.e., acetylcholine, dopamine, γ-aminobutyric acid (GABA), glutamate, histamine, melatonin, norepinephrine (NE), opioid, orexin, and serotonin (5-HT)) and nine modes of action (i.e., enzyme inhibitor, enzyme modulator, ion channel blocker, neurotransmitters releaser, positive allosteric modulator, receptor agonist, receptor antagonist, receptor partial agonist, and reuptake inhibitor). Hence, NbN use can help enrich the nomenclature and the elucidation of the relevant mechanism. Additionally, NbN use can help clinicians formulate informed decisions after confirmation of the "next pharmacological step" to be undertaken. NbN also includes five additional dimensions, namely, approved indications, efficacy and side effects, practical notes, neurobiology, and pregnancy. The second edition, revised in 2019 (NbN2R 2019), includes information on 136 compounds which encompass the majority of available psychotropic medications.

As shown in Table 1, "drugs for depression" are defined based on the NbN guidelines.

A History from Monoamine-Based Pharmacological Drugs to Esketamine

As shown in Fig. 2, a history of development of drugs for depression has been briefly described. Notably, the monoamine hypothesis of depression proposed nearly 60 years ago suggests that depleted levels of 5-HT, NE, and dopamine (DA) lead to the development of depressive disorders (Schildkraut, 1965). The monoamine hypothesis was supported by the observation of reserpine's effects on serotonin and catecholamine (i.e., DA and NE) and the pharmacological modes of action of drugs for depression (Hillhouse & Porter, 2015). In certain patients, depression is precipitated by the use of reserpine, a drug that was used as a pharmacological treatment for hypertension in the 1950s. Reserpine shows a property similar to vesicular monoamine transporter inhibitors, causing monoamine depletions in the brain (Muller et al., 1955). Additionally, while iproniazid was developed as a derivative of isoniazid in an original attempt to develop a new drug for tuberculosis, it was approved as a catalyst for a drug for depression (Fox & Gibas, 1953; López-Muñoz et al., 2014). Although iproniazid was approved as a new antitubercular compound in 1958 by the US Food and Drug Administration (US FDA), it was used as an off-label pharmacological treatment for major depression. Iproniazid exhibits a pharmacological mechanism similar to that of non-selective, irreversible monoamine oxidase inhibitors (MAOIs) and has been regarded as the first pharmacological treatment for depression (López-Muñoz et al., 2014). But it was withdrawn from a pharmacological market due to the marked side effects, including hypertensive crises and hepatic necrosis. Imipramine, following iproniazid approval and usage, was approved in 1959 by the FDA as an antidepressant drug (López-Muñoz et al. 2007, 2014; Pletscher, 1991). Furthermore, the reversible and selective MAO_A inhibitors (i.e., moclobemide and brofaromine) have been developed to improve the safety of MAO inhibitors (Lotufo-Neto et al., 2009).

Table 1 Drugs for depression based on the neuroscience-based nomenclature (NbN)

Indication-based	Pharmacology	Mode of action	Drugs
(TCA/TeCA)	Norepinephrine	Reuptake inhibitor (NET)	Desipramine, protriptyline, nortriptyline
	Norepinephrine, serotonin	Reuptake inhibitor (NET and SERT)	Lofepramine, amoxapine
	Serotonin, norepinephrine	Reuptake inhibitor (SERT and NET)	Imipramine, dosulepin, clomipramine
	Serotonin, norepinephrine	Reuptake inhibitor (SERT and NET), receptor antagonist (5-HT$_2$)	Amitriptyline
	Norepinephrine, serotonin	Reuptake inhibitor (NET and SERT), receptor antagonist (5-HT$_2$)	Doxepin
	Serotonin, dopamine	Receptor antagonist (5-HT$_2$ and D$_2$)	Trimipramine
(MAOI)	Serotonin, norepinephrine, dopamine	Enzyme inhibitor (MAO-A and MAO-B)	Isocarboxazid, phenelzine
		Reversible enzyme inhibitor (MAO-A)	Moclobemide
		Enzyme inhibitor (MAO-A and MAO-B), releaser (DA, NE)	Tranylcypromine
	Dopamine, norepinephrine, serotonin	Enzyme inhibitor (MAO-B and MAO-A)	Selegiline
(SSRI)	Serotonin	Reuptake inhibitor (SERT)	Citalopram, escitalopram, fluoxetine, fluvoxamine, paroxetine, sertraline
(SNRI)	Serotonin, norepinephrine	Reuptake inhibitor (SERT and NET)	Venlafaxine, duloxetine, milnacipran

Adopted from the European College of Neuropsychopharmacology (ECNP). Neuroscience-based Nomenclature, second edition, revised (NbN2R) [internet]. Utrecht, Netherlands: ECNP; 2019 [cited: Feb 17, 2021]. Available from: https://nbn2r.com, according to the Creative Commons license ECNP

MAO monoamine oxidase; *MAOI* monoamine oxidase inhibitor; *NET* norepinephrine transporter; *SERT* serotonin transporter; *SNRI* serotonin-norepinephrine reuptake inhibitor; *SSRI* selective serotonin reuptake inhibitor; *TCA* tricyclic antidepressant; *TeCA* tetracyclic antidepressant

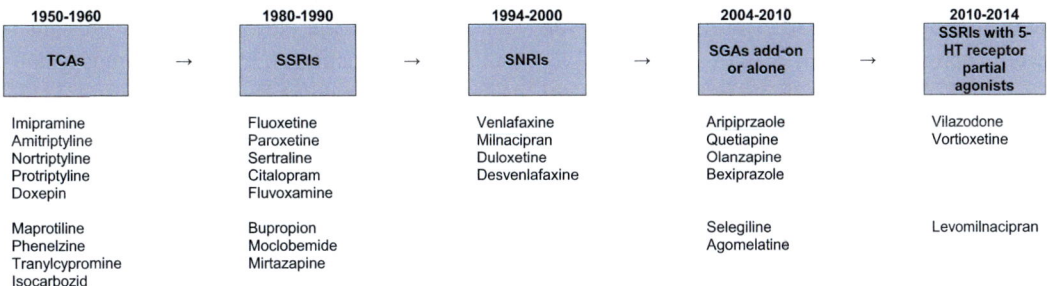

Fig. 2 A history of drugs for depression (reproduced from Shen, 2011a with permission to reprint from Taiwanese Society of Psychiatry). 5-HT serotonin; SGAs second-generation antidepressants; SNRIs serotonin and norepinephrine reuptake inhibitors antipsychotics; SSRIs selective serotonin reuptake inhibitors; *TCAs* tricyclic antidepressants

Although imipramine was developed as a derivative of promethazine to originally improve the pharmacological treatment for schizophrenia, it exhibited less remarkable antipsychotic properties. Imipramine has been established as the first drug in the class of tricyclic antidepressants

(TCAs), a class consisting of a diverse pharmacological profile including presynaptic norepinephrine reuptake transporter inhibition, presynaptic serotonin reuptake transporter inhibition, postsynaptic adrenergic receptor blockade, postsynaptic muscarinic receptor blockade, and postsynaptic histamine blockade (Cusack et al., 1994; Domino, 1999; Fangmann et al., 2008; Owens et al., 1997). In 1987, fluoxetine was approved as a drug for depression for selective serotonin reuptake inhibitor (SSRI) class by the US FDA (Wong et al., 1995, 2005). Based on the reports of decreased levels of 5-HT in the brain of patients with depressive suicides (Shaw et al., 1967), a role of 5-HT was reported in the pathogenetic mechanism of major depression in the late 1960s. Thus, the first report on the SSRI (fluoxetine) was published in 1974. Fluoxetine has been shown as a potent and selective serotonin reuptake inhibitor with relatively weak affinity for the norepinephrine transporter (Wong et al., 1975).

Following the development and approval of SSRIs, several atypical drugs for depression were developed. Bupropion was developed to improve the efficacy and safety of the treatment adopted for major depression. Bupropion is recognized as a dopamine-norepinephrine reuptake inhibitor (NDRI); hence, it has been classified as an "atypical" antidepressant drug that is markedly different from other antidepressant drugs. The immediate-release (IR), sustained-release (SR), and extended-release (XL) forms of bupropion were approved by the US FDA in 1989, 1996, and 2003, respectively (Fava et al., 2005; Stahl et al., 2005). Additionally, venlafaxine has been classified as an "atypical" drug for depression, based on its selective targets present on both 5-HT and NE transporters. The IR and extended-release (ER) forms of venlafaxine were approved by the US FDA in 1993 and 1997, respectively, for the treatment of major depression (Papakostas, 2008). Furthermore, vortioxetine is known as a novel "multimodal" drug that displays high binding affinity and complementary mechanisms of action for the 5-HT_{1A}, 5-HT_{1B}, 5-HT_{3A}, 5-HT_7, and serotonin transporters and shows considerable receptor affinity for the DA and NE transporters (Bang-Andersen et al., 2011; Mork et al., 2012).

In 2006, the Sequenced Treatment Alternative to Relieve Depression (STAR*D) study showed that several monoamine-based pharmacological drug classes that were established represented less than two-thirds of the remission rate for the treatment of major depression (Rush et al., 2006; Sinyor et al., 2010). Thus, the glutamatergic systems are examined as viable targets for the treatment of major depression; the noncompetitive N-methyl-D-aspartate (NMDA) receptor antagonist ketamine (esketamine), a drug which consistently exhibits rapid and sustained antidepressant actions, was approved by the US FDA in 2019 for the treatment of major depression (Chang et al., 2020). Ketamine was originally synthesized at Parke-Davis (Detroit, Michigan, USA) by Stevens in 1962. Ketamine is the derivative of phencyclidine and shows a combination of dissociative anesthetic action and hallucinogenic properties. Ketamine was originally approved for short-acting anesthetic use by the US FDA in the 1970s (Domino, 2010). Furthermore, ketamine was used as a recreational drug in the 1990s and was known as "special K." Hence, ketamine was grouped as a scheduled III nonnarcotic substance under the controlled substances act in 1999 (Drug Enforcement Administration, 2013). Based on accumulating evidence, the fast-acting and sustained antidepressant action elicited by ketamine is considered to be based on the mechanisms of action including the inhibition of the transmembrane domain (Yamamoto et al., 2015) in presynaptic and postsynaptic NMDA receptors (NMDARs) in GABAergic interneurons, the activation of postsynaptic α-amino-3-hydroxy-5-methyl-4-isoxazolepropionic acid receptors (AMPARs), and the brain-derived neurotrophic factor-tyrosine kinase B (BDNF-Trk B) signaling pathway. Moreover, an evidence exists that the gut microbiota are involved in the mechanism of antidepressant action exhibited through ketamine (Wang et al., 2020).

Neurogenesis and Antidepressant Action

In a theoretical framework for the pathogenesis of major depression, the neurogenesis theory has been proposed to address the limitations of the monoamine theory (Kim & Park, 2021; Park, 2019). The neurogenesis hypothesis postulates that major depressive disorder is triggered by the impairment of adult hippocampal neurogenesis, thus potentially explaining the lag in antidepressant action. A study by Jessberger and Kempermann (2003), using a mouse model, showed that immature hyperplastic neurons have to mature and are then preferentially recruited for dentate synaptic activity, thereby explaining the delayed action of antidepressants. As shown in Fig. 3, serotonergic drugs interact with postsynaptic 5-HT receptors via mediation by guanine nucleotide-binding proteins. Adenylate cyclase is stimulated by a G-protein, inducing the activation of cyclic adenosine monophosphate (cAMP) and resulting in increased levels of protein kinase. On the other hand, noradrenergic drugs interact with postsynaptic NE receptors. The coupling of adenylate cyclase with inhibitory G-proteins causes inhibition of adenylate cyclase. Since the secondary messenger systems of 5-HT and NE receptors are stimulatory and inhibitory G-proteins, respectively, decreased 5-HT and increased NE levels in the brain can be alleviated by increased levels of BDNF in the nuclei (Drazinix & Sazbo, 2017; Shen, 2011b). Antidepressants facilitate the functioning of new hippocampal cells, improve the incorporation of these newborn cells into the corresponding functional neural network, and increase the maturation and survival of these cells (Tanti & Belzung, 2013). While the norepinephrine system positively regulates early stages of adult neurogenesis, the serotonergic system positively regulates multiple processes of hippocampal neurogenesis, namely, proliferation, maturation, and survival. In addition, antidepressant actions are mediated by signaling pathways, including (a) the cAMP-phosphate kinase A (PKA)-cAMP-response element binding protein (CREB) pathway, (b) the Ras-mitogen-activated protein kinase A-CREB pathway, (c) the phospholipase Cγ (PLCγ)-Ca^{2+} pathway, and (d) the phosphatidylinositol 3-kinase (PI3K)-serine threonine kinase or protein kinase B (Akt) pathway (Eliwa et al., 2017). A final molecular pathway, involving the cAMP-PKA-CREB pathway, is upregulated. This pathway includes monoamine release, 5-HT_{1A} receptor activation, intercellular phosphorylation of CREB (pCREB), expression of BDNF, and binding of BDNF to the Trk B receptor (Brezun & Daszuta, 2000; Malberg & Blendy, 2005; Sairanen et al., 2005; Santarelli et al., 2003; Schmidt & Duman, 2007). Using an animal model, Nibuya et al. (1995) showed that the chronic use of antidepressants and electroconvulsive therapy (ECT) can increase the expression of TrkB and BDNF mRNA in the hippocampus and frontal lobe, respectively.

Studies have shown that the action of antidepressants is closely linked to the facilitation of BDNF-TrkB signaling. Antidepressant-induced neural plasticity also facilitates the activity-dependent reorganization of neural connections in response to environmental stimuli. In the alleviation of depressive symptoms, the rapid effects of ketamine are partially mediated by the activation of TrkB receptors, which results in structural plasticity changes (Rantamäki, 2017). In addition, antidepressant-mediated induction of neurogenesis can increase expression of a diverse variety of genes that upregulate vascular endothelial growth factor (VEGF), fibroblast growth factor (FGF), insulin-like growth factor (IGF), and other growth factors (Newton et al., 2003; Perera et al., 2007). Drugs for depression induce proliferation, maturation, and functional activation of newborn granule cell neurons, during a 2- to 4-week period, which corresponds with the therapeutic delay in the pharmacological activity of these drugs (Esposito et al., 2005; Perera et al., 2008). The mitotic effects of drugs for depression have been observed in the PFC and CA3 subfields of the hippocampus, but these effects have not been observed in the subventricular zone (SVZ) (Perera et al., 2007; Santarelli et al., 2003). In tree shrews treated with tianeptine, an increase in the rate of hippocampal neurogenesis was observed, with reversal of the stress-induced cortisol

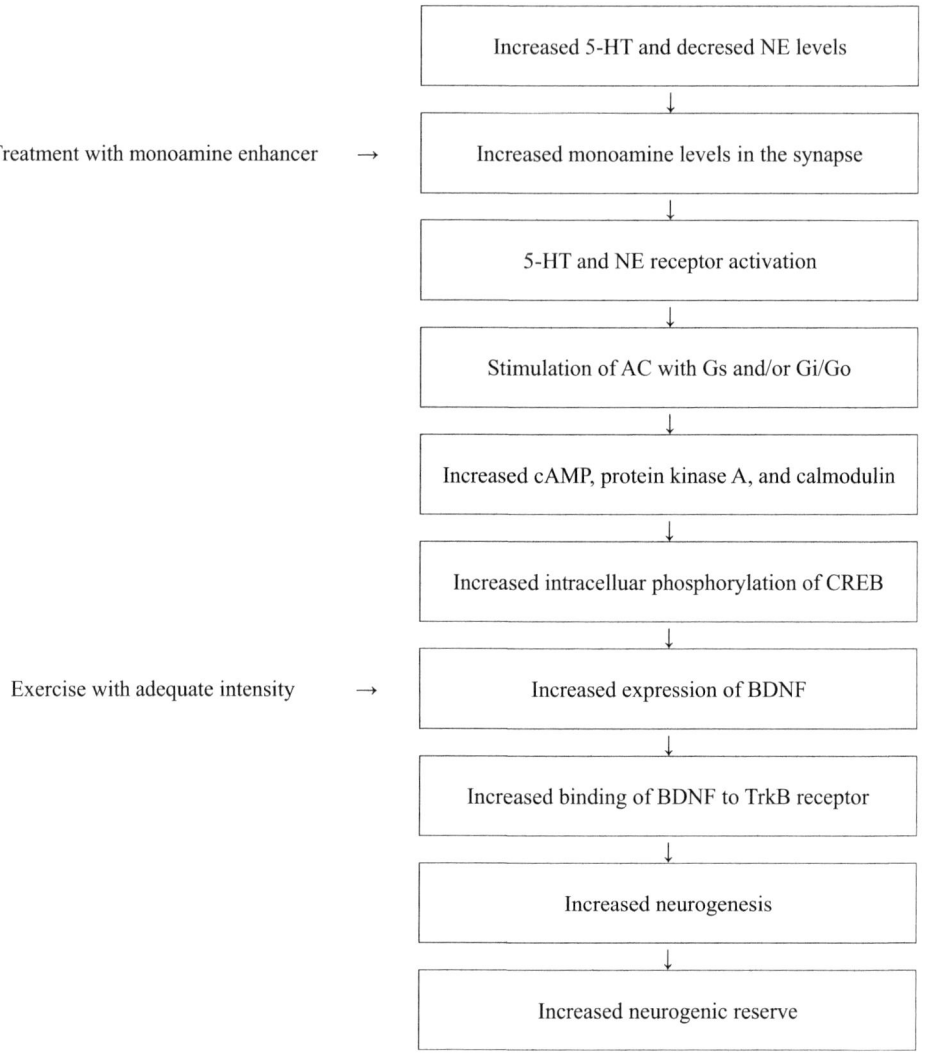

Fig. 3 A potential pathway explaining the relationship between antidepressant action and neurogenesis. 5-HT serotonin; AC adenylate cyclase; BDNF brain-derived neurotropic factor; cAMP cyclic adenosine monophosphate; CREB cAMP element-binding protein; TrkB tyrosine kinase B

response (Czeh et al., 2001). In addition, both the chronic uncontrolled stress (CUS) paradigm and the knockout of 5-HT$_{1A}$ receptors cause blockade of neurogenesis, as well as the loss of antidepressant efficacy (Santarelli et al., 2003; Wang et al., 2008). On the contrary, in the same strain of mice, the blockade of neurogenesis did not prevent environmental enrichment from improving performance in the novelty-suppressed feeding (NSF) test (Meshi et al., 2006). Although the relationship between adult neurogenesis and antidepressant action is complex and debatable, disrupted neurogenesis is associated with delayed efficacy of antidepressants, as shown using the NSF test (Santarelli et al., 2003). Additionally, with regard to the impact of hippocampal regional specificity on depression, it has been suggested that major depression is only affected by the more ventral regions of the hippocampus (Huckleberry et al., 2018). In terms of the action of antidepressants on hippocampal neurogenesis, in animal models, chronic treatment with monoamine enhancers increased the number of newborn neurons in the dorsal and ventral subregions. But agomelatine, a non-monoaminergic drug, was shown to increase neurogenesis only in the ventral subregion of the hippocampus (O'Leary & Cryan, 2014; Tanti & Belzung, 2013).

Exercise is regarded as a therapeutic option in many evidence-based treatment guidelines for depression (Park et al., 2014; Shen, 2016, 2020). Studies have consistently reported greater improvement of depressive symptoms in patients with major depression who exercise, compared with those who do not exercise (Mather et al., 2002; Singh et al., 1997, 2001, 2005). In addition, greater improvement of depressive symptoms was observed in sertraline-medicated major depression patients who exercised, than in such patients who did not (Babyak et al., 2000; Blumenthal et al., 1999, 2007; Herman et al., 2002). It can therefore be concluded that exercise can increase the level of BDNF in patients with major depression who are treated with antidepressants. As shown in Fig. 3, the underlying neurobiological mechanisms of the antidepressant effects of exercise are associated with increased hippocampal volume and increased levels of BDNF (Park, 2018; Shen, 2020). The antidepressant effect of exercise can also be explained by the regulation of growth factors, since exercise has anti-inflammatory effects, and inflammation can impair growth factor signaling in the brain (Kiecott-Glaser et al., 2015). Since exercise increases BDNF expression and neurogenesis, it can be used in treatment and prevention of depressive disorder (Harvey et al., 2018; Simon, 2018).

Taxonomy and its Relation to the Pharmacological Domain and Mode of Action

Historically, antidepressants have been grouped into the "first-generation" antidepressants including monoamine oxidase inhibitors (MAOIs) and tricyclic antidepressants (TCAs); "second-generation" antidepressants including selective serotonin receptor inhibitors (SSRIs), selective norepinephrine receptor inhibitors (NRIs),

tetracyclic antidepressant (TeCA), norepinephrine-dopamine reuptake inhibitors (NDRIs), and dopamine reuptake inhibitors (DRIs); and "third-generation" antidepressants including serotonin-norepinephrine reuptake inhibitors (SNRIs), mixed serotonin antagonists/reuptake inhibitors, and noradrenergic and specific serotonergic antidepressants (NaSSAs).

Structurally as another approach, antidepressants have been classified based on their chemical structure (e.g., TCAs) or presumed mechanism of action (e.g., SSRIs). As mentioned previously, based on the guidelines prescribed by NbN (ECNP, 2017; Moller et al., 2016; Frazer & Blier, 2016; Uchida, 2018; Zohar & Kasper, 2016), the drugs for depression can be defined using a combination of the pharmacological domain and mode of action. Thus, based on the NbN, drugs for depression are defined individually, but cannot be easily classified as subgroups. In this chapter, drugs for depression are classified in a practical manner through the combination of information available on the chemical structure and mechanism of action. NbN definitions for drugs for depression are additionally described. Table 2 summarizes the nomenclatures, indications, recommended doses, and clinical implications of antidepressants.

Selective Serotonin Reuptake Inhibitors (SSRIs)

SSRIs are defined as the inhibitors of 5-HT transporters (SERTs) based on the NbN guidelines (ECNP, 2017). These include citalopram, escitalopram, fluoxetine, fluvoxamine, paroxetine, and sertraline. Despite the incomplete understanding on the SSRI's mode of action, a consensus exists to have that greater availability of synaptic 5-HT and causes suppressed firing activity of 5-HT neurons within the raphé nuclei (Blier & Szabo, 2005; Sprouse et al., 2001). Briefly, SSRIs selectively block the reuptake of 5-HT through the inhibition of Na^+/K^+ adenosine triphosphatase (ATPase)-dependent transport at presynaptic G-protein receptor neurons (Shen, 2020). It is known that SSRIs increase 5-HT

Table 2 Nomenclatures, indications, recommended doses, and clinical implications of drugs for depression

Conventional nomenclature	Neuroscience-based nomenclature (NbN)	Indication	Dose	Clinical implication
(SSRI)				
Citalopram	Reuptake inhibitor (SERT)	Depressive disorder	10–60 mg/day	Selective to serotonin, lack of drug-drug interaction
Escitalopram	Reuptake inhibitor (SERT)	Depressive disorder	10–20 mg/day	Selective to serotonin, lack of drug-drug interaction
Fluoxetine	Reuptake inhibitor (SERT)	Depressive disorder, obsessive-compulsive disorder, panic disorder, eating disorder	20–80 mg/day	Low incidence of sexual dysfunction, monitoring drug-drug interaction
Fluvoxamine	Reuptake inhibitor (SERT)	Depressive disorder, obsessive-compulsive disorder	50–100 mg/day	Improving obsessive-compulsive symptoms
Paroxetine	Reuptake inhibitor (SERT)	Depressive disorder, obsessive-compulsive disorder, panic disorder, social anxiety disorder, generalized anxiety disorder, posttraumatic stress disorder	20–60 mg/day	Short half-life, side effects due to abrupt discontinuation, improving anxiety symptoms
Sertraline	Reuptake inhibitor (SERT)	Depressive disorder, obsessive-compulsive disorder, panic disorder, posttraumatic stress disorder, premenopausal dysphoric disorder, social anxiety disorder	50–200 mg/day	Improving anxiety and obsessive-compulsive symptoms, lack of drug-drug interaction
(SPARI)				
Vilazodone	Reuptake inhibitor (SERT), receptor partial agonist ($5\text{-}HT_{1A}$)	Depressive disorder	40 mg/day	Low incidence of weight gain and sexual dysfunction, lack of drug-drug interaction
Vortioxetine	Reuptake inhibitor (SERT), receptor partial agonist ($5\text{-}HT_{1A}$), receptor antagonist ($5\text{-}HT_3$)	Depressive disorder	10–20 mg/day	Low incidence of weight gain, preserving cognitive function, improving anxiety symptoms, lack of drug-drug interaction
(SARI)				
Trazodone	Receptor antagonist ($5\text{-}HT_{2A}$ and $5\text{-}HT_{1A}$)	Depressive disorder	150–300 mg/day	Improving insomnia, priapism
Nefazodone	Receptor antagonist ($5\text{-}HT_{2A}$ and $5\text{-}HT_{1A}$)	Depressive disorder	300–500 mg/day	Hepatic failure
(SNRI)				
Desvenlafaxine	Reuptake inhibitor (SERT and NET)	Depressive disorder	50 mg/day	Low incidence of gastrointestinal side effect

(continued)

Table 2 (continued)

Conventional nomenclature	Neuroscience-based nomenclature (NbN)	Indication	Dose	Clinical implication
Duloxetine	Reuptake inhibitor (SERT and NET)	Depressive disorder, generalized anxiety disorder, diabetic peripheral neuropathy, fibromyalgia	30–120 mg/day	More potent NET than SERT inhibition
Levomilnacipran	Reuptake inhibitor (SERT and NET)	Depressive disorder, fibromyalgia	40–120 mg/day	
Milnacipran	Reuptake inhibitor (SERT and NET)	Depressive disorder, fibromyalgia	50–100 mg/day	More potent NET than SERT inhibition
Venlafaxine	Reuptake inhibitor (SERT and NET)	Depressive disorder, panic disorder, social anxiety disorder, generalized anxiety disorder	75–375 mg/day	More potent SERT than NET inhibition
(NDRI)				
Bupropion	Reuptake inhibitor (NET and DAT), releaser (NE and DA)	Depressive disorder	150–300 mg/day	Low incidence of sexual dysfunction and weight gain, lack of drug-drug interaction
(NRI)				
Reboxetine	Reuptake inhibitor (NET)	Depressive disorder	8–10 mg/day	
(Melatonin receptor antagonist)				
Agomelatine	Agonist (MT_1 and MT_2), antagonist ($5\text{-}HT_{2C}$)	Depressive disorder	25–50 mg/day	Relatively safe in terms of drug-drug interaction
(NaSSA)				
Mirtazapine	Receptor antagonist (α_2, $5\text{-}HT_2$, and $5\text{-}HT_3$)	Depressive disorder	15–45 mg/day	Sedative effect, decreased gastrointestinal side effect
(SRE)				
Tianeptine	Unclear (glutamate)	Depressive disorder	25–50 mg/day	Relatively safe in terms of drug-drug interaction, used for the elderly
(MAOI)				
Isocarboxazid	Enzyme inhibitor (MAO-A and MAO-B)	Depressive disorder	20–60 mg/day	
Phenelzine	Enzyme inhibitor (MAO-A and MAO-B)	Depressive disorder	45–90 mg/day	
Selegiline	Enzyme inhibitor (MAO-B and MAO-A)	Depressive disorder	6–12 mg/day	
Tranylcypromine	Enzyme inhibitor (MAO-A and MAO-B), releaser (DA and NE)	Depressive disorder	20–60 mg/day	

(continued)

Table 2 (continued)

Conventional nomenclature (TCA/TeCA)	Neuroscience-based nomenclature (NbN)	Indication	Dose	Clinical implication
Amitriptyline	Reuptake inhibitor (SERT and NET), receptor antagonist ($5-HT_2$)	Depressive disorder	150–300 mg/day	Improving pain at low dose
Amoxapine	Reuptake inhibitor (SERT and NET)	Depressive disorder	200–300 mg/day	Blocking dopamine
Clomipramine	Reuptake inhibitor (SERT and NET)	Depressive disorder, obsessive-compulsive disorder	100–250 mg/day	Improving obsessive-compulsive symptoms
Desipramine	Reuptake inhibitor (NET)	Depressive disorder	150–300 mg/day	
Doxepin	Reuptake inhibitor (NET and SERT), receptor antagonist ($5-HT_2$)	Depressive disorder, anxiety disorder	150–300 mg/day	
Imipramine	Reuptake inhibitor (SERT and NET)	Depressive disorder, enuresis	150–300 mg/day	Improving enuresis
Protriptyline	Reuptake inhibitor (NET)	Depressive disorder	15–60 mg/day	
Trimipramine	Receptor antagonist ($5-HT_2$ and D_2)	Depressive disorder	150–300 mg/day	

MAOI monoamine oxidase inhibitor; *NaSSA* norepinephrine and specific serotonergic antidepressant; *NDRI* norepinephrine-dopamine reuptake inhibitor; *NET* norepinephrine transporter; *NRI* selective norepinephrine reuptake inhibitor; *SARI* serotonin antagonist/reuptake inhibitor; *SERT* serotonin transporter; *SNRI* serotonin-norepinephrine reuptake inhibitor; *SPARI* serotonin partial agonist/reuptake inhibitor; *SRE* serotonin reuptake enhancer; *SSRI* selective serotonin reuptake inhibitor; *TCA* tricyclic antidepressant; *TeCA* tetracyclic antidepressant

action at the synaptic cleft. But similar to the action of other drugs for depression, SSRIs display antidepressant action in a delayed manner. It is presumed that an increased availability of 5-HT by the SSRI cannot be directly related to its antidepressant action. Furthermore, since the sensitivity of somatodendritic/terminal $5-HT_{1A}$ autoreceptors can be decreased by repeated administration of an SSRI, in the context of the temporal association, it is possible that this effect can be directly related to the SSRI's antidepressant action.

Descriptions of Various SSRIs

The detailed descriptions for citalopram, escitalopram, fluoxetine, fluvoxamine, paroxetine, and sertraline are presented in the following:

- **Citalopram and Escitalopram**

  Citalopram which consists of two (R and S) enantiomers is known as a racemic citalopram with mild histamine and cytochrome P450 (CYP450) 2D6 inhibitory properties exhibited by the residue of the R enantiomer. Since citalopram is generally regarded as a well-tolerated SSRI, its application is promising in treating old adult patients with major depression. But, because of somewhat inconsistent therapeutic action at its low dose, an increase in the dose for optimization of therapeutic response may be required. It has been presumed that the reason is that the R enantiomer can interfere with the S enantiomer's inhibiting action at SERT. The potential interference of the R enantiomer may result in a reduced inhibition of SERT, with consequent reduced elevation synaptic 5-HT, which may reduce therapeutic action.

Escitalopram (the S enantiomer of citalopram) consists only of the S enantiomer of citalopram due to the removal of its R enantiomer. The use of escitalopram gives the same antidepressant action from the use of citalopram and the reduced burden of the liver. The antidepressant properties of escitalopram may be solely attributable to the active SERT properties. Owing to the removal of the R enantiomer of citalopram, escitalopram has less anti-histaminergic and CYP450 2D6 inhibitory properties. Escitalopram also has less inhibiting action at SERT because of the removal of the R enantiomer of citalopram which potentially interferes with the S enantiomer's action at SERT. Therefore, escitalopram may show efficacy at a lower dose. Thus, escitalopram is regarded as a better-tolerated SSRI with the fewest CYP450-mediated drug interactions among all five SSRIs.

- **Fluoxetine**

In addition to SERT inhibition, fluoxetine shows antagonistic action at $5-HT_{2C}$ receptors, leading to "disinhibition" (i.e., removal) of norepinephrine and dopamine release. Fluoxetine's blocking ability at $5-HT_{2C}$ receptors results in norepinephrine and dopamine disinhibition. Additionally, fluoxetine inhibits the activities of CYP450 2D6 and CYP450 3A4; these two CYP enzymes account for metabolism of over 80% of all psychotropic medications, greatly increasing the likelihood of a drug-drug interaction when fluoxetine is used. Fluoxetine presents with a long half-life, whereas its active metabolite demonstrates a comparatively longer half-life. The long half-life of fluoxetine may contribute to reduced incidence of sudden withdrawal symptoms. Furthermore, it is suggested that the conjunction of fluoxetine with a dopamine serotonin blocker (atypical antipsychotic (i.e., olanzapine)) is useful to increase the effectiveness in their antidepressant action for bipolar depression (López-Muñoz et al., 2018).

- **Fluvoxamine**

Fluvoxamine shows agonistic actions at σ_1 receptors, and such actions contribute to its

efficacy as an antianxiety drug, although the σ_1 actions have been poorly understood. Additionally, fluvoxamine is used as an agent in the treatment of psychotic and delusional depression. Since fluvoxamine is currently available as a controlled-release formulation, its once-daily administration is possible and recommended based on the clinical trials reporting favorable remission rates in anxiety disorder and OCD. Fluvoxamine is also available as an immediate-release formulation, and its twice-daily administration is required due to the existence of a shorter half-life. Fluvoxamine inhibits the activities of CYP450 1A2 and CYP450 3A4 (Shen, 1997). Fluvoxamine is more often used in treating obsessive-compulsive disorder and anxiety disorder rather than major depression, owing to the fact that it has never been approved as a drug for depression in the USA.

- **Paroxetine**

Paroxetine has both strong 5-HT reuptake inhibition properties and weak NE reuptake inhibition as well as anticholinergic action (M_1 muscarinic antagonism). Based on the anticholinergic action of paroxetine, it is more often used in the treatment of major depression patients with anxiety symptoms. What's more, since paroxetine is not only considered a potent inhibitor but also recognized as a substrate of CYP450 2D6 (Shen, 1997), serum levels of paroxetine can rapidly decline due to its short half-life when discontinued. Thus, the sudden discontinuation of paroxetine can contribute to its withdrawal symptoms, i.e., SSRI discontinuation symptoms. Since paroxetine inhibits nitric oxide synthase, it can lead to the exertion of sexual dysfunction side effects (Hsu & Shen, 1995; Shen & Hsu, 1995).

- **Sertraline**

Properties of sertraline include 5-HT reuptake inhibition, DA reuptake inhibition, and agonistic action at σ_1 receptors. Despite poor understanding of σ_1 actions, based on its agonistic action at σ_1 receptors, sertraline may be deemed useful in the treatment of anxiety symptoms and psychotic depression. Of all five SSRIs, only sertraline has weak dopamine transporter inhibitory actions mediated by its metabolite, desmethylsertraline (Shen, 1997). Similar to the use of bupropion or agomelatine (see section elsewhere in this chapter), sertraline's DA transporter inhibition can contribute to the improvement of several depressive symptoms in conjunction with SERT inhibition. Similarly, sertraline's DA reuptake inhibition can contribute to treating certain depressive symptoms, such as fatigue, lack of motivation, or anhedonia.

SSRI Side Effects

The side effects of SSRIs are related to the nonselective inhibition of norepinephrine and dopamine receptors. Although their improved selectivity relieves the risk of occurrence of side effects, including dry mouth, hypotension, and urinary retention, SSRIs present with various side effects. Most patients treated with SSRIs report sexual side effects, such as inhibition of sexual desire, sexual arousal (excitement), and orgasm. Sexual side effects warranting clinical assistance have been reported by 20–50% of those treated with SSRIs (Hsu & Shen, 1995; Shen & Hsu, 1995). It is presumed that SSRIs' activities at 5-HT$_2$ receptors in the brain and spinal cord can affect mental and physical aspects of arousal. Several counteracting therapies to these side effects have been clinically attempted. Nonetheless, even with success in redressing a single phase of sexual dysfunction, failure to redress others causes the complex interactions that occur during the sexual response cycle. Sexual side effects can be weakened and may disappear with improvement in depressive symptoms and with maintenance therapy of SSRIs. If sexual side effects persist, switching from SSRIs to bupropion or mirtazapine is recommended (Clayton & Shen, 1998). Additionally, the combination strategy of SSRIs with bupropion has been recommended (DeBattistia et al., 2000; Hirschfeld, 1999). The most common SSRI-induced sexual side effects are decreased sexual desire and delayed ejaculation rather than erection problems in men (Hsu & Shen, 1995) or decreased sexual desire and delayed orgasm rather than excitement with vasocongestion and lubrication

in women (Shen & Hsu, 1995). Sildenafil appears to be associated with an increase of sexual desire and helping achieve orgasm in male and female patients with SSRI-induced sexual side effects (Nurnberg, 2001; Shen et al., 1999). Moreover, it has been suggested that cyproheptadine, yohimbine, and dopamine-related medications (i.e., amantadine, amphetamine, and bromocriptine) may improve SSRI-induced sexual side effects (Jacobsen, 1992).

Nausea associated with vomiting or diarrhea may result from SSRIs' off-target stimulation of central and peripheral 5-HT$_3$ receptors. Several strategies have been proposed to decrease SSRI-induced gastrointestinal side effects. First, a gradual increase in dose is necessary to reduce gastrointestinal side effects (Shen, 2016). Second, administration of SSRIs concomitantly with food intake may reduce gastrointestinal side effects. Third, use of a 5-HT$_3$ antagonist (i.e., mirtazapine) can reduce gastrointestinal side effects.

Paroxetine alone among SSRIs may cause sedation because of its unique antihistamine properties. Thus, modafinil may be used for improvement of paroxetine-induced somnolence (DeBattistia et al., 2003). Conversely, since excessive synaptic serotonin may induce anxiety, SSRIs may induce an activation syndrome, particularly in individuals with the genetic variant of the short form of the serotonin transporter. This frequently occurs early in treatment. Similarly, SSRIs have been implicated in destabilization of bipolar disorder with an increase of both (hypo) mania and depression. Synaptic increase in any amine, including serotonin, may also result in insomnia (particularly middle insomnia), a complication that was experienced by 80% of patients receiving an SSRI in the STAR*D study. Trazodone, which blocks several postsynaptic serotonin receptors, can be used to treat SSRI-induced insomnia. SSRI-induced hyponatremia (<135 mmol/L) may occur due to the stimulation of ADH secretion mediated by 5-HT$_{2C}$, 5-HT$_{2A}$, and 5-HT$_7$ (but not 5-HT$_{1A}$) receptors. The clinical manifestations of SSRI-induced hyponatremia are relatively nonspecific. The symptoms of mild to moderate hyponatremia

(>120 mmol/L) include nausea, vomiting, malaise, headache, lethargy, confusion, obtundation, disorientation, pseudodepression, muscle weakness, and muscle cramps, whereas those of severe hyponatremia (<120 mmol/L) comprise depressed reflexes, extensor plantar responses, seizures, stupor, coma, respiratory arrest, and increased body weight. Female sex, age, and a history of co-medication usage (i.e., thiazide diuretics) are regarded as the main risk factors for SSRI-induced hyponatremia. A strong association between SSRIs and hospitalization has been reported due to hyponatremia exclusively related to its newly initiated treatment. An alternative treatment should be considered for individuals with clinically significant hyponatremia caused by newly initiated treatment with an SSRI. For patients presenting with ongoing antidepressant treatment, other causes should be explored and identified. In terms of its management, hyponatremia due to SSRI resolves after withdrawal of the causative drug and water restriction (Egger et al., 2006; Farmand et al., 2018).

Serotonin syndrome can result from the presence of toxic levels of synaptic 5-HT due to the administration of multiple serotonergic drugs including serotonin receptor agonists and 5-HT reuptake inhibitors. A diagnosis of serotonin syndrome can be made with the Hunter Serotonin Toxicity Criteria (sensitivity 84%, specificity 97%) (Beakley et al., 2015). These include (a) spontaneous clonus, (b) inducible clonus and agitation or diaphoresis, (c) ocular clonus and agitation or diaphoresis, (d) tremor and hyperrexia, (e) hypertonia, and (f) temperature > 38 °C and ocular clonus or inducible clonus (Boyer & Shannon, 2005; Dunkley et al., 2003). As CYP450 2D6, CYP450 3A4, and CYP450 2C19 catalyze the metabolism of serotonergic drugs (Shen, 1997), an increase to toxic levels of synaptic 5-HT over an unpredictable period can result from the use of concomitant drugs that decrease the activity of these enzymes. As MAO catalyzes the conversion and degradation from tryptophan to 5-HT, the use of MAOI can be regarded as one of the possible causes of serotonin syndrome (Scotton et al., 2019). The drugs commonly used in internal medicine that may contribute to the development of serotonin syndrome include analgesics (i.e., tramadol and fentanyl), antimigraine drugs, anticonvulsants (i.e., valproic acid), muscle relaxants (i.e., cyclobenzaprine), cough suppressants (i.e., dextromethorphan), antiemetics (i.e., ondansetron), antibiotics (i.e., linezolid), and antihistamine (i.e., chlorphenamine). The treatment of serotonin syndrome consists of discontinuation of all serotonergic drugs, administration of SERT receptor antagonists (i.e., cyproheptadine), provision of supportive care, and/or sedation with benzodiazepine.

SSRI discontinuation symptoms occur by sudden stopping or rapid dose reduction while administering SSRIs. The clinical manifestations of 5-HT withdrawal symptoms include dizziness, headache, nausea, diarrhea, insomnia, paresthesia, and others. SSRI discontinuation symptoms are often developed while administering SSRIs with a short half-life (i.e., paroxetine), whereas it is rarely developed while administering SSRIs with a long half-life (i.e., fluoxetine). SSRI discontinuation symptoms can be prevented by using SSRIs with a long half-life, by gradually reducing the dose of SSRIs, and by administering the SSRIs every other day before their discontinuation.

SSRI Overdose

SSRI overdose is considered a relatively safe condition, since few deaths occurring due to SSRI overdose have been reported. The most common symptoms of SSRI overdose comprise nausea, vomiting, tremor, somnolence, and others. With an administration at more than 75 times the usual dose of SSRI, more severe symptoms including cardiovascular disturbances, seizure, altered consciousness, and others can be manifested. The most common causes of deaths associated with SSRI overdose include seizure complications or cardiovascular disturbances (i.e., arrhythmia). The deaths associated with SSRI overdoses are commonly reported in combination with the consumption of TCA, alcohol, and other drugs. SSRI overdose can be recovered by gastric lavage and by providing supportive care in the emergency room. Monitoring of cardiac function and treatment for seizure is necessary in the condition of SSRI overdose combined with other drugs

(Dalfen & Stewart, 2001). Additionally, while SSRIs establish less severe drug-drug interactions compared to TCAs or MAOIs, most SSRIs inhibit the activity of CYP450. Since, among SSRIs, fluvoxamine shows relatively high inhibition action for CYP450 1A2, CYP450 2C, and CYP450 3A4 and mild inhibition action for CYP450 2D6 (Shen, 1997), its concomitant use with drugs establishing interactions with CYP450 enzymes should be carefully examined. While paroxetine and fluoxetine exert a significant impact on CYP450 2D6, sertraline and escitalopram do not demonstrate the establishment of interactions with CYP450 (Shen, 1997). It is a concern that SSRIs may increase TCA plasma concentrations. Fluvoxamine may increase the plasma concentration of clozapine, resulting in increased risk of occurrence of convulsion.

Serotonin Partial Agonist/Reuptake Inhibitors (SPARIs): Vilazodone and Vortioxetine

SPARIs are defined as reuptake inhibitors (SERT), receptor partial agonists (5-HT$_{1A}$), and/or receptor antagonists (5-HT$_3$) as per the NbN guidelines (ECNP, 2017).

• **Vilazodone**

Vilazodone is defined as a reuptake inhibitor (SERT) and receptor partial agonist (5-HT$_{1A}$) based on the NbN recommendations. Vilazodone was approved for major depression by the US FDA in 2011. It exerts partial agonistic actions at presynaptic 5-HT$_{1A}$ autoreceptors that lead to the activation of serotonin release through desensitization. Additionally, vilazodone increases serotonin release through the stimulation of postsynaptic 5-HT$_{1A}$ receptors. Furthermore, similar to that observed with SSRIs, vilazodone exhibits SERT blocking actions, resulting in an increased serotonin release at the synaptic cleft. But unlike SSRIs, vilazodone's actions at 5-HT$_{1A}$ receptors may not contribute to any delay in exhibition of favorable efficacy. Use

of a high dose of vilazodone may theoretically increase 5-HT levels more robustly than its standard dose and may demonstrate greater efficacy in certain patients and lesser tolerability in others. Daily doses of 50–80 mg for patients with treatment-resistant major depression or treatment-resistant OCD, comorbid depressive disorder and anxiety disorder, and other anxiety disorders can be considered. However, vilazodone does not show significant superiority to placebo in relapse prevention among patients with major depression (Khan et al., 2011; Choi et al., 2012; Durgam et al., 2018). Vilazodone presents with relatively favorable safety, with the exertion of only moderate-level side effects including nausea, vomiting, diarrhea, headache, insomnia, and others. Additionally, due to the demonstration of its partial agonistic actions at 5-HT$_{1A}$ autoreceptors, vilazodone exerts relatively less side effects of sexual dysfunction and weight gain in comparison with other serotonin reuptake-blocking drugs for depression. Since vilazodone is also studied in treatment for generalized anxiety disorder, superior efficacy has been reported in comparison with placebo (McCormack, 2015; Wang et al., 2016).

• **Vortioxetine**

Vortioxetine is defined as a reuptake inhibitor (SERT), receptor partial agonist (5-HT$_{1A}$), and receptor antagonist (5-HT$_3$) as per the NbN guidelines (ECNP, 2017). Vortioxetine was approved as a drug for depression by the FDA in 2013. It exerts an antidepressant action through multiple mechanisms of action (Trischler et al., 2014): Vortioxetine not only shows 5-HT reuptake inhibition actions but also exhibits partial agonistic actions at presynaptic 5-HT$_{1A}$ autoreceptors; additionally, it presents with activation actions at postsynaptic 5-HT$_{1A}$ receptors. Vortioxetine shows 5-HT$_3$ receptor antagonistic actions, leading to an increased NE and acetylcholine release. Moreover, vortioxetine exerts antidepressive and anxiolytic effects through antagonistic actions at 5-HT$_6$, 5-HT$_7$, and 5-HT$_{1D}$ receptors and through partial agonistic actions at 5-HT$_{1B}$ receptors (Mork et al., 2012). Furthermore, the use of an animal model showed that vortioxetine increases the levels of extracellular 5-HT, NE, and DA in the hippocampus and prefrontal cortex through regulation of monoamine (Leiser et al., 2014). Since vortioxetine presents with a long half-life (57 hours), its once-daily administration is recommended. The recommended maximum daily dose of vortioxetine is 10 mg for the poor metabolizers of CYP450 2D6. Lu AA34443 is a major metabolite of vortioxetine and exerts no pharmacological actions, although the half-life of Lu AA34443 is similar to that of vortioxetine. N-Hydroxy-piperazine demonstrates an inhibitory action on SERTs, although it is not a major metabolite of vortioxetine. In treating major depression, vortioxetine shows superior efficacy in comparison with placebo and demonstrates similar efficacy in comparison with other drugs for depression, as evidenced in clinical trials with a duration of 6–8 weeks. Additionally, the efficacy of vortioxetine in maintenance therapy has been evidenced (Thase et al., 2022). Furthermore, it has been proven that vortioxetine is efficacious in the treatment of not only depression but also anxiety and cognitive dysfunction, as highlighted in preclinical trials (Trischler et al., 2014). The European Medical Agency has even approved vortioxetine for the efficacy of cognitive improvement. But US FDA has not approved its efficacy for any cognitive benefit, considering that it is merely due to the improvement of concentration, which is one symptom of major depression of the *DSM-5*. Vortioxetine presents with relatively favorable safety and tolerability. In clinical trials conducted involving adult and elderly patients with major depression, it has been reported that those treated with 2.5–10 mg per day of vortioxetine present lower discontinuation rates due to the occurrence of side effects than those treated with other drugs for depression (Katona et al., 2012; Khin et al., 2011). But those treated with 15–20 mg per day of vortioxetine present higher discontinuation rates than those treated with duloxetine (Khin et al., 2011). The common side effects of vortioxetine include

nausea, vomiting, and headache (Baldwin et al., 2016). Remarkably, vortioxetine presents with a relatively low incidence of sexual dysfunction side effects in comparison with other antidepressant drugs (Alvarez et al., 2012). Furthermore, vortioxetine shows a relatively low incidence of insomnia, since it is presumed that it exerts less influence on the sleep-arousal cycle than that observed with the use of other drugs for depression.

Serotonin Antagonist/Reuptake Inhibitors (SARIs): Trazodone and Nefazodone

SARIs are defined as receptor antagonists (5-HT$_{2A}$ and 5-HT$_{1A}$) as per the NbN guidelines (ECNP, 2017). A receptor antagonist of 5-HT$_{2A}$ and 5-HT$_{1A}$, which has been classified as a 5-HT$_{2A/2C}$ antagonist and reuptake inhibitor (SARI) in the presumed mechanism of action, includes nefazodone and trazodone. Trazodone shows hypnotic and antidepressant actions at low doses and high doses, respectively. A low dose of trazodone exerts a hypnotic effect through the potent actions at 5-HT$_{2A}$, α_1, and H$_1$ receptors. A high dose of trazodone exhibits an antidepressant action through 5-HT reuptake inhibition that functions in synergy with 5-HT$_{2A}$ and 5-HT$_{2C}$ antagonism. Additionally, trazodone's antagonism of 5-HT$_2$ receptors stimulates 5-HT$_{1A}$ receptors. Moreover, nefazodone demonstrates strong 5-HT$_{2A}$ antagonism with 5-HT$_{2C}$ antagonism and weak 5-HT reuptake inhibition. m-Chlorophenylpiperazine (m-CPP), which is a main metabolite of trazodone and nefazodone, is considered a potent agonist of 5-HT$_{2C}$ receptor. The primary indication of trazodone and nefazodone is major depression. Several studies have shown that trazodone and nefazodone have equal efficacy to TCA in the treatment of major depression. In comparative clinical drug trials, trazodone has not been proven its clinical antidepressant efficacy is as good as venlafaxine (Cunningham et al., 1994) or bupropion (Weisler et al., 1994). Therefore, the inadequate efficacy of trazodone in treating patients with major depressive disorder is often seen either in the elderly (Goh et al., 2012) or in adolescent patients (Sultan & Courtney, 2017).

Trazodone and nefazodone also display anti-anxiety actions that precede antidepressant actions. It has been reported that trazodone presents with an equal efficacy to chlordiazepoxide in the treatment of generalized anxiety disorder. Treatment with a low dose (25–100 mg/day) of trazodone exerts sedative effects, whereas a low dose (less than 250 mg/day) of nefazodone shows anxiolytic actions. Additionally, nefazodone is deemed effective in the treatment of posttraumatic stress disorder.

In side effects, the incidence of nausea occurring due to trazodone and nefazodone usage is lower than that reported with the use of SSRIs. It has been reported that nausea is the most common cause reported for the discontinuation of nefazodone. It is well known that the incidence of nausea increases as the dose of trazodone increases. Furthermore, trazodone and nefazodone may induce dry mouth due to their α_1-adrenergic receptor inhibitions. Owing to its α_1-adrenergic receptor inhibition, trazodone may induce orthostatic hypotension in elderly patients. Priapism, which occurs in 1 in 6000 men treated with trazodone, may be fatal. An acute treatment of priapism involves an injection of α-adrenergic receptor agonist (i.e., epinephrine) into the penis. Since priapism is pharmacologically mediated through the adrenergic pathway, it is more rarely related to the effect of nefazodone rather than trazodone. Since trazodone and nefazodone present with relative wide therapeutic ranges, it is known that they are safe for usage even after the occurrence of an overdose. Moreover, 5-HT_2 antagonists are considered relatively safe in drug-drug interactions. However, it has been reported that nefazodone may induce liver failure (Aranda-Michel et al., 1999). Thus, nefazodone-induced liver failure is highlighted with a boxed warning by the US FDA. In augmentation of trazodone with antihypertensive drugs that may aggravate hypotension, a careful monitoring of blood pressure is necessary. Since 5-HT_2 antagonists exhibit similar effects to those observed with serotonin, theoretically, a combination of 5-HT_2 antagonist with MAOI may increase the risk of serotonin syndrome development. Nefazodone is deemed a potent inhibitor of CYP450 3A3/3A4, related to the metabolism of alprazolam, triazolam, ketoconazole, erythromycin, and carbamazepine. The withdrawal syndromes of trazodone and nefazodone are rarely reported. For an example, it has been reported that insomnia and other symptoms are developed after abrupt discontinuation of trazodone (Otani et al., 1994). It has also been reported that paresthesia and dizziness are developed after abrupt discontinuation of nefazodone (Bennazzi, 1998).

Serotonin-Norepinephrine Reuptake Inhibitors (SNRIs)

SNRIs are defined as reuptake inhibitors (SERT and NET) in the NbN guidelines (ECNP, 2017). SNRIs consist of venlafaxine, desvenlafaxine, duloxetine, milnacipran, and levomilnacipran and can be used as the primary treatment for major depressive disorder and anxiety disorder.

Descriptions of Various SNRIs

SNRIs can exert a therapeutic effect on chronic pain or somatic symptom disorder. They can be used as a secondary treatment option for patients who have remained nonresponsive to SSRIs or other drugs for depression. Additionally, about 33% of the patients who have shown failure to adequately respond to drugs for depression or electroconvulsive treatment may exhibit responses to venlafaxine treatment. Since SNRIs exert minimal effects on the CYP450 enzyme, drug interactions are rarely considered when using SNRIs (Shen, 1997). Owing to the relatively short half-life of SNRIs, they are used twice daily. Moreover, sustained-release (SR) formulations of SNRIs are used once daily. High-dose SNRI usage can result in elevated blood pressure. In mode of action, SNRIs inhibit the reuptake of norepinephrine and 5-HT at neuronal synapses. 5-HT reuptake inhibition is predominant with low-dose SNRI usage, whereas both norepinephrine and dopamine reuptake inhibition are predominant with high-dose SNRI usage. SNRIs exert a favorable therapeutic effect on major depression, and this finding is supported by several studies that have reported that SNRIs demonstrate quicker and better remission rate than SSRIs. Additionally, SNRIs exert a favorable therapeutic effect on anxiety disorders including generalized anxiety disorder. The recommended daily doses for venlafaxine, duloxetine, and milnacipran are 75–225 mg, 30–60 mg, and 25–100 mg, respectively. Since their side effects, including nausea, are common, prescription of the minimum starting doses is recommended. Moreover, careful monitoring of blood pressure is required, since increases in diastolic blood pressure can result from an increase in SNRI dosage. Half the general doses for SNRIs are recommended for patients with poor liver or kidney function. Unlike SSRIs, SNRIs present a dose-response relationship. Thus, low- and high-dose SNRIs are recommended for mild depression and severe or recurrent depression, respectively (Shelton et al., 2005a, b).

- **Venlafaxine**
  Venlafaxine is deemed a more potent NET inhibitor than SERT inhibitor. Furthermore, venlafaxine is known as the only drug for depression that induces the downregulation of β-adrenergic receptor-coupled cAMP level with a single dose. Thus, venlafaxine demonstrates a relatively quick onset of pharmacological action. Further, venlafaxine weakly binds to protein (27%). As venlafaxine poorly inhibits the activities of CYP450 enzymes (Shen, 1997), venlafaxine presents with a low possibility for the establishment of drug-drug interactions. Venlafaxine is effective for the treatment of major depression. It has been reported that venlafaxine is characterized by a greater possibility to aid the achievement of remission compared to TCA or SSRI (Einarson et al., 1999; Thase et al., 2001). Additionally, venlafaxine has been approved as the treatment of generalized anxiety disorder by the US FDA (Davidson et al., 1999). Moreover, venlafaxine can be effective in the treatment of neuropathic pain, fibromyalgia, and other chronic pain disorders (Kiayias et al., 2000). It has been

evidenced that venlafaxine 150–300 mg/day is effective in the treatment of child and adult ADHD. Like SSRIs, the side effects of venlafaxine include insomnia, nausea, and sexual dysfunction. A unique side effect of venlafaxine is hypertension through the activation of the noradrenergic system. The incidences of hypertension are 5% and 13% in the administration of its doses of less than 200 mg/day and more than 300 mg/day, respectively (Thase, 1998). Increased blood pressure can occur after the administration of high-dose (more than 120 mg/day) venlafaxine, a condition which rarely develops following the administration of low-dose venlafaxine. Thus, it is imperative that blood pressure should be carefully monitored during the first 2 months of venlafaxine administration. Deaths occurring due to venlafaxine overdose are rarely reported. It has been reported that seizure or serotonin syndrome may develop in case of an overdose occurring due to consumption of more than 10 g of venlafaxine. An overdose of venlafaxine can be managed with gastric lavage and with the provision of supportive care.

- **Desvenlafaxine**
  Desvenlafaxine is recognized as an active metabolite converted from venlafaxine that is a substrate for CYP450 2D6. Desvenlafaxine is characterized by greater NE transporter (NET) inhibition relative to serotonin SERT inhibition as compared to venlafaxine. The plasma levels of venlafaxine are nearly half those of desvenlafaxine after venlafaxine administration. As the use of a CYP450 2D6 inhibitor leads to a shift in the plasma levels toward more venlafaxine and less desvenlafaxine and as it reduces the relative extent of NET inhibition, the relative ratio of plasma levels of venlafaxine to desvenlafaxine can vary in concomitantly used drugs after venlafaxine administration. The ratio of these two drugs can also cause a variance in CYP450 2D6 genetic polymorphism because poor metabolizers can cause a shift in the plasma levels toward more venlafaxine and less desvenlafaxine. Based on such considerations, unpredictable NET inhibition can be

presented by considering a given dose of venlafaxine in a certain patient at a specific time, whereas more predictable NET inhibition can be presented by considering desvenlafaxine. It has been reported that desvenlafaxine is efficacious in reducing vasomotor symptoms in perimenopausal women, regardless of the presence of comorbidity in such women. It has been evidenced that desvenlafaxine is also efficacious in the treatment of major depression. According to one promising study, it has been suggested that SNRIs, especially desvenlafaxine, can be used in the treatment of vasomotor symptoms (Stahl, 2009, 2013b). Similar to those observed with venlafaxine, the side effects of desvenlafaxine include nausea, dizziness, sweating, constipation, appetite loss, sexual dysfunction, and increased blood pressure, among other effects. Desvenlafaxine is metabolized by the action of CYP450 3A4 to a less extent. Thus, dose control of desvenlafaxine is necessary for the elderly, renal failure patients, and hepatic failure patients.

- **Duloxetine**

   Duloxetine inhibits NE and 5-HT more strongly than venlafaxine. But it remains unknown whether the phenomenon of stronger inhibition implies that duloxetine is a better drug for depression than venlafaxine. Additionally, duloxetine is not active at the muscarinic receptor or histamine receptor. Duloxetine has been approved as a treatment option for major depression. Duloxetine is also useful in the treatment of physical symptoms and pain symptoms (Picerking et al., 2018). Moreover, duloxetine is used in the treatment of stress urinary incontinence (Norton et al., 2002). The common side effects of duloxetine comprise nausea, dry mouth, lethargy, sexual dysfunction, and others. Initial titration of duloxetine with low-dose strategy to improve patients' nauseation is advised if they start to receive duloxetine (Lee et al., 2012).

- **Milnacipran and Levomilnacipran**

   Milnacipran inhibits the reuptake of 5-HT and NE with a ratio of 1:1. Milnacipran is an SNRI used in the treatment of fibromyalgia. But milnacipran is not approved for the treatment of major depression in the USA, unlike in

Japan, Korea, Taiwan, and many other European countries such as France. Notably, its half-life is 8 hours. Since milnacipran is excreted by the kidney, its metabolism is affected by the severity of kidney disease. On the contrary, there exists no significant association with CYP450 enzymes.

   Levomilnacipran is an active enantiomer of the racemic drug of milnacipran. Unlike milnacipran, levomilnacipran inhibits the reuptake of 5-HT and NE with a ratio of less than 1:1 (Auclair et al., 2013). Thus, levomilnacipran is a relatively more potent NET inhibitor than SERT inhibitor; this pharmacodynamics property is more pronounced in improving the symptom of worry as seen in patients with generalized anxiety disorder. In 2013, the extended-release oral formula of levomilnacipran has been approved by the US FDA for treating patients with major depressive disorder (Montgomery et al., 2013).

SNRI Side Effects

In the side effects, SNRIs can induce gastrointestinal disturbance, sexual dysfunction, transient withdrawal symptoms, and others. Similar to the effects observed with SSRIs, SNRIs do not impact ECG activity and lower the seizure threshold. SNRIs rarely induce sedative effects and, at times, result in weight gain to a less remarkable extent. Additionally, high-dose SNRIs can result in high diastolic blood pressure. If hypertension occurs when using SNRIs, then the dose should be subjected to reduction or discontinuation, or switching should be considered. If high blood pressure occurs but if the antidepressant action is evident when using SNRIs, reduction in the SNRI doses should be considered in combination with the consumption of antihypertensive drugs. Moreover, although the overdose toxicity of SNRIs is rarely reported, it is speculated that SNRIs are safer than TCAs in the effects observed with an overdose.

SNRI Overdose

Side effects due to SNRI overdoses include drowsiness, tachycardia, grand mal seizure, and others. The overdose toxicity can be treated with supportive and symptomatic treatments. In drug-drug interactions, bupropion can increase blood venlafaxine concentration, and venlafaxine can increase blood haloperidol concentration. Since SNRIs exert no interfering effects on CYP450 enzymes and low protein-binding interactions, drug interactions are considered less significant when using SNRIs. If SNRIs are combined with an MAOI, the risk of serotonin syndrome development is increased. Thus, the combination of SNRI with MAOI should be avoided.

Norepinephrine-Dopamine Reuptake Inhibitor (NDRI): Bupropion

NDRI (i.e., bupropion) is defined as a reuptake inhibitor (NET and DAT) and releaser (NE and DA) in the NbN guidelines (ECNP, 2017). Bupropion's main mode of action is both NE reuptake inhibition and DA reuptake inhibition and is independently through a direct action on the 5-HT system. But bupropion is characterized by a less extent of NET and DA transporter (DAT) inhibition compared to that observed with SSRIs and SNRIs that demonstrate SERT inhibition. Thus, it implies that lower levels of NET and DAT inhibitory activities are necessary for agents to be used as drugs for depression compared to the use with SERT inhibition. This may be supported by observations that SNRI usage presents with markedly less NET inhibition rather than SERT inhibition, and monoamine releasers (psychostimulants, i.e., amphetamine, cocaine, and methylphenidate) are characterized by high degrees of DA inhibition. Furthermore, an antidepressant action of bupropion may result from low levels of combined NE and DA inhibition. Since bupropion shows less remarkable affinity to postsynaptic histamine, NE, 5-HT, DA, and acetylcholine receptors, it is presumed that bupropion use leads to weight gain, anticholinergic effects, or cardiovascular issues in side effects to a certain extent. Bupropion is available in the form of three formulations, including immediate release (IR), sustained release (SR), and extended/modified release (XR) for administration in regimens of thrice daily, twice daily, and once daily, respectively. Additionally, the times to achieve peak plasma levels of bupropion IR, SR, and XR formulations are 1–2 hours, 3 hours, and 5 hours,

respectively. Thus, bupropion SR and XR formulations are not only more convenient but are also more capable of reducing the seizure risk at peak plasma drug levels compared to its IR formulation. An antidepressant action of bupropion has been evidenced by several clinical trials. Bupropion shows greater efficacy in the treatment of moderate or severe major depression in comparison with placebo, and it exhibits similar efficacy in the treatment of major depression in comparison with SSRI and SNRI. The combination of bupropion with an existing SSRI or SNRI is considered a useful treatment option for major depression patients who have not shown responses to previously conducted serotonergic-focused treatment. The combination of bupropion and SSRI or SNRI has been described comprehensively in the section on treatment strategies for treatment-resistant major depression. Bupropion was also approved as a treatment option for nicotine addiction by the US FDA in 1997. Since bupropion can occupy DATs in the striatum and the nucleus accumbens, it can mitigate cravings but cannot be subjected to abuse. Additionally, bupropion is regarded as a treatment option for both children and adult patients with ADHD. Bupropion is an effective treatment strategy for patients with comorbidity of ADHD and depression, for those with comorbidity of ADHD and conduct disorder or substance use disorder, and for those with ADHD and tic symptoms induced by monoamine releasers (psychostimulants). But clinicians are encouraged to prescribe bupropion as an add-on drug if the patients' residual symptom shows fatigue or lack of energy; bupropion is discouraged to prescribe as the first-line drug if the patients show the predominant symptom of suicidal idea, phobic, panic, or obsessive-compulsive symptom.

Selective Norepinephrine Reuptake Inhibitor (NRI): Reboxetine and Atomoxetine

NRI (i.e., reboxetine) is defined as a reuptake inhibitor (NET) in the NbN guidelines (ECNP, 2017).

- **Reboxetine**

    Reboxetine increases the neurotransmission of NE through direct activation of the locus coeruleus. Prefrontal dopamine can be activated through an increased neurotransmission of NE. Thus, it is expected that selective norepinephrine reuptake inhibition may be considered effective in the treatment of fatigue, psychomotor retardation, and apathy, issues which are more closely associated with NE than 5-HT. Reboxetine use is popular in France, and other European countries, but it is not widespread in the USA. According to a double-blind placebo-controlled randomized trial (Eyding et al., 2010; Tanum, 2000), reboxetine shows a similar efficacy to desipramine or fluoxetine in the treatment of major depression. Reboxetine has been found to be superior to fluoxetine in the treatment of moderate depression or melancholic depression (Massana et al., 1999). Moreover, the open trial findings have shown that reboxetine is effective in the improvement of ADHD symptoms, when not responsive to methylphenidate, and in the reduction of binge-eating episode frequency in eating disorders (Sepede et al., 2012). Reboxetine has been therefore used off-label for ADHD, binge-eating disorder, panic disorder, and others. The initial dose of reboxetine is 8 mg/day, and its maximum dose is 10 mg/day. The common side effects of reboxetine include dry mouth, constipation, sexual dysfunction, and hypotension, among other effects. It has been known that reboxetine rarely exerts cardiovascular system-related side effects. While reboxetine is metabolized mainly through the action of CYP450 3A4, it is rarely implicated in drug-drug interactions. But a combination of reboxetine and MAOI increases the risk for serotonin syndrome development.

    By the way, atomoxetine that has been approved by the US FDA for the treatment of attention deficit/hyperactivity disorder (Gibson et al., 2006) belongs to this category of selective norepinephrine reuptake inhibitor. Most of the patients tolerate atomoxetine's side effects (mild decreased appetite, stomatic upset, dizziness, etc.) well (Gibson et al., 2006; Gau et al., 2007).

- **Atomoxetine**

    Atomoxetine was the first norepinephrine reuptake inhibitor marketed in the USA. It was also the first non-stimulant drug approved for the treatment of ADHD in the USA. But atomoxetine has not been approved for the treatment of major depression. Atomoxetine acts as a highly selective inhibitor of presynaptic NE receptors, but has low affinity for NE and DA receptors. In the treatment of ADHD, atomoxetine has a similar therapeutic effect to methylphenidate. Atomoxetine improves defiant behaviors, depressive symptoms, and anxiety symptoms, as well as ADHD symptoms, in ADHD patients with oppositional defiant disorder, tic disorder, depressive disorder, or anxiety disorder (Cheng et al., 2007; Garnock-Jone & Keating, 2009).

Melatonin Receptor Agonist: Agomelatine

A melatonin receptor agonist (i.e., agomelatine) is defined as an agonist (MT_1 and MT_2) and as an antagonist ($5-HT_{2C}$) in the NbN guidelines (ECNP, 2017). Agomelatine has been developed by the Servier pharmaceutical company and has been marketed with the brand name of Valdoxan in European countries. Since agomelatine is structurally similar to melatonin, agomelatine demonstrates an agonistic action at melatonin and exhibits an antagonistic action at $5-HT_{2C}$ receptors. Agomelatine has been approved as a treatment option for major depression by the European Medical Agency (EMA) in 2009, and it is the first drug for depression to be used among other melatonin-related medications. In Australia, agomelatine has been also approved in the Therapeutic Goods Administration (TGA) in 2010, but its payment has not been approved by the Pharmaceutical Benefits Advisory Committee (PBCA). Two pivotal clinical trials have shown the somewhat disappointing results that daily dose of 25 mg of agomelatine does not show significantly difference than placebo at all from week 2 to week 8. The daily dose of 50 mg has shown significant efficacy from week 2 to week 6 (Stahl et al., 2010; Zajecka et al., 2010). Servier and

Norvatis (the co-developer) in 2011 announced their giving up the pursuit of getting US FDA approval (Norman & Olver, 2019). Since agomelatine exerts less remarkable effects on rapid eye movement (REM) sleep, it increases non-REM sleep and shows a relatively rapid sedating effect in comparison with other drugs for depression. Additionally, based on its effect exerted on the serotonergic system, agomelatine presents with an awakening effect in daylight.

Agomelatine's bioavailability is considered low because of the exhibition of first-pass effect and as its half-life ranges from 1 to 2 hours. Since agomelatine demonstrates protein-binding property in blood with 95% affinity, doses should be carefully regulated for patients with liver cirrhosis, chronic kidney disease, and other diseases. Its combination with CYP450 1A2 inhibitors (i.e., fluvoxamine, ciprofloxacin) is contraindicated (Laudon & Fryman-Marom, 2014; Posadzki et al., 2018). Agomelatine improves the sleep disturbance-arousal cycle observed in major depression through the exhibition of significant improvements in sleep efficiency, slow-wave sleep (SWS), and the distribution of delta activity throughout the night, but with no change observed in the amount or latency of rapid eye movement (REM) sleep (Quera-Salva et al., 2010). Additionally, the effectiveness of agomelatine on anhedonia has been evidenced by an open-label, 8-week phase 4 trial involving 143 patients with major depression (Gargoloff et al., 2016). In an interesting article (Stahl, 2014), the profile of agonistic MT_1/MT_2 receptors and antagonistic $5\text{-}HT_{2C}$ receptor of agomelatine has been simplified as a bupropion-like drug with both NE and DA reuptake inhibitions. But elevated liver enzymes have been in 4.5% of patients taking daily agomelatine 50 mg (Zajecka et al., 2010).

Noradrenergic and Specific Serotonergic Antidepressant (NaSSA): Mirtazapine

NaSSA (i.e., mirtazapine) is defined as a receptor antagonist (α_2, $5\text{-}HT_2$, and $5\text{-}HT_3$) in the NbN guidelines (ECNP, 2017). Mirtazapine shows characteristics including rapid reduction of insomnia and anxiety and rapid improvement of depressive symptoms. Since mirtazapine is not significantly affected by CYP450-mediated metabolism, it presents with a benefit in the combination and augmentation strategies formulated with other drugs with 5-HT with or without NE reuptake inhibitors for depression (Chen et al., 2018). As mirtazapine acts as an antagonist at α_2 adrenergic receptor and as an inverse agonist at the $5\text{-}HT_{2C}$ receptor, it activates the adrenergic, serotonergic, and dopaminergic systems. Additionally, mirtazapine rarely induces nausea because of the antagonistic action at the $5\text{-}HT_3$ receptor and improves anxiety and insomnia, but increases appetite because of the antagonistic action at the $5\text{-}HT_{2C}$ receptor. Unlike an SSRI or an SNRI, mirtazapine does not cause remarkable inhibition in sexual function because of its antagonistic action at the $5\text{-}HT_{2A}$ receptor. Moreover, since mirtazapine acts as a strong inverse agonist at the H_1 receptor, it presents with a strong sedating effect. Owing to mirtazapine's relatively low affinity for α_1 adrenergic (implicating ejaculation function and sedation) and muscarinic (implicating dryness of the mouth and confusion) receptors, the effects of mirtazapine can be minimal at its clinical dose. But among certain patients, early severe adverse effects of mirtazapine may decrease drug compliance (Wooderson et al., 2011).

Mirtazapine can be effectively used in the treatment of elderly patients who present with main symptoms including insomnia, decreased appetite or weight, anxiety, agitation, and others. While older patients tend to show positive responses to mirtazapine, young or female patients tend to be intolerable to its early side effects. Although the initial dose of mirtazapine is generally 15 mg/day before sleep, it can be reduced to 3.75 mg/day or 7.5 mg/day for elderly patients or those who show a tendency to develop side effects. With gradual dose titration, most patients can tolerate the required dosage of 30 mg/day in a week. The maximum dose of mirtazapine is 45 mg/day. Moreover, the starting dose and maintenance dose should be reduced for older adult patients and comorbid hepatic or renal disease patients (Fava & Offidani, 2011).

The most common side effects are sedation or somnolence, weight gain, dizziness, dry mouth, constipation, and others. About half of the patients treated with mirtazapine have sedation or somnolence at the beginning of its intake. Because of the rapid adaptation to mirtazapine's inverse agonistic action at the H_1 receptor, sedation or somnolence does not persist after several weeks of use. Increased exercise and diet control are necessary to improve weight gain that may be deemed a secondary phenomenon of mirtazapine-associated increase in appetite. Additionally, an increase of more than 20% in cholesterol level can occur among more than 15% of the patients treated with mirtazapine, and a cholesterol level greater than 500 mg/dL can occur among 6% of those treated with mirtazapine. Moreover, blood dyscrasia including agranulocytosis occurs at the rate of less than 1 in 1000. Hypertension, orthostatic hypotension, blood vessel dilation combined with peripheral edema, dizziness, and other side effects may be caused in response to mirtazapine treatment. Side effects occurring due to mirtazapine overdose may include drowsiness, tachycardia, memory impairment, and others. Side effects occurring due to mirtazapine overdose can be managed with gastric lavage, cardiac monitoring, and provision of supportive care. In drug interactions, a significant inhibitory effect due to action of the CYP450 enzyme has not been reported. A combination of mirtazapine with MAOI should not be used, and mirtazapine should not be used within 14 days after the discontinuation of MAOI.

Serotonin Reuptake Enhancer (SRE): Tianeptine

Although tianeptine is conventionally classified as an SRE, thus far, the pharmacological domains of tianeptine are defined as glutamate and opioid. Its mode of action is not clearly defined in the NbN guidelines (ECNP, 2017). Tianeptine was developed in the 1960s. Despite its unclearly defined mode of action, unlike other classes of drugs for depression, tianeptine is considered a selective 5-HT reuptake enhancer in the central nervous system as well as an inhibitor for neural

dendrite atrophy occurring due to stress. In an animal model, tianeptine has been reported to increase the rate of hippocampal neurogenesis and to decrease stress-induced cortisol response (Czeh et al., 2001; Park, 2019), while the structure of tianeptine is similar to that of a TCA.

Tianeptine rarely induces worsened cognitive function, cardiovascular side effects, somnolence, weight change, and other side effects. Thus, tianeptine can be used for the elderly or for side effect-sensitive patients. Its common side effects comprise nausea, constipation, abdominal pain, headache, dizziness, and others. Tianeptine bypasses the first-pass effect of the liver and is rapidly excreted. Thus, significant drug interactions and significant clinical efficacy have never been reported. But several studies have reported that tianeptine exhibits an equal efficacy to other classes of drugs for depression including SSRIs in antidepressant and relapse prevention effects. Additionally, it has been reported that tianeptine exerts a therapeutic effect on anxiety and depression symptoms of patients with alcohol use disorder. A dose of 25–50 mg/day of tianeptine is recommended, and twice-daily or three times daily administration is necessary due to its relatively short half-life (Perahia et al., 2008).

Monoamine Oxidase Inhibitors (MAOIs)

Monoamine oxidase (MAO) has two subtypes, monoamine oxidase A (MAO-A) and monoamine oxidase B (MAO-B). In the NbN guidelines (ECNP, 2017), MAOIs are defining isocarboxazid and phenelzine as the enzyme inhibitors, MAO-A and MAO-B, respectively. Moclobemide is defined as a reversible enzyme inhibitor (MAO-A). Tranylcypromine is defined as an enzyme inhibitor (MAO-A and MAO-B) and releaser (DA and NE). In the past, MAOIs have been reported to exert a therapeutic effect on atypical depression, but this notion is no longer to have significant meaning in clinical psychiatry nowadays.

Monoamine oxidases (MAO) are a family of enzymes that catalyze the oxidation of monoamines, using oxygen to remove their amine group. MAO-A has selective effects on 5-HT and NE, whereas MAO-B exerts more significant effects on phenylethylamine. Both MAO-A and MAO-B show similar effects on dopamine and tyramine. Thus, inhibition of MAO-A presents with antidepressant actions. On the contrary, inhibition of MAO-B is not effective as a drug for depression because no direct effect is exerted on either 5-HT or norepinephrine metabolism. Furthermore, irreversible MAOI can induce hypertension through the inhibition of tyramine metabolism. Thus, patients consuming irreversible MAOIs (i.e., phenelzine and tranylcypromine) should not eat tyramine-containing food that can increase blood pressure. Nowadays, irreversible MAOIs are rarely used in the treatment of major depression because of the dietary restrictions (i.e., food intake restrictions). Irreversible MAOI is used as a symptomatic therapeutic agent for Parkinson's disease in certain cases. Thus, when administering a drug for depression for Parkinson's disease, the ascertainment of the concomitant use of MAOI should be considered.

Reversible inhibitor of MAO-A (RIMA) (i.e., moclobemide) is used as a drug for depression in clinical psychiatry. Although MAOI is not currently used as the first-line treatment option for major depression, RIMA can be considered a pharmacological treatment option for the patients who have not shown responses to other classes of drugs for depression. Since moclobemide is reversible and selective, food intake restrictions are not needed. Additionally, a starting dose of 150 mg/day moclobemide and twice-daily administration is recommended. In patients who have presented with insufficient treatment response, the therapeutic dose of 300–600 mg/day is recommended, and a dose of up to 900 mg/day may be considered. It is imperative to reduce 1/3 to 1/2 of the moclobemide dose for liver disease patients, whereas dose regulation is not necessary for renal disease patients. A dose of moclobemide should not be increased within the first week of its dosing, since its bioavailability is increased for that period (Youdim et al., 2006). Its common side effects are orthostatic hypotension, headache, insomnia, weight gain, sexual dysfunction, peripheral edema, somnolence, and others. Its serious adverse reactions are serotonin syndrome and hypertensive crisis. Since moclobemide establishes a significant interaction with other drugs, combinations with other drugs should be carefully used.

Monoamine Releasers (Tricyclic Antidepressants/Tetracyclic Antidepressants (TCAs/TeCAs))

In the NbN guidelines (ECNP, 2017), TCAs/TeCAs are defined as follows: Desipramine, protriptyline, and nortriptyline are reuptake inhibitors (NET). Lofepramine and amoxapine are defined as reuptake inhibitors (NET and SERT). Imipramine, dosulepin, and clomipramine are defined as reuptake inhibitors (SERT and NET). Amitriptyline is defined as a reuptake inhibitor (SERT and NET) and as a receptor antagonist ($5\text{-}HT_2$). Doxepin is defined as a reuptake inhibitor (NET and SERT) and as a receptor antagonist ($5\text{-}HT_2$). Trimipramine is defined as a receptor antagonist ($5\text{-}HT_2$ and D_2). The nomenclatures of TCAs and tetracyclic antidepressants (TeCAs) have been derived based on their own compound structure including three and four circular nucleus shapes, respectively. TCAs and TeCAs have been generally used in the 1970s. Since TCAs and TeCAs have been characterized by an increased exertion of cardiovascular system-related and anticholinergic side effects and by a severe overdose toxicity compared to other drugs for depression, their uses have been gradually decreased. But TCAs and TeCAs are currently used as an alternative strategy or combination strategy for moderate depression and in case of comorbidity of melancholic feature or psychotic feature. TCAs are classified by the number of methyl groups present on the nitrogen side chain. Desipramine and nortriptyline (secondary amine TCAs) are identified as the metabolites that lose one methyl group on a nitrogen side chain of imipramine and amitriptyline (tertiary amine TCAs), respectively. Amoxapine (TeCA), which has tricyclic nuclei with a modified piperazine side chain, is a derivative of loxapine (Stahl, 2013a).

TCAs pharmaceutically show functions through the reuptake inhibition of norepinephrine and serotonin and through competitive antagonistic actions at muscarinic (M_1), histaminergic (H_1), and α_1- and α_2-adrenergic receptors. Relative predominance of norepinephrine and serotonin reuptake inhibition actions are closely related to sedation and activation effects of TCAs, respectively. In the mode of pharmaceutical action, TCAs have been characterized by relatively weak 5-HT reuptake inhibition, with the exception of clomipramine. In vivo studies have shown that, with the exception of clomipramine, TCAs exhibit less remarkable 5-HT reuptake inhibition actions. Thus, with the exception of clomipramine, TCAs are considered extremely weak serotonergic antidepressants. Furthermore, TCAs and TeCAs inhibit dopamine reuptake to a certain extent. Among drugs for depression, only sertraline and bupropion demonstrate dopamine reuptake inhibition. But their actions are weak (Richelson, 2002; Owens et al., 1997; Sanderson et al., 2005). Most TCAs and TeCAs are primarily indicated for major depression by the US FDA. Additionally, doxepin and imipramine are indicated for anxiety and enuresis, respectively, by the US FDA. Clomipramine has been approved as a therapeutic medication for OCD by the FDA, whereas it is not approved as a medication for major depression. Although the indications have not been approved by the US FDA, several TCAs are used to treat insomnia (i.e., amitriptyline and doxepin), headache (i.e., amitriptyline, imipramine, and doxepin), chronic pain (i.e., doxepin and maprotiline), bulimia nervosa (i.e., imipramine and desipramine), and others. It has also been reported that imipramine and doxepin exert therapeutic effects on generalized anxiety disorder and indigestion, respectively. TCAs are partly metabolized by the action of CYP450 2D6, and it is known that 5%–7% of the general population lacks this enzymatic system. Additionally, CYP450 2D6 is best characterized by CYP450 which exhibits polymorphism in humans, and its three phenotypes include poor metabolizers (PMs), extensive metabolizers (EMs), and ultrarapid metabolizers (UMs). Since ethnic differences have been reported in CYP450 2D6 expression between Caucasians and Asians, the metabolism of TCAs is affected by this ethnic difference (Lee et al., 2008). Furthermore, the metabolism of TCAs can be affected by age and activation or inhibition of other medications.

A positive relationship exists between serum level and therapeutic response of TCAs in patients with endogenous depression. First, a sigmoid positive relationship has been noted between serum level and the therapeutic response of imipramine. The clinical response of imipramine increases in proportion to the rise in its plasma level up to 250 ng/mL, but with higher serum levels, the obtainment of a plateau denotes the achievement of the highest clinical response and the response does not increase further. It has been reported that depressed patients with imipramine plasma levels of less than 150 ng/mL, 150–225 ng/mL, and more than 225 ng/mL present with 30%, 67%, and 93% response rates, respectively (Glassman et al., 1977). Second, a curvilinear relationship exists between serum level and clinical response of nortriptyline. Thus, it is known that the therapeutic window of nortriptyline is 50–150 ng/mL of its plasma level.

- **Desipramine**

    Desipramine is classified as a TCA and is pharmacologically characterized by more potent NE reuptake inhibition than 5-HT reuptake inhibition. Since desipramine is considered a major metabolite of lofepramine, desipramine is similar in its pharmacological effect to lofepramine and demonstrates a longer duration of action than lofepramine. A primary indication of desipramine is major depression (i.e., moderate-to-severe depression) (Janowsky & Byerley, 1984; Kerihuel & Dreyfus, 1991). It has been reported that desipramine also exerts a therapeutic effect on bulimia nervosa, as evidenced by the findings of a double-blind, placebo-controlled study (McCann & Agras, 1990). Desipramine exerts lower sedative and anticholinergic side effects than that observed with the use of other TCAs. Desipramine is mainly metabolized by the action of CYP450 2D6 and CYP450 1A2 and its half-life is 24 hours. The starting dose of desipramine is 25–50 mg/day, and its therapeutic dose range is 150–300 mg/day. Maprotiline is a TCA and, in certain cases, it is classified as a TeCA. Its mode of action is selective norepinephrine reuptake inhibition, and clinical indications include major depression, chronic pain, and insomnia. In side effects, maprotiline may cause convulsion and exhibit common features of TCAs.

- **Maprotiline**

  Maprotiline is metabolized by the action of CYP450 2D6 and its half-life is approximately 51 hours. The initial dose of maprotiline is maintained at 75 mg/day for 2 weeks, and a gradual increase in its dose to 230 mg/day over a period of 6 weeks is required. It is recommended that the dose of maprotiline is maintained at less than 175 mg/day.

- **Nortriptyline**

  Nortriptyline is a TCA and is known as an inhibitor of selective norepinephrine reuptake. Additionally, the administration of a high dose of nortriptyline enhances serotonin signal transmission. As nortriptyline exhibits lower sedative effects and presents with a risk for orthostatic hypotension development compared to the effects observed with the use of other TCAs, it can be safely used for elderly patients. Based on an established dose-response curve, nortriptyline can present with a stable therapeutic response at its plasma concentration of 50–150 ng/mL. It is recommended that the starting dose of nortriptyline be 50 mg/day, which may be increased by 50–100 mg/day every week. Reduction of the dose of nortriptyline after a period of 3 weeks may help control symptoms. Furthermore, it is recommended that for the elderly patients, the starting dose of nortriptyline should be 25 mg/day, which may be increased by 50–100 mg/day every week up to a maximum of 150 mg/day. Nortriptyline is metabolized through the action of CYP450 2D6 and CYP450 3A4, and its half-life is 36 hours.

- **Protriptyline**

  Protriptyline is a TCA that is characterized by more potent NE reuptake inhibition than 5-HT reuptake inhibition. Its primary indication is major depression, and it can be adjunctively used for treating ADHD, cocaine use disorder, and others. It is recommended that, for young adult patients, the starting

dose of protriptyline should be 15 mg/day during the first week which may be increased by 5–10 mg/day every week up to a maximum of 60 mg/day. Moreover, it is recommended that, for the old adult patients, its starting dose should be 10 mg/day during the first week which may be increased up to a maximum of 30–40 mg. Protriptyline is metabolized by the action of CYP450 2D6, and its half-life is 74 hours.

- **Amoxapine**

  Amoxapine is classified as a TCA and is classified as a TeCA in certain cases. Its primary indication is major depression, and it can be used for comorbidity of anxiety and agitation. In the mode of action, amoxapine demonstrates the effects of $5-HT_{2A}$ receptor antagonism and D_2 receptor antagonism as well as norepinephrine reuptake inhibition. Thus, amoxapine can be used for the treatment of psychotic depression and bipolar depression (Anton & Sexauer, 1983; Jue et al., 1982). Amoxapine also exhibits DA receptor inhibition-related side effects including extrapyramidal side effects, galactorrhea, movement disorders, and others (Hayes & Kristoff, 1986). It is recommended that the starting dose of amoxapine should be 50 mg/day which may be increased to a maximum of 400 mg/day in gradual increments. Additionally, it is recommended that for old adult patients, the starting dose should be 25 mg which may be increased by 25 mg/day every week up to a maximum of 300 mg/day.

- **Doxepin**

  Doxepin is a TCA that presents with a sedative effect. It is characterized by H_1 receptor antagonism as well as norepinephrine reuptake inhibition and serotonin reuptake inhibition. Doxepin is used for the treatment of major depression comorbid with anxiety, insomnia, and psychotic symptoms (Anderson, 2000; Godfrey, 1996). The use of doxepin for patients with sleep disturbance is supported. Since doxepin shows high affinity for H_1 receptor at low doses, it can act as a selective antagonist for H_1 receptor when the dose is lowered. For the treatment of major depression, the starting dose of doxepin can be recommended as 25 mg/day, and its dose can be increased to a maximum of 300 mg/day. Furthermore, for the treatment of insomnia, its therapeutic dose can be recommended as 3–6 mg/day. In terms of side effect profiles, unlike those observed with the use of other TCAs, doxepin-related effect is characterized by a less remarkable presence of daytime sleepiness, somnolence, and tolerance due to long-term use (Roth et al., 2007; Singh & Becker, 2007; Stahl, 2008).

- **Mianserin**

    Mianserin is an α_2 receptor antagonist that increases norepinephrine and 5-HT release. The release of norepinephrine and 5-HT is increased by mianserin through receptor blockade rather than reuptake inhibition. Moreover, mianserin presents an antidepressant action, based on antagonism of 5-HT_{2A}, 5-HT_{2C}, 5-HT_3, and 5-HT_7 receptors, and exhibits a sedative action, based on an antagonism of the H_1 receptor. Moreover, mianserin can induce weight or appetite increase, based on antagonism of the 5-HT_{2C} receptor. Mianserin is classified as a TeCA and used in the treatment of depression, anxiety, and insomnia. The starting and maximum doses of mianserin are 30 mg/day and 90 mg/day, respectively. Compared to TCAs, mianserin induces less sexual dysfunction and cardiovascular system-related side effects. The most commonly reported side effects of mianserin are somnolence and weight or appetite increase. Mianserin rarely induces convulsion, blood dyscrasia, increased suicidal ideation, and other severe side effects (McAinsh & Cruickshank, 1990; Wakeling, 1983).

TCA Side Effects

TCA side effects occurring due to the antagonistic actions at the M_1, H_1, and α_1/α_2 receptors include anticholinergic effects (i.e., dry mouth, blurred vision, constipation, urinary incontinence, hypohidrosis, tachycardia, confusion, and memory impairment), sedation, weight gain, and orthostatic hypotension, respectively. Weight gain is a pharmacologically uncontrolled side effect of amitriptyline and doxepin. Moreover, in terms of TCA uses, careful consideration of cardiovascular system-related side effects may be necessary. Thus, before using TCAs, electrocardiography (ECG) may be done for patients with cardiovascular disease. Its therapeutic window is considered narrow, since both intracardiac conduction delay and arrhythmia occurring due to TCA overdose may cause death. Acute renal failure following amoxapine overdose has been also reported (Frendin & Swainson, 1985). Additionally, high therapeutic doses of several TCAs (i.e., maprotiline and amoxapine) may lead to the occurrence of neurological side effects including extrapyramidal symptoms and seizure. The neurological side effects of TCAs can be controlled by reducing dose and by preventing a gradual increase in dose. Especially, in the use of amoxapine, the risk of tardive dyskinesia needs to be warned (Revet et al., 2020). Furthermore, discontinuation of TCAs can cause withdrawal syndrome (i.e., dizziness, anergia, headaches, and nightmares) and anticholinergic rebound syndrome (i.e., nausea, vomiting, diarrhea, excessive salivation, sweating, anxiety, and delirium). Thus, in the context of discontinuation and reduction of TCA doses, their gradual tapering should be considered (Richelson, 2002; Owens et al., 1997; Sanderson et al., 2005).

Adjunctive Treatments with Dopamine-Serotonin Modulators for Depression

Dopamine-serotonin modulators have been used as an adjunctive treatment for depression in patients who did not show satisfactory treatment response on existing SNRIs or SSRIs or mirtazapine. Those drugs were approved for adjective treatment for depression by US FDA including aripiprazole in 2007, quetiapine in 2009, olanzapine in 2009, and brexpiprazole in 2015.

Aripiprazole (partial agonist and antagonist (dopamine and serotonin)) has two 6-week, double-blind, placebo-controlled studies (Berman et al., 2007) on the adjunctive treatment for major depression. The results showed that those

Table 3 Adjunctive use of a dopamine and serotonin modulator in major depressive disorder (expanded from Shen, 2011b)

DA and 5-HT modulator	Initial dose (mg)	Range of daily dose (mg)	Final average daily dose
Aripiprazole	2 or 5	2–20	11
Quetiapine	25	150–300	188
Olanzapine	6	6 or 12	9.6
Brexpiprazole	0.5 or 1	2–3	2

5-HT serotonin; *DA* dopamine

SNRI- or SSRI-medicated patients get significant improvement in depressive symptoms and response and remission rate with the daily dose range of 2–20 mg of aripiprazole (Table 3). The noticeable side effects of aripiprazole adjective treatment are akathisia, headache, restlessness, and fatigue (Berman et al., 2007).

Quetiapine (multimodal (dopamine, serotonin, and norepinephrine)) has one 6-week, double-blind, placebo-controlled study (McIntyre et al., 2006) on this adjunctive treatment for major depression. Two daily dosages of 150 mg and 300 mg of quetiapine were used. The results showed that both depressive symptom and remission rate have significantly more in patients with adjunctive treatment than those without. The average final quetiapine is 188 mg (Table 3). The side effects of quetiapine adjunctive treatment are sedation, somnolence, and lethargy (McIntyre et al., 2006).

Olanzapine (antagonist (dopamine and serotonin)) has three randomized, double-blind trials (Corya et al., 2006; Shelton et al. 2005a, b; Thase et al. 2007a, b) on this adjunctive treatment for major depression. The daily dose ranges are 6 mg or 12 mg of olanzapine. The results showed that both depressive symptom and remission rate have significantly more in patients with adjunctive treatment than those without. The daily dose ranges are 6 mg or 12 mg of olanzapine (Table 3).

Brexpiprazole (partial agonist and antagonist (dopamine and serotonin)) has two 6-week, double-blind, placebo-controlled studies with daily brexpiprazole of 1 mg and 3 mg (Thase et al., 2015a) and with daily brexpiprazole of 2 mg (Thase et al., 2015b) on this adjunctive treatment for major depression. The daily dose of 1 mg of brexpiprazole is not effective. But both daily doses of 2 mg and 3 mg of brexpiprazole significantly improve depressive symptom and remission rate, in patients with adjunctive treatment than those without. The initial dose of brexpiprazole is 2 mg or 3 mg (Table 3). The important side effects from this brexpiprazole are weight gain, akathisia, and headache (Thase et al., 2015a, b).

Novel Treatments for Depression

Ketamine/Esketamine

Ketamine is defined as an antagonist (glutamate) in the NbN guidelines (ECNP, 2017). It is an ionotropic glutamatergic NMDAR antagonist (Yamamoto et al., 2015). Ketamine is used as an adjunctive therapy for treatment-refractory or suicidal patients who continue to receive an SSRI, SNRI, or mirtazapine. Ketamine exhibits a rapid antidepressant action in patients with major depression. Reduction of depressive symptoms among patients with major depression can be achieved within 2 hours after an intravenous ketamine infusion, and the effects may be prolonged for more than 2 weeks (Autry et al., 2011). Studies based on an animal model showed that a rapid antidepressant action of ketamine is causally related to a rapid synthesis of BDNF. A rapid antidepressant action of ketamine is induced by blockade of the eukaryotic elongation factor 2 (eEF2) kinase, which occurs through an inactivation of eEF2 in antagonizing NMDAR (Autry et al., 2011; Kavalalil & Monteggia, 2012). An intravenous ketamine infusion reduces suicidality in patients with treatment-resistant major depression (Price et al., 2009). Furthermore, it has been reported that ketamine-based improvement of suicidal ideation among patients with major depression may or may not be related to the reduction of depression and anxiety symptoms (Ballard et al., 2014). About 50% of the major depression patients who have experienced suicidal ideation for more than 3 months present with rapid reduction and remission of suicidal ideation after subjection to an intravenous ketamine infusion of 0.50 mg/kg for 3 weeks and an infusion of 0.75 mg/kg for 3 weeks thereafter (Ionescu et al., 2016). Besides lithium and clozapine, ketamine has been considered a novel antisuicidal drug in patients with major depression.

Esketamine is an enantiomer of ketamine, and its relative binding affinity at the glutamatergic NMDAR is four times that of ketamine. Again, esketamine is also used as an adjunctive therapy for treatment-refractory or suicidal patients who continue to receive an SSRI, SNRI, or mirtazapine. Esketamine has been developed for the use in the form of a nasal spray. The antidepressive and antisuicidal effects of esketamine in patients with major depression have been also reported. Thus, esketamine has been designated as a "breakthrough therapy" by US FDA with the indication for treatment-resistant major depression in March 2021 (Molero et al., 2018). In a double-blind placebo-controlled randomized trial conducted for investigating the efficacy of an intranasal form of esketamine, a significant reduction in suicidal ideation and an improvement in depressive symptoms have been reported at 4 hours and 24 hours, respectively, after administration of esketamine at a dosage of 84 mg twice a week for a period of 4 weeks. No difference in the reduced suicidal ideation has been observed at 25 days after the administration of esketamine (Canuso et al., 2018). But it has been reported that ketamine or esketamine cannot decrease suicidal ideation in bipolar depression patients comorbid with suicidal ideation (Grunebaum et al., 2017). The most commonly reported side effects of esketamine include nausea, dizziness, dissociation, headache, and others (Molero et al., 2018). Although ketamine may increase a sense of dependency in the users, it has been recently reported that ketamine can reduce the recurrence rate of alcohol use disorder (McAndrew et al., 2017). This is a new finding of anticraving effect

to curb the desire for patients with alcohol use disorder (Shen, 2018). Since suicidal attempt and committing suicide are closely related in patients with major depression and as the prevalence of suicidal attempts is 10 times than that of committing suicide each year (Jeon et al., 2010a, b), esketamine use may lead to an important development in the realm of psychopharmacology in the future (Hashimoto et al., 2020).

Brexanolone

Brexanolone is a positive allosteric modulator (GABA) in the NbN guidelines (ECNP, 2017). Brexanolone, the first synthetic version of allopregnanolone, was approved as the drug of choice for the treatment of postpartum depression by the FDA in 2019 (Ali et al., 2021; Cristea & Naudet, 2019). Since serum allopregnanolone levels steadily increase throughout pregnancy and markedly decrease to normal serum levels in the postpartum period, it has been indicated that allopregnanolone may cause postpartum depression (Meltzer-Brody et al., 2018). As brexanolone is an exogenous analog of allopregnanolone and a neuroactive steroid, it binds to five-unit transmembrane $GABA_A$ receptors. Despite the poor understanding of its exact mode of action, it is presumed that the inhibitory action of GABA is enhanced through the binding of brexanolone to $GABA_A$ receptors. Thus, through the exhibition of enhanced inhibitory action, brexanolone reduces anxiety and depression symptoms and presents with side effects of sedation, drowsiness, and dizziness (Leader et al., 2019). Common administration of brexanolone entails continuous intravenous infusion over a period of 60 hours in inpatient facilities. A gradual increase in the dosage per hour from the starting dose of 30 µg/kg/h to a maintenance dose of 90 µg/kg/h is necessary until 52 hours, before being tapered off to 30 µg/kg/h by 60 hours (Kanes et al., 2017). Brexanolone presents with three major inactive metabolites with a total plasma clearance of 1 L/h/kg and shows an equal amount of excretion in the urine and feces; additionally, its half-life is 9 hours. Brexanolone is subjected to considerable non-CYP450-mediated   hepatic   metabolism through keto reduction, glucuronidation, and sulfation. Brexanolone's drug-drug interactions have been reported to have CNS depressants mainly due to its sedating effects. No other drug interaction associated with brexanolone has been documented (Abramowicz et al., 2019). Before its US FDA approval, the safety, tolerability, and efficacy of brexanolone were evaluated in a one open-label study and in three randomized placebo-controlled trials, using the reduction of mean score based on the Hamilton Depression Rating Scale (HAMD) as the primary outcome. A significant reduction of the HAMD score compared to that of placebo was observed in each of the randomized controlled trials conducted with an intravenous infusion administration of brexanolone over a period of 60 hours. But patients' access to brexanolone may be limited by the costs of medication (>USD 30,000 per 60-hour infusion) and owing to the necessity of admission to a medically supervised setting.

Zuranolone

Until now, zuranolone (SAGE-217) has not yet been defined in the NbN guidelines. Zuranolone is a neuroactive steroid and positive allosteric modulator of the $GABA_A$ receptor. It has been granted breakthrough therapy for postpartum depression by the US FDA. Zuranolone has been characterized by the pharmacological properties optimized for oral dosing and produced pharmacodynamic effects consistent with GABAergic activity in vivo following oral dosage and CNS exposure. Unlike brexanolone, zuranolone can be used in outpatient care settings (Frieder et al., 2019; Gunduz-Bruce et al., 2019). Furthermore, zuranolone shares a similar molecular profile with brexanolone. Zuranolone shows a serum half-life of 16–23 hours. In a phase 3, double-blind, randomized, outpatient, placebo-controlled trial (randomization 1:1 to placebo vs. zuranolone 30 mg, administered orally for 2 weeks) (Dligiannidis et al., 2021), it was reported that zuranolone improves the core symptoms of depression in patients with postpartum depression. Moreover, zuranolone is generally well-tolerated.

Mifepristone

So far, mifepristone has not yet been defined in the NbN guidelines. Mifepristone was approved by the FDA in 2016 as a drug to terminate early pregnancy by blocking progesterone action. An augmentation strategy of medication for depression with a dopamine serotonin blocker (second-generation (atypical) antipsychotic) is generally recommended for patients with psychotic depression in most pharmacological treatment guidelines. Psychotic depression is characterized by hypercortisolism compared to other depressive subtypes (Schatzberg & Rothchild, 1992), because psychotic depression presents with elevated levels of urinary free cortisol (Anton, 1987), elevated serum adrenocorticotropic hormone (ACTH), and cortisol levels (Poserner et al., 2000), as well as high rates of non-suppression on challenge with dexamethasone and high post-dexamethasone serum cortisol levels (Nelson & Davis, 1997; Schatzberg et al., 1983). Additionally, mifepristone is known as a potent reversible antagonist at glucocorticoid receptors. Based on the findings of a phase 3 trial, mifepristone has been shown to have a rapid onset of therapeutic actions in patients with psychotic depression. The strongest association between high mifepristone serum level and treatment response, followed by changes in adrenocorticotropin hormone and cortisol, has been evidenced based on a combined analysis of five double-blind phase 2 or 3 trial findings that helped evaluate efficacy and effectiveness of mifepristone treatment for the psychotic symptoms of psychotic depression. Furthermore, a dose of 1200 mg/day of mifepristone helps achieve its therapeutic serum levels (Block et al., 2018). But mifepristone has not been approved as a drug for psychotic depression.

Potential Future Developments of Drugs for Depression

Although many drugs have been used for the treatment of major depression, because of the unmet needs, it is required to develop the novel drugs for depression. Thus, NMDA, neuropeptide,

S.-C. Park and W. W.-D. Shen

histamine, and choline receptors are considered the potential targets for the development of novel drugs for depression. In addition, in the realm of new pharmacological treatment of major depression, one strategy focuses on the development of anti-glutamatergic agents. A different strategy is to focus on the potentiation of the cAMP response element-binding protein (CREB) activities, which is associated with cAMP-mediated gene transcription. The potentiation of the CREB cascade is based on the inhibition of phosphodiesterase-4 (PDE4) activity. Thus, it is expected that an inhibitor of PDE4 can help increase intracellular levels of CREB and may be considered as a medication for depression. Another strategy is to focus on direct enhancement of not only dopamine function but also on the increase in B-cell lymphoma protein 2 (Bcl-2) levels. It is thought that increased DA transmission in limbic structures (i.e., the nucleus accumbens) is partly related to its potential antidepressant actions (Shiloh et al., 2006). Moreover, oxidative stress has been regarded as one of the causative factors for depressive-like behaviors. Thus, antioxidant agents are investigated as the novel drug for depression to be beneficial in the treatment of depressive behaviors. Herein, it is expected that novel, effective, and safe drugs for depression would be developed based on the understanding of the pathophysiological implications in major depression (Thakare & Patel, 2015).

Sex, Ethnicity, and Racial Differences in Pharmacokinetics and Pharmacodynamics of Drugs for Depression

Sex

Although sex differences in the pharmacokinetics of drugs for depression have not been conclusively studied (Park et al. 2015a, b), several factors can affect them. First, since women have less gastric acid and slower gastric emptying than men do, they have slower gastrointestinal absorption. Second, women have a higher ratio of adipose tissue and lean body mass, which can affect the half-life of lipophilic medications for depression

(e.g., bupropion, trazodone). Third, a woman's volume of distribution might be affected by water retention associated with the menstrual cycle. Fourth, since estrogen is a substrate for both CYP450 3A4 and 1A2, hepatic enzyme function may be different between men and women (Bigos et al., 2009).

Several studies on the sex differences in the pharmacokinetics of drugs for depression have shown contradictory findings. For similar doses of drugs for depression, some studies showed that women have higher serum drug levels than men do, whereas other studies have reported no sex differences in the pharmacokinetics of drugs for depression. The discrepancy between studies might be explained by alterations in the hepatic metabolism of TCAs, which can be affected by a patient's use of oral contraceptive.

In addition, other sex differences in the pharmacodynamics of drugs for depression have been reported. Based on the reanalysis of data from a study comparing imipramine and sertraline for chronic depression, premenopausal women showed a better response to sertraline than to imipramine. On the contrary, the opposite response pattern has been reported for men (Kornstein et al., 2009). For the SSRIs, bupropion, and duloxetine, no sex differences exist in pharmacodynamics of drugs for depression (Kornstein et al., 2006; Papakostas et al., 2007). But more robust pain response to duloxetine and better improvement in anxiety and somatic symptoms have been reported for SSRIs in women compared to men. Based on a post hoc analysis of STAR*D, a better response to citalopram has been reported for women compared to men (Young et al., 2009).

Race and Ethnicity

As with sex differences, the potential effects of race and ethnicity on the pharmacokinetics and pharmacodynamics of drugs for depression have not been evaluated conclusively. In the reanalysis of data from the STAR*D, the poorest, intermediate, and strongest responses to citalopram have been reported in African, Hispanic, and Caucasian

Americans, respectively (Lesser et al., 2007). But the differences were explained mainly by the baseline demographic differences in the groups (Lesser et al., 2007). In a meta-analysis of studies (Lewis-Fernandez et al., 2006), no differences exist in efficacy to duloxetine between Caucasian and Hispanic patients. Since the Asian population tends to have the poor-metabolizer trait for CYP450 2C19, Asians can show some pharmacokinetic differences to drugs for depression (Roden et al., 2006; Wu-Chou et al., 2016), but no associations that exist between monoamine transporter polymorphisms and antidepressant responses have been reported in Asians in many pharmacogenetic studies (Nnadi et al., 2005). The observed differences between African and Caucasian Americans might be caused by differences in the ability to metabolize certain drugs for depression; a low prevalence of the CYP450 2D6 poor-metabolizer mutation has been reported in one genetic study of African Americans from the southern USA (Evans et al., 1993). A reanalysis of the STAR*D data reported that a particular allele of 5-HT2A which is six times more common in African Americans than that in Caucasian Americans is associated with a poor response to treatment (McMahon et al., 2006).

Symptom-Based Selection of Drugs for Depression

Evidence-based medicine has helped establish various clinical recommendations and give evidence for the pharmacological treatment of major depression. The specific class of drugs for depression and the optimal dosage of SSRIs have been suggested through systematic reviews and meta-analyses of studies for the treatment of major depression. A study pertaining to systematic review and network meta-analysis, involving 522 randomized, double-blind, placebo-controlled trials (116,477 participants) conducted for the acute treatment of adult patients with major depression, showed that vortioxetine has the highest levels of both efficacy (response rate) and acceptability (dropout rate) in head-to-head studies of 21 drugs for depression (Cipriani et al.,

2018). A study pertaining to systematic review and meta-analysis including 40 randomized, double-blind, placebo-controlled trials (10,039 participants) shows a relationship between the dose and efficacy of SSRIs for the treatment of major depression. Additionally, evidence exists for a relationship between high dosage and dropout rate occurring due to side effects of SSRIs. Thus, considering both efficacy and dropout rate, the optimal dosage of SSRIs for the treatment of major depression is estimated to be 250 mg of imipramine (Jakubovski et al., 2016). But despite the findings reported by Cipriani et al. (2018), it cannot be concluded that one treatment option is superior to other treatment options in which drugs for depression should be chosen in the context of such a complex clinical situation. Moreover, evidence-based treatment algorithms showed limitations that they cannot be used to recommend the specific selection of primary and secondary pharmacological treatment options for major depression (Park, 2020; Won et al., 2014). Additionally, a drug for depression can be chosen based on genetic testing. The prediction of high or low serum levels of the substrate drugs can be done based on the existence of several genetic forms of the CYP450 enzyme system (i.e., CYP450 2D6, 2C19, and 3A4). Thus, the side effects and therapeutic effects in certain people can be explained by such genetic testing approaches combined with therapeutic drug monitoring of drugs for depression. However, the genetic markers in psychopharmacology cannot be used to suggest a clinician's prescription for a specific individual with depressive disorder but can be used to potentially explain greater or lesser likelihood of response, nonresponse, or side effects. Furthermore, treatment responses are not defined based on the "all or none" phenomenon. While genetic information can influence a clinician's prescription for drugs for depression, it cannot dictate a single compelling choice. Several genes may assist in therapeutic decision-making for drugs for depression, as follows: SLC 6A4 variation, which is related to serotonin reuptake, may have potential therapeutic implications for poor response, slow response, and poor tolerability to SSRIs/SNRIs. 5-HT$_{2C}$ variation, related to the regulation of dopamine and norepinephrine release, may indicate poor response and poor tolerability to dopamine serotonin blockers (atypical antipsychotics). COMT Val variation, related to the regulation of dopamine levels in the prefrontal cortex and to the metabolism of dopamine and norepinephrine, may indicate reduced executive function. MFTHFR variation, related to the regulation of L-methylfolate levels and methylation, may indicate reduced executive function. Although genetic testing continues to expand in the realm of psychopharmacology, it remains limited in clinical practice (Stahl, 2009, 2013a, b).

The "next-generation treatments for mental disorders" proposed by Insel (2012) have been suggested according to the theoretical background transition from chemical imbalance to dysfunction of the neural circuit. According to the change from a disease-based approach to a symptom-based approach in the clinical treatment target, it is recommended that pharmacotherapy should be more symptom-targeted for the next-generation treatments. Furthermore, a machine learning-based model to understand the symptom clusters and their treatment responses of major depression has been identified (Chekroud et al., 2017). Using the data-driven approach, the novel symptom clusters have been identified, and prediction models for efficacy of drugs for depression have been established based on depressive symptoms evaluated using the self-report Quick Inventory of Depressive Symptomatology (QIDS-SR) (Rush et al., 2003) for 4039 major depression patients enrolled in the STAR*D study (Rush et al., 2008; Trivedi et al., 2006; Warden et al., 2007) and for 460 major depression patients enrolled in the Combining Medication to Enhance Depression Outcome (CO-MED) study (Rush et al., 2011). Hence, the depressive symptoms (QIDS-SR) are divided into three clusters including sleep/insomnia symptoms (i.e., early insomnia, middle insomnia, and late insomnia), core emotional symptoms (i.e., energy/fatigue, concentration/decision-making, loss of interest, depressed mood, and worthlessness), and atypical symptoms (i.e., psychomotor agitation, psychomotor retardation, suicidal ideation, and hypersomnia). For each symptom cluster, a new model has been

subjected to training through machine learning for patients receiving citalopram in the STAR*D study. Moreover, using the model with the best performance for the three symptom clusters in the CO-MED study, studies have been conducted to verify whether the response to drugs for depression can be generalized in a clinical study sample. As a result, it has been reported that core emotional symptom cluster can be predicted with above-chance performance for escitalopram with placebo and venlafaxine with mirtazapine. Additionally, a sleep/insomnia symptom cluster can be predicted with above-chance performance for escitalopram with bupropion. The three depressive symptom clusters are differentiated in their responses to different classes of drugs for depression. Thus, these findings can help support the choice of the optimal drugs for depression for a specific depressive symptom cluster in clinical settings.

Thus, the symptom-based approach can help individualize the pharmacological treatment portfolio for each patient with major depression more effectively than the evidence-based approach that defines the treatment options generally on a disease entity level. The symptom-based treatment algorithm for major depression, which has been proposed by Stahl (Stahl, 2009, 2013a, b), consists of the following five steps. Step 1: A list of specific symptoms that are presented by individual patients is deconstructed into diagnostic and associated symptoms of major depression. Step 2: These symptoms are matched with the neural circuits that theoretically mediate them. Step 3: Specific neurotransmitters and pharmacological mechanisms are targeted. Step 4: Neural circuit dysfunctions are matched with the psychopharmacological regulation of these circuits by neurotransmitters. Step 5: At a time, a unique treatment portfolio for each major depression patient is constructed. In turn, through the selection of treatment options that help target specific psychopharmacological mechanisms, the targeted symptoms are eliminated. Information on the treatment options with different mechanisms is added or replaced for persistent symptoms. The symptom-based treatment algorithm used for major depression aims to increase negative effect, to decrease positive effect, or to indicate all as therapeutic targets and contributes to the integration of the therapeutic process. In major depression, the increasing of negative effect is known to be mediated by dysregulation of 5-HT and norepinephrine. SSRI, SNRI, and others can be used as treatment options for the increase in negative effect. The decrease in positive effect is known to be mediated by dysregulation of DA and NE. Augmentation therapy involving the use of NDRI/NRI/SNRI and monoamine releaser (psychostimulant)/MAOI can be considered as a treatment option for the decrease in positive effect. On the other hand, to simultaneously treat the increase in negative effect and the decrease in positive effect, adoption of combination therapies, including SSRI and NDRI, SNRI and NDRI, SSRI/SNRI and monoamine releaser (psychostimulant), and α2 antagonist and SSRI/ SNRI/NDRI/MAOI, can be considered as a treatment option to target DA, NE, and 5-HT. Additionally, based on the findings reported from the STAR*D study, only two-thirds of patients with major depression have shown the achievement of remission after a year of treatment with four sequential drugs for depression during the treatment period of 12 weeks. Therefore, the symptom-based selection of drugs for depression is proposed as the construction strategy of a portfolio of multiple agents to treat all residual symptoms of major depression for the purpose of achieving sustained remission (Rush et al., 2006; Sinyor et al., 2010).

As shown in Table 4, the symptoms that are included in the diagnostic criteria for major depression can be approached using the symptom-based selection of drugs for depression (Shiloh et al., 2006; Stahl, 2009, 2013b). About 10%–30% of patients with major depression have a chronic course of depressive illness, even with adequate pharmacological treatment (Keller et al., 1984). In major depression patients who do not achieve remission, the most common residual symptoms are insomnia, fatigue, decreased concentration, and lack of interest (Stahl, 2013b). The treatment algorithm for residual symptoms is as follows: In the neural circuit dysfunction corresponding to each symptom, it is assured that decreased concentration,

loss of interest, fatigue, and insomnia can be reduced to the prefrontal cortex, nucleus accumbens, striatum or spinal cord, and hypothalamus, respectively. Furthermore, in the neurotransmitter corresponding to each symptom, decreased concentration, loss of interest, and fatigue, conditions that are mediated by the regulation of NE or DA and insomnia, can be mediated by the regulation of 5-HT, GABA, or histamine, respectively. Thus, the augmentation of NDRI/NRI/SNRI and the use of monoamine releasers (psychostimulants)/dopamine serotonin blockers (atypical antipsychotics)/5-HT$_{1A}$ agonists can be considered as treatment options to target decreased concentration, decreased interest, and fatigue. Furthermore, implementation of the use of medications for insomnia (hypnotics) or sedating drugs for depression and discontinuation of the use of activating drugs for depression can be considered as treatment options to target insomnia. Additionally, as shown in Table 4, the associated symptoms that are not included in the diagnostic criteria for major depression can also be approached using symptom-based selection (Shiloh et al., 2006; Stahl, 2009, 2013b). These symptoms include anxiety, pain, excessive sleep, sexual dysfunction, and vasomotor symptoms. In the context of the hypothetically malfunctioning brain circuits, anxiety is reduced to the ventral prefrontal cortex, and pain is reduced to the ventral prefrontal cortex, thalamus, and spinal cord. Moreover, sleepiness/hypersomnia is reduced to the basal forebrain, and sexual dysfunction is reduced to the nucleus accumbens. Vasomotor symptoms are reduced to the hypothalamus. In anxiety that is mediated by the regulation of 5-HT or GABA, the augmentation of SSRI/SNRI and the use of benzodiazepine/α2 antagonists/dopamine serotonin blockers (atypical antipsychotics) can be considered as treatment options. In pain that is mediated by the regulation of 5-HT and norepinephrine, the augmentation of SNRI and the use of an α2δ ligand (i.e., gabapentin) can be considered as treatment options. In sleepiness/hypersomnia that is mediated by the regulation of DA, NE, and histamine, the augmentation with monoamine releaser (psychostimulant) can be considered as a treatment option, and the discontinuation of antihistamine, antimuscarinic, and α1 blockers can be recommended. In sexual dysfunction that is

Table 4 Symptom-based selection of drugs for depression

Symptoms	Hypothetically malfunctioning neural circuits	Selected pharmacological mechanisms
Symptoms that are included in the diagnostic criteria of major depression		
Concentration problem	Prefrontal cortex	Augmentation of NDRI/NRI/SNRI and psychostimulants/atypical antipsychotics/5-HT$_{1A}$ agonists
Loss of interest	Nucleus accumbens	Augmentation of NDRI/NRI/SNRI and psychostimulants/atypical antipsychotics/5-HT$_{1A}$ agonists
Fatigue	Striatum and spinal cord	Augmentation of NDRI/NRI/SNRI and psychostimulants/atypical antipsychotics/5-HT$_{1A}$ agonists
Insomnia	Hypothalamus	Hypnotics or sedating antidepressants Discontinuation of activating antidepressants
Associated symptoms that are not included in the diagnostic criteria of major depression		
Anxiety	Ventral prefrontal cortex	Augmentation of SSRI/SNRI and benzodiazepine/α2 antagonists/atypical antipsychotics
Pain	Ventral prefrontal cortex, thalamus, and spinal cord	Augmentation of SNRI and α2δ ligand (i.e., gabapentin)
Sleepiness/hypersomnia	Basal forebrain	Augmentation with psychostimulant Discontinuation of antihistamine, antimuscarinic, and α1 blockers
Sexual dysfunction	Nucleus accumbens	MAOI or psychostimulants
Vasomotor symptoms	Hypothalamus	SNRI (i.e., desvenlafaxine)

mediated by the regulation of dopamine, the use of NDRI, α$_2$ antagonists, MAOI, or monoamine releasers (psychostimulants) may be considered as a treatment option, and the discontinuation of SSRIs or SNRIs is suggested. Finally, in vasomotor symptoms that are associated with the regulation of 5-HT and NE, the use of SNRI (i.e., desvenlafaxine) can be considered as a treatment option.

Strategies for Treatment-Resistant Major Depression

Treatment-resistant major depression refers to cases in those patients who present with no response to treatment involving the use of two different types of drugs for depression with adequate dose and duration (Dunner et al., 2006). In the definition of treatment-resistant major depression, the adequate dose denotes the dose of drug for depression equal to imipramine 200–300 mg or fluoxetine 20–40 mg, and the adequate duration denotes more than 4 weeks or more than 6 weeks of treatment period (Berlim & Turecki, 2007a). The definition of treatment-resistant major depression is further modified and stated as the first treatment failure to one adequate treatment trial, as the response rate decreases along with treatment trials (Dunner, 2013). High risk of non-remission or longer time to remission is related to the development of more severe symptoms, presence of melancholic or anxious features, psychiatric comorbidities, presence of psychotic symptoms, poor physical health, poor adherence with treatment, unemployment, and impairment in role functioning. Additionally, higher risk of non-remission is associated with age older than 40 years and longer duration of episodes (Rush et al., 2006). Moreover, longer time to remission is associated with treatment in specialty mental health settings, atypical features, previous suicidal attempts, and poor mental health status (Mojtabai, 2017). In the treatment strategy for treatment-resistant major depression, the staging model including optimization, switching, combination, and augmentation is discussed below (Berlim & Turecki, 2007b; Chen et al., 2018; Ruhe et al., 2012). Furthermore, evidence-based recommendations for combination and augmentation strategies have been suggested in Table 5 (Won et al., 2014).

Table 5 Evidence-based recommendations for combination and augmentation strategies for treatment-resistant major depression

	Strongly recommend	Weakly recommend	Weakly not recommend	Strongly not recommend
Combination agent of antidepressant	TCA/TeCA SSRI SNRI NaSSA (mirtazapine)	NDRI (bupropion) SARI (trazodone)		MAOI
Augmentation agent of antidepressant	DA and 5-HT modulator Lithium Benzodiazepine		Lamotrigine T$_3$ Methylphenidate Modafinil	Atomoxetine

5-HT serotonin; *DA* dopamine; *MAOI* monoamine oxidase inhibitor; *NaSSA* norepinephrine and specific serotonergic antidepressant; *NDRI* norepinephrine-dopamine reuptake inhibitor; *SNRI* serotonin-norepinephrine reuptake inhibitor; *SSRI* selective serotonin reuptake inhibitor; *TCA* tricyclic antidepressant; *TeCA* tetracyclic antidepressant

Optimization

The optimization of an initial pharmacotherapy can be considered the first treatment strategy, before using switching or other strategies for treatment-resistant major depression. The simple strategy for non-remission can be an extension of the treatment period of an initial pharmacotherapy by 2 or 4 weeks. The treatment period extension can contribute to an improvement of natural course of depressive episode. It has been reported that faster responses to drugs for depression are observed in depressed patients who exhibit the 5-HTTLPR L/L type in the promotor region polymorphism of the 5-HT gene. This finding gives a theoretical background for the period extension of initial pharmacotherapy. More than 6 weeks of initial pharmacotherapy can be deemed an optimal strategy, since a period of 4 weeks may be insufficient in treating certain patients with major depression (Schweizer et al., 1990). But remission is not achieved in patients with treatment-resistant depression despite subjection to pharmacotherapy for a period longer than 6 weeks (Wooderson et al., 2011). Based on the existence of a possible neurobiological mechanism, tolerance and reduced response rate can result from long-term antidepressant treatment (Andrews et al., 2011; Fava & Offidani, 2011). In optimizing dosage for drugs for depression, before the development of SSRI, the maximum dose maintenance of TCA and MAOI has been considered an important treatment strategy for treatment-resistant major

depression. Furthermore, in SSRI usage, a better response rate has been observed based on the maintenance of the maximum dose rather than usual dose. However, in SSRI usage, no differences exist between the efficacies of low dosage (i.e., fluoxetine 20 mg, paroxetine 20 mg, and sertraline 50 mg) and high dosage (i.e., fluoxetine 60 mg, paroxetine 50 mg, and sertraline 200 mg). Usually, the optimal dose of TCA cannot be achieved due to the exhibition of diverse acute and chronic side effects. For example, findings of a meta-analysis proposed a biphasic relationship of efficacy to serum level of nortriptyline, with a therapeutic window between 46 and 236 ng/mL (Chen et al., 2018; Ribero et al., 2000). Therefore, to evaluate the efficacy of high-dose drugs for depression treatment is necessary for patients who present with non-remission following low-dose treatment. Furthermore, the optimal doses of bupropion, mirtazapine, and other classes of drugs for depression are more undefined than those of SSRIs.

Switching

In the switching strategy, the across-class switching is more supported rather than the within-class switching in SSRI-resistant major depression, as evidenced by the results of a meta-analysis (Papakostas et al., 2008). On the contrary, in treatment-resistant major depression, the response or remission is not improved by switching the class of drugs for depression (Souery et al., 2011). Upon comparing the efficacies of drugs for depression, the efficacies of clomipramine, venlafaxine, and escitalopram are superior to those of other drugs for depression. It has been presumed that no association exists between efficacy and the class of drugs for depression (Montgomery et al., 2007; Won et al., 2014). But several recommendations on the within-class and across-class switches in the treatment of antidepressant-resistant major depression have been reported. Since SSRIs are initially considered as the treatment option for major depression, the switching strategy-based trials are mainly conducted for SSRI-resistant major depression. The recommendations about the switching strategy for SSRI-resistant major depression are as follows: First, the switching from one SSRI to another SSRI has been supported by several study findings (Joffe et al., 1996). An open trial has shown that 42–76% of the major depression patients who exhibit no response to one type of SSRI show responses to the 6-week switching strategy involving the use of another SSRI. Second, the switching from SSRI to TCA has been supported by several study findings (Peselow et al., 1989; Thase et al., 2002). A double-blind trial showed that 47% of the major depression patients ($n = 117$) that has demonstrated no response to sertraline shows responses to the 12-week switching strategy conducted with imipramine. Another trial showed that 73% of the major depression patients ($n = 15$) that has demonstrated no response to paroxetine exhibit responses to the switching strategy conducted with imipramine. However, in the switching from SSRI to TCA, various adverse effect profiles (i.e., cardiotoxicity) of TCA should be considered. Third, the switching from SSRI to SNRI has been supported by several study findings (De Montigny et al., 1999; Rosso et al., 2012). A study showed that about 30% of the antidepressant-resistant depression patients ($n = 84$) exhibit responses to a 12-week switching strategy conducted with venlafaxine. Another study showed that more than 50% of the SSRI-resistant major depression patients exhibit responses to the switching strategy conducted

with venlafaxine or duloxetine. Since the effects of duloxetine or milnacipran treatment on pain or fibromyalgia and depression have been reported, it is suggested that duloxetine or milnacipran can be used for treating major depression patients with antidepressant resistance associated with such physical symptoms. Moreover, there is less considerable evidence for adoption of the switching strategy from SSRI to DNRI (bupropion). Although the treatment responses have been reported after the switching from SSRI to bupropion (Rosso et al., 2012), less substantial evidence exists for validating the effect of bupropion treatment on SSRI-resistant major depression. Fifth, there is less remarkable evidence for the switching from SSRI to NaSSA (mirtazapine) (Baldomero et al., 2005; Fava et al., 2006). Although mirtazapine may be used as a second medication for depression for SSRI-resistant major depression, less considerable evidence exists for the comparative efficacy of mirtazapine and other drugs for depression.

Combination

Combination strategy is to add a second drug for depression to an existing drug for depression in the treatment of antidepressant-resistant major depression. As shown in Fig. 4, drugs for depression can be classified behaviorally and physiologically (Chen et al., 2018; Shen, 2011a, b, 2016) based on the neurotransmission of three monoamines (5-HT, NE, and DA).

Guidelines for the Combination Strategy
In the context of the combination strategy for drug for depression, the guidelines of choosing an add-on drug for depression are summarized:

- 5-HT is more closely related to impulse, with increased relation of NE to energy and with increased relation of DA to drive in behavior physiology (Healy & McMonagle, 1997).
- Both combined actions of 5-HT and NE mediate anxiety and irritability. Those of 5-HT and DA mediate sex, appetite, and aggression; and

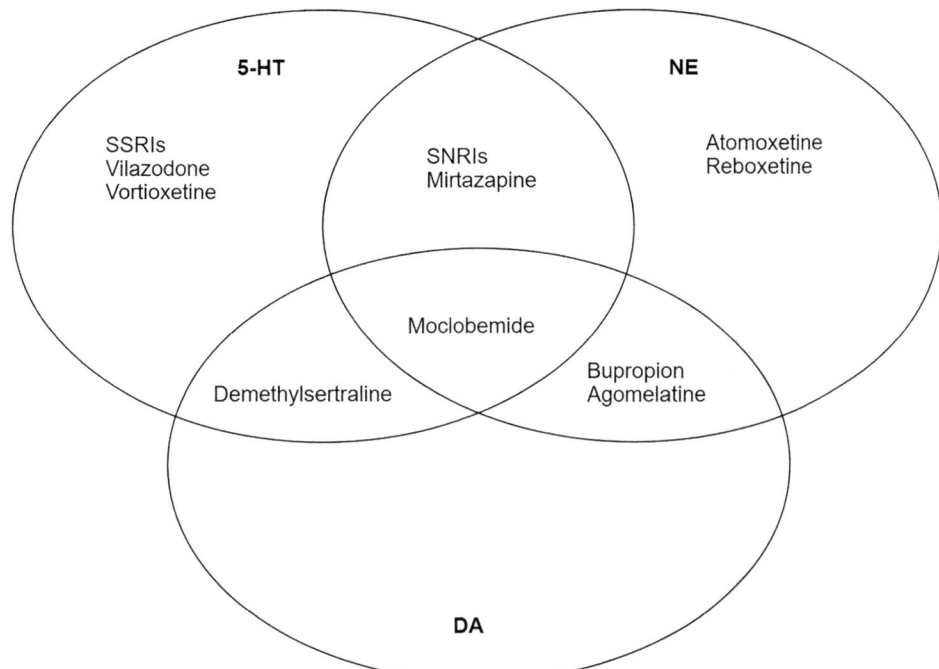

Fig. 4 The relationship between three monoamines and commonly used drugs for depression (reproduced with permission to reprint from Shen (2011b), aipei: Ho-Chi Publishing Company, Taipei, Taiwan; and from Chen et al. (2018) and Taiwanese Society of Psychiatry). 5-HT serotonin; DA dopamine; NE norepinephrine; SNRIs serotonin and norepinephrine reuptake inhibitors; SSRIs selective serotonin reuptake inhibitors

those of NE and DA mediate motivation (Healy & McMonagle, 1997).

- Higher and quicker remission rates are seen in patients with major depression treated with a dual-action drug for depression (i.e., SNRIs and mirtazapine) rather than a single-action drug for depression (i.e., SSRI) (Thase et al., 2001, 2007a, b, 2010). The same findings of higher and quicker remission rates between dual-action and single-action drugs for depression are also shown in patients with generalized anxiety disorder (Allgulander et al., 2001, 2004).
- Vilazodone and vortioxetine which show a property of 5-HT partial agonists improve tolerability for several side effects, such as sexual dysfunction (Chen et al., 2018).
- Adding a 5-HT-related drug for depression as a second drug for depression is necessary, if the major depression patients who are not treated with a 5-HT-related drug for depression are comorbid with anxiety symptoms (Chen et al., 2018).
- Adding a 5-HT-related drug for depression as a second drug for depression is necessary, if the patients who are treated with an adequate 5-HT-related drug for depression still experience any type of pain symptoms (i.e., migraine, diabetic peripheral neuropathy, gastrointestinal pain, muscle pain, or arthritis pain) (Chen et al., 2018).
- Adding a dopamine-related drug for depression as a second drug for depression is necessary, if the patients still have the predominant symptom of "fatigue or loss of energy," which is the sixth diagnostic criterion of major depression in *the Diagnostic and Statistical Manual of Mental Disorders, 5th edition (DSM-5)* (American Psychiatric Association, 2013; Chen et al., 2018).

Possible Combination Therapies

The possible combination strategies of drugs for depression have been recommended for several reasons:

- The combination of an SNRI or an SSRI and mirtazapine can be used. It has been reported that responding but non-remitted major depression patients treated with SNRI or SSRI usually present with five residual symptoms including sleeping disturbances, fatigue, loss of interest, guilt, and poor concentration (Nierenberg et al., 1999). Mirtazapine can be used as a second drug for depression to the existing SNRI or SSRI to treat the above-listed residual symptoms. Level 4 of the STAR*D study showed that a remission of 13.7% or more in major depression patients who have failed the three previous prospective medication trials is achieved following administration of a combination of venlafaxine and mirtazapine (McGrath et al., 2006). Furthermore, in the combination trial, using the mean doses of 210.3 mg for venlafaxine and 35.7 mg for mirtazapine, the lowest side effect profiles and a lack of dietary restrictions have been documented in patients. Thus, the combination of venlafaxine and mirtazapine, currently known as "California rocket fuel" (Hannan et al., 2007; Silva et al., 2016), has received recognition.

- The combination of an SNRI or an SSRI and bupropion or agomelatine can be used as follows. Level 2 of the STAR*D study showed that a remission of 29.7% or more in major depression patients who have used 50 mg/day or 100 mg/day of citalopram is achieved by introducing bupropion in a 10-week combination therapy (Rush et al., 2006). Since bupropion is involved in the neurotransmission of DA and NE, as depicted in Fig. 3, it is presumed that bupropion can be used for the patients presenting with the predominance of fatigue or loss of energy as the residual symptoms (Nierenberg et al., 1999). Similar to that observed with bupropion, agomelatine is also involved in the neurotransmission of DA and NE, as illustrated in Fig. 3. Thus, agomelatine can be used in a similar manner to bupropion in the context of combination therapy (Stahl, 2014). Both bupropion and agomelatine show favorable side effect profiles, in patients without weight gain.

- The combination of mirtazapine and bupropion can be used as follows. Bupropion can be used as an add-on drug for depression in major depression patients who have not achieved remission after treatment with adequate dosage of mirtazapine. The combination therapy of bupropion and mirtazapine can be used in patients with the predominance of residual symptoms including sleep disturbances and fatigue and loss of energy or those with intolerability to any SNRI- or SSRI-induced side effect of nausea and/or vomiting (Chen et al., 2018).

Augmentation

Augmentation strategy is to add an existing drug for depression with nondrug for depression in the treatment of treatment-resistant major depression. Mostly, drugs for psychosis have been used as an augmentation agent for psychotic depression or atypical depression. With the advent of dopamine serotonin blockers (second-generation (atypical) antipsychotic drugs) in the 1990s (Shen, 1999), because of a relatively favorable side effect profile, the use of dopamine serotonin blockers (atypical antipsychotics) as an augmentation or monotherapy agent for major depression has been supported by the trial findings (Park et al., 2015b). But dopamine serotonin blockers not effective in treating the two core symptoms of major depression (i.e., loss of interest and psychomotor retardation) (Nelson et al., 2008). A meta-analysis reported by Nelson and Papakotas (2009) has demonstrated that the overall response rates of dopamine serotonin blockers and placebo are 44.2% and 29.9%, respectively, indicating the efficacy of several dopamine serotonin blockers used as adjunctive therapy for treatment-resistant major depression. First, aripiprazole was approved as an option of medication for psychosis for an augmentation strategy of major depression in 2006 by the US FDA. Owing to its partial agonist activity at the D_2/D_3 receptors and 5-HT_{1A} receptors, it has been observed that aripiprazole can contribute to the synergistic effect of the medication for depression in the treatment of major depression (Nelson et al., 2008). In randomized controlled trials conducted for investigating the efficacy of aripiprazole as an add-on agent, a higher remission rate has been observed in the adjunctive aripiprazole group compared to the placebo group (Berman et al., 2007; Marcus et al., 2008). Furthermore, in a randomized controlled trial involving patients who reported inadequate response to drugs for depression (escitalopram, fluoxetine, paroxetine controlled release, sertraline, or venlafaxine extended release), higher response rate and remission rate have been observed in the aripiprazole group relative to the placebo group. Regarding side effects, although akathisia was observed in 18% of the patients of the aripiprazole group, conditions in 40% of the patients with akathisia resolved by the endpoint of the study. Furthermore, no significant differences have been found in the weight gain and serum levels of total cholesterol, triglyceride, high-density lipoprotein, low-density lipoprotein, and fasting glucose between the two groups (Berman et al., 2009). Thus, aripiprazole is an effective and tolerated option as augmentation therapy for treatment-resistant major depression (Chen et al., 2018). As shown in Table 3, an adjunctive use of a new serotonin-dopamine modulator (i.e., brexpiprazole) has been also evidenced in the treatment of major depression (Thase et al., 2015a, b; Mishra et al., 2022). Second, quetiapine was approved by the US FDA as an acute treatment option for bipolar depression, with a recommended dose of 300 mg/day. Furthermore, the antidepressant action of quetiapine in patients with schizophrenia spectrum disorders has been observed. Moreover, in a randomized controlled trial conducted for augmentation of treatment-resistant major depression, a greater improvement of symptoms was observed in the adjunctive use group with quetiapine (a target dose of 400 mg/day with a maximum dose of 800 mg/day) when compared to the use of lithium (Doree et al., 2007). In another randomized controlled trial, patients with an inadequate response to ongoing antidepressant treatment were allocated into three adjunctive intervention groups including quetiapine 150 mg/day, quetiapine 300 mg/day, and placebo. The overall remission rate was observed to be the highest in the quetiapine 300 mg/day group (52.7%), followed by the quetiapine 150 mg/day group (42.0%) and the placebo group (32.9%). Sedation and somnolence have been identified as the most frequently reported side effects of quetiapine (El-Khalili et al., 2010). Thus, quetiapine can be considered an effective and tolerable augmentation option for treatment-resistant major depression (Chen et al., 2018). Third, olanzapine augmentation of fluoxetine has also been approved by the US FDA for the treatment of treatment-resistant major depression. The starting doses of 5 mg/day of olanzapine and 20 mg/day of fluoxetine have been recommended. A randomized controlled study showed that higher remission rates for treatment-resistant major depression have been observed in the olanzapine/fluoxetine group compared to the olanzapine and fluoxetine group. Furthermore, greater improvement of symptoms is observed in the olanzapine/fluoxetine group when compared to the other groups (Corya et al., 2006). In another randomized controlled trial, greater improvement of symptoms is observed in the olanzapine/fluoxetine group compared to the fluoxetine-alone and olanzapine-alone groups. Additionally, a higher remission rate is observed in the olanzapine/fluoxetine group compared to the olanzapine group; however, it was not observed in comparison with the fluoxetine group. In the side effects documented, more cases of frequently increased appetite and weight gain were observed in the patients treated with olanzapine (both as monotherapy and in augmentation) (Shelton et al., 2001). Thus, olanzapine can be regarded as a relative tolerated option as augmentation therapy for treatment-resistant major depression. However, the possible risks of increased appetite, weight gain, and metabolic symptoms should be considered when treating treatment-resistant major depression (Chen et al., 2018). Fourth, an adjunctive use of risperidone was observed compared to placebo in five randomized controlled trials conducted for risperidone augmentation of treatment-resistant major depression. Moreover, higher remission rate has been observed for the adjunctive use group of risperidone compared to the placebo in a 4-week randomized controlled trial for 97 patents with treatment-resistant major depression (Keitner et al., 2009). Furthermore, in an open trial, the recurrence prevention period is prolonged in case of the risperidone augmentation group compared to the placebo group (Rapaport et al., 2006). Fifth, although no randomized controlled trial has been reported for ziprasidone augmentation, several open trials have reported that the treatment response of SSRI-resistant major depression could be enhanced by augmenting ziprasidone with SSRI compared to placebo (Papakostas et al., 2004). Further trials for the efficacy of ziprasidone augmentation for major depression have been done. Moreover, the efficacy of asenapine augmentation or paliperidone augmentation for major depression has been rarely studied.

Other psychotropic agents can be used as an augmentation agent for treatment-resistant major depression. First, lithium has been proposed as an effective augmentation agent for treatment-resistant major depression. It has been reported that, after the augmentation of lithium (at a dose of 900–1500 mg/day with a serum level of 0.5–1.3 meq/L), improved responses have been observed in 56% of the patients with treatment-resistant major depression (Price et al., 1986). Additionally, a study reported three recurrent major depressive episodes and the existence of a family history of major depression or bipolar disorder in a first-degree relative for consideration as predictors for higher response rate to lithium augmentation for treatment-resistant major depression. However, no significant correlation has been reported between efficacy of lithium augmentation and sex, age, suicidal attempts, psychotic symptoms, or atypical depressive symptoms (Sugawara et al., 2010). In level 3 of the STAR*D study, patients with inadequate response to the two antidepressant treatments were randomized and allocated into groups pertaining to the augmentation with lithium or T_3. Although the dose of lithium augmentation was 450–900 mg/day, it was 225–450 mg/day for the patients who were intolerant to the initial dose. No significant difference in remission rate between the lithium group (15.9%) and T_3 group (24.7%) has been observed. A greater proportion of the lithium group showed intolerance due to side effects compared with the T3 group. Thus, the efficacy of lithium treatment may be affected by the administration of an inadequate dose (Nierenberg et al., 2006). Despite the consistently reported efficacy of lithium augmentation, the risk of developing intoxication and the existence of a narrow therapeutic window should be considered for the patients treated with lithium (Chen et al., 2018). After the advent of dopamine serotonin blockers, augmentation with lithium was less commonly considered because of the availability of more tolerable agents for augmentation therapy for treatment-resistant major depression (Dunner, 2013). Second, thyroid hormones (i.e., T_3 and T_4) have been used as an augmenting agent for treatment-resistant major depression for over 50 years. Greater bioavailability in serum and greater affinity for thyroid receptors are presented by T_3 rather than T_4. In a T3 augmentation (25–50 μg/day) study involving patients who presented with inadequate response to a 6-week SSRI treatment regimen, the response rate at 2 weeks was 20%. All responsive patients were female, whereas none of the male patients were included in the responders (Cooper-Kazaz & Lerer, 2008). In level 3 of the STAR*D study, a dosage of 25–50 μg/day of T_3 was used as an augmenting agent in patients presenting with inadequate response to two antidepressant treatments. As compared with the augmentation group subjected to lithium treatment, a better tolerance rate was presented by the T_3 group (Nierenberg et al., 2006). Thus, T_3 can be proposed as a tolerated augmentation agent for treatment-resistant major depression, although the monitoring of thyroid function and the risk of developing T_3-related side effects (i.e., atrial flutter) should be considered. Third, medicines for bipolar disorder (mood stabilizers) including valproate, carbamazepine, lamotrigine, gabapentin, and topiramate may be used as an augmenting agent for treatment-resistant major depression. But less substantial evidence exists to support augmentation therapy with mood stabilizer. Moreover, in a randomized controlled trial, no significant difference is found in responses to the lamotrigine-fluoxetine group and placebo-fluoxetine group. Thus, medicines for bipolar disorder (mood stabilizers) may be regarded as an augmenting agent for treatment-resistant major depression. But less substantial evidence exists to suggest the use and the risk of developing sedative effects, and monitoring for serum level of medicines for bipolar disorder (mood stabilizers) should be considered.

Conclusion

Since serendipitous discoveries of iproniazid in 1957 in the USA and imipramine in 1958 in Switzerland (López-Muñoz et al., 2014), the progress for drugs for depression has come a long way in the past 65 years. Now the clinicians are equipped with various kinds of decent efficacious drugs for depression.

The priority in the past several decades was for the psychiatrists to look for new drugs for depression in treating patients with major depressive disorder. But the focus today is how to teach psychiatrists and patients with the correct knowledge how to diagnose depression and to use drugs for depression wisely and correctly.

The main mechanism of drugs for depression is to increase the level of BDNF of the brain to produce neurogenesis in the neurons (Cotman & Berchtold, 2002; Shen, 2016) through the use of drugs. Besides taking drugs for depression, exercise with adequate intensity can increase the expression of BDNF (Cotman & Berchtold, 2002; Shen, 2020). Please be sure to encourage the patients with depression to exercise, which does not produce intolerable side effect.

Acknowledgments The clinicians are advised to read the indications and correct dosages of the drug for depression in the package insert or a *Physicians' Reference Book* before prescribing the any medication.

Financial Support and Sponsorship The authors do not receive any support in writing this chapter.

Conflicts of Interest The authors declare no conflicts of interest in describing all the drugs in this chapter.

References

Abramowicz, M., Zuccotti, G., Pflomm, J. M., et al. (2019). Brexanolone (Zulresso) for postpartum depression. *JAMA, 322*, 73–74.

Ali, M., Aamir, A., Diwan, M. N., et al. (2021). Treating postpartum depression: What do you know about brexanolone? *Diseases, 9*, 52.

Allgulander, C., Hackett, D., & Salinas, E. (2001). Venlafaxine extended release (ER) in the treatment of generalised anxiety disorder: twenty-four-week placebo-controlled dose-ranging study. *Br J Psychiatry, 179*, 15–22.

Allgulander, C., Dahl, A. A., Austin, C., et al. (2004). Efficacy of sertraline in a 12-week trial for generalized anxiety disorder. *Am J Psychiatry, 161*, 1642–1649.

Alvarez, E., Perez, V., Dragheim, M., et al. (2012). A double-blind, randomized, placebo-controlled, active reference study of Lu AA21004 in patients with major depressive disorder. *International Journal of Neuropsychopharmacology, 15*, 589–600.

American Psychiatric Association. (2013). *Diagnostic and statistical manual of mental disorders* (5th ed., DSM-5). American Psychiatric Association.

Anderson, I. M. (2000). Selective serotonin reuptake inhibitors versus tricyclic antidepressants: A meta-analysis and efficacy and tolerability. *Journal of Affective Disorders, 58*, 19–36.

Andrews, P. W., Kornstein, S. G., Halberstadt, L. J., et al. (2011). Blue again: Pertubational effects of antidepressants suggest monoaminergic homeostasis in major depression. *Frontiers in Psychology, 2*, 1–24.

Anton, R. F. (1987). Urinary free cortisol in psychotic depression. *Biological Psychiatry, 22*, 24–34.

Anton, R. F., & Sexauer, J. D. (1983). Efficacy of amoxapine in psychotic depression. *The American Journal of Psychiatry, 140*, 1344–1347.

Aranda-Michel, J., Koehler, A., Bejarano, P. A., et al. (1999). Nefazodone-induced liver failure: Report of three cases. *Annals of Internal Medicine, 130*, 285–288.

Auclair, A. L., Martel, J. C., Assié, M. B., et al. (2013). Levomilnacipran (F2695), a norepinephrine-preferring SNRI: Profile in vitro and in models of depression and anxiety. *Neuropharmacology, 70*, 338–347.

Autry, A. E., Adachi, M., Nosyreva, E., Na, E. S., Los, M. F., Cheng, P. F., et al. (2011). NMDA receptor blockade at rest triggers rapid behavioral antidepressant response. *Nature, 475*, 91–95.

Babyak, M., Blumenthal, J. A., Herman, S., et al. (2000). Exercise treatment for major depression: Maintenance of therapeutic benefit at 10 months. *Psychosomatic Medicine, 62*, 633–638.

Baldomero, E. B., Ubago, J. G., Cercos, C. L., et al. (2005). Venlafaxine extended release versus conventional antidepressants in the remission of depressive disorder after previous antidepressant failure: ARGOS study. *Depression and Anxiety, 22*, 68–76.

Baldwin, D. S., Chrones, L., Florea, I., et al. (2016). The safety and tolerability of vortioxetine: Analysis of data from randomized placebo-controlled trials and open-label extension studies. *Journal of Psychopharmacology, 30*, 242–252.

Ballard, E. D., Ionescu, D. F., Vande Voort, J. L., et al. (2014). Improvement in suicidal ideation after ketamine infusion: Relationship reductions in depression and anxiety. *Journal of Psychiatric Research, 58*, 161–166.

Bang-Andersen, B., Ruhland, T., Jorgensen, M., et al. (2011). Discovery of 1-[2-(2,4-Dimethylsulfanyl)phenyl]piperazine (Lu AA21004): A novel multimodal compound for the treatment of major depressive disorder. *Journal of Medicinal Chemistry, 54*, 3206–3221.

Beakley, B. D., Kaye, A. M., & Kaye, A. D. (2015). Tramadol, pharmacology, side effects, and serotonin syndrome: A review. *Pain Physician, 18*, 395–400.

Bennazzi, F. (1998). Nefazodone withdrawal symptoms. *Canadian Journal of Psychiatry, 43*, 194–195.

Berlim, M. T., & Turecki, G. (2007a). What is the meaning treatment resistant/refractory major depression (TRD)? A systematic review of current randomized trials. *European Neuropsychopharmacology, 17*, 696–707.

Berlim, M. T., & Turecki, G. (2007b). Definition, assessment, and staging of treatment-resistant/refractory major depression: A review of current concepts and methods. *Canadian Journal of Psychiatry, 52*, 46–54.

Berman, R. M., Marcus, R. N., Swanink, R., et al. (2007). The efficacy and safety of aripiprazole as adjunctive therapy in major depressive disorder: A multicenter, randomized, double-blind, placebo-controlled study. *The Journal of Clinical Psychiatry, 68*, 843–853.

Berman, R. M., Fava, M., Thase, M. E., et al. (2009). Aripiprazole augmentation in major depressive disorder: A double-blind, placebo-controlled study in patients with inadequate response to antidepressants. *CNS Spectrums, 14*, 197–206.

Bigos, K. L., Pollock, B. G., Stankevich, B. A., et al. (2009). Sex differences in the pharmacokinetics and pharmacodynamics of antidepressants: An updated review. *Gender Medicine, 6*, 522–543.

Blier, P., & Szabo, S. T. (2005). Potential mechanisms of action of atypical antipsychotic medications in treatment-resistant depression and anxiety. *The Journal of Clinical Psychiatry, 66*(suppl), 30–40.

Block, T. S., Kushner, H., Kalin, N., et al. (2018). Combined analysis of mifepristone for psychotic depression: Plasma levels associated with clinical response. *Biological Psychiatry, 84*, 46–54.

Blumenthal, J. A., Babyak, M. A., Moore, K. A., et al. (1999). Effects of exercise training on older patients with major depression. *Archives of Internal Medicine, 159*, 2349–2356.

Blumenthal, J. A., Babyak, M. A., Doraiswamy, P. M., et al. (2007). Exercise and pharmacotherapy in the treatment of major depressive disorder. *Psychosomatic Medicine, 69*, 587–596.

Boyer, E. W., & Shannon, M. (2005). The serotonin syndrome. *New England Journal of Medicine, 352*, 1112–1120.

Brezun, J. M., & Daszuta, A. (2000). Serotonin may stimulate granule cell proliferation in the adult hippocampus, as observed in rats grafted with foetal raphe neurons. *The European Journal of Neuroscience, 12*, 391–399.

Canuso, C. M., Singh, J. B., Fedgchin, M., et al. (2018). Efficacy and safety of intranasal esketamine for the rapid reduction of symptoms of depression and suicidality in patients at imminent risk for suicide: Results of a double-blind, randomized, placebo-controlled study. *The American Journal of Psychiatry, 175*, 620–630.

Chang, L., Wei, Y., & Hashimoto, K. (2020). Antidepressant actions of ketamine and its two enantiomers. In K. Hashimoto, S. Ide, & K. Ikeda (Eds.), *Ketamine from abused drug to rapid-acting antidepressant* (pp. 105–126). Springer.

Chekroud, A. M., Gueorguieva, R., Krumholz, H., et al. (2017). Reevaluating the efficacy and predictability of antidepressant treatments: A symptom clustering approach. *JAMA Psychiatry, 74*, 370–378.

Chen, C. Y. A., Chiu, Y. H., & Shen, W. W. (2018). Drug augmentation for treatment-refractory major depressive disorder. *Taiwanese Journal of Psychiatry, 32*, 188–199.

Cheng, J. Y., Chen, R. Y., Ko, J. S., et al. (2007). Efficacy and safety of atomoxetine for attention-deficit/hyperactivity disorder in children and adolescents – Meta-analysis and meta-regression analysis. *Psychopharmacology, 194*, 197–209.

Choi, E., Zmarlicka, M., & Ehret, M. J. (2012). Vilazodone: a novel antidepressant. *American Journal of Health-System Pharmacy, 69*, 1551–1557.

Cipriani, A., Furukawa, T. A., Salanti, G., et al. (2018). Comparative efficacy and acceptability of 21 antidepressant drugs for the acute treatment of adults with major depressive disorder: A systematic review and network meta-analysis. *Lancet, 391*, 1357–1366.

Clayton, D. O., & Shen, W. W. (1998). Psychotropic drug-induced sexual function disorders: Diagnosis, incidence and management. *Drug Safety, 19*, 299–312.

Cooper-Kazaz, R., & Lerer, B. (2008). Efficacy and safety of triiodothyronine supplementation in patients with major depressive disorder treated with specific serotonin reuptake inhibitors. *The International Journal of Neuropsychopharmacology, 11*, 685–699.

Corya, S. A., Williamson, D., Sanger, T. M., et al. (2006). A randomized, double-blind comparison of olanzapine/fluoxetine combination, olanzapine, fluoxetine, and venlafaxine in the treatment of resistant depression. *Depression and Anxiety, 23*, 364–372.

Cotman, C. W., & Berchtold, N. C. (2002). Exercise: A behavioral intervention to enhance brain health and plasticity. *Trends in Neurosciences, 6*, 295–301.

Cristea, I. A., & Naudet, F. (2019). US Food and Drug Administration approval of esketamine and brexanolone. *Lancet, 6*, 975–977.

Cunningham, L. A., Borison, R. L., Carman, J. S., et al. (1994). A comparison of venlafaxine, trazodone, and placebo in major depression. *Journal of Clinical Psychopharmacology, 14*, 99–106.

Cusack, B., Nelson, A., & Richelson, E. (1994). Binding of antidepressants to human brain receptors; focus on newer generation compounds. *Psychopharmacology, 114*, 559–565.

Czeh, B., Michaelis, T., Watanabe, T., et al. (2001). Stress-induced changes in cerebral metabolites, hippocampal volume, and cell proliferation are prevented by antidepressant treatment with tianeptine. *Proceedings of the National Academy of Science of the USA, 98*, 12796–12801.

Dalfen, A. K., & Stewart, D. E. (2001). Who develops severe or fatal adverse drug reactions to selective serotonin reuptake inhibitors? *Canadian Journal of Psychiatry, 46*, 258–263.

Davidson, J. R., DuPont, R. L., Hedges, D., et al. (1999). Efficacy, safety, and tolerability of venlafaxine extended release and buspirone in outpatients with generalized anxiety disorder. *The Journal of Clinical Psychiatry, 60*, 528–535.

De Montigny, G., Silverston, P. J., Debonnel, G., et al. (1999). Venlafaxine for treatment resistant depression: A Canadian multi-center, open label trial. *Journal of Clinical Psychopharmacology, 19*, 400–406.

DeBattistia, C., Posener, J. A., Kalehzan, B. M., et al. (2000). Acute antidepressant effects of intravenous hydrocortisone and CRH in depressed patients: A double-blind, placebo-controlled study. *The American Journal of Psychiatry, 157*, 1334–1337.

DeBattistia, C., Dghramji, K., Menza, M. A., et al. (2003). Adjunctive modafinil for the short-term treatment of fatigue and sleepiness in patients with major depressive disorder: A preliminary double-blind, placebo-controlled study. *The Journal of Clinical Psychiatry, 64*, 1057–1064.

Dligiannidis, K. M., Meltzer-Brody, S., Gunduz-Bruce, H., et al. (2021). Effect of zuranolone vs placebo in postpartum depression: a randomized clinical trial. *JAMA Psychiatry*, e211559, in press.

Domino, E. F. (1999). History of modern psychopharmacology: A personal view with an emphasis on antidepressants. *Psychosomatic Medicine, 61*, 591–598.

Domino, E. F. (2010). Taming the ketamine tiger. *Anesthesiology, 113*, 678–684.

Doree, J. P., Des Rosier, J., Lew, V., et al. (2007). Quetiapine augmentation of treatment-resistant depression; a comparison of lithium. *Current Medical Research and Opinion, 23*, 333–341.

Drazinix, C. M., & Sazbo, S. T. (2017). Neurotransmitters and receptors in psychiatric disorder (chapter 2). In A. F. Schatzberg & C. B. Nemeroff (Eds.), *The American Psychiatric Association publishing textbook of psychopharmacology*. APA Publishing.

Drug Enforcement Administration. (2013). *Ketamine*. Department of Justice, Drug Enforcement Administration.

Dunkley, E. J., Isbister, G. K., Sibbritt, D., et al. (2003). The hunter serotonin toxicity criteria: Simple and accurate diagnostic decision rules for serotonin toxicity. *The Quarterly Journal of Medicine, 96*, 635–642.

Dunner, D. L. (2013). Treatment-resistant depression. *Taiwanese Journal of Psychiatry, 27*, 110–120.

Dunner, D. L., Rush, A. J., Russell, J. M., et al. (2006). Prospective, long-term, multicenter study of the naturalistic outcomes of patients with treatment-resistant depression. *The Journal of Clinical Psychiatry, 67*, 688–695.

Durgam, S., Gommoll, C., Migliore, R., et al. (2018). Relapse prevention in adults with major depressive disorder treated with vilazodone: A randomized, double-blind, placebo-controlled trial. *International Clinical Psychopharmacology, 33*, 304–311.

Egger, C., Muehlbacher, M., Nickel, M., et al. (2006). A review on hyponatremia associated with SSRIs, reboxetine and venlafaxine. *International Journal of Psychiatry in Clinical Practice, 10*, 17–26.

Einarson, T. R., Arikian, S. R., Casciano, J., et al. (1999). Comparison of extended-release venlafaxine, selective serotonin reuptake inhibitors, and tricyclic antidepressants in the treatment of depression: A meta-analysis of randomized controlled trials. *Clinical Therapeutics, 21*, 296–308.

Eliwa, H., Belzung, C., & Surget, A. (2017). Adult hippocampal neurogenesis: Is it the alpha and omega of antidepressant action? *Biochemical Pharmacology, 141*, 86–99.

El-Khalili, N., Joyce, M., Atkinson, S., et al. (2010). Extended-release quetiapine fumarate (quetiapine XR) as adjunctive therapy in major depressive disorder (MDD) in patients with an inadequate response to ongoing antidepressant treatment: A multicenter, randomized, double-blind, placebo-controlled study. *The International Journal of Neuropsychopharmacology, 13*, 917–932.

Esposito, M. S., Piatti, V. C., Laplagne, D. A., et al. (2005). Neuronal differentiation in the adult hippocampus recapitulates embryonic development. *The Journal of Neuroscience, 25*, 10074–10086.

European College of Neuropsychopharmacology (ECNP). (2017). *Neuroscience-based nomenclature, second edition-revised (NbN-2R)*. European College of Neuropsychopharmacology.

Evans, W. E., Relling, M. V., Rahman, A., et al. (1993). Genetic basis for a lower prevalence of deficient 2D6 oxidative drug metabolism phenotypes in black Americans. *Journal of Clinical Investigation, 91*, 2150–2154.

Eyding, D., Lelgmann, M., Grouven, U., et al. (2010). Reboxetine for acute treatment of major depression; systematic review and meta-analysis of published and unpublished placebo and selective serotonin reuptake inhibitor controlled trials. *BMJ, 341*, c4737.

Fangmann, P., Assion, H. J., Juckel, G., et al. (2008). Half a century of antidepressant drugs: On the clinical introduction of monoamine oxidase inhibitors, tricyclics, and tetracyclics. Part II: tricyclics and tetracyclics. *Journal of Clinical Psychopharmacology, 7*, 106–113.

Farmand, S., Lindh, J. D., Calissendorff, J., et al. (2018). Differences in associations of antidepressants and hospitalization due to hyponatremia. *The American Journal of Medicine, 31*, 56–63.

Fava, G. A., & Offidani, E. (2011). The mechanism of tolerance in antidepressant action. *Progress in Neuro-Psychopharmacology & Biological Psychiatry, 35*, 1593–1602.

Fava, M., Rush, A. J., Thase, M. E., et al. (2005). 15 years of clinical experience of bupropion HCI: From bupropion to bupropion SR to bupropion XL. *Primary Care Companion to the Journal of Clinical Psychiatry, 7*, 106–113.

Fava, M., Rush, A. J., Wisniewski, S. R., et al. (2006). A comparison of mirtazapine and nortriptyline following two consecutive failed medication treatment for depressed outpatients: A STAR*D report. *The American Journal of Psychiatry, 163*, 1161–1172.

Fox, H. H., & Gibas, J. T. (1953). Synthetic tuberculostats. VII monoalkyl derivatives of isonicotylhydrazine. *Journal of Organic Chemistry, 18*, 994–1002.

Frazer, A., & Blier, P. (2016). A neuroscience-based nomenclature (NbN) for psychotropic agents. *The International Journal of Neuropsychopharmacology, 19*, 1–2.

Frendin, T. J., & Swainson, C. P. (1985). Acute renal failure secondary to non-traumatic rhabdomyolysis following amoxapine overdose. *The New Zealand Medical Journal, 98*, 690–691.

Frieder, A., Fersh, M., Hainline, R., et al. (2019). Pharmacology of postpartum depression: Current approaches and novel drug development. *CNS Drugs, 33*, 265–282.

Gargoloff, P. D., Corral, R., Herbst, R., et al. (2016). Effectiveness of agomelatine on anhedonia in depressed patients: An outpatient, open-label, real-world study. *Human Psychopharmacology, 31*, 412–418.

Garnock-Jone, K. P., & Keating, G. M. (2009). Atomoxetine: A review of its use in attention-deficit hyperactivity disorder in children and adolescents. *Pediatric Drugs, 11*, 203–226.

Gau, S. S. F., Huang, Y. S., & Soong, W. T. (2007). A randomized, double-blind, placebo-controlled clinical trial on once-daily atomoxetine in Taiwanese children and adolescents with attention-deficit/hyperactivity disorder. *Journal of Child and Adolescent Psychopharmacology, 17*, 447–460.

Gibson, A. P., Bettinger, T. L., Patel, N. C., et al. (2006). Atomoxetine versus stimulants for treatment of attention deficit/ hyperactivity disorder. *Annals of Pharmacotherapy, 40*, 1134–1142.

Glassman, A. H., Perel, J. M., Shostak, M., et al. (1977). Clinical implications of imipramine plasma levels for depressive illness. *Archives of General Psychiatry, 34*, 197–204.

Godfrey, R. G. (1996). A guide to the understanding and use of tricyclic antidepressants in the overall management of fibromyalgia and other chronic pain syndromes. *Archives of Internal Medicine, 156*, 1047–1052.

Goh, K. K., Tam, W. K., & Shen, W. W. (2012). Trazodone is not as effective as other antidepressants in treating a patient with major depressive disorder. *Taiwanese Journal of Psychiatry, 26*, 311–315.

Grunebaum, M. F., Ellis, S. P., Keilp, J. G., et al. (2017). Ketamine versus midazolam in bipolar depression with suicidal thoughts: A pilot midazolam-controlled randomized clinical trial. *Bipolar Disorders, 19*, 176–183.

Gunduz-Bruce, H., Silber, C., Kaul, I., et al. (2019). Trial of SAGE-217 in patients with major depressive disorder. *New England Journal of Medicine, 381*, 903–911.

Hannan, N., Hmazah, Z., Akinpeloye, H. O., et al. (2007). Venlafaxine-mirtazapine combination in the treatment of persistent depressive illness. *Journal of Psychopharmacology, 21*, 161–163.

Harvey, S. B., Overland, S., Hatch, S. L., et al. (2018). Exercise and the prevention of depression: Results of the HUNT cohort study. *Americal Journal of Psychiaty, 175*, 28–36.

Hashimoto, K., Ide, S., & Ikeda, K. (2020). *Ketamine from abused drug to rapid-acting antidepressant*. Springer.

Hayes, R. E., & Kristoff, C. A. (1986). Adverse reactions to five new antidepressants. *Clinical Pharmacy, 5*, 471–480.

Healy, D., & McMonagle, T. (1997). The enhancement of social functioning as a therapeutic principle in the management of depression. *Journal of Psychopharmacology, 11*(supp 4), S25–S31.

Herman, S., Blumental, J. A., Babyak, M., et al. (2002). Exercise therapy for depression in middle-aged and older adults: Predictors of early dropout and treatment failure. *Health Psychology, 21*, 553–563.

Hillhouse, T. M., & Porter, J. H. (2015). A brief history of the development of antidepressant drugs: From monoamine to glutamate. *Experimental and Clinical Psychopharmacology, 23*, 1–21.

Hirschfeld, R. M. (1999). Management of sexual side effects of antidepressant therapy. *The Journal of Clinical Psychiatry, 60*(suppl 14), 27–30.

Hsu, J. H., & Shen, W. W. (1995). Male sexual side effects associated with antidepressants: A descriptive clinical study of 32 patients. *International Journal of Psychiatry in Medicine, 25*, 191–201.

Huckleberry, K. A., Shue, F., Copeland, T., et al. (2018). Dorsal and ventral hippocampal adult-born neurons contribute to context fear memory. *Neuropsychopharmacology, 43*, 2487–2496.

Insel, T. R. (2012). Next-generation treatments for mental disorders. *Science Translational Medicine, 4*, 155psc19.

Ionescu, D. F., See, M. B., & Pavone, K. J. (2016). Rapid and sustained reductions in current suicidal ideation following repeated doses of intervenous ketamine: Secondary analysis of an open-label study. *The Journal of Clinical Psychiatry, 77*, e719–e725.

Jacobsen, E. M. (1992). Fluoxetine-induced sexual dysfunction and an open trial of yohimbine. *The Journal of Clinical Psychiatry, 53*, 119–122.

Jakubovski, E., Varigonda, A. L., Freemantle, N., et al. (2016). Systematic review and meta-analysis: Dose-response relationship of selective serotonin reuptake inhibitors in major depressive disorder. *The American Journal of Psychiatry, 173*, 174–183.

Janowsky, D. S., & Byerley, B. (1984). Desipramine: an overview. *Journal of Clinical Psychiatry, 45*, 3–9.

Jeon, H. J., Lee, J. Y., Lee, Y. M., et al. (2010a). Lifetime prevalence and correlates of suicidal ideation, plan, and single and multiple attempts in a Korean nationwide study. *The Journal of Nervous and Mental Disease, 198*, 643–646.

Jeon, H. J., Lee, J. Y., Lee, Y. M., et al. (2010b). Unplanned versus planned suicide attempters, precipitants, methods, and an association with mental disorders in a Korea-based community sample. *Journal of Affective Disorders, 127*, 274–280.

Jessberger, S., & Kempermann, G. (2003). Adult-born hippocampal neurons mature into activity-dependent responsiveness. *The European Journal of Neuroscience, 18*, 2707–2712.

Joffe, R. T., Levitt, A. J., Sokolov, S. T., et al. (1996). Response to an open trial of a second SSRI in major depression. *The Journal of Clinical Psychiatry, 57*, 114–115.

Jue, S. G., Dawson, G. W., & Brogden, R. N. (1982). Amoxapine: A review of its pharmacology and efficacy in depressed states. *Drugs, 24*, 1–23.

Kanes, S., Colquhoun, H., Gunduz-Bruce, H., et al. (2017). Brexanolone (SAGE-547 injection) in post-partum depression: A randomized controlled trial. *Lancet, 390*, 480–489.

Katona, C., Hansen, T., & Olsen, C. K. (2012). A randomized, double-blind, placebo-controlled, duloxetine-referenced, fixed dose study comparing the efficacy and safety of Lu AA21004 in elderly patients with major depressive disorder. *International Clinical Psychopharmacology, 27*, 215–223.

Kavalalil, A. E., & Monteggia, L. M. (2012). Synaptic mechanism underlying rapid antidepressant action of ketamine. *The American Journal of Psychiatry, 169*, 1150–1156.

Keitner, G. I., Garlow, S. J., Ryan, C. E., et al. (2009). A randomized, placebo-controlled trial of risperidone augmentation for patients difficult-treat unipolar, non-psychotic major depression. *Journal of Psychiatric Research, 43*, 203–214.

Keller, M. B., Klerman, G. L., Lavori, P. W., et al. (1984). Long-term outcome of episodes of major depression: Clinical and public health significance. *JAMA, 252*, 788–792.

Kerihuel, J. C., & Dreyfus, J. F. (1991). Meta-analysis of the efficacy and tolerability of the tricyclic antidepressant lofepramine. *The Journal of International Medical Research, 19*, 183–201.

Khan, A., Cutler, A. J., Kajdasz, D. K., et al. (2011). A randomized, double-blind, placebo-controlled, 8-week study of vilazodone, a serotonergic agent for the treatment of major depressive disorder. *The Journal of Clinical Psychiatry, 72*, 441–447.

Khin, N. A., Chen, Y. F., Yang, Y., et al. (2011). Exploratory analyses of efficacy data from major depressive disorder trials submitted to the US Food and Drug Administration in support of new drug applications. *The Journal of Clinical Psychiatry, 72*, 464–472.

Kiayias, J. A., Vlachou, E. D., & Lakka-Papadodima, E. (2000). Venlafaxine HCl in the treatment of painful peripheral diabetic neuropathy. *Diabetes Care, 23*, 699.

Kiecott-Glaser, J. K., Derry, H. M., & Fagundes, C. P. (2015). Inflammation: Depression fans the flames and feasts on the heat. *The American Journal of Psychiatry, 172*, 1075–1091.

Kim, I. B., & Park, S. C. (2021). Neural circuitry-neurogenesis coupling model of depression. *International Journal of Molecular Sciences, 22*, 2468.

Kornstein, S. G., Wohlreich, M. M., & Mallinckrodt, et al. (2006). Duloxetine efficacy for major depressive disorder in male vs. female patients: Data from 7 randomized, double-blind, placebo-controlled trials. *The Journal of Clinical Psychiatry, 67*, 761–770.

Kornstein, S. G., Schatzberg, A. F., Thase, M. E., et al. (2009). Gender differences in treatment response to sertraline versus imipramine in chronic depression. *The American Journal of Psychiatry, 157*, 1445–1452.

Laudon, M., & Fryman-Marom, A. (2014). A therapeutic effect of melatonin receptor agonists on sleep and comorbid disorders. *International Journal of Molecular Sciences, 15*, 15924–15950.

Leader, L., O'Connell, M., & Vandenberg, A. (2019). Brexanolone for postpartum depression: Clinical evidence and practical considerations. *Pharmacotherapy, 39*, 1105–1112.

Lee, M. S., Kang, R. H., & Hahn, S. W. (2008). Pharmacogenetics of ethnic populations. In C. H. Ng, K. M. Lin, B. S. Singh, et al. (Eds.), *Ethno-psychopharmacology* (pp. 62–85). Cambridge University Press.

Lee, M. S., Ahn, Y. M., Chung, S. H., et al. (2012). The effect of initial duloxetine dosing strategy on nausea in Korean patients with major depressive disorder. *Psychiatry Investigation, 9*, 391–399.

Leiser, S. C., Pehrson, A. L., Robichaud, P. J., et al. (2014). Multimodal antidepressant vortioxetine increases frontal cortical oscillations unlike escitalopram and duloxetine: a quantitative EEG study in rats. *British Journal of Pharmacology, 171*, 666–675.

Lesser, I. M., Castro, D. B., Gaynes, B. N., et al. (2007). Ethnicity/race and outcome in the treatment of depression: Results from STAR*D. *Medical Care, 45*, 1043–1051.

Lewis-Fernandez, R., Blanco, C., Mallinckrodt, C. H., et al. (2006). Duloxetine in the treatment of major depressive disorder: Comparisons of safety and efficacy in US Hispanic and majority Caucasian patients. *The Journal of Clinical Psychiatry, 67,* 1379–1390.

López-Muñoz, F., Alamo, C., Juckel, G., et al. (2007). Half a century of antidepressant drugs: On the clinical introduction of monoamine oxidase inhibitor, tricyclics, and tetracyclics. Part I: Monoamine oxidase inhibitors. *Journal of Clinical Psychopharmacology, 27,* 555–559.

López-Muñoz, F., Shen, W. W., Alamo, C., et al. (2014). Serendipitous discovery of first two antidepressants. *Taiwanese Journal of Psychiatry, 28,* 67–70.

López-Muñoz, F., Shen, W. W., D'Ocon, P., et al. (2018). A history of the pharmacological treatment of bipolar disorder. *International Journal of Molecular Sciences, 19,* 2143–2181.

Lotufo-Neto, F., Trivedi, M., Thase, M., et al. (2009). Meta-analysis of the reversible inhibitors of monoamine oxidase type a moclobemide and brofaromine for the treatment of depression. *Neuropsychopharmacology, 20,* 226–227.

Malberg, J. E., & Blendy, J. A. (2005). Antidepressant action: To the nucleus and beyond. *Trends in Pharmacological Sciences, 26,* 631–638.

Marcus, R. N., McQuade, R. D., Carson, W. H., et al. (2008). The efficacy and safety of aripiprazole as adjunctive therapy in major depressive disorder: A second multicenter, randomized, double-blind, placebo-controlled study. *Journal of Clinical Psychopharmacology, 28,* 156–165.

Massana, J., Moller, H. J., Burrows, G. D., et al. (1999). Reboxetine: A double-blind comparison with fluoxetine in major depressive disorder. *International Clinical Psychopharmacology, 14,* 73–80.

Mather, A. S., Rodriguez, C., Guthrie, M. F., et al. (2002). Effects of exercise on depressive symptoms in older adults with poorly responsive depressive disorder: Randomized controlled trial. *The British Journal of Psychiatry, 180,* 411–415.

McAinsh, J., & Cruickshank, J. M. (1990). Beta-blockers and central nervous side effects. *Pharmacology & Therapeutics, 46,* 1630197.

McAndrew, A., Lawn, W., Stevens, T., et al. (2017). A proof-of-concept investigation into ketamine as a pharmacological treatment for alcohol dependence: Study protocol for a randomized controlled trial. *Trials, 18,* 159.

McCann, U. D., & Agras, W. S. (1990). Successful treatment of nonpurging bulimia nervosa with desipramine: A double-blind, placebo-controlled study. *The American Journal of Psychiatry, 147,* 1509–1513.

McCormack, P. L. (2015). Vilazodone: A review in major depressive disorder. *Drugs, 75,* 1915–1923.

McGrath, P. J., Stewart, J. W., Fava, M., et al. (2006). Tranylcypromine versus venlafaxine plus mirtazapine following three failed antidepressant medication trials for depression: A STAR*D report. *The American Journal of Psychiatry, 163,* 1531–1541.

McIntyre, A., Gendron, A., & Amanda McIntyre, A. (2006). Quetiapine adjunct to selective serotonin reuptake inhibitors or venlafaxine in patients with major depression, comorbid anxiety, and reuptake inhibitors or venlafaxine in patients with major depression, comorbid anxiety, and residual depressive symptoms: a randomized, placebo-controlled pilot study. *Depression and Anxiety, 24,* 487–494.

McMahon, F. J., Buervenich, S., Charney, D., et al. (2006). Tranylcypromine versus venlafaxine plus mirtazapine following three failed antidepressant medication trials for depression: A STAR*D report. *The American Journal of Psychiatry, 157,* 344–350.

Meltzer-Brody, S., Colquhoun, H., Riesenberg, R., et al. (2018). Brexanolone injection in post-partum depression: Two multicentre, double-blind, randomised, placebo-controlled, phase 3 trials. *Lancet, 392,* 1058–1070.

Meshi, D., Drew, M. R., Saxe, M., et al. (2006). Hippocampal neurogenesis is not required for behavioral effects of environmental enrichment. *Nature Neuroscience, 9,* 729–731.

Mishra, A., Sarangi, S. C., Maiti, R., et al. (2022). Efficacy and safety of adjunctive serotonin-dopamine activity modulators in major depression: A meta-analysis of randomized controlled trials. *Journal of Clinical Pharmacology, 62,* 721–732.

Mojtabai, R. (2017). Nonremission and time to remission among remitters in major depressive disorder: Revisiting STAR*D. *Depression and Anxiety, 34,* 1123–1133.

Molero, P., Rmos-Quiroga, J. A., Martin-Santos, R., et al. (2018). Antidepressant efficacy and tolerability of ketamine and esketamine: A critical review. *CNS Drugs, 32,* 411–420.

Moller, H.-J., Schmitt, A., & Falkai, P. (2016). Neuroscience-based nomenclature (NbN) to replace traditional terminology of psychotropic medications. *European Archives of Psychiatry and Clinical Neuroscience, 266,* 385–386.

Montgomery, S. A., Baldwin, D. S., Blier, P., et al. (2007). Which antidepressant have demonstrated superior efficacy? A review of the evidence. *International Clinical Psychopharmacology, 22,* 323–329.

Montgomery, S. A., Mansuy, L., Ruth, A., et al. (2013). Efficacy and safety of levomilnacipran sustained release in moderate to severe major depressive disorder: A randomized, double-blind, placebo-controlled, proof-of-concept study. *The Journal of Clinical Psychiatry, 74,* 363–369.

Mork, A., Pehrson, A., Brennum, L. T., et al. (2012). Pharmacological effects of Lu AA21004: A novel multimodal compound for the treatment of major depressive disorder. *The Journal of Pharmacology and Experimental Therapeutics, 340,* 666–675.

Muller, J. C., Pryor, W. W., Gibbons, J. E., et al. (1955). Depression and anxiety occurring Rauwolfia therapy. *JAMA, 69,* 392–399.

Muscholl, E. (1995). The evolution of experimental pharmacology as a biological science: The pioneering work of Buchheim and Schmiederg. *British Journal of Pharmacology, 116*, 2155–2159.

Nelson, J. C., & Davis, J. J. (1997). DST studies in psychotic depression: A meta-analysis. *The American Journal of Psychiatry, 154*, 1497–1503.

Nelson, J. C., & Papakotas, G. I. (2009). Atypical antipsychotic augmentation in major depressive disorder: A meta-analysis of placebo-controlled randomized trials. *The American Journal of Psychiatry, 166*, 980–991.

Nelson, J. C., Pikalov, A., & Berman, R. M. (2008). Augmentation treatment in major depressive disorder: Focus on aripiprazole. *Neuropsychiatric Disease and Treatment, 4*, 937–948.

Newton, S. S., Collier, E. F., Hunsberger, J., et al. (2003). Gene profile of electroconvulsive seizures: Induction of neurotrophic and angiogenic factors. *The Journal of Neuroscience, 23*, 10841–10851.

Nibuya, M., Morlnobu, S., & Duman, R. S. (1995). Regulation of BDNF and trkB mRNA in rat brain by chronic electroconvulsive seizure and antidepressant drug treatments. *The Journal of Neuroscience, 75*, 7539–7547.

Nierenberg, A. A., Keefe, B. R., Leslie, V. E., et al. (1999). Residual symptoms in depressed patients who respond acutely to fluoxetine. *The Journal of Clinical Psychiatry, 2*, 1086–1109.

Nierenberg, A. A., Fava, M., Trivedi, M. H., et al. (2006). A comparison of lithium and T3 augmentation following two failed medication treatments for depression: A STAR*D report. *The American Journal of Psychiatry, 163*, 1519–1530.

Nnadi, C. U., Goldberg, J. F., & Malhotra, A. K. (2005). Pharmacogenetics in mood disorder. *Current Opinion in Psychiatry, 18*, 33–39.

Norman, T. R., & Olver, J. S. (2019). Agomelatine for depression: Expanding the horizons? *Expert Opinion on Pharmacotherapy, 20*, 647–656.

Norton, P. A., Zinner, N. R., Yalcin, I., et al. (2002). Duloxetine versus placebo in the treatment of stress urinary incontinence. *American Journal of Obstetrics and Gynecology, 187*, 40–48.

Nurnberg, H. G. (2001). Managing treatment-emergent sexual dysfunction associated with serotonergic antidepressants: Before and after sildenafil. *Journal of Psychiatric Practice, 7*, 92–108.

Nutt, D. J. (2009). Beyond psychoanaleptics: Can improve antidepressant drug nomenclature? *Journal of Psychopharmacology, 23*, 343–345.

O'Leary, O. F., & Cryan, J. F. (2014). A ventral view on antidepressant action: Roles for adult hippocampal neurogenesis along the dorsoventral axis. *Trends in Pharmacological Sciences, 35*, 675–687.

Otani, K., Tanaka, O., Kaneko, S., et al. (1994). Mechanisms of the development of trazodone withdrawal symptoms. *International Clinical Psychopharmacology, 13*(Suppl), S84–S89.

Owens, M. J., Morgan, W. N., Plott, S. J., et al. (1997). Neurotransmitter receptor and transporter binding profile of antidepressants and their metabolites. *The Journal of Pharmacology and Experimental Therapeutics, 283*, 1305–1322.

Papakostas, G. I. (2008). Tolerability of modern antidepressants. *Journal of Clinical Psychiatry, 69*(suppl E1), 8–13.

Papakostas, G. I., Peterson, T. J., Nierenberg, A. A., et al. (2004). Ziprasidone augmentation of selective serotonin reuptake inhibitors (SSRIs) for SSRI-resistant major depressive disorder. *The Journal of Clinical Psychiatry, 65*, 217–221.

Papakostas, G. I., Kornstein, S. G., Clayton, A. H., et al. (2007). Relative antidepressant efficacy of bupropion and the selective serotonin reuptake inhibitors in major depressive disorder: Gender-age interactions. *International Clinical Psychopharmacology, 22*, 226–229.

Papakostas, G. I., Fava, M., & Thase, M. E. (2008). Treatment of SSRI-resistant depression: A meta-analysis comparing within-versus across-class switches. *Biological Psychiatry, 63*, 699–704.

Park, S. C. (2018). Antidepressive effects of exercise. *Journal of Korean Neuropsychiatric Association, 57*, 139–144.

Park, S. C. (2019). Neurogenesis and antidepressant actions. *Cell and Tissue Research, 377*, 95–106.

Park, S. C. (2020). Symptom-based selection of antidepressants. *Journal of the Korean Medical Association, 63*, 216–226.

Park, S. C., Oh, H. S., Oh, D. H., et al. (2014). Evidence-based non-pharmacological treatment guideline for depression in Korea. *Journal of Korean Medical Science, 29*, 12–22.

Park, S. C., Lee, M. S., Shinfuku, N., et al. (2015a). Gender differences in depressive symptom profiles and patterns of psychotropic drug usage in Asian patients with depression: Findings from the Research on Asian Psychotropic Prescription Patterns for Antidepressants (REAP-AD) study. *Aust N Z J Psychiatry, 49*, 833–841.

Park, S. C., Shinfuku, N., Maramis, M. M., et al. (2015b). Adjunctive antipsychotic prescriptions for outpatients with depressive disorders in Asia: The Research on Asian Psychotropic Prescription Patterns for Antidepressants (REAP-AD) study. *The American Journal of Psychiatry, 172*, 684–685.

Perahia, D. G., Quail, D., Desaiah, D., et al. (2008). Switching to duloxetine from selective serotonin reuptake inhibitor antidepressants: A multicenter clinical trial comparing 2 switching techniques. *The Journal of Clinical Psychiatry, 69*, 95–105.

Perera, T. D., Coplan, J. D., Lisanby, S. H., et al. (2007). Antidepressant-induced neurogenesis in the hippocampus of adult nonhuman primates. *The Journal of Neuroscience, 27*, 4894–4901.

Perera, T. D., Park, S., & Nemirovskaya, Y. (2008). Cognitive role of neurogenesis in depression and antidepressant treatment. *The Neuroscientist, 124*, 326–338.

Peselow, E. D., Filippi, A., Goodnick, P., et al. (1989). The short-and long-term efficacy of paroxetine HC. *Psychopharmacology Bulletin, 25*, 233–239.

Picerking, G., Macian, N., Delage, N., et al. (2018). Milnacipran poorly modulates pain in patients suffering from fibromyalgia: A randomized double-blind controlled study. *Drug Design, Development and Therapy, 12*, 2485–2496.

Pletscher, A. (1991). The discovery of antidepressants: A winding path. *Experientia, 47*, 4–8.

Posadzki, P. P., Bajpai, R., Kyaw, B. M., et al. (2018). Melatonin and health: An umbrella review of health outcomes and biological mechanisms of action. *BMC Medicine, 16*, 18.

Poserner, J. A., DeBattista, C., Williams, G. H., et al. (2000). 24-hours monitoring of cortisol and corticotropin secretion in psychotic and nonpsychotic major depression. *Archives of General Psychiatry, 57*, 755–760.

Price, L. H., Charney, D. S., & Heninger, G. R. (1986). Variability of response to lithium augmentation in refractory depression. *The American Journal of Psychiatry, 143*, 1387–1392.

Price, R. B., Nock, M. K., Charney, D. S., et al. (2009). Effects of intravenous ketamine on explicit and implicit measures of suicidality in treatment-resistant depression. *Biological Psychiatry, 66*, 522–526.

Quera-Salva, M.-A., Lemoine, P., & Guilleminault, C. (2010). Impact of the novel antidepressant agomelatine on disturbed sleep-wake cycle in depressed patients. *Human Psychopharmacology, 25*, 222–229.

Rantamäki, T. (2017). BDNF receptor TrkB as a target of antidepressant. In Y.-K. Kim (Ed.), *Major depressive disorder: Risk factors, characteristics and treatment options* (pp. 277–294). Nova Science Publisher.

Rapaport, M. H., Gharabawi, G. M., Canuso, C. M., et al. (2006). Effects of risperidone augmentation in patients with treatment-resistant depression: Results of open-label treatment followed by double-blind continuation. *Neuropsychopharmacology, 31*, 2505–2513.

Revet, A., Montastruc, F., Roussin, A., et al. (2020). Antidepressants and movement disorders: A postmarketing study in the world pharmacovigilance database. *BMC Psychiatry, 20*, 308.

Ribero, M. G., Pereira, E. L. A., Santos-Jesus, R., et al. (2000). Nortriptyline blood levels and clinical outcome: Meta-analysis of published studies. *Brazilian Journal of Psychiatry, 22*, 51–56.

Richelson, E. (2002). The clinical relevance of antidepressant interaction with neurotransmitter transmitter transporters and receptors. *Psychopharmacology Bulletin, 36*, 133–150.

Roden, D. M., Altman, R. B., Benewitz, N. L., et al. (2006). Pharmacogenomics: Challenges and opportunities. *Annals of Internal Medicine, 145*, 749–757.

Rosso, G., Rigardetto, S., Borrgetto, F., et al. (2012). A randomized, single-blind, comparison of duloxetine with bupropion in the treatment of SSRI-resistant major depression. *Journal of Affective Disorders, 136*, 172–176.

Roth, T., Rogowski, R., Hull, S., et al. (2007). Efficacy and safety of doxepin 1 mg, and 6 mg in adults with primary insomnia. *Sleep, 30*, 1555–1561.

Ruhe, H. G., van Rooijen, G., Spiker, J., et al. (2012). Staging methods for treatment resistant depression: A systematic review. *Journal of Affective Disorders, 137*, 35–45.

Rush, A. J., Trivedi, M. H., Ibrahim, H. M., et al. (2003). The 16-Item Quick Inventory of Depressive Symptomatology (QIDS), Clinician Rating (QIDS-C), and Self-Report (QIDS-SR): A psychometric evaluation in patients with chronic major depression. *Biological Psychiatry, 54*, 573–583.

Rush, A. J., Trivedi, M. H., Wisniewski, S. R., et al. (2006). Acute and longer-term outcomes in depressed outpatients requiring one or several treatment steps: A STAR*D report. *The American Journal of Psychiatry, 3*, 1905–1917.

Rush, A. J., Wisniewski, S. R., Warden, D., et al. (2008). Selecting among second-step antidepressant medication monotherapies: Predictive value of clinical, demographic, or first-step treatment features. *Archives of General Psychiatry, 65*, 870–880.

Rush, A. J., Trivedi, M. H., Stewart, J. W., et al. (2011). Combining medications to enhance depression outcomes (CO-MED): Acute and long-term outcomes of a single-blind randomized study. *The American Journal of Psychiatry, 168*, 689–701.

Sairanen, M., Lucas, G., Ernfos, P., et al. (2005). Brain-derived neurotrophic factor and antidepressant drugs have different but coordinated effects on neuronal turnover, proliferation, and survival in the adult dentate gyrus. *The Journal of Neuroscience, 25*, 1089–1094.

Sanderson, N. B., Armstrong, S. C., & Cozza, K. L. (2005). Med-psych drug-drug interaction update: An overview of psychotropic drug interactions. *Psychosomatics, 46*, 464–494.

Santarelli, L., Saxe, M., Gross, C., et al. (2003). Requirement of hippocampal neurogenesis for the behavioral effects of antidepressants. *Science, 301*, 805–809.

Schatzberg, A., & Rothchild, A. (1992). Psychotic (delusional) major depression: Should it be included as distinct syndrome in DSM-IV. *The American Journal of Psychiatry, 149*, 733–745.

Schatzberg, A., Rothchild, A., Stahl, J. B., et al. (1983). The dexamethasone suppression test: Identification of subtypes of depression. *The American Journal of Psychiatry, 154*, 1497–1503.

Schildkraut, J. J. (1965). The catecholamine hypothesis of affective disorders: A review of supporting evidence. *The American Journal of Psychiatry, 122*, 509–522.

Schmidt, H. D., & Duman, R. S. (2007). The role of neurotrophic factors in adult hippocampal neurogenesis, antidepressant treatments and animal models of depressive-like behavior. *Behavioural Pharmacology, 18*, 391–418.

Schweizer, E., Rickles, K., Amsterdam, J. D., et al. (1990). What constitute an adequate antidepressant trial for fluoxetine? *The Journal of Clinical Psychiatry, 51,* 8–11.

Scotton, W. J., Hill, L. J., Williams, A. C., et al. (2019). Serotonin syndrome: Pathophysiology, clinical features, management, and potential future directions. *International Journal of Tryptophan Research, 12,* 1178646919873925.

Sepede, G., Corbo, M., Fiori, F., et al. (2012). Reboxetine in clinical practice: a review. *Clinical Therapeutics, 163,* e255–e262.

Shaw, D. M., Eccleston, E. G., & Camps, F. E. (1967). 5-Hydroxytryptamine in the hind-brain of depressive suicides. *The British Journal of Psychiatry, 113,* 1407–1411.

Shelton, R. C., Tollefson, G. D., Tohen, M., et al. (2001). A novel augmentation strategy for treating resistant major depressive disorder. *The American Journal of Psychiatry, 158,* 131–134.

Shelton, R. C., Entsuah, R., Padmanabhan, S. K., et al. (2005a). Venlafaxine XR demonstrates higher rates of sustained remission compared to fluoxetine, paroxetine, or placebo. *International Clinical Psychopharmacology, 20,* 233–238.

Shelton, R. C., Williamson, D. J., Corya, S. A., et al. (2005b). Olanzapine/fluoxetine combination for treatment. Resistant depression: A controlled study of SSRI and nortriptyline resistance. *The Journal of Clinical Psychiatry, 66,* 1289–1297.

Shen, W. W. (1997). The metabolism of psychoactive drugs: A review of enzymatic biotransformation and inhibition. *Biological Psychiatry, 41,* 814–826.

Shen, W. W. (1999). A history of antipsychotic drug development. *Comprehensive Psychiatry, 40,* 407–414.

Shen, W. W. (2011a). Treating chronic insomnia: A wake-up call shattering old dogmas? *Taiwanese Journal of Psychiatry, 25,* 59–62.

Shen, W. W. (2011b). *Clinical psychopharmacology for the 21st century* (3rd ed.). Ho-Chi Publishing Company.

Shen, W. W. (2016). Antidepressant therapy. *Aino Journal (Osaka), 15,* 1–13.

Shen, W. W. (2018). Anticraving therapy for alcohol use disorder: A clinical review. *Neuropsychopharamcology Reports, 38,* 105–116.

Shen, W. W. (2020). Rehabilitative and habilitative perspectives of exercise in treating major depressive disorder. *Cognition & Rehabilitation (Osaka), 1,* 102–111.

Shen, W. W., & Hsu, J. H. (1995). Female sexual side effects associated with selective serotonin reuptake inhibitors: A descriptive clinical study of 33 patients. *International Journal of Psychiatry in Medicine, 25,* 239–248.

Shen, W. W., Urosevich, Z., & Clayton, D. O. (1999). Sildenafil in the treatment of female sexual dysfunction induced by selective serotonin reuptake inhibitors. *The Journal of Reproductive Medicine, 44,* 535–542.

Shiloh, R., Styjer, R., Weizman, A., et al. (2006). *Atlas of psychiatric pharmacotherapy.* Taylor & Francis.

Silva, J., Mota, J., & Azevedo, P. (2016). California rocket fuel: And what about being a first line treatment? *European Psychiatry, 33*(S1), S551.

Simon, G. (2018). Should psychiatrist write the exercise prescription for depression? *The American Journal of Psychiatry, 175,* 2–3.

Singh, H., & Becker, P. M. (2007). Novel therapeutic usage of low-dose doxepin hydrochloride. *Expert Opinion on Investigational Drugs, 16,* 1295–1305.

Singh, N. A., Clements, J. M., & Fiatarone, M. A. (1997). A randomized controlled trial of progressive resistant training in depressed elders. *The Journals of Gerontology. Series A, Biological Sciences and Medical Sciences, 52,* 27–35.

Singh, N. A., Clements, J. M., & Singh, M. A. (2001). The efficacy of exercise as a long-term antidepressant in elderly subjects: A randomized, controlled trial. *The Journals of Gerontology. Series A, Biological Sciences and Medical Sciences, 56,* 497–504.

Singh, N. A., Stavrinos, T. M., Scarbek, Y., et al. (2005). A randomized controlled trial of high versus low intensity weight training versus general practitioner care for clinical depression in older adults. *The Journals of Gerontology. Series A, Biological Sciences and Medical Sciences, 60,* 768–776.

Sinyor, M., Schaffer, A., & Levitt, A. (2010). The Sequenced Treatment Alternatives to Relieve Depression (STAR*D) trial: A review. *Canadian Journal of Psychiatry, 55,* 126–135.

Souery, D., Serretti, A., Calati, R., et al. (2011). Switching antidepressant class does not improve response or remission in treatment-resistant depression. *Journal of Clinical Psychopharmacology, 22,* 323–329.

Sprouse, J., Braselton, J., Reynolds, L., et al. (2001). Activation of postsynaptic 5-HT(1A) receptors by fluoxetine despite the loss of firing-dependent serotonergic input: Electrophysiological and neurochemical studies. *Synapse, 41,* 49–57.

Stahl, S. M. (2008). Selective histamine 1 antagonism: Novel hypnotic and pharmacological actions challenge classical notions of antihistamine. *CNS Spectrums, 13,* 855–865.

Stahl, S. M. (2009). *Stahl's illustrated antidepressants.* Cambridge University Press.

Stahl, S. M. (2013a). Classifying psychotropic drugs by mode of action not by target disorders. *CNS Spectrums, 18,* 113–117.

Stahl, S. M. (2013b). *Stahl's essential psychopharmacology* (4th ed.). Cambridge University Press.

Stahl, S. M. (2014). Mechanism of action of agomelatine: A novel antidepressant exploiting synergy between monoaminergic and melatonergic properties. *CNS Spectrums, 19,* 207–212.

Stahl, S. M., Pardko, J. F., Haight, B. R., et al. (2005). SNRIs: Their pharmacology, clinical efficacy, and tolerability in comparison with other classes of antidepressants. *CNS Spectrums, 10,* 732–747.

Stahl, S. M., Fava, M., Trivedi, M. H., et al. (2010). Agomelatine in the treatment of major depressive disorder: An 8-week, multicenter, randomized, placebo-controlled trial. *The Journal of Clinical Psychiatry, 71*, 616–626.

Sugawara, H., Sakamoto, K., Harada, T., et al. (2010). Predictors of efficacy in lithium augmentation for treatment-resistant depression. *Journal of Affective Disorders, 125*, 165–168.

Sultan, M. A., & Courtney, D. B. (2017). Adjunctive trazodone and depression outcome in adolescents treated with serotonin re-uptake inhibitors. *Journal of the Canadian Academy of Child and Adolescent Psychiatry, 26*, 233–240.

Tanti, A., & Belzung, C. (2013). Hippocampal neurogenesis: A biomarker for depression or antidepressant effect? Methodological considerations and perspective for future research. *Cell and Tissue Research, 354*, 203–219.

Tanum, L. (2000). Reboxetine: Tolerability and safety profile in patients with major depressive disorder. *Acta Psychiatrica Scandinavica, 402*, 37–40.

Thakare, V. N., & Patel, B. M. (2015). Potential targets for the development of novel antidepressant: Future perspectives. *CNS & Neurological Disorders Drug Targets, 14*, 270–281.

Thase, M. E. (1998). Effects of venlafaxine on blood pressure: A meta-analysis of original data from abrupt discontinuation of venlafaxine. *The Journal of Clinical Psychiatry, 59*, 502–508.

Thase, M. E., Entsuah, A. R., & Rudolph. (2001). Remission rates during treatment with venlafaxine or selective serotonin reuptake inhibitors. *The British Journal of Psychiatry, 178*, 234–241.

Thase, M. E., Rush, A. J., Howland, R. H., et al. (2002). Double-blind switch study of imipramine of sertraline treatment of antidepressant-resistant chronic depression. *Archives of General Psychiatry, 59*, 233–239.

Thase, M. E., Corya, S. A., Osuntokun, O., et al. (2007a). A randomized, double-blind comparison of olanzapine/fluoxetine combination, olanzapine, and fluoxetine in treatment-resistant major depressive disorder. *The Journal of Clinical Psychiatry, 68*, 224–236.

Thase, M. E., Nierenberg, A. A., Vrijland, R., et al. (2007b). Efficacy of duloxetine and selective serotonin reuptake inhibitors: Comparison as assessed by remission rates in patients with major depressive disorder. *International Clinical Psychopharmacology, 27*, 672–676.

Thase, M. E., Nierenberg, A. A., Vrijland, R., et al. (2010). Remission with mirtazapine and selective serotonin reuptake inhibitors: A meta-analysis of individual patient data from 15 controlled trials of acute phase treatment of major depressive disorder. *International Clinical Psychopharmacology, 25*, 189–198.

Thase, M. E., Youakim, J. M., Skuban, A., et al. (2015a). Adjunctive brexpiprazole 1 and 3 mg for patients with major depressive disorder following inadequate response to antidepressants: A phase 3, randomized, double-blind study. *The Journal of Clinical Psychiatry, 76*, 1232–1240.

Thase, M. E., Youakim, J. M., Skuban, A., et al. (2015b). Efficacy and safety of adjunctive brexpiprazole 2 mg in major depressive disorder: A phase 3, randomized, placebo-controlled study in patients with inadequate response to antidepressants. *The Journal of Clinical Psychiatry, 76*, 1224–1231.

Thase, M. E., Jacobsen, P. L., Hanson, E., et al. (2022). Vortioxetine 5, 10, and 20 mg significantly reduces the risk of relapse compared with placebo in patients with remitted major depressive disorder: The RESET study. *Journal of Affective Disorders, 303*, 123–130.

Trischler, L., Felice, D., Colle, R., et al. (2014). Vortioxetine for the treatment of major depressive disorder. *Expert Review of Clinical Pharmacology, 7*, 731–745.

Trivedi, M. H., Rush, A. J., Wisniewski, S. R., et al. (2006). Evaluation of outcomes with citalopram for depression using measurement-based care in STAR*D: Implications for clinical practice. *American Journal of Psychiatry, 163*, 28–40.

Uchida, H. (2018). Neuroscience-based nomenclature: What is it, why is it needed, and what comes next? *Psychiatry and Clinical Neurosciences, 72*, 50–51.

Wakeling, A. (1983). Efficacy and side effects of mianserin, a tetracyclic antidepressant. *Postgraduate Medical Journal, 59*, 229–231.

Wang, J. W., David, D. J., Monckton, J. E., et al. (2008). Chronic fluoxetine stimulates maturation and synaptic plasticity of adult-born hippocampal granule cells. *The Journal of Neuroscience, 28*, 1374–1384.

Wang, S. M., Han, C., Lee, S. J., et al. (2016). Vilazodone for the treatment of depression: An update. *Chonnam Medical Journal, 52*, 91–100.

Wang, Y., Xu, X., Luo, A., et al. (2020). The role of gut microbiota in the antidepressant effects of ketamine. In K. Hashimoto, S. Ide, & K. Ikeda (Eds.), *Ketamine from abused drug to rapid-acting antidepressant* (pp. 127–141). Springer.

Warden, D., Rush, A. J., Trivedi, M. H., et al. (2007). The STAR*D project results: A comprehensive review of findings. *Current Psychiatry Reports, 9*, 449–459.

Weisler, R. H., Johnston, J. A., Lineberry, C. G., et al. (1994). Comparison of bupropion and trazodone for the treatment of major depression. *Journal of Clinical Psychopharmacology, 14*, 170–179.

Won, E., Park, S. C., Han, K. M., et al. (2014). Evidence-based, pharmacological treatment guideline for depression in Korea, revised edition. *Journal of Korean Medical Science, 29*, 468–484.

Wong, D. T., Bymaster, F. P., Hornig, J. S., et al. (1975). A new selective inhibitor for uptake of serotonin into synaptosomes of rat brain: 3-(p-trifluoromethylphenoxy). N-methyl-3-phenylpropylamine. *The Journal of Pharmacology and Experimental Therapeutics, 193*, 804–811.

Wong, D. T., Bymaster, F. P., & Engleman, E. A. (1995). Prozac (fluoxetine, lilly 110140), the first selective serotonin uptake inhibitor and antidepressant drug;

twenty years since its first publication. *Life Sciences, 57,* 411–441.

Wong, D. T., Perry, K. W., & Bymaster, F. P. (2005). The discovery of fluoxetine hydrochloride (Prozac). *Nature Reviews Drug Discovery, 4,* 764–774.

Wooderson, S. C., Juruena, M. F., Fekadu, A., et al. (2011). Prospective evaluation of specialist inpatient treatment for refractory affective disorders. *Journal of Affective Disorders, 131,* 92–103.

World Health Organization (WHO). (2018). *Anatomical Therapeutic Chemical (ATC) classification system [internet].* WHO.

Wu-Chou, A. I., Liu, Y. L., & Shen, W. W. (2016). Genetic polymorphisms of cytochrome P 450 and antidepressants. In V. Srinivasan, F. López-Muñoz, D. De Berardis, et al. (Eds.), *Melatonin, neuroprotective agents and antidepressant therapy* (pp. 533–543). Springer India.

Yamamoto, H., Hagino, Y., Kasai, S., et al. (2015). Specific roles of NMDA receptor subunits in mental disorders. *Current Molecular Medicine, 15,* 193–205.

Youdim, N. B., Edmondson, D., & Tipton, K. F. (2006). The therapeutic potential of monoamine oxidase inhibitors. *Nature Reviews Neuroscience, 7,* 295–309.

Young, E. A., Kornstein, S. G., Marcus, S. M., et al. (2009). Sex differences in response to citalopram. *Nature Reviews Neuroscience, 49,* 786.

Zajecka, J., Schatzberg, A., Stahl, S., et al. (2010). Efficacy and safety of agomelatine in the treatment of major depressive disorder: A multicenter, randomized, double-blind, placebo-controlled trial. *Journal of Clinical Psychopharmacology, 2010*(30), 135–144.

Zohar, J., & Kasper, S. (2016). Neuroscience-based Nomenclature (NbN): a call for action. *The World Journal of Biological Psychiatry, 17,* 318–320.

Hiroyuki Uchida, Euitae Kim, L. Fredrik Jarskog, W. Wolfgang Fleischhacker, Gary Remington, and Jeffrey A. Lieberman

Contents

This chapter is an update from the 4th edition. Previous edition authors were Seiya Miyamoto, David B. Merrill, L. Fredrik Jarskog, W. Wolfgang Fleishhacker, Stephen R. Marder and Jeffrey A. Lieberman

H. Uchida (✉)
Department of Neuropsychiatry, Keio University School of Medicine, Tokyo, Japan
e-mail: hiroyuki_uchida@keio.jp

E. Kim
Department of Psychiatry, Seoul National University College of Medicine, Seoul, Korea
e-mail: euitae.kim@snu.ac.kr

L. F. Jarskog
Department of Psychiatry, University of North Carolina School of Medicine, Chapel Hill, NC, USA
e-mail: lars_jarskog@med.unc.edu

W. W. Fleischhacker
Department of Psychiatry, Psychotherapy and Psychosomatics, Medical University Innsbruck, Innsbruck, Austria
e-mail: wolfgang.fleischhacker@i-med.ac.at

G. Remington
Department of Psychiatry, University of Toronto, Toronto, ON, Canada
e-mail: Gary.Remington@camh.ca

J. A. Lieberman
Department of Psychiatry, New York State Psychiatric Institute, Columbia University College of Physicians and Surgeons, New York, NY, USA
e-mail: jlieberman@nyspi.columbia.edu

© Springer Nature Switzerland AG 2024
A. Tasman et al. (eds.), *Tasman's Psychiatry*,
https://doi.org/10.1007/978-3-030-51366-5_134

Abstract

Since the era of antipsychotic pharmacotherapy began with the discovery of the antipsychotic properties of chlorpromazine in 1952, many new medications for psychosis (antipsychotics) have become available. All currently approved medications for psychosis, except pimavanserin, remain dopaminergic agents. Medications for psychosis alleviate psychotic symptoms and prevent relapse into psychosis in the treatment of schizophrenia. Moreover, they are also indicated for a variety of other conditions, including bipolar disorder and treatment-resistant depression. Medications for psychosis sometimes cause various side effects while their side effect profiles greatly differ among individual drugs. Much effort has recently been devoted to developing drugs that exert their effects by modulating neural systems other than the dopaminergic system, and several drugs have demonstrated promising results in Phase II and III trials. In addition, individualized treatment for a specific patient based on their biological characteristics is one of the future goals of psychopharmacological treatment of schizophrenia. These goals will be achieved with precise biological characterization of this illness.

Keywords

Dopamine · Serotonin · Glutamate · AMPA · NMDA · Antipsychotic · Schizophrenia · Treatment-resistant depression

Introduction

The current nomenclature of psychotropic drugs, including antipsychotic drugs, is primarily based on clinical indications (Zohar et al., 2015). However, boundaries among various categories of psychotropic drugs are becoming blurred. For example, some "antipsychotics" are indicated not only for schizophrenia but also for bipolar disorder and treatment-resistant depression. Moreover, frequently used terms such as "atypical antipsychotics" and "second-generation antipsychotics (SGAs)" include a variety of drugs in the same categories despite their different mechanisms of action and pharmacological profiles. For the time being and until a definition based on mechanism of action (see below) is widely accepted, a practical approach has been adopted by differentiating first-generation antipsychotics (FGAs), introduced before clozapine, from new generation antipsychotics (NGAs), after the advent of clozapine. To overcome this lack of categorical clarity, the neuroscience-based nomenclature (NbN) has been proposed (Zohar et al., 2015); the NbN abandons the category name of "antipsychotic drugs" (temporarily referred to as "medications for psychosis" instead) and gives each drug a specific category name based on its presumed mechanism of action such as dopamine/serotonin/norepinephrine antagonist for risperidone, dopamine/serotonin antagonist for olanzapine, and dopamine receptor antagonist for amisulpride. While this challenging project is considered to be an important step in light of the limitations of the current nomenclature, further work is clearly necessary to foster its development and acceptance by researchers, patients, and their caregivers. Therefore, in this chapter, both the conventional terms and the NbN terms will be used.

A major advance in the pathophysiology of schizophrenia that has led to potential therapeutic strategies is the accumulation of novel, relevant findings in the glutamatergic system and an array of targets in its synaptic complex. The first support for this hypothesis was found in the acute psychotomimetic effects of noncompetitive antagonists of glutamate N-methyl-D-aspartate (NMDA) receptors, phencyclidine (Allen & Young, 1978), and subsequently ketamine (Krystal et al., 1994). Both have been shown to cause a variety of schizophrenia-like symptoms in healthy people, including positive and negative symptoms and cognitive impairment and worsen psychosis in patients with schizophrenia (Luby et al., 1962; Javitt & Zukin, 1991). These observations, along with subsequent nonhuman and human data, led to the NMDA hypofunction hypothesis of schizophrenia. Subsequent preclinical and clinical studies using proton magnetic resonance spectroscopy (^{1}H-MRS) identified increases in glutamate and Glx (glutamate plus glutamine) levels in the striatum and frontal and medial temporal cortices (Merritt et al., 2016; Schobel et al., 2013). Thus, NMDA receptor hypofunction and increased glutamatergic neurotransmission are thought to be associated with schizophrenia, and a number of drugs that modulate NMDA receptors and target metabotropic glutamate receptors have been developed and tested for the treatment of schizophrenia (Kantrowitz et al., 2020). This said, all currently approved antipsychotic drugs, except pimavanserin, remain dopaminergic agents.

The dopamine system is mostly normal in patients with schizophrenia; however, the prefrontal cortex is dysfunctional. This is believed to be a major cause of the negative symptoms of schizophrenia and probably involves glutamatergic dysfunction. When an individual is faced by a complex problem, activity of the prefrontal cortex is required. Dopamine from the midbrain is required to optimize that problem-solving ability of the frontal cortex. Consequently, in a situation of frontal cortical demand, there is a signal increasing dopamine delivery to the cortex. Since the midbrain dopaminergic cells are intact, they are able to deliver an increased dopamine signal to cortical tissues. However, since the frontal cortex is dysfunctional, it is not capable of utilizing the dopamine and continues to demand ever more dopamine. However, the limbic system does respond to the extra dopamine, and in the setting of ever-increasing dopamine delivery, it begins to display signs of psychosis. One of the common dopamine receptors in the limbic system is dopamine D_2, and consequently, its blockade may have antipsychotic effects.

This chapter presents an updated review of the history, pharmacology, uses, efficacy, and side effects of drugs for psychosis on the basis of currently available evidence. As well, it reviews the benefits and limitations of these drugs in the management of schizophrenia and other disorders.

History of Medications for Psychosis (Antipsychotics): Development

The era of antipsychotic pharmacotherapy began with the discovery of the antipsychotic properties of chlorpromazine in 1952 in Paris by Delay and Deniker (Delay et al., 1952a, b). Haloperidol was later discovered by Janssen in 1958 (Janssen et al., 1959) and a number of FGAs, which are also referred to as "conventional" or "typical" medications for psychosis (antipsychotics), were subsequently developed (Fig. 1). The historic discovery that antagonism of the dopamine D_2 receptor is strongly associated with clinical effects (Seeman et al., 1976) has contributed to further development of medications for psychosis that block D_2 transmission. For many patients with schizophrenia, FGAs are effective in treating positive symptoms of the disorder and in preventing psychotic relapse. However, some treated with FGAs remain symptomatic and are considered either treatment resistant or only partially responsive. Moreover, FGAs do not improve negative symptoms or cognitive impairments, which often determine the level of functioning. FGAs also cause a variety of side effects such as extrapyramidal side effects (EPS), both acutely and with long-term exposure. These adverse events, in many cases, lead to poor treatment adherence, which in turn leads to relapse and rehospitalization.

Clozapine, the prototypical "atypical" medication for psychosis (antipsychotic), was synthesized in 1956 in Berne, Switzerland, and underwent extensive clinical testing in the 1970s, until its development was halted in the United States and limited in other countries because of an increased incidence of agranulocytosis. Nonetheless, the

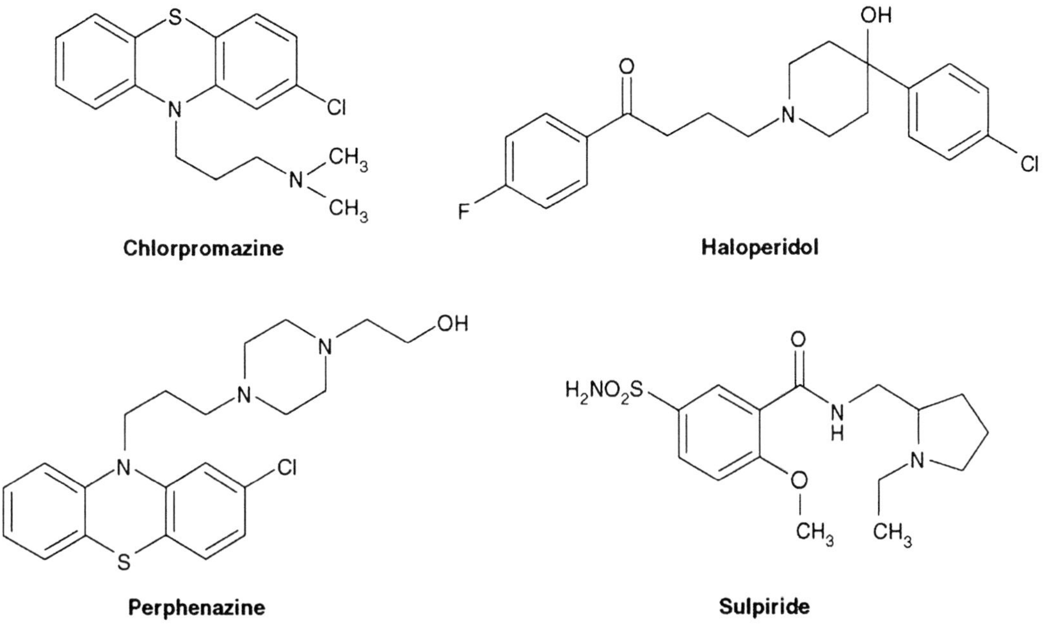

Chlorpromazine

Haloperidol

Perphenazine

Sulpiride

Fig. 1 Chemical structures of selected first-generation medications for psychosis (antipsychotics)

drug's superior efficacy ultimately led to its reintroduction in the United States in 1990 (Lieberman et al., 1989). The renaissance of clozapine was based on several advantages: superior efficacy to FGAs for treatment-refractory schizophrenia (Kane et al., 1988); possible advantages for negative, cognitive, and mood symptoms of schizophrenia (Lieberman, 1993); reduction in suicidal behavior (Meltzer et al., 2003); low liability for acute and chronic EPS; and an absence of prolactin elevation (Lieberman et al., 1989).

In the years that followed the reintroduction of clozapine, concerted research and development efforts were made to replicate the drug's therapeutic profile, while avoiding the associated risk of hematotoxicity. Although this goal was never fully realized, this initiative spawned the introduction of most currently available newer medications for psychosis (antipsychotics), which are henceforth referred to as NGAs (Fig. 2). Although none of these newer medications for psychosis have matched the singular effectiveness of clozapine, they have broadened the therapeutic repertoire available for the treatment of schizophrenia spectrum and other psychotic disorders.

Mechanism of Action of Medications for Psychosis

Dopamine Receptor Modulation

The dopamine D_2 receptor is regarded as the primary target associated with therapeutic antipsychotic effects and the attenuation of dopaminergic neural transmission via D_2 receptors in the mesolimbic pathway is thought to reduce psychotic symptoms (Seeman, 1992). Blockade of the D_2 receptor is also associated with the induction of EPS and prolactin elevation. All currently approved medications for psychosis (antipsychotics), with the exception, pimavanserin, have nanomolar affinity for the D_2 receptor and fully or partially block the actions of dopamine (Table 1).

Positron emission tomography (PET) studies in younger patients with schizophrenia have shown that striatal dopamine D_2 receptor blockade above 65% is associated with clinical response in the acute phase treatment of positive symptoms, while EPS are more likely to occur with a D_2 receptor blockade above 80% (Farde

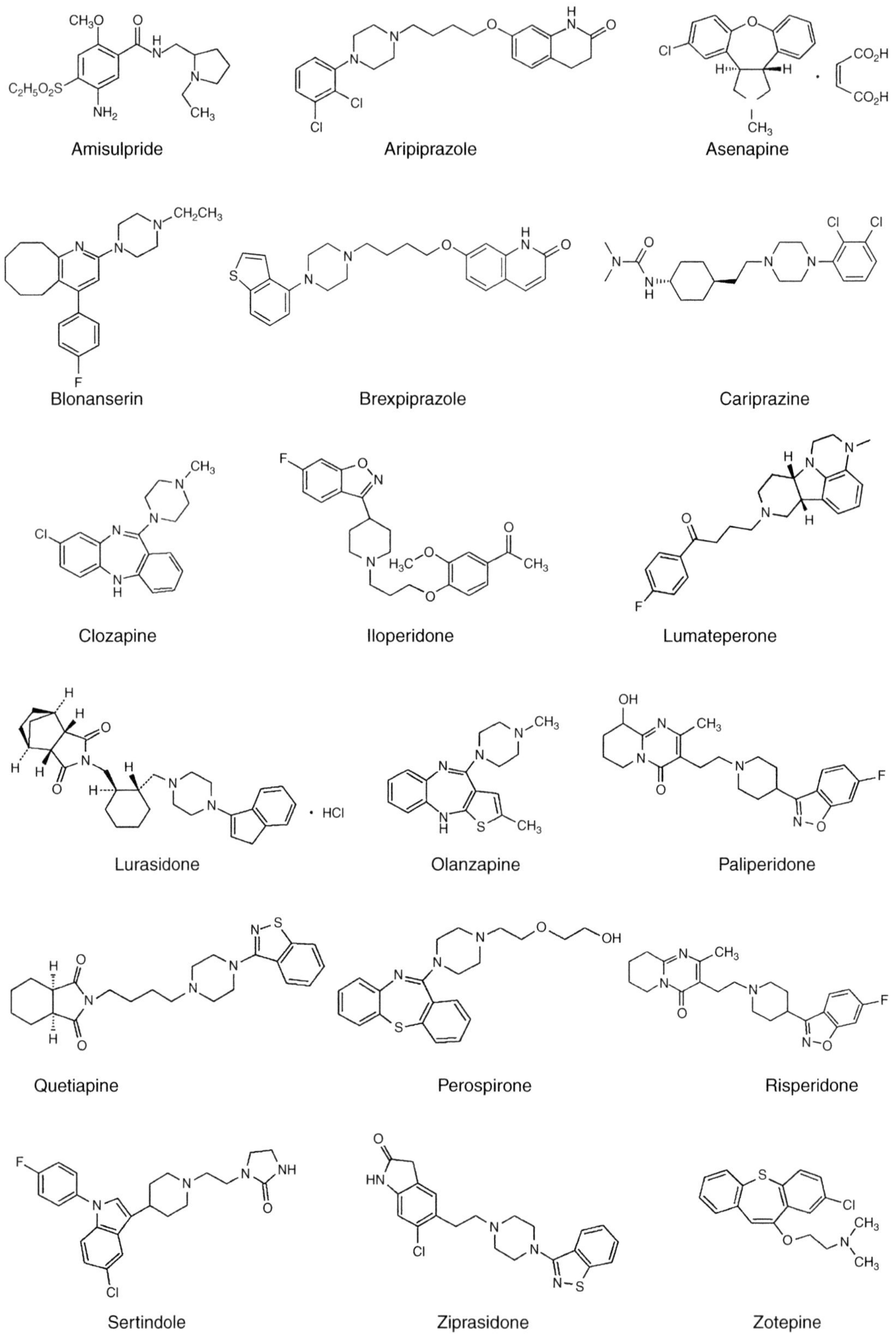

Fig. 2 Chemical structures of selected new generation medications for psychosis (antipsychotics)

et al., 1992; Kapur et al., 2000a; Uchida et al., 2011b). In addition, cognitive function is inversely related to >80% dopamine D_2 receptor occupancy levels (Sakurai et al., 2013). For relapse prevention, PET studies and clinical trials have suggested that dopamine D_2 receptor blockade with medications for psychosis could be maintained at a lower level (Uchida et al., 2008; Ozawa et al., 2019). On the other hand, a series of clinical PET studies in older adults with schizophrenia reported a lower therapeutic window of 50–60% of D_2 blockade for this specific population associated with the optimal balance of clinical response versus side effects (Graff-Guerrero et al., 2015; Uchida et al., 2014; Iwata et al., 2016). It should be noted that clozapine and quetiapine show less than 60% striatal D_2 receptor occupancy at therapeutically effective doses (Farde et al., 1992; Kapur et al., 2000b), suggesting that D_2 receptor antagonism alone cannot explain therapeutic efficacy. The low occupancy of striatal D_2 receptors by clozapine and quetiapine could account for their low EPS liability.

Newer medications for psychosis (antipsychotics) were originally termed "atypical antipsychotics" because of their low incidence of EPS at therapeutically effective doses. Although there is debate as to what constitutes "atypicality," the defining feature of this class of medications is the separation of the dose that results in a therapeutic effect from that which is associated with an increasing risk of EPS. The concept of "atypicality" is thought to be due to two properties: (1) their lower affinity for D_2 receptors and (2) their high affinity for serotonin 5-HT$_{2A}$ receptors. Meltzer et al. (1989) hypothesized that a relatively high affinity for 5-HT$_{2A}$ receptors relative to D_2 receptors is the critical feature of atypical drugs (Meltzer et al., 1989), whereas Kapur and Seeman (2001) proposed that low affinity for, and fast dissociation from, D_2 receptors may be the critical property for "atypicality" (Kapur & Seeman, 2001). Typical medications for psychosis generally bind more tightly than dopamine itself to the D_2 receptor and dissociate from it slowly. In contrast, some atypical medications for psychosis such as clozapine and quetiapine bind more loosely than dopamine to the D_2 receptor, with

dissociation constants (expressed as "k_{off}") that are higher than those for dopamine.

Partial D_2 agonism represents another strategy of attempting to attenuate dopaminergic neural transmission in the treatment of schizophrenia with fewer side effects, compared to full D_2 antagonism. Partial D_2 agonists have lower intrinsic activity at D_2 receptors than dopamine itself (nearly always <40% but >25%), allowing them to act as either functional agonists or antagonists, depending on synaptic dopamine levels and extent of receptor occupancy (Lieberman, 2004). Partial D_2 agonist activity, as reported for aripiprazole and brexpiprazole, appears to inhibit endogenous dopamine activity where it is high and when receptor occupancy is >90%. However, partial agonists will activate D_2 receptors when synaptic dopamine concentrations are low, or if receptor occupancy is <50%. Furthermore, since the intrinsic activity of currently approved antipsychotic partial agonists is nearly always equivalent to the tonic dopamine signal, between 25% and 40% the activity of dopamine itself, they maintain dopaminergic tone in the nigrostriatal and tuberoinfundibular pathways, thereby avoiding EPS and hyperprolactinemia normally associated with full D_2 antagonism.

Serotonin Receptor Modulation

The "dopamine–serotonin antagonism theory," conceived by Janssen et al. (1988) and popularized by Meltzer et al. (1989), assumes that a high ratio of serotonin 5-HT$_{2A}$ receptor to D_2 receptor blockade confers "atypicality" of medications for psychosis. Blockade of 5-HT$_{2A}$ receptors can increase dopaminergic transmission in the nigrostriatal pathway (Lucas & Spampinato, 2000), and thus can reduce the risk for EPS, and theoretically could improve negative symptoms and cognitive deficits in schizophrenia by increasing release of dopamine, acetylcholine, or both in the prefrontal cortex. This hypothesis applies to a greater or lesser extent to most NGAs (Table 1). There are, however, important limitations to this concept (Sharif et al., 2007). For example, amisulpride has no meaningful affinity for the

Table 1 Relative neurotransmitter receptor affinities of new generation medications for psychosis (antipsychotics)

Drug Receptor	Amisulpride	Aripiprazole	Asenapine	Blonanserin	Brexpiprazole	Cariprazine	Clozapine	Iloperidone
D_1	−		+++	−	+		+	+
D_2	+++	++++	+++	++++	++++	++++	+	+++
D_3	+++	++++	++++	++++	+++	++++	+	+++
D_4	−	++	+++	+	+++		++	++
$5\text{-}HT_{1A}$		+++	+++	+	++++	+++	+	+
$5\text{-}HT_{2A}$		+++	++++	++++	++++	++	+++	+++
$5\text{-}HT_{2C}$		++	++++	++	++	+	+++	
$5\text{-}HT_6$			++++	++	++		+++	++
$5\text{-}HT_7$	++	++	++++	+	+++		+++	++
α_1	−	++	+++	++	++++	+	+++	++++
α_2	−		+++	+	++		++	
H_1	−	++	+++	−	++	++	+++	+
M_1	−	−	−	−	−	−	+++	−

Receptor affinity values are based on each product information except for amisulpride (Abbas et al., 2009), blonanserin (Tenjin et al., 2013; Une & Kurumiya, 2007), paliperidone (Corena-McLeod, 2015), perospirone (Kato et al., 1990), risperidone (Corena-McLeod, 2015), sertindole (Juruena et al., 2011; Wang et al., 2013; Sanchez et al., 1991), and zotepine (Shobo et al., 2010)

Abbreviations: DA, dopamine; 5-HT, serotonin; NA, noradrenaline

Ki: ≥ 1001, minimal to none(−); $100 - <1000$, low affinity (+); $10 - <100$, moderate affinity (++); $1 - <10$, high affinity (+++); <1, very high affinity (++++)

Lumateperone	Lurasidone	Olanzapine	Paliperidone	Perospirone	Quetiapine	Risperidone	Sertindole	Ziprasidone	Zotepine
++		++	+	+	−	+			++
++	+++	++	+++	+++	+	+++	++++	+++	+++
		++	+++			++		+++	
++		++	++			++			
	+++		−	+++	+	+		+++	+
++++	++++	+++	++++	++++	+	++++	++++	++++	++++
−		++	++			++	++++	+++	+++
		+++	−			−			+++
	++++		+++			+++			+++
++		++	++	++	++	+++	+++	++	+++
	++		++	+	+	+++			++
−	−	+++	++	+++	++	++	−	++	++++
−	−	++	−	−	−	−	−	−	+

5-HT$_{2A}$ receptor, and all partial agonists have higher affinity to the D$_2$ receptor than to 5-HT$_{2A}$, yet clinically they have atypical profiles (Lieberman, 2004; Deeks & Keating, 2010). Risperidone and olanzapine exhibit high 5-HT$_{2A}$ receptor occupancy at doses that are not antipsychotic, and as doses of these drugs are increased beyond their usual therapeutic ranges, the risk for EPS increases despite maximum 5-HT$_{2A}$ receptor blockade (Kapur et al., 1999). Thus, high 5-HT$_{2A}$ affinity may contribute to the modulation of dopamine in the striatum and prefrontal cortex, but high 5-HT$_{2A}$ occupancy does not protect against the risk of EPS if D$_2$ receptor occupancy is greater than the EPS threshold (Sharif et al., 2007). Therefore, the 5-HT$_{2A}$/D$_2$ hypothesis does not satisfactorily explain "atypicality" (Kapur & Remington, 2001).

Some, but not all, atypical medications for psychosis (antipsychotics) are 5-HT$_{2C}$, 5-HT$_6$, and 5-HT$_7$ receptor antagonists as well as direct or indirect 5-HT$_{1A}$ receptor agonists (Table 1). There is some evidence that the combination of 5-HT$_{2A}$ and 5-HT$_{2C}$ receptor blockade is a more efficient way of increasing cortical dopamine release than either alone (Meltzer & Massey, 2011). Antagonism at 5-HT$_6$ and 5-HT$_7$ receptors may be useful for improving cognitive function (de Bruin & Kruse, 2015; Nikiforuk & Popik, 2013). Partial agonism of 5-HT$_{1A}$ receptors, resulting in activation and blockade of pre- and postsynaptic receptors, may contribute to the mechanism of action of some atypical medications for psychosis, including aripiprazole, asenapine, clozapine, perospirone, quetiapine, and ziprasidone. 5-HT$_{1A}$ partial agonist activity is thought to improve negative symptoms and cognitive impairment by enhancing prefrontal cortical dopamine release. In addition, 5-HT$_{1A}$ receptor agonists have been reported to possess antidepressant and anxiolytic properties, and to attenuate the EPS liability of D$_2$ antagonists (Miyamoto et al., 2012). However, this remains to be proven in clinical studies.

Following many years of serotonin research for the development of novel therapeutics, pimavanserin was approved by the FDA in 2016 as the first drug for psychosis without any significant antagonism of dopaminergic receptors. This drug is indicated for the treatment of hallucinations and delusions associated with Parkinson's disease psychosis and acts as an inverse agonist at serotonin 5-HT$_{2A}$ receptors and, to a lesser degree, at serotonin 5-HT$_{2C}$ receptors.

Modulation of Other Receptors

Most medications for psychosis (antipsychotics) interact with adrenergic, histaminergic, and muscarinic neurotransmitter systems and also with monoamine transporters (Table 1). Interactions with these receptor systems contribute to many common antipsychotic-induced side effects, but data also indicate the potential for previously unrecognized therapeutic benefits. For example, the blockade of H$_1$ receptors by medications for psychosis is probably related to weight gain and sedation, and α_1-adrenergic receptor blockade contributes to orthostatic hypotension and sedation. M$_1$ antagonist actions may cause central (e.g., cognitive impairment) and peripheral (e.g., constipation and dry mouth) anticholinergic adverse effects. In contrast, M$_1$ receptor agonism may be beneficial in treating cognitive dysfunction in schizophrenia (Ibrahim & Tamminga, 2012) and alpha or beta adrenergic blockade may reduce the risk of akathisia.

Effects on Intracellular Signal Transduction and Gene Expression

There is convincing evidence that antipsychotic action is associated with long-lasting adaptive modifications in neural functioning that involve changes in intracellular signal transduction and changes in gene expression in target neurons (Miyamoto et al., 2012). These changes appear to be initiated by the binding of medications for psychosis (antipsychotics) to various neurotransmitter receptors. However, the specific downstream effector molecules that are associated with therapeutic efficacy remain to be determined.

Attention has been focused on Akt (protein kinase B), glycogen synthase kinase-3 (GSK-3), and wingless (Wnt) signaling pathways, which have been associated with schizophrenia in many genetic and postmortem studies (Freyberg et al., 2010). Akt and GSK-3 play various roles in differentiation and development of neural cells, intracellular trafficking, apoptosis, and regulation of gene transcription. Thus, they may modulate synaptic plasticity of neurons. Accumulating evidence suggests that medications for psychosis may improve psychotic and affective symptoms, at least in part, through modulation of levels and activity of Akt-, GSK-3-, and Wnt-related intracellular signaling pathways (Freyberg et al., 2010; Karam et al., 2010).

DNA microarray expression profiling was used to identify candidate target genes that may be involved in the mechanism of action of medications for psychosis (antipsychotics). Several microarray studies in brains of rodents demonstrated that acute and chronic treatment with medications for psychosis, including clozapine, haloperidol, olanzapine, and risperidone, regulates the expression of genes involved in synaptic function, intracellular Ca^{2+} metabolism, cell survival, cell communication, immune response, nuclei acid metabolism, neural plasticity, signal transduction, ionic homeostasis, proteolysis, oxidative changes, metabolism and energy pathways, and neuronal growth factors (Kontkanen et al., 2002; Collins et al., 2014; de Bartolomeis et al., 2017). For example, immediate-early genes modulation by medications for psychosis is reported to control the early molecular processes of rapid synaptic plasticity induced by medications for psychosis (de Bartolomeis et al., 2017), and different medications for psychosis have been demonstrated to induce specific patterns of immediate-early gene expression (Matosin et al., 2016). Further research is warranted to clarify the nature and extent of influence exerted by medications for psychosis on neuronal gene expression and cellular function. This challenge is exemplified by lumateperone which indirectly modulates glutamatergic activity by increasing phosphorylation of the GluN2B subunit of NMDA receptors (Lieberman et al., 2016).

Neuroprotection

It is well documented that structural brain changes, such as ventricular enlargement and reduced gray and white matter volumes, may be present at the first psychotic episode of schizophrenia and possibly in the premorbid and prodromal phase (Lieberman et al., 2001; Pantelis et al., 2003). In addition, early longitudinal studies found that cortical gray matter loss may be progressive, especially early in the course of illness, and is associated with functional decline (Cahn et al., 2002; Ho et al., 2003). These findings raise the question of whether medications for psychosis (antipsychotics) could mitigate such pathophysiological progression in the early stages of illness. However, whether antipsychotic effects on the gray matter volume change are either protective or harmful remains uncertain (Goff et al., 2017). An early double-blind randomized controlled trial (RCT) demonstrated that long-term treatment with olanzapine, but not haloperidol, prevented progressive gray matter volume reductions in patients with first-episode psychosis (Lieberman et al., 2005b). More recently, a large multisite study by the Enhancing Neuro Imaging Genetics Through Meta-Analysis (ENIGMA) Consortium found that cortical thickness effect sizes were two to three times larger in patients receiving medications for psychosis relative to unmedicated ones (van Erp et al., 2018). Moreover, higher chlorpromazine dose equivalents were significantly correlated with thinner cortex in almost all regions. However, evidence linking brain volume changes to antipsychotic treatment is potentially confounded by the underlying illness and by uncertainty whether higher dosing of medications for psychosis (antipsychosis) is a response to, or a contributor to, progressive brain volume loss (Goff et al., 2017).

Although the precise mechanisms by which medications for psychosis (antipsychotics) may arrest or delay the pathomorphological process remain uncertain, there is in vitro and in vivo evidence that some newer medications for psychosis may exert neuroprotective effects through production of neurotrophic factors, attenuation of glutamate excitotoxicity, oxidative stress and apoptosis, and enhancement of neurogenesis and connectivity, all of which could provide a rationale for pharmacological intervention with medications for psychosis (Chen & Nasrallah, 2019).

New Generation Medications for Psychosis (Antipsychotics)

Medications for psychosis **(antipsychotics) have conventionally been classified as typical or** atypical medications for psychosis, and the concept of "atypicality" has promoted the research in elucidating mechanisms of antipsychotic action and side effects. However, the scientific rationale for such dichotomous classification is somewhat tenuous as both typical and atypical medications for psychosis drugs are considered heterogeneous. In this regard, the NbN may be more appropriate for better understanding of presumed mechanisms of actions of these drugs, although still preliminary.

Amisulpride

Amisulpride, a benzamide derivative, is a selective antagonist at D_3 and D_2 receptors, with little affinity for D_1-like or nondopaminergic receptors except for 5-HT_{7a} receptors (Scatton et al., 1997; Abbas et al., 2009). Amisulpride has higher affinity for D_3 than D_2 receptors (Schoemaker et al., 1997). Amisulpride has higher affinity to the short presynaptic D_2 receptor, so that at low doses, it preferentially blocks these presynaptic dopamine autoreceptors; postsynaptic dopamine receptor antagonism is apparent at higher doses. The effect of low doses on negative symptoms may be attributable to the presynaptic dopamine-releasing effect, which can have a psychomotor stimulatory action. Low doses also appear to block selectively temporal cortical D_2 and D_3 receptors, whereas higher doses result in considerable striatal D_2 and D_3 receptor occupancy, with associated EPS (Xiberas et al., 2001). Unfortunately, the benefits of low dose amisulpride are lost as the clinical dose in increased. Its propensity to elevate prolactin secretion even in the absence of EPS is related

to the fact that it is a substrate of the permeability glycoprotein and selectively antagonizes tuberoinfundibular D_2 receptors in a manner similar to that of risperidone (El-Mallakh & Watkins, 2019). Abbas et al. (2009) found that amisulpride is a potent 5-HT_{7A} receptor antagonist and that its potential antidepressant actions may require functional 5-HT_7 receptors in vivo.

Aripiprazole

Aripiprazole, a quinolinone derivative, is a partial dopamine agonist with high affinity for D_2 and D_3 receptors that acts on both postsynaptic D_2 receptors and presynaptic autoreceptors. The drug demonstrates properties of a functional agonist and antagonist in animal models of dopaminergic hypoactivity and hyperactivity, respectively (Kikuchi et al., 1995). It has been proposed that aripiprazole's activity at D_2 receptors is "functionally selective," meaning that the drug may activate a specific set of D_2 receptors, which are, in turn, coupled to particular G proteins that mediate its clinical effects (Shapiro et al., 2003). In addition, aripiprazole has partial agonist properties at D_4, 5-HT_{1A}, 5-HT_{2C}, 5-HT_7, and, to a much lesser extent (i.e., reduced intrinsic activity), 5-HT_{2A} receptors, so that it is functionally a 5-HT_{2A} antagonist. It also has modest affinity for α_1-adrenergic, H_1-histaminergic, and 5-HT_6 receptors, and no appreciable affinity for D_1 or muscarinic receptors. At present, it is unclear to what extent binding to receptors other than D_2 receptors contributes to the actions of aripiprazole, apart from their association with side effects.

Aripiprazole does not conform to the usual $5\text{-HT}_{2A}/D_2$ model of "atypicality." Its affinity for D_2 receptors exceeds that for 5-HT receptors by an order of magnitude (Shapiro et al., 2003), and PET studies indicate that aripiprazole occupies up to 90% of striatal D_2-like dopamine receptors at clinical doses in patients with schizophrenia (Mamo et al., 2007). In spite of this, aripiprazole causes minimal EPS, due to the partial (approximately 30%) activation of D_2 receptors.

Asenapine

Asenapine, a dibenzoxepinopyrrole derivative, has high affinity for a variety of serotonergic (5-HT_{1A}, 5-HT_{1B}, 5-HT_{2A}, 5-HT_{2B}, 5-HT_{2C}, 5-HT_5, 5-HT_6, and 5-HT_7), noradrenergic (α_1, α_{2A}, α_{2B}, and α_{2C}), histaminergic (H_1 and H_2), and dopaminergic (D_2, D_3, and D_4) receptors, but minimal affinity for muscarinic M_1 receptors (Bishara & Taylor, 2008; Shahid et al., 2009). Asenapine also acts as a partial agonist at 5-HT_{1A} receptors (Ghanbari et al., 2009). Preclinical studies show that asenapine potentiates both prefrontal dopaminergic and glutamatergic transmission (Frånberg et al., 2008), and it enhances cortical monoamine release at doses associated with antipsychotic activity (Frånberg et al., 2009).

Blonanserin

Blonanserin belongs to a series of 4-phenyl-2-(1-piperazinyl) pyridines and acts as an antagonist at D_2, D_3, and 5-HT_{2A} receptors. Its affinity for D_2 receptors is approximately six times greater than that for 5-HT_{2A} receptors so that there is no appreciable binding of $5HT_{2A}$ at clinical doses (Tenjin et al., 2013). Additionally, blonanserin has low affinity for 5-HT_{2C}, adrenergic α_1, histamine H_1, and muscarinic M_1 receptors. It also shows low inhibitory activity of neuronal reuptake of dopamine, 5-HT, and norepinephrine. A PET study showed that blonanserin occupies approximately 60–80% of striatal D_2 receptors with the clinically approved daily doses in patients with chronic schizophrenia (Tateno et al., 2013).

A preclinical study demonstrated that systematic administration of blonanserin increases extracellular levels of norepinephrine and dopamine, but not levels of 5-HT, glutamate, or γ-aminobutyric acid (GABA) in the prefrontal cortex (Ohoyama et al., 2011). It also enhances neuronal activity in the locus coeruleus and ventral tegmental area without affecting activity in the dorsal raphe nucleus or the mediodorsal thalamic nucleus (Ohoyama et al., 2011).

Brexpiprazole

Brexpiprazole acts as a partial agonist of 5-HT_{1A} receptor and dopamine D_2 and D_3 receptors (Citrome et al., 2015). Brexpiprazole has a lower intrinsic activity at D_2 receptor than aripiprazole, which may be associated with a lower risk of restlessness/akathisia. In addition to its intrinsic activity at D_2, its antagonism at 5-HT_{2A} and alpha 5-HT_{1B} receptors and partial agonism at 5-HT_{1A} receptors could be associated with fewer EPS, possibly due to enhanced downstream dopamine release in the striatum.

Cariprazine

Cariprazine is a partial agonist of dopamine D_3 and D_2 receptors with the higher affinity to D_3 receptor. Cariprazine also acts as a partial agonist of 5-HT_{1A} receptor and an antagonist of 5-HT_{2B} and 5-HT_{2A} receptors. Cariprazine is approved for the acute and maintenance treatment of schizophrenia and the acute treatment of manic, mixed, and depressive episodes of bipolar I disorder (Corponi et al., 2019). Post hoc analyses of clinical trial data suggest some efficacy of this drug in improving negative symptoms (Fleischhacker et al., 2019). Cariprazine is associated with no clinically significant changes in metabolic variables, prolactin levels, or QT intervals on the electrocardiogram (Citrome, 2016).

Clozapine

Clozapine is a tricyclic dibenzodiazepine derivative. It has relatively high affinity for 5-HT_{2A} receptors and lower affinity for D_2 receptors (Lieberman, 1993; Duncan et al., 1999). Clozapine has high antagonist affinity for D_4, H_1, M_1, and α_1 and α_2 adrenergic receptors (Bymaster et al., 1996). The drug exhibits weak partial agonist effects at 5-HT_{1A} receptors and at muscarinic M_1, M_2, and M_4 receptors, and the latter pharmacological properties have been hypothesized to contribute to antipsychotic activity and cognitive enhancement in animal models (Bymaster et al., 2002). N-desmethylclozapine, the major metabolite of clozapine, has potent partial agonist activity at the M_1 receptor and can potentiate NMDA currents in the hippocampus (Sur et al., 2003). It is possible that clozapine's unique therapeutic profile may be due, at least in part, to such M_1 receptor-mediated potentiation of NMDA receptor function.

Iloperidone

Iloperidone is a piperidinylbenzisoxazole derivative with relatively high affinity for 5-HT$_{2A}$, D$_2$, D$_3$, and α_1-adrenergic receptors, and moderate affinity for D$_4$, 5-HT$_6$, and 5-HT$_7$ receptors (Mauri et al., 2014). Iloperidone has low affinity for D$_1$, 5-HT$_{1A}$, and H$_1$ receptors, and has no appreciable affinity for muscarinic M$_1$ receptors. Similar to clozapine and olanzapine, it induces a depolarization blockade of A10 mesolimbic dopaminergic neurons, but not of A9 nigrostriatal neurons. Iloperidone increases dopamine and acetylcholine levels in the prefrontal cortex.

Lumateperone

Lumateperone is a butyrophenone-derivative compound and acts as an antagonist of 5-HT$_{2A}$ receptor and a dopamine D$_2$ receptor presynaptic partial agonist and postsynaptic antagonist, a D$_1$ receptor-dependent modulator of glutamate, and a serotonin reuptake inhibitor (Vyas et al., 2020). This drug simultaneously acts through the dopaminergic, serotonergic, and glutamatergic systems. Moreover, lumateperone indirectly modulates glutamatergic activity by increasing phosphorylation of the GluN2B subunit of NMDA receptors (Snyder et al., 2015). Additionally, it has affinity to the serotonin reuptake pump with about 50% receptor occupancy at therapeutic doses.

Lurasidone

Lurasidone is a benzoisothiazole derivative with very high affinity for D$_2$, 5-HT$_{2A}$, and 5-HT$_7$ receptors (Meyer et al., 2009; Ishibashi et al., 2010). It is also a partial agonist at 5-HT$_{1A}$ receptors and has moderate affinity for α_{2C} receptors. It has virtually no affinity for H$_1$ and muscarinic M$_1$ receptors, and minimal affinity for α_1, α_{2A}, D$_1$, D$_3$, and 5-HT$_{2C}$ receptors (Ishibashi et al., 2010). Preclinical studies in rodents showed that lurasidone was superior to certain SDAs in improving learning and memory impairment induced by the NMDA receptor antagonist MK-801 (Ishiyama et al., 2007; Enomoto et al., 2008). A PET study found that 80–160 mg of lurasidone resulted in >65% D$_2$ receptor occupancy in patients with schizophrenia or schizoaffective disorder (Potkin et al., 2014).

Olanzapine

Olanzapine, a thienobenzodiazepine derivative, is closely related in chemical structure to clozapine, and the two drugs share many receptor binding characteristics. Olanzapine has more potent antagonistic effects at 5-HT$_{2A}$ than at D$_2$ receptors, similarly to clozapine (Bymaster et al., 1997). However, olanzapine has substantially lower affinity for 5-HT$_{1A}$ and 5-HT$_7$ receptors in comparison with clozapine. A PET study found a striatal D$_2$ receptor occupancy of 71–80% in olanzapine's usual clinical dose range of 10–20 mg/day, and doses of 30 mg/day and higher were associated with greater than 80% D$_2$ occupancy in patients with schizophrenia (Kapur et al., 1998). Olanzapine produces a substantial blockade of muscarinic sites in vivo, as demonstrated by PET imaging (Raedler et al., 2000).

Paliperidone

Paliperidone, a benzisoxazole derivative, is 9-hydroxy-risperidone, the active metabolite of risperidone. The pharmacological properties of paliperidone are similar to those of risperidone with respect to receptor activity, potency, and onset and duration of action (van Beijsterveldt et al., 1994). However, paliperidone displays slightly less potent in vitro affinity for D$_1$, D$_2$,

5-HT_{1A}, 5-HT_{2A}, 5-HT_{2C}, α_1, α_2, and H_1 receptors than risperidone (Schotte et al., 1996). A paliperidone extended-release tablet (paliperidone ER), which is associated with small 24-hour peak-to-trough fluctuations in plasma concentration, undergoes limited hepatic metabolism and, consequently, may avoid significant hepatic-related drug-drug and drug-disease interactions (Kane et al., 2007). A PET study showed that paliperidone ER at 6–9 mg/day leads to a D_2 receptor occupancy of 70–80%, with the magnitude of the occupancy being similar between the striatum and temporal cortex (Arakawa et al., 2008).

Pimavanserin

Pimavanserin is indicated for the treatment of hallucinations and delusions associated with Parkinson's disease psychosis. Pimavanserin acts as an inverse agonist and antagonist preferentially at serotonin 5-HT_{2A} receptors and, to a lesser degree, at serotonin 5-HT_{2C} receptors (Sahli & Tarazi, 2018). This is the first FDA-approved drug for psychosis without any significant antagonism of dopaminergic receptors.

Perospirone

Perospirone is a benzisothiazole derivative with high affinity antagonism at 5-HT_{2A} and D_2 receptors (Onrust & McClellan, 2001). It displays partial 5-HT_{1A} agonist properties and some affinity for D_1, α_1-adrenergic, and H_1 receptors. The drug has no appreciable affinity for muscarinic receptors. Perospirone has at least four active metabolites, but all of them have lower affinity for D_2 and 5-HT_{2A} receptors than the parent drug. A PET study found that 16–48 mg of perospirone resulted in >70% D_2 receptor occupancy in patients (Arakawa et al., 2010).

Quetiapine

Quetiapine, a dibenzothiazepine derivative, has greater affinity for 5-HT_{2A} than for D_2 receptors. It also has some affinity for α_1-adrenergic and H_1 receptors and weak partial agonist effects at 5-HT_{1A} receptors. PET studies showed that quetiapine occupies 22–68% of striatal D_2 receptors and 48–70% of cortical 5-HT_2 receptors at therapeutic doses in patients with schizophrenia (Gefvert et al., 1998). Quetiapine produces transiently high striatal D_2 occupancy, but the effect lasts for only a few hours (Kapur et al., 2000b). Norquetiapine (N-desalkylquetiapine), a major active human metabolite of quetiapine, is a potent inhibitor of the norepinephrine transporter, enhancing noradrenaline and dopamine levels similarly to some medications for depression, and has partial 5-HT_{1A} agonist activity (Jensen et al., 2008). Quetiapine is available both as an immediate-release (IR) and as an extended-release (XR) formulation. The XR formulation has similar bioavailability also in terms of dopamine D2 receptor blockade compared with the IR formulation (Mamo et al., 2008).

Risperidone

Risperidone, a benzisoxazole derivative, has substantially higher affinity for 5-HT_{2A} and D_2 receptors than does clozapine. A PET study found that risperidone occupies 75–80% of striatal D_2 receptors and 78–88% of cortical 5-HT_2 receptors when administered to patients with schizophrenia at a dose of 6 mg/day (Farde et al., 1995). Despite high levels of D_2 receptor occupancy, moderate doses of risperidone (4–6 mg/day) exhibit a lower EPS risk than some FGAs. This may be due to its 5-HT_{2A} antagonist properties (Meltzer et al., 1989). However, at higher doses, risperidone consistently produces EPS (Marder & Meibach, 1994), indicating that 5-HT_{2A} receptor antagonism alone cannot eliminate EPS associated with substantial D_2 receptor blockade. PET studies showed no regional differences in D_2 receptor occupancy, but a study using $\text{L-}[B\text{-}^{11}C]\text{dopa}$ reported that risperidone may have stabilizing effects on levels of dopamine synthesis (Ito et al., 2009). Similar to clozapine, risperidone has relatively high affinity for α_1 and α_2 adrenergic receptors, but whether adrenergic receptor antagonism contributes to the therapeutic action of risperidone is unknown.

Sertindole

Sertindole is a phenylindole derivative with high affinity antagonism at $5-HT_{2A}$, $5-HT_{2C}$, α_1-adrenergic, and D_2-like receptors (Schotte et al., 1996). In conventional preclinical models, the receptor-binding profile of sertindole accurately predicted antipsychotic efficacy with low EPS liability (Arnt & Skarsfeldt, 1998). SPECT studies found that sertindole is similar to risperidone in occupying a high percentage of striatal D_2-like receptors and cortical $5-HT_{2A}$ receptors in patients with schizophrenia (Bigliani et al., 2000). Electrophysiologically, sertindole exhibits dose-dependent marked limbic selectivity.

Ziprasidone

Ziprasidone, a hydrochloride salt of a benzisothiazolylpiperazine analog, has potent $5-HT_{2A}$, D_2, and D_3 affinities and exhibits partial $5-HT_{1A}$ agonist properties (Ferris, 2000). Ziprasidone also has significant affinity for $5-HT_{1D}$, $5-HT_{2C}$, H_1, and α_1-adrenergic receptors (Arnt & Skarsfeldt, 1998), and it inhibits the reuptake of norepinephrine and 5-HT. The absence of weight gain with ziprasidone, which has relatively high $5-HT_{2C}$ and H_1 affinities, may be related to its partial agonism at $5-HT_{1A}$ receptors. PET studies revealed 60–75% striatal D_2/D_3 receptor occupancy in patients with schizophrenia at clinically effective doses (Mamo et al., 2004; Suzuki et al., 2013). Another PET study found that ziprasidone exerts a 10% higher D_2/D_3 receptor occupancy in extrastriatal compared with striatal regions (Vernaleken et al., 2008).

Zotepine

Zotepine is a dibenzothiepine analog of clozapine with high affinity for $5-HT_{2A}$, $5-HT_{2C}$, $5-HT_6$, $5-HT_7$, α_1-adrenergic, and H_1 receptors (Needham et al., 1996). It has modest affinity for D_1, D_2, D_3, and D_4 receptors. It is also a potent norepinephrine reuptake inhibitor. Preclinical studies suggested that low-dose zotepine increases dopaminergic neurotransmission, whereas at higher doses, the drug acts as a dopaminergic antagonist. SPECT studies indicated that zotepine occupies 68–78% of striatal D_2-like receptors at usual clinical doses in patients with schizophrenia (Barnas et al., 2001).

Conditions Treated with Medications for Psychosis (Antipsychotics) Other than Schizophrenia

At present, medications for psychosis are prescribed primarily for schizophrenia; however, increasingly NGAs are also being used for other psychiatric disorders. A proportion of these uses are empirically well supported, but only preliminary or moderate evidence exists for others. Furthermore, data on long-term safety of medications for psychosis in these populations are still insufficient and clearly needed.

Major Depression with Psychotic Features

Psychotic symptoms, such as delusions or hallucinations, have been observed in up to 25% of patients with major depressive disorder. Major depression with psychotic features is seen in patients of all ages and carries a high risk of short-term morbidity and suicide. Psychosis in the context of major depression often responds poorly to monotherapy with medication for depression and usually requires the use of adjuvant medications for psychosis (antipsychotics). According to a Cochrane Review regarding pharmacological treatment for psychotic depression, some evidence indicates that combination therapy with an antidepressant plus an antipsychotic is more effective than either treatment alone or placebo (Kruizinga et al., 2021). This seems also the case for relapse prevention; the STOP-PD II RCT demonstrated that continuing sertraline plus olanzapine compared with sertraline plus placebo reduced the risk of relapse over 36 weeks among remitted patients with psychotic depression (Flint et al., 2019). Increasingly, medications for psychosis have also been studied for use in non-psychotic depression.

Treatment-Resistant Depression

Currently, quetiapine, aripiprazole, olanzapine, and brexpiprazole are FDA approved for adjuvant use with medications for depression for the treatment of major depression that did not respond to monotherapy with medication for depression. Efficacy of monotherapy with a medication for psychosis (antipsychotic) for nonpsychotic depression has been inconsistent in the literature (Cutler et al., 2009; Papakostas et al., 2012; Suppes et al., 2016).

Bipolar Disorder

A number of medications for psychosis (antipsychotics) have FDA approval for the treatment of manic/mixed states of bipolar disorder. A meta-analysis of 68 RCTs (16,073 participants) that compared medications for psychosis and mood stabilizers for acute mania found that medications for psychosis as a class were significantly more effective than mood stabilizers and that risperidone, olanzapine, and haloperidol were the most effective individual agents (Cipriani et al., 2011). On the other hand, a recent network meta-analysis of 14,256 manic patients randomized to one of 18 active treatments or placebo found that all 12 antimanic drugs, including lithium and medications for psychosis, were comparable in improving manic symptoms except risperidone versus aripiprazole and versus valproate, both significantly favoring risperidone (Yildiz et al., 2015). A recent direct random-assignment 6-month comparison of lithium and quetiapine (Clinical Health Outcomes Initiative in Comparative Effectiveness) found the two drugs to be equal in efficacy, but that lithium was associated with reduced adverse effects (Nierenberg et al., 2016). To date, there have been no double-blind, controlled monotherapy studies of clozapine in the treatment of acute bipolar mania, but reports suggest that the drug may be beneficial in the management of treatment-refractory bipolar disorder (Aksoy Poyraz et al., 2015; Pacchiarotti et al., 2020).

Several medications for psychosis have demonstrated efficacy in the treatment of bipolar depression, which carries a greater overall burden of suicide risk, functional impairment, and duration of active symptomatology than does bipolar mania. Quetiapine, lurasidone, and lumateperone, as monotherapy, and the combination of olanzapine and fluoxetine have received FDA approval for bipolar depression. A small meta-analysis of seven RCTs for bipolar depression found that low and high doses of medications for psychosis were comparable in improving depression but the low doses were associated with reduced clinical harms (Bartoli et al., 2017). However, only lithium has demonstrated anti-suicide effects in patients with bipolar illness (Hawkins et al., 2021).

Medications for psychosis (antipsychotics) have also increasingly been studied for long-term maintenance treatment for bipolar disorder and a number of medications for psychosis including aripiprazole, olanzapine, quetiapine, and ziprasidone have received FDA approval for monotherapy treatment for this indication. However, given the potential risks regarding long-term use of medications for psychosis, including tardive dyskinesia and metabolic dysregulation (e.g., weight gain, dyslipidemia, and diabetes mellitus), the outcomes of studies such as the CHOICE direct comparative study, and documented poorer functional outcomes associated with increased use of medications for psychosis in bipolar patients (Sleem & El-Mallakh, 2021), caution should be exercised when choosing a medication for psychosis as a foundational treatment in bipolar disorder.

Tourette's Disorder

Tourette's disorder is a neurobehavioral illness characterized by motor and vocal tics that usually begin in childhood and may persist indefinitely. The pathophysiology of Tourette's disorder remains poorly understood.

When motor or vocal tics interfere with functioning, medications for psychosis (antipsychotics) can be effective in reducing their severity (McNaught & Mink, 2011). Haloperidol and pimozide are approved in many countries and, together with fluphenazine, have historically been used most commonly for moderate to severe tics. Increasingly, newer medications for psychosis have also been studied for tic suppression. Among them, efficacy of risperidone and aripiprazole have been well documented (Pringsheim et al., 2019). Given the risk of side effects with all medications for psychosis, these drugs should generally be reserved for patients with significant tic-related functional impairment or pain that is refractory to safer treatments.

Borderline Personality Disorder

Medications for psychosis (antipsychotics) are sometimes used in the treatment of borderline personality disorder (BPD) (Sakurai et al., 2020). Although psychotherapy is the mainstay of BPD treatment, medications for psychosis (antipsychotics) may be useful as adjunctive therapy in more severe cases. The available evidence is limited to smaller studies of individual agents. A Cochrane review found that no pharmacological therapy seems effective in specifically treating BPD pathology (Stoffers-Winterling et al., 2022). Based on the evidence for antisuicidal and anti-aggressive properties of clozapine, a role for this agent in severe cases of BPD has been proposed, but supporting data are very limited. Given the circumscribed utility of medications for psychosis for BPD, and their potential for causing adverse effects, most patients with this condition should not routinely be treated with medications for psychosis.

Substance-Related Disorders

A variety of substances, including amphetamines, cocaine, alcohol, cannabis, and phencyclidine, can cause psychotic symptoms during intoxication or drug withdrawal. In addition, substance-use disorders are common in patients with schizophrenia, and the presence of these disorders may negatively influence response to medications for psychosis (antipsychotics). Unfortunately, there is a dearth of data to guide treatment decisions for patients with substance-use comorbidities.

In general, the efficacy of medications for psychosis (antipsychotics) for substance-use disorders is not supported by empirical data (Chan et al., 2019a, b). On the other hand, medications for psychosis, including clozapine and risperidone, may have a role in reduction of substance use or craving when it is comorbid with schizophrenia (Krause et al., 2019). However, in the light of paucity of well-designed clinical trial data, no firm conclusions can be drawn. Notably, when a psychotic disorder is not present, medications for psychosis are not generally effective in the treatment of substance-use disorders.

Behavioral Disturbances in Patients with Dementia

Dementia, whether due to Alzheimer's disease (AD) or other causes, is frequently associated with behavioral disturbances and psychosis. While medications for psychosis (antipsychotics) are often used to manage these symptoms, concerns have emerged regarding an increased risk of cerebrovascular adverse events (e.g., stroke and transient ischemic episodes), the metabolic syndrome (MS), cognitive decline, and mortality with at least some NGAs. Because of these concerns, Health Canada issued an advisory in late 2002, and in the United States, the FDA issued its own public health advisory in April 2005, warning of an increased risk of mortality with NGAs in elderly patients with dementia. This warning was extended to all medications for psychosis in 2008. Consistent with the FDA's assessment, a meta-analysis found an increased relative risk of death with use of NGAs, relative to placebo treatment, in patients with dementia (Schneider et al., 2005), although absolute numbers were small.

Schneider et al. conducted a federally funded, multisite, randomized, double-blind, 36-week, placebo-controlled trial to determine the effectiveness of olanzapine, quetiapine, and risperidone, as compared with placebo, in 421 outpatients with AD and psychosis, aggression, or agitation (Schneider et al., 2006). The primary outcome of median time to discontinuation of treatment for any reason ranged from 5 to 8 weeks, with no significant differences among the four groups. In a network meta-analysis of 17 RCTs of aripiprazole, olanzapine, quetiapine, and risperidone in 5373 patients with AD or other types of dementia, no drug was consistently associated with better results than the others across all effectiveness and safety outcomes (Yunusa et al., 2019). It is important to note that these studies examined symptom improvement in patients, not the effect on the ease or quality of caregiving, which is one of the primary drivers of restarting medications for psychosis in this population (Aerts et al., 2019).

Together, these studies suggest modest therapeutic efficacy at best with the use of NGAs for psychosis and behavioral disturbances in patients with dementia, with probable small increases in the risk of cerebrovascular-related morbidity and mortality. The use of medications for psychosis (antipsychotics) in this population should be assessed on a case-by-case basis, giving judicious consideration to the risks and benefits of medications for psychosis and alternative treatment strategies. When medications for psychosis are used in elderly patients, lower dosages are indicated because of age-related declines in drug metabolism, vulnerability to drug–drug interactions, a high incidence of comorbid physical illness, and increased sensitivity to EPS due to age-related declines in the dopaminergic neurotransmission (Uchida et al., 2009a).

Parkinson's Disease and Huntington's Disease

Patients with Parkinson's disease (PD) may experience psychotic symptoms, which are often induced by treatment with dopaminergic agents. These symptoms may improve with reduction in the dopaminergic medication dose, but this usually worsens Parkinsonism and is often not tolerated. Although the literature in this area remains relatively sparse, several open-label and double-blind, placebo-controlled studies have found that clozapine and possibly quetiapine can offer substantial benefit to this patient population by decreasing psychotic symptoms without significantly worsening Parkinsonism (Seppi et al., 2011). In contrast, olanzapine and risperidone have produced mixed results regarding efficacy for PD-related psychosis, and more potential for EPS than clozapine or quetiapine. Recently, pimavanserin received FDA approval for the treatment of hallucinations and delusions associated with Parkinson's disease psychosis. Pimavanserin acts as an inverse agonist and antagonist preferentially at serotonin 5-HT_{2A} receptors and, to a lesser degree, at serotonin 5-HT_{2C} receptors.

Patients with psychiatric manifestations of Huntington's disease (HD) can sometimes benefit from medications for psychosis (antipsychotics). As with PD, the use of medications for psychosis (antipsychotics) in HD frequently worsens motor dysfunction (Buckley, 2001). Some medications for psychosis, including sulpiride and clozapine, may have modest beneficial effects in the management of motor and psychiatric symptoms of HD (Quinn & Marsden, 1984; van Vugt et al., 1997) although the evidence base is highly limited.

Behavioral Problems in Children and Adolescents

Attention-deficit hyperactivity disorder (ADHD), oppositional defiant disorder (ODD), and conduct disorder (CD) are common psychiatric conditions among children and adolescents. Some of the patients with these disorders demonstrate periods of hyperactivity, screaming, and agitation with combativeness. Several medications for psychosis (antipsychotics) have shown some efficacy in treating these conditions. In fact, risperidone and aripiprazole are currently FDA approved for irritability associated with autism. A meta-analysis of 11 RCTs of medications for psychosis found moderate-quality evidence that risperidone has a moderate-to-large effect on conduct problems and aggression in youth with subaverage IQ and ODD, CD, or disruptive behavior disorder not otherwise specified, with and without ADHD (Pringsheim et al., 2015). In addition, risperidone was found to have a moderate effect on disruptive and aggressive behavior in youth with average IQ and ODD or CD, with and without ADHD. On the other hand, data on the use of haloperidol, thioridazine, and quetiapine in aggressive youth with CD was considered to be of low or very-low quality. Similarly, a more recent Cochrane Review concluded that risperidone may reduce aggression and conduct problems in the short term in children and youths with disruptive behavior disorders and this intervention is associated with significant weight gain (Loy et al., 2017). Given the substantial risks associated with antipsychotic treatment in this vulnerable population, further controlled studies are required to determine the specific role of medications for psychosis in the treatment of nonpsychotic disorders among pediatric patients, and shared decision-making to consider carefully the risks and potential benefits should always accompany such treatment.

Effects of Medications for Psychosis (Antipsychotics) on Symptoms of Schizophrenia

Positive Symptoms

All medications for psychosis (antipsychotics) improve the positive symptoms of schizophrenia, including hallucinations and delusions.

In a network meta-analysis of 402 RCTs comparing efficacy of 32 medications for psychosis for the acute treatment of adults with multiepisode schizophrenia (117 studies and 31,179 participants for analysis of positive symptoms), the medication for psychosis with the largest standardized mean difference compared with placebo for reduction of positive symptoms was amisulpride followed by risperidone, clozapine, olanzapine, paliperidone, chlorpromazine, haloperidol, and asenapine (Huhn et al., 2019).

Three large long-term pragmatic studies have compared the effectiveness of NGAs with FGAs. The Clinical Antipsychotic Trials of Intervention Effectiveness (CATIE) study compared the effectiveness of the mid-potency FGA perphenazine (8–32 mg/day, mean dose 20.8 mg/day) with olanzapine (7.5–30 mg/day, mean dose 20.1 mg/day), quetiapine (200–800 mg/day,

mean dose 534.4 mg/day), risperidone (1.5–6.0 mg/day, mean dose 3.9 mg/day), and ziprasidone (40–160 mg/day, mean dose 112.8 mg/day) in 1493 chronic outpatients with schizophrenia for up to 18 months of treatment (Lieberman et al., 2005a). Results from the first phase of this study demonstrated that olanzapine had a significant advantage for "all-cause discontinuation," the primary outcome measure. The time to treatment discontinuation for lack of efficacy was significantly longer in the olanzapine group than in the FGA perphenazine group (hazard ratio 0.47), the quetiapine group (hazard ratio 0.41), the risperidone group (hazard ratio 0.45), or the ziprasidone group (hazard ratio 0.59). However, the other NGAs did not differ from perphenazine on the primary outcome measure and all five medications displayed comparable changes in Positive and Negative Syndrome Scale (PANSS) positive scores.

In a second large pragmatic trial, the European First Episode Schizophrenia Trial (EUFEST) (Kahn et al., 2008), almost 500 patients with first-episode schizophrenia were randomly allocated to open-label treatment with either low-dose haloperidol or one of four NGAs (amisulpride, olanzapine, quetiapine, or ziprasidone) for 1 year. Compared with haloperidol, all NGAs showed a significantly lower risk for early study discontinuation, the primary outcome variable, but there were no group differences in PANSS total scores.

In the Cost Utility of the Latest Antipsychotic Drugs in Schizophrenia Study (CUtLASS), in which FGAs and NGAs were compared as classes (most patients in the first group were treated with sulpiride and most in the second group with olanzapine), no significant differences in positive symptoms were observed between patients treated with FGAs compared with NGAs (Jones et al., 2006), although it needs to be stressed that the trial was terminated early due to recruitment difficulties.

In summary, CATIE, CUtLASS, and EUFEST provided some support for the superiority of NGAs over FGAs but no unequivocal evidence for any non-clozapine medication for psychosis (antipsychotic) over another on positive or overall symptoms.

Negative Symptoms

Negative symptoms of schizophrenia include blunted affect, emotional and social withdrawal, poverty of speech, anhedonia, avolition, and apathy. They may be divided into three subtypes that are often difficult to distinguish: (1) primary enduring (or deficit), (2) primary nonenduring, and (3) negative symptoms that are secondary to other causes, such as depression, positive symptoms, neurological side effects, substance misuse, and environmental factors (Buchanan & Gold, 1996; Murphy et al., 2006). Approximately 70% of patients with schizophrenia develop primary negative symptoms before the onset of positive symptoms (Häfner et al., 1992). The proportion of persistent primary negative symptoms may be 15–20% (Kirkpatrick et al., 2006; Buchanan, 2007; Bobes et al., 2010). They represent a core feature of the illness and may be associated with long-term disability and prolonged hospitalization (Buchanan & Gold, 1996).

In a network meta-analysis of 402 RCTs comparing efficacy of 32 medications for psychosis (antipsychotics) for the acute treatment of adults with multi-episode schizophrenia (132 studies and 32,015 participants for analysis of negative symptoms), the medication for psychosis with the largest standardized mean difference compared with placebo for reduction of negative symptoms was clozapine followed by zotepine, amisulpride, olanzapine, asenapine, perphenazine, risperidone, and paliperidone (Huhn et al., 2019).

There is debate as to whether the effects of medications for psychosis (antipsychotics) on negative symptoms are related to a reduction in primary negative symptoms, secondary negative symptoms, or both (Murphy et al., 2006). Path analyses have suggested that risperidone and olanzapine exert direct effects on primary negative symptoms, independent of changes in positive, depressive, or extrapyramidal symptoms (Möller, 1993; Tollefson & Sanger, 1997), but the statistical approaches were performed post hoc, and only the most efficacious doses of the NGAs were used. Post hoc analyses of the data from an RCT comparing cariprazine and risperidone found the superiority of cariprazine in the improvement of negative symptoms and no significant differences on pseudo-specificity measures such as positive symptoms, depression, and extrapyramidal symptoms, which may support a genuine treatment effect for cariprazine in negative symptom improvement (Fleischhacker et al., 2019). These questions will ultimately be answered when more objective measures of negative symptoms become standardized.

Cognitive Symptoms

Cognitive impairments are widely suggested as a core feature of schizophrenia and almost all patients with the illness have some degree of such deficits (Keefe et al., 2005). Patients typically perform one to two standard deviations below normal on a variety of neuropsychological measures, particularly those that assess attention, verbal skills, processing speed, and executive function (Woodward et al., 2005). Progressive decline in cognitive function has been documented to occur before the onset of psychosis and a wide range of cognitive deficits are usually present at the first psychotic break. After the initial onset, cognitive deficits tend to remain relatively stable, or to worsen slowly (Harvey et al., 1999) while the nature and course of cognition in late life is heterogeneous (Rajji & Mulsant, 2008).

Cognitive impairment associated with schizophrenia (CIAS) is consistently and robustly associated with negative symptoms (Galderisi et al., 2002). CIAS is also more strongly related to social and vocational functioning than positive symptoms and may substantially impair activities of daily living and quality of life (QOL) (Green, 1996; Harvey et al., 1998). Abnormalities in a number of neurotransmitter systems, most notably dopaminergic, glutamatergic, cholinergic, serotonergic, and GABAergic systems, are thought to be associated with CIAS (Galletly, 2009). Because CIAS is among the strongest predictors of functional outcome in schizophrenia (Green et al., 2000), the targeting of CIAS has become a major focus of treatment.

A number of investigators have investigated the effects of medications for psychosis (antipsychotics) on CIAS. A network meta-analysis of 35 RCTs comparing effects of NGAs on cognition in patients with schizophrenia found a decreased verbal working memory in patients treated with clozapine, olanzapine, quetiapine, and FGAs compared to ziprasidone and a positive effect of sertindole on executive function compared to clozapine, olanzapine, ziprasidone, and FGAs, as well as possible positive effect of clozapine and olanzapine on verbal fluency (Nielsen et al., 2015). It should be noted that this meta-analysis did not find any drug having a uniform cognitive profile.

It is also unclear whether the improvements observed with NGAs in previous studies represent true cognitive enhancement or only a relative reduction in EPS and anticholinergic-related cognitive effects, as compared with FGAs (Harvey & Keefe, 2001; Carpenter & Gold, 2002). This has led to a debate on whether lower doses of FGAs might show comparable efficacy to NGAs for CIAS (Mishara & Goldberg, 2004). Several studies with improved methodology suggest that low doses of haloperidol may, in fact, have a less deleterious impact on cognitive function (Green et al., 2002; Keefe et al., 2004).

Furthermore, some studies have found mild benefits in cognitive performance with certain FGAs (Heinrichs, 2007). The most compelling of these is the CATIE trial, in which the sole FGA included, perphenazine, yielded more improvement in cognition after 18 months than any Phase I NGAs (Keefe et al., 2007). Notably, the magnitude of cognitive improvement was small and probably not clinically significant for all agents assessed in this analysis. In contrast, the EUFEST investigators found no differences between haloperidol and a number of NGAs with respect to an improvement of cognitive functioning (Davidson et al., 2009).

It should also be acknowledged that the relationship between dosage of medications for psychosis (antipsychotics) or dopamine D_2 receptor blockade and cognitive impairment may not be linear as observed in both young adult and elderly patients with schizophrenia. Limited evidence suggests that excessive D_2 blockade beyond a threshold of approximately 75% may result in cognitive impairment in this population (Sakurai et al., 2013; Uchida et al., 2009c).

Taken together, the variable results of the studies reviewed here suggest that medications for psychosis (antipsychotics), when dosed properly, may yield at most modest improvements in CIAS. Neither class appears to be clearly superior to the other. Clearly, cognitive functioning in patients with schizophrenia cannot be normalized by medications for psychosis alone.

Affective Symptoms

Depressive symptoms are common in all phases of schizophrenia and are associated with poorer outcomes, impaired social and vocational functioning, lower QOL, and an increased risk of relapse and suicide (Burton, 2006). The prevalence of comorbid depression in first-episode schizophrenia has been reported to be 26% (Herniman et al., 2019). When patients with schizophrenia demonstrate depressive features, differential diagnostic considerations should include major depression, antipsychotic-induced side effects, demoralization, concurrent abuse of or sudden withdrawal from substances, schizoaffective disorder, and primary negative symptoms.

In a network meta-analysis of 402 RCTs comparing efficacy of 32 medications for psychosis (antipsychotics) for the acute treatment of adults with multi-episode schizophrenia (89 studies and 19,683 participants for analysis of depressive symptoms), the medications for psychosis with the largest standardized mean difference compared with placebo for reduction of depressive symptoms was sulpiride followed by clozapine, amisulpride, olanzapine, aripiprazole, cariprazine, paliperidone, and asenapine (Huhn et al., 2019). While medications for psychosis show some therapeutic effects in reducing depressive symptoms among patients with schizophrenia, non-psychopharmacological approach should be considered as well.

Suicidal Behavior

Suicidal behavior is common in schizophrenia. According to a meta-analysis of 35 observational studies with 16,747 patients with schizophrenia, the pooled lifetime prevalence of suicide attempts was 26.8% (Lu et al., 2019). Earlier age of onset and high-income countries were significantly associated with a higher prevalence of suicide attempts.

Among the NGAs, clozapine appears to possess particular antisuicidal properties (Meltzer, 2012). The International Suicide Prevention Trial (InterSePT), a double-blind RCT, compared suicidal behavior in 980 patients at relatively high risk for suicide, who were randomized to clozapine or olanzapine treatment and followed for up to 2 years (Meltzer et al., 2003). Suicidal behavior was significantly less common in patients treated with clozapine, as reflected by fewer patients attempting suicide, requiring hospitalizations or rescue interventions to prevent suicide, or requiring concomitant treatment with medication for depression, medications for anxiety, or soporifics. Although the InterSePT study has been criticized for certain methodological limitations, such as lack of blinding among patients and clinicians, the results were compelling enough to garner the FDA's approval of an indication unique to clozapine: reduction of the risk of recurrent suicidal behavior in patients with schizophrenia or schizoaffective disorder. Since statistically significant superiority of clozapine in reducing death by suicide, compared to other medications for psychosis (antipsychotics), was not confirmed by a recent meta-analysis probably due to limited available data, further investigations are warranted (Vermeulen et al., 2019).

Quality of Life

Quality of life (QOL) is now recognized as an important therapeutic outcome of antipsychotic treatment in schizophrenia. Patients with schizophrenia experience a significantly lower level of QOL than healthy individuals (Dong et al., 2019). Considerable evidence suggests that several factors can affect QOL, including severity of symptoms, presence of anxiety or depression, age, antipsychotic-induced side effects, sociodemographic variables, and

patients' subjective responses to medication (Hofer et al., 2004).

A meta-analysis of seven RCTs ($n = 1573$) comparing medications for psychosis (antipsychotics) and placebo found that quality of life might be better in drug-treated participants although the quality of evidence was considered low (Ceraso et al., 2020).

During Phase I of the CATIE trial, there were no significant differences between perphenazine and any NGAs, as measured by Quality of Life Scale (QLS) (Swartz et al., 2007) and by quality-adjusted life-year ratings that combined measures of symptoms and side effects (Rosenheck et al., 2006). Similarly, CUtLASS 1 found no statistical advantage for NGAs in terms of QOL over 1 year (Jones et al., 2006). In fact, patients treated with FGAs in CUt-LASS 1 demonstrated a trend toward greater improvements in QLS and symptom scores, and neither inadequate power nor patterns of drug discontinuation accounted for these results. CUtLASS 2, an extension of the National Health Service-funded, pragmatic, open, multicenter RCT, compared clozapine against other NGAs and found no statistically significant differences in QLS scores but did detect a trend ($p = 0.08$) toward higher scores with clozapine (Lewis et al., 2006a). In summary, as with other measures of response to medications for psychosis addressed above, the findings from the CATIE and CUt-LASS questioned the superiority of NGAs compared with FGAs in improving QOL (Hasan et al., 2013).

Treatment of Different Phases of Schizophrenia

Nearly all acute psychotic episodes in schizophrenia and schizoaffective disorder should be treated with medications for psychosis (antipsychotics), given the overwhelming evidence in support of their efficacy. Little evidence was found to support a negative effect of acute or maintenance antipsychotic treatment on outcomes, compared with withholding treatment (Goff et al., 2017).

Prodromal Stage of Schizophrenia

The prodromal phase of schizophrenia and other psychotic disorders has become a major focus of research, which has led to the development of early intervention services. Investigators have established clinically defined prodromal diagnostic criteria to identify "ultra-high risk" (UHR) subjects (Yung & McGorry, 1996; Miller et al., 2002). In a meta-analysis of 2502 UHR individuals, there was a mean transition risk of 18% at 6 months of follow-up, 22% at 1 year, 29% at 2 years, 32% at 3 years, and 36% at 3 years (Fusar-Poli et al., 2012). The main goals of early intervention are to reduce prodromal symptoms and to delay or prevent the progression to psychosis.

A number of interventions, including antipsychotic treatment, did not show clear advantages in preventing transition to psychosis although the evidence level is very low. The most promising evidence is available for psychotherapeutic measures (Devoe et al., 2020). Most trials were relatively short in duration and involved small numbers of subjects. Furthermore, the use of medications for psychosis (antipsychotics) in subjects who may not convert to psychosis has raised considerable ethical concerns given the potential of prescribing them to individuals who may not really profit from them, yet incur a risk of side effects.

Childhood Schizophrenia

Childhood-onset schizophrenia (COS) is a rare and severe form of the disorder, characterized by onset before age 13 years. COS usually continues to progress in adolescence and adulthood and has a poor prognosis. Children with schizophrenia experience predominantly negative symptoms, but positive symptoms, most commonly auditory hallucinations, also occur (Masi et al., 2006). Functional impairment may be marked.

Pharmacotherapy is usually necessary for the control of psychotic symptoms. However, compared with adult patients with schizophrenia, there have been fewer studies on the pharmacological treatment of patients with COS.

A network meta-analysis of 12 RCTs ($n = 2158$; 8–19 years old) including 8 medications for psychosis (antipsychotics) (aripiprazole, asenapine, paliperidone, risperidone, quetiapine, olanzapine, molindone, and ziprasidone) found comparable efficacy among medications for psychosis for acute treatment of children and adolescents with schizophrenia, except that efficacy appeared inferior for ziprasidone and unclear for asenapine (Pagsberg et al., 2017). Weight gain was most prominent among those treated with olanzapine; prolactin elevation was associated with risperidone, paliperidone, and olanzapine. Sedation or insomnia did not differ among these medications for psychosis.

Unfortunately, as with FGAs, pediatric patients appear to have a greater susceptibility than adults to NGA-induced side effects, particularly EPS, weight gain, and dysphoria, but also hyperprolactinemia and leukocyte dyscrasias (Masi et al., 2006). However, despite a higher rate of side effects, clozapine appears to be a uniquely beneficial second-line agent for the treatment of COS patients with refractory illness (Remschmidt & Theisen, 2012). Further studies are required to identify the best treatment options for COS.

First-Episode Schizophrenia

It is well known that patients with first-episode schizophrenia are more responsive and more sensitive to antipsychotic side effects than chronically ill patients (Hasan et al., 2012). Two meta-analyses of studies in patients with first-episode schizophrenia concluded that there is a significant association between duration of untreated psychosis (DUP) and outcomes such as positive symptoms, negative symptoms, and functionality (Marshall et al., 2005; Perkins et al., 2005). However, there is no robust evidence that available interventions are successful in reducing DUP during the first episode of psychosis (Oliver et al., 2018). Moreover, recent studies suggested that duration of active psychosis (DAP) after treatment during early phases is a better predictor of long-term negative symptoms and social functioning than DUP (Lyne et al., 2017, Pardo-de-Santayana et al., 2020).

According to a network meta-analysis of 19 RCTs including 12 medications for psychosis (antipsychotics) ($n = 2669$), amisulpride, olanzapine, ziprasidone, and risperidone were found to be more efficacious in terms of overall symptom reduction than haloperidol, but the evidence was very low to moderate quality (Zhu et al., 2017). In addition, little difference in efficacy was detected among NGAs. In contrast, there were more apparent drug differences in adverse effects; olanzapine was associated with at least one use of drugs to treat parkinsonian symptoms and quetiapine with less akathisia than haloperidol, aripiprazole, risperidone, and olanzapine. Molindone was superior to risperidone, haloperidol, and olanzapine in terms of weight gain, and superior to risperidone in terms of prolactin elevation. These findings suggest that choice of the medication for psychosis used in first-episode schizophrenia should be made on the basis of likely adverse effects (Taylor, 2017).

Late-Life Schizophrenia

Efficacy and safety of medications for psychosis (antipsychotics) have been primarily studied in younger patients (Kane et al., 1983), and very limited data are available for patients with late-life schizophrenia (LLS). Whereas accumulated evidence indicates the age-related increased sensitivity with respect to safety/tolerability of medications for psychosis, choice and dose of medications for psychosis for patients with LLS have not been systematically addressed in clinical trials. Most of the data on the effectiveness of medications for psychosis drugs derive from clinical trials in behavioral and psychological symptoms with dementia (BPSD) (Ballard et al., 2005; Schneider et al., 2006). To our knowledge, there have been only three double-blind randomized control trials comparing two medications for psychosis (antipsychotics) specifically among patients with LLS (Howanitz et al., 1999; Jeste et al., 2003; Kennedy et al., 2003), and no evidence is available on which medication for psychosis should be used first. Moreover, no published double-blind randomized control trials

are available with regard to dosing of medications for psychosis. The available literature on dosing of medications for psychosis in the elderly with schizophrenia mostly consists of cross-sectional studies of prescription surveys and expert consensus. Expert consensus guidelines generally recommend the use of a lower dose of medications for psychosis in this population, for example, a dose range of 1.25–3.5 mg/day for the elderly (Alexopoulos et al., 2004), which is compatible with the findings from positron emission tomography studies that examined dopamine D_2 receptor blockade with medications for psychosis in patients with LLS (Uchida et al., 2009a; Graff-Guerrero et al., 2015).

Treatment During the Acute Phase

Route of Administration

Medications for psychosis (antipsychotics) are available as tablets, liquid concentrates, orally dissolving formulations, sublingual preparations, topical formulations, inhalable formulations, short-acting intramuscular (IM) preparations, or long-acting injectable preparations. Except for emergency situations in which involuntary IM administration may be required for safety reasons, the choice of route of application should be adjusted to patient preference.

One disadvantage of oral administration is that ingestion is less dependably assured than with parenteral administration. For example, a patient may appear to accept oral medication but will actually "cheek" and later spit the drug. Liquid concentrates, orally disintegrating tablets, or sublingual preparations are available for some medications for psychosis (antipsychotics) and can mostly avoid this problem. Another disadvantage of oral administration of medications for psychosis is that pharmacokinetic factors such as hepatic disease or slow gastrointestinal absorption may increase the half-life and the time required to achieve steady-state concentration. A new formulation of inhaled loxapine was approved by the FDA in 2012. In addition, a transdermal patch of blonanserin was approved in Japan, and asenapine was approved in the United States in 2019. The availability of these new formulations can provide more options to patients.

Short-acting IM formulations are particularly useful in the treatment of severely disturbed patients who present a danger to themselves or others and may require medication over objection. Short-acting IM formulations of NGAs are currently available for aripiprazole, olanzapine, and ziprasidone (Table 2). The IM formulations of these drugs have demonstrated good tolerability and superior efficacy compared with placebo or low-dose active comparator medication in acutely agitated patients with schizophrenia (Brook et al., 2000; Breier et al., 2002; Andrezina et al., 2006). However, a disadvantage of IM administration is the risk of injury to the patient (or members of the treatment team), usually by needle stick or accompanying physical restraint. In addition, higher doses of high-potency drugs can lead to acute dystonia or akathisia, which may increase the patient's agitation. Moreover, the coercive treatment may have negative effects on the patient-physician relationship and on future treatment adherence.

Some medications for psychosis (antipsychotics) can also be administered intravenously. This practice is generally associated with a more rapid onset of clinical action than occurs with IM preparations, although there is no evidence from well-designed clinical trials to support this. However, intravenous use of medications for psychosis may increase the risk of cardiac arrhythmias and autonomic complications, and should be considered only after other options have failed or ruled out. Long-acting formulations are seldom prescribed for acute psychotic episodes because they take several weeks to months to reach steady-state levels and are eliminated very slowly.

Selection of Medications for Psychosis (Antipsychotic Agents)

When choosing the optimal medication for psychosis (antipsychotic) for each individual patient, clinicians need to assess which medication is likely to provide the most suitable combination of efficacy and safety/tolerability and is able to improve their QOL and social functioning. Primary considerations include the patient's prior response to medications for psychosis and side effect experience, adherence history, available formulations, the drug's pharmacology and safety profiles, potential for drug interactions, the presence of comorbid medical conditions, and the patient's stated preference (Buchanan et al., 2010; Hasan et al., 2012). The patient and, if available, significant others should be included in a discussion of these considerations, and informed consent should precede medication administration. In this context, shared decision-making (SDM) is expected to improve therapeutic alliance, treatment satisfaction, and medication adherence as demonstrated in one pragmatic RCT targeting inpatients with schizophrenia (Hamann et al., 2020). When selecting medications for psychosis (antipsychotics), the risk of both acute and chronic adverse effects should be weighed against the profound disability that can accompany severe mental illness and the potential for medications for psychosis to greatly ameliorate the lives of patients and their families. It is prudent for clinicians to document their reasoning when selecting medications for psychosis.

In a network meta-analysis of 402 RCTs ($n = 53{,}463$) comparing efficacy of 32 medications for psychosis (antipsychotics) for the acute treatment of adults with multi-episode schizophrenia, the medication for psychosis with the largest standardized mean difference compared with placebo for reduction of overall symptoms was clozapine followed by amisulpride, zotepine, thiotixene, olanzapine, risperidone, perphenazine, thioridazine, zuclopenthixol, paliperidone, and haloperidol (Huhn et al., 2019). It should be noted though that acutely aggressive and suicidal patients are usually excluded from clinical trials and severely ill patients are unlikely to be included because of lack of capacity of providing consent, which limits their generalizability. In addition, there has been increased concern over the safety profile of *NGAs* such as weight gain and metabolic abnormalities. These side effects are associated with potential long-term health risks in an already at-risk population and also decrease adherence to treatment regimens, which should be identified and addressed from the beginning of the treatment. At present, NGAs (excluding clozapine) are recommended in several guidelines as first-

Table 2 Features and properties of new generation medications for psychosis (antipsychotics)

	Routes of administration and available formulations	Recommended frequency	Approved dose ranges for schizophrenia (mg/day)	Mean elimination half-life (hours)	Enzymes mainly involved for metabolism
Amisulpride	Oral (Tab)	Twice daily	400–1200	12	Weakly metabolized (two inactive metabolites accounting for approximately 4% of the dose)
Aripiprazole	Oral (Tab, OD, OS), SAI	Once daily	10–30	75	CYP2D6, 3A4
	LAI	Every 4 weeks	300–400 mg/4 weeks	30–47 days	
Asenapine	Oral (sublingual)	Twice daily	10–20	24	CYP1A2; UGT1A4
Blonanserin	Oral (Tab)	Twice daily	8–24	68	CYP3A4
Brexpiprazole	Oral (Tab)	Once daily	2–4	91	CYP3A4, 2D6
Cariprazine	Oral (Cap)	Once daily	1.5–6	2–4 days for cariprazine; 1–3 weeks for didesmethyl cariprazine	CYP3A4, 2D6
Clozapine	Oral (Tab)	Once or twice daily	300–900	12–16	CYP1A2, 2D6, 3A4
Iloperidone	Oral (Tab)	Twice daily	12–24	18–26	CYP2D6, 3A4
Lumateperone	Oral (Cap)	Once daily	42	18	CYP3A4, 2C8, 1A2; UGT1A1, 1A4, 1C4
Lurasidone	Oral (Tab)	Once daily	40–160	18	CYP3A4
Olanzapine	Oral (Tab, OD), SAI	Once daily	5–20	30	CYP1A2, 2D6
	LAI	Every 2 or 4 weeks	150–300 mg/2 weeks	30 days	
Paliperidone	Oral (ER-Tab)	Once daily	3–12	23	59% of the dose excreted unchanged into urine; limited role for CYP2D6, 3A4
	LAI	Every 4 weeks	39–234 mg/4 weeks	25–49 days	
		Every 12 weeks	273–819 mg/12 weeks	84–139 days	
Perospirone	Oral (Tab)	Three times a day	12–48	5–8	CYP3A4
Quetiapine	Oral (Tab, ER-Tab)	Twice (ER: once) daily	150–750	7	CYP3A4
Risperidone	Oral (Tab, OD, OS)	Once or twice daily	0.5–16	15	CYP2D6
	LAI	Every 2 weeks	25–50 mg/2 weeks	3–6 days	
		Every 4 weeks	90–120 mg/4 weeks	9–11 days	

(continued)

Table 2 (continued)

	Routes of administration and available formulations	Recommended frequency	Approved dose ranges for schizophrenia (mg/day)	Mean elimination half-life (hours)	Enzymes mainly involved for metabolism
Sertindole	Oral (Tab)	Once daily	4–24	55–90	CYP2D6, 3A4
Ziprasidone	Oral (Tab with food), SAI	Twice daily	40–160	7	CYP3A4
Zotepine	Oral (Tab)	Three times daily	75–450	13–16	CYP3A4

Data are based on each product information
Abbreviations: CAP, capsule; CYP, cytochrome P450; ER, extended-release; LAI, long-acting injection; OD, orally disintegrating tablet; OS, oral solution; SAI, short-acting injection; Tab, tablet; UDT, uridine 5′-diphospho-glucuronosyl-transferases; AKR, aldoketoreductase

line agents for acute therapy for schizophrenia (Hasan et al., 2012; Osser et al., 2013). In addition, although the evidence level is considered low, choice of specific medications for psychosis is suggested by expert consensus depending on types of symptoms and treatment goals (Sakurai et al., 2021).

Dosage of Medications for Psychosis

The goal of pharmacological treatment is to maximize efficacy and minimize adverse effects with the lowest effective and best tolerated dose. Determination of the optimum dosage of medication for psychosis (antipsychotics) is complicated by the fact that both response and susceptibility to side effects may differ from patient to patient and may differ across phases of the illness (Buckley & Correll, 2008). For example, first-episode patients generally respond to lower doses of medication for psychosis than do patients with recurrent episodes (Takeuchi et al., 2019). Poor or partial responders may benefit from somewhat higher doses (Sakurai et al., 2016). The physiology of this variability is not well understood, but it seems likely that individual variations in pharmacokinetic as well as pharmacodynamic factors play a role. Dosage should be titrated in consideration of side effects and efficacy.

PET studies in younger patients with schizophrenia have shown that striatal dopamine D_2 receptor occupancy with medications for psychosis (antipsychotics) above 65% is associated with clinical response in the acute phase treatment of positive symptoms of schizophrenia, while EPS

are more likely to occur with a D_2 receptor occupancy above 80% (Farde et al., 1992; Kapur et al., 2000a; Uchida et al., 2011b). A pooled analysis of 12 PET studies found 60% D_2 receptor occupancy to achieve a 25% or greater symptom reduction and 72% for a 50% or greater symptom reduction (Uchida et al., 2011b). This range of 65–80% has been referred to as a "therapeutic window" for full D_2 antagonist drugs. Because of intrinsic activity of partial agonists, a higher receptor occupancy (>90%) is required for efficacy. Interestingly, selective partial agonism of the presynaptic (short) D_2 is associated with reduced dopamine release and D_2 receptor occupancy of 39–60% (Kim et al., 2012; Snyder et al., 2015; Vanover et al., 2019).

A recent meta-analysis of 26 studies involving 5618 patients with schizophrenia examined the dose-response relationship to seek optimal doses of medication for psychosis (antipsychotics) than in acute phase treatment (Takeuchi et al., 2020). Minimum effective dose (MED), defined as a dose that was significantly superior to placebo for the primary outcome in at least one identified study, was first determined as follows: aripiprazole (10 mg/day), asenapine (10 mg/day), clozapine (300 mg/day), haloperidol (4 mg/day), iloperidone (8 mg/day), lurasidone (40 mg/day), olanzapine (7.5 mg/day), paliperidone (3 mg/day), quetiapine (150 mg/day), risperidone (2 mg/day), sertindole (12 mg/day), and ziprasidone (40 mg/day). Both twofold and threefold MEDs were superior to one MED for effectiveness, and one MED was superior to twofold and threefold MEDs for side effects. These results suggest that twofold or threefold MED could be a target dose for patients with acute schizophrenia with close monitoring of side effects.

Maintenance Treatment

Therapeutic Goals

The goals of maintenance treatment are to prevent relapse, maintain long-term control of symptoms and behavior, improve functioning and QOL, consolidate remission, and promote recovery, with minimal adverse effects (Barnes et al., 2020; Hasan et al., 2013). Managing treatment adherence is an important part of maintenance treatment plans. There is substantial evidence from clinical trials that APDs can reduce the risk of relapse. In fact, a recent meta-analysis of 75 RCTs with a total of 9145 patients demonstrated the superiority of medications for psychosis (antipsychotics) over placebo in preventing relapse (Ceraso et al., 2020). Of note, psychosocial interventions, including family intervention, supported employment, assertive community treatment, social skills training, cognitive behavioral therapy, cognitive remediation, and social cognitive training, are recommended as important adjunctive treatments to antipsychotic pharmacological therapy (Barnes et al., 2020; Hasan et al., 2013). Pharmacological treatment should be tailored to the needs and preferences of the individual patient. Managing treatment adherence is an important part of maintenance treatment plans.

Selection of Medication for Psychosis (Antipsychotics)

Patient preference and drug experience including the patient's prior response and side effects are key determinants of medication selection.

Results from the CATIE phase 1 study demonstrated that olanzapine was superior to quetiapine and risperidone in terms of time to discontinuation for any cause, despite substantial metabolic side effects associated with olanzapine (Lieberman et al., 2005a). The FGA perphenazine and four NGAs displayed comparable changes in PANSS scores. Perphenazine was similar in efficacy, tolerability, cognition (Keefe et al., 2007), cost (Rosenheck et al., 2006), QOL (Swartz et al., 2007), and psychosocial functioning (Swartz et al., 2007) to the NGAs, suggesting that certain FGAs remain viable treatment options in chronic schizophrenia.

In CUtLASS, no significant differences in QOL, positive or negative symptoms, side effects, or patient satisfaction were observed between patients treated with FGAs and those treated with NGAs (Jones et al., 2006).

Unlike CATIE and CUtLASS, EUFEST found a clear advantage of the four NGAs (amisulpride, olanzapine, quetiapine, and ziprasidone), both as a group and individually, over haloperidol in terms of the risk for any-cause discontinuation, Clinical Global Impression scores, Global Assessment of Functioning scores (Kahn et al., 2008), and the proportions of response and remission (Boter et al., 2009). However, there were no group differences in PANSS total scores, the Calgary Depression Scale, QOL scores, or the improvement of cognitive functions (Davidson et al., 2009). On the other hand, an integrated analysis of CATIE and EUFEST data pointed towards an advantage of olanzapine compared to the other medications for psychosis (antipsychotics) used in these studies with respect to the effects of hostility on study discontinuation (Volavka et al., 2016).

In summary, CATIE, CUtLASS, and EUFEST did not provide unequivocal evidence to support the superiority of one medication for psychosis (antipsychotic) over another. To date, no single antipsychotic drug has emerged as superior in preventing relapse among all of the available medications for psychosis. Hence, the choice of medications for psychosis for maintenance treatment should rely on individual risk–benefit analysis and patient's preference, especially considering the individual safety profiles of the available medications. Although the evidence level is considered low, expert consensus suggests a choice of specific medications for psychosis for common clinical situations (Sakurai et al., 2021).

Medications for Psychosis (Antipsychotics): Dose and Dosing Interval

While medications for psychosis (antipsychotics) significantly reduce the risk of relapse, they are associated with a variety of dose-related serious adverse events, including motor (Uchida et al., 2011b) and cardiovascular side effects (Schneeweiss & Avorn, 2009; Ray et al., 2009b) as well as cognitive impairments (Sakurai et al., 2012). Therefore, it is ideal to minimize exposure to these drugs and to identify the lowest effective dose for each individual patient (Uchida et al., 2009b, 2011a). A meta-analysis of 13 studies with 1395 patients with schizophrenia found maintenance treatment with moderately low doses (50–100% of World Health Organization defined standard doses) was comparable in overall treatment failure or hospitalization to standard doses, whereas very low doses (less than 50% of standard doses) increased the risk of relapse (Uchida et al., 2011a). On the other hand, no significant differences were found in the rate of dropouts due to side effects between either standard dose versus low dose or very low dose. However, these results have to be interpreted in light of a small number of relevant studies.

Regular dosing with intervals of drug's half-life in blood has been the standard because this produces the high and continuous level of dopamine D_2 receptor blockade that is thought to maximize therapeutic response in the acute treatment phase. However, different dosing schedules have been tried to reduce dose-dependent antipsychotic side effects in the maintenance phase (Servonnet et al., 2020). An initial strategy was "targeted medications for psychosis (antipsychosis) dosing," in which antipsychotic treatment is resumed at the earliest signs of psychotic relapse following "drug holiday." This dosing method has been consistently reported to increase the risk of relapse and rehospitalization compared to regular dosing (Jolley & Hirsch, 1993; Schooler, 1991; Jolley et al., 1990; Gaebel, 1994; Gaebel et al., 2002). The second strategy is "extended, but regular dosing of medications for psychosis," in which medications for psychosis are regularly given but with longer intervals than usual such as every 2–3 days

for oral medications for psychosis (Remington & Kapur, 2010). While animal data have shown reductions of tardive dyskinesia as well as dopamine supersensitivity in support of this dosing regimen (Turrone et al., 2005; Samaha et al., 2008), human clinical trial data are still preliminary.

Adherence

Nonadherence can lead to psychotic relapse, rehospitalization, more frequent clinic and emergency room visits, and poorer long-term functioning. According to a recent meta-analysis of 19 studies ($n = 2184$) that examined medication adherence by using electronic adherence monitoring, adherence rate of medications for psychosis was 71.1% [from seven studies, $n = 256$, 95% confidence interval (CI) 58.0–84.1] while an 80% threshold is widely used to define satisfactory adherence (Yaegashi et al., 2020). Risk factors of nonadherence included younger age, poor insight into illness, cannabis abuse, and severe positive symptoms (El Abdellati et al., 2020). In addition, good adherence to antipsychotic treatment was associated with positive attitude toward medication of both patients and their family, family involvement, and illness insight.

Injectable Long-Acting Formulations

Long-acting injectable (LAI, also commonly referred to as depot) formulations allow for stable plasma concentrations of the active drug to remain at a therapeutic dose range for an extended period of time. Therefore, they are particularly advantageous for patients with adherence problems and for those with a history of severe relapse upon medication discontinuation. They also simplify compliance monitoring, as they are generally administered at scheduled intervals by a health care provider, and ameliorate the daily burden of taking oral medication for the patient (Rauch & Fleischhacker, 2013). In addition, benefits with respect to functioning and service engagement have been reported (Peters et al., 2019). Thus, LAI medications for psychosis (antipsychotics) are considered to be useful in the maintenance phase of treatment.

Meanwhile, there has been increasing evidence on the effectiveness of LAI medications for psychosis (antipsychotics) in the early disease stage (Schreiner et al., 2015; Huang et al., 2018; Kim et al., 2021). Nonadherence to treatment is more prominent in the earlier phases of schizophrenia (Lang et al., 2010) and contributes to an increased risk of persistent psychotic symptoms and relapse (Morken et al., 2008). Prolonged or recurrent psychotic episodes may be related to disease progression (Wyatt, 1991; Perkins et al., 2005; Anderson et al., 2014). Therefore, administration of LAI medications for psychosis may also be considered in early illness stages to improve long-term prognosis in terms of treatment adherence and prognosis. Indeed, a recent study regarding the use of LAI medications for psychosis in early schizophrenia has showed benefits in delaying time to hospitalization and improving social functioning (Kane et al., 2020; Kim et al., 2021).

As of January 2021, at least five FGAs and four NGAs are available in LAI depot formulations (Table 2). There are two LAI risperidone preparations. An intramuscular formulation is prepared by encapsulating the drug in biodegradable polymeric microspheres, while a subcutaneous formulation locks the active drug within a polymer that solidifies only after injection. LAI paliperidone palmitate was the first once-monthly depot NGA approved in the United States in 2009; its once-every-3-month depot preparation was approved in 2015, and its once-every-6-month depot preparation was approved in 2021. LAI olanzapine pamoate was also approved by the FDA in 2009. Its main safety concern is the risk of a post-injection syndrome of excessive sedation/delirium. Therefore, after receiving LAI olanzapine pamoate, a patient must be monitored by a health care provider for at least 3 h. LAI aripiprazole was approved by the FDA in 2013, and the related aripiprazole lauroxil was approved in 2015.

Results of meta-analyses of LAI medications for psychosis (antipsychotics) versus oral medications for psychosis have yielded mixed results (Misawa et al., 2016; Kishimoto et al., 2018; Peters et al., 2019). However, naturalistic cohort studies showed the advantages for LAI medications for psychosis over oral medications for

psychosis in relapse prevention in patients with schizophrenia (Grimaldi-Bensouda et al., 2012; Tiihonen et al., 2017). Specifically, Tiihonen et al. reported real-world effectiveness of LAI in a nationwide cohort of patients with schizophrenia, employing within-individual analyses (Tiihonen et al., 2017). In order to confirm real-world effectiveness of LAI medications for psychosis compared to oral medications for psychosis, large and long-term pragmatic studies are warranted. Moreover, current evidence is insufficient to recommend one specific LAI medication for psychosis over another.

Monotherapy Versus Polypharmacy

Medications for psychosis (antipsychotics) polypharmacy (i.e., a concurrent use of more than one medication for psychosis) is common in real-world clinical settings with prevalence rates varying widely (4–70%), depending on the setting and the patient population (Fleischhacker & Uchida, 2014). The current evidence suggests some clinical benefits of medications for psychosis polypharmacy such as better symptom (Ortiz-Orendain et al., 2017) and a reversal of metabolic side effects with concomitant use of aripiprazole (Mizuno et al., 2014). Also, a recent nationwide cohort study also found the lowest rehospitalization rates in patients with schizophrenia taking combination medications for psychosis, primarily clozapine paired with other agents or LAIs paired with other agents (Tiihonen et al., 2019). However, most of the data from randomized clinical trials on medications for psychosis polypharmacy are very heterogeneous (Correll et al., 2009) and derive from short-term studies, limiting the assessment of long-term efficacy and safety. Therefore, the interpretation of findings in the literature should be made conservatively in light of potential burden of additional side effects caused by medications for psychosis polypharmacy. One small meta-analysis examined whether patients receiving medications for psychosis polypharmacy should switch to medications for psychosis monotherapy or stay on polypharmacy and found a significant difference in study discontinuation due to all causes in favor of staying on medications for psychosis polypharmacy when compared to converting polypharmacy to monotherapy (Matsui et al., 2019). On the other hand, since many patients appear to tolerate the switch and clinical worsening due to the regimen change from polypharmacy to monotherapy has been reported to be reversed by introducing the previous regimen, converting medications for psychosis polypharmacy to monotherapy could be a valid and reasonable treatment option. Thus, medications for psychosis polypharmacy may work for some clinically difficult conditions; however, it should be the exception rather than the rule.

Treatment Resistance

One-fifth to one-third of all patients with schizophrenia experience significant psychotic symptoms despite adequate trials of pharmacological treatment and are considered "treatment-resistant schizophrenia" (TRS) (Lieberman, 1999). Additional subsets of patients may be treatment intolerant, treatment noncompliant, or slow to respond. The possibility that upregulation of the dopamine system in response to dopamine blocking agents may lead to dopamine supersensitivity psychosis (Chouinard et al., 2017) in a manner similar to tardive dyskinesia is another possible consideration. The Treatment Response and Resistance in Psychosis (TRRIP) Working Group Consensus Guidelines on Diagnosis and Terminology proposed the following criteria to define treatment-resistance as minimum requirement: at least moderate severity determined by a standardized rating scale for ≥ 12 weeks; at least moderate functional impairment measured using a validated scale; ≥ 2 past adequate treatment episodes (i.e., ≥ 600 mg/day of chlorpromazine equivalent dose for ≥ 6 weeks) with different medications for psychosis (antipsychotics); and $\geq 80\%$ of prescribed doses taken, confirmed with at least two sources (Howes et al., 2017).

Clozapine

Clozapine is the only medication to consistently demonstrate efficacy against psychotic symptoms in well-defined patient populations with TRS, and it is the only medication for psychosis (antipsychotic) approved by the FDA for this indication. The drug's singular efficacy in TRS was first conclusively demonstrated in a pivotal study involving 268 patients who were prospectively established to be treatment resistant using narrow criteria (Kane et al., 1988). About 30% of patients receiving clozapine responded at 6 weeks, compared with 7% treated with chlorpromazine. This superiority of clozapine has been replicated in subsequent clinical trials. According to a meta-analysis of 21 studies, 40.1% of patients with treatment-resistant schizophrenia responded to clozapine, with a mean reduction in PANSS of 22.0 points, a reduction of 25.8% (95% CI, 24.7–26.9%) from baseline (Siskind et al., 2017).

Other Medications for Psychosis (Antipsychotics)

The use of other medications for psychosis (antipsychotics) for TRS has been explored, with largely disappointing results. One early RCT suggested a possible role for high-dose olanzapine in the management of TRS (Meltzer et al., 2008), but a preponderance of evidence from four RCTs has not shown this strategy to be as effective as treatment with clozapine (Duggan et al., 2005). The CUt-LASS 2 trial compared the effectiveness of clozapine with that of other NGAs (amisulpride, olanzapine, quetiapine, and risperidone) in 136 patients with schizophrenia with clinician-defined poor response to two or more medications for psychosis (Lewis et al., 2006a, b). Clozapine showed a statistically significant advantage over other NGAs with respect to the total PANSS score, but not with respect to QOL, which was the primary outcome in this study, or associated costs of care over 1 year.

Because of clozapine's superior efficacy relative to other medications for psychosis (antipsychotics), monotherapy with this agent represents the "gold standard" for treatment of patients with TRS. Unfortunately, the increased risk of serious side effects compared to other agents combined with the requirement for regular blood monitoring render it unsuitable as a first-line intervention. Trials of other dopamine serotonin blockers before proceeding to clozapine are therefore still state of the art, but clinicians should consider clozapine early in the management of treatment resistance to other medications for psychosis. Pharmacoepidemiological studies have shown that in many patients, years go by before they are started on clozapine (John et al., 2018). Certain patients may respond preferentially to a single agent of this class and potentially avoid the burden of treatment with clozapine. Clinicians should not, however, withhold clozapine from patients who have failed therapeutic trials with two or more other agents.

Even with clozapine monotherapy of adequate dosage and duration, one-third to two-thirds of people with TRS still have persistent positive symptoms (Cipriani et al., 2009). One strategy commonly employed in the management of such patients is to augment clozapine with other psychotropic medications. The evidence supporting this practice, however, is very limited. A meta-analysis of 46 studies suggested the use of aripiprazole, fluoxetine, and sodium valproate as augmentation agents for total psychosis symptoms and memantine for negative symptoms (Siskind et al., 2018). However, limitations clearly included short follow-up periods and poor study quality. Indeed, no drug has consistently and repeatedly demonstrated efficacy as an adjuvant to clozapine in TRS. Further controlled studies are needed to clarify the potential of other psychotropic agents to serve as adjuvants in TRS. Lastly, there is some evidence that electroconvulsive therapy may lead to symptom improvement, especially when added to clozapine (Petrides et al., 2015).

Use of Plasma Levels of Medications for Psychosis (Antipsychotics)

Therapeutic drug monitoring (TDM), the measurement of serum or plasma drug levels for dose optimization, is, if used appropriately, a valid tool for patient-matched pharmacotherapy (Hiemke et al., 2018). Due to genetic and nongenetic factors such as age, gender, comorbid illness, nutrition, concomitant medications, adherence, and smoking, it is well known that large inter- and intraindividual variability in plasma drug levels exists among patients

treated with the same antipsychotic dose (Ng et al., 2009). Genetic factors, particularly genetic polymorphisms of the drug-metabolizing enzymes, are major causes of these interindividual differences. In the light of such wide inter- and intraindividual differences, TDM of medications for psychosis (antipsychotics) is useful in certain situations. For example, before deciding that an agent is ineffective despite an adequate trial duration at a sufficient dose, it is important to determine whether lack of efficacy may be due to such factors as pharmacokinetic variables (e.g., cytochrome P450 oxidase polymorphisms) or medication nonadherence. A low plasma level at a normally therapeutic drug dose may prompt the clinician to raise the dose, investigate etiologies of poor absorption or accelerated metabolism, or address medication compliance. A high plasma level, on the other hand, may suggest lowering the dose, because adverse effects may be obscuring therapeutic effects.

Drug Interactions and Medications for Psychosis (Antipsychotics)

Medications for psychosis (antipsychotics) are subject to drug–drug interactions with other psychotropic drugs or with medications used in the treatment of comorbid psychiatric or physical illnesses, or for controlling adverse effects (Spina & de Leon, 2007; Patteet et al., 2012). Most pharmacokinetic interactions with medications for psychosis occur at the metabolic level and usually involve changes in the activity of drug-metabolizing enzymes involved in their biotransformation, such as cytochrome P450 (CYP) monooxygenases and uridine diphosphate glucuronosyltransferases (UGTs) (Spina & de Leon, 2007). The major CYP isoenzymes responsible for metabolism of NGAs are CYP1A2, CYP2C9, CYP2C19, CYP2D6, and CYP3A4. A notable exception is amisulpride, which is primarily excreted in the urine and shows no effect on the activity of the main human CYP isoenzymes (Spina & de Leon, 2007). Moreover, CYP isoenzymes play only a limited role in the metabolism of paliperidone (Patteet et al., 2012).

When medications are metabolized by the same CYP system, drug–drug interactions can occur. Plasma levels of NGAs are affected by medications that induce or inhibit CYP isoenzymes. These interactions usually involve medications for depression and antiepileptic drugs. Table 2 summarizes clinically significant pharmacokinetic drug interactions affecting NGAs.

There are other common drug–drug interactions that may be clinically important. Valproic acid increases mean plasma quetiapine levels by 77% via an unclear mechanism (Aichhorn et al., 2006). Medications for psychosis (antipsychotics) can antagonize the effects of dopamine agonists and levodopa. Medications for psychosis may also enhance the effects of central nervous system depressants such as analgesics, anxiolytics, and hypnotics. Coadministration of clozapine and benzodiazepines can cause lethargy, ataxia, loss of consciousness, and, rarely, respiratory arrest in the first 24–48 h after the first clozapine dose (Spina & de Leon, 2007).

There is also evidence that medications for psychosis (antipsychotics) and their active metabolites are, to various extents, substrates or inhibitors of the efflux transporter of the blood–brain barrier, permeability-glycoprotein (P-glycoprotein), which could limit their absorption or entry into the brain (see the section "Endocrine and Sexual Effects").

Knowledge of the interaction profiles of individual medications for psychosis is useful in choosing an appropriately effective drug that is less likely to interfere with other medications. In addition, avoidance of unnecessary polypharmacy and the readiness to adjust antipsychotic dosage based on clinical response, and possibly plasma drug concentration, will help to minimize adverse drug interactions (Spina & de Leon, 2007).

Medications for Psychosis (Antipsychotics) and Pregnancy

Treatment with medications for psychosis (antipsychotics) during pregnancy is likely to serve a protective function for women with schizophrenia, as discontinuation of medication for psychosis clearly increases the risk of relapse. Such relapse may pose risks to the fetus and mother, including stillbirth, prematurity, small size for gestational age, difficulties with parenting, and loss of custody (Nilsson et al., 2002; Yaeger et al., 2006). On the other hand, antipsychotic treatment during pregnancy can increase the risk of gestational diabetes mellitus (Kucukgoncu et al., 2020).

Data on the impact of in utero exposure to medications for psychosis (antipsychotics) on congenital malformations seem mixed. According to a meta-analysis of 12 studies including 1782 patients with schizophrenia who took NGAs during pregnancy and 1,322,749 controls, the use of NGAs during the first trimester of pregnancy was associated with an increased risk for major congenital malformations with an odds ratio of 2.03; however, no specific pattern of malformations was found. An increased risk was also detected for preterm births (odds ratio, 1.85) (Terrana et al., 2015). On the other hand, a nationwide Medicaid database including 1,360,101 pregnant women with a live-born infant reported that the relative risks for overall and cardiac malformations, after adjustment for confounders, were not significantly increased with 1.05 and 1.06, respectively, for NGAs (Huybrechts et al., 2016).

With regard to long-term motor development following in utero exposure to medications for psychosis (antipsychotics), several cohort studies (Johnson et al., 2012; Mortensen et al., 2003; Hurault-Delarue et al., 2016; Shao et al., 2015) consistently showed a deficit in motor functioning in the first 9 months of life, but which appeared to spontaneously resolve thereafter (Poels et al., 2018).

The risk profile of medications for psychosis (antipsychotics) in pregnancy needs to be carefully considered. As the ethical and practical challenges of conducting placebo-controlled trials in this area render it difficult to distinguish risks posed by medications for psychosis (antipsychotics) from those due to underlying illness and associated behaviors, the field has to rely on evidence from observational and cohort studies.

The long-term risks of antipsychotic treatment during breastfeeding are similarly unknown, and there are insufficient data to ensure that the benefits of breastfeeding outweigh the risks of drug exposure to the infant.

Whenever possible, patients should obtain a prepregnancy consultation to discuss with their clinician(s), partner, and family the known and potential risks and benefits of antipsychotic treatment during pregnancy. In negotiating this difficult decision, emphasis should be placed on optimizing the mother's health and ability to parent (Yaeger et al., 2006). Social and other non-pharmacological supports should be maximized during pregnancy and in the postpartum. If medication for psychosis is used, monotherapy at the lowest effective dose may help to minimize risks to the fetus.

Adverse Effects

Acute Extrapyramidal Symptoms (Dystonia, Parkinsonism, and Akathisia)

Medications for psychosis (antipsychotics)-induced EPS can occur acutely or after chronic treatment. Virtually all medications for psychosis are capable of producing EPS, although they vary considerably in their propensity to do so. As a group, NGAs produce less EPS compared to FGAs; however, the frequency of EPS among individual NGAs is heterogeneous. In a network meta-analysis of 136 RCTs ($n = 24,991$) comparing medications for psychosis and placebo for the acute treatment of adults with multi-episode schizophrenia, the following drugs were significantly less likely to cause EPS than haloperidol with regard to EPS risk, starting with the best: clozapine, perazine, sertindole, placebo, olanzapine, quetiapine, asenapine, aripiprazole, thioridazine, amisulpride, iloperidone, brexpiprazole, paliperidone, ziprasidone, risperidone, lurasidone, zotepine, and chlorpromazine (Huhn et al., 2019). Drugs that possess a high potency with respect to the D_2 receptor generally carry the highest risk of EPS. Mid- and low-potency drugs pose less risk.

Commonly occurring acute EPS include Parkinsonism, dystonia, and akathisia. Acute EPS typically develop within hours to weeks of initiation of medications for psychosis. These movement disorders are dose dependent and almost always reversible, remitting after the offending agent has been discontinued. The increasing use of NGAs is believed to have substantially reduced the burden of acute EPS.

Medication-induced Parkinsonism usually occurs within days to weeks of antipsychotic treatment initiation and is characterized by rigidity, tremor, and bradykinesia. Care should be taken to distinguish these patients from those with depression, catatonia, negative symptoms of schizophrenia, or Parkinson's disease. Medication-induced Parkinsonism occurs most commonly with high-potency medications for psychosis (antipsychotics), especially when anticholinergic medication is not administered concurrently. Other risk factors include older age, higher dose, a history of Parkinsonism, and underlying basal ganglia pathology.

Dystonia presents as sustained muscular contraction, with contorting, twisting, or abnormal posturing affecting mainly the muscles of the head and neck, but sometimes the trunk or extremities. It tends to be sudden in onset, with 95% of cases occurring in the first 96 h of antipsychotic treatment (van Harten et al., 1999). Dystonia is often dramatic in presentation and may be highly distressing and painful for patients. Laryngeal dystonia is potentially fatal, as it may compromise the airway. Dystonia sometimes occurs with perceptual, mainly visual, alteration (Uchida et al., 2003). Risk factors for acute dystonia include men, young age, use of high-potency medications for psychosis (antipsychotics), higher dose, and IM administration (Ayd, 1961).

Akathisia is characterized by both subjective and objective somatic restlessness. Patients with akathisia typically experience an inner sensation of restlessness or an irresistible urge to move various parts of their bodies. The disorder appears objectively as psychomotor agitation, such as continuous pacing, rocking from foot to foot, or the inability to sit still. Akathisia usually begins within hours to days of administration of medications for psychosis (antipsychotics) (Casey, 1993). It can be extremely distressing for patients and may contribute to medication nonadherence and self-injurious behavior.

Patients should be monitored for EPS at weekly intervals during initiation medications for psychosis (antipsychotics) and until their medication dose has been stable for at least 2 weeks. The treatment of acute EPS depends on the specific side effect.

The initial treatment of Parkinsonism is to lower the dose of medications for psychosis (antipsychotics), since doses above the EPS threshold are unlikely to yield additional clinical benefit (Farde et al., 1992; Kapur et al., 2000a; Uchida et al., 2011b). If symptoms persist, switching to a medication for psychosis associated with a lower risk of EPS should be considered. Alternatively, adding an anticholinergic drug may be efficacious, although these drugs carry their own risks of adverse effects, including dry mouth, constipation, urinary retention, blurred vision, cognitive impairment, and abuse, and therefore should be used at the possible lowest dose.

Acute dystonia usually responds rapidly to treatment with an anticholinergic (e.g., benztropine) or an antihistaminergic (e.g., diphenhydramine) agent, especially when administered parenterally. Oral anticholinergic medication should then be continued for at least a few days to prevent recurrence of dystonia.

Initial treatment options for akathisia include lowering the dose of the medication for psychosis (antipsychotic) and switching to a medication for psychosis with a lower risk of akathisia. If symptoms persist, or if psychotic symptoms necessitate a higher dose of the medication, a trial of a β-adrenergic antagonist (e.g., 30–90 mg/day of propranolol) or a benzodiazepine may be attempted; anticholinergic agents, in contrast, are generally less effective for akathisia (Casey, 1993; Miller & Fleischhacker, 2000).

Tardive Dyskinesia and Related Syndromes

Tardive dyskinesia (TD) is a repetitive, involuntary, hyperkinetic movement disorder caused by months to years of exposure to medications for psychosis (antipsychotics). It can be associated with reduced quality of life, difficulty with certain motor tasks, as well as lead to embarrassment and social withdrawal (Citrome & Saklad, 2020). TD is characterized most commonly by orofacial movements, such as lip smacking, chewing, and tongue protrusion, and also by choreoathetoid movements of the neck, limbs, trunk, and occasionally the diaphragm (Casey, 1999). The abnormal movements of TD usually increase with emotional arousal and are absent during sleep. Dyskinetic movements generally begin either while the patient is receiving a medication for psychosis or within a few weeks of discontinuing a medication for psychosis.

Less common than the choreoathetoid movements described above are variants of TD known as tardive dystonia, tardive akathisia, and tardive parkinsonism. These syndromes resemble their acute counterparts, but by definition, they emerge after longer-term antipsychotic treatment. Tardive dystonia and tardive akathisia can be particularly disabling and often persist despite discontinuation of medications for psychosis (antipsychotics). More than one variant of TD may be present simultaneously.

A meta-analysis regarding medications for psychosis (antipsychotics)-induced tardive dyskinesia in patients with schizophrenia underscores the superiority of NGAs (21%) over FGAs (30%) with respect to prevalence and incidence rates (Carbon et al., 2017). When dyskinetic movements emerge, a neurological evaluation should be considered to rule out other etiologies. The clinician should also consider reducing the dose of the medication for psychosis or switching to an agent with lower potential for causing TD, although the evidence is still insufficient (Bergman et al., 2018). Either of these approaches may paradoxically cause an initial worsening of TD, but in many cases, the movements will return to baseline and further improve with time. Clozapine may have the lowest potential for causing the condition (Casey, 1999).

Two novel reversible vesicular monoamine transporter 2 (VMAT2) inhibitors, valbenazine and deutetrabenazine, received FDA approval for the treatment of TD in 2017. These VMAT-2 inhibitors are effective in treating TD, both acutely and long-term, with no evidence for increased risk of depression or suicide [unlike tetrabenazine which may increase suicide ideation in Huntington Disease (Artukoglu et al., 2020)]. They are thought to act by inhibiting the transporter that regulates monoamine (particularly dopamine, but also noradrenaline, serotonin, and histamine) uptake from the cytoplasm to the synaptic vesicle for storage and release. In addition, vitamin E, vitamin B_6, and amantadine may also be effective for the treatment of TD (Artukoglu et al., 2020) although they are off-label.

Neuroleptic Malignant Syndrome

Experts differ on the core diagnostic features of neuroleptic malignant syndrome (NMS), but most agree that the condition is characterized by the triad of rigidity, hyperthermia ($>38\ °C$), and autonomic instability (especially tachycardia and blood pressure lability), occurring in association with the use of medications for psychosis (antipsychotics). In addition, elevated serum creatine kinase level, leukocytosis, and/or alteration in level of consciousness are often present. The disorder usually develops during the initial weeks of antipsychotic treatment and can be sudden in onset. NMS is a medical emergency that should be distinguished from such differential diagnostic considerations as catatonia, serotonin syndrome, sepsis, endocrinopathy, and heat stroke (Fleischhacker et al., 1990). If not diagnosed and treated promptly, death can occur in 5–20% of cases.

NMS is relatively rare, occurring in 0.1–0.2% of patients treated with medications for psychosis (antipsychotics) (Schneider et al., 2020; Lao et al., 2020). The risk of NMS with NGAs is lower than that with FGAs (Schneider et al., 2020). Preexisting organic pathologies of the central nervous system, lithium treatment, infection/exsiccosis, and the withdrawal of medication with anticholinergic properties or alcohol were reported to be other risk factors.

If NMS is suspected, the offending medications for psychosis (antipsychotics) should be discontinued immediately, and the patient should be transferred to a medical setting where supportive treatment can be initiated (Strawn et al., 2007). Treatment guidelines suggest concrete therapy recommendations such as benzodiazepines, dantrolene, bromocriptine, amantadine, intensive care, and/or electroconvulsive therapy (ECT), however, with high heterogeneity (Schönfeldt-Lecuona et al., 2020). After several weeks of recovery, a medication for psychosis may be gradually reintroduced. Generally, NGAs with a low D_2 affinity are selected, and the dose is titrated slowly while observing for signs of NMS recurrence.

Endocrine and Sexual Effects

Many FGAs and NGAs can elevate serum prolactin levels by antagonizing the tonic inhibitory actions of dopamine on lactotrophic cells in the pituitary. In a network meta-analysis of 90 RCTs ($n = 21,569$) comparing medications for psychosis (antipsychotics) and placebo for the acute treatment of adults with multi-episode schizophrenia, olanzapine, asenapine, lurasidone, sertindole, haloperidol, amisulpride, risperidone, and paliperidone were associated with significantly elevated prolactin levels (mean difference range 4.47–48.51 ng/mL) (Huhn et al., 2019). Although some patients develop partial tolerance to the pituitary effect of medications for psychosis after several weeks, most have chronic serum prolactin elevation so long as antipsychotic treatment is continued (Haddad & Wieck, 2004). A few medications for psychosis are associated with increases in serum prolactin that appear to exceed striatal D_2 receptor occupancy. Risperidone and paliperidone are the best examples of this. This may be due to the observation that they have a high affinity to the P-glycoprotein and have to be dosed to overwhelm this protein in the blood–brain barrier. However, since the lactotrophs in the anterior pituitary are outside the blood–brain barrier and are not protected by the P-glycoprotein, they experience a significantly higher level of these drugs (El-Mallakh & Watkins 2019).

Women experience, on average, significantly greater elevation in prolactin than do men during long-term treatment with the same dose of medication for psychosis (antipsychotic) (Haddad & Wieck, 2004). Hyperprolactinemia may manifest differently in men and women, and there is great individual variation in the prolactin level at which symptoms appear (Haddad & Wieck, 2004). In women, prolactin elevation may lead to menstrual disturbances, including anovulatory cycles and infertility, menses with abnormal luteal phases, or frank amenorrhea and hypoestrogenemia (Malik et al., 2011). Women may also experience decreased libido, impaired arousal, anorgasmia, and, at least theoretically, increased long-term risk of osteoporosis. An association between use of medications for psychosis and risk of breast cancer has still been in debate (Hicks et al., 2020; Pottegård et al., 2018). The major effects of hyperprolactinemia in men are loss of libido, hypospermatogenesis, erectile or ejaculatory deficits, and occasionally galactorrhea or priapism. Medications for psychosis -induced sexual side effects can be particularly bothersome for some patients, whereas for others, sexual dysfunction associated with schizophrenia itself may improve with antipsychotic treatment (Haddad & Wieck, 2004).

Clinicians should screen for signs and symptoms of hyperprolactinemia prior to and during treatment with prolactin-elevating medications for psychosis (antipsychotics) (Riecher-Rössler et al., 2013). If symptomatic hyperprolactinemia is present, consideration should be given to lowering the dose of the medication for psychosis or switching to a "prolactin-sparing" agent. Other options include adjunctive use of metformin (Wu et al., 2012) and aripiprazole (Kelly et al., 2018).

Weight Gain and Obesity

Obesity can negatively affect self-image, impair social adjustment, and reduce medication compliance (Marder et al., 2004). Furthermore, it can have serious deleterious effects on health and life expectancy via a number of disease processes, including hypertension, coronary artery disease, osteoarthritis, type 2 diabetes mellitus (T2DM), stroke, sleep apnea, and certain cancers. Approximately half of all people with schizophrenia meet criteria for obesity, representing a relative risk of nearly two when compared with the general population (Newcomer, 2007). Genetic and lifestyle factors may contribute to this phenomenon, but in large measure, medications for psychosis (antipsychotics)-induced weight gain are responsible. In a network meta-analysis of 116 RCTs ($n = 28,317$) comparing medications for psychosis and placebo for the acute treatment of adults with multi-episode schizophrenia, about half of the medications for psychosis examined caused significantly more weight gain than placebo with mean differences as follows: zotepine (3.21 kg), olanzapine (2.78 kg), sertindole (2.47 kg), chlorpromazine (2.37 kg), iloperidone (2.18 kg), quetiapine (1.94 kg), clozapine (1.89 kg), paliperidone (1.49 kg), and risperidone (1.44 kg) (Huhn et al., 2019). Another network meta-analysis of 83 studies (study durations, 2–13 weeks) comparing 18 different medications for psychosis (18,750 patients) with placebo (4210 patients), there was no evidence of weight gain with ziprasidone, haloperidol, fluphenazine, aripiprazole, lurasidone, cariprazine, amisulpride, or flupenthixol when compared with placebo while weight gain was observed with brexpiprazole, risperidone, paliperidone, quetiapine, iloperidone, sertindole, olanzapine, zotepine, and clozapine. Ranking on the basis of degree of weight gain, clozapine (3.01 kg) was identified as causing the most weight gain, followed by zotepine (2.80 kg), olanzapine (2.73 kg), and sertindole (2.37 kg) (Pillinger et al., 2020).

The mechanism of action underlying medications for psychosis (antipsychotics)-induced weight gain has been hypothesized to involve changes in appetite and satiety mediated by H_1 receptor antagonism (Newcomer, 2007). Consistent with this hypothesis, medications for psychosis possessing a high affinity for H_1 receptors, such as clozapine, olanzapine, and zotepine, are associated with considerable weight-gain potential. Agents with little H_1 affinity, such as aripiprazole, haloperidol, and ziprasidone, tend to be more weight neutral. Antagonism of $5-HT_{2C}$ receptors may also contribute to medications for psychosis-induced weight gain.

Recognizing the considerable potential for obesity to impact negatively on morbidity and mortality, regular monitoring of the body mass index (BMI), calculated as weight in kilograms divided by the square of the height in meters (kg/m^2), for every patient with schizophrenia is recommended. Waist circumference monitoring is also recommended, as central obesity is a risk factor for diabetes, hypertension, and dyslipidemia, independent of BMI. A waist circumference of >102 cm (>40 in) in men or > 88 cm (>35 in) in women is indicative of abdominal obesity (Newcomer, 2007). Clinicians should encourage exercise and a healthy diet in all patients to prevent initial weight gain and obesity, and perform lifestyle intervention for obese patients (Cooper et al., 2016). The relative weight gain potential of the various medications for psychosis (antipsychotics) should be a consideration in medication selection. In terms of adjunctive treatments for weight gain associated with medications for psychosis, metformin has the most robust evidence base, is associated with ~3 kg weight loss compared to placebo in 12–24 week trials (Mizuno et al., 2014), and has been endorsed by several schizophrenia treatment guidelines including the recently published American Psychiatric Association guidelines (American Psychiatric Association, 2021). In addition, adjunctive aripiprazole is recommended as a possible intervention for weight gain associated with clozapine and olanzapine (Cooper et al., 2016). Adjunctive metformin is also considered to attenuate or reduce weight gain following medications for psychosis (Mizuno et al., 2014).

Diabetes

The prevalence of T2DM is twice as high among patients with schizophrenia as in the general population with a pooled relative risk of 1.82 (Stubbs et al., 2015), placing these patients at increased risk of coronary artery disease, stroke, peripheral vascular disease, retinopathy, nephropathy, neuropathy, and ketoacidosis. Accumulating data indicate that some NGAs increase this burden. A network meta-analysis of 37 studies comparing 16 different medications for psychosis (antipsychotics) (10,681

patients) with placebo (3032 patients), there was no evidence of change in glucose concentrations with amisulpride, asenapine, sertindole, ziprasidone, brexpiprazole, quetiapine, risperidone, paliperidone, aripiprazole, haloperidol, cariprazine, and iloperidone. Ranking on the basis of degree of associated glucose alteration identified clozapine the worst, followed by zotepine (Huhn et al., 2019). There is evidence that some individuals are also at risk for insulin-dependent diabetes as some NGAs may directly induce apoptosis of pancreatic β cells (Ozasa et al. 2013).

Measurement of plasma glucose level before starting a new medication for psychosis (antipsychotic) and regular follow-up assessments should be performed. If fasting glucose levels cannot be obtained, hemoglobin A_{1c} levels are acceptable (Marder et al., 2004). Elevation of either of these levels should prompt a referral to a primary care provider or a specialist physician for management. Clinicians should inquire and educate patients about the symptoms of new-onset diabetes, which include weight loss, polyuria, polydipsia, and blurred vision.

Dyslipidemia

Elevated lipid levels, particularly low-density lipoprotein (LDL) and triglyceride (TG) levels, are associated with coronary artery disease and myocardial infarction. Compared with placebo, a network meta-analysis of 24 studies comparing 24 different medications for psychosis (antipsychotics) (7439 patients) with placebo (2419 patients) found no strong evidence of change in LDL levels with ziprasidone, lurasidone, risperidone, paliperidone, aripiprazole, and brexpiprazole; a decrease with cariprazine; and increases with quetiapine and olanzapine (Pillinger et al., 2020). As for TG level change, no strong evidence was found with brexpiprazole, lurasidone, sertindole, cariprazine, ziprasidone, aripiprazole, risperidone, paliperidone, amisulpride, haloperidol, and iloperidone; and increases with quetiapine, olanzapine, zotepine, and clozapine [34 studies; 15 medications for psychosis (10,965 patients) and placebo (3021 patients)] (Pillinger et al., 2020).

Because patients with schizophrenia are, as a group, at increased risk for coronary artery disease, lipid levels should be monitored more frequently than in the general population. Measurement of lipid levels before starting a new medication for psychosis (antipsychotic) and regular follow-up assessments should be performed. If lipid abnormalities are present, the clinician should refer the patient to a primary care provider or a specialist physician. If this is not possible, the clinician should advise the patient to reduce dietary fat intake and should be prepared to prescribe a lipid-lowering drug in the event that cholesterol levels do not normalize with this intervention (Marder et al., 2004). Patients with schizophrenia taking medications for psychosis (antipsychotics) appear to respond to lipid-lowering agents in a manner consistent with the general population (Vincenzi et al., 2013). Adding or switching to drugs with safer profile may modestly improve lipid levels (Chen et al., 2012), although when switching medications for psychosis there is also the risk that psychopathology may worsen so these types of changes must be done carefully using a shared decision-making process with the patient.

Cardiovascular Effects

Common cardiovascular effects of medications for psychosis (antipsychotics) include orthostatic hypotension, sinus tachycardia, and electrocardiogram (ECG) changes.

Orthostatic hypotension occurs most frequently with low-potency medications for psychosis (antipsychotics) including clozapine, as a result of their high α_1-adrenergic affinities, but may also be seen with other agents. Orthostatic hypotension is most likely to occur early in treatment and with dose titration. Patients should be assessed regularly during these times for vital sign changes consistent with orthostatic hypotension and for symptoms such as dizziness, light-headedness, presyncope, and syncope. Elderly patients are particularly vulnerable to this side effect, which may predispose to falls and related injuries. Most patients develop tolerance to the orthostatic effects of medications

for psychosis over a period of weeks. Gradual dose titration often reduces the risk of orthostatic hypotension. Management strategies include advising patients to slowly shift body slowly as well as reducing or dividing the medication dose or switching to a lower α_1-adrenergic affinity medication for psychosis.

Sinus tachycardia may result from the anticholinergic effects of medications for psychosis (antipsychotics) or be secondary to orthostatic hypotension. Tachycardia can occur with any medication for psychosis but is especially common with clozapine, affecting approximately 25% of patients treated with the drug (Lieberman et al., 1989). Although an increase in resting heart rate may be well tolerated in most patients, caution should be exercised when treating elderly patients and those with preexisting heart disease. If tachycardia is sustained or accompanied by symptoms of cardiac ischemia, an ECG should be obtained and a serum marker of myocardial damage (e.g., troponin or creatine kinase MB) should be considered.

Abnormalities in cardiac electrophysiology, as reflected in ECG changes, have been observed with most medications for psychosis (antipsychotics) (Pillinger et al., 2020). Prolongation of the QTc interval, especially beyond 500 ms, is of particular concern because it increases the risk of torsades de pointes (TdP), a ventricular tachyarrhythmia that can lead to sudden death. TdP has been hypothesized to account for the increased risk of sudden cardiac death observed in users of FGAs and NGAs (Ray et al., 2009a). Among them, sertindole, amisulpride, ziprasidone, iloperidone, risperidone, olanzapine, and quetiapine are reported to increase the QTc interval compared to placebo (Huhn et al., 2019). A baseline ECG is recommended prior to treatment with QTc-prolonging medications for psychosis (e.g., ziprasidone) in patients with any of the aforementioned risk factors, in elderly patients, and in those who are taking other drugs known to prolong the QTc interval (Marder et al., 2004). An ECG is also indicated if the patient presents with syncope or other symptoms associated with a prolonged QTc interval (Marder et al., 2004). The risk of QTc prolongation may be decreased by using the lowest effective dose, correcting hypokalemia and hypomagnesemia, and minimizing concomitant exposure to other QTc-prolonging drugs.

Clozapine is associated with an increased risk of myocarditis and cardiomyopathy. A meta-analysis of 28 studies of 258,961 people exposed to clozapine reported event rates of myocarditis and cardiomyopathy of 0.007 and 0.006, respectively (Siskind et al., 2020). Myocarditis typically occurs during the first 2 months of clozapine treatment and manifests with nonspecific signs and symptoms, including fever, tachycardia, chest pain, dyspnea, flu-like symptoms, eosinophilia, elevated cardiac enzyme levels, and ECG changes. No single finding is pathognomonic, and diagnosis is usually guided by a preponderance of clinical evidence, often including echocardiography and/or cardiac magnetic resonance imaging. If untreated, progression to fulminant myocarditis may be rapid, with an attendant mortality rate as high as 50%. If the condition is diagnosed early, however, drug discontinuation and supportive treatment frequently lead to spontaneous recovery. Routine screening for clozapine-induced myocarditis, such as with serial measurements of cardiac enzymes during the first month of treatment, has been proposed, but large-scale prospective studies are needed to establish whether this would result in earlier detection of the disorder. Clinicians should also be vigilant for the development of clozapine-associated dilated cardiomyopathy, which may manifest after months to years of clozapine treatment and may itself be associated with significant morbidity and mortality.

Gastrointestinal Effects

The anticholinergic effects of medications for psychosis (antipsychotics) include xerostomia (dry mouth) and constipation. These are relatively commonly encountered with clozapine and low-potency medications for psychosis but can also occur with other agents. Because they are frequently dose related, xerostomia and constipation may improve with dose reduction. Elderly patients are particularly susceptible to anticholinergic effects, owing to reduced cholinergic function. Since constipation is

often neither recognized nor reported to psychiatrists by patients, active evaluation of constipation may be necessary (Koizumi et al., 2013).

Patients with xerostomia should be advised to maintain hydration, use sugarless gum or sugarless hard candy, and obtain regular dental care. Constipation can usually be treated with stool softeners and laxatives. In rare instances, paralytic ileus occurs in association with clozapine or haloperidol, in which case medication for psychosis (antipsychotics) should be discontinued and medical or surgical consultation sought promptly. Paradoxically, clozapine may also cause sialorrhea (excess salivation). This may be treated with a few drops of 1% atropine solution, either directly applied sublingually or mixed in a glass of water and administered as a mouthwash on an as-needed basis.

Respiratory Effects

Medication for psychosis (antipsychotics) have been reported to increase the risk of aspiration pneumonia due to the impairment of swallowing and cough reflexes, sedation, excessive salivation, and changes in pharyngeal and laryngeal muscle tones (Dzahini et al., 2018). Although not definitive, decreased level of substance P, which plays a critical role in these reflexes, due to the use of medication for psychosis (Ishida et al., 2011), may be associated with impaired swallowing reflex. In a meta-analysis of 14 studies that reported the association between use of medications for psychosis and incidence of pneumonia, the risk of pneumonia was found to be significantly increased by both FGAs and NGAs (risk ratio 1.69, 95%CI 1.34–2.15; risk ratio 1.93, 95%CI 1.55–2.41, respectively) (Dzahini et al., 2018). The risk of pneumonia may become higher as the dose of medications for psychosis increases (Haga et al., 2018).

Inhaled loxapine may induce respiratory difficulties in people with asthma or emphysema, and there is a *Risk Evaluation and Mitigation Strategy* (*REMS*) that needs to be implemented when this formulation is used.

Hepatic Effects

Asymptomatic elevations of liver enzyme levels occur in up to one-third of patients taking medications for psychosis (antipsychotics), usually within 6 weeks of drug initiation (Barnes & McPhillips, 1999; Burns, 2001; Marwick et al., 2012). These abnormalities typically remain stable or resolve with continued treatment (Marwick et al., 2012). Clinically significant medication for psychosis-induced hepatic effects are uncommon, with rates varying among individual agents. Occasionally, symptomatic hepatotoxicity (cholestatic or hepatitic) may be associated with medications for psychosis (Burns, 2001). In these cases, the offending medication should be promptly discontinued. Patients taking a medication for psychosis who develop nausea, abdominal pain, and jaundice should have their liver function evaluated. Since medication-induced jaundice is infrequent, other etiologies should be ruled out before the cause is judged to be antipsychotic treatment.

Hematological Effects

Medication for psychosis (antipsychotics) may cause blood dyscrasias, including leukopenia, neutropenia, agranulocytosis, and eosinophilia. Among them, clozapine-induced agranulocytosis (defined as an absolute neutrophil count less than $500/mm^3$) is the best known side effect although neutropenia may occur with other medications for psychosis (antipsychotics) with approximately the same frequency (Myles et al., 2019). Early agranulocytosis-related fatalities led to the withdrawal of clozapine in some European countries and severe restrictions on its use in others. With the advent of mandatory, systematic absolute neutrophil count monitoring, clozapine has been approved for the treatment-resistant schizophrenia. The incidence of clozapine-associated neutropenia defined as an absolute neutrophil count of $<1500/\mu$ L was 3.8% and severe neutropenia ($<500/\mu$L) 0.9%, and the peak incidence of severe neutropenia occurred at 1 month of exposure and declined to negligible levels after 1 year of treatment (Myles et al., 2018). If agranulocytosis does occur, the clinician should discontinue clozapine treatment, monitor for signs of infection, and consider bone marrow aspiration; if granulopoiesis is deficient, protective isolation is advisable. Treatment with granulocyte colony stimulating factor may be considered, as this agent can restore granulopoietic function and reduce recovery time. It should also be noted that transient neutropenia with spontaneous remission is a common side effect of clozapine (22%) (Hummer et al., 1994).

Ocular Effects

Cataracts are ocular lens opacities that can impair vision. Cataracts have also been found in beagles treated with supratherapeutic doses of quetiapine but not in other nonhuman animal species, including monkeys. Some case reports exist of lens opacities developing in quetiapine-treated humans; however, most of these patients had known risk factors for cataracts (Marder et al., 2004). In addition, more recent data do not find evidence of the association between use of medications for psychosis and cataract (Pakzad-Vaezi et al., 2013; Chou et al., 2016; Chu et al., 2017). Still, because patients with schizophrenia often have known risk factors for cataracts (e.g., diabetes, hypertension, and poor nutrition), regular ocular evaluations are advisable. Clinicians should also inquire about visual changes and should ask specifically about the quality of distance vision and blurred vision (Marder et al., 2004). If changes are present, referral for ophthalmological evaluation is warranted.

Other Side Effects

Sedation can occur with virtually any medication for psychosis (antipsychotic) but is particularly common with low-potency medications for psychosis, including clozapine, olanzapine, quetiapine, and zotepine. This side effect is typically most pronounced during dose titration, as the majority of patients develop some tolerance with continued administration of medication for psychosis. Significant sedation may persist in some individuals, however, especially in those treated with clozapine. While the sedation that accompanies initiation of medications for psychosis may be beneficial in agitated patients, persistent sedation can be socially and vocationally disabling. This effect can sometimes be mitigated by lowering the dose, by consolidating divided doses into a single evening dose, or by switching to a less sedating agent. In spite of the high prevalence of medication for psychosis-induced sedation, pharmacological treatments have not been well studied.

Medication for psychosis (antipsychotics), in particular clozapine, can lower the seizure threshold. The incidence for seizures with medications for psychosis other than clozapine is as low as 0.03–0.05% while the clozapine-associated seizure rate is 0.18% (Druschky et al., 2019). Findings on the relationship between clozapine dose or concentration and incidence of seizures have been inconsistent (Williams & Park, 2015; Haring et al., 1994).

Future Directions in Drug Research and Treatment

Research of medications for psychosis (antipsychotics) has two primary aims: to increase our understanding of existing medications and to develop novel agents with greater effectiveness. To date, many trials of medications for psychosis have been funded, conducted, and/or reported by representatives of the pharmaceutical industry. Several government-funded, long-term, large-scale trials, conducted under "real-world" conditions and involving representative patient samples, have extended the information provided in these reports (Lieberman et al., 2005a; Jones et al., 2006; Kahn et al., 2008). Moreover, recent methodology of network meta-analysis enables direct and indirect comparisons of medication for psychosis in a comprehensive manner (Huhn et al., 2019; Pillinger et al., 2020), which has provided relevant information to guide physicians.

Future research on existing medications for psychosis (antipsychotics) is likely to evaluate outcomes that have received relatively less attention, historically, than has short-term reduction in positive symptoms of schizophrenia. These outcomes include impact on negative and cognitive symptoms, relapse prevention, improvement in social and vocational functioning, suicide prevention, and reduction in family and caregiver burden. Early intervention strategies, targeting prodromal and high-risk patients, are also increasingly an area of intensive investigation (Fleischhacker & Simma, 2012). While most non-clozapine agents demonstrate approximately equal efficacy for positive symptoms, it is possible that certain agents may have specific utility in other symptom domains or at particular stages of the illness. In addition, long-term safety especially with regard to metabolic side effects and cardiovascular events warrant further investigations.

Potential between-drug differences in symptom- and stage-specific efficacy, as well as known differences in side effect profiles, suggest that a priority for future research should be the individualization of antipsychotic treatment. At present, one cannot reliably predict which medication for psychosis (antipsychotic) is best suited for any given patient. However, it is likely that pharmacogenetic/pharmacogenomic advances will eventually allow clinicians to select medications based on the genetic variability across patients (Malhotra et al., 2012). In fact, several gene variants related to antipsychotic response and adverse effects have been reported, and several commercial pharmacogenomic tests have become available (Yoshida & Müller, 2020). In addition, neuroimaging (e.g., positron emission tomography, structural magnetic resonance imaging, functional magnetic resonance imaging, diffusion tensor imaging, and magnetic resonance spectroscopy) and electrophysiological (e.g., event-related potentials) studies are becoming increasingly sophisticated, providing more detailed information about brain morphological changes, pathways, and circuits involved in the pathophysiology of schizophrenia (Egerton et al., 2018; Iwata et al., 2019; Miyazaki et al., 2020). It is hoped that individual differences in these variables may ultimately serve as predictors of treatment response.

The second aim of antipsychotic research is to develop novel agents with greater effectiveness. Pimavanserin is one good example; this is the first FDA-approved drug for psychosis without any significant antagonism of dopaminergic receptors although it is not indicated for schizophrenia. Moreover, several novel non-dopaminergic medications for psychosis (antipsychotics) are currently in Phase I, II, or III clinical trials. Emerging evidence of glutamatergic dysregulation in schizophrenia has led to research on a number of agents with direct or indirect effects on the glutamate system. Similarly, the central cholinergic system is being investigated for its potential role in the cognitive deficits observed in schizophrenia. Compounds that enhance cholinergic activity have been hypothesized to improve these deficits (Brannan et al., 2021). Similarly, trace amines, which were first studied in schizophrenia in the 1970s by Richard Wyatt at the National Institute of Mental Health, appear to be finally reaching clinical use (Koblan et al., 2020). These and other agents merit further investigation and will likely engender a wealth of research in the near future. Lastly, the methodology of clinical trials is being scrutinized. As the number of failed trials is increasing and placebo responses are increasing in schizophrenia trials, reexamination of design elements for future trials is clearly warranted (Gopalakrishnan et al., 2020).

Conclusions

Despite recent advances in pharmacotherapeutic options, schizophrenia and other psychotic disorders are pathophysiologically complex and are generally associated with persistent debilitating symptoms for which there are no known cures. Consequently, many therapeutic needs remain unmet. A number of efficacy and effectiveness studies have clarified the heterogeneous profiles of medications for psychosis (antipsychotics). To date, clozapine is the only evidence-based treatment for refractory patients and none of the currently approved antipsychotics are better than clozapine for the treatment of schizophrenia. However, the pharmacological basis of clozapine's

superiority, or even its precise difference from other antipsychotic drugs, is still not known. Therapeutic outcome could be improved by more effective use of existing pharmacological and non-pharmacological treatments and by ongoing development of better novel therapies. Successful treatment outcomes are often challenged by the remarkable heterogeneity of therapeutic response and adverse effects between patients and between different illness stages. In the absence of a clearly superior class of medication, risk–benefit analysis, clinical judgment, and shared decision-making should be relied upon in an effort to pair patients with medications for psychosis that are likely to be safe and effective. Individualized treatment for a specific patient based on their biological characteristics is one of the future goals of psychopharmacological treatment of schizophrenia. In addition, novel agents with potential efficacy for a variety of symptoms of schizophrenia and psychotic disorders are being explored and are eagerly awaited. These goals will be achieved with precise biological characterization of this illness.

Conflict of Interest Statements Dr. Uchida has received grants from Eisai, Otsuka Pharmaceutical, Dainippon-Sumitomo Pharma, and Meiji-Seika Pharma; speaker's honoraria from Otsuka Pharmaceutical, Dainippon-Sumitomo Pharma, Eisai, and Meiji-Seika Pharma; and advisory panel payments from Dainippon-Sumitomo Pharma within the past 3 years.

Dr. Euitae Kim has received grants from Otsuka and participated in advisory/speaker meetings organized by Janssen, Otsuka, and Bukwang Pharm Company.

Dr. Jarskog has received grant support from Boeringer Ingelheim, Otsuka, and Corcept, and has consulted for Signant Health and UpToDate within the past 3 years.

Dr. Fleischhacker has received research grants from Lundbeck and Otsuka. He has received honoraria for educational programs from Janssen, Richter, and Recordati, and advisory board honoraria from Angelini, Otsuka, Boehringer-Ingelheim, and Dainippon/Sumitomo.

Dr. Remington has received research and advisory board support from HLS Therapeutics, as well as consultant fees from Mitsubishi Tanabe Pharma Corporation.

Dr. Lieberman neither accepts nor receives any personal financial remuneration for consulting, speaking, or research activities from any pharmaceutical, biotechnology, or medical device companies. He receives support administered through Columbia University and the Research Foundation for Mental Hygiene in the form of funding and medication supplies for investigator initiated research from Denovo, Taisho, Sunovion, and Genentech, and for company sponsored Phase II, III, and IV studies from Alkermes, Allergan, and Boehringer Ingelheim. However, none of this research support contributes to his institutional compensation. He is a consultant to or member of the advisory board of Intracellular Therapies, Lilly, Pierre Fabre, Pear Therapeutics, and Gilgamesh Therapeutics for which he receives no remuneration. He is a paid consultant for Signant, a clinical research services organization, and holds a patent from Repligen that neither has nor currently yields any royalties.

References

Abbas, A. I., Hedlund, P. B., Huang, X. P., et al. (2009). Amisulpride is a potent 5-HT7 antagonist: Relevance for antidepressant actions in vivo. *Psychopharmacology (Berl), 205*, 119–128.

Aerts, L., Cations, M., Harrison, F., Jessop, T., Shell, A., Chenoweth, L., & Brodaty, H. (2019). Why deprescribing antipsychotics in older people with dementia in long-term care is not always successful: Insights from the HALT study. *International Journal of Geriatric Psychiatry, 34*(11), 1572–1581.

Aichhorn, W., Marksteiner, J., Walch, T., et al. (2006). Influence of age, gender, body weight and valproate comedication on quetiapine plasma concentrations. *International Clinical Psychopharmacology, 21*, 81–85.

Aksoy Poyraz, C., Turan, Ş., Demirel, Ö. F., et al. (2015). Effectiveness of ultra-rapid dose titration of clozapine for treatment-resistant bipolar mania: Case series. *Therapeutic Advances in Psychopharmacology, 5*, 237–242.

Alexopoulos, G. S., Streim, J., Carpenter, D., et al. (2004). Using antipsychotic agents in older patients. *The Journal of Clinical Psychiatry, 65*(Suppl 2), 5–99; discussion 100–102; quiz 103–4.

Allen, R. M., & Young, S. J. (1978). Phencyclidine-induced psychosis. *The American Journal of Psychiatry, 135*, 1081–1084.

American Psychiatric Association (Ed.). (2021). *Practice guideline for the treatment of patients with schizophrenia* (3rd ed.). American Psychiatric Association.

Anderson, K. K., Voineskos, A., Mulsant, B. H., et al. (2014). The role of untreated psychosis in neurodegeneration: A review of hypothesized mechanisms of neurotoxicity in first-episode psychosis. *Canadian Journal of Psychiatry, 59*, 513–517.

Andrezina, R., Josiassen, R. C., Marcus, R. N., et al. (2006). Intramuscular aripiprazole for the treatment of acute agitation in patients with schizophrenia or schizoaffective disorder: A double-blind, placebo-controlled comparison with intramuscular haloperidol. *Psychopharmacology (Berl), 188*, 281–292.

Arakawa, R., Ito, H., Takano, A., et al. (2008). Dose-finding study of paliperidone ER based on striatal and extrastriatal dopamine D2 receptor occupancy in patients with schizophrenia. *Psychopharmacology (Berl), 197*, 229–235.

Arakawa, R., Ito, H., Takano, A., et al. (2010). Dopamine D2 receptor occupancy by perospirone: A positron emission tomography study in patients with schizophrenia and healthy subjects. *Psychopharmacology (Berl), 209*, 285–290.

Arnt, J., & Skarsfeldt, T. (1998). Do novel antipsychotics have similar pharmacological characteristics? A review of the evidence. *Neuropsychopharmacology, 18*, 63–101.

Artukoglu, B. B., Li, F., Szejko, N., & Bloch, M. H. (2020). Pharmacologic Treatment of Tardive Dyskinesia: A Meta-Analysis and Systematic Review. *J Clin Psychiatry, 81*(4), 19r12798. https://doi.org/10.4088/JCP.19r12798. PMID: 32459404

Ayd, F. J., Jr. (1961). A survey of drug-induced extrapyramidal reactions. *JAMA, 175*, 1054–1060.

Ballard, C., Margallo-Lana, M., Juszczak, E., et al. (2005). Quetiapine and rivastigmine and cognitive decline in Alzheimer's disease: Randomised double blind placebo controlled trial. *BMJ, 330*, 874.

Barnas, C., Quiner, S., Tauscher, J., et al. (2001). In vivo (123)I IBZM SPECT imaging of striatal dopamine 2 receptor occupancy in schizophrenic patients. *Psychopharmacology (Berl), 157*, 236–242.

Barnes, T. R., & Mcphillips, M. A. (1999). Critical analysis and comparison of the side-effect and safety profiles of the new antipsychotics. *The British Journal of Psychiatry. Supplement, 38*, 34–43. PMID: 10884898.

Barnes, T. R., Drake, R., Paton, C., et al. (2020). Evidence-based guidelines for the pharmacological treatment of schizophrenia: Updated recommendations from the British Association for Psychopharmacology. *Journal of Psychopharmacology, 34*, 3–78.

Bartoli, F., Dell'osso, B., Crocamo, C., et al. (2017). Benefits and harms of low and high second-generation antipsychotics doses for bipolar depression: A meta-analysis. *Journal of Psychiatric Research, 88*, 38–46.

Bergman, H., Rathbone, J., Agarwal, V., et al. (2018). Antipsychotic reduction and/or cessation and antipsychotics as specific treatments for tardive dyskinesia. *Cochrane Database of Systematic Reviews, 2*, Cd000459.

Bigliani, V., Mulligan, R. S., Acton, P. D., et al. (2000). Striatal and temporal cortical D2/D3 receptor occupancy by olanzapine and sertindole in vivo: A [123I] epidepride single photon emission tomography (SPET) study. *Psychopharmacology (Berl), 150*, 132–140.

Bishara, D., & Taylor, D. (2008). Upcoming agents for the treatment of schizophrenia: Mechanism of action, efficacy and tolerability. *Drugs, 68*, 2269–2292.

Bobes, J., Arango, C., Garcia-Garcia, M., et al. (2010). Prevalence of negative symptoms in outpatients with schizophrenia spectrum disorders treated with antipsychotics in routine clinical practice: Findings from the CLAMORS study. *The Journal of Clinical Psychiatry, 71*, 280–286.

Boter, H., Peuskens, J., Libiger, J., et al. (2009). Effectiveness of antipsychotics in first-episode schizophrenia and schizophreniform disorder on response and remission: An open randomized clinical trial (EUFEST). *Schizophrenia Research, 115*, 97–103.

Brannan, S. K., Sawchak, S., Miller, A. C., Lieberman, J. A., Paul, S. M., & Breier, A. (2021). Muscarinic cholinergic receptor agonist and peripheral antagonist for schizophrenia. *The New England Journal of Medicine, 384*(8), 717–726.

Breier, A., Meehan, K., Birkett, M., et al. (2002). A double-blind, placebo-controlled dose-response comparison of intramuscular olanzapine and haloperidol in the treatment of acute agitation in schizophrenia. *Archives of General Psychiatry, 59*, 441–448.

Brook, S., Lucey, J. V., & Gunn, K. P. (2000). Intramuscular ziprasidone compared with intramuscular haloperidol in the treatment of acute psychosis. Ziprasidone I.M. Study Group. *The Journal of Clinical Psychiatry, 61*, 933–941.

Buchanan, R. W. (2007). Persistent negative symptoms in schizophrenia: An overview. *Schizophrenia Bulletin, 33*, 1013–1022.

Buchanan, R. W., & Gold, J. M. (1996). Negative symptoms: Diagnosis, treatment and prognosis. *International Clinical Psychopharmacology, 11*(Suppl 2), 3–11.

Buchanan, R. W., Kreyenbuhl, J., Kelly, D. L., et al. (2010). The 2009 schizophrenia PORT psychopharmacological treatment recommendations and summary statements. *Schizophrenia Bulletin, 36*, 71–93.

Buckley, P. F. (2001). Broad therapeutic uses of atypical antipsychotic medications. *Biological Psychiatry, 50*, 912–924.

Buckley, P. F., & Correll, C. U. (2008). Strategies for dosing and switching antipsychotics for optimal clinical management. *The Journal of Clinical Psychiatry, 69*(Suppl 1), 4–17.

Burns, M. J. (2001). The pharmacology and toxicology of atypical antipsychotic agents. *Journal of Toxicology. Clinical Toxicology, 39*, 1–14.

Burton, S. (2006). Symptom domains of schizophrenia: The role of atypical antipsychotic agents. *Journal of Psychopharmacology, 20*, 6–19.

Bymaster, F. P., Calligaro, D. O., Falcone, J. F., et al. (1996). Radioreceptor binding profile of the atypical antipsychotic olanzapine. *Neuropsychopharmacology, 14*, 87–96.

Bymaster, F. P., Rasmussen, K., Calligaro, D. O., et al. (1997). In vitro and in vivo biochemistry of olanzapine: A novel, atypical antipsychotic drug. *The Journal of Clinical Psychiatry, 58*(Suppl 10), 28–36.

Bymaster, F. P., Felder, C., Ahmed, S., et al. (2002). Muscarinic receptors as a target for drugs treating schizophrenia. *Current Drug Targets. CNS and Neurological Disorders, 1*, 163–181.

Cahn, W., Hulshoff Pol, H. E., Lems, E. B., et al. (2002). Brain volume changes in first-episode schizophrenia: A

1-year follow-up study. *Archives of General Psychiatry, 59*, 1002–1010.

Carbon, M., Hsieh, C. H., Kane, J. M., et al. (2017). Tardive dyskinesia prevalence in the period of second-generation antipsychotic use: A meta-analysis. *The Journal of Clinical Psychiatry, 78*, e264–e278.

Carpenter, W. T., & Gold, J. M. (2002). Another view of therapy for cognition in schizophrenia. *Biological Psychiatry, 51*, 969–971.

Casey, D. E. (1993). Neuroleptic-induced acute extrapyramidal syndromes and tardive dyskinesia. *The Psychiatric Clinics of North America, 16*, 589–610.

Casey, D. E. (1999). Tardive dyskinesia and atypical antipsychotic drugs. *Schizophrenia Research, 35*(Suppl), S61–S66.

Ceraso, A., Lin, J. J., Schneider-Thoma, J., et al. (2020). Maintenance treatment with antipsychotic drugs for schizophrenia. *Cochrane Database of Systematic Reviews, 8*, Cd008016.

Chan, B., Freeman, M., Kondo, K., et al. (2019a). Pharmacotherapy for methamphetamine/amphetamine use disorder-a systematic review and meta-analysis. *Addiction, 114*, 2122–2136.

Chan, B., Kondo, K., Freeman, M., et al. (2019b). Pharmacotherapy for cocaine use disorder-a systematic review and meta-analysis. *Journal of General Internal Medicine, 34*, 2858–2873.

Chen, A. T., & Nasrallah, H. A. (2019). Neuroprotective effects of the second generation antipsychotics. *Schizophrenia Research, 208*, 1–7.

Chen, Y., Bobo, W. V., Watts, K., et al. (2012). Comparative effectiveness of switching antipsychotic drug treatment to aripiprazole or ziprasidone for improving metabolic profile and atherogenic dyslipidemia: A 12-month, prospective, open-label study. *Journal of Psychopharmacology, 26*, 1201–1210.

Chou, P. H., Chu, C. S., Lin, C. H., et al. (2016). Use of atypical antipsychotics and risks of cataract development in patients with schizophrenia: A population-based, nested case-control study. *Schizophrenia Research, 174*, 137–143.

Chouinard, G., Samaha, A. N., Chouinard, V. A., Peretti, C. S., Kanahara, N., Takase, M., & Iyo, M. (2017). Antipsychotic-induced dopamine supersensitivity psychosis: Pharmacology, criteria, and therapy. *Psychotherapy and Psychosomatics, 86*(4), 189–219.

Chu, C. S., Chou, P. H., Chen, Y. H., et al. (2017). Association between antipsychotic drug use and cataracts in patients with bipolar disorder: A population-based, nested case-control study. *Journal of Affective Disorders, 209*, 86–92.

Cipriani, A., Boso, M., & Barbui, C. (2009). Clozapine combined with different antipsychotic drugs for treatment resistant schizophrenia. *Cochrane Database of Systematic Reviews, 3*, Cd006324.

Cipriani, A., Barbui, C., Salanti, G., et al. (2011). Comparative efficacy and acceptability of antimanic drugs in acute mania: A multiple-treatments meta-analysis. *Lancet, 378*, 1306–1315.

Citrome, L. (2016). Cariprazine for the treatment of schizophrenia: A review of this dopamine D3-preferring

D3/D2 receptor partial agonist. *Clinical Schizophrenia & Related Psychoses, 10,* 109–119.

Citrome, L., & Saklad, S. R. (2020). Revisiting Tardive Dyskinesia: Focusing on the Basics of Identification and Treatment. *J Clin Psychiatry, 81*(2): TV18059AH3C. https://doi.org/10.4088/JCP. TV18059AH3C. PMID: 32078259.

Citrome, L., Stensbøl, T. B., & Maeda, K. (2015). The preclinical profile of brexpiprazole: What is its clinical relevance for the treatment of psychiatric disorders? *Expert Review of Neurotherapeutics, 15,* 1219–1229.

Collins, C. M., Wood, M. D., & Elliott, J. M. (2014). Chronic administration of haloperidol and clozapine induces differential effects on the expression of Arc and c-Fos in rat brain. *Journal of Psychopharmacology, 28,* 947–954.

Cooper, S. J., Reynolds, G. P., Barnes, T., et al. (2016). BAP guidelines on the management of weight gain, metabolic disturbances and cardiovascular risk associated with psychosis and antipsychotic drug treatment. *Journal of Psychopharmacology, 30,* 717–748.

Corena-Mcleod, M. (2015). Comparative pharmacology of risperidone and paliperidone. *Drugs R D, 15,* 163–174.

Corponi, F., Fabbri, C., Bitter, I., et al. (2019). Novel antipsychotics specificity profile: A clinically oriented review of lurasidone, brexpiprazole, cariprazine and lumateperone. *European Neuropsychopharmacology, 29,* 971–985.

Correll, C. U., Rummel-Kluge, C., Corves, C., et al. (2009). Antipsychotic combinations vs monotherapy in schizophrenia: A meta-analysis of randomized controlled trials. *Schizophrenia Bulletin, 35,* 443–457.

Cutler, A. J., Montgomery, S. A., Feifel, D., et al. (2009). Extended release quetiapine fumarate monotherapy in major depressive disorder: A placebo- and duloxetine-controlled study. *The Journal of Clinical Psychiatry, 70,* 526–539.

Davidson, M., Galderisi, S., Weiser, M., et al. (2009). Cognitive effects of antipsychotic drugs in first-episode schizophrenia and schizophreniform disorder: A randomized, open-label clinical trial (EUFEST). *The American Journal of Psychiatry, 166,* 675–682.

De Bartolomeis, A., Buonaguro, E. F., Latte, G., et al. (2017). Immediate-early genes modulation by antipsychotics: Translational implications for a putative gateway to drug-induced long-term brain changes. *Frontiers in Behavioral Neuroscience, 11,* 240.

De Bruin, N. M., & Kruse, C. G. (2015). 5-HT6 receptor antagonists: Potential efficacy for the treatment of cognitive impairment in schizophrenia. *Current Pharmaceutical Design, 21,* 3739–3759.

Deeks, E. D., & Keating, G. M. (2010). Blonanserin: A review of its use in the management of schizophrenia. *CNS Drugs, 24,* 65–84.

Delay, J., Deniker, P., Harl, et al. (1952a). [N-dimethylamino-prophylchlorophenothiazine (4560 RP) therapy of confusional states]. *Annales Medico-Psychologiques (Paris), 110,* 398–403.

Delay, J., Deniker, P., & Harl, J. M. (1952b). [Therapeutic use in psychiatry of phenothiazine of central elective

action (4560 RP)]. *Annales Medico-Psychologiques (Paris), 110,* 112–117.

Devoe, D. J., Farris, M. S., Townes, P., & Addington, J. (2020). Interventions and Transition in Youth at Risk of Psychosis: A Systematic Review and Meta-Analyses. *J Clin Psychiatry, 81*(3):17r12053. https://doi.org/10.4088/JCP.17r12053. PMID: 32433834.

Dong, M., Lu, L., Zhang, L., et al. (2019). Quality of life in schizophrenia: A meta-analysis of comparative studies. *The Psychiatric Quarterly, 90,* 519–532.

Druschky, K., Bleich, S., Grohmann, R., et al. (2019). Seizure rates under treatment with antipsychotic drugs: Data from the AMSP project. *The World Journal of Biological Psychiatry, 20,* 732–741.

Duggan, L., Fenton, M., Rathbone, J., et al. (2005). Olanzapine for schizophrenia. *Cochrane Database of Systematic Reviews, 2,* Cd001359. https://doi.org/10.1002/14651858.CD001359.pub2. PMID: 15846619.

Duncan, G. E., Zorn, S., & Lieberman, J. A. (1999). Mechanisms of typical and atypical antipsychotic drug action in relation to dopamine and NMDA receptor hypofunction hypotheses of schizophrenia. *Molecular Psychiatry, 4,* 418–428.

Dzahini, O., Singh, N., Taylor, D., et al. (2018). Antipsychotic drug use and pneumonia: Systematic review and meta-analysis. *Journal of Psychopharmacology, 32,* 1167–1181.

Egerton, A., Broberg, B. V., Van Haren, N., et al. (2018). Response to initial antipsychotic treatment in first episode psychosis is related to anterior cingulate glutamate levels: A multicentre (1)H-MRS study (OPTiMiSE). *Molecular Psychiatry, 23,* 2145–2155.

El Abdellati, K., De Picker, L., & Morrens, M. (2020). Antipsychotic treatment failure: A systematic review on risk factors and interventions for treatment adherence in psychosis. *Frontiers in Neuroscience, 14,* 531763.

El-Mallakh, R. S., & Watkins, J. (2019). Prolactin elevations and the permeability glycoprotein. *Primary Care Companion CNS Disorders, 21*(3), pii: 18nr02412.

Enomoto, T., Ishibashi, T., Tokuda, K., et al. (2008). Lurasidone reverses MK-801-induced impairment of learning and memory in the Morris water maze and radial-arm maze tests in rats. *Behavioural Brain Research, 186,* 197–207.

Farde, L., Nordström, A. L., Wiesel, F. A., et al. (1992). Positron emission tomographic analysis of central D1 and D2 dopamine receptor occupancy in patients treated with classical neuroleptics and clozapine. Relation to extrapyramidal side effects. *Archives of General Psychiatry, 49,* 538–544.

Farde, L., Nyberg, S., Oxenstierna, G., et al. (1995). Positron emission tomography studies on D2 and 5-HT2 receptor binding in risperidone-treated schizophrenic patients. *Journal of Clinical Psychopharmacology, 15,* 19s–23s.

Ferris, P. (2000). Ziprasidone. *Current Opinion in CPNS Investigational Drugs, 2,* 58–70.

Fleischhacker, W. W., & Simma, A. M. (2012). Managing the prodrome of schizophrenia. *Handbook of*

Experimental Pharmacology, 212, 125–134. https://doi.org/10.1007/978-3-642-25761-2_5. PMID: 23129330.

Fleischhacker, W. W., & Uchida, H. (2014). Critical review of antipsychotic polypharmacy in the treatment of schizophrenia. *The International Journal of Neuropsychopharmacology, 17*, 1083–1093.

Fleischhacker, W. W., Unterweger, B., Kane, J. M., et al. (1990). The neuroleptic malignant syndrome and its differentiation from lethal catatonia. *Acta Psychiatrica Scandinavica, 81*, 3–5.

Fleischhacker, W., Galderisi, S., Laszlovszky, I., et al. (2019). The efficacy of cariprazine in negative symptoms of schizophrenia: Post hoc analyses of PANSS individual items and PANSS-derived factors. *European Psychiatry, 58*, 1–9.

Flint, A. J., Meyers, B. S., Rothschild, A. J., et al. (2019). Effect of continuing olanzapine vs placebo on relapse among patients with psychotic depression in remission: The STOP-PD II randomized clinical trial. *JAMA, 322*, 622–631.

Frånberg, O., Wiker, C., Marcus, M. M., et al. (2008). Asenapine, a novel psychopharmacologic agent: Preclinical evidence for clinical effects in schizophrenia. *Psychopharmacology (Berl), 196*, 417–429.

Frånberg, O., Marcus, M. M., Ivanov, V., et al. (2009). Asenapine elevates cortical dopamine, noradrenaline and serotonin release. Evidence for activation of cortical and subcortical dopamine systems by different mechanisms. *Psychopharmacology (Berl), 204*, 251–264.

Freyberg, Z., Ferrando, S. J., & Javitch, J. A. (2010). Roles of the Akt/GSK-3 and Wnt signaling pathways in schizophrenia and antipsychotic drug action. *The American Journal of Psychiatry, 167*, 388–396.

Fusar-Poli, P., Bonoldi, I., Yung, A. R., et al. (2012). Predicting psychosis: Meta-analysis of transition outcomes in individuals at high clinical risk. *Archives of General Psychiatry, 69*, 220–229.

Gaebel, W. (1994). Intermittent medication – An alternative? *Acta Psychiatrica Scandinavica. Supplementum, 382*, 33–38.

Gaebel, W., Janner, M., Frommann, N., et al. (2002). First vs multiple episode schizophrenia: Two-year outcome of intermittent and maintenance medication strategies. *Schizophrenia Research, 53*, 145–159.

Galderisi, S., Maj, M., Mucci, A., et al. (2002). Historical, psychopathological, neurological, and neuropsychological aspects of deficit schizophrenia: A multicenter study. *The American Journal of Psychiatry, 159*, 983–990.

Galletly, C. (2009). Recent advances in treating cognitive impairment in schizophrenia. *Psychopharmacology (Berl), 202*, 259–273.

Gefvert, O., Bergström, M., Långström, B., et al. (1998). Time course of central nervous dopamine-D2 and 5-HT2 receptor blockade and plasma drug concentrations after discontinuation of quetiapine (Seroquel) in patients with schizophrenia. *Psychopharmacology (Berl), 135*, 119–126.

Ghanbari, R., El Mansari, M., Shahid, M., et al. (2009). Electrophysiological characterization of the effects of asenapine at 5-HT(1A), 5-HT(2A), alpha(2)-adrenergic and D(2) receptors in the rat brain. *European Neuropsychopharmacology, 19*, 177–187.

Goff, D. C., Falkai, P., Fleischhacker, W. W., et al. (2017). The long-term effects of antipsychotic medication on clinical course in schizophrenia. *The American Journal of Psychiatry, 174*, 840–849.

Gopalakrishnan, M., Zhu, H., Farchione, T. R., Mathis, M., Mehta, M., Uppoor, R., & Younis, I. (2020). The Trend of Increasing Placebo Response and Decreasing Treatment Effect in Schizophrenia Trials Continues: An Update From the US Food and Drug Administration. *J Clin Psychiatry, 81*(2):19r12960. https://doi.org/10.4088/JCP.19r12960. PMID: 32141721.

Graff-Guerrero, A., Rajji, T. K., Mulsant, B. H., et al. (2015). Evaluation of antipsychotic dose reduction in late-life schizophrenia: A prospective dopamine D2/3 receptor occupancy study. *JAMA Psychiatry, 72*, 927–934.

Green, M. F. (1996). What are the functional consequences of neurocognitive deficits in schizophrenia? *The American Journal of Psychiatry, 153*, 321–330.

Green, M. F., Kern, R. S., Braff, D. L., et al. (2000). Neurocognitive deficits and functional outcome in schizophrenia: Are we measuring the "right stuff"? *Schizophrenia Bulletin, 26*, 119–136.

Green, M. F., Marder, S. R., Glynn, S. M., et al. (2002). The neurocognitive effects of low-dose haloperidol: A two-year comparison with risperidone. *Biological Psychiatry, 51*, 972–978.

Grimaldi-Bensouda, L., Rouillon, F., Astruc, B., et al. (2012). Does long-acting injectable risperidone make a difference to the real-life treatment of schizophrenia? Results of the Cohort for the General study of Schizophrenia (CGS). *Schizophrenia Research, 134*, 187–194.

Haddad, P. M., & Wieck, A. (2004). Antipsychotic-induced hyperprolactinaemia: Mechanisms, clinical features and management. *Drugs, 64*, 2291–2314.

Häfner, H., Riecher-Rössler, A., Maurer, K., et al. (1992). First onset and early symptomatology of schizophrenia. A chapter of epidemiological and neurobiological research into age and sex differences. *European Archives of Psychiatry and Clinical Neuroscience, 242*, 109–118.

Haga, T., Ito, K., Sakashita, K., et al. (2018). Risk factors for pneumonia in patients with schizophrenia. *Neuropsychopharmacology Reports, 38*, 204–209.

Hamann, J., Holzhüter, F., Blakaj, S., et al. (2020). Implementing shared decision-making on acute psychiatric wards: A cluster-randomized trial with inpatients suffering from schizophrenia (SDM-PLUS). *Epidemiology and Psychiatric Sciences, 29*, e137.

Haring, C., Neudorfer, C., Schwitzer, J., et al. (1994). EEG alterations in patients treated with clozapine in relation to plasma levels. *Psychopharmacology (Berl), 114*, 97–100.

Harvey, P. D., & Keefe, R. S. (2001). Studies of cognitive change in patients with schizophrenia following novel antipsychotic treatment. *The American Journal of Psychiatry, 158*, 176–184.

Harvey, P. D., Howanitz, E., Parrella, M., et al. (1998). Symptoms, cognitive functioning, and adaptive skills in geriatric patients with lifelong schizophrenia: A comparison across treatment sites. *The American Journal of Psychiatry, 155*, 1080–1086.

Harvey, P. D., Silverman, J. M., Mohs, R. C., et al. (1999). Cognitive decline in late-life schizophrenia: A longitudinal study of geriatric chronically hospitalized patients. *Biological Psychiatry, 45*, 32–40.

Hasan, A., Falkai, P., Wobrock, T., et al. (2012). World Federation of Societies of Biological Psychiatry (WFSBP) guidelines for biological treatment of schizophrenia, part 1: Update 2012 on the acute treatment of schizophrenia and the management of treatment resistance. *The World Journal of Biological Psychiatry, 13*, 318–378.

Hasan, A., Falkai, P., Wobrock, T., et al. (2013). World Federation of Societies of Biological Psychiatry (WFSBP) guidelines for biological treatment of schizophrenia, part 2: Update 2012 on the long-term treatment of schizophrenia and management of antipsychotic-induced side effects. *The World Journal of Biological Psychiatry, 14*, 2–44.

Hawkins, E. M., Coryell, W., Leung, S., Parikh, S. V., Weston, C., Nestadt, P., Nurnberger, J. I., Jr., Kaplin, A., Kumar, A., Farooqui, A. A., El-Mallakh, R. S., & National Network of Depression Centers Suicide Prevention Task Group. (2021). Effects of somatic treatments on suicidal ideation and completed suicides. *Brain and Behavior: A Cognitive Neuroscience Perspective, 11*(11), e2381.

Heinrichs, R. W. (2007). Cognitive improvement in response to antipsychotic drugs: Neurocognitive effects of antipsychotic medications in patients with chronic schizophrenia in the CATIE trial. *Archives of General Psychiatry, 64*, 631–632.

Herniman, S. E., Allott, K., Phillips, L. J., et al. (2019). Depressive psychopathology in first-episode schizophrenia spectrum disorders: A systematic review, meta-analysis and meta-regression. *Psychological Medicine, 49*, 2463–2474.

Hicks, B. M., Busby, J., Mills, K., et al. (2020). Post-diagnostic antipsychotic use and cancer mortality: A population based cohort study. *BMC Cancer, 20*, 804.

Hiemke, C., Bergemann, N., Clement, H. W., et al. (2018). Consensus guidelines for therapeutic drug monitoring in neuropsychopharmacology: Update 2017. *Pharmacopsychiatry, 51*, 9–62.

Ho, B. C., Andreasen, N. C., Nopoulos, P., et al. (2003). Progressive structural brain abnormalities and their relationship to clinical outcome: A longitudinal magnetic resonance imaging study early in schizophrenia. *Archives of General Psychiatry, 60*, 585–594.

Hofer, A., Kemmler, G., Eder, U., et al. (2004). Quality of life in schizophrenia: The impact of psychopathology, attitude toward medication, and side effects. *The Journal of Clinical Psychiatry, 65*, 932–939.

Howanitz, E., Pardo, M., Smelson, D. A., et al. (1999). The efficacy and safety of clozapine versus chlorpromazine in geriatric schizophrenia. *The Journal of Clinical Psychiatry, 60*, 41–44.

Howes, O. D., Mccutcheon, R., Agid, O., et al. (2017). Treatment-resistant schizophrenia: Treatment Response and Resistance in Psychosis (TRRIP) working group consensus guidelines on diagnosis and terminology. *The American Journal of Psychiatry, 174*, 216–229.

Huang, M., Yu, L., Pan, F., et al. (2018). A randomized, 13-week study assessing the efficacy and metabolic effects of paliperidone palmitate injection and olanzapine in first-episode schizophrenia patients. *Progress in Neuro-Psychopharmacology & Biological Psychiatry, 81*, 122–130.

Huhn, M., Nikolakopoulou, A., Schneider-Thoma, J., et al. (2019). Comparative efficacy and tolerability of 32 oral antipsychotics for the acute treatment of adults with multi-episode schizophrenia: A systematic review and network meta-analysis. *Lancet, 394*, 939–951.

Hummer, M., Kurz, M., Barnas, C., et al. (1994). Clozapine-induced transient white blood count disorders. *The Journal of Clinical Psychiatry, 55*, 429–432.

Hurault-Delarue, C., Damase-Michel, C., Finotto, L., et al. (2016). Psychomotor developmental effects of prenatal exposure to psychotropic drugs: A study in EFEMERIS database. *Fundamental & Clinical Pharmacology, 30*, 476–482.

Huybrechts, K. F., Hernández-Díaz, S., Patorno, E., et al. (2016). Antipsychotic use in pregnancy and the risk for congenital malformations. *JAMA Psychiatry, 73*, 938–946.

Ibrahim, H. M., & Tamminga, C. A. (2012). Treating impaired cognition in schizophrenia. *Current Pharmaceutical Biotechnology, 13*, 1587–1594.

Ishibashi, T., Horisawa, T., Tokuda, K., et al. (2010). Pharmacological profile of lurasidone, a novel antipsychotic agent with potent 5-hydroxytryptamine 7 (5-HT7) and 5-HT1A receptor activity. *The Journal of Pharmacology and Experimental Therapeutics, 334*, 171–181.

Ishida, T., Uchida, H., Suzuki, T., et al. (2011). Plasma substance P level in patients with schizophrenia: A cross-sectional study. *Psychiatry and Clinical Neurosciences, 65*, 526–528.

Ishiyama, T., Tokuda, K., Ishibashi, T., et al. (2007). Lurasidone (SM-13496), a novel atypical antipsychotic drug, reverses MK-801-induced impairment of learning and memory in the rat passive-avoidance test. *European Journal of Pharmacology, 572*, 160–170.

Ito, H., Takano, H., Takahashi, H., et al. (2009). Effects of the antipsychotic risperidone on dopamine synthesis in human brain measured by positron emission tomography with L-[beta-11C]DOPA: A stabilizing effect for dopaminergic neurotransmission? *The Journal of Neuroscience, 29*, 13730–13734.

Iwata, Y., Nakajima, S., Caravaggio, F., et al. (2016). Threshold of dopamine D2/3 receptor occupancy for hyperprolactinemia in older patients with schizophrenia. *The Journal of Clinical Psychiatry, 77*, e1557–e1563.

Iwata, Y., Nakajima, S., Plitman, E., et al. (2019). Glutamatergic neurometabolite levels in patients with ultra-treatment-resistant schizophrenia: A cross-sectional 3T proton magnetic resonance spectroscopy study. *Biological Psychiatry, 85*, 596–605.

Janssen, P. A., Van De Westeringh, C., Jageneau, A. H., et al. (1959). Chemistry and pharmacology of CNS depressants related to 4-(4-hydroxy-phenylpiperidino) butyrophenone. I. Synthesis and screening data in mice. *Journal of Medicinal and Pharmaceutical Chemistry, 1*, 281–297.

Janssen, P. A., Niemegeers, C. J., Awouters, F., et al. (1988). Pharmacology of risperidone (R 64 766), a new antipsychotic with serotonin-S2 and dopamine-D2 antagonistic properties. *The Journal of Pharmacology and Experimental Therapeutics, 244*, 685–693.

Javitt, D. C., & Zukin, S. R. (1991). Recent advances in the phencyclidine model of schizophrenia. *The American Journal of Psychiatry, 148*, 1301–1308.

Jensen, N. H., Rodriguiz, R. M., Caron, M. G., et al. (2008). N-desalkylquetiapine, a potent norepinephrine reuptake inhibitor and partial 5-HT1A agonist, as a putative mediator of quetiapine's antidepressant activity. *Neuropsychopharmacology, 33*, 2303–2312.

Jeste, D. V., Barak, Y., Madhusoodanan, S., et al. (2003). International multisite double-blind trial of the atypical antipsychotics risperidone and olanzapine in 175 elderly patients with chronic schizophrenia. *The American Journal of Geriatric Psychiatry, 11*, 638–647.

John, A. P., Ko, E. K. F., & Dominic, A. (2018). Delayed initiation of clozapine continues to be a substantial clinical concern. *Canadian Journal of Psychiatry, 63*, 526–531.

Johnson, K. C., Laprairie, J. L., Brennan, P. A., et al. (2012). Prenatal antipsychotic exposure and neuromotor performance during infancy. *Archives of General Psychiatry, 69*, 787–794.

Jolley, A. G., & Hirsch, S. R. (1993). Continuous versus intermittent neuroleptic therapy in schizophrenia. *Drug Safety, 8*, 331–339.

Jolley, A. G., Hirsch, S. R., Morrison, E., et al. (1990). Trial of brief intermittent neuroleptic prophylaxis for selected schizophrenic outpatients: Clinical and social outcome at two years. *BMJ, 301*, 837–842.

Jones, P. B., Barnes, T. R., Davies, L., et al. (2006). Randomized controlled trial of the effect on quality of life of second- vs first-generation antipsychotic drugs in schizophrenia: Cost utility of the latest antipsychotic drugs in schizophrenia study (CUtLASS 1). *Archives of General Psychiatry, 63*, 1079–1087.

Juruena, M. F., De Sena, E. P., & De Oliveira, I. R. (2011). Sertindole in the management of schizophrenia. *Journal of Central Nervous System Disease, 3*, 75–85.

Kahn, R. S., Fleischhacker, W. W., Boter, H., et al. (2008). Effectiveness of antipsychotic drugs in first-episode schizophrenia and schizophreniform disorder: An open randomised clinical trial. *Lancet, 371*, 1085–1097.

Kane, J. M., Rifkin, A., Woerner, M., et al. (1983). Low-dose neuroleptic treatment of outpatient schizophrenics. I. Preliminary results for relapse rates. *Archives of General Psychiatry, 40*, 893–896.

Kane, J., Honigfeld, G., Singer, J., et al. (1988). Clozapine for the treatment-resistant schizophrenic. A double-blind comparison with chlorpromazine. *Archives of General Psychiatry, 45*, 789–796.

Kane, J., Canas, F., Kramer, M., et al. (2007). Treatment of schizophrenia with paliperidone extended-release tablets: A 6-week placebo-controlled trial. *Schizophrenia Research, 90*, 147–161.

Kane, J. M., Schooler, N. R., Marcy, P., et al. (2020). Effect of long-acting injectable antipsychotics vs usual care on time to first hospitalization in early-phase schizophrenia: A randomized clinical trial. *JAMA Psychiatry, 77*, 1–8.

Kantrowitz, J. T., Grinband, J., Goff, D. C., et al. (2020). Proof of mechanism and target engagement of glutamatergic drugs for the treatment of schizophrenia: RCTs of pomaglumetad and TS-134 on ketamine-induced psychotic symptoms and pharmacoBOLD in healthy volunteers. *Neuropsychopharmacology, 45*, 1842–1850.

Kapur, S., & Remington, G. (2001). Dopamine D(2) receptors and their role in atypical antipsychotic action: Still necessary and may even be sufficient. *Biological Psychiatry, 50*, 873–883.

Kapur, S., & Seeman, P. (2001). Does fast dissociation from the dopamine d(2) receptor explain the action of atypical antipsychotics?: A new hypothesis. *The American Journal of Psychiatry, 158*, 360–369.

Kapur, S., Zipursky, R. B., Remington, G., et al. (1998). 5-HT2 and D2 receptor occupancy of olanzapine in schizophrenia: A PET investigation. *The American Journal of Psychiatry, 155*, 921–928.

Kapur, S., Zipursky, R. B., & Remington, G. (1999). Clinical and theoretical implications of 5-HT2 and D2 receptor occupancy of clozapine, risperidone, and olanzapine in schizophrenia. *The American Journal of Psychiatry, 156*, 286–293.

Kapur, S., Zipursky, R., Jones, C., et al. (2000a). Relationship between dopamine D(2) occupancy, clinical response, and side effects: A double-blind PET study of first-episode schizophrenia. *The American Journal of Psychiatry, 157*, 514–520.

Kapur, S., Zipursky, R., Jones, C., et al. (2000b). A positron emission tomography study of quetiapine in schizophrenia: A preliminary finding of an antipsychotic effect with only transiently high dopamine D2 receptor occupancy. *Archives of General Psychiatry, 57*, 553–559.

Karam, C. S., Ballon, J. S., Bivens, N. M., et al. (2010). Signaling pathways in schizophrenia: Emerging targets and therapeutic strategies. *Trends in Pharmacological Sciences, 31*, 381–390.

Kato, T., Hirose, A., Ohno, Y., et al. (1990). Binding profile of SM-9018, a novel antipsychotic candidate. *Japanese Journal of Pharmacology, 54*, 478–481.

Keefe, R. S., Seidman, L. J., Christensen, B. K., et al. (2004). Comparative effect of atypical and conventional antipsychotic drugs on neurocognition in first-episode psychosis: A randomized, double-blind trial of olanzapine versus low doses of haloperidol. *The American Journal of Psychiatry, 161*, 985–995.

Keefe, R. S., Eesley, C. E., & Poe, M. P. (2005). Defining a cognitive function decrement in schizophrenia. *Biological Psychiatry, 57*, 688–691.

Keefe, R. S., Bilder, R. M., Davis, S. M., et al. (2007). Neurocognitive effects of antipsychotic medications in patients with chronic schizophrenia in the CATIE trial. *Archives of General Psychiatry, 64*, 633–647.

Kelly, D. L., Powell, M. M., Wehring, H. J., et al. (2018). Adjunct aripiprazole reduces prolactin and prolactin-related adverse effects in premenopausal women with psychosis: Results from the DAAMSEL clinical trial. *Journal of Clinical Psychopharmacology, 38*, 317–326.

Kennedy, J. S., Jeste, D., Kaiser, C. J., et al. (2003). Olanzapine vs haloperidol in geriatric schizophrenia: Analysis of data from a double-blind controlled trial. *International Journal of Geriatric Psychiatry, 18*, 1013–1020.

Kikuchi, T., Tottori, K., Uwahodo, Y., et al. (1995). 7-(4-[4-(2,3-Dichlorophenyl)-1-piperazinyl]butyloxy)-3,4-dihydro-2(1H)-quinolinone (OPC-14597), a new putative antipsychotic drug with both presynaptic dopamine autoreceptor agonistic activity and postsynaptic D2 receptor antagonistic activity. *The Journal of Pharmacology and Experimental Therapeutics, 274*, 329–336.

Kim, E., Howes, O. D., Kim, B. H., et al. (2012). Predicting brain occupancy from plasma levels using PET: Superiority of combining pharmacokinetics with pharmacodynamics while modeling the relationship. *Journal of Cerebral Blood Flow and Metabolism, 32*, 759–768.

Kim, S., Kim, S., Koh, M., Choi, G., Kim, J. J., Paik, I. H., Kim, S. H., Choi, Y. S., Lee, Y., Suh, J., Takeuchi, H., Uchida, H., & Kim, E. (2021). Effects of Long-Acting Injectable Paliperidone Palmitate on Clinical and Functional Outcomes in Patients With Schizophrenia Based on Illness Duration. *J Clin Psychiatry, 82*(1), 20m13446. https://doi.org/10.4088/JCP.20m13446. PMID: 33434958

Kirkpatrick, B., Fenton, W. S., Carpenter, W. T., Jr., et al. (2006). The NIMH-MATRICS consensus statement on negative symptoms. *Schizophrenia Bulletin, 32*, 214–219.

Kishimoto, T., Hagi, K., Nitta, M., et al. (2018). Effectiveness of long-acting injectable vs oral antipsychotics in patients with schizophrenia: A meta-analysis of prospective and retrospective cohort studies. *Schizophrenia Bulletin, 44*, 603–619.

Koblan, K. S., Kent, J., Hopkins, S. C., Krystal, J. H., Cheng, H., Goldman, R., & Loebel, A. (2020). A non-D2-receptor-binding drug for the treatment of schizophrenia. *The New England Journal of Medicine, 382*(16), 1497–1506.

Koizumi, T., Uchida, H., Suzuki, T., et al. (2013). Oversight of constipation in inpatients with schizophrenia: A cross-sectional study. *General Hospital Psychiatry, 35*, 649–652.

Kontkanen, O., Törönen, P., Lakso, M., et al. (2002). Antipsychotic drug treatment induces differential gene expression in the rat cortex. *Journal of Neurochemistry, 83*, 1043–1053.

Krause, M., Huhn, M., Schneider-Thoma, J., et al. (2019). Efficacy, acceptability and tolerability of antipsychotics in patients with schizophrenia and comorbid substance use. A systematic review and meta-analysis. *European Neuropsychopharmacology, 29*, 32–45.

Krystal, J. H., Karper, L. P., Seibyl, J. P., et al. (1994). Subanesthetic effects of the noncompetitive NMDA antagonist, ketamine, in humans. Psychotomimetic, perceptual, cognitive, and neuroendocrine responses. *Archives of General Psychiatry, 51*, 199–214.

Kucukgoncu, S., Guloksuz, S., Celik, K., et al. (2020). Antipsychotic exposure in pregnancy and the risk of gestational diabetes: A systematic review and meta-analysis. *Schizophrenia Bulletin, 46*, 311–318.

Lang, K., Meyers, J. L., Korn, J. R., et al. (2010). Medication adherence and hospitalization among patients with schizophrenia treated with antipsychotics. *Psychiatric Services, 61*, 1239–1247.

Lao, K. S. J., Zhao, J., Blais, J. E., et al. (2020). Antipsychotics and risk of neuroleptic malignant syndrome: A population-based cohort and case-crossover study. *CNS Drugs, 34*, 1165–1175.

Lewis, S. W., Barnes, T. R., Davies, L., et al. (2006a). Randomized controlled trial of effect of prescription of clozapine versus other second-generation antipsychotic drugs in resistant schizophrenia. *Schizophrenia Bulletin, 32*, 715–723.

Lewis, S. W., Davies, L., Jones, P. B., et al. (2006b). Randomised controlled trials of conventional antipsychotic versus new atypical drugs, and new atypical drugs versus clozapine, in people with schizophrenia responding poorly to, or intolerant of, current drug treatment. *Health Technology Assessment, 10*, iii–iv, ix–xi, 1–165.

Lieberman, J. A. (1993). Understanding the mechanism of action of atypical antipsychotic drugs. A review of compounds in use and development. *The British Journal of Psychiatry. Supplement, 22*, 7–18. Erratum in: Br J Psychiatry 1994 May;164:709. PMID: 7906527.

Lieberman, J. A. (1999). Pathophysiologic mechanisms in the pathogenesis and clinical course of schizophrenia. *The Journal of Clinical Psychiatry, 60*(Suppl 12), 9–12.

Lieberman, J. A. (2004). Dopamine partial agonists: A new class of antipsychotic. *CNS Drugs, 18*, 251–267.

Lieberman, J. A., Kane, J. M., & Johns, C. A. (1989). Clozapine: Guidelines for clinical management. *The Journal of Clinical Psychiatry, 50*, 329–338.

Lieberman, J. A., Perkins, D., Belger, A., et al. (2001). The early stages of schizophrenia: Speculations on pathogenesis, pathophysiology, and therapeutic approaches. *Biological Psychiatry, 50*, 884–897.

Lieberman, J. A., Stroup, T. S., Mcevoy, J. P., et al. (2005a). Effectiveness of antipsychotic drugs in patients with chronic schizophrenia. *The New England Journal of Medicine, 353*, 1209–1223.

Lieberman, J. A., Tollefson, G. D., Charles, C., et al. (2005b). Antipsychotic drug effects on brain morphology in first-episode psychosis. *Archives of General Psychiatry, 62*, 361–370.

Lieberman, J. A., Davis, R. E., Correll, C. U., et al. (2016). ITI-007 for the treatment of schizophrenia: A 4-week randomized, double-blind, controlled trial. *Biological Psychiatry, 79*, 952–961.

Loy, J. H., Merry, S. N., Hetrick, S. E., et al. (2017). Atypical antipsychotics for disruptive behaviour disorders in children and youths. *Cochrane Database of Systematic Reviews, 8*, Cd008559.

Lu, L., Dong, M., Zhang, L., et al. (2019). Prevalence of suicide attempts in individuals with schizophrenia: A meta-analysis of observational studies. *Epidemiology and Psychiatric Sciences, 29*, e39.

Luby, E. D., Gottlieb, J. S., Cohen, B. D., et al. (1962). Model psychoses and schizophrenia. *The American Journal of Psychiatry, 119*, 61–67.

Lucas, G., & Spampinato, U. (2000). Role of striatal serotonin2A and serotonin2C receptor subtypes in the control of in vivo dopamine outflow in the rat striatum. *Journal of Neurochemistry, 74*, 693–701.

Lyne, J., Joober, R., Schmitz, N., et al. (2017). Duration of active psychosis and first-episode psychosis negative symptoms. *Early Intervention in Psychiatry, 11*, 63–71.

Malhotra, A. K., Zhang, J. P., & Lencz, T. (2012). Pharmacogenetics in psychiatry: Translating research into clinical practice. *Molecular Psychiatry, 17*, 760–769.

Malik, P., Kemmler, G., Hummer, M., et al. (2011). Sexual dysfunction in first-episode schizophrenia patients: Results from European First Episode Schizophrenia Trial. *Journal of Clinical Psychopharmacology, 31*, 274–280.

Mamo, D., Kapur, S., Shammi, C. M., et al. (2004). A PET study of dopamine D2 and serotonin 5-HT2 receptor occupancy in patients with schizophrenia treated with therapeutic doses of ziprasidone. *The American Journal of Psychiatry, 161*, 818–825.

Mamo, D., Graff, A., Mizrahi, R., et al. (2007). Differential effects of aripiprazole on D(2), 5-HT(2), and 5-HT(1A) receptor occupancy in patients with schizophrenia: A triple tracer PET study. *The American Journal of Psychiatry, 164*, 1411–1417.

Mamo, D. C., Uchida, H., Vitcu, I., et al. (2008). Quetiapine extended-release versus immediate-release formulation: A positron emission tomography study. *The Journal of Clinical Psychiatry, 69*, 81–86.

Marder, S. R., & Meibach, R. C. (1994). Risperidone in the treatment of schizophrenia. *The American Journal of Psychiatry, 151*, 825–835.

Marder, S. R., Essock, S. M., Miller, A. L., et al. (2004). Physical health monitoring of patients with schizophrenia. *The American Journal of Psychiatry, 161*, 1334–1349.

Marshall, M., Lewis, S., Lockwood, A., et al. (2005). Association between duration of untreated psychosis and outcome in cohorts of first-episode patients: A systematic review. *Archives of General Psychiatry, 62*, 975–983.

Marwick, K. F., Taylor, M., & Walker, S. W. (2012). Antipsychotics and abnormal liver function tests: Systematic review. *Clinical Neuropharmacology, 35*, 244–253.

Masi, G., Mucci, M., & Pari, C. (2006). Children with schizophrenia: Clinical picture and pharmacological treatment. *CNS Drugs, 20*, 841–866.

Matosin, N., Fernandez-Enright, F., Lum, J. S., et al. (2016). Molecular evidence of synaptic pathology in the CA1 region in schizophrenia. *NPJ Schizophrenia, 2*, 16022.

Matsui, K., Tokumasu, T., Takekita, Y., et al. (2019). Switching to antipsychotic monotherapy vs. staying on antipsychotic polypharmacy in schizophrenia: A systematic review and meta-analysis. *Schizophrenia Research, 209*, 50–57.

Mauri, M. C., Paletta, S., Maffini, M., et al. (2014). Clinical pharmacology of atypical antipsychotics: An update. *EXCLI Journal, 13*, 1163–1191.

Mcnaught, K. S., & Mink, J. W. (2011). Advances in understanding and treatment of Tourette syndrome. *Nature Reviews. Neurology, 7*, 667–676.

Meltzer, H. Y. (2012). Clozapine: Balancing safety with superior antipsychotic efficacy. *Clinical Schizophrenia & Related Psychoses, 6*, 134–144.

Meltzer, H. Y., & Massey, B. W. (2011). The role of serotonin receptors in the action of atypical antipsychotic drugs. *Current Opinion in Pharmacology, 11*, 59–67.

Meltzer, H. Y., Matsubara, S., & Lee, J. C. (1989). Classification of typical and atypical antipsychotic drugs on the basis of dopamine D-1, D-2 and serotonin2 pKi values. *The Journal of Pharmacology and Experimental Therapeutics, 251*, 238–246.

Meltzer, H. Y., Alphs, L., Green, A. I., et al. (2003). Clozapine treatment for suicidality in schizophrenia: International Suicide Prevention Trial (InterSePT). *Archives of General Psychiatry, 60*, 82–91.

Meltzer, H. Y., Bobo, W. V., Roy, A., et al. (2008). A randomized, double-blind comparison of clozapine and high-dose olanzapine in treatment-resistant patients with schizophrenia. *The Journal of Clinical Psychiatry, 69*, 274–285.

Merritt, K., Egerton, A., Kempton, M. J., et al. (2016). Nature of glutamate alterations in schizophrenia: A meta-analysis of proton magnetic resonance spectroscopy studies. *JAMA Psychiatry, 73*, 665–674.

Meyer, J. M., Loebel, A. D., & Schweizer, E. (2009). Lurasidone: A new drug in development for schizophrenia. *Expert Opinion on Investigational Drugs, 18*, 1715–1726.

Miller, C. H., & Fleischhacker, W. W. (2000). Managing antipsychotic-induced acute and chronic akathisia. *Drug Safety, 22*, 73–81.

Miller, T. J., Mcglashan, T. H., Rosen, J. L., et al. (2002). Prospective diagnosis of the initial prodrome for schizophrenia based on the Structured Interview for Prodromal Syndromes: Preliminary evidence of interrater reliability and predictive validity. *The American Journal of Psychiatry, 159*, 863–865.

Misawa, F., Kishimoto, T., Hagi, K., et al. (2016). Safety and tolerability of long-acting injectable versus oral antipsychotics: A meta-analysis of randomized controlled studies comparing the same antipsychotics. *Schizophrenia Research, 176*, 220–230.

Mishara, A. L., & Goldberg, T. E. (2004). A meta-analysis and critical review of the effects of conventional neuroleptic treatment on cognition in schizophrenia: Opening a closed book. *Biological Psychiatry, 55*, 1013–1022.

Miyamoto, S., Miyake, N., Jarskog, L. F., et al. (2012). Pharmacological treatment of schizophrenia: A critical review of the pharmacology and clinical effects of current and future therapeutic agents. *Molecular Psychiatry, 17*, 1206–1227.

Miyazaki, T., Nakajima, W., Hatano, M., et al. (2020). Visualization of AMPA receptors in living human brain with positron emission tomography. *Nature Medicine, 26*, 281–288.

Mizuno, Y., Suzuki, T., Nakagawa, A., et al. (2014). Pharmacological strategies to counteract antipsychotic-induced weight gain and metabolic adverse effects in schizophrenia: A systematic review and meta-analysis. *Schizophrenia Bulletin, 40*, 1385–1403.

Möller, H. J. (1993). Neuroleptic treatment of negative symptoms in schizophrenic patients. Efficacy problems and methodological difficulties. *European Neuropsychopharmacology, 3*, 1–11.

Morken, G., Widen, J. H., & Grawe, R. W. (2008). Non-adherence to antipsychotic medication, relapse and rehospitalisation in recent-onset schizophrenia. *BMC Psychiatry, 8*, 32.

Mortensen, J. T., Olsen, J., Larsen, H., et al. (2003). Psychomotor development in children exposed in utero to benzodiazepines, antidepressants, neuroleptics, and anti-epileptics. *European Journal of Epidemiology, 18*, 769–771.

Murphy, B. P., Chung, Y. C., Park, T. W., et al. (2006). Pharmacological treatment of primary negative symptoms in schizophrenia: A systematic review. *Schizophrenia Research, 88*, 5–25.

Myles, N., Myles, H., Xia, S., et al. (2018). Meta-analysis examining the epidemiology of clozapine-associated neutropenia. *Acta Psychiatrica Scandinavica, 138*, 101–109.

Myles, N., Myles, H., Xia, S., et al. (2019). A meta-analysis of controlled studies comparing the association between clozapine and other antipsychotic medications and the development of neutropenia. *The Australian and New Zealand Journal of Psychiatry, 53*, 403–412.

Needham, P. L., Atkinson, J., Skill, M. J., et al. (1996). Zotepine: Preclinical tests predict antipsychotic efficacy and an atypical profile. *Psychopharmacology Bulletin, 32*, 123–128.

Newcomer, J. W. (2007). Antipsychotic medications: Metabolic and cardiovascular risk. *The Journal of Clinical Psychiatry, 68*(Suppl 4), 8–13.

Ng, W., Uchida, H., Ismail, Z., et al. (2009). Clozapine exposure and the impact of smoking and gender: A population pharmacokinetic study. *Therapeutic Drug Monitoring, 31*, 360–366.

Nielsen, R. E., Levander, S., Kjaersdam Telléus, G., et al. (2015). Second-generation antipsychotic effect on cognition in patients with schizophrenia – A meta-analysis of randomized clinical trials. *Acta Psychiatrica Scandinavica, 131*, 185–196.

Nierenberg, A. A., McElroy, S. L., Friedman, E. S., Ketter, T. A., Shelton, R. C., Deckersbach, T., McInnis, M. G., Bowden, C. L., Tohen, M., Kocsis, J. H., Calabrese, J. R., Kinrys, G., Bobo, W. V., Singh, V., Kamali, M., Kemp, D., Brody, B., Reilly-Harrington, N. A., Sylvia, L. G., Shesler, L. W., Bernstein, E. E., Schoenfeld, D., Rabideau, D. J., Leon, A. C., Faraone, S., & Thase, M. E. (2016 Jan). Bipolar CHOICE (Clinical Health Outcomes Initiative in Comparative Effectiveness): A pragmatic 6-month trial of lithium versus quetiapine for bipolar disorder. *The Journal of Clinical Psychiatry, 77*(1), 90–99.

Nikiforuk, A., & Popik, P. (2013). Amisulpride promotes cognitive flexibility in rats: The role of 5-HT7 receptors. *Behavioural Brain Research, 248*, 136–140.

Nilsson, E., Lichtenstein, P., Cnattingius, S., et al. (2002). Women with schizophrenia: Pregnancy outcome and infant death among their offspring. *Schizophrenia Research, 58*, 221–229.

Ohoyama, K., Yamamura, S., Hamaguchi, T., et al. (2011). Effect of novel atypical antipsychotic, blonanserin, on extracellular neurotransmitter level in rat prefrontal cortex. *European Journal of Pharmacology, 653*, 47–57.

Oliver, D., Davies, C., Crossland, G., et al. (2018). Can we reduce the duration of untreated psychosis? A systematic review and meta-analysis of controlled interventional studies. *Schizophrenia Bulletin, 44*, 1362–1372.

Onrust, S. V., & Mcclellan, K. (2001). Perospirone. *CNS Drugs, 15*, 329–337. discussion 338.

Ortiz-Orendain, J., Castiello-De Obeso, S., Colunga-Lozano, L. E., et al. (2017). Antipsychotic combinations for schizophrenia. *Cochrane Database of Systematic Reviews, 6*, Cd009005.

Osser, D. N., Roudsari, M. J., & Manschreck, T. (2013). The psychopharmacology algorithm project at the Harvard South Shore Program: An update on schizophrenia. *Harvard Review of Psychiatry, 21*, 18–40.

Ozasa, R., Okada, T., Nadanaka, S., Nagamine, T., Zyryanova, A., Harding, H., Ron, D., & Mori, K. (2013). The antipsychotic olanzapine induces apoptosis in insulin-secreting pancreatic β cells by blocking PERK-mediated translational attenuation. *Cell Structure and Function, 38*(2), 183–195. https://doi.org/10.1247/csf.13012. Epub 2013 Jun 28. Erratum in: Cell Structure and Function. 2013;38(2):227. Erratum in: Cell Structure and Function. 2014;39(1):21.

Ozawa, C., Bies, R. R., Pillai, N., et al. (2019). Model-guided antipsychotic dose reduction in schizophrenia: A pilot, single-blind randomized controlled trial. *Journal of Clinical Psychopharmacology, 39*, 329–335.

Pacchiarotti, I., Anmella, G., Colomer, L., et al. (2020). How to treat mania. *Acta Psychiatrica Scandinavica, 142*, 173–192.

Pagsberg, A. K., Tarp, S., Glintborg, D., et al. (2017). Acute antipsychotic treatment of children and adolescents with schizophrenia-Spectrum disorders: A systematic review and network meta-analysis. *Journal of the American Academy of Child and Adolescent Psychiatry, 56*, 191–202.

Pakzad-Vaezi, K. L., Etminan, M., & Mikelberg, F. S. (2013). The association between cataract surgery and atypical antipsychotic use: A nested case-control study. *American Journal of Ophthalmology, 156*, 1141–1146.e1.

Pantelis, C., Velakoulis, D., Mcgorry, P. D., et al. (2003). Neuroanatomical abnormalities before and after onset of psychosis: A cross-sectional and longitudinal MRI comparison. *Lancet, 361*, 281–288.

Papakostas, G. I., Vitolo, O. V., Ishak, W. W., et al. (2012). A 12-week, randomized, double-blind, placebo-controlled, sequential parallel comparison trial of ziprasidone as monotherapy for major depressive disorder. *The Journal of Clinical Psychiatry, 73*, 1541–1547.

Pardo-De-Santayana, G., Vázquez-Bourgon, J., Gómez-Revuelta, M., et al. (2020). Duration of active psychosis during early phases of the illness and functional outcome: The PAFIP 10-year follow-up study. *Schizophrenia Research, 220*, 240–247.

Patteet, L., Morrens, M., Maudens, K. E., et al. (2012). Therapeutic drug monitoring of common antipsychotics. *Therapeutic Drug Monitoring, 34*, 629–651.

Perkins, D. O., Gu, H., Boteva, K., et al. (2005). Relationship between duration of untreated psychosis and outcome in first-episode schizophrenia: A critical review and meta-analysis. *The American Journal of Psychiatry, 162*, 1785–1804.

Peters, L., Krogmann, A., Von Hardenberg, L., et al. (2019). Long-acting injections in schizophrenia: A 3-year update on randomized controlled trials published January 2016–March 2019. *Current Psychiatry Reports, 21*, 124.

Petrides, G., Malur, C., Braga, R. J., et al. (2015). Electroconvulsive therapy augmentation in clozapine-resistant schizophrenia: A prospective, randomized study. *The American Journal of Psychiatry, 172*, 52–58.

Pillinger, T., Mccutcheon, R. A., Vano, L., et al. (2020). Comparative effects of 18 antipsychotics on metabolic function in patients with schizophrenia, predictors of metabolic dysregulation, and association with psychopathology: A systematic review and network meta-analysis. *Lancet Psychiatry, 7*, 64–77.

Poels, E. M. P., Schrijver, L., Kamperman, A. M., et al. (2018). Long-term neurodevelopmental consequences of intrauterine exposure to lithium and antipsychotics: A systematic review and meta-analysis. *European Child & Adolescent Psychiatry, 27*, 1209–1230.

Potkin, S. G., Keator, D. B., Kesler-West, M. L., et al. (2014). D2 receptor occupancy following lurasidone treatment in patients with schizophrenia or schizoaffective disorder. *CNS Spectrums, 19*, 176–181.

Pottegård, A., Lash, T. L., Cronin-Fenton, D., et al. (2018). Use of antipsychotics and risk of breast cancer: A Danish nationwide case-control study. *British Journal of Clinical Pharmacology, 84*, 2152–2161.

Pringsheim, T., Hirsch, L., Gardner, D., et al. (2015). The pharmacological management of oppositional behaviour, conduct problems, and aggression in children and adolescents with attention-deficit hyperactivity disorder, oppositional defiant disorder, and conduct disorder: A systematic review and meta-analysis. Part 2: Antipsychotics and traditional mood stabilizers. *Canadian Journal of Psychiatry, 60*, 52–61.

Pringsheim, T., Holler-Managan, Y., Okun, M. S., et al. (2019). Comprehensive systematic review summary: Treatment of tics in people with Tourette syndrome and chronic tic disorders. *Neurology, 92*, 907–915.

Quinn, N., & Marsden, C. D. (1984). A double blind trial of sulpiride in Huntington's disease and tardive dyskinesia. *Journal of Neurology, Neurosurgery, and Psychiatry, 47*, 844–847.

Raedler, T. J., Knable, M. B., Jones, D. W., et al. (2000). In vivo olanzapine occupancy of muscarinic acetylcholine receptors in patients with schizophrenia. *Neuropsychopharmacology, 23*, 56–68.

Rajji, T. K., & Mulsant, B. H. (2008). Nature and course of cognitive function in late-life schizophrenia: A systematic review. *Schizophrenia Research, 102*, 122–140.

Rauch, A. S., & Fleischhacker, W. W. (2013). Long-acting injectable formulations of new-generation antipsychotics: A review from a clinical perspective. *CNS Drugs, 27*, 637–652.

Ray, W. A., Chung, C. P., Murray, K. T., et al. (2009a). Atypical antipsychotic drugs and the risk of sudden cardiac death. *The New England Journal of Medicine, 360*, 225–235.

Ray, W. A., Chung, C. P., Murray, K. T., et al. (2009b). Atypical antipsychotic drugs and the risk of sudden cardiac death. *New England Journal of Medicine, 360*, 225–235.

Remington, G., & Kapur, S. (2010). Antipsychotic dosing: How much but also how often? *Schizophrenia Bulletin, 36*, 900–903.

Remschmidt, H., & Theisen, F. (2012). Early-onset schizophrenia. *Neuropsychobiology, 66*, 63–69.

Riecher-Rössler, A., Rybakowski, J. K., Pflueger, M. O., et al. (2013). Hyperprolactinemia in antipsychotic-naive patients with first-episode psychosis. *Psychological Medicine, 43*, 2571–2582.

Rosenheck, R. A., Leslie, D. L., Sindelar, J., et al. (2006). Cost-effectiveness of second-generation antipsychotics and perphenazine in a randomized trial of treatment for chronic schizophrenia. *The American Journal of Psychiatry, 163*, 2080–2089.

Sahli, Z. T., & Tarazi, F. I. (2018). Pimavanserin: Novel pharmacotherapy for Parkinson's disease psychosis. *Expert Opinion on Drug Discovery, 13*, 103–110.

Sakurai, H., Bies, R., Stroup, S., et al. (2012). Dopamine D2 receptor occupancy and cognition in schizophrenia: Analysis of the CATIE data. *Schizophrenia Bulletin, 39*, 564–574.

Sakurai, H., Bies, R. R., Stroup, S. T., et al. (2013). Dopamine D2 receptor occupancy and cognition in schizophrenia: Analysis of the CATIE data. *Schizophrenia Bulletin, 39*, 564–574.

Sakurai, H., Suzuki, T., Bies, R. R., et al. (2016). Increasing versus maintaining the dose of olanzapine or risperidone in schizophrenia patients who did not respond to a modest dosage: A double-blind randomized controlled trial. *The Journal of Clinical Psychiatry, 77*, 1381–1390.

Sakurai, H., Uchida, H., Kato, M., et al. (2020). Pharmacological management of depression: Japanese expert consensus. *Journal of Affective Disorders, 266*, 626–632.

Sakurai, H., Yasui-Furukori, N., Suzuki, T., Uchida, H., Baba, H., Watanabe, K., Inada, K., Kikuchi, Y. S., Kikuchi, T., Katsuki, A., Kishida, I., & Kato, M. (2021). Pharmacological Treatment of Schizophrenia: Japanese Expert Consensus. *Pharmacopsychiatry, 54* (2), 60–67. https://doi.org/10.1055/a-1324-3517. Epub 2021 Jan 12. PMID: 33434943; PMCID: PMC7946533.

Samaha, A. N., Reckless, G. E., Seeman, P., et al. (2008). Less is more: Antipsychotic drug effects are greater with transient rather than continuous delivery. *Biological Psychiatry, 64*, 145–152.

Sanchez, C., Arnt, J., Dragsted, N., et al. (1991). Neurochemical and in vivo pharmacological profile of sertindole, a limbic-selective neuroleptic compound. *Drug Development Research, 22*, 239–250.

Scatton, B., Claustre, Y., Cudennec, A., et al. (1997). Amisulpride: From animal pharmacology to therapeutic action. *International Clinical Psychopharmacology, 12*(Suppl 2), S29–S36.

Schneeweiss, S., & Avorn, J. (2009). Antipsychotic agents and sudden cardiac death – How should we manage the risk? *New England Journal of Medicine, 360*, 295–296.

Schneider, L. S., Dagerman, K. S., & Insel, P. (2005). Risk of death with atypical antipsychotic drug treatment for dementia: Meta-analysis of randomized placebo-controlled trials. *JAMA, 294*, 1934–1943.

Schneider, L. S., Tariot, P. N., Dagerman, K. S., et al. (2006). Effectiveness of atypical antipsychotic drugs in patients with Alzheimer's disease. *The New England Journal of Medicine, 355*, 1525–1538.

Schneider, M., Regente, J., Greiner, T., et al. (2020). Neuroleptic malignant syndrome: Evaluation of drug safety data from the AMSP program during 1993-2015.

European Archives of Psychiatry and Clinical Neuroscience, 270, 23–33.

Schobel, S. A., Chaudhury, N. H., Khan, U. A., et al. (2013). Imaging patients with psychosis and a mouse model establishes a spreading pattern of hippocampal dysfunction and implicates glutamate as a driver. *Neuron, 78*, 81–93.

Schoemaker, H., Claustre, Y., Fage, D., et al. (1997). Neurochemical characteristics of amisulpride, an atypical dopamine D2/D3 receptor antagonist with both presynaptic and limbic selectivity. *The Journal of Pharmacology and Experimental Therapeutics, 280*, 83–97.

Schönfeldt-Lecuona, C., Kuhlwilm, L., Cronemeyer, M., et al. (2020). Treatment of the neuroleptic malignant syndrome in international therapy guidelines: A comparative analysis. *Pharmacopsychiatry, 53*, 51–59.

Schooler, N. R. (1991). Maintenance medication for schizophrenia: Strategies for dose reduction. *Schizophrenia Bulletin, 17*, 311–324.

Schotte, A., Janssen, P. F., Gommeren, W., et al. (1996). Risperidone compared with new and reference antipsychotic drugs: In vitro and in vivo receptor binding. *Psychopharmacology (Berl), 124*, 57–73.

Schreiner, A., Aadamsoo, K., Altamura, A. C., et al. (2015). Paliperidone palmitate versus oral antipsychotics in recently diagnosed schizophrenia. *Schizophrenia Research, 169*, 393–399.

Seeman, P. (1992). Dopamine receptor sequences. Therapeutic levels of neuroleptics occupy D2 receptors, clozapine occupies D4. *Neuropsychopharmacology, 7*, 261–284.

Seeman, P., Lee, T., Chau-Wong, M., et al. (1976). Antipsychotic drug doses and neuroleptic/dopamine receptors. *Nature, 261*, 717–719.

Seppi, K., Weintraub, D., Coelho, M., et al. (2011). The Movement Disorder Society Evidence-Based Medicine Review Update: Treatments for the non-motor symptoms of Parkinson's disease. *Movement Disorders, 26*(Suppl 3), S42–S80.

Servonnet, A., Uchida, H., & Samaha, A. N. (2020). Continuous versus extended antipsychotic dosing in schizophrenia: Less is more. *Behavioural Brain Research, 401*, 113076.

Shahid, M., Walker, G. B., Zorn, S. H., et al. (2009). Asenapine: A novel psychopharmacologic agent with a unique human receptor signature. *Journal of Psychopharmacology, 23*, 65–73.

Shao, P., Ou, J., Peng, M., et al. (2015). Effects of clozapine and other atypical antipsychotics on infants development who were exposed to as fetus: A post-hoc analysis. *PLoS One, 10*, e0123373.

Shapiro, D. A., Renock, S., Arrington, E., et al. (2003). Aripiprazole, a novel atypical antipsychotic drug with a unique and robust pharmacology. *Neuropsychopharmacology, 28*, 1400–1411.

Sharif, Z., Miyamoto, S., & Lieberman, J. A. (2007). Pharmacotherapy of schizophrenia. In D. R. Sibley, I. Hanin, M. Kuhar, & P. Skolnick (Eds.), *Handbook of contemporary neuropharmacology*. Wiley.

Shobo, M., Kondo, Y., Yamada, H., et al. (2010). Norzotepine, a major metabolite of zotepine, exerts atypical antipsychotic-like and antidepressant-like actions through its potent inhibition of norepinephrine reuptake. *The Journal of Pharmacology and Experimental Therapeutics, 333*, 772–781.

Siskind, D., Siskind, V., & Kisely, S. (2017). Clozapine response rates among people with treatment-resistant schizophrenia: Data from a systematic review and meta-analysis. *Canadian Journal of Psychiatry, 62*, 772–777.

Siskind, D. J., Lee, M., Ravindran, A., et al. (2018). Augmentation strategies for clozapine refractory schizophrenia: A systematic review and meta-analysis. *The Australian and New Zealand Journal of Psychiatry, 52*, 751–767.

Siskind, D., Sidhu, A., Cross, J., et al. (2020). Systematic review and meta-analysis of rates of clozapine-associated myocarditis and cardiomyopathy. *The Australian and New Zealand Journal of Psychiatry, 54*, 467–481.

Sleem, A., & El-Mallakh, R. S. (2021). Advances in the psychopharmacotherapy of bipolar disorder type I. *Expert Opinion on Pharmacotherapy, 22*(10), 1267–1290.

Snyder, G. L., Vanover, K. E., Zhu, H., et al. (2015). Functional profile of a novel modulator of serotonin, dopamine, and glutamate neurotransmission. *Psychopharmacology (Berl), 232*, 605–621.

Spina, E., & De Leon, J. (2007). Metabolic drug interactions with newer antipsychotics: A comparative review. *Basic & Clinical Pharmacology & Toxicology, 100*, 4–22.

Stoffers-Winterling, J. M., Storebø, O. J., Pereira Ribeiro, J., Kongerslev, M. T., Völlm, B. A., Mattivi, J. T., Faltinsen, E., Todorovac, A., Jørgensen, M. S., Callesen, H. E., Sales, C. P., Schaug, J. P., Simonsen, E, & Lieb, K. (2022). Pharmacological interventions for people with borderline personality disorder. *Cochrane Database Syst Rev, 11*(11), CD012956. https://doi.org/10.1002/14651858.CD012956.pub2. PMID: 36375174; PMCID: PMC9662763.

Strawn, J. R., Keck, P. E., Jr., & Caroff, S. N. (2007). Neuroleptic malignant syndrome. *The American Journal of Psychiatry, 164*, 870–876.

Stubbs, B., Vancampfort, D., De Hert, M., et al. (2015). The prevalence and predictors of type two diabetes mellitus in people with schizophrenia: A systematic review and comparative meta-analysis. *Acta Psychiatrica Scandinavica, 132*, 144–157.

Suppes, T., Silva, R., Cucchiaro, J., et al. (2016). Lurasidone for the treatment of major depressive disorder with mixed features: A randomized, double-blind, placebo-controlled study. *The American Journal of Psychiatry, 173*, 400–407.

Sur, C., Mallorga, P. J., Wittmann, M., et al. (2003). N-desmethylclozapine, an allosteric agonist at muscarinic 1 receptor, potentiates N-methyl-D-aspartate receptor activity. *Proceedings of the National Academy of Sciences of the United States of America, 100*, 13674–13679.

Suzuki, T., Graff-Guerrero, A., Uchida, H., et al. (2013). Dopamine $D_{2/3}$ occupancy of ziprasidone across a day: A within-subject PET study. *Psychopharmacology (Berl), 228*, 43–51.

Swartz, M. S., Perkins, D. O., Stroup, T. S., et al. (2007). Effects of antipsychotic medications on psychosocial functioning in patients with chronic schizophrenia: Findings from the NIMH CATIE study. *The American Journal of Psychiatry, 164*, 428–436.

Takeuchi, H., Siu, C., Remington, G., et al. (2019). Does relapse contribute to treatment resistance? Antipsychotic response in first- vs. second-episode schizophrenia. *Neuropsychopharmacology, 44*, 1036–1042.

Takeuchi, H., Mackenzie, N. E., Samaroo, D., et al. (2020). Antipsychotic dose in acute schizophrenia: A meta-analysis. *Schizophrenia Bulletin, 46*, 1439–1458.

Tateno, A., Arakawa, R., Okumura, M., et al. (2013). Striatal and extrastriatal dopamine D2 receptor occupancy by a novel antipsychotic, blonanserin: A PET study with [11C]raclopride and [11C]FLB 457 in schizophrenia. *Journal of Clinical Psychopharmacology, 33*, 162–169.

Taylor, D. (2017). Choice of antipsychotic to treat first-episode schizophrenia. *Lancet Psychiatry, 4*, 653–654.

Tenjin, T., Miyamoto, S., Ninomiya, Y., et al. (2013). Profile of blonanserin for the treatment of schizophrenia. *Neuropsychiatric Disease and Treatment, 9*, 587–594.

Terrana, N., Koren, G., Pivovarov, J., et al. (2015). Pregnancy outcomes following in utero exposure to second-generation antipsychotics: A systematic review and meta-analysis. *Journal of Clinical Psychopharmacology, 35*, 559–565.

Tiihonen, J., Mittendorfer-Rutz, E., Majak, M., et al. (2017). Real-world effectiveness of antipsychotic treatments in a Nationwide cohort of 29 823 patients with schizophrenia. *JAMA Psychiatry, 74*, 686–693.

Tiihonen, J., Taipale, H., Mehtälä, J., et al. (2019). Association of antipsychotic polypharmacy vs monotherapy with psychiatric rehospitalization among adults with schizophrenia. *JAMA Psychiatry, 76*, 499–507.

Tollefson, G. D., & Sanger, T. M. (1997). Negative symptoms: A path analytic approach to a double-blind, placebo- and haloperidol-controlled clinical trial with olanzapine. *The American Journal of Psychiatry, 154*, 466–474.

Turrone, P., Remington, G., Kapur, S., et al. (2005). Continuous but not intermittent olanzapine infusion induces vacuous chewing movements in rats. *Biological Psychiatry, 57*, 406–411.

Uchida, H., Suzuki, T., Tanaka, K. F., et al. (2003). Recurrent episodes of perceptual alteration in patients treated with antipsychotic agents. *Journal of Clinical Psychopharmacology, 23*, 496–499.

Uchida, H., Mamo, D. C., Kapur, S., et al. (2008). Monthly administration of long-acting injectable risperidone and striatal dopamine D2 receptor occupancy for the management of schizophrenia. *The Journal of Clinical Psychiatry, 69*, 1281–1286.

Uchida, H., Mamo, D. C., Mulsant, B. H., et al. (2009a). Increased antipsychotic sensitivity in elderly patients: Evidence and mechanisms. *The Journal of Clinical Psychiatry, 70*, 397–405.

Uchida, H., Pollock, B. G., Bies, R. R., et al. (2009b). Predicting age-specific dosing of antipsychotics. *Clinical Pharmacology and Therapeutics, 86*, 360–362.

Uchida, H., Rajji, T. K., Mulsant, B. H., et al. (2009c). D2 receptor blockade by risperidone correlates with attention deficits in late-life schizophrenia. *Journal of Clinical Psychopharmacology, 29*, 571–575.

Uchida, H., Suzuki, T., Takeuchi, H., et al. (2011a). Low dose vs standard dose of antipsychotics for relapse prevention in schizophrenia: Meta-analysis. *Schizophrenia Bulletin, 37*, 788–799.

Uchida, H., Takeuchi, H., Graff-Guerrero, A., et al. (2011b). Dopamine D2 receptor occupancy and clinical effects: A systematic review and pooled analysis. *Journal of Clinical Psychopharmacology, 31*, 497–502.

Uchida, H., Suzuki, T., Graff-Guerrero, A., et al. (2014). Therapeutic window for striatal dopamine D(2/3) receptor occupancy in older patients with schizophrenia: A pilot PET study. *The American Journal of Geriatric Psychiatry, 22*, 1007–1016.

Une, T., & Kurumiya, S. (2007). Pharmacological profile of blonanserin. *Japanese Journal of Clinical Psychopharmacology, 10*, 1263–1272.

Van Beijsterveldt, L. E., Geerts, R. J., Leysen, J. E., et al. (1994). Regional brain distribution of risperidone and its active metabolite 9-hydroxy-risperidone in the rat. *Psychopharmacology (Berl), 114*, 53–62.

Van Erp, T. G. M., Walton, E., Hibar, D. P., et al. (2018). Cortical brain abnormalities in 4474 individuals with schizophrenia and 5098 control subjects via the enhancing neuro imaging genetics through meta analysis (ENIGMA) consortium. *Biological Psychiatry, 84*, 644–654.

Van Harten, P. N., Hoek, H. W., & Kahn, R. S. (1999). Acute dystonia induced by drug treatment. *BMJ, 319*, 623–626.

Van Vugt, J. P., Siesling, S., Vergeer, M., et al. (1997). Clozapine versus placebo in Huntington's disease: A double blind randomised comparative study. *Journal of Neurology, Neurosurgery, and Psychiatry, 63*, 35–39.

Vanover, K. E., Davis, R. E., Zhou, Y., Ye, W., Brašic, J. R., Gapasin, L., Saillard, J., Weingart, M. E., Mates, S., & Wong, D. F. (2019). Dopamine D₂ receptor occupancy of lumateperone (ITI-007): A Positron Emission Tomography Study in patients with schizophrenia. *Neuropsychopharmacology, 44*(3), 598–605.

Vermeulen, J. M., Van Rooijen, G., Van De Kerkhof, M. P. J., et al. (2019). Clozapine and long-term mortality risk in patients with schizophrenia: A systematic review and meta-analysis of studies lasting 1.1–12.5 years. *Schizophrenia Bulletin, 45*, 315–329.

Vernaleken, I., Fellows, C., Janouschek, H., et al. (2008). Striatal and extrastriatal D2/D3-receptor-binding

properties of ziprasidone: A positron emission tomography study with [18F]Fallypride and [11C]raclopride (D2/D3-receptor occupancy of ziprasidone). *Journal of Clinical Psychopharmacology, 28*, 608–617.

Vincenzi, B., Borba, C. P., Gray, D. A., et al. (2013). An exploratory study examining lipid-lowering medications in reducing fasting serum lipids in schizophrenia patients treated with atypical antipsychotics. *Annals of Clinical Psychiatry, 25*, 141–148.

Volavka, J., Van Dorn, R. A., Citrome, L., et al. (2016). Hostility in schizophrenia: An integrated analysis of the combined Clinical Antipsychotic Trials of Intervention Effectiveness (CATIE) and the European First Episode Schizophrenia Trial (EUFEST) studies. *European Psychiatry, 31*, 13–19.

Vyas, P., Hwang, B. J., & Brašic, J. R. (2020). An evaluation of lumateperone tosylate for the treatment of schizophrenia. *Expert Opinion on Pharmacotherapy, 21*, 139–145.

Wang, S. M., Han, C., Lee, S. J., et al. (2013). Asenapine, blonanserin, iloperidone, lurasidone, and sertindole: Distinctive clinical characteristics of 5 novel atypical antipsychotics. *Clinical Neuropharmacology, 36*, 223–238.

Kruizinga, J., Liemburg, E., Burger, H., Cipriani, A., Geddes, J., Robertson, L., Vogelaar, B., & Nolen, W. A. (2021). Pharmacological treatment for psychotic depression. *Cochrane Database Syst Rev, 12*(12), CD004044. https://doi.org/10.1002/14651858.CD004044.pub5. PMID: 34875106; PMCID: PMC8651069.

Williams, A. M., & Park, S. H. (2015). Seizure associated with clozapine: Incidence, etiology, and management. *CNS Drugs, 29*, 101–111.

Woodward, N. D., Purdon, S. E., Meltzer, H. Y., et al. (2005). A meta-analysis of neuropsychological change to clozapine, olanzapine, quetiapine, and risperidone in schizophrenia. *The International Journal of Neuropsychopharmacology, 8*, 457–472.

Wu, R. R., Jin, H., Gao, K., et al. (2012). Metformin for treatment of antipsychotic-induced amenorrhea and weight gain in women with first-episode schizophrenia: A double-blind, randomized, placebo-controlled study. *The American Journal of Psychiatry, 169*, 813–821.

Wyatt, R. J. (1991). Neuroleptics and the natural course of schizophrenia. *Schizophrenia Bulletin, 17*, 325–351.

Xiberas, X., Martinot, J. L., Mallet, L., et al. (2001). In vivo extrastriatal and striatal D2 dopamine receptor blockade by amisulpride in schizophrenia. *Journal of Clinical Psychopharmacology, 21*, 207–214.

Yaegashi, H., Kirino, S., Remington, G., et al. (2020). Adherence to Oral antipsychotics measured by electronic adherence monitoring in schizophrenia: A systematic review and meta-analysis. *CNS Drugs, 34*, 579–598.

Yaeger, D., Smith, H. G., & Altshuler, L. L. (2006). Atypical antipsychotics in the treatment of schizophrenia during pregnancy and the postpartum. *The American Journal of Psychiatry, 163*, 2064–2070.

Yildiz, A., Nikodem, M., Vieta, E., et al. (2015). A network meta-analysis on comparative efficacy and all-cause discontinuation of antimanic treatments in acute bipolar mania. *Psychological Medicine, 45*, 299–317.

Yoshida, K., & Müller, D. J. (2020). Pharmacogenetics of antipsychotic drug treatment: Update and clinical implications. *Molecular Neuropsychiatry, 5*, 1–26.

Yung, A. R., & Mcgorry, P. D. (1996). The initial prodrome in psychosis: Descriptive and qualitative aspects. *The Australian and New Zealand Journal of Psychiatry, 30*, 587–599.

Yunusa, I., Alsumali, A., Garba, A. E., et al. (2019). Assessment of reported comparative effectiveness and safety of atypical antipsychotics in the treatment of behavioral and psychological symptoms of dementia: A network meta-analysis. *JAMA Network Open, 2*, e190828.

Zhu, Y., Krause, M., Huhn, M., et al. (2017). Antipsychotic drugs for the acute treatment of patients with a first episode of schizophrenia: A systematic review with pairwise and network meta-analyses. *Lancet Psychiatry, 4*, 694–705.

Zohar, J., Stahl, S., Moller, H. J., et al. (2015). A review of the current nomenclature for psychotropic agents and an introduction to the Neuroscience-based Nomenclature. *European Neuropsychopharmacology, 25*, 2318–2325.

Medications for Bipolar Disorder

Philip B. Mitchell

Contents

Abstract

This chapter reviews current evidence for medications in the treatment of patients with bipolar disorder. It will overview the historic "mood stabilizers" such as lithium, valproate, and carbamazepine, and also lamotrigine and the medications for psychosis (antipsychotics). This pragmatic approach is cognizant of the major shift in prescribing patterns for bipolar disorder which has occurred internationally over the last two decades, with a dramatic increase in the use of medications for psychosis for this condition.

P. B. Mitchell (✉)
Discipline of Psychiatry and Mental Health, University of New South Wales, Sydney, NSW, Australia
e-mail: p.mitchell@unsw.edu.au

© Springer Nature Switzerland AG 2024
A. Tasman et al. (eds.), *Tasman's Psychiatry*,
https://doi.org/10.1007/978-3-030-51366-5_135

A wide range of treatments have now been shown to be effective for the acute treatment of mania, with it now evident that medications for psychosis are to be preferred in the short-term control of this urgent presentation. Bipolar depression remains a therapeutic challenge and remains an area in urgent need of new and more effective treatments. Maintenance therapy is critical in view of the high rates of relapse and disability with this condition. While lithium remains the gold standard treatment option for prevention of mania, lamotrigine has been shown to be effective in preventing depressive recurrences. Further, many of the dopamine serotonin blockers (atypical antipsychotics) have also been found to be useful, particularly for manic relapse, with long-acting injectables having a demonstrable advantage. Finally, growing evidence is confirming that optimal treatment comprises effective mood-stabilizing pharmacotherapy concurrent with evidence-based psychological therapies.

Keywords

Bipolar disorder · Mania · Depression · Treatment · Pharmacotherapy · Psychological · Lithium · Valproate · Divalproex · Lamotrigine · Carbamazepine · Antipsychotic · Antidepressant · Olanzapine · Risperidone · Quetiapine · Lurasidone · Paliperidone · Aripiprazole · Asenapine · Lurasidone · Cariprazine

Introduction

Bipolar disorder affects about 1.0–1.5% of the population (Merikangas et al., 2011; Mitchell et al., 2013). It is highly disabling, accounting for over 1% of all years lost due to disability (YLDs) due to health condition globally, because of its early onset, severity, and highly recurrent nature (Ferrari et al., 2016). Furthermore, it is associated with one of the highest risks of suicide of all the psychiatric disorders (Miller & Black, 2020). This is a challenging and complex condition to treat, as management needs to address both acute care of manic and depressive episodes, as well as maintenance treatment with its focus on relapse prevention.

This chapter overviews the evidence for medications in the treatment of patients with bipolar disorder. It overviews the historic "mood stabilizers" such as lithium, valproate, and carbamazepine, and also lamotrigine and the medications for psychosis (antipsychotics). This pragmatic approach is cognizant of the major shift in prescribing patterns for bipolar disorder which has occurred internationally over the last two decades. A recent US study examining the 20-year period 1997–2016 (Rhee et al., 2020) reported both a major increase in the prescription of medications for psychosis (from use at 12% to 51% of outpatient psychiatric visits) and a concomitant reduction in the use of lithium and anticonvulsants (from 62% to 26% of visits).

In addition to reviewing the evidence for medications for bipolar disorder, the roles of adjunctive psychotherapies and other relevant medications (such as medications for depression) are addressed. For an excellent historical overview of pharmacological treatments for bipolar disorder, see Lopez-Munoz et al. (2018). Table 1 lists FDA-approved drugs for the treatment of bipolar disorder in the United States.

Where relevant, this chapter will use the neuroscience-based nomenclature (NbN) for classification of psychotropic medications. It should be noted that the NbN system uses the term "medications for bipolar disorder" rather than "mood stabilizers." This chapter also uses NbN terminology for specific medications (Zohar et al., 2015).

Medications for Acute Mania

General Considerations

Mania (with or without mixed features) is characterized by increased activity, racing thoughts, elevated/euphoric mood, irritability, impulsivity, lability, reduced need for sleep, and marked

Table 1 Approved medications in the United States for bipolar disorder

Drug	Acute mania	Mixed	Acute depression	Maintenance
Aripiprazole	×	×	×	
Asenapine	×	×		
Carbamazepine-ER	×	×		
Cariprazine	×	×		
Chlorpromazine	×			
Divalproex-DR	×			
Divalproex-ER	×	×		
Lamotrigine				×
Lithium	×			×
Lumateperone	×			
Lurasidone			×	
Olanzapine	×	×		×
Olanzapine–fluoxetine			×	
Quetiapine	×		×	×[a]
Risperidone	×	×		×[b]
Ziprasidone	×	×		×[a]

[a]Combination with lithium or valproate
[b]Long-acting injectable only

functional impairment or distress. Psychotic experiences such as paranoid or grandiose delusions and hallucinations (particularly auditory or more rarely, visual) are common. Hospitalization is often necessary when attempting to control acute bipolar mania, though some less severely unwell patients may be able to be managed in the outpatient or community setting.

The first step in choosing the initial pharmacotherapy for mania should involve assessment of the severity of symptoms, which may guide the need for monotherapy versus combination therapy. In addition, consideration should be given to maintenance therapy planning even in acutely manic patients, given the recurrent nature of bipolar disorder.

A seminal multi-treatments meta-analysis of medications for the acute treatment of mania (Cipriani et al., 2011) found that haloperidol, risperidone, olanzapine, lithium, quetiapine, aripiprazole, carbamazepine, asenapine, valproate, and ziprasidone were significantly more effective than placebo, whereas gabapentin, lamotrigine, and topiramate were not. Haloperidol was found to be more effective than most of the other agents but is not widely used in contemporary practice in view

of its propensity for causing severe extrapyramidal adverse effects. Risperidone and olanzapine had a very similar profile of comparative efficacy, being more effective than valproate and ziprasidone. Olanzapine, risperidone, and quetiapine were associated with significantly fewer discontinuations than lithium. However, a recent head-to-head comparison of lithium and quetiapine found that quetiapine was associated with more frequent, intense, and impairing side effects than lithium (Nierenberg et al., 2016).

Overall, this important meta-analysis found that medications for psychosis were significantly more effective than lithium and the anticonvulsants in the acute management of mania. The authors recommended that risperidone, olanzapine, and haloperidol should be considered preferred first options for the acute treatment of manic episodes (Cipriani et al., 2011).

With respect to the management of acute mania with mixed features, data from adequately powered treatment trials is limited, but the existing evidence suggests the effectiveness of the dopamine serotonin blockers (atypical antipsychotics) and valproate (Chakrabarty et al., 2020).

Lithium (NbN – Mode of Action: Enzyme Modulator; Displaces Intracellular Sodium)

Since the introduction of lithium for the treatment of "psychotic excitement" (Cade, 1949; Mitchell, 1999; Mitchell & Hadzi-Pavlovic, 1999) and later confirmation of its antimanic effect by the team led by Mogens Schou (Schou et al., 1954), this medication has been the keystone of bipolar disorder therapy. It is still regarded as the "gold standard" treatment for this condition (McIntyre et al., 2020). It received approval for use in mania by the FDA (U.S. Food and Drug Administration) in 1970, after a series of National Institute of Mental Health (NIMH) trials were done to satisfy FDA requirements, as the pharmaceutical industry did not pursue the studies because lithium could not be patented. The short-term efficacy of lithium in the treatment of acute mania is supported by at least five placebo-controlled trials and was confirmed by the meta-analysis of Cipriani et al. (2011). Combined treatment of lithium with dopamine serotonin blockers (atypical antipsychotics) or benzodiazepines is frequently necessary for most manic patients.

Pharmacokinetics

Lithium is absorbed completely within 8 h of oral intake, with peak plasma levels reached between 1 and 3 h after ingestion. Absorption is not affected by the presence of food, making it preferable for some patients experiencing gastric irritation to take lithium immediately after meals. Distribution of lithium after absorption into the bloodstream is rapid, with the majority eventually entering the intracellular space (intracellular accumulation is key to the mechanism of action of lithium which accumulates preferentially in more active neurons).

On average, the brain concentration of lithium is lower than the plasma concentration with the half-life of lithium in the brain being longer than that in blood (Plenge et al., 1994). A recent 7-Tesla MRI brain scan study of 21 euthymic bipolar disorder patients on maintenance lithium treatment (Stout et al., 2020) found a strong correlation between plasma and brain lithium concentrations. Furthermore, that study found that lithium was distributed widely across brain regions, but there was a well-defined significant cluster corresponding closely to the left hippocampus that was characterized by high lithium content. This specific anatomical finding was of particular interest in view of the major role of the hippocampus in emotion processing and regulation, and the consistent atrophy of the hippocampus in untreated patients with bipolar disorder. Cellular and modeling studies reveal that lithium is not distributed evenly within the cytosol but has concentrations that are higher near sodium channels and in postsynaptic terminals.

Renal clearance is the primary route of lithium elimination. Age-related changes necessitate more careful use of lithium in the elderly, as elimination may be significantly decreased, lending to a greater risk for elevated or toxic levels. Lithium is reabsorbed from the proximal tubule, so clinicians should be aware that lithium elimination may be substantially increased by certain medications, whereas some diuretics that act at the distal tubule may lead to unwanted lithium retention (Table 2).

Mechanism of Action

Lithium's mechanism of action – in both the acute management of mania and in maintenance treatment – is poorly understood. Recent research into molecular targets for lithium, including transcriptomic and epigenetic studies, has been well summarized by Bellivier and Marie-Claire (2018). Other studies have examined the effect of lithium on circadian rhythms (Sawai et al., 2019; Sanghani et al., 2020). More promising have been genetic studies of predictors of response to lithium. Hou et al. (2016), using data from the International Consortium on Lithium Genetics, found that response to lithium was associated with two genes for long noncoding RNA. Amare et al., using the same dataset, found that response to lithium was associated with polygenic risk scores for major depression (Amare et al., 2020) and schizophrenia, as well as genes for HLA antigens and inflammatory markers (Amare et al., 2018).

Clinical Application

Although lithium is effective for mania, clinical improvement is relatively slow, with an initial onset of therapeutic response generally occurring no sooner than 7–14 days. For this reason, concurrent use of dopamine-serotonin blockers (atypical antipsychotics) is usually required to gain acute control of any disturbed behavior. Adverse effects are detailed in Table 3. These effects may be less apparent during acute care with lithium than during maintenance treatment.

Pretreatment procedures and testing are recommended before beginning therapy with lithium: general medical history, physical examination and weight, full blood count, creatinine and eGFR (estimated glomerular filtration rate), thyroid function studies, calcium, and pregnancy testing for women of childbearing age. Lithium carries a teratogenicity category D warning,

Table 2 Potential interactions with lithium

Increase lithium levels	Decrease lithium levels	Adverse reactions
Angiotensin-converting enzyme inhibitors	Acetazolamide	Iodides (hypothyroidism)
	Alkalinizing agents	Calcium channel blockers
NSAIDs, including ibuprofen, indomethacin	Caffeine	
COX-2 inhibitors, including celecoxib	Carbonic anhydrase inhibitors	ECT
	Laxatives	Succinylcholine
Thiazide diuretics	Osmotic diuretics	Methyldopa
Metronidazole	Theobromine diuretic Theophylline Urea Xanthine preparations	

Table 3 Adverse effects of lithium

System affected	Adverse effect(s)	Treatment/comments
Cardiovascular	Bradycardia, syncope, AV block, EKG changes, sick-sinus syndromes	Infrequent or rare events; EKG monitoring; do not use lithium with sick sinus; discontinuation of lithium may be needed
Central nervous system	Mental dullness, headache, memory troubles, muscle weakness	Commonly seen, especially at start of therapy. Dose reduction may help
Dermatological	Acne, hair loss, psoriasis, skin reactions	Standard dermatological treatments, avoid or discontinue lithium if severe (particularly if pre-existing psoriasis)
Endocrine	Goiter, elevated TSH, hypothyroidism, hyperparathyroidism with hypercalcemia	Thyroid problems common with long-term therapy. Can be treated with exogenous L-thyroxine
Gastrointestinal	Nausea, GI pain/distress, vomiting, diarrhea, frequent loose stools	More common at start of therapy but may indicate toxicity; check lithium levels, take lithium with food or change preparation, lower/slower dosing
Hematological	Leukocytosis	Common, predominant neutrophilia, reversible
Metabolic	Weight gain	Avoid caloric beverages, adopt healthy diet, increase physical activity
Neurological	Postural tremor, confusion, ataxia	Fine tremor is frequent side effect, dosage reduction and/or beta-blockers may help. Other effects should prompt lithium level check
Ophthalmic	Blurred vision, nystagmus, eye pain	Rare. Reduce dose or discontinue lithium
Renal	Polydipsia, polyuria, reduced concentration capacity	Non-thiazide diuretics with caution, lower dosage, take dose once daily at bedtime, decrease dietary protein (exclude diabetes insipidus)

meaning it has known teratogenesis. Electrocardiography should be obtained for patients over the age of 40 years.

For treatment of acute mania, lithium should be dosed to achieve therapeutic plasma concentrations (sampled 12 h after the last dose) of between 0.8 and 1.2 mEq/L, avoiding toxic effects more commonly seen with levels at or above 1.5 mEq/L. A routine initial dosing regimen in physically healthy patients is the administration of lithium carbonate 300 mg three or four times daily, with trough plasma level determination on day 4 or day 5 of the treatment, targeting the therapeutic range and watching for any signs of toxicity. A lower starting dose and slower titration are recommended in older patients and those with impaired renal function. When used to treat milder mania, lower starting doses (300 mg BID) and lower blood levels (0.6–0.8 mEq/L) may sufficiently stabilize mood. Lower doses and levels of lithium may be sufficient when combination pharmacotherapy strategies are used, such as lithium plus an anticonvulsant or medication for psychosis. These lower doses may greatly reduce the frequency of troublesome side effects during early treatment and thus enhance adherence.

The most common acute side effects of lithium are nausea, vomiting, tremor, diarrhea, polydipsia, and polyuria. During acute treatment, gastrointestinal disturbances and tremor can be mitigated by slower dosing strategies. These adverse effects typically subside within 1–2 weeks, although on occasion they can persist. Worsening of any of these adverse effects, as well as the development of bradycardia, syncope, confusion, or ataxic gait should prompt the clinician to immediately measure lithium blood levels and consider changes in dosing or treatment alternatives. Psoriasis, acne, and other skin/hair reactions may also occur.

Lithium toxicity may be heralded by emergence or aggravation of any of the above adverse effects, progressing to include one or more other symptoms such as coarsening tremor, sluggishness, slurred speech, parkinsonism, hyperreflexia, myoclonic twitches, and confusion. Early recognition of toxicity is critical as is investigation of the cause (Table 4). The possibility of drug interactions with lithium must always be considered,

Table 4 Potential causes of high lithium levels and/or toxicity

Excess dosing
Overdose
Laboratory error, or less than 10–12 h between last drug dose and blood sampling

Increase in lithium absorption
    Diarrhea
    Vomiting

Drug interactions reducing renal clearance (NSAIDs, thiazides)

Dehydration
    Fever/illness
    Hot weather/heat exhaustion
    Intense physical activity/profuse sweating
    Sauna/steam room

Renal insufficiency or intercurrent renal disease
Sodium deficiency (low-caloric or low-sodium diet)

particularly with thiazide (and possibly loop) diuretics, angiotensin-converting enzyme (ACE) inhibitors, and nonsteroidal anti-inflammatory drugs (NSAIDs).

Anticonvulsants

Early reports of the beneficial behavioral effects associated with the use of anticonvulsants in patients with bipolar disorder, particularly for mania, date back to the 1960s (see Lopez-Munoz et al., 2018). There is not a distinct class effect for anticonvulsants in the management of bipolar disorder as not all anticonvulsants have been shown to have efficacy (for example, there is no evidence from controlled trials for the efficacy of gabapentin or topiramate). The likely explanation for this variability for anticonvulsants in bipolar disorder relates to the complex differences in the mechanisms of action for these drugs (Muzina et al., 2005). In line with drugs that address abnormal physiology, all effective anticonvulsants inhibit sodium channels in an activity-dependent manner, just as lithium enters cells preferentially through open sodium channels in an activity-dependent manner (El-Mallakh & Huff, 2001).

Here, anticonvulsants that have been demonstrated to be antimanic in randomized, double-blind, placebo-controlled studies with adequate sample size are considered. Note that there is no evidence that lamotrigine is an effective antimanic agent (its acute and prophylactic efficacy in bipolar depression is discussed later in this chapter).

Divalproex/Sodium Valproate (NbN – Pharmacology: Voltage-Dependent Sodium Channel Antagonist)

Divalproex sodium is comprised of both sodium valproate and valproic acid in a 1:1 molar relationship. With valproate first reported to have antimanic properties by the French psychiatrist Lambert in the 1970s (López-Muñoz et al., 2018), divalproex sodium became the first anticonvulsant to be approved for bipolar mania by the FDA in 1995, after a large-scale, randomized, double-blind parallel-group study found divalproex to be equivalent to lithium in superiority over placebo for the management of acute mania (Bowden et al., 1994). In this study, efficacy of divalproex appeared to be independent of prior responsiveness to lithium. An extended-release (ER) preparation of divalproex subsequently received FDA approval for acute manic or mixed episodes of bipolar disorder (Bowden et al., 2006).

Divalproex is recommended as one of the first-line acute treatment option for mania in most evidence-based practice guidelines, either as monotherapy or in combination with medications for psychosis for more severe manic episodes, though – as noted above – the meta-analysis of Cipriani et al. (2011) ranked a number of medications for psychosis ahead of this agent.

Pharmacokinetics

Following oral administration of divalproex, its absolute bioavailability quickly approaches 100%; the ER preparation bioavailability is closer to 90%. This suggests that a slightly higher dose of the ER form is needed in order to reach bioequivalence with the immediate-release preparation (8–20% higher dose for ER). Peak plasma concentrations are achieved within 3–5 h, although up to 17 h may be needed to reach the peak concentration for the ER form. The mean terminal half-life is about 12–16 h for either preparation, with steady-state conditions usually being achieved within 3–4 days.

Valproate is metabolized almost entirely by the liver via the cytochrome P450 2D6 system. Mitochondrial beta-oxidation is the other major metabolic pathway. Valproate does not induce its own metabolism. It inhibits drug oxidation and may increase serum levels of other oxidatively metabolized drugs, such as phenytoin, phenobarbital, and monoamine enhancers (tricyclic antidepressants). Coadministration with other microsomal enzyme-inducing drugs, such as carbamazepine, will decrease plasma valproic acid levels. Inhibitors of the P450 system, such as selective serotonin reuptake inhibitors (SSRIs), can increase levels of valproic acid. Toxicity can also be induced when divalproex is co-prescribed with other drugs that may be highly protein bound.

Mechanism of Action

While the mechanisms of action of valproate in bipolar disorder is poorly understood, it has been proposed that it acts via inhibition of voltage sensitive sodium channels which appears via the most active glutamatergic and GABAergic pathways.

Clinical Application

Before initiating acute treatment of mania with divalproex, a general medical history with special attention to any past problems of a hepatic or hematological nature should be obtained, in addition to baseline liver profile and complete blood cell count. Pregnancy screening should be conducted before commencement, particularly given its FDA "black box" warning of teratogenicity category D with neural tube defects and reduced cognitive function in children exposed to it. In view of this, and the potential for causing polycystic ovarian syndrome, its use should be avoided in young women of childbearing potential.

Divalproex can be dosed with either an oral loading or standard titration strategy. For acute management of severely ill manic patients, divalproex oral loading with a therapeutic starting dose of 20–30 mg/kg per day is a commonly used approach. Oral loading typically achieves serum concentrations of valproic acid >50 mg/L by day 2, which represents the low end of the therapeutic range of serum concentrations for this agent of

50–125 mg/L. It is important to note that in post hoc analyses of randomized trials, a serum level of ≥80 mg/L is necessary to achieve symptom reduction. Similar results can be obtained by using the ER preparation. This oral loading strategy for divalproex has been demonstrated to lead to a more rapid antimanic effect than standard titration divalproex (Hirschfeld et al., 2003). Lower, divided doses are initially recommended (250 mg TID or 500 mg BID) for patients with less severe mania or in elderly patients.

Adverse effects seen during initial therapy with divalproex are usually mild, transient, and easily managed. Gastrointestinal distress and sedation are the most common side effects during acute treatment, although other dose-related effects including tremor and benign hepatic transaminase elevations are occasionally encountered. Reduced dosage or slower upward titration can be helpful, and in addition, the use of divalproex sodium formulation or ER divalproex instead of valproic acid may also lessen these side effects. Tremor may be minimized through dose reduction or concomitant beta-blocker medication. During acute treatment with divalproex, clinicians should consider the possibility of rare but potentially serious adverse effects such as irreversible hepatic failure, hemorrhagic pancreatitis, or hyperammonemic encephalopathy when patients experience severe abdominal distress, confusion, or delirium. Hematological adverse effects may occur, particularly thrombocytopenia.

Divalproex commonly displaces otherwise highly protein-bound drugs from their protein-binding sites, requiring greater awareness of clinically significant drug interactions. Divalproex can also inhibit the metabolism of other drugs cleared by the liver. Most notably, it inhibits lamotrigine metabolism by 50%, resulting in a critical need to initiate lamotrigine at 50% lower doses when coadministered with divalproex.

Carbamazepine (NbN – Pharmacology: Glutamate; Mode of Action: Voltage-Gated Sodium and Calcium Channel Blocker)

The initial reports of the antimanic efficacy of carbamazepine came from Okuma in Japan in the early 1970s (Lopez-Munoz et al., 2018).

There were approximately 20 double-blind studies examining the efficacy of carbamazepine and its 10-keto analogue oxcarbazepine in the treatment of mania prior to the completion of the seminal first large, randomized, double-blind, placebo-controlled parallel trial of carbamazepine monotherapy using a beaded, extended-release formulation. That trial led to its approval for mania by the FDA (Muzina et al., 2002; Weisler et al., 2004, 2005). A post hoc analysis also found benefit for patients with mixed states.

Pharmacokinetics

Carbamazepine is 80% bioavailable after oral administration, with immediate and beaded release formulations bioequivalent. Nearly 80% is protein bound with the primary route of hepatic metabolism via the cytochrome P450 3A4 system to its active epoxide form. With initial administration of the ER form, the half-life averages between 35 and 40 h, but with repeated administration, this decreases to 12–17 h due to autoinduction. Inhibitors of the P450 3A4 will increase carbamazepine levels whereas inducers can decrease levels.

Clinical Application

Before treatment with carbamazepine, a general history and physical examination should be performed, including screening for any history of liver disease or blood dyscrasias. Routine laboratory tests including complete blood count, liver profile, and electrolytes with creatinine should be obtained. More frequent monitoring is indicated in the elderly or in any individual on carbamazepine who develops fever, easy bruising or bleeding, weakness, or infection. Further, the possibility of a current or potential pregnancy in young women should be excluded, as it is a category D teratogenic agent, with the major problems being neural tube defects, reduced cognitive ability, cardiovascular and urinary tract anomalies, and cleft palate.

There is no clear target serum level of carbamazepine for acute mania, though levels used in the management of epilepsy are used as a guide. Immediate-release carbamazepine therapy may be started at a total divided daily dose of 200–600 mg, with incremental increases of 200–1000 mg/day followed by careful monitoring of blood levels, side effects, and clinical efficacy. The beaded, ER form may be started at 400 mg/day and increased as tolerated up to 1600 mg/day. Many factors can affect carbamazepine blood levels beyond direct dosing, including autoinduction of metabolism and significant drug interactions.

During acute treatment with carbamazepine, the most commonly observed side effects are nausea, fatigue, blurred vision, diplopia, and ataxia. More gradual initial titration of the dose may minimize these effects. Liver transaminase elevations, rash, hyponatremia, and blood dyscrasias may complicate acute treatment. Rash develops in about 10% of patients, requiring cessation of this medication.

Antipsychotics (NbN: Medications for Psychosis)

Medications for psychosis (antipsychotic drugs) are considered in greater detail elsewhere in this handbook. The chemistry, pharmacology, adverse effects, and precautions associated with the use of these drugs for schizophrenia apply also to bipolar disorder. Particular attention should be paid to the metabolic risks associated with the dopamine-serotonin blockers (atypical antipsychotics), along with associated monitoring and safety requirements. While there is strong evidence from meta-analyses (as detailed above) for the efficacy of haloperidol in acute mania, this agent is not widely used in contemporary practice due to its propensity for causing severe extrapyramidal symptoms. For this reason, this section will focus on the dopamine-serotonin blockers for which there is a strong evidence base for efficacy in acute mania and widespread use in clinical practice (Rhee et al., 2020).

Randomized, placebo-controlled trials have demonstrated the efficacy of eight dopamine-serotonin blockers or dopamine partial agonists, specifically, olanzapine, risperidone, quetiapine, ziprasidone, aripiprazole, asenapine, paliperidone, and cariprazine. These trials have led to FDA approval of the majority of these for the management of acute bipolar mania (with or without

psychotic features). Only paliperidone has not received approval for bipolar mania. With the exception of quetiapine, all approved agents are also approved for mixed states in bipolar disorder. As detailed above, a multiple-treatment meta-analysis including six approved dopamine serotonin blockers, haloperidol, lithium, valproate, carbamazepine, and other anticonvulsants concluded that risperidone, olanzapine, and haloperidol should be considered the preferred medications for the acute treatment of bipolar mania (Cipriani et al., 2011).

Olanzapine (NbN – Pharmacology: Dopamine and Serotonin; Mode of Action: Antagonist at D2 and 5-HT2 Receptors)

Olanzapine became the first dopamine-serotonin blocker (atypical antipsychotic) to be indicated for the treatment of mania associated with bipolar disorder when it received approval by the FDA in 2000, based on data from two large placebo-controlled trials (Tohen et al., 1999, 2000). Further, in manic patients only partially responsive to either lithium or divalproex, the use of adjunctive olanzapine led to significantly faster antimanic response and a greater reduction in symptom severity than placebo augmentation (Tohen et al., 2002).

The antimanic dose of olanzapine is between 15 and 20 mg/day, although lower doses may be effective when used adjunctively with other medications for bipolar disorder. During initial treatment of a severe manic episode in clinical practice, off-label doses of 30–40 mg/day may be required, which can be safely reduced once the acute episode has resolved.

Olanzapine is also available in a short-acting intramuscular (IM) formulation approved for use in the treatment of agitation associated with schizophrenia or bipolar I mania. The IM use of olanzapine is of particular utility in the hospitalized manic patient with severe agitation and can be administered as an initial 10 mg dose and repeated if needed in 2 h and again 4 h later, up to a maximum daily dose of 30 mg. This maximum dosing may lead to substantial orthostasis, so must be employed with great caution. Concomitant administration of IM lorazepam should be avoided as this combination may lead to pronounced sedation and respiratory depression.

Olanzapine binds with high affinity to dopaminergic, serotonergic, adrenergic, muscarinic, and histaminergic receptors. It has a low risk for causing extrapyramidal side effects. The major side effects are sedation, weight gain, and metabolic syndrome. The exact mechanism of action as an antimanic agent remains unknown.

Risperidone (NbN – Pharmacology: Dopamine, Serotonin, and Norepinephrine; Mode of Action: Antagonist at D2, 5-HT2, and Norepinephrine Alpha-1 and Minimally Alpha-2 Receptors)

Substantial evidence supports the efficacy of risperidone in acute mania. An initial investigator-initiated study compared risperidone with haloperidol and lithium, demonstrating that all three agents significantly reduced manic symptoms (Segal et al., 1998). Two subsequent industry-sponsored multicenter, double-blind, placebo-controlled monotherapy studies of risperidone for acute mania found response rates greater than that seen in placebo-treated subjects (Hirschfeld et al., 2004; Khanna et al., 2005). Two trials of risperidone added to medications for bipolar disorder (lithium, valproate, or carbamazepine) for 3 weeks also showed an advantage of risperidone over placebo (Sachs et al., 2002; Yatham et al., 2003). The risperidone data are consistent with observations made from the olanzapine studies – combined dopamine-serotonin blocker (atypical antipsychotic) and traditional mood stabilizer therapy leads to greater improvement in mania than monotherapy (Smith et al., 2007). The mean modal antimanic doses for risperidone as monotherapy or add-on therapy ranged from 4 to 6 mg/day, with the lower end of the range used when combined with another medication for bipolar disorder.

Risperidone is a potent $5HT_{2A}$ and D_2 antagonist. It has less affinity for D_3 and D_4 receptors than clozapine. It also blocks H_1, α_1, and α_2 but not acetylcholine receptors. Doses of risperidone above 6 mg/day are known to produce >80%

blockade of D_2 receptors in the basal ganglia, resulting in an increased likelihood of extrapyramidal side effects. Risperidone levels are high in the anterior pituitary because that part of the brain's circulation lacks permeability glycoprotein which actively removes risperidone from neuropil (El-Mallakh & Watkins, 2009), consequently it can dramatically increase prolactin levels. As with olanzapine, its exact mechanism of therapeutic action in bipolar disorder is not known, although medications for psychosis (antipsychotics), in general, may exert antimanic actions, at least in part, via D_2 receptor antagonism. Central dopaminergic hyperactivity has been postulated as one component of the pathophysiology of mania (Gerner et al., 1976).

Paliperidone (NbN – Pharmacology: Dopamine, Serotonin, and Norepinephrine; Mode of Action: Antagonist at D2, 5-HT2, and Norepinephrine Alpha-1 and Alpha-2 Receptors)

Paliperidone is a metabolite of risperidone. Paliperidone extended release oral formulation (paliperidone-ER) has been approved by the FDA for the treatment of schizophrenia and schizoaffective disorder, but not bipolar disorder. Findings from randomized controlled trials have been inconsistent. Its efficacy and safety as monotherapy in the acute treatment of mania and mixed episodes have been studied with two 3-week randomized, double-blind, placebo-controlled trials. In the first study, eligible patients were treated with paliperidone-ER 3–12 mg/day, quetiapine 400–800 mg/day, or placebo for 3 weeks (Vieta et al., 2010); paliperidone-ER was found to be superior to placebo in reducing manic symptoms. A second study compared a fixed dosing schedule with paliperidone-ER 3, 6, and 12 mg/day and placebo (Berwaerts et al., 2012a). At the end of 3 weeks of study, only paliperidone 12 mg/day was superior to placebo in reducing manic symptoms. In a third trial, paliperidone-ER 3–12 mg/day adjunctive therapy was not superior to placebo adjunctive therapy to lithium or valproate during a 6-week treatment of acute mania (Berwaerts et al., 2011).

Quetiapine (NbN – Pharmacology: Dopamine and Serotonin; Mode of Action: Antagonist at D2, 5-HT2 Receptors; Its Metabolite, Norquetiapine, is an Inhibitor of the Norepinephrine Transporter and Alpha-1 Receptor)

Unlike other acute mania trials, quetiapine mania studies excluded mixed states. Accordingly, quetiapine is FDA approved only for the treatment of acute manic episodes associated with bipolar disorder. Two quetiapine monotherapy studies over 12 weeks demonstrated greater efficacy than placebo in acute mania (McIntyre et al., 2005; Bowden et al., 2005). Quetiapine combined with lithium or divalproex in a 3-week, multicenter, randomized controlled trial resulted in greater response rates compared with those receiving medication for bipolar disorder plus placebo (Sachs et al., 2004). In both monotherapy and adjunctive therapy indications for bipolar I mania, the recommended dose for quetiapine is 600 mg/day, following a BID titration schedule over 5 days after starting with 50 mg BID.

Quetiapine is an antagonist at multiple neurotransmitter receptor sites in the brain, including $5HT_{1A}$, $5HT_2$, D_1, D_2, histamine H_1, acetylcholine M_1, α_1, and α_2. It has no appreciable affinity at benzodiazepine receptors. Its dopamine antagonism may be responsible for its antimanic effects. Rapid disassociation from the D_2 receptor is thought to contribute to quetiapine's minimal propensity for inducing extrapyramidal side effects or prolactin elevation. Antagonism of H_1 and M_1 receptors may explain somnolence observed with quetiapine use.

Ziprasidone (NbN – Pharmacology: Dopamine and Serotonin; Mode of Action: Antagonist at D2 and 5-HT2 Receptors)

Ziprasidone monotherapy has demonstrated rapid onset of antimanic activity leading to significantly greater response rates compared with placebo in two double-blind, randomized controlled trials (Keck et al., 2003a). FDA regulatory approval was given in 2004 for acute manic or mixed episodes associated with bipolar disorder, with or without psychotic features.

Similarly to the other dopamine-serotonin blockers (atypical antipsychotics), ziprasidone has a high affinity for dopamine, serotonin, and α-adrenergic receptors with only a moderate affinity for histaminic receptors. Like the majority of dopamine-serotonin blockers, ziprasidone increases the risk for somnolence (Gao et al., 2013), which is believed to be due to the blockade of histamine receptors.

Ziprasidone may be started at 40 mg twice daily (given with meals to enhance absorption) and increased up to 160 mg/day within the first week, with an average dose of around 120 mg/day in study responders. Ziprasidone is available in a short-acting IM form that can help to control associated agitation rapidly with 10 mg IM every 2 h (or 20 mg every 4 h) as needed up to 40 mg/day.

Aripiprazole (NbN – Pharmacology: Dopamine and Serotonin; Mode of Action: Partial Agonist at D2 and 5-HT1A Receptors, and Antagonist at 5-HT2 Receptors)

Aripiprazole has been investigated in two randomized placebo controlled trials of acute bipolar mania, which led to FDA approval in 2004 for monotherapy treatment of acute manic or mixed episodes associated with bipolar disorder. In the first 3-week placebo-controlled monotherapy trial, aripiprazole produced a significant reduction in manic symptoms compared with placebo by day 4, and this statistically significant difference was maintained throughout the trial (Keck et al., 2003b). These results were confirmed by a second placebo-controlled study by Sachs et al. (2006).

Compared with the other approved dopamine-serotonin blockers (atypical antipsychotics), aripiprazole possesses a novel mechanism of action, appearing to mediate its primary therapeutic effects through partial agonism at the D_2 receptor with nearly 30% intrinsic activity. Partial agonism is also seen at the $5HT_{1A}$ receptor, and like other dopamine-serotonin blockers, aripiprazole displays antagonism at the $5HT_{2A}$ receptor, although it, and all other D2 partial agonists, have less affinity at $5HT_{2A}$ than D2 receptors. Aripiprazole has moderate affinity for histamine and α-adrenergic

4192 P. B. Mitchell

receptors and these receptors are usually not adequately blocked at doses ≤ 20 mg/day, and no appreciable affinity for cholinergic muscarinic receptors.

Treatment of manic or mixed episodes in bipolar disorder with aripiprazole may commence with 15–30 mg/day, with responders requiring an average daily dose near 30 mg/day. An IM formulation of aripiprazole was approved to treat agitation associated with schizophrenia or bipolar mania. In 2015, the company discontinued this formulation.

Asenapine (NbN – Pharmacology: Dopamine, Serotonin, and Norepinephrine; Mode of Action: Antagonist at D2, 5-HT2, and Norepinephrine Alpha-2 Receptors)

Asenapine is approved by the FDA for acute mania and mixed episodes, based on findings from two 3-week, randomized, double-blind, placebo-controlled studies. In both studies, asenapine was flexibly dosed with 5 or 10 mg BID and olanzapine 5–20 mg/day was used as an active comparator. In the first study, both asenapine and olanzapine were significantly superior to placebo in reducing manic symptoms and response rates (McIntyre et al., 2009). In the second study, both asenapine and olanzapine were significantly superior to placebo in reducing manic symptoms; however, only olanzapine was superior to placebo in response and remission rates (McIntyre et al., 2010). Asenapine adjunctive therapy to a medication for bipolar disorder was also significant superior to placebo adjunctive therapy to a mood stabilizer in reducing manic symptoms.

Consistent with other dopamine-serotonin blockers (atypical antipsychotics), asenapine has a higher affinity for 5-HT_{2A} than D_2 receptors. However, asenapine has a relatively higher affinity for and antagonistic activity at a broad range of serotonin, dopamine, α-adrenergic, and histamine receptors (Chwieduk & Scott, 2011). Asenapine has no appreciable affinity for muscarinic receptors and a low affinity for β-adrenergic receptors, histamine H_3 receptors, and 5-HT_3 receptors.

Asenapine is the only medication for psychosis available in a sublingual formulation. For monotherapy, asenapine should be initiated at 10 mg twice daily, with a final dose of 5–10 mg twice daily as recommended based on tolerability. A starting dose of 5 mg twice daily is recommended as adjunctive therapy to lithium or valproate. The sublingual tablet should not be crushed, chewed, or swallowed, but placed under the tongue until dissolved completely. Eating or drinking should be avoided for at least 10 min after administration. Somnolence and a bitter taste are some of the most commonly reported side effects.

Cariprazine (NbN – Pharmacology: Dopamine and Serotonin; Mechanism of Action: Partial Agonist at D2, D3, and 5-HT_{1A} Receptors, and Antagonist at 5-HT_{2A} Receptors)

Cariprazine was approved by the FDA in 2015 for the management of acute mania with or without mixed features. It has been shown to be effective in acute mania or mixed episodes in three randomized controlled trials at doses of 3–12 mg daily (Saraf et al., 2019). This drug is a dopamine D2/D3 partial agonist with a 10-fold greater affinity for D3 over D2 receptors (Saraf et al., 2019); nonetheless, it is dosed to achieve >90% D2 receptor occupancy, so the differences in affinity at D3 are of questionable importance. Common adverse effects include akathisia, headache, constipation, and nausea. Akathisia has been reported in about one-third of patients; parkinsonism may also occur.

Dosing, Tolerability, and Safety Issues with Dopamine Serotonin Blockers (Atypical Antipsychotics)

Table 5 provides dosing information for the use of dopamine serotonin blockers (atypical antipsychotics) for the management of acute bipolar mania.

Of principal concern with longer term use is the risk for metabolic derangements, notably weight gain, type 2 diabetes, and lipid elevations. Prior to initiating therapy for mania with any dopamine-serotonin blocker (atypical antipsychotic), baseline determination of fasting glucose, weight, waist circumference, and lipid profile is recommended.

Table 5 Dosing of dopamine serotonin blockers (atypical antipsychotics) for acute mania

Atypical	Starting daily dose (mg)	Target daily dose (mg)
Olanzapine	5–10	15–20
Risperidone	2–4	4–6
Quetiapine	100	400–800
Ziprasidone	80	120–160
Aripiprazole	15–20	25–30
Asenapine	20	10–20
Cariprazine	1.5	6

Benzodiazepines: Clonazepam (NbN – Pharmacology: γ-Aminobutyric Acid; Mechanism of Action: Potentiating γ-Aminobutyric Acid Action)

Older randomized controlled studies have demonstrated the efficacy of clonazepam and its superiority to lithium in the short-term management of acute manic symptoms (Curtin & Schulz, 2004). These agents probably hyperpolarize the presumed depolarized resting potential of central nervous system cells of manic individuals. This treatment option has been displaced by use of dopamine-serotonin antagonists and dopamine partial agonists, but may be a safer short-term option in some patients.

Medications for Acute Bipolar Depression

General Considerations

Acute bipolar depression presents substantial diagnostic and treatment challenges for clinicians. As depressive episodes are the most common initial presentation of bipolar disorder, it may take years for the diagnosis of bipolar disorder to become apparent. In terms of phenomenology, bipolar depression is more likely than major depressive disorder (MDD) to be characterized by atypical features (hypersomnia and hyperphagia), psychomotor retardation, and psychotic features (Mitchell et al., 2008; Frankland et al., 2018; McIntyre et al., 2020). In general, bipolar depression is more difficult to treat than MDD, with significant limitations in the extant treatment options (Nierenberg et al., 2015).

The place of medications for depression (antidepressants) in the management of bipolar depression requires some comment here. A special task force of the International Society for Bipolar Disorders on the role of medications for depression recommended that the use of medications for depression for bipolar depression was appropriate, as long as this use was adjunctive to a medication for bipolar disorder (Pacchiarotti et al., 2013). The evidence of their efficacy in bipolar depression is inconsistent, with pivotal studies reporting both positive (Gijsman et al., 2004) and negative (Sachs et al., 2007) treatment outcomes. Interestingly, however, neither study reported significant increases in rates of rapid cycling or induction of mania. Consistent with that, a large Swedish registry study confirmed no increase in rates of relapse with antidepressants in bipolar disorder patients also prescribed a concurrent medication for bipolar disorder (Viktorin et al., 2014).

A recent meta-analysis (Bahji et al., 2019) has confirmed the efficacy of electroconvulsive therapy (ECT) in bipolar depression, finding that response rates and speed of response are higher than for patients with MDD.

The only agents with FDA approval for acute treatment of bipolar depression are combined fluoxetine/olanzapine (OFC), quetiapine, lurasidone, lumateperone and cariprazine.

Lithium (NbN – Mode of Action; Enzyme Modulator, Displaces Intracellular Sodium)

There is very limited evidence for lithium in the acute management of bipolar depression. There have only been 8 small, placebo-controlled trials comprising a total of 116 bipolar patients (Young et al., 2010). All studies employed crossover designs, with seven of the eight trials reporting lithium to be superior to placebo.

Although not well studied, most recommendations for the use of lithium in the acute management of depression associated with bipolar disorder are to dose to a minimum serum lithium level of 0.8 mEq/L. There is scant evidence available with which to compare the efficacy of lithium relative to standard medications for depression. In a double-blind, placebo-controlled study of bipolar depressed patients on lithium, the addition of either paroxetine or imipramine was not effective in alleviating depressive symptoms in patients with serum lithium levels greater than 0.8 mEq/L (Nemeroff et al., 2001). In a trial focused on the efficacy of quetiapine in the acute treatment of bipolar depression, there was a lithium monotherapy arm (Young et al., 2010). Lithium was associated with numerically greater but not statistically significant improvement in depressive symptoms compared with placebo throughout the 8-week study period. Post hoc analyses of lithium at concentrations of ≥ 0.8 or <0.8 mEq/L also revealed no significant difference from placebo in the improvement of depressive symptoms.

Anticonvulsants

With the exception of lamotrigine, the antidepressant effects of anticonvulsants in bipolar disorder have been poorly studied and there is very limited supportive evidence for their use. Most of the attention to date has been centered on divalproex/valproate and lamotrigine, with very few controlled data regarding the use of other anticonvulsants in acute bipolar depression.

Divalproex/Valproate (NbN – Pharmacology: Glutamate; Mode of Action: Blockade of Voltage-Dependent Sodium Channel)

While there have been no large-scale randomized controlled trials of divalproex/valproate in acute bipolar depression, there is some very limited evidence for efficacy from several small randomized

trials. In an 8-week, double-blind, placebo-controlled, randomized clinical trial of 25 mostly male outpatients with acute bipolar I depression, divalproex was significantly more likely to reduce symptoms of both depression and anxiety (Davis et al., 2005). In the largest study of divalproex in the acute treatment of patients with bipolar I depression ($n = 20$) or II ($n = 34$) who were naive to mood-stabilizer therapy (Muzina et al., 2011), divalproex treatment produced statistically significant improvement in depressive symptoms compared with placebo from week 3 onwards.

A meta-analysis with 142 patients found that response and remission rates were significantly greater for divalproex than placebo (Bond et al., 2010). In terms of adverse effects, subjects receiving divalproex-ER compared with those receiving placebo reported more frequent nausea, increased appetite, weight gain, diarrhea, fatigue, dry mouth, and stomach cramps.

Lamotrigine (NbN – Pharmacology: Glutamate; Mode of Action: Voltage-Gated Sodium Channel Blocker)

There have been five randomized trials of lamotrigine monotherapy in the acute management of bipolar depression. Only one demonstrated efficacy, but that efficacy was not for the primary outcome measure. In the first study, Calabrese et al. (1999) randomly assigned 195 subjects with acute bipolar I depression to receive lamotrigine 50 or 200 mg/day or placebo for 7 weeks. There was no significant benefit on the primary outcome measure – the Hamilton Depression Rating Scale (HAMD). However, in terms of the secondary outcome measure (the MADRS), patients in the lamotrigine 200 mg/day group demonstrated significant improvement compared with placebo. Subsequent trials for acute bipolar depression did not confirm that first study (Calabrese et al., 2008). However, in a meta-analysis of these 5 studies, which included a total of 1072 participants, more patients treated with lamotrigine than placebo responded to treatment as measured with two different depression rating scales (Geddes et al., 2009). Moreover, the significant difference between lamotrigine and placebo was related to baseline depression severity. In the group of patients with a baseline Hamilton Depression Rating Scale (HAMD) total score of >24, the efficacy of lamotrigine was superior to that of placebo. However, in the group of patients with a baseline HAMD total score of ≤24, the efficacy of lamotrigine was not superior to that of placebo.

Lamotrigine adjunctive therapy to lithium has been found to be superior to placebo when used adjunctive to lithium in the treatment of bipolar I and II depression (van der Loos et al., 2009). A higher percentage of patients responded to add-on lamotrigine compared with placebo.

A head-to-head 7-week comparison study of OFC versus lamotrigine in the acute treatment of bipolar depression did not find a significant difference between the two treatments (Brown et al., 2006). In a 16-week single-blind comparison study of lamotrigine with lithium monotherapy in the acute treatment of bipolar II depression, both lamotrigine (200 mg/day) and lithium (0.6–1.2 mEq/L) significantly reduced depressive symptoms from baseline to the endpoint, and there were no significant difference between the two groups (Suppes et al., 2008).

With regard to adverse effects, lamotrigine is generally well tolerated, with the main safety issue being rash, and in particular the development of Stevens-Johnson syndrome. In the aforementioned trials, the incidence of rash did not differ between the placebo and treatment arms. Despite this, rash is generally recognized as the side effect most likely to complicate the drug's clinical use significantly. In early epilepsy trials, rash led to hospitalization and treatment discontinuation, or Stevens–Johnson syndrome, in 0.3% of adults treated with lamotrigine. In the bipolar population, severe rashes were also reported (see the Maintenance section later). It is well documented that the risk of rash is heightened in children less than 12 years old, by the coadministration of valproic acid, or by exceeding the recommended initial dose or rate of dose escalation of lamotrigine. To reduce the risk for severe rashes, standard titration schedules (see the section "Medications for Maintenance Treatment of Bipolar Disorder") should be followed. During postmarketing, rare cases of aseptic meningitis were reported.

Atypical Antipsychotics (NbN: Dopamine-Serotonin Blockers)

Olanzapine–Fluoxetine Combination (OFC) and Olanzapine (NbN – Olanzapine – Pharmacology: Dopamine and Serotonin; Mode of Action: Antagonist at D2 and 5-HT$_2$ Receptors; Fluoxetine – Drug for Depression – Acts on Serotonin; Mechanism of Action: Inhibitor of Serotonin Transporter)

The first controlled study of a dopamine serotonin blocker (atypical antipsychotic) for the treatment of bipolar depression was completed by Tohen et al. (2003b). This 8-week, placebo-controlled, parallel-group trial randomized patients with bipolar I depression to olanzapine, OFC, or placebo (using a 4:1:4 randomization). Treatment with olanzapine was initiated at 5 mg/day and could be increased to 20 mg/day in 5 mg increments. OFC was initiated at 6 and 25 mg/day (olanzapine 6 mg with fluoxetine 25 mg) but could be increased to 6 and 50 or 12 and 50 mg/day after at least 1 day at each dose. Both olanzapine and OFC demonstrated significantly greater and sustained mean symptomatic improvements over placebo. The response rate was significantly higher in the OFC-treated group than the olanzapine-treated group, both of which were higher than the rate of response for patients taking placebo.

Both OFC and olanzapine were relatively well tolerated compared with placebo. Adverse events that occurred more frequently in patients treated with either olanzapine or OFC included somnolence, weight gain, increased appetite, dry mouth, asthenia, and diarrhea. Olanzapine and OFC caused significantly higher weight gain than placebo.

Quetiapine (NbN – Pharmacology: Dopamine and Serotonin; Mode of Action: Antagonist at D2, 5-HT2 Receptors; Its Metabolite is an Inhibitor of the Norepinephrine Transporter and Alpha-1)

Completion of two large, multicenter, double-blind, placebo-controlled, parallel-group 8-week studies of quetiapine monotherapy led to its approval by the FDA as the second pharmacological agent for acute bipolar depression (Calabrese et al., 2005; Thase et al., 2006). These studies included subjects experiencing either bipolar I or II depression and did not exclude rapid cycling. After meeting enrolment criteria similar to the aforementioned OFC trial, subjects were randomized to quetiapine 300 mg/day, quetiapine 600 mg/day, or placebo. Both studies confirmed significant antidepressant benefits associated with quetiapine over placebo, regardless of bipolar I or II status or the presence of rapid cycling. There was no added benefit with the higher 600 mg dose of quetiapine.

Two large international studies also found that quetiapine 300 and 600 mg/day were superior to placebo in reducing depressive symptom in bipolar I or II depression (McElroy et al., 2010; Young et al., 2010). Quetiapine extended-release form (quetiapine-XR) 300 mg/day was also found to be superior to placebo in reducing depressive symptoms in patients with bipolar I or II depression (Suppes et al., 2010).

There was minimal difference in efficacy between fixed-dose 600 and 300 mg/day, but there was a higher rate of significant side effects in the quetiapine immediate-release (IR) 600 mg group than that in the 300 mg/day group (Gao et al., 2011). The FDA approved quetiapine-IR 300 mg/day for the acute treatment of bipolar depression. Patients treated with quetiapine had significantly higher rates of weight gain, dry mouth, sedation, somnolence, dizziness, and constipation than those treated with placebo.

Aripiprazole (NbN – Pharmacology: Dopamine and Serotonin; Mode of Action: Partial Agonist at D2 and 5-HT$_{1A}$ Receptors, and Antagonist at 5-HT$_2$ Receptors)

Two large, randomized, double-blind, placebo-controlled trials did not show superiority of aripiprazole over placebo in reducing depressive symptoms in bipolar depression at study end at 8 weeks (Thase et al., 2008).

Ziprasidone (NbN – Pharmacology: Dopamine and Serotonin; Mode of Action: Antagonist at D2 and 5-HT$_2$ Receptors)

Two large, randomized, double-blind, placebo-controlled studies (Sachs et al., 2011; Lombardo et al., 2012) of ziprasidone as monotherapy or adjunctive therapy to medications for bipolar disorder in bipolar depression did not demonstrate efficacy of this agent.

Lurasidone (NbN – Pharmacology: Dopamine and Serotonin; Mode of Action: Antagonist at D2 and 5-HT$_2$ Receptors)

Lurasidone was the third treatment to be approved by the FDA for bipolar depression. It has high affinity for 5-HT$_{2A}$ and 5HT$_7$ receptors that exceed the affinity at D$_2$ (antagonist effects), moderate affinity for 5HT$_{1A}$ (partial agonist effect), and α_{2C} receptors (antagonist effect) but these receptors are not recruited at moderate doses ≤40 mg daily, and no appreciable affinity for H$_1$ and M$_1$ receptors.

In a monotherapy study, subjects meeting DSM-IV-TR criteria for bipolar I depression, with or without rapid cycling, with moderate depressive severity were randomized to a 6-week, double-blind treatment with either lurasidone 20–60 mg/day (average in the mid 30 s), lurasidone 80–120 mg/day (average in the mid 80 s), or placebo (Loebel et al., 2014a). Lurasidone treatment resulted in significantly greater depressive symptom reduction at endpoint for both the lurasidone 20–60 mg/day group and the 80–120 mg/day group versus placebo. Responder rates were significantly higher for lurasidone 20–60 mg/day and 80–120 mg/day compared with placebo. The number needed to treat (NNT) for response with monotherapy was 5 (for both lower and higher dose groups), and for remission was 6 for lower dose and 7 for higher dose (Ali et al., 2020). In a similarly designed Japanese study, efficacy was seen at the low dose (20–60 mg/day), but not the high dose (80–120 mg/day) (Kato et al., 2020). These data suggest that there may be a therapeutic window at doses between 20 and 60 mg daily, and 80 mg daily should not be exceeded since efficacy drops with high doses.

In an adjunctive therapy study, lurasidone 20–120 mg/day adjunctive to either lithium or valproate was superior to placebo adjunctive therapy in the treatment of moderate-to-severe bipolar I depression (Loebel et al., 2014b). Responder rates were significantly higher for the lurasidone than for the placebo group.

Lurasidone was well tolerated with no significant difference in severe side effects compared to placebo. The most frequently reported adverse events during monotherapy were nausea, headache, and akathisia. Minimal changes in weight, lipids, and measures of glycemic control were observed.

Risperidone (NbN – Pharmacology: Dopamine, Serotonin, and Norepinephrine; Mode of Action: Antagonist at D2, 5-HT$_2$, and Norepinephrine Alpha-1 and Alpha-2 Receptors)

A double-blind adjunctive trial of risperidone given singly or in combination with paroxetine in a small sample of patients with bipolar I or II disorder ($n = 30$) who were receiving a stable dose of medication for bipolar disorder demonstrated that risperidone, risperidone plus paroxetine, and paroxetine were all modestly effective when added to a medication for bipolar disorder (Shelton & Stahl, 2004). There have been no reported large randomized trials of risperidone in bipolar depression as either monotherapy or adjunctive to lithium or valproate.

Cariprazine (NbN – Pharmacology: Dopamine and Serotonin; Mechanism of Action: Partial Agonist at D2, D3, and 5-HT$_{1A}$ Receptors, and Antagonist at 5-HT$_{2A}$ Receptors)

More recently, cariprazine has also been found to be effective in the acute treatment of bipolar depression as well as mania, leading to its approval for this indication by the FDA. This has been demonstrated in three randomized controlled phase 3 trials (Durgam et al., 2016; Earley

et al., 2019, 2020) with this effect being confirmed in a multi-treatments meta-analysis (Bahji et al., 2020). These studies involved a total of 1383 patients. The most frequent adverse effects were akathisia, restlessness, nausea, fatigue, increased appetite, dizziness, sedation, and dry mouth; akathisia was the main adverse effect responsible for discontinuation (Saraf et al., 2019).

Lumateperone (NbN – Pharmacology: Dopamine and Serotonin; Mechanism of Action: Postsynaptic D2 Antagonist, Presynaptic D2 Partial Agonist, and 5HT$_{2A}$ Antagonist)

Lumateperone was approved for bipolar depression in December 2021. It was approved for depression in both type I and type II bipolar illness, and both as a single agent (Calabrese et al., 2021) and as adjunctive therapy with lithium and divalproex. The monotherapy study included a total of 386 patients and lumateperone 42 mg daily was associated with a placebo-subtracted difference of −4.6 points improvement on the MADRS ($P < 0.0001$). In the adjunctive study, there were 348 patients and a placebo-subtracted difference of −2.4 points on the MADRS ($P = 0.021$ vs. lithium or divalproex plus placebo).

Lumateperone was well tolerated and the adverse effect profiles were remarkably similar in the two studies. Both studies reported somnolence (13% vs. 3% in placebo for both), dizziness (8% vs. placebo 4%, monotherapy; 11% vs. 2%, adjunctive), nausea (8% vs. placebo 3%, monotherapy; 9% vs. 4%, adjunctive), and dry mouth (5% vs. placebo 1%, for both).

Medications for Maintenance Treatment of Bipolar Disorder

General Considerations

Preventing, or reducing the likelihood of, further bipolar episodes through effective maintenance treatment is critical in order to improve the poor long-term outcomes that exist for individuals with this condition. Lithium was the first demonstrated effective maintenance agent, with the seminal studies being undertaken during the 1960s and 1970s. Subsequently, some of the anticonvulsants and most of the dopamine-serotonin blockers (atypical antipsychotics) have also been found to be effective preventive agents.

Efficacy of Maintenance Treatments

A recent network meta-analysis of maintenance treatments for bipolar disorder (Kishi et al., 2020a) found the following agents to be more effective than placebo in reducing recurrence/relapse *for any mood episode*: aripiprazole (oral monotherapy; oral combined with valproate or lamotrigine; long-acting injectable), asenapine, lithium (monotherapy; combined with oxcarbazepine or valproate), olanzapine, quetiapine, risperidone (long-acting injectable only), and valproate. Carbamazepine and paliperidone were not found to be effective. *For prevention of manic episodes*, all treatments were more effective than placebo except for carbamazepine and lamotrigine. *For prevention of depressive episodes*, effective treatments were aripiprazole (only in combination with valproate), lamotrigine (monotherapy and combination with valproate), lithium, olanzapine, and quetiapine.

In a further meta-analytic paper from the same group, Kishi et al. (2020b) reported that the recurrence rate of any mood episode at 6 months (the primary outcome) was significantly lower in the maintenance group than in the discontinuation group. The maintenance group also exhibited significantly fewer depressive and manic/hypomanic/mixed episode recurrence rates. Maintaining drug treatment during clinically stable BD prevented recurrence for up to 24 months. Discontinuation of medications for at least 1 month significantly increased recurrence risk; however, just less than half of patients who discontinued drugs for 6 months did not experience recurrence.

Rapid cycling bipolar disorder (defined by the occurrence of at least four mood episodes over 12 months) remains a particular challenge, with no replicated robust placebo-controlled evidence of efficacy with any agent (Bauer et al., 2018).

The comparative efficacy of mood stabilizing treatments in the maintenance phase has been a major focus of research. Hayes et al. (2016a) compared rates of monotherapy treatment failure in individuals prescribed lithium, valproate, olanzapine, or quetiapine in a UK population-based cohort study using electronic health records. Five thousand and eighty-nine patients with bipolar disorder were prescribed lithium, valproate, olanzapine, or quetiapine as monotherapy. Treatment failure was defined as time to stopping medication or add-on of another medication for bipolar disorder, medication for psychosis, medications for depression, or benzodiazepine. In unadjusted analyses, the duration of successful monotherapy was longest in individuals treated with lithium. Treatment failure had occurred in 75% of those prescribed lithium by 2.05 years, compared to 0.76 years for those prescribed quetiapine, 0.98 years for those prescribed valproate, and 1.13 years for those prescribed olanzapine. Lithium's superiority remained in a propensity score matched analysis. Lithium appeared to be more successful as a monotherapy maintenance treatment than valproate, olanzapine, or quetiapine.

The comparative efficacy of maintenance treatments for bipolar disorder in preventing psychiatric, cardiovascular, and all-cause hospitalization was investigated using a Finish dataset for over 18,000 patients admitted for bipolar disorder between 1987 and 2012 and followed up for a mean of 7 years (Lähteenvuo et al., 2018). That study found that 9721 patients (54.0%) had at least 1 psychiatric rehospitalization. Risperidone long-acting injection, gabapentin, perphenazine long-acting injection, and lithium carbonate were associated with the lowest risk of psychiatric rehospitalization. Concerning all-cause hospitalization, lithium was associated with the lowest risk. The most frequently used antipsychotic treatment, quetiapine, showed only modest effectiveness in terms of risk of psychiatric rehospitalization and risk of all-cause hospitalization. Long-acting injections were associated with substantially better outcomes compared with identical oral medications for psychosis (antipsychotics). Results from sensitivity analyses showed consistent beneficial effects only for lithium, and for long-acting injections compared with their oral counterparts. These large, long-term studies offer compelling evidence for use of lithium as a first-line agent.

The advantage of long-acting injectable forms of antipsychotics over oral medications for psychosis in preventing recurrences in bipolar disorder was also reported by Keramatian et al. (2019). Undertaking a systematic review, that group found that long-acting injectable dopamine-serotonin blockers and partial dopamine agonists (second-generation antipsychotics), particularly risperidone and aripiprazole, may be a safe and effective alternative to oral medications in the management of bipolar disorder.

Using a clinically pragmatic methodology, Kessing et al. (2018) undertook a systematic literature search of non-randomized controlled observational studies on lithium monotherapy vs treatment with another maintenance medication for bipolar disorder in monotherapy. They found that in 8 out of 9 identified studies, including a total of almost 14,000 patients, maintenance lithium monotherapy was associated with greater rates of improved outcome compared with another medication for bipolar disorder in monotherapy, including valproate, lamotrigine, unspecified anticonvulsants, carbamazepine/lamotrigine, or medications for psychosis olanzapine, quetiapine, and unspecified medications for psychosis (antipsychotics).

In practice, combination therapy is common in the management of bipolar disorder. In a recent study (Fung et al., 2019), 43% of patients were prescribed three or more medications at any time (complex polypharmacy), while 25% received three or more treatments for the majority of time. While such multiple prescribing was not associated with more side effects, it did correlate with poor adherence, suggesting that multiple prescribing may be counterproductive for many patients. Further, multiple medication prescription is often associated with sedation, cognitive effects, and weight gain.

Prevention of Suicide

Another focus in the maintenance treatment of bipolar disorder is the relative capacity of medications to prevent suicide attempts and completed suicide. There has been particular interest in reports of lithium reducing risk of suicide and whether this effect is specific to that treatment or a nonspecific effect of medications for bipolar disorder more broadly. Hayes et al. (2016b) compared rates of self-harm, unintentional injury, and suicide in patients with bipolar disorder who were prescribed lithium, valproate sodium, olanzapine, or quetiapine in a nationally representative UK sample of more than 14,000 subjects using electronic health data. Those taking lithium had significantly reduced self-harm and unintentional injury rates compared to the other treatments. While the suicide rate was also lower in the lithium group than for other treatments, there were too few events to allow for accurate estimates. The authors interpreted the results of the study as supporting the hypothesis that lithium reduces impulsive aggression in addition to stabilizing mood.

A recent systematic review of suicidal behavior with various treatments for bipolar disorder (del Matto et al., 2020) examined 18 prospective (number of participants: 153,786), 10 retrospective (number of participants: 61,088), and 16 ecological studies (total sample: 2062). Most of the observational studies reported a reduction in suicide in patients with mood disorders who were prescribed lithium. All studies which investigated treatment duration reported that long-term lithium provided more benefits than short-term lithium in suicide risk.

Other studies have investigated the comparative effect on suicide rates with other medications, in particular valproate. Chen et al. (2019) reported on a meta-analysis of six placebo-controlled studies of valproate in bipolar disorder, finding no significant difference between valproate and placebo in the incidence rates of suicide attempts or completed suicides.

Adverse Effects

The comparative adverse effects of the common maintenance therapies for bipolar disorder were studied using nationally representative UK electronic health records by Hayes et al. (2016c). The study included patients prescribed lithium ($n = 2148$), valproate ($n = 1670$), olanzapine ($n = 1477$), or quetiapine ($n = 1376$). Compared to patients prescribed lithium, those taking valproate, olanzapine, and quetiapine had reduced rates of chronic kidney disease (stage 3 or more severe) following adjustment for relevant confounding factors. Hypothyroidism was reduced in those taking valproate and olanzapine compared to those taking lithium. Rates of new onset hyperthyroidism and hypercalcemia were also reduced relative to lithium. However, rates of greater than 15% weight gain (an increase of 2 points of body mass index) on valproate, olanzapine, and quetiapine were higher than in individuals prescribed lithium, as were rates of hypertension in the olanzapine treated group. In a direct head-to-head comparison of lithium and quetiapine, quetiapine was associated with more frequent, intense, and impairing side effects than lithium (Nierenberg et al., 2016).

Teratogenesis and Adverse Pregnancy Outcomes

While there is an undoubted increase in rates of teratogenesis with the anticonvulsants valproate and carbamazepine, there has been uncertainty concerning the effects of lithium on the developing fetus. In one of the largest studies to address this issue, Patorno et al. (2017) studied a US cohort of over 1.3 million pregnancies in women who were enrolled in Medicaid and who delivered a live-born infant between 2000 and 2010, with a focus on the 663 infants exposed to lithium. The adjusted risk ratio for cardiac malformations among infants exposed to lithium as compared with unexposed infants was significantly increased at 1.65, with this effect being dose-related (i.e., 1.1 for doses less than 600 mg, compared to 3.2 for more than 900 mg). The prevalence of right ventricular outflow tract obstruction defects was significantly increased at 0.60% among lithium-exposed infants versus 0.18% among unexposed infants (adjusted risk ratio, 2.66). Overall, use of lithium during the first trimester was associated with an increased risk of cardiac malformations, including Ebstein's anomaly. However, the magnitude of this effect was smaller than had been previously postulated.

Unlike the other anticonvulsants, growing evidence is indicating that lamotrigine is not a teratogenic agent. For example, a prospective Israeli study of 218 women prescribed lamotrigine for either neurological or psychiatric disorders (Diav-Citrin et al., 2017) found no increase in teratogenic effects compared to over 800 women not prescribed lamotrigine or any known teratogenic agents.

Concerning the potential effect of medications for bipolar disorder on adverse pregnancy outcomes (not teratogenesis), Cohen et al. (2019) examined Medicaid data on almost 1.5 million pregnancies, including over 10,000 on lithium or an anticonvulsant medication for bipolar disorder. The authors found no relationship between medication use and preeclampsia, placental abruption, growth restriction, and preterm birth after controlling for confounding by indication.

Adherence and Undertreatment

One of the major limitations of medications for bipolar disorder is the consistently high reported poor rates of adherence, with relevant causative factors including side effects and difficulties in accepting the diagnosis and need for treatment. A number of approaches have been recently proposed. For example, Sajatovic et al. (2018) reported on a randomized controlled trial of a customized adherence enhancement program specifically designed for poorly adherent bipolar disorder patients, and which targeted individual adherence barriers. Compared to a structured bipolar disorder psychoeducation program, the adherence intervention led to significantly improved medication adherence, functioning, and mental health resource use. Pakpour et al. (2017) have also reported on the efficacy of a program using a multifaceted intervention that included motivational interviewing and psychoeducation in improving medication adherence among patients with bipolar disorder. Related to this are worrying reports of inadequate treatment rates in various ethnic groups such as Hispanic (Salcedo et al., 2017) populations.

Adjunctive Psychological Therapies

While the focus of this section is on pharmacological maintenance treatments for bipolar disorder, some comment on the efficacy of adjunctive psychological treatments is apposite, as most treatment guidelines propose such adjunctive therapies as one component of optimal management for this disorder. In a recent component network meta-analysis (Miklowitz et al., 2021), the authors identified 39 randomized clinical trials of structured psychological treatments adjunct to medications for bipolar disorder, with 3863 participants. They found that manualized psychological treatments were associated with lower recurrence rates than control treatments. Psychoeducation with guided practice of illness management skills in a family or group format was associated with reducing recurrences versus the same strategies in an individual format. They also found, with less certainty, that family or conjoint therapy and interpersonal therapy were associated with stabilizing depressive symptoms compared with treatment as usual.

Lithium (NbN – Mode of Action; Enzyme Modulator, Displaces Intracellular Sodium)

Standard dosing to achieve blood levels of lithium between 0.8 and 1.0 mEq/L or higher is generally recommended over "low" dosing to levels between 0.4 and 0.6 mEq/L. Maintenance at lithium levels in this low range has been associated with a 2.6-fold greater risk of relapse compared with standard dosing into the higher range (Keller et al., 1992). Furthermore, patients maintained in the higher range are less likely to experience subsyndromal episodes. However, lithium side effects were found to be more frequent when dosing to achieve concentrations of 0.8 and 1.0 mEq/L was employed. The clinician therefore needs to balance potential efficacy against significant adverse effects as the latter may lead to poor adherence.

The seminal study of the adverse effects of lithium was the meta-analysis of McKnight et al. (2012) which included 385 studies. The findings are particularly pertinent to patients on long-term maintenance use of lithium. On average, glomerular filtration rate was found to be reduced, as was urinary concentrating ability. The risk of renal failure was low, with only 0.5% of patients received renal replacement therapy. The prevalence of clinical hypothyroidism was increased. Lithium treatment was associated with increased blood calcium and parathyroid hormone levels, with the authors recommending routine measurement of calcium concentrations before and after initiation of treatment. Patients receiving lithium gained more weight. There was no significant increased risk of congenital malformations, alopecia, or skin disorders (though the methodology may have reduced confidence in those latter findings which have been repeatedly reported in clinical studies).

In terms of renal effects, lithium-induced nephrogenic diabetes insipidus should be considered in patients with increased thirst and polyuria where the 24-hour urine volume exceeds 2.5 L. This condition is due to a lack of responsiveness to anti-diuretic hormone (ADH). Lithium can also cause chronic tubulointerstitial nephritis resulting in renal insufficiency as detailed above. Helpful clinical guidance on preventing and managing renal complications of lithium has been published in recent years (Kripalani et al., 2009; Schoot et al., 2020). Prevention of renal effects is diminished by once daily administration, using the clinically lowest effective concentrations, and avoiding lithium toxicity. For those with renal impairment, a close working relationship with a nephrologist is recommended. Lithium-related decline in glomerular function is associated with an increase in renal microcysts and normal sized kidneys (Ali & El-Mallakh, 2021). The absence of microcysts is strongly suggestive that the renal dysfunction is not lithium-related.

Further, neurotoxicity during maintenance treatment with lithium is not uncommon and is often related to concurrent medications, as detailed above in acute mania. Lithium-induced neurotoxicity typically occurs with chronic accumulation rather than following acute overdose. A recent important case series report (Hlaing et al., 2020) highlighted the protracted nature of lithium neurotoxicity long after the lithium concentrations return to the therapeutic range, with its severity correlating poorly with lithium concentrations, which normalize quickly.

Anticonvulsants

Lamotrigine (NbN – Pharmacology: Glutamate; Mode of Action: Voltage-Gated Sodium Channel Blocker)

Two 18-month maintenance trials confirmed the efficacy of lamotrigine in bipolar disorder, (Bowden et al., 2003; Calabrese et al., 2003). These studies led to approval of lamotrigine by the FDA as a maintenance treatment for bipolar I disorder.

In recently manic or hypomanic patients with bipolar I disorder, both lithium and lamotrigine were superior to placebo at delaying the time until additional pharmacotherapy was required for treatment of a mood episode. Lamotrigine demonstrated significantly greater efficacy for prolonging the time to a depressive episode compared with lithium. In recently depressed patients with bipolar I disorder, both lamotrigine and lithium were significantly superior to placebo at delaying the time to intervention for any mood episode. Lamotrigine and lithium were superior to placebo on analysis of overall survival in the study, and lamotrigine, but not lithium, was superior to placebo in terms of delaying the time to intervention for depressive episodes. These studies established that lamotrigine was significantly more effective than placebo in preventing relapse in patients with bipolar disorder while demonstrating primary effectiveness against depression. In a later pooled analysis of these two studies, Goodwin et al. (2004) found that while both prevented recurrence of any mood episode, lamotrigine was more effective in preventing depressive than manic recurrence.

In terms of the mechanism of action of lamotrigine in bipolar disorder, this is poorly understood (de Miranda et al., 2019). As discussed in the review of Miranda, preclinical studies have indicated that neurotransmitter systems, especially serotoninergic, noradrenergic, and glutamatergic, as well as non-neurotransmitter pathways such as inflammation and oxidative processes might play a role in lamotrigine's antidepressant effects; however, all of these effects are small and indirect. It is much more likely that inhibition of sodium entry in an activity-dependent fashion, which is a shared feature of all effective anticonvulsant mood-stabilizing agents, including lamotrigine, is central to the action of these agents.

No laboratory tests are required before initiating lamotrigine, although routine physical examination, basic baseline chemistries, and pregnancy testing are advisable. A history of hypersensitivity to lamotrigine is the only contraindication to use. The risk of rash must be discussed before initiating therapy. Calabrese et al. (2002) reviewed data on 3153 patients treated with lamotrigine in mood disorder trials and compared rash rates with those for 1056 placebo-treated patients, finding that rates of benign rash were 8.3% and 6.4% in lamotrigine and placebo-treated patients, respectively.

Any lamotrigine-treated patient developing any rash that cannot readily be explained by a known other cause, such as contact dermatitis, should immediately discontinue lamotrigine and notify their treating physician prior to resuming therapy. In unclear or difficult cases, dermatological consultation should be obtained before restarting or rechallenging with lamotrigine.

Following a careful and gradual dosage titration schedule reduces the risk of rash. Lamotrigine should be initiated at 25 mg/day with the first dosage increase to 50 mg/day 2 weeks later. Lamotrigine may be increased to 100 mg/day in week 5 and to 200 mg/day in week 6. The presence of enzyme-inducing drugs, such as carbamazepine, requires doubling of the lamotrigine dose while concomitant administration with valproate necessitates a 50% reduction in lamotrigine dosing. Concomitant prescription of estrogen-containing oral contraceptives may lead to enzyme induction and increased metabolism of lamotrigine, which may require an increase in lamotrigine dose. The obverse should also be considered; lamotrigine causes a modest level of auto-induction of CYP 3A4 which is greater in Asians (Hussein & Posner, 1997), and effectiveness of low estrogen oral contraceptive agents may be reduced, and this needs to be communicated to those prescribing the birth control agent.

The most common side effects observed with lamotrigine are dizziness, ataxia, somnolence, headache, diplopia, blurred vision, nausea, and vomiting. These side effects are typically mild and rarely lead to treatment discontinuation. Decreasing the dose or slowing the rate of dose escalation may alleviate these side effects.

Divalproex/Valproate (NbN – Pharmacology: Glutamate; Mode of Action: Voltage-Gated Sodium Channel Blocker)

This anticonvulsant is commonly used in maintenance therapy for bipolar disorder, although it does not carry a labeled indication for maintenance in the United States or elsewhere.

Bowden et al. (2000) conducted the only large-scale double-blind, randomized, controlled maintenance study in bipolar I disorder, which followed patients from an index manic episode through a subsequent 52-week maintenance phase. Subjects were randomized into the placebo-controlled, double-blind, parallel-group maintenance study and placed into divalproex, lithium, or placebo groups. Study drugs were dosed to serum trough concentrations of between 71 and 126 µg/mL for valproate and between 0.8 and 1.2 mEq/L for lithium. No significant difference was found in the time to recurrence of any mood episode during maintenance therapy between the three groups. It should be noted that this study did not recruit an enriched sample which is the practice for maintenance studies done subsequent to this study.

In a comparator study of olanzapine versus divalproex over a 47-week period of initial acute treatment of mania or mixed state followed by maintenance treatment of 251 patients with bipolar disorder (Tohen et al., 2003a), there were no significant overall differences between the two treatments in the rates of symptomatic remission or subsequent relapse over the entire 47-week period.

The relative efficacy of lithium and valproate (alone and in combination with lithium) in the maintenance treatment of bipolar disorder was investigated in a large randomized open-label multisite UK study (the BALANCE trial) by Geddes et al. (2010). Hazard ratios for the primary outcome indicated that the combination of lithium

and valproate was more effective in delaying time to recurrence than was valproate monotherapy though the combination was not significantly better than lithium monotherapy. Lithium monotherapy was slightly significantly more effective than valproate monotherapy.

As detailed in the introduction to maintenance treatments above, the recent multi-treatments meta-analysis of Kishi et al. (2020a) supported the efficacy of valproate monotherapy. Overall, despite some important negative findings, valproate is in widespread clinical use and recommended in most treatment guidelines.

Carbamazepine (NbN – Pharmacology: Glutamate; Mode of Action: Voltage-Gated Sodium and Calcium Channel Blocker)

As detailed above, the recent meta-analysis of Kishi et al. (2020) found no evidence for the efficacy of carbamazepine in maintenance treatment. This is consistent with an earlier meta-analysis of medications for bipolar disorder in the prevention of recurrent affective disorders (Davis et al., 1999). Expert opinion is that carbamazepine is inferior to lithium in the maintenance treatment of bipolar disorder (Muzina et al., 2002). There is some evidence supporting its use in combination with other medications for bipolar disorder for maintenance, particularly with lithium.

Atypical Antipsychotics (NbN – Dopamine-Serotonin Blockers and Dopamine Partial Agonists)

The dopamine-serotonin blockers and dopamine partial agonists (atypical antipsychotics) which have received FDA approval for maintenance treatment of bipolar disorder are olanzapine (oral), aripiprazole (oral and long-acting injection), quetiapine, ziprasidone, and risperidone (long-acting injectable formulation). Both quetiapine and ziprasidone received FDA approval specifically as adjunctive therapy (in combination with lithium or valproate), not as monotherapy.

Olanzapine (NbN – Pharmacology: Dopamine and Serotonin; Mode of Action: Antagonist at D2 and 5-HT$_2$ Receptors)

An 18-month placebo-controlled study evaluating the preventive efficacy of olanzapine adjunctive to either lithium or valproate found that those on the combination therapy sustained symptomatic remission longer than those on lithium or valproate monotherapy (Tohen et al., 2004). In a head-to-head comparison, olanzapine mono-therapy was significantly more effective than lith-ium in preventing manic relapse/recurrence (Tohen et al., 2005). Olanzapine and lithium were comparable in preventing relapse or recur-rence of depression. In a placebo-controlled, dou-ble-blind study to examine the efficacy of olanzapine monotherapy, the median time to relapse into any mood episode, mania or depression was significantly longer among patients continuing to receive olanzapine than in placebo-treated patients (Tohen et al., 2006). However, olanzapine prevented relapse into mania more effectively than into depression. Those treated with olanzapine had significantly higher rates of weight increase, dry mouth, increased appetite, somnolence, and fatigue than those treated with placebo.

Olanzapine is also available in a long-acting injectable form (Jann & Penzak, 2018), though this is not FDA-approved for bipolar disorder. The occurrence of a postinjection delirium/seda-tion syndrome (and some reported deaths) with the long-acting form in about 1.5% of patients, with the consequent regulatory requirement for a period of monitored observation postinjection, has limited its use clinically.

Aripiprazole (NbN – Pharmacology: Dopamine and Serotonin; Mode of Action: Partial Agonist at D2 and 5-HT$_{1A}$ Receptors, and Antagonist at 5-HT$_2$ Receptors)

A 26-week randomized placebo-controlled study of aripiprazole monotherapy supported the mood-stabilizing effect of this dopamine serotonin blocker (atypical antipsychotic) (Keck et al., 2006). Aripiprazole was superior to placebo in delaying the time to relapse, although this effect was found for manic prophylaxis but not for depression. Adverse events with an incidence of ≥5% and at least twice that of the placebo group included tremor and akathisia.

Aripiprazole once-monthly injection was approved by the FDA for maintenance mono-therapy for bipolar I disorder in 2017 (Calabrese et al., 2017b).

Quetiapine Adjunctive to Lithium or Divalproex (NbN – Pharmacology: Dopamine and Serotonin; Mode of Action: Antagonist at D2, 5-HT$_2$ Receptors; Its Metabolite Is an Inhibitor of the Norepinephrine Transporter)

There have been two large studies of quetiapine therapy adjunctive to lithium or divalproex in the maintenance treatment of bipolar I disorder (Suppes et al., 2009; Vieta et al., 2008). Patients with bipolar I disorder were randomly assigned to receive either quetiapine (400–800 mg/day) or placebo plus lithium (0.5–1.2 mEq/L) or divalproex (50–125 µg/mL) for 104 weeks. Both studies showed that the combination therapy was significantly superior to lithium or divalproex plus placebo in increasing the time to recurrence of any mood event. Secondary analyses showed that adjunctive quetiapine was significantly more effective than placebo in increasing the time to recurrence of both mania and depression. Seda-tion was more common in those receiving quetiapine. Increases in weight, body mass index, HbA$_{1c}$, and triglyceride in the quetiapine group were significantly higher than those receiv-ing placebo.

Ziprasidone Adjunctive to Lithium or Valproate (NbN – Pharmacology: Dopamine and Serotonin; Mode of Action: Antagonist at D2 and 5-HT$_2$ Receptors)

A 6-month randomized, placebo-controlled study of ziprasidone adjunctive to medication for bipolar disorder in patients with bipolar I disorder showed that ziprasidone was superior to placebo in preventing mood episode relapse (Bowden et al., 2010). Time to intervention for a mood episode was significantly longer for ziprasidone than placebo. Post hoc analyses showed that ziprasidone was superior to placebo in preventing manic, but not depressive relapse. Patients on ziprasidone were more likely to have elevated prolactin levels.

Risperidone (NbN – Pharmacology: Dopamine, Serotonin, and Norepinephrine; Mode of Action: Antagonist at D2, 5-$_{HT2}$, and Norepinephrine Alpha-2 Receptors)

Risperidone is one of the dopamine-serotonin blockers (atypical antipsychotics) with a long-acting intramuscular injection form which has been FDA-approved for use in bipolar disorder. Slow release of risperidone occurs with gradual hydrolysis of the polymer beginning about 3 weeks after initial injection. It is estimated that the 50 mg injectable dose given every 2 weeks is roughly equivalent to up to 4 mg of oral risperidone given daily. The main role of this formulation is for individual patients with mania prone to frequent relapse or recurrence due to poor oral medication adherence.

With regard to the supporting studies, risperidone long-acting injectable therapy was studied as an adjunctive therapy to treatment as usual (MacFadden et al., 2009). The time to relapse was significant longer in patients receiving adjunctive long-acting risperidone injection than those receiving adjunctive placebo. Common adverse events were tremor, insomnia, muscle rigidity, weight gain, and hypokinesia. The second study randomly assigned patients to placebo injection or long-acting injectable risperidone (12.5–50 mg biweekly) for up to 24 months. The time to recurrence for any mood episode was significantly longer in the risperidone group, with a significant effect on manic but not depressive relapse.

Paliperidone (NbN – Pharmacology: Dopamine, Serotonin, and Norepinephrine; Mode of Action: Antagonist at D2, 5-HT2, and Norepinephrine Alpha-2 Receptors)

There has been one large 24-month randomized placebo-controlled trial of paliperidone in the maintenance treatment of bipolar I disorder (Berwaerts et al., 2012b). The time to recurrence of any mood symptoms was significantly longer with paliperidone-ER than placebo. However, findings were mixed. During the first year, the NNT was 8 with paliperidone-ER relative to placebo for preventing any mood symptoms, but at the end of the second year, there was no significant difference between paliperidone-ER and placebo in preventing recurrence of any mood symptoms. The time to recurrence of manic symptoms was significantly longer in the paliperidone group versus the placebo group. There was no significant effect on recurrence of depressive symptoms.

Paliperidone is also available in monthly, trimonthly, and hexamonthly long-acting injectable forms – paliperidone palmitate. There have been no controlled trials of these long-acting depot formulation in bipolar disorder. Once monthly paliperidone palmitate has been demonstrated to reduce episodes of psychosis, depression, and mania in individuals with schizoaffective disorder compared to placebo (Fu et al., 2015).

Lurasidone (NbN – Pharmacology: Dopamine and Serotonin; Mode of Action: Antagonist at D2 and 5-HT$_2$ Receptors)

In the only published report of lurasidone as a maintenance treatment (Calabrese et al., 2017a), the addition of lurasidone to lithium or valproate – compared with adjunctive lithium – did not significantly reduce relapse into any mood episode. There have been no reports of lurasidone monotherapy for maintenance treatment for bipolar disorder.

Cariprazine (NbN – Pharmacology: Dopamine and Serotonin; Mechanism of Action: Partial Agonist at D2, D3, and 5-HT$_{1A}$ Receptors, and Antagonist at 5-HT$_{2A}$ Receptors)

There have been no published trials of cariprazine as a maintenance treatment for bipolar disorder. This is unfortunate since most acute studies in bipolar patients (depression and mania) end before the drug has reached steady state, raising a major question regarding long-term safety, efficacy, and appropriate long-term dosing.

Conclusion

A wide range of treatments have now been shown to be effective for the *acute treatment of mania*, with it now being evident that medications for psychosis (antipsychotics) are to be preferred in the short-term control of this urgent presentation. *Bipolar depression* remains a therapeutic challenge and remains an area in urgent need of new and more effective treatments. *Maintenance therapy* is critical in view of the high rates of relapse and disability with this condition. While lithium remains the gold standard treatment option for prevention of mania, lamotrigine has been shown to be effective in preventing depressive recurrences. Further, many of the dopamine-serotonin blockers and dopamine partial agonists (atypical antipsychotics) have also been found to be useful, particularly for manic relapse, with long-acting injectables having a demonstrable advantage. Finally, growing evidence is confirming that optimal treatment comprises effective pharmacotherapy concurrent with evidence-based psychological therapies.

References

Ali, Z., & El-Mallakh, R. S. (2021). Lithium nephropathy and renal microcysts. *Current Psychiatry, 20*(6), 34–38, 50.

Ali, Z., Tegin, C., & El-Mallakh, R. S. (2020). Evaluating lurasidone as a treatment option for bipolar disorder. *Expert Opinion in Pharmacotherapy, 21*(3), 253–260. https://doi.org/10.1080/14656566.2019.1695777

Amare, A. T., Schubert, K. O., Hou, L., et al. (2018). Association of polygenic score for schizophrenia and HLA antigen and inflammation genes with response to lithium in bipolar affective disorder: A genome-wide association study. *JAMA Psychiatry, 75*, 65–74.

Amare, A. T., Schubert, K. O., Hou, L., et al. (2020). Association of polygenic score for major depression with response to lithium in patients with bipolar disorder. *Molecular Psychiatry.* Online ahead of print.

Bahji, A., Hawken, E. R., Sepehry, A. A., et al. (2019). ECT beyond unipolar major depression: Systematic review and meta-analysis of electroconvulsive therapy in bipolar depression. *Acta Psychiatrica Scandinavica, 139*, 214–226.

Bahji, A., Ermacora, D., Stephenson, C., et al. (2020). Comparative efficacy and tolerability of pharmacological treatments for the treatment of acute bipolar depression: A systematic review and network meta-analysis. *Journal of Affective Disorders, 269*, 154–184.

Bauer, M., Andreassen, O. A., Geddes, J. R., et al. (2018). Areas of uncertainties and unmet needs in bipolar disorders: Clinical and research perspectives. *Lancet Psychiatry, 5*, 930–939.

Bellivier, F., & Marie-Claire, C. (2018). Molecular signatures of lithium treatment: Current knowledge. *Pharmacopsychiatry, 51*, 212–219.

Berwaerts, J., Lane, R., Nuamah, I. F., et al. (2011). Paliperidone extended-release as adjunctive therapy to lithium or valproate in the treatment of acute mania: A randomized, placebo-controlled study. *Journal of Affective Disorders, 129*, 252–260.

Berwaerts, J., Xu, H., Nuamah, I., et al. (2012a). Evaluation of the efficacy and safety of paliperidone extended-release in the treatment of acute mania: A randomized, double-blind, dose-response study. *Journal of Affective Disorders, 136*, e51–e60.

Berwaerts, J., Melkote, R., Nuamah, I., et al. (2012b). A randomized, placebo-and active-controlled study of paliperidone extended-release as maintenance treatment in patients with bipolar I disorder after an acute manic or mixed episode. *Journal of Affective Disorders, 138*, 247–258.

Bond, D. J., Lam, R. W., & Yatham, L. N. (2010). Divalproex sodium versus placebo in the treatment of acute bipolar depression: A systematic review and meta-analysis. *Journal of Affective Disorders, 124*, 228–234.

Bowden, C. L., Brugger, A. M., Swann, A. C., et al. (1994). Efficacy of divalproex vs. lithium and placebo in the treatment of mania. The Depakote Mania Study Group. *JAMA, 271*, 918–924.

Bowden, C. L., Calabrese, J. R., McElroy, S. L., et al. (2000). A randomized, placebo-controlled 12-month trial of divalproex and lithium in treatment of outpatients with bipolar I disorder. Divalproex Maintenance Study Group. *Archives of General Psychiatry, 57*, 481–489.

Bowden, C. L., Calabrese, J. R., Sachs, G., et al. (2003). A placebo-controlled 18-month trial of lamotrigine and lithium maintenance treatment in recently manic or hypomanic patients with bipolar I disorder. *Archives of General Psychiatry, 60,* 392–400.

Bowden, C. L., Grunze, H., Mullen, J., et al. (2005). A randomized, double-blind, placebo-controlled efficacy and safety study of quetiapine or lithium as monotherapy for mania in bipolar disorder. *Journal of Clinical Psychiatry, 66,* 111–121.

Bowden, C. L., Swann, A. C., Calabrese, J. R., et al. (2006). A randomized, placebo-controlled, multicenter study of divalproex sodium extended release in the treatment of acute mania. *Journal of Clinical Psychiatry, 67,* 1501–1510.

Bowden, C. L., Vieta, E., Ice, K. S., et al. (2010). Ziprasidone plus a mood stabilizer in subjects with bipolar I disorder: A 6-month, randomized, placebo-controlled, double-blind trial. *Journal of Clinical Psychiatry, 71,* 130–137.

Brown, E. B., McElroy, S. L., Keck, P. E., Jr., et al. (2006). A 7-week, randomized, double-blind trial of olanzapine/fluoxetine combination versus lamotrigine in the treatment of bipolar I depression. *Journal of Clinical Psychiatry, 67,* 1025–1033.

Cade, J. F. (1949). Lithium salts in the treatment of psychotic excitement. *Medical Journal of Australia, 2,* 349–352.

Calabrese, J. R., Bowden, C. L., Sachs, G. S., et al. (1999). A double-blind placebo-controlled study of lamotrigine monotherapy in outpatients with bipolar I depression. Lamictal 602 Study Group. *Journal of Clinical Psychiatry, 60,* 79–88.

Calabrese, J. R., Sullivan, J. R., Bowden, C. L., et al. (2002). Rash in multicenter trials of lamotrigine in mood disorders: Clinical relevance and management. *Journal of Clinical Psychiatry, 63,* 1012–1019.

Calabrese, J. R., Bowden, C. L., Sachs, G., et al. (2003). A placebo-controlled 18-month trial of lamotrigine and lithium maintenance treatment in recently depressed patients with bipolar I disorder. *Journal of Clinical Psychiatry, 64,* 1013–1024.

Calabrese, J. R., Keck, P. E., Jr., MacFadden, W., et al. (2005). A randomized, double-blind, placebo-controlled trial of quetiapine in the treatment of bipolar I or II depression. *American Journal of Psychiatry, 162,* 1351–1360.

Calabrese, J. R., Huffman, R. F., White, R. L., et al. (2008). Lamotrigine in the acute treatment of bipolar depression: Results of five double-blind, placebo-controlled clinical trials. *Bipolar Disorders, 10,* 323–333.

Calabrese, J. R., Pikalov, A., Streicher, C., et al. (2017a). Lurasidone in combination with lithium or valproate for the maintenance treatment of bipolar I disorder. *European Neuropsychopharmacology, 27,* 865–876.

Calabrese, J. R., Sanchezm, R., Jin, N., et al. (2017b). Efficacy and safety of aripiprazole once-monthly in the maintenance treatment of bipolar I disorder: A double-blind, placebo-controlled, 52-week randomized withdrawal study. *Journal of Clinical Psychiatry, 78*(3), 324–331. https://doi.org/10.4088/JCP.16m11201

Calabrese, J. R., Durgam, S., Satlin, A., et al. (2021). Efficacy and safety of lumateperone for major depressive episodes associated with bipolar I or bipolar II disorder: A phase 3 randomized placebo-controlled trial. *The American Journal of Psychiatry, 178*(12), 1098–1106. https://doi.org/10.1176/appi.ajp.2021.20091339

Chakrabarty, T., Keramatian, K., & Yatham, L. N. (2020). Treatment of mixed features in bipolar disorder: An updated view. *Current Psychiatry Reports, 22,* 15.

Chen, T. Y., Kamali, M., Chu, C. S., et al. (2019). Divalproex and its effect on suicide risk in bipolar disorder: A systematic review and meta-analysis of multinational observational studies. *Journal of Affective Disorders, 245,* 812–818.

Chwieduk, C. M., & Scott, L. J. (2011). Asenapine: A review of its use in the management of mania in adults with bipolar I disorder. *CNS Drugs, 25,* 251–267.

Cipriani, A., Barbui, C., Salanti, G., et al. (2011). Comparative efficacy and acceptability of antimanic drugs in acute mania: A multiple-treatments meta-analysis. *The Lancet, 378,* 1306–1315.

Cohen, J. M., Huybrechts, K. F., Patorno, E., et al. (2019). Anticonvulsant mood stabilizer and lithium use and risk of adverse pregnancy outcomes. *Journal of Clinical Psychiatry, 80,* 18m12572.

Curtin, F., & Schulz, P. (2004). Clonazepam and lorazepam in acute mania: A Bayesian meta-analysis. *Journal of Affective Disorders, 78*(3), 201–208. https://doi.org/10.1016/S0165-0327(02)00317-8

Davis, J. M., Janicak, P. G., & Hogan, D. M. (1999). Mood stabilizers in the prevention of recurrent affective disorders: A meta-analysis. *Acta Psychiatrica Scandinavica, 100,* 406–417.

Davis, L. L., Bartolucci, A., & Petty, F. (2005). Divalproex in the treatment of bipolar depression: A placebo-controlled study. *Journal of Affective Disorders, 85,* 259–266.

de Miranda, A. S., de Miranda, A. S., & Teixeira, A. L. (2019). Lamotrigine as a mood stabilizer: Insights from the pre-clinical evidence. *Expert Opinion on Drug Discovery, 14,* 179–190.

Del Matto, L., Muscas, M., Murru, A., et al. (2020). Lithium and suicide prevention in mood disorders and in the general population: A systematic review. *Neuroscience and Biobehavioral Reviews, 116,* 142–153.

Diav-Citrin, O., Shechtman, S., Zvi, N., et al. (2017). Is it safe to use lamotrigine during pregnancy? A prospective comparative observational study. *Birth Defects Research, 109,* 1196–1203.

Durgam, S., Earley, W., Lipschitz, A., et al. (2016). An 8-week randomized, double-blind, placebo-controlled evaluation of the safety and efficacy of cariprazine in patients with bipolar I depression. *American Journal of Psychiatry, 173,* 271–281.

Earley, W., Burgess, M. V., Rekeda, L., et al. (2019). Cariprazine treatment of bipolar depression: A randomized double-blind placebo-controlled phase 3 study. *American Journal of Psychiatry, 176,* 439–448.

Earley, W. R., Burgess, M. V., Khan, B., et al. (2020). Efficacy and safety of cariprazine in bipolar I depression: A double-blind, placebo-controlled phase 3 study. *Bipolar Disorders, 22*, 372–384.

El-Mallakh, R. S., & Huff, M. O. (2001). Mood stabilizers and ion regulation. *Harvard Review of Psychiatry, 9*(1), 23–32. https://doi.org/10.1080/10673220127873

El-Mallakh, R. S., & Watkins, J. (2009). Prolactin elevations and the permeability glycoprotein. *Primary Care Companion CNS Disorders, 21*(3) pii: 18nr02412.

Ferrari, A. J., Stockings, E., Khoo, J. P., et al. (2016). The prevalence and burden of bipolar disorder: Findings from the Global Burden of Disease Study 2013. *Bipolar Disorders, 18*, 440–450.

Frankland, A., Roberts, G., Holmes-Preston, E., et al. (2018). Clinical predictors of conversion to bipolar disorder in a prospective longitudinal familial high-risk sample: Focus on depressive features. *Psychological Medicine, 48*, 1713–1721.

Fu, D. J., Turkoz, I., Simonson, R. B., et al. (2015). Paliperidone palmitate once-monthly reduces risk of relapse of psychotic, depressive, and manic symptoms and maintains functioning in a double-blind, randomized study of schizoaffective disorder. *Journal of Clinical Psychiatry, 76*(3), 253–262. https://doi.org/10.4088/JCP.14m09416

Fung, V. C., Overhage, L. N., Sylvia, L. G., et al. (2019). Complex polypharmacy in bipolar disorder: Side effect burden, adherence, and response predictors. *Journal of Affective Disorders, 257*, 7–22.

Gao, K., Kemp, D. E., Fein, E., et al. (2011). Number needed to treat to harm for discontinuation due to adverse events in the treatment of bipolar depression, major depressive disorder, and generalized anxiety disorder with atypical antipsychotics. *Journal of Clinical Psychiatry, 72*, 1063–1071.

Gao, K., Pappadopulos, K., Karayal, O. N., et al. (2013). Risk for adverse events and discontinuation due to adverse events of ziprasidone monotherapy relative to placebo in the acute treatment of bipolar depression, mania, and schizophrenia. *Journal of Clinical Psychopharmacology, 33*, 425–431.

Geddes, J. R., Calabrese, J. R., & Goodwin, G. M. (2009). Lamotrigine for treatment of bipolar depression: Independent meta-analysis and meta-regression of individual patient data from five randomized trials. *British Journal of Psychiatry, 94*, 4–9.

Geddes, J. R., Goodwin, G. M., Rendell, J., et al. (2010). Lithium plus valproate combination therapy versus monotherapy for relapse prevention in bipolar I disorder (BALANCE): A randomized open-label trial. BALANCE Investigators and Collaborators. *The Lancet, 375*, 385–395.

Gerner, R. H., Post, R. M., & Bunney, W. E., Jr. (1976). A dopaminergic mechanism in mania. *American Journal of Psychiatry, 133*, 1177–1180.

Gijsman, H. J., Geddes, J. R., Rendell, J. M., et al. (2004). Antidepressants for bipolar depression: A systematic review of randomized, controlled trials. *American Journal of Psychiatry, 161*, 1537–1547.

Goodwin, G. M., Bowden, C. L., Calabrese, J. R., et al. (2004). A pooled analysis of 2 placebo-controlled 18-month trials of lamotrigine and lithium maintenance in bipolar I disorder. *Journal of Clinical Psychiatry, 65*, 432–441.

Hayes, J. F., Marston, L., Walters, K., et al. (2016a). Lithium vs. valproate vs. olanzapine vs. quetiapine as maintenance monotherapy for bipolar disorder: A population-based UK cohort study using electronic health records. *World Psychiatry, 15*, 53–58.

Hayes, J. F., Pitman, A., Marston, L., et al. (2016b). Self-harm, unintentional injury, and suicide in bipolar disorder during maintenance mood stabilizer treatment: A UK population-based electronic health records study. *JAMA Psychiatry, 73*, 630–637.

Hayes, J. F., Marston, L., Walters, K., et al. (2016c). Adverse renal, endocrine, hepatic, and metabolic events during maintenance mood stabilizer treatment for bipolar disorder: A population-based cohort study. *PLoS Medicine, 13*, e1002058.

Hirschfeld, R. M., Baker, J. D., Wozniak, P., et al. (2003). The safety and early efficacy of oral-loaded divalproex versus standard-titration divalproex, lithium, olanzapine, and placebo in the treatment of acute mania associated with bipolar disorder. *Journal of Clinical Psychiatry, 64*, 841–846.

Hirschfeld, R. M., Keck, P. E., Jr., Kramer, M., et al. (2004). Rapid antimanic effect of risperidone monotherapy: A 3-week multicenter, double-blind, placebo-controlled trial. *American Journal of Psychiatry, 161*, 1057–1065.

Hlaing, P. M., Isoardi, K. Z., Page, C. B., et al. (2020). Neurotoxicity in chronic lithium poisoning. *Internal Medicine Journal, 50*, 427–432.

Hou, L., Heilbronner, U., Degenhardt, F., et al. (2016). Genetic variants associated with response to lithium treatment in bipolar disorder: A genome-wide association study. *The Lancet, 387*, 1085–1093.

Hussein, Z., & Posner, J. (1997). Population pharmacokinetics of lamotrigine monotherapy in patients with epilepsy: Retrospective analysis of routine monitoring data. *British Journal of Clinical Pharmacology, 43*(5), 457–465. https://doi.org/10.1046/j.1365-2125.1997.00594.x

Jann, M. W., & Penzak, S. R. (2018). Long-acting injectable second-generation antipsychotics: An update and comparison between agents. *CNS Drugs, 32*, 241–257.

Kato, T., Ishigooka, J., Miyajima, M., et al. (2020). Double-blind, placebo-controlled study of lurasidone monotherapy for the treatment of bipolar I depression. *Psychiatry and Clinical Neurosciences, 74*(12), 635–644. https://doi.org/10.1111/pcn.13137

Keck, P. E., Jr., Versiani, M., Potkin, S., et al. (2003a). Ziprasidone in the treatment of acute bipolar mania: A three-week, placebo-controlled, double-blind, randomized trial. *American Journal of Psychiatry, 160*, 741–748.

Keck, P. E., Jr., Marcus, R., Tourkodimitris, S., et al. (2003b). A placebo-controlled, double-blind study of the efficacy and safety of aripiprazole in patients with acute bipolar mania. *American Journal of Psychiatry, 160*, 1651–1658.

Keck, P. E., Jr., Calabrese, J. R., McQuade, R. D., et al. (2006). A randomized, double-blind, placebo-controlled 26-week trial of aripiprazole in recently manic patients with bipolar I disorder. *Journal of Clinical Psychiatry, 67*, 626–637.

Keller, M. B., Lavori, P. W., Kane, J. M., et al. (1992). Subsyndromal symptoms in bipolar disorder. A comparison of standard and low serum levels of lithium. *Archives of General Psychiatry, 49*, 371–376.

Keramatian, K., Chakrabarty, T., & Yatham, L. N. (2019). Long-acting injectable second-generation/atypical antipsychotics for the management of bipolar disorder: A systematic review. *CNS Drugs, 33*, 431–456.

Kessing, L. V., Bauer, M., Nolen, W. A., et al. (2018). Effectiveness of maintenance therapy of lithium vs other mood stabilizers in monotherapy and in combinations: A systematic review of evidence from observational studies. *Bipolar Disorders, 20*, 419. https://doi.org/10.1111/bdi.12623. Online ahead of print.

Khanna, S., Vieta, E., Lyons, B., et al. (2005). Risperidone in the treatment of acute mania: Double-blind, placebo-controlled study. *British Journal of Psychiatry, 187*, 229–234.

Kishi, T., Ikuta, T., Matsuda, Y., et al. (2020a). Mood stabilizers and/or antipsychotics for bipolar disorder in the maintenance phase: A systematic review and network meta-analysis of randomized controlled trials. *Molecular Psychiatry*. https://doi.org/10.1038/s41380-020-00946-6. Online ahead of print.

Kishi, T., Matsuda, Y., Sakuma, K., et al. (2020b). Recurrence rates in stable bipolar disorder patients after drug discontinuation v. drug maintenance: A systematic review and meta-analysis. *Psychological Medicine, 13*, 1–9. https://doi.org/10.1017/S0033291720003505. Online ahead of print.

Kripalani, M., Shawcross, J., Reilly, J., et al. (2009). Lithium and chronic kidney disease. *British Medical Journal, 339*, b2452.

Lähteenvuo, M., Tanskanen, A., Taipale, H., et al. (2018). Real-world effectiveness of pharmacologic treatments for the prevention of rehospitalization in a Finnish nationwide cohort of patients with bipolar disorder. *JAMA Psychiatry, 75*, 347–355.

Loebel, A., Cucchiaro, J., Silva, R., et al. (2014a). Lurasidone monotherapy in the treatment of bipolar I depression: A randomized, double-blind, placebo-controlled study. *American Journal of Psychiatry, 171*(2), 160–168.

Loebel, A., Cucchiaro, J., Silva, R., et al. (2014b). Lurasidone as adjunctive therapy with lithium or valproate for the treatment of bipolar I depression: A randomized, double-blind, placebo-controlled study. *American Journal of Psychiatry, 171*(2), 169–177.

Lombardo, I., Sachs, G., Kolluri, S., et al. (2012). Two 6-week, randomized, double-blind, placebo-controlled studies of ziprasidone in outpatients with bipolar I depression: Did baseline characteristics impact trial outcome? *Journal of Clinical Psychopharmacology, 32*, 470–478.

López-Muñoz, F., Shen, W. W., D'Ocon, P., et al. (2018). A history of the pharmacological treatment of bipolar disorder. *International Journal of Molecular Sciences, 19*, 2143.

Macfadden, W., Alphs, L., Haskins, J. T., et al. (2009). A randomized, doubleblind, placebo-controlled study of maintenance treatment with adjunctive risperidone long-acting therapy in patients with bipolar I disorder who relapse frequently. *Bipolar Disorders, 11*, 827–839.

McElroy, S. L., Weisler, R. H., Chang, W., et al. (2010). A double-blind, placebo-controlled study of quetiapine and paroxetine as monotherapy in adults with bipolar depression (EMBOLDEN II). *Journal of Clinical Psychiatry, 71*, 163–174.

McIntyre, R. S., Brecher, M., Paulsson, B., et al. (2005). Quetiapine or haloperidol as monotherapy for bipolar mania – A 12-week, double-blind, randomized, parallel-group, placebo-controlled trial. *European Neuropsychopharmacology, 15*, 573–585.

McIntyre, R. S., Cohen, M., Zhao, J., et al. (2009). A 3-week, randomized, placebo-controlled trial of asenapine in the treatment of acute mania in bipolar mania and mixed states. *Bipolar Disorders, 11*, 673–686.

McIntyre, R. S., Cohen, M., Zhao, J., et al. (2010). Asenapine in the treatment of acute mania in bipolar I disorder: A randomized, double-blind, placebo-controlled trial. *Journal of Affective Disorders, 122*, 27–38.

McIntyre, R. S., Berk, M., Brietzke, E., et al. (2020). Bipolar disorders. *Lancet, 396*, 1841–1856.

McKnight, R. F., Adida, M., Budge, K., et al. (2012). Lithium toxicity profile: A systematic review and meta-analysis. *Lancet, 379*, 721–728.

Merikangas, K. R., Jin, R., He, J. P., et al. (2011). Prevalence and correlates of bipolar spectrum disorder in the world mental health survey initiative. *Archives of General Psychiatry, 68*, 241–251.

Miklowitz, D. J., Efthimiou, O., Furukawa, T. A., et al. (2021). Adjunctive psychotherapy for bipolar disorder: A systematic review and component network meta-analysis. *JAMA Psychiatry, 78*, 141–150.

Miller, J. N., & Black, D. W. (2020). Bipolar disorder and suicide: A review. *Current Psychiatry Reports, 22*, 6.

Mitchell, P. B. (1999). On the 50th anniversary of John Cade's discovery of the anti-manic effect of lithium. *Australian and New Zealand Journal of Psychiatry, 33*, 623–628.

Mitchell, P. B., & Hadzi-Pavlovic, D. (1999). John Cade and the discovery of lithium treatment for manic depressive illness. *Medical Journal of Australia, 171*, 262–264.

Mitchell, P. B., Goodwin, G. M., Johnson, G. F., et al. (2008). Diagnostic guidelines for bipolar depression: A probabilistic approach. *Bipolar Disorders, 10*, 144–152.

Mitchell, P. B., Johnston, A. K., Frankland, A., et al. (2013). Bipolar disorder in a national survey using the World Mental Health Version of the Composite International Diagnostic Interview: The impact of differing diagnostic algorithms. *Acta Psychiatrica Scandanivica, 127*, 381–393.

Muzina, D. J., El-Sayegh, S., & Calabrese, J. R. (2002). Antiepileptic drugs in psychiatry–focus on randomized controlled trials. *Epilepsy Research, 50*, 195–202.

Muzina, D. J., Elhaj, O., Gajwani, P., et al. (2005). Lamotrigine and antiepileptic drugs as mood stabilizers in bipolar disorder. *Acta Psychiatrica Scandinavica. Supplementum, 111*(426), 21–28.

Muzina, D. J., Gao, K., Kemp, D. E., et al. (2011). Acute efficacy of divalproex sodium versus placebo in mood stabilizer-naive bipolar I or II depression: A double-blind, randomized, placebo-controlled trial. *Journal of Clinical Psychiatry, 72*, 813–819.

Nemeroff, C. B., Evans, D. L., Gyulai, L., et al. (2001). Double-blind, placebo-controlled comparison of imipramine and paroxetine in the treatment of bipolar depression. *American Journal of Psychiatry, 158*, 906–912.

Nierenberg, A. A., McIntyre, R. S., & Sachs, G. S. (2015). Improving outcomes in patients with bipolar depression: A comprehensive review. *Journal of Clinical Psychiatry, 76*, e10.

Nierenberg, A. A., McElroy, S. L., Friedman, E. S., Ketter, T. A., Shelton, R. C., Deckersbach, T., et al. (2016). Bipolar CHOICE (Clinical Health Outcomes Initiative in Comparative Effectiveness): a pragmatic 6-month trial of lithium versus quetiapine for bipolar disorder. *Journal of Clinical Psychiatry, 77*(1), 90–99. https://doi.org/10.4088/JCP.14m09349

Pacchiarotti, I., Bond, D. J., Baldessarini, R. J., et al. (2013). The International Society for Bipolar Disorders (ISBD) task force report on antidepressant use in bipolar disorders. *American Journal of Psychiatry, 170*, 1249–1262.

Pakpour, A. H., Modabbernia, A., Lin, C.-Y., et al. (2017). Promoting medication adherence among patients with bipolar disorder: A multicenter randomized controlled trial of a multifaceted intervention. *Psychological Medicine, 47*, 2528–2539.

Patorno, E., Huybrechts, K. F., Bateman, B. T., et al. (2017). Lithium use in pregnancy and the risk of cardiac malformations. *New England Journal of Medicine, 376*, 2245–2254.

Plenge, P., Stensgaard, A., Jensen, H. V., et al. (1994). 24-hour lithium concentration in human brain studied by Li-7 magnetic resonance spectroscopy. *Biological Psychiatry, 36*, 511–516.

Rhee, T. G., Olfson, M., Nierenberg, A. A., et al. (2020). 20-year trends in the pharmacologic treatment of bipolar disorder by psychiatrists in outpatient care settings. *American Journal of Psychiatry, 177*, 706–715.

Sachs, G. S., Grossman, F., Ghaemi, S. N., et al. (2002). Combination of a mood stabilizer with risperidone or haloperidol for treatment of acute mania: A double-blind, placebo-controlled comparison of efficacy and safety. *American Journal of Psychiatry, 159*, 1146–1154.

Sachs, G., Chengappa, K. N., Suppes, T., et al. (2004). Quetiapine with lithium or divalproex for the treatment of bipolar mania: A randomized, doubleblind, placebo-controlled study. *Bipolar Disorders, 6*, 213–223.

Sachs, G., Sanchez, R., Marcus, R., et al. (2006). Aripiprazole in the treatment of acute manic or mixed episodes in patients with bipolar I disorder: A 3-week placebo-controlled study. *Journal of Psychopharmacology, 20*, 536–546.

Sachs, G. S., Nierenberg, A. A., Calabrese, J. R., et al. (2007). Effectiveness of adjunctive antidepressant treatment for bipolar depression. *New England Journal of Medicine, 356*, 1–12.

Sachs, G. S., Ice, K. S., Chappell, P. B., et al. (2011). Efficacy and safety of adjunctive oral ziprasidone for acute treatment of depression in patients with bipolar I disorder: A randomized, double-blind, placebo-controlled trial. *Journal of Clinical Psychiatry, 72*, 1413–1422.

Sajatovic, M., Tatsuoka, C., Cassidy, K. A., et al. (2018). A 6-month, prospective, randomized controlled trial of customized adherence enhancement versus bipolar-specific educational control in poorly adherent individuals with bipolar disorder. *Journal of Clinical Psychiatry, 79*, 17m12036.

Salcedo, S., McMaster, K. J., & Johnson, S. L. (2017). Disparities in treatment and service utilization among Hispanics and non-Hispanic whites with bipolar disorder. *Journal of Racial and Ethnic Health Disparities, 4*, 354–363.

Sanghani, H. R., Jagannath, A., Humberstone, T., et al. (2020). Patient fibroblast circadian rhythms predict lithium sensitivity in bipolar disorder. *Molecular Psychiatry, 26*, 5252. https://doi.org/10.1038/s41380-020-0769-6. Online ahead of print.

Saraf, G., Pinto, J. V., & Yatham, L. N. (2019). Efficacy and safety of cariprazine in the treatment of bipolar disorder. *Expert Opinion in Pharmacotherapy, 20*, 2063–2072.

Sawai, Y., Okamoto, T., Muranaka, Y., et al. (2019). In vivo evaluation of the effect of lithium on peripheral circadian clocks by real-time monitoring of clock gene expression in near-freely moving mice. *Scientific Reports, 9*, 10909.

Schoot, T. S., Molmans, T. H. J., Grootens, T. P., et al. (2020). Systematic review and practical guideline for the prevention and management of the renal side effects of lithium therapy. *European Neuropsychopharmacology, 31*, 16–32.

Schou, M., Juel-Nielsen, N., Stromgren, E., et al. (1954). The treatment of manic psychoses by the administration of lithium salts. *Journal of Neurology, Neurosurgery and Psychiatry, 17*, 250–260.

Segal, J., Berk, M., & Brook, S. (1998). Risperidone compared with both lithium and haloperidol in mania: A double-blind randomized controlled trial. *Clinical Neuropharmacology, 21*, 176–180.

Shelton, R. C., & Stahl, S. M. (2004). Risperidone and paroxetine given singly and in combination for bipolar depression. *Journal of Clinical Psychiatry, 65*, 1715–1719.

Smith, L. A., Cornelius, V., Warnock, A., et al. (2007). Acute bipolar mania: A systematic review and meta-analysis of co-therapy vs. monotherapy. *Acta Psychiatrica Scandinavica, 115*, 12–20.

Stout, J., Hozer, F., Coste, A., et al. (2020). Accumulation of lithium in the hippocampus of patients with bipolar disorder: A lithium-7 magnetic resonance imaging study at 7 Tesla. *Biological Psychiatry, 88*, 426–433.

Suppes, T., Marangell, L. B., Bernstein, I. H., et al. (2008). A single blind comparison of lithium and lamotrigine for the treatment of bipolar II depression. *Journal of Affective Disorders, 111*, 334–443.

Suppes, T., Vieta, E., Liu, S., et al. (2009). Maintenance treatment for patients with bipolar I disorder: Results from a North American study of quetiapine in combination with lithium or divalproex (trial 127). *American Journal of Psychiatry, 166*, 476–488.

Suppes, T., Datto, C., Minkwitz, M., et al. (2010). Effectiveness of the extended release formulation of quetiapine as monotherapy for the treatment of acute bipolar depression. *Journal of Affective Disorders, 121*, 106–115.

Thase, M. E., Macfadden, W., Weisler, R. H., et al. (2006). Efficacy of quetiapine monotherapy in bipolar I and II depression: A double-blind, placebo-controlled study (the BOLDER II study). *Journal of Clinical Psychopharmacology, 26*, 600–609.

Thase, M. E., Jonas, A., Khan, A., et al. (2008). Aripiprazole monotherapy in nonpsychotic bipolar I depression: Results of 2 randomized, placebo-controlled studies. *Journal of Clinical Psychopharmacology, 28*, 13–20.

Tohen, M., Sanger, T. M., McElroy, S. L., et al. (1999). Olanzapine versus placebo in the treatment of acute mania. Olanzapine HGEH Study Group. *American Journal of Psychiatry, 156*, 702–709.

Tohen, M., Jacobs, T. G., Grundy, S. L., et al. (2000). Efficacy of olanzapine in acute bipolar mania: A double-blind, placebo-controlled study. The Olanzapine HGGW Study Group. *Archives of General Psychiatry, 57*, 841–849.

Tohen, M., Chengappa, K. N., Suppes, T., et al. (2002). Efficacy of olanzapine in combination with valproate or lithium in the treatment of mania in patients partially nonresponsive to valproate or lithium monotherapy. *Archives of General Psychiatry, 59*, 62–69.

Tohen, M., Ketter, T. A., Zarate, C. A., et al. (2003a). Olanzapine versus divalproex sodium for the treatment of acute mania and maintenance of remission: A 47-week study. *American Journal of Psychiatry, 160*, 1263–1271.

Tohen, M., Vieta, E., Calabrese, J., et al. (2003b). Efficacy of olanzapine and olanzapine-fluoxetine combination in the treatment of bipolar I depression. *Archives of General Psychiatry, 60*, 1079–1088.

Tohen, M., Chengappa, K. N., Suppes, T., et al. (2004). Relapse prevention in bipolar I disorder: 18-month comparison of olanzapine plus mood stabilizer v. mood stabilizer alone. *British Journal of Psychiatry, 184*, 337–345.

Tohen, M., Greil, W., Calabrese, J. R., et al. (2005). Olanzapine versus lithium in the maintenance treatment of bipolar disorder: A 12-month, randomized, double-blind, controlled clinical trial. *American Journal of Psychiatry, 162*, 1281–1290.

Tohen, M., Calabrese, J. R., Sachs, G. S., et al. (2006). Randomized, placebo-controlled trial of olanzapine as maintenance therapy in patients with bipolar I disorder responding to acute treatment with olanzapine. *American Journal of Psychiatry, 163*, 247–256.

van der Loos, M. L., Mulder, P. G., Hartong, E. G., et al. (2009). Efficacy and safety of lamotrigine as add-on treatment to lithium in bipolar depression: A multicenter, double-blind, placebo-controlled trial. *Journal of Clinical Psychiatry, 70*, 223–231.

Vieta, E., Suppes, T., Eggens, I., et al. (2008). Efficacy and safety of quetiapine in combination with lithium or divalproex for maintenance of patients with bipolar I disorder (international trial 126). *Journal of Affective Disorders, 109*, 251–263.

Vieta, E., Nuamah, I. F., Lim, P., et al. (2010). A randomized, placebo-and active-controlled study of paliperidone extended release for the treatment of acute manic and mixed episodes of bipolar I disorder. *Bipolar Disorders, 12*, 230–243.

Viktorin, A., Lichtenstein, P., Thase, M. E., et al. (2014). The risk of switch to mania in patients with bipolar disorder during treatment with an antidepressant alone and in combination with a mood stabilizer. *American Journal of Psychiatry, 171*, 1067–1073.

Weisler, R. H., Kaiali, A. H., & Ketter, T. A. (2004). A multicenter, randomized, double-blind, placebo-controlled trial of extended-release carbamazepine capsules as monotherapy for bipolar disorder patients with manic or mixed episodes. *Journal of Clinical Psychiatry, 65*, 478–484.

Weisler, R. H., Keck, P. E., Jr., Swann, A. C., et al. (2005). Extended-release carbamazepine capsules as monotherapy for acute mania in bipolar disorder: A multicenter, randomized, double-blind, placebo-controlled trial. *Journal of Clinical Psychiatry, 66*, 323–330.

Yatham, L. N., Grossman, F., Augustyns, I., et al. (2003). Mood stabilizers plus risperidone or placebo in the treatment of acute mania. International, double-blind, randomized controlled trial. *British Journal of Psychiatry, 182*, 141–147.

Young, A. H., McElroy, S. L., Bauer, M., et al. (2010). A double-blind, placebo-controlled study of quetiapine and lithium monotherapy in adults in the acute phase of bipolar depression (EMBOLDEN I). *Journal of Clinical Psychiatry, 71*, 150–162.i.

Zohar, J., Stahl, S., Moller, H. J., Blier, P., Kupfer, D., Yamawaki, S., … Nutt, D. (2015). A review of the current nomenclature for psychotropic agents and an introduction to the Neuroscience-based Nomenclature. *European Neuropsychopharmacology, 25*(12), 2318–2325.

Zui Narita, Michiko Fujimoto, Elizabeth Winter, Paul Nestadt, and Akira Sawa

Contents

Previous edition authors were Tammy C. Saah, Steven J. Garlow, Mark Hyman Rapaport, Deidre M. Edwards, Kathryn L. Hale and Rachel E. Maddux

Z. Narita
Department of Behavioral Medicine, National Institute of Mental Health, National Center of Neurology and Psychiatry, Tokyo, Japan

M. Fujimoto
Department of Psychiatry, Osaka University Graduate School of Medicine, Osaka, Japan
e-mail: mfujimoto@psy.med.osaka-u.ac.jp

E. Winter
Department of Veterans Affairs, Office of Inspector General, Baltimore, MD, USA

P. Nestadt
Departments of Psychiatry, Johns Hopkins University School of Medicine, Baltimore, MD, USA

Departments of Mental Health, Johns Hopkins Bloomberg School of Public Health, Baltimore, MD, USA
e-mail: pn@jhmi.edu

A. Sawa (✉)
Departments of Psychiatry, Johns Hopkins University School of Medicine, Baltimore, MD, USA

Departments of Mental Health, Johns Hopkins Bloomberg School of Public Health, Baltimore, MD, USA

Departments of Neuroscience, Johns Hopkins University School of Medicine, Baltimore, MD, USA

Departments of Biomedical Engineering, Johns Hopkins University School of Medicine, Baltimore, MD, USA

Departments of Genetic Medicine, Johns Hopkins University School of Medicine, Baltimore, MD, USA

Departments of Pharmacology, Johns Hopkins University School of Medicine, Baltimore, MD, USA
e-mail: asawa1@jhmi.edu

© Springer Nature Switzerland AG 2024
A. Tasman et al. (eds.), *Tasman's Psychiatry*,
https://doi.org/10.1007/978-3-030-51366-5_136

Abstract

There has been an exponential increase in the number of medications demonstrated to be effective for the treatment of anxiety and anxiety disorders. This chapter provides a brief review of the history and evolution of psychopharmacologic strategies. Data for efficacy and adverse effects of these medications for different anxiety disorders are presented based on clinical trial reports so that the reader can compare and contrast the advantages and disadvantages of different classes of medications for treatment of each anxiety disorder described in the International Classification of Diseases and the Diagnostics and Statistical Manual of Mental Disorders over the past three decades. Specifically, this chapter describes current evidence of medications for anxiety disorders for five disorders: generalized anxiety disorder, social anxiety disorder, panic disorder, posttraumatic stress disorder, and obsessive-compulsive disorder. For some drugs, this chapter includes the "Special Populations" sections that show evidence specific to children, adolescents, and elderly patients if available. This chapter also provides a table that summarizes the efficacy of each drug for these five disorders. The other table summarizes pharmacological properties of medications commonly used to treat anxiety disorders. At present, clinicians have a wide array of medications available for the treatment of anxiety and anxiety disorders. The breadth of accessible treatment options greatly facilitates our ability to help patients. Nevertheless, appropriate diagnosis and rapport are still the foundations of any pharmacological intervention that is made with the patients.

Keywords

SSRI · SNRI · TCA · MAOI · Benzodiazepine · General anxiety disorder · Social anxiety disorder · Panic disorder · Posttraumatic stress disorder · Obsessive-compulsive disorder

Introduction

There has been an exponential increase in the number of medications demonstrated to be effective for the treatment of anxiety and anxiety disorders (Tables 1 and 2). This chapter provides a brief review of the history and evolution of these psychopharmacologic strategies. The efficacy and adverse effects data from clinical trials with different disorders are presented so that the reader can compare and contrast the advantages and disadvantages of different classes of medications for the treatment of each anxiety disorder described in the International Classification of Diseases (ICD-11) and the Diagnostics and Statistical Manual of Mental Disorders (DSM-5) over the past three decades (DSM-III, DSM-III-R, DSM-IV, DSM-IV-TR, DSM-5). As the definitions of specific anxiety disorders continue to evolve, additional research will be conducted to develop more selective and efficacious treatments (Table 2).

In the late nineteenth century, management of anxiety progressed from alcohol, bromides, and opiates to barbiturates, which were developed in the early twentieth century. Barbiturates were effective in decreasing anxiety symptoms, but they were addictive and lethal in overdose. There was continued development of medications for anxiety disorders such as meprobamate (effective but addictive and lethal in overdose) and the H_1 antihistamine hydroxyzine, which was less effective against the symptoms of anxiety than meprobamate and barbiturates and very sedating. The next major advance in anxiolytic therapy was Klein and Fink's work (1962) demonstrating that some monoamine enhancers (tricyclic antidepressants; TCAs) with varying degrees of inhibition of the reuptake of serotonin and norepinephrine and antagonism of muscarinic cholinergic, H_1 histamine, and α_1-adrenergic receptors, were useful in the treatment of panic disorder. This was rapidly followed by studies reporting that monoamine oxidase inhibitors (MAOIs), which increase the amounts of norepinephrine, serotonin, and dopamine available for release, could be used to treat panic disorder.

However, the major advancement in the field of medications for anxiety disorders in the 1960s

Table 1 Efficacy of psychotropic medications for the treatment of anxiety disorders

	Generalized anxiety disorder	Social anxiety disorder	Panic disorder	Posttraumatic stress disorder	Obsessive–compulsive disorder
Proven efficacy	SSRIs	SSRIs	SSRIs	SSRIs	SSRIs
	Venlafaxine Duloxetine	MAOIs/RIMAs	Venlafaxine		Clomipramine
	Monoamine Enhancers (TCAs)	Benzodiazepines	Mirtazapine		
	Trazodone	Venlafaxine	MAOIs		
	Benzodiazepines		Monoamine Enhancers (TCAs)		
	Buspirone		Benzodiazepines		
Some evidence	Nefazodone	Mirtazapine	Duloxetine	Mirtazapine	MAOIs
	Mirtazapine	Bupropion	Reboxetine	Venlafaxine	Venlafaxine
	Pregabalin	Clonazepam[a]	Nefazodone	Duloxetine	Topiramate
	Risperidone	Nefazodone	Vortioxetine	MAOIs	Lamotrigine
	Olanzapine	Gabapentin	Pagoclone	TCAs	Pregabalin
	Tiagabine	Pregabalin	Gabapentin	Nefazodone	Pindolol[a]
	Topiramate	Topiramate	Tiagabine	Trazodone	Amisulpiride
	Lesopitron	Levetiracetam	Valproic acid	Propranolol	Olanzapine[a]
	Riluzole	Olanzapine	Olanzapine[a]	Tiagabine	Quetiapine
	Opipramol Hydroxyzine		Levetiracetam	Topiramate Lamotrigine Risperidone Olanzapine Fluphenazine	Risperidone[a] Aripiprazole Memantine
				Aripiprazole Asenapine Prazosin	
Not effective		Monoamine Enhancers (TCAs)	Bupropion	Bupropion	Bupropion
		Buspirone	Bupropion		Mirtazapine
		Pindolol	Trazodone		Monoamine Enhancers (TCAs)
		Valproic acid			Trazodone Buspirone Haloperidol
No data	MAOIs Bupropion	Trazodone			Nefazodone

[a]Listed medication augmented with SSRI

was the development and approval of benzodiazepines, a group of agents that bind to a site on the γ-aminobutyric acid (GABA) receptor complex (primarily GABA$_A$) and allosterically enhance GABA-mediated inhibition. These agents were safer than barbiturates and meprobamate, had a rapid onset of action (so patients felt better quickly), and had a broad spectrum of efficacy ranging from non-pathological situational anxiety to anxiety disorders. Benzodiazepines with different absorption times and half-lives were developed and have been valuable not only for treating anxiety and anxiety disorders but also for treating seizure disorders, insomnia, and alcohol withdrawal. Unfortunately, like alcohol, opiates, bromides, and barbiturates, prolonged use of

Table 2 Summary of pharmacological properties of medications commonly used to treat anxiety

Agent	Mechanism of Action	Onset of Action	Titration	Abuse Liability	Need for Discontinuation Taper	Potential for Withdrawal Syndrome	Probability of Lethality in Overdose
Sertraline	SSRI	Delayed (in 2 wk)	Yes	Very low	Yes, but not mandatory	Very low	Low
Paroxetine	SSRI	Delayed (in 2 wk)	Yes	Very low	Yes	Moderate	Low
Fluvoxamine	SSRI	Delayed (in 2 wk)	Yes	Very low	Yes, but not mandatory	Very low	Low
Fluoxetine	SSRI	Delayed (in 2 wk)	Yes	Very low	No	Lowest	Low
Citalopram	SSRI	Delayed (in 2 wk)	Yes	Very low	Yes, but not mandatory	Very low	Low
Escitalopram	SSRI	Delayed (in 2 wk)	Sometimes	Very low	Yes, probably not mandatory	Very low	Low
Venlafaxine	SNRI	Delayed (in 2 wk)	Yes	Very low	Yes	Moderate	Low
Duloxetine	SNRI	Delayed (in 2 wk)	Yes	Very low	Yes	Moderate	Low
Mirtazapine	α_2-Antagonist	Delayed	Yes	Very low	Yes	Moderate	Low
Nefazodone	$5HT_2$ blockade	Delayed	Yes	Very low	Yes	Moderate	Low
Bupropion	Norepinephrine?	Delayed	Yes	Very low	Yes	Low	Low
Monoamine Enhancers (TCAS)	Reuptake blockade	Delayed (2 wk)	Yes	Very low	Yes	Moderate	Moderate–high
MAOIs	Monoamine inhibition	Delayed (2 wk)	Yes	Very low	Yes	Moderate	Moderate–high
Buspirone	$5HT_{1A}$ partial agonist	Delayed (2 wk)	Yes	Very low	Yes	Low–moderate	Low
Clonazepam	Modulates $GABA_A$ receptor	Rapid	Yes	Moderate	Yes	Moderate–high	Low
Alprazolam	Modulates $GABA_A$ receptor	Very rapid	Yes	Moderate	Yes	High	Low
Alprazolam XR	Modulates $GABA_A$ receptor	Rapid	Yes	Moderate	Yes	Moderate–high	Low

(continued)

Table 2 (continued)

Agent	Mechanism of Action	Onset of Action	Titration	Abuse Liability	Need for Discontinuation Taper	Potential for Withdrawal Syndrome	Probability of Lethality in Overdose
Lorazepam	Modulates $GABA_A$ receptor	Very rapid	Yes	Moderate	Yes	Moderate–high	Low
Diazepam	Modulates $GABA_A$ receptor	Rapid	Yes	Moderate	Yes	Moderate	Low
Beta-blockers	Blocks β-adrenergic receptors	Rapid	Sometimes	Low	No (acute use)	No (acute use)	Low–moderate
Gabapentin	Not fully known but GABAergic	Moderate(d)	Yes	Low	Yes	Low–moderate	Low
Pregabalin	Not fully known but GABAergic	Rapid	Yes	Low	Yes	Low–moderate	Low
Aripiprazole	D_2, D_3, $5\text{-}HT_{1A}$ and $5\text{-}HT_{2A}$ blockade	Delayed (2 wk)	Yes	Unknown	Yes	Unknown	Low–moderate
Risperidone	$5\text{-}HT_2$ and D_2 blockade	Rapid	Probably	Low	Yes	Low	Low
Haloperidol	D_2 blockade	Rapid	Yes	Low	Yes	Low	Low
Olanzapine	$5\text{-}HT_2$ and D_2 blockade	Rapid	Probably	Low	Yes	Low	Low
Quetiapine	$5\text{-}HT_2$ and D_2 blockade	Rapid	Yes	Low	Yes	Low	Low

benzodiazepines leads to tolerance, potential physiological dependence and withdrawal, and possible addiction. The next major class of agents approved was azapyrones, of which buspirone is the best known. This agent is a partial agonist at $5-HT_{1A}$ autoreceptors (with 50% intrinsic activity) and an α_2-adrenergic antagonist.

There was a cascade of GABA enhancers (anxiolytic) research in the 1990s. As a class, selective serotonin reuptake inhibitors (SSRIs) were demonstrated to be efficacious treatments for most of the anxiety disorders described in the DSM-5. Although these agents have a delayed onset when contrasted with benzodiazepines, they have a broader spectrum of action, no problems with dependence, and are much less likely to cause withdrawal symptoms. The 1990s also saw the approval of venlafaxine as a treatment for generalized anxiety disorder. One of the most intriguing evolutions in medications for anxiety (anxiolytics) development has been the ongoing research investigating the use of anticonvulsant medications and dopamine serotonin blockers (second-generation antipsychotics) for the treatment of anxiety disorders.

This chapter describes current evidence of medications for anxiety (anxiolytics) for five disorders: generalized anxiety disorder, social anxiety disorder, panic disorder, posttraumatic stress disorder, and obsessive-compulsive disorder. For some drugs, "Special Populations" sections are included, which describe evidence specific to children, adolescents, and elderly patients if available. "Clinical Vignette," "Future Directions," and "Conclusion" are also included in the last portion of this chapter.

A General Approach to Using Medication with Anxious Patients

A complete medical and psychiatric history is essential for making an appropriate diagnostic formulation, while also demonstrating an interest in the patient's situation and the "reality" of their experience. Patients seeking treatment for anxiety require reassurance that their physicians hear their concerns and have the ability to provide relief. Formulating an appropriate differential diagnosis is critical to the success of any psychopharmacological intervention.

The diagnosis suggests the class of medication to be trialed and the length of pharmacotherapy. Potential differential diagnoses for patients with anxiety disorders include the following: adjustment disorders secondary to life stressors, anxiety disorders secondary to a medical condition, symptoms of anxiety secondary to a medical condition, anxiety secondary to alcohol or substance abuse or dependence, generalized anxiety disorder (GAD), panic disorder (PD), social anxiety disorder (SAD), specific phobia, posttraumatic stress disorder (PTSD), and obsessive-compulsive disorder (OCD).

Discussing the diagnostic formulation with the patient is an important intervention that validates the patient's concerns and facilitates the patient's commitment to the treatment plan. Forming this therapeutic alliance is crucial because patients with anxiety disorders may be reticent to take medication due to ruminations about side effects and a greater likelihood to experience somatic preoccupations and heightened somatic sensitivity. A collaborative approach where physicians and patients form a "team" to monitor both the potential benefits and the liabilities of any medication intervention empowers the patient and enhances adherence. A clinical pearl is "to start low and go slow" when initiating pharmacological treatment for patients with anxiety disorders. Interestingly, although patients with anxiety disorders frequently require more gradual initial titration schedules, patients usually need greater maintenance dosages of medications for depression than dosages commonly used to treat major depressive disorder.

Generalized Anxiety Disorder (GAD)

Selective Serotonin Reuptake Inhibitors

A number of trials have been carried out to assess the efficacy of SSRIs for the treatment of patients with GAD, with several studies demonstrating that paroxetine is an effective short-term and continuous treatment for patients with GAD. Pollack et al. (2001) reported an 8-week randomized, double-blind, placebo-controlled, flexible-dosage trial investigating the efficacy of paroxetine in 326 adult outpatients with GAD. Inclusion criteria both at screen and baseline, following a 1–week single-blind, placebo run-in, were a Hamilton Anxiety Rating Scale (HAM-A) total score of ≥ 20 and a score of ≥ 2 on HAM-A items 1 (anxious mood) and 2 (tension). Patients assigned to paroxetine started treatment at 10 mg/day for the first week and were titrated to 20 mg/day by week 2. After the second week, patients tolerating the medication could have their medication increased every 7 days by 10 mg/day to a maximum of 50 mg/day (mean daily dose 26.8 ± 7.5 mg/day). Response to treatment was defined by a score of (1) "very much improved" or (2) "much improved" on the CGI-Global Improvement Scale. A HAM-A total score of 7 or less defined remission. About 74% of paroxetine patients met the response criteria versus 55.6% of placebo-treated patients, and 43% of paroxetine-treated patients met remission criteria compared with 26.3% of placebo-treated patients.

Rickels et al. (2003) described an 8-week randomized, double-blind, placebo-controlled trial of paroxetine 20 or 40 mg/day to treat GAD in 566 outpatients with DSM-IV-TR diagnosis of GAD and a score of ≥ 20 on the HAM-A. The primary outcome measure was the change from baseline in total score on the HAM-A. Response was defined as a rating of "very much improved" or "much improved" on the CGI. Remission was defined as a HAM-A total score of ≤ 7. Both the 20 and 40 mg/day paroxetine-treated groups demonstrated statistically significant reductions in HAM-A total scores compared with the placebo group. Response rates were 62% and 68% for the 20 and 40 mg/day paroxetine-treated groups, respectively, compared with 46% for the placebo group. Remission rates were 30% and 36% of patients in the 20 and 40 mg/day paroxetine groups, respectively, compared with 20% given placebo.

Stocchi et al. (2003) reported a 24-week double-blind GAD study contrasting paroxetine (20–50 mg/day) and placebo in 652 adult outpatients with a DSM-IV-TR GAD diagnosis and a CGI-S score of ≥ 4. Patients whose CGIS scores decreased by at least 2 points to ≤ 3 after 8 weeks of double-blind paroxetine were randomly assigned to

double-blind treatment with paroxetine ($n = 278$) or placebo ($n = 288$) for an additional 24 weeks. The primary efficacy measure was the proportion of the patients relapsing (an increase in CGI-S score of at least 2 points to a score of ≥ 4 or withdrawal due to lack of efficacy) during double-blind treatment. Significantly fewer paroxetine-treated subjects relapsed during the 24-week double-blind phase (10.9% vs. 39.9%; $p > 0.001$). Subjects randomized to placebo discontinuation were almost five times more likely to relapse than those continuously on double-blind paroxetine [estimated hazard ratio $= 0.213$; 95% confidence interval (CI) $= 0.1$–0.3; $p < 0.001$].

Escitalopram has also been extensively studied as a treatment for GAD. Davidson et al. (2004) conducted an 8-week randomized, double-blind, placebo-controlled, multicenter, flexible-dose study of escitalopram (10 mg/day for the first 4 weeks and then flexibly dosed from 10 to 20 mg/day) of 315 US outpatients who met DSM-IV-TR criteria for GAD. The mean change from baseline to week 8 in the HAM-A was used as the primary efficacy variable. Mean changes from baseline to week 8 on the HAM-A total score using a last-observation-carried-forward approach were -11.3 for escitalopram and -7.4 for placebo. Response rates at week 8 were 68% for escitalopram and 41% for placebo among completers, and 58% for escitalopram and 38% for placebo looking at the last-observation-carried-forward values. There was no statistical difference in reported adverse events in the escitalopram-treated group (8.9%) compared with the placebo-treated group (5.1%).

Davidson et al. (2005) also examined the safety and efficacy of long-term, open-label treatment of GAD with escitalopram. Three 8-week, double-blind, placebo-controlled trials of nearly identical design were conducted with escitalopram in GAD (DSM-IV-TR criteria). Patients completing these trials were given the option of entering a 24-week, open-label, flexible-dose trial of escitalopram (10–20 mg/day). A total of 299 of 526 patients (56.8%) completed 24 weeks of open-label treatment. The mean HAM-A score at baseline of open-label treatment was 13.1. Long-term escitalopram treatment led to continued improvement in anxiety and quality-of-life measures. Of those completing 24 weeks of treatment, 92.0% were responders (CGI $\leq$ 2). The mean HAM-A score in the completer analysis was 6.9, and the mean HAM-A score in the last observation carried forward analysis was 9.2 at endpoint. Although uncontrolled, these data mirror the improvement observed in double-blind continuation studies with other SSRIs.

Allgulander et al. (2004) performed the first large, global multicenter, placebo-controlled, flexible-dose study of sertraline (50–150 mg/day) compared with placebo in GAD outpatients. After a 1-week single-blind placebo lead-in, patients were randomly assigned to 12 weeks of double-blind treatment with placebo ($n = 188$) or flexible doses of sertraline ($n = 182$). Sertraline-treated subjects showed significantly greater improvement (mean = 11.7 points) over placebo patients (mean = 8.0) on the HAM-A at week 4 (primary outcome measure).

Brawman-Mintzer et al. (2006) performed a similar study comparing sertraline (50–200 mg/day) with placebo in outpatients with GAD for 10 weeks. The primary efficacy measure was the change from baseline in HAM-A total score. Response was defined as a $\geq$50% decrease in HAM-A total score at endpoint. Sertraline produced a statistically significant reduction in anxiety symptoms, as measured by the HAM-A total changes scores ($p = 0.032$). Response rates were 59.2% for the sertraline group compared with 48.2% for the placebo group.

Slee et al. (2019) conducted a systematic review and network meta-analysis on randomized trials in adult outpatients with GAD. The primary outcomes were the change of the HAM-A and tolerability. In this study, 22 different drugs or placebo for 25,441 patients in 89 trials were evaluated. Escitalopram ($n = 1581$) was more efficacious than placebo in large cumulative samples of patients without increased discontinuation (mean difference: –2.45; 95% CI –3.27 to –1.63). Sertraline ($n = 485$; mean difference, –2.88; 95% CI, –4.17 to –1.59) and fluoxetine ($n = 264$; mean difference, –2.43; 95% CI, –3.74 to –1.16) were also efficacious although the findings were limited by small sample sizes. Paroxetine ($n = 1862$) also showed a significant effect (mean difference,

–2.29; 95% CI, –3.11 to –1.47) but was poorly tolerated compared with placebo.

Special Populations–Children, Adolescents, and Elderly Patients

The efficacy and safety of sertraline treatment for children with GAD were investigated in a 9-week, double-blind, placebo-controlled study by Rynn et al. (2001). Twenty-two children and adolescents, aged 5–17 years, who met DSM-IV-TR criteria for GAD and had a HAM-A rating of $\geq$16, participated in this study. All patients underwent a 2–3-week pre-evaluation period, followed by a 9-week double-blind treatment phase. Sertraline was initiated at 25 mg/day for the first week, and then titrated to 50 mg/day for weeks 2–9. Mean HAM-A scores for the sertraline group decreased from 20.6 $\pm$ 3.6 at baseline to 7.8 $\pm$ 5.7 at week 9, whereas those receiving placebo decreased from 23.3 $\pm$ 4.0 at baseline to 21.0 $\pm$ 7.8 at week 9. Mean CGI-Severity scores improved significantly for the sertraline-treated group (4.0 baseline to 2.4 weeks) compared to the placebo-treated group (4.0 baseline to 3.9 weeks). Importantly, no statistically significant differences were found in adverse events between the two groups. Only two patients in the placebo-treated group and one patient in the sertraline group dropped out prior to study completion. These data suggest that low doses of sertraline may be effective and well tolerated for children with GAD.

Strawn et al. (2020) performed an 8-week double-blinded, randomized, placebo-controlled trial of escitalopram in adolescents with GAD. In this study, 51 pediatric patients were treated with escitalopram of 15 mg/day (flexible titration to 20 mg/day) or placebo (escitalopram, $n = 26$; placebo, $n = 25$). The escitalopram group showed a greater mean change on the Pediatric Anxiety Rating Scale than the placebo group (–8.65 $\pm$ 1.3 vs. –3.52 $\pm$ 1.1) and superior improvement on CGI scores, with similar adverse events between the two groups.

Lenze et al. (2005) examined the efficacy and tolerability of citalopram in 34 subjects with anxiety disorders aged 60 years or older (30 subjects meeting DSM-IV-TR criteria for GAD). Subjects were randomly assigned to treatment with citalopram or placebo for 8 weeks. Response was defined as a score of 1 (very much improved) or 2 (much improved) on the CGI-I scale or a 50% reduction in the HAM-A scale score. Ten (67%) of the 15 citalopram-treated subjects diagnosed with GAD met response criteria versus 4 (27%) of the 15 placebo-treated subjects with GAD. Later, Lenze et al. (2009) investigated escitalopram as a treatment for GAD in geriatric subjects; 85 patients were randomized to 10–20 mg/day escitalopram and 92 patients to placebo. After 12 weeks of treatment, the escitalopram-treated patients showed greater improvement in anxiety symptoms by CGI-I in observed cases, but not in intention-to-treat analysis. In summary, data from randomized, placebo-controlled treatment trials suggest that SSRIs may be useful in the treatment of adults, children, and elderly patients with GAD. The medication dosages used to treat individuals with GAD are similar to those used to treat major depressive disorder.

Serotonin–Norepinephrine Reuptake Inhibitors (Venlafaxine, Duloxetine)

There have been several placebo-controlled multicenter studies demonstrating that venlafaxine XR (extended release) is an effective treatment of GAD. Davidson et al. (1999) performed an 8-week placebo-controlled trial of 75 and 150 mg/day of venlafaxine XR contrasted with 30 mg/day of buspirone. Venlafaxine XR reduced HAM-A scores significantly more than either placebo or 30 mg/day of buspirone. In an additional study by Rickels et al. (2000), patients with GAD were randomized to three doses of venlafaxine XR, namely, 75, 150, or 225 mg/day; all three doses were more effective in decreasing HAM-A scores than placebo. Results from a 6-month, fixed-dose study of venlafaxine XR (37.5, 75, or 150 mg/day) versus placebo suggested that patients treated with venlafaxine did significantly better than patients treated with placebo. Secondary analysis from this study suggested that the two higher doses of venlafaxine were also associated with significant improvement in social functioning (Haskins et al., 1999).

Gelenberg et al. (2000) compared the 6-month efficacy and safety of flexible doses of venlafaxine XR in adult outpatients with GAD. A total of 251 adults who met DSM-IV-TR criteria for GAD without comorbid major depressive disorder met the following criteria: (1) a screen and baseline score of at least 18 on the HAM-A, (2) at least 2 on items 1 and 2 of the HAM-A, (3) a total score of at least 9 on the Raskin Depression Scale at screen and baseline, and (4) a Covi Anxiety Scale score greater than the total score on the Raskin. Patients were started on venlafaxine XR at 75 mg/day or placebo during week 1. After day 8, the dose could be increased to 150 mg/day of venlafaxine XR or placebo. At day 15, increases were again allowed to the maximum 225 mg/day or placebo. A patient was considered a "responder" if the HAM-A score decreased by at least 40% from baseline or if the CGI score was 1 (very much improved) or 2 (much improved). By week 2, 42% of patients in the venlafaxine XR group were categorized as responders compared with 21% of placebo patients. During weeks 6–28, response rates were 69% for venlafaxine XR and 42–46% for placebo. Adverse events led to treatment discontinuation for 17% of the placebo-treated subjects and 26% of the venlafaxine-treated subjects.

Stahl et al. (2007) performed a post hoc analysis of five placebo-controlled, double-blind, randomized studies examining venlafaxine XR efficacy specifically on psychic and somatic symptoms of GAD from 8 weeks up to 6 months. They found that within the first 8 weeks of treatment, responders showed improvement in anxious mood, tension, behavior, muscular pain, cardiovascular irregularities, and respiratory symptoms. After extended periods of treatment, the last symptoms to respond were insomnia, fear, sensory/pain, gastrointestinal distress, and autonomic symptoms. They observed that continuing treatment with venlafaxine XR up to 6 months further improved early-responding symptoms and also targeted late-responding symptoms.

US Food and Drug Administration (FDA) granted approval for duloxetine as acute treatment of GAD in February 2007, and granted approval for maintenance treatment in November 2009.

Endicott et al. (2007) performed three randomized, double-blind, placebo-controlled studies examining the efficacy of duloxetine versus placebo in acute treatment of GAD. All three trials included patients fulfilling criteria for moderate GAD as defined by scores on the Hospital Anxiety & Depression Scale anxiety subscore ≥ 10 and CGI-Severity of Illness ≥ 4. The first study included 168 patients treated with duloxetine 60 mg/day, 170 patients with duloxetine 120 mg/day, and 175 patients with placebo. Patients entered a 1-week single-blind placebo lead-in, followed by 9 weeks of double-blind acute therapy, and a 2-week discontinuation period. Study 2 included a 10-week flexible-dose (60–120 mg/day) trial of duloxetine ($n = 168$) or placebo ($n = 159$) followed by a 2-week discontinuation period. Study 3 randomized patients to either 10 weeks of duloxetine 60–120 mg/day ($n = 162$), venlafaxine 75–225 mg/day ($n = 161$), or placebo ($n = 161$), followed by a 2-week discontinuation period. The primary outcome measure was the Sheehan Disability Scale (SDS) change in score from baseline to study completion. SDS global functioning scores improved significantly in duloxetine-treated patients. In study 1, scores decreased by 7.8 points in the 60 mg/day group, 7 points in the 120 mg/day group, and 3.8 points in the placebo group; study 2 scores improved by 5.8 points in the duloxetine group compared with a decrease of 3.1 points in placebo-treated patients; study 3 revealed an 8-point decrease in SDS in duloxetine-treated patients compared with a 5.4-point decrease in placebo-treated patients. SDS scores in work, social life, and family/home responsibility showed significant improvement across all three studies and duloxetine-treated patients had greater percentages of SDS global functioning scores in the normative range (normative range defined as a score of ≤ 5).

Hartford et al. (2007) conducted a multicenter, randomized, double-blind, parallel-group study investigating the efficacy of duloxetine versus venlafaxine XR and placebo in GAD. After initial screening, patients were randomized to 10 weeks of treatment with flexible doses of duloxetine (60–120 mg/day; $n = 162$), venlafaxine XR (75–225 mg/day; $n = 164$), or placebo ($n = 161$), followed by a 2-week taper to discontinuation of all medications. The primary outcome measure was the mean change in HAM-A scores from baseline to discontinuation. The duloxetine arm displayed a 46% improvement, with significant changes noted at week 1. Patients on venlafaxine showed a 50% improvement from baseline, with significant changes noted in week 2. Placebo-treated patients showed a 37% improvement. Overall discontinuation rates were not significantly different; however, adverse-event-related discontinuation was statistically greater among the patients treated with duloxetine (14.2%) and venlafaxine XR (11%) compared with placebo (1.9%).

A study by Rynn et al. (2008) demonstrated efficacy of duloxetine for acute treatment of GAD. A total of 327 subjects were randomized to flexible-dose, progressive-titration duloxetine (60–120 mg/day, $n = 168$) or placebo ($n = 159$). Total HAM-A score changes (primary outcome) for duloxetine-treated patients were reduced by an average of 36% compared with a placebo group mean reduction of 25%. Separation between duloxetine and placebo groups on the primary outcome measure was noted as early as 2 weeks into treatment. Duloxetine-treated patients had greater discontinuation rates due to adverse reactions, mainly nausea, dizziness, and somnolence.

Wu et al. (2011) reexamined the efficacy, safety, and tolerability of duloxetine monotherapy in GAD in a double-blind, placebo-controlled study. This study consisted of a screening period, active treatment for 15 weeks, and a 2-week taper. The primary outcome measure was the Hospital Anxiety and Depression Scale Anxiety subscale (HADS-A) and the secondary outcome measure was the HAM-A. Patients were randomly assigned to duloxetine 60 mg/day ($n = 108$) or placebo ($n = 102$). By week 7 of active treatment, responders were maintained on duloxetine 60 mg/day ($n = 67$, 62%) and nonresponders had duloxetine dose increased to 120 mg/day ($n = 41$, 38%). It was found that 72.5% of duloxetine-treated patients completed the 15 weeks of active treatment compared with 75.9% of placebo-treated patients. Discontinuation due to adverse events was significantly elevated in the duloxetine group at 12% versus 2.9% in the placebo group. Somnolence, nausea, and dizziness were more frequently reported with duloxetine. Duloxetine-treated patients demonstrated a significant improvement in the primary outcome measure with a 51.4% improvement from baseline to endpoint compared with 36.4% in the placebo group.

Li et al. (2018) conducted a meta-analysis to investigate the efficacy of duloxetine during short-term treatment in adults with GAD. The primary efficacy outcome was mean change in the HADS-A. Change in the HADS-A from baseline to the follow-up was reported in four trials (duloxetine, $n = 547$; placebo, $n = 627$). Duloxetine showed a significantly greater effect than placebo (mean difference, 2.32; 95% CI, 1.77 to 2.88).

Slee et al. (2019) performed a network-meta-analysis of randomized controlled trials to investigate the effect of serotonin–norepinephrine reuptake inhibitors in patients GAD. They verified that venlafaxine ($n = 2275$) and duloxetine ($n = 1355$) had significantly greater effects on the HAM-A with relatively good acceptability (mean difference, -3.13; 95% CI, -4.13 to -2.13).

Davidson et al. (2008) examined duloxetine treatment in preventing relapse of GAD. A total of 887 patients were treated with flexible-dose (60–120 mg/day) duloxetine for 26 weeks. Responders from that phase ($\geq 50\%$ HAM-A reduction to score of ≤ 11 for the last 2 weeks) entered the double-blind phase and were assigned to duloxetine ($n = 204$) or placebo ($n = 201$). Placebo-treated patients had a 41.8% relapse rate, defined as a ≥ 2 point increase in illness severity scales or discontinuation of medication secondary to lack of efficacy, compared with duloxetine-treated patients, who had a 13.7% relapse rate. Bodkin et al. (2011) examined predictors of relapse in patients who responded to treatment with duloxetine for GAD in a 6-month double-blind, randomized trial. Patients who displayed a good response to duloxetine for 6 months were randomized to continue duloxetine ($n = 216$) or a blind taper to placebo ($n = 213$). They determined that residual anxious mood (HAM-A), pain severity while awake (VAS pain), and low psychosocial function were associated with the greatest risk of relapse. Piero and Locati (2011) evaluated the efficacy of duloxetine versus escitalopram in a 6-month trial. HAMA, HAM-D, CGI, and GAF scores were assessed at baseline, 1 month into treatment, 3 months into treatment, and at the end of the treatment period. Although both medications showed significant improvements in all measures, duloxetine had greater remission and response rates and greater improvement in the HAM-A somatic subscale than escitalopram 1 month into treatment.

In summary, doses of venlafaxine XR as low as 37.5 mg/day and as high as 225 mg/day are effective in decreasing symptoms of anxiety for patients with GAD. Studies indicate that venlafaxine XR is an effective short- and long-term treatment for GAD. Side effects appeared to be mild and tended to decrease in number and intensity over the course of treatment. Nausea, dry mouth, and somnolence were the most commonly reported side effects. Duloxetine has been shown to be effective in decreasing symptoms of generalized anxiety in doses 60 mg/day to 120 mg/day, and can be used for both short-term and maintenance treatment for GAD. Duloxetine is well tolerated, with the most common side effects being somnolence, dizziness, and nausea.

Special Populations–Children, Adolescents, and Elderly Patients

The efficacy and safety of venlafaxine XR for children and adolescents with GAD were investigated by Rynn et al. (2007) in two randomized, double-blind, placebo-controlled trials. Patients received either flexible doses of venlafaxine XR ($n = 157$) or placebo ($n = 163$) for an 8-week study period. The primary outcome measure was the change from baseline to completion on a selected portion of the Schedule for Affective Disorders and Schizophrenia for School-Age Children. In a pooled analysis, the venlafaxine XR-treated patients displayed significantly improved scores on the primary outcome measure compared with placebo-treated patients, -17.4 versus -12.7. Significant adverse events noted in the venlafaxine group were changes in height, weight, blood pressure, pulse, and cholesterol levels.

Strawn et al. (2015) conducted a randomized placebo-controlled study to examine the effect of 10-week flexibly dosed duloxetine (30–120 mg daily) in youth aged 7 through 17 years with a primary diagnosis of GAD (duloxetine, $n = 135$; placebo, $n = 137$). The primary outcome measure was the Pediatric Anxiety Rating Scale. The duloxetine group showed significantly greater improvement from baseline to 10 weeks compared to the placebo group (-9.7 vs. -7.1; Cohen's d, 0.5). Alaka et al. (2014) evaluated the effect of 10-week duloxetine in patients aged 65 or older with GAD in a randomized placebo-controlled trial (duloxetine, $n = 151$; placebo, $n = 140$). The primary outcome measure was the HAM-A. At week 10, duloxetine showed a significantly greater effect than placebo (-15.9 vs. -11.7) but adverse events occurred twice as frequently in the duloxetine treated group.

Monoamine Enhancers (Tricyclic Antidepressants) and Monoamine Oxidase Inhibitors

A variety of monoamine enhancers (TCAs) have been demonstrated to be effective treatments for GAD (Liebowitz et al., 1988; Hoehn-Saric et al., 1993). However, the side effects (including anticholinergic activity, weight gain, daytime sedation, and orthostatic hypotension) and difficulty of titrating the dosage of these medications have made their use uncommon. In 2000, Zohar and Westenberg reviewed the use of monoamine enhancers (TCAs) in GAD, citing multiple double-blind studies demonstrating comparable efficacy between benzodiazepines and imipramine. Imipramine showed greater efficacy in reducing somatic and hyperarousal symptoms associated with anxiety, as well as obsessionality. In one open-label study (Wingerson et al., 1992) of clomipramine, patients experienced significant improvement in anxiety symptoms, but half discontinued the medication due to intolerable side effects. Schmitt et al. (2005) performed a review and meta-analysis comparing efficacy and tolerability of medications for depression against placebo for treatment of generalized anxiety

disorder, including imipramine. Imipramine had the second lowest number needed to treat, but the dearth of studies directly comparing imipramine to other medications for depression made comparison of effect sizes impossible. Interestingly, the studies included in the review suggested that imipramine was tolerated as well as newer medications for depression including paroxetine and venlafaxine. Currently, there are no monoamine oxidase inhibitors (MAOIs) FDA approved for use in GAD. However, given MAOIs action of increasing the synaptic concentration of dopamine, serotonin, and norepinephrine, it is possible that MAOIs may have efficacy in cases of refractory GAD in which SSRIs, SNRIs, or other more typical treatment regimens have been unsuccessful.

Other Medications for Depression (Trazodone, Nefazodone, Mirtazapine, Bupropion, Vilazodone, Vortioxetine)

Some older studies suggest that trazodone is an effective treatment for GAD (Rickels et al., 1993). There has been one open-label study suggesting that nefazodone may be efficacious for patients with GAD (Hedges et al., 1996). Mirtazapine has serotonergic and antihistaminergic effects. There have been only two open-label trials of mirtazapine in GAD (Strawn et al., 2018a). In the first, mirtazapine was associated with improvement in HAM-A scores over the 8-week duration of treatment. In the second, mirtazapine 30 mg/day was associated with improvement in HAMA-A scores over the 12-week treatment period, and by the end of the study, nearly 80% of participants had responded to treatment.

There is only one, small randomized controlled trial for bupropion in GAD (Bystritsky et al., 2008); 24 patients were randomly assigned to either bupropion XL (extended release) 150–300 mg/day or escitalopram 10–20 mg/day for 12 weeks. Bupropion and escitalopram showed similar efficacy and tolerability profiles.

Vilazodone 40 mg/day (but not 20 mg/day) was shown to be effective in reducing HAM-A total score change from baseline when compared to placebo in a multi-center, double-blind, parallel group, fixed dose trial (Gommoll et al., 2015). A flexible dose, multi-center, double-blind, randomized control trial (Durgam et al., 2016) demonstrated that both vilazodone 20 mg/day and 40 mg/day were statistically superior to placebo on all measures of anxiety and functional disability. Adverse events occurred at twice the rate of placebo, nausea, diarrhea, dizziness, fatigue, and erectile dysfunction being most common.

A 2015 meta-analysis of vortioxetine (Pae et al., 2015) suggested that vortioxetine was significantly more effective than placebo in reducing HAM-A scores from baseline with discontinuation similar between the groups. However, a second meta-analysis 4 years later (Qin et al., 2019) showed no significant difference between multiple doses of vortioxetine and placebo, but acknowledged there continue to be limited randomized controlled trials with the medication and its efficacy in GAD.

Benzodiazepine Medications

While benzodiazepines were the mainstay of anxiety treatment for many years, much of their study has been broadly for "anxiety disorders," and not GAD, which was not defined until DSM-III. Alprazolam is the only benzodiazepine FDA approved for use in GAD, although diazepam and clonazepam are sometimes preferred due to their longer half-lives. There have been many studies demonstrating that benzodiazepines as a class are more effective than a placebo in the treatment of GAD (Greenblatt et al., 1983). In general, the response rate from these benzodiazepine studies is in the 60–80% range (Rickels et al., 1983; Uhlenhuth et al., 1998). However, long-term benzodiazepine use may cause symptoms consistent with CNS depression, like fatigue, dizziness, increased reaction time, impaired driving skills, decreased cognitive functioning, and increased fall risk in elderly adults. Long-term use may also lead to physiological dependency, especially in individuals with a history of or predisposition to substance use disorders. Current prescribing guidelines suggest that benzodiazepines should not be utilized for longer than 3–6 months.

Rapaport et al. (2006) investigated combination therapy with alprazolam ODT and SSRI/SNRI compared with SSRI/SNRI monotherapy. A total of 129 subjects with a primary diagnosis of GAD were randomized 1:1 to 8 weeks of open-label treatment with alprazolam or no augmentation. The primary efficacy measure was time to response, defined as a $\geq 50\%$ decrease from baseline in HAM-A total score. There was no statistically significant difference between treatment groups in time to response. The results revealed that at weeks 2 and 4 CGI-I scores demonstrated that combination therapy (SSRI/SNRI + alprazolam) was associated with earlier improvement in response than SSRI/SNRI monotherapy (week 2, 42.9% vs. 21.1%, $p < 0.01$; week 4, 63.3% vs. 41.3%, $p < 0.05$). Similarly, subject-rated measures of global improvement (PGI) suggested that combination therapy ameliorated symptoms earlier than monotherapy (week 2, 41.9% vs. 20.4%, $p < 0.05$; week 3, 50.8% vs. 30.2%, $p < 0.05$; week 6, 70.0% vs. 42.8%, $p < 0.01$) of concern, approximately one-third of patients in this trial could not be tapered off alprazolam ODT during the specified taper phase, a finding that requires more investigation.

Special Populations–Children and Adolescents

Benzodiazepines for GAD have not been widely studied in children and adolescents. In one study (Simeon et al., 1992), children (mean age 12.6) treated with alprazolam did not show significant improvement on CGI scores. The most common side effects of benzodiazepines in children and adolescents are irritability, drowsiness, and dry mouth.

Buspirone

Buspirone has been shown to be more effective than placebo in decreasing HAM-A scores in a large number of placebo-controlled trials (Rickels et al., 1982, 1988; Laakmann et al., 1998). It is thought that buspirone may be particularly effective against psychological symptoms such as worry, tension, irritability, and apprehension, but less effective in ameliorating somatic symptoms because of its activation of locus coeruleus (Riblet et al., 1982; Rickels et al., 1982).

Data from placebo-controlled and comparative studies against benzodiazepines suggest that the onset of action of buspirone usually takes at least 2 weeks (Connor & Davidson, 1998; Laakmann et al., 1998). However, clinical experience suggests that buspirone treatment may take 3–4 weeks before one sees a truly beneficial effect in patients with GAD. Buspirone also requires multiple daily administrations to be effective. Although a meta-analysis suggested that buspirone may be given twice per day for the treatment of anxiety and GAD, the majority of studies and the current labeling for buspirone suggest that it should be administered three times per day (Sramek et al., 1999). Discontinuation of buspirone has not been associated with the development of withdrawal symptoms; however, over time, patients discontinued from pharmacotherapy tend to relapse (Rickels et al., 1988). Mokhber et al. (2010) conducted an 8-week randomized single-blind trial. In this study, 46 elderly patients with GAD were randomly assigned to sertraline ($n = 21$; 50–100 mg/day) or buspirone ($n = 25$; 10–15 mg/day). The primary outcome measure was the HAM-A. Both groups showed improvements in the HAM-A throughout the study period. After 2 and 4 weeks, buspirone was significantly superior to sertraline, but after 8 weeks, this difference did not reach statistical significance.

In summary, although there are considerable data suggesting that buspirone is an effective treatment for GAD, most of these data are from acute trials. It is clear that buspirone's time to onset is delayed, in contrast to the benzodiazepines, and that it may not be as useful as benzodiazepines for the treatment of GAD patients with primarily somatic symptoms.

Special Populations–Children and Adolescents

Strawn et al. (2018b) reviewed randomized controlled trials of buspirone in children and adolescents. Neither fixed-dose nor flexible dose trials separated buspirone from placebo, while buspirone arms had twice the drop-outs due to lightheadedness.

Anticonvulsant Medication

Two different anticonvulsants have been studied for the treatment of GAD: pregabalin and tiagabine. Pande et al. (2003) reported a 6-week, double-blind, four-armed placebo-controlled comparison of pregabalin (150 and 600 mg/day) and lorazepam (6 mg/day). This study had three phases: a 1-week placebo lead-in phase, a 4-week double-blind treatment phase, and a 1-week taper phase. The primary efficacy measure was the change in HAM-A tools scores over 4 weeks. In this study, both pregabalin groups and the lorazepam groups significantly differentiated from placebo in terms of decreases in HAM-A total scores. However, pregabalin subjects had a lower dropout rate than lorazepam subjects. Feltner et al. (2003) performed an identical study in which a 600 mg/day pregabalin group and a 6 mg/day lorazepam group both separated from placebo on the HAM-A, but the 150 mg/day pregabalin group did not. Rickels et al. (2005) undertook a five-arm, double-blind, placebo-controlled study in outpatients with GAD. In this 6-week study, three doses of pregabalin (300, 450, and 600 mg/day) were contrasted with alprazolam 1.5 mg/day and placebo. All three pregabalin groups and the alprazolam group caused statistically significantly greater changes in HAM-A total scores and the HAM-A psychic anxiety subscale compared with placebo.

Feltner et al. (2008) reported a study on the long-term efficacy of pregabalin 450 mg/day compared with placebo. A total of 338 GAD patients were treated with pregabalin in an initial open-label 8-week course and were then randomized to continue pregabalin (450 mg/day) or placebo and were treated for an additional 24 weeks. Pregabalin-treated patients had significantly lower rates of relapse (42%) at endpoint compared with placebo (65%). Time to relapse was also significantly longer for pregabalin-treated patients. Adverse events were low in both groups, at 6.0% for the pregabalin group and 2.4% for the placebo group. Hadley et al. (2012) examined the benefit of switching GAD patients on long-term benzodiazepine treatment to pregabalin versus placebo, with the primary outcome measure being the ability to remain benzodiazepine free and secondary measures being mean HAM-A score changes and Physician Withdrawal Checklist (PWC) endpoint scores. They found that numerically but not statistically more patients tapered off alprazolam and onto pregabalin remained benzodiazepine free compared with those receiving placebo. The pregabalin-treated patients also had significantly greater HAM-A score reductions and lower endpoint mean PWC scores.

Montgomery et al. (2006) investigated the efficacy of two doses of pregabalin versus venlafaxine and placebo for the treatment of moderate to severe GAD (HAM-A ≥ 20) in 421 patients. The primary outcome measure was the change in HAM-A mean total score. In three active treatment groups, pregabalin (400 and 600 mg/day) and venlafaxine (75 mg/day) caused significantly greater decreases in HAM-A total scores than placebo. The active groups did not differentiate in other measures. The discontinuation rates were slightly higher for the venlafaxine group; 20.4% versus 6.2% for the 400 mg/day pregabalin group, 13.6% for pregabalin 600 mg/day, and 9.9% for the placebo group. A similar study by Kasper et al. (2009) investigated the efficacy of pregabalin and venlafaxine XR compared with placebo. This double-blind trial was conducted over 8 weeks. Patients were assigned to pregabalin ($n = 121$) at doses of 150–600 mg/day, venlafaxine XR ($n = 125$) 75–225 mg/day, and placebo ($n = 128$). Significant decreases in the primary outcome measure, HAM-A scores from baseline to endpoint, were noted in the pregabalin arm (-14.5) compared with the venlafaxine XR (-12.0) and placebo (-11.7) arms. Treatment with pregabalin also demonstrated rapid onset of action, with the pregabalin arm separating from the other two by day 4. Pregabalin-treated patients reported similar rates of severe adverse events to placebo, with venlafaxine XR patients reporting significantly higher rates of adverse events.

A study by Rickels et al. (2012) investigated pregabalin as an adjunctive treatment in GAD patients who had partial response to therapeutic doses of SSRI/SNRI. A total of 356 patients were randomized to either pregabalin (150–600 mg/day) or placebo for 8-week augmentation of their existing SSRI/SNRI regimen. There were significantly greater decreases in HAM-A scores from baseline to endpoint in the pregabalin group compared with placebo (-7.6 vs. -6.4), with responder rates significantly higher for pregabalin patients (47.5% vs. 35.2%). This suggests that pregabalin augmentation may be an appropriate strategy for partial responders on an SSRI/SNRI.

Generoso et al. (2017) conducted a systematic review and meta-analysis that includes placebo-controlled trials. They used Hedges' g to evaluate the effect of pregabalin on anxiety symptoms in patients with GAD. Eight randomized placebo-controlled trials ($n = 2229$) were included in the meta-analysis. A random-effects model demonstrated that pregabalin was superior than placebo (Hedges' g, 0.37; 95% CI, 0.30–0.44). No significant difference between pregabalin and placebo was shown in terms of the dropout rates. The authors concluded that pregabalin was comparable to benzodiazepines in clinical effects, and had lower dropout rates than benzodiazepine.

In summary, there have been multiple positive studies demonstrating that pregabalin, in a dose range from 150 to 600 mg/day, is effective in the treatment of GAD. Despite this rather impressive array of data, pregabalin is not approved by the FDA for this indication at present.

Pollack et al. (2005) reported a large, 8-week, randomized, double-blind, flexible-dose, placebo-controlled study investigating tiagabine (4–16 mg/day) versus placebo. Tiagabine did not differentiate from placebo in the planned primary outcome analysis, which was a last-observation-carried-forward analyses of the HAM-A, but it did differentiate from placebo when a mixed models' repeated measures analysis was performed. It was well tolerated and was not associated with problems with sexual functioning, weight, or sleep-related side effects. There have been no studies of tiagabine in children or adolescents with GAD.

Medications for Psychosis (Antipsychotics)

Since 2005, antipsychotic treatment of GAD has been explored as both an adjunct to other first-line agents and also for monotherapy management. Pollack et al. (2006) openly treated individuals with fluoxetine. Those who did not respond to fluoxetine entered into a double-blind phase where they were randomized to receive either olanzapine (20 mg/day) or placebo. In this small study, olanzapine augmentation significantly improved CGI-I scores and caused a significant reduction in HAM-A total scores. The rates of both response and remission were much greater in the olanzapine-treated group than in the placebo-treated group. However, olanzapine was associated with a significant increase in weight.

Brawman-Mintzer et al. (2005) investigated the efficacy of risperidone augmentation for treatment-resistant GAD. In this trial, individuals who continued to have a HAM-A score of ≥18 or a CGI-S score of moderate or greater after 4 weeks of anxiolytic treatment were randomized to receive risperidone (0.5–1.5 mg/day). Risperidone augmentation caused a statistically significant decrease in HAM-A total scores. Aripiprazole was investigated in a small, open-label augmentation trial of patients on therapeutic doses of a medicine for depression (antidepressant) and still reporting HAM-A total scores of ≥14 or CGI scores of ≥4 by Menza et al. (2007). The mean dose of aripiprazole at the endpoint was 13.9 mg/day. Patients with aripiprazole augmentation displayed significant decreases in HAM-A and CGI scores at 6 weeks, compared with placebo; 5 out of 9 aripiprazole patients were responders and 1 was considered to be a remitter by week 6. It was well tolerated by patients, with only weight gain (average 10 lb) remaining after discontinuation of medication.

Bandelow et al. (2010) investigated quetiapine XR (extended release) as monotherapy for GAD treatment. A total of 873 patients were enrolled in a double-blind, placebo-controlled study of either quetiapine XR 50 mg/day, quetiapine XR 150 mg/day, paroxetine 20 mg/day, or placebo over a 10-week trial (8 weeks of active treatment and 2 weeks of drug taper to discontinuation). Primary endpoint measurement was the change in HAM-A score from baseline to week 8. Both paroxetine and quetiapine XR patients displayed significant overall decreases in HAM-A scores, and also improvement on the psychic subscale score compared with placebo. Mean HAM-A score changes were −13.95 for quetiapine XR 50 mg/day, −15.96 for quetiapine XR 150 mg/day, −14.45 for paroxetine 20 mg/day, and −12.30 for placebo. Patients receiving quetiapine XR 150 mg/day were the only group to report significantly reduced HAM-A somatic subscale scores. Remission rates (HAM-A scores of ≤7) were significantly higher in the quetiapine XR 150 mg/day (42.6%) and paroxetine (38.8%) groups compared with placebo and low–dose quetiapine. Patients on quetiapine XR reported less sexual dysfunction compared with placebo and paroxetine, and did not report higher rates of EPS than those receiving paroxetine. These findings were replicated by three large-scale studies of similar design. The study by Merideth et al. (2012) examined quetiapine XR compared with escitalopram and placebo, and those of Katzman et al. (2011) and Khan et al. (2011) investigated quetiapine XR at doses of up to 300 mg/day versus placebo without SSRI/SNRI comparator. All four studies demonstrated quetiapine XR to be efficacious in monotherapy treatment of GAD. Despite these positive studies, the FDA did not approve quetiapine for GAD secondary to the risks of adverse events including metabolic consequences and QT prolongation outweighing potential therapeutic benefit.

A study by Lohoff et al. (2010) investigated ziprasidone for both augmentation and monotherapy in treatment-resistant GAD patients, defined as nonresponders to standard antidepressant therapy. No statistically significant differences were found in HAM-A scores for monotherapy ziprasidone compared with placebo. However, further investigation into ziprasidone efficacy for GAD is warranted.

Treatment Conclusions

People with GAD suffer from a combination of both somatic symptoms and psychic anxiety. There has been a wealth of clinical trials that demonstrate myriad classes of agents are effective in the short-term treatment of GAD. Several SSRIs and venlafaxine XR have received FDA approval as a short-term treatment for GAD. Duloxetine has received FDA approval for acute and maintenance treatment. The longer-term maintenance studies clearly suggest that patients continue to improve and are more likely to achieve remission from their symptoms of GAD if they are continued at full dose of their treatment over a period of at least 1 year. There also are data suggesting that discontinuation of pharmacotherapy is associated with the recurrence of illness.

Individuals have difficulty with the initial upward titration of medications for depression, so it is not unreasonable to consider adding a benzodiazepine for a short period of time to ease the transition and provide interim relief of symptoms. Unfortunately, it is also apparent that there is a subgroup of individuals with GAD who have a more refractory course. There are limited data available to suggest the most appropriate augmentation or switch therapies for patients with treatment-refractory GAD. However, the literature suggests that both acute augmentation and monotherapy maintenance with low doses of dopamine serotonin blockers (atypical antipsychotic) and anticonvulsant medications may be beneficial in relieving residual symptoms of GAD.

Social Anxiety Disorder

Selective Serotonin Reuptake Inhibitors

SSRIs have emerged as first-line treatment for SAD. Most of the efficacy data are derived from multicenter, double-blind trials of paroxetine, sertraline, and fluvoxamine. The first study to demonstrate that an SSRI was efficacious in treating SAD was a 12-week trial of 183 patients randomized to either paroxetine (20–50 mg/day) or placebo. About 55% of patients randomized to paroxetine improved significantly, versus 24% of those randomized to placebo (Stein et al., 1998).

These results were replicated in a second, large multicenter study for SAD (Baldwin et al., 1999).

Multiple large-scale studies of escitalopram for acute treatment and relapse prevention of SAD have been published (Lader et al., 2004; Kasper et al., 2005; Montgomery et al., 2005). Kasper et al. (2005) recruited 358 adult outpatients for their 12-week, multisite, randomized, placebo-controlled, flexible-dosage study. A total of 290 participants completed the study, with 145 patients enrolled in each treatment group at the study's endpoint. Although escitalopram was significantly better than placebo on all primary and secondary efficacy measures, it also had a higher incidence of headache, nausea, and sexual side effects than placebo, which should be taken into consideration. Among completers, escitalopram was significantly better than placebo on all efficacy measures, even with an extremely high placebo response rate of 39%. Lader et al. (2004) enrolled 837 participants in a 24-week, multisite, randomized study where patients received placebo ($n = 166$), 5 mg escitalopram ($n = 167$), 10 mg escitalopram ($n = 167$), 20 mg escitalopram ($n = 170$), or 20 mg paroxetine ($n = 169$). Escitalopram 20 mg was significantly more effective than placebo or paroxetine 20 mg on five of six symptom dimensions of the Liebowitz Social Anxiety Scale (LSAS). Baldwin et al. (2016) conducted a meta-analysis to examine the effect of escitalopram of various doses in patients with SAD by including placebo-controlled trials. The primary outcome measure was the estimated treatment difference in the LSAS total score at week 12. Two randomized controlled trials were eligible for the meta-analysis of 10-mg and 20-mg escitalopram. Both 10-mg ($n = 724$) and 20-mg ($n = 712$) escitalopram treatments showed significantly greater effects than placebo (mean difference, -4.56; 95% CI, -8.14 to -0.99 and mean difference, -10.10; 95% CI, -13.67 to -6.53, respectively).

Liebowitz et al. (2003) conducted a study enrolling 415 outpatients with severe generalized social anxiety in a 12-week, double-blind study of sertraline (50–200 mg/day) versus placebo. Among completers, sertraline was significantly more effective than placebo in causing improvement on a wide array of measures, including the

LSAS, CGI-I, Q-LES-Q, and SDS. In a subgroup analysis, 87 subjects were identified as having Quality of Life Enjoyment and Satisfaction Questionnaire (Q-LES-Q) scores more than two standard deviations below mean normative community values. About 37% of these subjects randomized to sertraline treatment had restoration of Q-LES-Q scores back to the normal range versus 11% of placebo-treated subjects.

In a study of fluvoxamine treatment for 82 patients with SAD, 42% of patients randomized to fluvoxamine responded to treatment (mean dose 202 mg/day), compared with 24% of patients who received placebo (Stein et al., 1999). In another large double-blind study, Westenberg et al. (2004) enrolled 300 patients in a 12-week, multicenter study comparing 100–300 mg/day fluvoxamine controlled release (CR) with placebo. Fluvoxamine CR led to a greater improvement on the primary outcome measures compared with placebo. Ratings of sexual functioning were performed and were not statistically significant between the groups. (For more details on social anxiety, see ▶ Chap. 21, "Gene-Environment Interactions.") Liu et al. (2018) performed a meta-analysis to evaluate the effect of fluvoxamine in randomized placebo-controlled design. The analysis showed that fluvoxamine was more efficacious than placebo on the LSAS ($n = 736$; mean difference, 11.90; 95% CI, 8.09 to 15.79) and the CGI-S ($n = 650$; mean difference, 0.52; 95% CI, 0.33 to 0.72). Participants randomized to fluvoxamine also showed higher response rates compared with placebo ($n = 1001$; odds ratio, 1.71; 95% CI, 1.30–2.24).

The use of sertraline for the longer-term treatment of SAD was investigated in a two-part study where subjects entered a 20-week, double-blind, placebo-controlled study that was followed by a 24-week, double-blind extension for first-phase treatment responders (Van Ameringen et al., 2001). A total of 204 adult outpatients with SAD were randomly assigned 2:1 to sertraline or placebo. Following a 1-week placebo run-in period, patients received an initial dose of 50 mg/day of sertraline or placebo. After 4 weeks, the dose could be increased by 50 mg/day every 3 weeks up to a maximum of 200 mg/day. In the event of intolerable side effects, doses could be reduced to a minimum of 50 mg/day. The mean dose of sertraline was 146.7 mg/day. Primary efficacy measures included the CGI-Improvement Scale, the patient rated Marks Fear Questionnaire (FQ), and the physician rated Brief Social Anxiety Disorder Scale. At the study endpoint, 53% of sertraline-treated patients compared with 29% of placebo-treated patients were (2) "much improved" or (1) "very much improved" as measured by the CGI-Improvement Scale. The mean score on the FQ decreased from 23.07 at baseline to 15.54 at endpoint for the sertraline-treated group, whereas the placebo-treated group decreased from 21.72 at baseline to 19.38 at endpoint. The mean score on the Brief Social Anxiety Disorder Scale for the sertraline-treated group decreased from 47.48 at baseline to 31.18 at endpoint, versus 45.72 to 37.23 for the placebo-treated group. Fifty sertraline responders and 15 placebo responders entered the double-blind, placebo-controlled relapse-prevention phase of the study. Patients who were randomly assigned to continue sertraline treatment received a mean daily dose of 148.0 mg/day. About 88% of sertraline-treated patients, 40% of placebo-switched patients, and 40% of the initial placebo-responder patients completed the study; 4% of the subjects who received sertraline maintenance relapsed compared with 36% of the placebo-switch group.

In summary, SSRIs are effective and generally well-tolerated first-line medications for patients seeking treatment for SAD, especially those presenting with comorbid depression. Maintenance therapy trials with escitalopram and sertraline confirmed the continued benefit of maintenance pharmacotherapy.

Special Populations–Children and Adolescents

Isolan et al. (2007) investigated the efficacy and safety of escitalopram for children and adolescents with SAD in an open-label, placebo-controlled trial. Twenty patients were treated with escitalopram 5–20 mg/day for 12 weeks. The primary outcome measure was change from baseline to endpoint in the CGI-I, and 65% of patients responded to treatment by this measure.

Serotonin–Norepinephrine Reuptake Inhibitors (Venlafaxine, Duloxetine)

The first case reports and open-label studies investigating the efficacy of venlafaxine in SAD were published in the mid-1990s (Kelsey, 1995). More recently, multiple double-blind studies examining the efficacy of venlafaxine ER in SAD have been published (Allgulander et al., 2004; Rickels et al., 2004; Liebowitz et al., 2005). Rickels et al. (2004) conducted a 12-week, double-blind, placebo-controlled study of venlafaxine extended release (XR) in 272 outpatients with generalized SAD. Among 172 completers, venlafaxine ER was significantly more effective than placebo, as demonstrated by LSAS, CGI-I, and CGI-S scores. These results were supported by the double-blind, placebo-controlled, flexible-dose trial of Liebowitz et al. (2005), which examined the effects of venlafaxine XR (75–225 mg/day) in 440 patients with SAD. Venlafaxine ER treatment caused a significant improvement over placebo on the LSAS (primary outcome measure), with response rates being 44% in venlafaxine-treated patients versus 30% in placebo-treated patients, where response was defined as a CGI-I score of 1 or 2. In addition, remission rates were significantly greater in the venlafaxine XR group than in the placebo group (20% vs. 7%, where remission was an LSAS score of ≤ 30). Allgulander et al. (2004) conducted a similarly designed study, examining the efficacy of 75–255 mg/day venlafaxine XR, 20–50 mg/day paroxetine, or placebo for 12 weeks. Treatment with venlafaxine XR demonstrated a significant improvement over placebo on all efficacy variables, although no significant differences were observed between the venlafaxine XR and paroxetine groups. Even so, both active treatments led to remission in less than half of patients, suggesting that short-term, 12-week studies may not be long enough to attain the medications' full therapeutic effects with a chronic illness such as SAD.

In a six-week open-label trial of duloxetine 60 mg/day followed by an 18-week randomized trial of either duloxetine 60 mg/day or 120 mg/day, Simon et al. (2010) found that patients experienced a significant improvement on the LSAS. The effect was maintained at 24 weeks, but there was no significant difference between the doses.

Special Populations–Children
March et al. (2007) studied the efficacy of venlafaxine ER in pediatric SAD in a study in which 293 patients were randomized to either placebo or venlafaxine XR (37.5–225 mg/day) for 16 weeks. Venlafaxine XR-treated patients had a significantly greater improvement on the CGI-I compared with placebo (56% vs. 37%), but three venlafaxine-treated patients reported treatment-emergent suicidality, with no attempts or suicides.

Monoamine Enhancers (Tricyclic Antidepressant Medication) and Monoamine Oxidase Inhibitors

In general, monoamine enhancers (TCAs) have not been found to be effective for the treatment of SAD. MAOIs have been shown to be superior to placebo treatment for SAD (Gelernter et al., 1991; Versiani et al., 1992). The mean daily dose of the drug in these trials ranged between 60 and 90 mg/day. Although surpassed by the SSRIs as first-line agents, there is no doubt that first-generation MAOIs are effective. Nardi et al. (2010b) investigated the efficacy and dose-response relationship of the first-generation MAOI tranylcypromine. They found that 60 mg/day tranylcypromine was more efficacious than 30 mg/day, as determined by significantly lower LSAS scores (mean change -35.0 vs. -17.9). They also found a twofold greater change in the social anxiety symptom scale with the higher 60 mg/day dose. Unfortunately, the risk of hypertensive crisis and the need for patients to follow a tyramine-free diet make this class of drugs unappealing for the majority of patients.

The promising early results derived from efficacy studies of first-generation MAOIs spawned the study of reversible monoamine oxidase inhibitors (RIMAs) for this indication. RIMAs have significantly less risk of hypertensive crisis, and there is no need for a tyramine-restricted diet. They also have a more tolerable side effect profile. Brofaromine and moclobemide are the two most studied RIMAs. However, there is conflicting evidence about their efficacy for the treatment of SAD. Of four published placebo-controlled multi-center moclobemide studies, three were positive (International Multicenter Clinical Trial Group, 1997; Noyes et al., 1997; Blanco & Liebowitz, 1998; Schneier et al., 1998; Stein et al., 2002b). Two controlled studies of brofaromine in SAD showed superior efficacy to placebo (van Vliet et al., 1993; Fahlen et al., 1995). However, brofaromine is no longer under development in the United States and moclobemide is not available there.

Other Medications for Depression (Nefazodone, Mirtazapine, Bupropion, Vilazodone, Vortioxetine)

Muehlbacher et al. (2005) conducted a double-blind, placebo-controlled trial of mirtazapine treatment for SAD patients. A total of 66 women enrolled in the 10-week, fixed-dose study of 30 mg/day mirtazapine or placebo. After 10 weeks of treatment, significant differences were noted between mirtazapine and placebo on the SPIN and the LSAS. Subsequent studies by Schutters et al. (2010) on 60 patients with SAD failed to demonstrate efficacy of monotherapy with mirtazapine (30–45 mg/day) in the treatment of SAD. However, Schutters et al. (2011) later demonstrated that paroxetine treatment was efficacious in the same population of patients who failed mirtazapine monotherapy, and that augmentation with mirtazapine significantly decreased rates of sexual dysfunction compared with paroxetine monotherapy (38% vs. 50%).

Emmanuel et al. (1991, 2000) published two papers investigating bupropion treatment for SAD. In the more recent work, a small 12-week, open-label, flexible-dose study, the efficacy of bupropion SR in the treatment of SAD was evaluated. Eighteen subjects with DSM-IV-TR diagnosed SAD were seen weekly for the first 4 weeks, then biweekly for the next 8 weeks. Bupropion SR (slow release) was initiated at 100 mg/day and titrated to a maximum of 200 mg twice daily, depending on response to treatment. Ten subjects completed the 12-week treatment period; 5 of the 10 were responders. The mean dose of bupropion SR was 366 ± 68 mg/day. The mean CGI-I scores changed from 4.3 ± 0.7 at baseline to 2.5 ± 0.9 at week 12, and the mean Brief Social Phobia Scale (BSPS) scores decreased from 42.2 ± 13 at baseline to 21.0 ± 11 at week 12.

Van Ameringen et al. (2007) found no significant difference between nefazodone and placebo in scores on the CGI or LSAS. A small randomized trial of vilazodone by Careri et al. (2015) demonstrated significant improvement over placebo on the LSAS and the CGI. A 2017 Cochrane Review (Williams et al., 2017) reported an ongoing randomized controlled trial of vortioxetine in the treatment of comorbid major depressive disorder and social anxiety disorder.

Benzodiazepine Medication

The benzodiazepines, clonazepam, and alprazolam have been shown to be efficacious in treating SAD (Gelernter et al., 1991; Dubuff et al., 1995; Nardi & Perna, 2006). Pollack et al. (2014) conducted a multicenter, 12-week double-blind randomized controlled trial. In this study, patients with the LSAS greater than 50 were enrolled. The authors compared sertraline plus clonazepam ($n = 63$) vs. sertraline plus placebo ($n = 59$). Although response rate did not significantly differ (sertraline plus clonazepam: 27%, sertraline plus placebo: 17%), sertraline plus clonazepam was superior to placebo on the improvement of the LSAS (27 ± 24.4 vs. 16 ± 23.0). The limitations of clonazepam and alprazolam are the same as when used in any indication. Owing to the potential for abuse and drug withdrawal, their use must be monitored carefully. This is a particularly problematic issue in SAD because of the high rate of comorbid substance abuse. Benzodiazepines may be best suited for patients with situational and performance anxiety on an as-needed basis.

Buspirone

Buspirone showed promise during open-label studies of SAD but did not differentiate from placebo in controlled trials (van Vliet et al., 1997). Data from open-label trials also suggest that buspirone may be used as an adjuvant with SSRIs when patients exhibit only partial response to monotherapy (Van Ameringen et al., 1996).

Beta-Blocker Medication

Stein et al. (2001) conducted a double-blind, placebo-controlled crossover study of pindolol potentiation of paroxetine for the treatment of SAD. Pindolol was selected as an adjunct to SSRIs because it is a 5-HT_{1A} autoreceptor antagonist in addition to being a β-adrenergic receptor blocker. It is postulated that antagonism of the 5-HT_{1A} autoreceptor would stimulate presynaptic release of serotonin. In fact, pindolol has been shown to accelerate the antidepressant response of SSRIs in some studies (Perez et al., 1997; Zanardi et al., 1997). Fourteen patients who were less than "very much improved" on CGI-I ratings after 10 weeks of maximally tolerated dose of paroxetine were randomized to receive 5 mg of pindolol TID or placebo for 4 weeks. After 4 weeks, subjects were tapered and crossed over to receive the other agent (pindolol or placebo) for another 4-week period. Pindolol augmentation was not more effective than placebo augmentation in this study. A 2017 Cochrane Review (Williams et al., 2017) did not find sufficient evidence to recommend use of beta-blockers for either generalized social anxiety disorder or performance anxiety.

Anticonvulsant Medication

Anticonvulsants have been evaluated for efficacy and safety in patients with SAD. Pande et al. (2004) reported a 10-week, double-blind, placebo-controlled trial of two fixed doses of pregabalin (150 or 600 mg/day). The 600 mg/day dose of pregabalin significantly reduced total scores on the LSAS when contrasted with placebo. The 600 mg/day dose was associated with a significant reduction in total fear, total avoidance, social fear, and social avoidance as measured on the brief social phobia scores. These results were replicated in a later study by Feltner

et al. (2011), showing that 600 mg/day significantly improved LSAS scores compared with placebo, 300 mg/day, and 450 mg/day. Improvement with the 600 mg/day dose was noted by week 1 of treatment, and it was well tolerated with dizziness and somnolence being the most commonly reported adverse events. The same group carried out a longer-term study (Greist et al., 2011) monitoring the efficacy and safety of pregabalin in preventing relapse in SAD. A total of 153 patients who had met responder criteria to 450 mg/day of pregabalin over 10 weeks of treatment were randomized to placebo or continuing equivalent dose of pregabalin. Pregabalin-treated patients had significantly delayed times to relapse and greater maintenance of improvements on the LSAS and Marks Fear Questionnaire. Dizziness and increased infection were reported with pregabalin, but were otherwise tolerated well.

Pande et al. (1999) conducted a study in which 69 patients with SAD were randomized for 14 weeks to receive either gabapentin (900–3600 mg/day) or placebo. In this study, 32% of patients randomized to flexible-dose gabapentin were considered responders versus 14% of patients receiving placebo. A 2010 12-week randomized controlled trial of levetiracetam versus placebo (Stein et al., 2010) was unable to demonstrate significant between-group differences on the LSAS. Further studies, especially maintenance studies, are still called for in order to determine the potential benefits of anticonvulsant therapy for patients with SAD.

Medication for Psychosis (Antipsychotics)

A small double-blind pilot study randomized 12 people to either olanzapine or placebo (Barnett et al., 2002). Treatment began at 5 mg/day and was titrated up to 20 mg/day. The olanzapine arm showed significantly greater improvement on the Brief Social Phobia Scale and the Social Phobia Inventory; patients taking olanzapine had more drowsiness and dry mouth. In 2007, Vaishnavi et al. randomized 15 patients to either quetiapine (up to 400 mg/day) or placebo for 8 weeks. Although there were no significant differences on the primary outcomes, there was a trend towards greater improvement on the Brief Social Phobia Scale and the CGI in the quetiapine arm than in the placebo arm.

Treatment Conclusions

Medications for depression (except for monoamine enhancers), benzodiazepines, and anticonvulsants have all been demonstrated to be effective pharmacotherapies for patients with SAD. Since all of the medications for depression s can cause activation, it is important to start the initial titration at a low dose and titrate the dose up over a period of weeks to months. There is some evidence to suggest that short-term usage of a benzodiazepine when initiating a stimulating medication for depression may allow patients to tolerate the medication for depression more easily and may facilitate more rapid titration of the medication. Although these data are limited, this approach is widely practiced in many clinical settings.

SAD frequently requires not only pharmacotherapy, but also some form of adjunctive cognitive behavioral therapy in order to facilitate decreasing anxious behaviors. For individuals who cannot tolerate traditional medication for depression and who are fearful of benzodiazepines, there are some data supporting the use of anticonvulsants. Although none of the anticonvulsants is currently approved for treatment of SAD, gabapentin and pregabalin have been reported to be efficacious for patients with SAD. As had been previously discussed with GAD, some individuals are not fully responsive to first or second treatment intervention. There is a paucity of evidence-based data to guide further intervention, although many patients do seem to respond best to either combination therapy.

Panic Disorder

Selective Serotonin Reuptake Inhibitors

SSRIs are generally accepted as a first-line treatment for PD. The major advantage of these agents is their tolerability compared with other options and thus longer-term acceptance by patients. There is currently evidence that fluoxetine, sertraline, paroxetine, fluvoxamine, citalopram, and escitalopram are effective in the acute treatment of PD. Fluoxetine, sertraline, paroxetine, citalopram, and escitalopram are approved for PD in the USA and Europe, whereas fluvoxamine is only used off-label.

There are two published large multicenter, double-blind, placebo-controlled studies that investigated the efficacy of fluoxetine versus placebo. Michelson et al. (1998) conducted a large multicenter, double-blind, placebo-controlled trial investigating the efficacy of 10 or 20 mg/day fluoxetine versus placebo. The results suggested that fluoxetine reduced panic attack frequency, global distress, and agoraphobic distress. Patients with a CGI-Improvement score of (1) "very much improved" or (2) "much improved" at the end of the acute trial were entered into a 24-week continuation study (Michelson et al., 1998). During this study, patients either continued on active treatment or received placebo. Patients randomized to fluoxetine continued to improve during the course of 24 weeks whereas patients who were randomized to placebo demonstrated a marked exacerbation in symptoms and were more likely to meet criteria for a recurrence of PD (Michelson et al., 1999).

Michelson et al. (2001) also conducted a 12-week, randomized, placebo-controlled trial of fluoxetine (10–60 mg/day) conducted at nine sites in Europe. A total of 180 adult patients who met DSM-IV-TR criteria for PD and had a minimum of four full panic attacks in the month preceding entry into the study, and two full panic attacks during the 2-week baseline evaluation period, were randomized to treatment. Patients also had to have at least moderate symptom severity, as measured with the Panic Disorder Severity Scale (PDSS), and a CGI-Severity rating of 4 or greater. Fluoxetine was initiated at 10 mg/day for week 1 and increased to 20 mg/day until week 6. Patients who had not achieved a CGI-Severity score of 2 or less were increased to 40 mg/day at week 6. Fluoxetine could be further increased to 60 mg/day. The final mean daily dose of fluoxetine was only 29.6 mg/day. Fluoxetine was statistically superior to placebo at both the 6-week point and at the study endpoint. Fluoxetine-treated patients also had a statistically significant improvement on the Sheehan Disability Scale (SDS). Reports of adverse events among the placebo and fluoxetine groups were similar and low.

Four large multicenter trials investigating the efficacy of sertraline in the treatment of PD have been published. Two studies were placebo-controlled, fixed-dose trials and the other two were placebo-controlled, flexible-dose studies. In the flexible-dose sertraline trials, patients were started on 25 mg/day for the first week and then the dose was escalated to 50 mg/day. Sertraline-treated patients experienced significant reductions in panic attack frequency, CGI-Improvement and Severity Scales, the PDSS, the Patient Global Evaluation (PGE) Rating Scale, and the Q-LES-Q (Londborg et al., 1998; Pollack et al., 1998; Rapaport et al., 1998).

Sertraline was also found to be efficacious in fixed dose, with the 150 mg/day dose seeming to be the most effective dose in these studies (Dubuff et al., 1995). There is one large multicenter long-term trial, by Rapaport et al. (2001), in which individuals who initially entered 10-week treatment trials were then offered 1 year of open-label treatment of sertraline and responders were subsequently entered into a double-blind, 26-week placebo discontinuation trial. Clinically significant deterioration was examined over the course of the double-blind phase; deterioration was specified as (1) relapse, (2) discontinuation due to insufficient clinical response, and (3) exacerbation of panic symptomatology. At the end of 26 weeks of double-blind treatment, patients randomized to placebo were twice as likely to discontinue owing to lack of efficacy as patients treated with sertraline. Similarly, 13% of sertraline-treated patients experienced acute exacerbation of panic symptoms versus 30% of placebo-treated patients. These data demonstrated that maintenance treatment with sertraline was associated with continued improvement and protected patients from recurrence (Rapaport et al., 2001). The mean daily dose of sertraline was 112.1 mg/day.

A head-to-head comparison trial was conducted by Bandelow et al. (2004), comparing paroxetine with sertraline. A total of 225 patients with PD were randomly assigned in a double-blind study to paroxetine 40–60 mg/day or sertraline 50–150 mg/day. Patients were assessed with the Panic and Agoraphobia Scale (PAS) scores, CGI-I (responders defined as those with a score of ≤ 2), and panic attack frequency. At the endpoint, paroxetine and sertraline were associated with equivalent levels of improvement on the PAS total score, and also all secondary measures. About 78% of patients taking paroxetine and 82% of patients taking sertraline were CGI-I responders. Paroxetine and sertraline had equivalent efficacy in the acute treatment of PD, but sertraline treatment resulted in less weight gain and less clinical worsening during the taper phase.

There have been a number of multicenter, double-blind, placebo-controlled studies demonstrating the efficacy of paroxetine for the treatment of PD. Ballenger et al. (1998) reported that 40 mg/day of paroxetine was more effective than placebo in reducing panic attacks and the symptoms of PD. In a large multicenter European trial, paroxetine (20–60 mg/day) was contrasted with clomipramine (50–150 mg/day) and placebo. Both active treatments were more effective than placebo in reducing the number of panic attacks. The paroxetine group also seemed to have a more rapid onset of action and reported fewer side effects (Lecrubier et al., 1997). Patients who completed the study were entered into a 9-month extension study; subjects treated with paroxetine and clomipramine demonstrated continued improvement over time (Lecrubier & Judge, 1997).

A head-to-head comparison trial of paroxetine and citalopram was conducted by Perna et al. (2001). Fifty-eight patients with PD were randomly assigned to one of the two active treatments (both 10–50 mg/day) in a single-blind design. Patients were assessed by the PASS, the SDS, and the FQ at baseline, day 7, and day 60. Responders were defined by a reduction of at least 50% from baseline on both PASS and SDS global scores at day 60. At the endpoint, 84% of patients receiving paroxetine and 86% of patients receiving citalopram responded. The PASS total scores decreased from 9.3 ± 4.9 at baseline to 3.5 ± 4.6 at day 60 for the paroxetine-treated group, and from 8.7 ± 3.7 to 3.1 ± 3.0 for the citalopram-treated group. Only one patient from each group dropped out because of side effects.

Sheehan et al. (2005) reported a pooled analysis that combined three identical trials investigating the efficacy and tolerability of controlled-release paroxetine (paroxetine CR) treatment of adults with PD. Subjects were randomly assigned to paroxetine CR (25–75 mg/day; $n = 444$) or placebo ($n = 445$) in these double-blind, 10-week clinical trials. The primary outcome measure used was the percentage of patients who were free of panic attacks in the 2 weeks prior to endpoint. About 63% of paroxetine CR-treated subjects were panic free compared with 53% of placebo-treated subjects ($p < 0.005$). Adverse events leading to study withdrawal were minimal and occurred in 11% of the paroxetine CR group and 6% of the placebo group.

Several smaller placebo-controlled trials and one large multicenter placebo-controlled acute trial have been published demonstrating that citalopram is an effective treatment of PD. In the large multicenter study of citalopram, placebo, and clomipramine, 475 patients were randomized to 8 weeks of treatment. The patients received three flexible doses of citalopram and one of clomipramine. All three doses of citalopram (10–60 mg/day) and clomipramine (60–90 mg/day) were more effective than placebo in decreasing panic attacks to zero and decreasing phobic symptoms. The 20–30 mg/day dose of citalopram seems to be the most effective treatment (Wade et al., 1997). Patients who in the physician's judgment were responders in the acute trial were offered continued double-blind treatment for 10 months. In this double-blind continuation study, citalopram was demonstrated to be more effective than placebo in decreasing phobic avoidance and interpersonal sensitivity (Leinonen et al., 1999).

Fluvoxamine has been reported to be more effective than placebo for the treatment of PD (Hoehn-Saric et al., 1993). Asnis et al. (2001) performed an 8-week, double-blind, parallel-group study comparing fluvoxamine with placebo in 188 patients with DSM-III-R defined PD (with or without agoraphobia). Patients in this study were required to have at least one panic attack per week for at least 4 weeks. Patients were randomized to double-blind treatment if they had an average weekly panic attack severity score of 25 (number of attacks × severity 0–10) and had at least one full panic attack in the final week of placebo washout. Treatment was initiated at 50 mg/day of fluvoxamine or placebo. The dose was titrated upwards to 150 mg/day during the first 2 weeks. Thereafter, it was titrated to 300 mg/day as needed, based on side effects and efficacy. Fluvoxamine was significantly superior to placebo at the endpoint on three of four primary outcome measures: proportion of patients free from full panic attacks, percentage reduction in frequency of full panic attacks per week, and PD severity. Fluvoxamine was superior to placebo with respect to a global improvement in CGI-Severity Scale scores, 64% versus 42%.

Pollack et al. (2003) extended the work of Goddard et al. (2001) by investigating the acute and continuation efficacy of combination treatment with benzodiazepines and SSRIs. In a double-blind, three-arm study, 60 patients with PD were randomized to paroxetine and placebo (PP), paroxetine co-administered with clonazepam followed by a tapered benzodiazepine discontinuation phase (PC-D), or paroxetine plus clonazepam with ongoing combination treatment (PC-M). All treatment groups demonstrated significant improvement by the endpoint (PP 68.2%; PC-M 60.0%, PC-D 77.8%). However, the results should be interpreted with caution, since this was a moderate-sized study and there was a high rate of early termination (60 subjects at baseline with only 34 study completers). Combined treatment with paroxetine and clonazepam resulted in more rapid response than with the SSRI alone, but there was no differential benefit beyond the initial few weeks of therapy. This suggests that combined treatment followed by benzodiazepine taper may provide early benefit while avoiding the potential adverse consequences of long-term combination therapy.

Dannon et al. (2007) conducted a 12-month, naturalistic, comparison study of citalopram, fluoxetine, fluvoxamine, and paroxetine. A total of 200 patients with PD were randomized equally to one of four SSRIs at comparable doses (citalopram mean dose 28.2 mg/day, fluoxetine

29.3 mg/day, fluvoxamine 187.2 mg/day, paroxetine 28.2 mg/day), with all groups reaching the final SSRI dose by week 4. A total of 127 patients completed the 12 months of treatment, with good response across all medications. Paroxetine had the highest rates of retention at 76%, followed by citalopram at 68%, fluvoxamine at 60%, and fluoxetine at 50%. The primary endpoint was the Panic Self-Questionnaire administered at baseline and monthly. In the first month of treatment, paroxetine and citalopram had lower rates of panic attacks; however, by month 3, efficacy was similar among all four medications.

Stahl et al. (2003) conducted a 10-week, randomized, double-blind, placebo-controlled trial of escitalopram and citalopram. A total of 366 patients were randomly assigned to escitalopram, citalopram, or placebo. Escitalopram treatment showed significant improvement compared with placebo in Panic and Agoraphobia Scale, CGI-I, CGI-S, CGI-Phobic Avoidance, HAM-A, PGE, and anticipatory anxiety duration scores. Citalopram treatment showed significant improvement compared with placebo in Panic and Agoraphobia Scale, CGI-I, CGI-S, CGI-Phobic Avoidance, PGE, and Quality of Life Enjoyment and Satisfaction Questionnaire scores. Choi et al. (2012) performed a 24-week open-label multicenter trial of escitalopram for 119 patients with PD. The total Panic Disorder Severity Scale (PDSS) and Sheehan Disability Scale scores were improved significantly over the 24 weeks of treatment.

In summary, the majority of data suggest that SSRIs as a class are effective in the treatment of PD. One of the advantages of SSRIs is that they tend to be fairly well tolerated, in contrast to some of the other treatments available for PD. Although a few individuals may have some initial problems with restlessness and increased anxiety, data suggest that starting at lower doses such as 25 mg/day of sertraline or 10 mg/day of paroxetine may decrease the risk of medication for depression "jitteriness." This particular adverse effect may be related to excessive synaptic serotonin levels (Abell & El-Mallakh, 2021) that occur when SSRIs are given to individuals that have the genetic variant of the short form of the serotonin transporter, which is associated with reduced

expression of the protein (El-Mallakh & Ali, 2019). Patients may certainly experience other types of side effects such as headaches, nausea, and diarrhea. However, most of these side effects diminish with time and are well tolerated, particularly if the patients are informed of the possibility of these transient side effects.

Serotonin–Norepinephrine Reuptake Inhibitors

There is currently evidence that venlafaxine, duloxetine, and reboxetine are effective in the treatment of PD. Venlafaxine is approved for PD in the USA and Europe, whereas duloxetine and reboxetine are only for off-label use. There have been several double-blind, placebo-controlled studies (Pollack et al., 1996; Bradwejn et al., 2005) suggesting that venlafaxine may be an effective treatment for patients with PD. Bradwejn et al. (2005) studied 361 adult outpatients with PDs. Subjects were randomly assigned to receive venlafaxine ER (75–225 mg/day) or placebo for up to 10 weeks in a double-blind study. Venlafaxine was associated with a lower mean number of panic attacks (venlafaxine −5, placebo −3.7), higher response (venlafaxine 68.1%, placebo 55.4%), and remission rates (venlafaxine 35.6%, placebo 24.4%) in CGI-I. A similar study was conducted by Liebowitz et al. (2009) that did not demonstrate significant efficacy of venlafaxine compared with placebo on the primary outcome measure, the percentage of panic-free patients according to the Sheehan Panic and Anticipatory Anxiety Scale (52% vs. 43%).

Pollack et al. (2007) were the first to report a large-scale, controlled trial of venlafaxine XR, paroxetine, and placebo. A total of 664 non-depressed adult outpatients who met DSM-IV-TR criteria for PD (with or without agoraphobia) were randomly assigned to 12 weeks of treatment with placebo, or fixed doses of venlafaxine XR (75 or 150 mg/day) or paroxetine 40 mg/day. The primary efficacy measure was the percentage of patients free from full-symptom panic attacks, assessed with the PAAS. Response was defined as a CGI-I rating of very much improved or much improved; remission was defined as a CGI-S rating of not at all ill or borderline ill and no PAAS full-symptom panic attacks. Measures of depression, anxiety, phobic fear and avoidance, anticipatory anxiety, functioning, and quality of life were also assessed. Intention-to-treat analysis showed that mean improvement on most measures was greater with venlafaxine XR and paroxetine than with placebo. No significant differences were observed between active treatment groups. Panic-free rates at the endpoint ranged from 54% of those treated with venlafaxine XR and 61% of those treated with paroxetine compared with 35% for placebo. Approximately 75% of patients given active treatment were responders and nearly 45% achieved remission. The placebo response rate was approximately 55% and the remission rate approximately 25%.

Ferguson et al. (2007) examined the long-term efficacy of venlafaxine ER in preventing relapse of responders; 291 patients underwent 12-week open-label treatment with venlafaxine ER, with response defined as ≤ 1 panic attack per week during the last 2 weeks of open-label treatment. A total of 169 responders were enrolled in the double-blind phase of venlafaxine versus placebo treatment for 26 weeks. Time to relapse was the primary outcome, with relapse defined as ≥ 2 full-symptom panic attacks per week for two consecutive weeks or until discontinuation of medication. Those treated with venlafaxine ER had significantly longer times to relapse, with 76.4% of venlafaxine-treated patients remaining panic attack free at the end of the study compared with 55% in the placebo group. Overall, relapse rates were 22.5% for venlafaxine and 50% for placebo. Venlafaxine-treated patients also showed significant improvements in quality of life and disability.

Simon et al. (2009) conducted an 8-week, open-label, flexible dose prospective trial of duloxetine (60–120 mg/day) for 17 patients with PD. Duloxetine treatment resulted in significant improvement in PDSS, supporting duloxetine as a potential drug for panic disorder.

There have been a few studies suggesting that reboxetine is effective in the treatment of PD. Versiani et al. (2002) conducted a multicenter, randomized, placebo-controlled, parallel-group, double-blind clinical trial in Brazil and Italy. A total of 82 patients with DSM-III-R diagnosis of PD with or without agoraphobia and who had experienced at least four panic attacks in the month preceding their admission to the study were included in the study. Participants had an initial 7-day washout period and then were randomly assigned to receive 6–8 mg/day of reboxetine or placebo for 8 weeks. Primary parameters of efficacy were the mean number of total panic attacks symptoms (spontaneous and situational) per week [as measured by the Sheehan Panic Attack and Anxiety Scale (SPAAS)], the global scores of severity and phobic symptomatology (Phobia Scale), and the score on the CGI. Efficacy measurements were determined by comparing baseline and last assessment scores. The study revealed that there was a significant reduction in the mean number of panic attacks and phobic symptoms in the reboxetine group compared with the placebo group. Consistent with the reduction of both major panic attacks and improvement in phobic symptomatology, significantly greater improvement in mean CGI scores for severity of illness was observed for patients in the reboxetine group (5.2 at baseline vs. 2.5 at last assessment) compared with those in the placebo group (5.0 at baseline vs. 3.8 at last assessment; $p = 0.0002$).

Monoamine Enhancers (Tricyclic Antidepressant Medication) and Monoamine Oxidase Inhibitors

Imipramine was the first monoamine enhancer (TCA) observed to treat panic attacks and PD effectively and is one of the best-studied monoamine enhancers (TCA) (Klein & Fink, 1962; Klein, 1964; Boyer, 1995). Double-blind, placebo-controlled trials of desipramine and nortriptyline have also been performed (Munjack et al., 1988; Kalus et al., 1991; Lydiard et al., 1993); however, the monoamine enhancer (TCA) clomipramine has been more widely studied. There is currently evidence that imipramine and clomipramine are effective in the treatment of PD. Clomipramine is approved for PD in Europe but not in the

USA. Imipramine is used off-label for PD. There are a large number of double-blind, placebo-controlled trials demonstrating clomipramine's efficacy as a treatment for PD (den Boer, 1998; Calliard et al., 1999). Interestingly, there are a number of both placebo-controlled and crossover design studies suggesting that clomipramine may be more efficacious than imipramine in the treatment of PD (Cassano et al., 1988; Modigh et al., 1992; Sasson et al., 1999).

Longitudinal data suggest that patients who achieve complete remission with imipramine treatment will maintain their remission with continuous treatment (Mavissakalian & Perel, 1992b). Mavissakalian and Perel (1992a) demonstrated that patients who met stringent remission criteria for at least 6 months could sustain their panic-free status even after their imipramine dosage was decreased by 50%. This finding is particularly important because such a dosage reduction might increase tolerability and thus the adherence of patients to therapy. Longer-term adherence is important because Mavissakalian and Perel (2001) found that the longer periods of panic-free maintenance therapy were associated with a greater likelihood of remaining in remission after discontinuation of imipramine therapy Eighteen patients participated in the second year of a double-blind extension study: seven patients in the placebo group continued on placebo, seven of the imipramine group were switched to placebo, and four continued with imipramine. Two patients from the imipramine-placebo switch group relapsed and two patients in the placebo continuation group relapsed. None of the imipramine continuation group subjects relapsed. This suggests that there may actually be an advantage of continued prophylaxis beyond the first year of maintenance imipramine therapy.

Monoamine enhancers (TCAs), unfortunately, are associated with a significant side-effect burden. The jitteriness associated with initiation of monoamine enhancers (TCAs) is frequently frightening and may be intolerable for some patients with PD. It requires that monoamine enhancers (TCAs) be started at very low doses and titrated up very slowly. Other problems with monoamine enhancers (TCAs) that decrease adherence include anticholinergic side effects, orthostatic hypotension, sexual dysfunction, and weight gain. In a naturalistic follow-up study by Noyes et al. (1989), the majority of patients discontinued monoamine enhancers (TCA) treatment because of side effects (Noyes et al., 1996). Although monoamine enhancers (TCAs) have been widely used for the treatment of PD, their side-effect profile and slow time to onset of action make them a difficult class of medication for many patients to tolerate.

The first-generation MAOIs such as phenelzine and tranylcypromine have been demonstrated to be effective in placebo-controlled trials (Sheehan et al., 1980) although neither phenelzine nor tranylcypromine is formally approved for use in PD. Nardi et al. (2010b) found that both 30 and 60 mg/day doses of tranylcypromine were efficacious in treating PD, with a 69.6% reduction in panic symptoms in the 30 mg/day group and 74.8% in the 60 mg/day group. There are also a variety of smaller studies demonstrating that the RIMAs moclobemide and brofaromine are effective in the treatment of PD (Bakish, 1992; van Vliet et al., 1993; Kruger & Dahl, 1999; Tiller et al., 1999). Unfortunately, it is unlikely that either of the newer reversible and selective MAOIs will be available in the United States. Therefore, our pharmacotherapy is limited to the older irreversible and nonselective agents.

The irreversible and nonselective agents have problems with dietary restrictions, weight gain, insomnia, sexual dysfunction, medication interactions, and orthostasis. They can be lethal in overdose and hypertensive crises can be precipitated by tyramine-rich foods and by some commonly sold over-the-counter cold medicines. Despite these concerns, many expert clinicians feel that MAOIs may be a useful treatment for severely ill patients with PD, particularly those with comorbid PD and depressive disorder.

Other Medications for Depression (Bupropion, Mirtazapine, Nefazodone, Trazodone, Vortioxetine)

There have been a few double-blind trials of bupropion and trazodone for the treatment of

PD, but neither was found to be more effective than placebo (Sheehan et al., 1983; Charney et al., 1986). Mirtazapine, nefazodone, and vortioxetine are all used off-label for PD. Ribeiro et al. (2001) compared mirtazapine and fluoxetine in a randomized double-blind, flexible-dose trial of 27 patients with DSM-IV-TR diagnosed PD. Participants were required to have a minimum of three panic attacks during the 2 weeks before enrollment into the study and a score of >18 on the HAM-A. Following a 1-week, single-blind, placebo run-in, patients were randomly assigned to mirtazapine 15 mg/day or fluoxetine 10 mg/day. At week 3, doses could be raised to 30 mg/day of mirtazapine or 20 mg/day of fluoxetine according to clinical response. Fourteen patients were treated with mirtazapine and 13 with fluoxetine. Three patients treated with mirtazapine and two treated with fluoxetine dropped out because of side effects. Among completers, the mean daily dose of mirtazapine was 18.3 ± 1.3 mg/day and that of fluoxetine was 14.0 ± 1.0 mg/day. HAM-A scores decreased from 15.7 ± 10.0 at baseline to 10.7 ± 11.2 at endpoint for the mirtazapine-treated subjects and from 28.8 ± 6.5 at baseline to 11.8 ± 7.5 at endpoint for the fluoxetine-treated subjects. In both groups, the number of panic attacks decreased from three at baseline to zero at endpoint, suggesting that both medications eliminated spontaneous panic attacks.

Nefazodone is a combined serotonin receptor antagonist and serotonin reuptake inhibitor (SARI). There have been a few open-label trials of nefazodone for PD showing early promise. DeMartinis et al. (1996) conducted an 8-week open trial of nefazodone (200–600 mg/day) for 14 patients with PD. Nefazodone significantly improved in panic attack severity, phobic avoidance, HAM-D, HAM-A, CGI-Severity, and Sheehan Disability Scale scores. Bystritsky et al. (1999) et al. conducted a 12-week open-label trial of nefazodone (50–400 mg/day) for 10 patients with PD. Nefazodone significantly improved in PDSS and CGI-scale. Papp et al. (2000) conducted a 12-week open-label trial with nefazodone for 19 patients with PD. Intent-to treat response rate of nefazodone was 37%

(7/19). Further study is required to clarify the benefit of nefazodone in PD.

Vortioxetine is a novel type of medication for depression working as a 5-HT_3 receptor antagonist and 5-HT_{1A} receptor agonist. Shah and Northcutt (2018) conducted an open-label, flexible dose trial of vortioxetine for 27 patients with PD and found significant improvement in PDSS.

Benzodiazepine Medication

Alprazolam and clonazepam are high-potency benzodiazepines that have been approved by the FDA and are widely used in the treatment of PD. Alprazolam has a relatively short half-life, requires frequent dosing (up to four times per day), and may cause significant discontinuation problems in some patients. In the Cross-National Collaborative Panic Disorder Study, alprazolam at a mean dose of 5.7 mg/day was found to be more effective than placebo in decreasing the number of panic attacks and allowing patients to become free from panic attacks (Ballenger et al., 1988). In phase 2 of the study (Cross-National Collaborative Panic Study, Second Phase Investigators, 1992), 1168 PD patients were randomized to receive alprazolam, imipramine, or placebo. Both imipramine and alprazolam were more effective in treating PD than placebo.

Rickels and Schweizer have published a number of papers comparing alprazolam with imipramine and placebo. In a report of an acute 8-week study, they found that alprazolam and imipramine were equally effective in treating PD but there were a large number of dropouts in the imipramine-treatment group (Rickels & Schweizer, 1998). During a 6-month maintenance phase, 62% of alprazolam-treated patients completed the study and were panic free, in contrast to only 26% of imipramine-treated and placebo-treated patients. A recent meta-analysis showed that benzodiazepine medication was more effective in reducing the number of panic attacks than monoamine enhancers (TCAs) (risk ratio = 1.13; 95%CI = 1.01–1.27) (Offidani et al., 2013). In general, alprazolam has been demonstrated in both short- and long-term studies to be effective. There has been considerable concern about the risk of dependency and also difficulty in discontinuing alprazolam in some patients. Additionally, evidence that benzodiazepine use may contribute to falls, motor vehicle accidents, and the onset of dementia has all limited their use, particularly among older patients (Islam et al., 2016; Woolcott et al., 2009; Hemmelgarn et al., 1997).

Although clonazepam has been used for the treatment of PD for a long time, it was not until the 1990s that two large, definitive, multicenter, double-blind, placebo-controlled studies demonstrated clonazepam's efficacy as monotherapy for patients with PD. In a large fixed-dose study, 413 patients were randomized to receive one of five doses of clonazepam (0.5, 1.0, 2.0, 3.0, 4.0 mg/day) or placebo (Rosenbaum et al., 1997). There was a 3-week dosage escalation period followed by 6 weeks of fixed-dose therapy and then a 7-week discontinuation phase. Doses of clonazepam greater than 1 mg/day were effective in decreasing the number of panic attacks and were well tolerated. Some exacerbation of symptoms was experienced by patients during the double-blind discontinuation phase, but patients with these symptoms did not approach the initial baseline level of dysfunction. In a flexible-dose study by Moroz and Rosenbaum (1999), 222 patients were randomized to clonazepam treatment and 216 patients received placebo. Patients randomized to clonazepam had a decrease in the number of panic attacks and an improvement in CGI-Severity scores. Clonazepam was well tolerated but there was some worsening of symptoms during discontinuation.

Nardi et al. (2011) investigated the short-term efficacy of clonazepam compared with paroxetine in an 8-week trial of 120 patients. They observed a significantly reduced frequency of panic attacks in clonazepam-treated patients by week 4 compared with paroxetine-treated patients (92.06% vs. 70.16%). Both treatments demonstrated efficacy by week 8. The same group extended their investigation to examine the long-term efficacy of both treatments over 34 months at target doses of clonazepam 2 mg/day and paroxetine 40 mg/day (Nardi et al., 2012). Patients with a good response to the initial 8-week trial were continued on the same treatment, and partial responders were switched to combination treatment with both medications. Although both treatments had similar efficacy in long-term maintenance of PD, clonazepam had significant, albeit small, improvements in CGI-I scores and gave fewer adverse events (28.9% vs. 70.6%) than paroxetine. Freire et al. (2017) conducted 6-year posttreatment follow-up with clonazepam and paroxetine for 120 patients with PD and found lower recurrence with clonazepam treatment than with paroxetine treatment according to CGI-S scores.

Goddard et al. (2001) looked at early co-administration of clonazepam with sertraline for PD in a double-blind trial of 50 patients. Patients received open-label sertraline for 12 weeks (target dose 100 mg/day) and were randomized to receive either 0.5 mg/day of clonazepam three times per day or placebo for the first 4 weeks of the trial. The response rates were markedly different at the end of week 1, with 41% of sertraline/clonazepam-treated patients responding to treatment versus 4% of sertraline/placebo-treated patients. At week 3, 63% of sertraline/clonazepam patients responded versus 32% of sertraline/placebo patients.

Katzenick et al. (2006) investigated combination therapy with alprazolam ODT and SSRI/SNRI compared with SSRI/SNRI monotherapy. A total of 245 subjects diagnosed with PD, with or without agoraphobia, were randomized to 8 weeks of open-label treatment with alprazolam. The primary efficacy measure was time to response, a $\geq 50\%$ decrease from baseline in HAM-A total score. Secondary measures included change from baseline in HAM-A total score, CGI-I, and PGI scales. A statistically significant difference between treatment groups in time to response was not found. Analyses revealed that combination treatment was statistically significantly more likely to show an earlier response than monotherapy ($p < 0.05$). Combination treatment also demonstrated that a significantly higher proportion of subjects met response criteria than those receiving monotherapy (35% vs. 22%, $p < 0.05$). Additionally, combination treatment demonstrated statistically significant earlier improvement than monotherapy on the mean change from baseline in total

HAM-A, CGI-I, and PGI scores. Both treatment regimens were generally well tolerated.

Concern about withdrawal on discontinuation of benzodiazepines may have impacted clinician prescribing practices. However, Nardi et al. (2010a) determined that a slow discontinuation of clonazepam, 0.25 mg/week, did not result in major withdrawal symptoms or exacerbation of panic attacks in 73 patients with PD who had been receiving clonazepam treatment for at least 3 years; 68.9% of patients were clonazepam-free within 4 months of tapering, and 26% required a further 3 months of tapering to be discontinued entirely off the benzodiazepine. There were no significant adverse effects associated with longer-term clonazepam treatment.

Other benzodiazepines have also been demonstrated to be effective in the treatment of PD. In an 8-week, double-blind, placebo-controlled trial, alprazolam (4.9 mg/day) and diazepam (40 mg/day) were equally effective and superior to placebo for the treatment of PD (Noyes et al., 1996). In a smaller comparative study, lorazepam was as effective as alprazolam in the treatment of PD (Schweizer et al., 1988). Chlordiazepoxide, clorazepate, and oxazepam have also been studied for PD in small clinical trials (Sartori and Singewald 2019). In summary, it is clear that two high-potency benzodiazepines, alprazolam and clonazepam, are effective in the treatment of PD. Unfortunately, there are concerns about dependency and adverse effects associated with the use of benzodiazepine medication, including contributions to dementia and delirium, accidents, and increasing recognition of the role that benzodiazepines play in the opioid overdose epidemic.

Buspirone

In general, buspirone is not thought to be an effective monotherapy for PD.

Anticonvulsant Medication

Anticonvulsants are not approved for PD in the USA or Europe. There are published case reports, case series, and small studies investigating at least five different anticonvulsant medications. Sandford et al. (2001) conducted a small randomized, double-blind, placebo-controlled crossover trial of pagoclone versus placebo in 16 patients with PD. Pagoclone is believed to act as a partial agonist at the $GABA_A$ benzodiazepine receptor. Patients were randomized to receive either pagoclone 0.1 mg TID during weeks 1 and 2 or pagoclone 0.1 mg TID during weeks 4 and 5. There was a decrease in the number of full panic attacks from baseline among pagoclone-treated patients but not among placebo-dosed patients; however, this difference did not reach statistical significance.

Zwanzger et al. (2001) investigated the putative anxiolytic properties of tiagabine in a clinical case series. Four patients meeting DSM-IV-TR criteria for PD were treated. All patients reported a reduction in panic attacks and improvement in anxiety levels within 4 weeks, although one patient discontinued treatment because of side effects. Sheehan et al. (2007) further investigated tiagabine efficacy in PD in 28 patients, but did not find a clinically significant improvement. Zwanzger et al. (2009) again investigated tiagabine efficacy in a subsequent study with 19 patients randomized to either tiagabine or placebo. No statistical significance was found in clinical symptom improvements between treatments.

There are some scattered reports suggesting that the valproic acid and gabapentin may be useful, particularly in either atypical or treatment-resistant patients with PD (Keck et al., 1993; Woodman & Noyes, 1994). A small study by Masdrakis et al. (2010) did not demonstrate significant benefit of lamotrigine in the treatment of panic disorder with agoraphobia.

Papp (2006) reported the results of an open-label, flexible-dose study of levetiracetam treatment for 18 subjects meeting criteria for PD with or without agoraphobia. Eighteen patients completed this 3-month, open-label study and 85% of them were rated very much or much improved on the CGI-I. Panic attack frequency, total HAM-A scores, and CGI-S scores decreased significantly during these 12 weeks of open-label treatment. The median dose of levetiracetam used in the study was 1138 mg/day. The most common side effects were sedation, headaches, and some irritability.

Medications for Psychosis (Antipsychotics)

There are no approved medications for psychosis (antipsychotics) for PD. A study of olanzapine augmentation of SSRI-resistant PD was conducted by Sepede et al. (2006). In this study of 31 adult outpatients with PD who were unresponsive to standard SSRI treatment, subjects were given 5 mg/day of olanzapine. Twenty-six patients completed this 12-week, open-label study. By week 12, 82% of patients were defined as responders and demonstrated significant improvement on all the rating scales. Fifteen patients achieved remission status (no panic attacks and a HAM-A total score of <7). Common side effects in this 12-week study were weight gain and drowsiness.

Prosser et al. (2009) studied the efficacy of risperidone monotherapy compared with SSRI treatment in PD. Fifty-six patients were randomized to receive low-dose risperidone (0.125–1 mg/day) or paroxetine 30 mg/day for an 8-week trial. Retention rates were low for the study, with 39.4% dropout in the risperidone group and 60.9% in the paroxetine group. Both risperidone and paroxetine were found to be efficacious in reducing panic attack occurrence and severity; however, no statistical significance was noted by the endpoint on any scale in risperidone-treated patients compared with paroxetine-treated patients.

Following some case reports, which showed the efficacy of quetiapine in PD, Goddard et al. (2015) conducted an 8-week double-blind, placebo-controlled, randomized trial of quetiapine extended release with 43 patients with SSRI-resistant PD. The study showed that quetiapine extended release treatment was not statistically superior to placebo in the treatment x time interaction although there was improvement in PDSS total scores across the 8-week trial.

Novel Treatments

Several alternative treatments for PD are being investigated. Suvorexant is an FDA-approved insomnia agent that functions as an orexin receptor antagonist. The clinical trial of suvorexant for PD is in phase 4 (Sartori & Singewald, 2019). Cannabidiol, a major phytocannabioids, has been in phase 3 clinical trial in treating symptoms of GAD, SAD, PD, or agoraphobia in adults recently (Sartori & Singewald, 2019). One meta-analysis showed no statically significant differences of efficacy between propranolol and benzodiazepines for PD in the short-term (mean difference = 1.58, 95%CI = −2.33–5.50) (Steenen et al., 2016).

Treatment Conclusions

Although panic attacks are a cardinal feature of PD, many patients suffer from PD but rarely have spontaneous panic attacks. These individuals are plagued by problems with anxiety sensitivity and anticipatory anxiety. In general, patients with PD, with or without agoraphobia, tend to be exquisitely sensitive to their body's somatic cues, and therefore they are remarkably sensitive to all of the potential somatic effects of pharmacotherapy. Clinicians treating patients with PD must help them gain a sense of control and mastery as pharmacotherapy is initiated. One approach to ensuring that this occurs is to enlist the patient as an active participant in monitoring both the potential positive and negative effects of pharmacotherapy. There are currently a host of effective short-term treatments for PD, including the SRIs, SNRIs, MAOIs, monoamine enhancers (TCAs), and benzodiazepines. SSRIs and SNRIs are generally accepted as first-line treatments for PD. However, if one does not gradually titrate a patient with PD up on these medications, one can be certain of noncompliance owing to their side-effect burden. A second pharmacological technique that has been effective for patients who suffer from PD is to co-administer a small dose of a benzodiazepine such as clonazepam or alprazolam with a medication for depression, and then withdraw the benzodiazepine over a period of several weeks, after the patient has reached a therapeutic dosage of the medication for depression.

Panic attacks may be effectively controlled relatively early in the course of treatment; however, anticipatory anxiety and somatic sensitivity take significantly longer to resolve. Some form of behavioral intervention, even a brief one in a medication management session, can facilitate more rapid improvement in anticipatory anxiety and phobic behavior. Aside from the widely used augmentation of medications for depression with benzodiazepines, there are very few data guiding our choice of second- and third-line approaches to the treatment of patients with PD.

Posttraumatic Stress Disorder

Selective Serotonin Reuptake Inhibitors

There have been several open-label and double-blind, placebo-controlled studies demonstrating that SSRIs are effective for the treatment of PTSD. Open-label trials with the SSRIs currently available suggest that each may be effective in decreasing the core symptoms of PTSD. Three of the SSRIs have been studied in a double-blind, placebo-controlled fashion. Paroxetine and sertraline are approved for PTSD in the USA and Europe, whereas fluoxetine, citalopram, escitalopram, and fluvoxamine are prescribed for off-label use. Fluoxetine was the first of the SSRIs to be studied in controlled trials. In two small studies, civilian patients treated with fluoxetine demonstrated significant improvement compared with placebo treatment. In one of these trials, van der Kolk et al. (1994) also included a military veteran subgroup, and fluoxetine decreased symptoms of PTSD in this cohort. Martenyi et al. (2002) conducted a double-blind, randomized study comparing the effects of fluoxetine and placebo on relapse prevention in 131 subjects meeting DSM-IV-TR criteria for PTSD. The subjects underwent 12 weeks of acute treatment, and patients who responded were re-randomized and continued in a 24-week relapse-prevention phase with fluoxetine or placebo. Subjects in the fluoxetine/fluoxetine group were found to be less likely to relapse than those in the fluoxetine/placebo group.

Sertraline and paroxetine have been approved by the FDA for the treatment of PTSD. Two large placebo-controlled, acute-treatment trials of sertraline for noncombat-related PTSD have been published. Davidson et al. (2001) and Brady et al. (2000) found that sertraline was more effective than placebo treatment in decreasing the Clinician Administered Posttraumatic Stress Disorder Scale 2 (CAPS-2), the patient-rated Impacts of Events Scale, and the CGI-Severity and Improvement Rating Scales. Both of the studies were flexible-dose studies in which 50–200 mg/day of sertraline was administered. Londborg et al. (2001) reported open-label continuation data for individuals who had participated in these acute trials (Davidson et al., 2001). They demonstrated that patients who responded during the acute phase not only maintained their response but also continued to improve after 6 months of continuation treatment. About 54% of individuals who did not acutely respond to SSRI treatment during the initial 12-week trial had a significant decrease in their CAPS-2 scale scores by the end of 6 months of open-label treatment. Davidson et al. (2001) reported the results of the placebo discontinuation phase of this study. Individuals who responded to 6 months of open-label therapy were rerandomized to receive double-blind treatment with sertraline or placebo. Placebo discontinuation was associated with a significant risk of relapse (26% vs. 5%) and reemergence of the core symptoms of PTSD.

Rapaport et al. (2002) investigated the effects of sertraline treatment on quality of life in the samples described in the studies by Londborg et al. (2001) and Davidson et al. (2001). By the end of the 12-week acute phase, 58% of sertraline responders had achieved Q-LES-Q total scores within 10% of community norms and had shown significant improvement on the Medical Outcomes Study 36-Item Short Form Health Survey (SF-36) subscale scores. Continuation of treatment resulted in a further 20% improvement in quality-of-life and psychosocial measures. During the double-blind maintenance phase, randomization to placebo was associated with significant worsening of Q-LES-Q scores and quality of life.

These findings from Davidson et al. (2001), Londborg et al. (2001), and Rapaport et al. (2002) suggest that some patients with PTSD will require sustained SSRI treatment, possibly for years, in order to protect them from exacerbation of PTSD. Zohar et al. (2002) conducted a double-blind, placebo-controlled study of sertraline (50–200 mg/day) in military veterans with PTSD. Forty-two patients were randomized to 10 weeks of sertraline or placebo treatment. Primary efficacy measures included the CAPS-2 and CGI-I. Although the sertraline treatment group demonstrated improved scores on the CAPS-2, no statistically significant difference was observed. At the study endpoint, mean CGI-I scores were lower in the sertraline-treated group compared with placebo (2.4 ± 0.3 for sertraline and 3.4 ± 0.3 for placebo). CGI-I responders rates were 53% for sertraline and 20% for placebo and combined CGI-I and CAPS-2 responder rates (≥30% reduction in baseline CAPS-2 score) were 41% for sertraline and 20% for placebo. Sertraline was well tolerated and may be an effective treatment for patients with combat-related PTSD, although larger studies are needed to confirm these results. A study by Robb et al. (2010) showed no significant benefit with sertraline over placebo in pediatric PTSD, so although sertraline has proven efficacy in adults it has not translated equally to the child and adolescent population.

Paroxetine is also approved by the FDA as an acute treatment for PTSD. There has been one fixed-dose study demonstrating that both 20 and 40 mg/day of paroxetine were more effective than placebo for the treatment of PTSD (Marshall et al., 2001). In this study, patients treated with paroxetine had significantly greater improvement in the CAPS-2 and were more likely to have CGI-Improvement Scale scores of (2) "much improved" or (1) "very much improved." In the 12-week flexible-dose study, 20–50 mg of paroxetine treatment was more effective than placebo treatment in decreasing the CAPS-2 scale, the Davidson Trauma Scale, and the SDS (Tucker et al., 2001). Marshall et al. (2007) replicated these results in a follow-up study in which they found paroxetine (maximum 60 mg/day) was significantly more efficacious than placebo. After a 10-week treatment period, 66.7% of paroxetine-treated patients were rated as responders on the CGI-I compared with 27.3% of placebo-treated patients. Those receiving paroxetine also had

significant improvements on the CAPS, Dissociative Experiences Scale (DES), and self-reported Inventory of Interpersonal Problems (IIP) scores. After the initial 10-week treatment period, subjects maintained on paroxetine for another 12 weeks continued to show significant and ongoing improvement in all measures compared with those receiving placebo.

English et al. (2006) conducted an 8-week open trial of citalopram for PTSD ($N = 18$). The result demonstrated the moderate effect of citalopram in the PTSD treatment in CAPS, HAM-A, Global Assessment of Function, and CGI-rating scales. However meta-analysis failed to show the significant difference between 8 and 12 weeks of citalopram treatment and placebo for PTSD (effect size $= 0.18, 95\% \text{ CI} = -0.56$–0.91) (Lee et al. 2016).

Escitalopram has also been studied for PTSD. Qi et al. (2017) treated 45 patients with high doses of escitalopram (40 mg/day) and demonstrated reduced CAPS scores at 4 weeks. In a 56-week randomized controlled trial of escitalopram in 198 subjects by Zohar et al. (2018), the treated participants showed improvement in CAPS, Pittsburgh Sleep Quality Inventory, and measures of depression and global illness severity.

Fluvoxamine has a weaker evidence base. Escalona et al. (2002) conducted a 14-week open-label study of fluvoxamine in 15 veterans with PTSD and found significant improvement in total CAPS scores, but no replications or follow-ups have yet been attempted. In conclusion, data from open-label studies and double-blind, placebo-controlled studies suggest that SSRIs, as a class, are effective in the treatment of PTSD. In addition, longer-term research confirms the continued benefit of extended SSRI treatment.

Serotonin–Norepinephrine Reuptake Inhibitors

Venlafaxine and duloxetine are often used off-label for PTSD. Davidson et al. (2006b) compared venlafaxine ER, sertraline, and placebo in a 12-week, double-blind, randomized, flexible-dose, multicenter trial of adult outpatients. Patients met DSM-IV-TR PTSD criteria of at least 6 months' duration and had a score >60 on the Clinician Administered PTSD Scale (CAPS-SX-17). Subjects were randomly assigned to receive placebo, venlafaxine ER (37.5–300 mg/day, mean maximum dose 225 mg/day), or sertraline (25–200 mg/day, mean maximum dose 151 mg/day). Mean changes in CAPS-SX-17 scores were −41.8, −39.4, and −33.9 for venlafaxine ER, sertraline, and placebo, respectively. Week 12 remission rates were 30.2, 24.3, and 19.6% for the venlafaxine XR, sertraline, and placebo groups, respectively. Davidson et al. (2006a) also conducted a 6-month, double-blind, placebo-controlled, multisite trial of venlafaxine ER with 329 adult outpatients. Patients were randomly assigned to placebo or venlafaxine ER (37.5–300 mg/day, mean maximum dose 221.5 mg/day) for 24 weeks. Mean changes from baseline to CAPS-SX-17 total scores at endpoint were −51.7 for venlafaxine ER and −42.9 for placebo. The results from these studies suggest that venlafaxine ER is an effective short- and long-term (6-month) treatment for PTSD.

Two studies investigated the efficacy of duloxetine in the treatment of PTSD. Villarreal et al. (2010) enrolled 20 military PTSD patients in a 12-week open-label trial of duloxetine (mean dose 81 mg/day). In the 15 patients who completed the full 12-week trial, significant improvements were found in total CAPS score changes from baseline to endpoint. Significant improvements were also found in secondary measures of CAPS subscales, sleep measures, and HAM-D depression ratings. About 45% of patients were responders by the end of the trial, with most improvement noted by week 2 of treatment. The most common adverse events were constipation, diarrhea, and nausea, with two patients developing tachycardia. Walderhaug et al. (2010) conducted a study of duloxetine in treatment-resistant PTSD with comorbid major depressive disorder (MDD). Twenty-one combat veterans with PTSD and MDD were given duloxetine 60–120 mg/day for 8 weeks. Significant improvements were noted in both PTSD and MDD symptoms. Notably, duloxetine was efficacious in decreasing nightmares and improving sleep. Future double-blind, placebo-controlled trials would help

to clarify further the role of duloxetine in the treatment of PTSD.

Monoamine Enhancers (Tricyclic Antidepressant Medication) and Monoamine Oxidase Inhibitors

Monoamine enhancers (TCAs) and MAOIs are sometimes used off-label for PTSD. There have been several small double-blind, placebo-controlled trials suggesting that both amitriptyline and imipramine are more effective than placebo in decreasing signs and symptoms of PTSD. However, after the advent of SSRIs that have significant safety and tolerability advantages, additional studies with monoamine enhancers (TCAs) including clomipramine were not pursued (Davidson et al., 1990; Kosten et al., 1991).

There is little literature suggesting that MAOIs are more effective than placebo in the treatment of PTSD. The two best-studied MAOIs are the irreversible nonselective MAOI phenelzine and the reversible selective MAOI brofaromine. Placebo-controlled studies by Davidson et al. (1987) and Kosten et al. (1991) suggested that phenelzine is more effective than placebo in decreasing symptoms of PTSD. Their findings are supported by the work of Lerer et al. (1987), who demonstrated that phenelzine was a beneficial treatment for veterans with PTSD. As summarized by Stein et al. (2000), there have been several studies investigating the efficacy of brofaromine in the treatment of PTSD. International studies with brofaromine demonstrated that it was more effective than placebo, while the initial work performed in the United States differentiated brofaromine from placebo treatment. Although failed studies are not uncommon, this has unfortunately led to discontinuation of the research program with brofaromine in the United States.

Other Medications for Depression (Bupropion, Mirtazapine, Nefazodone, Trazodone)

Bupropion, Mirtazapine, Nefazodone, and Trazodone have all been considered for off-label use in

PTSD. Davis et al. (2004) conducted a 12-week, double-blind, placebo-controlled study of nefazodone's effects on chronic PTSD. Forty-one individuals were enrolled in this study, of whom all but one suffered from combat-related PTSD. Patients were given either placebo or 100–600 mg of nefazodone twice daily. After 12 weeks, the subjects showed a significant improvement in CAPS total score, and also the CAPS hyperarousal criterion and HAM-D scores. No significant between-group differences were found on the other study efficacy measures; however, it should be noted that this study had a high attrition rate that may have affected its results.

Two studies compared the efficacies of nefazodone and sertraline (Saygin et al., 2002; McRae et al., 2004), with mixed results. Saygin et al. (2002) recruited 60 outpatients suffering from PTSD relating to an earthquake trauma and randomized them to flexible doses of nefazodone (200–400 mg/day) or sertraline (50–100 mg/day) for 5 months. Both medications produced similar improvements in symptoms, as measured by CGI scores, but sertraline scores on the eight-item Treatment-Outcome Posttraumatic Stress Disorder Scale (TOP-8) reached significance over the scores for nefazodone-treated patients. In a shorter, 12-week study, McRae et al. (2004) randomized 37 patients to either 25–100 mg sertraline given twice daily or 50–300 mg nefazodone given twice daily. Both medications significantly reduced PTSD symptoms, with no statistical difference between treatment groups on outcome measures, which included the 17-item CAPS-2 severity score, CGI-I, Davidson Trauma Scale (DTS), TOP-8, SDS, MADRS, HAM-A, and Pittsburgh Sleep Quality Index (PSQI). These trials suggest that nefazodone and sertraline may be equally effective for the short-term treatment of PTSD symptoms, with sertraline separating from nefazodone in longer continuation treatment.

There have also been six open-label studies investigating the efficacy of nefazodone in PTSD. The overall results of these six trials were summarized in a paper by Hidalogo et al. (1999), where they suggested that nefazodone treatment decreased the core symptoms of PTSD, improved sleep, and decreased symptoms of anger. Among

these six open-label studies, three focused on patients who were veterans with chronic PTSD. Both Hertzberg et al. (1998) and Davis et al. (2000) found that open-label nefazodone treatment decreased core symptoms of PTSD. Zisook et al. (2000) demonstrated that open-label nefazodone treatment was helpful for patients who were refractory to SSRI treatment for PTSD. In all of the open-label studies, nefazodone was titrated to what was considered to be a clinically effective dose, in the 400–600 mg/day range. Subsequently, the FDA revised nefazodone's labeling to include a warning of rare cases of liver failure in patients receiving this medication, but none of the aforementioned trials found any adverse hepatic events during the course of the studies.

Kim et al. (2005) investigated the effectiveness of mirtazapine during an open-label, 24-week continuation treatment study ($n = 12$). The results suggested that mirtazapine may be effective in short- and long-term treatment of PTSD. Mirtazapine was also studied in an 8-week, placebo-controlled, double-blind trial of 29 subjects who met DSM-IV-TR criteria for PTSD. Approximately 65% of mirtazapine-treated subjects met response criteria versus 20% for the placebo-treatment group.

Chung et al. (2004) conducted a 6-week, flexible-dose study comparing mirtazapine and sertraline in 95 outpatient and 18 inpatient Korean veterans who met DSM-IV-TR criteria for PTSD. Efficacy measures included the CAPS-2, HAM-D-17, and CGI-I. Response was defined as a $\geq$30% decrease in CAPS-2 total severity, a $\geq$50% decrease in total HAM-D-17 score, and a CGI-I score of <3. The mean daily dose of mirtazapine was 34.1 mg/day and that of sertraline was 101.5 mg/day. The mirtazapine-treated group showed a statistically significantly higher response rate than the sertraline-treated group on the CAPS-2 total score (88% vs. 69%). The HAM-D-17 total score and the CGI-I score decreased in both treatment groups, with no significant difference between the groups.

Schneier et al. (2015) ran a 24-week, placebo-controlled trial of combined treatment with an SSRI and mirtazapine in 36 patients with PTSD.

They found that the combination of sertraline with mirtazapine significantly improved PTSD symptoms compared to sertraline with placebo, in both remission rates and reduced depression severity. Recently, Davis et al. (2020) conducted an 8-week, randomized, placebo-controlled study of mirtazapine in 78 patients with PTSD. However, the mirtazapine-treated group failed to show the efficacy of mirtazapine monotherapy on PTSD in Structured Interview for Posttraumatic Stress Disorder scores.

Becker et al. (2007) conducted an 8-week placebo-controlled trial of bupropion, which failed to show the superiority of bupropion to reduce PTSD symptoms compared to placebo as measured by CAPS. Trazodone has shown some promise in treating PTSD associated nightmares in a pair of small studies (Brownlow et al., 2015; Detweiler et al., 2016).

Beta-Blocker Medication

Beta-blocker medication is not FDA approved for the treatment of PTSD, but has been used clinically. Results from two pilot studies that investigated a course of propranolol among individuals who had recently experienced an acute traumatic event indicated that propranolol was effective in decreasing PTSD symptoms and severity. Pittman et al. (2002) gave 40 mg of propranolol four times per day for 10 days (with initial dosing beginning within 6 h following the acute traumatic event) and found the dosing regimen to be effective in reducing PTSD symptoms 1 month later. Similarly, Vaiva et al. (2003) studied the efficacy of propranolol prescribed shortly after trauma exposure in the prevention of PTSD. Eleven patients were given 40 mg of propranolol three times per day for 7 days and the results were compared with those for eight patients who refused the propranolol but agreed to participate in the study. PTSD rates measured 2 months after the trauma were higher in the refusal group (3/8) than the propranolol-treated group (1/11), as were the levels of PTSD symptoms. Menzies (2012) reported comparable findings in a study of 36 patients treated with either one 40 mg dose of

regular formula propranolol or two doses of propranolol (one 40 mg dose of regular formula, followed 2 h later by one 60 mg dose of long-acting propranolol) after reactivation of their trauma memory. Some 91.7% of patients responded to propranolol on either dosing schedule with a Patient Global Improvement Scale (PGIS) mean score of 8.4/10, indicating "major improvement/back to normal self."

However, a study conducted by Hoge et al. (2012) did not find evidence that propranolol (up to 240 mg/day) was beneficial in preventing PTSD symptoms in 41 emergency room patients presenting with acute psychological trauma, who were monitored for up to 13 weeks post-trauma. Brunet et al. (2018) conducted a 6-week, randomized-controlled trial of pre-reactivation propranolol therapy for 60 patients with PTSD. Propranolol administered before a brief memory reactivation session improved PTSD symptoms compared to placebo as measured by patient self-report scales. Further larger-scale studies would be helpful in clarifying efficacy of propranolol in acute psychological trauma and PTSD.

Anticonvulsant Medication

No anticonvulsant medication is currently approved for the treatment of PTSD. Three studies investigated the efficacy of tiagabine with PTSD patients (Taylor, 2003; Connor et al., 2006; Davidson et al., 2007). Taylor (2003) studied a case series of seven patients where 8 mg/day tiagabine was added to the patients' existing medication regimens. During the 16-week trial, six of the seven patients improved markedly after 2 weeks of augmentation therapy. Tiagabine alleviated the severity of nightmares and avoidance, and arousal characteristics of PTSD. Complementing these results, Connor et al. (2006) enrolled 26 PTSD patients in 12 weeks of open-label treatment with 2–16 mg tiagabine daily. The 18 subjects who showed minimal response on the CGI-I scale were then randomized to 12 weeks of double-blind treatment with tiagabine or placebo, either being tapered off their medication or remaining at the same dose as they had taken at the end of the open-label phase. Among the eight completers, symptoms decreased by 68% in tiagabine-treated patients and 54% in placebo-treated patients, with equal relapse rates in the two groups. Davidson et al. (2007) conducted a large, double-blind, placebo-controlled study with 232 patients. Patients were randomized to tiagabine (titrated to a maximum dose of 16 mg/day) or placebo during a 12-week study period. No significant differences were found in the total CAPS score change from baseline to endpoint between tiagabine and placebo-treated patients. This suggests that tiagabine is not efficacious in treating PTSD.

Berlant pioneered the use of topiramate as a monotherapy or an adjuvant treatment for PTSD. In his published case series (Berlant, 2001), topiramate seemed to be particularly useful in improving sleep, decreasing nightmares, and decreasing intrusive thoughts. More recently, he published two open-label studies on the same topic, examining topiramate's effects on PTSD patients with and without hallucinations (Berlant & van Kammen, 2002; Berlant, 2004). Berlant and van Kammen (2002) found higher response rates in non-hallucinatory patients, who experienced a definite reduction in nightmares or intrusions. In the second study, only nonhallucinatory PTSD patients were recruited. Among the 33 patients enrolled in this 12-week, open-label study, 70% were responders, as reflected by a 30% reduction in PTSD Checklist-Civilian Version (PCL-C) scores. Although these results are promising, both studies were open-label and had high attrition rates.

Subsequently, Tucker et al. (2007) reported a double-blind, placebo-controlled trial with topiramate, in which 40 civilian PTSD patients were randomized to either topiramate (titrated up to the maximally tolerated dose or 400 mg/day) or placebo for a 12-week study period. The primary outcome measure, the change from baseline to endpoint in total CAPS score, was not statistically significant between the two groups. A significant decrease was found in the CAPS cluster B scores for re-experiencing symptoms in the topiramate group (74.9% vs. 50.2%) and also the Treatment Outcome PTSD scale (68% vs. 41/6%). This study, however, was not powered adequately. Lindley et al. (2007) conducted a comparable study, also showing no statistically significant improvement in endpoint CAPS score between topiramate (with patients receiving up to 200 mg/day) and placebo treatment. This study also had a high attrition rate and hence was inadequately powered. Yeh et al. (2010) carried out a similarly designed study supporting topiramate as efficacious in PTSD treatment, in which 35 civilian PTSD patients were randomized in a double-blind placebo-controlled study to topiramate (up to 200 mg/day) or placebo. CAPS total scores at the endpoint for the topiramate group showed a significant decrease (-57.78 vs. -32.41), and CAPS cluster B scores also showed significant improvements with topiramate treatment. Further research is needed to clarify topiramate's effects on PTSD.

To date, there has been one double-blind, placebo-controlled study of lamotrigine for the treatment of PTSD (Hertzberg et al., 1999). In this study, 15 patients were randomized in a 2:1 ratio to either lamotrigine or placebo. Five of the 10 patients who received lamotrigine had a significant decrease in their global PTSD Scale score as compared with one of four patients who received placebo. Lamotrigine-treated patients showed improvements in avoidance, numbing, and re-experiencing symptoms. Further studies are necessary to determine whether anticonvulsant treatment, either as a monotherapy or as an adjuvant with medications for depression, will be beneficial for patients with PTSD.

There is no supporting evidence to suggest that divalproex is beneficial in treating PTSD as a monotherapy, as shown by studies by Davis et al. (2008) and Hamner et al. (2009). Also, no significant improvement has been shown in the pediatric PTSD population, according to a study by Steiner et al. (2007). Baniasadi et al. (2014) conducted a randomized controlled trial of pregabalin for 37 male patients with PTSD and found a significant treatment effect, but this small study has not been replicated.

Medications for Psychosis (Antipsychotics)

No medication for psychosis (antipsychotic) is approved for the treatment of PTSD. With the development of the dopamine serotonin blockers (atypical antipsychotics), there has been a reemergence of interest in the possible use of medications for psychosis as either a primary or adjuvant treatment for PTSD. Risperidone has been examined as an adjunctive agent in the treatment of PTSD. Reich et al. (2004) studied 21 adult female outpatients who had a history of childhood abuse. In this 8-week, double-blind, placebo-controlled study of risperidone (0.5–8 mg/day), risperidone-treated patients experienced a significant reduction in the total CAPS-2 and all subscales except for the avoidant subscale (nine patients took concomitant medications during this study).

Four studies have explicitly examined risperidone as an augmenting agent in PTSD patients' medication regimens (Hamner et al., 2003; Bartzokis et al., 2004; Rothbaum et al., 2008; Krystal et al., 2011). Hamner et al. (2003) randomized 37 combat veterans with PTSD to a 5-week, flexible-dose, double-blind, placebo-controlled trial of 1–6 mg/day risperidone augmentation to a preexisting medication for depression/psychosis regimen. Outcome measures included the Positive and Negative Syndrome Scale (PANSS) and the CAPS. Among the 22 patients who completed the study, risperidone-treated patients showed a significantly greater decrease in psychotic symptoms, as measured by PANSS. Both groups showed a substantial decline in CAPS scores, with no significant difference between treatment groups. Bartzokis et al. (2004) recruited 65 patients with severe combat-related PTSD for a double-blind, placebo-controlled trial. These patients participated in a 5-week residential program followed by 3 months of outpatient follow-up. Both treatment groups showed significant improvements during the initial inpatient phase, but by week 16, risperidone treatment separated from placebo on the CAPS-Total, CAPS-D, HAM-A, and PANSS. This particular study suggested risperidone augmentation helps to decrease intrusive thoughts and markedly improves symptoms of PTSD. Rothbaum et al. (2008) conducted a 16-week study of risperidone augmentation of sertraline in PTSD, in which 45 patients were treated with open-label sertraline

for 8 weeks. Non-remitters ($n = 25$) were entered into double-blind placebo-controlled augmentation with risperidone (0.5–2 mg/day) or placebo for a further 8 weeks. PTSD and related symptoms improved in both treatment groups, with no significant difference in total CAPS score changes between the two groups. The risperidone group, however, had significant improvement in the Davidson Trauma Scale (DTS) sleep item. Krystal et al. (2011) conducted a large study with 296 veterans diagnosed with combat-related PTSD. Patients were maintained on their pre-study medication regimen and were randomized to receive risperidone (from 1 mg/day titrated up to 4 mg/day) or placebo for 24 weeks. Both groups improved on all measures, with no significant difference between the two treatment conditions. Risperidone was not superior to placebo as an augmenting agent for PTSD symptoms, depression, or anxiety. The data suggest that risperidone may not be useful in treating PTSD.

Five studies have examined olanzapine's efficacy in PTSD. Stein et al. (2002a) studied 19 male war veterans with military-related PTSD in a 12-week, double-blind, placebo-controlled study of 10–20 mg/day olanzapine versus placebo. Outcome measures included CAPS, PSQI, and the Center for Epidemiologic Studies Depression Scale (CES-D). Olanzapine-treated patients experienced a significantly greater reduction in PTSD symptoms, depressive symptoms, and sleep disturbance. Response rates as measured by the CGI-I were fairly low (30%) among olanzapine patients, and they were not statistically superior to scores for placebo patients. In another small double-blind, placebo-controlled study of olanzapine, 11 patients completed the study and there was no difference in the response rates between the two groups (Butterfeld et al., 2001). Pivac et al. (2004) conducted an open-label trial of olanzapine versus fluphenazine. In this study, 55 male war veterans with combat-related PTSD received 5–10 mg/day olanzapine or fluphenazine for 6 weeks. Both medications helped alleviate symptoms of PTSD, but olanzapine was more efficacious than fluphenazine at reducing negative symptoms, avoidance, and arousal. As expected, fluphenazine caused more extrapyramidal symptoms and olanzapine caused greater weight gain. Petty et al. (2001) reported an open-label study of 48 individuals who were started on open-label olanzapine treatment. Thirty patients completed the 8-week trial and tolerated olanzapine well. These patients showed clinically significant improvement in their CAPS-2 scores, HAM-D scores, and Brief Psychiatric Rating Scale (BPRS) scores. Carey et al. (2012) conducted a double-blind, placebo-controlled, flexible-dose study with 28 patients with noncombat-related chronic PTSD. Patients were randomized to olanzapine (maximum dose 15 mg/day) or placebo for 8 weeks. Olanzapine-treated patients showed a significantly greater improvement in total CAPS score (57.7% vs. 23.7%) compared with the placebo group. Six out of 14 patients treated with olanzapine reported significant weight gains, in the range 6–10 kg over the course of the study.

Aripiprazole has demonstrated solid results in a small number of trials. Villarreal et al. (2007) conducted a 12-week open-label monotherapy trial with 22 PTSD patients with both military and civilian backgrounds. Aripiprazole monotherapy significantly improved total CAPS scores (primary outcome measure), and also all CAPS subscales and anxiety and depression scales. However, there was a high dropout rate of eight patients, thought to be due to intolerability from a high starting dose of aripiprazole (10 mg/day). By the endpoint, 14 patients were classified as responders and two as remitters. Robert et al. (2009) enrolled 17 military veterans with PTSD in a 12-week open-label, flexible-dose augmentation trial of aripiprazole. Significant improvements in total CAPS scores from baseline to endpoint and also in the CAPS-B (reexperiencing) and CAPS-C (avoidance/numbing) subscales were reported; 53% of patients were responders by the study endpoint. Patients most commonly reported gastrointestinal symptoms, psychomotor agitation, and sedation. An open-label study by Youssef et al. (2012) investigated the efficacy of aripiprazole in combat-related PTSD. Ten veterans received aripiprazole (5–30 mg/day) for 12 weeks, and the overall change in CAPS total score from baseline to endpoint was measured.

Patients showed a significant improvement in CAPS total scores. A recent systematic review by Britnell et al. (2017), which included these trials, concluded that while there is some evidence so far. Further large-scale studies would be helpful in further substantiating the efficacy of aripiprazole in PTSD treatment.

Pilkinton et al. (2016) conducted a 12-week of asenapine open-label trial including 18 patients with PTSD. In this study, asenapine treatment improved PTSD symptoms on the CAPS, suggesting the possibility of asenapine as a novel treatment of PTSD. Well-designed, adequately powered studies of dopamine serotonin blockers (second-generation antipsychotics) are needed to clarify further the utility of these medications as monotherapy or augmentation agents for the treatment of PTSD.

Novel Treatments

Several alternative treatments for PTSD are being investigated. Raskind et al. (2003) compared prazosin and placebo in a 20-week, double-blind, crossover study of 10 Vietnam combat veterans diagnosed with chronic PTSD and severe trauma-related nightmares. Prazosin is a centrally active adrenergic antagonist commonly used in treating hypertension. Primary efficacy measures included two items from the CAPS (recurrent distressing dreams item and difficulty falling/staying asleep item), and the change in overall PTSD severity and functional status according to the CGI. Prazosin was shown to be more effective than placebo at the study endpoint in decreasing symptoms of recurrent distressing dreams (3.6 vs. 6.7, respectively, $p < 0.001$) and with difficulty falling/staying asleep (4.0 vs. 7.1, respectively, $p < 0.01$). At the study endpoint, prazosin displayed an average of 57.3 on the CAPS compared with 86.5 for the placebo group ($p < 0.01$). Prazosin was also associated with improved global functioning compared with placebo (prazosin mean CGI score 2.0 compared with placebo mean CGI score 4.5, $p < 0.01$). Prazosin was well tolerated. Raskind et al. (2007) reported a larger efficacy study including 40 veterans in a double-blind, placebo-controlled investigation of prazosin. Patients were randomized to placebo or prazosin, which was initiated at 1 mg/day and flexibly titrated over 28 days up to 15 mg/day. The final dose or placebo was then maintained for a further 8 weeks. The primary outcome measure was the CAPS, and secondary outcome measures were the Nightmare Frequency Questionnaire-Revised (NFQ), the PTSD Dream Rating Scale (PDRS), and the HAM-D. Prazosin had significant efficacy over placebo in reducing trauma-related nightmares, improving sleep quality, and increasing the amount of normal dreaming. The CAPS total score was also significantly improved on prazosin.

Taylor et al. (2008) also conducted a placebo-controlled trial of prazosin in 13 noncombat-related PTSD patients. This study revealed similar findings, with significant improvement in sleep duration by 94 min, increased REM sleep duration, reduced trauma-related nightmares, reduced distressed awakenings, improved CGI-I scores, and PDRS scores trending more towards normal dreaming. Byers et al. (2010) conducted the largest prazosin trial with 237 veteran PTSD patients, comparing prazosin with quetiapine in treatment of night PTSD symptoms. Both treatment groups showed similar reductions in night symptoms (nightmares, decreased REM sleep, poor quality sleep, distressed awakenings). However, patients receiving prazosin were more likely to continue the medication long-term compared with quetiapine. In a meta-analysis including 429 participants with PTSD, Reist et al. (2020) demonstrated a statistically significant improvement of overall PTSD scores including insomnia (standardized mean differences $= -0.31$, 95%CI $= -0.62$–0.01), following treatment with prazosin. These results suggest that prazosin is beneficial as a treatment for nighttime symptoms of PTSD.

A small study by Heresco-Levy et al. (2002) investigated the partial NMDAR agonist D-cycloserine (DCS) as a treatment for PTSD. DCS, an NMDA receptor partial agonist, is being investigated as a treatment in a number of diagnoses. In a double-blind, placebo-controlled crossover trial, 11 subjects who met DSM-IV-TR criteria for PTSD were treated with 50 mg/day DCS or placebo for 12 weeks. The CAPS was the primary efficacy measure. DCS therapy significantly decreased scores on the numbing and avoidance clusters of the CAPS but not total scores or other subscales. Accumulated data have not shown a benefit of DCS monotherapy on PTSD (Baker et al., 2018).

Heresco-Levy et al. (2009) then looked at the full NMDAR agonist D-serine (DSR) as either monotherapy or adjunctive treatment for chronic PTSD. DSR at 30 mg/kg/day was found to be significantly superior to placebo in reducing HAMA and Mississippi Scale for Combat-Related PTSD (MISS) scores, but did not show significant benefit over placebo in total CAPS scores. Later, Litz et al. (2012) conducted a randomized, double-blind, placebo-controlled trial of DCS as augmentation for exposure therapy, but did not find benefit with the use of DCS. In fact, their data suggested that DCS adjunctive treatment may result in poorer outcomes with exposure therapy, as indicated by a significant reduction in symptom improvement with DCS treatment compared with placebo.

Pollack et al. (2011) investigated the non-benzodiazepine medication for insomnia c eszopiclone in 23 PTSD patients. Eszopiclone treatment was associated with improvements in Short PTSD Rating Interview (SPRINT) scores, CAPS total scores, and Pittsburgh Sleep Quality Index (PSQI) scores. Recently, Dowd et al. (2020) conducted a 12-week, double-blind, randomized controlled study of eszopiclone for 25 patients with PTSD. They were unable to detect a significant difference between eszopiclone and placebo on CAPS, Short PTSD Rating Interview, or PSQI.

The novel medication for depression ketamine, an NMDA antagonist, has been evaluated for use in PTSD in a handful of clinical trials. The most recent, conducted by Feder et al. (2021), randomized 30 subjects to 2 weeks of either intravenous ketamine or a midazolam control and found significantly improved response in the ketamine group, with 67% response compared to 20% response in the midazolam control group. Ketamine response was maintained for a median of 4 weeks.

Despite this wealth of research, PTSD remains suboptimally treated in a large number of patients

(Watson, 2019), highlighting the ongoing need for more effective treatments. The research community has responded with many ongoing novel approaches that include, among others, the use of cannabinoids and psychedelics, and glucocorticoid receptors (https://clinicaltrials.gov/ct2/show/NCT01739335).

Treatment Conclusions

SSRIs and SNRIs remain the mainstay of pharmacological treatment for the core condition of PTSD, with adjunctive medicines often chosen to target specific patient presentations or comorbidities. Patients who have had PTSD for a long time and present with comorbid psychiatric syndromes can be difficult to treat. They generally respond very slowly to treatment and their care is frequently confounded by their overall complex psychiatric presentation. The key to treating such individuals successfully is to be clear about the target symptoms and to give the patient sustained treatment because improvement many times will require months of therapy. There are some promising smaller studies suggesting that such patients may do well with an augmentation approach.

Because of the high proportions of patients with PTSD who suffer from comorbid psychiatric conditions, such as depression, substance use disorders, and other anxiety disorders, clinicians will often find themselves utilizing treatments that address multiple conditions simultaneously. Nightmares associated with PTSD may be addressed with prazosin or less commonly with trazodone. Excessive irritability or paranoia may respond to dopamine serotonin blockers (second generation neuroleptics). Loss of sleep or appetite may indicate a role for mirtazapine. Residual neuropathic pain, related to the index trauma or not, may respond best to duloxetine. Substance use disorders must be addressed when present and comorbid depression cannot be ignored. Often the medication regimen will need to change throughout the course of illness, as paired with psychotherapy and support.

Obsessive–Compulsive Disorder

Selective Serotonin Reuptake Inhibitors

Large, well-designed, double-blind, placebo-controlled trials have demonstrated that fluoxetine, paroxetine, fluvoxamine, citalopram, escitalopram, and sertraline are effective acute treatments for OCD (Carpenter et al., 1996; Koponen et al., 1997; Mundo et al., 1997; Thomsen, 1997; Cartwright & Hollander, 1998; Greist & Jefferson, 1998; Todorov et al., 2000). There have been two large 12-week, parallel-group, double-blind, placebo-controlled, multicenter trials of fluoxetine in fixed doses of 20, 40, and 60 mg/day versus placebo (Tollefson et al., 1994a, b). Response was defined as a decrease of 35% or greater on the Yale-Brown Obsessive-Compulsive Scale (Y-BOCS). The three fluoxetine doses each had response rates of 33%. There was some suggestion that the 40 and 60 mg/day doses acted more rapidly and facilitated greater improvement in Y-BOCS scores. Montgomery et al. (1993) previously reported that 40 and 60 mg/day of fluoxetine, but not 20 mg/day, was more effective than placebo. One study looked at nonresponders of fluoxetine monotherapy to investigate the efficacy of augmentation strategies versus increasing fluoxetine dose (Diniz et al., 2011). A total of 54 patients were treated with the highest recommended (60 mg/day) or tolerated dose of fluoxetine for 8 weeks. Nonresponders were randomized to receive higher dose fluoxetine/placebo (≤80 mg/day fluoxetine), fluoxetine/clomipramine (≤40 mg/day, ≤75 mg/day), or fluoxetine/quetiapine (≤40 mg/day, ≤200 mg/day) for 12 weeks. Y-BOCS scores from baseline to endpoint were significantly improved with fluoxetine/placebo and fluoxetine/clomipramine compared with fluoxetine/quetiapine, suggesting that OCD nonresponders may benefit from either higher doses of fluoxetine, if tolerated, or clomipramine augmentation of mid-range dosed fluoxetine.

Although there are data available from a number of placebo-controlled, multicenter trials contrasting paroxetine with placebo, only one 12-week fixed-dose trial (20, 40, and 60 mg/day) has been published (Ballenger et al., 1998). In this study, all patients were initially started at 20 mg/day, and the 40 and 60 mg/day groups were titrated to these doses by the end of week 2. The 40 and 60 mg/day groups demonstrated statistically significant decreases on the Y-BOCS and the CGI-Severity Scale. The greatest improvement seemed to be present in the 60 mg/day group.

There have been several smaller trials and one large 10-week, double-blind, placebo-controlled, multicenter trial of fluvoxamine (100–300 mg/day) versus placebo in the treatment of OCD (Perse et al., 1987; Goodman et al., 1989; Jenike et al., 1990; Greist et al., 1995c). In the large study, Greist et al. (1995c) reported that 40% of fluvoxamine-treated patients were responders on the CGI-Improvement Scale compared with 15% of placebo-treated patients. The fluvoxamine-treated patients also had a significantly greater decrease in the mean Y-BOCS scores (Greist et al., 1995c). These studies suggest that fluvoxamine is effective in treating OCD.

In addition, Pallanti et al. (2002) conducted an open-label study of intravenous citalopram infusion in 39 treatment-resistant OCD adult patients. Thirty-eight patients completed the 21-day study of 20–80 mg intravenous citalopram. Twenty-three patients (59%) had a ≥25% decrease in their Y-BOCS scores. These patients were then given oral citalopram and continued to improve daily during this 63-day phase. Although intravenous citalopram was rapidly effective in this study, an increased incidence of cardiovascular adverse events was noted. Furthermore, placebo-controlled, double-blind studies should be performed to extend these results. Montgomery et al. (2001) performed a 401-patient, multisite, double-blind, placebo-controlled study of 20, 40, and 60 mg/day doses of citalopram. In contrast to previous studies during the 12-week study, 20 mg was found to be an effective dose, as measured by the change in Y-BOCS scores from baseline to study endpoint. All three doses were significantly superior to placebo in treating both obsessions and compulsions. There are data from single-blind comparator trials of citalopram versus paroxetine, fluvoxamine, and fluoxetine that suggest that

there is no significant difference in Y-BOCS scores between these medications (Mundo et al., 1997; Alaghband-Rad & Hakimshooshtary, 2009). These data are consistent with other open-label studies with adults and children suggesting that both monotherapy and mono-amine enhancers (TCA) adjunctive therapy with citalopram may decrease symptoms of OCD (Koponen et al., 1997; Thomsen, 1997; Marazziti et al., 2008).

Several studies have investigated escitalopram efficacy in OCD treatment. Stein et al. (2007) conducted a large, double-blind, placebo-con-trolled, active comparator study of escitalopram. A total of 466 adult patients were randomized to escitalopram 10 mg/day ($n = 116$), escitalopram 20 mg/day ($n = 116$), paroxetine 40 mg/day ($n = 119$), or placebo ($n = 115$) for a 24-week trial. Primary outcome measure was the Y-BOCS total score change from baseline to week 12, with remission defined as a Y-BOCS score of ≤10. All treatments were efficacious compared with placebo; however, escitalopram 20 mg/day was superior to the others in total Y-BOCS score improvement and in remission rates. Higher dose escitalopram was also associated with signifi-cantly earlier response, separating from the other treatments as early as week 6. Rabinowitz et al. (2008) treated 67 OCD patients with escitalopram 20 mg/day for 4 weeks. Those who were not classified as responders (Y-BOCS score reduction ≥25%) by week 4 were then continued at a higher dose of escitalopram for a further 12 weeks. Sixty-four patients entered the second phase of the trial, and escitalopram was flexibly titrated as tolerated up to 50 mg/day (the mean dose at the endpoint was 33.8 mg/day). By the endpoint, significant improvements in Y-BOCS scores were found compared with baseline, with 80% of patients classified as responders.

In a large 12-week, double-blind, placebo-controlled, multicenter, fixed-dose study of sertra-line (50, 100, and 200 mg/day), all three doses caused a significant reduction in Y-BOCS scores (Greist et al., 1995a). The higher doses showed selectively greater improvements on all measures but this did not reach statistical significance.

There have been two other double-blind, placebo-controlled, multicenter, flexible-dose trials of sertraline versus placebo in the treatment of OCD (Chouinard et al., 1990; Kronig et al., 1999). In both of these trials, sertraline was more efficacious than placebo on standard outcome measures. Greist et al. (1995a) also demonstrated the ability of sertraline to maintain improvement after 12 weeks of double-blind, placebo-controlled treatment during an additional 40-week double-blind fixed-dose trial of sertraline (50, 100, and 200 mg/day) (Greist et al., 1995b). At the 52-week endpoint, mean scores for all four outcome measures, Y-BOCS, the NIMH Global Obsessive-Compulsive Scale, and CGI-Severity and Improvement Scales, revealed significantly greater improvement for the sertraline group versus the placebo group (Noyes et al., 1997). Fifty-one of the patients who participated in the double-blind, placebo-controlled trial of sertraline were entered into a 2-year sertraline open-label study. Outcome data suggested that efficacy was not only maintained, but improvement continued; first-year mean Y-BOCS scores were 11.4 compared with second year mean Y-BOCS scores of 3.2.

Koran et al. (2002) performed a relapse-prevention study in which patients who had achieved a sustained response during a 52-week single-blind treatment phase were randomized to an additional 28 weeks of double-blind treatment with either sertraline or placebo. This 80-week study was conducted at 21 sites in the United States. Following a 1-week washout, 649 patients entered 16 weeks of single-blind treatment with flexible doses of sertraline. Patients were titrated up on a 50 mg incremental schedule to a maximum 200 mg/day depending on response and tolerability. Response was defined as a decrease of 25% from baseline on the Y-BOCS, and a CGI-Improvement Scale score of ≤ 3. Those who met responder criteria continued to receive single-blind sertraline treatment for an additional 36 weeks. Data from the single-blind trial revealed consistent improvement throughout the 52 weeks of treatment. Mean Y-BOCS scores decreased from 26.1 at baseline to 15.9 at the end of week 16 and to 10.3 at the end of week 52. Symptomatic improvement was associated with improvement in quality of life as measured by Q-LES-Q scores from 60% at baseline to 73.7% at the end of week 16 and to 78.2% at the end of week 52 (the mean Q-LES-Q score for a population of normal subjects is 80%). At the end of week 52, patients who continued to meet responder criteria were randomized to the 28-week double-blind discontinuation trial. Those patients randomly assigned to sertraline treatment were continued on the same daily dose at week 52. Patients assigned to placebo had their dose decreased by 50 mg/day every 3 days until they were taking only placebo tablets. Continued sertraline treatment was associated with sustained improvement during the 28-week relapse-prevention phase. Sertraline treatment was also significantly more effective than placebo in decreasing study discontinuation due to relapse or insufficient clinical response and acute exacerbation of obsessive-compulsive symptoms.

Skapinakis et al. (2016) conducted a systematic review and network meta-analysis of randomized controlled trials to investigate pharmacological investigations for management of OCD. The primary outcome measure was the YBOCS. All types of SSRIs showed a greater effect than placebo, as follows: fluoxetine (mean difference, -3.46; 95% CI, -5.27 to -1.58; 6 trials and 633 patients); fluvoxamine (mean difference, -3.60; 95% CI, -5.29 to -1.95; 13 trials and 521 patients); paroxetine (mean difference, -3.42; 95% CI, -5.10 to -1.61; 8 trials and 902 patients); sertraline (mean difference, -3.50; 95% CI, -5.30 to -1.63; 7 trials and 565 patients); citalopram (mean difference, -3.49; 95% CI, -5.62 to -1.31; 2 trials and 311 patients); and escitalopram (mean difference, -3.48; 95% CI, -5.61 to -1.23; 1 trial and 226 patients). A class effect for all SSRIs was significant as well (mean difference, -3.49; 95% CI, -5.12 to -1.81; 37 trials and 3158 patients). These data suggest that SSRIs are an effective first-line pharmacotherapy for the treatment of OCD in adults. Discontinuing SSRI treatment is associated with an exacerbation of symptoms and worsening in quality of life. (For more details on OCD, see ▶ Chap. 143, "Collaborative Care and Geriatric Psychiatry.")

Serotonin–Norepinephrine Reuptake Inhibitors–Venlafaxine, Duloxetine

Hollander et al. (2003a) conducted an open-label trial of venlafaxine and venlafaxine XR treatment of OCD. There were 39 patients in this trial, of whom 74% had not responded to at least one previous SSRI trial. The CGI-I was used to assess global improvement. About 69% of the patients were classified as responders (CGI-I of 1 or 2) by the end of the 18 months of treatment. A network meta-analysis of randomized controlled trials performed by Skapinakis et al. (2016) examined the effect of venlafaxine on the YBOCS in patients with OCD. The effect of venlafaxine did not reach statistical significance when compared with placebo potentially due to a small sample (mean difference, −3.22; 95% CI, −8.26 to 1.88; 2 trials and 98 patients). Future placebo-controlled studies with venlafaxine XR seem warranted.

A 2015 open-label study of duloxetine by Dougherty et al. demonstrated that duloxetine (up to 120 mg/day) resulted in statistically significant improvement on YBOCS scores for all 12 study completers; 8 participants dropped out of the study, most for adverse effects including nausea, fatigue, sexual dysfunction, and headache. Of study completers, 58.3% met criteria for full medication response. Mowla et al. (2016), in a small double-blinded study, randomized 46 patients to receive either 8 weeks of duloxetine (20–60 mg/day) or sertraline (50–200 mg/day) in addition to their regular OCD treatment after having failed a 12-week trial of fluvoxamine, fluoxetine, or citalopram. Both groups showed statistically significant improvement in YBOCS scores from baseline, but no difference between the treatment groups.

Monoamine Enhancers (Tricyclic Antidepressant Medication) and Monoamine Oxidase Inhibitors

There have been several meta-analyses comparing clomipramine and SSRIs (Cox et al., 1993; Piccinelli et al., 1995; Stein et al., 1995; Abramowitz, 1997; Greist, 1998; Kobak et al., 1998). In each case, clomipramine was found to be significantly more effective than the SSRIs. There has been some concern that this conclusion may be biased because many of the clomipramine studies were performed in patients who had not been previously treated with an effective pharmacotherapy, whereas patients in the later studies of the SSRIs had failed treatment with clomipramine or other agents. There is also concern that differences in study design might bias these results in favor of clomipramine (Greist, 1998). There have been double-blind comparison studies of clomipramine against fluoxetine, fluvoxamine, sertraline, and paroxetine (Freeman et al., 1994; Koran et al., 1996; Lopez-Ibor et al., 1996; Zohar & Judge, 1996; Bisserbe et al., 1997; Milanfranchi et al., 1997). None of these studies had the power to differentiate clomipramine from the SSRI in question. Therefore, findings of no difference between the two groups are not surprising. However, if one takes the entire body of evidence into account, analyses suggest that clomipramine is at least as effective as the SSRIs and may, in some instances, be more efficacious (Greist, 1998; Todorov et al., 2000). A recent meta-analysis performed by Skapinakis et al. (2016) evaluated the effect of clomipramine on the YBOCS in patients with OCD. Clomipramine was superior to placebo (mean difference, −4.72; 95% CI, −6.85 to −2.60; 13 trials and 831 patients).

Although there have been several studies with MAOIs, the results have been equivocal at best (Jenike et al., 1990; Vallejo et al., 1992). The two other strategies for enhancing serotonergic neurotransmission have involved manipulating the ratio of clomipramine to its metabolite desmethylclomipramine. One strategy involved the use of intravenous clomipramine in order to decrease first-pass hepatic metabolism and thus increase the ratio of clomipramine to desmethylclomipramine. These studies found that intravenous clomipramine reduces symptoms of OCD more rapidly than oral clomipramine (Warneke, 1984, 1985, 1989; Koran et al., 1994, 1997). However, over time, patients treated with intravenous clomipramine did not show a greater response than patients treated with oral clomipramine. The second strategy combined clomipramine with an SSRI (fluvoxamine) that inhibits its hepatic metabolism. This, again, shifts the ratio of clomipramine to desmethylclomipramine in favor of clomipramine.

Other Medications for Depression (Trazodone, Mirtazapine, Bupropion, Agomelatine)

In a small double-blind, placebo-controlled study with trazodone, Pigott et al. (1992b) found no significant differences between pre-treatment and post-treatment scores on OCD scales. Mirtazapine treatment was studied in a 10-week open-label trial of 10 adult outpatients who met DSM-IV-TR criteria for OCD for at least 1 year and had a Y-BOCS score >18 (Koran et al., 2002). Mirtazapine was started at a dose of 15 mg/day, then increased to 30 mg/day after 4 days and to 45 mg/day at the end of the second week. OCD severity was measured with the Y-BOCS and the CGI-Severity Scale. In this study, mean Y-BOCS scores decreased from 28.7 at baseline to 23.7 at week 10. Only two of the 10 subjects met responder criteria. These results suggest that mirtazapine may not be a useful treatment for subjects with OCD. Bupropion has also been studied in patients with OCD. Vulink et al. (2005) found that bupropion was an ineffective treatment for 12 patients with OCD in an open-label, fixed-dose study. Two patients responded (≥25% reduction in Y-BOCS score) and eight patients deteriorated during the 8-week study. Agomelatine has been studied as augmentation for OCD treatment. A 2014 case series (Tzavellas et al., 2014) showed that 12 patients who failed a 16-week trial of an SSRI had statistically significant improvement in YBOCS scores after addition of agomelatine to their regular treatment.

Benzodiazepine Medication

Pigott et al. (1992a) conducted a controlled trial of clonazepam augmentation in OCD patients treated with clomipramine or fluoxetine. Augmentation with clonazepam was found to be more effective than placebo in decreasing global measures of anxiety but not Y-BOCS scores. However, a 2003 double-blind, placebo-controlled trial of clonazepam as OCD monotherapy (Hollander, 2003b) did not demonstrate any significant difference between clonazepam and placebo on CGI, YBOCS, or HAM-A scores.

Buspirone

Harvey and Balon (1995) reviewed double-blind augmentation studies of buspirone used for patients with treatment-resistant OCD, noting that none demonstrated any advantage over placebo treatment.

Beta-Blocker Medication

Pindolol augmentation was evaluated in one small double-blind study of patients with treatment-resistant OCD. In this two-phase study performed by Dannon et al. (2000), 23 treatment-resistant patients were prospectively treated with 60 mg/day of paroxetine. Nonresponse was defined as a 25% improvement on the Y-BOCS score during this open-label treatment phase. Sixteen patients were randomized to receive double-blind augmentation therapy with either pindolol 2.5 mg TID or matching placebo for 6 weeks. The primary outcome measure was a change in Y-BOCS score and secondary measures included changes in HAM-A scales total scores and MADRS total scores. A significant decrease in Y-BOCS score was noted, but there were no changes compared with placebo in the MADRS or the HAM-A rating scores. This suggests that pindolol might have a unique role in decreasing symptoms in OCD, but is consistent otherwise with findings in both MDD and in SAD, where pindolol augmentation did not cause a significant change in ratings.

Anticonvulsant Medication

Anticonvulsant medications have undergone preliminary testing as adjunctive agents in OCD treatment. Rubio et al. (2006) examined topiramate augmentation to current medications for depression (antidepressants) in 12 treatment-resistant patients during a 12-week study. Topiramate was titrated from 25 to 400 mg/day as tolerated, and 10 of the 12 subjects were treatment responders with a $\geq 30\%$ reduction in their Y-BOCS scores. Berlin et al. (2011) conducted another augmentation study with 36 OCD patients who were treated with the highest tolerated SSRI dose for 12 weeks, then maintained on that dose for at least 6 weeks. They were randomized to receive topiramate (up to 400 mg/day as tolerated) or placebo in conjunction with their SSRI regimen for a 12-week, double-blind study. Those receiving topiramate had a significant (5.38 points) improvement on the Y-BOCS compulsion scale compared with placebo (0.6 points). No significant improvement was seen on the obsessions scale. Overall, topiramate augmentation appeared to have efficacy in treating compulsions, but it was poorly tolerated and led to both dose reduction and medication discontinuation.

Two studies examined augmentation with pregabalin in treatment-resistant OCD (Oulis et al., 2011; Nicola et al., 2011). Oulis et al. (2011) investigated augmentation of current SSRI regimens in a small group of 10 treatment-resistant (defined as failure to achieve a $\geq 35\%$ improvement on Y-BOCS) OCD patients who had already failed an adequate trial augmentation with medication for psychosis (antipsychotics). Patients received pregabalin augmentation at 225–675 mg/day for 8 weeks of open-label treatment and a 4-week follow-up. Significant benefit was found in overall Y-BOCS total score changes from baseline to endpoint. Nicola et al. (2011) found similar benefit with pregabalin augmentation of an SSRI in a study of 12 OCD patients over a 16-week open-label trial. The pregabalin dose ranged from 150 to 450 mg/day. There was a 26% overall reduction in Y-BOCS total scores by the endpoint, with significant improvements in both the obsessions and compulsions subcategories and also other secondary measures, HAM-A, HAM-D, and GAF. These results suggest that augmentation of an SSRI with pregabalin may be beneficial in the treatment of OCD. Mowla and Ghaedsharaf (2020) conducted

a 12-week double-blind placebo-controlled trial to evaluate the effect of pregabalin as an augmentation to sertraline in patients with treatment-resistant OCD. The YBOCS decreased from 26.13 ± 7.03 to 8.81 ± 3.47 in the pregabalin group ($n = 28$), and 26.85 ± 4.34 to 17.63 ± 4.22 in the placebo group ($n = 28$). A 35% decrease in the YBOCS was shown in 57.14% of the pregabalin group and 7.14% of the placebo group. Throughout the study, pregabalin was well tolerated.

A double-blind study by Bruno et al. (2012) randomized 33 OCD patients on standing SSRIs to lamotrigine 100 mg/day or placebo for 12 weeks. By the trial endpoint, significant improvements were noted in total Y-BOCS scores and also both obsessions and compulsions sub-scores in the lamotrigine group. Further large-scale studies are necessary to substantiate this potential benefit.

Farnia et al. (2018) examined the effect of gabapentin as an augmentation to fluoxetine in an 8-week randomized controlled trial. In this study, the YBOCS did not decrease over time and no group difference was observed (fluoxetine + gabapentin: $n = 33$, fluoxetine + placebo: $n = 33$).

A recent network meta-analysis conducted by Zhou et al. (2019) investigated the effect of augmentation agents to SSRIs in patients with treatment-resistant OCD. The primary outcome measure was the YBOCS. A network meta-analysis revealed that topiramate and lamotrigine had greater effects than placebo (mean difference, -6.05; 95% CI, -10.89 to -1.20 and mean difference, -6.07; 95% CI, -11.60 to -0.50, respectively). Topiramate showed an increased risk of side-effect discontinuation compared with placebo (odds ratio, 3.67; 95% CI, 1.01–13.34).

Medications for Psychosis (Antipsychotics)

Patients with OCD frequently do not respond to monotherapy with SSRIs or clomipramine. Preliminary studies suggest that treatment with dopamine blockers (typical neuroleptics) for patients with tics may be beneficial (McDougle, 1997). Open-label case series suggest that the newer dopamine serotonin blockers (atypical neuroleptics) risperidone and olanzapine may be useful in augmenting clinical response in treatment-refractory OCD patients (McDougle, 1997; Weiss et al., 1999). Pfanner et al. (2000) investigated the use of risperidone as an adjunct to SSRI therapy in an 8-week open-label trial of 20 patients with refractory OCD. Patients in this study had a DSM-IV-TR diagnosis of OCD with a duration of illness of at least 2 years, had demonstrated less than a 25% improvement with an SSRI after 6 months of treatment, and had a baseline Y-BOCS score of 30. Risperidone was added at 1 mg/day and titrated up to 3 mg/day over 2 weeks. The mean Y-BOCS score decreased by 26% over the course of the trial, from 36.1 at baseline to 24.8 at week 8. At the endpoint, eight patients had a CGI-Improvement score of 1 (very much improved), nine patients had a score of 2 (much improved), and only three patients had a score of 3 (minimally improved).

Li et al. (2005) published the results of a 16-patient, 9-week, crossover design study investigating the efficacy of risperidone and haloperidol augmentation of SSRI-resistant OCD. Individuals were randomized to different sequences of the three augmentation strategies (placebo–risperidone–haloperidol). Each active treatment sequence was followed by a 1-week washout phase in order to minimize carryover effects between augmentation strategies. Twelve individuals completed all three arms of this study. In this study, risperidone was better tolerated than haloperidol. Neither risperidone nor haloperidol differentiated placebo augmentation on Y-BOCS scale scores or other outcome variables. Simpson et al. (2013) examined the effect of risperidone as an augmenting agent to SSRIs in a randomized controlled design. In this study, 100 were randomized to either of three groups (risperidone, $n = 40$; exposure and ritual prevention, $n = 40$; and placebo, $n = 20$). At eight weeks, the risperidone group did not show a greater reduction in the YBOCs than the placebo group, and was inferior to the exposure and ritual prevention group.

Pessina et al. (2009) conducted a study of aripiprazole augmentation of SSRIs in treatment-resistant OCD. Twelve patients entered an open-label, flexible-dose (5–20 mg/day), 12-week study of aripiprazole. Significant improvement was found in Y-BOCS mean total scores by the endpoint. Muscatello et al. (2011) investigated augmentation of either SSRIs or clomipramine with aripiprazole in a double-blind, placebo-controlled trial in 30 patients. Patients were randomized to placebo or aripiprazole 15 mg/day for 16 weeks. Significant improvement was noted with aripiprazole augmentation over placebo in Y-BOCS total score and obsessions/compulsions sub-scores at the endpoint. Similar results were found in studies by Ak et al. (2011) and Sayyah et al. (2012). However, in a comparison trial by Selvi et al. (2011), aripiprazole was found to be inferior to risperidone in augmentation of treatment-resistant OCD. Forty-one patients who had failed 12 weeks of SSRI monotherapy were randomized to risperidone 3 mg/day or aripiprazole 15 mg/day for 8 weeks, and 72.2% of risperidone-treated patients showed improvement on the Y-BOCS obsession scores, which was significant compared with 50% of aripiprazole-treated patients.

Marazziti et al. (2005) added olanzapine to patients' previous medication regimes at a dose of 2.5 mg/day, titrating the medication up to 10 mg/day within 2 weeks. A significant number of patients showed at least a 35% reduction in total Y-BOCS score, which was maintained during 1 year of follow-up. Maina et al. (2008) investigated augmentation of treatment-resistant OCD with olanzapine or risperidone. Ninety-six patients with OCD were treated for 16 weeks with open-label SSRI at adequate doses. Fifty patients were considered to have failed SSRI treatment as defined by less than a 35% reduction in YBOCS scores and were randomized to 8 weeks of single-blind treatment with risperidone (1–3 mg/day) or olanzapine (2.5–10 mg/day). Both groups showed similar responses to treatment according to total Y-BOCS scores, with no significant difference being noted between the two medications. Risperidone was most often associated with amenorrhea and olanzapine with weight gain. There have been at least four double-blind, placebo-controlled augmentation trials of quetiapine augmentation in SSRI-resistant patients with OCD.

Denys et al. (2004) randomized 40 SSRI non-responding subjects with OCD to either quetiapine up to 300 mg/day or placebo for 8 weeks. About 40% of the quetiapine-treated cohort were classified as responders compared with 10% of the placebo-treated subjects. Fineberg et al. (2006) performed a combined analysis of the three published double-blind, placebo-controlled quetiapine augmentation studies. They suggested that there seemed to be a clinically significant signal associated with quetiapine augmentation. Kordon et al. (2008) conducted a 12-week, double-blind trial with 40 SSRI-resistant OCD patients, randomizing them to either quetiapine 400–600 mg/day or placebo. No significant benefit of quetiapine augmentation was found. Diniz et al. (2010) did find a significant improvement in treatment-resistant OCD symptoms with quetiapine augmentation of an SSRI compared with clomipramine augmentation. Vulink et al. (2009) also investigated quetiapine augmentation of SSRI, but excluded treatment-resistant OCD patients. They found statistically significant benefit with quetiapine augmentation (300–450 mg/day) in a 12-week trial of 66 patients on citalopram 60 mg/day over placebo, according to total YBOCS score from baseline to endpoint. However, large-scale placebo-controlled studies are clearly indicated in order to confirm and extend these findings.

A preliminary study by Metin et al. (2003) investigated the efficacy of amisulpride augmentation (200–600 mg) in 20 OCD patients who had not responded to SSRI treatment. In this open-label trial, 35% of patients reported significant improvement with combination therapy and 60% of patients reported moderate improvement.

Matsunaga et al. (2009) conducted a long-term trial of dopamine serotonin blockers (atypical antipsychotic) augmentation of SSRI in treatment-resistant OCD patients. A total of 137 patients received 12 weeks of treatment with fluvoxamine or paroxetine. Forty-four patients with less than 10% improvement in Y-BOCS scores with a CGI-I score of 3 or 4 were considered to be treatment resistant and were randomized to receive risperidone 1–5 mg/day, quetiapine 25–100 mg/day, or olanzapine 1–10 mg/day for at least 6 months. This study defined responders as those with a ≥25% improvement in Y-BOCS total

score and a CGI-I score of 1 or 2. No significant therapeutic benefit was seen with dopamine serotonin blockers (atypical antipsychotics) augmentation in these treatment-resistant patients, but these subjects did report significantly increased appetite and weight gain.

Zhou et al. (2019) examined the effect of medications for psychosis (antipsychotics) as augmenting agents to SSRIs by performing a network meta-analysis. The mean change in YBOCS score from baseline to follow-up was used as the measurement of the primary outcome. A random-effects model revealed that risperidone (mean difference, -4.47; 95% CI, -8.75 to -0.17) and aripiprazole (mean difference, -5.14; 95% CI, -9.95 to -0.28) had significantly greater effects than placebo. In summary, studies suggest that augmentation with a variety of dopamine serotonin blockers (atypical antipsychotic agents) may result in a substantial benefit for a subset of patients with treatment-refractory OCD.

Novel Treatments

Ghaleiha et al. (2013) conducted a randomized controlled trial to examine the effect of memantine for augmenting fluvoxamine. The authors enrolled 42 patients with OCD with the YBOCS of 21 or greater. Patients were evaluated by using the YBOCS at every two weeks for eight weeks. A significant effect for time x treatment interaction was demonstrated ($p = 0.006$), suggesting that the memantine group had a greater effect than placebo. Haghighi et al. (2013) evaluated the effect of adjuvant memantine to a standard SSRI or clomipramine in a 12-week randomized controlled trial. The time by group interaction showed that Y-BOCS scores significantly decreased over time in the memantine group compared with the placebo group ($p = 0.005$). Another randomized controlled trial was conducted by Modarresi et al. (2018) to investigate the effect of memantine in patients with SRI-refractory OCD. In this study, 32 patients were randomized to either 20-mg memantine or placebo augmentation and were evaluated at baseline and every 4 weeks for 12 weeks. After post-hoc analyses, a significant interaction between time and group for the reduction of the YBOCS at weeks 8 and 12 (both $p < 0.001$). The mean reduction of the YBOCS the memantine and placebo groups were respectively 17.2% versus -0.8% ($p < 0.001$) at 8 weeks and 40.9% versus -0.3% ($p < 0.001$) at 12 weeks. Zhou et al. (2019) evaluated the effect of memantine as an augmenting agent to SSRIs by performing a network meta-analysis. A random effects model verified that memantine was superior to placebo in the YBOCS improvement (mean difference, -8.94; 95% CI, -14.42 to -3.42). Notably, this effect was larger than those of the aforementioned agents that showed significant effects in this study, i.e., topiramate, lamotrigine, risperidone, and aripiprazole.

Odansetron and granisetron are 5-HT_3 receptor antagonists that have been investigated in OCD patients. Pallanti et al. (2009) treated 14 patients on stable regimens of SRIs and medications for psychosis (antipsychotics) in a 12-week, open-label trial of odansetron 0.5 mg twice daily; 64.3% of patients were classified as responders, with an average 23.2% reduction in Y-BOCS total score. Soltani et al. (2010) conducted an 8-week, double-blind, placebo-controlled trial in 42 patients with OCD. Patients were randomized to odansetron 4 mg/fluoxetine 20 mg or placebo/fluoxetine 20 mg. The odansetron-treated patients had significantly improved Y-BOCS total scores by week 8 compared with placebo. Askari et al. (2012) investigated granisetron augmentation of fluvoxamine, 42 patients being randomized to granisetron 1 mg every 12 h or placebo in a double-blind 8-week trial. Significantly more granisetron-treated patients achieved a response compared with placebo (90% vs. 35%). However, further studies are needed to investigate both odansetron and granisetron efficacy in OCD.

Two studies scrutinized D-cycloserine (DCS) as an adjunct to exposure therapy in OCD treatment. DCS was originally an anti-tuberculosis medication and does not have anxiolytic properties; however, DCS has been shown to increase response to conditioned fear extinction learning, the philosophy underlying exposure therapy. Kushner et al. (2007) found in a double-blind trial that 125 mg of DCS administered 2 h prior to each exposure therapy session decreased the total number of sessions required and also significantly decreased obsession-related distress versus placebo. However, another study by Storch et al. (2007) failed to show significant benefit with DCS augmentation of exposure therapy. A 2014 meta-analysis of randomized, double-blind, placebo-controlled trials (Rodrigues et al., 2014) that included four trials for OCD found that DCS enhanced exposure treatment for anxiety disorders when given close to the actual exposure and in low doses. A more recent meta-analysis from 2017 (Mataix-Cols et al., 2017) of 1047 patients across 21 clinical trials demonstrated that patients who received DCS had a statistically significant improvement from pre-treatment to post-treatment measures, but not on pre-treatment to mid-treatment, or pre-treatment to follow-up measures. The authors concluded that DCS had a small augmentation effect on exposure therapy.

Special Populations–Children and Adolescents

Paroxetine has been studied as a treatment for OCD in children. Geller et al. (2004) performed a multisite, double-blind, placebo-controlled, flexible-dose study randomizing 203 patients to receive either 10–50 mg/day paroxetine or placebo for 10 weeks. A total of 145 patients completed the study, with 33/98 (33.7%) treatment responders among patients receiving paroxetine in contrast to 25/105 (23.8%) responders among patients receiving placebo. The Children's Yale-Brown Obsessive-Compulsive Scale (CY-BOCS) scores demonstrated a significant decrease for the paroxetine-treated group compared with the placebo-treated group. However, the two treatment groups did not differ in terms of CGI-S, CGI-I, and Global Assessment of Functioning (GAF) ratings.

Riddle et al. (2001) conducted a 10-week, double-blind, placebo-controlled, multisite study investigating the efficacy of fluvoxamine in childhood OCD. Out of 120 individuals who were randomized, 24/57 (42.1%) treated with fluvoxamine and 17/63 (26%) treated with placebo were responders (defined as a $\geq25\%$ decrease in CY-BOCS scores).

Another study evaluated the efficacy of long-term sertraline treatment of children and adolescents aged 6–18 years with OCD (Cook et al., 2001). A total of 137 patients were assigned to 12 weeks of double-blind, placebo-controlled treatment with subsequent open-label sertraline treatment (50–200 mg/day) for 52 weeks. At the study endpoint, the mean daily sertraline dose was 108 mg/day for children (6–12 years) and 132 mg/day for adolescents (13–18 years). Outcome measures included the CY-BOCS, the NIMH Obsessive-Compulsive Scale, and the CGI-Severity and Improvement Scales. Mean CY-BOCS scores changed from 22.8 at study entry to 17.0 at the completion of the double-blind phase and to 10.8 at the 52-week endpoint. The mean CGI-Severity scores declined from 4.6 to 3.7 and to 2.7, respectively. Only 12% of patients discontinued because of side effects.

Masi et al. (2010) examined aripiprazole augmentation of SSRI or clomipramine monotherapy in 39 adolescents with treatment-resistant OCD. Patients were given aripiprazole 5–20 mg/day in conjunction with their pretrial SSRI or clomipramine dose. About 59% of patients were classified as responders by the treatment endpoint with a CGI-I of 1 or 2 and a CGI-S of 3 or less. It was found that 62.5% of these patients with comorbid tic disorder/Tourette syndrome improved after starting aripiprazole. In another trial by Grant et al. (2007), six treatment-resistant OCD pediatric patients were given riluzole in an open-label 12-week trial. Four patients reached response rates, with reductions of $\geq$46% in Y-BOCS scores by the endpoint and improvements on the CGI-I scale. Riluzole was well tolerated. Future large-scale, placebo-controlled trials would help to clarify the efficacy of these medications in the pediatric population.

Treatment Conclusions

The SRIs and clomipramine are the only FDA-approved treatments for OCD. These agents require individualized titration and there are many instances where patients end up on much higher doses than those used in the treatment of mood disorders. Although this clinical observation is not unambiguously supported by existing clinical trial data, there seems to be merit to this approach. Patients who have comorbid tic disorders with OCD may be more responsive to the combination of medications for psychosis with a serotonergic agent. There is a growing body of literature supporting off-label usage of numerous medication classes for augmentation strategies. As the mechanisms of OCD are better understood, there will be increasing knowledge about novel psychopharmacological targets for medication. Studies suggest that the combination of pharmacotherapies with behavioral modification may be the optimal approach for the care of these patients.

Clinical Vignette
Ms. K is a 46-year-old Hispanic female who presents with a chief complaint of recurrent major depressive disorder.

Ms. K has had multiple brief treatments for depressive disorder but always discontinues her treatment. She reports that she was physically abused by her older brother while in adolescence and 2 years ago was involved in a car-jacking. Although Ms. K was not harmed during the car-jacking, she developed a depressive disorder immediately following the incident. When questioned further, it is clear that Ms. K has recurrent nightmares about this event, has flashbacks, startles easily, avoids driving on all streets except main arteries and the freeway, and has complained of feeling distant from her husband and children. During the interview, Ms. K breaks down and is tearful. She discusses how ashamed she feels for being "weak" and having these pervasive symptoms. When the concept of posttraumatic stress disorder is introduced to Ms. K, she is relieved to know that she is not the odd person out.

Ms. K was started on sertraline 25 mg/day and this dose was gradually tapered up by 25 mg/week to 100 mg/day.

(continued)

She had a gradual reduction in her intrusive thoughts and nightmares that began during the second week of treatment. Ms. K's depressive symptoms also began to resolve by week 4. Since Ms. K was not having side effects and still was symptomatic at 6 weeks, the dosage of sertraline was increased to a 150 mg/day. Over the next 8 weeks, Ms. K showed continued improvement in both her depressive symptoms and many of her symptoms of posttraumatic stress disorder. She became more active with her family and affectionate with her husband, and she began to expand the areas where she felt comfortable driving. Over time, Ms. K continued to demonstrate improvements in symptoms. She is currently maintained on 150 mg/day of sertraline and has been able to return to work and most of her normal activities over a period of 2 years.

Future Directions

The development of future pharmacological treatments for anxiety disorders may take new and exciting avenues. There is a growing body of preclinical evidence suggesting that basic fear involves the interaction between three of the 13 nuclei of the amygdala with decreased activation of the medio-prefrontal cortex and concomitant increased activation of the hypothalamus, the noradrenergic system, and specific areas related to fear such as the peri-aqueductal gray region of the brain. This has led investigators to explore the use of NMDA antagonists in an attempt to inhibit the development of fear responses, and other compounds such as NMDA receptor agonists, which might be used to facilitate new learning and thereby decrease fear responses. For example, there is ongoing research investigating whether the effects of psychotherapy may be potentiated using D-cycloserine, a tuberculosis medication with partial mechanism as an NMDA receptor

agonist. This research may be the harbinger of a variety of new pharmacological approaches targeted at specific receptors and secondary messenger systems. Pilot studies employing medication augmentation strategies are currently under way for SAD and PTSD.

Conclusion

In closing, clinicians have a wide array of medications available for the treatment of anxiety and anxiety disorders. The breadth of accessible treatment options greatly facilitates our ability to help patients. Safe and effective treatments exist for everything from short-term treatment of non-pathological anxiety to previously intractable anxiety disorders such as OCD. However, the most important therapeutic agents that a clinician possesses are still sound clinical skills and judgment. Appropriate diagnosis and rapport are the foundations of any pharmacological intervention that made with our patients.

References

Abell, S. R., & El-Mallakh, R. S. (2021). Serotonin-mediated anxiety: How to recognize and treat it. *Current Psychiatry, 20*(11), 37–40.

Abramowitz, J. S. (1997). Effectiveness of psychological and pharmacological treatments for obsessive-compulsive disorder: A quantitative review. *Journal of Consulting and Clinical Psychology, 65*(1), 44–52.

Ak, M., Bulut, S. D., Bozhurt, A., et al. (2011). Aripiprazole augmentation of serotonin reuptake inhibitors in treatment-resistant obsessive-compulsive disorder: A 10-week open-label study. *Advances in Therapy, 28*(4), 341–348.

Alaghband-Rad, J., & Hakimshooshtary, M. (2009). A randomized controlled clinical trial of citalopram versus fluoxetine in children and adolescents with obsessive-compulsive disorder (OCD). *European Journal of Child and Adolescent Psychiatry, 18*, 131–135.

Alaka, K. J., Noble, W., Montejo, A., Dueñas, H., Munshi, A., Strawn, J. R., Lenox-Smith, A., Ahl, J., Bidzan, L., Dorn, B., & Ball, S. (2014). Efficacy and safety of duloxetine in the treatment of older adult patients with generalized anxiety disorder: A randomized, double-blind, placebo-controlled trial. *International Journal of Geriatric Psychiatry, 29*(9), 978–986.

Allgulander, C., Mangano, R., Zhang, J., et al. (2004). Efficacy of venlafaxine ER in patients with social anxiety disorder: A double-blind, placebo-controlled, parallel-group comparison with paroxetine. *Human Psychopharmacology, 19*, 387–396.

Askari, N., Moin, M., Sanati, M., et al. (2012). Granisetron adjunct to fluvoxamine for moderate to severe obsessive-compulsive disorder: A randomized, double-blind, placebo-controlled trial. *CNS Drugs, 26*(1), 883–892.

Asnis, G. M., Hameedi, F. A., Goddard, A. W., et al. (2001). Fluvoxamine in the treatment of panic disorder: A multi-center, double-blind, placebo-controlled study in outpatients. *Psychiatric Research, 103*(1), 1–14.

Baker, J. F., Cates, M. E., & Luthin, D. R. (2018). D-cycloserine in the treatment of posttraumatic stress disorder. *Mental Health Clinician, 7*(2), 88–94.

Bakish, D. (1992). Reversible monoamine-a inhibitors in panic disorder. *Clinical Neuropharmacology, 15*(Suppl 1, Pt A), 432A–433A.

Baldwin, D., Bobes, J., Stein, D. J., et al. (1999). Paroxetine in social phobia/social anxiety disorder. Randomised, double-blind, placebo-controlled study. Paroxetine Study Group. *British Journal of Psychiatry, 175*, 120–126.

Baldwin, D. S., Asakura, S., Koyama, T., Hayano, T., Hagino, A., Reines, E., & Larsen, K. (2016). Efficacy of escitalopram in the treatment of social anxiety disorder: A meta-analysis versus placebo. *European Neuropsychopharmacology: The Journal of the European College of Neuropsychopharmacology, 26*(6), 1062–1069.

Ballenger, J. C., Burrows, G. D., DuPont, R. L. J., et al. (1988). Aprazolam in panic disorder and agoraphobia: Results from a multicenter trial. I. Efficacy in short-term treatment. *Archives of General Psychiatry, 45*(5), 413–422.

Ballenger, J. C., Wheadon, D. E., Steiner, M., et al. (1998). Double-blind, fixed-dose, placebo-controlled study of paroxetine in the treatment of panic disorder. *American Journal of Psychiatry, 155*(1), 36–42.

Bandelow, B., Behnke, K., Lenoir, S., et al. (2004). Sertraline versus paroxetine in the treatment of panic disorder: An acute, double-blind noninferiority comparison. *Journal of Clinical Psychiatry, 65*, 405–413.

Bandelow, B., Chouinard, G., Bobes, J., et al. (2010). Extended-release quetiapine fumarate (quetiapine XR): A once-daily monotherapy effective in generalized anxiety disorder. Data from a randomized, double-blind, placebo- and active-controlled study. *International Journal of Neuropsychopharmacology, 13*, 305–320.

Baniasadi, M., Hosseini, G., Bordbar, M. R. F., et al. (2014). Effect of pregabalin augmentation in treatment of patients with combat-related chronic posttraumatic stress disorder: A randomized controlled trial. *Journal of Psychiatric Practice, 20*(6), 419–427.

Barnett, S. D., Kramer, M. L., Casat, C. D., Connor, K. M., & Davidson, J. R. (2002). Efficacy of olanzapine in social anxiety disorder: A pilot study. *Journal of Psychopharmacology, 16*(4), 365–368.

Bartzokis, G., Lu, P., Turner, J., et al. (2004). Adjunctive risperidone in the treatment of chronic combat-related posttraumatic stress disorder. *Biological Psychiatry, 57*, 474–479.

Becker, M. E., Hertzberg, M. A., Scott, D., Moore, S. D., et al. (2007). A placebo-controlled trial of bupropion SR in the treatment of chronic posttraumatic stress disorder. *Journal of Clinical Psychopharmacology, 27*, 193–197.

Berlant, J. L. (2001). Topiramate in posttraumatic stress disorder: Preliminary clinical observations. *Journal of Clinical Psychiatry, 62*(Suppl 17), 60–63.

Berlant, J. L. (2004). Prospective open-label study of add-on and monotherapy topiramate in civilians with chronic nonhallucinatory posttraumatic stress disorder. *BMC Psychiatry, 4*, 24.

Berlant, J. L., & van Kammen, D. P. (2002). Open-label topiramate as primary or adjunctive therapy in chronic civilian posttraumatic stress disorder: A preliminary report. *Journal of Clinical Psychiatry, 63*, 15–20.

Berlin, H. A., Koran, L. M., Jenke, M. A., et al. (2011). Double-blind, placebo-controlled trial of topiramate augmentation in treatment-resistant obsessive-compulsive disorder. *Journal of Clinical Psychiatry, 72*(5), 716–721.

Bisserbe, J. C., Lane, R. M., & Flament, M. F. (1997). A double-blind comparison of sertraline and clomipramine in outpatients with obsessive-compulsive disorder. *European Psychiatry, 12*, 82–93.

Blanco, C., & Liebowitz, M. R. (1998). Dimensional versus categorical response to moclobemide in social phobia. *Journal of Clinical Psychopharmacology, 18*(4), 344–346.

Bodkin, J. A., Allgulander, C., Llorica, P. M., et al. (2011). Predictors of relapse in a study of duloxetine treatment for patients with generalized anxiety disorder. *Human Psychopharmacology: Clinical and Experimental, 26*, 258–266.

Boyer, W. (1995). Serotonin uptake inhibitors are superior to imipramine and alprazolam in alleviating panic attacks: A meta-analysis. *International Clinical Psychopharmacology, 10*(1), 45–49.

Bradwejn, J., Ahokas, A., Stein, D. J., et al. (2005). Venlafaxine extended-release capsules in panic disorder. *British Journal of Psychiatry, 187*, 352–359.

Brady, K., Pearlstein, T., Asnis, G. M., et al. (2000). Efficacy and safety of sertraline treatment of posttraumatic stress disorder: A randomized controlled trial. *JAMA, 283*(14), 1837–1844.

Brawman-Mintzer, O., Knapp, R. G., & Nieter, P. L. (2005). Adjunctive risperidone in generalized anxiety disorder: A double-blind, placebo-controlled study. *Journal of Clinical Psychiatry, 66*(10), 1321–1325.

Brawman-Mintzer, O., Knapp, R. G., Rynn, M., et al. (2006). Sertraline treatment for generalized anxiety disorder: A randomized, double-blind, placebo-controlled study. *Journal of Clinical Psychiatry, 67*(6), 874–881.

Britnell, S. R., Jackson, A. D., Brown, J. N., & Capehart, B. P. (2017). Aripiprazole for post-traumatic stress disorder: A systematic review. *Clinical Neuropharmacology, 40*(6), 273–278.

Brownlow, J. A., Harb, G. C., & Ross, R. J. (2015). Treatment of sleep disturbances in post-traumatic stress disorder: A review of the literature. *Current Psychiatry Reports, 17*(6), 41.

Brunet, A., Daniel Saumier, D., Liu, A., et al. (2018). Reduction of PTSD symptoms with pre-reactivation propranolol therapy: A randomized controlled trial. *The American Journal of Psychiatry, 175*(5), 427–433.

Bruno, A., Mico, U., Pandolfo, G., et al. (2012). Lamotrigine augmentation of serotonin reuptake inhibitors in treatment-resistant obsessive-compulsive disorder: A double-blind, placebo-controlled study. *Journal of Psychopharmacology, 26*(11), 1456–1462.

Butterfeld, M., Becker, M., Connor, K., et al. (2001). Olanzapine in the treatment of posttraumatic stress disorder: A pilot study. *International Clinical Psychopharmacology, 16*(4), 197–203.

Byers, M. G., Allison, K. M., Wendel, C. S., et al. (2010). Prazosin versus quetiapine for nighttime posttraumatic stress disorder symptoms in veterans. *Journal of Clinical Psychopharmacology, 30*(2), 225–229.

Bystritsky, A., Rosen, R., Suri, R., et al. (1999). Pilot open-label study of nefazodone in panic disorder. *Depression and Anxiety, 10*(3), 137–139.

Bystritsky, A., Kerwin, L., Feusner, J. D., & Vapnik, T. (2008). A pilot controlled trial of bupropion XL versus escitalopram in generalized anxiety disorder. *Psychopharmacology Bulletin, 41*(1), 46–51.

Calliard, V., Roullion, F., & Viwl, J. (1999). Comparative effects of low and high doses of clomipramine and placebo in panic disorder: A double-blind controlled study. *Acta Psychiatrica Scandinavica, 99*, 51–58.

Careri, J. M., Draine, A. E., Hanover, R., & Liebowitz, M. R. (2015). A 12-week double-blind, placebo-controlled, flexible-dose trial of vilazodone in generalized social anxiety disorder. *The Primary Care Companion for CNS Disorders, 17*(6). https://doi.org/10.4088/PCC.15m01831

Carey, P., Suliman, S., Ganesan, K., et al. (2012). Olanzapine monotherapy in posttraumatic stress disorder: Efficacy in a randomized, double-blind, placebo-controlled study. *Human Psychopharmacology: Clinical and Experimental, 27*, 386–391.

Carpenter, L. L., McDougle, C. J., Epperson, C. N., et al. (1996). A risk-benefit assessment of drugs used in the management of obsessive-compulsive disorder. *Drug Safety, 15*(2), 116–134.

Cartwright, C., & Hollander, E. (1998). SSRIs in the treatment of obsessive-compulsive disorder. *Depression and Anxiety, 8*(Suppl 1), 105–113.

Cassano, G. B., Petracca, A., Perugi, G., et al. (1988). Clomipramine for panic disorder: I. The first

10 weeks of a long-term comparison with imipramine. *Journal of Affective Disorders, 14*(2), 123–127.

Charney, D. S., Woods, S. W., Goodman, W. K., et al. (1986). Drug treatment of panic disorder: The comparative efficacy of imipramine, alprazolam, and trazodone. *Journal of Clinical Psychiatry, 47*(12), 580–586.

Choi, K. W., Woo, J. M., Kim, Y. R., et al. (2012). Long-term escitalopram treatment in Korean patients with panic disorder: A prospective, naturalistic, open-label, multicenter trial. *Clinical Psychopharmacology and Neuroscience, 10*(1), 44–48.

Chouinard, G., Goodman, W., Greist, J., et al. (1990). Results of a double-blind placebo controlled trial of a new serotonin uptake inhibitor, sertraline, in the treatment of obsessive-compulsive disorder. *Psychopharmacology Bulletin, 26*(3), 279–284.

Chung, M. Y., Min, K. H., Jun, Y. J., et al. (2004). Efficacy and tolerability of mirtazapine and sertraline in Korean veterans with posttraumatic stress disorder: A randomized open label study. *Human Psychopharmacology: Clinical and Experimental, 10*, 489–494.

Connor, K. M., & Davidson, J. R. (1998). Generalized anxiety disorder: Neurobiological and pharmacotherapeutic perspectives. *Biological Psychiatry, 44*(12), 1286–1294.

Connor, K. M., Davidson, J. R., Weisler, R., et al. (2006). Tiagabine for posttraumatic stress disorder: Effects of open-label and double-blind discontinuation treatment. *Psychopharmacology, 184*, 21–25.

Cook, E. H., Wagner, K. D., March, J. S., et al. (2001). Long-term sertraline treatment of children and adolescents with obsessive-compulsive disorder. *Journal of the American Academy of Child and Adolescent Psychiatry, 40*(10), 1175–1181.

Cox, B. J., Swinson, R. P., Morrison, B., et al. (1993). Clomipramine, fluoxetine, and behavior therapy in the treatment of obsessive-compulsive disorder: A meta-analysis. *Journal of Behavior Therapy and Experimental Psychiatry, 24*(2), 149–153.

Cross-National Collaborative Panic Study, Second Phase Investigators. (1992). Drug treatment of panic disorder. Comparative efficacy of alprazolam, imipramine, and placebo. *British Journal of Psychiatry, 160*, 191–202. (Erratum published in *British Journal of Psychiatry*, 1992, 161, 724).

Dannon, P. N., Sasson, Y., Hirschmann, S., et al. (2000). Pindolol augmentation in treatment-resistant obsessive compulsive disorder: A double-blind placebo controlled trial. *European Neuropsychopharmacology, 10*(3), 165–169.

Dannon, P. N., Iancu, I., Lowengrub, K., et al. (2007). A naturalistic long-term comparison study of selective serotonin reuptake inhibitors in the treatment of panic disorder. *Clinical Neuropharmacology, 30*(6), 326–334.

Davidson, J., Walker, J. I., & Kilts, C. (1987). A pilot study of phenelzine in the treatment of posttraumatic stress disorder. *British Journal of Psychiatry, 150*, 252–255.

Davidson, J., Kudler, H., Smith, R., et al. (1990). Treatment of posttraumatic stress disorder with amitriptyline and placebo. *Archives of General Psychiatry, 47*(3), 259–266.

Davidson, J. R., DuPont, R. L., Hedges, D., et al. (1999). Efficacy, safety, and tolerability of venlafaxine extended release and buspirone in outpatients with generalized anxiety disorder. *Journal of Clinical Psychiatry, 60*(8), 528–535.

Davidson, J., Pearlstein, T., Londborg, P., et al. (2001). Efficacy of sertraline in preventing relapse of posttraumatic stress disorder: Results of a 28-week double-blind, placebo-controlled study. *American Journal of Psychiatry, 158*(12), 1974–1981.

Davidson, J., Bose, A., Korotzer, A., et al. (2004). Escitalopram in the treatment of generalized anxiety disorder: Double-blind, placebo controlled, flexible-dose study. *Depression and Anxiety, 19*, 234–240.

Davidson, J., Bose, A., & Wang, Q. (2005). Safety and efficacy of escitalopram in the long-term treatment of generalized anxiety disorder. *Journal of Clinical Psychiatry, 66*(11), 1441–1446.

Davidson, J., Baldwin, D., Stein, D. J., et al. (2006a). Treatment of posttraumatic stress disorder with venlafaxine extended release. *Archives of General Psychiatry, 63*, 1158–1165.

Davidson, J., Rothbaum, B. O., Tucker, P., et al. (2006b). Venlafaxine extended release in posttraumatic stress disorder. *Journal of Clinical Psychopharmacology, 26*(3), 259–267.

Davidson, J. R., Brady, K., Mellman, T. A., et al. (2007). The efficacy and tolerability of tiagabine in adult patients with posttraumatic stress disorder. *Journal of Clinical Psychopharmacology, 27*(1), 85–88.

Davidson, J., Allgulander, C., Pollack, M. H., et al. (2008). Efficacy and tolerability of duloxetine in elderly patients with generalized anxiety disorder: A pooled analysis of four randomized, double-blind, placebo-controlled studies. *Human Psychopharmacology, 23*(6), 519–526.

Davis, L. L., Nugent, A. L., Murray, J., et al. (2000). Nefazodone treatment for chronic posttraumatic stress disorder: An open trial. *Journal of Clinical Psychopharmacology, 20*(2), 159–164.

Davis, L. L., Jewell, M. E., Ambrose, S., et al. (2004). A placebo-controlled study of nefazodone for the treatment of chronic posttraumatic stress disorder. *Journal of Clinical Psychopharmacology, 24*(3), 291–297.

Davis, L. L., Davidson, J. R., Ward, L. C., et al. (2008). Divalproex in the treatment of posttraumatic stress disorder: A randomized, double-blind, placebo-controlled trial in a veteran population. *Journal of Clinical Psychopharmacology, 28*(1), 84–88.

Davis, L. L., Pilkinton, P., Lin, C., et al. (2020). A randomized, placebo-controlled trial of mirtazapine for the treatment of posttraumatic stress disorder in veterans. *The Journal of Clinical Psychiatry, 81*(6), 20m13267.

DeMartinis, N. A., Schweizer, E., & Rickels, K. (1996). An open-label trial of nefazodone in high comorbidity panic disorder. *The Journal of Clinical Psychiatry, 57*(6), 245–248.

den Boer, J. A. (1998). Pharmacotherapy of panic disorder: Differential efficacy from a clinical viewpoint. *Journal of Clinical Psychiatry, 59*(Suppl 8), 30–36.

Denys, D., de Geus, F., van Megen, H., et al. (2004). A double-blind, randomized, placebo-controlled trial of quetiapine addition in patients with obsessive-compulsive disorder refractory to serotonin reuptake inhibitors. *Journal of Clinical Psychiatry, 65*, 1040–1048.

Detweiler, M. B., Bhuvaneshwar, P., Joseph, C., Jennifer, S. B., Jonna, G. D., & Brian W. L. (2016). Treatment of Post-Traumatic Stress Disorder Nightmares at a Veterans Affairs Medical Center. *Journal of Clinical Medicine Research 5*(12). https://doi.org/10.3390/jcm5120117.

Diniz, J. B., Shavitt, R. G., Pereira, C. A. B., et al. (2010). Quetiapine versus clomipramine in the augmentation of selective serotonin reuptake inhibitors for the treatment of obsessive-compulsive disorder: A randomized, open-label trial. *Journal of Psychopharmacology, 24*(3), 297–307.

Diniz, J. B., Shavitt, R. G., Fossaluza, V., et al. (2011). A double-blind, randomized, controlled trial of fluoxetine plus quetiapine or clomipramine versus fluoxetine plus placebo for obsessive-compulsive disorder. *Journal of Clinical Psychopharmacology, 31*(6), 763–768.

Dougherty, D. D., Corse, A. K., Chou, T., Duffy, A., Arulpragasam, A. R., Deckersbach, T., Jenike, M. A., & Keuthen, N. J. (2015). Open-label study of duloxetine for the treatment of obsessive-compulsive disorder. *The International Journal of Neuropsychopharmacology, 18*(2), pyu062.

Dowd, S. M., Zalta, A. K., Burgess, H. J., et al. (2020). Double-blind randomized controlled study of the efficacy, safety and tolerability of eszopiclone *vs* placebo for the treatment of patients with post-traumatic stress disorder and insomnia. *World Journal of Psychiatry, 10*(3), 21–28.

Dubuff, E., Ferguson, J., & Londborg, P. (1995). Double-blind comparison of three fixed doses of sertraline and placebo in patients with panic disorder [abstract]. Presented at the Eighth Congress of the European College of Neuropsychopharmacology, Venice.

Durgam, S., Gommoll, C., Forero, G., Nunez, R., Tang, X., Mathews, M., & Sheehan, D. V. (2016). Efficacy and safety of vilazodone in patients with generalized anxiety disorder: A randomized, double-blind, placebo-controlled, flexible-dose trial. *Journal of Clinical Psychiatry, 77*(12), 1687–1694.

El-Mallakh, R. S., & Ali, Z. (2019). Therapeutic implications of the serotonin transporter gene in depression. *Biomarkers in Neuropsychiatry, 1*, 100004.

Emmanuel, N. P., Lydiard, R. B., & Ballenger, J. C. (1991). Treatment of social phobia with bupropion. *Journal of Clinical Psychopharmacology, 11*(4), 276–277.

Emmanuel, N. P., Brawman-Mintzer, O., Morton, W. A., et al. (2000). Bupropion-SR in treatment of social phobia. *Depression and Anxiety, 12*(2), 111–113.

Endicott, J., Russell, J. M., Raskin, J., et al. (2007). Duloxetine treatment for role functioning improvement in generalized anxiety disorder: Three independent studies. *Journal of Clinical Psychiatry, 68*, 518–524.

English, B. A., Jewell, M., Jewell, G., et al. (2006). Treatment of chronic posttraumatic stress disorder in combat veterans with citalopram: An open trial. *Journal of Clinical Psychopharmacology, 26*(1), 84–88.

Escalona, R., Canive, J. M., Calais, L. A., et al. (2002). Fluvoxamine treatment in veterans with combat-related post-traumatic stress disorder. *Depression and Anxiety, 15*, 29–33.

Fahlen, T., Nilsson, H. L., Borg, K., et al. (1995). Social phobia: The clinical efficacy and tolerability of the monoamine oxidase-A and serotonin uptake inhibitor brofaromine. A double-blind placebo-controlled study. *Acta Psychiatrica Scandinavica, 92*(5), 351–358.

Farnia, V., Gharehbaghi, H., Alikhani, M., Almasi, A., Golshani, S., Tatari, F., Davarinejad, O., Salemi, S., Sadeghi Bahmani, D., Holsboer-Trachsler, E., & Brand, S. (2018). Efficacy and tolerability of adjunctive gabapentin and memantine in obsessive compulsive disorder: Double-blind, randomized, placebo-controlled trial. *Journal of Psychiatric Research, 104*, 137–143.

Feder, A., Costi, S., Rutter, S. B., Collins, A. B., Govindarajulu, U., Jha, M. K., Horn, S. R., Kautz, M., Corniquel, M., Collins, K. A., & Bevilacqua, L. (2021). A randomized controlled trial of repeated ketamine administration for chronic posttraumatic stress disorder. *American Journal of Psychiatry, 178*, 193.

Feltner, D. E., Crockatt, J. G., Dubovsky, S. J., et al. (2003). A randomized, doubleblind, placebo-controlled, fixed-dose, multicenter study of pregabalin in patients with generalized anxiety disorder. *Journal of Clinical Psychopharmacology, 23*(3), 240–249.

Feltner, D., Wittchen, H., Kavoussi, R., et al. (2008). Long-term efficacy of pregabalin in generalized anxiety disorder. *International Clinical Psychopharmacology, 23*(1), 18–28.

Feltner, D. E., Liu-Dumaw, M., Schweizer, E., et al. (2011). Efficacy of pregabalin in generalized social anxiety disorder: Results of a double-blind, placebo-controlled, fixed-dose study. *International Clinical Psychopharmacology, 26*(4), 213–220.

Ferguson, J. M., Khan, A., Mangano, R., et al. (2007). Relapse prevention of panic disorder in adult outpatient responders to treatment with venlafaxine extended release. *Journal of Clinical Psychiatry, 68*(1), 58–68.

Fineberg, N., Stein, D., Premkumar, P., et al. (2006). Adjunctive quetiapine for serotonin reuptake inhibitor-resistant obsessive-compulsive disorder: A meta-analysis of randomized controlled treatment trials. *International Clinical Psychopharmacology, 21*(6), 337–343.

Freeman, C. P., Trimble, M. R., Deakin, J. F., et al. (1994). Fluvoxamine versus clomipramine in the treatment of obsessive-compulsive disorder: A multicenter, randomized, double-blind, parallel group comparison. *Journal of Clinical Psychiatry, 55*(7), 301–305.

Freire, R. C., Amrein, R., Mochcovitch, M. D., et al. (2017). A 6-year posttreatment follow-up of panic disorder patients: Treatment with clonazepam predicts lower recurrence than treatment with paroxetine. *Journal of Clinical Psychopharmacology, 37*, 429–434.

Gelenberg, A. J., Lydiard, R. B., Rudolph, R. L., et al. (2000). Efficacy of venlafaxine extended-release capsules in nondepressed outpatients with generalized anxiety disorder: A 6-month randomized controlled trial. *JAMA, 283*(23), 3082–3088.

Gelernter, C. S., Uhde, T. W., Cimbolic, P., et al. (1991). Cognitive-behavioral and pharmacological treatments of social phobia. A controlled study. *Archives of General Psychiatry, 48*(10), 938–945.

Geller, D., Wagner, K., Emslie, G., et al. (2004). Paroxetine treatment in children and adolescents with obsessive-compulsive disorder: A randomized, multicenter, double-blind, placebo-controlled trial. *Journal of the American Academy of Child and Adolescent Psychiatry, 43*(11), 1387–1396.

Generoso, M. B., Trevizol, A. P., Kasper, S., Cho, H. J., Cordeiro, Q., & Shiozawa, P. (2017). Pregabalin for generalized anxiety disorder: An updated systematic review and meta-analysis. *International Clinical Psychopharmacology, 32*(1), 49–55.

Ghaleiha, A., Entezari, N., Modabbernia, A., Najand, B., Askari, N., Tabrizi, M., Ashrafi, M., Hajiaghaee, R., & Akhondzadeh, S. (2013). Memantine add-on in moderate to severe obsessive-compulsive disorder: Randomized double-blind placebo-controlled study. *Journal of Psychiatric Research, 47*(2), 175–180.

Goddard, A. W., Brouette, T., Almai, A., et al. (2001). Early coadministration of clonazepam with sertraline for panic disorder. *Archives of General Psychiatry, 58*(7), 681–686.

Goddard, A. W., Mahmud, W., Medlock, C., et al. (2015). A controlled trial of quetiapine XR coadministration treatment of SSRI-resistant panic disorder. *Annals of General Psychiatry, 14*, 26.

Gommoll, C., Durgam, S., Mathews, M., Forero, G., Nunez, R., Tang, X., & Thase, M. E. (2015). A double-blind, randomized, placebo-controlled, fixed-dose phase III study of vilazodone in patients with generalized anxiety disorder. *Depression and Anxiety, 32*(6), 451–459.

Goodman, W. K., Price, L. H., Rasmussen, S. A., et al. (1989). Efficacy of fluvoxamine in obsessive-compulsive disorder. A double-blind comparison with placebo. *Archives of General Psychiatry, 46*(1), 36–44.

Grant, P., Lougee, L., Hirschtritt, M., et al. (2007). An open-label trial of riluzole, a glutamate antagonist, in children with treatment-resistant obsessive-compulsive disorder. *Journal of Child and Adolescent Psychopharmacology, 17*(6), 761–767.

Greenblatt, D. J., Shader, R. I., & Abernethy, D. R. (1983). Drug therapy. Current status of benzodiazepines. *New England Journal of Medicine, 309*(7), 410–416.

Greist, J. H. (1998). The comparative effectiveness of treatments for obsessive-compulsive disorder. *Bulletin of the Menninger Clinic, 62*(4, Suppl A), A65–A81.

Greist, J., Chouinard, G., DuBoff, E., et al. (1995a). Double-blind parallel comparison of three dosages of sertraline and placebo in outpatients with obsessive-compulsive disorder. *Archives of General Psychiatry, 52*, 289–295.

Greist, J. H., Jefferson, J. W., Kobak, K. A., et al. (1995b). A 1 year doubleblind placebo-controlled fixed dose study of sertraline in the treatment of obsessive-compulsive disorder. *International Clinical Psychopharmacology, 10*(2), 57–65.

Greist, J. H., Jenike, M. A., & Robinson, D. (1995c). Efficacy of fluvoxam-ine in obsessive-compulsive disorder: Results of a multicentre, doubleblind, placebo-controlled trial. *European Journal of Clinical Research, 7*, 195–204.

Greist, J. H., & Jefferson, J. W. (1998). Pharmacotherapy for Obsessive-Compulsive Disorder. *The British Journal of Psychiatry.* Supplement, no. 35, 64–70.

Greist, J. H., Lui-Dumaw, M., Schweizer, E., et al. (2011). Efficacy of pregabalin in preventing relapse in patients with generalized social anxiety disorder: Results of a double-blind, placebo-controlled 26-week study. *International Clinical Psychopharmacology, 26*(5), 243–251.

Hadley, S. J., Mandel, F. S., & Schweizer, E. (2012). Switching from long-term benzodiazepine therapy to pregabalin in patients with generalized anxiety disorder: A double-blind, placebo-controlled trial. *Journal of Psychopharmacology, 26*(4), 461–470.

Haghighi, M., Jahangard, L., Mohammad-Beigi, H., Bajoghli, H., Hafezian, H., Rahimi, A., Afshar, H., Holsboer-Trachsler, E., & Brand, S. (2013). In a double-blind, randomized and placebo-controlled trial, adjuvant memantine improved symptoms in inpatients suffering from refractory obsessive-compulsive disorders (OCD). *Psychopharmacology, 228*(4), 633–640.

Hamner, M., Faldowski, R., Ulmer, H., et al. (2003). Adjunctive risperidone treatment in posttraumatic stress disorder: A preliminary controlled trial of effects on comorbid psychotic symptoms. *International Clinical Psychopharmacology, 18*(1), 1–18.

Hamner, M. B., Faldowski, R. A., Robert, S., et al. (2009). A preliminary controlled trial of divalproex in posttraumatic stress disorder. *Annals of Clinical Psychiatry, 21*(2), 89–94.

Hartford, J., Kornstein, S., Liebowitz, M., et al. (2007). Duloxetine as an SNRI treatment for generalized anxiety disorder: Results from a placebo and active-controlled trial. *International Clinical Psychopharmacology, 22*(3), 167–174.

Harvey, K. V., & Balon, R. (1995). Augmentation with buspirone: A review. *Annals of Clinical Psychiatry, 7*, 143–147.

Haskins, J. T., Aguiar, L., & Pallay, A. (1999). Double-blind, placebo controlled study of once-daily venlafaxine XR (V-XR) in outpatients with generalized anxiety disorder. Presented at the annual meeting of the Anxiety Disorders Association of America, San Diego, CA.

Hedges, D. W., Reimherr, F. W., Strong, R. E., et al. (1996). An open trial of nefazodone in adult patients with generalized anxiety disorder. *Psychopharmacology Bulletin, 32*(4), 671–676.

Hemmelgarn, B., Suissa, S., Huang, A., et al. (1997). Benzodiazepine use and the risk of motor vehicle crash in the elderly. *JAMA, 278*, 27–31.

Heresco-Levy, U., Kremer, I., Javitt, D. C., et al. (2002). Pilot-controlled trial of D-cycloserine for the treatment of posttraumatic stress disorder. *International Journal of Neuropsychopharmacology, 5*, 301–307.

Heresco-Levy, U., Vass, A., Bloch, B., et al. (2009). Pilot controlled trial of D-serine for the treatment of posttraumatic stress disorder. *International Journal of Neuropsychopharmacology, 12*, 1275–1282.

Hertzberg, M. A., Feldman, M. E., Beckham, J. C., et al. (1998). Open trial of nefazodone for combat-related posttraumatic stress disorder. *Journal of Clinical Psychiatry, 59*(9), 460–464.

Hertzberg, M. A., Butterfeld, M. I., Feldman, M. E., et al. (1999). A preliminary study of lamotrigine for the treatment of posttraumatic stress disorder. *Biological Psychiatry, 45*(9), 1226–1229.

Hidalogo, R., Hertzberg, M. A., Mellman, T., et al. (1999). Nefazodone in posttraumatic stress disorder: Results from six open-label trials. *International Clinical Psychopharmacology, 14*(2), 61–68.

Hoehn-Saric, R., McLeod, D. R., & Hipsley, P. A. (1993). Effect of fluvoxamine on panic disorder. *Journal of Clinical Psychopharmacology, 13*(5), 321–326.

Hoge, E. A., Worthington, J. J., Nagurney, J. T., et al. (2012). Effect of acute posttrauma propranolol on PTSD outcome and physiological responses during script-driven imagery. *CNS Neuroscience and Therapeutics, 18*, 21–27.

Hollander, E., Friedberg, J., Wasserman, S., et al. (2003a). Venlafaxine in treatment-resistant obsessive-compulsive disorder. *Journal of Clinical Psychiatry, 64*, 546–550.

Hollander, E., Kaplan, A., & Stahl, S. M. (2003b). A double-blind, placebo-controlled trial of clonazepam in obsessive-compulsive disorder. *The World Journal of Biological Psychiatry, 4*(1), 30–34.

International Multicenter Clinical Trial Group. (1997). Moclobemide in social phobia: A double-blind, placebo-controlled clinical study. *European Archives of Psychiatry and Clinical Neuroscience, 247*(2), 71–80.

Islam, M. M., Iqbal, U., Walther, B., Atique, S., Dubey, N. K., Nguyen, P. A., Poly, T. N., Masud, J. H., Li, Y. C., & Shabbir, S. A. (2016). Benzodiazepine use and risk of dementia in the elderly population: A systematic review and meta-analysis. *Neuroepidemiology, 47*(3–4), 181–191.

Isolan, L., Pheula, G., Salum, G. A., et al. (2007). An open-label trial of escitalopram in children and adolescents with social anxiety disorder. *Journal of Child and Adolescent Psychopharmacology, 17*(6), 751–759.

Jenike, M. A., Baer, L., & Greist, J. H. (1990). Clomipramine versus fluoxetine in obsessive-compulsive disorder: A retrospective comparison of side effects and efficacy. *Journal of Clinical Psychopharmacology, 10*(2), 122–124.

Kalus, O., Asnis, G. M., Rubinson, E., et al. (1991). Desipramine treatment in panic disorder. *Journal of Affective Disorders, 21*(4), 239–244.

Kasper, S., Stein, D., Loft, H., et al. (2005). Escitalopram in the treatment of social anxiety disorder: Randomized, placebo-controlled, flexible-dosage study. *British Journal of Psychiatry, 186*, 222–226.

Kasper, S., Herman, B., Nivoli, G., et al. (2009). Efficacy of pregabalin and venlafaxine-XR in generalized anxiety disorder: Results of a double-blind, placebo-controlled 8-week trial. *International Clinical Psychopharmacology, 24*(2), 87–96.

Katzenick, D. J., Saidi, J., Vanelli, M. R., et al. (2006). Time to response in panic disorder in a naturalistic setting: Combination therapy with alprazolam orally disintegrating tablets and serotonin reuptake inhibitors compared to serotonin reuptake inhibitors alone. *Psychiatry, 12*(3), 39–49.

Katzman, M. A., Brawman-Mintzer, O., Reyes, E. B., et al. (2011). Extended release quetiapine fumarate (quetiapine XR) monotherapy as maintenance treatment for generalized anxiety disorder: A long-term, randomized, placebo-controlled trial. *International Clinical Psychopharmacology, 26*(1), 11–24.

Keck, P. E., Jr., Taylor, V. E., Tugrul, K. C., et al. (1993). Valproate treatment of panic disorder and lactate-induced panic attacks. *Biological Psychiatry, 33*(7), 542–546.

Kelsey, J. E. (1995). Venlafaxine in social phobia. *Psychopharmacology Bulletin, 31*(4), 767–771.

Khan, A., Joyce, M., Atkinson, S., et al. (2011). A randomized, double-blind study of once-daily extended release quetiapine fumarate (quetiapine XR) monotherapy in patients with generalized anxiety disorder. *Journal of Clinical Psychopharmacology, 31*(4), 418–428.

Kim, W., Pae, C. U., Chae, J. H., et al. (2005). The effectiveness of mirtazapine in the treatment of posttraumatic stress disorder: A 24-week continuation therapy. *Psychiatry and Clinical Neurosciences, 59*(6), 743–747.

Klein, D. (1964). Delineation of two-drug responsive anxiety syndromes. *Psychopharmacologia, 5*, 397–408.

Klein, D., & Fink, M. (1962). Psychiatric reaction patterns to imipramine. *American Journal of Psychiatry, 119*, 432–438.

Kobak, K. A., Greist, J. H., Jefferson, J. W., et al. (1998). Behavioral versus pharmacological treatments of obsessive-compulsive disorder: A meta-analysis. *Psychopharmacology, 136*(3), 205–216.

Koponen, H., Lepola, U., Leinonen, E., et al. (1997). Citalopram in the treatment of obsessive-compulsive disorder: An open pilot study. *Acta Psychiatrica Scandinavica, 96*(5), 343–346.

Koran, L. M., Faravelli, C., & Pallanti, S. (1994). Intravenous clomipramine for obsessive-compulsive disorder. *Journal of Clinical Psychopharmacology, 14*(3), 216–218.

Koran, L. M., McElroy, S. L., Davidson, J. R., et al. (1996). Fluvoxamine versus clomipramine for obsessive-compulsive disorder: A double-blind comparison. *Journal of Clinical Psychopharmacology, 16*(2), 121–129.

Koran, L. M., Sallee, F. R., & Pallanti, S. (1997). Rapid benefit of intravenous pulse loading of clomipramine in obsessive-compulsive disorder. *American Journal of Psychiatry, 154*(3), 396–401.

Koran, L. M., Hackett, E., Rubin, A., et al. (2002). Efficacy of sertraline in the long-term treatment of obsessive-compulsive disorder. *American Journal of Psychiatry, 159*(1), 88–95.

Kordon, A., Wahl, K., Koch, N., et al. (2008). Quetiapine addition to serotonin reuptake inhibitors in patients with severe obsessive-compulsive disorder. *Journal of Clinical Psychopharmacology, 28*(5), 550–554.

Kosten, T. R., Frank, J. B., Dan, E., et al. (1991). Pharmacotherapy for posttraumatic stress disorder using phenelzine or imipramine. *Journal of Nervous and Mental Disease, 179*(6), 366–370.

Kronig, M. H., Apter, J., Asnis, G., et al. (1999). Placebo-controlled, multicenter study of sertraline treatment for obsessive-compulsive disorder. *Journal of Clinical Psychopharmacology, 19*(2), 172–176.

Kruger, M. B., & Dahl, A. A. (1999). The efficacy and safety of moclobemide compared to clomipramine in the treatment of panic disorder. *European Archives of Psychiatry and Clinical Neuroscience, 249*(Suppl 1), S19–S24.

Krystal, J. H., Rosenheck, R. A., Cramer, J. A., et al. (2011). Adjunctive risperidone treatment for antidepressant-resistant symptoms of chronic military service-related PTSD. *JAMA, 306*(5), 493–502.

Kushner, M. G., Kim, S. W., Donahue, C., et al. (2007). D-Cycloserine augmented exposure therapy for obsessive-compulsive disorder. *Biological Psychiatry, 62*, 835–838.

Laakmann, G., Schule, C., Lorkowski, G., et al. (1998). Buspirone and lorazepam in the treatment of generalized anxiety disorder in outpatients. *Psychopharmacology, 136*(4), 357–366.

Lader, M., Stender, K., Burger, V., et al. (2004). Efficacy and tolerability of escitalopram in 12- and 24-week treatment of social anxiety disorder: Randomised, double-blind, placebo-controlled, fixed-dose study. *Depression and Anxiety, 19*, 241–248.

Lecrubier, Y., & Judge, R. (1997). Long-term evaluation of paroxetine, clomipramine and placebo in panic disorder. Collaborative Paroxetine Panic Study Investigators. *Acta Psychiatrica Scandinavica, 95*(2), 153–160.

Lecrubier, Y., Bakker, A., Dunbar, G., et al. (1997). A comparison of paroxetine, clomipramine and placebo in the treatment of panic disorder. Collaborative Paroxetine Panic Study Investigators. *Acta Psychiatrica Scandinavica, 95*(2), 145–152.

Lee, D. J., Carla, W. S., Jonathan P. W., Meena, V., Ann, M. R., & Charles, W. H. (2016). Psychotherapy Versus Pharmacotherapy for Posttraumatic Stress Disorder: Systemic Review And Meta-analyses to Determine First-line Treatments. *Depression and Anxiety, 33*(9), 792–806.

Leinonen, E., Skarstein, J., Behnke, K., et al. (1999). Efficacy and tolerability of mirtazapine versus citalopram: A double-blind, randomized study in patients with major depressive disorder. Nordic Antidepressant Study Group. *International Clinical Psychopharmacology, 14*(6), 329–337.

Lenze, E. J., Mulsant, B. H., Shear, M. K., et al. (2005). Efficacy and tolerability of citalopram in the treatment of late-life anxiety disorders: Results from an 8-week randomized, placebo-controlled trial. *American Journal of Psychiatry, 162*(1), 146–150.

Lenze, E. J., Rollman, B. L., Shear, M. K., et al. (2009). Escitalopram for older adults with generalized anxiety disorder: A randomized controlled trial. *JAMA, 301*(3), 295–303.

Lerer, B., Bleich, A., Kotler, M., et al. (1987). Posttraumatic stress disorder in Israeli combat veterans. Effect of phenelzine treatment. *Archives of General Psychiatry, 44*(11), 976–981.

Li, X., May, R., Tolbert, L., et al. (2005). Risperidone and haloperidol augmentation of serotonin reuptake inhibitors in refractory obsessive-compulsive disorder: A crossover study. *Journal of Clinical Psychiatry, 66*, 736–743.

Li, X., Zhu, L., Zhou, C., Liu, J., Du, H., Wang, C., & Fang, S. (2018). Efficacy and tolerability of short-term duloxetine treatment in adults with generalized anxiety disorder: A meta-analysis. *PLoS One, 13*(3), e0194501.

Liebowitz, M. R., Fyer, A. J., Gorman, J. M., et al. (1988). Tricyclic therapy of the DSM-III anxiety disorders: A review with implications for further research. *Journal of Psychiatric Research, 22*(Suppl 1), 7–31.

Liebowitz, M. R., DeMartinis, N. A., Weihs, K., et al. (2003). Efficacy of sertraline in severe generalized social anxiety disorder: Results of a double-blind, placebo-controlled study. *Journal of Clinical Psychiatry, 64*(4), 785–792.

Liebowitz, M. R., Gelenberg, A. J., & Munjack, D. (2005). Venlafaxine extended release vs. placebo and paroxetine in social anxiety disorder. *Archives of General Psychiatry, 62*, 190–198.

Liebowitz, M. R., Asnis, G., Mangano, R., et al. (2009). A double-blind, placebo-controlled, parallel-group, flexible-dose study of venlafaxine extended release capsules in adult outpatients with panic disorder. *Journal of Clinical Psychiatry, 70*(40), 550–561.

Lindley, S. E., Carlson, E. B., & Hill, K. (2007). A randomized, double-blind, placebo-controlled trial of augmentation topiramate for chronic combat-related

posttraumatic stress disorder. *Journal of Clinical Psychopharmacology, 27*(6), 677–681.

Litz, B. T., Salters-Pedneault, K., Steenkamp, M. M., et al. (2012). A randomized placebo-controlled trial of D-cycloserine and exposure therapy for posttraumatic stress disorder. *Journal of Psychiatric Research, 46*, 1184–1190.

Liu, X., Li, X., Zhang, C., Sun, M., Sun, Z., Xu, Y., & Tian, X. (2018). Efficacy and tolerability of fluvoxamine in adults with social anxiety disorder: A meta-analysis. *Medicine, 97*(28), e11547.

Lohoff, F. M., Etemad, B., Mandos, L. A., et al. (2010). Ziprasidone treatment of refractory generalized anxiety disorder. *Journal of Clinical Psychopharmacology, 30*(2), 185–189.

Londborg, P. D., Wolkow, R., Smith, W. T., et al. (1998). Sertraline in the treatment of panic disorder. A multisite, double-blind, placebo-controlled, fixed-dose investigation. *British Journal of Psychiatry, 173*, 54–60.

Londborg, P. D., Hegel, M. T., Goldstein, S., et al. (2001). Sertraline treatment of posttraumatic stress disorder: Results of 24 weeks of open-label continuation treatment. *Journal of Clinical Psychiatry, 65*(5), 325–331.

Lopez-Ibor, J. J., Jr., Saiz, J., Cottraux, J., et al. (1996). Double-blind comparison of fluoxetine versus clomipramine in the treatment of obsessive-compulsive disorder. *European Neuropsychopharmacology, 6*(2), 111–118.

Lydiard, R. B., Morton, W. A., Emmanuel, N. P., et al. (1993). Preliminary report: Placebo-controlled, double-blind study of the clinical and metabolic effects of desipramine in panic disorder. *Psychopharmacology Bulletin, 29*(2), 183–188.

Maina, G., Pessina, E., Albert, U., et al. (2008). 8-week, single-blind, randomized trial comparing risperidone versus olanzapine augmentation of serotonin reuptake inhibitors in treatment-resistant obsessive–compulsive disorder. *European Neuropsychopharmacology, 18*, 364–372.

Marazziti, D., Pfanner, C., Dell'Osso, B., et al. (2005). Augmentation strategy with olanzapine in resistant obsessive compulsive disorder: An Italian longterm open-label study. *Journal of Psychopharmacology, 19*, 392–394.

Marazziti, D., Golia, F., Consoli, G., et al. (2008). Effectiveness of long-term augmentation with citalopram to clomipramine in treatment-resistant OCD patients. *CNS Spectrum, 13*(11), 971–976.

March, J. S., Entusah, A. R., Rynn, M., et al. (2007). A randomized controlled trial of venlafaxine ER versus placebo in pediatric social anxiety disorder. *Biological Psychiatry, 62*, 1149–1154.

Marshall, R., Beebe, K., Oldham, M., et al. (2001). Efficacy and safety of paroxetine treatment for chronic PTSD: A fixed dose, placebo controlled study. *American Journal of Psychiatry, 158*(12), 1982–1988.

Marshall, R. D., Lewis-Fernandez, R., Blanco, C., et al. (2007). A controlled trial of paroxetine for chronic

PTSD, dissociation, and interpersonal problems in mostly minority adults. *Depression and Anxiety, 24*, 77–84.

Martenyi, F., Brown, E. B., Zhang, H., et al. (2002). Fluoxetine v. placebo in prevention of relapse in post-traumatic stress disorder. *Journal of Mental Science, 181*(4), 315–320.

Masdrakis, V. G., Papadimitriou, G. N., & Oulis, P. (2010). Lamotrigine administration in panic disorder with agoraphobia. *Clinical Neuropharmacology, 33*(3), 126–128.

Masi, G., Pranner, C., Millepiedi, S., et al. (2010). Aripiprazole augmentation in 39 adolescents with medication-resistant obsessive-compulsive disorder. *Journal of Clinical Psychopharmacology, 30*(6), 688–693.

Mataix-Cols, D., Fernández de la Cruz, L., Monzani, B., Rosenfield, D., Andersson, E., Pérez-Vigil, A., Frumento, P., de Kleine, R. A., Difede, J., Dunlop, B. W., Farrell, L. J., Geller, D., Gerardi, M., Guastella, A. J., Hofmann, S. G., Hendriks, G. J., Kushner, M. G., Lee, F. S., Lenze, E. J., Levinson, C. A., McConnell, H., Otto, M. W., Plag, J., Pollack, M. H., Ressler, K. J., Rodebaugh, T. L., Rothbaum, B. O., Scheeringa, M. S., Siewert-Siegmund, A., Smits, J. A. J., Storch, E. A., Ströhle, A., Tart, C. D., Tolin, D. F., van Minnen, A., Waters, A. M., Weems, C. F., Wilhelm, S., Wyka, K., Davis, M., Rück, C., the DCS Anxiety Consortium, Altemus, M., Anderson, P., Cukor, J., Finck, C., Geffken, G. R., Golfels, F., Goodman, W. K., Gutner, C., Heyman, I., Jovanovic, T., Lewin, A. B., McNamara, J. P., Murphy, T. K., Norrholm, S., & Thuras, P. (2017). D-Cycloserine augmentation of exposure-based cognitive behavior therapy for anxiety, obsessive-compulsive, and posttraumatic stress disorders: A systematic review and meta-analysis of individual participant data. *JAMA Psychiatry, 74*(5), 501–510.

Matsunaga, H., Nagata, T., Hayashida, K., et al. (2009). A long-term trial of the effectiveness and safety of atypical antipsychotic agents in augmenting SSRI-refractory obsessive-compulsive disorder. *Journal of Clinical Psychiatry, 70*(6), 863–868.

Mavissakalian, M., & Perel, J. M. (1992a). Clinical experiments in maintenance and discontinuation of imipramine therapy in panic disorder with agoraphobia. *Archives of General Psychiatry, 49*(4), 318–323.

Mavissakalian, M., & Perel, J. M. (1992b). Protective effects of imipramine maintenance treatment in panic disorder with agoraphobia. *American Journal of Psychiatry, 149*(8), 1053–1057.

Mavissakalian, M. R., & Perel, J. M. (2001). 2nd year maintenance and discontinuation of imipramine in panic disorder with agoraphobia. *Annals of Clinical Psychiatry, 13*(2), 63–67.

McDougle, C. J. (1997). Update on pharmacologic management of OCD: Agents and augmentation. *Journal of Clinical Psychiatry, 58*(Suppl 12), 11–17.

McRae, A., Brady, K., Mellman, T., et al. (2004). Comparison of nefazodone and sertraline for the treatment of posttraumatic stress disorder. *Depression and Anxiety, 19*, 190–196.

Menza, M., Dobkin, R. D., & Marin, H. (2007). An open-label trial of aripiprazole augmentation for treatment-resistant generalized anxiety disorder. *Journal of Clinical Psychopharmacology, 22*(2), 207–209.

Menzies, R. P. D. (2012). Propranolol, traumatic memories, and amnesia: A study of 36 cases. *Journal of Clinical Psychiatry, 73*(1), 129–130.

Merideth, C., Cutler, A. J., She, F., et al. (2012). Efficacy and tolerability of extended release quetiapine fumarate monotherapy in the acute treatment of generalized anxiety disorder: A randomized, placebo controlled and active-controlled study. *International Clinical Psychopharmacology, 27*(1), 40–54.

Metin, O., Yazici, K., Tot, S., et al. (2003). Amisulpride augmentation in treatment resistant obsessive-compulsive disorder: An open trial. *Human Psychopharmacology, 18*, 463–467.

Michelson, D., Lydiard, R. B., Pollack, M. H., et al. (1998). Outcome assessment and clinical improvement in panic disorder: Evidence from a randomized controlled trial of fluoxetine and placebo. The Fluoxetine Panic Disorder Study Group. *American Journal of Psychiatry, 155*(11), 1570–1577.

Michelson, D., Pollack, M., Lydiard, R. B., et al. (1999). Continuing treatment of panic disorder after acute response: Randomised, placebo-controlled trial with fluoxetine. The Fluoxetine Panic Disorder Study Group. *British Journal of Psychiatry, 174*, 213–218.

Michelson, D., Allgulander, C., Dantendorfer, K., et al. (2001). Efficacy of usual antidepressant dosing regimens of fluoxetine in panic disorder: Randomised, placebo-controlled trial. *British Journal of Psychiatry, 179*, 514–518.

Milanfranchi, A., Ravagli, S., Lensi, P., et al. (1997). A double-blind study of fluvoxamine and clomipramine in the treatment of obsessive-compulsive disorder. *International Clinical Psychopharmacology, 12*(3), 131–136.

Modarresi, A., Sayyah, M., Razooghi, S., Eslami, K., Javadi, M., & Kouti, L. (2018). Memantine augmentation improves symptoms in serotonin reuptake inhibitor-refractory obsessive-compulsive disorder: A randomized controlled trial. *Pharmacopsychiatry, 51*(6), 263–269.

Modigh, K., Westberg, P., & Eriksson, E. (1992). Superiority of clomipramine over imipramine in the treatment of panic disorder: A placebo-controlled trial. *Journal of Clinical Psychopharmacology, 12*(4), 251–261.

Mokhber, N., Azarpazhooh, M. R., Khajehdaluee, M., Velayati, A., & Hopwood, M. (2010). Randomized, single-blind, trial of sertraline and buspirone for treatment of elderly patients with generalized anxiety disorder. *Psychiatry and Clinical Neurosciences, 64*(2), 128–133.

Montgomery, S. A., McIntyre, A., Osterheider, M., et al. (1993). A doubleblind, placebo-controlled study of fluoxetine in patients with DSM-III-R obsessive-compulsive disorder. The Lilly European OCD Study Group. *European Neuropsychopharmacology, 3*(2), 143–152.

Montgomery, S., Kasper, S., Stein, D., et al. (2001). Citalopram 20 mg, 40 mg, and 60 mg are all effective and well tolerated compared with placebo in obsessive-compulsive disorder. *International Clinical Psychopharmacology, 16*(2), 75–86.

Montgomery, S. A., Nil, R., Durr-Pal, N., et al. (2005). A 24-week randomized, double-blind, placebo-controlled study of escitalopram for the prevention of generalized social anxiety disorder. *Journal of Clinical Psychiatry, 66*(10), 1270–1278.

Montgomery, S. A., Tobias, K., Zornberg, G. L., et al. (2006). Efficacy and safety of pregabalin in the treatment of generalized anxiety disorder: A 6-week, multicenter, randomized, double-blind, placebo-controlled comparison of pregabalin and venlafaxine. *Journal of Clinical Psychiatry, 67*(5), 771–782.

Moroz, G., & Rosenbaum, J. F. (1999). Efficacy, safety, and gradual discontinuation of clonazepam in panic disorder: A placebo-controlled, multicenter study using optimized dosages. *Journal of Clinical Psychiatry, 60*(9), 604–612.

Mowla, A., & Ghaedsharaf, M. (2020). Pregabalin augmentation for resistant obsessive-compulsive disorder: A double-blind placebo-controlled clinical trial. *CNS Spectrums, 25*(4), 552–556.

Mowla, A., Boostani, S., & Dastgheib, S. A. (2016). Duloxetine augmentation in resistant obsessive-compulsive disorder: A double-blind controlled clinical trial. *Journal of Clinical Psychopharmacology, 36*(6), 720–723.

Muehlbacher, M., Nickel, M., Nickel, C., et al. (2005). Mirtazapine treatment of social phobia in women: A randomized, double-blind, placebo-controlled study. *Journal of Clinical Psychopharmacology, 25*, 580–583.

Mundo, E., Bianchi, L., & Bellodi, L. (1997). Efficacy of fluvoxamine, paroxetine, and citalopram in the treatment of obsessive-compulsive disorder: A single-blind study. *Journal of Clinical Psychopharmacology, 17*(4), 267–271.

Munjack, D. J., Usigli, R., Zulueta, A., et al. (1988). Nortriptyline in the treatment of panic disorder and agoraphobia with panic attacks. *Journal of Clinical Psychopharmacology, 8*(3), 204–207.

Muscatello, M. R., Bruno, A., Pandolfo, G., et al. (2011). Effect of aripiprazole augmentation of serotonin reuptake inhibitors or clomipramine in treatment-resistant obsessive-compulsive disorder. *Journal of Clinical Psychopharmacology, 31*(2), 174–179.

Nardi, A., & Perna, G. (2006). Clonazepam in the treatment of psychiatric disorders: An update. *International Clinical Psychopharmacology, 21*(3), 131–142.

Nardi, A. E., Freire, R. C., Valenca, A. M., et al. (2010a). Tapering clonazepam in patients with panic disorder after at least 3 years of treatment. *Journal of Clinical Psychopharmacology, 30*(3), 290–293.

Nardi, A. E., Lopes, F. L., Valenca, A. M., et al. (2010b). Double-blind comparison of 30 and 60 mg tranylcypromine daily in patients with panic disorder comorbid with social anxiety disorder. *Psychiatry Research, 175*, 260–265.

Nardi, A. E., Valenca, A. M., Freire, R. C., et al. (2011). Psychopharmacotherapy of panic disorder: 8-week randomized trial with clonazepam and paroxetine. *Brazilian Journal of Medical and Biological Research, 44*(4), 366–373.

Nardi, A. E., Freire, R. C., Mochcovitch, M. D., et al. (2012). A randomized, naturalistic, parallel-group study for the long-term treatment of panic disorder with clonazepam or paroxetine. *Journal of Clinical Psychopharmacology, 32*(1), 120–126.

Nicola, M. D., Tedeschi, D., Martinotti, G., et al. (2011). Pregabalin augmentation in treatment-resistant obsessive-compulsive disorder. *Journal of Clinical Psychopharmacology, 31*(5), 675–677.

Noyes, R., Jr., Garvey, M. J., Cook, B. L., et al. (1989). Problems with tricyclic antidepressant use in patients with panic disorder or agoraphobia: Results of a naturalistic follow-up study. *Journal of Clinical Psychiatry, 50*, 163–169.

Noyes, R., Jr., Burrows, G. D., Reich, J. H., et al. (1996). Diazepam versus alprazolam for the treatment of panic disorder. *Journal of Clinical Psychiatry, 57*(8), 349–355.

Noyes, R., Jr., Moroz, G., Davidson, J. R., et al. (1997). Moclobemide in social phobia: A controlled dose-response trial. *Journal of Clinical Psychopharmacology, 17*(4), 247–254.

Offidani, E., Guidi, J., Tomba, E., et al. (2013). Efficacy and tolerability of benzodiazepines versus antidepressants in anxiety disorders: A systematic review and meta-analysis. *Psychotherapy and Psychosomatics, 82*, 355–362.

Oulis, P., Mourikis, I., & Konstantakopoulos, G. (2011). Pregabalin augmentation in treatment-resistant obsessive-compulsive disorder. *International Clinical Psychopharmacology, 26*(4), 221–224.

Pae, C. U., Wang, S. M., Han, C., Lee, S. J., Patkar, A. A., Masand, P. S., & Serretti, A. (2015). Vortioxetine, a multimodal antidepressant for generalized anxiety disorder: A systematic review and meta-analysis. *Journal of Psychiatric Research, 64*, 88–98.

Pallanti, S., Quercioli, L., & Koran, L. M. (2002). Citalopram intravenous infusion in resistant obsessive-compulsive disorder: An open trial. *Journal of Clinical Psychiatry, 63*(9), 796–801.

Pallanti, S., Bernardi, S., Antonini, S., et al. (2009). Ondansetron augmentation in treatment-resistant obsessive-compulsive disorder: A preliminary, singleblind, prospective study. *CNS Drugs, 23*(12), 1047–1055.

Pande, A. C., Greiner, M., Adams, J. B., et al. (1999). Placebo-controlled trial of the CCK-B antagonist, CI-988, in panic disorder. *Biological Psychiatry, 46*(6), 860–862.

Pande, A. C., Crockatt, J. G., Feltner, D. E., et al. (2003). Pregabalin in generalized anxiety disorder: A placebo-controlled trial. *American Journal of Psychiatry, 160*(3), 533–540.

Pande, A. C., Feltner, D., Jefferson, J., et al. (2004). Efficacy of the novel anxiolytic pregabalin in social anxiety disorder: A placebo-controlled, multicenter study. *Journal of Clinical Psychopharmacology, 24*(2), 141–149.

Papp, L. C. (2006). Safety and efficacy of levetiracetam for patients with panic disorder: Results of an open-label, fixed-flexible dose study. *Journal of Clinical Psychiatry, 67*(10), 1573–1576.

Papp, L. A., Coplan, J. D., Martinez, J. M., et al. (2000). Efficacy of open-label nefazodone treatment in patients with panic disorder. *Journal of Clinical Psychopharmacology, 20*(5), 544–546.

Perez, V., Gilaberte, I., Faries, D., et al. (1997). Randomised, double-blind, placebo-controlled trial of pindolol in combination with fluoxetine antidepressant treatment. *The Lancet, 349*(9065), 1594–1597.

Perna, G., Bertani, A., Caldirola, D., et al. (2001). A comparison of citalopram and paroxetine in the treatment of panic disorder: A randomized, singleblind study. *Pharmacopsychiatry, 34*(3), 85–90.

Perse, T. L., Greist, J. H., Jefferson, J. W., et al. (1987). Fluvoxamine treatment of obsessive-compulsive disorder. *American Journal of Psychiatry, 144*(12), 1543–1548.

Pessina, E., Albert, U., Bogetto, F., et al. (2009). Aripiprazole augmentation of serotonin reuptake inhibitors in treatment-resistant obsessive-compulsive disorder: A 12-week open-label preliminary study. *International Clinical Psychopharmacology, 24*(5), 265–269.

Petty, F., Brannan, S., Casada, J., et al. (2001). Olanzapine treatment for posttraumatic stress disorder: An open-label study. *International Clinical Psychopharmacology, 16*(6), 331–337.

Pfanner, C., Marazziti, D., Dell'Osso, L., et al. (2000). Risperidone augmentation in refractory obsessive-compulsive disorder: An open-label study. *International Clinical Psychopharmacology, 15*(5), 297–301.

Piccinelli, M., Pini, S., Bellantuono, C., et al. (1995). Efficacy of drug treatment in obsessive-compulsive disorder. A meta-analytic review. *British Journal of Psychiatry, 166*(4), 424–443.

Piero, A., & Locati, E. (2011). An open, non-randomized comparison of escitalopram and duloxetine for the treatment of subjects with generalized anxiety disorder. *Human Psychopharmacology: Clinical and Experimental, 26*, 63–71.

Pigott, T. A., L'Heureux, F., & Rubenstein, C. S. (1992a). A controlled trial of clonazepam augmentation in OCD patients treated with clomipramine or fluoxetine. New

Research Program and Abstracts of the 145th Meeting of the American Psychiatric Association, Abstract 144.

Pigott, T. A., L'Heureux, F., Rubenstein, C. S., Bernstein, S. E., Hill, J. L., & Murphy, D. L. (1992b). A double-blind, placebo controlled study of trazodone in patients with obsessive-compulsive disorder. *Journal of Clinical Psychopharmacology, 12*(3), 156–162.

Pilkinton, P., Berry, C., Norrholm, S., et al. (2016). An open label pilot study of adjunctive asenapine for the treatment of posttraumatic stress disorder. *Psychopharmacology Bulletin, 46*(2), 8–17.

Pittman, R. K., Sanders, K. M., Zusman, R. M., et al. (2002). Pilot study of secondary prevention of posttraumatic disorders with propranolol. *Biological Psychiatry, 51*, 189–192.

Pivac, N., Kovacic-Kozaric, D., & Muck-Seler, D. (2004). Olanzapine versus fluphenazine in an open trial in patients with psychotic combat-related posttraumatic stress disorder. *Psychopharmacology, 175*, 451–456.

Pollack, M. H., Worthington, J. J., III, Otto, M. W., et al. (1996). Venlafaxine for panic disorder: Results from a double-blind, placebo-controlled study. *Psychopharmacology Bulletin, 32*(4), 667–670.

Pollack, M. H., Otto, M. W., Worthington, J. J., et al. (1998). Sertraline in the treatment of panic disorder: A flexible-dose multicenter trial. *Archives of General Psychiatry, 55*(11), 1010–1016.

Pollack, M. H., Zaninelli, R., Goddard, A., et al. (2001). Paroxetine in the treatment of generalized anxiety disorder: Results of a placebo-controlled, flexible-dosage trial. *Journal of Clinical Psychiatry, 62*(5), 350–357.

Pollack, M. H., Simon, N. M., Worthington, J. J., et al. (2003). Combined paroxetine and clonazepam treatment strategies compared to paroxetine monotherapy for panic disorder. *Journal of Psychopharmacology, 17*(3), 276–282.

Pollack, M. H., Roy-Byrne, P. P., Van Ameringen, M., et al. (2005). The selective GABA reuptake inhibitor tiagabine for the treatment of generalized anxiety disorder: Results of a placebo-controlled study. *Journal of Clinical Psychiatry, 66*(11), 1401–1408.

Pollack, M. H., Simon, N. M., Zalta, A. K., et al. (2006). Olanzapine augmentation of fluoxetine for refractory generalized anxiety disorder: A placebo controlled study. *Biological Psychiatry, 59*, 211–215.

Pollack, M. H., Lepola, U., Koponen, H., et al. (2007). A double-blind study of the efficacy of venlafaxine extended-release, paroxetine, and placebo in the treatment of panic disorder. *Depression and Anxiety, 24*(1), 1–14.

Pollack, M. H., Hoge, E. A., Worthington, J. J., et al. (2011). Eszopiclone for the treatment of posttraumatic stress disorder and associated insomnia: A randomized, double-blind, placebo-controlled trial. *Journal of Clinical Psychiatry, 72*(7), 892–897.

Pollack, M. H., Van Ameringen, M., Simon, N. M., Worthington, J. W., Hoge, E. A., Keshaviah, A., & Stein, M. B. (2014). A double-blind randomized controlled trial of augmentation and switch strategies for refractory social anxiety disorder. *The American Journal of Psychiatry, 171*(1), 44–53.

Prosser, J. M., Yard, S., Steele, A., et al. (2009). A comparison of low-dose risperidone to paroxetine in the treatment of panic attacks: A randomized, single-blind study. *BMC Psychiatry, 9*, 25.

Qi, W., Martin, G., & Arieh, S. (2017). Efficacy and Tolerability of High-Dose Escitalopram in Posttraumatic Stress Disorder. *Journal of Clinical Psychopharmacology, 37*(1), 89–93.

Qin, B., Huang, G., Yang, Q., Zhao, M., Chen, H., Gao, W., & Yang, M. (2019). Vortioxetine treatment for generalised anxiety disorder: A meta-analysis of anxiety, quality of life and safety outcomes. *BMJ Open, 9*(11), e033161.

Rabinowitz, I., Baruch, Y., & Barak, Y. (2008). High-dose escitalopram for the treatment of obsessive-compulsive disorder. *International Clinical Psychopharmacology, 23*(1), 49–53.

Rapaport, M. H., Wolkow, R. M., & Clary, C. M. (1998). Methodologies and outcomes from the sertraline multicenter flexible-dose trials. *Psychopharmacology Bulletin, 34*(2), 183–189.

Rapaport, M. H., Wolkow, R., Rubin, A., et al. (2001). Sertraline treatment of panic disorder: Results of a long-term study. *Acta Psychiatrica Scandinavica, 4*(104), 289–298.

Rapaport, M. H., Endicott, J., & Clary, C. M. (2002). Posttraumatic stress disorder and quality of life: Results across 64 weeks of sertraline treatment. *Journal of Clinical Psychiatry, 63*, 59–65.

Rapaport, M. H., Skarky, S. B., Katzelnick, D. J., et al. (2006). Time to response in generalized anxiety disorder in a naturalistic setting: Combination therapy with alprazolam orally disintegrating tablets and serotonin reuptake inhibitors compared to serotonin reuptake inhibitors alone. *Psychiatry, 3*(12), 50–59.

Raskind, M. A., Peskind, E. R., Kanter, E. D., et al. (2003). Reduction of nightmares and other PTSD symptoms in combat veterans by prazosin: A placebo-controlled study. *American Journal of Psychiatry, 160*(2), 371–373.

Raskind, M. A., Peskind, E. R., Joff, D. J., et al. (2007). A parallel group placebo controlled study of prazosin for trauma nightmares and sleep disturbance in combat veterans with posttraumatic stress disorder. *Biological Psychiatry, 61*, 928–934.

Reich, D., Winternitz, S., Hennen, J., et al. (2004). A preliminary study of risperidone in the treatment of posttraumatic stress disorder related to childhood abuse in women. *Journal of Clinical Psychiatry, 65*, 1601–1606.

Reist, C., Streja, E., Tang, C. C., et al. (2020). Prazosin for treatment of post-traumatic stress disorder: A systematic review and meta-analysis. *CNS Spectrums, 26*, 338–344.

Ribeiro, L., Busnello, J. V., Kauer-Sant'Anna, M., et al. (2001). Mirtazapine versus fluoxetine in the treatment of panic disorder. *Brazilian Journal of Medical and Biological Research, 34*(10), 1303–1307.

Riblet, L. A., Taylor, D. P., Eison, M. S., et al. (1982). Pharmacology and neurochemistry of buspirone. *Journal of Clinical Psychiatry, 43*(12, Pt 2), 11–18.

Rickels, K., & Schweizer, E. (1998). Panic disorder: Long-term pharmacotherapy and discontinuation. *Journal of Clinical Psychopharmacology, 18*(6, Suppl 2), 12S–18S.

Rickels, K., Weisman, K., Norstad, N., et al. (1982). Buspirone and diazepam in anxiety: A controlled study. *Journal of Clinical Psychiatry, 43*(12, Pt 2), 81–86.

Rickels, K., Case, W. G., Downing, R. W., et al. (1983). Long-term diazepam therapy and clinical outcome. *JAMA, 250*(6), 767–771.

Rickels, K., Schweizer, E., Csanalosi, I., et al. (1988). Long-term treatment of anxiety and risk of withdrawal. Prospective comparison of clorazepate and buspirone. *Archives of General Psychiatry, 45*(5), 444–450.

Rickels, K., Downing, R., Schweizer, E., et al. (1993). Antidepressants for the treatment of generalized anxiety disorder. A placebo-controlled comparison of imipramine, trazodone, and diazepam. *Archives of General Psychiatry, 50*(11), 884–895.

Rickels, K., Pollack, M. H., Sheehan, D. V., et al. (2000). Efficacy of extended-release venlafaxine in nondepressed outpatients with generalized anxiety disorder. *American Journal of Psychiatry, 157*(6), 968–974.

Rickels, K., Zaninelli, R., McCafferty, J., et al. (2003). Paroxetine treatment of generalized anxiety disorder: A double-blind, placebo-controlled study. *American Journal of Psychiatry, 160*(4), 749–756.

Rickels, K., Mangano, R., & Khan, A. (2004). A double-blind, placebo-controlled study of a flexible dose of venlafaxine ER in adult outpatients with generalized social anxiety. *Journal of Clinical Psychopharmacology, 24*(5), 488–496.

Rickels, K., Pollack, M. H., Feltner, D. E., et al. (2005). Pregabalin for treatment of generalized anxiety disorder. *Archives of General Psychiatry, 62*, 1022–1030.

Rickels, K., Shiovitz, T. M., Ramey, T. S., et al. (2012). Adjunctive therapy with pregabalin in generalized anxiety disorder patients with partial response to SSRI or SNRI treatment. *International Clinical Psychopharmacology, 27*(3), 142–150.

Riddle, M., Reeve, E., Yaryura-Tobias, J., et al. (2001). Fluvoxamine for children and adolescents with obsessive-compulsive disorder: A randomized, controlled, multicenter trial. *Journal of the American Academy of Child and Adolescent Psychiatry, 40*, 222–229.

Robb, A. D., Cueva, J. E., Sporn, J., et al. (2010). Sertaline treatment of children and adolescents with post-traumatic stress disorder: A double-blind, placebo-controlled trial. *Journal of Child and Adolescent Psychopharmacology, 20*(6), 463–471.

Robert, S., Hamner, M. B., Durkalski, V. L., et al. (2009). An open-label assessment of aripiprazole in the treatment of PTSD. *Psychopharmacology Bulletin, 42*(1), 69–80.

Rodrigues, H., Figueira, I., Lopes, A., Gonçalves, R., Mendlowicz, M. V., Coutinho, E. S., & Ventura, P. (2014). Does D-cycloserine enhance exposure therapy for anxiety disorders in humans? A meta-analysis. *PLoS One, 9*(7), e93519.

Rosenbaum, J. F., Moroz, G., & Bowden, C. L. (1997). Clonazepam in the treatment of panic disorder with or without agoraphobia: A dose-response study of efficacy, safety, and discontinuance. Clonazepam Panic Disorder Dose-Response Study Group. *Journal of Clinical Psychopharmacology, 17*(5), 390–400.

Rothbaum, B. O., Killeen, T. K., Davidson, J. R. T., et al. (2008). Placebo-controlled trial of risperidone augmentation for selective serotonin reuptake inhibitor-resistant civilian posttraumatic stress disorder. *Journal of Clinical Psychiatry, 69*(4), 520–525.

Rubio, G., Jimenez-Arriero, M., Martinez-Gras, I., et al. (2006). The effects of topiramate adjunctive treatment added to antidepressants in patients with resistant obsessive-compulsive disorder. *Journal of Clinical Psychopharmacology, 26*, 341–344.

Rynn, M. A., Siqueland, L., & Rickels, K. (2001). Placebo-controlled trial of sertraline in the treatment of children with generalized anxiety disorder. *American Journal of Psychiatry, 158*(12), 2008–2014.

Rynn, M. A., Riddle, M. A., Yeung, P. P., et al. (2007). Efficacy and safety of extended-release venlafaxine in the treatment of generalized anxiety disorder in children and adolescents: Two placebo-controlled trials. *American Journal of Psychiatry, 164*(2), 290–300.

Rynn, M., Russell, J., Erickson, J., et al. (2008). Efficacy and safety of du-loxetine in the treatment of generalized anxiety disorder: A flexible-dose, progressive-titration, placebo-controlled trial. *Depression and Anxiety, 25*, 182–189.

Sandford, J. J., Forshall, S., Bell, C., et al. (2001). Crossover trial of pagoclone and placebo in patients with DSM-IV panic disorder. *Journal of Psychopharmacology, 15*(3), 205–208.

Sartori, S. B., & Singewald, N. (2019). Novel pharmacological targets in drug development for the treatment of anxiety and anxiety-related disorders. *Pharmacology & Therapeutics, 204*, 107402.

Sasson, Y., Iancu, I., Fux, M., et al. (1999). A double-blind crossover comparison of clomipramine and desipramine in the treatment of panic disorder. *European Neuropsychopharmacology, 9*(3), 191–196.

Saygin, M., Sungur, M., Sabol, E., et al. (2002). Nefazodone versus sertraline in treatment of posttraumatic stress disorder. *Psychopharmacology Bulletin, 12*(1), 1–5.

Sayyah, M., Sayyah, M., Boostani, H., et al. (2012). Effects of aripiprazole augmentation in treatment-resistant obsessive-compulsive disorder (a double-blind clinical trial). *Depression and Anxiety, 29*, 850–854.

Schmitt, R., Gazalle, F. K., Lima, M. S., Cunha, A., Souza, J., & Kapczinski, F. (2005). The efficacy of antidepressants for generalized anxiety disorder: A systematic review and meta-analysis. *Brazilian Journal of Psychiatry, 27*(1), 18–24.

Schneier, F. R., Goetz, D., Campeas, R., et al. (1998). Placebo-controlled trial of moclobemide in social phobia. *Journal of Mental Science, 172*, 70–77.

Schneier, F. R., Campeas, R., Carcamo, J., et al. (2015). Combined mirtazapine and SSRI treatment of PTSD: A placebo-controlled trial. *Depression and Anxiety, 32*(8), 570–579.

Schutters, S. I. J., van Megen, H. J., Van Veen, J. F., et al. (2010). Mirtazapine in generalized social anxiety disorder: A randomized, doubleblind, placebo-controlled study. *International Clinical Psychopharmacology, 25*(5), 302–304.

Schutters, S. I. J., van Megen, H. J., Van Veen, J. F., et al. (2011). Paroxetine augmentation in patients with generalised social anxiety disorder, non-responsive to mirtazapine or placebo. *Human Psychopharmacology: Clinical and Experimental, 26*, 72–76.

Schweizer, E., Fox, I., Case, G., et al. (1988). Lorazepam vs. alprazolam in the treatment of panic disorder. *Psychopharmacology Bulletin, 24*(2), 224–227.

Selvi, Y., Atli, A., Aydin, A., et al. (2011). The comparison of aripiprazole and risperidone augmentation in selective serotonin reuptake inhibitor-refractory obsessive-compulsive disorder: A single-blind, randomized study. *Human Psychopharmacology: Clinical and Experimental, 26*, 51–57.

Sepede, G., Berardis, D. D., Gambi, F., et al. (2006). Olanzapine augmentation in treatment-resistant panic disorder. *Journal of Clinical Psychopharmacology, 26*(1), 45–49.

Shah, A., & Northcutt, J. (2018). An openlabel, flexible dose adaptive study evaluating the efficacy of vortioxetine in subjects with panic disorder. *Annals of General Psychiatry, 17*, 19.

Sheehan, D. V., Ballenger, J., & Jacobsen, G. (1980). Treatment of endogenous anxiety with phobic, hysterical, and hypochondriacal symptoms. *Archives of General Psychiatry, 37*(1), 51–59.

Sheehan, D. V., Davidson, J., Manschreck, T., et al. (1983). Lack of efficacy of a new antidepressant (bupropion) in the treatment of panic disorder with phobias. *Journal of Clinical Psychopharmacology, 3*(1), 28–31.

Sheehan, D. V., Burnham, D. B., Iyengar, M. K., et al. (2005). Efficacy and tolerability of controlled-release paroxetine in the treatment of panic disorder. *Journal of Clinical Psychiatry, 66*, 34–40.

Sheehan, D. V., Sheehan, K. H., Raj, B. A., et al. (2007). An open-label study of tiagabine in panic disorder. *Psychopharmacology Bulletin, 40*(3), 32–40.

Simeon, J. G., Ferguson, H. B., Knott, V., Roberts, N., Gauthier, B., Dubois, C., & Wiggins, D. (1992). Clinical, cognitive, and neurophysiological effects of alprazolam in children and adolescents with overanxious and avoidant disorders. *Journal of the American Academy of Child and Adolescent Psychiatry, 31*(1), 29–33.

Simon, N. M., Kaufman, R. E., Hoge, E. A., et al. (2009). Open-label support for duloxetine for the treatment of panic disorder. *CNS Neuroscience & Therapeutics, 15*(1), 19–23.

Simon, N. M., Worthington, J. J., Moshier, S. J., Marks, E. H., Hoge, E. A., Brandes, M., Delong, H., & Pollack, M. H. (2010). Duloxetine for the treatment of generalized social anxiety disorder: A preliminary randomized trial of increased dose to optimize response. *CNS Spectrums, 15*(7), 367–373.

Simpson, H. B., Foa, E. B., Liebowitz, M. R., Huppert, J. D., Cahill, S., Maher, M. J., McLean, C. P., Bender, J., Marcus, S. M., Williams, M. T., Weaver, J., Vermes, D., Van Meter, P. E., Rodriguez, C. I., Powers, M., Pinto, A., Imms, P., Hahn, C.-G., & Campeas, R. (2013). Cognitive-behavioral therapy vs risperidone for augmenting serotonin reuptake inhibitors in obsessive-compulsive disorder: A randomized clinical trial. *JAMA Psychiatry, 70*(11), 1190–1199.

Skapinakis, P., Caldwell, D. M., Hollingworth, W., Bryden, P., Fineberg, N. A., Salkovskis, P., Welton, N. J., Baxter, H., Kessler, D., Churchill, R., & Lewis, G. (2016). Pharmacological and psychotherapeutic interventions for management of obsessive-compulsive disorder in adults: A systematic review and network meta-analysis. *The Lancet. Psychiatry, 3*(8), 730–739.

Slee, A., Nazareth, I., Bondaronek, P., Liu, Y., Cheng, Z., & Freemantle, N. (2019). Pharmacological treatments for generalised anxiety disorder: A systematic review and network meta-analysis. *Lancet (London, England), 393*(10173), 768–777.

Soltani, F., Sayyah, M., Feizy, F., et al. (2010). A double-blind, placebo-controlled pilot study of ondansetron for patients with obsessive-compulsive disorder. *Human Psychopharmacology: Clinical and Experimental, 25*, 509–513.

Sramek, J. J., Hong, W. W., Hamid, S., et al. (1999). Meta-analysis of the safety and tolerability of two dose regimens of buspirone in patients with persistent anxiety. *Depression and Anxiety, 9*(3), 131–134.

Stahl, S. M., Gergel, I., Li, D., et al. (2003). Escitalopram in the treatment of panic disorder: A randomized, double-blind, placebo-controlled trial. *The Journal of Clinical Psychiatry, 64*(11), 1322–1327.

Stahl, S. M., Ahmed, S., & Haudiquet, V. (2007). Analysis of the rate of improvement of specific psychic and somatic symptoms of general anxiety disorder during long-term treatment with venlafaxine ER. *CNS Spectrum, 12*(9), 703–711.

Steenen, S. A., Wijk, A. J., Heijden, G. J. M. G., et al. (2016). Propranolol for the treatment of anxiety disorders: Systematic review and meta-analysis. *Journal of Psychopharmacology, 30*(2), 128–139.

Stein, D. J., Spadaccini, E., & Hollander, E. (1995). Meta-analysis of pharmacotherapy trials for obsessive-compulsive disorder. *International Clinical Psychopharmacology, 10*(1), 11–18.

Stein, M. B., Liebowitz, M. R., Lydiard, R. B., et al. (1998). Paroxetine treatment of generalized social phobia (social anxiety disorder): A randomized controlled trial. *JAMA, 280*(8), 708–713.

Stein, M. B., Fyer, A. J., Davidson, J. R., et al. (1999). Fluvoxamine treatment of social phobia (social anxiety disorder): A double-blind, placebo-controlled study. *American Journal of Psychiatry, 156*(5), 756–760.

Stein, D. J., Seedat, S., van der Linden, G. J., et al. (2000). Selective serotonin reuptake inhibitors in the treatment of posttraumatic stress disorder: A meta-analysis of randomized controlled trials. *International Clinical Psychopharmacology, 15*(Suppl 2), S31–S39.

Stein, M. B., Sareen, J., Hami, S., et al. (2001). Pindolol potentiation of paroxetine for generalized social phobia: A double-blind, placebo-controlled, crossover study. *American Journal of Psychiatry, 158*(10), 1725–1727.

Stein, M., Kline, N., & Matloff, J. (2002a). Adjunctive olanzapine for SSRI-resistant combat-related PTSD: A double-blind, placebo-controlled study. *American Journal of Psychiatry, 159*(10), 1777–1779.

Stein, D., Cameron, A., Amrein, R., et al. (2002b). Moclobemide is effective and well tolerated in the long-term pharmacotherapy of social anxiety disorder with or without comorbid anxiety disorder. *International Clinical Psychopharmacology, 17*(4), 161–170.

Stein, D. J., Andersen, E. W., Tonnoir, B., et al. (2007). Escitalopram in obsessive-compulsive disorder: A randomized, placebo-controlled, paroxetine-referenced, fixed-dose, 24-week study. *Current Medical Research and Opinion, 23*(4), 701–711.

Stein, M. B., Ravindran, L. N., Simon, N. M., Liebowitz, M. R., Khan, A., Brawman-Mintzer, O., Lydiard, R. B., & Pollack, M. H. (2010). Levetiracetam in generalized social anxiety disorder: A double-blind, randomized controlled trial. *The Journal of Clinical Psychiatry, 71*(5), 627–631.

Steiner, H., Saxena, K. S., Carrion, V., et al. (2007). Divalproex sodium for the treatment of PTSD and conduct disordered youth: A pilot randomized controlled clinical trial. *Child Psychiatry and Human Development, 38*, 183–193.

Stocchi, F., Nordera, G., Jokinen, R. H., et al. (2003). Efficacy and tolerability of paroxetine for the long-term treatment of generalized anxiety disorder. *Journal of Clinical Psychiatry, 64*(3), 250–258.

Storch, E. A., Merlo, L. J., Begtson, M., et al. (2007). D-Cycloserine does not enhance exposure-response prevention therapy in obsessive-compulsive disorder. *International Clinical Psychopharmacology, 22*(4), 230–237.

Strawn, J. R., Welge, J. A., Wehry, A. M., Keeshin, B., & Rynn, M. A. (2015). Efficacy and tolerability of antidepressants in pediatric anxiety disorders: A systematic review and meta-analysis. *Depression and Anxiety, 32*(3), 149–157.

Strawn, J. R., Geracioti, L., Rajdev, N., Clemenza, K., & Levine, A. (2018a). Pharmacotherapy for generalized anxiety disorder in adult and pediatric patients: An evidence-based treatment review. *Expert Opinion on Pharmacotherapy, 19*(10), 1057–1070.

Strawn, J. R., Mills, J. A., Cornwall, G. J., Mossman, S. A., Varney, S. T., Keeshin, B. R., & Croarkin, P. E. (2018b). Buspirone in children and adolescents with anxiety: A review and Bayesian analysis of abandoned randomized controlled trials. *Journal of Child and Adolescent Psychopharmacology, 28*(1), 2–9.

Strawn, J. R., Mills, J. A., Schroeder, H., Mossman, S. A., Varney, S. T., Ramsey, L. B., Poweleit, E. A., Desta, Z., Cecil, K., & DelBello, M. P. (2020). Escitalopram in adolescents with generalized anxiety disorder: A double-blind, randomized, placebo-controlled study. *The Journal of Clinical Psychiatry, 81*(5), 20m13396.

Taylor, F. (2003). Tiagabaine for posttraumatic stress disorder: A case series of 7 women. *Journal of Clinical Psychiatry, 64*, 1421–1425.

Taylor, F. B., Martin, P., Thompson, C., et al. (2008). Prazosin effects on objective sleep measures and clinical symptoms in civilian trauma posttraumatic stress disorder: A placebo-controlled study. *Biological Psychiatry, 63*(6), 629–632.

Thomsen, P. H. (1997). Child and adolescent obsessive-compulsive disorder treated with citalopram: Findings from an open trial of 23 cases. *Journal of Child and Adolescent Psychopharmacology, 7*(3), 157–166.

Tiller, J. W., Bouwer, C., & Behnke, K. (1999). Moclobemide and fluoxetine for panic disorder. International Panic Disorder Study Group. *European Archives of Psychiatry and Clinical Neuroscience, 249*(Suppl 1), S7–S10.

Todorov, C., Freeston, M. H., & Borgeat, F. (2000). On the pharmacotherapy of obsessive-compulsive disorder: Is a consensus possible? *Canadian Journal of Psychiatry. Revue Canadienne de Psychiatrie, 45*(3), 257–262.

Tollefson, G., Birkett, M., & Koran, L. (1994a). Continuation treatment of OCD: Double blind and open label experience with fluoxetine. *Journal of Clinical Psychiatry, 55*(Suppl 10), 69–76.

Tollefson, G. D., Rampey, A. H., Jr., Potvin, J. H., et al. (1994b). A multicenter investigation of fixed-dose fluoxetine in the treatment of obsessive-compulsive disorder. *Archives of General Psychiatry, 51*(7), 559–567. (Erratum published in *Archives of General Psychiatry, 1994, 51*(11), 864).

Tucker, P., Zaninelli, R., Yehuda, R., et al. (2001). Paroxetine in the treatment of chronic posttraumatic stress disorder: Results of a placebo-controlled, flexible-dosage trial. *Journal of Clinical Psychiatry, 62*(11), 860–868.

Tucker, P., Trautman, R. P., Wyatt, D. B., et al. (2007). Efficacy and safety of topiramate monotherapy in civilian posttraumatic stress disorder: A randomized, double-blind, placebo-controlled study. *Journal of Clinical Psychiatry, 68*(2), 201–206.

Tzavellas, E., Karaiskos, D., Ilias, I., Liappas, I., & Paparrigopoulos, T. (2014). Agomelatine augmentation in obsessive compulsive disorder: A preliminary report. *Psychiatriki, 25*(3), 179–184.

Uhlenhuth, E. H., Alexander, P. E., Dempsey, G. M., et al. (1998). Medication side effects in anxious patients: Negative placebo responses? *Journal of Affective Disorders, 47*(1–3), 183–190.

Vaishnavi, S., Alamy, S., Zhang, W., Connor, K. M., & Davidson, J. R. (2007). Quetiapine as monotherapy for social anxiety disorder: A placebo-controlled study. *Progress in Neuro-Psychopharmacology & Biological Psychiatry, 31*(7), 1464–1469.

Vaiva, G., Ducrocq, F., Jezequel, K., et al. (2003). Immediate treatment with propranolol decreases posttraumatic stress disorder two months after trauma. *Biological Psychiatry, 54*, 947–949.

Vallejo, J., Olivares, J., Marcos, T., et al. (1992). Clomipramine versus phenelzine in obsessive-compulsive disorder. A controlled clinical trial. *British Journal of Psychiatry, 161*, 665–670.

Van Ameringen, M., Mancini, C., & Wilson, C. (1996). Buspirone augmentation of selective serotonin reuptake inhibitors (SSRIs) in social phobia. *Journal of Affective Disorders, 39*(2), 115–121.

Van Ameringen, M. A., Lane, R. M., Walker, J. R., et al. (2001). Sertraline treatment of generalized social phobia: A 20-week, double-blind, placebo-controlled study. *American Journal of Psychiatry, 158*(2), 275–281.

Van Ameringen, M., Mancini, C., Oakman, J., Walker, J., Kjernisted, K., Chokka, P., Johnston, D., Bennett, M., & Patterson, B. (2007). Nefazodone in the treatment of generalized social phobia: A randomized, placebo-controlled trial. *The Journal of Clinical Psychiatry, 68*(2), 288–295.

van der Kolk, B. A., Dreyfuss, D., Michaels, M., et al. (1994). Fluoxetine in posttraumatic stress disorder. *Journal of Clinical Psychiatry, 55*(12), 517–522.

van Vliet, I. M., Westenberg, H. G., & den Boer, J. A. (1993). MAO inhibitors in panic disorder: Clinical effects of treatment with brofaromine. A double blind placebo controlled study. *Psychopharmacology, 112*(4), 483–489.

van Vliet, I. M., den Boer, J. A., Westenberg, H. G., et al. (1997). Clinical effects of buspirone in social phobia: A double-blind placebo-controlled study. *Journal of Clinical Psychiatry, 58*(4), 164–168.

Versiani, M., Nardi, A. E., Mundim, F. D., et al. (1992). Pharmacotherapy of social phobia. A controlled study with moclobemide and phenelzine. *British Journal of Psychiatry, 161*, 353–360.

Versiani, M., Cassano, G., Perugi, G., et al. (2002). Reboxetine, a selective norepinephrine reuptake inhibitor, is an effective and well-tolerated treatment for panic disorder. *Journal of Clinical Psychiatry, 63*(1), 31–37.

Villarreal, G., Calais, L. A., Canive, J. M., et al. (2007). Prospective study to evaluate the efficacy of aripiprazole as a monotherapy in patients with severe chronic posttraumatic stress disorder: An open trial. *Psychopharmacology Bulletin, 40*(2), 6–18.

Villarreal, G., Canive, J. M., Calais, L. A., et al. (2010). Duloxetine in military posttraumatic stress disorder. *Psychopharmacology Bulletin, 43*(3), 26–34.

Vulink, N. C., Denys, D., & Westenberg, H. G. (2005). Bupropion for patients with obsessive-compulsive disorder: An open-label, fixed-dose study. *Journal of Clinical Psychiatry, 66*, 228–230.

Vulink, N. C., Denys, D., Fluitman, S. B., et al. (2009). Quetiapine augments the effect of citalopram in non-refractory obsessive-compulsive disorder: A randomized, double-blind, placebo-controlled study of 76 patients. *Journal of Clinical Psychiatry, 70*(7), 1001–1008.

Wade, A. G., Lepola, U., Koponen, H. J., et al. (1997). The effect of citalopram in panic disorder. *British Journal of Psychiatry, 170*, 549–553.

Walderhaug, E., Kasserman, S., Aikins, D., et al. (2010). Effects of duloxetine in treatment-refractory men with posttraumatic stress disorder. *Pharmacopsychiatry, 43*(2), 45–49.

Warneke, L. B. (1984). The use of intravenous chlorimipramine in the treatment of obsessive-compulsive disorder. *Canadian Journal of Psychiatric Nursing, 29*(2), 135–141.

Warneke, L. B. (1985). Intravenous chlorimipramine in the treatment of obsessional disorder in adolescence: Case report. *Journal of Clinical Psychiatry, 46*(3), 100–103.

Warneke, L. (1989). Intravenous chlorimipramine therapy in obsessive-compulsive disorder. *Canadian Journal of Psychiatric Nursing, 34*(9), 853–859.

Watson, P. (2019). PTSD as a public mental health priority. *Current Psychiatry Reports, 21*(7), 61.

Weiss, E. L., Potenza, M. N., McDougle, C. J., et al. (1999). Olanzapine addition in obsessive-compulsive disorder refractory to selective serotonin reuptake inhibitors: An open-label case series. *Journal of Clinical Psychiatry, 60*(8), 524–527.

Westenberg, H., Stein, D., Yang, H., et al. (2004). A double-blind placebo-controlled study of controlled release fluvoxamine for the treatment of generalized social anxiety disorder. *Journal of Clinical Psychopharmacology, 24*(1), 49–55.

Williams, T., Hattingh, C. J., Kariuki, C. M., Tromp, S. A., van Balkom, A. J., Ipser, J. C., & Stein, D. J. (2017). Pharmacotherapy for social anxiety disorder (SAnD). *The Cochrane Database of Systematic Reviews, 10*(10), CD001206. https://doi.org/10.1002/14651858. CD001206.pub3

Wingerson, D., Nguyen, C., & Roy-Byrne, P. P. (1992). Clomipramine treatment for generalized anxiety disorder (letter). *Journal of Clinical Psychopharmacology, 12*, 214–215.

Woodman, C. L., & Noyes, R., Jr. (1994). Panic disorder: Treatment with valproate. *Journal of Clinical Psychiatry, 55*(4), 134–136.

Woolcott, J. C., Richardson, K. J., Wiens, M. O., et al. (2009). Meta-analysis of the impact of 9 medication classes on falls in elderly persons. *Archives of Internal Medicine, 169*, 1952–1960.

Wu, W., Wang, G., Ball, S. G., et al. (2011). Duloxetine versus placebo in the treatment of patients with generalized anxiety disorder in China. *Chinese Medical Journal, 124*(20), 3260–3268.

Yeh, M. S., Mari, J. J., Costa, M. C. P., et al. (2010). A double-blind randomized controlled trial to study the efficacy of topiramate in a civilian sample of PTSD. *CNSN Euroscience and Therapeutics, 17*, 305–310.

Youssef, N. A., Marx, C. E., Bradford, D. W., et al. (2012). An open-label pilot study of aripiprazole for male and female veterans with chronic posttraumatic stress disorder who respond suboptimally to antidepressants. *International Clinical Psychopharmacology, 27*(4), 191–196.

Zanardi, R., Artigas, F., Franchini, L., et al. (1997). How long should pindolol be associated with paroxetine to improve the antidepressant response? *Journal of Clinical Psychopharmacology, 17*(6), 446–450.

Zhou, D.-D., Zhou, X.-X., Li, Y., Zhang, K.-F., Lv, Z., Chen, X.-R., Wan, L.-Y., Wang, W., Wang, G.-M., Li, D.-Q., Ai, M., & Kuang, L. (2019). Augmentation agents to serotonin reuptake inhibitors for treatment-resistant obsessive-compulsive disorder: A network meta-analysis. *Progress in Neuro-Psychopharmacology & Biological Psychiatry, 90*, 277–287.

Zisook, S., Chentsova-Dutton, Y. E., Smith-Vaniz, A., et al. (2000). Nefazodone in patients with treatment-refractory posttraumatic stress disorder. *Journal of Clinical Psychiatry, 61*(3), 203–208.

Zohar, J., & Judge, R. (1996). Paroxetine versus clomipramine in the treatment of obsessive–compulsive disorder. OCD Paroxetine Study Investigators. *British Journal of Psychiatry, 169*(4), 468–474.

Zohar, J., & Westenberg, H. G. (2000). Anxiety disorders: A review of tricyclic antidepressants and selective serotonin reuptake inhibitors. *Acta Psychiatrica Scandinavica. Supplementum, 403*, 39–49.

Zohar, J., Amital, D., Miodownik, C., et al. (2002). Double-blind placebo-controlled pilot study of sertraline in military veterans with posttraumatic stress disorder. *Journal of Clinical Psychopharmacology, 22*(2), 190–195.

Zohar, J., Fostick, L., Juven-Wetzler, A., et al. (2018). Secondary prevention of chronic PTSD by early and short-term administration of escitalopram: A prospective randomized, placebo-controlled, double-blind trial. *The Journal of Clinical Psychiatry, 79*(2), 16m10730.

Zwanzger, P., Baghai, T. C., Schule, C., et al. (2001). Tiagabine improves panic and agoraphobia in panic disorder patients. *Journal of Clinical Psychiatry, 62*(8), 656–657.

Zwanzger, P., Eser, D., Nothdurfter, C., et al. (2009). Effects of the GABA-reuptake inhibitor tiagabine on panic and anxiety in patients with panic disorder. *Pharmacopsychiatry, 42*, 266–269.

Contents

This chapter is an update from the 4th edition. Previous edition authors were David N. Neubauer, Jared Minkel, and Andrew D. Krystal

K. Mishima (✉)
Department of Neuropsychiatry, Akita University
Graduate School of Medicine, Akita-city, Akita, Japan
e-mail: mishima@med.akita-u.ac.jp

© Springer Nature Switzerland AG 2024
A. Tasman et al. (eds.), *Tasman's Psychiatry*,
https://doi.org/10.1007/978-3-030-51366-5_137

Abstract

Insomnia is a common complaint in patients with psychiatric disorders. Pharmacologic approaches to the treatment of insomnia are often recommended in combination with sleep hygiene education and cognitive behavioral therapy. Currently, the insomnia treatment medications include benzodiazepines, nonbenzodiazepines, melatonin receptor agonist, dual orexin receptor antagonists, and histamine H_1 receptor antagonist. All have demonstrated efficacy in treating one or more insomnia symptoms and have been shown to have satisfactory safety characteristics in clinical trials with populations of normal subjects and individuals suffering with chronic insomnia. Therefore, the most recently approved insomnia treatment medications no longer have an implied limitation on their duration of use, and the indications specifically note efficacy for sleep onset and sleep maintenance. All of benzodiazepines, nonbenzodiazepines, and dual orexin receptor antagonists are categorized as Schedule IV controlled substances; however, ramelteon and doxepin have no abuse liability and are classed as a nonscheduled medication. Although various medications for depression, psychosis, and other psychotropic agents are prescribed with the primary intention of improving sleep onset, maintenance, and quality, generally, there is limited insomnia efficacy evidence in non-psychiatric populations, and there are significant safety concerns with all of these medications. Traditionally, insomnia caused by some underlying disease, such as a psychiatric disorder, has been referred to as secondary insomnia. However, it has become clear that in many cases insomnia can persist independently, even after the causative illness has resolved. The concept of comorbid insomnia as an alternative to secondary insomnia has been proposed, and it is appropriate to treat comorbid insomnia from the earliest stages of onset and recurrence of psychiatric disorders.

Keywords

Insomnia · Benzodiazepines · Nonbenzodiazepines · Melatonin receptor agonist · Dual orexin receptor antagonists · Histamine H_1 receptor antagonist · Sedating medications · OTC medications

Introduction

Insomnia is a common complaint in the general adult population, and the prevalence is significantly greater in patients with psychiatric disorders. Approximately 30% of adults experience occasional insomnia and about 10% suffer with chronic symptoms (National Institutes of Health, 2005). Insomnia may occur idiopathically or may arise with distressing circumstances, psychological conditioning, environmental factors, jet lag, shift work schedules, medication effects, and medical, psychiatric, and sleep disorders. Insomnia among psychiatric patients may be influenced by particular disorders, the severity of the illness, and the treatment approaches. Individuals with mood and anxiety disorders are especially likely to experience insomnia; however, most psychiatric disorders increase the risk of sleep disturbances. Often insomnia is the earliest symptom heralding a depressive episode (Baglioni et al., 2011), and it may remain a residual symptom following effective treatment of the mood disturbance (Perlis et al., 1997; Nierenberg et al., 1999; McClintock et al., 2011).

The relationship between sleep and psychiatric illness has many different dimensions. In addition to psychiatric illness increasing the likelihood of insomnia, the presence of persistent insomnia increases the future risk of relapses or the new onset of mood, anxiety, and substance abuse disorders. Several psychotropic medications include insomnia as common side effects. Certain psychiatric disorders, particularly major depression, may be associated with changes in the distribution and amount of sleep stages. There are commonalities among neurotransmitters known to regulate sleep, mood, and anxiety. Strategic manipulations of sleep may have antidepressant effects, while sleep deprivation can precipitate mania in bipolar disorder patients.

Although insomnia may simply describe the inadequate quantity or quality of sleep when one has the opportunity to be sleeping, insomnia nosologies typically require the presence of difficulty falling asleep, staying asleep, or excessively early awakening, in conjunction with daytime consequences (American Psychiatric Association, 2013). These daytime effects may include fatigue, poor concentration, irritability, and excessive worry about sleep. Insomnia episodes may be acute or chronic and may or may not be associated with comorbid medical, psychiatric, or other sleep disorders. Chronic insomnia has been associated with increased absenteeism, decreased quality of life, higher healthcare costs, and a significant societal economic burden.

Insomnia treatment generally addresses the context of the sleep disturbance, with the goal of optimizing the management of comorbid disorders and creating an environment conducive for sleep to improve. Typically, this requires a comprehensive evaluation to identify associated conditions and specific factors that may undermine the experience of good-quality sleep. Treatment approaches may involve a variety of general sleep hygiene recommendations, in addition to specific behavioral recommendations. Psychotherapeutic interventions may be valuable for selected patients. Cognitive behavioral therapy for insomnia has been shown to be an effective modality for chronic insomnia subjects. Pharmacologic approaches to the treatment of insomnia are often recommended in combination with these educational, psychotherapeutic, and behavioral modalities (Smith et al., 2002).

Sleep-Wake Cycle Regulation

Under normal conditions, sleep and wakefulness are strongly influenced by a network of central nervous system processes that promote sleepiness or arousal according to time-dependent mechanisms. Key brainstem arousal pathways include acetylcholine-producing cell groups activating the cortex through thalamic relay neurons and monoaminergic cell groups with projections to hypothalamic and basal forebrain regions. These latter arousal mechanisms include noradrenergic and serotonergic, and also hypothalamic histaminergic pathways. The hypothalamic ventrolateral preoptic nucleus (VLPO) has been identified as a critical sleep-active cell group that inhibits the stimulation from the monoaminergic arousal pathways. Inhibitory γ-aminobutyric acid (GABA) and galanin-related VLPO activity serve as a switch that stabilizes the states of sleep and wakefulness, which are further modulated and reinforced by the neuropeptide orexin. The hypothalamic suprachiasmatic nucleus (SCN) maintains 24-h periodicity through a complex transcription–translation feedback loop involving multiple genes. The primary input to the SCN is retinal photoperiod information, which entrains the circadian rhythm to the environmental day–night cycle (Saper et al., 2001).

Homeostatic and circadian process interactions have been postulated to explain the normal human pattern of daytime and evening wakefulness for approximately 16 h alternating with about 8 h of nighttime sleep. The homeostatic process represents the overall balance of sleep and waking whereby a sleep drive increases with continued wakefulness. The homeostatic sleep drive is low in the morning following nighttime sleep but then increases steadily until nighttime sleep again occurs to reverse the process and reduce the sleep drive. In spite of the increased homeostatic evening sleep drive, people generally are the most awake and alert during the evening hours. This is due to the circadian arousal that peaks in the evening and opposes the homeostatic sleepiness. As bedtime approaches, the circadian arousal declines and allows sleep onset to occur due to the unopposed homeostatic sleep drive. The SCN directs the pineal gland activity in producing and releasing the hormone melatonin. The SCN contains a high concentration of melatonin receptors. Agonist activity at these SCN melatonin receptors reduces their firing rate, which contributes to the arousal decline and subsequent sleep onset. Essentially, the homeostatic process determines the sleep need, while the circadian process optimizes sleep to occur at nighttime (Richardson, 2005).

Historical Overview

A wide assortment of substances and medications have been employed in the attempt to promote sleep. Fermented beverages and opium have been available for millennia. Sedating concoctions, such as opium-containing laudanum, were very commonly used for ailments including insomnia from the seventeenth through the nineteenth centuries. Numerous patent medicine compounds were promoted as sleep aids. Chloral hydrate and paraldehyde became available in the mid-nineteenth century. Barbiturates and related compounds frequently were recommended for insomnia from the early twentieth century until recent decades. Meprobamate, methyprylon, ethchlorvynol, glutethimide, barbiturates, and methaqualone were commonly prescribed for insomnia in the mid-1900s. A group of chemically related compounds, the benzodiazepine medications for insomnia (hypnotics), became the predominant selection for pharmacologic insomnia management beginning in the 1970s due to their improved safety profile. The non-benzodiazepine medications for insomnia (hypnotics), sharing basic pharmacodynamic features with benzodiazepines, but having different chemical structures, were launched in the 1990s, primarily to reduce the risk of physical dependence, by having significantly shorter half-lives. Melatonin receptor agonists and dual orexin receptor antagonists, sleep medications with new mechanisms of action, followed and continue to be developed at the present time.

Compounds currently taken with the intention of treating insomnia include bedtime alcohol, unregulated complementary and alternative substances, over-the-counter (OTC) antihistamines, sedating prescription medications (e.g., medications for depression and medications for psychosis) used on an off-label basis, and US Food and Drug Administration (FDA)-approved insomnia treatment medications. A useful historical generalization is that efficacy has always been present with many older substances being capable of inducing sleep. However, earlier remedies were often associated with serious safety problems, including lethality and dependence. The predominant evolutionary theme of insomnia pharmacotherapy has been the trend of improved safety. Current treatment options represent significant progress regarding tolerability, adverse events, and abuse liability.

FDA-Approved Insomnia Treatment Medications

Currently, the insomnia treatment medications approved by the FDA that are in common use include a variety of agents that enhance GABAergic inhibition by binding to the benzodiazepine binding site on the GABA$_A$ receptor complex (benzodiazepines and nonbenzodiazepines), a melatonin receptor agonist (ramelteon), dual orexin receptor antagonists, and a histamine H$_1$ receptor antagonist (doxepin). Table 1 shows the available doses, approximate elimination half-lives, and US Drug Enforcement Administration (DEA) schedule for each of these medications. All have demonstrated efficacy in treating one or more insomnia symptoms and have been shown to have satisfactory safety characteristics in clinical trials with populations of normal subjects and individuals suffering with chronic insomnia. Generally, insomnia treatment medications have been evaluated with objective sleep laboratory and outpatient subjective assessments. These two types of evaluations provide

complementary efficacy information. In recent years, there has been a trend toward longer-term efficacy and safety trials and primary outcome measures reflecting sleep-onset latency and sleep maintenance. These directions have paralleled the approved prescribing guidelines.

Previously, all of the insomnia treatment medications were approved with indications for short-term use; however, four new medications approved by the FDA since 2005 (eszopiclone, ramelteon, zolpidem extended-release, and low-dose doxepin) have had no implied limitation on their duration of use. Accordingly, these may be prescribed for as long as medically appropriate. Recent approvals have also included language specifying sleep onset and/or sleep maintenance indications. This information can help direct appropriate medication selection

Table 1 FDA-approved insomnia treatment medications

Medication	Available doses (mg)	Elimination half-life (hours)	DEA schedule	Active metabolite
Benzodiazepines and nonbenzodiazepines				
Immediate-release benzodiazepines				
• Estazolam	1, 2	8–24	IV	Yes
• Flurazepam	15, 30	48–120	IV	Yes
• Quazepam	7.5, 15	48–120	IV	Yes
• Temazepam	7.5, 15, 22.5, 30	8–20	IV	No
• Triazolam	0.125, 0.25	2–4	IV	No
Immediate-release nonbenzodiazepines				
• Eszopiclone	1, 2, 3	5–7	IV	No
• Zaleplon	5, 10	1	IV	No
• Zolpidem	5, 10	1.5–2.4	IV	No
Modified-release nonbenzodiazepine				
• Zolpidem ER	6.25, 12.5	2.8–2.9	IV	No
Alternative-delivery nonbenzodiazepines				
• Zolpidem oral spray (Zolpimist); gastric absorption	5, 10	1.5–2.4	IV	No
• Zolpidem sublingual (Edular); gastric absorption	5, 10	1.5–2.5	IV	No
• Zolpidem sublingual (Intermezzo); oral and gastric absorption	1.75, 3.5	1.5–2.5	IV	No
Melatonin receptor agonist				
• Ramelteon	8	1–2.6	None	Yes
Dual orexin receptor antagonist				
• Suvorexant	5, 10, 15, 20	8–19	IV	No
• Lemborexant	5, 10	50	IV	No
• Daridorexant	25, 50	6–10	IV	No
Histamine H$_1$ receptor antagonist				
• Doxepin (Silenor)	3, 6	15	None	No

for individual patient insomnia patterns. Abuse liability is reflected in the DEA scheduling. All of the benzodiazepines and nonbenzodiazepines are categorized as Schedule IV controlled substances due to a relatively low risk of abuse; however, ramelteon and doxepin have no abuse liability and are classed as a nonscheduled medication.

A systematic review of 170 trials (36 interventions, 47,950 patients) and network meta-analysis of 154 trials (30 interventions, 44,089 patients) examining the benefits and harms of sleep medications for insomnia in adults found that the nonbenzodiazepine eszopiclone and the dual orexin receptor antagonist lemborexant had good profiles in both acute and long-term treatment. However, eszopiclone can cause adverse events, and data on the safety of lemborexant are inconclusive. Many approved drugs (benzodiazepines, daridorexant, suvorexant, trazodone, etc.) are effective in the acute treatment of insomnia, but they are either poorly tolerated or there is a lack of information on their long-term effects. Melatonin and the melatonin receptor agonist ramelteon have shown no clear benefit (De Crescenzo et al., 2022).

Benzodiazepine and Nonbenzodiazepine Medications for Insomnia (Hypnotics)

The benzodiazepines and nonbenzodiazepines have been the primary medications prescribed for the treatment of insomnia for over 40 years. This class includes the benzodiazepines, a group of medications with a shared chemical structure, and newer compounds that do not have the benzodiazepine structure but act at the same binding site on the $GABA_A$ receptor complex and are referred to as nonbenzodiazepines. The typical benzodiazepine structure is a seven-membered diazepine ring fused with a benzene ring.

The benzodiazepine medications for insomnia (hypnotics) were approved by the FDA in the 1970s and 1980s. All of the nonbenzodiazepines have been approved since the 1990s and have unique structures without benzodiazepine rings. There has been a trend toward short-to-intermediate half-life medications for insomnia, in contrast to earlier long-acting medications. Until recently, all of these medications for insomnia have been immediate-release formulations; however, at least one controlled-release medication for insomnia has now become available (zolpidem ER). Several nonbenzodiazepines employ formulations by which zolpidem is absorbed through the oral mucosa.

The benzodiazepine and nonbenzodiazepine medications for insomnia function as positive allosteric modulators of GABA responses at the $GABA_A$ receptor complex, which is a five-subunit, ligand-gated, transmembrane structure with a central chloride ion channel (see Fig. 1). GABA is the primary inhibitory neurotransmitter in the central nervous system, and it influences the intracellular-extracellular negative chloride ion balance. In humans, the $GABA_A$ receptor complex most typically comprises two α, two β, and one γ glycoprotein subunits. A benzodiazepine recognition site exists at the interface of α and γ subunits. When a benzodiazepine agonist is present at this site and GABA activates the receptor, a greater number of chloride ions are able to enter the cell due to an increase in the frequency of ion channel opening. The result is greater hyperpolarization and enhanced inhibitory activity. The action of all of

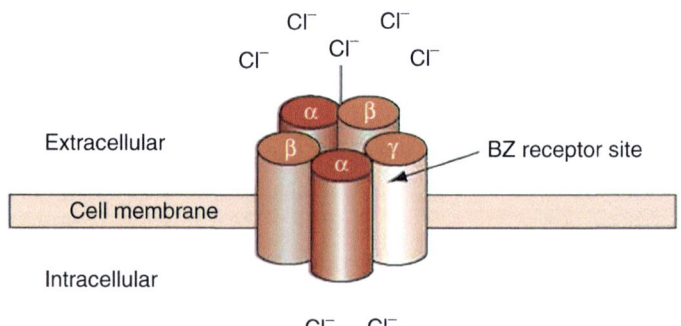

Fig. 1 $GABA_A$ receptor complex with benzodiazepine recognition site

the benzodiazepine receptor agonists can be reversed by the benzodiazepine receptor antagonist, flumazenil. Although the wide distribution of GABA$_A$ receptors may lead to global sedating effects, targeted action at VLPO GABAergic neurons likely plays a key role in the hypnotic effect of the medications in this class. Various benzodiazepine medications may also be employed as sedatives, medications for anxiety, muscle relaxants, and anticonvulsants due to binding to GABA$_A$ receptors in specific regions of the brain. Other medications and substances influencing consciousness by affecting the function of the GABA$_A$ receptor include barbiturates, propofol, and ethanol.

Multiple subtypes have been identified for the glycoprotein subunits constituting the GABA$_A$ receptor complex (Bateson, 2006). The α subunit, which has at least six subtypes, serves a key role in the pharmacodynamic effects of this receptor. The most common configuration in the central nervous system is $\alpha1\beta2\gamma2$. Benzodiazepines are relatively nonselective among the α subunits, with affinity for the 1, 2, 3, and 5 α subunit subtypes. The non-benzodiazepine medications for insomnia have greater selectivity for the α1 subtype, which is associated with sedation, but also amnesia and ataxia (Mohler et al., 2002). This subtype selectivity may contribute to improved tolerability and decreased discontinuation effects among the non-benzodiazepine agents.

Pharmacokinetic properties vary widely among these benzodiazepine receptor agonist medications for insomnia. All of these medications approved for the treatment of insomnia are absorbed relatively rapidly and may enhance sleep onset. The elimination half-life and dose determine the relative duration of sedating effects during the desired sleep period, but also potential undesired sedation or impairment during waking hours. Generally, tolerance to the sleep-enhancing effects of the benzodiazepine receptor agonist medications for insomnia has not been demonstrated in controlled clinical trials.

The benzodiazepine and nonbenzodiazepine medications for insomnia generally are well tolerated. Adverse effects may include somnolence, headache, dizziness, nausea, diarrhea, and anterograde amnesia. Rarely patients may exhibit sleep walking or confused behaviors within a few hours after taking a dose of medication for insomnia. Longer elimination half-lives and dosages increase the risk for next-day residual daytime effects and associated impairment. When prescribing to the elderly, it should be noted that long-term use of benzodiazepine receptor agonists, including medications for anxiety and medications for insomnia, has been noted to potentially increase the risk of developing dementia (Penninkilampi & Eslick, 2018). Rebound insomnia, characterized by worsened sleep for several nights relative to the pre-treatment baseline insomnia, may occur upon sudden discontinuation of the medications in this class. Tapering the dose may reduce the magnitude of this effect. Recrudescence of the original insomnia symptoms also may occur with discontinuation of medications for insomnia. New withdrawal symptoms rarely will occur with rapid discontinuation of shorter-acting benzodiazepines (Roehrs & Roth, 2006).

Benzodiazepine Medications for Insomnia (Hypnotics)

Generally, benzodiazepines have sedative, antianxiety, muscle-relaxant, anticonvulsant, and amnestic properties. The five benzodiazepine medications for insomnia (hypnotics) currently available in the United States are estazolam, flurazepam, temazepam, triazolam, and quazepam. Their elimination half-lives range from a few hours (triazolam) to a few days (flurazepam and quazepam). Estazolam, flurazepam, and quazepam have active metabolites. Benzodiazepine medications for insomnia not available in the United States include flunitrazepam, loprazolam, lormetazepam, and nitrazepam. Flunitrazepam is considered an abused substance in the United States.

Nonbenzodiazepine Medications for Insomnia (Hypnotics)

There are six nonbenzodiazepine medications for insomnia (hypnotic) available in the United States: eszopiclone, zaleplon, zolpidem, zolpidem extended-release, orally disintegrating zolpidem, and transorally absorbed zolpidem. The elimination half-lives range from the ultrashort zaleplon at 1 h to eszopiclone at approximately 6 h in adults. Zopiclone is available in selected countries outside the United States. Eszopiclone and zolpidem extended-release are approved by the FDA, with indications for treating insomnia associated with difficulty with sleep onset and sleep maintenance, and both have no implied limitation on their duration of use. A transoral zolpidem is indicated for helping people return to sleep after middle-of-the-night awakenings if there are four or more hours left in the night (Roth et al., 2013).

Whereas the clinical trials for benzodiazepine medications for insomnia have been relatively short term, nightly and non-nightly dosing studies with the nonbenzodiazepine medications for insomnia have been conducted for extended periods. These have included placebo-controlled, double-blind subjective efficacy assessments for 6 months of treatment and open-label safety and tolerability studies for 1 year of use. It has been demonstrated that efficacy is maintained without the development of tolerance and that adverse events remain relatively infrequent with extended use (Krystal et al., 2003; Perlis et al., 2004; Ancoli-Israel et al., 2005; Roth et al., 2005; Erman et al., 2006a).

Prescribing Guidelines

The benzodiazepine and nonbenzodiazepine medications for insomnia generally should be taken when going to bed. Doses taken prior to bedtime may risk psychomotor impairment, confusion, and amnesia occurring before sleep occurs. Patients should avoid hazardous activities after taking the medication for insomnia and should be prepared to spend at least 7–8 h in bed following ingestion of the medication. The exception is the very short-acting zaleplon, which should no longer be active 4 h following ingestion. Only transoral zolpidem is specifically indicated for middle-of-the-night dosing, although the short half-life of zaleplon and several prior studies suggest the utility of this agent also for use for middle-of-the-night awakenings.

Lower doses generally are recommended for older adults and patients with hepatic impairment. The concomitant use of other CNS depressants, including alcohol and various psychotropic medications, may lead to excessive sedation, in addition to cognitive and psychomotor impairment. There is a relatively low risk of pharmacokinetic interaction with these approved medications for insomnia, as none are significant inducers or inhibitors of hepatic metabolic isozymes. Concomitant medications will rarely inhibit the metabolism of the medications for insomnia and increase their pharmacokinetic area under the curve and the peak serum concentration (Hesse et al., 2003). All of the nonbenzodiazepine medications for insomnia are pregnancy category C and all of the benzodiazepine medications for insomnia are pregnancy category X.

Clinical Considerations

The benzodiazepine and nonbenzodiazepine medications for insomnia have durations of action ranging from very short to very long. Accordingly, medication selections can be made to optimize insomnia efficacy for individual patients while minimizing the risk of undesired daytime effects. The newer nonbenzodiazepine medications for insomnia have been popular choices owing to the short to moderately short half-lives and the generally positive tolerability and safety profiles. Although primarily intended for bedtime dosing and nighttime sleep, they may be employed later during the night or when schedules require daytime sleep. Caution regarding the duration of action of medications for insomnia is necessary when a period of less than 8 h is available for sleep. Patients taking longer acting medications for insomnia should avoid hazardous daytime activities until they can assess potential impairment from the medications. The long half-life benzodiazepine medications for insomnia can accumulate with nightly dosing so that daytime blood levels increase and reach a steady state only after 1 week or longer (Vermeeren, 2004). The tendency for women to metabolize zolpidem more slowly than men has led the FDA to recommend that women initially be prescribed the lower available dose to minimize the risk for next-morning residual sleepiness or impairment.

Medications for insomnia may be taken nightly or on an as-needed basis, which may be episodic, frequent, or rare. Patients may begin with nightly use and transition to intermittent dosing. Placebo-controlled clinical trials with both zolpidem tablet formulations have demonstrated continued efficacy for up to 6 months of non-nightly use (Walsh et al., 2000; Perlis et al., 2004; Erman et al., 2006a; Krystal et al., 2008). Rebound insomnia was not evident on discontinuation from intermittent use. Insomnia is a chronic disorder, and some patients clearly benefit from chronic treatment. The formal indications for two recently approved medications for insomnia, eszopiclone and zolpidem extended-release, do not include any implied restriction on the duration of use.

The potential for abuse or dependence of benzodiazepine receptor agonist medications has been a longstanding concern. Although these medications for insomnia do share some degree of abuse liability, the risk is very low with insomnia patients who do not have substance abuse comorbidity. Most patients taking medications for insomnia use them for up to a few weeks. Longer term bedtime use at therapeutic doses generally reflects therapy-seeking behavior, rather than abuse. Dose escalation of medications for insomnia is rare (Walsh et al., 2005; Roehrs & Roth, 2006).

Melatonin Receptor Agonists

Several investigational agents have targeted the sleep-promoting and circadian rhythm-entraining activity associated with the SCN and the effects of endogenous melatonin. One melatonin receptor agonist, ramelteon, is currently available; however, others are in the developmental phase.

Ramelteon is the single S-enantiomer of a tricyclic indan derivative. It was developed to target the sleep-enhancing role performed by the melatonin receptors in the SCN. Ramelteon is unique in being a non-sedating, sleep-promoting medication with a novel mechanism of action. It does not interact to a significant extent with other neurotransmitter systems, including GABA and the monoamines. The FDA approval is for the treatment of insomnia characterized by difficulty with sleep onset. There is no implied limitation for the duration that it may be used. Clinical trials have demonstrated efficacy in improving sleep onset in both objective and subjective assessments in adult and older adult subjects (Seiden et al., 2005; Zammit et al., 2005).

Mechanism of Action

Ramelteon attenuates the circadian arousal signal that normally is present in the evening (see Fig. 2). Over time, it may improve sleep onset through stabilization of the circadian system. It is a selective agonist for the MT_1 and MT_2 melatonin receptor subtypes, which are present in high concentrations in the SCN. Agonists at the MT_1 subtype promote sleep onset by decreasing the firing rate of selected SCN neurons, whereas the MT_2 subtype appears to have a central role in the reinforcement and entrainment of the circadian rhythm that strongly influences the timing of the sleep-wake cycle (Kato et al., 2005). Animal studies have shown greater potency in comparison with exogenous melatonin (Miyamoto et al., 2004; Yukuhiro et al., 2004). Human studies have shown a relatively flat dose-response relationship (Erman et al., 2006b).

Pharmacokinetics

Ramelteon is rapidly absorbed and reaches a peak concentration within 1 h. A high-fat meal will decrease absorption and may decrease efficacy. The elimination half-life is under 3 h. Ramelteon is metabolized primarily through the CYP1A2 isozyme and secondarily through the CYP2C subfamily and CYP3A4 isozymes. There is one active metabolite (M-II), which has an elimination half-life of 2–5 h and the same pharmacodynamic characteristics as the parent compound, but at markedly reduced potency (Takeda Pharmaceuticals North America, 2005; Karim et al., 2006).

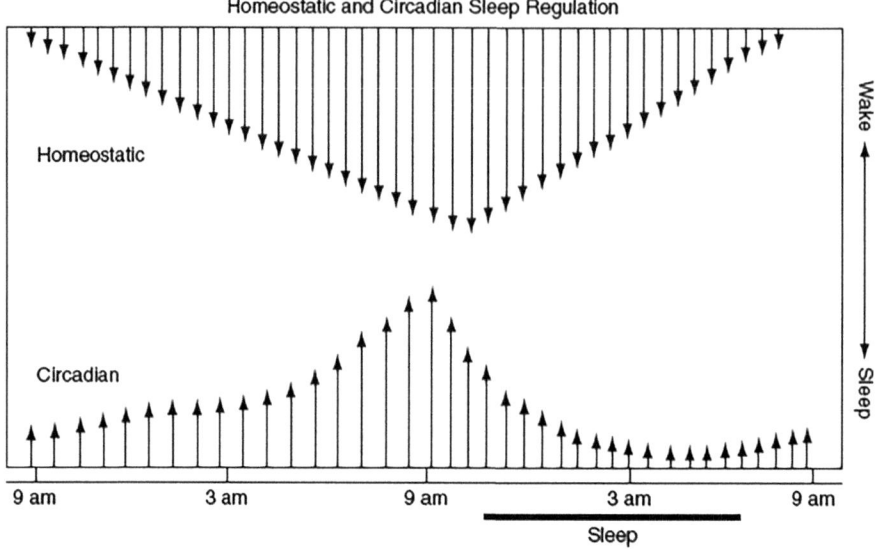

Fig. 2 Normal homeostatic and circadian sleep-wake cycle influences during a 24-h period

Prescribing Guidelines

Ramelteon is available in a single 8 mg strength, which is the dose recommended for adults and older adults, and in patients with mild-to-moderate chronic obstructive pulmonary disease (COPD), sleep apnea, and hepatic impairment. It should not be co-administered with fluvoxamine, which may inhibit the metabolism of ramelteon. It may be taken approximately 30 min prior to bedtime, as there are no psychomotor or cognitive impairments associated with the medication. However, patients should avoid hazardous activities after taking the medication owing to the anticipated sleepiness. It may be used on a chronic basis and is classed as nonscheduled by the DEA (Takeda Pharmaceuticals North America, 2005).

Clinical Considerations

Ramelteon may be beneficial for patients with difficulty falling asleep and remaining asleep during the early part of the night. It may not help with symptoms of sleep maintenance or early morning awakening. Patients taking ramelteon may not experience the medication's maximum benefit in reducing the sleep-onset latency until they have taken it for several nights or, perhaps, a few weeks. They should be informed that their experience with ramelteon will not cause the presleep sedation that they may have noted with other sedating OTC or prescription medications.

Ramelteon generally is well tolerated. The adverse events reported by at least 2% more of ramelteon-treated subjects compared with placebo-treated subjects during the clinical trials included somnolence, dizziness, and fatigue (Takeda Pharmaceuticals North America, 2005). It is not associated with any discontinuation effects, such as withdrawal symptoms or rebound insomnia. Ramelteon has been assessed for up to 1 year of nightly use in an open-label study (DeMicco et al., 2006).

Ramelteon has been shown to have a complete absence of abuse liability in studies with animals and humans. In a double-blind, crossover study, sedative abusing subjects were given ramelteon (16, 80, and 160 mg), placebo, and an active control (triazolam). Measures of drug likeability, behavioral and cognitive performance, and ataxia all showed no differentiation between ramelteon and placebo, whereas high-dose triazolam was associated with impairment and elevated likeability (Johnson et al., 2006).

Dual Orexin Receptor Antagonists

This new class of medications for insomnia suppresses wakefulness by inhibiting the binding of the wake-promoting neuropeptides orexin A and orexin B to orexin type 1 receptors (Ox1R) and orexin type 2 receptors (Ox2R). Suvorexant was approved in 2014 and lemborexant in 2019 in the United States and Japan; daridorexant was approved in the United States and the EU in 2022, becoming the first orexin receptor antagonist available in the EU.

Benzodiazepine and nonbenzodiazepine medications for insomnia are recommended for short-term use due to the potential risk of abuse or dependence, whereas dual orexin receptor antagonists have a relatively low risk and are therefore listed as schedule IV controlled substances in the United States.

In clinical trials, suvorexant (Kuriyama & Tabata, 2017), lemborexant (Moline et al., 2021; Murphy et al., 2017), and daridorexant (Mignot et al., 2022) have demonstrated efficacy in adult patients with insomnia in both objective and subjective assessments.

Mechanism of Action

The orexinergic nucleus contains about 70,000 neurons in humans and is localized in the lateral hypothalamus. It facilitates and stabilizes the daytime neural activity of the arousal-promoting nucleus groups, and is involved in maintaining wakefulness, with Ox2R in particular being more closely related to physiological sleep than Ox1R (Jacobson et al., 2022). Orexin receptors are exclusively Ox1R in the locus coeruleus (noradrenergic), whereas only Ox2R is expressed in the tuberomammillary nucleus (histaminergic). Both receptors are expressed in the dorsal raphe nucleus (serotonergic), the initiating nucleus of cholinergic nerves localized in the lateral dorsal tegmental nucleus, and the pedunculopontine tegmental nucleus. Although lemborexant, suvorexant, and daridorexant inhibit both Ox1R and Ox2R, lemborexant has comparatively stronger inhibitory activity (binding affinity) for Ox2R, which is more closely related to physiological sleep regulation. Orexin neuropeptide is released predominantly during the day and is not active at night. Thus, orexin receptor blockade with orexin antagonists simply enforces normal physiology.

Pharmacokinetics

The elimination half-life is 8–19 h for suvorexant, about 50 h for lemborexant, and 6–10 h for daridorexant. The PK profile of lemborexant is biphasic, and the elimination phase is gradual, so the elimination half-life is calculated to be longer (Beuckmann et al., 2017).

The receptor occupancy for the hypnotic effects of dual orexin receptor antagonists is estimated to be 65–80%, which is higher than that of benzodiazepine and nonbenzodiazepine medications for insomnia (26–29%); thus, comparisons of elimination half-lives between different classes of sleep medications are not clinically meaningful (Gotter et al., 2013). Dual orexin receptor antagonists are estimated to have a duration of action of 6–8 h and belong to the short-acting class.

Prescribing Guidelines

In two practice guidelines by the American Academy of Sleep Medicine (Sateia et al., 2017) and the American College of Physicians (Qaseem et al., 2016), suvorexant is recommended for chronic insomnia in adults. Lemborexant and daridorexant are not mentioned in the guidelines because they were launched after the guidelines were issued, but a recent systematic review and network meta-analysis found that both agents are useful in the acute treatment of insomnia, with lemborexant showing a good profile in long-term treatment (De Crescenzo et al., 2022). Daridorexant is the first insomnia medication to have been evaluated as showing effectiveness in improving not only nighttime symptoms but also daytime functioning (Mignot et al., 2022).

Clinical Considerations

Dual orexin receptor antagonists are mainly metabolized by CYP3A4, and concomitant use with drugs that inhibit CYP3A4 requires caution. The effect of sleep onset may be weakened by dietary influences. Since dual orexin receptor antagonists also inhibit P-glycoprotein, caution should be exercised when they are used in patients who are currently using digoxin.

Orexin receptor antagonists may affect the reward system and cause drug preference reactions. Dual orexin receptor antagonists have a relatively low risk of drug preference compared to $GABA_A$ receptor agonists, but they may produce significant drug preference at high doses compared to placebo. However, they are listed as a DEA Schedule IV drug with relatively low risk of abuse, in the United States.

Histamine H_1 Antagonists

The most commonly used agents in this class are diphenhydramine and doxylamine, available over the counter, often in combination with analgesics, such as acetaminophen. The evidence base related to these medications is limited. There are no published placebo controlled trials of doxylamine and several relatively small studies indicate that diphenhydramine appears to improve sleep maintenance more than sleep onset (Rickels et al., 1983; Meuleman et al., 1987; Kudo & Kurihara, 1990; Morin et al., 2005; Glass et al., 2008). Evidence for the efficacy of low-dose doxepin, a medication that is approved for the treatment of depression at higher doses, is more substantial.

Mechanism of Action

Antihistamines were generally developed to treat allergies, but also have sleep-enhancing effects due to antagonism of histamine H_1 receptors. Low-dose doxepin has demonstrated effects in the last third of the night, including the final hour of an 8-h sleep period, but has minimal adverse effects related to daytime drowsiness, even shortly after waking (Roth et al., 2007; Scharf et al., 2008; Krystal et al., 2010, 2011, 2013). Although the reasons for this unique profile have not been definitively identified, there is reason to believe that doxepin in 3–6 mg doses blocks the wake-promoting effects of histamine which are maximal at the end of the night; however, after waking, either additional histamine release overcomes the H_1 blockade or the activation of other parallel wake-promoting systems such as hypocretin/orexin and norepinephrine support wakefulness despite H_1 blockade (Krystal et al., 2013). Doxylamine and diphenhydramine are relatively nonselective in their pharmacologic effects, with significant cholinergic receptor antagonism along with H_1 antagonism. Low-dose doxepin, in contrast, is a highly selective H_1 antagonist (Krystal et al., 2013). Doxepin is also FDA approved for the treatment of depression at dosages ranging from 75 to 300 mg, where it has broad pharmacologic effects including inhibition of serotonin reuptake and anticholinergic effects.

Prescribing Guidelines

Doxepin is effective for the treatment of insomnia at dosages of 3–6 mg. Like diphenhydramine, low-dose doxepin has been shown to have stronger benefits for sleep maintenance than for sleep onset (Krystal, 2009; Krystal et al., 2013). Notably, it is the only medication currently available with demonstrated benefits in the last third of the night, including the final hour of an 8-h sleep period, with minimal adverse effects in the morning after waking (Roth et al., 2007; Scharf et al., 2008; Krystal et al., 2010, 2011, 2013).

Clinical Considerations

Nonselective antihistamines have adverse effects associated with cholinergic antagonism, including blurred vision, dry mouth, constipation, urinary retention, and delirium. In addition, dizziness, weight gain, and daytime sedation have been reported. Owing to these effects, H_1 antagonists (other than doxepin, which has not shown these adverse effects) should be avoided in patients with dementia, urinary retention, and narrow-angle glaucoma who may be at increased risk of complications due to cholinergic antagonism. Diphenhydramine has been associated with relatively infrequent agitation and insomnia. Although strong evidence is lacking, case reports have raised a concern that doxylamine may, in rare cases, be associated with coma and rhabdomyolysis (Koppel et al., 1987). Doxepin in the 3–6 mg range has not been associated with anticholinergic side effects, most likely due to its H_1 selectivity. In addition, weight gain is not associated with doxepin, although it is a common side effect of less selective antihistamines (Krystal et al., 2013). The abuse potential of the antihistamines is negligible, making them attractive for patients with a history of substance use disorders. Finally, the beneficial effects on allergies make them particularly well suited to patients with insomnia and co-occurring nasal congestion and/or allergies.

Prescription Medication Off-Label Use

A wide assortment of sedating medications without formal indications for the treatment of insomnia are prescribed with the primary intention of improving sleep onset, maintenance, and quality (Roehrs & Roth, 2004). These are mostly medications for depression but also include medications for psychosis and other psychotropic agents. A survey of prescribing practices during 2002 showed that three of the top four medications prescribed in association with insomnia were sedating medications for depression (Walsh, 2004). Historically, a decline in benzodiazepine prescriptions for insomnia during the 1980s and 1990s coincided with a significant rise in the use of trazodone (Walsh & Schweitzer, 1999).

Several factors have likely contributed to the widespread off-label use of these medications. Although these sedating agents may be selected for psychiatric patients with prominent insomnia symptoms, often they are prescribed for individuals with no psychiatric comorbidity. Since insomnia is frequently associated with depressive disorders, it may seem that medications for depression would always be beneficial. The appeal of these medications may be enhanced by the lack of abuse liability and an absence of restrictions on their duration of use. Until recently, all FDA-approved insomnia treatment medications were DEA Schedule IV controlled substances and were indicated for short-term use. In some cases, cost may also have been an issue owing to the availability of generic formulations of older medications for depression.

Generally, there is limited insomnia efficacy evidence in non-psychiatric populations, and there are significant safety concerns with all of these medications. Many clinical studies are short term, have few subjects, are not placebo controlled, have only subjective outcome measures, and do not incorporate dose-response analyses. Moreover, risk-benefit considerations have not been well delineated for insomnia populations without psychiatric comorbidity. Common to all of these medications is the risk of undesired daytime sedation and related impairment due to the relatively long elimination half-lives, generally in the range of 15–30 h. Although prominent withdrawal effects are unlikely, rebound insomnia may occur with sudden discontinuation.

Sedating Medications for Depression

Medications for depression with sedating properties are among the most widely prescribed drugs for insomnia patients with and without psychiatric comorbidity (Walsh, 2004). In mood and anxiety disorder patients, they may be employed as a single agent to promote both improved sleep and other psychiatric symptomatology. Often this is associated with residual daytime sedation. More commonly, these agents are prescribed at a relatively low dose, sometimes in combination with other psychiatric medications. The medications for depression most commonly recommended for the treatment of insomnia have been the monoamine enhancers (tricyclic antidepressants) amitriptyline and doxepin, trazodone, and mirtazapine. Nefazodone previously had been used to enhance sleep but currently is prescribed infrequently owing to an increased risk for liver failure. The majority of sleep-related studies with sedating medications for depression have been with populations of depressed patients, although some have been done with insomnia subjects and healthy individuals with no sleep complaints.

Most available medications for depression have varying degrees of serotonin (5-HT) and norepinephrine (NE) presynaptic reuptake inhibition and assorted postsynaptic receptor interactions. Although these are thought to be fundamental to the antidepressant effects, sedation likely results from the postsynaptic antagonist activity of medications on 5-HT, NE, and histamine receptors.

Monoamine Enhancers (Tricyclic Antidepressants)

Monoamine enhancers (tricyclic antidepressants) include both tertiary and secondary amines. The tertiary amine tricyclics, which tend to be more sedating, include amitriptyline, doxepin, imipramine, and trimipramine. Doxepin has been shown to decrease sleep-onset latency, increase total sleep time, decrease the amount of rapid eye movement (REM) sleep, and delay the onset of the first REM episode (Roth et al., 1982; Hajak et al., 2001). The monoamine enhancers (tricyclic antidepressants) may exacerbate periodic limb movements in vulnerable individuals and are associated with a wide range of adverse effects. Potential toxicity includes cardiac arrhythmias and seizures. Anticholinergic effects may include blurred vision, dry mouth, constipation, urinary hesitancy and retention, confusion, and delirium. Antihistaminic effects may include sedation, increased appetite, and weight gain. Hypotension may result from postsynaptic α_1 adrenergic receptor antagonist activity. Amitriptyline is strongly anticholinergic, and doxepin is strongly antihistaminic. As noted earlier, recent studies have confirmed the sleep-enhancing effects of ultra-low-dose doxepin at strengths less than 10 mg.

Trazodone

Although trazodone is widely prescribed to treat insomnia, there is minimal clinical trial evidence to support its efficacy (Montgomery et al., 1983;

Saletu-Zyhlarz et al., 2002; Mendelson, 2005). The doses often range from 300 to 600 mg; however, it generally is prescribed for insomnia in the range of 25–150 mg. Studies have reported improved sleep latency, total sleep time, and sleep efficiency. Trazodone has postsynaptic antagonist activity at 5-HT_{1A}, 5-HT_{1C}, 5-HT_2, histamine H_1, and α_1 adrenergic receptors, and also weak presynaptic serotonin reuptake inhibition. It has an elimination half-life of approximately 5–9 h and is metabolized to a serotonergic metabolite, m-chlorophenylpiperazine (mCPP). Residual sedation is a common complaint. Orthostatic hypotension may occur due to the α_1 antagonism. Priapism is a rare adverse event. Case reports suggest that trazodone may contribute to ventricular arrhythmias and the development of serotonin syndrome (Janowsky et al., 1983; Rao, 1997).

Mirtazapine

Mirtazapine is commonly associated with sedation, which may be beneficial for nighttime sleep, but undesired during waking hours. The elimination half-life is approximately 20–40 h. It has presynaptic antagonist activity at α_2 autoreceptors and postsynaptic antagonistic effects at 5-HT_1, 5-HT_2, histamine H_1, and α_1 adrenergic receptors. The sedating property likely results from a combination of these actions. Clinical studies have demonstrated improvements in sleep-onset and maintenance parameters in normal subjects and patients with major depression (Asian et al., 2002; Winokur et al., 2003). Mirtazapine also has been shown to increase slow-wave sleep (Schmid et al., 2006). It has been suggested that mirtazapine may be more sedating at lower doses due to the balance of serotonergic and histaminic effects relative to noradrenergic activity. Adverse effects may include increased appetite and weight gain.

Sedating Medications for Psychosis

Patients with psychiatric disorders often report improved nighttime sleep when taking antipsychotic medications. Among the dopamine serotonin blockers (atypical antipsychotics), clozapine, olanzapine, and quetiapine are especially sedating. Although the clinical studies are limited in size and number, they have examined the subjective and polysomnographic sleep effects of sedating medications for psychosis in populations of psychiatric patients, insomnia subjects, and healthy subjects without sleep complaints. Studies have reported improvements in sleep latency, total sleep time, and sleep efficiency (Salin-Pascual et al., 1999; Cohrs et al., 2004). There have been variable effects on specific sleep stages. Beneficial sleep effects may result from combinations of antagonist activity at histaminic and multiple subtypes of 5-HT receptors. Both dopamine blockers (typical) and dopamine serotonin blockers (atypical) medications for psychosis are associated with a wide spectrum of potential adverse effects. The dopamine serotonin blockers (atypical antipsychotics) increase the risk of weight gain and impaired glucose utilization. Undesired sleep effects may include excessive sedation, sleep fragmentation, and increased periodic limb movements during sleep. The medications for psychosis are employed optimally as sleep-promoting agents within the confines of the approved psychiatric indications.

Miscellaneous Prescription Medications

Aside from approved insomnia treatment medications, medications for depression, medications for psychosis, and antihistamines, various other psychotropic agents occasionally are prescribed for the treatment of insomnia. Among these are barbiturates, chloral hydrate, cyclobenzaprine, γ-hydroxybutyrate (GHB), and assorted anticonvulsants. All may have sedating effects but are associated with significant safety concerns, including toxicity and potential lethality.

Chloral Hydrate

Currently, chloral hydrate most commonly is used for its acute sedating effects in preparation for medical tests and procedures. It rarely is recommended for the treatment of insomnia

owing to toxicity and the development of tolerance and dependence. Chloral hydrate is available in oral capsule, syrup, and suppository formulations. It is very rapidly metabolized to trichloroethanol, which modulates the $GABA_A$ receptor complex. The elimination half-life is approximately 5–10 h. Chloral hydrate has been shown to improve sleep onset and continuity, but tolerance to the beneficial effects evolves with repeated doses. Discontinuation may result in withdrawal symptoms. Adverse effects may include unpleasant taste, gastrointestinal distress, respiratory depression, hypotension, and ataxia. It can be lethal in overdose.

γ-Hydroxybutyrate

GHB is a sedating, abused substance, which simultaneously is classed by the DEA as Schedule I and III. As sodium oxybate (Xyrem), GHB is approved by the FDA for the treatment of the cataplexy and excessive daytime sleepiness in narcolepsy patients and is distributed through a single centralized pharmacy. Clinical trials have shown that it enhances nighttime slow-wave sleep. Typically, it is given in a liquid form in two nighttime doses. The approximate elimination half-life is less than 1 h. The pharmacodynamic actions have not been fully elucidated; however, there are effects on dopamine activity and the $GABA_B$ receptor. The improvements to nighttime sleep have led some practitioners to prescribe GHB on an off-label basis for insomnia patients who have not benefited from standard treatments. Potential adverse effects include nocturnal enuresis, confusion, mental depression, respiratory depression, coma, and death. GHB should not be combined with alcohol or other CNS depressants.

Anticonvulsants

Several anticonvulsant medications have been observed to have sedating properties that may be beneficial for nighttime sleep but represent an undesired effect during the daytime. Clinical trials assessing insomnia treatment efficacy have been performed with selected anticonvulsants, and some are occasionally prescribed with the primary intention of improving sleep. Two examples are gabapentin (Neurontin) and tiagabine (Gabitril).

Gabapentin is indicated for the treatment of epilepsy and postherpetic neuralgia. The mechanism of action is unknown. The observed sedating effects have led to clinical use with insomnia patients and controlled clinical trials (Bateson, 2006). It has been reported to increase slow-wave sleep in normal subjects (Foldvary-Schaefer et al., 2002). Adverse effects include dizziness, somnolence, fatigue, and ataxia.

Tiagabine presynaptically inhibits GABA reuptake through binding at the GAT-1 transporter. The result is increased GABA availability at the synapse. It is indicated as adjunct therapy in the treatment of partial seizures. Clinical studies have demonstrated increases in slow-wave sleep (Walsh et al., 2006). Adverse effects include dizziness, somnolence, nervousness, and tremor. An FDA warning noted that tiagabine has been associated with new-onset seizures and status epilepticus in individuals without epilepsy.

Antihypertensives

Prazosin is an antihypertensive agent that was found to have benefits for sleep, primarily in patients with posttraumatic stress disorder (PTSD) who experience sleep disturbance due to chronic nightmares and hyperarousal (Taylor & Raskind, 2002). At present, prazosin is the only sleep-enhancing agent shown to reduce sleep impairment due to nightmares. Patients have reported significant improvements in nightmare frequency, insomnia severity, and PTSD symptom severity (measured by self-report), and studies using home-PSG have reported improvements in total sleep time, REM sleep time, and REM duration (Raskind et al., 2003, 2007; Taylor et al., 2008; Germain et al., 2012). However, a more recent 26-week clinical trial conducted in 304 veterans with PTSD plus frequent nightmares did not support the efficacy of Prazosin (Raskind et al., 2018). Further validation is needed.

A pharmacologic mechanism to explain the benefits on nightmares has not yet been established (Krystal & Davidson, 2007), but prazosin is known to be an α_1 adrenoreceptor antagonist. The recommended starting dose is 1 mg, which is slowly titrated up to an effective dose, generally found to be in the 2–6 mg range (Taylor et al., 2008). Studies have found 15–20 mg to be an upper limit for prazosin. Although studies to date have not been large enough to define the rates of adverse events relative to placebo definitively, the most common symptoms reported include orthostatic hypotension (most likely to occur the morning after a dose increase), dizziness, nasal congestion, initial insomnia, dry mouth, sweating, depression, and lower extremity edema (Raskind et al., 2003, 2007; Taylor et al., 2008; Germain et al., 2012).

OTC Medications

OTC antihistamine sleep aids are regulated substances that are available without a prescription. Most of these first-generation, centrally acting antihistamine products contain diphenhydramine hydrochloride; however, some have diphenhydramine citrate or doxylamine succinate as the active ingredients. They are marketed as single agents or in combination with acetaminophen or aspirin and are approved for occasional sleeplessness for people having difficulty falling asleep. None are specifically indicated for the treatment of insomnia. There are very limited efficacy data on the treatment of sleep disturbances (Kudo & Kurihara, 1990). In the United States, OTC diphenhydramine is available in doses up to 50 mg and doxylamine up to 25 mg.

Sleep may be promoted by a sedating action resulting from postsynaptic H_1 receptor blockade that inhibits the stimulating effect of histamine in the mammillary nucleus of the posterior hypothalamus. Normally, these histaminergic neurons promote wakefulness through extensive projections to the cerebral cortex and brainstem (Tashiro & Yanai, 2007). The OTC sleep aids are rapidly absorbed and typically reach a peak blood level within 1 h. The relatively long half-lives of these compounds (approximately 8 h) may lead to residual next-morning sedation and a complaint of grogginess. Tolerance to the sedating effects may develop (Richardson et al., 2002). Adverse effects may include paradoxical stimulation and restlessness and thickening of bronchial secretions. Postsynaptic muscarinic receptor blockade may produce anticholinergic effects, such as blurred vision, urinary hesitance and retention, and confusion and delirium. Elderly patients and those concomitantly on medications with anticholinergic effects are especially vulnerable for these adverse reactions. Additionally, there may be additive effects with other CNS depressants.

Complementary and Alternative Substances

Numerous unregulated compounds are marketed as sleep aids. Limited efficacy evidence supports the claims, and safety data generally are lacking. Common ingredients in these homeopathic and herbal preparation sleep aids are valerian, hops, kava, passion flower, skullcap, and lavender. Generally, these compounds include large numbers of individual chemicals. Therefore, specific active ingredients and mechanisms of action are difficult to determine. Valerian preparations have been the best studied in this category. Improved sleep parameters have been reported in some studies; however, the overall findings remain inconclusive (Bent et al., 2006).

Melatonin preparations have been used extensively in the attempt to improve sleep. Endogenous melatonin has a central role in the regulation of the sleep-wake cycle. Since endogenous melatonin normally is elevated during the nighttime when humans sleep, theoretically supplementing the body's melatonin could enhance sleep onset, duration, and quality. Melatonin is synthesized primarily in the pineal gland. Exogenous melatonin is rapidly absorbed and has an elimination half-life of less than 1 h. Melatonin has been studied across a wide dose range. Meta-analyses of clinical trials examining the efficacy of melatonin for various insomnia populations have failed to provide consistent positive findings. However, it appears to be safe for short-term use. Potential long-term risks are unknown. Although bedtime

use to enhance sleep onset may have limited value, melatonin may be beneficial in the treatment of circadian rhythm disorders, particularly the delayed sleep phase syndrome.

Insomnia in Psychiatric Patients

In recent years, the relationship between psychiatric disorders and insomnia has come to be appreciated as circular and synergistic. Psychiatric illnesses, particularly anxiety and mood disorders, have long been recognized as frequent causes of insomnia. The *Diagnostic and Statistical Manual of Mental Disorders*, fifth edition (DSM-5) diagnostic criteria formalize this association in some cases (American Psychiatric Association, 2013). Clinical experience suggests that almost all patients with mood and anxiety disorders have sleep disturbances either chronically or surrounding exacerbations of their psychiatric illnesses. However, it has become evident that insomnia also increases the risk of future relapse or the development of new onset mood, anxiety, and substance abuse disorders. This association can promote a downward spiral of symptom severity and quality of life for patients that may further complicate treatment efforts. On the other hand, the close relationship between insomnia, depression, and anxiety symptoms can be viewed as an opportunity for targeted approaches that may provide significant benefits for patients.

In addition to major depression and dysthymic disorder, insomnia often occurs with bipolar disorder during depressive and manic episodes. Although manic patients may report a decreased need for sleep, many others feel distressed due to an inability to sleep.

Various sociodemographic and disease factors have been suggested as clinical predictors of bipolar relapse, including age, gender, childhood maltreatment, number of previous mood episodes (and/or rapid cycling), residual depressive symptoms, comorbid physical (e.g., thyroid dysfunction) or psychiatric (e.g., anxiety) disorders, nicotine, alcohol, obesity, number of psychotropic number of medications, medication adherence, and other sociodemographic and disease factors,

K. Mishima

but the contribution of each of these factors is small. In this context, sleep-wake cycle and circadian rhythm disorders have shown promise as clinical predictors. Indeed, a number of research reports suggest that sleep and circadian rhythm disturbances during remission, and sleep symptoms such as insomnia and hypersomnia, are associated with bipolar relapse (Meyrel et al., 2022).

Insomnia is especially problematic for anxiety disorder patients with panic disorder, PTSD, generalized anxiety disorder, and social phobia. Patients with panic disorder often will experience distressing panic episodes that awaken them from sleep. They may evolve considerable anticipatory anxiety about going to sleep, which ultimately may lead to sleep insufficiency and increased anxiety. PTSD patients often experience poor sleep quality and vivid nightmares (Green, 2003). Patients with generalized anxiety disorder often experience chronic anxiety that affects them throughout the night, with resulting difficulty falling asleep and repeated awakenings. Social phobia patients report significantly worse sleep quality and difficulty falling asleep in comparison with healthy control subjects (Stein et al., 1993).

Treating the insomnia symptoms of patients with concurrent psychiatric disorders is a two-pronged approach. Specific therapeutic interventions addressing the primary psychiatric condition may include psychotherapeutic, behavioral, and pharmacologic strategies. Optimizing the treatment of the underlying disorder should ultimately improve sleep. However, many commonly prescribed medications for depression, such as the selective serotonin reuptake inhibitors (SSRIs), rarely improve the sleep symptoms rapidly, and some patients develop insomnia as a side effect of the medication. The use of sedating medications for depression to treat depression and insomnia is limited by daytime sedation and other adverse effects.

General approaches to insomnia are those applicable to a broad range of patients and include sleep hygiene and behavioral interventions, cognitive behavioral therapy, and insomnia treatment medications. These therapies may be employed concurrently, with specific treatment strategies targeting the psychiatric disorders. There are several advantages to this two-pronged approach. First, there is

greater choice among medications for the psychiatric symptoms, rather than restricting the options to sedating agents. There also is a greater opportunity for flexibility in the dosage, timing, and duration of use of medications targeting different symptoms. Second, medications for insomnia may provide immediate relief and, subsequently, decreased distress and improved quality of life. Medication for insomnia can offset the stimulating effect of some medications for depression. Finally, these general insomnia treatment approaches can directly address the perpetuating factors that reinforce chronic insomnia.

In clinical practice, medications for insomnia often are prescribed concomitantly with medications for depression for patients with mood and anxiety disorders. Pharmacokinetic and pharmacodynamic studies of fluoxetine and sertraline combined with zolpidem have been performed in healthy, nondepressed women (Allard et al., 1998, 1999). No clinically significant interactions were identified. In another clinical study, patients prescribed an SSRI concurrently with zolpidem or a placebo were assessed. The study population included individuals successfully treated for depression with an SSRI, but who were complaining of persistent insomnia. The patients treated with medications for insomnia reported significantly improved sleep and daytime functioning (Asnis et al., 1999). Eszopiclone was studied in a placebo-controlled trial in depressed patients treated with fluoxetine. The patients in the eszopiclone-fluoxetine group demonstrated rapid improvement in sleep parameters and also a greater magnitude of the antidepressant effect (Fava et al., 2006).

Therapeutic intervention for sleep disorders comorbid with psychiatric disorders is generally delayed. Traditionally, insomnia caused by some underlying disease, such as a psychiatric disorder, or a physical disease that causes pain or itching, has been referred to as secondary insomnia. However, it has become clear that in many cases insomnia can persist independently, even after the original causative illness has resolved (McClintock et al., 2011). As a result, the 2005 NIH Consensus Statement proposed the concept of comorbid insomnia as an alternative to secondary insomnia

(NIH-Consensus-Statement, 2005). The concept of secondary insomnia, in which insomnia caused by mental illness resolves with psychiatric symptoms, may delay treatment, and it is appropriate to treat comorbid insomnia from the beginning. Based on this concept of comorbid insomnia, the International Classification of Sleep Disorders (ICSD-3), revised in 2014, eliminated the concept of secondary insomnia and changed classification from being based on the cause to the duration of illness (chronic insomnia disorder and short-term insomnia disorder) (ICSD-3, 7593).

Although it is important to recognize and manage the insomnia symptoms that may result from psychiatric illnesses, it also is critical to evaluate psychiatric patients for other potential causes of their sleep disturbances. These may include the stimulating effects of psychotropic and other medications, medical disorders, underlying primary sleep disorders, circadian rhythm disorders, irregular schedules, and maladaptive habits and routines. Sleep apnea patients can present solely with insomnia complaints. Restless legs syndrome and periodic limb movements, which can be exacerbated by most medications for depression, can result in difficulty falling asleep and repeated awakenings. Agoraphobic or socially withdrawn patients may spend excessive time at home, sleep at irregular times, and avoid the photoperiod that normally reinforces the sleep-wake cycle.

Conclusion

A wide assortment of substances and medications have been recommended for individuals suffering with insomnia. Compared with older insomnia treatment medications, the current-generation medications maintain efficacy but have significantly improved safety profiles. The diversity of approved agents allows the selection of the medication to be customized for individual patients, depending on their clinical histories. Recently, there have been significant advances in both pharmacodynamic and pharmacokinetic approaches in promoting improved sleep. The most recently approved insomnia treatment medications no longer have an implied limitation on their duration of

use, and the indications specifically note efficacy for sleep onset and sleep maintenance. New pharmacologic approaches continue to be developed.

References

Allard, S., Sainati, S., Roth-Schechter, B., et al. (1998). Minimal interaction between fluoxetine and multiple-dose zolpidem in healthy women. *Drug Metabolism and Disposition, 26*, 617–622.

Allard, S., Sainati, S. M., & Roth-Schechter, B. F. (1999). Coadministration of short-term zolpidem with sertraline in healthy women. *Journal of Clinical Pharmacology, 39*, 184–191.

American Psychiatric Association. (2013). *Diagnostic and statistical manual of mental disorders* (DSM-5) (5th ed.). American Psychiatric Publishing.

Ancoli-Israel, S., Richardson, G. S., Mangano, R. M., et al. (2005). Long-term use of sedative hypnotics in older patients with insomnia. *Sleep Medicine, 6*, 107–113.

Asian, S., Isik, E., & Cosar, B. (2002). The effects of mirtazapine on sleep: A placebo controlled, double-blind study in young healthy volunteers. *Sleep, 25*, 677–679.

Asnis, G. M., Chakraburtty, A., DuBoff, E. A., et al. (1999). Zolpidem for persistent insomnia in SSRI-treated depressed patients. *Journal of Clinical Psychiatry, 60*, 668–676.

Baglioni, C., Battagliese, G., Feige, B., et al. (2011). Insomnia as a predictor of depression: A meta-analytic evaluation of longitudinal epidemiological studies. *Journal of Affective Disorders, 135*(1–3), 10–19.

Bateson, A. N. (2006). Further potential of the GABA receptor in the treatment of insomnia. *Sleep Medicine, 7*(Suppl 1), S3–S9.

Bent, S., Padula, A., Moore, D., et al. (2006). Valerian for sleep: A systematic review and meta-analysis. *American Journal of Medicine, 119*, 1005–1012.

Beuckmann, C. T., Suzuki, M., Ueno, T., et al. (2017). In vitro and in silico characterization of lemborexant (E2006), a novel dual orexin receptor antagonist. *The Journal of Pharmacology and Experimental Therapeutics, 362*(2), 287–295.

Cohrs, S., Rodenbeck, A., Guan, Z., et al. (2004). Sleep-promoting properties of quetiapine in healthy subjects. *Psychopharmacology, 174*, 421–429.

Connor, K., Budd, K., Snavely, D., et al. (2012). Efficacy and safety of suvorexant, an orexin receptor antagonist, in patients with primary insomnia: A 3-month phase 3 trial (trial #1). *Journal of Sleep Research, 21*(SI), 97.

De Crescenzo, F., D'Alò, G. L., Ostinelli, E. G., et al. (2022). Comparative effects of pharmacological interventions for the acute and long-term management of insomnia disorder in adults: A systematic review and network meta-analysis. *Lancet, 400*(10347), 170–184.

DeMicco, M., Wang-Weigand, S., & Zhang, J. (2006). Long-term therapeutic effects of ramelteon treatment in adults with chronic insomnia: A 1 year study. *Sleep, 29*(Abstract Suppl), A234.

Erman, M., Krystal, A., Zammit, G., et al. (2006a). Zolpidem extended-release 12.5 mg, taken for 24 weeks "as needed" up to 7 nights/week, improves subjective measures of therapeutic global impression, sleep onset, and sleep maintenance in patients with chronic insomnia. *International Journal of Neuropsychopharmacology, 9*(Suppl 1), S256.

Erman, M., Seiden, D., Zammit, G., et al. (2006b). An efficacy, safety, and dose-response study of ramelteon in patients with chronic primary insomnia. *Sleep Medicine, 7*, 17–24.

Fava, M., McCall, W. V., Krystal, A., et al. (2006). Eszopiclone co-administered with fluoxetine in patients with insomnia coexisting with major depressive disorder. *Biological Psychiatry, 59*, 1052–1060.

Foldvary-Schaefer, N., De Leon, S. I., Karafa, M., et al. (2002). Gabapentin increases slow-wave sleep in normal adults. *Epilepsia, 43*, 1493–1497.

Germain, A., Richardson, R., Moul, D. E., et al. (2012). Placebo-controlled comparison of prazosin and cognitive-behavioral treatments for sleep disturbances in US military veterans. *Journal of Psychosomatic Research, 72*(2), 89–96.

Glass, J. R., Sproule, B. A., Herrmann, N., et al. (2008). Effects of 2-week treatment with temazepam and diphenhydramine in elderly insomniacs: A randomized, placebo-controlled trial. *Journal of Clinical Psychopharmacology, 28*(2), 182–188.

Gotter, A. L., Winrow, C. J., Brunner, J., et al. (2013). The duration of sleep promoting efficacy by dual orexin receptor antagonists is dependent upon receptor occupancy threshold. *BMC Neuroscience, 14*(90), 90.

Green, B. (2003). Posttraumatic stress disorder: Symptom profiles in men and women. *Current Medical Research and Opinion, 19*, 200–204.

Hajak, G., Rodenbeck, A., Voderholzer, U., et al. (2001). Doxepin in the treatment of primary insomnia: A placebo-controlled, double-blind, polysomnographic study. *Journal of Clinical Psychiatry, 62*, 453–463.

Herring, W. J., Snyder, E., Budd, K., et al. (2012a). Orexin receptor antagonism for treatment of insomnia: A randomized clinical trial of suvorexant. *Neurology, 79*(23), 2265–2274.

Herring, W., Snyder, E., Paradis, E., et al. (2012b). Long term safety and efficacy of suvorexant in patients with primary insomnia. *Sleep, 35*, A217.

Hesse, L. M., von Moltke, L. L., & Greenblatt, D. J. (2003). Clinically important drug interactions with zopiclone, zolpidem and zaleplon. *CNS Drugs, 17*, 513–532.

ICSD-3. (2014). *International classification of sleep disorders* (3rd ed.). American Academy of Sleep Medicine.

Jacobson, L. H., Hoyer, D., & de Lecea, L. (2022). Hypocretins (orexins): The ultimate translational neuropeptides. *Journal of Internal Medicine, 291*(5), 533–556.

Janowsky, D., Curtis, G., Zisook, S., et al. (1983). Ventricular arrhythmias possibly aggravated by trazodone. *American Journal of Psychiatry, 140*, 796–797.

Johnson, M. W., Suess, P. E., & Griffiths, R. R. (2006). Ramelteon: A novel hypnotic lacking abuse liability and sedative adverse effects. *Archives of General Psychiatry, 63*, 1149–1157.

Karim, A., Tolbert, D., & Cao, C. (2006). Disposition kinetics and tolerance of escalating single doses of ramelteon, a high-affinity MT1 and MT2 melatonin receptor agonist indicated for treatment of insomnia. *Journal of Clinical Pharmacology, 46*, 140–148.

Kato, K., Hirai, K., Nishiyama, K., et al. (2005). Neurochemical properties of ramelteon (TAK-375), a selective MT1/MT2 receptor agonist. *Neuropharmacology, 48*, 301–310.

Koppel, C., Tenczer, J., & Ibe, K. (1987). Poisoning with over-the-counter doxylamine preparations: An evaluation of 109 cases. *Human Toxicology, 6*(5), 355–359.

Krystal, A. D. (2009). A compendium of placebo-controlled trials of the risks/benefits of pharmacological treatments for insomnia: The empirical basis for US clinical practice. *Sleep Medicine Reviews, 13*(4), 265–274.

Krystal, A. D., & Davidson, J. (2007). The use of prazosin for the treatment of trauma nightmares and sleep disturbance in combat veterans with posttraumatic stress disorder. *Biological Psychiatry, 61*(8), 925–927.

Krystal, A. D., Walsh, J. K., Laska, E., et al. (2003). Sustained efficacy of eszopiclone over 6 months of nightly treatment: Results of a randomized, doubleblind, placebo-controlled study in adults with chronic insomnia. *Sleep, 26*, 793–799.

Krystal, A. D., Erman, M., Zammit, G. K., et al. (2008). Long-term efficacy and safety of zolpidem extended-release 12.5 mg, administered 3 to 7 nights per week for 24 weeks, in patients with chronic primary insomnia: A 6-month, randomized, double-blind, placebo-controlled, parallel-group, multicenter study. *Sleep, 31*(1), 79–90.

Krystal, A. D., Durrence, H. H., Scharf, M., et al. (2010). Efficacy and safety of doxepin 1 mg and 3 mg in a 12-week sleep laboratory and outpatient trial of elderly subjects with chronic primary insomnia. *Sleep, 33*(11), 1553–1561.

Krystal, A. D., Lankford, A., Durrence, H. H., et al. (2011). Efficacy and safety of doxepin 3 and 6 mg in a 35-day sleep laboratory trial in adults with chronic primary insomnia. *Sleep, 34*(10), 1433–1442.

Krystal, A. D., Richelson, E., & Roth, T. (2013). Review of the histamine system and the clinical effects of H_1 antagonists: Basis for a new model for understanding the effects of insomnia medications. *Sleep Medicine Reviews, 17*(4), 263–272.

Kudo, Y., & Kurihara, M. (1990). Clinical evaluation of diphenhydramine hydrochloride for the treatment of insomnia in psychiatric patients: A double-blind study. *Journal of Clinical Pharmacology, 30*, 1041–1048.

Kuriyama, A., & Tabata, H. (2017). Suvorexant for the treatment of primary insomnia: A systematic review and meta-analysis. *Sleep Medicine Reviews, 35*, 1–7.

Leibenluft, E., Albert, P. S., Rosenthal, N. E., et al. (1996). Relationship between sleep and mood in patients with rapid-cycling bipolar disorder. *Psychiatry Research, 63*, 161–168.

McClintock, S. M., Husain, M. M., Wisniewski, S. R., et al. (2011). Residual symptoms in depressed outpatients who respond by 50% but do not remit to antidepressant medication. *Journal of Clinical Psychopharmacology, 31*(2), 180–186.

Mendelson, W. B. (2005). A review of the evidence for the efficacy and safety of trazodone in insomnia. *Journal of Clinical Psychiatry, 66*, 469–476.

Meuleman, J., Nelson, R. C., & Clark, R. L. (1987). Evaluation of temazepam and diphenhydramine as hypnotics in a nursing-home population. *Drug Intelligence and Clinical Pharmacy, 21*(9), 716–720.

Meyrel, M., Scott, J., & Etain, B. (2022). Chronotypes and circadian rest-activity rhythms in bipolar disorders: A meta-analysis of self- and observer rating scales. *Bipolar Disorders, 24*(3), 286–297.

Mignot, E. (2013). The perfect hypnotic? *Science, 340*(6128), 36–38.

Mignot, E., Mayleben, D., Fietze, I., et al. (2022). Safety and efficacy of daridorexant in patients with insomnia disorder: Results from two multicentre, randomised, double-blind, placebo-controlled, phase 3 trials. *Lancet Neurology, 21*(2), 125–139.

Miyamoto, M., Nishikawa, H., Doken, Y., et al. (2004). The sleep-promoting action of ramelteon (TAK-375) in freely moving cats. *Sleep, 27*, 1319–1325.

Mohler, H., Fritschy, J. M., & Rudolph, U. (2002). A new benzodiazepine pharmacology. *Journal of Pharmacology and Experimental Therapeutics, 300*, 2–8.

Moline, M., Zammit, G., Cheng, J. Y., et al. (2021). Comparison of the effect of lemborexant with placebo and zolpidem tartrate extended release on sleep architecture in older adults with insomnia disorder. *Journal of Clinical Sleep Medicine, 17*(6), 1167–1174.

Montgomery, I., Oswald, I., Morgan, K., et al. (1983). Trazodone enhances sleep in subjective quality but not in objective duration. *British Journal of Clinical Pharmacology, 16*, 139–144.

Morin, C., Koetter, U., Bastien, C., et al. (2005). Valerian–hops combination and diphenhydramine for treating insomnia: A randomized placebo-controlled clinical trial. *Sleep, 28*(11), 1465–1471.

Murphy, P., Moline, M., Mayleben, D., et al. (2017). Lemborexant, a dual orexin receptor antagonist (DORA) for the treatment of insomnia disorder: Results from a Bayesian, adaptive, randomized, double-blind, placebo-controlled study. *Journal of Clinical Sleep Medicine, 13*(11), 1289–1299.

National Institutes of Health. (2005). National Institutes of Health State of the Science conference statement on manifestations and management of chronic insomnia in adults, June 13–15, 2005. *Sleep, 28*, 1049–1057.

Nierenberg, A. A., Keefe, B. R., Leslie, V. C., et al. (1999). Residual symptoms in depressed patients who respond acutely to fluoxetine. *Journal of Clinical Psychiatry, 60*, 221–225.

NIH Consensus Statement. (2005). NIH state-of-the-science conference statement on manifestations and management of chronic insomnia in adults. *NIH Consensus and State-of-the-Science Statements, 22*(2), 1–30.

Penninkilampi, R., & Eslick, G. D. (2018). A systematic review and meta-analysis of the risk of dementia associated with benzodiazepine use, after controlling for protopathic bias. *CNS Drugs, 32*(6), 485–497.

Perlis, M. L., Giles, D. E., Buysse, D. J., et al. (1997). Which depressive symptoms are related to which sleep electroencephalographic variables? *Biological Psychiatry, 42*, 904–913.

Perlis, M. L., McCall, W. V., Krystal, A. D., et al. (2004). Long-term, non-nightly administration of zolpidem in the treatment of patients with primary insomnia. *Journal of Clinical Psychiatry, 65*, 1128–1137.

Qaseem, A., Kansagara, D., Forciea, M. A., et al. (2016). Management of chronic insomnia disorder in adults: A clinical practice guideline from the American College of Physicians. *Annals of Internal Medicine, 165*(2), 125–133.

Rao, R. (1997). Serotonin syndrome associated with trazodone. *International Journal of Geriatric Psychiatry, 12*, 129–130.

Raskind, M. A., Peskind, E. R., Kanter, E. D., et al. (2003). Reduction of nightmares and other PTSD symptoms in combat veterans by prazosin: A placebo-controlled study. *American Journal of Psychiatry, 160*(2), 371–373.

Raskind, M. A., Peskind, E. R., Hoff, D. J., et al. (2007). A parallel group placebo controlled study of prazosin for trauma nightmares and sleep disturbance in combat veterans with posttraumatic stress disorder. *Biological Psychiatry, 61*(8), 928–934.

Raskind, M. A., Peskind, E. R., Chow, B., et al. (2018). Trial of prazosin for post-traumatic stress disorder in military veterans. *The New England Journal of Medicine, 378*(6), 507–517.

Richardson, G. S. (2005). The human circadian system in normal and disordered sleep. *Journal of Clinical Psychiatry, 66*, 3–9; quiz, 42–43.

Richardson, G. S., Roehrs, T. A., Rosenthal, L., et al. (2002). Tolerance to daytime sedative effects of H_1 antihistamines. *Journal of Clinical Psychopharmacology, 22*, 511–515.

Rickels, K., Morris, R. J., Newman, H., et al. (1983). Diphenhydramine in insomniac family practice patients: A double-blind study. *Journal of Clinical Pharmacology, 23*(5–6), 234–242.

Roehrs, T., & Roth, T. (2004). 'Hypnotic' prescription patterns in a large managed-care population. *Sleep Medicine, 5*, 463–466.

Roehrs, T. A., & Roth, T. (2006). Safety of insomnia pharmacotherapy. *Sleep Medicine Clinics, 1*, 399–407.

Roth, T., Zorick, F., Wittig, R., et al. (1982). The effects of doxepin HCl on sleep and depression. *Journal of Clinical Psychiatry, 43*, 366–368.

Roth, T., Walsh, J. K., Krystal, A., et al. (2005). An evaluation of the efficacy and safety of eszopiclone over 12 months in patients with chronic primary insomnia. *Sleep Medicine, 6*, 487–495.

Roth, T., Rogowski, R., Hull, S., et al. (2007). Efficacy and safety of doxepin 1 mg, 3 mg, and 6 mg in adults with primary insomnia. *Sleep, 30*(11), 1555–1561.

Roth, T., Krystal, A., Steinberg, F. J., et al. (2013). Novel sublingual low-dose zolpidem tablet reduces latency to sleep onset following spontaneous middle-of-the-night awakening in insomnia in a randomized, doubleblind, placebo-controlled, outpatient study. *Sleep, 36*(2), 189–196.

Saletu-Zyhlarz, G. M., Abu-Bakr, M. H., Anderer, P., et al. (2002). Insomnia in depression: Differences in objective and subjective sleep and awakening quality to normal controls and acute effects of trazodone. *Progress in Neuro-Psychopharmacology and Biological Psychiatry, 26*, 249–260.

Salin-Pascual, R. J., Herrera-Estrella, M., Galicia-Polo, L., et al. (1999). Olanzapine acute administration in schizophrenic patients increases delta sleep and sleep efficiency. *Biological Psychiatry, 46*, 141–143.

Saper, C. B., Chou, T. C., & Scammell, T. E. (2001). The sleep switch: Hypothalamic control of sleep and wakefulness. *Trends in Neuroscience, 24*, 726–731.

Sateia, M. J., Buysse, D. J., Krystal, A. D., et al. (2017). Clinical practice guideline for the pharmacologic treatment of chronic insomnia in adults: An American Academy of Sleep Medicine Clinical Practice Guideline. *Journal of Clinical Sleep Medicine, 13*(2), 307–349.

Scharf, M., Rogowski, R., Hull, S., et al. (2008). Efficacy and safety of doxepin 1 mg, 3 mg, and 6 mg in elderly patients with primary insomnia: A randomized, double-blind, placebo-controlled crossover study. *Journal of Clinical Psychiatry, 69*(10), 1557–1564.

Schmid, D. A., Wichniak, A., Uhr, M., et al. (2006). Changes of sleep architecture, spectral composition of sleep EEG, the nocturnal secretion of cortisol, ACTH, GH, prolactin, melatonin, ghrelin, and leptin, and the DEX-CRH test in depressed patients during treatment with mirtazapine. *Neuropsychopharmacology, 31*, 832–844.

Seiden, D., Zee, P., Weigand, S., et al. (2005). Double-blind, placebo-controlled outpatient clinical trial of ramelteon for the treatment of chronic insomnia in an elderly population. *Sleep, 28*(Abstract Suppl), A228.

Smith, M. T., Perlis, M. L., Park, A., et al. (2002). Comparative meta-analysis of pharmacotherapy and behavior therapy for persistent insomnia. *American Journal of Psychiatry, 159*, 5–11.

Stein, M. B., Kroft, C. D., & Walker, J. R. (1993). Sleep impairment in patients with social phobia. *Psychiatry Research, 49*, 251–256.

Sun, H., Kennedy, W., Wilbraham, D., et al. (2013). Phase II randomized, 4-way crossover, double-blind, placebo-controlled, multi-center, dose-finding trial with esmirtazapine in patients with primary insomnia. *Sleep, 36*, 259–267.

Takeda Pharmaceuticals North America. (2005). *Rozerem prescribing information*. Takeda Pharmaceuticals North America.

Tashiro, M., & Yanai, K. (2007). Molecular imaging of histamine receptors in the human brain. *Brain and Nerve, 59*, 221–231.

Taylor, F., & Raskind, M. A. (2002). The α_1-adrenergic antagonist prazosin improves sleep and nightmares in civilian trauma posttraumatic stress disorder. *Journal of Clinical Psychopharmacology, 22*(1), 82–85.

Taylor, F. B., Martin, P., Thompson, C., et al. (2008). Prazosin effects on objective sleep measures and clinical symptoms in civilian trauma PTSD: A placebo-controlled study. *Biological Psychiatry, 63*(6), 629–632.

Vermeeren, A. (2004). Residual effects of hypnotics: Epidemiology and clinical implications. *CNS Drugs, 18*, 297–328.

Walsh, J. K. (2004). Drugs used to treat insomnia in 2002: -Regulatory-based rather than evidence-based medicine. *Sleep, 27*, 1441–1442.

Walsh, J. K., & Schweitzer, P. K. (1999). Ten-year trends in the pharmacological treatment of insomnia. *Sleep, 22*, 371–375.

Walsh, J. K., Roth, T., Randazzo, A., et al. (2000). Eight weeks of non-nightly use of zolpidem for primary insomnia. *Sleep, 23*, 1087–1096.

Walsh, J. K., Roehrs, T., & Roth, T. (2005). Pharmacologic treatment of primary insomnia. In M. H. Kryger, T. Roth, & W. C. Dement (Eds.), *Principles and practice of sleep medicine* (4th ed., pp. 749–760). Elsevier Saunders.

Walsh, J. K., Zammit, G., Schweitzer, P. K., et al. (2006). Tiagabine enhances slow wave sleep and sleep maintenance in primary insomnia. *Sleep Medicine, 7*, 155–161.

Winokur, A., DeMartinis, N. A., III, McNally, D. P., et al. (2003). Comparative effects of mirtazapine and fluoxetine on sleep physiology measures in patients with major depression and insomnia. *Journal of Clinical Psychiatry, 64*, 1224–1229.

Young, D. M. (1995). Psychiatric morbidity in travelers to Honolulu, Hawaii. *Comprehensive Psychiatry, 36*, 224–228.

Yukuhiro, N., Kimura, H., Nishikawa, H., et al. (2004). Effects of ramelteon (TAK-375) on nocturnal sleep in freely moving monkeys. *Brain Research, 1027*, 59–66.

Zammit, G., Roth, T., Erman, M., et al. (2005). Double-blind, placebo-controlled polysomnography and outpatient trial to evaluate the efficacy and safety of ramelteon in adult patients with chronic insomnia. *Sleep, 28*(Abstract Suppl), A228–A229.

Therapeutic Use of Dopamine Enhancers (Stimulants)

Alessandro Zuddas and Sara Carucci

Contents

This chapter is an update from the 4th edition. Previous edition authors were Jonathan Posner, Laurence L. Greenhill and J. Craig Nelson. Dr. Alessandro Zuddas died before the publication of the work was completed.

A. Zuddas · S. Carucci (✉)
Department of Biomedical Sciences, Section Neuroscience and Clinical Pharmacology, University of Cagliari, Cagliari, Italy

Child and Adolescent Neuropsychiatry Unit, "A. Cao" Paediatric Hospital, Cagliari, Italy
e-mail: sara.carucci@unica.it

© Springer Nature Switzerland AG 2024
A. Tasman et al. (eds.), *Tasman's Psychiatry*,
https://doi.org/10.1007/978-3-030-51366-5_138

Abstract

According to international guidelines, dopamine enhancers (stimulants) are recommended as the first-line pharmacologic treatment for ADHD. Their efficacy on ADHD core and related symptoms has been confirmed since the middle of last century and thousands of studies have been published so far on the topic. Within the present chapter, the molecular mechanism of action of dopamine enhancers as well as evidence of their efficacy, safety, and tolerability are discussed. Indications for their use in ADHD during childhood and adulthood as well as in comorbidities in particular sub-populations will be discussed. Other indications, including Binge Eating Disorder, Depression and Excessive Day Sleepiness in Narcolepsy for which dopamine enhancers have been approved or currently studied are also reported.

Keywords

ADHD · Methylphenidate · Amphetamine · Lisdexamfetamine · Modafinil · Efficacy · Safety · Adverse effects · Indications · Comorbidities

Monoamine Enhancers (Stimulant Medications) for the Treatment of ADHD in Children, Adolescents, and Adults

Introduction

Following the original report by Charles Bradley (1937) of an amphetamine that paradoxically "calmed" hyperactive children, methylphenidate (MPH) was first synthesized in 1944 by Leandro Panizzon in the Chemische Industrie Basel (CIBA) laboratories in Basel: while he was not impressed by the results of a self-administration experiment, his wife Marguerite ("*Rita*" for short) noticed its cognitive effects on her tennis playing. MPH was then registered in 1950, with indications for "fatigue" and "confusion"; soon after, in 1954, MPH was marketed under the trade-name of Ritalin®, in acknowledgment of Rita (Wenthur, 2016). The US FDA approved MPH in child psychiatry in 1961 and in 1977 the results of a randomized double blind trial clearly showed its efficacy on a cohort of children with attention deficit hyperactivity disorder (Barkley, 1977); these results were confirmed in the following years by a very large number of clinical studies: in 2021, more than 3000 clinical trials of MPH were referenced in PubMed (www.ncbi.nlm.nih. gov/pubmed).

There is now an increased recognition that ADHD is an impairing condition that often persists into adulthood: stimulant medications (methylphenidate, dextroamphetamine and derivates) and "noradrenergic" medications (atomoxetine and guanfacine) are the most effective medications for ADHD; their use is well established and consistently recommended in evidence-based clinical guidelines across the world (Wolraich et al., 2019).

Molecular Mechanism of Action of Dopamine Enhancers (Stimulant Medications)

Amfetamines (International nonpropriety name, also known as amphetamines which is derived from the chemical name α-methylphenethylamine) and methylphenidate both enhance the impact of dopamine (DA) and norepinephrine (NE) neurotransmission, by blocking their reuptake in the respective monoamine transporters (Arnsten & Pliszka, 2011; Volkow et al., 2002). Amfetamines also increase the catecholamine release from synaptic vesicles (Robertson et al., 2009). The therapeutic effects of stimulants on behavior and attention appear to be related to the enhanced neurotransmission of these catecholamines, especially in the prefrontal cortex (Arnsten, 2011).

Racemic **amfetamine** contains equal amounts of d- and l-amfetamine isomer. The amfetamine *d* isomer (*d*-amfetamine – DEX) inhibits reuptake via both membrane transporters and, at higher concentration, vesicular transporters. DEX also increases basal DA and NE release by binding to the vesicular monoamine uptake 2-transporter (VMAT2) and inducing reverse transport through the plasma membrane into the Cytoplasm. In vitro, affinity of DEX is higher for norepinephrine transporters (NET; mainly in prefrontal cortex) than for DA (in the striatum), and much lower for serotonin. The amfetamine *l*-isomer is 13-fold less potent than the *d*-isomer in inhibiting the accumulation of DA or NA into vesicles, and 5–7 time less potent in inhibiting the synaptic membrane transport (Easton et al., 2007). **Methylphenidate** (MPH) shows a similar potency in inhibiting synaptic re-uptake of DA and a slightly less potency in inhibiting NE re-uptake, but it is 40- and 70-fold less potent than DEX at inhibiting vesicular accumulation of DA or NA. Racemic MPH consists of both *d*- and *l*-threo-enantiomers in a 50/50 ratio. The *d*-threo-enantiomer is pharmacologically more active than the l-threo-enantiomer (10-fold for norepinephrine reuptake; 10- to 40-fold for dopamine reuptake (Easton et al., 2007; Heal & Pierce, 2006)) (Fig. 1).

Animal studies have shown that therapeutic doses of methylphenidate (and atomoxetine)

may improve prefrontal cortex (PFC) function by increasing endogenous NE stimulation of α2-receptors and DA stimulation of D1-receptors in rats and monkeys. These drugs may produce an inverted U dose response, whereby higher doses may actually impair PFC cognitive performance. In addition to the PFC, stimulant medications might induce their therapeutic effects through actions in striatum (e.g., caudate), and in the

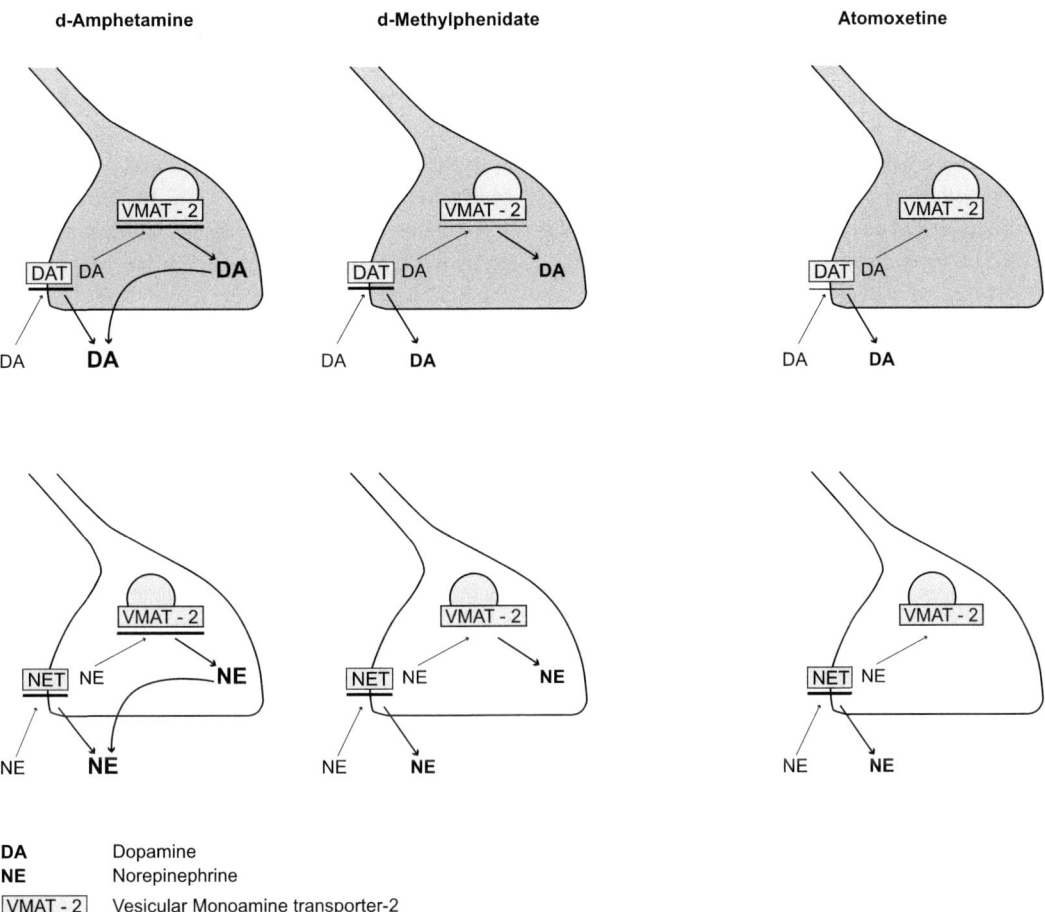

DA	Dopamine
NE	Norepinephrine
VMAT - 2	Vesicular Monoamine transporter-2
DAT	Dopamine transporter
NET	Norepinephrine transporter
——	Inhibition of transporter
⟶	Width of the arrow indicate the degree of inhibition of monoamine transporter

Fig. 1 Mechanisms of action of stimulant medications (based on Easton et al., 2007). The amfetamine *d*-isomer (*d*-Amfetamine) prevents extra-neuronal DA or NA from entering the neuron by inhibiting their reuptake via both membrane transporters and, at higher concentration, via vesicular transporters. *d*-Amfetamine also increases basal DA and NE release by binding to the Vesicular Monoamine Uptake 2-Transporter (VMAT2); the higher cytoplasmic pool of DA or NA induces reverse transport through the plasma membrane into the cytoplasma. Methylphenidate (MPH) shows a similar potency in inhibiting synaptic re-uptake of DA and a slightly less potency in inhibiting NE re-uptake, but it is 40- and 70-fold less potent than DEX at inhibiting vesicular accumulation of DA or NA

posterior association cortices. PET studies have shown that therapeutic doses of stimulant medications may induce DA receptors activation in human striatum, corresponding to a small but significant increase in DA release measured in rodent striatum (Arnsten & Pliszka, 2011). During cognitive tasks MPH has been shown to increase cerebral blood flow in dorsolateral prefrontal and posterior parietal cortices in healthy controls (Mehta et al., 2000) and in the prefrontal cortex in adults with ADHD (Schweitzer et al., 2004); it appears to decrease metabolic activation of task-irrelevant brain regions, thus focusing activation of task-relevant regions and improving performance. MPH significantly increases DA only when given with the academic task but not when given with the neutral task, with the same failing to increase DA levels when performed with placebo. In the evaluation of self-reports of saliency, MPH-treated subjects rated the salient task as interesting and exciting or motivating compared to placebo-treated subjects (Volkow et al., 2008).

More recently, stimulants have been suggested to be able to modulate functional connectivity: in attention and motivation networks, methylphenidate normalizes activation and functional connectivity deficits in medication-naïve children with ADHD during a rewarded continuous performance task and it normalizes frontocingulate underactivation during error processing (Rubia et al., 2011); in adolescents with ADHD, stimulant medications demonstrated effects on the *functional connectivity* of frontoparietal brain networks, with beneficial effects on working memory performance. During inhibitory tasks, children with *ADHD* exhibit a raised motivational threshold at which task-relevant stimuli become sufficiently salient to deactivate the Default Mode Network (DMN): treatment with MPH normalizes this threshold, rendering their pattern of task-related DMN deactivation indistinguishable from that of typically developing children (Liddle et al., 2011). It should be noted, however, that improvements in cognition following methylphenidate do not necessarily correlate with behavioral improvements in ADHD symptoms (Bedard et al., 2015; Coghill et al., 2007).

A more recent meta-analysis (Pereira-Sanchez et al., 2021), however, showed that this evidence is not compelling. Some recent studies found an increase in "intrinsic functional connectivity (IFC)," between one of the core posterior hubs, the posterior cingulate cortex, and the lateral parietal cortex; other network-focused studies, however, observed stimulant associated decreases in mean IFC of multiple networks or increased within-community segregation in multiple networks, including DMN.

Moreover, studies that included typically developing controls reported "normalizing" changes with medication; true normalization should include statistically reliable changes between conditions: in these studies, however, normalization was reported as disappearance (i.e., loss of a significant difference, which can just reflect noisy data) of some imaging group differences between patients and control participants when patients were on treatment. Significant methodological differences make these results controversial and functional Magnetic Resonance Imaging (fMRI) research has been a focus of criticism and skepticism regarding its reliability, transparency, and potential for translation to the clinical realm (Pereira-Sanchez et al., 2021).

Efforts in promoting best practices in conducting and reporting neuroimaging studies are ongoing, improving acquisitions technics and analytic methods, as well as more reproducible and transparent reporting and multiple well-powered independent replications. Broader application of multisite harmonized efforts (the ABCD study may be considered an example [Bjork et al., 2017]) may provide large-scale, deeply phenotyped samples with data openly available to investigators may address the neuroscience of treatments for ADHD and other disorders.

Atypical Dopamine Enhancers (Stimulants)

Modafinil and its derivatives are considered atypical stimulants (Tanda et al., 2021). Modafinil exists as a racemic mixture of S- and R-enantiomers. Of note, the R-enantiomer is thought to be the source of modafinil's psychotropic properties and is marketed independently as armodafinil. Modafinil is known to be a weak inhibitor of dopamine reuptake (with affinity 6 to

100 lower to dopamine transporter DAT compared to methylphenidate [Schmitt & Reith, 2011]), acting on different conformation of the dopamine transporter: classical stimulants have higher affinity for the DAT when it is in a conformation open to the extracellular compared to the intracellular space, whereas affinity of modafinil for the DAT is similar for both outward and inward conformations (Loland et al., 2012). This lack of pronounced selectivity of binding to DAT conformations has been reported to be responsible of the unique in vivo pharmacological profile of modafinil: in animal models, modafinil lack the selectivity between *Shell* and *Core* regions of the *Nucleus Accumbens*, typically observed after cocaine administration, with in turn is considered the mechanism of drug reinforcement and abuse potential (Mereu et al., 2017).

Modafinil has little to no in vivo affinity for the serotonin (5HT) or norepinephrine (NE) transporters, although elevated concentrations of NE and 5HT in the prefrontal cortex and hypothalamus have been observed following modafinil administration, possibly as an indirect effect of increased extracellular dopamine (Volkow et al., 2009; Wisor, 2013). Additionally, modafinil has been postulated to increase signaling in the hypothalamic orexin and histamine neurotransmitter pathways (Ishizuka et al., 2012), and animal studies have also suggested a glutamatergic effect (Mahler et al., 2014).

Modafinil is approved for narcolepsy and sleep disorders: its use as a potential pharmacological treatment for Psychostimulant Use Disorder (PSUD) has been explored (Mereu et al., 2017; Tanda et al., 2021). It has been shown effective in ameliorating ADHD symptoms in children, adolescents (Kahbazi et al., 2009), and adult patients (Arnold et al., 2014). The potential advantage of modafinil is its very low propensity for causing euphoric effects associated with traditional psychostimulants.

Pharmacokinetics and Specific Formulations

A large number of different stimulant preparations and formulations have been approved in the last decades, although availability varies by country. These include: immediate release, intermediate duration, and longer-acting stimulants, a transdermal patch, and a long-lasting liquid, with further novel formulations and delivery systems in development (Table 1).

Absorption of **amfetamine** is rapid, with peak plasma levels occurring about 3 h after oral administration. Children eliminate amfetamine faster than adults, the elimination half-life of DEX being about 1 h shorter in 6–12 year old children (average 9 h) than in adults (average about 10 h). Food (e.g., high fat diet) can delay absorption, but does not have a significant clinical impact. Acidification of urines increases urinary excretion of amfetamines. Ingestions of acidic substances, such as fruit juices, may lower absorption, whereas gastrointestinal alkalinizing agents, such as sodium bicarbonate, will increase absorption.

Amfetamine undergoes liver metabolism where it is oxidized to form 4-hydroxy-amfetamine, alpha-hydroxy-amfetamine, or norephedrine. P450 enzymes metabolize a significant proportion of DEX on first liver passage. Norephedrine and 4-hydroxy-amfetamine are both active and they are subsequently oxidized to form 4-hydroxy-norephedrine. Following oral administration, approximately 96% of the oral dose is found in the urine, only 0.3% recovered in the feces over a period of 120 h.

As suggested by the pharmacokinetic profile, the onset of action of amfetamines is rapid, within 1 h after administration. With immediate release preparations, the duration of action varies from 4 to 6 h, which is slightly longer than for MPH, but typically requires at least bid administration to ensure adequate coverage. Mixtures of different d-amfetamine and *dl*-amfetamine salts formulated for immediate or sustained release are available in the United States but not in Europe.

An extended release formulation of **mixed amfetamine salts** (Adderall XR®) is a racemic mixture of dextro- and levo-isomers of amfetamine salts, which provides 50% immediate release MAS and 50% delivered at a second pulse 4–6 h later. Adderall XR® (Shire) 20 mg provides comparable plasma concentrations to Adderall

Table 1 Stimulant medications approved for the pharmacological treatment of ADHD[a]

Stimulant	Molecular mechanism of action	Generic names (brand names) and formulations	Duration
Amphetamines	Dopamine (DA) and norepinephrine (NE) neurotransmission enhancement, by blocking their reuptake in the respective monoamine transporters Increase of catecholamine release from synaptic vesicles by inhibition of VMAT-2 Reverse transport through the plasma membrane into the cytoplasma	*Amphetamine and dextroamphetamine mixed salts*:	
		Tablet (Adderall®)	4–8 h
		Extended-release capsules (Adderall XR®)	8–12 h
		Amphetamine extended-release:	
		Oral suspension liquid (Adzenys ER)	
		Oral suspension capsule (Dyanavel® XR)	8–12 h
		Orally disintegrating tablet (Adzenys XR™)	9–12 h
		Amphetamine sulfate:	
		Tablet (Evekeo®)	4–6 h
		Orally disintegrating tablet (Evekeo ODT™)	4–6 h
		Methamphetamine:	
		Tablet (Desoxyn®)	4–8 h
		Dextroamphetamine sulfate:	
		Extended-release tablet (Dexedrine®)	6–9 h
		Liquid (Procentra®)	4–8 h
		Tablet (Zenzedi®)	4–8 h
		Mixed amphetamine salts:	
		Extended-release capsule (Mydayis™)	16 h
Lisdexamfetamine	It is a pro-drug of dextroamphetamine. After oral administration, it is rapidly absorbed from the gastrointestinal tract and converted to dextroamphetamine, responsible for the drug's activity	*Lisdexamfetamine dimesylate*:	
		Capsule (Vyvanse®)	10–12 h
		Chewable tablet (Vyvanse®)	8–12 h
Methylpheinidate	Dopamine (DA) and norepinephrine (NE) increase neurotransmission, by blocking DA and NE transporters Redistribution of VMAT-2	*Methylphenidate hydrochloride extended-release*:	
		Capsule (Adhansia XR™; Aptensio XR™; Jornay PM™; Metadate CD®; Ritalin LA®, Medikinet®, Equasym®)	7–+12 h
		Tablet (Concerta®; Metadate ER®; Methylin® ER; Ritalin SR®)	7–12 h
		Chewable tablet (QuilliChew ER™)	8–12 h
		Liquid (Quillivant XR®)	8, 10, 12 h
		Methylphenidate hydrochloride:	
		Tablet (Ritalin®)	3–5 h
		Liquid (Methylin®)	3–5 h
		Chewable tablet (Methylin®)	3–5 h
		Methylphenidate:	
		Transdermal patch (Daytrana®)	10–12 h
		Orally disintegrating tablet (Cotempla™ XR)	8–12 h

(continued)

Table 1 (continued)

Stimulant	Molecular mechanism of action	Generic names (brand names) and formulations	Duration
		Dexmethylphenidate hydrochloride:	
		Tablet (Focalin®)	3–5 h
		Extended-release capsule (Focalin XR®)	12 h

[a]FDA stimulant medications approved for the treatment of ADHD. Revised from Children and Adults with Attention-Deficit/Hyperactivity Disorder (CHADD), January 2020

immediate release 10 mg bid administered 4 h apart. Its administration results in ascending plasma levels of amfetamine up to a peak at about 7 h after dosing. This is followed by a gradual decline that results in a low but detectable plasma levels 24 h after dosing.

Lisdexamfetamine (LDX) is dextroamfetamine that has been covalently attached to the essential amino acid, L-lysine. Lisdexamfetamine itself is not pharmacodynamically active and neither does it result in high dextroamfetamine levels when injected or snorted: for these reasons, it is considered to have a lower abuse potential compared to conventional stimulants.

LDX is water soluble: after oral administration, it is rapidly absorbed from the gastrointestinal tract. Absorption is thought to be mediated by the high-capacity PEPT1 transporter. Following oral administration, the amide linkage between the two molecules is enzymatically hydrolyzed releasing the active DEX. Most of this hydrolysis takes place within the red blood cells. Food does not affect the observed AUC and Cmax of DEX but prolongs Tmax by approximately 1 h compared to orally administered dexamfetamine (from 3.8 h at fasted state to 4.7 h after a high fat meal).

The Tmax of lisdexamfetamine dimesylate is approximately 1 h following an oral dose: the Tmax of dexamfetamine is approximately 3.5 h. Pharmacokinetics of d-amfetamine after single-dose LDX oral administration is linear: in adults, there is no accumulation of DEX at steady state, nor accumulation of lisdexamfetamine dimesylate after once-daily dosing for seven consecutive days. The plasma elimination half-life of lisdexamfetamine averaged less than 1 h in studies in volunteers. The

mean half-life of d-amfetamine is 11 h, although there is intra-individual variability (Boellner et al., 2010; Ermer et al., 2010; McGough et al., 2003).

DEX generated by hydrolyzed LDX do not undergo to the extensive liver first pass metabolism observed after orally administered DEX, this probably helps to explain the differences in oral dose/efficacy ratio observed with the two medications. Therefore, it is *not* possible to predict what dose of LDX will be required for a patient previously stabilized on DEX and vice versa.

Oral **Methylphenidate** (MPH) is rapidly absorbed from the gastrointestinal tract, with peak plasma concentrations occurring about 1.5–3 h after administration. Food can delay the time to maximum plasma concentration from 1.5 h when fasting to 2.5 h after a heavy breakfast: it is usually recommended to give the medication just before breakfast. MPH is primarily metabolized through a single step pathway involving de-esterification by carboxylesterase 1A (CES1) to ritalinic acid. Ritalinic acid has no clinically significant pharmacological activity and is excreted in the urine. Variants in the CES-1 gene can lead to wide variations in MPH metabolism (Zhu et al., 2008).

D*l-threo*-methylphenidate undergoes enantioselective metabolism in the liver, which results in marked differences in the plasma concentrations of its isomers. The elimination plasma steady half-life of *d-threo*-methylphenidate is about 3–3.5 h. Because of this short-half life, steady state for methylphenidate is not achieved during regular treatment and there is usually no carry over from 1 day into the next, although there is a theoretical possibility of steady-state developing where high doses of extended release preparations are used or when

MPH is prescribed to those with poor metabolism. Neither of these possibilities has been properly studied and clinical relevance is uncertain. Metabolism is similar in school-age children and adults, and for both there is a wide individual variability in peak plasma concentrations (Markowitz & Patrick, 2008).

Laboratory school classroom studies suggest a close relationship between pharmacokinetic (PK) profile and pharmacodynamic (PD) properties (Swanson et al., 2004). Optimal clinical effect appears associated with rapidly increasing levels in the morning followed by a steadily increasing plasma levels across the rest of the day: preparations with bi- or tri-modal release systems (immediate release followed by one or two extended-release phases) ensure an initial sharp plasma peak occurring about 1.5 h after dosing, followed by a second peak several hours after, and then a gradual decline. Concerta XL® (Janssen-Cilag), Matoride XL® (Sandoz), Equasym XL® (Shire), Medikinet Retard® (Medice), and Ritalin LA® (Novartis) all provide a mixture of immediate and extended-release MPH; they differ in the mechanisms of the delayed-release system and in the proportion of immediate release to delayed release methylphenidate. Effects on behavior of these different formulations parallel their different PK profile (concentrations in the blood) over time (Banaschewski et al., 2006). For Concerta XL and Matoride XL effects lasts 10–12 h, Equasym XL®, Ritalin LA, and Medikinet Retard last between 6 and 8 h.

Linear dose effects were found for OROS MPH and also for dex-methylphenidate (Stein et al., 2011). Similarly, dose-dependent efficacy of Mixed Salt Amfetamines in childhood ADHD has been demonstrated versus placebo in a large, parallel group study. For extended release MAS, duration of the behavioral and cognitive effects was demonstrated for 12 h (Biederman et al., 2002).

Long-acting formulations of methylphenidate, decreasing the risk of forgetting or that children may be ashamed to take medication at school, have been associated with significantly better adherence (Gajria et al., 2014).

More recently, new long-acting formulations of methylphenidate have been developed (or are currently under development) in order to overcoming possible challenges in swallowing tablets/capsules, an important issue for a subgroup of children with ADHD, and decreasing the potential for adverse effects related to dosage peak that may occur when the child chew the tablet or opened and sprinkled over applesauce or suspended in water the capsule content, leading to incomplete delivery or to a release of the full dose more rapid than intended (Childress & Berry, 2010).

Transdermal patches, long-acting liquid, chewable, or orally disintegrating formulations, as well as novel delivery platform allowing methylphenidate to be administered at nighttime but with symptomatic control staring in the morning at wakening have been recently approved in the USA and in some other countries.

In particular, transdermal patches (Daytrana®) are prepared in order to allow about 12 h of effect if worn for 9 h preparations with different concentrations [10 mg/9 h (1.1 mg/h), 15 mg/9 h (1.6 mg/h), 20 mg/9 h (2.2 mg/h), 30 mg/9 h (3.3 mg/h)] available in the USA. MPH released by the patch do not undergo to extensive first pass liver metabolism observed with orally administered MPH: l-threo-methylphenidate is not rapidly metabolized as after oral administration and could be detectable in the plasma; it has been suggested, but never confirmed, that l-threo-methylphenidate may be associated with the development of adverse events and/or tolerance.

Oral solutions of immediate-release MPH have been available for many years in the USA; and Dexedrine in the UK. Quillivant XR® is a recent liquid long-lasting formulation of MPH: it is available as a suspension of 5 mg/ml, with 20% immediate release and 80% extended release. Peak plasma levels occur approximately 5 h after dosing with effects marketed to last for up to 12 h.

Similarly, with a chewable tablet formulation (Quilliches ER®), containing 30% immediate release and 70% extended release MPH, formulated as 15% MPH chloride salt and 85% as MPH ionically bound to a group of sodium polystyrene sulfone particles, peak levels were measured after 4.2 h and after 6 h with a kinetics equal to 2 doses of IR administered 6 h apart (Abbas et al., 2016). With an orally disintegrating tablet (ODT) formulation (Cotempla®) clinical effect was evident at 1 h and lasted through 12 h post-dose (Childress et al., 2017, 2021b).

A specific formulation (HLD 200, Jornay PM® approved by US FDA in August 2018 and commercially available in the US market since June 2019) has been designed as a once-daily, evening-dosed, delayed-release, and extended-release methylphenidate (DR/ER-MPH), in order to provide therapeutic effect beginning upon awakening and lasting into the evening: after the treatment-optimization phase in a randomized, double-blind, multicenter, placebo-controlled, phase 3 trial, an optimized dose administered at 8: 00 PM, significantly improved functional impairment versus placebo in the early morning and in the late afternoon/evening (Pliszka et al., 2017; Wilens et al., 2021).

Finally, other formulations have been developed in order to extend the therapeutic duration of short-acting drugs such as MPH and, in same time, protecting the required high drug loads from tampering and improper extraction. Different multilayer coating formulation have been developed: Aptensio XR® has been approved in the USA (Childress et al., 2021a; S. B. Wigal et al., 2015) and investigated also in children aged 4–6; ORADUR technology has been recently successfully tested in Taiwan (Huang et al., 2021).

These differing delivery profiles provide the clinician with increased options when choosing which preparation to use to meet the needs of particular patients, as well as a more flexible and sensitive individualized adjustment while retaining the benefits of an ER preparation. It should be noted, however, that PK profiles may show considerable inter-individual variation and caution should be observed when generalizing from aggregated profiles to individual patient cases. The onset of action, in particular, can be delayed or may be attenuate in the afternoon, requiring the concomitant administration of a low dose of immediate release preparation. Also just because two preparations have very similar profiles at the group level (e.g., Concerta XL and Matoride XL) this does not mean that individual patients will experience them as identical and care should be taken when switching from one to the other.

Interaction with Other Drugs

Stimulants show little interference with the metabolism and pharmacokinetics of other medications. There is, however, a potential to inhibit the metabolism of anticonvulsants such as phenobarbital, phenytoin and primidone, and of monoamine enhancers (tricyclic antidepressants). Stimulants may potentiate stimulating effects of other drugs on the cardiovascular or central nervous system; they also may increase pressor response to vasopressor agents. Thus, caution should be used when combining with other sympathomimetic agents, including atomoxetine, as there is the potential for tachycardia, tremor, and nervousness.

Four cases of sudden death were reported in children with ADHD taking concurrent administration of MPH and clonidine, but no causal links between MPH treatment and cardiovascular adverse events or the occurrence of ECG abnormalities have been established (Tourette's Syndrome Study Group 2002).

Stimulants are appropriately contraindicated in concomitant use – or use within the last 2 weeks – with monoamine oxidase (MAO) inhibitors.

Clinical Efficacy and Effectiveness on ADHD

Efficacy of Dopamine Enhancers (Stimulants) on ADHD Core Symptoms

ADHD medications are within the most effective psychiatric pharmacological treatments with effect sizes clearly larger than other psychotropic drugs and even higher than several common internal medicine medications, placing them in the range found with antibiotic for infection (Leucht et al., 2012). Since Bradley's discovery in 1937 and the publication of several case reports showing that behavioral problems in children could be improved by treatment with amphetamines, interest in psychostimulants progressively increased, in particular after the first controlled trials performed in the '60. Currently AMP and MPH use is well established in the treatment of ADHD and usually recommended in evidence-based clinical guidelines across the world (Banaschewski et al., 2006; Coghill & Seth, 2015; Cortese et al., 2013b; Graham et al., 2011; Pliszka, 2007; Taylor et al., 2004; Wolraich et al., 2019). Their response rate is estimated around 70%, rising to 95% when subjects considered as non-responders are treated with a second different drug (Hodgkins et al., 2012).

In several countries, however, amphetamine is within controlled substances classified along with drugs like heroin, which has inevitably retarded or prevented its adoption for the treatment of children.

Lisdexamfetamine, a long-acting amphetamine pro-drug was licensed in the USA in 2007, and later in several European countries in 2013 for the treatment of ADHD children and adolescents with an inadequate clinical response to previous MPH. LDX has also been approved in several European countries for the treatment of adults suffering from ADHD, including ADHD adults not older than 55 years in Switzerland from 2014 and all adults in Denmark, Sweden, and the United Kingdom from 2015 (Siffel et al., 2020). The last NICE guidelines in the United Kingdom recommend LDX as a first-line pharmacological treatment for ADHD in adult subjects (NICE, 2018).

The efficacy of stimulant medications in reducing the ADHD core symptoms in the short-term is supported by numerous placebo-controlled randomized trials that confirm their effectiveness in the short term, with effect sizes ranging between 0.8 and 1.1 in children and adolescents (Banaschewski et al., 2006; Faraone & Buitelaar, 2010).

The efficacy of LDX in relieving the symptoms of ADHD has been demonstrated in a series of pivotal randomized controlled trials in North America and Europe (Adler et al., 2008; Biederman et al., 2007; Coghill et al., 2013; Findling et al., 2011), with very large effect size (1.80) compared to placebo on ADHD symptoms. Evidence from trials of at least 12-month duration indicates that the safety and tolerability profile of LDX is similar to those of other stimulants in people with ADHD (Coghill et al., 2014b; Findling et al., 2008; Weisler et al., 2009). The randomized withdrawal after 26 weeks of open label treatment indicated the benefit of continued LDX, with treatment emerging adverse events

generally consistent with those associated with the MPH preparation, used as the active comparator (Coghill et al., 2014a).

A Cochrane review including 38 parallel-group trials and 147 cross-over trials (Storebø et al., 2015) questioned the quality of the evidence, concluding that the effect sizes of methylphenidate in managing and reducing ADHD symptoms should be considered uncertain; a more recent comprehensive Network Meta-Analysis (Cortese et al., 2018) questioned this assumption and showed that all medications approved for ADHD are statistically significantly superior to placebo. This Network Meta-Analysis (NMA) included 133 double-blind, randomized, controlled trials (RCTs), 81 of which were performed in children/adolescents, 51 in adults, and one in both, with an average duration of 7 weeks. All ADHD medications were effective in decreasing the severity of inattention and disruptive symptoms as rated by clinicians, with the largest effect sizes for amphetamines and methylphenidate. In the 81 RCTs performed in children and adolescents, efficacy data with a follow up closest to 12 weeks, showed an effect size of **1.02** (95% CI 1.19–0.85) for amphetamines, and of **0.78** (95% CI 0.93–0.62) for methylphenidate. In the 51 RCTs performed in the adult population effect sizes resulted slightly lower compared to those found in children and adolescents both for amphetamines **0.79** (95% CI 0.99–0.58), and methylphenidate **0.49** (95% CI 0.64–0.35).

Within the same NMA, considering that lisdexamfetamine (LDX) is differently metabolized from other amphetamines, with possible differences in its efficacy and tolerability, the authors separated lisdexamfetamine from the other compounds and performed a post-hoc analysis showing that LDX was less well tolerated in children and adolescents compared to placebo (OR 2.69, 95% CI 1.40–5.16), whereas tolerability of the other amphetamines was slightly better (1.83, 0.84–4.02); in adults, the opposite pattern emerged (lisdexamfetamine vs placebo:, 2.74, 0.80–9.30; other amphetamines vs. placebo, 3.66, 1.36–9.87).

Taking into account the evidence on both efficacy and safety, Cortese's network meta-analysis supports methylphenidate as the preferred first-choice medication for the short-term treatment of ADHD in children and adolescents, while amphetamines appear to be indicated as the first choice in adults. Although, at a group level amphetamines resulted more efficacious than methylphenidate, data from 6 crossover randomized controlled trials (RCTs) including 174 subjects (L. E. Arnold, 2000), showed that, at the patient level, approximately 41% (72/174) of participants had a comparable good response to both amphetamines and methylphenidate, 28% showed a better response to amphetamines, 16% had a better response to methylphenidate, while the rest (15%) did not respond to either medication (Cortese, 2020).

Efficacy of Dopamine Enhancers (Stimulants) on ADHD Associated Features and Quality of Life

As discussed above, stimulants rapidly and significantly contribute to ameliorate the ADHD core symptoms reducing restlessness, inattentiveness, and impulsiveness. It is currently widely accepted, however, that ADHD treatment should be focused not only to the control of patients' symptoms, but also to obtain an improvement in global functioning and quality of life.

Health-related quality of life (HRQoL) can be described as the impact that a disorder exerts on the individual's perception of his position in life, considering the culture and value systems in which he live, and in relation to his goals, expectations, standards, and concerns (Adamo et al., 2015; Coghill et al., 2017b; "The World Health Organization Quality of Life assessment (WHOQOL): position paper from the World Health Organization," 1995).

By definition, ADHD interferes with a subject's academic, occupational, familiar and social functioning; the nature of the impairment usually vary from patient to patient and with age, encompassing diverse outcomes including underachievement at school or at work, unemployment or low income, substance abuse, smoking, early pregnancy, problems with the law, divorce, or acquiring sexually transmitted disease ("Optimizing clinical outcomes across domains of life in

adolescents and adults with ADHD," 2011). Clinical trials and observational studies using questionnaire-based instruments report that subjects with ADHD usually are severely compromised in their quality of life. A systematic review (D. R. Coghill et al., 2017b) investigating the Health Related Quality of Life (HRQoL) and the functional outcomes in randomized placebo-controlled trials of medications for ADHD reported that, within the total 34 included studies, effect sizes of ADHD medications vs. placebo in children and adolescents were >0.5 in 7/9 and 4/8 studies for HRQoL and functioning, respectively. Effect sizes were typically larger for stimulants than for non-stimulants and the larger effect sizes for HRQoL and/or functional outcomes (≥ 0.8) were found in short-term studies of long-acting stimulant medications (LDX, OROS-MPH and TD-MPH).

The domains with more significant improvement, as measured by multi-domain instruments, were related to achievement/school, risk-taking and interpersonal relations. The same review reported effect sizes >0.5 in only 1/6 studies for QoL and 1/8 studies for functioning in trials performed in adults. Such a noticeable difference can be explained by the diversity of instruments used to assess HRQoL and functional impairment in the clinical trials of ADHD medications and in particular by the fact that almost all HRQoL and functional outcomes were parent-rated in child/adolescent studies and self-rated in adult studies (Coghill et al., 2017b). It is not clear at the moment if changes in HRQOL are mediated by symptom changes, changes in functional impairment, or other factors. One study reported that improvement of HRQoL due to pharmacological treatment persisted following medication withdrawal, even when symptom severity increased (Banaschewski et al., 2014), indicating that potential cause/effect relationships between medication use, symptoms and functional impairment reduction, and HRQOL require further investigation (Zuddas et al., 2018).

Other important features associated to ADHD comprise various neurocognitive deficits including both executive functions (response inhibition, working memory, attentional set shifting, planning, etc.) and non-executive functions (basic storage aspects of memory, timing, reaction time, and reaction time variability), as well as motivational factors such as delay aversion and decision-making (Coghill et al., 2014c).

The cognitive deficits associated with ADHD generally co-occur with the ADHD core symptoms; however, they are not necessarily causally related to each other, rather they run in parallel. It is therefore important to understand if the pharmacological treatments can be helpful in improving these associated cognitive deficits. Coghill et al. (2014c) in their systematic review and meta-analysis evidenced that methylphenidate was superior to placebo in all the investigated domains with the highest effect sizes for reaction time variability (SMD 0.62, 95% CI: 0.90–0.34) and non-executive memory (SMD 0.60, 95% CI: 0.79–0.41) followed by response inhibition (SMD 0.41, 95% CI: 0.55–0.27), executive memory (SMD 0.26, 95% CI: 0.39–0.13), and reaction time (SMD 0.24, 95% CI: 0.33–0.15).

A recent randomized, placebo-controlled discontinuation study examined the effects of methylphenidate on executive functioning in children and adolescents with ADHD after long-term use (Rosenau et al., 2021). Ninety-four subjects (ages 8–18 years) who used MPH beyond 2 years were either assigned to 7 weeks of continued treatment with 36 or 54 mg of extended-release methylphenidate or to gradual withdrawal over 3 weeks to placebo for 4 weeks. Working memory, response inhibition, attentional flexibility, and psychomotor speed were compared between both groups showing that the discontinuation group made more errors on working memory, with no significant differences on the other domains. The authors therefore conclude that methylphenidate can have a beneficial effect on working memory also in the long-term use with implications for clinical and pharmacological management decisions (Rosenau et al., 2021).

In a 2-year open-label study of LDX, no potentially clinically significant deteriorations on cognitive functions were observed from baseline over the 2 years of the study. Clinically significant improvements were observed at 6 months and at the end of treatment on the CANTAB tasks

"*Delayed Matching to Sample (DMS),*" and "*Spatial Working Memory (SWM)*" (Coghill et al., 2018).

Evidence from observational, pharmacoepidemiologic, and registry studies indicate that ADHD medications can have a positive impact on many other important aspects of patients' lives. Increased academic achievement and decreased absenteeism at school have been found from a population-based study in 370 ADHD children: longer duration of treatment predicted decreased absenteeism rates and children treated with stimulants were 1.8 times less likely to subsequently be retained a grade (Barbaresi et al., 2007). A recent systematic review and meta-analysis also confirmed that methylphenidate can improve math productivity and accuracy and can increase reading speed, although academic improvements resulted minimal compared to symptom improvements and were substantially limited to math (Kortekaas-Rijlaarsdam et al., 2019).

In terms of physical well-being and health, it is well known that children with ADHD are more prone to accidental injury and trauma compared with healthy peers (Dalsgaard et al., 2015), which may lead to increased mortality (Dalsgaard et al., 2015), and it is also worth noting that even parents of ADHD subjects have higher risks of burn injury, fracture, and traumatic brain injury than fathers and mothers of children without ADHD (Li et al., 2021). Stimulant medications have been related to a reduced risk of emergency admission to hospital for trauma, accidents, and unintentional injuries in ADHD children, adolescents, and adults (Brunkhorst-Kanaan et al., 2021; Ruiz-Goikoetxea et al., 2018). Liou and colleagues (2018) reported a preventive effect of ADHD medication for the risk of traumatic brain injury (TBI) in ADHD children, adolescents, and young adults.

Using a self-controlled case series study design, Man et al. (2015) found that the relative incidence of trauma-related ED admissions was lower during exposed periods to MPH compared with nonexposed periods with no differences according to gender. There are also several studies showing a beneficial effect of stimulant medication on driving performance in adolescents and young adults with ADHD. Biederman and colleagues (2012) evidenced that driving performance were significantly ameliorated by lisdexamfetamine compared to placebo. Cox and colleagues (2012) reported extended-release methylphenidate to be more effective in improving driving performance in comparison to extended release mixed amphetamine salts and a study by Sobanski et al. (2008) found better driving performance in ADHD drivers medicated with methylphenidate vs. unmedicated ADHD drivers.

Importantly ADHD medications have been associated also with lower risk of depression (Chang et al., 2016) and suicidal events (Chen et al., 2014), and decreased rates of substance abuse (Chang et al., 2014) and criminality (Chang et al., 2019; Lichtenstein et al., 2012). In periods during which patients were receiving medications, these studies showed a significant decrease in negative outcomes although, also in these cases, effect sizes are generally somewhat smaller than those found for symptom reduction per se.

Efficacy in Specific Clinical ADHD Subgroups

Knowledge about the effectiveness of stimulants in ADHD was derived by clinical trials firstly conducted on children and, more recently, on adolescents and adults. Knowledge has also been expanded in specific subpopulations, including pre-schoolers and subjects suffering from ADHD plus comorbid disorders.

Preschoolers

MPH is not approved for ADHD under the age of 6 years, while amphetamine/dextroamphetamine is approved for use down to age 3 years, although data supporting this indication are limited.

A comprehensive review published in 2011 investigated the role of medications in pre-schoolers (Kaplan & Adesman, 2011). Several studies reported that treatment efficacy in younger children was comparable to that found for older children, with significant improvements in structured situations (Connor, 2002); however, different methodology across studies did not allow to draw firm conclusions. To address the lack of a precise methodology around safety and efficacy data in pre-schoolers, the NIMH funded a randomized, placebo-controlled, multisite MPH trial entitled Preschool ADHD Treatment Study (PATS). After completing a parent-training phase, about 165 families were randomized into a double-blind, placebo-controlled, crossover, dose-optimization study. Prior to entry into the double-blind phase of the protocol, all children received medication treatment during an open safety phase lasting 1 week. In the following double-blind phase, children were on placebo or active medication (1.25, 2.5, 5, or 7.5 mg) three times daily, with their dose switched weekly. By comparing MPH vs placebo, MPH resulted significantly efficacious, however the magnitude of its effect was lower than that typically observed in school-aged children, with an increased frequency and severity of adverse events, i.e., greater mood lability, irritability, decreased appetite, and sleep problems (T. Wigal et al., 2006). Treatment discontinuation was observed in about 11% of cases, a rate significantly higher than 2% of school-age children with ADHD included into the MTA Study (Greenhill et al., 2006). Furthermore, continuous treatment for about 9–10 months was associated with a slight but detectable decrease in height and weight growth (Swanson et al., 2007).

An open, pharmacokinetic, pilot study conducted in a small subset of the PATS sample suggested that pre-schoolers may experience a significantly lower MPH clearance than do school-age children, leading to higher, sustained MPH plasma levels, even when corrected for weight (Wigal et al., 2006). This pharmacokinetic difference could explain why pre-schoolers with ADHD are more sensitive to the side effects of MPH.

Children with Autism Spectrum Disorder (ASD)

ADHD symptoms can be present in about 80% of children with ASD and comorbidity of ADHD in the context of the autism spectrum disorders (ASD) is very common, with 24–83% of ASD children meeting criteria for a formal diagnosis of ADHD (Frazier et al., 2001; Simonoff et al., 2013).

A number of open studies conducted in the 1970s and 1980s showed that stimulant medication, particularly dextroamphetamine, resulted in increased stereotypical movements in children with ASD (Birmaher et al., 1988; Campbell, 1975; Di Martino et al., 2004). More recent studies report that the response rate to stimulants in the ASD population is lower than in subjects without ASD with an higher discontinuation rate due to severe adverse events (Posey et al., 2004). A multi-site double-blind trial performed in children with autism or other PDD and significant ADHD symptoms treated with MPH immediate release (7.5–25 mg/day t.i.d) showed that the compound was effective in reducing ADHD symptoms severity, although a positive responses was only seen in about 50% of cases (Greenhill et al., 2006), instead of the 70% usually seen in non-ASD children with ADHD. Tolerability was also lower, with a proportion of children unable to tolerate the side effects of MPH of about 18% compared to that observed in normal children (generally less than 5% (T. Wigal et al., 2006)). Other studies, however, reported that treatment with stimulants was effective in the ASD population with ADHD in a similar manner to that presenting with ADHD alone (Santosh et al., 2006) and methylphenidate appeared to improve some social behaviors and self-regulation of children with ASD (Jahromi et al., 2009).

A meta-analysis published in 2013 showed MPH to be effective (effect size 0.67) for treating ADHD symptoms in children with pervasive developmental disorders; however, again, the authors evidenced that its relatively lower tolerability must be considered with particular attention in this population (Reichow et al., 2013), keeping in mind that irritability, aggression, self-harm behaviors, emotional lability, and dysphoria are frequently reported in children with ASD treated with psychostimulants (Cortese et al., 2012; Di Martino et al., 2004; Ghuman et al., 2009), and, in turn, they may further impair their social interactions.

A more recent review by the Cochrane group (Sturman et al., 2017) reported, within a low-quality evidence, that high-dose methylphenidate (0.43–0.60 mg/kg/dose) was significantly effective on hyperactivity, as rated both by teachers and parents, but not on inattention. Adverse events in terms of ASD core symptom worsening or benefits on social interaction were not reported.

ADHD and Anxiety

Early studies suggested that children with comorbid anxiety or internalizing symptoms have a lower benefit by stimulant medications on ADHD symptoms; however, several more recent, high quality studies, including the MTA study, confirmed that subjects with ADHD and comorbid anxiety show a robust clinical response to stimulants, questioning previous evidence. A recent meta-analysis indicates that stimulant medications at the group level have a beneficial effect on anxiety (Coughlin et al., 2015). Methylphenidate appears also to be effective in increasing positive social interactions and social behavior and to reduce social anxiety (Patin & Hurlemann, 2015). Its efficacy in improving emotion processing as well as its correlated effect in improving faces and facial emotions recognition may explain the observed improvement in social interaction (Sturman et al., 2017).

Tics and Tourette's Syndrome

While Tourette's syndrome was once considered an absolute contraindication to stimulants, several more recent studies have shown that stimulants are quite effective in controlling ADHD in the context of Tourette's disorder. A small double-blind study found that a relatively high dose of stimulants (MPH up to 45 mg b.i.d or DEX up to 22.5 mg b.i.d. for body weight more than 30 Kg) may increase tic severity during the first week of treatment, but after 3 weeks of treatment with MPH in 17 out 20 patient, tics severity was similar to placebo (only 9 out 20 with DEX; (Castellanos et al., 1997)).

Two recent meta-analyses showed that ADHD medications do not worsen tic severity (Bloch et al., 2009; Cohen et al., 2015): alpha-2 agonists and atomoxetine demonstrated statistically significant improvement in tic symptoms with treatment (ES: 0.74), whereas both methylphenidate and desipramine demonstrated improvement in tic symptoms at trend levels (ES: 0.28) (Bloch et al., 2009).

In the view of this evidence, the EMA no longer consider tics to be an absolute contraindication for the use of stimulants in EU (European Medicines Agency, 2010). Another recent review analyzing 8 RCTs with 510 children with ADHD and chronic Tic Disorder evidenced an ADHD core symptoms improvement with all medications including methylphenidate and dextroamphetamine, although evidence have been rated as low-quality for MPH to very low-quality for DEX. Tic symptoms improved in children treated with methylphenidate, and with a combination of methylphenidate and clonidine, while high-doses of dextroamphetamine appeared to worsen tics in one study. This review confirms previous findings that while stimulants do not worsen tics significantly in most people, they may exacerbate tics at an individual level. In this case, alpha agonists or atomoxetine may represent an alternative pharmacological option (Osland et al., 2018).

Substance Abuse Disorder (SUD)

ADHD and SUDs are frequently comorbid in clinical settings, with a prevalence rate of SUDs two to four times higher in people with ADHD than in the general population (Biederman, 1998; Cumyn et al., 2009). Both the severity and persistence of ADHD symptoms may affect the course of adolescent substance abuse predicting a greater risk for negative outcomes (Dalsgaard et al., 2014) and a reduced effectiveness of treatment (B. S. Molina & Pelham, 2003). Considering their misuse potential, stimulants could be considered potentially dangerous, putatively leading to the development of SUDs: the existing evidence in literature, however, do not confirm this risk, showing on the contrary, that treating ADHD young patients with MPH may reduce their risk for SUD in later life with benefits of MPH treatment outweighing its risks (Chamakalayil et al., 2020).

Studies using data from large databases and registries showed that medication treatment of ADHD, mostly with stimulants, was associated with a significantly reduced risk of SUD outcomes (Chang et al., 2014; Quinn et al., 2017; Steinhausen & Bisgaard, 2014) and that within individuals this reduction in risk of SUD outcomes is related to periods of adherence to the medication prescribed (Quinn et al., 2017; Steinhausen & Bisgaard, 2014). In a long-term follow-up (18 months) of 60 male adult patients with ADHD and comorbid severe SUD receiving compulsory inpatient treatment (39 patients receiving pharmacological treatment for ADHD and 30 untreated), the group that received pharmacological treatment for ADHD exhibited fewer substance abuse relapses, received more frequently voluntary treatments in accordance with a rehabilitation plan, required less frequent compulsory care, were more frequently accommodated in supportive housing or a rehabilitation center, and displayed a higher employment rate than the non-treated group (Bihlar Muld et al., 2015).

Because of the lower DAT availability and occupancy, MPH may result less efficacious in subjects with SUD, especially ADHD patients addicted to cocaine (Martinez et al., 2007; Skoglund et al., 2016). In these cases, Lisdexamfetamine dimesylate (LDX) may represent another safe and effective option for adults with ADHD (Adler et al., 2017; Cortese et al., 2018; Soutullo et al., 2013) also considering its lower misuse potential related to its biological mechanism of enzymatic hydrolysis before to obtain the active compound D-amphetamine (Adler et al., 2017).

In any case, ADHD patients and SUD comorbidity should be closely monitored for possible side effects and potential misuse/diversion of stimulants.

Disruptive Behavioral Problems and Conduct Disorder with Aggression

Aggression is a critical area of clinical concern and stimulants have been extensively studied for the management of aggression. ADHD children show a frequent comorbidity with Oppositional Defiant Disorder (ODD) and/or Conduct Disorder (CD): stimulants efficacy in ameliorating comorbid disruptive behavior including aggression and irritability has been largely demonstrated within systematic reviews and meta-analyses with a medium to large effect size on aggression in pediatric ADHD samples (Connor et al., 2002; Pappadopulos et al., 2006; Pringsheim et al., 2015).

Connor et al. (2002) performed a meta-analysis including 28 RCT (21 on MPH, 5 on amphetamine "AMP" and 2 on pemoline "PEM") in order to investigate the effects of stimulant medications on covert and overt aggression in ADHD children with comorbid ODD/CD. Overt aggression was defined as "aggression resulting in a direct confrontation with the environment" (physical assault, verbal threats, oppositional and defiant behavior, conduct problems, rage attacks, and irritability), whereas covert aggression was intended as "aggression that is furtive and hidden from the environment" (cheating, lying, stealing, and fire-setting). The overall weighted mean effect size was 0.84 for overt aggression and 0.69 for covert aggression in children with ADHD. Effect sizes based on overall ratings were comparable for studies of MPH ($d = 0.80$) and AMP ($d = 0.83$), and dose was not significantly associated with effect size. Stimulants resulted less effective for overt aggression in subjects with a comorbid diagnosis of CD.

The review of Pappadopulos et al. (2006) confirmed that stimulants exert a medium to large effect on pediatric aggression with a mean effect size of 0.78. This slightly lower effect size compared to that of 0.84 found in the above meta-analysis by Connor et al. (2002), has been interpreted by the authors to be related to the statistical influence of the included studies published after 2002.

Pringsheim et al. (2015) substantially performed an update of the Connor et al. (2002) review. The authors included 12 papers published between 2002 and 2013 meeting the same inclusion criteria (i.e., placebo controlled randomized trials examining stimulant effects on aggression-related behaviors within the context of ADHD children and adolescents): of the identified studies 11 were on MPH both immediate release (IR) and long acting formulation, and 1 on lisdexamfetamine (LDX), including a total of 1681 participants. ODD or CD comorbidities were present in between 44% and 93% of subjects. Psychostimulants determined a significant benefit compared to placebo with an effect size of 0.84 (95% CI 0.59–1.10) on teachers' measures and 0.55 (95% CI 0.36–0.73) on parent-rated oppositional, behavior, conduct problems, and aggression.

Using an open label stimulant monotherapy optimization protocol, Blader et al. (2010) showed that in children whose aggressive behavior develops in the context of ADHD and of ODD or CD, a systematic and well-monitored titration of stimulant monotherapy result in clinically relevant reductions in the levels of aggression and prevent the need for additional medications.

Pelham et al. investigated the efficacy of a long-acting mixed amphetamine salt compared to methylphenidate, in a crossover placebo-controlled trial with seven treatment arms using different daily dosages (three MPH, three AMP and one placebo) in 21 patients (aged 6–12 years) with a primary diagnosis of ADHD with or without ODD ($N = 14$) or CD ($N = 5$). No differences in aggression were reported between the two medications, but AMP resulted more effective than methylphenidate on ADHD symptoms (Pelham et al., 1999).

A recent meta-analysis also shows that MPH is clearly an important therapeutic option for comorbid disruptive behavioral problems and aggression in patients with ADHD and conduct disorder (Balia et al., 2018). In this view, psychostimulants should be considered as preferable to medications for psychosis (antipsychotics) in managing aggression and disruptive behaviors owing to fewer adverse effects and a better safety profile (Balia et al., 2018).

Long-Term Efficacy

As discussed above, short-term studies on stimulant medications have robustly confirmed their benefit and efficacy. The long-term effects of stimulants, on the other hand, have been less well investigated and the possibility that their effects may last only as long as the patients continue to be on medication is still matter of discussion.

The majority of RCTs examining stimulant efficacy have a short follow-up time, generally not longer than 3 months (Cortese et al., 2018); the longer ones rarely address the full developmental period from childhood through adolescence (Maia et al., 2017; Taylor, 2019). A few studies investigating the stimulants response after one or more years of treatment by using a discontinuation design found that, after 8 months of treatment, about the 80% of children with

ADHD relapsed when switched from MPH or LDX (D. R. Coghill et al., 2014a) to placebo. Other long-term trials have shown that stimulant effects can be maintained over periods ranging from 12 months (Abikoff et al., 2004; Gillberg et al., 1997) to 24 months (Abikoff et al., 2004).

In a recent double-blind RCT discontinuation study in which subjects treated with methylphenidate for an average of 4.5 years were assigned to randomly continue or discontinue medication, continuation resulted in an on-going benefit with respect to ADHD symptoms (Matthijssen et al., 2019), although effects sizes were smaller compared to the short-term trials. This could be explained as due to different treatment or personal characteristics, to the possible decreased effectiveness of the medication over time, to an inadequate dose adjustment, or the overrepresentation in the study of participants with a mild or resolving presentation of ADHD (Cortese, 2020).

Difficulties in overcoming bias and confounding, in managing costs and ethical issues related to long-term trials made RCTs on long-term effects of ADHD medications extremely challenging (Taylor, 2019).

In order to assess the relative effectiveness of different treatment modalities on long-term outcomes of ADHD in school-age children, the National Institute of Mental Health (NIMH) MTA Study was carried out in the 90s. It still represents one of the most valid sources of information on the long-term effects of MPH. Within this large-scale, random-allocation, non-blind trial, 579 children with ADHD, combined subtype, aged between 7 and 9 years, were recruited at seven US sites and a comparison was made between careful medication management (MedMgt), intensive behaviorally oriented psychosocial therapy (Beh), a combination of these two (Comb), and a community comparison group (CC) that received treatment as usual (usually medication) ("A 14-month randomized clinical trial of treatment strategies for attention-deficit/hyperactivity disorder. The MTA Cooperative Group. Multimodal Treatment Study of Children with ADHD," 1999).

All four treatment groups were reported to have, over time, a significant reduction of ADHD symptoms, with significant differences in degrees of change among groups (MTA Cooperative Group, 1999). Over the period of 14 months of the trial, a significant better outcome was observed in those who had received the "carefully titrated and managed" medication compared to those enrolled in the *Behavioral intervention* group (Beh) or *Community Control* group (CC). Overall, the careful medication management resulted superior to the CC, even though two-thirds of CC subjects were taking ADHD medications prescribed by their personal doctors while enrolled into the study. The outcomes of the *Combined treatment* group (Comb) and *Medication Management* group (MedMgt) treatments did not differ significantly on any measure of ADHD symptoms. However, the combination of behavioral therapy and medication did have some benefits: better control of aggressive behavior at home, improved overall sense of satisfaction of parents, and possibly reduced medication dosage.

At the 2-year follow-up, the benefits persisted; at the 36 months follow-up, however, a time when parents and children were free to choose the actual treatment, all four original groups showed a similar outcome (Jensen et al., 2007). Finally at the 8- and 16-year follows-up no significant advantage was observed for any of the four groups that had originally received the intensive therapies (B. S. G. Molina et al., 2009; Swanson et al., 2017).

Different explanations are possible: the effects of more intensive therapy can disappear when intensive treatment is stopped; after the 14-month point, the groups were no longer as initially randomized and self-selection of patients to treatments at the end of the randomization phase may have led to similar outcome (many children assigned to behavioral intervention started medication, and a significant percentage of those on intensive medication management actually withdrew medication); even for those who remained in the condition of original assignment, the level of fidelity to the detailed protocol of therapy would not have been the same (Taylor, 2019). The most favorable overall development was found in children initially randomized to the MTA medication regime, whether or not they were taking medication at 36 months, thus suggesting some long-lasting benefit for early treated ADHD children.

These findings were confirmed by a major review of 111 studies (4777 children) (T. Spencer et al., 1996a) and meta-analyses (Charach et al., 2011).

Likewise, debate persists over whether the stimulant-induced reductions in ADHD symptoms alleviate the long-term negative outcomes associated with the disorder (Biederman et al., 2009; B. S. G. Molina et al., 2009).

A narrative review and analysis published by Buitelaar et al. (2015) aimed to better understand the relapse of ADHD symptoms upon discontinuation of medication in ADHD subjects who have responded to medication treatment and to explore differences among medications in maintaining treatment response. Randomized withdrawal studies of dexmethylphenidate hydrochloride (d-MPH), methylphenidate modified-release (MPH-LA), lisdexamphetamine dimesylate (LDX), guanfacine extended-release (GXR), and atomoxetine (ATX) in both children/adolescents and adults with ADHD were reviewed. The percentage of relapse was significantly higher and the time-to-relapse significantly shorter with placebo compared to active treatment in patients who were previously stable on 5 weeks to 1 year of active treatment, suggesting clinically significant benefit with continued long-term pharmacotherapy.

Currently, there is a lack of evidence on how long it is appropriate to continue treatment with stimulants for responding subjects and which can be the outcome after a discontinuation following a long-term medication treatment. The maintenance of response and the potential relapse upon discontinuation represents important issues to consider in the management of ADHD-medicated subjects (Buitelaar et al., 2015).

Safety, Tolerability, and Management of the Adverse Events of Dopamine Enhancers (Stimulants)

Stimulants are generally safe and well tolerated; adverse effects range from mild to moderate severity and in most case are transitory, reversible,

and easily manageable by the prescriber (Cortese et al., 2013b; Graham et al., 2011). Common side effects of these dopaminergic agents include sleep problems and insomnia, irritability, appetite and weight loss, headache, increased blood pressure (BP), tachycardia, stomach and abdominal pain.

Most of the adverse effects tend to remit within the first months of treatment, with no significant differences between normal and poor metabolizers (Banaschewski et al., 2006). More severe adverse reactions, including psychotic, neurological symptoms, or allergic reactions, have rarely been observed as a consequence of stimulant medications (Cortese et al., 2013b; Graham et al., 2011).

Despite these information, in 2007 the EMA's Committee for Medicinal Products for Human Use (CHMP) reviewed MPH extensive safety data in order to answer to public concerns over cardiovascular (BP and HR increase and arrhythmias) and cerebrovascular risks (migraine, cerebrovascular accidents, stroke, cerebral infarction, vasculitis, and ischemia). In 2009, the CHMP concluded that the benefit of MPH outweighed the risks in children over 6 years with a formal diagnosis of ADHD, nevertheless CHMP recommended to provide safety information across all EU members for a standardized prescription approach. The CHMP further concluded that research was needed on the long-term effects of MPH (Inglis et al., 2016); therefore, as a result of these recommendations, the Attention Deficit Hyperactivity Disorder Drugs Use Chronic Effects (ADDUCE) consortium was established including experts in the fields of ADHD in order to confirm MPH safety (http://www.adhd-adduce.org).

Cardiovascular Effects

MPH and AMP are sympathomimetic agents that increase noradrenergic and dopaminergic transmission; based on their mechanism of action, there is a potential for an adverse effect on cardiac functioning with an increasing of heart rate (HR) and blood pressure (BP) (Hammerness et al., 2011a; Volkow et al., 2003). Although some authors suggest that MPH and AMP induce only minor increases in BP and HR and do not statistically or clinically significantly alter the QTc trait, with sudden death remaining an extremely rare event (Stiefel & Besag, 2010), data from individual studies suggests that beside small elevations of BP ($\leq$5 mmHg) and HR ($\leq$10 beats/min) at a group level with MPH (Hammerness et al., 2011a) and AMP (Coghill et al., 2014b), around 5–15% of subjects may experience greater increases in HR or BP or may report a cardiovascular adverse event during ADHD medication treatment (Hammerness et al., 2011a).

The 10-year follow-up of the MTA study (Vitiello et al., 2012) found no significant effect of methylphenidate on either systolic or diastolic BP; however, the use of stimulants was associated with a higher HR at follow-up after 3 and 8 years. A recent meta-analysis by Hennissen et al. (2017) including 18 trials with data from 5837 ADHD children and adolescents (80.7% boys) treated with MPH, AMP, or ATX and an average treatment duration of 28.7 weeks, revealed that all three medications were associated with a small, but statistically significant increase in Systolic BP (SBP). Compared to AMP, MPH did not show a significant impact on Diastolic BP (DBP) and HR. Discontinuation of treatment due to any cardiovascular effect was observed in only the 2% of patients. In the majority of patients, the cardiovascular effects resolved spontaneously or with medication dose changes. Another meta-analysis performed in ADHD adult patients showed that compared to placebo, patients randomized to stimulant medications reported a small, but significant, increases in SBP (+2.0 mmHg) and HR (+5.7 bpm), with no effect on DBP (Mick et al., 2013). No changes in ECG parameters, including PR, QRS, and QT intervals, have been associated with the use of MPH and ATX (Hammerness et al., 2011b). The comparative tolerability of oral medications for ADHD in children and adolescents ($n = 11,018$) and adults (5362) was also estimated in a recent network meta-analysis (NMA). Within the results of this NMA, the systolic blood pressure appeared to be slightly increased with the use of amphetamines (SMD 0.09, 95% CI 0.01–0.18) in children and adolescents, and with use of methylphenidate (0.17, 0.05–0.30) in adults, compared with placebo. In terms of diastolic blood pressure, the use of amphetamines (0.21, 0.12–0.31) and methylphenidate (0.24, 0.14–0.33) in children and adults, and methylphenidate (0.20, 0.08–0.32) only in adults, resulted also significantly associated with a correlated medication slight increase compared to placebo (Cortese et al., 2018).

The available epidemiological studies, summarized by Hammerness et al. (2011a), do not show any significant association between ADHD drugs and serious cardiovascular events. A recent large study of 1,200,438 children and young adults between 2 and 24 years found no evidence that ADHD drugs increased the risk of serious cardiovascular events (Cooper et al., 2011). Similarly, several large registry studies of adults (Habel et al., 2011), and children aged 3 to 17 years (Schelleman et al., 2011), suggest that ADHD medication, when medically supervised, was not associated with an increased risk of severe cardiovascular events.

The recent published results from the ADDUCE 24 months prospective study comparing three cohorts (756 medication-naive ADHD nearly to start methylphenidate treatment (MPH group), 391 medication-naive ADHD subjects who did not intend to start any ADHD medication (no-MPH group), and a control group without ADHD, $n = 263$) showed that pulse rate and systolic and diastolic blood pressure were higher in the MPH group compared with the no-MPH group after 24 months of treatment (Man et al., 2023).

Furthermore, in a cross-sectional study including 162 ADHD adolescents and young adults treated with MPH for >2 years and compared to 71 same age population (12–25 years), ADHD subjects who had never received methylphenidate, the 24-h systolic blood pressure (SBP) and heart rate (HR), although significantly higher during daytime in medicated individuals, resulted similar in both groups during the night; the Left Ventricular Mass measured by echocardiography did not also differ between the two groups, reassuring from possible risks of persistent long-term adverse effects on the cardiovascular system (Buitelaar et al., 2022).

In summary, serious unexpected cardiac or psychiatric adverse events associated with stimulants have been known for years and are extremely rare. However, it is a concern that even small, but persistent, BP and/or HR increases may amplify the risk for serious cardiovascular events in the long term: it is therefore important to understand whether these cardiac effects of ADHD medication seen at the group level may translate into clinically relevant adverse event in some individual patients (Hennissen et al., 2017).

Management of Cardiovascular Effects

Current guidelines (Cortese et al., 2013b) recommend that patients being considered for ADHD medication should have an extensive clinical assessment, including identification of any known heart disease, any history of syncope with exercise, and any family history of sudden unexpected death under the age of 40 years. HR and BP should be taken at baseline, after each dose change and repeated every 3–6 months. Both should be measured in the patient at rest. If the first measure is elevated (pulse rate > 100 bpm, systolic or diastolic BP $>$95th percentile), it should be repeated at least twice, within 10 min of interval. It is important using age-adapted cuff sizes for BP measurement and height-adjusted BP percentiles as proposed by the European Society of Hypertension (Parati et al., 2008).

Routine blood tests, EKGs, and echocardiograms are not indicated before initiating stimulant treatment for a patient with unremarkable medical history and physical examination. If the patient has a history of hypertension and/or complains of syncope, arrhythmias, or chest pain, then the clinician should refer the patient for consultation with a (pediatric) cardiologist, since these symptoms may indicate hypertrophic cardiomyopathy, which has been associated with SUD. Where persistent tachycardia (resting heart rate > 110 bpm) or a history suggestive of arrhythmia or familial risk is identified, it is appropriate to request a 24-h ECG. This should be read and reported by an experienced pediatric cardiologist (Cortese et al., 2013b).

Effects on Growth and Pubertal Maturation

One of the most common side effects of ADHD medications is appetite loss, with consequent weight loss and height gain reduction occurring after extended use (Cortese et al., 2013b; Storebø et al., 2018). Although many studies have measured the effects of treatment with stimulant medications on growth, a clear consensus has not been reached on whether the changes in growth are specifically related to stimulants or to other causes such as the condition of ADHD itself.

The mechanism by which growth is affected by treatment with stimulants has not yet been completely clarified. Appetite loss and the consequent reduction in caloric intake represent the obvious probable cause of the growth slowdown (Vitiello, 2008). Other possible mechanisms could include medication effects on hepatic and/or CNS growth factors and direct effects on cartilage (Faraone et al., 2008).

The clinical significance of medication-related reductions in height during development has also been questioned (Vitiello, 2008), and the key question is whether children treated with medication obtain their expected height as adults, or not (Swanson et al., 2017). Studies providing longitudinal data suggest the height deficit is approximately 1 cm/year during the first 3 years of treatment, while other data suggest that these effects tend to attenuate over time and ultimate adult growth parameters are generally not affected (Peyre et al., 2013). Other authors reported that the height or weight changes might be instead a specific feature of ADHD (Hanć & Cieślik, 2008; T. J. Spencer et al., 1996b; Swanson et al., 2007). While, on average, the effects of stimulants on growth are reported to be modest, a substantial discrepancy with some children completely unaffected, and others experiencing a significant growth suppression, has been noted (Faraone et al., 2008).

Swanson and the MTA Cooperative Group (Swanson et al., 2017) re-examined children's physical growth and revealed that the "*New medicated subgroup*" was, at the 36 months follow-up point, 3.04 cm shorter and 2.71 kg lighter than the "*Not medicated group*." During the follow-up observation into adulthood, the prolonged use of MPH in the ADHD group resulted in an average height of 1.29 ± 0.55 cm shorter than the control group ($p < 0.01$, $d = 0.21$). Within the treated sample, adherence to drug treatment was defined as consistent, inconsistent, or negligible: participants belonging to the consistent or inconsistent pattern were 2.55 ± 0.73 cm shorter than the subgroup with the negligible pattern ($p < 0.0005$, $d = 0.42$) (Swanson et al., 2017).

A recent systematic review and meta-analysis including 18 studies with data on 4868 ADHD children and adolescents showed that MPH was associated with a small statistically significant pre-post treatment difference and a consequent possible minimal clinical impact both for height (SMD = 0.27, 95% CI 0.16–0.38) and weight (SMD = 0.33, 95% CI 0.22–0.44) Z scores, with a prominent impact on weight during the first 12 months and on height within the first 24–30 months (Carucci et al., 2021).

In a 2 year open label study of LDX in children and adolescents with ADHD, LDX was associated with a significant small although transient (prominently within the first 36 weeks of treatment) reduction in mean weight, height, and body mass index z-scores and a small increase in the proportion of participants in the lowest weight and BMI categories. Within the same study, there was no evidence of delayed onset of puberty (T. Banaschewski et al., 2018).

The recent naturalistic, longitudinal, controlled ADDUCE study found that there was little evidence of a MPH effect on growth in the medicated population compared to the No-MPH group in the long-term treatment (24 months). Only weight velocity showed an initial slowing at 6 months in the MPH group; however, no group differences were found with respect to BMI at any time point (Man et al., 2023).

Data on the impact of MPH on timing of puberty are currently limited and contrasting. Spencer et al. (1996a), in their study examining 124 ADHD male children and adolescents and a matched control group, did not detect any obvious influence of methylphenidate using a self-staging questionnaire for pubertal maturation. The same result was found in a similar study performed on 124 ADHD girls (Biederman et al., 2003). A recent analysis from the follow up MTA study confirmed the absence of evidence that stimulants can significantly impact the timing of puberty.

Within this study 342 ADHD subjects and a non-ADHD control group ($n = 159$) completed the self-report Tanner staging at the 36-month follow-up assessment. No statistically significant differences in Tanner stages of pubertal development were found between the ADHD and non-ADHD groups at the age of assessment (between 10 and 14 years of age). Further comparisons were made accordingly the medication status comparing 61 ADHD participants who were *medicated* with other ADHD subjects who were *never* ($n = 56$), *newly* ($n = 74$) and *inconsistently* ($n = 116$) medicated with stimulants. No differences on pubertal maturation were found also among these four ADHD medication subgroups (Greenfield et al., 2014). Poulton et al. (2013) reported a delay in pubertal maturation in the long term (after 3 years of continuous treatment with stimulants) in adolescents aged 14 to 16 years. No significant difference in the stage of puberty was found for the sample of boys aged 12.0–13.9 years, suggesting that medication may delay the rate of maturation during puberty but not the onset of puberty (Poulton et al. 2013).

Management of Dopamine Enhancers' (Stimulants)effects on Growth

In order to manage appetite loss, prevent growth suppression, and ensure an adequate growth pattern to subjects in the developmental age, current guidelines (Cortese, 2020; Cortese et al., 2013b) recommend:

- Monitoring appetite, weight, height, and BMI regularly. Height should be measured every 6 months in children and young people. Weight should be measured every 3 months in children ≤ 10 year old; at 3 and 6 months after starting treatment in children >10 year old and young people, and every 6 months thereafter and in adults.
- Differentiating between pre-treatment eating problems and medication-induced eating problems.
- Giving medication after meals, rather than before, if weight loss is of clinical concern; encouraging the use of high-calorific snacks and late-evening meals; planning drug

holidays: i.e., discontinuing medication on weekends or school vacations to allow for catch-up growth;

- Reducing the dose or switching to an alternative class or formulation.
- Referring to a pediatric endocrinologist/growth specialist if height and weight values are below critical thresholds.
- If weight changes in an adult as a result of ADHD pharmacologic treatment, change medication.
- Complaints of upset stomach or nausea may benefit from a change of formulation (i.e., switching from oral to transdermal MPH). Otherwise, the problem can be treated symptomatically with antacid tablets or by changing to a sustained-release MPH formulation, which is absorbed more slowly.

Tics

As stimulant medications can increase the dopaminergic activity in the basal ganglia, from a theoretical point of view; they can contribute to exacerbate tic severity (Albin, 2006). In the past, a history of chronic or episodic tics or a family history of Tourette's disorder has been considered a contraindication to MPH use. However, ADHD and tic disorders are frequently comorbid in the same individual and treatment of these cases with stimulants can result in an improvement of both the ADHD symptoms and the tics, whereas in other cases the tics can worsen.

A meta-analysis, including 9 DB-RPC trials, examining the efficacy of medications for ADHD in children with comorbid tic (n subjects = 477) concluded that MPH generally does *not* worsen tic severity in the short term, although it is possible that MPH may worsen tics in individual cases (Bloch et al., 2009). The Cochrane database systematic review by Pringsheim and Steeves (2011) concluded the same. The recent publication by Krinzinger et al. (2019) including five comparative studies, four non-comparative studies, and five case reports on tics and/or other dyskinesias as a potential adverse outcome of MPH treatment indicated that MPH is generally well tolerated in subjects without a history of tic but overall caution is indicated in those with comorbid tics or tic

disorder especially for the management in the long term.

Management of Tics

The management of tics in ADHD subjects should include the observation of the intensity of tics over a 3-month period before taking any decision regarding ADHD medication. High levels of monitoring and supervision are required, and if the tics worsen over a prolonged period of time (so as not to confuse a medication related increase with the regular waxing of tics), the stimulant should be suspended. It is critical to document the type and severity of tics *before* starting treatment, in order to establish a baseline against which to assess treatment-associated changes. If tics are related to the ADHD treatment, a dose reduction or substitution with a different molecule should be considered; if the previous measures are not effective, consider to add a D2 blocker for the control of tics or to stop medication (Cortese, 2020).

Sleep Problems

ADHD subjects often experience sleep problems and it is therefore important to rule out primary sleep disorders that may mimic or exacerbate ADHD. Both AMP and MPH may increase the latency to sleep onset, shorten sleep duration, and decrease sleep efficiency (Kidwell et al., 2015). Current evidence regarding sleep disorders (Krinzinger et al., 2019) show contrasting results, with three non-comparative studies indicating that MPH is safe/well-tolerated in this regard, as does one case study. Two studies concluded that the relationship between sleep disorders and ADHD is complex, while four comparative studies are unclear. One large comparative study (Cortese et al., 2015) indicates that atomoxetine may cause fewer sleep AEs than MPH.

Management of Sleep Problems

It is suggested to assess sleep at baseline and to continue to monitor it throughout treatment. When behavioral measures (i.e., sleep hygiene) are insufficient and medicated subjects are displaying a good medication response, a review of the possible causes of sleep problems should be performed: treating restless legs syndrome if present; considering changes in the time of dosing, with most of the medication given early in the day and, if stimulants cause a rebound effect, adding small doses of short-acting stimulants in the evening. Should sleep problems persist, an alternative stimulant class or switch drug formulations should be considered, as well as adding an evening dose of melatonin (Cortese, 2020; S. Cortese et al., 2013a).

Seizures, Epilepsy, or EEG Abnormalities

Current published studies do not indicate evidence for seizures as an adverse event of MPH treatment in children with no prior history (Krinzinger et al., 2019). Although more research is needed to confirm the safety of long-term stimulants in children and young people at risk of seizures, current evidence supports the use of MPH for the treatment of ADHD in patients with well-controlled epilepsy and even in those with infrequent seizures (Torres et al., 2008).

Management of Epilepsy and Seizures

When epilepsy is poorly controlled, the frequency of seizures should be carefully monitored; if their frequency increases, or if seizures are new or worsening, ADHD medication should be reviewed and medication that might be contributing to the seizures stopped. ADHD medication should be cautiously reintroduced, if it is unlikely to be the cause of the seizures.

Psychosis and Psychotic Like Symptoms

Drugs for ADHD have rarely been associated with psychosis and/or psychotic-like symptoms as a potential adverse outcome. A review of the literature suggests the absence of any significant association between MPH treatment for ADHD and psychosis. Two studies (Cortese et al., 2015; Paternite et al., 1999) provided evidence that MPH reduces risk of psychotic-like symptoms, and one study (Hechtman et al., 2004) that it reduces the risk hospitalization for psychosis. However, some open-label trial extension studies reported discontinuations due to psychotic symptoms and two comparative studies (Cherland & Fitzpatrick, 1999; Shyu et al., 2015) suggest the

need for caution. One of these (Shyu et al., 2015) was a large cohort study that specifically studied psychotic disorders as a potential adverse outcome of MPH treatment. A long-term safety and efficacy study of LDX in ADHD children and adolescents found psychiatric treatment-related adverse events as infrequent: psychotic symptoms have been reported only for one subject in 191 subjects that completed the study (D. R. Coghill et al., 2017a).

In contrast, some positive clinical experience (drugs for ADHD being helpful) have been reported for managing ADHD symptoms in the context of a psychotic disorder. A single study found that medication with MPH in childhood could show a protective effect by reducing schizotypic features in adults. Analyzing electronic medical records, a population study (K. K. Man et al., 2016) showed that of 20,586 patients (mean age at beginning of observation 6.95 years) who were prescribed with MPH and followed-up for a mean of about 10 years, only 103 reported an incident psychotic event (0.5%). On average, each participant was exposed to MPH for 2.17 years. The overall incidence of psychotic events during the MPH exposure period was 6.14 per 10,000 patient-years. No increased risk was found during MPH-exposed compared with non-exposed periods; however, an increased risk was found during the pre-exposure period, which may be because of an association between psychotic events and the behavioral and attentional symptoms that led to psychiatric assessment and initiation of MPH treatment.

Management of Psychotic Symptoms

If symptoms occur with a therapeutic dose of ADHD medication, reduce the dose or discontinue the drug. Once psychotic symptoms resolve, consider rechallenge with ADHD medication.

Suicidal Ideation and Behaviors

Suicide-related events during treatment with drugs for ADHD have been extensively investigated. Long-term MPH treatment appears inversely related to suicidal behavior.

Three large comparative cohorts and two smaller studies provided in fact evidence in favour of MPH regarding suicidal behavior (Krinzinger et al., 2019). A recent population-based study, analyzing electronic medical records, showed that the incidence of suicide attempts was higher in the period immediately *before* the start of MPH treatment, remaining elevated immediately after the start of MPH treatment but returning to baseline levels during continuation of the treatment. This may reflect emerging psychiatric symptoms that trigger medical consultations, resulting in MPH treatment (K. K. C. Man et al., 2017). Population-based studies indicate that medication for ADHD is associated with a reduced long-term risk (that is, 3 years later) for depression (Chang et al., 2016) and a potential protective effect on suicidal behaviour (Chen et al., 2014). Although the good tolerability in terms of suicidal thoughts and behaviors, individual patients being treated with medications for ADHD should be observed for the emergence of suicidal ideation, as part of routine monitoring.

Irritability, Mood Liability, and Anxiety

Irritability, mood liability, anxiety, hostility, and explosive out-bursts may be observed in 5–10% of children taking ADHD medication, although evidence regarding irritability/emotional reactivity outcomes of long-term MPH treatment is limited. One large comparative study (Cortese et al., 2015) and four smaller non-comparative studies provided evidence in favor of MPH regarding irritability/emotional reactivity, while a small comparative study of pre-schoolers and a case report are unclear (Krinzinger et al., 2019).

It should be taken in account that among children whose aggressive behavior develops in the context of ADHD with oppositional defiant disorder or conduct disorder, systematic, well-monitored titration of mono-therapy often reduces aggression considerably, thus averting the need for additional medications (Blader et al., 2010).

Depression

Low mood and depressive symptoms have been reported as a potential adverse outcome of long-term MPH treatment: nevertheless, several studies provide evidence in favour of a protective effect of MPH on the appearance of low mood/depression.

These include three comparative studies in children and young adults: two large cohort studies with sample sizes >1000 and a RCT and three non-comparative studies (Krinzinger et al., 2019) as well as the recent ADDUCE longitudinal prospective 24 months study (Man et al., 2023).

Substance Misuse, Overuse, and Substance Use Disorder (SUD) Risk

DA-releasing drugs have the potential for misuse and can be diverted by patients or families to this end (Wilens et al., 2008). The extended-release formulations of drugs for ADHD are less prone to diversion because the once-daily administration reduces the requirement of taking medication during school hours and makes parental supervision easier to enforce, and because, according to their physical characteristics, extended-release formulations are also less easily crushed into powder for injection or snorting. Non-dopaminergic medications (that is, ATX) are another option with low abuse potential.

Concerns have been raised that therapeutic use of ADHD medication may result in "sensitization" and possibly increase the risk for substance use disorder (SUD) later in life. Recent evidence suggests that the medical use of CNS stimulants markedly increased in the last years: in a USA adult cohort of approximately 30,000 subjects, about a 80% increasing of adult stimulant prescription was found from 2013 to 2018; stimulant use occurred in a population reporting multiple neurological and mental disorders, prevailing in female, younger, and white race/ethnicity individuals (Moore et al., 2021). The risk of overuse was recently examined also in a Sweden cohort, showing a positive association with previous diagnosis of alcohol and drug misuse, higher age, and previous use of ADHD medication (Bjerkeli et al., 2018). Overall, these data could be correlated to a higher formulation of diagnosis and treatment in the adult population (Chung et al., 2019), however, might suggest any contribution of pharmacological issues (i.e., "sensitization") in explaining the rising overuse of CNS stimulants. Further studies should take into account the concomitant role of other biological, cognitive, sociocultural, and diagnostic biases sustaining this phenomenon, considering that also non-medical use of CNS stimulants has dramatically increased in the last few years (Carlier et al., 2019).

Moreover, ADHD is associated with impulsivity and conduct disturbances, and represents itself a risk factor for SUD. A huge Swedish study of over half a million people found a strong association (more than threefold) between ADHD and subsequent drug use disorders after adjusting for some confounders (i.e., sex and parental education) (Sundquist et al., 2015). Anyway, naturalistic follow-up studies do not support the contention that drugs for ADHD increase the risk for SUD. Among the others, a Swedish registry study found the use of ADHD medication associated with decreased rates of substance use in patients with ADHD (Chang et al., 2014).

In general, it is possible to conclude that ADHD medication could expose to a greater risk of misuse and overuse but remains widely recognized as a protective factor against the risk of generic SUD both for stimulants and for other substances.

See Table 2 for a summary of the monitoring and managing strategies for major adverse events.

Clinical Implication and Recommendation

In general, continuous treatment with medications for ADHD is recommended; for patients who have shown treatment response for a short time, treatment should be continued as long as it remains clinically effective (NICE, 2018).

In patients who are stable, treatment should be reviewed annually, assessing clinical need, benefits, and side effects. In patients who did not have severe problems at baseline (e.g., legal problems, history of severe impulsivity) that could become worse upon treatment discontinuation, the need for continuing treatment should be periodically assessed (NICE, 2018).

When making treatment decisions about patients who are responding well to treatment, it is important to set accurate expectations after discontinuation. Patients who discontinue should be periodically evaluated to assess possible initial relapse symptoms.

Other Indications of Dopamine Enhancers (Stimulants)

Excessive Daytime Sleepiness in Narcolepsy

Narcolepsy is a chronic and disabling neurologic disorder, affecting about 1 in 2000 people worldwide with an onset generally occurring before the age of 16 years (Kornum et al., 2017; Szabo et al., 2019).

The *International Classification of Sleep Disorders–Third Edition* (ICSD-3) includes two main types of the disorder: a Narcolepsy Type 1 (NT1) and a Type 2 (NT2) (American Academy of Sleep Medicine (AASM), 2014). Both types of narcolepsy criteria include (1) chronic daily excessive sleepiness lasting ≥ 3 months; and (2) a polysomnographic testing indicating REM sleep onset of <15 min on night-time sleep, a mean sleep latency time of 8 min on the multiple sleep latency test (MSLT), and two or more sleep onset REM periods (SOREMPs). For a diagnosis of **NT1** also **cataplexy**, and/or reduced cerebrospinal fluid (CSF) levels of hypocretin 1 (orexin A) have to be present. **NT2 narcolepsy** include the same polysomnographic criteria as NT1 but without cataplexy or with a known CSF Hcrt1/orexin-A level < 110 pg/mL (AASM, 2014).

NT1 is probably caused by a deficiency of hypocretin (orexin) signalling, due to a selective loss of hypocretin-producing neurons in the hypothalamus. The mechanism underlying this loss is potentially the result of an autoimmune-related neuronal death mediated by genetic and environmental factors (Kornum et al., 2017; Szabo et al., 2019).

Symptoms of narcolepsy include excessive daytime sleepiness (EDS) (Szabo et al., 2019), cataplexy (an involuntary loss of muscle tone during wakefulness, typically evoked by strong emotions, present in up to 60% of patients), disturbed night-time sleep, hypnagogic and hypnopompic hallucinations, and sleep paralysis (Kornum et al., 2017). EDS has its origins in hypocretin deficiency as the lack of hypocretin reduces excitatory signalling to neurons involved in synthesis of the norepinephrine (NE), dopamine (DA), serotonin (5-hydroxytryptamine

Table 2 Strategies for the management of major adverse effects during treatment with stimulants[a]

Adverse event	Monitoring	Management strategies
Cardiovascular	Perform an extensive clinical assessment, including identification of any heart disease, history of syncope with exercise, and any family history of sudden death under the age of 40 years Measure HR and BP after each dose change and every 3–6 month (standardized charts recommended) EKG and blood tests not mandatory	In case of hypertension and/or complains of syncope, arrhythmias, or chest pain, refer the patient for consultation with a (pediatric) cardiologist If sustained tachycardia (>110 beats/mina t rest), arrhythmia, or SBP >95th percentile (or a clinically significant increase) on two occasions, reduce dose and refer patient to a specialist

Growth and pubertal maturation	Measure height, weight and BMI baseline and every 6 month in children and adolescents with standardized charts
	Measure weight every 3 month in children ≤10 year; at 3 month and 6 month after starting treatment in children >10 year, and every 6 month thereafter or more often if needed
	Differentiate between pre-treatment eating problems and medication-induced eating problems
		If weight loss is a clinical concern:
		    Give medication after meals, rather than before
		    Encourage the use of high-calorific snacks and late evening meals
		    Plan drug holidays
		    Reduce the dose or switch to a different class or formulation
		Refer to a pediatric endocrinologist/growth specialist if height and weight values are below critical thresholds

Gastro-intestinal symptoms	Assess and monitor the type and severity of gastro-intestinal symptoms and their possible relation to stimulants medication	Stomach pain or nausea, may benefit from a change of formulation (i.e., from oral to transdermal, or to a long release formulation) or from antacid tablets
Tics	Document the type and severity of tics before starting treatment in order to establish a baseline level Observe the intensity of tics for 3 months before taking any decision regarding pharmacological treatment	Consider a dose reduction or substitution with a different molecule (e.g., guanfacine, atomoxetine, clonidine) If these measure are not effective, consider to add a D2 blocker or to stop medication

Sleep problems	Assess sleep at baseline and continue to monitor it throughout treatment
	Suggest behavioral measures (sleep hygiene)
	Polysomnography indicated in case of suspicion of sleep-breathing disorder, episodic nocturnal phenomena, limb movements, and unexplained excessive daytime sleepiness
		Review the possible causes of sleep problems:
		    Treat restless legs syndrome if present
		    Consider changing the time of dosing
		    If stimulants cause a rebound effect, add small doses of short-acting stimulants in the evening
		    If sleep problems persist consider an alternative stimulant class or switch drug formulations
		    Consider adding melatonin

Seizures, epilepsy or EEG abnormalities	When epilepsy is poorly controlled, carefully monitor the frequency of seizures	In case of new or worsening seizures, review medication and stop any drug that may contribute to the seizures Cautiously reintroduce ADHD medication if it is not cause of the seizures
Psychosis and psychotic like symptoms	Document the absence of psychotic symptoms before starting treatment and monitor them over the course of drug treatment	In case of psychosis or psychotic symptoms reduce or discontinue the drug Once psychotic symptoms resolve, consider a rechallenge test

[a]Recommendations from the National Institute for Health and Care Excellence (NICE) guidelines and the European ADHD Guidelines Group

[5-HT]), and histamine neurotransmitters, known for their wake promoting properties (Kornum et al., 2017; Szabo et al., 2019). It is quite common that sleepiness in narcolepsy is not recognized as related to a specific neurological disorder and it is often misinterpreted as related to attention deficit hyperactivity disorder (ADHD), particularly in children, where narcolepsy can appear as attention problems, hyperactive behaviors, and emotional lability leading to a wrong diagnosis of ADHD (Scammell, 2015).

The main goals of narcolepsy treatment include improvement of nocturnal sleep, decrease of cataplexy episodes, reduction of abnormal dream phenomena, improvement of alertness and sleepiness, together with improvement of quality of life and safety. However, currently only a scarce number of pharmacological treatments have gained so far specific regulatory approval for the treatment of this disorder and the majority of medications are able to manage only a few symptoms (Szabo et al., 2019). In this scenario, amphetamines, stimulants, and wake-promoting agents appear to be especially useful in managing EDS (Abad & Guilleminault, 2017; Kornum et al., 2017; Szabo et al., 2019) by their action of increasing the release or inhibiting the reuptake of NE and DA.

Modafinil and Armodafinil are considered first-line treatments for the management of EDS.

Modafinil

Modafinil is approved for the treatment of narcolepsy in the USA and Europe. Its mechanism of action results in an increase of DA in the CNS by blocking the DA transporter (Volkow et al., 2009). Modafinil also blocks the reuptake of noradrenaline by the noradrenergic terminals on sleep-promoting neurons in the ventrolateral preoptic area, partially explaining its wake-promoting. Modafinil reaches its maximal plasma levels within 2 h; it is metabolized in the liver into two inactive forms. Its half-life is between 9 and 14 h while the steady state is reached after 2–4 days. When administered at a dosage between 200 and 400 mg once or twice a day it is significantly effective in reducing EDS ("Randomized trial of modafinil as a treatment for the excessive daytime somnolence of narcolepsy: US Modafinil in Narcolepsy Multicenter Study Group," 2000). Common adverse effects include irritability, palpitation, sweating, headache, tremor, nausea, and (rarely) skin rash. Due to his mechanism of action, modafinil has a low risk for dependency, development of tolerance, and impulsive behavior.

Armodafinil

The R-enantiomer of modafinil is approved only in the USA, for the treatment of narcolepsy. Armodafinil has a half-life and adverse effect profile similar to modafinil; however, its plasma concentration is longer lasting, resulting in a more prolonged effect of the drug. Its efficacy has been recognized when administered at the doses of 150–250 mg once per day (Harsh et al., 2006).

Solriamfetol

It is one of the most recently approved treatments for EDS associated with narcolepsy in the USA. It is a dopamine and norepinephrine reuptake inhibitor, with an half-life of 6–7.6 h, and a steady plasma level of about 3 days: solriamfetol showed a significant improvement in sleep latency at daily doses of 150 and 300 mg. Headache, nausea, decreased appetite, nasopharyngitis, dry mouth, and anxiety have been reported as the most commonly adverse events in a Phase 3 clinical trial (Thorpy et al., 2017).

Classical Dopamine Enhancers (Stimulants)

Second-line agents for EDS include methylphenidate and amphetamines (dextroamphetamine or amphetamine–dextromethamphetamine combination, or amphetamine sulfate). These stimulants increase the release of noradrenaline, dopamine, and serotonin and inhibit the reuptake of amines by dopamine transporter. Wakefulness may, therefore, be promoted by increased amine signaling through direct effects on the cortex or via subcortical pathways.

Amphetamines have been used in the context of narcolepsy since the 1930s. At lower doses, amphetamines act by releasing NE whereas, higher doses of amphetamines also inhibit the reuptake transporters (Morón et al., 2002). Adderall (immediate release) is a mixed amphetamine salt that is approved in the USA for the treatment of narcolepsy. D-amphetamine (e.g., Dexedrine), lisdexamphetamine (Vyvanse®), and other amphetamine medications exert similar effects to those of Adderall and are used off-label.

For patients who are non-responders to modafinil/armodafinil and who have been using amphetamines for their EDS, switching to an amphetamine extended-release formulations or to an intermediate-release formulation of methylphenidate may be useful.

Similar to amphetamines, methylphenidate exerts its action by increasing DA and NE release with minimal effects on catecholamine storage. As a consequence of its easily absorption and the rapid passage through the blood-brain barrier, MPH has a rapid wake-promoting effect. A dose between 10 and 60 mg/day has been shown to be significantly effective in improving EDS (Szabo et al., 2019).

A recent population-based, retrospective cohort study in Taiwan showed that the risk of hospitalization for Motor Vehicle Accidents Injuries (MVAI) among subjects with narcolepsy was significantly higher than those without narcolepsy; however, the use of modafinil or methylphenidate, as monotherapy or combined treatment, was associated with a lower risk of MVAI in the narcolepsy cohort (Tzeng et al., 2019).

Due to the possible occurrence of tolerance, many physicians avoid prescribing these medications or tend to prescribe doses that are too small to manage the sleepiness of narcolepsy. It is important to know that narcoleptic subject are often resistant to addiction, and that they generally need doses larger than common people in order to manage their ESD (Turner, 2019). Prescribing amphetamine-based stimulants to people with narcolepsy at doses that are not appropriate for them represents a great disservice. It is important to remember that without adequate stimulant medication patients' ability to participate safely and productively in society is seriously limited (Turner, 2019)

Bing Eating Disorder

Binge eating disorder (BED) is a diagnostic category of eating disorders introduced in 2013 in the *Diagnostic and Statistical Manual of Mental Disorders* (*DSM-5;* APA, 2013). BED diagnostic criteria include episodes of binge eating (i.e., eating large amounts of food while experiencing subjective feelings of loss of control) occurring at least 1 day a week over the past 3 months, marked distress about the binge eating without inappropriate compensatory behaviors, distinguishing this disorder from bulimia nervosa (Reas & Grilo, 2015). BED is the most common eating disorder with a prevalence rate about 2–5% in adults; it is quite equally distributed in both sexes and across different ethnic and minority groups (Marques et al., 2011).

Typically, BED symptoms begin during the developmental age, but the disorder can also be seen in the elderly with longer duration of illness reported (Kornstein et al., 2016). BED is also frequently associated with obesity, and it is also frequently comorbid with medical and psychiatric illnesses (mainly depression and anxiety but also substance use disorder, borderline personality disorder, and bipolar disorder [R. C. Kessler et al., 2013]). Despite its high prevalence rate and its frequent comorbidity with clinical and psychiatric problems, diagnosis of BED is commonly under recognized or mistreated.

A recent meta-analysis (Brownley et al., 2016) concluded that the preferred treatment for BED should include cognitive-behavior therapy (CBT) and medications: lisdexamfetamine (LDX), second-generation medications for depression (antidepressants), sibutramine (discontinued in many countries mainly due to cardiovascular risk), and topiramate.

Apart from LDX, that in 2015 received the approval from the US FDA for moderate/severe BED treatment in adults, no other medications have been approved and currently the other pharmacotherapy options for BED are currently limited to an "off label" use. LDX, a classical ADHD medication, was tested because high levels of impulsivity occur in BED, suggesting that this disorder can be related to a dysfunction of dopamine and/or norepinephrine neurotransmission (R. M. Kessler et al., 2016). LDX received the FDA approval for BED after an intense development program including three short-term trials: one dose-finding Phase II study, two dose-optimization Phase III studies (McElroy et al., 2015b, 2016), and two long-term trials, aiming to investigate its safety and efficacy (Gasior et al., 2017; Hudson et al., 2017). LDX was shown to be effective in reducing binge-eating episodes per week at doses between 50 and 70 mg/day (McElroy et al., 2015b). The recommended starting dose is 30 mg/day with increments of 20 mg/day to achieve the recommended target dose of 50 to 70 mg/day (Appolinario et al., 2019). When examining secondary outcome measures LDX also resulted in an increased weight loss compared to placebo (Reas & Grilo, 2015); however, LDX does not have a formal indication for weight loss.

Considering the strong association between dopamine and eating behaviour, also methylphenidate was investigated for BED. In a randomized study where 22 subjects treated with MPH at flexible dosage ranging from 18–72 mg/day were compared with 27 patients on CBT, both groups presented a comparable remission of binge episodes with a significant difference in BMI at Week 12 for the methylphenidate group (Quilty et al., 2019).

At last, although not a psychostimulant per se, also armodafinil, acting as indirect dopamine receptor agonist, was associated to a significant reduction in binge eating frequency, BMI and obsessive symptoms, at a mean dosage at endpoint evaluation of 216.7 (43.9) mg (McElroy et al., 2015a).

Depression

Major depressive disorder is one of the most impairing mental illnesses with high rates of partial or no response to medications for depression (antidepressants), often requiring more intensive treatment including off label treatments or an augmentation strategy (Kolar, 2018).

Psychostimulants have been considered a pharmacological option in the past, both as a monotherapy or as an augmentation treatment, and are currently commonly prescribed off label to treat depression in adults with MDD (Olfson et al., 2014). According to their pharmacodynamic profile, the use of psychostimulants can target depressive symptoms such as fatigue, apathy, and cognitive difficulties, making these medications

generally more acceptable than medications for psychosis (antipsychotics) (Kolar, 2018).

The acute (24–48 h) effect of stimulants on depressed mood has been well established within several controlled studies, although other placebo-controlled trials in primary or major depression indicated a lack of efficacy of stimulants in the long term (4–6 week) treatment (Satel & Nelson, 1989). Benefits on decreased energy, lack of concentration, decreased alertness, and daytime sleepiness are not long lasting and disappear when stimulants are discontinued.

A recent systematic review and meta-analysis investigating the efficacy of psychostimulants in adults with unipolar or bipolar depression analyzed 21 randomized, placebo-controlled clinical trials on armodafinil, amphetamine, dextroamphetamine, lisdexamfetamine, methylphenidate, or modafinil. The treatment with psychostimulants resulted associated with a statistically significant improvement in depressive symptoms compared to placebo (OR, 1.41; 95% CI, 1.13–1.78; $p = 0.003$) both among subjects with Major Depressive Disorder (MDD) and Bipolar Depression (BD). The highest clinically significant improvements were found for ar/modafinil and dextroamphetamine and subgroup analyses revealed that adjunctive therapy, but not monotherapy, was associated with a clinically significant improvement in depressive symptoms (McIntyre et al., 2017).

In 2013, the systematic review and meta-analysis performed by Goss and coll. (2013), including 4 MDD and 2 BD RCTs, modafinil resulted significantly effective in improving overall depression scores and determined a good remission rate both in MDD and bipolar depression. Modafinil also showed a significant positive effect on fatigue symptoms and result well tolerated.

Currently, however, it can be concluded that data are not sufficient to affirm that psychostimulants can have a putative role as adjunctive or monotherapy for the treatment of depression as most published studies have significant methodological limitations. A favorable hypothesis at the moment can be that psychostimulants may be helpful if appropriately tested in selected domains of psychopathology (e.g., cognitive emotional processing), rather than as "broad-spectrum" medications for depression (antidepressants) (McIntyre et al., 2017).

References

Abad, V. C., & Guilleminault, C. (2017). New developments in the management of narcolepsy. *Nature and Science Sleep, 9*, 39–57. https://doi.org/10.2147/nss.s103467

Abbas, R., Palumbo, D., Walters, F., Belden, H., & Berry, S. A. (2016). Single-dose pharmacokinetic properties and relative bioavailability of a novel methylphenidate extended-release chewable tablet compared with immediate-release methylphenidate chewable tablet. *Clinical Therapeutics, 38*(5), 1151–1157. https://doi.org/10.1016/j.clinthera.2016.02.026

Abikoff, H., Hechtman, L., Klein, R. G., Weiss, G., Fleiss, K., Etcovitch, J., ... Pollack, S. (2004). Symptomatic improvement in children with ADHD treated with long-term methylphenidate and multimodal psychosocial treatment. *Journal of the American Academy of Child and Adolescent Psychiatry, 43*(7), 802–811. https://doi.org/10.1097/01.chi.0000128791.10014.ac

Adamo, N., Seth, S., & Coghill, D. (2015). Pharmacological treatment of attention-deficit/hyperactivity disorder: Assessing outcomes. *Expert Review of Clinical Pharmacology, 8*(4), 383–397. https://doi.org/10.1586/17512433.2015.1050379

Adler, L. A., Goodman, D. W., Kollins, S. H., Weisler, R. H., Krishnan, S., Zhang, Y., & Biederman, J. (2008). Double-blind, placebo-controlled study of the efficacy and safety of lisdexamfetamine dimesylate in adults with attention-deficit/hyperactivity disorder. *The Journal of Clinical Psychiatry, 69*(9), 1364–1373.

Adler, L. A., Lynch, L. R., Shaw, D. M., Wallace, S. P., O'Donnell, K. E., Ciranni, M. A., ... Faraone, S. V. (2017). Effectiveness and duration of effect of open-label lisdexamfetamine dimesylate in adults with ADHD. *Journal of Attention Disorders, 21*(2), 149–157. https://doi.org/10.1177/1087054713485421

Albin, R. L. (2006). Neurobiology of basal ganglia and Tourette syndrome: Striatal and dopamine function. *Advances in Neurology, 99*, 99–106.

American Academy of Sleep Medicine. (2014). International classification of sleep disorders, 3rd ed. Darien, IL: American Academy of Sleep Medicine.

American Psychiatric Association. (2013). *Diagnostic and statistical manual of mental disorders* (5th ed.). American Psychiatric Association.

Appolinario, J. C., Nardi, A. E., & McElroy, S. L. (2019). Investigational drugs for the treatment of binge eating disorder (BED): An update. *Expert Opinion on Investigational Drugs, 28*(12), 1081–1094. https://doi.org/10.1080/13543784.2019.1692813

Arnold, L. E. (2000). Methyiphenidate vs. amphetamine: Comparative review. *Journal of Attention Disorders,* *3*(4), 200–211. https://doi.org/10.1177/108705470000300403

Arnold, V. K., Feifel, D., Earl, C. Q., Yang, R., & Adler, L. A. (2014). A 9-week, randomized, double-blind, placebo-controlled, parallel-group, dose-finding study to evaluate the efficacy and safety of modafinil as treatment for adults with ADHD. *Journal of Attention Disorders, 18*(2), 133–144. https://doi.org/10.1177/1087054712441969

Arnsten, A. F. T., & Pliszka, S. R. (2011). Catecholamine influences on prefrontal cortical function: Relevance to treatment of attention deficit/hyperactivity disorder and related disorders. *Pharmacology, Biochemistry & Behavior, 99*(2), 211–216.

Arnsten, A. F. T. (2011). Catecholamine influences on dorsolateral prefrontal cortical networks. *Biol Psychiatry, 69*(12), e89–99.

Balia, C., Carucci, S., Coghill, D., & Zuddas, A. (2018). The pharmacological treatment of aggression in children and adolescents with conduct disorder. Do callous-unemotional traits modulate the efficacy of medication? *Neuroscience and Biobehavioral Reviews, 91*, 218–238. https://doi.org/10.1016/j.neubiorev.2017.01.024

Banaschewski, T., Coghill, D., Santosh, P., Zuddas, A., Asherson, P., Buitelaar, J., ... Taylor, E. (2006). Long-acting medications for the hyperkinetic disorders. A systematic review and European treatment guideline. *European Child & Adolescent Psychiatry, 15*(8), 476–495.

Banaschewski, T., Johnson, M., Lecendreux, M., Zuddas, A., Adeyi, B., Hodgkins, P., ... Coghill, D. R. (2014). Health-related quality of life and functional outcomes from a randomized-withdrawal study of long-term lisdexamfetamine dimesylate treatment in children and adolescents with attention-deficit/hyperactivity disorder. *CNS Drugs, 28*(12), 1191–1203. https://doi.org/10.1007/s40263-014-0193-z

Banaschewski, T., Johnson, M., Nagy, P., Otero, I. H., Soutullo, C. A., Yan, B., ... Coghill, D. R. (2018). Growth and puberty in a 2-year open-label study of lisdexamfetamine dimesylate in children and adolescents with attention-deficit/hyperactivity disorder. *CNS Drugs, 32*(5), 455–467. https://doi.org/10.1007/s40263-018-0514-8

Barbaresi, W. J., Katusic, S. K., Colligan, R. C., Weaver, A. L., & Jacobsen, S. J. (2007). Modifiers of long-term school outcomes for children with attention-deficit/hyperactivity disorder: Does treatment with stimulant medication make a difference? Results from a population-based study. *Journal of Developmental & Behavioral Pediatrics, 28*(4), 274–287.

Barkley, R. A. (1977). The effects of methylphenidate on various types of activity level and attention in hyperkinetic children. *Journal of Abnormal Child Psychology, 5*(4), 351–369.

Bedard, A. C., Stein, M. A., Halperin, J. M., Krone, B., Rajwan, E., & Newcorn, J. H. (2015). Differential impact of methylphenidate and atomoxetine on sustained attention in youth with attention-deficit/hyperactivity disorder. *Journal of Child Psychology and Psychiatry, 56*(1), 40–48. https://doi.org/10.1111/jcpp.12272

Biederman, J. (1998). Attention-deficit/hyperactivity disorder: A life-span perspective. *Journal of Clinical Psychiatry, 59*(Suppl 7), 4–16.

Biederman, J., & Fried, R. (2012). The effects of lisdexamfetamine dimesylate on the driving performance of young adults with ADHD. *European Neuropsychopharmacology, 22*, S436.

Biederman, J., Lopez, F. A., Boellner, S. W., & Chandler, M. C. (2002). A randomized, double-blind, placebo-controlled, parallel-group study of SLI381 (Adderall XR) in children with attention-deficit/hyperactivity disorder. *Pediatrics, 110*(2 Pt 1), 258–266.

Biederman, J., Faraone, S. V., Monteaux, M. C., Plunkett, E. A., Gifford, J., & Spencer, T. (2003). Growth deficits and attention-deficit/hyperactivity disorder revisited: Impact of gender, development, and treatment. *Pediatrics, 111*(5 Pt 1), 1010–1016. https://doi.org/10.1542/peds.111.5.1010

Biederman, J., Krishnan, S., Zhang, Y., McGough, J. J., & Findling, R. L. (2007). Efficacy and tolerability of lisdexamfetamine dimesylate (NRP-104) in children with attention-deficit/hyperactivity disorder: A phase III, multicenter, randomized, double-blind, forced-dose, parallel-group study. *Clinical Therapeutics, 29*(3), 450–463.

Biederman, J., Monteaux, M. C., Spencer, T., Wilens, T. E., & Faraone, S. V. (2009). Do stimulants protect against psychiatric disorders in youth with ADHD? A 10-year follow-up study. *Pediatrics, 124*(1), 71–78. https://doi.org/10.1542/peds.2008-3347

Bihlar Muld, B., Jokinen, J., Bolte, S., & Hirvikoski, T. (2015). Long-term outcomes of pharmacologically treated versus non-treated adults with ADHD and substance use disorder: A naturalistic study. *Journal of Substance Abuse Treatment, 51*, 82–90. https://doi.org/10.1016/j.jsat.2014.11.005

Birmaher, B., Quintana, H., & Greenhill, L. L. (1988). Methylphenidate treatment of hyperactive autistic children. *Journal of the American Academy of Child & Adolescent Psychiatry, 27*(2), 248–251.

Bjerkeli, P. J., Vicente, R. P., Mulinari, S., Johnell, K., & Merlo, J. (2018). Overuse of methylphenidate: An analysis of Swedish pharmacy dispensing data. *Clinical Epidemiology, 10*, 1657–1665.

Bjork, J. M., Straub, L. K., Provost, R. G., & Neale, M. C. (2017). The ABCD study of neurodevelopment: Identifying neurocircuit targets for prevention and treatment of adolescent substance abuse. *Current Treatment Options in Psychiatry, 4*(2), 196–209. https://doi.org/10.1007/s40501-017-0108-y

Blader, J. C., Pliszka, S. R., Jensen, P. S., Schooler, N. R., & Kafantaris, V. (2010). Stimulant-responsive and stimulant-refractory aggressive behavior among children with ADHD. *Pediatrics, 126*(4), e796–e806. https://doi.org/10.1542/peds.2010-0086

Bloch, M. H., Panza, K. E., Landeros-Weisenberger, A., & Leckman, J. F. (2009). Meta-analysis: Treatment of attention-deficit/hyperactivity disorder in children with comorbid tic disorders. *Journal of the American Academy of Child and Adolescent Psychiatry, 48*(9), 884–893. https://doi.org/10.1097/CHI.0b013e3181b26e9f

Boellner, S. W., Stark, J. G., Krishnan, S., & Zhang, Y. (2010). Pharmacokinetics of lisdexamfetamine dimesylate and its active metabolite, d-amphetamine, with increasing oral doses of lisdexamfetamine dimesylate in children with attention-deficit/hyperactivity disorder: A single-dose, randomized, open-label, crossover study. *Clinical Therapeutics, 32*(2), 252–264. https://doi.org/10.1016/j.clinthera.2010.02.011

Brownley, K. A., Berkman, N. D., Peat, C. M., Lohr, K. N., Cullen, K. E., Bann, C. M., & Bulik, C. M. (2016). Binge-eating disorder in adults: A systematic review and meta-analysis. *Annals of Internal Medicine, 165*(6), 409–420. https://doi.org/10.7326/m15-2455

Brunkhorst-Kanaan, N., Libutzki, B., Reif, A., Larsson, H., McNeill, R. V., & Kittel-Schneider, S. (2021). ADHD and accidents over the life span – A systematic review. *Neuroscience and Biobehavioral Reviews, 125*, 582–591. https://doi.org/10.1016/j.neubiorev.2021.02.002

Buitelaar, J., Asherson, P., Soutullo, C., Colla, M., Adams, D. H., Tanaka, Y., . . . Upadhyaya, H. (2015). Differences in maintenance of response upon discontinuation across medication treatments in attention-deficit/hyperactivity disorder. *European Neuropsychopharmacology, 25*(10), 1611–1621. https://doi.org/10.1016/j.euroneuro.2015.06.003

Buitelaar, J. K., van de Loo-Neus, G. H. H., Hennissen, L., Greven, C. U., Hoekstra, P. J., Nagy, P., Ramos-Quiroga, A., Rosenthal, E., Kabir, S., Man, K. K. C., Ic, W., & Coghill, D. (2022). ADDUCE consortium. Long-term methylphenidate exposure and 24-hours blood pressure and left ventricular mass in adolescents and young adults with attention deficit hyperactivity disorder. *Eur Neuropsychopharmacol, 64*, 63–71. https://doi.org/10.1016/j.euroneuro.2022.09.001

Campbell, M. (1975). Pharmacotherapy in early infantile autism. *Biological Psychiatry, 10*(4), 399–423.

Carlier, J., Giorgetti, R., Vari, M. R., et al. (2019). Use of cognitive enhancers: Methylphenidate and analogs. *European Review for Medical and Pharmacological Sciences, 23*, 3–15.

Carucci, S., Balia, C., Gagliano, A., Lampis, A., Buitelaar, J. K., Danckaerts, M., . . . Zuddas, A. (2021). Long term methylphenidate exposure and growth in children and adolescents with ADHD. A systematic review and meta-analysis. *Neuroscience and Biobehavioral Reviews, 120*, 509–525. https://doi.org/10.1016/j.neubiorev.2020.09.031

Castellanos, F. X., Giedd, J. N., Elia, J., Marsh, W. L., Ritchie, G. F., Hamburger, S. D., & Rapoport, J. L. (1997). Controlled stimulant treatment of ADHD and comorbid Tourette's syndrome: Effects of stimulant and dose. *Journal of the American Academy of Child and Adolescent Psychiatry, 36*(5), 589–596. https://doi.org/10.1097/00004583-199705000-00008

Chamakalayil, S., Strasser, J., Vogel, M., Brand, S., Walter, M., & Dürsteler, K. M. (2020). Methylphenidate for attention-deficit and hyperactivity disorder in adult patients with substance use disorders: Good clinical practice. *Frontiers in Psychiatry, 11*, 540837. https://doi.org/10.3389/fpsyt.2020.540837

Chang, Z., Lichtenstein, P., Halldner, L., D'Onofrio, B., Serlachius, E., Fazel, S., . . . Larsson, H. (2014). Stimulant ADHD medication and risk for substance abuse. *Journal of Child Psychology and Psychiatry, 55*(8), 878–885. https://doi.org/10.1111/jcpp.12164

Chang, Z., D'Onofrio, B. M., Quinn, P. D., Lichtenstein, P., & Larsson, H. (2016). Medication for attention-deficit/hyperactivity disorder and risk for depression: A nationwide longitudinal cohort study. *Biological Psychiatry, 80*(12), 916–922. https://doi.org/10.1016/j.biopsych.2016.02.018

Chang, Z., Ghirardi, L., Quinn, P. D., Asherson, P., D'Onofrio, B. M., & Larsson, H. (2019). Risks and benefits of attention-deficit/hyperactivity disorder medication on behavioral and neuropsychiatric outcomes: A qualitative review of pharmacoepidemiology studies using linked prescription databases. *Biological Psychiatry, 86*(5), 335–343. https://doi.org/10.1016/j.biopsych.2019.04.009

Charach, A., Yeung, E., Climans, T., & Lillie, E. (2011). Childhood attention-deficit/hyperactivity disorder and future substance use disorders: Comparative meta-analyses. *Journal of the American Academy of Child and Adolescent Psychiatry, 50*(1), 9–21. https://doi.org/10.1016/j.jaac.2010.09.019

Chen, Q., Sjölander, A., Runeson, B., D'Onofrio, B. M., Lichtenstein, P., & Larsson, H. (2014). Drug treatment for attention-deficit/hyperactivity disorder and suicidal behaviour: Register based study. *BMJ, 348*, g3769. https://doi.org/10.1136/bmj.g3769

Cherland, E., & Fitzpatrick, R. (1999). Psychotic side effects of psychostimulants: A 5-year review. *Canadian Journal of Psychiatry – Revue Canadienne de Psychiatrie, 44*(8), 811–813.

Childress, A. C., & Berry, S. A. (2010). The single-dose pharmacokinetics of NWP06, a novel extended-release methylphenidate oral suspension. *Postgraduate Medicine, 122*(5), 35–41. https://doi.org/10.3810/pgm.2010.09.2199

Childress, A. C., Kollins, S. H., Cutler, A. J., Marraffino, A., & Sikes, C. R. (2017). Efficacy, safety, and tolerability of an extended-release orally disintegrating methylphenidate tablet in children 6–12 years of age with attention-deficit/hyperactivity disorder in the laboratory classroom setting. *Journal of Child and Adolescent Psychopharmacology, 27*(1), 66–74. https://doi.org/10.1089/cap.2016.0002

Childress, A. C., Foehl, H. C., Newcorn, J. H., Faraone, S. V., Levinson, B., & Adjei, A. L. (2021a). Long-term treatment with extended-release methylphenidate treatment in children aged 4 to <6 years. *Journal of the American Academy of Child and Adolescent Psychiatry.* https://doi.org/10.1016/j.jaac.2021.03.019

Childress, A. C., Kollins, S. H., Cutler, A. J., Marraffino, A., & Sikes, C. R. (2021b). Open-label dose optimization of methylphenidate extended-release orally disintegrating tablet in a laboratory classroom study of children with attention-deficit/hyperactivity disorder. *Journal of Child and Adolescent Psychopharmacology, 31*(5), 342–349. https://doi.org/10.1089/cap.2020.0142

Chung, W., Jiang, S. F., Paksarian, D., Nikolaidis, A., Castellanos, F. X., Merikangas, K. R., & Milham, M. P. (2019). Trends in the prevalence and incidence of attention-deficit/hyperactivity disorder among adults and children of different racial and ethnic groups. *JAMA Network Open, 2*(11), e1914344. https://doi.org/10.1001/jamanetworkopen.2019

Coghill, D., & Seth, S. (2015). Effective management of attention-deficit/hyperactivity disorder (ADHD) through structured re-assessment: The Dundee ADHD Clinical Care Pathway. *Child and Adolescent Psychiatry and Mental Health, 9*, 52. https://doi.org/10.1186/s13034-015-0083-2

Coghill, D. R., Rhodes, S. M., & Matthews, K. (2007). The neuropsychological effects of chronic methylphenidate on drug-naive boys with attention-deficit/hyperactivity disorder. *Biological Psychiatry, 62*(9), 954–962. https://doi.org/10.1016/j.biopsych.2006.12.030

Coghill, D., Banaschewski, T., Lecendreux, M., Soutullo, C., Johnson, M., Zuddas, A., . . . Squires, L. (2013). European, randomized, phase 3 study of lisdexamfetamine dimesylate in children and adolescents with attention-deficit/hyperactivity disorder. *European Neuropsychopharmacology, 23*(10), 1208–1218. https://doi.org/10.1016/j.euroneuro.2012.11.012

Coghill, D. R., Banaschewski, T., Lecendreux, M., Johnson, M., Zuddas, A., Anderson, C. S., . . . Squires, L. A. (2014a). Maintenance of efficacy of lisdexamfetamine dimesylate in children and adolescents with attention-deficit/hyperactivity disorder: Randomized-withdrawal study design. *Journal of the American Academy of Child and Adolescent Psychiatry, 53*(6), 647–657. e641. https://doi.org/10.1016/j.jaac.2014.01.017

Coghill, D. R., Caballero, B., Sorooshian, S., & Civil, R. (2014b). A systematic review of the safety of lisdexamfetamine dimesylate. *CNS Drugs, 28*(6), 497–511. https://doi.org/10.1007/s40263-014-0166-2

Coghill, D. R., Seth, S., Pedroso, S., Usala, T., Currie, J., & Gagliano, A. (2014c). Effects of methylphenidate on cognitive functions in children and adolescents with attention-deficit/hyperactivity disorder: Evidence from a systematic review and a meta-analysis. *Biological Psychiatry, 76*(8), 603–615. https://doi.org/10.1016/j.biopsych.2013.10.005

Coghill, D. R., Banaschewski, T., Nagy, P., Otero, I. H., Soutullo, C., Yan, B., . . . Zuddas, A. (2017a). Long-term safety and efficacy of lisdexamfetamine dimesylate in children and adolescents with ADHD: A phase IV, 2-year, open-label study in Europe. *CNS Drugs, 31*(7), 625–638. https://doi.org/10.1007/s40263-017-0443-y

Coghill, D. R., Banaschewski, T., Soutullo, C., Cottingham, M. G., & Zuddas, A. (2017b). Systematic review of quality of life and functional outcomes in randomized placebo-controlled studies of medications for attention-deficit/hyperactivity disorder. *European Child & Adolescent Psychiatry, 26*(11), 1283–1307. https://doi.org/10.1007/s00787-017-0986-y

Coghill, D. R., Banaschewski, T., Bliss, C., Robertson, B., & Zuddas, A. (2018). Cognitive function of children and adolescents with attention-deficit/hyperactivity disorder in a 2-year open-label study of lisdexamfetamine dimesylate. *CNS Drugs, 32*(1), 85–95. https://doi.org/10.1007/s40263-017-0487-z

Cohen, S. C., Mulqueen, J. M., Ferracioli-Oda, E., Stuckelman, Z. D., Coughlin, C. G., Leckman, J. F., & Bloch, M. H. (2015). Meta-analysis: Risk of tics associated with psychostimulant use in randomized, placebo-controlled trials. *Journal of the American Academy of Child and Adolescent Psychiatry, 54*(9), 728–736. https://doi.org/10.1016/j.jaac.2015.06.011

Connor, D. F. (2002). Preschool attention deficit hyperactivity disorder: A review of prevalence, diagnosis, neurobiology, and stimulant treatment. *Journal of Developmental and Behavioral Pediatrics, 23*(1 Suppl), S1–S9.

Connor, D. F., Glatt, S. J., Lopez, I. D., Jackson, D., & Melloni, R. H., Jr. (2002). Psychopharmacology and aggression. I: A meta-analysis of stimulant effects on overt/covert aggression-related behaviors in ADHD. *Journal of the American Academy of Child and Adolescent Psychiatry, 41*(3), 253–261.

Cooper, W. O., Habel, L. A., Sox, C. M., Chan, K. A., Arbogast, P. G., Cheetham, T. C., ... Ray, W. A. (2011). ADHD drugs and serious cardiovascular events in children and young adults. *The New England Journal of Medicine, 365*(20), 1896–1904. https://doi.org/10.1056/NEJMoa1110212

Cortese, S. (2020). Pharmacologic treatment of attention deficit-hyperactivity disorder. *The New England Journal of Medicine, 383*(11), 1050–1056. https://doi.org/10.1056/NEJMra1917069

Cortese, S., Castelnau, P., Morcillo, C., Roux, S., & Bonnet-Brilhault, F. (2012). Psychostimulants for ADHD-like symptoms in individuals with autism spectrum disorders. *Expert Review of Neurotherapeutics, 12*(4), 461–473. https://doi.org/10.1586/ern.12.23

Cortese, S., Brown, T. E., Corkum, P., Gruber, R., O'Brien, L. M., Stein, M., ... Owens, J. (2013a). Assessment and management of sleep problems in youths with attention-deficit/hyperactivity disorder. *Journal of the American Academy of Child and Adolescent Psychiatry, 52*(8), 784–796.

Cortese, S., Holtmann, M., Banaschewski, T., Buitelaar, J., Coghill, D., Danckaerts, M., ... European, A. G. G. (2013b). Practitioner review: Current best practice in the management of adverse events during treatment with ADHD medications in children and adolescents. *Journal of Child Psychology & Psychiatry & Allied Disciplines, 54*(3), 227–246.

Cortese, S., Panei, P., Arcieri, R., Germinario, E. A., Capuano, A., Margari, L., ... Curatolo, P. (2015). Safety of methylphenidate and atomoxetine in children with attention-deficit/hyperactivity disorder (ADHD): Data from the Italian national ADHD registry. *CNS Drugs, 29*(10), 865–877. https://doi.org/10.1007/s40263-015-0266-7

Cortese, S., Adamo, N., Del Giovane, C., Mohr-Jensen, C., Hayes, A. J., Carucci, S., ... Cipriani, A. (2018). Comparative efficacy and tolerability of medications for attention-deficit hyperactivity disorder in children, adolescents, and adults: A systematic review and network meta-analysis. *Lancet Psychiatry, 5*(9), 727–738. https://doi.org/10.1016/s2215-0366(18)30269-4

Coughlin, C. G., Cohen, S. C., Mulqueen, J. M., Ferracioli-Oda, E., Stuckelman, Z. D., & Bloch, M. H. (2015). Meta-analysis: Reduced risk of anxiety with psychostimulant treatment in children with attention-deficit/hyperactivity disorder. *Journal of Child and Adolescent Psychopharmacology, 25*(8), 611–617. https://doi.org/10.1089/cap.2015.0075

Cox, D. J., Davis, M., Mikami, A. Y., Singh, H., Merkel, R. L., & Burket, R. (2012). Long-acting methylphenidate reduced collision rates of young adult drivers with attention-deficit/hyperactivity disorder. *Journal of Clinical Psychopharmacology, 32*(2), 225–230.

Cumyn, L., French, L., & Hechtman, L. (2009). Comorbidity in adults with attention-deficit hyperactivity disorder. *Canadian Journal of Psychiatry, 54*(10), 673–683. https://doi.org/10.1177/070674370905401004

Dalsgaard, S., Mortensen, P. B., Frydenberg, M., & Thomsen, P. H. (2014). ADHD, stimulant treatment in childhood and subsequent substance abuse in adulthood – A naturalistic long-term follow-up study. *Addictive Behaviors, 39*(1), 325–328. https://doi.org/10.1016/j.addbeh.2013.09.002

Dalsgaard, S., Østergaard, S. D., Leckman, J. F., Mortensen, P. B., & Pedersen, M. G. (2015). Mortality in children, adolescents, and adults with attention deficit hyperactivity disorder: A nationwide cohort study. *Lancet, 385*(9983), 2190–2196. https://doi.org/10.1016/s0140-6736(14)61684-6

Di Martino, A., Melis, G., Cianchetti, C., & Zuddas, A. (2004). Methylphenidate for pervasive developmental disorders: Safety and efficacy of acute single dose test and ongoing therapy: An open-pilot study. *Journal of Child & Adolescent Psychopharmacology, 14*(2), 207–218.

Easton, N., Steward, C., Marshall, F., Fone, K., & Marsden, C. (2007). Effects of amphetamine isomers, methylphenidate and atomoxetine on synaptosomal and synaptic vesicle accumulation and release of dopamine and noradrenaline in vitro in the rat brain. *Neuropharmacology, 52*(2), 405–414. https://doi.org/10.1016/j.neuropharm.2006.07.035

Ermer, J., Homolka, R., Martin, P., Buckwalter, M., Purkayastha, J., & Roesch, B. (2010). Lisdexamfetamine dimesylate: Linear dose-proportionality, low intersubject and intrasubject variability, and safety in an open-label single-dose pharmacokinetic study in healthy adult volunteers. *Journal of Clinical Pharmacology, 50*(9), 1001–1010. https://doi.org/10.1177/0091270009357346

European Medicines Agency. (2010). *Guideline on clinical investigation of medicinal products for the treatment of attention deficit hyperactivity disorder.* http://www.ema.europa.eu/docs/en_GB/document_library/Scientific_guideline/2010/08/WC500095686.pdf

Faraone, S. V., & Buitelaar, J. (2010). Comparing the efficacy of stimulants for ADHD in children and adolescents using meta-analysis. *European Child & Adolescent Psychiatry, 19*(4), 353–364. https://doi.org/10.1007/s00787-009-0054-3

Faraone, S. V., Biederman, J., Morley, C. P., & Spencer, T. J. (2008). Effect of stimulants on height and weight: A review of the literature. *Journal of the American Academy of Child & Adolescent Psychiatry, 47*(9), 994–1009.

Findling, R. L., Childress, A. C., Krishnan, S., & McGough, J. J. (2008). Long-term effectiveness and safety of lisdexamfetamine dimesylate in school-aged children with attention-deficit/hyperactivity disorder. *CNS Spectrums, 13*(7), 614–620.

Findling, R. L., Childress, A. C., Cutler, A. J., Gasior, M., Hamdani, M., Ferreira-Cornwell, M. C., & Squires, L. (2011). Efficacy and safety of lisdexamfetamine dimesylate in adolescents with attention-deficit/hyperactivity disorder. *Journal of the American Academy of Child and Adolescent Psychiatry, 50*(4), 395–405. https://doi.org/10.1016/j.jaac.2011.01.007

Frazier, J. A., Biederman, J., Bellordre, C. A., Garfield, S. B., Geller, D. A., Coffey, B. J., & Faraone, S. V. (2001). Should the diagnosis of attention-deficit/hyperactivity disorder be considered in children with pervasive developmental disorder? *Journal of Attention Disorders, 4*(4), 203–211. https://doi.org/10.1177/108705470100400402

Gajria, K., Lu, M., Sikirica, V., Greven, P., Zhong, Y., Qin, P., & Xie, J. (2014). Adherence, persistence, and medication discontinuation in patients with attention-deficit/hyperactivity disorder – A systematic literature review. *Neuropsychiatric Disease and Treatment, 10*, 1543–1569. https://doi.org/10.2147/ndt.s65721

Gasior, M., Hudson, J., Quintero, J., Ferreira-Cornwell, M. C., Radewonuk, J., & McElroy, S. L. (2017). A phase 3, multicenter, open-label, 12-month extension safety and tolerability trial of lisdexamfetamine dimesylate in adults with binge eating disorder. *Journal of Clinical Psychopharmacology, 37*(3), 315–322. https://doi.org/10.1097/jcp.0000000000000702

Ghuman, J. K., Aman, M. G., Lecavalier, L., Riddle, M. A., Gelenberg, A., Wright, R., ... Fort, C. (2009). Randomized, placebo-controlled, crossover study of methylphenidate for attention-deficit/hyperactivity disorder symptoms in preschoolers with developmental disorders. *Journal of Child and Adolescent Psychopharmacology, 19*(4), 329–339. https://doi.org/10.1089/cap.2008.0137

Gillberg, C., Melander, H., vonKnorring, A. L., Janols, L. O., Thernlund, G., Hagglof, B., ... Kopp, S. (1997). Long-term stimulant treatment of children with attention-deficit hyperactivity disorder symptoms – A randomized, double-blind, placebo-controlled trial. *Archives of General Psychiatry, 54*(9), 857–864.

Goss, A. J., Kaser, M., Costafreda, S. G., Sahakian, B. J., & Fu, C. H. (2013). Modafinil augmentation therapy in unipolar and bipolar depression: a systematic review and meta-analysis of randomized controlled trials. *J Clin Psychiatry, 74*(11), 1101–1107. https://doi.org/10.4088/JCP.13r08560

Graham, J., Banaschewski, T., Buitelaar, J., Coghill, D., Danckaerts, M., Dittmann, R. W., ... Taylor, E. (2011). European guidelines on managing adverse effects of medication for ADHD. *European Child and Adolescent Psychiatry, 20*(1), 17–37.

Greenfield, B., Hechtman, L., Stehli, A., & Wigal, T. (2014). Sexual maturation among youth with ADHD and the impact of stimulant medication. *European Child & Adolescent Psychiatry, 23*(9), 835–839. https://doi.org/10.1007/s00787-014-0521-3

Greenhill, L., Kollins, S., Abikoff, H., McCracken, J., Riddle, M., Swanson, J., ... Cooper, T. (2006). Efficacy and safety of immediate-release methylphenidate treatment for preschoolers with ADHD. *Journal of the American Academy of Child and Adolescent Psychiatry, 45*(11), 1284–1293. https://doi.org/10.1097/01.chi.0000235077.32661.61

Habel, L. A., Cooper, W. O., Sox, C. M., Chan, K. A., Fireman, B. H., Arbogast, P. G., ... Selby, J. V. (2011). ADHD medications and risk of serious cardiovascular events in young and middle-aged adults. *JAMA, 306*(24), 2673–2683.

Hammerness, P. G., Perrin, J. M., Shelley-Abrahamson, R., & Wilens, T. E. (2011a). Cardiovascular risk of stimulant treatment in pediatric attention-deficit/hyperactivity disorder: Update and clinical recommendations. *Journal of the American Academy of Child and Adolescent Psychiatry, 50*(10), 978–990. https://doi.org/10.1016/j.jaac.2011.07.018

Hammerness, P. G., Surman, C. B. H., & Chilton, A. (2011b). Adult attention-deficit/hyperactivity disorder treatment and cardiovascular implications. *Current Psychiatry Reports, 13*(5), 357–363.

Hanć, T., & Cieślik, J. (2008). Growth in stimulant-naive children with attention-deficit/hyperactivity disorder using cross-sectional and longitudinal approaches. *Pediatrics, 121*(4), e967–e974. https://doi.org/10.1542/peds.2007-1532

Harsh, J. R., Hayduk, R., Rosenberg, R., Wesnes, K. A., Walsh, J. K., Arora, S., ... Roth, T. (2006). The efficacy and safety of armodafinil as treatment for adults with excessive sleepiness associated with narcolepsy. *Current Medical Research and Opinion, 22*(4), 761–774. https://doi.org/10.1185/030079906x100050

Heal, D. J., & Pierce, D. M. (2006). Methylphenidate and its isomers – Their role in the treatment of attention-deficit hyperactivity disorder using a transdermal delivery system. *CNS Drugs, 20*(9), 713–738. https://doi.org/10.2165/00023210-200620090-00002

Hechtman, L., Abikoff, H., Klein, R. G., Weiss, G., Respitz, C., Kouri, J., ... Pollack, S. (2004). Academic achievement and emotional status of children with ADHD treated with long-term methylphenidate and multimodal psychosocial treatment. *Journal of the American Academy of Child and Adolescent Psychiatry, 43*(7), 812–819. https://doi.org/10.1097/01.chi.0000128796.84202.eb

Hennissen, L., Bakker, M. J., Banaschewski, T., Carucci, S., Coghill, D., Danckaerts, M., ... Buitelaar, J. K. (2017). Cardiovascular effects of stimulant and non-stimulant medication for children and adolescents with ADHD: A systematic review and meta-analysis of trials of methylphenidate, amphetamines and atomoxetine. *CNS Drugs, 31*(3), 199–215. https://doi.org/10.1007/s40263-017-0410-7

Hodgkins, P., Shaw, M., Coghill, D., & Hechtman, L. (2012). Amfetamine and methylphenidate medications for attention-deficit/hyperactivity disorder: Complementary treatment options. *European Child & Adolescent Psychiatry, 21*(9), 477–492.

Huang, Y. S., Yeh, C. B., Chen, C. H., Shang, C. Y., & Gau, S. S. (2021). A randomized, double-blind, placebo-controlled, two-way crossover clinical trial of ORADUR-methylphenidate for treating children and adolescents with attention-deficit/hyperactivity disorder. *Journal of Child and Adolescent Psychopharmacology, 31*(3), 164–178. https://doi.org/10.1089/cap.2020.0104

Hudson, J. I., McElroy, S. L., Ferreira-Cornwell, M. C., Radewonuk, J., & Gasior, M. (2017). Efficacy of lisdexamfetamine in adults with moderate to severe binge-eating disorder: A randomized clinical trial. *JAMA Psychiatry, 74*(9), 903–910. https://doi.org/10.1001/jamapsychiatry.2017.1889

Inglis, S. K., Carucci, S., Garas, P., Häge, A., Banaschewski, T., Buitelaar, J. K., ... Coghill, D. C. (2016). Prospective observational study protocol to investigate long-term adverse effects of methylphenidate in children and adolescents with ADHD: The Attention Deficit Hyperactivity Disorder Drugs Use Chronic Effects (ADDUCE) study. *BMJ Open, 6*(4), e010433. https://doi.org/10.1136/bmjopen-2015-010433

Ishizuka, T., Murotani, T., & Yamatodani, A. (2012). Action of modafinil through histaminergic and orexinergic neurons. *Vitamins and Hormones, 89*, 259–278. https://doi.org/10.1016/b978-0-12-394623-2.00014-7

Jahromi, L. B., Kasari, C. L., McCracken, J. T., Lee, L. S., Aman, M. G., McDougle, C. J., ... Posey, D. J. (2009). Positive effects of methylphenidate on social communication and self-regulation in children with pervasive developmental disorders and hyperactivity. *Journal of Autism and Developmental Disorders, 39*(3), 395–404. https://doi.org/10.1007/s10803-008-0636-9

Jensen, P. S., Arnold, L. E., Swanson, J. M., Vitiello, B., Abikoff, H. B., Greenhill, L. L., ... Hur, K. (2007). 3-year follow-up of the NIMH MTA study. *Journal of the American Academy of Child and Adolescent Psychiatry, 46*(8), 989–1002.

Kahbazi, M., Ghoreishi, A., Rahiminejad, F., Mohammadi, M. R., Kamalipour, A., & Akhondzadeh, S. (2009). A randomized, double-blind and placebo-controlled trial of modafinil in children and adolescents with attention deficit and hyperactivity disorder. *Psychiatry Research, 168*(3), 234–237. https://doi.org/10.1016/j.psychres.2008.06.024

Kaplan, A., & Adesman, A. (2011). Clinical diagnosis and management of attention deficit hyperactivity disorder in preschool children. *Current Opinion in Pediatrics, 23*(6), 684–692. https://doi.org/10.1097/MOP.0b013e32834cbbba

Kessler, R. C., Berglund, P. A., Chiu, W. T., Deitz, A. C., Hudson, J. I., Shahly, V., ... Xavier, M. (2013). The prevalence and correlates of binge eating disorder in the World Health Organization World Mental Health Surveys. *Biological Psychiatry, 73*(9), 904–914. https://doi.org/10.1016/j.biopsych.2012.11.020

Kessler, R. M., Hutson, P. H., Herman, B. K., & Potenza, M. N. (2016). The neurobiological basis of binge-eating disorder. *Neuroscience and Biobehavioral Reviews, 63*, 223–238. https://doi.org/10.1016/j.neubiorev.2016.01.013

Kidwell, K. M., Van Dyk, T. R., Lundahl, A., & Nelson, T. D. (2015). Stimulant medications and sleep for youth with ADHD: A meta-analysis. *Pediatrics, 136*(6), 1144–1153. https://doi.org/10.1542/peds.2015-1708

Kolar, D. (2018). Addictive potential of novel treatments for refractory depression and anxiety. *Neuropsychiatric Disease and Treatment, 14*, 1513–1519. https://doi.org/10.2147/ndt.s167538

Kornstein, S. G., Kunovac, J. L., Herman, B. K., & Culpepper, L. (2016). Recognizing binge-eating disorder in the clinical setting: A review of the literature. *The Primary Care Companion for CNS Disorders, 18*(3). https://doi.org/10.4088/PCC.15r01905

Kornum, B. R., Knudsen, S., Ollila, H. M., Pizza, F., Jennum, P. J., Dauvilliers, Y., & Overeem, S. (2017). Narcolepsy. *Nature Reviews. Disease Primers, 3*, 16100. https://doi.org/10.1038/nrdp.2016.100

Kortekaas-Rijlaarsdam, A. F., Luman, M., Sonuga-Barke, E., & Oosterlaan, J. (2019). Does methylphenidate improve academic performance? A systematic review and meta-analysis. *European Child & Adolescent Psychiatry, 28*(2), 155–164. https://doi.org/10.1007/s00787-018-1106-3

Krinzinger, H., Hall, C. L., Groom, M. J., Ansari, M. T., Banaschewski, T., Buitelaar, J. K., ... Liddle, E. B. (2019). Neurological and psychiatric adverse effects of long-term methylphenidate treatment in ADHD: A map of the current evidence. *Neuroscience and*

Biobehavioral Reviews, 107, 945–968. https://doi.org/10.1016/j.neubiorev.2019.09.023

Leucht, S., Hierl, S., Kissling, W., Dold, M., & Davis, J. M. (2012). Putting the efficacy of psychiatric and general medicine medication into perspective: Review of meta-analyses. *The British Journal of Psychiatry, 200*(2), 97–106. https://doi.org/10.1192/bjp.bp.111.096594

Li, D. J., Chen, Y. L., Chen, Y. Y., Hsiao, R. C., Lu, W. H., & Yen, C. F. (2021). Increased risk of traumatic injuries among parents of children with attention deficit/hyperactivity disorder: A nationwide population-based study. *International Journal of Environmental Research and Public Health, 18*(7). https://doi.org/10.3390/ijerph18073586

Lichtenstein, P., Halldner, L., Zetterqvist, J., Sjölander, A., Serlachius, E., Fazel, S., . . . Larsson, H. (2012). Medication for attention deficit-hyperactivity disorder and criminality. *The New England Journal of Medicine, 367*(21), 2006–2014. https://doi.org/10.1056/NEJMoa1203241

Liddle, E. B., Hollis, C., Batty, M. J., Groom, M. J., Totman, J. J., Liotti, M., . . . Liddle, P. F. (2011). Task-related default mode network modulation and inhibitory control in ADHD: Effects of motivation and methylphenidate. *Journal of Child Psychology & Psychiatry & Allied Disciplines, 52*(7), 761–771.

Liou, Y. J., Wei, H. T., Chen, M. H., Hsu, J. W., Huang, K. L., Bai, Y. M., . . . Chen, T. J. (2018). Risk of traumatic brain injury among children, adolescents, and young adults with attention-deficit hyperactivity disorder in Taiwan. *The Journal of Adolescent Health, 63*(2), 233–238. https://doi.org/10.1016/j.jadohealth.2018.02.012

Loland, C. J., Mereu, M., Okunola, O. M., Cao, J., Prisinzano, T. E., Mazier, S., . . . Newman, A. H. (2012). R-modafinil (armodafinil): A unique dopamine uptake inhibitor and potential medication for psychostimulant abuse. *Biological Psychiatry, 72*(5), 405–413.

Mahler, S. V., Hensley-Simon, M., Tahsili-Fahadan, P., LaLumiere, R. T., Thomas, C., Fallon, R. V., . . . Aston-Jones, G. (2014). Modafinil attenuates reinstatement of cocaine seeking: Role for cystine-glutamate exchange and metabotropic glutamate receptors. *Addiction Biology, 19*(1), 49–60. https://doi.org/10.1111/j.1369-1600.2012.00506.x

Maia, C. R., Cortese, S., Caye, A., Deakin, T. K., Polanczyk, G. V., Polanczyk, C. A., & Rohde, L. A. (2017). Long-term efficacy of methylphenidate immediate-release for the treatment of childhood ADHD. *Journal of Attention Disorders, 21*(1), 3–13. https://doi.org/10.1177/1087054714559643

Man, K. K., Chan, E. W., Coghill, D., Douglas, I., Ip, P., Leung, L. P., . . . Wong, I. C. (2015). Methylphenidate and the risk of trauma. *Pediatrics, 135*(1), 40–48. https://doi.org/10.1542/peds.2014-1738

Man, K. K., Coghill, D., Chan, E. W., Lau, W. C., Hollis, C., Liddle, E., . . . Wong, I. C. (2016). Methylphenidate and the risk of psychotic disorders and hallucinations in children and adolescents in a large health system. *Translational Psychiatry, 6*(11), e956. https://doi.org/10.1038/tp.2016.216

Man, K. K. C., Coghill, D., Chan, E. W., Lau, W. C. Y., Hollis, C., Liddle, E., . . . Wong, I. C. K. (2017). Association of risk of suicide attempts with methylphenidate treatment. *JAMA Psychiatry, 74*(10), 1048–1055. https://doi.org/10.1001/jamapsychiatry.2017.2183

Man, K. K. C., Häge, A., Banaschewski, T., Inglis, S. K., Buitelaar, J., Carucci, S., Danckaerts, M., Dittmann, R. W., Falissard, B., Garas, P., Hollis, C., Konrad, K., Kovshoff, H., Liddle, E., McCarthy, S., Neubert, A., Nagy, P., Rosenthal, E., Sonuga-Barke, E. J. S., Zuddas, A., Wong, I. C. K., Coghill, D. (2023). ADDUCE Consortium. Long-term safety of methylphenidate in children and adolescents with ADHD: 2-year outcomes of the Attention Deficit Hyperactivity Disorder Drugs Use Chronic Effects (ADDUCE) study. *Lancet Psychiatry, 10*(5), 323–333. https://doi.org/10.1016/S2215-0366(23)00042-1

Markowitz, J. S., & Patrick, K. S. (2008). Differential pharmacokinetics and pharmacodynamics of methylphenidate enantiomers: Does chirality matter? *Journal of Clinical Psychopharmacology, 28*(3 Suppl 2), S54–S61.

Marques, L., Alegria, M., Becker, A. E., Chen, C. N., Fang, A., Chosak, A., & Diniz, J. B. (2011). Comparative prevalence, correlates of impairment, and service utilization for eating disorders across US ethnic groups: Implications for reducing ethnic disparities in health care access for eating disorders. *The International Journal of Eating Disorders, 44*(5), 412–420. https://doi.org/10.1002/eat.20787

Martinez, D., Narendran, R., Foltin, R. W., Slifstein, M., Hwang, D. R., Broft, A., . . . Laruelle, M. (2007). Amphetamine-induced dopamine release: Markedly blunted in cocaine dependence and predictive of the choice to self-administer cocaine. *The American Journal of Psychiatry, 164*(4), 622–629. https://doi.org/10.1176/ajp.2007.164.4.622

Matthijssen, A. M., Dietrich, A., Bierens, M., Kleine Deters, R., van de Loo-Neus, G. H. H., van den Hoofdakker, B. J., et al. (2019). Continued benefits of methylphenidate in ADHD after 2 years in clinical practice: A randomized placebo-controlled discontinuation study. *The American Journal of Psychiatry, 176*(9), 754–762. https://doi.org/10.1176/appi.ajp.2019.18111296

McElroy, S. L., Guerdjikova, A. I., Mori, N., Blom, T. J., Williams, S., Casuto, L. S., & Keck, P. E., Jr. (2015a). Armodafinil in binge eating disorder: A randomized, placebo-controlled trial. *International Clinical Psychopharmacology, 30*(4), 209–215. https://doi.org/10.1097/yic.0000000000000079

McElroy, S. L., Hudson, J. I., Mitchell, J. E., Wilfley, D., Ferreira-Cornwell, M. C., Gao, J., . . . Gasior, M. (2015b). Efficacy and safety of lisdexamfetamine for treatment of adults with moderate to severe binge-eating disorder: A randomized clinical trial. *JAMA*

Psychiatry, 72(3), 235–246. https://doi.org/10.1001/jamapsychiatry.2014.2162

McElroy, S. L., Hudson, J., Ferreira-Cornwell, M. C., Radewonuk, J., Whitaker, T., & Gasior, M. (2016). Lisdexamfetamine dimesylate for adults with moderate to severe binge eating disorder: Results of two pivotal phase 3 randomized controlled trials. *Neuropsychopharmacology, 41*(5), 1251–1260. https://doi.org/10.1038/npp.2015.275

McGough, J. J., Biederman, J., Greenhill, L. L., McCracken, J. T., Spencer, T. J., Posner, K., . . . Swanson, J. M. (2003). Pharmacokinetics of SLI381 (ADDERALL XR), an extended-release formulation of Adderall. *Journal of the American Academy of Child and Adolescent Psychiatry, 42*(6), 684–691. https://doi.org/10.1097/01.chi.0000046850.56865.cb

McIntyre, R. S., Lee, Y., Zhou, A. J., Rosenblat, J. D., Peters, E. M., Lam, R. W., . . . Jerrell, J. M. (2017). The efficacy of psychostimulants in major depressive episodes: A systematic review and meta-analysis. *Journal of Clinical Psychopharmacology, 37*(4), 412–418. https://doi.org/10.1097/jcp.0000000000000723

Mehta, M. A., Owen, A. M., Sahakian, B. J., Mavaddat, N., Pickard, J. D., & Robbins, T. W. (2000). Methylphenidate enhances working memory by modulating discrete frontal and parietal lobe regions in the human brain. *The Journal of Neuroscience, 20*(6), RC65.

Mereu, M., Chun, L. E., Prisinzano, T. E., Newman, A. H., Katz, J. L., & Tanda, G. (2017). The unique psychostimulant profile of ($\pm$)-modafinil: Investigation of behavioral and neurochemical effects in mice. *The European Journal of Neuroscience, 45*(1), 167–174. https://doi.org/10.1111/ejn.13376

Mick, E., McManus, D. D., & Goldberg, R. J. (2013). Meta-analysis of increased heart rate and blood pressure associated with CNS stimulant treatment of ADHD in adults. *European Neuropsychopharmacology, 23*(6), 534–541.

Molina, B. S., & Pelham, W. E., Jr. (2003). Childhood predictors of adolescent substance use in a longitudinal study of children with ADHD. *Journal of Abnormal Psychology, 112*(3), 497–507. https://doi.org/10.1037/0021-843x.112.3.497

Molina, B. S. G., Hinshaw, S. P., Swanson, J. M., Arnold, L. E., Vitiello, B., Jensen, P. S., . . . Houck, P. R. (2009). The MTA at 8 years: Prospective follow-up of children treated for combined-type ADHD in a multisite study. *Journal of the American Academy of Child and Adolescent Psychiatry, 48*(5), 484–500.

Moore, T. J., Wirtz, P. W., Kruszewski, S. P., & Alexander, G. C. (2021). Changes in medical use of central nervous system stimulants among US adults, 2013 and 2018: A cross-sectional study. *BMJ Open,* (11), e048528. https://doi.org/10.1136/bmjopen-2020-048528

Morón, J. A., Brockington, A., Wise, R. A., Rocha, B. A., & Hope, B. T. (2002). Dopamine uptake through the norepinephrine transporter in brain regions with low levels of the dopamine transporter: Evidence from

knock-out mouse lines. *The Journal of Neuroscience, 22*(2), 389–395. https://doi.org/10.1523/jneurosci.22-02-00389.2002

National Institute for Health and Care Excellence. (2018). Attention deficit hyperactivity disorder: diagnosis and management (NG87). nice.org.uk/guidance/ng87

Olfson, M., Kroenke, K., Wang, S., & Blanco, C. (2014). Trends in office-based mental health care provided by psychiatrists and primary care physicians. *The Journal of Clinical Psychiatry, 75*(3), 247–253. https://doi.org/10.4088/JCP.13m08834

Optimizing clinical outcomes across domains of life in adolescents and adults with ADHD. (2011). *The Journal of Clinical Psychiatry, 72*(7), 1008–1014. https://doi.org/10.4088/JCP.10063ah1

Osland, S. T., Steeves, T. D., & Pringsheim, T. (2018). Pharmacological treatment for attention deficit hyperactivity disorder (ADHD) in children with comorbid tic disorders. *Cochrane Database of Systematic Reviews, 6*(6), Cd007990. https://doi.org/10.1002/14651858.CD007990.pub3

Pappadopulos, E., Woolston, S., Chait, A., Perkins, M., Connor, D. F., & Jensen, P. S. (2006). Pharmacotherapy of aggression in children and adolescents: Efficacy and effect size. *Journal of Canadian Academy of Child and Adolescent Psychiatry, 15*(1), 27–39.

Parati, G., Stergiou, G. S., Asmar, R., Bilo, G., de Leeuw, P., Imai, Y., . . . Mancia, G. (2008). European Society of Hypertension guidelines for blood pressure monitoring at home: A summary report of the second international consensus conference on home blood pressure monitoring. *Journal of Hypertension, 26*(8), 1505–1526. https://doi.org/10.1097/HJH.0b013e328308da66

Paternite, C. E., Loney, J., Salisbury, H., & Whaley, M. A. (1999). Childhood inattention-overactivity, aggression, and stimulant medication history as predictors of young adult outcomes. *Journal of Child & Adolescent Psychopharmacology, 9*(3), 169–184.

Patin, A., & Hurlemann, R. (2015). Social cognition. *Handbook of Experimental Pharmacology, 228*, 271–303. https://doi.org/10.1007/978-3-319-16522-6_10

Pelham, W. E., Aronoff, H. R., Midlam, J. K., Shapiro, C. J., Gnagy, E. M., Chronis, A. M., . . . Waxmonsky, J. (1999). A comparison of ritalin and adderall: Efficacy and time-course in children with attention-deficit/hyperactivity disorder. *Pediatrics, 103*(4), e43.

Pereira-Sanchez, V., Franco, A. R., Vieira, D., de Castro-Manglano, P., Soutullo, C., Milham, M. P., & Castellanos, F. X. (2021). Systematic review: Medication effects on brain intrinsic functional connectivity in patients with attention-deficit/hyperactivity disorder. *Journal of the American Academy of Child and Adolescent Psychiatry, 60*(2), 222–235. https://doi.org/10.1016/j.jaac.2020.10.013

Peyre, H., Hoertel, N., Cortese, S., Acquaviva, E., Limosin, F., & Delorme, R. (2013). Long-term effects of ADHD medication on adult height: Results from the NESARC. *Journal of Clinical Psychiatry, 74*(11), 1123–1124.

A. Zuddas and S. Carucci

Pliszka, S. R. (2007). Pharmacologic treatment of attention-deficit/hyperactivity disorder: Efficacy, safety and mechanisms of action. *Neuropsychology Review, 17*(1), 61–72.

Pliszka, S. R., Wilens, T. E., Bostrom, S., Arnold, V. K., Marraffino, A., Cutler, A. J., . . . Newcorn, J. H. (2017). Efficacy and safety of HLD200, delayed-release and extended-release methylphenidate, in children with attention-deficit/hyperactivity disorder. *Journal of Child and Adolescent Psychopharmacology, 27*(6), 474–482. https://doi.org/10.1089/cap.2017.0084

Posey, D. J., McDougle, C. J., Aman, M. G., Arnold, L. E., Scahill, L., McCracken, J. T., . . . Network, R. A. (2004). A randomized, double-blind, placebo-controlled, crossover trial of methylphenidate in children with hyperactivity associated with pervasive developmental disorders. *Neuropsychopharmacology, 29*(Suppl 1), S142–S143.

Poulton, A. S., Melzer, E., Tait, P. R., Garnett, S. P., Cowell, C. T., Baur, L. A., & Clarke, S. (2013). Growth and pubertal development of adolescent boys on stimulant medication for attention deficit hyperactivity disorder. *Medical Journal of Australia, 198*(1), 29–32.

Pringsheim, T., & Steeves, T. (2011). Pharmacological treatment for attention deficit hyperactivity disorder (ADHD) in children with comorbid tic disorders. *Cochrane Database of Systematic Reviews, 4*, CD007990. https://doi.org/10.1002/14651858.CD007990.pub2

Pringsheim, T., Hirsch, L., Gardner, D., & Gorman, D. A. (2015). The pharmacological management of oppositional behaviour, conduct problems, and aggression in children and adolescents with attention-deficit hyperactivity disorder, oppositional defiant disorder, and conduct disorder: A systematic review and meta-analysis. Part 1: Psychostimulants, alpha-2 agonists, and atomoxetine. *Canadian Journal of Psychiatry, 60*(2), 42–51. https://doi.org/10.1177/0706743715 6000202

Quilty, L. C., Allen, T. A., Davis, C., Knyahnytska, Y., & Kaplan, A. S. (2019). A randomized comparison of long acting methylphenidate and cognitive behavioral therapy in the treatment of binge eating disorder. *Psychiatry Research, 273*, 467–474. https://doi.org/10.1016/j.psychres.2019.01.066

Quinn, P. D., Chang, Z., Hur, K., Gibbons, R. D., Lahey, B. B., Rickert, M. E., . . . D'Onofrio, B. M. (2017). ADHD medication and substance-related problems. *The American Journal of Psychiatry, 174*(9), 877–885. https://doi.org/10.1176/appi.ajp.2017.16060686

Randomized trial of modafinil as a treatment for the excessive daytime somnolence of narcolepsy: US Modafinil in Narcolepsy Multicenter Study Group. (2000). *Neurology, 54*(5), 1166–1175. https://doi.org/10.1212/wnl.54.5.1166

Reas, D. L., & Grilo, C. M. (2015). Pharmacological treatment of binge eating disorder: Update review and synthesis. *Expert Opinion on Pharmacotherapy, 16*(10), 1463–1478. https://doi.org/10.1517/14656566.2015.1053465

Reichow, B., Volkmar, F. R., & Bloch, M. H. (2013). Systematic review and meta-analysis of pharmacological treatment of the symptoms of attention-deficit/hyperactivity disorder in children with pervasive developmental disorders. *Journal of Autism and Developmental Disorders, 43*(10), 2435–2441. https://doi.org/10.1007/s10803-013-1793-z

Robertson, S. D., Matthies, H. J., & Galli, A. (2009). A closer look at amphetamine-induced reverse transport and trafficking of the dopamine and norepinephrine transporters. *Molecular Neurobiology, 39*(2), 73–80. https://doi.org/10.1007/s12035-009-8053-4

Rosenau, P. T., Openneer, T. J. C., Matthijssen, A. M., van de Loo-Neus, G. H. H., Buitelaar, J. K., van den Hoofdakker, B. J., . . . Dietrich, A. (2021). Effects of methylphenidate on executive functioning in children and adolescents with ADHD after long-term use: A randomized, placebo-controlled discontinuation study. *Journal of Child Psychology and Psychiatry.* https://doi.org/10.1111/jcpp.13419

Rubia, K., Halari, R., Mohammad, A. M., Taylor, E., & Brammer, M. (2011). Methylphenidate normalizes frontocingulate underactivation during error processing in attention-deficit/hyperactivity disorder. *Biological Psychiatry, 70*(3), 255–262. https://doi.org/10.1016/j.biopsych.2011.04.018

Ruiz-Goikoetxea, M., Cortese, S., Aznarez-Sanado, M., Magallón, S., Alvarez Zallo, N., Luis, E. O., . . . Arrondo, G. (2018). Risk of unintentional injuries in children and adolescents with ADHD and the impact of ADHD medications: A systematic review and meta-analysis. *Neuroscience and Biobehavioral Reviews, 84*, 63–71. https://doi.org/10.1016/j.neubiorev.2017.11.007

Santosh, P. J., Baird, G., Pityaratstian, N., Tavare, E., & Gringras, P. (2006). Impact of comorbid autism spectrum disorders on stimulant response in children with attention deficit hyperactivity disorder: A retrospective and prospective effectiveness study. *Child: Care, Health and Development, 32*(5), 575–583. https://doi.org/10.1111/j.1365-2214.2006.00631.x

Satel, S. L., & Nelson, J. C. (1989). Stimulants in the treatment of depression: A critical overview. *The Journal of Clinical Psychiatry, 50*(7), 241–249.

Scammell, T. E. (2015). Narcolepsy. *The New England Journal of Medicine, 373*(27), 2654–2662. https://doi.org/10.1056/NEJMra1500587

Schelleman, H., Bilker, W. B., Strom, B. L., Kimmel, S. E., Newcomb, C., Guevara, J. P., . . . Hennessy, S. (2011). Cardiovascular events and death in children exposed and unexposed to ADHD agents. *Pediatrics, 127*(6), 1102–1110.

Schmitt, K. C., & Reith, M. E. (2011). The atypical stimulant and nootropic modafinil interacts with the dopamine transporter in a different manner than classical cocaine-like inhibitors. *PLoS One, 6*(10), e25790. https://doi.org/10.1371/journal.pone.0025790

Schweitzer, J. B., Lee, D. O., Hanford, R. B., Zink, C. F., Ely, T. D., Tagamets, M. A., ... Kilts, C. D. (2004). Effect of methylphenidate on executive functioning in adults with attention-deficit/hyperactivity disorder: Normalization of behavior but not related brain activity. *Biological Psychiatry, 56*(8), 597–606. https://doi.org/10.1016/j.biopsych.2004.07.011

Shyu, Y. C., Yuan, S. S., Lee, S. Y., Yang, C. J., Yang, K. C., Lee, T. L., & Wang, L. J. (2015). Attention-deficit/hyperactivity disorder, methylphenidate use and the risk of developing schizophrenia spectrum disorders: A nationwide population-based study in Taiwan. *Schizophrenia Research, 168*(1–2), 161–167. https://doi.org/10.1016/j.schres.2015.08.033

Siffel, C., Page, M., Maxwell, T., Thun, B., Kolb, N., Rosenlund, M., ... Keja, J. (2020). Patterns of lisdexamfetamine dimesylate use in children, adolescents, and adults with attention-deficit/hyperactivity disorder in Europe. *Journal of Child and Adolescent Psychopharmacology, 30*(7), 439–447. https://doi.org/10.1089/cap.2019.0173

Simonoff, E., Taylor, E., Baird, G., Bernard, S., Chadwick, O., Liang, H., ... Jichi, F. (2013). Randomized controlled double-blind trial of optimal dose methylphenidate in children and adolescents with severe attention deficit hyperactivity disorder and intellectual disability. *Journal of Child Psychology and Psychiatry, 54*(5), 527–535. https://doi.org/10.1111/j.1469-7610.2012.02569.x

Skoglund, C., Brandt, L., Almqvist, C., D'Onofrio, B. M., Konstenius, M., Franck, J., & Larsson, H. (2016). Factors associated with adherence to methylphenidate treatment in adult patients with attention-deficit/hyperactivity disorder and substance use disorders. *Journal of Clinical Psychopharmacology, 36*(3), 222–228. https://doi.org/10.1097/jcp.0000000000000501

Sobanski, E., Sabljic, D., Alm, B., Skopp, G., Kettler, N., Mattern, R., & Strohbeck-Kuhner, P. (2008). Driving-related risks and impact of methylphenidate treatment on driving in adults with attention-deficit/hyperactivity disorder (ADHD). *Journal of Neural Transmission, 115*(2), 347–356.

Soutullo, C., Banaschewski, T., Lecendreux, M., Johnson, M., Zuddas, A., Anderson, C., ... Coghill, D. R. (2013). A post hoc comparison of the effects of lisdexamfetamine dimesylate and osmotic-release oral system methylphenidate on symptoms of attention-deficit hyperactivity disorder in children and adolescents. *CNS Drugs, 27*(9), 743–751. https://doi.org/10.1007/s40263-013-0086-6

Spencer, T., Biederman, J., Wilens, T., Harding, M., Odonnell, B. A. D., & Griffin, S. (1996a). Pharmacotherapy of attention-deficit hyperactivity disorder across the life cycle. *Journal of the American Academy of Child and Adolescent Psychiatry, 35*(4), 409–432. https://doi.org/10.1097/00004583-199604000-00008

Spencer, T. J., Biederman, J., Harding, M., O'Donnell, D., Faraone, S. V., & Wilens, T. E. (1996b). Growth deficits in ADHD children revisited: Evidence for disorder-associated growth delays? *Journal of the American Academy of Child & Adolescent Psychiatry, 35*(11), 1460–1469.

Stein, M. A., Waldman, I. D., Charney, E., Aryal, S., Sable, C., Gruber, R., & Newcorn, J. H. (2011). Dose effects and comparative effectiveness of extended release dexmethylphenidate and mixed amphetamine salts. *Journal of Child and Adolescent Psychopharmacology, 21*(6), 581–588. https://doi.org/10.1089/cap.2011.0018

Steinhausen, H. C., & Bisgaard, C. (2014). Substance use disorders in association with attention-deficit/hyperactivity disorder, co-morbid mental disorders, and medication in a nationwide sample. *European Neuropsychopharmacology, 24*(2), 232–241. https://doi.org/10.1016/j.euroneuro.2013.11.003

Stiefel, G., & Besag, F. M. C. (2010). Cardiovascular effects of methylphenidate, amphetamines and atomoxetine in the treatment of attention-deficit hyperactivity disorder. *Drug Safety, 33*(10), 821–842.

Storebø, O. J., Ramstad, E., Krogh, H. B., Nilausen, T. D., Skoog, M., Holmskov, M., ... Gluud, C. (2015). Methylphenidate for children and adolescents with attention deficit hyperactivity disorder (ADHD). *Cochrane Database of Systematic Reviews,* (11), Cd009885. https://doi.org/10.1002/14651858.CD009885.pub2

Storebø, O. J., Pedersen, N., Ramstad, E., Kielsholm, M. L., Nielsen, S. S., Krogh, H. B., ... Gluud, C. (2018). Methylphenidate for attention deficit hyperactivity disorder (ADHD) in children and adolescents – Assessment of adverse events in non-randomised studies. *Cochrane Database of Systematic Reviews, 5*(5), Cd012069. https://doi.org/10.1002/14651858.CD012069.pub2

Sturman, N., Deckx, L., & van Driel, M. L. (2017). Methylphenidate for children and adolescents with autism spectrum disorder. *Cochrane Database of Systematic Reviews, 11*(11), Cd011144. https://doi.org/10.1002/14651858.CD011144.pub2

Sundquist, J., Ohlsson, H., Sundquist, K., & Kendler, K. S. (2015). Attention-deficit/hyperactivity disorder and risk for drug use disorder: A population-based follow up and co-relative study. *Psychological Medicine, 45*(5), 977–983.

Swanson, J. M., Wigal, S. B., Wigal, T., Sonuga-Barke, E., Greenhill, L. L., Biederman, J., ... Hatch, S. J. (2004). A comparison of once-daily extended-release methylphenidate formulations in children with attention-deficit/hyperactivity disorder in the laboratory school (the Comacs study). *Pediatrics, 113*(3 Pt 1), e206–e216.

Swanson, J. M., Elliott, G. R., Greenhill, L. L., Wigal, T., Arnold, L. E., Vitiello, B., ... Volkow, N. D. (2007). Effects of stimulant medication on growth rates across 3 years in the MTA follow-up. *Journal of the American Academy of Child and Adolescent Psychiatry, 46*(8), 1015–1027. https://doi.org/10.1097/chi.0b013e3180686d7e

Swanson, J. M., Arnold, L. E., Molina, B. S. G., Sibley, M. H., Hechtman, L. T., Hinshaw, S. P., et al. (2017).

Young adult outcomes in the follow-up of the multi-modal treatment study of attention-deficit/hyperactivity disorder: Symptom persistence, source discrepancy, and height suppression. *Journal of Child Psychology and Psychiatry, 58*(6), 663–678. https://doi.org/10.1111/jcpp.12684

Szabo, S. T., Thorpy, M. J., Mayer, G., Peever, J. H., & Kilduff, T. S. (2019). Neurobiological and immunogenetic aspects of narcolepsy: Implications for pharmacotherapy. *Sleep Medicine Reviews, 43*, 23–36. https://doi.org/10.1016/j.smrv.2018.09.006

Tanda, G., Hersey, M., Hempel, B., Xi, Z. X., & Newman, A. H. (2021). Modafinil and its structural analogs as atypical dopamine uptake inhibitors and potential medications for psychostimulant use disorder. *Current Opinion in Pharmacology, 56*, 13–21. https://doi.org/10.1016/j.coph.2020.07.007

Taylor, E. (2019). ADHD medication in the longer term. *Zeitschrift für Kinder- und Jugendpsychiatrie und Psychotherapie, 47*(6), 542–546. https://doi.org/10.1024/1422-4917/a000664

Taylor, E., Dopfner, M., Sergeant, J., Asherson, P., Banaschewski, T., Buitelaar, J., ... Zuddas, A. (2004). European clinical guidelines for hyperkinetic disorder – First upgrade. *European Child and Adolescent Psychiatry, Supplement, 13*(1), I/7–I/30.

The MTA Cooperative Group. Multimodal Treatment Study of Children with ADHD. (1999). A 14-month randomized clinical trial of treatment strategies for attention-deficit/hyperactivity disorder. *Archives of General Psychiatry, 56*(12), 1073–1086. https://doi.org/10.1001/archpsyc.56.12.1073

The World Health Organization Quality of Life assessment (WHOQOL): Position paper from the World Health Organization. (1995). *Social Science & Medicine, 41*(10), 1403–1409. https://doi.org/10.1016/0277-9536(95)00112-k

Thorpy, M. J., Dauvilliers, Y., Shapiro, C., Mayer, G., Corser, B. C., Chen, D., ... Emsellem, H. (2017). 0675 a randomized, placebo-controlled, phase 3 study of the safety and efficacy of JZP-110 for the treatment of excessive sleepiness in patients with narcolepsy. *Sleep, 40*(Suppl_1), A250–A250. https://doi.org/10.1093/sleepj/zsx050.674

Torres, A. R., Whitney, J., & Gonzalez-Heydrich, J. (2008). Attention-deficit/hyperactivity disorder in pediatric patients with epilepsy: Review of pharmacological treatment. *Epilepsy & Behavior, 12*(2), 217–233. https://doi.org/10.1016/j.yebeh.2007.08.001

Tourette's Syndrome Study Group. (2002). Treatment of ADHD in children with tics: A randomized controlled trial. *Neurology, 58*(4), 527–536. https://doi.org/10.1212/wnl.58.4.527

Turner, M. (2019). The treatment of narcolepsy with amphetamine-based stimulant medications: A call for better understanding. *Journal of Clinical Sleep Medicine, 15*(5), 803–805. https://doi.org/10.5664/jcsm.7788

Tzeng, N. S., Hsing, S. C., Chung, C. H., Chang, H. A., Kao, Y. C., Mao, W. C., ... Chien, W. C. (2019). The risk of hospitalization for motor vehicle accident injury in narcolepsy and the benefits of stimulant use: A nationwide cohort study in Taiwan. *Journal of Clinical Sleep Medicine, 15*(6), 881–889. https://doi.org/10.5664/jcsm.7842

Vitiello, B. (2008). Understanding the risk of using medications for attention deficit hyperactivity disorder with respect to physical growth and cardiovascular function. *Child & Adolescent Psychiatric Clinics of North America, 17*(2), 459–474. xi.

Vitiello, B., Elliott, G. R., Swanson, J. M., Arnold, L. E., Hechtman, L., Abikoff, H., ... Gibbons, R. (2012). Blood pressure and heart rate over 10 years in the multimodal treatment study of children with ADHD. *American Journal of Psychiatry, 169*(2), 167–177.

Volkow, N. D., Wang, G.-J., Fowler, J. S., Logan, J., Franceschi, D., Maynard, L., ... Swanson, J. M. (2002). Relationship between blockade of dopamine transporters by oral methylphenidate and the increases in extracellular dopamine: Therapeutic implications. *Synapse, 43*(3), 181–187.

Volkow, N. D., Wang, G. J., Fowler, J. S., Molina, P. E., Logan, J., Gatley, S. J., ... Swanson, J. M. (2003). Cardiovascular effects of methylphenidate in humans are associated with increases of dopamine in brain and of epinephrine in plasma. *Psychopharmacology, 166*(3), 264–270. https://doi.org/10.1007/s00213-002-1340-7

Volkow, N. D., Fowler, J. S., Wang, G.-J., Telang, F., Logan, J., Wong, C., Ma, J., Pradhan, K., Benveniste, H., & Swanson, J. M. (2008). Methylphenidate decreased the amount of glucose needed by the brain to perform a cognitive task. *PLoS One, 3*(4), e2017. https://doi.org/10.1371/journal.pone.0002017

Volkow, N. D., Fowler, J. S., Logan, J., Alexoff, D., Zhu, W., Telang, F., ... Apelskog-Torres, K. (2009). Effects of modafinil on dopamine and dopamine transporters in the male human brain: Clinical implications. *JAMA, 301*(11), 1148–1154. https://doi.org/10.1001/jama.2009.351

Weisler, R., Young, J., Mattingly, G., Gao, J., Squires, L., & Adler, L. (2009). Long-term safety and effectiveness of lisdexamfetamine dimesylate in adults with attention-deficit/hyperactivity disorder. *CNS Spectrums, 14*(10), 573–585.

Wenthur, C. J. (2016). Classics in chemical neuroscience: Methylphenidate. *ACS Chemical Neuroscience, 7*(8), 1030–1040. https://doi.org/10.1021/acschemneuro.6b00199

Wigal, T., Greenhill, L., Chuang, S., McGough, J., Vitiello, B., Skrobala, A., ... Stehli, A. (2006). Safety and tolerability of methylphenidate in preschool children with ADHD. *Journal of the American Academy of Child and Adolescent Psychiatry, 45*(11), 1294–1303. https://doi.org/10.1097/01.chi.0000235082.63156.27

Wigal, S. B., Nordbrock, E., Adjei, A. L., Childress, A., Kupper, R. J., & Greenhill, L. (2015). Efficacy of methylphenidate hydrochloride extended-release capsules

(Aptensio XR™) in children and adolescents with attention-deficit/hyperactivity disorder: A phase III, randomized, double-blind study. *CNS Drugs, 29*(4), 331–340. https://doi.org/10.1007/s40263-015-0241-3

Wilens, T. E., Adler, L. A., Adams, J., Sgambati, S., Rotrosen, J., Sawtelle, R., . . . Fusillo, S. (2008). Misuse and diversion of stimulants prescribed for ADHD: A systematic review of the literature. *Journal of the American Academy of Child & Adolescent Psychiatry, 47*(1), 21–31.

Wilens, T. E., Faraone, S. V., Hammerness, P. G., Pliszka, S. R., Uchida, C. L., DeSousa, N. J., . . . Newcorn, J. H. (2021). Clinically meaningful improvements in early morning and late afternoon/evening functional impairment in children with ADHD treated with delayed-release and extended-release methylphenidate. *Journal of Attention Disorders*, 10870547211020073. https://doi.org/10.1177/10870547211020073

Wisor, J. (2013). Modafinil as a catecholaminergic agent: Empirical evidence and unanswered questions. *Frontiers in Neurology, 4*, 139. https://doi.org/10.3389/fneur.2013.00139

Wolraich, M. L., Hagan, J. F., Jr., Allan, C., Chan, E., Davison, D., Earls, M., . . . Zurhellen, W. (2019). Clinical practice guideline for the diagnosis, evaluation, and treatment of attention-deficit/hyperactivity disorder in children and adolescents. *Pediatrics, 144*(4). https://doi.org/10.1542/peds.2019-2528

Zhu, H. J., Patrick, K. S., Yuan, H. J., Wang, J. S., Donovan, J. L., DeVane, C. L., . . . Markowitz, J. S. (2008). Two CES1 gene mutations lead to dysfunctional carboxylesterase 1 activity in man: Clinical significance and molecular basis. *American Journal of Human Genetics, 82*(6), 1241–1248. https://doi.org/10.1016/j.ajhg.2008.04.015

Zuddas, A., Banaschewski, T., Coghill, D., & Stein, M. A. (2018). ADHD treatment: Psychostimulants. In Banaschewski, D. Coghill, & A. Zuddas (Eds.), *Oxford textbook of attention deficit hyperactivity disorder* (pp. 379–292). Oxford University Press.

Cognitive Enhancers and Treatments for Alzheimer's Disease

D. P. Devanand and R. Fremont

Contents

This chapter is an update from the 4th edition. Previous edition author was D. P. Devanand

D. P. Devanand (✉)
Division of Geriatric Psychiatry, New York State
Psychiatric Institute, College of Physicians and Surgeons
of Columbia University, New York, NY, USA
e-mail: dpd3@cumc.columbia.edu

R. Fremont
Department of Psychiatry, Icahn Mount Sinai School of
Medicine, New York, NY, USA
e-mail: rachel.fremont@mssm.edu

© Springer Nature Switzerland AG 2024
A. Tasman et al. (eds.), *Tasman's Psychiatry*,
https://doi.org/10.1007/978-3-030-51366-5_139

Abstract

Alzheimer's disease is a devastating yet common neurodegenerative illness affecting over six million people in the United States alone. Unfortunately, there are currently no available therapies that can modify the trajectory of this disorder and fully prevent progression of this illness. However, there are seven medications that are approved by the U.S. Food and Drug Administration (FDA) to treat the symptoms of Alzheimer's disease. These medications include memantine, donepezil, galantamine, rivastigmine, aducanumab, lecanemab, and a medication that combines memantine and donepezil. In this chapter, the neuropathology of Alzheimer's disease is discussed. A detailed description of the medications approved for Alzheimer's disease and an overview of good prescribing practices for these medications is also provided. Further, a summary of other treatments and interventions that are not approved by the FDA but have some evidence of efficacy in addressing the symptoms of Alzheimer's disease are provided. Although there is no cure for Alzheimer's disease, there are treatments that can improve the lives of patients with this disorder.

Keywords

Alzheimer's disease · Dementia · Acetylcholinesterase Inhibitors · AchEIs · Memantine · Donepezil · Amyloid · Tau · Aducanumab · Lecanemab

History and Background

Prior to the 1970s, various medications proposed for the treatment of dementia included monoamine releasers (psychostimulants), vasodilators, ergoloids, and medication "cocktails." In the United States, only one medication, ergoloid mesylates (dihydroergotoxin mesylate), was approved for an ill-defined condition, senile mental decline. By the late 1980s, the advent of research-based diagnostic criteria for the dementia of Alzheimer's disease (AD) (McKhann et al., 1984), an initial understanding of its underlying pathology, and the development of mechanism-based pharmacological therapeutics provided the framework for clinical trials to exploit new treatment strategies.

DSM-5 has broad categories, Mild Neurocognitive Disorder and Major Neurocognitive Disorder, that encompass a broad range of patients with cognitive symptoms and deficits (American Psychiatric Association, 2013). Dementia is defined as being a Major Neurocognitive Disorder whereas mild cognitive impairment (MCI) is considered a Mild Neurocognitive Disorder. MCI was originally defined on the basis of a verbal episodic memory deficit that markedly increases the risk of transitioning to a clinical diagnosis of dementia over time (Petersen et al., 1995). In research clinics, a diagnosis of MCI typically leads to a conversion rate to dementia, primarily AD, of about 12–15% per year (Petersen et al., 2001). However, MCI is not necessarily a predementia stage of AD, since many people who fulfill criteria do not progress to dementia and some improve to not being cognitively impaired (Devanand et al., 2008).

During the 1990s, research-and consensus-based criteria emerged to address vascular dementia (Chui et al., 1992; Roman et al., 1993), dementia with Lewy bodies (McKeith et al., 1996), frontotemporal dementia syndrome, and Parkinson's disease dementia. Subsequently, National Institute on Aging-Alzheimer's Association Criteria have been developed that also include biomarkers in addition to clinical criteria (McKhann et al., 2011; Albert et al., 2011). While neuroimaging biomarkers are used in research settings including clinical trials, recent developments suggest that blood-based bio-markers eventually may replace expensive brain imaging procedures in the diagnostic process (Blennow & Zetterberg, 2018).

Regulatory and Methodological Issues

The US Food and Drug Administration (FDA) utilizes de facto guidelines for establishing that a drug has "antidementia efficacy." These require that (1) clinical trials are double-blind and placebo-controlled, (2) patients fulfill criteria for a primary dementia such as AD [e.g., DSM-5 or NIA criteria (McKhann et al., 2011), and (3) appropriate efficacy instruments are used. Further, a putative antidementia drug must show efficacy at improving memory or retarding its deterioration, since memory impairment is one of the primary features of dementia. The drug also must have an effect, determined independently from neuropsychological assessment, using a global clinical measure or a measure of functional incapacity in order to address the concern that drug-related memory improvement may be observed with psychometric testing but not be clinically meaningful. The FDA guidelines that patients fulfill established criteria for a primary dementia is an attempt to avoid the inclusion of ill-defined dementia syndromes in clinical trials, to obtain sample homogeneity, and to achieve expert consensus that the medication is effective in a specific group of patients with an illness such as AD.

Most clinical trials of antidementia drugs undertaken in the United States to seek regulatory approval for mild/moderate AD have used the Alzheimer's Disease Assessment Scale – Cognitive Subscale (ADAS-Cog) (Mohs et al., 1997) as the index of cognitive change and a version of a clinical global impression of change (CGIC) or the clinical dementia rating scale (CDR) as the "global" clinical measure (Schneider et al., 1997). The ADAS-Cog includes the following measures: word recall, naming, commands, constructional and ideational praxis, orientation, word recognition, spoken language, comprehension of spoken language, word finding, and recall of test instructions. The CGIC is usually rendered

as a seven-point, ordinal scale upon which a clinician indicates their impression of clinical change from "no change" to "minimal," "moderate," or "marked" improvement or worsening. The CDR is a scale used by clinicians based on six domains of cognitive and functional ability. The CDR scores are 0 for normal, 0.5 for very mild dementia (often includes amnestic MCI), 1 for mild dementia, 2 for moderate dementia, and 3 for severe dementia. Trials conducted in people with moderate and severe dementia have used other cognitive measures, such as the Severe Impairment Battery (SIB) (Saxton et al., 1990) and/or the Alzheimer's Disease Cooperative Study – Activities of Daily Living (ADCS-ADL) Scale to assess aspects of daily functioning (Galasko et al., 1997).

This chapter focuses on the dementia of AD. Many of the issues raised will become relevant to other dementias or cognitive impairment syndromes with further research.

Typical Inclusion Criteria for AD Clinical Trials

These include the presence of probable AD (NINCDS-ADRDA Work Group criteria) (McKhann et al., 1984) and NIA-AA criteria; a specified Mini Mental State Examination (MMSE) (Folstein et al., 1975) score (e.g., typically between 12 and 26 for "mild to moderate AD"); a score of less than 4 on the Modified Hachinski Scale (Rosen et al., 1980); a computed tomography (CT) or magnetic resonance imaging (MRI) scan consistent with AD with no evidence of significant focal lesions; generally good physical health other than the dementia (confirmed by medical history, physical examination, neurological examination, ECG, and laboratory tests); normal blood pressure; and having a reliable caregiver to participate in clinical evaluations. Some research studies also require a positive PET amyloid scan or CSF amyloid Beta and tau/phospho tau levels consistent with AD.

Exclusion criteria typically consist of a history of other psychiatric or neurological disorders, of a stroke or CT/MRI evidence of a stroke, significant concurrent physical illness, or abnormal laboratory findings. Therefore, dementia populations studied in clinical trials consist mainly of mildly-to-moderately impaired outpatients living at home with their families, who are otherwise medically healthy and lack significant behavioral symptoms. Many trials of potential disease-modifying drugs have expanded inclusion criteria to allow the concurrent use of approved antidementia therapies, for example, cholinesterase inhibitors (CheIs) and memantine.

Etiopathology and Implications for Treatment

This chapter offers a survey of treatments for AD ranging from approved therapies to some agents that have been investigated in depth or are in clinical trials. The survey is organized according to the underlying rationale for the therapy. Certain approaches of theoretical interest will also be mentioned, since the field is changing quickly. The review begins with a schematic summary of the pathobiology of AD, in order to provide a theoretical framework within which to present the wide array of potential therapies.

Neuropathology of AD

The definitive diagnosis of AD is based on the clinical dementia syndrome together with certain neuropathologic features first described by Alois Alzheimer in 1907. These include neuritic *plaques* (primarily amyloid) and neurofibrillary *tangles* (primarily abnormally phosphorylated tau protein) spread diffusely through the cerebral cortex and hippocampus (Khachaturian, 1985; Mirra et al., 1991) (Fig. 1). Plaques, found outside the neuron, are spherical structures possessing a central core of amyloid protein surrounded by distended abnormal neuronal dendrites and small axons. Tangles consist of bundles of filaments in cell bodies, axons, and dendrites. These histologic abnormalities are associated with loss of synaptic density and neuronal death. Much has been learned in recent years about the chain of events leading to these defining features of AD, and this has led to a number of drug development strategies.

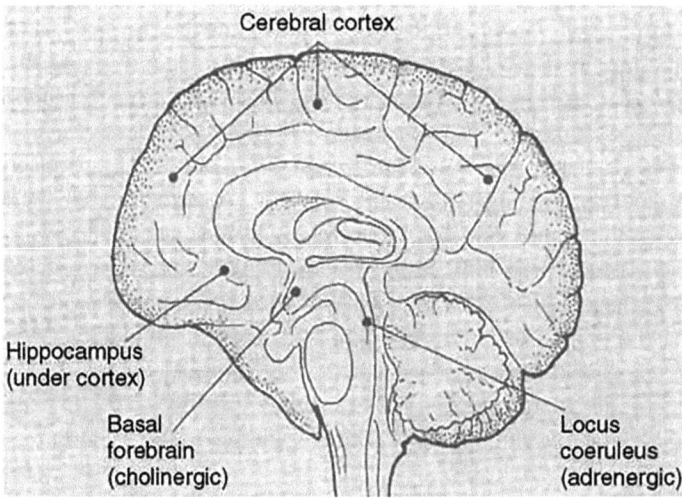

Fig. 1 The right half of a human brain (facing left) showing some of the areas affected by AD. The hippocampus is affected early in the course of the disease, which accounts for short-term memory losses. As the disease progresses, the cerebral cortex is increasingly involved, causing the cognitive and personality defects. Regions that are the sources of transmitters to wide areas of the brain also form neurofibrillary tangles, including the cholinergic basal forebrain, the noradrenergic locus coeruleus, and the serotoninergic dorsal raphe (not shown). (Source: Tariot et al. (1993). Reproduced with permission of John Wiley & Sons, Ltd.)

Dominant Theories of AD Pathobiology

The "amyloid cascade" hypothesis posits that dysregulation in amyloid precursor protein (APP) occurs early in the illness and leads to increased production of amyloid beta protein (pathological peptide $A\beta_{1-42}$) and other $A\beta$ peptide fragments. This in turn leads to a number of downstream events including the development of extracellular amyloid plaques that may accelerate the formation of neurofibrillary pathology (tangles) as well as synaptic loss, and cell death. This hypothesis is supported by the fact that a macromolecular form of $A\beta$ is toxic by itself. The exact reasons why $A\beta$ leads to synaptic dysfunction and cell death are not fully elucidated, but important insights have been gleaned. Proteolytic cleavage of APP can occur via three different secretases, α, β, and γ. When α-secretase cleaves APP, generation of pathogenic $A\beta$ species is precluded. When α-secretase activity is diminished, or that of β-secretase is increased, the generation of pathogenic $A\beta$ species is enhanced, a process also involving subsequent cleavage by γ-secretase. The discovery of presenilin 1 and 2 mutations in familial AD kindreds and their central role in aberrant APP processing further strengthens the amyloid hypothesis.

Another theory is the tau (a protein that stabilizes microtubules which, when defective, is thought to lead to CNS dysfunction) and tangle hypothesis. Plaques and tangles in AD can be separated in time and space, and it is actually the extent of tangle, not plaque, burden that correlates with the overall severity of dementia. Further, in mice genetically modified to overexpress familial APP mutations, there is relatively little neuronal death, abnormal phosphorylation of tau, or tangle burden, suggesting that aberrant APP processing may not be sufficient to account for the molecular and pathological phenotype of AD. Finally, there are certain mutations in the gene coding for tau that have been identified; these people will develop the clinical phenotype of dementia with a tangle-only pathological phenotype. There are, therefore, non-amyloidogenic pathways for the development of dementia supported by evolving cellular and molecular studies of the evolution of tangles.

Much of this work focuses on aberrant kinase phosphorylation of tau, with consequences that would be expected to include synaptic dysfunction, impaired cellular signaling, and vulnerability to a variety of stressors including $A\beta$ accumulation.

A Pathogenic Cascade Hypothesis

For purposes of organizing emerging treatment strategies, a broad, hypothetical schematic view of the evolution of AD is displayed in Fig. 2. The disease begins relatively early in life, and the times of onset of cognitive impairment and dementia are influenced by a variety of interacting factors such as aging; genetic susceptibility or resistance; environmental susceptibility factors such as toxin exposure, head injury, or smoking; possible exogenous protective factors such as exposure to certain medications; and endogenous susceptibility or protective factors such as menopause, diet, exercise, degree of educational attainment, major depression, chronic psychological stress, or even lack of social connectedness. One theory is that the balance between stress and resilience tips at some point that varies from person to person, initiating a series of molecular events such as oxidative stress, excitotoxicity, altered APP processing, or early apoptosis. Compensatory mechanisms may protect against cell injury for some period of time, such as alterations in DNA repair, and enzyme upregulation. These stress responses eventually break down, leading to neuronal injury in common cellular signaling pathways and eventual loss of synapses, neurotransmitter failure, and cell death. The molecular phenotype may precede both the neuropathology and clinical phenotype by many years, perhaps decades; and the clinical manifestations may not emerge until there is significant cellular damage.

The neurodegenerative process is asynchronous, meaning that viable neurons coexist with damaged or dead cells, and are highly variable both within an individual and among individuals over time. The fact that both inherited and sporadic forms of AD share many clinical and pathological features supports the hypothesis that there is a convergence of biochemical abnormalities downstream from the various initiating factors.

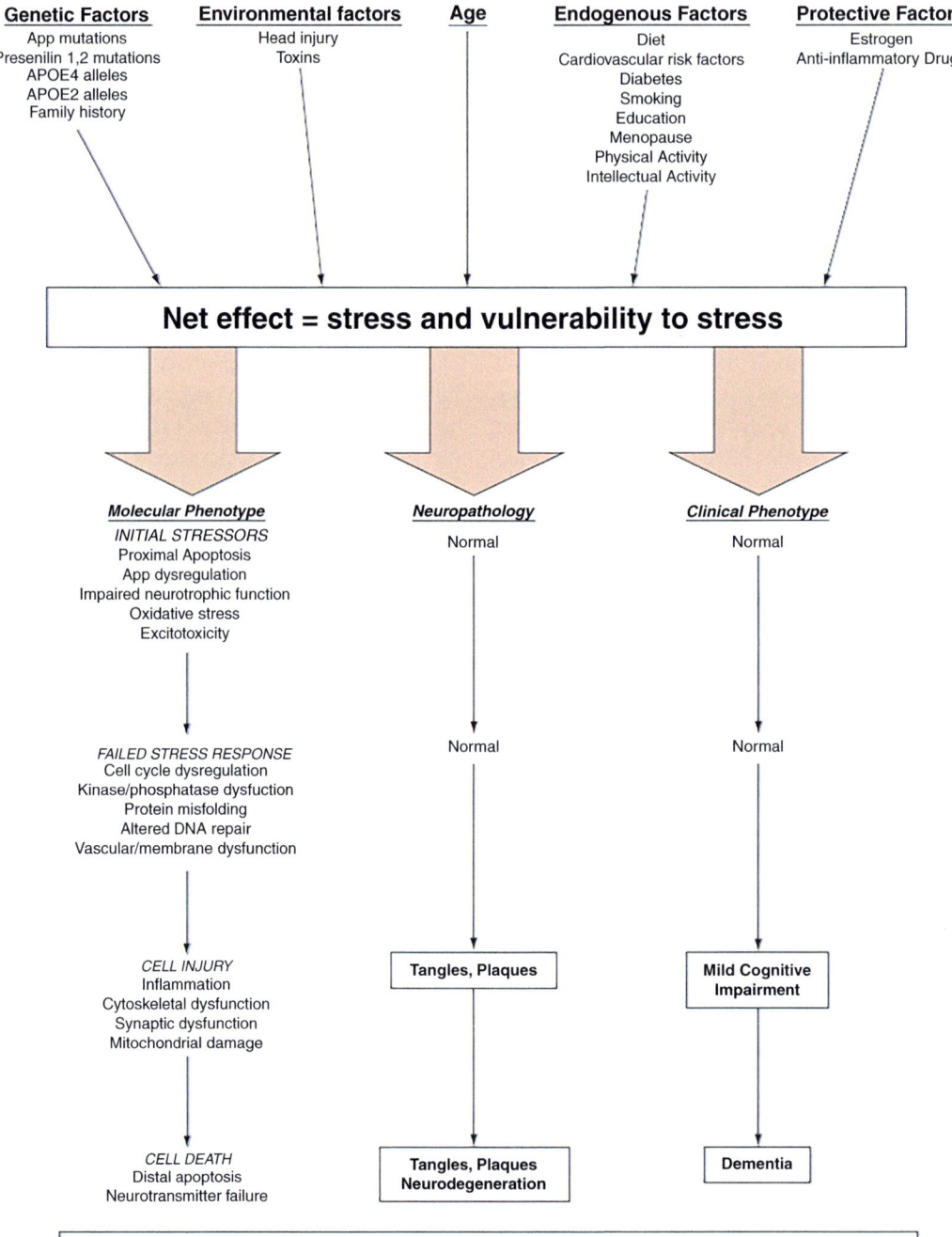

Fig. 2 The evolution of AD based on the stress hypothesis. (Modified from Tariot and Federoff (2003). Reproduced with permission of Lippincott Williams & Wilkins)

Each aspect of this cascade offers a potential therapeutic target. For instance, neurotransmitter-based therapies might afford symptomatic relief without altering the pathological outcome. Impaired signaling has been identified in cholinergic and glutamatergic pathways, and restorative therapies targeting cholinergic and glutamatergic neurotransmission have been approved as treatments for AD, though with limited short-term and no long-term disease-modifying effects. Interventions that decrease pathogenic $A\beta$ production or fibrillization of $A\beta$ or decrease tau protein phosphorylation may prevent neuronal degeneration, and treatments with neuroprotective properties might result in enhanced cellular resistance and slower decline in clinical deterioration. These hopes have yet to be realized despite numerous large clinical trials.

Treatment Paradigms

Approaches to the treatment of AD can be grouped into several conceptual categories (Table 1). One approach attempts to treat the associated features of the illness such as sleep-wake cycle, appetitive disruption, or behavioral disorders such as aggression, psychosis, agitation, depression, anxiety, or apathy. This approach, although clinically of considerable importance, is not the focus of this chapter. A second approach addresses the cognitive signs of the illness such as memory, language, praxis, attention, and orientation. Most neurotransmitter-based therapies have this goal in mind. A third approach aims to slow the rate of progression of the illness, preserving patients' quality of life or autonomy. A fourth approach is primary prevention, delaying the time to onset of illness. Success with this approach could have considerable impact. For example, delaying the onset of AD by 5 years would halve its incidence (Jorm et al., 1987; Brookmeyer et al., 1998). Indeed, important modifiable risk factors including physical activity and diet have already been identified (Scarmeas et al., 2009). The identification of at-risk and preclinical AD populations is becoming important for targeting primary and secondary prevention clinical trials in AD (Pillai & Cummings, 2013).

Neurotransmitter-Based Approaches

Activity of cerebral cortical choline acetyltransferase (ChAT), the key enzyme in acetylcholine synthesis (Bowen et al., 1976; Davies &

Table 1 Conceptualized treatment strategies for patients with cognitive impairment

Symptomatic and/or Restorative

Targets: Impaired cognition, depression, psychosis, agitation, aggression, anxiety, insomnia

Examples: CheIs, various cholinergic agonists, medications for depression (antidepressants), medications for psychosis (antipsychotics), medications for bipolar disorder (mood stabilizers), antianxiety agents, medications for insomnia (hypnotics), NMDA and AMPA receptor modulators, angiotensin-converting enzyme inhibitors

Neurotrophic factors: NGF, brain-derived neurotrophic factors, estrogens

Note: Some substances may have symptomatic or restorative effects but are unproven. These may include Hydergine and neotropics such as piracetam

Pathophysiologically Directed

Targets: Underlying pathophysiology of neurodegeneration, including inflammation, production of oxidizing free radicals, excitatory amino acids

Examples: Anti-inflammatory agents, calcium channel blockers, NMDA and AMPA receptor modulators. Transplantation of hormonally active tissues, or NGF gene therapy using viral vectors have been undertaken experimentally

Etiologically Directed

Targets: β-Amyloid formation or hyperphosphorylated tau protein

Examples: Modulators of APP expression, β-and γ-secretase inhibitors, inhibitors of β-amyloid protein aggregation or deposition, active or passive immunization with antibodies to β-amyloid

Note: The interventions listed above include a few that are available and marketed, but most have not been demonstrated effective, and some conceptual treatments are not yet developed

Maloney, 1976; Perry et al., 1977), is reduced, along with the loss of cholinergic cell bodies in the nucleus basalis (Whitehouse et al., 1982; Arendt et al., 1983). There are correlations between cortical ChAT reduction or nucleus basalis cell reduction and plaque densities in cortical areas (Arendt et al., 1985) (Fig. 3). The cholinergic deficit correlates with the decline in scores on the Blessed-Roth Dementia Rating Scale (Perry et al., 1978). Over time, it became clear that many other neuronal systems also are affected in this disorder.

Variable involvement of noradrenergic systems, including the nucleus locus coeruleus (nLC), occurs in AD (see Fig. 3) (Bondareff et al., 1982). Significantly greater neuronal loss in the nLC occurred in AD patients who showed

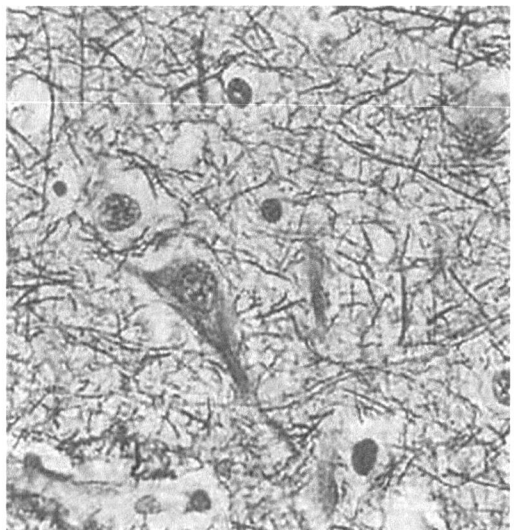

Fig. 3 Human brain tissue in AD, showing neurofibrillary tangles. (Courtesy US Department of Energy/Science Photo Library)

clinical manifestations of depression before death than in those who did not (Zubenko & Moossy, 1988; Zweig et al., 1988). Similarly, demented patients with major depression had a tenfold reduction in norepinephrine compared with non-depressed demented patients (Zubenko et al., 1990). These findings suggest that many symptoms in AD may be modulated through nLC function, as they are involved in primary depression and anxiety. Agents that modify catecholamine levels broadly have been trialed in some dementias (Fremont et al., 2020).

Neurotransmitter perturbations in AD provide the most accessible targets for therapeutic interventions. Since cholinergic deficits represent the most consistent neurotransmitter depletion and appear early in the disease process (Francis et al., 1999), cholinergic manipulation has been a major focus of applied clinical pharmacological research.

Cholinergic Treatment Approaches
The primary implication of the cholinergic hypothesis is that potentiation of central cholinergic function should improve the cognitive impairment associated with AD. This simple "neurotransmitter replacement" rationale has been validated by the consistent effects of CheIs across trials (see Fig. 4).

Other cholinergic treatment approaches, including precursor loading, direct cholinergic receptor stimulation, and indirect cholinergic stimulation, have proven ineffective, effective but too toxic, or have not been completely developed.

Theoretically, cholinergic therapies may have effects beyond the short-term symptomatic improvement of cognition and may modify the pathogenic processes of the illness (Thai et al., 1997). For example, activation of M_1 muscarinic receptors can stimulate secretion of APPs via the α-secretase pathway such that there is a decrease in the production of toxic and insoluble β-amyloid peptides, thus decreasing the formation of amyloid plaques and promoting the normal processing of APP (Nitsch et al., 1992; Inestrosa et al., 1996; Muller et al., 1997). These effects have been reported with some cholinergic agonists and CheIs (Buxbaum et al., 1992; Lahiri et al., 1992; Nitsch et al., 1992; Haroutunian et al., 1997), but remain to be proven in clinical trials (Buxbaum et al., 1992; Haroutunian et al., 1997; Lahiri et al., 2000).

Precursor Loading
Early investigations focused on acetylcholine precursors, including choline and phosphatidylcholine (lecithin), in attempts to augment acetylcholine synthesis. Numerous trials of cholinergic precursors, most with small sample sizes, generally failed to improve cognitive performance in patients with AD; only 10 of 43 such trials reported any positive effect (Jorm, 1986).

Muscarinic Agonists and Antagonists
The rationale for the use of direct cholinergic agonists rests on the fact that postsynaptic M_1 cholinergic receptors are relatively intact in AD and that presynaptic M_2 cholinergic receptors, which are decreased in AD, regulate acetylcholine release. Cholinergic agonists studied in the past included bethanechol, oxotremorine, pilocarpine, RS–86, and arecoline.

Clinical trials results of numerous cholinergic agonists have been disappointing owing to doubtful efficacy and high incidences of muscarinic-related adverse effects. The muscarinic agonists included alvameline (LU 25-109), arecoline,

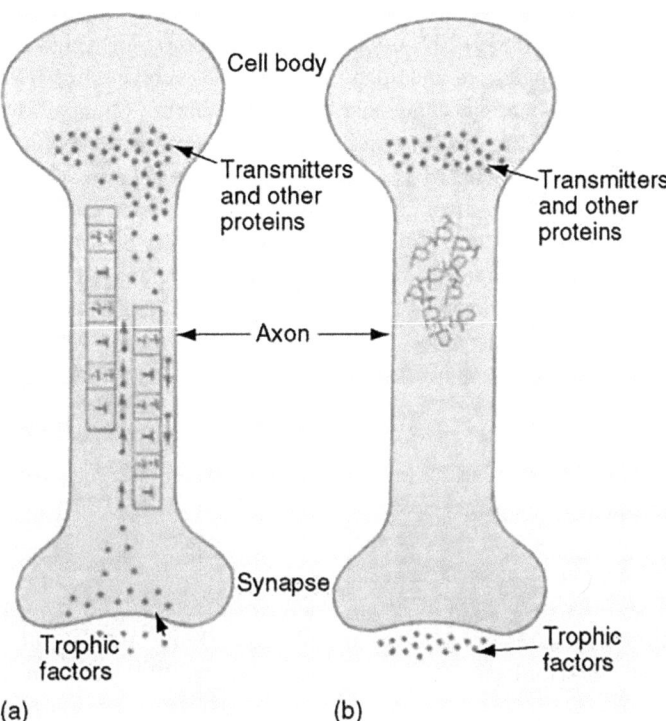

Fig. 4 (**a**) A normal neuron. The microtubule-associated protein tau (τ) promotes the assembly of tubulin (T) into tubular microtubules. The microtubules serve as the highways for transporting transmitters and other proteins from the cell body to the synapse, and also of trophic factors (necessary for survival of the neuron) from the synapse back to the cell body. In AD (**b**), tau is abnormally phosphorylated (represented by Ps attached to τ). This destroys tau's ability to promote the polymerization of tubulin into microtubules, robbing the cell of the major highways for transporting many proteins from the cell body to the synapse and from the synapse back to the cell body. Proteins pile up in the cell body unable to be moved. It is presumed that trophic substances pile up outside the synapse, also unable to be transported. The abnormal phosphorylation of tau is associated with its self-aggregation into the neurofibrillary tangles. (Source: Tariot et al. (1993). Reproduced with permission of John Wiley & Sons, Ltd.)

cevimeline (AF-102B), milameline, sabcomeline (SB 202026), SR 46559, talsaclidine, and xanomeline. The failure of these muscarinic agonists with predominant M_1 activity, albeit each with a somewhat different receptor profile, suggests that direct stimulation of muscarinic receptors may not be sufficient to treat AD. Also, nonselective, direct receptor agonists may be relatively ineffective because they provide a prolonged and tonic cholinergic stimulation of postsynaptic cells; or perhaps they actually inhibit acetylcholine release through their action on presynaptic M_2 receptors. By comparison, the relative efficacy of Chels may reflect their additional actions, such as nicotinic stimulation or nonenzymatic actions.

Other Indirect Cholinergic-Enhancing Strategies

Various methods of indirect cholinergic enhancement included DuP 996 (linopirdine), which enhances the presynaptic, potassium-mediated release of acetylcholine, ondansetron, a 5-HT_3 antagonist, which may act as a cholinergic facilitator, enhancing acetylcholine release, and various partial nicotine agonists. Despite interesting preclinical rationales, these drugs have not shown clinical effectiveness (Rockwood et al., 1997; Dysken et al., 2002).

Nicotinic Agonists

Nicotinic acetylcholine receptors (nAChRs) are reduced in the hippocampus and the temporal and frontal cortex in patients with AD compared with age-matched controls (Nordberg, 2001). Nicotine can improve memory, cognition, and attention in AD patients, and a nicotinic antagonist produces the opposite effect (Newhouse et al., 2001). In a study of 74 patients with MCI who were randomized, double-blind, to nicotine dermal patch 15 mg daily or placebo for 6 months, nicotine showed an advantage in measures of attention and processing speed and to a lesser extent memory. There were no differences in the clinician's global improvement measure; safety and tolerability of nicotine were excellent (Newhouse et al., 2012).

Out of concern that nicotine usefulness is limited by the extent of its peripheral and parasympathetic effects, efforts have also focused on the development of a selective CNS nicotinic agonist, active at the $\alpha_4\beta_2$ subtype of nAChRs, because of high densities of these receptors in brain regions involved in memory processes and AD (Nordberg, 2001).

Ispronicline (TC-1734, AZ 3480) is a partial nAChR agonist with a high affinity and selectivity for the $\alpha_4\beta_2$ subtype. Initial Phase 1 data showed good tolerability (Dunbar et al., 2006), but in a Phase 2 comparative study neither TC-1734 nor donepezil met the primary outcome measure. The α_7 nAChR subtype received particular attention (Mazurov et al., 2006) but trials of α_7 nicotinic agonists including R3487/MEM 3454, SSR 180711C, PH-399733, and GTS-21 were abandoned largely related to lack of proven efficacy.

Cholinesterase Inhibitors

Modest, significant, and reliable improvements in memory were observed in early trials of the acetylcholinesterase (AChE) inhibitor physostigmine (Christie et al., 1981; Davis & Mohs, 1982; Mohs et al., 1985; Stern et al., 1987), but it caused adverse reactions and required frequent dosing (Table 2). Early success with dosage individualization (Jorm, 1986) led to multicenter trials of the

Table 2 FDA-approved drugs to improve cognitive function in Alzheimer's disease: Dosages[a] of marketed CheIs

Drug	How supplied	Initial dosage	Maintenance dosage	Comments
Tacrine	10, 20, 30, and 40 mg capsules	10 mg QID	30 or 40 mg QID	120–160 mg/day are efficacious doses. Reversible direct hepatotoxicity in about 1–3 patients, requiring initial biweekly transaminase monitoring and dose titration. Not commonly used
Donepezil	5, 10, 23 mg tablets and orally dispersible tablets	5 mg QD	5–23 mg QD	5 and 10 mg are both effective doses; 10 mg may be somewhat more efficacious in some trials; 23 mg may be marginally more efficacious but is also associated with more side effects
Donepezil	23 mg tablets	23 mg QD	23 mg QD	Indicated for moderate-to-severe AD. Slightly superior to donepezil 10 mg in cognitive outcomes but not in function, and associated with more than double the incidence of gastrointestinal side effects
Rivastigmine	1.5, 3, 4.5, and 6 mg capsules, transdermal patch	1.5 mg BID	3, 4.5, or 6 mg BID	Effective dosage range 3–6 mg BID. Doses of 4.5 mg BID may be optimal. May be taken with food
Galantamine	4, 8, and 12 mg tablets, 4 mg/mL solution, 8, 12, and 24 mg extended-release capsules	4 mg BID, 8 mg QD	8 or 12 mg BID, 16 or 24 mg QD	Effective dosage range is 16–24 mg/day; 16 mg/day is the modal optimal dose

Note: The FDA-approved indication for these drugs is "indicated for the treatment of mild to moderate dementia of the Alzheimer's type," except for donepezil, which is approved for mild, moderate, and severe AD
[a]Initial dosages should be maintained for at least 2 and preferably 4–6 weeks before increasing. Adverse events may occur with dosage titration

CheIs tacrine, velnacrine, and sustained-release physostigmine (Davis et al., 1992; Antuono, 1995; Thal et al., 1996), and subsequently provided the rationale and justification for larger clinical trials and eventual approval and marketing of several CheIs.

Tacrine was the first CheI to be approved for the treatment of AD, but it is no longer actively marketed, primarily because of liver toxicity. The newer CheIs donepezil, rivastigmine, and galantamine are in clinical use. Other CheIs include phenserine, huperzine A, ZT-1 (a prodrug of huperzine A), ladostigil, and bisnorcymserine. Phenserine is a long-acting CheI that may also reduce Aβ levels (Haroutunian et al., 1997), but a large-scale trial was negative (Becker & Greig, 2012). Huperzine A is an extract from club moss commonly used as an herbal remedy in Asia and available in the United States as a dietary supplement. A large-scale NIH-funded placebo-controlled trial using huperzine A was negative (Rafii et al., 2011).

Mechanisms of Cholinesterase Inhibition

By inhibiting the actions of AChE, CheIs effectively increase the amount of ACh available for intrasynaptic cholinergic receptor stimulation. Butyrylcholinesterase (BuChE), another central enzyme that hydrolyzes ACh, is also targeted for AD therapy. AChE-positive neurons diffusely project to the cortex and are thought to modulate cortical processing and responses to new stimuli; BuChE-positive neurons project principally to the frontal cortex and might have a role in influencing attention, executive function, emotional memory, and behavior (Lane et al., 2006).

An AChE inhibitor can work at either of two sites on AChE, an ionic subsite or a catalytic esteratic subsite, to prevent the interaction between ACh and AChE. Tacrine and donepezil act at the ionic subsite. Rivastigmine acts at the catalytic esteratic subsite (Enz & Floersheim, 1997). Galantamine acts at both the ionic site and catalytic binding site (Lane et al., 2006).

Donepezil and galantamine are relatively selective for AChE (Brufani & Filocamo, 1997), whereas tacrine and rivastigmine inhibit both AChE and BuChE. Binding to the AChE sites

may be either reversible or irreversible, and may be competitive or noncompetitive with acetylcholine. Galantamine is an example of a competitive CheI, competing with acetylcholine for AChE.

In addition, AChE is present in a few molecular forms containing one (monomeric G1), two (dimeric G2), or four (tetrameric G4) catalytic subunits. The G1 and G4 forms are present in the human brain, varying in proportion in different brain regions (Atack et al., 1986). The tetrameric G4 is located on the presynaptic membranes within the cholinergic synaptic cleft; the monomer G1 is found on postsynaptic membranes. Whereas G4 is decreased along with the neuronal loss of presynaptic cholinergic neurons, postsynaptic cholinergic receptor neurons and G1 ACh are not decreased significantly with AD or aging (Enz & Floersheim, 1997). Similar to AChE, BuChE exists in G1, G2, and G4 molecular forms, with a preponderance of the G4 isoform in the brain (Arendt et al., 1992). The BuChE monomeric G1 increases by 30–60% in the AD brain whereas the BuChE tetrameric G4 isoform decreases or remains at the same level. Hence, investigations of CheI should consider the molecular form-specific characteristics of AChE and BuChE inhibition. Rivastigmine is a CheI that is highly selective for the postsynaptic G1 monomer form of AChE, whereas galantamine is less so and donepezil is not selective.

A summary of pharmacokinetics and pharmacodynamics is presented in Table 3.

Individual Cholinesterase Inhibitors: Clinical Studies, Dosing, and Adverse Effects

Despite the slight variations in the mode of action of the three clinically available CheIs (Table 4) for treatment of AD, there is no evidence of meaningful difference among them with respect to efficacy (Birks, 2006).

Tacrine

Tacrine is a noncompetitive reversible inhibitor of AChE. It binds near the catalytically active site of the AChE molecule to inhibit enzyme activity, and also has other actions (Adem et al., 1990). Two clinical trials resulted in FDA approval of tacrine in 1993 (Farlow et al., 1992; Knapp et al., 1994). Tacrine is no longer actively marketed

Table 3 Pharmacodynamic and pharmacokinetics of marketed CheIs

Drug	Pharmaco-dynamics	Absorption	Bioavail-ability (%)	Peak Plasma (hours)	Elimination Half-life (hours)	Protein Binding (%)	Metabolism/ Comments
Tacrine	Noncompetitive, reversible CheI, both butyryl and acetyl-CheI, also multiple other actions	Delayed by food	17	1–2	2–4	55	Via 1A2. nonlinear pharmacokinetics: hepatoxicity requires regular monitoring of serum alanine aminotransferases
Donepezil	Noncompetitive, reversible acetyl-CheI	Not affected by food	100	3–4	70	96	Via 2D6, 3A4. Nonlinear pharmacokinetics at 10 mg/day
Rivastigmine	Noncompetitive CheI, both butyryl-and acetyl-CheI, may differentially effect different acetyl-ChEs	Delayed by food	40	1.4–2.6	<5	40	Hydrolysis by esterases and excreted in urine (nonhepatic). Duration of cholinesterase inhibition longer than plasma half-life. Nonlinear pharmacokinetics
Galantamine	Competitive, reversible CheI, modulates nicotine receptors	Delayed by food	90	1	7	18	Via 2D6, 3A4

Note: Pharmacodynamic effects of some ChIs are longer than their elimination half-lives. Drugs that inhibit or induce the cytochrome enzymes might be expected to increase or decrease blood levels. For the most part, however, drug interactions with donepezil, rivastigmine, and galantamine have not been clinical problems

owing to a high incidence of hepatotoxicity and the frequent dosing schedule.

Dosing. The recommended starting dose is 10 mg QID to be maintained for 6 weeks, while serum transaminase levels are monitored every other week. If tolerated, the dose is then increased to 20 mg QID. After 6 weeks, the dosage should be increased to 30 mg QID, again with biweekly monitoring, and then, if tolerated, to 40 mg QID for the next 6 weeks.

Donepezil

Donepezil is a long-acting piperidine-based highly selective and reversible AChE inhibitor. Two Phase 3 clinical trials (Rogers et al., 1998a, b) resulted in FDA approval for mild to moderate stages of AD in 1996. Subsequent randomized clinical trials included a trial of duration 6 months (Burns et al., 1999), a Scandinavian study of 12 months (Winblad et al., 2001), a study of nursing-home patients (Tariot et al., 2001), a trial assessing functional decline (Mohs et al., 2001), and a 6-month trial in which a large proportion were more severely impaired than in previous trials (Feldman et al., 2001).

Two 24-week clinical trials conducted in nursing-home patients in Sweden (Winblad et al., 2006) and outpatients in Japan (Homma et al., 2000) with severe AD (MMSE 1–10) provided sufficient cognitive and global efficacy for the FDA to approve donepezil for this indication in 2006. Donepezil is the only CheI indicated for severe AD (Winblad et al., 2009).

Earlier concerns about cost-effectiveness in the United Kingdom (Courtney et al., 2004) have become less relevant because donepezil, galantamine, and rivastigmine (other than the transdermal patch) are now generic and inexpensive.

Table 4 Key published placebo-controlled, randomized CheI clinical trials[a]

Study	Duration (weeks)	No.	Age (years)	Dose (mg/day)	Completers (%)
Tacrine					
Knapp et al. (1994)	26	663	73	120	32
				160	28
				P	68
Donepezil					
Rogers et al. (1998b)	24	473	73	5	85
				10	68
				P	80
Burns et al. (1999)	24	818	72	5	78
				10	74
				P	80
Tariot et al. (2001)	24	208	86	10	74
				p	82
Feldman et al. (2001)	24	290	74	10	67
Winblad et al. (2001)	52	286	73	P	67
Rivastigmine					
Corey-Bloom et al. (1998)	26	699	75	1–4	85
				6–12	65
				P	84
Rösler et al. (1999)	26	725	72	1–4	86
				6–12	67
				P	87
Galantamine					
Tariot et al. (2000)	20	978	77	8	76
				16	78
				24	78
				P	84
Raskind et al. (2000)	24	636	75	24	68
				32	58
				P	81
Wilcock et al. (2000)	24	653	72		

Note: Rounded to two figures. All trials had 58–64% female subjects except for tacrine (Knapp et al., 1994: 53%), and donepezil nursing home trial (Tariot et al., 2001: 82%). Dropouts are for all reasons, to avoid bias, not just those attributed to side effects

[a]All trials included only patients with probable AD (NINCDS-ADRDA criteria) or dementia of Alzheimer's type (DSM-IV-TR criteria), and generally with baseline MMSE scores between 10 and 26 inclusive, except for the galantamine trials that used narrower ranges of 10–22 and 11–24

Donepezil has shown some positive effects in patients with MCI (Salloway et al., 2004; Petersen et al., 2005; Doody et al., 2009), but efficacy has not been consistent or robust enough to lead to FDA approval for this indication. A significant proportion of MCI patients do not progress clinically to AD, and this large subgroup decreases the likelihood of finding a therapeutic effect with a CheI.

Dosing/Formulation. Donepezil is initiated at 5 mg/day and then increased to 10 mg/day after 4–6 weeks. Raising the dose earlier increases the risk for cholinergic adverse events. Doses of 5 or 10 mg/day are effective; 10 mg tends to be somewhat more effective and to have more adverse effects than 5 mg when the various trials as a group are evaluated. Donepezil is also available as an oral disintegrating tablet, avoiding the need to swallow tablets.

In 2010, the FDA approved a higher daily dose of donepezil (23 mg/day) for the treatment of AD in the moderate-to-severe stages based on positive results from a large, global, Phase 3 clinical trial that compared switching to donepezil 23 mg/day with continuing treatment on donepezil 10 mg/day. In that trial, no benefit was seen in the co-primary endpoint of global functioning; however, donepezil 23 mg/day provided a small but significant improvement in the cognitive endpoint compared with donepezil 10 mg/day (Christensen, 2012). In a safety study of 1434 patients receiving donepezil 23 mg daily versus 10 mg daily, the most common adverse events were nausea, vomiting, and diarrhea, which were more common on the higher dose (donepezil 23 mg/day, 11.8, 9.2, 8.3%, respectively; donepezil 10 mg/day, 3.4, 2.5, 5.3%, respectively). However, there were no differences in serious adverse events in the two groups (Farlow et al., 2011). A 1-year extension study with donepezil 23 mg/day did not reveal any increase in adverse events beyond the initial weeks of the study (Tariot et al., 2012).

Rivastigmine

Rivastigmine is a pseudo-irreversible, selective AChE subtype inhibitor. Although it also inhibits BuChE, it is relatively selective for the postsynaptic G1 monomer form of AChE in areas of the cortex and hippocampus. After binding to AChE, the carbamate portion of rivastigmine is slowly hydrolyzed, cleaved, conjugated to a sulfate, and excreted. Thus, its metabolism is essentially extrahepatic and is unlikely to have significant pharmacokinetic interactions.

Four Phase 3 randomized, placebo-controlled, 26-week-long clinical trials were completed of similar design, but differing mainly in dosing methods. In two published trials, doses were titrated weekly during the first 7 weeks to one of two dosage ranges, 1–4 or 6–12 mg/day, and dose decreases were not permitted, possibly contributing to lesser tolerability and seemingly more side effects during these stages of treatment (Corey-Bloom et al., 1998; Rösler et al., 1999). Some results of the third trial were included in secondary reports (Birks et al., 2000; Schneider, 2002).

A 6-month, multicenter trial of transdermal rivastigmine showed that the patch (4.6 mg/24 h) provided similar benefits to oral rivastigmine at lower doses but with fewer side effects, and with similar side-effect burden at higher doses (Winblad et al., 2007). The rivastigmine transdermal patch appears to have better tolerability than rivastigmine capsules, with fewer gastrointestinal adverse events and discontinuations because of these adverse events (Sadowsky et al., 2010). Simple daily rotation of patch location appears to reduce the frequency of skin reactions.

Dosing/Formulation. The recommended starting dose of oral rivastigmine is 1.5 mg BID, taken with meals, increasing to 3 mg BID after a minimum of 2 weeks of treatment if the initial dose is well tolerated. Subsequent increases to 4.5 and then 6 mg BID should be based on good tolerability with the previous dose and may be considered after a minimum 2-week treatment interval. Higher daily doses, averaging about 9–10 mg, were associated with better efficacy than lower doses. Rivastigmine is available as an oral solution. Transdermal rivastigmine is started with an initial 4.6 mg patch per day for at least 4 weeks before the dose is raised to a 9.5 mg patch per day maintenance dose, based on good tolerability with the previous dose.

Galantamine

Galantamine, an alkaloid originally extracted from Amaryllidaceae *(Galcinthus woronowii,* the Caucasian snowdrop), but now synthesized, is a reversible, competitive inhibitor of AChE with relatively little BuChE activity (Harvey, 1995). Competitive inhibitors compete with ACh at AChE binding sites, so their inhibition is, theoretically, dependent on intrasynaptic ACh. Another characteristic is its allosteric modulation of nicotinic receptor sites, thus possibly enhancing cholinergic transmission by presynaptic nicotinic stimulation (Maelicke et al., 2001).

Several multicenter trials involving over 2400 subjects (Raskind et al., 2000; Tariot et al., 2000; Wilcock et al., 2000; Rockwood et al., 2001) have been published. The results of two trials indicated that treatment with either 24 or 32 mg/day galantamine improved cognition, clinician's global assessment of change, and activities of daily living (ADL) scores, with lesser adverse effects at the lower dose. The results of a third trial showed that daily doses of 16 or 24 mg were effective and 8 mg was not. In patients with mild-to-moderate AD who have shown cognitive benefits from up to 5 months of galantamine treatment, continuing therapy maintains previously achieved benefit compared with patients who show more disease progression after galantamine is discontinued (Gaudig et al., 2011), consistent with studies of discontinuation of donepezil.

Dosing/Formulation. Initial dosing of the original formulation of galantamine is 4 mg BID, and should be raised to 8 mg BID after 2–4 weeks. For patients who are tolerating the medication but not responding, the dose can be raised to 12 mg BID after a further 4 weeks. The FDA approved galantamine in April 2001 under the trade name Reminyl, which was later changed to Razadyne. An extended-release formulation of galantamine (Razadyne ER) was subsequently approved by the FDA for once-daily use (Brodaty et al., 2005). The extended-release formation of galantamine is started at 8 mg/day and increased to 16 mg/day after 4 weeks. After 4 weeks, the dose can be further increased to 24 mg/day based upon clinical benefit and tolerability.

Effects of Cholinesterase Inhibitors on Daily Functional Activities

Across most studies, donepezil treatment leads to a small improvement in daily functioning compared to placebo (Gauthier et al., 2010). In a prospective, nonrandomized, multicenter study in a routine clinical setting that included 784 AD patients treated with donepezil, rivastigmine, or galantamine, after 6 months of CheI treatment 49% and 74% of patients showed improvement/ no change in Instrumental Activities of Daily Living (IADL) and Physical Self Maintenance Scale (PSMS) score, respectively. The improved/ unchanged patients exhibited better cognitive status at baseline (Wattmo et al., 2012).

Adverse Effects of Cholinesterase Inhibitors

Adverse events of the marketed CheIs are summarized in Table 5. Significant cholinergic side effects can occur in up to about 25% of patients receiving higher doses. Often they are related to

Table 5   Adverse effects of CheIs: Summary of adverse event data in placebo-controlled, randomized clinical trials[a]

Drug	Adverse Events
Tacrine	Nausea, vomiting, diarrhea, dyspepsia, myalgia, anorexia, dizziness, confusion, insomnia, rare agranulocytosis Approximately 50% of patients will develop direct, reversible hepatotoxicity manifested by elevated transaminases Drug interactions may include increased cholinergic effects with bethanacol; increased plasma tacrine levels with cimetidine or fluvoxamine. This may occur by inhibition of P450 1A2. The association of tacrine with haloperidol may increase Parkinsonism and tacrine increases theophylline concentration
Donepezil	Nausea, diarrhea, insomnia, vomiting, muscle cramps, fatigue, anorexia, dizziness, abdominal pain, myasthenia, rhinitis, weight loss, anxiety, syncope (2 vs. 1%). Rare risk of bradycardia, prolonged QTc interval
Rivastigmine	Nausea, vomiting, anorexia, dizziness, abdominal pain, diarrhea, malaise, fatigue, asthenia, headache, sweating, weight loss, somnolence, syncope (3 vs. 2%). Rarely, severe vomiting with esophageal rupture
Galantamine	Nausea, vomiting, diarrhea, anorexia, weight loss, abdominal pain, dizziness, tremor, syncope (2 vs. 1%)

[a]The method of obtaining adverse events and their reporting vary among trials. Adverse event estimates vary widely among the CheIs from study to study and thus relative adverse event rates among drugs are difficult to estimate. Cholinergic side effects generally occur early and are related to initiating or increasing medication. They tend to be mild and self-limited. Medications should be restarted at the lowest doses after temporarily stopping. See prescribing information referenced in Table 2

the initial titration of medication. Patients tend rapidly to become tolerant to the adverse events when they occur. Nausea, diarrhea, vomiting, and weight loss are the most common side effects of the CheIs. Few trials have directly compared CheIs. The proportion of patients losing greater than 7% of their baseline weight varies from approximately 10 to 24% at the higher doses and from 2 to 10% in the placebo-treated patients The rate of anorexia at the highest dosage of CheIs is

8–25% versus 3–10% with placebo. One study comparing donepezil and rivastigmine showed more frequent adverse events associated with rivastigmine during the titration phase but similar frequencies with rivastigmine and donepezil in the maintenance phase (Bullock et al., 2005). Two clinical trials testing galantamine in MCI showed an increased mortality rate in the galantamine group (1.5%) compared with placebo (0.4%), leading to an FDA alert. However, two clinical trials testing donepezil and one with rivastigmine in MCI did not show an increase in mortality compared with placebo (Salloway et al., 2004; Petersen et al., 2005; Feldman et al., 2007). A donepezil trial in vascular dementia also showed an increased incidence of deaths in one trial (10 vs. 0) but not in two others (Kavirajan & Schneider, 2007).

Because of the actions of CheIs, these drugs require caution when used in patients with significant asthma, significant chronic obstructive pulmonary disease, cardiac conduction defects, active peptic ulcers, or clinically significant bradycardia (Hernandez et al., 2009; Kröger et al., 2012). Appropriate considerations are also involved in general anesthesia since CheIs may prolong the effects of succinylcholine-type drugs.

Donepezil

The most common gastrointestinal side effects of donepezil include nausea, vomiting, diarrhea, and anorexia. Some patients develop muscle cramps, headache, dizziness, syncope, flushing, insomnia, nightmares, weakness, drowsiness, fatigue, and agitation. Weight loss of >7% of baseline occurred at twice the rate of placebo in a nursing home study, but not in outpatient trials. Adverse effects occurred at higher rates when the titration from 5 to 10 mg was made in 1 week compared with 6 weeks.

Rivastigmine

Adverse effects are primarily gastrointestinal and occurred in the high-dose (6–12 mg/day) group. They occurred mainly during dose escalation and led to withdrawal in one study in 23% of the high-dose group, 7% of the low-dose group, and 7% of the placebo group.

Adverse effects in the higher dose group were sweating, fatigue, asthenia, weight loss, malaise, dizziness, somnolence, nausea, vomiting, anorexia, and flatulence. In the maintenance phase, dizziness, nausea, vomiting, dyspepsia, and sinusitis occurred more in the 6–12 mg/day group than in the placebo group.

General precautions with CheIs (as indicated in the prescribing information)

By increasing the central and peripheral cholinergic stimulation, CheIs may:

1. Increase gastric acid secretion, increasing the risk for GI bleeding especially in patients with ulcer disease or those taking anti-inflammatories
2. Produce bradycardia, especially in patients with sick sinus or other supraventricular conduction delay, leading to syncope, falls, and possible injury
3. Exacerbate obstructive pulmonary disease
4. Cause urinary outflow obstruction
5. Increase risk for seizures
6. Prolong the effects of succinylcholine-type muscle relaxants

Galantamine

Principal adverse effects are nausea, vomiting, diarrhea, anorexia, weight loss, abdominal pain, dizziness, and tremor. Again, adverse events were more frequent earlier in the course of treatment and during the dosage titration from 16 to 24 mg/day and higher. In one trial, the effective dosage of 16 mg/day was not associated with overall greater adverse events than the placebo-treated group. Doses of 16 mg/day when titrated over a 4-week period were best tolerated.

Glutamatergic Therapies

The N-methyl-D-aspartate (NMDA) receptor, a glutamate receptor subtype, has important effects on learning and memory. Stimulation by the excitatory amino acid glutamate results in long-term potentiation of neuronal activity basic to memory formation (Cotman et al., 1988). There appears to be a decrease in cerebral cortical and hippocampal NMDA receptors in AD. Glycine, acting at an adjoining glycine-B receptor, modulates the effects of glutamate.

Memantine

Memantine is an uncompetitive NMDA antagonist that binds to the NMDA receptor-operated cation channels and may act by inhibiting calcium ion influx, thus modulating excitotoxicity. Memantine also acts as an uncompetitive antagonist at the 5-HT$_3$ receptor but the clinical significance of this is unknown (Rammes et al., 2001). The FDA approved memantine for moderate-to-severe AD in 2003. Three randomized, placebo-controlled trials have shown benefits over placebo in moderate or severe AD on several measures. Two studies, one of which evaluated participants who were being treated with donepezil, showed significant improvements on the SIB (which assesses attention, orientation, language, memory, and social interactions), the modified ADCS-ADL Scale, and the Clinician's Interview-Based Impression of Change Plus Caregiver Input (CIBIC-Plus) global score (Reisberg et al., 2003; Tariot et al., 2004). A fourth 6-month trial in moderate-to-severe AD did not show significant effects in favor of memantine.

The third study, conducted in nursing homes in Latvia, showed a significant advantage of memantine over placebo in the Behavioral Rating Scale for Geriatric patients (BGP), which assessed day-to-day functioning and a global impression of change (Winblad & Poritis, 1999). Another 6-month placebo-controlled trial in moderate-to-severe AD outpatients did not show advantages for memantine (Van Dyck et al., 2007).

Three trials evaluated memantine in mild-to-moderate AD. Two of the three trials failed to show advantages for memantine on cognitive, global, functional, or behavioral outcomes (www.forestclinicaltrials.com; unpublished data, 99,679, data on file, H. Lundbeck A/S, 2004; Peskind et al., 2006). According to the Cochrane Database of Systemic Reviews, pooled data on memantine in mild-to-moderate AD show a marginal beneficial effect at 6 months that was clinically insignificant with no effect on behavior or ADLs (McShane et al., 2006) and the drug is not FDA approved for mild AD.

Dosing

Memantine is available in tablet and oral solution form. Treatment is initiated with 5 mg/daily for 1 week and increased by 5 mg/daily in divided doses to a maintenance dose of 10 mg twice per day. The nursing home trial in Latvia showed efficacy with 10 mg/day doses. An extended-release version with once-daily dosing (7, 14, 21, 28 mg) is also available.

Adverse Effects

Significant side effects can include dizziness, constipation, headache, and confusion. Some studies show an increase in agitation and others show improvement in agitation. Generally, memantine is well tolerated and is thought to have low potential for drug interactions.

Combined Cholinesterase Inhibitor and Memantine

In a study of 295 AD patients who were receiving donepezil and were then randomized to one of four treatment conditions, there was a clear benefit for donepezil alone compared with placebo, a smaller benefit for memantine compared with placebo, and lack of superior efficacy with donepezil plus memantine compared with donepezil or memantine alone (Howard et al., 2012). However, subsequent studies have suggested that coadministration of memantine and CheIs may have a synergistic effect (Gauthier & Molinuevo, 2013). Meta-analyses show that combination therapy results in significant benefits in behavior and global outcomes for patients with mild to severe AD when compared to CHeI monotherapy (Matsunaga et al., 2014; Schmidt et al., 2015). There are conflicting data on whether adding memantine to a CheI improves cognition (Tariot et al., 2004). Still, this combination is often used clinically and in 2014, a fixed combination of memantine and donepezil in once-a-day capsule form was approved by the FDA for moderate to severe AD (Namzaric). Namzaric comes in extended release capsules containing 7, 14, 21, or 28 mg of memantine combined with 10 mg donepezil. Instructions for administering this medication are to start patients with a single capsule of 7 mg memantine/10 mg donepezil and then up titrate in 7 mg memantine increments to a maximum dose of 28 mg memantine/10 mg donepezil. The recommended maximum dose for patients with renal impairment is 14 mg memantine/10 mg donepezil. Side effects are identical for those listed above for donepezil and memantine. There does not appear to be reduced tolerability or increased side effects when combining these medications (Owen, 2016; Riverol et al., 2011).

Treatment Approach with FDA-Approved Compounds

Optimal duration of treatment with approved CheI compounds and memantine is unknown but overall efficacy extends at least 9–12 months and in one study up to 3 years (Burns et al., 2007) based on the clinical trials and open-label extension studies. In clinical trials, medication effects are often stabilization or lack of worsening of symptoms or cognitive function relative to placebo-treated patients who continue to decline. Therefore, the clinical observation of minimal or no clinical worsening may be sufficient reason to continue medication treatment if patients are tolerating therapy.

Effect on Behavior

The evidence that CheIs may improve behavior is based on case series and secondary analyses of efficacy trials (Kaufer et al., 1996; Raskind et al., 1997). Patients enrolled into CheI trials were selected largely on the basis of their ability to cooperate; they were not generally agitated, psychotic, or depressed. In two trials, there were significant effects on hallucinations or aberrant motor behavior items and in one trial each on agitation or apathy. One prospective clinical trial with donepezil in patients with AD and significant psychopathology was designed as a withdrawal trial (Holmes et al., 2004) in which patients received open-label donepezil for 3 months and were then randomized to either placebo or donepezil. The placebo group showed relative worsening of behavior. There was no effect on behavior in any of the three donepezil trials in severe AD; and a trial targeting patients with severe AD and significant behavioral disorder also found no significant effect with donepezil on behavioral symptoms (Howard et al., 2007). Another study showed no differences among donepezil, placebo, and *Melissa officinalis* (lemon balm) oil in the treatment of agitation in AD (Burns et al., 2011). Memantine showed beneficial effects on behavior in one (Tariot et al., 2004) of three trials in moderate-to-severe AD (Reisberg et al., 2003), and inconsistent and non-robust effects in patients with mild-to-moderate AD (McShane et al., 2006).

Other Neurotransmitter-Based Approaches

Degeneration in central catecholaminergic systems in AD and their important role in brain-related functions provide the rationale for pharmacological enhancement of these systems. The general strategies employed are analogous to those used with cholinergic agents: precursor loading, inhibition of degradative enzymes, and use of agonists. Studies of dopamine precursors (e.g., with tyrosine, L-dopa) and agonists (e.g., clonidine, guanfacine, amantadine, bromocriptine) have largely been negative (Schneider & Tariot, 1997).

At a dose of up to 10 mg/day, selegiline relatively selectively inhibits MAO-B. The overall effect of selegiline at this dose may be to increase CNS levels of dopamine and some trace neurotransmitters such as phenylethylamine without affecting norepinephrine levels. Selegiline is currently marketed for the treatment of Parkinson's disease, for which it has demonstrated effects in maintaining motor function (Parkinson Study Group, 1993). Selegiline also may have potential to preserve surviving neurons (Tatton, 1993). Elevated levels of MAO-B, found in the brains of patients with AD, might lead to elevated levels of neurotoxins such as 6-hydroxydopamine, quinones, and free radicals. Inhibition of monoamine oxidase (MAO) on a chronic basis could theoretically retard progression of the illness based on antitoxic mechanisms. Finally, selegiline may have neuronal "rescue" effects, based on neurotrophic-like properties not involving MAO inhibition.

An early multicenter trial found that selegiline, 5 mg BID, both alone and combined with vitamin E, 1000 IU BID, was associated with maintenance of ADLs and survival in the community compared with placebo in moderately to severely impaired AD patients but with no improvement in cognition (Sano et al., 1997). Other clinical trials with selegiline have been performed with variable results and insufficient evidence to justify routine clinical use in AD (Birks & Flicker, 2003).

Rasagiline, an irreversible inhibitor of MAO-B approved in 2006 by the FDA as initial monotherapy or adjunct therapy to levodopa for the treatment of Parkinson's disease (Nayak & Henchcliffe, 2008). Safinamide is another medication approved by the FDA as an add-on treatment for Parkinson's disease patients with motor fluctuations that has multiple mechanisms of action including reversible inhibition of MAO-B (Blair & Dhillon, 2017). Other agents with effects on MAO-B including Ladostigil which combines MAO-B inhibition with cholinesterase inhibition in a single molecule and is neuroprotective in neuronal cell cultures (Youdim, 2013) is currently being investigated for Parkinson's disease, Alzheimer's disease, and mild cognitive impairment, without fair tolerability but without clear benefit thus far (Schneider et al., 2019).

Histamine

Cortical histamine levels are significantly reduced in the brains of patients with AD (Panula et al., 1998). The H_3 receptor is a presynaptic inhibitory autoreceptor that diminishes histamine release. Preclinical evidence suggests that H_3 histamine receptor antagonists improve performance in attention and memory tasks whereas H_3 receptor agonists produce a decline in cognitive function (Witkin & Nelson, 2004).

Dimebon (latrepirdine), which has H_3 antagonist properties, has been approved in Russia since the early 1980s. After an initial publication that showed an advantage for dimebon over placebo in cognition and several other measures in an 11-site study conducted in Russia (Doody et al., 2008), a subsequent international trial failed to show differentiation of dimebon from placebo on any measure. Further development of dimebon and other H_3 antagonists remains uncertain.

Somatostatin

Somatostatin is a neuropeptide that is believed to play a role in learning and memory (Schettini, 1991; Matsuoka et al., 1995). Some studies have shown a correlation between impairment in somatostatin-mediated neurotransmission in the brain and AD (Davies et al., 1980; Bissette & Myers, 1992). FK960 is a synthetic compound that enhances somatostatin release and has been shown to improve memory in animal models of amnesia (Yamazaki et al., 1996; Matsuoka & Aigner, 1997). FK962 was developed to optimize further the pharmacological action and pharmacokinetic properties of FK**960**, and was shown to be beneficial in animals (Tokita et al., 2005). This compound completed Phase 1 and 2 testing, but showed no efficacy in AD.

Growth Hormone-Releasing Hormone (GHRH)

In 152 adults (66 patients with MCI and 86 cognitively intact older subjects), 20 weeks of GHRH administration had favorable effects on cognition in both adults with MCI and healthy older adults (Baker et al., 2012). This approach remains untested in patients with AD.

Treatments with Other Mechanisms of Action

Neurotrophic Factors

Nerve Growth Factor (NGF)

NGF, a trophic factor for cholinergic neurons, may target forebrain cholinergic neurons that release the majority of acetylcholine in the cortex and hippocampus. Because NGF does not cross the blood–brain barrier, it was directly infused into brain ventricles in a study that was terminated prematurely owing to significant adverse events including weight loss and pain. Additional indicators of toxicity come from animal studies of intraventricular NGF showing abnormal axonal sprouting of sympathetic nerves and migration of Schwann cells into the subpial space surrounding the brainstem and spinal cord forming a dense sheath (Tuszynski, 2002).

Subsequently, a trial introducing the actual NGF gene was performed (Tuszynski et al., 2005) by engineering the patient's own fibroblasts to produce human NGF, and then stereotactically inject the cells into the nucleus basalis. The first two patients had serious brain hemorrhages, resulting in death in one patient due to movement during the injection procedure. These patients were alert but sedated for the procedure. There were no surgical complications in the following six patients who received general anesthesia. Patients were followed for 18–24 months. One patient developed two syncopal events, possibly related to the procedure. Two of the six subjects showed improved MMSE scores, one showed no decline, two showed minor decline, and one had a large decline. Compared with historical controls, rates of decline in the ADAS-cog also showed improvement. Positron emission tomography (PET) scans in four patients showed significant increases in FDG uptake in most cortical regions receiving cholinergic input from the nucleus basalis. Brain autopsy in one patient who died showed survival of the implanted autologous fibroblasts with little evidence of inflammation. In situ hybridization for messenger RNA encoded NGF showed robust NGF expression by the transplanted fibroblasts. Immuno-cytochemistry showed sprouting of cholinergic axons in and around the transplant site. The absence of controls and the small number of patients precluded conclusive evidence; however, this trial served as an impetus for studies in stem cell and gene therapy but these strategies have yet to demonstrate efficacy in controlled clinical trials.

Other Trophic Factors

Cerebrolysin
Cerebrolysin is a synthetic compound comprised of neurotrophic peptides from the porcine brain that has been found to reduce amyloid burden and improve synaptic plasticity in transgenic mice (Rockenstein et al., 2003). It is thought to regulate APP maturation and transport to sites where $A\beta$ protein is generated (Rockenstein et al., 2006). In an initial placebo-controlled trial, small improvements in ADAS-cog scores and global assessments of severity were seen (Ruether et al., 2001). A second randomized trial of 192 patients did not find significant effects on the ADAS-cog but did on global assessment (Panisset et al., 2002); a third observed improvements on ADLs (Muresanu et al., 2002). Overall, the evidence for efficacy in dementia is not convincing (Allegri & Guekht, 2012).

Estrogens
Estrogens may have cholinergic neurotrophic and neuroprotective effects and may enhance cognitive function (Simpkins et al., 1994). In ovariectomized rats, estradiol replacement enhanced learning, reversed the decrease in neuronal choline uptake and ChAT caused by ovariectomy, and prevented the ovariectomy-associated decline in NGF and in brain-derived neurotrophic factor mRNA.

A beneficial role for estrogen in AD, cognitive function, mood, and aging is suggested by observations of an inverse relationship of estrogen replacement therapy dose and duration with dementia diagnoses on death certificates (Paganini-Hill & Henderson, 1994), and by preliminary trials suggesting a cognitive enhancing effect of estradiol, estrone, and conjugated estrogens in AD (Ohkura et al., 1994; Asthana et al., 2001).

Clinical trials of conjugated equine estrogens (CEE) to improve cognition in both hysterectomized and nonhysterectomized women with AD have not led to success, and indeed women treated with CEE were somewhat worse in both cognition and safety, including 5% who developed deep vein thrombosis (Henderson et al., 2000; Mulnard et al., 2000). The Women's Health Initiative Memory Study (WHIMS), a randomized, double blind, placebo-controlled clinical trial, showed that postmenopausal CEE supplementation (both CEE alone and CEE plus progestin) increased the risk of all cause dementia and did not prevent MCI in healthy women aged 65 years or older

(Shumaker et al., 2003). CEE plus progestin was associated with increased risk for heart disease, stroke, pulmonary embolism, and breast cancer.

Metabolic Enhancement

In view of regional decreases in glucose utilization and abnormal oxidative metabolism, drugs have been employed with the aim of correcting these abnormalities, including ergot alkaloids and nootropics (Olin et al., 2001). Although hydergine was once one of the most frequently prescribed medications in the world, it is now rarely prescribed. Numerous clinical trials in various elderly patient groups have not confirmed its efficacy. Another ergoloid derivative available in Europe, nicergoline was associated with significant improvement in some areas such as orientation and attention (Fioravanti & Flicker, 2001), but no results are available from systematic clinical trials in AD. Nicergoline, dihydroergotamine, and other ergolines can induce chronic pleural thickening, fibrosis, and effusion.

Nootropics

Piracetam, oxiracetam, pramiracetam, aniracetam, CI 933, and BMY 21502 are pyrrolidone derivatives of γ-aminobutyric acid (GABA). Although they do not appear to have GABA-like effects, they are postulated to have neuroprotective properties. These drugs may also enhance the CNS microcirculation by reducing platelet activity, by reducing adherence of red blood cells to vessel walls, and may stimulate central cholinergic activity. These diverse effects are termed nootropic to indicate the class of drugs that are structurally related to piracetam and may improve learning and memory (Nicholson, 1990). Double blind, multicenter, controlled studies have shown mixed results with piracetam in the treatment of dementia in the elderly (Vernon & Sorkin, 1991; Flicker & Evans, 2001). Pramiracetam and oxiracetam have been evaluated in large-scale multicenter studies without significant clinical efficacy in dementia.

Amyloid-Based Interventions

Significant efforts have been made toward developing disease-modifying anti-amyloid therapies, but all have failed to date in short-to intermediate-term clinical trials in AD.

Active Immunotherapy

Active immunotherapy introduces an antigen that triggers an immune response. Investigations with AN 1792 in transgenic mice exhibiting CNS pathology similar to AD showed reduced development of amyloid plaques and decreased cognitive decline (Janus et al., 2000). The Phase 2 trial of AN 1792 in humans was terminated early owing to the development of a T-cell immune-mediated aseptic meningoencephalitis in 18 of 300 subjects (6%) who were randomized to the study drug. A total of 59 subjects (19.7%) developed an antibody response. No significant differences were found between antibody responder and placebo groups for ADAS-cog, Disability Assessment for Dementia, Clinical Dementia Rating, MMSE, or CGIC.

A subset ($n = 11$) of antibody responders had cerebrospinal fluid (CSF) analysis showing no change in Aβ levels, but a decrease in tau (Gilman et al., 2005), suggesting the possibility of a reduced rate of cellular degeneration in antibody responders. A series of autopsies of antibody responders showed a markedly reduced number of amyloid plaques compared with patients treated with placebo or active vaccine who did not show high antibody titers (Nicoll et al., 2003; Orgogozo, 2006), thus suggesting the possibility that Aβ immunotherapy may reduce plaque burden. Paired MRI scans of antibody responders showed a greater decrease in brain volume and greater ventricular enlargement compared with placebo with no correlation with cognitive decline. The reasons for these findings are unclear, but the authors postulated the role of amyloid removal and fluid shifts. In 2019, a 15-year post-mortem neuropathological follow-up of patients from this trial demonstrated that amyloid burden was reduced in people who received the vaccine but most patients had progressed clinically to severe dementia (Nicoll et al., 2019).

A second active vaccine, ACC-001, was developed to induce a highly specific antibody response to Aβ. Work with ACC-001 was suspended in 2008 due to the development of vasculitis in a patient and exploratory analysis showed no benefit of the vaccine (Pasquier et al., 2016). CAD 106, or amilomotide, is another active vaccine that has shown a favorable safety profile and acceptable antibody response (Winblad et al., 2012; Farlow et al., 2015); efficacy remains to be established. This vaccine is based on the first six N-terminal amino acids of Aβ, which is believed to stimulate human B cells without a T-cell response, thereby possibly avoiding the adverse effects seen with AN 1792. Other vaccines currently in the trial phase include ABvac40, which is in phase 2, and UB-311 (Vaz & Silvestre, 2020).

Passive Immunotherapy

Antibodies are delivered directly, averting the need to mount an immune response; however, antibodies have to be given frequently and have a relatively short lifespan (less than 1 month) compared with active immunization.

The anti-Aβ antibody m266 restores hippocampal cholinergic dysfunction and impaired learning in APP transgenic mice (Bales et al., 2006). Solanezumab is a humanized version of the m266 monoclonal antibody that binds to the Aβ 16–23 amino acid region. Data obtained from two Phase 3 human studies in AD showed that it did not improve cognition significantly, but pooled data from the two studies restricted to mild AD patients showed evidence of some efficacy (Doody et al., 2014). A further phase 3 trial of mild AD cases showed that Solanezumab had no impact on disease progression (Honig et al., 2018).

AAB-001 (bapineuzumab) is a monoclonal antibody to Aβ that showed initial promise with evidence of amyloid reduction on PET scanning (Rinne et al., 2010), but subsequent Phase 3 trials did not show superior efficacy to placebo. Tau and phospho tau were reduced without change in Aβ in CSF (Blennow et al., 2012), and there was concern about vascular changes including microvascular hemorrhages (Sperling et al., 2012).

Unfortunately, Phase 3 trials failed to show clinical improvement (Salloway et al., 2014). Other monoclonal antibodies that have been trialed include ponezumab, gantenermab, and crenezumab, all of whose development was halted (Landen et al., 2017; Ostrowitzki et al., 2017). Other similar medications include aducanumab, lecanemab, and donanemab.

Despite early studies suggesting that aducanumab decreased Aβ aggregates and may slow clinical progression in AD (Sevigny et al., 2016), phase 3 clinical trials of aducanumab were cancelled in 2019 at the recommendation of the drug safety monitoring board following a futility analysis (Fillit & Green, 2021). Aducanumab reduced brain amyloid as measured by PET amyloid radiotracer imaging by 20–25% across two studies but did not show clear clinical superiority to placebo on cognitive and functional measures in these two studies (Fillit & Green, 2021). However, the company that developed aducanumab subsequently submitted an application to the FDA for approval of the drug based on preliminary data from the halted trials, which is currently pending (Fillit & Green, 2021). While still controversial, aducanumab, marketed as Aduhelm, received approval from the FDA for the treatment of Alzheimer's disease in June 2021 through the accelerated approval pathway (Dhillon, 2021). The accelerated approval pathway provides patients with access to medications where there may be an expectation of clinical benefit but without definitive proof of this benefit. In this case, accelerated approval was based on the drug's effect on the reduction of amyloid beta plaque, which has been documented in at least one post mortem case (Plowey et al., 2022). While reduction of amyloid beta plaque might be expected to predict clinical benefit, it has not yet been definitively proven to do so. Therefore, the accelerated approval pathway for aducanumab will require the company to verify clinical benefit in a post-approval trial and if the sponsor cannot prove benefit, the FDA may withdraw approval of the drug. Lecanemab is a humanized monoclonal antibody that binds with high affinity to amyloid-beta protofibrils which are thought to be more toxic to neurons. Lecanemab received accelerated approval from the FDA in 2023, and results of an 18-month long multicenter, double-blind phase 3 trial involving 1795 individuals between the ages of 50–90 with early Alzheimer's disease with evidence of amyloid were recently published (van Dyck et al., 2023). This study demonstrated that lecanemab reduced markers of amyloid and resulted in "moderately less decline on measures of cognition and function than placebo" at 18 months. Similar to aducanumab, lecanemab will not be fully covered by Medicare until it receives traditional FDA approval. Donanemab targets a pyroglutamated form of $A\beta$ that aggregates in amyloid plaques (Demattos et al., 2012). There is currently an ongoing phase 2 clinical trial evaluating efficacy and tolerability of donanemab (Clinical trial: NCT03367403).

Human intravenous immunoglobulin (IVIg), polyclonal antibodies which contain antibodies against $A\beta$, have been studied for therapeutic benefit in AD. In a small pilot study, IVIg was administered monthly for 6 months to five patients with AD. The total $A\beta$ levels in the CSF decreased by 30.1%, serum $A\beta$ increased by 233%, and ADASc and MMSE scores essentially stabilized (Dodel et al., 2004). A second open-label pilot study reported similar findings (Relkin et al., 2005). However, a large-scale, randomized, double-blind, placebo-controlled trial did not show efficacy in cognitive measures although there was indication that $A\beta$ was reduced in the blood and brain of patients who had the apolipoprotein E e4 genotype (Relkin et al., 2017).

Secretase Inhibitors

Both β-secretase (β-site APP cleavage enzyme; BACE) and γ-secretase abnormally cleave APP, an early action in the cascade of pathological events. Both enzymes have been targeted for therapeutics as anti-amyloid disease-modifying treatments; however, the development of this class of compounds has been challenging. BACE inhibitors are large molecules that have difficulty penetrating the blood-brain barrier and the receptor site is difficult to bind (Walker et al., 2005).

The current γ-secretase inhibitors also block the Notch signaling protein, a physiologically essential substrate that acts as an important intermediate in many cellular functions, most notably the regulation and differentiation of cells in the gastrointestinal tract and immune system. Interference with Notch can result in untoward biological consequences and gastrointestinal adverse events, and is of concern in the development of some γ-secretase inhibitors. In a Phase 3 trial of the γ-secretase inhibitor semagacestat in AD, there was no difference from placebo in cognition and a worsening in function at the higher dose of drug (Doody et al., 2013). A Phase 2 trial of the γ-secretase inhibitor avagacestat was abandoned due to poor tolerability (Coric et al., 2012). Verubecestat, a potent γ-secretase inhibitor, was shown to reduce CSF AB in AD patients but a Phase 3 trial showed no benefit in cognition (Egan et al., 2018). A modulator of γ-secretase, tarenflurbil was also developed, but a Phase 3 study showed no clinical benefit (Green et al., 2009).

Selective $A\beta_{42}$-Lowering Agents

A subset of nonsteroidal anti-inflammatory drugs (NSAIDs) may have direct $A\beta_{42}$ lowering properties. In contrast to γ-secretase inhibitors, selective $A\beta_{42}$-lowering agents (SALAs) do not interfere with physiologically essential substrates such as Notch because they do not bind to the active site. The SALAs allosterically modulate γ-secretase, which alters the enzymatic site of action on APP to produce shorter, less toxic $A\beta$ fragments (Beher et al., 2004).

The R-enantiomer of flurbiprofen [(R)-flurbiprofen, tarenflurbil) has no significant cyclo-oxygenase inhibition, which is a target of (S)-flurbiprofen and other NSAIDs. Preclinical animal studies showing improved spatial memory led to a Phase 1 trial in normal volunteers (Galasko et al., 2004). A 12-month Phase 2 trial of 207 participants with mild-to-moderate AD showed no overall treatment effect (Wilcock et al., 2006) and a Phase 3 trial showed no improvement in cognition and ADLs compared with placebo (Green et al., 2009).

Two nitric oxide (NO)-releasing derivatives of flurbiprofen, HCT 1026 and NCX 2216, possess anti-amyloidogenic activity in AD models (Gasparini et al., 2005). These drugs consist of an NO-donating moiety coupled to flurbiprofen. When in the presence of cells or tissues, NO is slowly released, mimicking the physiological levels of NO. This NO release is thought to improve the tolerability of the parent drug flurbiprofen (Wallace et al., 2004). The lack of further development of these derivatives may have been influenced by the failure of the parent drug tarenflurbil.

Anti-Aggregation Agents

After APP has been cleaved into Aβ fragments, glycosaminoglycans (GAG) bind to the Aβ protein that permits polymerization into amyloid plaques. Agents that interfere with this aggregation process may be therapeutic. Tramiprosate is a GAG mimetic that binds to Aβ at GAG binding sites, thus reducing Aβ aggregation by thwarting GAG binding (Gervais et al., 2001). A Phase 2 study of 58 patients with mild-to-moderate AD demonstrated tolerability over a 3-month period at three different doses, and 42 subjects entered an open-label phase for 17 months. The most frequent adverse events were nausea, vomiting, and diarrhea. A dose-dependent reduction in CSF Aβ_{42} levels, but not Aβ_{40}, was observed after 3 months; suggesting that the drug affects brain amyloid. There were no significant differences in cognition after 3 months (Aisen et al., 2006). A multicenter Phase 3 trial in 1052 patients involving two doses failed to show efficacy (Aisen et al., 2011).

A second anti-aggregation agent, scyllo-cyclohexanehexol (AZD-103), also showed salutary effects in preclinical studies (McLaurin et al., 2006). Fibril formation due to aggregated Aβ is strongly facilitated by phosphatidylinositol lipids. Scyllo-cyclohexanehexol is a phosphatidylinositol lipid derivative that competes with the phosphatidylinositol lipids for binding to Aβ and interferes with Aβ fibril assembly. A Phase 2 trial in 353 patients with AD randomized to one of three doses of scyllo-cyclohexanehexol or placebo failed to show efficacy for any drug dose (Salloway et al., 2011).

Overall, anti-amyloid clinical trials have not generated a single agent with proven efficacy though a few have shown small effects in subgroups of patients without demonstrating clear and consistent efficacy across larger clinical trials, e.g., aducanumab. Another limitation is that these treatments have been associated with amyloid-related imaging abnormalities (ARIA) that may indicate cerebrovascular injury.

Metal Chelators

The fibrillization of Aβ is influenced by heavy metal ions, leading to the hypothesis that heavy metal chelators can reduce polymerization. Clioquinol was used for parasitic gastrointestinal disease but removed from the market because it was linked to a rare form of optic nerve damage called subacute myelo-optico-neuropathy (SMON) (Tateishi, 2000). It has copper and zinc chelation properties, and it reduced brain amyloid in transgenic mice (Cherny et al., 2001). A small Phase 2 study in 36 patients with moderately severe AD did not show evidence of cognitive benefit but did show a decline in plasma Aβ levels (Ritchie et al., 2003). Further development of this drug for AD was halted owing to concerns regarding toxicity and impurities in the clinical formulation. PBT-2 is another chelator with similar properties to clioquinol. In a Phase 2 trial, PBT-2 showed a small clinical benefit on some executive functions but not other cognitive measures, with no impact on amyloid biomarkers (Faux et al., 2010). An iron chelator deferiprone was shown to lower brain iron levels and improve motor performance in a Phase 2 study in Parkinson's disease (Devos et al., 2014) and there is a Phase 2 study underway to determine if deferiprone can slow progression in patients with prodromal or mild AD (clinical trial: NCT03234686).

Statins

Epidemiological studies suggested an association between cholesterol-lowering therapy with HMG-CoA reductase inhibitors and a lower risk of AD (Jick et al., 2000; Wolozin et al., 2000). The mechanism underlying this possible relationship is uncertain, but could be related to cholesterol reduction and mediation of inflammatory effects. Another proposed method of action suggests that statins may activate the α-secretase pathway, shifting Aβ processing away from β-and

γ-secretase. The result of increased α-secretase activity translates to increased levels of a less toxic form of Aβ and lower levels of Aβ_{42}.

In a randomized, placebo-controlled, double-blind trial of 44 patients over 26 weeks, simvastatin did not significantly alter CSF levels of Aβ overall, although Aβ_{40} levels were decreased in a post hoc subgroup with mild AD (Simons et al., 2002). The results of a randomized, placebo-controlled trial of atorvastatin in mild-to-moderate AD in 67 patients over 12 months were not statistically significant (Sparks et al., 2005). Simvastatin did not differ from placebo in a large study of patients with AD (Sano et al., 2011) and atorvastatin was ineffective in patients receiving donepezil (Feldman et al., 2010). However, one recent study has suggested that use of statins may be especially beneficial for AD patients homozygous for the Apolipoprotein E4 variant (Geifman et al., 2017) and a recent meta-analysis of statin use in AD suggests that they may slow deterioration of neuropsychiatric symptoms and improve activities of daily living (Xuan et al., 2020).

Curcumin

Curcumin is the principal curcuminoid in the widely used Indian spice turmeric. It is a potent antioxidant and has been investigated for possible protective roles in AD inflammatory processes. Curcumin inhibits Aβ_{40} aggregation in a dose-dependent manner in vitro (Ono et al., 2004) and reduces plaque burden in mice by disaggregating Aβ and preventing fibril and oligomer formation (Yang et al., 2005). The clinical study of curcumin is limited partly by its poor solubility and poor bioavailability in the brain (Chin et al., 2013).

Docosahexaenoic Acid (DHA)

Epidemiological studies suggest a reduced risk of AD associated with fish consumption (Kalmijn et al., 1997; Barberger-Gateau et al., 2002, Zhang et al., 2016). DHA is one of the major omega-3 fatty acids found in fish oil, an integral component of neural membrane phospholipids, and the major polyunsaturated fatty acid in the brain (Lauritzen et al., 2001). In a multicenter Phase 3 randomized, placebo-controlled trial in 402 patients with mild-to-moderate AD, DHA was not superior to placebo on any cognitive or functional measure (Quinn et al., 2010). In 2016, a Cochrane review found no convincing evidence for the efficacy of supplementation with omega-3 polyunsaturated fatty acids in the treatment of AD (Burckhardt et al., 2016). Still, clinical trials involving DHA and other omega-3 fatty acids are ongoing with one recent study suggesting that higher doses of DHA may be needed than have previously been used (Arellanes et al., 2020).

Peroxisome Proliferator-Activated Receptor-γ Agonists

Attention has been placed on insulin-modifying therapy in view of the association between AD with elevated plasma insulin levels and reduced insulin sensitivity. In animals, insulin was found to facilitate Aβ release from intraneuronal compartments and interfere with the degradation of Aβ (Leissring et al., 2003). Insulin may also contribute to cerebral inflammation because increased plasma insulin levels in response to a glucose challenge are associated with elevated inflammatory markers (Hak et al., 2001). There was initial evidence of efficacy with no serious adverse events from a pilot trial of intranasal insulin in a mixed sample of MCI and AD patients (Craft et al., 2012). However, a larger randomized clinical trial of adults with MCI and AD found no cognitive or functional benefit of intranasal insulin over placebo (Craft et al., 2020). MK-677 is a compound that induces secretion of insulin-like growth factor-1 (IGF-1). In a 1-year study in 563 patients with AD, there were no therapeutic advantages for MK-677 over placebo.

Peroxisome proliferator-activated receptor-γ (PPAR-γ) agonists improve insulin sensitivity, and it has been hypothesized that PPAR-γ agonists may protect neurons from the presumed harmful effects of insulin resistance. Rosiglitazone is a PPAR-γ agonist that showed initial promise in clinical trials (Watson et al., 2005; Risner et al., 2006). However, in a Phase 3 clinical trial involving 693 patients with mild-to-moderate AD, two different doses (Rosiglitazone XR 2 and 8 mg) failed to show superiority over placebo in cognitive and functional outcomes (Gold et al., 2010). A small pilot study in patients with amnestic mild cognitive impairment showed that patients on metformin, whose mechanism of action includes PPAR-γ activation, performed better on the selective reminding test than patients on placebo (Luchsinger et al., 2016). A subsequent pilot study in AD patients suggested metformin use was associated with improved executive function compared to patients taking placebo (Koenig et al., 2017) and an epidemiological study showed that diabetic patients over the age of 65 on metformin had a lower risk of dementia than those on sulfonylureas (Orkaby et al., 2017). Still, a recent systematic review suggested that based on current data, no particular antidiabetic drug could be recommended in AD or MCI (Munoz-Jimenez et al., 2020).

Tau–Based Therapies

Tau-based therapies have been proposed because neurofibrillary tangles with abnormally phosphorylated tau proteins are a cardinal pathological feature of AD. LMTX, a derivative of methylene blue, is thought to block the polymerization of tau and appeared to be promising in early studies in AD patients (Wischik et al., 2015). However, subsequent phase 3 trials did not show conclusive evidence of efficacy in AD patients (Gauthier et al., 2016b). Subsequent analysis of this trial and another brought up the possibility that low dose LMTX may be beneficial in mild AD and an ongoing clinical trial exploring this possibility is currently in progress (Clinical trial identifier: NCT03446001). In addition to the development of novel molecules targeting Tau pathology, a number of groups have started developing Tau immunotherapies as a possible treatment for AD and other tauopathies.

Active Tau Immunotherapy
In vitro experiments demonstrated that tau oligomerization could be prevented using the monoclonal antibody DC8E8 (Konsetkova et al., 2014). It was then shown that immunization of transgenic tauopathy rats with a vaccine based on the sequence that DCE8E8 bound to resulted in decreased levels of pathological tau and improved behavior (Kontsekova et al., 2014). Based on

these results, AADvac1 was developed and a phase 1 trial in patients with mild to moderate AD showed acceptably safety and immunogenicity (Novak et al., 2017). A larger phase 2 trial is reported as completed but results have not yet been published (clinical trial: NCT02579252). Another active vaccine against the pSer396/404 epitope of tau ACI-35 is currently in phase 1 testing (clinical trial: ISRCTN13033912).

Passive Tau Immunotherapy
Multiple pharmaceutical companies have invested in passive tau immunotherapy agents. One of the ones whose development is furthest along is semorinemab, an anti-tau IgG4 antibody. Phase 1 studies of semorinemab suggested good tolerability and safety but a subsequent phase 2 trial for AD met none of the primary or secondary endpoints and did not show evidence for efficacy (Press release, 2020). Gosuranemab (BMS-986168 or BIIB092) is another humanized IgG4 monoclonal anti-tau antibody. A randomized single ascending dose study of gosuranemab reported no serious or severe adverse reactions (Qureshi et al., 2018). A phase 2 trial of gosuranemab is underway (clinical rial id# NCT03352557). Notably, a phase 2 trial of this same medication for progressive supranuclear palsy (PSP), another tauopathy, showed that the medication had no efficacy on primary outcomes (Biogen). Tilavonemab (ABBV-8E12) is another anti-tau antibody currently in development. A Phase 1 trial showed adequate safety (West et al., 2017). Phase 2 trials utilizing this drug for the treatment of AD are ongoing but similar to gosuranemab, a Phase 2 trial for PSP showed no efficacy. There are numerous other passive immunization strategies in the pipeline including strategies based on antisense oligonucleotides and zinc finger nuclease technology.

Currently, it is unclear whether tau-targeted therapies will hold the key to improving treatment for AD.

Valproate and Lithium

Some studies suggest that lithium and valproic acid have anti-amyloid and neuroprotective properties (Tariot et al., 2002). Valproic acid protects cultured rat hippocampal neurons against Aβ and glutamate-induced injury, possibly owing to stabilization of intracellular calcium levels (Mark et al., 1995). Glycogen synthase kinases (GSK)-3alpha/beta also play an important role in signaling downstream effects of ER stress, including the phosphorylation of tau. Valproic acid also inhibits GSK-**3** and may protect against cell damage due to ER stress (Kim et al., 2005). However, in a multicenter Phase 3 clinical trial, valproate did not attenuate the progression of AD or the likelihood of developing agitation over time (Tariot et al., 2011). Interestingly, tideglusib, an irreversible inhibitor of GSK-3beta, was also shown to produce no significant improvement in AD patients in a large phase 2 clinical trial (Lovestone et al., 2015).

Among its many actions, lithium inhibits GSK-3 and this is its postulated mechanism of action to treat patients with AD. A few reports suggest that chronic lithium use reduces the risk of dementia (Nunes et al., 2007; Kessing et al., 2008), but other data show increased dementia risk with lithium use (Dunn et al., 2005). A placebo-controlled, single-blind lithium trial showed no cognitive effects in patients with AD (Hampel et al., 2009), but a recent Brazilian trial of lithium in 45 patients with MCI showed a small advantage for lithium ($n = 24$) over placebo ($n = 21$) in attention and other cognitive domains (Forlenza et al., 2011). In these studies, tolerability was good at low lithium oral doses leading to blood levels in the 0.2–0.8 mmol/L range, which is lower than the 0.6–1.6 mmol/L blood level range used in patients with bipolar disorder. Large-scale Phase 3 trials using lithium in AD have not yet been reported.

Inflammatory Processes

Inflammation in the brain is a feature of AD (Aisen & Davis, 1994). Immune/inflammatory reactions may be established by reactive microglia surrounding senile plaques and astrocyte proliferation, and inflammatory cytokines are produced such as tumor necrosis factor alpha, interleukin-1 (IL-1),

α_2-macroglobulin, and α_1-antichymotrypsin. IL-1 and IL-6 promote the synthesis of β-APP.

Anti-inflammatory Medications

Epidemiological evidence supports the use of NSAIDs as preventatives of AD. However, clinical trial data have been discouraging after an initial 44-patient placebo-controlled trial of indomethacin (Rogers et al., 1993). Low-dose prednisone (10 mg) did not prove effective in a 1-year placebo-controlled trial by the ADCS (Aisen et al., 2000). There have been clinical trial failures with the cyclooxygenase-2 inhibitors celecoxib and rofecoxib, and also with the nonspecific cyclooxygenase inhibitor naproxen (Aisen et al., 2002). In the ADAPT trial of 2071 persons followed for 2–4 years, there was a short-term increased risk of dementia with the NSAIDs naproxen and celecoxib compared with placebo but a lowered risk with longer term follow-up (Breitner et al., 2011). Overall, the efficacy of anti-inflammatory agents in AD has yet to be demonstrated.

Antioxidants

Many studies have reported evidence of increased levels of oxidative or free radical damage to neurons in AD in addition to normal aging. Vitamin E has antioxidant properties and has been evaluated in a few trials in AD patients. In a trial comparing vitamin E, 1000 IU BID, and selegiline, 5 mg BID, in outpatients with moderate-to-severe dementia (Sano et al., 1997), overall both drugs were effective in maintaining patients' ADLs and prolonging survival in the community by about 5–7 months. There was no effect on cognition, however. A subsequent trial in 769 patients with MCI comparing vitamin E with donepezil and placebo did not find efficacy with vitamin E (Petersen et al., 2005). However, a 2014 study suggested that patients with mild to moderate AD had a slowing of disease progression when taking Vitamin E (Dysken et al., 2014). Still, a recent Cochrane review suggests limited benefit of Vitamin E in AD. Another antioxidant, sodium selenate, was shown to lessen deterioration of brain structures in a Phase 2 trial (Malpas et al., 2016) and a subsequent study suggested increased selenium levels were associated with improved cognitive performance (Cardoso et al., 2019). N-acetylcysteine (NAC) can improve lipid antioxidant defenses and a small clinical trial of NAC showed no significant improvement on primary cognitive measures but some improvement in a sub-analysis over 6 months of treatment (Adair et al., 2001). Antioxidant use does not change CSF biomarkers of AD (Galasko et al., 2012).

Ginkgo Biloba

Ginkgo biloba extract (GbE may exert neuroprotective effects under conditions of hypoxia preventing neuronal cell death (Spinnewyn, 1992), inhibit the toxic effects of β-amyloid (Bastianetto et al., 2000), and act as a free radical scavenger (Dorman et al., 1992). In aged animals, oral GbE treatment leads to increases in the densities of hippocampal muscarinic acetylcholine receptors and cortical 5-HT$_{1A}$ receptors (Huguet et al., 1994) and enhances high-affinity choline uptake into hippocampal synaptosomes (Kristofikova et al., 1992). GbE has been trialed numerous times for AD but multiple meta-analyses of GbE trials for dementia and AD specifically have suggested that while some individual studies suggest cognitive improvement with this natural medicine, results are not consistent enough to support its effectiveness and safety and more research is needed (Yang et al., 2016; Yuan et al., 2017).

A GbE preparation (EGb761 or Tanakan) is approved in multiple European countries for the treatment of disturbances of cerebral function related to dementia syndromes. Several clinical trials have been conducted in patients with dementia, including AD (Weitbrecht & Jansen, 1986; Hofferberth, 1994; Le Bars et al., 1997; Maurer et al., 1997) and mixed populations with patients suffering from either vascular dementia or dementia of the Alzheimer type (Haase et al., 1996; Kanowski et al., 1996). A 6-month placebo-controlled trial involving over 500 patients with mild-to-moderate AD showed no differences between 240 mg/day, 120 mg/day, and placebo in cognitive or global outcomes (Schneider et al., 2005). A large-scale, long-term prevention trial in cognitively intact ($n = 2587$) and MCI ($n = 482$) subjects showed no reduction in the development of dementia or AD (DeKosky et al., 2008). However, a meta-analysis in 2015 suggested that EGb761 at 240 mg/day may stabilize or slow decline in cognition and may be especially helpful for dementia patients with neuropsychiatric symptoms (Tan et al., 2015).

Anti-Excitotoxic Therapies

Calcium Channel Blockers

The process of neuronal death in aging and in AD may be mediated by an increase in intracellular free calcium, which activates various destructive enzymes (such as proteases, endonucleases, and phospholipases) and disrupts intracellular processes. Neuronal calcium levels are regulated by specific proteins known as L-type calcium channels. In principle, blocking the increase in intracellular free calcium may retard these mechanisms and thus slow progression of disease.

Although there have been initial efforts with some compounds such as MEM 1003, nimodipine is the calcium channel blocker that has been tested most extensively in patients with dementia. In one trial, the low-dose nimodipine group (30 mg TID) showed less deterioration on several memory tests over a 10–12-week treatment period than the placebo or high-dose nimodipine (60 mg TID) group (Tollefson, 1990). Another trial showed significant cognitive effects for nimodipine compared with dihydroergotoxin mesylates (Hydergine) and placebo (Kanowski et al., 1988). It is not clear, however, whether the study subjects fulfilled criteria for possible AD.

Although nimodipine has shown some benefits in AD clinical trials (Fritze & Waiden, 1995), a meta-analysis of dementia trials concluded that although there may be evidence for short-term benefits, there was no justification for its use as a long-term antidementia agent (López-Arrieta & Birks, 2002). Nimodipine is used to treat cognitive impairment and dementia in several European countries, but has not been approved by the FDA for dementia.

Combined Therapies

The literature on combining CheIs with memantine has been discussed. Physicians should be skeptical also of combining available anti-inflammatories or hormones with CheIs, first because of the lack of demonstrated efficacy of these drugs, and second because of the additive adverse events, especially gastrointestinal events. Clinical experience suggests that the combination of CheIs with vitamin E or *Ginkgo biloba* does not appear to worsen adverse effects, but there is no efficacy evidence for this practice either, or for the clear efficacy of the latter drugs. Selegiline has been combined with CheIs in small pilot studies suggesting an additive effect, but this remains to be demonstrated unequivocally. Additional strategies employing combination therapies in AD are likely to emerge.

Diet and Exercise

In 1880, elderly community-dwelling subjects in northern Manhattan who ate a Mediterranean-type diet and had high physical activity were less likely to develop AD than subjects who had a poor diet with no physical activity (Scarmeas et al., 2009). Another study of 1410 subjects in France did not show such a clear effect. Since these data come from epidemiological studies without structural brain imaging information, it is unclear if these effects of diet and exercise protect against vascular dementia, AD, or both. Clinical trials with diet and exercise in AD have led to mixed results. Indeed, a recent systematic review of randomized controlled trials involving the Mediterranean-type diet showed mostly non-significant results (Radd-Vagenas et al., 2018). Other trials involving the dietary approaches to stop hypertension (DASH) diet and modifications of this diet are currently being trialed (Cremonini et al., 2019). In 2015, a multi-domain intervention study including exercise, cognitive training, and vascular risk monitoring in over a thousand at-risk older adults found that patients in the intervention group had statistically improved cognition compared to controls (Ngandu et al., 2015). The clinical importance of this statistical difference remains unclear. Another trial of 1680 people looked at at-risk older adults

randomized to receive standard of care or multi-domain intervention that included nutrition, physical activity, and cognitive training with or without omega 3 polyunsaturated fatty acid supplementation. They found no difference between any of the arms, suggesting there was no benefit from any of the interventions (Andrieu et al., 2017). To date, the role of life-style modifications in preventing or slowing AD remains unclear.

Souvenaid, a proprietary nutritional supplement with DHA, vitamins, and other components, some of which are believed to be precursors for phosphatides in cell membranes, showed a small advantage over placebo in a composite neuropsychological test battery score in 259 patients studied at 6 months (Scheltens et al., 2012). However, at 12 weeks, placebo showed a small advantage over Souvenaid in the same study. Further, a larger-scale double blind randomized controlled trial suggested no benefit of Souvenaid compared to placebo in AD (Shah et al., 2013). In a study of 946 patients with AD, compared with usual care, a health and nutrition training program did not lead to cognitive or functional improvement over 1 year, although malnutrition improved in the intervention group (Salva et al., 2011). In a controlled trial, adding physical, cognitive, social, and sensory stimulation to donepezil did not lead to greater improvement (Andersen et al., 2012). An exercise program can improve physical function in AD (Pitkälä et al., 2013), but a positive therapeutic effect on cognition has not been established in clinical trials. Even in positive studies, the magnitude of the impact of diet and exercise is very small and it appears that very large samples are needed to show a statistically significant effect. Therefore, a clinically manifest improvement in cognition in individual subjects is unlikely with diet and exercise alone, although these interventions clearly have advantages in improving physical health overall.

Vitamin D Deficiency

Vitamin D levels have been shown to be low in patients with AD, particularly in northern latitudes with low sun exposure (Wilkins et al., 2006), and high vitamin D levels may be associated with improved cognition in patients with AD

(Oudshoom et al., 2008). While some studies of the relationship between vitamin D level and cognitive decline have been less conclusive (Schneider et al., 2014), a recent meta-analysis suggested that the overall evidence supports an association between vitamin D deficiency with dementia and AD (Chai et al., 2019). In a sample of 43 white patients with AD, those who took memantine plus vitamin D for 6 months had a statistically and clinically relevant gain in cognition compared with the memantine-only or vitamin D-only groups, underlining possible synergistic and potentiating benefits of the combination (Annweiler et al., 2012). However, a larger randomized double-blind placebo-controlled study of over 2000 women 65 and older who were randomized to either take calcium with 400 IU vitamin D or placebo found no difference in the development of MCI or AD over a mean follow-up period of 7.8 years (Rossom et al., 2012). Therefore, it is still unclear whether vitamin D supplementation and specifically high dose vitamin D supplementation is helpful in older adults at risk for dementia or AD patients.

Summary

After the success, although modest, of CheIs and memantine, there was great optimism that treatments that target specific pathogenic pathways, particularly the amyloid pathway based on the amyloid cascade hypothesis and amyloid brain pathology, would result in major advances in the treatment of AD. Unfortunately, the results of clinical trials with these agents have been negative. These findings have led some to question the validity of the amyloid hypothesis (Herrup, 2010) and others to suggest that the clinical failure of β-amyloid-lowering agents does not mean that the hypothesis itself is incorrect but rather that manipulating β-amyloid directly is an unrealistic strategy for therapeutic intervention, given the complex role of β-amyloid in neuronal physiology (Teich & Arancio, 2012). A more direct (and possibly simpler) approach to AD therapeutics is to rescue synaptic dysfunction directly by tackling the mechanisms by which elevated levels of β-amyloid

disrupt synaptic physiology (Teich & Arancio, 2012). Another key issue is that amyloid pathology may begin several years to decades before clinical manifestations of the disease, and toxic β-amyloid levels may have caused irreversible damage to downstream cellular pathways by the time of clinical presentation. Based on this view (Jack et al., 2010), subjects with abnormalities identified by amyloid brain imaging are selected for intervention in these prevention trials. Anti-tau therapy trials have also focused either on prevention in patients known to be at high risk (with AD or PSP) or those patients who are currently symptomatic.

The results from studies of vitamins, diet, exercise, and related approaches have either been negative or have shown a marginal effect at best. Although such strategies remain popular among large segments of the population, they are unlikely to have any sustained therapeutic effects in patients with established AD, although they can have beneficial effects on overall physical and mental health.

Currently available FDA-approved treatments for AD (donepezil, rivastigmine, galantamine, and memantine) decrease symptoms and do not decelerate or prevent the progression of the disease. These therapies demonstrate modest but consistent benefit for cognition, global status, and functional ability. The search for disease-modifying interventions continues, and approaches that target new mechanisms and pathways may eventually lead to therapeutic results superior to those achieved to date. New avenues include the testing of repurposed drugs, e.g., the antiviral drug valacyclovir based on the infectious disease hypothesis of AD (NCT 03282916), but as with all other trials in this field only a clear, positive result is likely to have an impact on clinical use to modify the course of this progressive neurodegenerative disease.

Acknowledgments This work was supported in part by grants NIA R01AG17761 and NIA R01AG040093.

References

Adair, J. C., Knoefel, J. E., & Morgan, N. (2001). Controlled trial of N-acetylcysteine for patients with probable Alzheimer's disease. *Neurology, 57*(8), 1515–1517.

Adem, A., Mohammed, A. K., & Winblad, B. (1990). Multiple effects of tetrahydroaminoacridine on the cholinergic system: Biochemical and behavioural aspects. *Journal of Neural Transmission Parkinsons Disease and Dementia Section, 2,* 113–128.

Aisen, P. S., & Davis, K. L. (1994). Inflammatory mechanisms in Alzheimer's disease: Implications for therapy. *American Journal of Psychiatry, 151,* 1105–1113.

Aisen, P. S., Davis, K. L., Berg, J. D., et al. (2000). A randomized controlled trial of prednisone in Alzheimer's disease. Alzheimer's Disease Cooperative Study. *Neurology, 54,* 588–593.

Aisen, P. S., Schafer, K., Grundman, M., et al. (2002). Results of a multicenter trial of rofecoxib and naproxen in Alzheimer's disease. *Neurobiology of Aging, 23,* S429.

Aisen, P. S., Saumier, D., Briand, R., et al. (2006). A Phase II study targeting amyloid-beta with 3APS in mild-to-moderate Alzheimer disease. *Neurology, 67*(10), 1757–1763.

Aisen, P. S., Gauthier, S., Ferris, S. H., et al. (2011). Tramiprosate in mild-to-moderate Alzheimer's disease – A randomized, double-blind, placebo-controlled, multi-centre study (the Alphase Study). *Archives of Medical Science, 7*(1), 102–111.

Albert, M. S., DeKosky, S. T., Dickson, D., et al. (2011). The diagnosis of mild cognitive impairment due to Alzheimer's disease: Recommendations from the National Institute on Aging-Alzheimer's Association workgroups on diagnostic guidelines for Alzheimer's disease. *Alzheimer's & Dementia, 7*(3), 270–279.

Allegri, R. F., & Guekht, A. (2012). Cerebrolysin improves symptoms and delays progression in patients with Alzheimer's disease and vascular dementia. *Drugs Today, 48*(Suppl A), 25–41.

American Psychiatric Association. (2013). *Diagnostic and statistical manual of mental disorders* (DSM-5) (5th ed.). American Psychiatric Publishing.

Andersen, F., Viitanen, M., Halvorsen, D. S., et al. (2012). The effect of stimulation therapy and donepezil on cognitive function in Alzheimer's disease. A community based RCT with a two-by-two factorial design. *BMC Neurology, 12,* 59.

Andrieu, S., Guyonnet, S., Coley, N., et al. (2017). Effect of long-term omega 3 polyunsaturated fatty acid supplementation with or without multidomain intervention on cognitive function in elderly adults with memory complaints (Mapt): A randomised, placebo-controlled trial. *The Lancet Neurology, 16*(5), 377–389.

Annweiler, C., Herrmann, F. R., Fantino, B., et al. (2012). Effectiveness of the combination of memantine plus vitamin D on cognition in patients with Alzheimer disease: A pre-post pilot study. *Cognitive and Behavioral Neurology, 25*(3), 121–127.

Antuono, P. G. (1995). Effectiveness and safety of velnacrine for the treatment of Alzheimer's disease. A double-blind, placebo-controlled study. Mentane Study Group. *Archives of Internal Medicine, 155,* 1766–1772.

Arellanes, I. C., Choe, N., Solomon, V., et al. (2020). Brain delivery of supplemental docosahexaenoic acid (Dha): A randomized placebo-controlled clinical trial. *eBioMedicine, 59,* 102883.

Arendt, T., Bigl, V., Arendt, A., et al. (1983). Loss of neurons in the nucleus basalis of Meynert in Alzheimer's disease, paralysis agitans and Korsakoff's disease. *Acta Neuropathologica, 61,* 101–108.

Arendt, T., Bigl, V., Tennstedt, A., et al. (1985). Neuronal loss in different parts of the nucleus basalis is related to neuritic plaque formation in cortical target areas in Alzheimer's disease. *Neuroscience, 14,* 1–14.

Arendt, T., Bruckner, M. K., Lange, M., et al. (1992). Changes in acetylcholinesterase and butyrylcholinesterase in Alzheimer's disease resemble embryonic development – A study of molecular forms. *Neurochemistry International, 21*(3), 381–396.

Asthana, S., Baker, L. D., Craft, S., et al. (2001). High-dose estradiol improves cognition for women with AD: Results of a randomized study. *Neurology, 57,* 605–612.

Atack, J. R., Perry, E. K., Bonham, J. R., et al. (1986). Molecular forms of acetylcholinesterase and butyrylcholinesterase in the aged human central nervous system. *New England Journal of Medicine, 47,* 263–277.

Baker, L. D., Barsness, S. M., Borson, S., et al. (2012). Effects of growth hormonereleasing hormone on cognitive function in adults with mild cognitive impairment and healthy older adults: Results of a controlled trial. *Archives of Neurology, 69*(11), 1420–1429.

Bales, K. R., Tzavara, E. T., Wu, S., et al. (2006). Cholinergic dysfunction in a mouse model of Alzheimer disease is reversed by an anti-A beta antibody. *Journal of Clinical Investigation, 116*(3), 825–832.

Barberger-Gateau, P., Letenneur, L., Deschamps, V., et al. (2002). Fish, meat, and risk of dementia: Cohort study. *BMJ, 325*(7370), 932–933.

Bastianetto, S., Ramassamy, C., Dore, S., et al. (2000). The *Ginkgo biloba* extract (EGb 761) protects hippocampal neurons against cell death induced by beta-amyloid. *European Journal of Neuroscience, 12,* 1882–1890.

Becker, R. E., & Greig, N. H. (2012). Was phenserine a failure or were investigators misled by methods? *Current Alzheimer Research, 9,* 1174–1181.

Beher, D., Clarke, E. E., Wrigley, J. D., et al. (2004). Selected non-steroidal anti-inflammatory drugs and their derivatives target gamma-secretase at a novel site. Evidence for an allosteric mechanism. *Journal of Biological Chemistry, 279*(42), 43419–43426.

Birks, J. (2006). Cholinesterase inhibitors for Alzheimer's disease. *Cochrane Database of Systematic Reviews,* (1), CD005593.

Birks, J., & Flicker, L. (2003). Selegiline for Alzheimer's disease. *Cochrane Database of Systematic Reviews,* (1), CD000442.

Birks, J., Grimley Evans, J., Iakovidou, V., et al. (2000). Rivastigmine for Alzheimer's disease. *Cochrane Database of Systematic Reviews,* (4), CD001191.

Bissette, G., & Myers, B. (1992). Somatostatin in Alzheimer's disease and depression. *Life Sciences, 51,* 1389–1410.

Blair, H. A., & Dhillon, S. (2017). Safinamide: A review in Parkinson's disease. *CNS Drugs, 31*(2), 169–176.

Blennow, K., & Zetterberg, H. (2018). Biomarkers for Alzheimer's disease: Current status and prospects for the future. *Journal of Internal Medicine, 284*(6), 643–663.

Blennow, K., Zetterberg, H., Rinne, J. O., et al. (2012). Effect of immunotherapy with bapineuzumab on cerebrospinal fluid biomarker levels in patients with mild to moderate Alzheimer disease. *Archives of Neurology, 69*(8), 1002–1010.

Bondareff, W., Mountjoy, C. Q., & Roth, M. (1982). Loss of neurons of origin of the adrenergic projection to cerebral cortex (nucleus locus coeruleus) in senile dementia. *Neurology, 32*, 164–168.

Bowen, D. M., Smith, C. B., White, P., et al. (1976). Neurotransmitter-related enzymes and indices of hypoxia in senile dementia and other abiotrophies. *Brain, 99*, 459–496.

Breitner, J. C., Baker, L. D., Montine, T. J., et al. (2011). Extended results of the Alzheimer's Disease Anti-inflammatory Prevention Trial. *Alzheimer's & Dementia, 7*(4), 402–411.

Brodaty, H., Corey-Bloom, J., Potocnik, F. C., et al. (2005). Galantamine prolonged-release formulation in the treatment of mild to moderate Alzheimer's disease. *Dementia and Geriatric Cognitive Disorders, 20*(2–3), 120–132.

Brookmeyer, R., Gray, S., & Kawas, C. (1998). Projections of Alzheimer's disease in the United States and the public health impact of delaying disease onset. *American Journal of Public Health, 88*, 1337–1342.

Brufani, M., & Filocamo, L. (1997). Rational design of new acetylcholinersterase inhibitors. In R. Becker (Ed.), *Alzheimer's disease: From molecular biology to therapy* (pp. 171–177). Birkhauser.

Bullock, R., Touchon, J., Bergman, H., et al. (2005). Rivastigmine and donepezil treatment in moderate to moderately severe Alzheimer's disease over a 2-year period. *Current Medical Research and Opinions, 21*(8), 1317–1327.

Burckhardt, M., Herke, M., Wustmann, T., Watzke, S., Langer, G., & Fink, A. (2016). Omega-3 fatty acids for the treatment of dementia. *Cochrane Database of Systematic Reviews*, (4), CD009002.

Burns, A., Rossor, M., Hecker, J., et al. (1999). The effects of donepezil in Alzheimer's disease – Results from a multinational trial. *Dementia and Geriatric Cognitive Disorders, 10*, 237–244.

Burns, A., Gauthier, S., & Perdomo, C. (2007). Efficacy and safety of donepezil over 3 years: An open-label, multicentre study in patients with Alzheimer's disease. *International Journal of Geriatric Psychiatry, 22*(8), 806–812.

Burns, A., Perry, E., Holmes, C., et al. (2011). A double-blind placebo-controlled randomized trial of *Melissa officinalis* oil and donepezil for the treatment of agitation in Alzheimer's disease. *Dementia and Geriatric Cognitive Disorders, 31*(2), 158–164.

Buxbaum, J. D., Oishi, M., Chen, H. I., et al. (1992). Cholinergic agonists and interleukin 1 regulate processing and secretion of the Alzheimer beta/A4 amyloid protein precursor. *Proceedings of the National Academy of Sciences of the United States of America, 89*, 10075–10078.

Cardoso, B. R., Roberts, B. R., Malpas, C. B., et al. (2019). Supranutritional sodium selenate supplementation delivers selenium to the central nervous system: Results from a randomized controlled pilot trial in Alzheimer's disease. *Neurotherapeutics, 16*(1), 192–202.

Chai, B., Gao, F., Wu, R., et al. (2019). Vitamin D deficiency as a risk factor for dementia and Alzheimer's disease: An updated meta-analysis. *BMC Neurology, 19*(1), 284.

Cherny, R. A., Atwood, C. S., Xilinas, M. E., et al. (2001). Treatment with a copper-zinc chelator markedly and rapidly inhibits beta-amyloid accumulation in Alzheimer's disease transgenic mice. *Neuron, 30*, 665–676.

Chin, D., Huebbe, P., Pallauf, K., et al. (2013). Neuroprotective properties of curcumin in Alzheimer's disease – Merits and limitations. *Current Medicinal Chemistry, 20*(32), 3955–3985.

Christensen, D. D. (2012). Higher-dose (23 mg/day) donepezil formulation for the treatment of patients with moderate-to-severe Alzheimer's disease. *Postgraduate Medicine, 124*(6), 110–116.

Christie, J. E., Shering, A., Ferguson, J., et al. (1981). Physostigmine and arecoline: Effects of intravenous infusions in Alzheimer presenile dementia. *British Journal of Psychiatry, 138*, 46–50.

Chui, H. C., Victoroff, J. I., Margolin, D., et al. (1992). Criteria for the diagnosis of ischemic vascular dementia proposed by the State of California Alzheimer's Disease Diagnostic and Treatment Centers. *Neurology, 42*, 473–480.

Corey-Bloom, J., Anand, R., Veach, J., et al. (1998). A randomized trial evaluating the efficacy and safety of ENA 713 (rivastigmine tartrate): A new acetylcholinesterase inhibitor, in patients with mild to moderately severe Alzheimer's disease. *International Journal of Geriatric Psychopharmacology, 1*, 55–65.

Coric, V., van Dyck, C. H., Salloway, S., et al. (2012). Safety and tolerability of the γ-secretase inhibitor avagacestat in a phase 2 study of mild to moderate Alzheimer disease. *Archives of Neurology, 69*(11), 1430–1440.

Cotman, C. W., Monaghan, D. T., & Ganong, A. H. (1988). Excitatory amino acid neurotransmission: NMDA receptors and Hebb-type synaptic plasticity. *Annual Review of Neuroscience, 11*, 61–80.

Courtney, C., Farrell, D., Gray, R., et al. (2004). Long-term donepezil treatment in 565 patients with Alzheimer's disease (AD2000): Randomised double blind trial. *The Lancet, 363*(9427), 2105–2115.

Craft, S., Baker, L. D., Montine, T. J., et al. (2012). Intranasal insulin therapy for Alzheimer disease and amnestic mild cognitive impairment: A pilot clinical trial. *Archives of Neurology, 69*(1), 29–38.

Craft, S., Raman, R., Chow, T. W., et al. (2020). Safety, efficacy, and feasibility of intranasal insulin for the treatment of mild cognitive impairment and Alzheimer disease dementia: A randomized clinical trial. *JAMA Neurology, 77*(9), 1099–1109.

Cremonini, A. L., Caffa, I., Cea, M., Nencioni, A., Odetti, P., & Monacelli, F. (2019). Nutrients in the prevention of Alzheimer's disease. *Oxidative Medicine and Cellular Longevity, 2019*, 9874159.

Davies, P., & Maloney, A. J. (1976). Selective loss of central cholinergic neurons in Alzheimer's disease (Letter). *The Lancet, ii*, 1403.

Davies, P., Katzman, R., & Terry, R. D. (1980). Reduced somatostatin-like immunoreactivity in cerebral cortex from cases of Alzheimer senile dementia. *Nature, 288*, 279–280.

Davis, K. L., & Mohs, R. C. (1982). Enhancement of memory processes in Alzheimer's disease with multiple-dose intravenous physostigmine. *American Journal of Psychiatry, 139*, 1421–1424.

Davis, K. L., Thai, L. J., Gamzu, E. R., et al. (1992). A double-blind, placebo-controlled multicenter study of tacrine for Alzheimer's disease. The Tacrine Collaborative Study Group. *New England Journal of Medicine, 327*, 1253–1259.

DeKosky, S. T., Williamson, J. D., Fitzpatrick, A. L., et al. (2008). Ginkgo biloba for prevention of dementia: A randomized controlled trial. *JAMA, 300*(19), 2253–2262.

Demattos, R. B., Lu, J., Tang, Y., et al. (2012). A plaque-specific antibody clears existing β-amyloid plaques in Alzheimer's disease mice. *Neuron, 76*(5), 908–920.

Devanand, D. P., Liu, X., Pradhaban, G., et al. (2008). Combining early markers strongly predicts conversion from mild cognitive impairment to Alzheimer's disease. *Biological Psychiatry, 64*, 871–879.

Devos, D., Moreau, C., Devedjian, J. C., et al. (2014). Targeting chelatable iron as a therapeutic modality in Parkinson's disease. *Antioxidants & Redox Signaling, 21*(2), 195–210.

Dhillon, S. (2021). Aducanumab: First approval. *Drugs, 81*(12), 437–1443. https://doi.org/10.1007/s40265-021-01569-z. Erratum in: Drugs. 2021;81(14):1701.

Dodel, R. C., Du, Y., Depboylu, C., et al. (2004). Intravenous immunoglobulins containing antibodies against beta-amyloid for the treatment of Alzheimer's disease. *Journal of Neurology, Neurosurgery and Psychiatry, 75*(10), 1472–1474.

Doody, R. S., Gavrilova, S. I., Sano, M., et al. (2008). Effect of dimebon on cognition, activities of daily living, behaviour, and global function in patients with mild-to-moderate Alzheimer's disease: A randomised, double-blind, placebo-controlled study. *The Lancet, 372*(9634), 207–215.

Doody, R. S., Ferris, S. H., Salloway, S., et al. (2009). Donepezil treatment of patients with MCI: A 48-week randomized, placebo-controlled trial. *Neurology, 72*(18), 1555–1561.

Doody, R. S., Raman, R., Farlow, M., et al. (2013). A phase 3 trial of semagacestat for treatment of Alzheimer's disease. *New England Journal of Medicine, 369*(4), 341–350.

Doody, R. S., Thomas, R. G., Farlow, M., et al. (2014). Phase 3 trials of solanezumab for mild-to-moderate Alzheimer's disease. *New England Journal of Medicine, 370*(4), 311–321.

Dorman, D. C., Cote, L. M., & Buck, W. B. (1992). Effects of an extract of *Gingko biloba* on bromethalin-induced cerebral lipid peroxidation and edema in rats. *American Journal of Veterinary Research, 53*, 138–142.

Dunbar, G., Demazieres, A., Monreal, A., et al. (2006). Pharmacokinetics and safety profile of ispronicline (TC-1734), a new brain nicotinic receptor partial agonist, in young healthy male volunteers. *Journal of Clinical Pharmacology, 46*(7), 715–726.

Dunn, N., Holmes, C., & Mullee, M. (2005). Does lithium therapy protect against the onset of dementia? *Alzheimer Disease and Associated Disorders, 19*, 20–22.

Dysken, M., Kuskowski, M., Love, S., et al. (2002). Ondansetron in the treatment of cognitive decline in Alzheimer dementia. *American Journal of Geriatric Psychiatry, 10*, 212–215.

Dysken, M. W., Sano, M., Asthana, S., et al. (2014). Effect of vitamin E and memantine on functional decline in Alzheimer disease: The TEAM-AD VA cooperative randomized trial. *Journal of the American Medical Association, 311*(1), 33–44.

Egan, M. F., Kost, J., Tariot, P. N., et al. (2018). Randomized trial of verubecestat for mild-to-moderate Alzheimer's disease. *New England Journal of Medicine, 378*(18), 1691–1703.

Enz, A., & Floersheim, P. (1997). Cholinesterase inhibitors: An overview of their mechanisms of action. In R. Becker (Ed.), *Alzheimer's disease: From molecular biology to therapy* (pp. 211–215). Birkhauser.

Farlow, M., Gracon, S. I., Hershey, L. A., et al. (1992). A controlled trial of tacrine in Alzheimer's disease. The Tacrine Study Group. *JAMA, 268*, 2523–2529.

Farlow, M., Veloso, F., Moline, M., et al. (2011). Safety and tolerability of donepezil 23 mg in moderate to severe Alzheimer's disease. *BMC Neurology, 11*, 57.

Farlow, M. R., Andreasen, N., Riviere, M.-E., et al. (2015). Long-term treatment with active Aβ immunotherapy with CAD106 in mild Alzheimer's disease. *Alzheimer's Research & Therapy, 7*(1), 23.

Faux, N. G., Ritchie, C. W., Gunn, A., et al. (2010). PBT2 rapidly improves cognition in Alzheimer's disease: Additional Phase II analyses. *Journal of Alzheimer's Disease, 20*(2), 509–516.

Feldman, H., Gauthier, S., Hecker, J., et al. (2001). A 24-week, randomized, double-blind study of donepezil in moderate to severe Alzheimer's disease. *Neurology, 57*, 613–620.

Feldman, H. H., Ferris, S., Winblad, B., et al. (2007). Effect of rivastigmine on delay to diagnosis of Alzheimer's disease from mild cognitive impairment: The InDDEx study. *Lancet Neurology, 6*(6), 501–512.

Feldman, H. H., Doody, R. S., Kivipelto, M., et al. (2010). Randomized controlled trial of atorvastatin in mild to moderate Alzheimer disease: LEADe. *Neurology, 74*(12), 956–964.

Fillit, H., & Green, A. (2021). Aducanumab and the FDA – Where are we now? *Nature Reviews Neurology.* Published online January 13, 2021, 1–2.

Fioravanti, M., & Flicker, L. (2001). Efficacy of Nicergoline in dementia and other age associated forms of cognitive impairment. *Cochrane Database of Systematic Reviews, (4), CD003159.*

Flicker, L., & Evans, J. G. (2001). Piracetam for dementia or cognitive impairment. *Cochrane Database of Systematic Reviews, (2), CD001011.*

Folstein, M. F., Folstein, S. E., & McHugh, P. R. (1975). Mini-mental state: A practical method for grading the cognitive state of patients for the clinician. *Journal of Psychiatric Research, 12*, 189–198.

Forlenza, O. V., Diniz, B. S., Radanovic, M., et al. (2011). Disease-modifying properties of long-term lithium treatment for amnestic mild cognitive impairment: Randomized controlled trial. *British Journal of Psychiatry, 198*, 351–356.

Francis, P. T., Palmer, A. M., Snape, M., et al. (1999). The cholinergic hypothesis of Alzheimer's disease: A review of progress. *Journal of Neurology, Neurosurgery and Psychiatry, 66*, 137–147.

Fremont, R., Manoochehri, M., Armstrong, N. M., et al. (2020). Tolcapone treatment for cognitive and behavioral symptoms in behavioral variant frontotemporal dementia: A placebo-controlled crossover study. *Journal of Alzheimer's Disease, 75*(4), 1391–1403.

Fritze, J., & Waiden, J. (1995). Clinical findings with nimodipine in dementia: Test of the calcium hypothesis. *Journal of Neural Transmission, 46*(Suppl), 439–453.

Galasko, D., Bennett, D., Sano, M., et al. (1997). An inventory to assess activities of daily living for clinical trials in Alzheimer's disease. The Alzheimer's Disease Cooperative Study. *Alzheimer Disease and Associated Disorders, 11*, S33–S39.

Galasko, D., Graff-Radford, N., Murphy, M. P., et al. (2004). Safety, tolerability, pharmacokinetics and $A\beta$ levels following short-term administration of R-flurbiprofen in healthy elderly individuals: A phase I study. In *Presented at the 9th international conference on Alzheimer's disease and related disorders*, Philadelphia, PA.

Galasko, D. R., Peskind, E., Clark, C. M., et al. (2012). Antioxidants for Alzheimer disease: A randomized clinical trial with cerebrospinal fluid biomarker measures. *Archives of Neurology, 69*(7), 836–841.

Gasparini, L., Ongini, E., Wilcock, D., et al. (2005). Activity of flurbiprofen and chemically related anti-inflammatory drugs in models of Alzheimers disease. *Brain Research Reviews, 48*, 400–408.

Gaudig, M., Richarz, U., Han, J., et al. (2011). Effects of galantamine in Alzheimer's disease: Double-blind withdrawal studies evaluating sustained versus interrupted treatment. *Current Alzheimer Research, 8*(7), 771–780.

Gauthier, S., & Molinuevo, J. L. (2013). Benefits of combined cholinesterase inhibitor and memantine treatment in moderate-severe Alzheimer's disease. *Alzheimer's & Dementia, 9*(3), 326–331.

Gauthier, S., Lopez, O. L., Waldemar, G., et al. (2010). Effects of donepezil on activities of daily living: Integrated analysis of patient data from studies in mild, moderate and severe Alzheimer's disease. *International Psychogeriatrics, 22*(6), 973–983.

Geifman, N., Brinton, R. D., Kennedy, R. E., Schneider, L. S., & Butte, A. J. (2017). Evidence for benefit of statins to modify cognitive decline and risk in Alzheimer's disease. *Alzheimer's Research & Therapy, 9*(1), 10.

Gervais, F., Chalifour, R., Garceau, D., et al. (2001). Glycosaminoglycan mimetics: A therapeutic approach to cerebral amyloid angiopathy. *Amyloid, 8*(Suppl 1), 28–35.

Gilman, S., Koller, M., Black, R. S., et al. (2005). AN1792 (QS-21)-201 study Team. Clinical effects of Abeta immunization (AN1792) in patients with AD in an interrupted trial. *Neurology, 64*(9), 1553–1562.

Gold, M., Alderton, C., Zvartau-Hind, M., et al. (2010). Rosiglitazone monotherapy in mild-to-moderate Alzheimer's disease: Results from a randomized, double-blind, placebo-controlled phase III study. *Dementia and Geriatric Cognitive Disorders, 30*(2), 131–146.

Green, R. C., Schneider, L. S., Amato, D. A., et al. (2009). Effect of tarenflurbil on cognitive decline and activities of daily living in patients with mild Alzheimer disease: A randomized controlled trial. *JAMA, 302*(23), 2557–2564.

Haase, J., Halama, P., & Horr, R. (1996). Effectiveness of brief infusions with *Ginkgo biloba* special extract EGb 761 in dementia of the vascular and Alzheimer-type. *Zeitschrift für Gerontologie und Geriatrie, 29*, 302–309. (in German).

Hak, A. E., Pols, H. A., Stehouwer, C. D., et al. (2001). Markers of inflammation and cellular adhesion molecules in relation to insulin-resistance in nondiabetic elderly: The Rotterdam study. *Journal of Clinical Endocrinology and Metabolism, 86*, 4398–4405.

Hampel, H., Ewers, M., Bürger, K., et al. (2009). Lithium trial in Alzheimer's disease: A randomized, single-blind, placebo-controlled, multicenter 10-week study. *Journal of Clinical Psychiatry, 70*(6), 922–931.

Haroutunian, V., Greig, N., Pei, X. F., et al. (1997). Pharmacological modulation of Alzheimer's β-amyloid precursor protein levels in the CSF of rats with forebrain cholinergic system lesions. *Brain Research. Molecular Brain Research, 46*(1–2), 161–168.

Harvey, A. L. (1995). The pharmacology of galanthamine and its analogues. *Pharmacology and Therapeutics, 68*, 113–128.

Henderson, V. W., Paganini-Hill, A., Miller, B. L., et al. (2000). Estrogen for Alzheimer's disease in women:

Randomized, double-blind, placebo-controlled trial. *Neurology, 54*, 295–301.

Hernandez, R. K., Farwell, W., Cantor, M. D., et al. (2009). Cholinesterase inhibitors and incidence of bradycardia in patients with dementia in the veterans affairs New England healthcare system. *Journal of the American Geriatrics Society, 57*(11), 1997–2003.

Herrup, K. (2010). Re-imagining Alzheimer's disease: An age-based hypothesis. *Journal of Neuroscience, 30*, 16755–16762.

Hofferberth, B. (1994). The efficacy of EGb 761 in patients with senile dementia of the Alzheimer type, a double-blind, placebo-controlled study on different levels of investigation. *Human Psychopharmacology, 9*, 215–222.

Holmes, C., Wilkinson, D., Dean, C., et al. (2004). The efficacy of donepezil in the treatment of neuropsychiatric symptoms in Alzheimer disease. *Neurology, 63*(2), 214–219.

Homma, A., Takeda, M., Imai, Y., et al. (2000). Clinical efficacy and safety of donepezil on cognitive and global function in patients with Alzheimer's disease. A 24-week, multicenter, double-blind, pleacebo-controlled study in Japan. E2020 Study Group. *Dementia and Geriatric Cognitive Disorders, 11*(6), 299–313.

Honig, L. S., Vellas, B., Woodward, M., et al. (2018). Trial of solanezumab for mild dementia due to Alzheimer's disease. *The New England Journal of Medicine, 378*(4), 321–330.

Howard, R. J., Juszocak, E., Ballard, C. G., et al. (2007). Donepezil for the treatment of agitation in Alzheimer's disease. *New England Journal of Medicine, 357*(14), 1382–1392.

Howard, R., McShane, R., Lindesay, J., et al. (2012). Donepezil and memantine for moderate-to-severe Alzheimer's disease. *New England Journal of Medicine, 366*(10), 893–903.

Huguet, F., Drieu, K., & Piriou, A. (1994). Decreased cerebral 5-HT1A receptors during ageing: Reversal by *Ginkgo biloba* extract (EGb 761). *Journal of Pharmacy and Pharmacology, 46*, 316–318.

Inestrosa, N. C., Alvarez, A., Perez, C. A., et al. (1996). Acetylcholinesterase accelerates assembly of amyloid-beta-peptides into Alzheimer's fibrils: Possible role of the peripheral site of the enzyme. *Neuron, 16*, 881–891.

Jack, C. R., Jr., Knopman, D. S., Jagust, W. J., et al. (2010). Hypothetical model of dynamic biomarkers of the Alzheimer's pathological cascade. *Lancet Neurology, 9*(1), 119–128.

Janus, C., Pearson, J., McLaurin, J., et al. (2000). Aβ peptide immunization reduces behavioural impairment and plaques in a model of Alzheimer's disease. *Nature, 408*, 979–982.

Jick, H., Zornberg, G. L., Jick, S. S., et al. (2000). Statins and the risk of dementia. *The Lancet, 356*(9242), 1627–1631.

Jorm, A. F. (1986). Effects of cholinergic enhancement therapies on memory function in Alzheimer's disease:

A meta-analysis of the literature. *Australian and New Zealand Journal of Psychiatry, 20*, 237–240.

Jorm, A. F., Korten, A. E., & Henderson, A. S. (1987). The prevalence of dementia: A quantitative integration of the literature. *Acta Psychiatrica Scandinavica, 76*, 465–479.

Kalmijn, S., Launer, L. J., Ott, A., et al. (1997). Dietary fat intake and the risk of incident dementia in the Rotterdam Study. *Annals of Neurology, 42*(5), 776–782.

Kanowski, S., Fischof, P., & Hiersemenzel, R. (1988). Wirksamkeitsnachweis von Neotropika am Beispiel von Nimodipin-ein Beitrag zur entwicklung geeigneter klinischer Prufmodelle. *Zeitschrift fur Gerontopsychologie und Psychiatrie, 1*, 35–44.

Kanowski, S., Herrmann, W. M., Stephan, K., et al. (1996). Proof of efficacy of the ginkgo biloba special extract EGb 761 in outpatients suffering from mild to moderate primary degenerative dementia of the Alzheimer-type or multi-infarct dementia. *Pharmacopsychiatry, 29*, 47–56.

Kaufer, D. I., Cummings, J. L., & Christine, D. (1996). Effect of tacrine on behavioral symptoms in Alzheimer's disease: An open-label study. *Journal of Geriatric Psychiatry and Neurology, 9*, 1–6.

Kavirajan, H., & Schneider, L. N. (2007). Efficacy and adverse effects of cholinesterase inhibitors and memantine in vascular dementia: A meta-analysis of randomized controlled trials. *Lancet Neurology, 6*(9), 782–792.

Kessing, L. V., Söndergård, L., Forman, J. L., et al. (2008). Lithium treatment and risk of dementia. *Archives of General Psychiatry, 65*, 1331–1335.

Khachaturian, Z. S. (1985). Diagnosis of Alzheimer's disease. *Archives of Neurology, 42*, 1097–1105.

Kim, A. J., Shi, Y., Austin, R. C., et al. (2005). Valproate protects cells from ER stress-induced lipid accumulation and apoptosis by inhibiting glycogen synthase kinase-3. *Journal of Cell Science, 118*(Pt 1), 89–99.

Knapp, M. J., Knopman, D. S., Solomon, P. R., et al. (1994). A 30-week randomized controlled trial of high-dose tacrine in patients with Alzheimer's disease. The Tacrine Study Group. *JAMA, 271*(13), 985–991.

Koenig, A. M., Mechanic-Hamilton, D., Xie, S. X., et al. (2017). Effects of the insulin sensitizer metformin in Alzheimer disease: Pilot data from a randomized placebo-controlled crossover study. *Alzheimer Disease and Associated Disorders, 31*(2), 107–113.

Kontsekova, E., Zilka, N., Kovacech, B., Novak, P., & Novak, M. (2014). First-in-man tau vaccine targeting structural determinants essential for pathological tau–tau interaction reduces tau oligomerisation and neurofibrillary degeneration in an Alzheimer's disease model. *Alzheimer's Research & Therapy, 6*(4), 44.

Kristofikova, Z., Benesova, O., & Tejkalova, H. (1992). Changes of high-affinity choline uptake in the hippocampus of old rats after long term administration of two nootropic drugs (tacrine and *Ginkgo biloba* extract). *Dementia, 3*, 304–307.

Kröger, E., Berkers, M., Carmichael, P. H., et al. (2012). Use of rivastigmine or galantamine and risk of adverse cardiac events: A database study from The Netherlands. *American Journal of Geriatric Pharmacotherapy, 10*(6), 373–380.

Lahiri, D. K., Nall, C., & Farlow, M. R. (1992). The cholinergic agonist carbachol reduces intracellular beta-amyloid precursor protein in PC 12 and C6 cells. *Biochemistry International, 28*, 853–860.

Lahiri, D. K., Farlow, M. R., Hintz, N., et al. (2000). Cholinesterase inhibitors, beta-amyloid precursor protein and amyloid beta-peptides in Alzheimer's disease. *Acta Neurologica Scandinavica, 176*(Suppl), 60–67.

Landen, J. W., Cohen, S., Billing, C. B., et al. (2017). Multiple-dose ponezumab for mild-to-moderate Alzheimer's disease: Safety and efficacy. *Alzheimer's & Dementia (New York, N.Y.), 3*(3), 339–347.

Lane, R. M., Potkin, S. G., & Enz, A. (2006). Targeting acetylcholinesterase and butyrylcholinesterase in dementia. *International Journal of Neuropsychopharmacology, 9*, 101–124.

Lauritzen, L., Hansen, H. S., Jorgensen, M. H., et al. (2001). The essentiality of long chain n-3 fatty acids in relation to development and function of the brain and retina. *Progress in Lipid Research, 40*(1–2), 1–94.

Le Bars, P. L., Katz, M. M., Berman, N., et al. (1997). A placebo-controlled, double-blind, randomized trial of an extract of *Ginkgo biloba* for dementia. North American EGb Study Group. *JAMA, 278*, 1327–1332.

Leissring, M. A., Farris, W., Chang, A. Y., et al. (2003). Enhanced proteolysis of beta-amyloid in APP transgenic mice prevents plaque formation, secondary pathology, and premature death. *Neuron, 40*, 1087–1093.

López-Arrieta, J. M., & Birks, J. (2002). Nimodipine for primary degenerative, mixed and vascular dementia. *Cochrane Database of Systematic Reviews*, (3), CD000147.

Lovestone, S., Boada, M., Dubois, B., et al. (2015). A Phase II trial of tideglusib in Alzheimer's disease. *Journal of Alzheimer's Disease, 45*(1), 75–88.

Luchsinger, J. A., Perez, T., Chang, H., et al. (2016). Metformin in amnestic mild cognitive impairment: Results of a pilot randomized placebo controlled clinical trial. *Journal of Alzheimer's Disease, 51*(2), 501–514.

Maelicke, A., Samochocki, M., Jostock, R., et al. (2001). Allosteric sensitization of nicotinic receptors by galantamine, a new treatment strategy for Alzheimer's disease. *Biological Psychiatry, 49*, 279–288.

Malpas, C. B., Vivash, L., Genc, S., et al. (2016). A Phase IIa randomized control trial of vel015 (Sodium selenate) in mild-moderate Alzheimer's disease. *Journal of Alzheimer's Disease, 54*(1), 223–232.

Mark, R. J., Ashford, J. W., Goodman, Y., et al. (1995). Anticonvulsants attenuate amyloid beta-peptide neurotoxicity, Ca^{2+} deregulation, and cytoskeletal pathology. *Neurobiology of Aging, 16*(2), 187–198.

Matsunaga, S., Kishi, T., & Iwata, N. (2014). Combination therapy with cholinesterase inhibitors and memantine for Alzheimer's disease: A systematic review and meta-analysis. *The International Journal of Neuropsychopharmacology, 18*(5), pyu115.

Matsuoka, N., & Aigner, T. G. (1997). FK960 [*N*-(4-acetyl-1-piperazinyl)-*p*-fluorobenzamide monohydrate], a novel potential antidementia drug, improves visual recognition memory in rhesus monkeys: Comparison with physostigmine. *Journal of Pharmacology and Experimental Therapeutics, 280*, 1201–1209.

Matsuoka, N., Yamazaki, M., & Yamaguchi, I. (1995). Changes in brain somatostatin in memory deficient rats: Comparison with cholinergic markers. *Neuroscience, 66*, 617–626.

Maurer, K., Ihl, R., Dierks, T., et al. (1997). Clinical efficacy of *Ginkgo biloba* special extract EGb 761 in dementia of the Alzheimer type. *Journal of Psychiatric Research, 31*, 645–655.

Mazurov, A., Hauser, T., & Miller, C. H. (2006). Selective alpha7 nicotinic acetylcholine receptor ligands. *Current Medicinal Chemistry, 13*(13), 1567–1584.

McKeith, I. G., Galasko, D., Kosaka, K., et al. (1996). Consensus guidelines for the clinical and pathologic diagnosis of dementia with Lewy bodies (DLB). Report of the Consortium on DLB International Workshop. *Neurology, 47*, 1113–1124.

McKhann, G., Drachman, D., Folstein, M., et al. (1984). Clinical diagnosis of Alzheimer's disease. Report of the NINCDS-ADRDA Work Group under the auspices of Department of Health and Human Services Task Force on Alzheimer's Disease. *Neurology, 34*, 939–944.

McKhann, G. M., Knopman, D. S., Chertkow, H., et al. (2011). The diagnosis of dementia due to Alzheimer's disease: Recommendations from the National Institute on Aging-Alzheimer's Association workgroups on diagnostic guidelines for Alzheimer's disease. *Alzheimer's & Dementia, 7*(3), 263–269.

McLaurin, J., Kierstead, M. E., Brown, M. E., et al. (2006). Cyclohexanehexol inhibitors of Abeta aggregation prevent and reverse Alzheimer phenotype in a mouse model. *Nature Medicine, 12*(7), 801–808.

McShane, R., Areosa Sastre, A., & Minakaran, N. (2006). Memantine for dementia. *Cochrane Database of Systematic Reviews*, (2), CD003154.

Mirra, S. S., Heyman, A., McKeel, D., et al. (1991). The Consortium to Establish a Registry for Alzheimer's Disease (CERAD). Part II. Standardization of the neuropathologic assessment of Alzheimer's disease. *Neurology, 41*, 479–486.

Mohs, R. C., Davis, B. M., Johns, C. A., et al. (1985). Oral physostigmine treatment of patients with Alzheimer's disease. *American Journal of Psychiatry, 142*, 28–33.

Mohs, R. C., Knopman, D., Petersen, R. C., et al. (1997). Development of cognitive instruments for use in clinical trials of antidementia drugs: Additions to the Alzheimer's Disease Assessment Scale that broaden its scope. The Alzheimer's Disease Cooperative

Study. *Alzheimer Disease and Associated Disorders, 11*, S13–S21.

Mohs, R. C., Doody, R. S., Morris, J. C., et al. (2001). 1-year, placebo-controlled preservation of function survival study of donepezil in AD patients. *Neurology, 57*, 481–488.

Muller, D., Mendla, K., Farber, S. A., et al. (1997). Muscarinic M1 receptor agonists increase the secretion of the amyloid precursor protein ectodomain. *Life Sciences, 60*, 985–991.

Mulnard, R. A., Cotman, C. W., Kawas, C., et al. (2000). Estrogen replacement therapy for treatment of mild to moderate Alzheimer disease: A randomized controlled trial. Alzheimer's Disease Cooperative Study. *JAMA, 283*, 1007–1015.

Muñoz-Jiménez, M., Zaarkti, A., García-Arnés, J. A., & García-Casares, N. (2020). Antidiabetic drugs in Alzheimer's disease and mild cognitive impairment: A systematic review. *Dementia and Geriatric Cognitive Disorders, 49*(5), 423–434.

Muresanu, D. F., Rainer, M., & Moessler, H. (2002). Improved global function and activities of daily living in patients with AD: A placebo-controlled clinical study with the neurotrophic agent Cerebrolysin. *Journal of Neural Transmission, 62*(Suppl), 277–285.

Nayak, L., & Henchcliffe, C. (2008). Rasagiline in treatment of Parkinson's disease. *Neuropsychiatric Disease and Treatment, 4*(1), 23–32.

Newhouse, P. A., Potter, A., Kelton, M., et al. (2001). Nicotinic treatment of Alzheimer's disease. *Biological Psychiatry, 49*(3), 268–278.

Newhouse, P., Kellar, K., Aisen, P., et al. (2012). Nicotine treatment of mild cognitive impairment: A 6-month double-blind pilot clinical trial. *Neurology, 78*(2), 91–101.

Ngandu, T., Lehtisalo, J., Solomon, A., et al. (2015). A 2 year multidomain intervention of diet, exercise, cognitive training, and vascular risk monitoring versus control to prevent cognitive decline in at-risk elderly people (Finger): A randomised controlled trial. *Lancet, 385*(9984), 2255–2263.

Nicholson, C. D. (1990). Pharmacology of nootropics and metabolically active compounds in relation to their use in dementia. *Psychopharmacology, 101*, 147–159.

Nicoll, J. A., Wilkinson, D., Holmes, C., et al. (2003). Neuropathology of human Alzheimer disease after immunization with amyloid-beta peptide: A case report. *Nature Medicine, 9*(4), 448–452.

Nicoll, J. A. R., Buckland, G. R., Harrison, C. H., et al. (2019). Persistent neuropathological effects 14 years following amyloid-β immunization in Alzheimer's disease. *Brain, 142*(7), 2113–2126.

Nitsch, R. M., Slack, B. E., Wurtman, R. J., et al. (1992). Release of Alzheimer amyloid precursor derivatives stimulated by activation of muscarinic acetylcholine receptors. *Science, 258*, 304–307.

Nordberg, A. (2001). Nicotinic receptor abnormalities of Alzheimer's disease: Therapeutic implications. *Biological Psychiatry, 49*(3), 200–210.

Novak, P., Schmidt, R., Kontsekova, E., et al. (2017). Safety and immunogenicity of the tau vaccine AADvac1 in patients with Alzheimer's disease: A randomised, double-blind, placebo-controlled, phase 1 trial. *The Lancet Neurology, 16*(2), 123–134.

Nunes, P. V., Forlenza, O. V., & Gattaz, W. F. (2007). Lithium and risk for Alzheimer's disease in elderly patients with bipolar disorder. *British Journal of Psychiatry, 190*, 359–360.

Ohkura, T., Isse, K., Akazawa, K., et al. (1994). Evaluation of estrogen treatment in female patients with dementia of the Alzheimer type. *Endocrine Journal, 41*, 361–371.

Olin, J., Schneider, L., Novit, A., et al. (2001). Hydergine for dementia. *Cochrane Database of Systematic Reviews*, (2), CD000359.

Ono, K., Hasegawa, K., Naiki, H., et al. (2004). Curcumin has potent anti-amyloidogenic effects for Alzheimer's beta-amyloid fibrils in vitro. *Journal of Neuroscience Research, 75*(6), 742–750.

Orgogozo, J. M. (2006). Vaccination treatment of AD (abstract S5-04-04). *Alzheimer's & Dementia, 2*, S94.

Orkaby, A. R., Gaziano, J. M., Djousse, L., & Driver, J. A. (2017). Statins for primary prevention of cardiovascular events and mortality in older men. *Journal of the American Geriatrics Society, 65*(11), 2362–2368.

Ostrowitzki, S., Lasser, R. A., Dorflinger, E., et al. (2017). A phase III randomized trial of gantenerumab in prodromal Alzheimer's disease. *Alzheimer's Research & Therapy, 9*(1), 95.

Oudshoorn, C., Mattace-Raso, F. U., van der Velde, N., et al. (2008). Higher serum vitamin D_3 levels are associated with better cognitive test performance in patients with Alzheimer's disease. *Dementia and Geriatric Cognitive Disorders, 25*(6), 539–543.

Owen, R. T. (2016). Memantine and donepezil: A fixed drug combination for the treatment of moderate to severe Alzheimer's dementia. *Drugs of Today (Barcelona, Spain), 52*(4), 239–248.

Paganini-Hill, A., & Henderson, V. W. (1994). Estrogen deficiency and risk of Alzheimer's disease in women. *American Journal of Epidemiology, 140*, 256–261.

Panisset, M., Gauthier, S., Moessler, H., et al. (2002). Cerebrolysin in Alzheimer's disease: A randomized, double-blind, placebo-controlled trial with a neurotrophic agent. *Journal of Neural Transmission, 109*(7–8), 1089–1104.

Panula, P., Rinne, J., Kuokkanen, K., et al. (1998). Neuronal histamine deficit in Alzheimer's disease. *Neuroscience, 82*(4), 993–997.

Parkinson Study Group. (1993). Effects of tocopherol and deprenyl on the progression of disability in early Parkinson's disease. *New England Journal of Medicine, 328*(3), 176–183.

Pasquier, F., Sadowsky, C., Holstein, A., et al. (2016). Two phase 2 multiple ascending-dose studies of vanutide cridificar (ACC-001) and qs-21 adjuvant in mild-to-moderate Alzheimer's disease. *Journal of Alzheimer's Disease, 51*(4), 1131–1143.

Perry, E. K., Perry, R. H., Blessed, G., et al. (1977). Necropsy evidence of central cholinergic deficits in senile dementia. *The Lancet, i*, 189.

Perry, E. K., Tomlinson, B. E., Blessed, G., et al. (1978). Correlation of cholinergic abnormalities with senile plaques and mental test scores in senile dementia. *The British Medical Journal, ii*, 1457–1459.

Peskind, E. R., Potkin, S. G., Pomara, N., et al. (2006). Memantine treatment in mild to moderate Alzheimer disease: A 24-week randomized, controlled trial. *American Journal of Geriatric Psychiatry, 14*(8), 704–715.

Petersen, R. C., Smith, G. E., Ivnik, R. J., et al. (1995). Apolipoprotein E status as a predictor of the development of Alzheimer's disease in memory-impaired individuals. *JAMA, 273*(16), 1274–1278.

Petersen, R. C., Stevens, J. C., Ganguli, M., et al. (2001). Practice parameter: Early detection of dementia: Mild cognitive impairment (an evidence-based review). Report of the Quality Standards Subcommittee of the American Academy of Neurology. *Neurology, 56*, 1133–1142.

Petersen, R. C., Thomas, R. G., Grundman, M., et al. (2005). Alzheimer's Disease Cooperative Study Group. Vitamin E and donepezil for the treatment of mild cognitive impairment. *New England Journal of Medicine, 352*, 2379–2388.

Pillai, J. A., & Cummings, J. L. (2013). Clinical trials in predementia stages of Alzheimer disease. *Medical Clinics of North America, 97*(3), 439–457.

Pitkälä, K. H., Pöysti, M. M., Laakkonen, M. L., et al. (2013). Effects of the Finnish Alzheimer disease exercise trial (FINALEX): A randomized controlled trial. *JAMA Internal Medicine, 173*(10), 894–901.

Plowey, E. D., Bussiere, T., Rajagovindan, R., et al. (2022). Alzheimer disease neuropathology in a patient previously treated with aducanumab. *Acta Neuropathologica, 144*(1), 143–153. https://doi.org/10.1007/s00401-022-02433-4

Quinn, J. F., Raman, R., Thomas, R. G., et al. (2010). Docosahexaenoic acid supplementation and cognitive decline in Alzheimer disease: A randomized trial. *JAMA, 304*(17), 1903–1911.

Qureshi, I. A., Tirucherai, G., Ahlijanian, M. K., Kolaitis, G., Bechtold, C., & Grundman, M. (2018). A randomized, single ascending dose study of intravenous BIIB092 in healthy participants. *Alzheimer's & Dementia (New York, N.Y.), 4*, 746–755.

Radd-Vagenas, S., Duffy, S. L., Naismith, S. L., Brew, B. J., Flood, V. M., & Fiatarone Singh, M. A. (2018). Effect of the Mediterranean diet on cognition and brain morphology and function: A systematic review of randomized controlled trials. *The American Journal of Clinical Nutrition, 107*(3), 389–404.

Rafii, M. S., Walsh, S., Little, J. T., et al. (2011). A phase II trial of huperzine A in mild to moderate Alzheimer disease. *Neurology, 76*(16), 1389–1394.

Rammes, G., Rupprecht, R., Ferrari, U., et al. (2001). The *N*-methyl-D-aspartate receptor channel blockers memantine, MRZ 2/579 and other amino-alkyl-cyclohexanes antagonise 5-HT$_3$ receptor currents in cultured HEK-293 and N1E-115 cell systems in a non-competitive manner. *Neuroscience Letters, 306*(1–2), 81–84.

Raskind, M. A., Sadowsky, C. H., Sigmund, W. R., et al. (1997). Effect of tacrine on language, praxis, and non-cognitive behavioral problems in Alzheimer disease. *Archives of Neurology, 54*, 836–840.

Raskind, M., Peskind, E. R., Wessel, T., et al. (2000). Galantamine in Alzheimer's disease – A 6-month, randomized, placebo-controlled trial with a 6-month extension. *Neurology, 54*, 2261–2268.

Reisberg, B., Doody, R., Stoffler, A., et al. (2003). Memantine in moderate-to-severe Alzheimer's disease. *New England Journal of Medicine, 348*(14), 1333–1341.

Relkin, N., Szabo, P., Adamiak, B., et al. (2005). Intravenous immunoglobulin (IVIg) treatment causes dose-dependent alterations in β-amyloid $(A\beta)$ levels and anti-Aβ antibody titers in plasma and cerebrospinal fluid (CSF) of Alzheimer's disease (AD) patients. *Neurology, 64*(Suppl 1), A144.

Relkin, N. R., Thomas, R. G., Rissman, R. A., et al. (2017). A phase 3 trial of IV immunoglobulin for Alzheimer disease. *Neurology, 88*(18), 1768–1775.

Rinne, J. O., Brooks, D. J., Rossor, M. N., et al. (2010). ^{11}C-PiB PET assessment of change in fibrillar amyloid-beta load in patients with Alzheimer's disease treated with bapineuzumab: A phase 2, double-blind, placebo-controlled, ascending-dose study. *Lancet Neurology, 9*(4), 363–372.

Risner, M. E., Saunders, A. M., Altman, J. F., et al. (2006). Efficacy of rosiglitazone in a genetically defined population with mild-to-moderate Alzheimer's disease. *Pharmacogenomics Journal, 6*(4), 246–254.

Ritchie, C. W., Bush, A. I., Mackinnon, A., et al. (2003). Metal-protein attenuation with iodochlorhydroxyquin (clioquinol) targeting Abeta amyloid deposition and toxicity in Alzheimer disease: A pilot phase 2 clinical trial. *Archives of Neurology, 60*, 1685–1691.

Riverol, M., Slachevsky, A., & López, O. L. (2011). Efficacy and tolerability of a combination treatment of memantine and donepezil for Alzheimer's disease: A literature review evidence. *European Neurological Journal, 3*(1), 15–19.

Rockenstein, E., Adame, A., Mante, M., et al. (2003). The neuroprotective effects of Cerebrolysin in a transgenic model of Alzheimer's disease are associated with improved behavioral performance. *Journal of Neural Transmission, 110*(11), 1313–1327.

Rockenstein, E., Torrance, M., Mante, M., et al. (2006). Cerebrolysin decreases amyloid-beta production by regulating amyloid protein precursor maturation in a transgenic model of Alzheimer's disease. *Journal of Neuroscience Research, 83*(7), 1252–1261.

Rockwood, K., Beattie, B. L., Eastwood, M. R., et al. (1997). A randomized, controlled trial of linopirdine in the treatment of Alzheimer's disease. *Canadian Journal of Neurological Sciences, 24*, 140–145.

Rockwood, K., Mintzer, J., Truyen, L., et al. (2001). Effects of a flexible galantamine dose in Alzheimer's disease: A randomised, controlled trial. *Journal of Neurology, Neurosurgery and Psychiatry, 71*, 589–595.

Rogers, J., Kirby, L. C., Hempelman, S. R., et al. (1993). Clinical trial of indomethacin in Alzheimer's disease. *Neurology, 43*, 1609–1611.

Rogers, S. L., Doody, R. S., Mohs, R. C., et al. (1998a). Donepezil improves cognition and global function in Alzheimer's disease: A 15-week, double blind, placebo-controlled study. Donepezil Study Group. *Archives of Internal Medicine, 158*, 1021–1031.

Rogers, S. L., Farlow, M. R., Doody, R. S., et al. (1998b). A 24-week, double-blind, placebo-controlled trial of donepezil in patients with Alzheimer's disease. Donepezil Study Group. *Neurology, 50*, 136–145.

Roman, G. C., Tatemichi, T. K., Erkinjuntti, T., et al. (1993). Vascular dementia: Diagnostic criteria for research studies. Report of the NINDS-AIREN International Workshop. *Neurology, 43*, 250–260.

Rosen, W. G., Terry, R. D., Fuld, P. A., et al. (1980). Pathological verification of ischemic score in differentiation of dementias. *Annals of Neurology, 7*, 486–488.

Rösler, M., Anand, R., Cicin-Sain, A., et al. (1999). Efficacy and safety of rivastigmine in patients with Alzheimer's disease: International randomised controlled trial. *BMJ, 318*, 633–638.

Rossom, R. C., Espeland, M. A., Manson, J. E., et al. (2012). Calcium and vitamin D supplementation and cognitive impairment in the women's health initiative. *Journal of the American Geriatrics Society, 60*(12), 2197–2205.

Ruether, E., Husmann, R., Kinzler, E., et al. (2001). A 28-week, double-blind, placebo-controlled study with Cerebrolysin in patients with mild to moderate Alzheimer's disease. *International Clinical Psychopharmacology, 16*(5), 253–263. [erratum appears in *International Clinical Psychopharmacology,* 2001, *16*(6), 372].

Sadowsky, C. H., Farlow, M. R., Meng, X., et al. (2010). Safety and tolerability of rivastigmine transdermal patch compared with rivastigmine capsules in patients switched from donepezil: Data from three clinical trials. *International Journal of Clinical Practice, 64*(2), 188–193.

Salloway, S., Ferris, S., Kluger, A., et al. (2004). Efficacy of donepezil in mild cognitive impairment: A randomized placebo-controlled trial. *Neurology, 63*, 651–657.

Salloway, S., Sperling, R., Keren, R., et al. (2011). A phase 2 randomized trial of ELND005, scyllo-inositol, in mild to moderate Alzheimer disease. *Neurology, 77*(13), 1253–1262.

Salloway, S., Sperling, R., Fox, N. C., et al. (2014). Two phase 3 trials of bapineuzumab in mild-to-moderate Alzheimer's disease. *New England Journal of Medicine., 370*(4), 322–333.

Salvá, A., Andrieu, S., Fernandez, E., et al. (2011). Health and nutrition promotion program for patients with dementia (NutriAlz): Cluster randomized trial. *Journal of Nutrition, Health and Aging, 15*(10), 822–830.

Sano, M., Ernesto, C., Thomas, R. G., et al. (1997). A controlled trial of selegiline, alpha-tocopherol, or both as treatment for Alzheimer's disease. The Alzheimer's Disease Cooperative Study. *New England Journal of Medicine, 336*, 1216–1222.

Sano, M., Bell, K. L., Galasko, D., et al. (2011). A randomized, double-blind, placebo-controlled trial of simvastatin to treat Alzheimer disease. *Neurology, 77*(6), 556–563.

Saxton, J., McGonigle-Gibson, K. L., Swihart, A. A., et al. (1990). Assessment of the severely impaired patient: Description and validation of a new neuropsychological test battery. *Psychological Assessment, 2*, 298–303.

Scarmeas, N., Luchsinger, J. A., Schupf, N., et al. (2009). Physical activity, diet, and risk of Alzheimer disease. *JAMA, 302*(6), 627–637.

Scheltens, P., Twisk, J. W., Blesa, R., et al. (2012). Efficacy of Souvenaid in mild Alzheimer's disease: Results from a randomized, controlled trial. *Journal of Alzheimers Disease, 31*(1), 225–236.

Schettini, G. (1991). Brain somatostatin: Receptor-coupled transducting mechanisms and role in cognitive functions. *Pharmacological Research, 23*(3), 203–214.

Schmidt, R., Hofer, E., Bouwman, F. H., et al. (2015). EFNS-ENS/EAN guideline on concomitant use of cholinesterase inhibitors and memantine in moderate to severe Alzheimer's disease. *European Journal of Neurology, 22*(6), 889–898.

Schneider, L. (2002). Rivastigmine. In N. Qizilbash, L. Schneider, H. Chui, et al. (Eds.), *Evidence-based dementia practice* (pp. 499–509). Blackwell Science.

Schneider, L., & Tariot, P. N. (1997). Cognitive enhancers for Alzheimer's disease. In A. Tasman, J. Kay, & J. A. Lieberman (Eds.), *Psychiatry* (Vol. 2, pp. 1685–1701). WB Saunders.

Schneider, L. S., Olin, J. T., Doody, R. S., et al. (1997). Validity and reliability of the Alzheimer's Disease Cooperative Study – Clinical global impression of change. The Alzheimer's Disease Cooperative Study. *Alzheimer Disease and Associated Disorders, 11*, S22–S32.

Schneider, L. S., DeKosky, S. T., Farlow, M. R., et al. (2005). A randomized, double-blind, placebo-controlled trial of two doses of Ginkgo biloba extract in dementia of the Alzheimer's type. *Current Alzheimer Research, 2*(5), 541–551.

Schneider, A. L. C., Lutsey, P. L., Alonso, A., et al. (2014). Vitamin D and cognitive function and dementia risk in a biracial cohort: The ARIC Brain MRI Study. *European Journal of Neurology, 21*(9), 1211–1218, e69–70.

Schneider, L. S., Geffen, Y., Rabinowitz, J., et al. (2019). Low-dose ladostigil for mild cognitive impairment: A phase 2 placebo-controlled clinical trial. *Neurology, 93*(15), e1474–e1484.

Sevigny, J., Chiao, P., Bussière, T., et al. (2016). The antibody aducanumab reduces Aβ plaques in Alzheimer's disease. *Nature, 537*(7618), 50–56.

Shah, R. C., Kamphuis, P. J., Leurgans, S., et al. (2013). The S-Connect study: Results from a randomized, controlled trial of Souvenaid in mild-to-moderate Alzheimer's disease. *Alzheimer's Research & Therapy, 5*(6), 59.

Shumaker, S. A., Legault, C., Rapp, S. R., et al. (2003). Estrogen plus progestin and the incidence of dementia and mild cognitive impairment in postmenopausal women: The Women's Health Initiative Memory Study: A randomized controlled trial. *JAMA, 289*, 2651–2662.

Simons, M., Schwärzler, F., Lütjohann, D., et al. (2002). Treatment with simvastatin in normocholesterolemic patients with Alzheimer's disease: A 26-week randomized, placebo-controlled, double-blind trial. *Annals of Neurology, 52*(3), 346–350.

Simpkins, J. W., Singh, M., & Bishop, J. (1994). The potential role for estrogen replacement therapy in the treatment of the cognitive decline and neurodegeneration associated with Alzheimer's disease. *Neurobiology of Aging, 15*(Suppl 2), S195–S197.

Sparks, D. L., Sabbagh, M. N., Connor, D. J., et al. (2005). Atorvastatin for the treatment of mild to moderate Alzheimer disease: Preliminary results. *Archives of Neurology, 62*(5), 753–757.

Sperling, R., Salloway, S., Brooks, D. J., et al. (2012). Amyloid-related imaging abnormalities in patients with Alzheimer's disease treated with bapineuzumab: A retrospective analysis. *Lancet Neurology, 11*(3), 241–249.

Spinnewyn, B. (1992). *Ginkgo biloba* extract (EGb 761) protects against delayed neuronal death in gerbil. In Y. Christen, J. Costentin, & M. Lacour (Eds.), *Effects of Ginkgo Biloba Extract (EGb 761) on the Central Nervous System* (pp. 113–118). Elsevier.

Stern, Y., Sano, M., & Mayeux, R. (1987). Effects of oral physostigmine in Alzheimer's disease. *Annals of Neurology, 22*, 306–310.

Tan, M.-S., Yu, J.-T., Tan, C.-C., et al. (2015). Efficacy and adverse effects of ginkgo biloba for cognitive impairment and dementia: A systematic review and meta-analysis. *Journal of Alzheimer's Disease, 43*(2), 589–603.

Tariot, P. N., & Federoff, H. J. (2003). Current treatment for Alzheimer's disease and future prospects. *Alzheimer Disease and Associated Disorders, 17*(Suppl 4), S105–S113.

Tariot, P., Schneider, L., & Coleman, P. D. (1993). Treatment of Alzheimer's disease: Glimmers of hope? *Chemistry and Industry, 20*, 801–807.

Tariot, P. N., Solomon, P. R., Morris, J. C., et al. (2000). A 5-month, randomized, placebo-controlled trial of galantamine in AD. The Galantamine USA-10 Study Group. *Neurology, 54*, 2269–2276.

Tariot, P. N., Cummings, J. L., Katz, I. R., et al. (2001). A randomized, double-blind, placebo-controlled study of the efficacy and safety of donepezil in patients with Alzheimer's disease in the nursing home setting. *Journal of the American Geriatrics Society, 49*(12), 1590–1599.

Tariot, P. N., Loy, R., Ryan, J. M., et al. (2002). Mood stabilizers in Alzheimer's disease: Symptomatic and

neuroprotective rationales. *Advanced Drug Delivery Reviews, 54*(12), 1567–1577.

Tariot, P. N., Farlow, M. R., Grossberg, G. T., et al. (2004). Memantine treatment in patients with moderate to severe Alzheimer disease already receiving donepezil: A randomized controlled trial. *JAMA, 291*(3), 317–324.

Tariot, P. N., Schneider, L. S., Cummings, J., et al. (2011). Chronic divalproex sodium to attenuate agitation and clinical progression of Alzheimer disease. *Archives of General Psychiatry, 68*(8), 853–861.

Tariot, P., Salloway, S., Yardley, J., et al. (2012). Long-term safety and tolerability of donepezil 23 mg in patients with moderate to severe Alzheimer's disease. *BMC Research Notes, 5*, 283.

Tateishi, J. (2000). Subacute myelo-optico-neuropathy: Clioquinol intoxication in humans and animals. *Neuropathology, 20*(Suppl), S20–S24.

Tatton, W. (1993). "Tropic-like" reduction of nerve cell death by deprenyl without monoamine oxidase inhibition. *Neurology Forum, 4*, 3–10.

Teich, A. F., & Arancio, O. (2012). Is the amyloid hypothesis of Alzheimer's disease therapeutically relevant? *Biochemical Journal, 446*(2), 165–177.

Thai, L. J., Carta, A., Doody, R., et al. (1997). Prevention protocols for Alzheimer disease. Position paper from the international working group on harmonization of dementia drug guidelines. *Alzheimer Disease and Associated Disorders, 11*, 46–49.

Thal, L. J., Schwartz, G., Sano, M., et al. (1996). A multicenter double-blind study of controlled-release physostigmine for the treatment of symptoms secondary to Alzheimer's disease. Physostigmine study group. *Neurology, 47*, 1389–1395.

Tokita, K., Inoue, T., Yamazaki, S., et al. (2005). FK962, a novel enhancer of somatostatin release, exerts cognitive-enhancing actions in rats. *European Journal of Pharmacology, 527*(1–3), 111–120.

Tollefson, G. D. (1990). Short-term effects of the calcium channel blocker nimodipine (Bay-e-9736) in the management of primary degenerative dementia. *Biological Psychiatry, 27*, 1133–1142.

Tuszynski, M. H. (2002). Growth-factor gene therapy for neurodegenerative disorders. *Lancet Neurology, 1*(1), 51–57.

Tuszynski, M. H., Thal, L., Pay, M., et al. (2005). A phase 1 clinical trial of nerve growth factor gene therapy for Alzheimer disease. *Nature Medicine, 11*(5), 551–555.

van Dyck, C. H., Tariot, P. N., Meyers, B., et al. (2007). A 24-week randomized, controlled trial of memantine in patients with moderate-to-severe Alzheimer disease. *Alzheimer Disease and Associated Disorders, 21*(2), 136–143.

van Dyck, C. H., Swanson, C. J., Aisen, P., et al. (2023). Lecanemab in early alzheimer's disease. *The New England Journal of Medicine, 388*(1), 9–21.

Vaz, M., & Silvestre, S. (2020). Alzheimer's disease: Recent treatment strategies. *European Journal of Pharmacology, 887*, 173554.

Vernon, M. W., & Sorkin, E. M. (1991). Piracetam. An overview of its pharmacological properties and a review of its therapeutic use in senile cognitive disorders. *Drugs and Aging, 1*, 17–35.

Walker, L. C., Ibegbu, C. C., Todd, C. W., et al. (2005). Emerging prospects for the disease-modifying treatment of Alzheimer's disease. *Biochemical Pharmacology, 69*(7), 1001–1008.

Wallace, J. L., Muscara, M. N., de Nucci, G., et al. (2004). Gastric tolerability and prolonged prostaglandin inhibition in the brain with a nitric oxide-releasing flurbiprofen derivative, NCX-2216 [3-[4-(2-fluoro-alpha-methyl-[1,1'-biphenyl]-4-acetyloxy)-3-methoxyphenyl]-2-propenoic acid 4-nitrooxy butyl ester]. *Journal of Pharmacology and Experimental Therapeutics, 309*, 626–633.

Watson, G. S., Cholerton, B. A., Reger, M. A., et al. (2005). Preserved cognition in patients with early Alzheimer disease and amnestic mild cognitive impairment during treatment with rosiglitazone: A preliminary study. *American Journal of Geriatric Psychiatry, 13*(11), 950–958.

Wattmo, C., Wallin, A. K., & Minthon, L. (2012). Functional response to cholinesterase inhibitor therapy in a naturalistic Alzheimer's disease cohort. *BMC Neurology, 12*, 134.

Weitbrecht, W. U., & Jansen, W. (1986). Primary degenerative dementia: Therapy with *Ginkgo biloba* extract. Placebo-controlled double-blind and comparative study. *Fortschritte der Medizin, 104*, 199–202. (in German).

West, T., Hu, Y., Verghese, P. B., et al. (2017). Preclinical and clinical development of abbv-8e12, a humanized anti-tau antibody, for treatment of Alzheimer's disease and other tauopathies. *The Journal of Prevention of Alzheimer's Disease, 4*(4), 236–241.

Whitehouse, P. J., Price, D. L., Struble, R. G., et al. (1982). Alzheimer's disease and senile dementia: Loss of neurons in the basal forebrain. *Science, 215*, 1237–1239.

Wilcock, G. K., Lilienfeld, S., & Gaens, E. (2000). Efficacy and safety of galantamine in patients with mild to moderate Alzheimer's disease: Multicentre randomised controlled trial. Galantamine International-1 Study Group. *BMJ, 321*, 1445–1449.

Wilcock, G. K., Black, S. E., Haworth, J., et al. (2006). Efficacy and safety of MPC-7869 (R-flurbiprofen), a selective Aβ42-lowering agent, in Alzheimer's disease (AD): Results of a 12-month phase 2 trial and 1-year follow-on study. *Alzheimer's and Dementia, 2*(3, Suppl 1), S81–S82.

Wilkins, C. H., Sheline, Y. I., Roe, C. M., et al. (2006). Vitamin D deficiency is associated with low mood and worse cognitive performance in older adults. *American Journal of Geriatric Psychiatry, 14*(12), 1032–1040.

Winblad, B., & Poritis, N. (1999). Memantine in severe dementia: Results of the 9M-Best Study (Benefit and efficacy in severely demented patients during treatment with memantine). *International Journal of Geriatric Psychiatry, 14*, 135–146.

Winblad, B., Bonura, M. L., Rossini, B. M., et al. (2001). Nicergoline in the treatment of mild-to-moderate Alzheimer's disease: A European multicentre trial. *Clinical Drug Investigation, 21*, 621–632.

Winblad, B., Kilander, L., Eriksson, S., et al. (2006). Donepezil in patients with severe Alzheimer's disease: Double-blind, parallel-group, placebo-controlled study. *The Lancet, 367*(9516), 1057–1065. [erratum appears in *The Lancet*, **2006**, *367(9527)*, 1980].

Winblad, B., Cummings, J., Andreasen, N., et al. (2007). A six-month double-blind, randomized, placebo-controlled study of a transdermal patch in Alzheimer's disease-rivastigmine patch versus capsule. *International Journal of Geriatric Psychiatry, 22*(5), 456–467.

Winblad, B., Black, S. E., Homma, A., et al. (2009). Donepezil treatment in severe Alzheimer's disease: A pooled analysis of three clinical trials. *Current Medical Research Opinion, 25*(11), 2577–2587.

Winblad, B., Andreasen, N., Minthon, L., et al. (2012). Safety, tolerability, and antibody response of active Aβ immunotherapy with CAD106 in patients with Alzheimer's disease: Randomised, double-blind, placebo-controlled, first-in-human study. *Lancet Neurology, 11*(7), 597–604.

Wischik, C. M., Staff, R. T., Wischik, D. J., et al. (2015). Tau aggregation inhibitor therapy: An exploratory phase 2 study in mild or moderate Alzheimer's disease. *Journal of Alzheimer's Disease, 44*(2), 705–720.

Witkin, J. M., & Nelson, D. L. (2004). Selective histamine H3 receptor antagonists for treatment of cognitive deficiencies and other disorders of the central nervous system. *Pharmacology and Therapeutics, 103*(1), 1–20.

Wolozin, B., Kellman, W., Ruosseau, P., et al. (2000). Decreased prevalence of Alzheimer disease associated with 3-hydroxy-3-methylglutaryl coenzyme A reductase inhibitors. *Archives of Neurology, 57*(10), 1439–1443.

Xuan, K., Zhao, T., Qu, G., Liu, H., Chen, X., & Sun, Y. (2020). The efficacy of statins in the treatment of Alzheimer's disease: A meta-analysis of randomized controlled trial. *Neurological Sciences, 41*(6), 1391–1404.

Yamazaki, M., Matsuoka, N., Maeda, N., et al. (1996). FK960 *N*-(4-acetyl-1-piperazinyl)-*p*-fluorobenzamide monohydrate ameliorates the memory deficits in rats through a novel mechanism of action. *Journal of Pharmacology and Experimental Therapeutics, 279*, 1157–1173.

Yang, F., Lim, G. P., Begum, A. N., et al. (2005). Curcumin inhibits formation of Aβ oligomers and fibrils and binds plaques and reduces amyloid in vivo. *Journal of Biological Chemistry, 280*(7), 5892–5901.

Yang, G., Wang, Y., Sun, J., Zhang, K., & Liu, J. (2016). Ginkgo biloba for mild cognitive impairment and Alzheimer's disease: A systematic review and meta-analysis of randomized controlled trials. *Current Topics in Medicinal Chemistry, 16*(5), 520–528.

Youdim, M. B. (2013). Multi target neuroprotective and neurorestorative anti-Parkinson and anti-Alzheimer drugs ladostigil and m30 derived from rasagiline. *Experimental Neurobiology, 22*(1), 1–10.

Yuan, Q., Wang, C.-W., Shi, J., & Lin, Z.-X. (2017). Effects of Ginkgo biloba on dementia: An overview of systematic reviews. *Journal of Ethnopharmacology, 195*, 1–9.

Zhang, Y., Chen, J., Qiu, J., Li, Y., Wang, J., & Jiao, J. (2016). Intakes of fish and polyunsaturated fatty acids and mild-to-severe cognitive impairment risks: A dose-response meta-analysis of 21 cohort studies.

The American Journal of Clinical Nutrition, 103(2), 330–340.

Zubenko, G. S., & Moossy, J. (1988). Major depression in primary dementia. Clinical and neuropathologic correlates. *Archives of Neurology, 45*, 1182–1186.

Zubenko, G. S., Moossy, J., & Kopp, U. (1990). Neurochemical correlates of major depression in primary dementia. *Archives of Neurology, 47*, 209–214.

Zweig, R. M., Ross, C. A., Hedreen, J. C., et al. (1988). The neuropathology of aminergic nuclei in Alzheimer's disease. *Annals of Neurology, 24*, 233–242.

Pharmacological Treatment of Substance Use Disorders

132

A. Benjamin Srivastava, Frances R. Levin, and Edward V. Nunes

Contents

This chapter is an update from the 4th edition. Previous edition authors were Wilfrid Noel Raby, John J. Mariani, Frances R. Levin and Edward V. Nunes

A. B. Srivastava (✉) · F. R. Levin · E. V. Nunes
Department of Psychiatry, Columbia University College of Physicians and Surgeons/New York State Psychiatric Institute, New York, NY, USA
e-mail: Benjamin.Srivastava@nyspi.columbia.edu; frl2@cumc.columbia.edu; nunesed@nyspi.columbia.edu

© Springer Nature Switzerland AG 2024
A. Tasman et al. (eds.), *Tasman's Psychiatry*,
https://doi.org/10.1007/978-3-030-51366-5_140

Abstract

Substance use disorders (SUDs) represent a significant public health challenge, with morbidity and mortality from overdoses and other use-related causes increasing substantially during the COVID-19 pandemic. Thus, a working knowledge of treatment of substance use disorders is prudent for any psychiatrist. The United States Food and Drug Administration (FDA) has approved medications for the treatment of opioid, alcohol, and tobacco use disorders; however, evidence-based pharmacotherapy exists for many of the major classes of substances. This chapter will detail mechanistic approaches to the treatment of SUDs, evidence for the use of pharmacotherapy for the treatment of SUDs including efficacy and safety, and important clinical caveats including optimal management of SUDs and co-occurring psychiatric disorders. Important nuances regarding the treatment of withdrawal, relapse prevention, harm reduction, and how different medications may be appropriate for a given purpose will be described. Additionally, avenues for medication development will be discussed, suggesting ways in which pharmacotherapy for SUDs can be improved.

Keywords

Alcohol · Opioids · Cannabis · Stimulants · Intoxication · Withdrawal · Abstinence

It has been said, that the disuse of spirits should be gradual; but my observations authorize me to say, that persons who have been addicted: to them, should abstain from them suddenly and entirely. "Taste not, handle not, touch not," should be inscribed upon every vessel that contains spirits in the house of a man, who wishes to be cured of habits of intemperance. To obviate, for a while, the debility which arises from the sudden abstraction of the stimulus of spirits, laudanum, or bitters infused in water, should be taken, and perhaps a larger quantity of beer or wine than is consistent with the strict rules of temperate living; By the temporary use of these substitutes for spirits, I have never known the transition to sober habits, to be attended with any bad effects, but often with permanent health of body, and peace of mind.

– Benjamin Rush, An Inquiry into the Effect of Ardent Spirits upon the Human Body and Mind, with an Account of the Means of Preventing and of the Remedies for Curing Them. (Rush, 1823)

In 1784, Benjamin Rush, a physician and one of the Founding Fathers of the United States first published *An Inquiry into the Effect of Ardent Spirits upon the Human Body and Mind, with an Account of the Means of Preventing and of the Remedies for Curing Them*, one of the first treatises exploring the "disease concept" of alcoholism. Rush advocated abstinence, but his suggestion decreasing the use of "ardent spirits" with beer or wine as substitutes is prescient; medications exploiting similar pharmacotherapeutic mechanisms as the substance itself may be useful in attenuating substance use, an axiom of substance use disorders pharmacotherapy. Clinicians now have access to effective medications, especially for alcohol, nicotine, and opioid use disorders, while medications for cannabis and for cocaine and other stimulants are under development. This chapter seeks to instruct clinicians on the use of available agents approved by the US Food and Drug Administration (FDA) to treat substance use disorders. It also highlights significant findings in medication development for marijuana and stimulants, including cocaine and methamphetamine. The first section gives an overview of general pharmacological mechanisms or strategies for the treatment of substance use disorders. This is followed by sections detailing medication treatments for each of the major addictive substances, including alcohol, opioids, nicotine, stimulants, and cannabis (Rush, 1823).

Pharmacological Principles of Medication Treatment for Substance Use Disorders

The receptor systems at which addictive drugs exert their effects are generally well known. Accordingly, pharmacological strategies employ medications that bind to the same receptors, acting either as agonist, partial agonist, or antagonist. A related key principle is that addictive drugs function as reinforcers. All addictive drugs function as positive reinforcers which promote the development of habitual use. Addictive drugs also function as negative reinforcers in that unpleasant withdrawal symptoms develop after cessation of drug-taking. Resuming use of the drug relieves withdrawal. Pavlovian conditioning also develops, where cues, either environmental (e.g., persons, places, things) or internal (e.g., feeling states, sensations elicited by small doses of the drug) trigger craving for the drug and drug use. One way or another, medication strategies seek to address these mechanisms, reducing the rewarding effects of drugs, reducing withdrawal, or reducing craving (Koob & Volkow, 2016).

Agonist Replacement: Treatment with an agonist seeks to safely substitute for the addictive drug while minimizing impairment. The most time-honored and successful examples are methadone maintenance for opioid use disorder and various nicotine replacement products for nicotine use disorder. Methadone is the prototype and displays the pharmacokinetic features of an effective agonist: (1) slow absorption by the oral route, which reduces the rewarding "high" that results from rapid drug absorption by injection use, for example, and (2) long half-life, producing relatively steady blood levels and thus minimizing withdrawal symptoms. The effectiveness of methadone also depends on advancing gradually to a high dose, which induces tolerance in the opioid system. This results in both limited agonist effects and thus little or no impairment and "tolerance blockade" such that the patient no longer experiences agonist effects from taking typical opioids. The clinician can gauge whether methadone maintenance is working if the patient says "I took some heroin and nothing happened, I was wasting my money." Agonists are also effective for treating withdrawal symptoms. Methadone or other opioid agonists, and nicotine replacement, are used in this way. Barbiturates or benzodiazepines can be thought of as indirect agonists with respect to alcohol and are the treatments of choice for alcohol withdrawal. However, chronic use of such agents for alcohol dependence, as occurs with methadone for opiate dependence, has not proven effective (e.g., Charnoff et al., 1963) suggesting limits to this approach.

High Affinity Partial Agonists: Partial agonists also substitute for the addictive drug while exerting less agonist effect. Slow kinetics with long half-life in combination with high receptor affinity and slow dissociation from receptors results in long duration of action and less tendency for withdrawal. The high affinity for the receptor also results in blocking the effects is the addictive drug, which generally has lower receptor affinity. Examples in the pharmacopeia are buprenorphine for opioid use disorder and varenicline for tobacco use disorder.

Antagonists: An antagonist binds the receptor with high affinity, blocking the effects of the addictive drug. Naltrexone is the prototype antagonist effective for treatment of opioid use disorder. A patient on naltrexone with adequate blood level will typically report feeling nothing if they attempt taking an opioid. Naltrexone is also effective for some patients with alcohol use disorder, based presumably on an indirect effect of the opioid system on the reinforcing effects of alcohol.

Other Mechanisms: A variety of other medications indirectly modulate the receptor systems impacted by addictive substances. Examples include anticonvulsants which modulate GABA or NMDA receptors and are used to treat alcohol use disorder and have been tested for cocaine use disorder. Alpha-2 agonists, which attenuate sympathetic nervous system activity, are effective for reducing the symptoms of opioid withdrawal by blocking the component of adrenergic overdrive. Alpha-2 agonists have also been tested for their putative effects in attenuating the physiological response to stress, given that stress increases the rewarding effects of addictive drugs and promotes drug use. The effect of naltrexone for alcohol use disorder is thought to involve opioid receptor modulation of the inhibitory effect of GABA neurons on dopamine neurons in the ventral striatum and mediate the reward signal. Blockade of opioid receptors by naltrexone increases GABA inhibition on dopamine outflow in response to addictive substances, reducing their reward effects. Accordingly, naltrexone has been found to be effective for treatment of alcohol use disorder and has also been tested with mixed results for treatment of stimulant and nicotine use disorders. A person with alcohol use disorder for whom naltrexone works will say that alcohol no longer produces the same compulsion to binge drink; they may also note experiencing less craving in response to alcohol-associated cues, suggesting naltrexone interfering with Pavlovian conditioning around alcohol. Disulfiram, in a class by itself, makes the ingestion of alcohol highly unpleasant by inhibiting the metabolism of alcohol in the liver promoting the buildup of the toxic metabolite acetaldehyde. In operant conditioning terms, disulfiram converts alcohol into a punishment.

Alcohol

Pharmacotherapy for alcohol use disorder (AUD) can be divided into two categories: medication for the treatment of alcohol withdrawal and medication for reduction in drinking. Alcohol principally acts as a gamma-aminobutyric acid (GABA)-A receptor agonist and an n-methyl-d-aspartate (NMDA) receptor antagonist. Chronic alcohol intake results in down-regulation of GABA-A receptors and upregulation of NMDA receptors (Kosten & O'Connor, 2003). Thus, in individuals with chronic drinking patterns, a characteristic constellation of symptoms presents, termed the *alcohol withdrawal syndrome,* may develop. These symptoms may be mild or absent in many chronic drinkers but may also be quite severe and even life-threatening. Insomnia, agitation, tremor, and tachycardia may begin within a few hours and last for 24–48 h. If withdrawal is allowed to continue, increasing autonomic instability, visual, auditory, or tactile hallucinations with a clear

sensorium, and generalized tonic clonic seizures can occur 12–48 h after drinking ends. Alcohol withdrawal delirium, also known as delirium tremens, can accompany more severe alcohol withdrawal states, typically presents in the first week after cessation of drinking, and encompasses symptoms of altered consciousness, perceptual disturbances, disorientation, autonomic instability, sweating, and confusion (Pelic & Myrick, 2003; Schuckit, 2014). Monitoring for the presence and severity of alcohol withdrawal, including withdrawal delirium, and prompt pharmacological intervention are important, as severe autonomic instability can lead to cardiovascular collapse and death (Schuckit, 2014).

Medically Supervised Withdrawal

The goal of medically supervised withdrawal (e.g., detoxification) is to allow the discontinuation of alcohol use safely while protecting the patient against the consequences of central nervous system (CNS) excitability. The medically supervised management of alcohol withdrawal can occur in a variety of inpatient and outpatient settings. It is important to emphasize that the treatment of withdrawal is only the beginning phase of treatment and that longer term treatment for AUD is necessary beyond simply transitioning the patient from a state of physiological dependence on alcohol.

During the initial assessment of a patient with AUD, the presence and severity of withdrawal signs and symptoms need to be assessed. A quantitative and reliable method to do this is the use of validated scales, such as the revised Clinical Institute Assessment for Alcohol (CIWA-Ar). This practical scale evaluates sweating, nausea and vomiting, tremor, agitation, tactile, visual or auditory disturbances, headache, and orientation, all of which are assessed on a scale of 0–7, except orientation (0–4). This scale has been extensively validated in clinical settings (Sullivan et al., 1989), with a score of 10 being a good cut-off to separate those who will and will not require medications. Hence patients who score less than 10 and do not have a history of withdrawal seizures or withdrawal delirium, and do not have medical illnesses, usually can stop drinking safely with supportive treatment not necessarily aided by pharmacological intervention (Nuss et al., 2004). Patients with CIWA-Ar scores of 15 or higher require medication treatment (Foy et al., 1988). Patients with scores between 8 and 15 (moderate severity withdrawal) benefit symptomatically from medication treatment (Mayo-Smith 1997), but it is up to the clinician's discretion.

If the patient can stop drinking with a CIWA-Ar score of less than 10, sufficient psychosocial support (family, friends, housing, adequate nutrition), and daily contact with the patient by a clinician most often indicate that outpatient detoxification can be attempted. In outpatient situations, the clinician may opt to provide the patient with a benzodiazepine at low to moderate dosage to alleviate withdrawal anxiety and insomnia. Benzodiazepines that are slowly absorbed and more slowly eliminated are preferred as they may have less abuse potential (e.g., lorazepam 1 mg three or four times per day; clonazepam 1 mg three times per day; chlordiazepoxide 25 mg twice per day). If the patient is malnourished, 100 mg of thiamine (vitamin B_1) daily will protect the patient from developing Wernicke's encephalopathy, characterized by the acute onset of ataxia, nystagmus, and ophthalmoplegia, or from developing a sixth cranial nerve palsy. When in doubt, an intramuscular dose of thiamine followed by daily multivitamins is prudent, given the potentially serious sequelae of Wernicke's encephalopathy.

When the CIWA-Ar score is greater than 15 (severe withdrawal), then medications are required to protect patients from withdrawal symptoms. Blood alcohol levels tend to be elevated in withdrawal-prone patients, although this cannot always be relied upon as a predictor of withdrawal, given the rapid elimination of alcohol from the system. Benzodiazepines (Amato et al., 2010), barbiturates, and anticonvulsants (Minozzi et al., 2010) form the essential armamentarium available to the clinician.

With the CIWA-Ar, monitoring of vital signs, electrolytes, and nutritional status, the effect of administered medications can be readily assessed.

Benzodiazepines remain first-rank medications to treat alcohol withdrawal, since they both effectively treat withdrawal symptoms and protect against the advent of withdrawal seizures. Benzodiazepines are positive allostatic modulators of the GABA-A receptor, increasing the frequency of channel opening. Generally, benzodiazepines with long half-lives (e.g., chlordiazepoxde, diazepam, clonazepam) are preferred. For patients with hepatic impairment, lorazepam, oxazepam, and temazepam are preferred. Barbiturates, including phenobarbital, are also positive allostatic modulators of the GABA-A receptor, which act by prolonging the duration of chloride channel opening, increasing the risk for oversedation. However, phenobarbital still remains a time-honored and effective medication, and recent studies have shown that in general medical (Nisavic et al., 2019) and surgical trauma patients (Nejad et al., 2020) barbiturates may be safe and effective.

Several protocols exist; they involve either fixed-dose or symptom-induced dosing (Saitz et al., 1994). Long-acting benzodiazepines (e.g., diazepam, chlordiazepoxide, or clonazepam) are preferable to shorter acting agents. Table 1 illustrates fixed-dose and symptom-induced dosing for at-risk patients (Romach & Sellers, 1991; Mayo-Smith, 1997; Daeppen et al., 2002). Two randomized trials comparing symptom-triggered dosing versus fixed dose (Saitz et al., 1994; Daeppen et al., 2002) concluded that administering benzodiazepines when symptomatic shortened the duration of treatment and decreased the amount of medication used, without compromising the safety and comfort of patients.

Anticonvulsants and Other Adjunct Medications

Benzodiazepines or phenobarbital remain the gold standard for treatment of alcohol withdrawal. However, there has been great interest in developing nonbenzodiazepine treatment alternatives for alcohol withdrawal, since in the outpatient setting benzodiazepines present the risk of misuse and have additive sedative and respiratory

Table 1 Approaches to moderate or severe alcohol withdrawal syndrome

CIWA-Ar ≥10 (moderate) or ≥ 20 (severe) Fixed Dose Method	
Normal liver enzyme levels diazepam	Elevated liver enzyme levels oxazepam
1. 20 mg PO q 1 h for three consecutive doses 2. CIWA-Ar assessment after third dose 3. If CIWA-Ar <10, then monitor vital signs, CIWA-Ar q 2 h 4. If CIWA-Ar ≥ 10, diazepam 20 mg PO, repeat CIWA-Ar q 1 h 5. Administer diazepam 20 mg PO q 1 h until CIWA-Ar <10 6. Continue diazepam 20 mg q 6 h for 48 h; CIWA-Ar, vital signs, q 6 h	1. 30 mg PO q 6 h for 4 consecutive doses, then 15 mg for 8 doses 2. CIWA-Ar assessment 30 min after each dose 3. If CIWA-Ar between 10 and 15, give 15 mg oxazepam PO prn; if CIWA-Ar >15, give 30 mg oxazepam PO prn
Symptom-Triggered Dosing Depending on liver enzyme status, diazepam 10 mg doses can be substituted for oxazepam 15 mg doses 1. Vital signs, CIWA-Ar assessment every 6 h 2. If CIWA-Ar score between 10 and 15, give oxazepam 15 mg PO. If CIWA-Ar score > 15, give oxazepam 30 mg 3. Monitor CIWA-Ar every 30 min, continue to give oxazepam until CIWA-Ar <10 4. Patients are monitored for 72 h	

depressant effects when combined with alcohol. The evaluation of anticonvulsant drugs has yielded some promising results (Ait-Daoud et al., 2006; Leggio et al., 2008), particularly carbamazepine (Bjorkqvist et al., 1976; Malcolm et al., 1989) and gabapentin (Mariani et al., 2006; Myrick et al., 2009). These agents offer an advantage over benzodiazepines for mild-to-moderate cases of alcohol withdrawal treated in the outpatient setting. However, for the treatment of moderate-to-severe alcohol withdrawal in the inpatient setting, anticonvulsant medications have not been proven to prevent the development of withdrawal delirium, and benzodiazepines remain the treatment of choice (Schuckit, 2014).

Maintenance Treatment

Although some individuals remit from AUD without specific treatment and do not relapse, it is often a chronic-relapsing illness, calling for long-term treatment. Treatment needs to be individualized based on the severity of the disorder and characteristics of the patient. Deciding on the appropriate setting of treatment is an essential aspect of treating planning for a patient with AUD. The first consideration is safety; for patients in severe alcohol withdrawal or who have unstable co-occurring medical or psychiatric

conditions, an inpatient setting is indicated. For the majority of patients who do not have acute safety or medical concerns, outpatient treatment is usually appropriate and well accepted. Typically, patients will seek the least restrictive treatment option available. The range of treatment setting options includes outpatient to inpatient or residential treatment, and available treatments include behavioral treatments, medication, and mutual support groups (Witkiewitz et al., 2019).

Pharmacotherapy for Alcohol Use Disorder

Several medications have been shown to be effective at preventing relapse in abstinent patients or promoting reductions in alcohol use. These medications have different pharmacological and clinic characteristics, which need to be taken into consideration when developing a treatment plan for an individual with AUD.

Disulfiram

Disulfiram blocks the enzyme acetaldehyde dehydrogenase, a key enzyme in the metabolic pathway of alcohol that transforms acetaldehyde derived from alcohol into acetate. When alcohol is consumed – even in small quantity – the inhibition of the enzyme causes the accumulation of acetaldehyde, which leads to increasingly unpleasant symptoms as the blood concentration of acetaldehyde increases. This reaction, known as the disulfiram–alcohol reaction, involves malaise, nausea, perhaps vomiting, restlessness, tachycardia, hypotension, sweating, and dyspnea. In severe reactions there may be respiratory depression, cardiovascular collapse, arrhythmias, myocardial infarction, acute congestive heart failure, unconsciousness, convulsions, and death. It usually begins 20–30 min after drinking, and will continue as long as alcohol is being metabolized. The strength of the reaction varies according to the amount of alcohol ingested. Severe reactions are generally restricted to patients with underlying medical vulnerabilities, and thus patients with cardiovascular or renal or other serious systemic illnesses should not be treated with disulfiram in most cases. In otherwise healthy individuals, the disulfiram reaction, while highly unpleasant, is not usually harmful. Detailed informed consent regarding the risks of disulfiram is important at the outset of treatment. In the case of a severe disulfiram reaction, supportive measures to restore blood pressure and treat shock should be instituted in hospital.

Disulfiram renders the aldehyde dehydrogenase inactive by irreversibly binding to it. In clinical terms, no tolerance to disulfiram develops with ongoing use: the longer a patient takes disulfiram, the greater the proportion of enzymes blocked, and the more sensitive to alcohol the patient becomes. Hence even discontinuing disulfiram for 1–2 days will not change the strength of the reaction; normal alcohol metabolism returns up to 2 weeks after stopping disulfiram, although this may vary. There are also patients who appear less sensitive to the disulfiram reaction; this may represent poor medication adherence or pharmacogenomic variability in sensitivity to disulfiram, and higher disulfiram doses may be required for some patients (e.g., 500 mg daily). An important caution regarding disulfiram is that the patient must abstain from alcohol use for at least 12 h prior to initiating therapy.

Patient selection and the environment in which the medication is given are most important. Patients with a strong motivation to stop drinking, but who feel they lack the self-determination to carry it through, can benefit the most from disulfiram, especially if it can be given in a supervised setting. The main limitation of disulfiram is that poor compliance with the medication can limit therapeutic benefit, with a recognized need for supervised administration to achieve maximum benefit. Meta-analytic review has found that supervised disulfiram treatment is associated with improved abstinence and less alcohol consumption in open label studies but not in blinded studies (Skinner et al., 2014). An effective strategy is to involve a significant other in monitoring daily intake of disulfiram, and Network Therapy (Galanter, 1993) is a useful blueprint for involving significant others. The medication monitor can be a family member or close friend, or it could be a clinician or clinic where the patient attends regularly. Disulfiram is a powerful treatment if the patient takes the medication. Patient, physician, and the designated medication monitor need to agree to a plan. The designated monitor should witness and confirm that the patient has taken the medication each day, verbally reinforce and encourage adherence, and report to the treating clinician if the medication is not taken. Non-adherence should trigger a clinic or office visit to review the plan and seek to bolster the patient's motivation.

Disulfiram is administered at a dose of 250–500 mg per day. Although disulfiram is usually well tolerated, monitoring of liver enzymes should be considered in vulnerable individuals as cases of hepatotoxicity have been reported (Björnsson et al., 2006). Hepatotoxicity is considered idiosyncratic, and baseline elevated liver function tests, irrespective of the cause, are not contraindications to starting disulfram treatment. The disulfiram reaction is vasodilatory and a stress on the cardiovascular system. Accordingly, disulfiram is relatively contraindicated in patients with significant cardiovascular, renal disease, or other systemic illnesses including congestive heart failure, liver disease, neuropathy, and chronic hypotension (LiverTox, 2012). Acute psychosis is a rare side effect of disulfiram, presumably because in addition to its effects on liver enzymes, disulfiram inhibits a number of other enzymes, including dopamine beta hydroxylase, inhibition of which increases brain dopamine levels. Accordingly, for patients with vulnerability to psychosis, weigh the risks versus potential benefits and monitor for worsening mental state (Nunes & Quitkin, 1987). Patients can carry an information card to alert emergency personnel of its use.

Naltrexone

Blockade of opioid receptors with naltrexone is an effective treatment for some patients with alcohol use disorder. The rewarding, calming, and euphoric effects of alcohol appear to be at least in part due to the activation of the opioid system. This is supported by reports that alcohol increases plasma levels of β-endorphins, especially in individuals with a strong family history of AUD (Gianoulakis et al., 1996). Opioid receptors on GABAergic neurons in the ventral striatum exert an inhibitory effect on GABA release, which in turn releases striatal dopamine neurons from inhibition by GABA. Naltrexone, by blocking these receptors may increase striatal GABA, thus inhibiting dopamine release in response to alcohol. This is consistent with clinical experience that patients for whom naltrexone works report diminished pleasure from alcohol and/or diminished cue induced craving (Anton, 2008). Naltrexone has been shown to decrease drinking (O'Malley et al., 1992; Volpicelli et al., 1994; Stromberg et al., 2001) and the number of relapses during a course of treatment (Garbutt et al., 1999). Three meta-analyses have corroborated the efficacy of naltrexone to treat AUD (Kranzler & Van Kirk, 2001; Streeton & Whelan, 2001; Rösner et al., 2010). A review by Pettinati et al. (2006) analyzed 29 published trials of opioid antagonists for AUD and found that naltrexone may have greatest effectiveness in reducing heavy or excessive drinking, as opposed to inducing complete abstinence. This concurs with the observation that naltrexone may be of most benefit for those who derived a greater sense of reward from drinking alcohol excessively. These findings suggest that naltrexone is appropriate for actively drinking patients as a measure to reduce heavy drinking, and also for abstinent patients to reduce the likelihood that any alcohol use will result in a return to heavy drinking. Some patients respond well to naltrexone,

while others report no improvement. Reasons for variability in response have been proposed but have been largely inconsistent in prospective studies (Garbutt et al., 2016). For example, retrospective data suggested that the functional polymorphism rs1799971 in the mu-opioid receptor gene OPRM1 was associated with reduced risk to relapse, but this was not confirmed prospectively (Oslin et al., 2015).

The usual dose of oral naltrexone is 50–100 mg per day. Higher doses (i.e., >300 mg) were associated with reversible liver toxicity, though this is uncommon at standard doses (LiverTox, 2012). In addition, clinicians should ascertain that the patient is not using opioids, as administration of naltrexone would precipitate withdrawal. Common side effects encountered with naltrexone may include nausea, gastric upset, and diarrhea, although these are self-limited and usually resolve after a few days or weeks. As with any oral medication, compliance is the key to successful treatment, and it is worth inquiring candidly from patients concerning their routine for taking the medication. Naltrexone can also be dosed on a three times per week schedule (e.g., 100 mg on Monday, 100 mg on Wednesday, 150 mg on Friday), which may make monitoring more convenient, for example in conjunction with three times per week clinic attendance. Shortcomings with oral naltrexone adherence have stimulated interest in long-acting formulations. Engaging significant others to monitor naltrexone-taking, as in the strategy described above for disulfiram, may also be considered.

Extended-release naltrexone (380 mg per unit) is a formulation in which naltrexone is incorporated into polymer microspheres for intramuscular gluteal injection. It produces adequate blood levels for about 1 month. It can be administered monthly by a healthcare professional, as part of a comprehensive care program for AUD. In addition to the safeguards mentioned for oral naltrexone, one must also monitor the injection site, as local reactions such as hyperemia and induration can occur, especially if the medication is injected repeatedly at the same site, and rotation of the injection site is important. Clinical trials have shown naltrexone to be well tolerated and effective at reducing drinking days, extending time to a first drinking day, and better abstinence rates compared with placebo (Kranzler et al., 2004; Garbutt et al., 2005; O'Malley et al., 2007). Injectable naltrexone should be particularly considered for patients with a history of relapse or with uneven adherence to daily oral medication, and could, for example, be administered to patients immediately after completion of a detoxification (inpatient or outpatient) in an effort to reduce relapse in the post-detoxification period.

Acamprosate

Acamprosate, an NMDA receptor modulator, has been widely studied in Europe and the United States, and is approved by the FDA for the treatment of AUD. A meta-analysis demonstrated a benefit in greater reduction in drinking in already abstinent patients (Jonas et al., 2014). Multi-site trials conducted in the United States where patients were not required to achieve pretrial abstinence have not shown benefit (Anton et al., 2006; Mason et al., 2006). Acamprosate appears to work by normalizing the dysregulation of NMDA-mediated glutamatergic neurotransmission that occurs during chronic alcohol consumption and withdrawal, and thus attenuates one of the physiological mechanisms that may prompt relapse, presumably by reducing craving, anxiety, and insomnia (Mason and Heyser., 2010). The COMBINE study, a multi-site trial comparing naltrexone and acamprosate to placebo among outpatients with alcohol use disorder, found no benefit of acamprosate compared to placebo, while naltrexone was found effective (Anton et al., 2006). The emerging consensus in the literature is that acamprosate is a good medication choice for individuals with AUD who have a goal of abstinence from alcohol and will have a period of pretreatment abstinence, ideally in a controlled therapeutic environment, and that the therapeutic benefit of acamprosate is primarily on reducing relapse risk. Acamprosate is well tolerated, and can be started at a treatment dose of 666 mg three times per day, although doses of up to approximately 3 g per day have been used (Anton et al., 2006), preferably once the patient has achieved abstinence. Diarrhea, headache, and abdominal pain are the most frequently experienced side effects.

Non-FDA Approved Medications

Topiramate

The anticonvulsant agent topiramate has been shown to be effective in reducing heavy drinking, presumably by attenuating alcohol-induced reward in the nucleus accumbens (Johnson et al., 2003, 2007). A meta-analysis showed that topiramate was associated with more abstinence and lower binge-drinking frequency (Blodgett et al., 2014). A potential clinical advantage of topiramate is the ability to achieve a clinical benefit without first achieving pretreatment abstinence. Topiramate utility is similar to that of naltrexone; it is a good medication choice for individuals who have moderation goals as opposed to abstinence, and are actively drinking at the time of treatment initiation. Cognitive side effects, including reductions in verbal fluency and working memory, are common. Further, topiramate is an inhibitor of carbonic anhydrase, which can result in metabolic acidosis (Knapp et al., 2008).

Baclofen

Baclofen is a GABA-B agonist that is FDA approved as an anti-spasmodic agent. A recent meta-analysis showed the baclofen was associated with a longer duration of abstinence, and heavy baseline drinking was associated with a larger treatment effect. Notably, doses greater than 60 mg/day did not improve abstinence (Pierce et al., 2018). However, a recent study showed significant sex differences: women showed improvements in heavy drinking and abstinence when given 30 mg baclofen, whereas men did not. Neither group showed statistically significant improvements in heavy drinking or abstinence when taking 90 mg (Garbutt et al., 2021). Baclofen has not been associated with improvement in other drinking outcomes, including binge drinking and percentage of abstinent days (Kranzler & Soyka, 2018).

Gabapentin

Gabapentin is a voltage gated calcium channel blocker that is FDA approved for the treatment of epilepsy and neuropathic pain. A recent meta-analysis showed that gabapentin is associated with a reduction in heavy drinking when compared to placebo (Kranzler et al., 2019). Recent studies have shown that gabapentin may also be helpful for individuals with AUD with a history of alcohol withdrawal (Anton et al., 2020), and high doses* (3600 mg) may yield improved outcomes in both heavy drinking and abstinence (Mariani et al., 2021a).

Summary

Alcohol use disorder is a common substance use disorder that can be treated with a variety of behavioral and medication therapies. The management of alcohol withdrawal is often the first stage of treatment, but in most cases is not a sufficient intervention to result in prolonged abstinence and should be used as a prelude to further counseling or medication treatment. Counseling, behavioral treatments, and 12-step (e.g., Alcoholics Anonymous) participation are cornerstones of treatment, but medications including disulfiram, naltrexone, acamprosate, and others may be helpful in achieving or maintaining abstinence or reducing severity. Treatment may be delivered in inpatient, residential, and outpatient settings, combining behavioral and medication therapies in combinations tailored individually to the needs of the patient. Other than disulfiram, which exerts a large effect on preventing drinking if taken consistently, the medications for alcohol use disorder exert small to medium sized effects, with number-needed-to-treat (NNT) estimates in the range of 5 to 7 or greater (Garbutt et al., 2005; Anton et al., 2006). Thus, only some patients can be expected to respond, and if one medication trial fails, alternative medications should be considered.

Opioids

Medically Supervised Withdrawal

Opioid withdrawal may begin anywhere from 4 to 24 h after the last dose of opioid drug or medication, depending on the half-life. Typically, it will begin with yawning, rhinorrhea, tearing,

tachycardia, back and limb pain, and abdominal cramps, leading to restlessness, irritability, muscle cramps, and shaking chills. As withdrawal appears to follow closely the rate of clearance of opioids from receptors, first symptoms emerge sooner with a short-acting opioid such as heroin, and later with methadone. The treatment of opioid withdrawal targets specific pathophysiological mechanisms implicated in the syndrome, mainly sympathetic overdrive and absence of opioid receptor stimulation, with α_2-adrenergic receptor agonists and μ-opioid receptor agonists being the cornerstone of treatment (Srivastava et al., 2020). Opioid receptor agonists produce more complete resolution of opioid withdrawal symptoms. Methadone is the prototype given its favorable characteristics of long half-life, although other opioid agonists can be used as well. The partial agonist buprenorphine also relieves withdrawal and has advantages of long duration of action and lower tolerance.

α_2-Adrenergic Agonist-Based Treatment

The use of this class of medication is predicated on its inhibition or reduction of the opioid withdrawal symptoms caused by sympathetic nervous system hyperactivity (insomnia, tachycardia, sweating, nausea, malaise). Other symptoms such as craving for opioids, muscle aches, and pain are less impacted, however (Jasinsky et al., 1985). Hence, clonidine is not usually used by itself but is more often combined with a benzodiazepine (for anxiety), a nonsteroidal anti-inflammatory drug (NSAID) or acetominophen (for muscle pains), and medications for gastrointestinal distress. In general, clonidine-based treatment of withdrawal is not as effective as detoxification methods based on agonists (methadone or buprenorphine), but it remains a useful component of the armamentarium.

Clonidine and lofexidine belong to this class of medications, and although not FDA approved for this purpose, clonidine is well known in the medical community as a useful medication to treat opioid withdrawal symptoms and to attempt detoxification

(Kleber, 2007). Lofexidine was FDA approved for the treatment of opioid withdrawal in 2018. Lofexidine is said to exert comparatively less hypotensive and sedative side effects, a potential advantage over clonidine, although comparative trials are lacking. Thus, sufficient dosages can be given to reduce withdrawal symptoms with less concern about producing iatrogenic hypotension, although keeping the patient hydrated and monitoring vital signs remain important. This appears to be due to the high affinity of lofexidine to only one of the three subtypes of α_2-adrenergic receptors, as opposed to the high affinity of clonidine to all receptor subtypes (Herman & O'Brien, 1997). Since opioid withdrawal may involve nausea, vomiting, and diarrhea, with consequent volume depletion, attention to hydration (oral or parenteral if necessary) is always important during opioid withdrawal.

Medically supervised withdrawal using clonidine or lofexidine can begin with abrupt cessation of opioid intake, with a transfer onto clonidine the same day, preferably performed in an inpatient setting. Symptoms of withdrawal begin to diminish usually within 30 min, with a peak effect reached after 2–3 h. The blood pressure should be checked before each dose, and when lower than approximately 85/55 mmHg the dose should be held, and the patient encouraged to remain supine and to drink copious fluids. Clonidine detoxification should not be attempted in pregnant patients and should be used with caution on patients with cardiovascular disease.

Buprenorphine

Buprenorphine is a μ-opioid receptor partial agonist that is FDA approved for the treatment of opioid use disorder. Compared with α_2-recpetor agonists, treatment with buprenorphine is associated with less severe withdrawal and greater treatment retention (Ling et al., 2005; Srivastava et al., 2020). Given its partial agonist properties, if buprenorphine is given too close in time to the last doses of other opioids, particularly with slowly eliminated agents such as methadone, withdrawal can be precipitated because a partial agonist is displacing full agonists from receptors.

For this reason, when initiating buprenorphine, it is important to wait until a patient is clearly developing withdrawal symptoms before starting buprenorphine, and a greater delay is prudent when the patient has been taking a long-acting opioid such as methadone. After withdrawal signs and symptoms have begun to emerge, a test dose of 2 mg sublingual (SL) can be administered. If withdrawal symptoms begin to improve, or at least do not worsen, then another 2 mg can be given 1 h later, followed by 4 mg, for a total of 8–16 mg SL on the first day. The most accepted method involves a 3-day protocol, which provides good control of withdrawal symptoms with few side effects. After initiation of buprenorphine, the first-day dose is repeated on the second day and decreased to 4–8 mg on the third day.

Some patients experience residual symptoms and may be treated with ancillary medications including clonidine/lofexidine, non-steroidal anti-inflammatory medications for muscle cramps, bismuth subsalicylate for diarrhea, and prochlorperazine or ondansetron for nausea, and sedative/medications for insomnia for anxiety and sleep. Combining buprenorphine with non-opioid medications may be particularly useful when the goal is to initiate naltrexone. The non-opioid medications will reduce the dose and duration of buprenorphine needed to keep the patient comfortable, an advantage in preparing the system to tolerate an opioid antagonist (Sigmon et al. 2012).

Methadone

Methadone is a full μ-opioid receptor agonist that is comparable to buprenorphine in terms of withdrawal treatment completion (Srivastava et al., 2020). Initial dosing is typically based on the patient's use history; for a patient with unknown or low tolerance, 10 mg may be an appropriate dose, whereas for a patient with higher tolerance, 20 mg–30 mg may be appropriate. In any case, the patient should not receive more than 40 mg on the first day, since methadone is a powerful agonist, and respiratory depression is a significant risk if the starting dose is too high. Methadone can be tapered by (1) 10 mg/day for 3 days and then 2 mg/day for 4 days or (2) 5 mg per day.

Medically Supervised Withdrawal from Opioids and Risk of Relapse and Overdose

Medically supervised withdrawal (i.e., detoxification) from opioids should not be considered an end for most patients but rather as a bridge to maintenance medication treatment with methadone, buprenorphine, or naltrexone. It is important to use the treatments outlined above to control acute opioid withdrawal symptoms. Withdrawal is highly aversive and drives patients back to opioids. Prompt effective treatment can help foster a treatment alliance. However, the risk of relapse to opioid use after completion of medically supervised withdrawal is very high (Nunes et al., 2018), and this includes elevated risk of overdose and death which are increased after discharge from controlled settings where detoxification has taken place (Binswanger et al., 2007; Ravndal & Amyndsen, 2010). Tolerance is reversed after detoxification and therefore, individuals are more sensitive to the agonist effects of opioids including respiratory depression and arrest. This risk has been heightened with the advent into the illicit drug supply of the high potency opioid agonist fentanyl and its analogues. After withdrawal symptoms have been controlled, starting one of the maintenance medication strategies outlined below, namely, methadone, buprenorphine, or extended-release naltrexone, should be considered the treatment of choice for most patients. A minority of patients will succeed in remaining abstinent after detoxification without a maintenance medication, but clinicians are not able to accurately predict which patients can safely follow this course, and the risk of getting it wrong is overdose and death. Patients should be educated and warned about overdose risk, and patients and their significant others should receive naloxone kits that can be used to reverse overdoses along with education on how to use them (Winhusen et al., 2020). It is important to have a strong aftercare plan in place including both maintenance medications and psychosocial treatment such as either residential treatment, an intensive outpatient program, outpatient counseling, and 12-step meeting participation.

Maintenance Treatments

Methadone

Methadone maintenance was introduced in 1965 by Dole and Nyswander and represented a major advance in the treatment of opioid use disorder. It allowed many chronic opioid-dependent patients who had previously been refractory to psychosocial treatments to become abstinent and lead productive lives. Methadone maintenance treatment is tightly regulated in the United States by federal regulations and can only be administered at specially licensed clinics, according to criteria summarized below. In principle, methadone maintenance is indicated for any patient who has failed to maintain abstinence and relapsed after one or more attempts at "drug-free" psychosocial treatment. According to the regulations, to be eligible for methadone maintenance, an individual must show evidence of at least 1 year of physiological dependence, demonstrated by at least a positive urine toxicology, and signs and symptoms of opioid withdrawal. In addition, the following individuals can be eligible:

- Individuals from penal institutions who meet the criteria prior to incarceration (risk of relapse is high in such individuals despite the fact that they may have been abstinent for some time in prison).
- Pregnant women (methadone maintenance is the treatment of choice in pregnant opioid-dependent patients; since it is most likely to eliminate or minimize drug use, it appears to be relatively safe in pregnancy, and it prevents withdrawal symptoms, which can be harmful to the fetus).
- Patients previously on methadone maintenance for up to 2 years after treatment is terminated.
- An individual younger than 18 years, after two documented attempts at medically supervised

withdrawal from opioids or drug-free treatment at least 1 week apart. A parent or legal guardian must sign *Form FDA-325 Consent to Methadone Maintenance.*

Individual states are permitted to make these criteria more stringent, but not less so. Federal regulations prohibit daily doses greater than 120 mg without first obtaining permission from state authorities. Some patients, particularly those who are rapid metabolizers of methadone, may require higher doses; for patients who continue to use opioids despite methadone doses in the 100–120 mg range, drawing of peak and trough methadone blood levels should be considered to assess rate of metabolism, and permission should be sought to attempt higher doses if appropriate.

The reason for effectiveness of methadone in maintenance treatment lies in its pharmacokinetics and in the phenomenon of tolerance blockade. Methadone displays the prototypical pharmacokinetic profile of an effective agonist substitution agent, with relatively slow absorption, and long half-life. It is absorbed from the stomach and reaches peak plasma levels in about 4 h, which contrasts with the rapid spikes in blood level associated with injection, inhalation, or insufflation of heroin. It becomes tightly bound to proteins in all tissues, including the brain. After 3–4 weeks of daily intake, the methadone concentration reaches a steady-state level. Further, with chronic administration at adequate doses (in the 60–100 mg per day range), methadone induces marked tolerance in the brain opioid system, such that other agonists are effectively blocked. Thus, after 1–2 weeks on methadone, as the dose is titrated upwards, patients will begin to report that they no longer get high when they take heroin, and heroin use often ceases completely at that point. Because of tolerance, intoxication and sedation are also avoided, so that methadone maintenance should not impair functioning, allowing patients to work and otherwise function normally in society. If sedation persists as a side effect, this suggests that the dose is too high, or that the patient does not tolerate methadone well. Estimating the initial dose for maintenance is similar to that for withdrawal (see above), and the

practice of observing the patient for 1–2 h is worth repeating here. The maintenance dose should aim at eliminating any sign of withdrawal, not produce any euphoria or sedation, and prevent the use of any illegal opioids (as evidenced by urine toxicology and report from the patient). What this dose will be varies considerably from patient to patient, but a national survey reported that better retention was obtained with doses in excess of 60 mg per day (Hargreaves, 1983), and that the target dose range is usually 60–100 mg (Mattick et al., 2009). The methadone dose needs to titrated upwards gradually over the first several weeks to give tolerance time to set in. Advancing the dose too quickly can result in overdose due to the long half-life resulting in increasing blood level before the patient has developed tolerance. Generally, the dose is started lower, at 20–30 mg per day, and gradually increased over 1–2 weeks into the target range. The concern here is that since methadone is a full agonist, and a patient's level of tolerance may vary, too high an initial dose of methadone could lead to overdose, particularly if the patient continues to use illicit opioids on top of the methadone at the outset. However, it is important to recognize that, as in many other instances in psychopharmacology, the most common reason for treatment failure with methadone maintenance is inadequate dosage, and every effort should be made to advance the dose to the target range, or beyond as tolerated, until abstinence is achieved.

The public health benefits of methadone maintenance have been documented thoroughly; to list three: methadone has reduced the spread of HIV among intravenous users of heroin (Barthwell et al., 1989; Novick et al., 1990), methadone use has decreased needle sharing with all the health ramifications that ensue from it (Ball et al., 1988), and methadone reduces the risk of opioid overdose, an otherwise common cause of death among opioid abusers. Also, there are the considerable benefits of the reduction in personal suffering from achievement of abstinence and the benefits of getting patients back to normal social functioning. Retention in treatment has been shown to be greater on methadone compared to buprenorphine (Hser et al., 2014). Thus, methadone remains the treatment of choice for many patients. A limitation is that methadone is restricted to specially licensed clinics, and methadone clinics are not available in many communities.

Special Management Issues During Methadone Maintenance

Three situations that may be encountered while on methadone maintenance warrant special mention: (a) the patient who continues to use drugs despite apparently adequate methadone doses; (b) the patient who needs surgery; and (c) the pregnant patient. They are outlined in the following.

Continued Drug Use During Methadone Maintenance. The optimal response to methadone maintenance is full remission of opioid use disorder (i.e., abstinence confirmed by regular urine toxicologies) and of any other substance use problems. Anything short of that should warrant clinical attention and ongoing efforts to achieve full remission. Full coverage of this topic is beyond the scope of this overview chapter, but several main points should be highlighted. First, methadone maintenance treatment almost always reduces the quantity of illicit opioids used, usually substantially, even if complete abstinence is not achieved. Although this does not represent remission, it is often a clinically significant improvement, suggesting that methadone treatment should be continued and augmented, and not abandoned. Second, ongoing opioid use should prompt efforts at increased counseling and behavioral treatment, increased methadone dosage, and/or blood level studies to determine if the patient is a rapid metabolizer. The effectiveness of methadone maintenance has been shown to depend upon concurrent counseling (McLellan et al., 1993). Specific behavioral methods, such as voucher incentives, exchangeable for goods and services, contingent on drug-free urine samples, have been shown to increase abstinence among methadone-maintained patients (Pierce et al., 2006). Nonopioid substance use is also common since methadone does not have a specific pharmacological effect on nonopioid drug use. Alcohol, cannabis, cocaine, and benzodiazepines are frequently seen. Again, the presence of these other substance problems should not prompt discontinuation of a methadone treatment that is otherwise succeeding in reducing opioid use. Instead, specific treatment efforts for each of these substances should be mounted, including both counseling and behavioral therapies, in addition to the pharmacological approaches described elsewhere in this chapter.

Surgery or Pain Control During Methadone Maintenance. In the case of surgery requiring general anesthesia, the opioid-dependent patient may require larger than expected doses of opioid analgesics for pain control because of preexisting tolerance. Tolerance also means that methadone is not an effective analgesic in patients maintained on methadone (Kreek, 1979). If the patient requires analgesia, the maintenance dose of methadone does not contribute to analgesia. Efforts should be made to control such pain with nonopioid medications or nerve blocks. Other opioids can be carefully added to control pain, but need to be administered at the usual frequency and dosages, at least initially, to avoid an overdose. The patient on methadone maintenance therapy should not receive any partial agonist medications such as buprenorphine, or full antagonists such as naltrexone that would induce a sudden withdrawal. For the whole perioperative period, the patient should receive their usual methadone opioid dose.

Buprenorphine

A major advance in the treatment of opioid use disorder was the 2002 approval by the FDA of buprenorphine for office-based treatment of illegal and medical opioid dependence. From a doctor who is qualified to prescribe buprenorphine, patients can receive treatment in total confidentiality in a range of clinic-or office-based settings. The need is large: it has been estimated that there are between 810,000 and 1 million chronic users of heroin, and only 200,000 patients receiving methadone maintenance treatment (McCance-Katz, 2004). With buprenorphine, the goal is to expand the availability of opioid use disorder treatment to primary care settings, beyond the capability of the methadone clinic system.

Prior to 2023, to dispense buprenorphine for the treatment of opioid use disorder, physicians who held a DEA license to prescribe controlled substances were required to meet certain

Table 2 Previously required physician qualification to prescribe Buprenorphine for opioid use disorder: Drug Abuse Treatment Act of 2000 (As Amended in H.R. 6344, 2006). Following the 2023 Consolidated Appropriations Act (P.L. 117-328), the waiver requirement and limits on number of patients treated were removed

Qualified Physicians must complete a Waiver Notification form certifying the following:

Possession of valid state license to practice medicine

Valid DEA registration

Commitment of physician or group practice to treat no more than 30 patients *if the physician has submitted the waiver notification less than 1 year ago, or is a new physician submitting notification*. Physicians who have submitted their notification at least 1 year ago *can treat up to 100 patients*

Capacity for physician to refer for ancillary services

Satisfaction of training and experience criteria, such as added qualification in addiction medicine (American Board of Medical Specialties). Complete list of criteria can be viewed at http://www.buprenorphine.samhsa.gov

requirements as specified in the Drug Abuse Treatment Act of 2000 (see Table 2), mainly completion of a brief training course. The 2023 Consolidated Appropriations Act (P.L. 117–328) removed the waiver requirement and limits on number of patients allowed. Courses are offered online or by seminars sponsored by professional organizations such as American Academy of Addiction Psychiatry (AAAP) and the American Society of Addiction Medicine (ASAM). Physicians who are Board Certified in Addiction Psychiatry or have passed the ASAM certifying examination automatically qualify. Subsequently, the physician submits a notification request to the Center for Substance Abuse Treatment of the Substance Abuse and Mental Health Services Administration (CSAT/SAMHSA). After verification with the DEA for an established number, a unique DEA number solely for the purpose of prescribing buprenorphine will be assigned. The physician may opt to be listed on a physician locator on the SAMHSA's website as a provider of buprenorphine treatment.

As previously described in the section on medically supervised withdrawal, buprenorphine is a high-affinity partial agonist at the μ-opioid receptor, binding the receptor but only partially activating it, while blocking access to the receptor by other opioids. Advantages include greater safety, including, importantly, low risk of respiratory depression, lower abuse liability due to the ceiling on agonist effects, and muted withdrawal effects due to slow dissociation from the receptor. Compared with ascending oral doses of methadone, SL buprenorphine displays a ceiling – it produces less

agonist effect and does not produce any further agonist effect beyond a certain dose. This holds across agonist effects including, importantly, respiratory depression (Walsh et al., 1994). Buprenorphine is available as 2, 4, and 8 mg tablets for SL administration, as either buprenorphine alone or buprenorphine–naloxone combinations. Buprenorphine has been available for a number of years by injection for analgesic purposes, but this parenteral formulation has been diverted and abused (Auriacombe et al., 2004). To counteract any potential for parenteral abuse, a 4:1 combination of buprenorphine and naloxone (2/0.5 and 8/2.0 mg tablets) for SL administration has also been produced. As naloxone is poorly absorbed SL, when buprenorphine–naloxone is placed under the tongue the predominant effect emanates from buprenorphine, an effect that can last as long as 96 h (Walsh et al., 1994); if an opioid-dependent individual attempts to dissolve the buprenorphine–naloxone combination for illicit injection purposes, naloxone will cause precipitated withdrawal, which is a strong deterrent, in addition to blocking most of the effect of buprenorphine, thus diminishing its reinforcing effect (Stoller et al., 2001; Comer & Collins, 2002).

Induction Procedure for Buprenorphine

The goal of induction is to establish the dose of buprenorphine that will eliminate all withdrawal symptoms, decrease or abolish the use of other opiates, and minimize cravings, with few side effects. Of note, most inductions can be accomplished at home and do not require in-person visits (Gunderson et al., 2010), which was greatly expanded during the COVID-19 pandemic (Tofighi et al., 2021). As in all of medicine, selecting the patients most suited for this treatment is key. Induction onto buprenorphine first requires a medical and psychiatric evaluation. Many psychiatrists will want to establish a relationship with a medical clinic to obtain a report from a recent medical examination that includes blood chemistry with LFTs and hematology results. However, the priority should be to initiate buprenorphine, while medical follow-up can be arranged once the patient is stable on buprenorphine. A room dedicated to induction, with vital sign monitoring equipment and comfortable chairs, is also useful. Buprenorphine was reported to be associated with increased liver enzymes (Petry et al., 2000), although this likely reflected co-occurring disorders such a viral hepatitis, since a large subsequent safety study with random assignment to buprenorphine or methadone maintenance found no evidence hepatotoxicity (Saxon et al., 2013). Apart from hypersensitivity to buprenorphine itself, which is rare, there are no absolute medical contraindications to buprenorphine treatment. However, the individual seeking treatment must meet criteria for opioid use disorder, as evidenced by the history of increasing use, tachyphylaxis requiring ever increasing dosage, and evidence of withdrawal, or opioid use by injection, or evidence of past opioid use disorder in the face of imminent risk of relapse.

Prior to receiving the first SL dose of either buprenorphine or buprenorphine/naloxone, an individual seeking treatment should display signs of mild-to-moderate withdrawal. This is important because in the absence of withdrawal signs, the partial agonist action of buprenorphine may displace full opioids such as heroin still present on receptors precipitating withdrawal. Precipitated opioid withdrawal can be severe causing profound discomfort. An objective scale such as the Objective Opioid Withdrawal Scale (OOWS) can assist the physician in estimating the extent of withdrawal (Handlesman et al., 1987), or the Clinical Opiate Withdrawal Scale (Wesson & Ling, 2003). Using these scales, a minimum score of 3 on the OOWS or 13 on the COWS is recommended to proceed with the induction. This can usually be achieved by instructing the patient to abstain from all opioids from around 9 p.m. on the evening prior to the induction until the scheduled time to present at the office, optimally in the morning, and preferably not on a Friday. If at the scheduled time the minimum withdrawal scores are not obtained, the patient should wait or return home until the next day, with the admonition to avoid taking any opiates, which would only delay the induction further. For patients unable to abstain for long enough to present to the office with some level of withdrawal symptoms, brief hospitalization for buprenorphine initiation should be considered.

Day 1 of Buprenorphine Induction. Either buprenorphine alone or the buprenorphine–naloxone combination can be used for short-acting opioids. A first dose of 2/0.5 mg or 4/1 mg of buprenorphine-naloxone can be used. It is important to tell the patient to keep the tablet under the tongue until the tablet has completely dissolved, and only then swallow. The patient should then be monitored for the next 2 h, during which some relief from withdrawal is experienced, usually within 30–45 min. If so, another dose similar to the first can be given, up to 8 mg for the first day, although clinicians can prescribe up to 16 mg if clinically indicated. If, however, the patient reports worsening of withdrawal symptoms shortly after receiving buprenorphine, this suggests a precipitated withdrawal (this eventuality can be avoided by ensuring that objective withdrawal signs and symptoms are documented prior to induction). In this case, the clinician may consider giving another dose of buprenorphine to favor the agonist effect and relieve withdrawal, or else may opt to delay continuing the induction until the next day, or treat specific withdrawal symptoms, as outlined in Table 3.

Table 3 Ancillary medications for attenuation of opioid withdrawal during initiation of naltrexone or Buprenorphine

Clonidine/lofexidine	Hypertension, tachycardia, anxiety
Acetaminophen/ibuprofen	Muscular aches and pain
Dicyclomine	Abdominal cramps
Loperamide	Diarrhea
Diphenhydramine/trazodone	Insomnia

For patients dependent on long-acting opiates such as methadone, the induction process may be more difficult due to a higher level of tolerance and methadone remaining on receptors even as withdrawal symptoms begin to manifest. According to the ASAM Practice Guidelines for the Use of Medications in the Treatment of Addiction Involving Opioid Use, patients on methadone should decrease their dosage to 30–40 mg per day or less 1 week before buprenorphine induction is scheduled (Kampman & Jarvis, 2015). At least 24 h should precede the beginning of the induction process, allowing for withdrawal signs and symptoms to emerge and be tabulated with the OOWS or COWS. In this instance, buprenorphine alone is preferred, starting at 2 mg, adding 2 mg every 1–2 h, usually not exceeding 8–16 mg during the first day.

For patients with a past history of opiate use disorder who are now abstinent but seeking relapse-prevention treatment, withdrawal symptoms are usually not present. They should be given a 2/0.5 mg dose, with a very gradual, slower rate of dosage increase aimed at decreasing cravings and psychological dependence.

Before departing at the end of the first day, patients should be advised that they may experience some withdrawal, for which some of the medications mentioned above may be used.

Day 2 of Buprenorphine Induction. When patients return, clinicians should inquire about withdrawal problems since leaving the office, and about any use of opiates, preferably verified by urine toxicology. If withdrawal signs are noted, the buprenorphine dose should be increased to reach 12–16 mg per day, in 2–4 mg increments. Sedation would suggest excessive dosing, in which event the dose should be lowered. Should the patient continue to present with withdrawal symptoms or cravings, the dose may be increased further on a third day of induction. The maximum approved daily dose is 32 mg, in divided doses.

Buprenorphine Maintenance

Buprenorphine treatment is currently best conceived as a maintenance treatment. Data from clinical trials shows that buprenorphine discontinuation is strongly associated with relapse (Fiellin

et al., 2014; Woody et al., 2008; Sigmon et al., 2013). Newer, longer-acting forms are also available in implantable and injectable formulations (Rosenthal et al., 2016; Lofwall et al., 2018; Haight et al., 2019). Methadone may be superior to buprenorphine on the key outcome of retention in treatment and should be considered if a patient does not respond well to buprenorphine or drops out of treatment. The largest comparative effectiveness trial to date found methadone superior to buprenorphine on retention (Hser et al., 2014). Meta analyses have suggested that when compared with methadone, treatment retention and reduction in opioid use are similar at high fixed doses; however, when flexible dosing is used, treatment retention is superior with methadone (Mattick et al., 2014). Generally, though buprenorphine is effective, high dropout rates (>50% by 6 months) are a problem acoss trials (Mattick et al., 2014). One solution may be a longer acting (monthly), injectable formulations of buprenorphine, which has been approved for use in the United States (Haight et al., 2019), although the first comparative trial of sublingual versus injectable buprenorphine found the injection superior to sublingual on some measures of opioid abstinence but no difference in retention (Lofwall et al., 2018).

Opioid Antagonist Treatment: Naltrexone

Naltrexone is an antagonist at the μ-opioid receptor, and blocks the action of endogenous and exogenous opioids at this receptor. In patients with opioid use disorder, naltrexone at adequate blood levels is highly effective, blocking the effects of heroin or other opioids. Patients may "test the blockade" periodically, but use of opioid is typically low while naltrexone is in force (Nunes et al., 2020). The problem with naltrexone is that it is easy for patients to discontinue, after which risk of relapse is high. The advent of the extended-release injection formulation has improved adherence to some degree, with dropout rates similar or only slightly worse than buprenorphine, but high dropout nonetheless (Lee et al., 2018). Thus, a clear plan and concurrent behavioral treatment is needed to encourage adherence.

A major challenge to extended-release naltrexone treatment is induction. Naltrexone cannot be initiated straightaway in patients physiologically dependent on opioids, because, due to its high receptor affinity, it would displace other opioids from the receptors and precipitate withdrawal. The traditional guidance, reflected in the prescribing information for extended-release injection naltrexone (Vivitrol), is that a patient must be abstinent for 7–10 days prior to initiation of naltrexone, to avoid precipitating withdrawal (Srivastava et al., 2020). It has been shown that initiation of naltrexone can be accomplished in a shorter period with methods involving more aggressive use of non-opioid medications, including clonidine and clonazepam to attenuate withdrawal (Bisaga et al., 2018). However, more rapid initiation requires intensive clinical monitoring, and may be more successful on an inpatient basis, particularly for patients with heroin use (Shulman et al., 2022).

Two large clinical trials have shown that after successful initiation of medication naltrexone produces similar retention in treatment compared to buprenorphine (Tanum et al., 2017; Lee et al., 2018), although overall retention was inferior when all randomized patients are included due to failure to initiate naltrexone in many patients (Lee et al., 2018). High rates of dropout after initiation (i.e., failure to continue with subsequent monthly injections) highlight the need for a clear plan and behavioral strategies to encourage adherence.

In recently detoxified individuals, low grade withdrawal symptoms (fatigue, nausea, low appetite, insomnia) may occur for several weeks after the initiation of naltrexone, but do not usually persist and patients can be reassured that these will resolve. Aside from instances in which patients already have elevated liver enzymes or liver failure and who should not receive naltrexone, a review of several studies failed to demonstrate that naltrexone causes elevation of liver enzymes (O'Brien & Cornish, 1999).

In contrast to methadone and buprenorphine, no dependence develops with naltrexone: it provides no reward, and does not cause withdrawal. Oral naltrexone is generally not recommended given poor adherence (Blanco & Volkow, 2019).

The monthly extended-release injection formulation has improved adherence and treatment retention when compared with oral naltrexone (Sullivan et al., 2019). Even longer acting subcutaneous implants of naltrexone have shown promise and are available outside the USA (Hulse et al., 2004; Krupitsky, 2012), but are not yet FDA approved.

Opioid Use Disorder and Pregnancy

The pregnant patient with opioid use disorder presents with special challenges and can contribute to adverse consequences such as fetal growth restriction, low birth weight, and neonatal abstinence syndrome (NAS). NAS presents with signs and symptoms ranging from failure to thrive to hyperreflexia, tremor, diarrhea, and dehydration (Senay, 1999). A recent Cochrane review showed that buprenorphine and methadone are similar in terms of efficacy and safety for pregnant women with OUD (Minozzi et al., 2020). In terms of prevention of NAS, Jones and colleagues showed than in a randomized controlled trial of 175 pregnant women with OUD that in neonates whose mothers were exposed to buprenorphine required significant less morphine for NAS treatment, had a significantly shorter treatment duration for NAS, and had a significantly shorter length of hospital stay (Jones et al., 2010). For NAS treatment, Methadone, buprenorphine, morphine, and phenobarbital have been used, and a recent meta-analysis showed that treatment with either methadone or buprenorphine, compared with morphine, resulted in shorter lengths of stay. However, methadone and buprenorphine have not been compared head-to-head in a randomized controlled trial (Disher et al., 2019).

Treatment of Co-Occurring Depression or PTSD and Opioid Use Disorder

Major depression is common among opioid-dependent subjects and has been associated with poor outcomes. Several studies suggest that monoamine enhancers are effective for the treatment of depression in opioid-dependent patients who are on methadone maintenance, and that such treatment is associated with improvement in drug use (Woody et al., 1975; Nunes et al., 1998). However, other trials, particularly trials of serotonin reuptake inhibitors, have been negative (for a reviews, see Nunes & Levin, 2004 and Nunes et al., 2004). Posttraumatic stress disorder (PTSD) is also common among patients with opioid use disorder, often associated with depression, and associated with worse outcome (Hien et al., 2000). A reasonable recommendation would be that major depression and other co-occurring mental disorders should be identified and treated among opioid-dependent patients once the opioid use disorder has been stabilized with methadone, buprenorphine, or naltrexone. It is important to take a careful history in an effort to identify a clear-cut major depressive syndrome, as opposed to depressive symptoms that may reflect toxic effects of drugs.

Summary

The FDA approval in 1992 of buprenorphine for the treatment of opioid use disorder brought forth an excellent medication that can be used flexibly across a range of settings, including primary care, mental health clinics, emergency departments, and general hospitals. It can be used for medically supervised withdrawal, to help transition to naltrexone treatment, or as a maintenance tool if abstinence-based treatment is not successful. Clonidine and lofexidine are good ancillary medications to help control withdrawal symptoms. Methadone continues to be a well-used medication for maintenance treatment, although its availability is restricted to specially licensed clinics with arduous attendance requirements. Extended-release naltrexone is an effective alternative to prevent relapse among patients who have been detoxified and achieved abstinence. Dropout rates are a serious problem and the risk of relapse, overdose, and overdose death re-emerge after medication is discontinued. Dropout rates are particularly high for buprenorphine and naltrexone, and behavioral strategies need to be put in place in an effort to encourage adherence.

Tobacco Use Disorder

Nicotine Replacement Therapy

Nicotine is the primary psychoactive compound in tobacco products, and the most basic treatment is nicotine replacement therapy (NRT). NRT is purified nicotine that is available in multiple forms (patches, gum, lozenges, spray, inhalers) to alleviate cravings and symptoms of physical dependence (Prochaska and Benowitz, 2019). All forms have shown similar efficacy in terms of smoking cessation rates (Lindson et al., 2019). Patches, which are generally absorbed more slowly, are generally recommended for 16–24 h, and use of lozenges, gum, and inhalers is recommended every 1–2 h (Prochaska and Benowitz, 2019). Combination NRT including a patch and another form has been shown to be more effective than a single NRT formulation in terms of quit rates (Lindson et al., 2019).

Bupropion

In 1997; the FDA approved sustained-release bupropion for tobacco use disorder. This derived from a serendipitous observation that there was an antismoking effect among US veterans receiving bupropion to treat depression, regardless of an antidepressant response. Its efficacy at inhibiting smoking was demonstrated in a pivotal clinical trial against placebo, during which the 150 and 300 mg doses demonstrated better efficacy than placebo, with better smoking cessation rates achieved with the highest dose (44.2% for 300 mg, 38.6% for 150 mg, 19.9% for placebo) (Hurt et al., 1997). The mechanism is thought to be linked to its blockade of dopamine and norepinephrine reuptake, and to some antagonism of nicotinic acetylcholine receptor (Ascher et al., 1995; Slemmer et al., 2000). The simplicity of its use and the good tolerability have made bupropion a first-line agent against nicotine use disorder.

Treatment with bupropion does not require that patients quit smoking at the inception of treatment. Rather, clinicians are advised to set a quit date approximately 1 week after bupropion is begun, although many smokers prefer to reduce their smoking gradually as their cravings decrease. While there continues to be a contraindication to the use of bupropion in patients with a preexisting history of seizures, the rate of new-onset seizures with bupropion is about 0.3%, and is mostly associated with the use of the immediate-release formulation at a dose exceeding 450 mg. The most common side effects reported are headaches, agitation, and insomnia. An emerging consensus supports the use of bupropion for at least 6 months after smoking cessation (Steinberg et al., 2006), and success at 6 months appears to be related to increased age, continued use of medication, and longer time to the first morning cigarette. Bupropion in combination with NRT increases quit rates beyond those of either drug alone (Stead et al., 2012).

Varenicline

The newest FDA-approved medication for tobacco use disorder is the nicotine receptor antagonist varenicline. Nicotine receptors come in pentameric arrangements of alpha (2–10) and beta (2–4) subunits with extensive brain distribution, and the one targeted by varenicline is the $\alpha_4\beta_2$ subtype. Animal studies indicate that β_2 receptor subtypes are involved in nicotine reinforcement, but not in withdrawal (Besson et al., 2006). Varenicline is a high affinity partial agonist. This means that while varenicline binds with high affinity to the $\alpha_4\beta_2$ nicotinic receptor, it stimulates less receptor-mediated activity than nicotine itself. This is important because of the rapid desensitization characteristic of nicotinic cholinergic receptors. It also binds the receptor with greater affinity than nicotine, so that it blocks the effects of nicotine if, for example, a patient smokes while on the medication. The end result of these pharmacological properties is the blockage of nicotine-induced dopamine release in the meso-limbic system, where neuronal systems related to reinforcement are located. This blockade is the proposed mechanism for the effect of varenicline in tobacco smokers. Smokers for

whom varenicline is effective will sometimes say they simply "forgot to smoke," indicating the absence of withdrawal symptoms or craving, and if they do smoke will often report feeling no effects, indicating the blockade of receptors.

Varenicline has been shown to have been more effective than either bupropion or single NRT and is comparable to combined NRT (Cahill et al., 2013). To reduce the incidence of nausea, the recommended titration is 0.5 mg per day with food for the first 3 days, then 0.5 mg twice per day for the next 4 days, followed by 1 mg twice per day. The treatment is then continued for a first period of 12 weeks. If the patient achieves abstinence, clinicians may consider continuing treatment for an additional 12 weeks to encourage long-term abstinence.

After FDA approval, anecdotal reports of treatment emergent neuropsychiatric side effects including depression, psychosis, and suicide, particularly among patients with psychiatric illness, emerged, prompting a black box warning. However, a large, multi-national randomized clinical trial (termed "EAGLES") involving 8000 smokers both with and without psychiatric illness found no evidence of increased treatment emergent neuropsychiatric symptoms (Anthenelli et al., 2016), leading the way for removal of the black box warning. Notably, as a secondary outcome, the EAGLES trial showed improved smoking cessation rates in varenicline compared with bupropion and NRT. Nonetheless, as a precaution, before initiating varenicline, patients should be instructed to be aware of and report side effects, including any changes in mental state.

Stimulants: Cocaine and Amphetamines

This section on stimulants will begin to discuss addiction disorders for which no FDA-approved treatment exists. The euphoric and stimulant effects of cocaine are primarily modulated through blockade of reuptake of dopamine and norepinephrine into the pre-synaptic neuron through blockade of the norepinephrine transporter (NET) and dopamine transporter, whereas amphetamines also release of dopamine and norepinephrine directly into the synaptic terminal via inhibition of vesicular monoamine transporter (VMAT). Thus, similarly to agonist therapy for opioids and tobacco, prescription monoamine releasers have been some of the most promising agents for the treatment of cocaine use disorder. A recent meta-analysis showed that prescription amphetamines, when given in robust doses, is associated with improved abstinence (Tardelli et al., 2020). Particularly robust is evidence for high dose mixed amphetamine salts (60 or 90 mg/day) in the treatment of co-morbid cocaine use disorder and attention deficit hyperactivity disorder (ADHD); mixed amphetamine salts, when combined with cognitive behavioral therapy (CBT), reduced symptoms for both disorders (Levin et al., 2015). For methamphetamine and amphetamine use disorder specifically, low quality evidence indicates that methylphenidate may reduce use (Chan et al., 2019). Other medications including bupropion, topiramate, disulfiram, glutamate modifying agents (i.e., ketamine), doxazosin have shown preliminary and/or mixed findings and thus require further study (Brandt et al., 2021). Recently, the combination of extended-release injection naltrexone and bupropion has been shown to be modestly effective compared to placebo for treatment of severe methamphetamine use disorder (Trivedi et al., 2021).

Cannabis

Currently, there are no FDA-approved medications to treat cannabis use disorder, although there are several leads that appear promising. Human laboratory studies have studied numerous potential pharmacotherapies to alleviate withdrawal symptoms. Medications evaluated include lithium, divalproex, nefazadone, baclofen, mirtazapine, oral tetrahydrocannabinol (THC), combined lofexidine and oral THC, quetiapine, nabilone (a synthetic CB-1 receptor agonist), and extended-release zolpidem (Brezing & Levin, 2018). To date, the medications that have shown

the most promise in reducing withdrawal symptoms are oral THC alone and in combination with lofexidine and nabilone. The combination pharmacotherapy may be superior to either agent alone. Another approach studied in the laboratory is to initiate abstinence and then observe whether a medication reduces self-administration of marijuana. Using this paradigm, mirtazapine, THC alone and in combination with lofexidine, nabilone, and acute and chronic dosing of naltrexone have been studied. To date, the combination therapy and chronic dosing of naltrexone, and nabilone have shown promise in reducing self-administration of recently abstinent heavy marijuana users.

Despite the promise shown in human laboratory stories, comparatively fewer medications have shown promise in clinical trials. Mirtazepine, GABA-A agonists, gabapentin, and CB-1 agonists have shown promise for attenuating cannabis withdrawal symptoms (Brezing & Levin, 2018), and n-acetyl-cysteine, a glutamate modulator, has shown improved abstinence in adolescents (Gray et al., 2012), though this effect was not seen in adults (Gray et al., 2017). A recent study also showed that quetiapine treatment was associated with a reduction from heavy to moderate (though not light) use (Mariani et al., 2021b).

Additionally, a recent pilot randomized controlled trial showed that varenicline, compared with placebo, improved abstinence measures in participants with cannabis use disorder (McRae-Clark et al., 2021).

Treatment of Co-occurring Major Depressive Disorder and Other Co-occurring Disorders

Other psychiatric disorders, including major depressive disorder, bipolar disorder, post-traumatic stress disorder (PTSD), and attention deficit hyperactivity disorder (ADHD) co-occur at elevated rates among people with substance use disorders, indicating that having one disorder elevates the risk of having the others. A number of clinical trials and observational studies have examined the impact of treating co-occurring psychiatric disorders on the outcome of a substance use disorder, and vice versa – the impact of reduction in substance use or abstinence on the outcome of mood or other disorders. In general, the findings suggest that treatment of the substance use disorder is a first priority often improving symptoms of depression or other disorders, but that careful diagnosis and pharmacotherapy of the co-occurring psychiatric disorders improves outcome and should be considered in treatment planning for patients with substance use disorders.

Co-occurring major depressive disorder has been studied most extensively (Nunes & Weiss, 2019). Much of what presents as a depressive syndrome among patients actively using alcohol or other drugs, responds to abstinence or psychosocial counseling, as reflected in observational studies and high placebo response rates in many placebo-controlled trials of medications for depression. This suggests much of the depression is substance induced and the main treatment is treatment of the substance use disorder. However, in trials that selected samples with a lower placebo response rate, either by including only patients still depressed after detoxification on an inpatient unit, or with independent depression established by clinical history, medications for depression do appear effective, improving both mood and substance use outcome (Nunes & Levin, 2004). Trials have similarly suggested adequate treatment of bipolar disorder with lithium or anticonvulsants improves both mood syndromes and co-occurring substance use (Nunes & Weiss, 2019).

Trauma is very common in the histories of patients with substance use disorders, as is post-traumatic stress disorder (PTSD). PTSD is notably common among patients with opioid use disorder, often associated with depression, and associated with worse outcomes (Hien et al., 2000). A thorough history is important to uncover and characterize trauma and its sequelae, which can include depression and other disorders in addition to PTSD. Behavioral interventions are the cornerstone of treatment for PTSD and are effective in patients with co-occurring substance use disorders. Sertraline was shown to be helpful in combination with the behavioral intervention Seeking Safety, for reducing PTSD symptoms among patients with PTSD and substance use disorders (Hien et al., 2015).

Attention Deficit Hyperactivity Disorder (ADHD) has its onset in childhood and appears to be a risk factor for development of substance use disorders in adolescence and adulthood. The prevalence of ADHD among adults seeking treatment for substance use disorders is in the 15% to 20% range. A thorough history, with exploration of the developmental and school history is important to making the diagnosis (Levin & Upadhaya, 2006). Several studies suggest that treatment of ADHD with robust doses of monoamine releasers among patients with co-occurring stimulant or other substance disorders is effective at both improving ADHD symptoms and reducing substance use (Konstenius et al., 2014; Levin et al., 2015). Monoamine releasers carry a risk of misuse, although clinical trials using extended-release formulations of methylphenidate or mixed amphetamine salts and have observed little misuse (Winhusen et al., 2011, Levin & Mariani, 2019). Treatments for ADHD that are not monoamine releasers, such as atomoxetine, have been tried but trials are limited and the effects appear weaker; although may have utility in reducing heavy drinking episodes among adults with ADHD and alcohol use disorder (Wilens et al., 1995).

Summary

Like the pharmacology of psychiatric disorders in general, the pharmacological treatment of substance use disorders is rapidly advancing, driven by active research. A number of FDA-approved medications for addictive disorders are available, including methadone, buprenorphine, and naltrexone for opioid use disorder, disulfiram, naltrexone, and acamprosate for alcohol use disorder, nicotine replacement (patch, gum, etc.), bupropion, and varenicline for smoking cessation. Other potentially promising medications are emerging, notably monoamine releasers for cocaine use disorder, and data are accumulating to support their use in substance use disorders. More tools are being developed to target brain adaptive systems altered by drugs of abuse. Clinicians can expect that in the coming years, more medications will be at their disposal to assist patients in recovery from substance use disorders.

Acknowledgments The authors wish to acknowledge the support from the National Institute on Drug Abuse.

References

Ait-Daoud, N., Malcolm, R. J., Jr., & Johnson, B. A. (2006). An overview of medications for the treatment of alcohol withdrawal and alcohol dependence with an emphasis on the use of older and newer anticonvulsants. *Addictive Behaviors, 31*(9), 1628–1649.

Amato, L., Minozzi, S., Vecchi, S., et al. (2010). Benzodiazepines for alcohol withdrawal. *Cochrane Database of Systematic Reviews, 3,* CD005063.

Anthenelli, R. M., Benowitz, N. L., West, R., St. Aubin, L., McRae, T., Lawrence, D., Ascher, J., Russ, C., Krishen, A., & Evins, A. E. (2016). Neuropsychiatric safety and efficacy of varenicline, bupropion, and nicotine patch in smokers with and without psychiatric disorders (EAGLES): A double-blind, randomised, placebo-controlled clinical trial. *Lancet, 387,* 2507–2520.

Anton, R. F. (2008). Naltrexone for management of alcohol use disorder. *New England Journal of Medicine, 359,* 715–721.

Anton, R. F., O'Malley, S. S., Ciraulo, D. A., et al. (2006). Combined pharmacotherapy and behavioral interventions for alcohol dependence. *JAMA, 295,* 2003–2017.

Anton, R. F., Latham, P., Voronin, K., et al. (2020). Efficacy of gabapentin for the treatment of alcohol use disorder in patients with alcohol withdrawal symptoms: A randomized clinical trial. *JAMA Internal Medicine, 180,* 728–736.

Ascher, J. A., Cole, J. O., Colin, J. N., et al. (1995). Bupropion: A review of its mechanism of antidepressant activity. *Journal of Clinical Psychiatry, 56,* 395–401.

Auriacombe, M., Fatseas, M., Dubernet, J., et al. (2004). French field experience with buprenorphine. *American Journal on Addictions, 13*(Suppl 1), S17–S28.

Ball, J. C., Lange, W. R., & Myers, E. (1988). Reducing the risk of AIDS through methadone maintenance treatment. *Journal of Health and Social Behavior, 29,* 214–226.

Barthwell, A., Senay, E., & Marks, R. (1989). Patients successfully maintained with methadone escaped human immunodeficiency virus infection. *Archives of General Psychiatry, 46,* 957–958.

Besson, M., David, V., Suarez, S., et al. (2006). Genetic dissociation of two behaviors associated with nicotine addiction: beta-2 containing nicotinic receptors are involved in nicotine reinforcement but not in withdrawal. *Psychopharmacology, 187,* 189–199.

Binswanger, I. A., Stern, M. F., & Devo, R. A. (2007). Release from prison- a high risk death for former inmates. *New England Journal of Medicine, 356,* 157–165.

Bisaga, A., Mannelli, P., Yu, M., Nangia, N., Graham, C. E., Tompkins, D. A., Kosten, T. R., Akerman, S. C., Silverman, B. L., & Sullivan, M. A. (2018). Outpatient transition to extended-release injectable naltrexone for patients with opioid use disorder: A phase 3 randomized trial. *Drug and Alcohol Dependence, 187,* 171–178.

Bjorkqvist, S. E., Isohanni, M., Makela, R., et al. (1976). Ambulant treatment of alcohol withdrawal symptoms with carbamazepine: A formal multicenter double-blind comparison with placebo. *Acta Psychiatrica Scandinavica, 53*(5), 333–342.

Björnsson, E., Nordlinder, H., & Olsson, R. (2006). Clinical characteristics and prognostic markers in disulfiram-induced injury. *Journal of Hepatology, 44*(4), 791–797.

Blanco, C., & Volkow, N. D. (2019). Management of opioid use disorder in the USA: Present status and future directions. *Lancet, 393,* 1760–1772.

Blodgett, J. C., Del Re, A. C., Maisel, N. C., & Finney, J. W. (2014). A meta-analysis of topiramate's effects for individuals with alcohols use disorders. *Alcoholism: Clinical and Experimental Research, 38*(6), 1481–1488.

Brandt, L., Chao, T., Comer, S. D., & Levin, F. R. (2021). Pharmacotherapeutic strategies for treating cocaine use disorder-what do we have to offer? *Addiction, 116*(4), 694–710.

Brezing, C. A., & Levin, F. R. (2018). The current state of pharmacological treatments for cannabis use disorder and withdrawal. *Neuropsychopharmacology, 43,* 173–194.

Cahill, K., Stevens, S., Perera, R., & Lancaster, T. (2013). Pharmacological interventions for smoking cessation: An overview and network meta-analysis. *Cochrane Database of Systematic Reviews,* CD009329.

Chan, B., Kondo, K., Freeman, M., Ayers, C., Montgomery, J., & Kansagara, D. (2019). Pharmacotherapy for cocaine use disorder— A systematic review and meta-analysis. *Journal of General Internal Medicine, 34,* 2858–2873.

Charnoff, S. M., Kissen, D., & reed, J.I. (1963). An evaluation of various psychotherapeutic agents in the long term treatment of chronic alcoholism: Results of a double blind study. *The American Journal of the Medical Sciences, 246,* 172–179. https://doi.org/10.1097/00000441-196308000-00006

Comer, S. D., & Collins, E. D. (2002). Self-administration of intravenous buprenorphine and the buprenorphine/naloxone combination by recently detoxified heroin abusers. *Journal of Pharmacology and Experimental Therapeutics, 303,* 695–703.

Daeppen, J. B., Gache, P., Landry, U., et al. (2002). Symptom-triggered vs. fixed-schedule doses of benzodiazepine for alcohol withdrawal. *Archives*

Internationales de Pharmacodynamie et de Thérapie, 162, 1117–1121.

Disher, T., Gullickson, C., Singh, B., et al. (2019). Pharmacological treatments for neonatal abstinence syndrome: A systematic review and network meta-analysis. *JAMA Pediatrics, 173*(3), 234–243.

Dole, V. P., & Nyswander, M. (1965). A medical treatment for diacetylmorphine (heroin) addiction. A clinical trial with methadone hydrochloride. *JAMA, 193*, 646–650.

Fiellin, D. A., Schottenfeld, R. S., Cutter, C. J., Moore, B. A., Barry, D. T., & O'Connor, P. G. (2014). Primary care-based buprenorphine taper vs maintenance therapy for prescription opioid dependence: A randomized clinical trial. *JAMA Internal Medicine, 174*, 1947–1954.

Foy, A., March, S., & Drinkwater, V. (1988). Use of an objective clinical scale in the assessment and management of alcohol withdrawal in a large general hospital. *Alcoholism, Clinical and Experimental Research, 12*(3), 360–364.

Galanter, M. (1993). Network therapy for addiction: A model for office practice. *American Journal of Psychiatry, 150*, 28–36.

Garbutt, J. C., West, S. L., & Carey, T. S. (1999). Pharmacological treatment of alcohol dependence: A review of the evidence. *JAMA, 282*, 1318–1325.

Garbutt, J. C., Kranzler, H. R., & O'Malley, S. S. (2005). Efficacy and tolerability of long-acting injectable naltrexone for alcohol dependence: A randomized controlled trial. *JAMA, 293*, 1617–1625.

Garbutt, J. C., Kampov-Polevoy, A. B., Kalka-Juhl, L. S., & Gallop, R. J. (2016). Association of the sweet lining phenotype and craving for alcohol with the response to naltrexone treatment in alcohol dependence: A randomized clinical trial. *JAMA Psychiatry, 73*(10), 1056–1063.

Garbutt, J. C., Kampov-Polevoy, A. B., Pedersen, C., Stansbury, M., Jordan, R., Willing, L., & Gallop, R. J. (2021). Efficacy and tolerability of baclofen in a U.S. community population with alcohol use disorder: a dose-response, randomized, controlled trial. *Neuropsychopharmacology, 46*(13), 2250–2256.

Gianoulakis, C., Krishnan, B., & Thavundavil, J. (1996). Enhanced sensitivity of pituitary beta-endorphin to ethanol in subject at high risk of alcoholism. *Archives of General Psychiatry, 53*, 250–257.

Colfax, G. N., Glenn-Milo, S., Moupali, D., et al. (2011). Mirtazapine to reduce metamphetamine use. *Archives of General Psychiatry, 68*(11), 1168–1175.

Gray, K. M., Carpenter, M. J., Baker, N. L., et al. (2012). A double-blind randomized controlled trial of *N*-acetylcysteine in cannabis-dependent adolescents. *American Journal of Psychiatry, 169*, 805–812.

Gray, K. M., Sonne, S. C., McClure, E. A., Ghitza, U. E., Matthews, A. G., McRae-Clark, A. L., et al. (2017). A randomized placebo-controlled trial of N-acetylcysteine for cannabisuse disorder in adults. *Drug and Alcohol Dependence, 177*, 249–257.

Gunderson, E. W., Wang, X., Fiellin, D. A., et al. (2010). Unobserved versus observed office buprenorphine/naloxone induction: A pilot randomized clinical trial. *Addiction Behavior, 35*, 537–540.

Haight, B. R., Learned, S. M., Laffont, C. M., et al. (2019). Efficacy and safety of a monthly buprenorphine depot injection for opioid use disorder: A multicentre, randomised, double-blind, placebo-controlled, phase 3 trial. *Lancet, 393*, 778–790.

Handlesman, L., Cochrane, K. J., Aronson, M. J., et al. (1987). Two new rating scales for opiate withdrawal. *American Journal of Drug and Alcohol Abuse, 13*, 293–308.

Hargreaves, W. (1983). Methadone dosage and duration of treatment for maintenance treatment. In J. A. Cooper, A. Altman, B. Brown, et al. (Eds.) *Research on the treatment of narcotic addiction: State of the art* (DHHS publication no ADM-83–1281, pp. 19–91). Rockville: National Institute on Drug Abuse.

Herman, B. H., & O'Brien, C. P. (1997). Clinical medications development for opiate addiction: Focus on non-opioids and opioids antagonists for the amelioration of opiate withdrawal symptoms and relapse prevention. *Seminars in Neuroscience, 9*, 158–172.

Hien, D. A., Nunes, E., Levin, F. R., & Fraser, D. (2000). Posttraumatic stress disorder and short-term outcome in early methadone treatment. *Journal of Substance Abuse Treatment, 19*(1), 31–37.

Hien, D. A., Levin, F. R., Ruglass, L. M., et al. (2015). Combining seeking safety with sertraline for PTSD and alcohol use disorders: A randomized controlled trial. *Journal of Consulting and Clinical Psychology, 83*, 359–369.

Hser, Y. I., Saxon, A. J., & Huang, D. (2014). Treatment retention among patients randomized to buprenorphine/naloxone compared to methadone in a multi-site trial. *Addiction, 109*, 79–87.

Hulse, G. K., Arnold-Reed, D. E., O'Neil, G., et al. (2004). Blood naltrexone and 6-beta-naltrexol levels following naltrexone implant: Comparing two naltrexone implants. *Addiction Biology, 9*, 59–65.

Hurt, R. D., Sachs, D. P., Glover, E. D., et al. (1997). A comparison of sustained-release bupropion and placebo for smoking cessation. *New England Journal of Medicine, 337*, 1195–1202.

Jasinsky, D. R., Johnson, R. E., Kocher, T. R., et al. (1985). Clonidine in morphine withdrawal: Differential effects on signs and symptoms. *Archives of General Psychiatry, 42*, 1063–1066.

Johnson, B. A., Ait-Daoud, N., Bowder, C. L., et al. (2003). Oral topiramate for treatment of alcohol dependence: A randomised controlled trial. *The Lancet, 361*, 1677–1685.

Johnson, B. A., Rosenthal, N., Capece, J. A., et al. (2007). Topiramate for treating alcohol dependence: A randomized controlled trial. *JAMA, 298*(14), 1641–1651.

Jonas, D. E., Amrick, H. R., Feltner, C., et al. (2014). Pharmacotherapy for adults with alcohol use disorders in outpatient settings: A systematic review and meta-analysis. *JAMA, 311*(18), 1889–1900.

Jones, H. E., Kaltenbach, K., Heil, S. H., Stine, S. M., Coyle, M. G., Arria, A. M., O'Grady, K. E., Selby, P., Martin, P. R., & Fischer, G. (2010). Neonatal abstinence syndrome after methadone or buprenorphine exposure. *The New England Journal of Medicine, 363*(24), 2320–2331.

Kampman, K., & Jarvis, M. (2015). American Society of Addiction Medicine (ASAM) National Practice Guideline for the Use of Medications in the Treatment of Addiction Involving Opioid Use. *Journal of Addiction Medicine, 9*(5), 358–367. https://doi.org/10.1097/ADM.0000000000000166

Kleber, H. D. (2007). Pharmacologic treatments for opioid dependence: Detoxification and maintenance options. *Dialogues in Clinical Neuroscience, 4*, 455–470.

Knapp, C., Ciraulo, D. A., Sarid-Segal, O., et al. (2008). Zonisamide, topiramate, and levetiracetam efficacy and neuropsychological effects in alcohol use disorder. *Journal of Clinical Psychopharmacology, 1*, 34–42.

Konstenius, M., Jayaram-Lindström, N., Guterstam, J., Beck, O., Philips, B., & Franck, J. (2014). Methylphenidate for attention deficit hyperactivity disorder and drug relapse in criminal offenders with substance dependence: a 24-week randomized placebo-controlled trial. *Addiction, 109*(3), 440–449. https://doi.org/10.1111/add.12369. Epub 2013 Dec 1. PMID: 24118269; PMCID: PMC4226329.

Koob, G. F., & Volkow, N. D. (2016). Neurobiology of addiction : A neurocircuitry analysis. *Lancet, 3*, 760–773.

Kosten, T. R., & O'Connor, P. G. (2003). Management of drug and alcohol withdrawal. *New England Journal of Medicine, 348*, 1786–1795.

Kranzler, H. R., & Soyka, M. (2018). Diagnosis and pharmacotherapy of alcohol use disorder: A review. *JAMA, 320*(8), 815–824.

Kranzler, H. R., & Van Kirk, J. (2001). Efficacy of naltrexone and acamprosate for alcoholism treatment: A meta-analysis. *Alcoholism, Clinical and Experimental Research, 25*, 1335–1341.

Kranzler, H. R., Wesson, D. R., & Billot, L. (2004). Naltrexone depot for treatment of alcohol dependence: A multi-center, randomized, placebo-controlled trial. *Alcoholism, Clinical and Experimental Research, 28*, 1051–1059.

Kranzler, H. R., Feinn, R., Morris, P., & Hartwell, E. E. (2019). A meta-analysis of the efficacy of gabapentin for treating alcohol use disorder. *Addiction, 114*(9), 1547–1555. https://doi.org/10.1111/add.14655. Epub 2019 Jun 5. PMID: 31077485; PMCID: PMC6682454.

Kreek, M. J. (1979). Methadone in treatment: Physiological and pharmacological issues. In R. L. Dupont, A. Goldstein, & J. O'Donnell (Eds.), *Handbook on drug abuse* (pp. 57–86). National Institute on Drug Abuse.

Krupitsky, E., Zvartau, E., Blokhina, E., et al. (2012). Randomized trial of long-acting sustained-release naltrexone implant vs oral naltrexone or placebo for preventing relapse to opioid dependence. *American Journal of Psychiatry, 69*, 973–981.

Lee, J. D., Nunes, E. V., Jr., Novo, P., et al. (2018). Comparative effectiveness of extended-release naltrexone versus buprenorphine-naloxone for opioid relapse prevention (X:BOT): A multicentre, open-label, randomised controlled trial. *Lancet, 391*, 309–318.

Leggio, L., Kenna, G. A., & Swift, R. M. (2008). New developments for the pharmacological treatment of alcohol withdrawal syndrome. A focus on non-benzodiazepine GABAergic medications. *Progress in Neuropsychopharmacology and Biological Psychiatry, 32*(5), 1106–1117.

Levin, F. R., & Mariani, J. J. (2019). Co-occurring substance use disorder and attention deficit hyperactivity disorder. In S. C. Miller, D. A. Fiellin, R. N. Rosenthal, & R. Saitz (Eds.), *The ASAM principles of addiction medicine* (6th ed.). Lippincott Williams & Wilkins (Wolters Kluwer).

Levin, F. R., & Upadhaya, H. P. (2006). Diagnosing ADHD in adults with substance use disorder: DSM-IV criteria and differential diagnosis. *Journal of Clinical Psychiatry, 68*, e18.

Levin, F. R., Mariani, J. J., Specker, S., Mooney, M., Mahony, A., Brooks, D. J., et al. (2015). Extended-release mixed amphetamine salts vs placebo for comorbid adult attention-deficit/hyperactivity disorder and cocaine use disorder. *JAMA Psychiatry, 72*, 593–602.

Lindson, N., Chepkin, S. C., Ye, W., Fanshawe, T. R., Bullen, C., & Hartmann-Boyce, J. (2019). Different doses, durations and modes of delivery of nicotine replacement therapy for smoking cessation. *Cochrane Database of Systematic Reviews, 4*, CD013308.

Ling, W., Amass, L., Shoptaw, S., et al. (2005). A multicenter randomized trial of buprenorphine-naloxone versus clonidine for opioid detoxification: Findings from the National Institute on Drug Abuse Clinical Trials Network. *Addiction, 100*, 1090–1100.

LiverTox: Clinical and Research Information on Drug-Induced Liver Injury [Internet]. Bethesda: National Institute of Diabetes and Digestive and Kidney Diseases; 2012-. Disulfiram. [Updated 2018 Jan 5]. Available from: https://www.ncbi.nlm.nih.gov/books/NBK548103/

Lofwall, M. R., Walsh, S. L., Nunes, E. V., et al. (2018). Weekly and monthly subcutaneous buprenorphine depot formulations vs daily sublingual buprenorphine with naloxone for treatment of opioid use disorder: A randomized clinical trial. *JAMA Internal Medicine, 178*, 764–773.

Malcolm, R., Ballenger, J. C., Sturgis, E. T., et al. (1989). Double-blind controlled trial comparing carbamazepine to oxazepam treatment of alcohol withdrawal. *American Journal of Psychiatry, 146*, 617–621.

Mariani, J. J., Rosenthal, R. N., Tross, S., et al. (2006). A randomized, open-label, controlled trial of gabapentin and phenobarbital in the treatment of alcohol withdrawal. *American Journal on Addictions, 15*, 76–84.

Mariani, J. J., Pavlicova, M., Choi, C. J., et al. (2021a). Quetiapine treatment for cannabis use disorder. *Drug and Alcohol Dependence, 218*, 108366.

Mariani, J. J., Pavlicova, M., Basaraba, C., et al. (2021b). Pilot randomized placebo-controlled clinical trial of high-dose gabapentin for alcohol use disorder. *Alcoholism: Clinical and Experimental Research, 45*, 1639–1652.

Mason, B. J., & Heyser, C. J. (2010). Acamprosate: A prototypic neuromodulator in the treatment of alcohol dependence. *CNS & Neurological Disorders Drug Targets, 9*(1), 23–32.

Mason, B. J., Goodman, A. M., Chabac, S., & Lehert, P. (2006). Effect of oral acamprosate on abstinence in patients with alcohol dependence in a double-blind, placebo-controlled trial: The role of patient motivation. *Journal of Psychiatric Research, 40*(5), 383–393.

Mattick, R. P., Breen, C., Kimber, J., & Davoli, M. (2009). Methadone maintenance therapy versus no opioid replacement therapy for opioid dependence. *Cochrane Database of Systematic Reviews, 3*, CD002209.

Mattick, R. P., Breen, C., Kimber, J., & Davoli, M. (2014). Buprenorphine maintenance versus placebo or methadone maintenance for opioid dependence. *Cochrane Database of Systematic Reviews, 2*, CD002207.

Mayo-Smith, M.F. (for the American Society of Addiction Medicine). (1997). Pharmacological management of alcohol withdrawal: A meta-analysis and evidence-based practice guidelines. *JAMA, 278*, 144–151.

McCance-Katz, E. F. (2004). Office-based buprenorphine treatment for opioid-dependent patients. *Harvard Review of Psychiatry, 12*, 321–338.

McLellan, A. T., Arndt, I. O., Metzger, D. S., et al. (1993). The effects of psychosocial services in substance abuse treatment. *JAMA, 269*, 1953–1959.

McRae-Clark, A. L., Gray, K. M., Baker, N. L., Sherman, B. J., Squeglia, L., Sahlem, G. L., Wagner, A., & Tomko, R. (2021). Varenicline as a treatment for cannabis use disorder: A placebo-controlled pilot trial. *Drug Alcohol Depend, 229*(Pt B), 109111. https://doi.org/10.1016/j.drugalcdep.2021.109111. Epub 2021 Sep 28. PMID: 34655945; PMCID: PMC8665036.

Minozzi, S., Amato, L., Vecchi, S., et al. (2010). Anticonvulsants for alcohol withdrawal. *Cochrane Database of Systematic Reviews, 3*, CD005064.

Minozzi, S., Amato, L., Jahanfar, S., Bellisario, C., Ferri, M., & Davoli, M. (2020). Maintenance agonist treatments for opiate-dependent pregnant women (2020). *Cochrane Database of Systematic Reviews, 11*(11), CD006318.

Myrick, H., Malcolm, R., Randall, P. K., et al. (2009). A double-blind trial of gabapentin versus lorazepam in the treatment of alcohol withdrawal. *Alcoholism, Clinical and Experimental Research, 33*(9), 1582–1588.

Nejad, S., Nisavic, M., Larentzakis, A., Dijkink, S., Chang, Y., Levine, A. R., de Moya, M., & Velmahos, G. (2020). Phenobarbital for acute alcohol withdrawal management in surgical trauma patients-A retrospective comparison study. *Psychosomatics, 61*(4), 327–335. https://doi.org/10.1016/j.psym.2020.01.008. Epub 2020 Feb 8. PMID: 32199629.

Nisavic, M., Nejad, S. H., Isenberg, B. M., Bajwa, E. K., Currier, P., Wallace, P. M., Velmahos, G., & Wilens, T. (2019). Use of phenobarbital in alcohol withdrawal management – A retrospective comparison study of phenobarbital and benzodiazepines for acute alcohol withdrawal management in general medical patients. *Psychosomatics, 60*(5), 458–467. https://doi.org/10.1016/j.psym.2019.02.002. Epub 2019 Feb 14. PMID: 30876654.

Novick, D. M., Joseph, H., Croxson, T. S., et al. (1990). Absence of antibody to human immunodeficiency virus in long-term, socially rehabilitated methadone maintenance patients. *Archives of Internal Medicine, 150*, 97–99.

Nunes, E. V., & Levin, F. R. (2004). Treatment of depression in patients with alcohol dependence or other drug dependence: A meta-analysis. *JAMA, 291*, 1887–1896.

Nunes, E. V., & Quitkin, F. (1987). Disulfiram and bipolar affective disorder. *Journal of Clinical Psychopharmacology, 7*, 284.

Nunes, E. V., & Weiss, R. W. (2019). Co-occurring mood and substance use disorders. In S. C. Miller, D. A. Fiellin, R. N. Rosenthal, & R. Saitz (Eds.), *The ASAM principles of addiction medicine* (6th ed.). Lippincott Williams & Wilkins (Wolters Kluwer).

Nunes, E. V., Quitkin, F. M., Donovan, S. J., et al. (1998). Imipramine treatment of opiate-dependent patients with depressive disorders. A placebo-controlled trial. *Archives of General Psychiatry, 55*, 153–160.

Nunes, E. V., Sullivan, M. A., & Levin, F. R. (2004). Treatment of depression in patients with opiate dependence. *Biological Psychiatry, 56*, 793–802.

Nunes, E. V., Gordon, M., & Friedmann, P. D. (2018). Relapse to opioid use disorder after inpatient treatment: Protective effect of injection naltrexone. *Journal of Substance Abuse Treatment, 85*, 49–55.

Nunes, E. V., Bisaga, A., & Kruptisky, E. (2020). Opioid use dropout from extended-release naltrexone in a controlled trial: Implications for mechanism. *Addiction, 115*, 239–246.

Nuss, M. A., Elmicki, D. M., Dunsworth, T. S., et al. (2004). Utilizing CIWA-Ar to assess use of benzodiazepine in patients vulnerable to alcohol withdrawal syndrome. *West Virginia Medical Journal, 100*, 21–25.

O'Brien, C. P., & Cornish, J. W. (1999). Opioids: Antagonists and partial agonists. In M. Galenter & H. D. Kleber (Eds.), *Textbook of substance abuse treatment* (pp. 281–294). American Psychiatric Press.

O'Malley, S. S., Jaffe, A. J., Chang, G., et al. (1992). Naltrexone and coping skills therapy for alcohol dependence. A controlled study. *Archives of General Psychiatry, 49*, 881–887.

O'Malley, S. S., Garbutt, J. C., Gastfriend, D. R., et al. (2007). Efficacy of extended-release naltrexone in alcohol-dependent patients who are abstinent before treatment. *Journal of Clinical Psychopharmacology, 27*(5), 507–512.

Oslin, D. W., Leong, S. H., Lynch, K. G., et al. (2015). Naltrexone vs placebo for the treatment of alcohol dependence: A randomized clinical trial. *JAMA Psychiatry, 72*(5), 430–437.

Pelic, P., & Myrick, H. (2003). Who is at greatest risk for delirium tremens? *Current Psychiatry, 2*, 14–18.

Petry, N. M., Bickel, W. K., Piasecki, D., et al. (2000). Elevated liver enzyme levels in opioid-dependent patients with hepatitis treated with buprenorphine. *American Journal on Addictions, 9*, 265–269.

Pettinati, H. M., O'Brien, C. P., Rabinowitz, A. R., et al. (2006). The status of naltrexone in the treatment of alcohol dependence: Specific effects on heavy drinking. *Journal of Clinical Psychopharmacology, 26*, 610–625.

Pierce, J. M., Petry, N. M., Stitzer, M. L., et al. (2006). Effects of lower-cost incentives on stimulant abstinence in methadone maintenance treatment: A National Drug Abuse Treatment Clinical Trials Network study. *Archives of General Psychiatry, 63*, 201–208.

Pierce, M., Sutterland, A., Beraha, E. M., Morley, K., & van den Brink, W. (2018). Efficacy, tolerability, and safety of low-dose and high-dose baclofen in the treatment of alcohol dependence: A systematic review and meta-analysis. *European Neuropsychopharmacolgy, 28*(7), 795–806.

Prochaska, J. J., & Benowitz, N. L. (2019). Current advances in research in treatment and recovery: Nicotine addiction. *Science Advances, 5*(10), eaay9763. https://doi.org/10.1126/sciadv.aay9763. PMID: 31663029; PMCID: PMC6795520.

Ravndal, E., & Amyndsen, E. J. (2010). Mortality among drug users after discharge from inpatient treatment: An 8-year prospective study. *Drug and Alcohol Dependence, 108*, 65–69.

Romach, M. K., & Sellers, E. M. (1991). Management of the alcohol withdrawal syndrome. *Annual Review of Medicine, 42*, 323–340.

Rosenthal, R. N., Lofwall, M. R., Kim, S., Chen, M., Beebe, K. L., & Vocci, F. J. (2016). Effect of buprenorphine implants on illicit opioid use among abstinent adults with opioid dependence treated with sublingual buprenorphine: A randomized clinical trial. *JAMA, 316*, 282–290.

Rösner, S., Hackl-Herrwerth, A., Leucht, S., et al. (2010). Opioid antagonists for alcohol dependence. *Cochrane Database of Systematic Reviews, 12*, CD001867.

Rush, B. (1823). *An inquiry into the effect of ardent spirits upon the human body and mind, with an account of the means of preventing and of the remedies for curing them* (8th ed.). James Loring.

Saitz, R., Mayo-Smith, M. F., Roberts, M. S., et al. (1994). Individualized treatment for alcohol withdrawal. A randomized double-blind controlled trial. *JAMA, 272*, 519–523.

Saxon, A. J., Ling, W., Hillhouse, M., Thomas, C., Hasson, A., Ang, A., Doraimani, G., Tasissa, G., Lokhnygina, Y., Leimberger, J., Bruce, R. D., McCarthy, J., Wiest,

K., McLaughlin, P., Bilangi, R., Cohen, A., Woody, G., & Jacobs, P. (2013). Buprenorphine/naloxone and methadone effects on laboratory indices of liver health: A randomized trial. *Drug and Alcohol Dependence, 128*(1–2), 71–76.

Schuckit, M. A. (2014). Recognition and management of withdrawal delirium (delirium tremens). *The New England Journal of Medicine, 371*(22), 2109–2113. https://doi.org/10.1056/NEJMra1407298. PMID: 25427113.

Senay, E. C. (1999). Opioids: Methadone maintenance. In M. Galanter & H. D. Kleber (Eds.), *Textbook of substance abuse treatment* (2nd ed., pp. 271–279). American Psychiatric Press.

Shulman, M., Hu, M. C., Sullivan, M. A., Akerman, S. C., Fratantonio, J., Barbieri, V., Nunes, E. V., & Bisaga, A. (2022). Patient characteristics associated with initiation of XR-naltrexone for opioid use disorder in clinical trials. *Drug and Alcohol Dependence, 233*, 109343.

Sigmon, S. C., Dunn, K. E., Saulsgiver, K., et al. (2013). A randomized, double-blind evaluation of buprenorphine taper duration in primary prescription opioid abusers. *JAMA Psychiatry, 70*, 1347–1354.

Sigmon, S. C. C., Bisaga, A., Nunes, E. V., et al. (2012). Opioid detoxification and naltrexone induction strategies: recommendations for clinical practice. *American Journal of Drug and Alcohol Abuse, 38*, 187–199.

Skinner, M. D., Lahmek, P., Pham, H., & Aubin, H. J. (2014). Disulfram efficacy in the treatment of alcohol dependence: A meta-analysis. *PLoS One, 9*, e87366.

Slemmer, J. E., Martin, B. R., & Damaj, M. I. (2000). Bupropion is a nicotinic antagonist. *Journal of Pharmacology and Experimental Therapeutics, 295*, 321–327.

Srivastava, A. B., Mariani, J. J., & Levin, F. R. (2020). New directions in the treatment of opioid withdrawal. *Lancet, 395*, 1938–1948.

Stead, L. F., Perera, R., Bullen, C., Mant, D., Hartmann-Boyce, J., Cahill, K., & Lancaster, T. (2012). Nicotine replacement therapy for smoking cessation. *Cochrane Database of Systematic Reviews, 11*, CD000146.

Steinberg, M. B., Foulds, J., Richardson, D. L., et al. (2006). Pharmacotherapy and smoking cessation at tobacco dependence clinic. *Preventive Medicine, 42*, 114–119.

Stoller, K., Walsh, S. L., Bigelow, G. E., et al. (2001). Effects of buprenorphine/naloxone in opioid-dependent humans. *Addiction, 96*, 230–242.

Streeton, C., & Whelan, G. (2001). Naltrexone, a relapse prevention maintenance treatment for alcohol dependence: A meta-analysis of randomized controlled trials. *Alcohol and Alcoholism, 36*, 544–552.

Stromberg, M. F., Mackler, S. A., & Volpicelli, J. R. (2001). Effect of acamprosate and naltrexone, alone or in combination, on ethanol consumption. *Alcohol, 23*, 109–116.

Sullivan, J. T., Sykora, K., Schneiderman, J., et al. (1989). Assessment of alcohol withdrawal: The revised clinical institute withdrawal assessment for alcohol scale (CIWA-Ar). *British Journal of Addiction, 84*, 1353–1357.

Sullivan, M. A., Rothenberg, J. L., Vosburg, S. K., et al. (2006). Predictors of retention in naltrexone maintenance for opioid dependence: Analysis of a stage 1 trial. *American Journal on Addictions, 15*, 150–159.

Sullivan, M. A., Bisaga, A., Pavlicova, M., Carpenter, K. M., Choi, C. J., Levin, F. R., Mariani, J. J., & Nunes, E. V. (2019). A randomized trial comparing extended-release injectable suspension and oral naltrexone, both combined with behavioral therapy, for the treatment of opioid use disorder. *American Journal of Psychiatry, 176*, 129–137.

Tanum, L., Solli, K. K., Latif, Z. E., et al. (2017). Effectiveness of injectable extended-release naltrexone vs daily buprenorphine-naloxone for opioid dependence: A randomized clinical noninferiority trial. *JAMA Psychiatry, 74*, 1197–1205.

Tardelli, V. S., Bisaga, A., Arcadepani, F. B., Gerra, G., Levin, F. R., & Fidalgo, T. M. (2020). Prescription psychostimulants for the treatment of stimulant use disorder: A systematic review and meta-analysis. *Psychopharmacology, 237*, 2233–2255.

Tofighi, B., McNeely, J., & Walzer, D. (2021, 2021). A telemedicine buprenorphine clinic to serve New York City: Initial evaluation of the NYC public hospital system's initiative to expand treatment access during the COVID-19 pandemic. *Journal of Addiction Medicine*. Online ahead of print.

Trivedi, M. H., Walker, R., & Ling, W. (2021). Bupropion and naltrexone in methamphetamine use disorder. *New England Journal of Medicine, 384*, 140–153.

Volpicelli, J. R., Alterman, A. L., & Hayashida, M. (1994). Naltrexone in the treatment of alcohol dependence. *Archives of General Psychiatry, 51*, 335–336.

Walsh, S. L., Preston, K. L., Stitzer, M. L., et al. (1994). Clinical pharmacology of buprenorphine: Ceiling effects at high doses. *Clinical Pharmacology and Therapeutics, 55*, 569–580.

Wesson, D. R., & Ling, W. (2003). The clinical opiate withdrawal scale (COWS). *Journal of Psychoactive Drugs, 35*, 253–259.

Wilens, T. E., Prince, J. B., & Biederman, J., et al. (1995) Attention deficit hyperactivity disorders in adult. *Psychiatric Services, 46*, 761–763, 765.

Winhusen, T. M., Lewis, D. F., Riggs, P. D., et al. (2011). Subjective effects, misuse, and adverse effects of osmotic-release methylphenidate treatment in adolescent substance abusers with attention-deficit/hyperactivity disorder. *Journal of Child and Adolescent Psychopharmacology, 21*, 455–463.

Winhusen, T., Walley, A., & Fanucchi, L. C. (2020). The opioid-overdose reduction continuum of care approach (ORCCA): Evidence-based practices in the HEALing Communities Study. *Drug and Alcohol Dependence, 217*, 108325.

Witkiewitz, K., Litten, R. Z., & Leggio, L. (2019). Advances in the science and treatment of alcohol use disorder. *Science Advances, 5*, eaax4043.

Woody, G. E., O'Brien, C. P., & Rickels, K. (1975). Depression and anxiety in heroin addicts: A placebo-controlled study of doxepin in combination with methadone. *American Journal of Psychiatry, 132*(4), 447–450.

Woody, G. E., Poole, S. A., Subramaniam, G., et al. (2008). Extended vs short-term buprenorphine-naloxone for treatment of opioid-addicted youth: A randomized trial. *JAMA, 300*, 2003–2011.

Medical Use of Cannabinoids and Psychedelic Compounds

Antonio Inserra, Danilo De Gregorio, and Gabriella Gobbi

Contents

A. Inserra · G. Gobbi (✉)
Neurobiological Psychiatry Unit, Department of
Psychiatry, McGill University, Montreal, QC, Canada
e-mail: antonio.inserra@mcgill.ca;
gabriella.gobbi@mcgill.ca

D. De Gregorio
Neurobiological Psychiatry Unit, Department of
Psychiatry, McGill University, Montreal, QC, Canada

Division of Neuroscience, Vita-Salute San Raffaele
University, Milan, Italy
e-mail: danilo.degregorio@mail.mcgill.ca;
degregorio.danilo@hsr.it

© Springer Nature Switzerland AG 2024
A. Tasman et al. (eds.), *Tasman's Psychiatry*,
https://doi.org/10.1007/978-3-030-51366-5_141

Abstract

Mounting evidence suggests safety and tolerability of cannabis-derived and psychedelic drugs as potential novel therapeutics for psychiatric, as well as sleep and pain disorders. Evidence concerning the therapeutic efficacy of these compounds remains controversial, although some promising preliminary results are available for them in specific disorders. For example, CBD is approved as medication for treatment-refractory epilepsy, while MDMA and psilocybin are being tested in phase 3 clinical trials respectively for treatment-refractory PTSD and MDD. Despite encouraging preliminary results, further preclinical and clinical trials are required to validate these findings. Most importantly, more systematic research is required to assess the potential long-term side effects that might arise following the use of cannabinoids and psychedelic compounds in psychiatric settings.

Keywords

Cannabinoids · Psychedelics · Psychiatric disorders · Sleep disorders · Pain disorders

General Introduction

In the last 10 years, the medical use of cannabinoid (delta-9-tetrahydrocannabinol-THC and cannabidiol-CBD) and psychedelic compounds (psilocybin, lysergic acid diethylamide-LSD, and 3,4-Methylenedioxymethamphetamine-MDMA) has gained popularity as a potential novel treatment of psychiatric Disorders despite having never received the approval from the drug regulatory agencies and having never passed the required clinical trials of Phase 1, 2, and 3 to test their safety and efficacy in larger populations. This unprecedented phenomenon leaves many unresolved questions in the field of clinical pharmacology since although several lines of evidence suggest that cannabinoids and Psychedelics may represent valid therapeutic tools in psychiatry, the evidence available is not conclusive. Therefore, more research is needed to fully understand the real efficacy and full spectrum of side effects of these compounds.

The purpose of this chapter is to offer a review of the data that support the use of medical cannabis and psychedelic compounds for the treatment of mental illness, while taking into account that –

with the exception of CBD for treatment-resistant epilepsy in children – none of these drugs has been approved as treatment for neuropsychiatric diseases by government agencies including the Food and Drug Administration (FDA), Health Canada, and the European Medicines Agency.

Surprisingly, medicinal cannabis and psyche-delics did not follow the classical medical phar-macological development as other commercially available drugs. There are multiple reasons for such unconventional drug development path. Firstly, these compounds were classified as Schedule 1 (in USA) or in homolog classes in other countries, thus preventing pharmaceutical companies from developing them abiding to the standards requested by governmental agencies. Secondly the public opinion has accepted the use of medical cannabis and psychedelics without enough scientific evidence, on the basis that they derive from plants. This reasoning appears some-what surprising, given that some 40 of the most essential medicines currently used to treat the most prevalent diseases (including cancer and cardiovascular diseases) also derive from plants; however, these FDA-approved plant-derived drugs (e.g., vincristine, digitoxin, ergotamine, and theophylline) were required to go through rigorous clinical trials before being deemed safe and efficacious for use in large populations by governmental agencies.

If a rational approach to medicinal cannabis and psychedelics is applied, these compounds should first be assessed in systematic clinical stud-ies to determine their efficacy in treating specific diseases (such as depression or pain) and their safety compared to standard treatments (e.g., approved medications for depression, analgesics). The recommended dosages of these plant-derived medicines, their side effects, *possible* drug–drug interactions and the required duration of treatment should be known. Lastly, it's important to know if more vulnerable populations (e.g., pregnant or breastfeeding women, youths, and elderly populations) can use these medicines as well. Therefore, a more rational, scientific approach to

medicinal cannabis and psychedelics is required. That is, to start asking the right clinical questions and answering them with clinical trials, as is done for any other plant-derived medicines (Gobbi, 2019).

Delta-9-tetrahydrocannabinol (THC)

THC Mechanism of Action

Delta-9-tetrahydrocannabinol (THC) is the main psychoactive compound of *Cannabis sativa* and *Cannabis indica* and, among a plethora of over 100, it is the most common phytocannabinoid in cannabis drug chemotypes (Hanuš et al., 2016). THC displays moderate affinity and low relative intrinsic activity (partial agonism) for the canna-binoid receptors CB1 (Ki = 40.7 nM) and CB2 (Ki = 36 nM) (Grassin-Delyle et al., 2014; Pertwee et al., 2010b), an interaction which under-lies its activities in modulating mood, pain, appe-tite, and sleep (Hanuš et al., 2016). THC also possesses anti-inflammatory (Evans, 1991), anti-oxidant and neuroprotective (Hampson et al., 1998), and bronchodilating (Williams et al., 1976) properties. While THC stimulates both CB1 and CB2 receptors, the role of these two proteins is distinct. Stimulation of CB1 receptors by THC leads to several behavioral effects in laboratory animals including hypothermia, sup-pression of locomotor activity, as well as anti-nociception (Martin et al., 1991). Animal studies showed that acute THC treatment as well as CB1 receptor stimulation produce an antidepressant effect (Bambico et al., 2007, 2012) whereas the chronic exposure to THC or the CB1 receptor agonist WIN55,212 during adolescence leads to the development of anxiety and certain features of depressive-like phenotype (Bambico et al., 2010; De Gregorio et al., 2020). On the one hand, the "high" or euphoria produced by the THC seems to be caused by its interaction with the CB1 receptor. Moreover, CB2 receptor stimulation is associated with pain relief and ant-inflammatory activities

(Pacher & Mechoulam, 2011). CB1 receptor stimulation inhibits forskolin-stimulated adenylate cyclase (AC) and leads to the inhibition N-, Q-, L-type calcium channels while stimulating the release of G proteins to activate inwardly rectifying potassium channel (Howlett et al., 2002) . On the other hand, it has been demonstrated that THC activates K+ channels and inhibition of calcium (Ca2+) channels, and this involves non-CB1, non-TRPV1 but G-protein-coupled receptors (O'Sullivan et al., 2005; Console-Bram et al. 2012). This receptor signaling also stimulates the activity of Mitogen-activated protein kinases (MAPKs). MAPK pathways are often activated by G-protein-coupled receptors (GPCRs) and can alter the activity of extracellular signal-regulated kinases (ERK) 1/2, p38 MAP kinase, and/or ERK5 proteins. The stimulation of their activity also contributes to the control of cell growth and metabolism.

The localization of the CB1 receptor is widespread in the brain, and its distribution parallels the known pharmacological actions of THC. Given that CB1 receptors are expressed especially in the limbic system, they represent a potential therapeutic target for the treatment of mental illness and pain (Herkenham et al., 1990; Hill et al., 2009a; Russo, 2016). THC-mediated CB2 receptor stimulation leads to inhibition of forskolin-stimulated AC activation and stimulating MAP kinases but lacks the ion channels-modulating effects of CB1. CB2 is localized mainly in cells of the immune system, such as bone marrow, thymus, spleen, tonsils, T and B lymphocytes, monocytes, and mast cells. Given that the maximal effect of THC at the CB receptor proteins is below that of synthetic cannabinoids such as nabilone or HU-210, THC is classified as a partial agonists because other ligands or cannabinoids are much more potent at cannabinoid receptors (Matsuda et al., 1990; Pacher et al., 2006).

Noteworthy, THC interacts with a wide variety of proteins including receptors, ion channels, and enzymes. For example, micromolar concentrations of THC can activate GPR18, GPR55, peroxisome proliferator-activated receptor gamma (PPARγ) nuclear receptors, as well as the transient receptor potential channel (TRP) A1, TRPV2, TRPV3, and TRPV4 while potentiating glycine-ligated ion channels (important for pain relief) and the activity of β-adrenoceptors (Hong & Liu, 2017). THC can either block or activate the non-cannabinoid receptor GPR55, depending on the experimental conditions.

The interaction of THC with the serotonergic (5-HT) neurotransmission is thought to be crucial for its effects. Indeed, acute intravenous administration of THC (0.1–1.5 mg/kg, i.p.) induces complex responses of 5-HT neurons, with some neurons excited, others inhibited, and some others unaffected. Repeated (5 days) but not acute THC administration enhances the 5-HT_{1A} receptor-mediated tonic neuronal activity in the hippocampus (Bambico et al., 2012). Similarly, THC acts through the CB1 receptor to modulate the dopaminergic system (Diana et al., 1998), an effect involved in its addictive effects and related withdrawal syndrome. Importantly, THC is able to allosterically modulate the μ- and δ-opioid receptors (Lichtman et al., 2001; Pertwee et al., 2010a). Indeed, studies using whole brain or cortical membranes found that application of THC leads to an increase in the dissociation of μ- and δ-opioid receptors specific ligands (Kathmann et al., 2006).

CBD is also an allosteric modulator of the μ- and δ-opioid receptors (Kathmann et al., 2006; Vaysse et al., 1987), thus suggesting that the interaction of THC with opioid receptors is a noncompetitive one. Indeed, clinically, when THC and morphine are co-administered, ¼ the dose of morphine (a μ- and δ-opioid receptors agonist) is required to reach significant reductions in pain (Naef et al., 2003), leading to the hypothesis that this reduction in morphine dose is due to the noncompetitive interaction of the THC with the opioid receptors.

Additionally, THC inhibits T-type calcium ($\text{Ca}_{\text{V}}3$) voltage-gated ion channels, conductance in Na^+ voltage-gated ion channels(−), and conductance in gap junctions between cells at concentrations between 1 and 10 μM. THC also interacts with a variety of enzymes such as monoamine oxidase, phospholipase, lysophosphatidylcholineacyl transferase, lipoxygenase, $\text{Na}^+\text{-K}^+\text{-ATPase}$, and $\text{Mg}^{2+}\text{-ATPase}$ (Evans, 1991; Pertwee, 1988; Pertwee & Cascio, 2014; Yamaori et al., 2011).

Medical Use of THC

Generally speaking, it is very difficult to reach a conclusion about the medical usefulness and effectiveness of THC since there are no systematic studies, and the published randomized controlled trials (RCTs) employ THC from various sources, different concentrations, doses, and administration route. In the present chapter, data analyzed with a systematic review and meta-analysis by Whiting and colleagues (2015) and by Black and colleagues (2019) were took as reference for clinical evidence.

Pain

Substantial evidence suggests that cannabis is an effective treatment for some chronic pain conditions in adults, even if the available clinical trials addressed different pain conditions employing different THC formulations and administration routes. A meta-analysis across 7 trials that evaluated nabiximols (a combination of THC CBD) and 1 trial that evaluated inhaled THC suggested that plant-derived cannabinoids increase the odds for improvement of pain by approximately 40% versus the control condition (OR = 1.41, 95% CI = 0.99, 2.00; 8 trials). The effects did not differ significantly across pain conditions, although it was not clear that there was adequate statistical power to test for such differences. Abrams and colleagues also indicated that cannabis reduced pain versus a placebo (OR = 3.43, 95% CI = 1.03, 11.48) (Abrams et al., 2007).

Whiting and colleagues reported other 28 randomized trials in patients with chronic pain (2454 participants). Twenty-two of these trials evaluated plant-derived cannabinoids (nabiximols, 13 trials; plant flower that was smoked or vaporized, 5 trials; THC oramucosal spray, 3 trials; oral THC, 1 trial), while 5 trials evaluated synthetic THC (i.e., nabilone) (Whiting et al., 2015). All but 1 of the selected primary trials used a placebo control, while the remaining trial used an active comparator (amitriptyline). The medical condition underlying the chronic pain was most often related to a neuropathy (17 trials); other

conditions included cancer pain, multiple sclerosis, rheumatoid arthritis, musculoskeletal issues, and chemotherapy-induced pain.

It is worth noting that the effect size for inhaled cannabis is consistent with a separate review of five trials of the effect of inhaled cannabis on neuropathic pain (Andreae et al., 2015). The pooled ORs from these trials reached 3.22 for pain relief versus placebo (OR 3.22, 95% CI = 1.59, 7.24) tested across 9 THC concentrations. There was also some evidence of a dose-dependent effect in these studies.

Two additional investigations studied the effects of cannabis flowers on acute pain (Wallace et al., 2015; Wilsey et al., 2016). One found a dose-dependent effect of vaporized cannabis flower on spontaneous pain, with the high dose (7% THC) showing the strongest effect size (Wallace et al., 2015). The other study found that vaporized cannabis flower reduced pain but did not identify a significant dose-dependent effect (Wilsey et al., 2016). These two studies are in agreement with the previous reviews by Whiting (Whiting et al., 2015) and Andreae (Andreae et al., 2015), which suggest a reduction in pain after cannabis administration.

Anorexia Nervosa

Pharmacological interventions in the treatment of anorexia nervosa have not been promising to date. Andries and colleagues (2014) conducted a prospective, double-blind, crossover RCT with a small sample of 24 women with anorexia nervosa of at least 5 years duration attending psychiatric and somatic therapy. In addition to their standard psychotherapy and nutritional interventions, the participants received dronabinol (which is an isomer of THC) 2.5 mg twice daily for 4 weeks or placebo for 4 weeks, and were randomly assigned to two treatment sequences (dronabinol/placebo or placebo/dronabinol). The primary outcome was weight change assessed weekly. The secondary outcome was change in Eating Disorder Inventory-2 (EDI-2) scores. The participants had a significant weight gain of 1.00 kg (95% CI 0.40–1.62) during dronabinol therapy versus

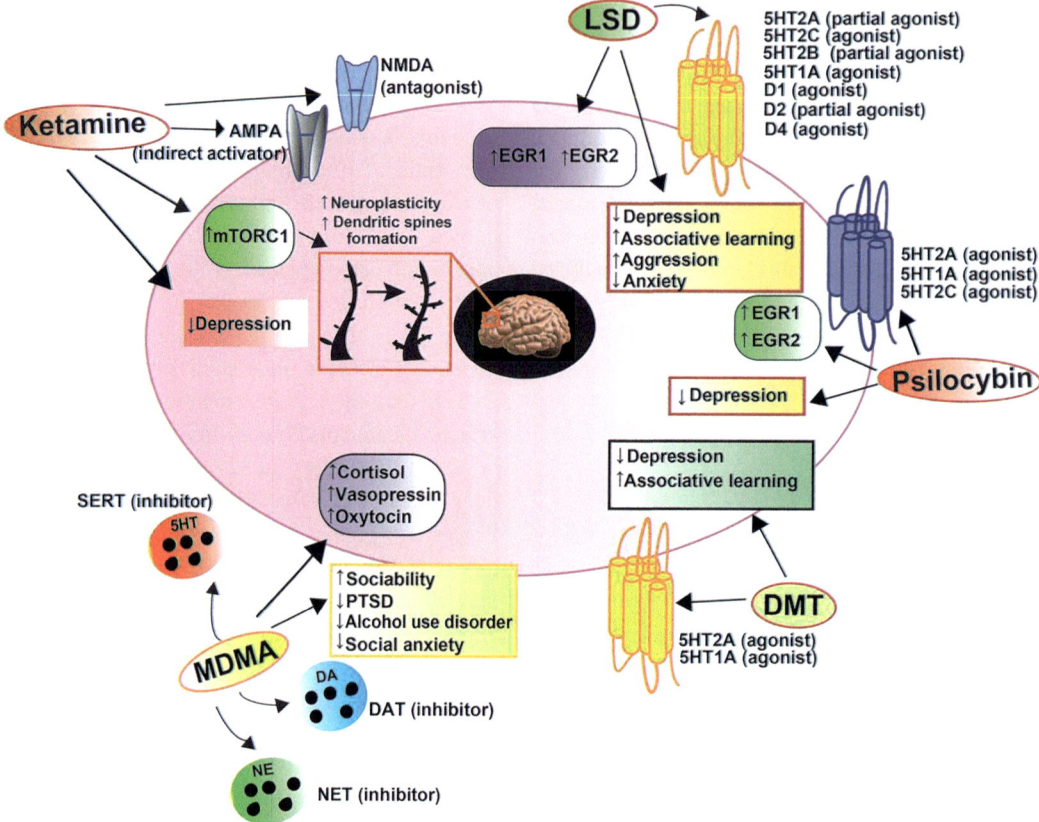

Fig. 1 Schematic overview of the main pharmacological targets of LSD, psilocybin, DMT, MDMA, and ketamine, the signaling cascades involved, hormonal modulation, as well as main behavioral outcomes following their administration in both animals and humans. Abbreviations are reported in the main text. (With permission from De Gregorio et al., 2021, J. Neuroscience)

0.34 kg (95% CI: −0.14 to 0.82) during placebo (t = 2.26, df = 23, p = 0.03) (Fig. 1). The average weekly change in body weight was 0.25 kg during dronabinol therapy, i.e., 0.17 kg (95% CI 0.01 to 0.32, t = 2.26, df = 22, p = 0.03) above placebo. Dronabinol significantly predicted weight gain in a multiple linear regression including EDI-2 body dissatisfaction score and leptin. EDI-2 subscale scores showed no significant changes over time (Andries et al., 2014). However, this study was in a very small sample, therefore its generalization to the population should be approached with caution. No concomitant behavioral changes were observed, suggesting that there was no real effect on the participants' attitudinal and behavioral traits related to eating disorders. However, the study was limited by small sample size and short treatment duration. Therefore, larger clinical trials are required.

Cannabis has long been thought to have an orexigenic effect, increasing food intake (Abel, 1975). Small residential studies conducted in the 1980s found that inhaled cannabis increased caloric intake by 40%, with most of the increase occurring as snacks and not during meals (Foltin et al., 1988). Accordingly, some evidence exists that oral cannabinoids might be able to enhance weight gain in patients with the HIV-associated wasting syndrome and anorexia nervosa. Indeed, in some countries (i.e., Canada), this is an indication for the use of THC in medicine. Clinical studies in this area have generally been small and short-lived and may not have yet identified the optimal dose of cannabinoid required. In one study in HIV patients, both dronabinol and inhaled cannabis increased weight significantly compared to placebo. Hence, the results of the clinical trials in AIDS wasting and cancer-associated anorexia-cachexia syndrome demonstrating little to no impact on appetite and weight were somewhat unexpected. There have not been any randomized controlled trials studying the effect of plant-derived cannabis on appetite and weight with weight as the primary endpoint. No benefit has been demonstrated in cancer-associated anorexia-cachexia syndrome.

Depression

No specific RCTs have evaluated the effects of THC on depression. However, some of the published studies evaluated the effects of depression as a secondary outcome in other diseases (such as pain and multiple sclerosis). Five RCTs (634 participants) enrolled patients for other conditions (chronic pain or multiple sclerosis with spasticity) and reported on depressive symptoms. Three studies (nabiximols, dronabinol) showed no effect using validated symptom scales. One study that evaluated three doses of nabiximols found increased depressive symptoms at the highest dose (11–14 sprays/day), but no difference compared to placebo at lower doses (Portenoy et al., 2012; Whiting et al., 2015). Black and colleagues performed a metanalysis comparing THC-CBD with either active comparator or placebo in 12 RCT (mostly in multiple sclerosis, MS), finding substantial heterogeneity across studies ($I^2 = 67\%$) and minimal effect sizes (Black et al., 2019). However, none of these studies included an RCT performed exclusively in patients with a primary diagnosis of major depression.

Sleep Disorder

At present, there are no published clinical trials that evaluated the effects of cannabinoids in patients with primary chronic insomnia, although a few clinical trials with different concentration of THC:CBD are ongoing.

THC has been reported to improve short-term sleep outcomes in patients with sleep disturbance associated with obstructive sleep apnea, fibromyalgia, chronic pain, or multiple sclerosis. Two RCTs (54 participants in total) evaluated cannabinoids (nabilone, dronabinol) for the treatment of sleep disorders. A trial with high risk of bias including 22 patients diagnosed with obstructive sleep apnea showed greater benefits of dronabinol (maximum dose of 10 mg daily) compared to baseline on supine apnea hypopnea index (mean difference from baseline(-14.1 ± 17.5; $p = 0.007$)) at 3 weeks follow-up (Prasad et al., 2013). No modifications in sleep architecture or serious adverse events were noted. A crossover trial in 32 patients with fibromyalgia and insomnia found improvements for nabilone 0.5 mg daily compared with 10 mg amitriptyline in insomnia (OR $-$ 3.25, 95% CI $= -5.26, -1.24$) and greater sleep restfulness (OR 0.48, 95% CI $= 0.01, 0.95$) at 2 weeks follow-up.

Nineteen trials (3231 participants total) enrolled patients with other conditions (chronic pain or multiple sclerosis) and reported on sleep outcomes. Nabiximols (13 studies), THC/CBD capsules (2 studies), smoked THC (2 studies), and dronabinol or nabilone were compared to placebo. Sleep outcomes were assessed at 2–15 weeks post-randomization. Eleven of the 19 trials were judged to have a high risk of bias, 6 had an uncertain risk of bias, and the other 2 were judged to have a low risk of bias. Despite this, the authors performed the analysis including the suspect studies. The meta-analysis found greater improvements with cannabinoids in sleep quality among 8 trials (weighted mean difference (WMD -0.58, 95% CI $= -0.87, -0.29$) and sleep disturbance among 3 trials (WMD -0.26, 95% CI $= -0.52, 0.00$) (Whiting et al., 2015).

Anxiety

At present, there are no RCTs investigating the anxiolytic effects of THC as a primary outcome measure. Four randomized controlled trials (232 participants) enrolled patients with chronic pain and reported also outcome in anxiety symptoms. The cannabinoids studied were: dronabinol, 10–20 mg daily; nabilone, maximum dose of 2 mg daily; and nabiximols, maximum dose of 4–48 sprays/day. Outcomes were assessed from 8 h to 6 weeks after randomization. These trials suggested greater short-term benefit with cannabinoids than a placebo on self-reported anxiety symptoms (Frank et al., 2008; Narang et al., 2008; Rog et al., 2005; Skrabek et al., 2008).

A metanalysis of the seven different studies evaluating anxiety symptoms after treatment with THC-CBD vs. placebo found that THC-CBD led to a significantly greater reductions in anxiety symptoms than placebo (OR -0.25, 95% CI = −0.49, −0.01), but with a heterogeneity of $I^2 =$ 65%, thus with a very low GRADE rating (Black et al., 2019). Again, most of the studies were performed in populations where the anxiety was secondary to other diseases.

PTSD

A double-blind, crossover RCT (Jetly et al., 2015) conducted with Canadian male military personnel with trauma-related nightmares also receiving standard treatments for PTSD was carried out with nabilone 0.5 mg–3.0 mg or placebo for 7 weeks. Nightmares, global clinical state, and general well-being were improved more with nabilone than with placebo ($p < 0.05$). There was no effect on sleep quality and quantity. Global clinical state was rated as very much improved or much improved for 7 of 10 subjects in the nabilone treatment period and 2 of 10 subjects in the placebo treatment period. One trial examined the effects of four different types of cannabis with varying THC and CBD content on PTSD symptoms in 76 veterans (ClinicalTrials.gov identifier NCT02759185). Another Canadian study aims to evaluate different formulations of THC and CBD in 42 adults with PTSD (ClinicalTrials.gov identifier NCT02517424). However, the results of these trials have not been published yet.

Psychosis

One small RCT reported the effect of THC in psychosis and found no significant change in the positive symptoms, but a worsening in the negative symptoms, as well as in cognitive functions (D'Souza et al., 2005).

Ongoing clinical trials are evaluating the use of THC in different kind of pain, substance use disorders, Alzheimer's disease, OCD (see clinicaltrials. gov).

CBD

CBD Mechanism of Action

Cannabidiol (CBD) is the main non-addictive phytocannabinoid. Even though CBD shows very low affinity for cannabinoid receptors, it possess the unique ability to antagonize CB1 at very low concentrations in the presence of THC (Thomas et al., 2007). This observed antagonism could be related to the ability of CBD to act as a negative allosteric modulator at CB1 receptors, reducing the euphoric effects of THC (Laprairie et al., 2015). Accordingly, CBD attenuates the intoxicating effects of cannabis, such as tachycardia, anxiety, hunger, and sedation in rats and humans (Nicholson et al., 2004; Russo & Guy, 2006; Russo, 2011).

The lack of CB1 receptor agonism by CBD is likely linked to its low capacity to produce the "high" like THC and to its low addiction liability. Noteworthy, CBD is included in a specific ratio of 1:1 with THC in the approved pharmaceutical preparation known as Sativex®, which has been studied in many controlled clinical trials (Flachenecker et al., 2014; Sastre-Garriga et al., 2011; Wade et al., 2010). Intriguingly, in vitro experiments reported that CBD isomerizes to THC under acidic conditions, but there is no evidence that directly supports that this actually in humans (Deiana et al., 2012; Golombek et al., 2020; Grotenhermen et al., 2017; Nahler et al., 2017; Russo, 2017).

CBD acts primarily as an agonist at TRPV1 (Bisogno et al., 2001; Iannotti et al., 2014) and 5-HT$_{1A}$ receptors (De Gregorio et al., 2019b; Russo et al., 2005). Moreover, CBD enhances adenosine receptor signaling (Carrier et al., 2006). The main effects elicited by CBD include anticonvulsive, anti-inflammatory, antioxidant, and antipsychotic action. These outcomes highlight the neuroprotective properties of CBD and support its potential role in the treatment of a plethora of neurological and neurodegenerative disorders, including Parkinson, Alzheimer and Huntington disease, Amyotrophic Lateral Sclerosis, Multiple Sclerosis and epilepsy (Hofmann & Frazier, 2013; Lago & Fernandez-Ruiz, 2007; Martín-Moreno et al., 2011; Scuderi et al., 2009).

Medical use of CBD

The only CBD formulation approved by the FDA (June 2018) is for the treatment of seizures associated with Lennox-Gastaut syndrome (LGS) or Dravet syndrome, two rare and particularly difficult-to-treat forms of epilepsy. This particular CBD is a purified form of CBD, extracted from *Cannabis sativa*, and has received FDA approval after extensive Phase III studies, which have been published in peer-reviewed medical journals (Devinsky et al., 2017, 2018a, b; Thiele et al., 2018).

There is limited evidence of the clinical effects of CBD in psychiatric disorders. The results available from completed RCTs show that CBD has some clinical indications in substance abuse disorders, psychosis, and anxiety. However further RCTs are required to better evaluate its efficacy and safety (Bonaccorso et al., 2019). Khoury and colleagues recently published a systematic review on the role of CBD in psychiatry (Khoury et al., 2019). After analyzing 609 articles, they found six case reports, seven randomized clinical trials, and 21 registered clinical trials, with a total of 201 subjects included. They then classified the level of evidence following criteria A (the highest) to C2 (the lowest), following the WFSBP task force standards (Bandelow et al., 2008). Their results indicated that, while CBD has no efficacy in major depressive and bipolar disorders, the level of evidence for cannabis withdrawal is B, cannabis addiction is C2, and treatment of positive symptoms in schizophrenia and anxiety in social anxiety disorder is C1. The most frequently reported side effects were sedation and dizziness without any severe adverse events.

It has to be pointed out that the use of CBD raises various safety concerns, especially with long-term use.

A new regulatory pathway would benefit consumers by providing safeguards to minimize risks related to CBD products. Some risk management tools could include clear labels, prevention of contaminants, CBD content limits, and measures such as minimum purchase age to mitigate the risk of ingestion by children. In addition, a new pathway could provide access and oversight for certain CBD-containing products for animals.

The current FDA's foods and dietary supplement authorities provide only limited tools for managing many of the risks associated with CBD products. Under the law, any substance, including CBD, must meet specific safety standards to be lawfully marketed as a dietary supplement or food additive (https://www.fda.gov/news-events/press-announcements/fda-concludes-existing-regulatory-frameworks-foods-and-supplements-are-not-appropriate-cannabidiol).

Anxiety

One randomized trial with a high risk of bias compared a single 600 mg dose of CBD to placebo in 24 participants with generalized social anxiety disorder. CBD was associated with a greater improvement on the anxiety factor of a 100-point visual analogue mood scale (mean difference from baseline -16.52, $p < 0.05$) compared to placebo (mean difference from baseline -35, $p = 0.018$) during a simulated public speaking test (Bergamaschi et al., 2011). Another small clinical trial also reports that CBD is more efficacious than placebo in social anxiety disorder (Crippa et al., 2011). Moreover, a recent clinical trial showed that CBD administered for 12 weeks on a fixed-flexible schedule titrated up to 800 mg/ reduced anxiety severity and has an adequate safety profile in 31 young people with treatment-resistant anxiety disorders.

Sleep

Even if CBD oil is often used off-label for improving sleep disorders, there are no RCT demonstrating its effect on sleep. One RCT in healthy volunteers did not find any difference between the placebo and CBD group in sleep-wake cycle regulation or sleep architecture (Linares et al., 2018).

Pain

There are currently RCTs being performed with CBD alone (Boyaji et al., 2020), while other RCTs employing the THC:CBD mixture nabiximol (THC/CBD in a 1:1 ratio) are reported above. However, few recent published clinical trials highlighted the inefficacy of CBD in certain features of neuropathic pain. Indeed, utilization of topical CBD in supplement to multimodal analgesia did not reduce pain or opioid consumption, or improve sleep scores following total knee arthroplasty TKA (Berger et al., 2022). Moreover, twenty healthy volunteers were included in a randomized, placebo-controlled, double-blinded, crossover study assessing pain intensities (using numeric rating scale), secondary hyperalgesia (von Frey filament), and allodynia (dry cotton swab). The oral application of 800-mg CBD failed to show a significant effect (Schneider et al., 2022).

Migraine

There are currently no cannabis-based drugs approved for use in migraine. However, different clinical studies indicate that phytocannabinoids might have a beneficial effect on the onset and duration of migraine headaches in adults.

According to these studies, oral drops of THC and CBD formulation significantly reduced nausea and vomiting associated with migraine attacks after 6 months of use along with the frequency of migraine headaches per month. THC and CBD drops MC were 51% more effective in reducing migraines than non-cannabis products [3]. However, well-designed experimental studies that assess the effectiveness and safety for treating migraine in adults are needed to support this hypothesis since one of the main problem associated with those studies are the were the lack of specification in the dosage either

the percentage of THC and CBD used. Currently, new clinical trials are ongoing to further assess the potential use of CBD and THC oil formulation as prophylactic treatment (NCT03972124) for acute (NCT05427630, NCT04360044) or chronic (NCT05337033, NCT04989413) migraine treatment.

Substance Abuse Disorder

A few studies have reported the efficacy of CBD or nabiximol in treating substance abuse disorders. A reduction of overall symptoms of cannabis withdrawal was observed after nabiximol compared to placebo (Allsop et al., 2014). Similarly, CBD attenuated symptoms of nicotine withdrawal scores compared to placebo (Lintzeris et al., 2019). Other two RCTs in patients with cannabis use disorder reported that nabixomol did not produce any significant effects (Lintzeris et al., 2019; Trigo et al., 2016, 2018).

Psychotic Disorders

In one study, CBD was reported not to significantly improve total symptoms (positive and negative) compared to placebo (Boggs et al., 2018), while in another, a lower level of positive symptoms and improvement in the CGI was observed in the CBD group (McGuire et al., 2018). CBD vs. an active comparator also improved symptoms of schizophrenia and the improvement was linked to an increase in anandamide levels (Leweke et al., 2012). Presently, several ongoing clinical trials are evaluating the efficacy of CBD on generalized anxiety, social anxiety, insomnia and OCD.

Other Endocannabinoids: N-Palmitoylethanolamide (PEA)

N-Palmitoylethanolamide (PEA) is a fatty acid amide belonging to the class of the N-acylethanolamine (NAE) that has been isolated from soy lecithin, eggs yolk, and peanut meal (Ganley et al., 1958). PEA is currently clinically employed as food-supplement for its neuroprotective (Raso et al., 2014), anti-neuroinflammatory (Solorzano et al., 2009), and analgesic (Calignano et al., 1998) effects. PEA selectively activates the peroxisome proliferator-activated receptor-alpha (PPAR-α) in vitro, which in turn heterodimerizes with 9-cis-retinoic acid receptor (RXR). This dimer then interacts with specific DNA sequences in the promoter regions of selective genes, known as peroxisome proliferator response elements (PPREs), thus stimulating anti-inflammatory gene expression (Daynes & Jones, 2002).

PEA displays a weak affinity for CB1 and CB2 receptors (Ho et al., 2008), and is thus unlikely to interact meaningfully with these receptors when taken along with the higher affinity THC. Moreover, PEA has agonist activity toward GPR55 (Baker et al., 2006), a non-CB1/CB2 cannabinoid receptor. Nevertheless, PEA can influence signaling through the CB2 receptor pathway (when examined alone): PEA reduces the enzymatic degradation of the endogenous ligand for the CB receptors N-arachidonylethanolamine (anandamide or AEA) by acting as a false substrate for fatty acid amide hydrolase (FAAH), which is the primary catabolic enzyme of AEA. This induces an increase in AEA concentrations which can in turn modulate inflammation via CB2, an effect which is also indirectly mediated by PEA (Facci et al., 1995). In addition, PEA acts directly as an allosteric modulator of TRPV1 channels, potentiating the activation of the channel by AEA and 2-arachidonylglycerol (2-AG), the other primary endogenous ligands for the cannabinoid receptors (Di Marzo et al., 2001). Recent studies have indicated that PEA is able to increase the levels of CB2 mRNA and protein as a result of PPAR-α activation (Guida et al., 2017). Indeed, after acute and repeated administration of the FAAH inhibitor URB597, the levels of PEA increased exponentially, producing an antidepressant-like effect (Gobbi et al., 2005). This suggests that PEA is also metabolized directly through FAAH.

Ghazizadeh-Hashemi and colleagues performed the first study in humans investigating the antidepressant potential of PEA (600 mg twice daily) as adjunctive treatment to standard antidepressant therapy with citalopram. The effects of PEA were evaluated in a 6-weeks double-blind, placebo-controlled RCT in 58 MDD patients (predominantly males) with a baseline Hamilton Depression Rating Scale (HAM-D) score P$\geq$19. The group of patients receiving add-on PEA showed significantly greater improvement in HAM-D scores throughout the entire trial period (weeks 2, 4, and 6), in the absence of additional side effects. Interestingly, in the PEA group there was a more rapid antidepressant effect, and the response rate at week 6 was 100% compared to 74% in the placebo group (Ghazizadeh-Hashemi et al., 2018). Overall, due to multifactorial pharmacological targets, PEA plays a protective role in situations of biochemical (Hill et al., 2009b) and neuro-inflammatory (Kim & Won, 2017) stress. For more detailed discussions on the potential of PEA in psychiatry and depression, please see the review (De Gregorio et al., 2019a).

Medical Psychedelic Drugs

Between 1950 and 1970, psychedelic research boomed across the globe as a potential novel breakthrough approach in psychiatry. Thousands of individuals received psychedelic compounds in clinical settings for a range of conditions spanning psychiatric disorders such as MDD, PTSD, and substance and alcohol abuse disorders, neurodevelopmental disorders such as autism spectrum disorder, and even as an empathogenic tool in couple psychotherapy (Inserra et al., 2021a). Despite encouraging results, psychedelic compounds were placed in Schedule I by the DEA and "The war on drugs" was launched. After psychedelic drugs were placed in Schedule I by the FDA and equivalent drug administrations worldwide, the research investigating the potential application of these compounds for the treatment of psychiatric disorders declined drastically.

In the last 30 years however, psychedelic compounds have been re-scrutinized under the lens of contemporary psychiatric and ethical standard, generating preliminary evidence that they might be useful pharmacotherapeutic and psychotherapy-enhancing agents as well as empathogenic and entheogenic agents. Starting with the pioneering clinical trials performed by Strassman et al. in which volunteers were given dimethyltryptamine (DMT) i.v. to assess safety (Strassman & Qualls, 1994; Strassman et al., 1994), eliciting what some of the volunteers have described as the most meaningful as well as challenging experiences of their lives (Strassman, 2001), the so-called "psychedelic renaissance" had begun. Since then, also due to the current state-of-the-art biomedical and imaging research, our understanding of the therapeutic mechanisms elicited by psychedelic compounds has increased drastically; however, although the knowledge generated allows for a glimpse of the biological and neurobiological mechanisms involved, a complete understanding of the molecular pathways engaged by psychedelics and the neurobiological mechanisms responsible for therapeutic improvement is still lacking.

The most important neurobiological effects elicited by psychedelics that induce anxiolytic and antidepressant effects in clinical trials employing psychedelics, appear to require the entrainment of neurogenesis and neuronal and synaptic plasticity, the regulation of neurotransmitter release and reuptake, and the regulation of pathway related to neuroimmunomodulation via the epigenetic regulation of gene expression, an important yet underexplored dimension in brain function and behavior.

A nasal spray containing the dissociative anesthetic and NMDA receptor antagonist ketamine was recently approved as a first in a new class of medications for depression (antidepressant drugs) by the FDA, creating new hopes for improvements in medications for depression (Kim et al., 2019). Psilocybin was reported to induce large-size antidepressant effects in several RCTs recently published (Davis et al., 2020). Other compounds such as MDMA and psilocybin are undergoing Phase III clinical trials for treatment-resistant PTSD and MDD, respectively, while Phase 1 clinical trials are investigating LSD and psilocybin as empathy-enhancing agents. Despite their name, the pharmacology of serotonergic classic psychedelics is not limited to 5-HT receptors but has multiple layers of complexity that involve several receptor families, homo and heteroreceptor complexes, and compound-specific biased intracellular cascades.

Psilocybin

Mechanism of Action

Psilocybin (4-phosphoryloxy-N,N-dimethyltryptamine) is a substituted indolealkylamine found in psilocybin-containing mushrooms and truffles (Hofmann et al., 1958). Psilocybin is an agonist at human serotonin receptors 5-HT$_{1A}$, 5-HT$_{2A}$, 5-HT$_{2B}$, and 5-HT$_{2C}$ receptors on dopaminergic and GABAergic neurons of the striatum, which are thought to be involved in the psilocybin-induced modulation of striatal activity. Indeed, blocking the 5-HT$_{2A}$ receptor nullifies the modulatory effects of psilocybin on mood states and emotional face recognition. Other effects however, such as the bias toward the processing of positive emotions, appear to be mediated by serotonergic receptors other than 5-HT$_{2A}$, possibly 5-HT$_{2B}$, 5-HT$_{2C}$, or 5-HT$_{1A}$ receptors.

Brain imaging studies have shown that psilocybin occupies up to 72% of all 5-HT$_{2A}$ receptors at clinically relevant doses inducing a hypermetabolic state in frontolateral and frontomedial cortices. These effects are accompanied by a decrease in default mode network (DMN) connectivity, a phenomenon which might be part of the neural substrates of mystical experiences, "ego death," and a transcendental sense of unity (Carhart-Harris et al., 2017). A recent study reported that psilocybin profoundly affects the claustrum, an understudied brain area which has the highest density of 5-HT$_{2A}$ receptor throughout the whole brain. More specifically, psilocybin decreased claustral connectivity with areas involved in emotion, memory, and attention (Barrett et al., 2020).

Potentially secondary to a 5-HT$_{2A}$ mediated activation of the HPA axis, psilocybin acutely increases the circulating levels of the stress-responsive adrenocorticotropic hormone (ACTH), cortisol, prolactin, and thyroid-stimulating hormone (Hasler et al., 2004). Concerning neurogenesis, recent studies found that psilocybin modulates neurogenesis-related gene expression in the rodent PFC and hippocampus (Jefsen et al., 2020). Previous studies found that while at lower doses a trend was observable toward increased neurogenesis, at higher doses neurogenesis was significantly decreased, suggesting that different dosages can have different outcomes on neuroplasticity and neuronal function overall (Catlow et al., 2013). Preclinical studies also point toward a facilitating effect of psilocybin over fear extinction at low doses, suggesting that it might hold potential for the treatment of PTSD (Catlow et al., 2013).

Clinical Trials

The therapeutic potential applications of psilocybin in psychiatry have revolved around its use coupled to supportive behavioral interventions as a medication for anxiety for individuals affected by a potentially life-threatening condition (Griffiths et al., 2016), and as an antidepressant for MDD (Davis et al., 2020). Two meta-analyses are available, which have reported on the outcomes of clinical trials investigating the therapeutic potential of synthetic psilocybin (0.2–0.4 mg/kg) for treatment-resistant MDD (Goldberg et al., 2020; Vargas et al., 2020). All studies included pre-intervention visits, support during the administration of psilocybin or placebo, and post-intervention support. The first meta-analysis found that by taking into consideration three randomized, double-blind, placebo-controlled pre-post between-group effect size for a total of 97 individuals treated, clinically but not statistically significant within-group pre-post and pre-follow-up improvements on anxiety [ES = 0.82, 95% CI (0.40, 1.23), $I^2 = 0\%$] and depression were reported [ES = 0.83, 95% CI (0.39, 1.26), $I^2 = 0\%$] (Goldberg et al., 2020). Importantly, no serious or persistent adverse events were reported and transient adverse reactions were similar across studies and resolved within hours of a psilocybin administration (Goldberg et al., 2020).

The second meta-analysis considered three clinical studies for a total of 92 patients treated (Vargas et al., 2020). This study found that the psilocybin group was significantly favored compared to the control group when considering BDI depression scores [WMD −4.589, 95% CI = −4.207, −0.971, $I^2 = 0\%$, $p = 0.002$] and STAI-traits anxiety scores [WMD −5.906, 95% CI = −7.852, −3.960, $I^2 = 0$, $p = 0.001$]. Interestingly, the reduction is not dose-dependent (Vargas et al., 2020). The improvements reported remained significant for a period of 38 to 189 days in BDI and in 14 to 189 days in STAI-Trait (Vargas et al., 2020).

Recently one RCT found that psilocybin at a single dose of 25 mg, but not 10 mg or 1 mg, reduced depression scores significantly more than a 1 mg dose over a period of 3 weeks; the difference between the 25 mg group and 1 mg group was -6.6 (95% confidence interval [CI], -10.2 to -2.9; P<0.001) (Goodwin et al., 2022).

Similarly, another RCT found that two separate doses of 25 mg of psilocybin 3 weeks apart followed by 6 weeks of daily placebo (psilocybin group) improved depression similarly to 6 weeks of daily dose of the SSRI escitalopram (20 mg). The mean scores on the QIDS-SR-16 at baseline were 14.5 in the psilocybin group and 16.4 in the escitalopram group. The mean (±SE) changes in the scores from baseline to week 6 were -8.0 ± 1.0 points in the psilocybin group and -6.0 ± 1.0 in the escitalopram group, for a between-group difference of 2.0 points (95% confidence interval [CI], -5.0 to 0.9) (P=0.17) (Carhart-Harris et al., 2021).

Importantly psilocybin (25 mg, two administrations) was reported to decrease the percentage of heavy drinking days during 32 weeks of follow-up. The percentage of heavy drinking days during the 32-week double-blind period was 9.7% for the psilocybin group and 23.6% for the diphenhydramine group, a mean difference of 13.9%; (95% CI, 3.0−24.7; $F_{1,86} = 6.43$; $P = .01$), (Bogenschutz et al., 2022).

Psilocybin can induce mystical and peak experiences, which seem to mediate at least partially the therapeutic improvements observed. The fact that the patients treated can experience these states, highlights the fact that ad-hoc training should be provided to the individuals willing to deliver this type of therapy should it be approved for use in the population. Importantly, psilocybin has been reported to increase brain-measured emotional responses, suggesting that it might increase the connection with one's own emotions (Roseman et al., 2018). Interestingly, psilocybin use has been associated with increased nature relatedness and a decrease in authoritarian political views (Lyons & Carhart-Harris, 2018). Aside from MDD and as an end-of-life therapy, therapeutic potential has been suggested for the treatment of OCD (Leonard & Rapoport, 1987; Moreno et al., 2006). Ongoing clinical trials are investigating the safety and feasibility of psilocybin for alcohol and substance use disorder, anorexia nervosa, cognitive impairments, Alzheimer's disease, as well as cluster, migraine, and post-traumatic headache [for a list of clinical trials, doses and regimens employed, and available results, please consult the extensive literature review (Inserra et al., 2021a)]. Together with LSD, psilocybin is one of the most commonly used substances for "microdosing" (Rosenbaum et al., 2020), a practice which consists in taking up to one tenth of a full dose, which is becoming widespread among the population based on claims of therapeutic efficacy for a range of diseases and psychological enhancement. These claims, however, are not based on scientific evidence and should therefore be taken with caution until more studies are available on the therapeutic efficacy and potential side effects of such an approach (Kuypers et al., 2019; Preller, 2019).

The main limitation of the clinical trials so far performed is the small sample size often due to economic constraints, which can increase bias potential. However, these studies provide preliminary evidence that indeed psilocybin could prove useful for the treatment of MDD, as well as for the treatment of the anxiety, distress, and depression associated with a potentially life-threatening illness.

Many clinical trials are ongoing or planned to study the effects of psilocybin in 1) depression, 2) PTSD, 3) obsessive-compulsive disorder, 4) autism spectrum disorder, 5) bipolar disorder, 6) alcohol and substance use disorder, 7) psychological and existential distress in palliative care, 8) anorexia nervosa and binge-eating disorder 9) chronic pain and phantom limb pain, 10) migraines and concussion headache, 11) co-occuring major depressive disorder and borderline personality disorder, 12) depression and anxiety in Parkinson's disease, 13) depression and burnout suffered by frontline work in the COVID pandemic, 14) mild cognitive impairment or early Alzheimer's disease and 15) to taper off opioid medications.

Lysergic Acid Diethylamide (LSD)

Mechanism of Action

Lysergic acid diethylamide (LSD) is a semisynthetic ergosterol that can be derived from the naturally occurring ergot alkaloid lysergic acid, which is contained in the rye parasite Claviceps purpurea. LSD is a partial agonist at the 5-HT_{2A}, 5-HT_{1A}, 5-HT_{2B}, 5-HT_{2C}, 5-HT_{1B}, 5-HT_{1D}, 5-HT_{1E}, 5-HT_6, and 5-HT_7 receptors. Evidence for an interaction at serotonin transporter (SERT) remains controversial given that while *in vitro* studies reported no SERT interaction (Rickli et al., 2016), *in vivo* studies report decreased effects of LSD in genetically modified *Sert* knockout mice (Krall et al., 2008; Kyzar et al., 2016).

Interestingly, and differently from the effects of other ligands at serotonin receptors, the 5-HT_{2A} and 5-HT_{2B} receptors "trap" LSD in their binding pocket, thus contributing to the slow dissociation rate of LSD and its long psychoactive effects, which can last for up to 12–16 h (Wacker et al., 2017). The resulting conformational changes of these receptors lead to strong functional selectivity for β-arrestin signaling over Gq signaling.

Similarly to the mechanism of action of the widely used SSRIs, LSD desensitizes the postsynaptic 5-HT_{1A} receptor, acutely decreasing 5-HT/DRN neurons firing and burst activity, while repeated administration increases serotonergic firing and elicits anxiolytic effects via a mechanism mediated by 5-HT_{1A} (De Gregorio et al., 2022), D2, and TAAR receptors (Artigas et al., 1996; Blier et al., 1993; De Gregorio et al., 2016). LSD also interacts with the glutamatergic system, an effect which results in the release of glutamate in deep cortical layers via a mechanism involving the AMPA and NMDA receptor, as well as the metabotropic glutamate 2 (mGlu2) receptor. Concerning the latter, it appears that a 5-HT$_{2A}$-mGlu2 heteroreceptor complex is required for LSD to induce psychotic-like symptoms via inducing psychedelic-unique patterns of transcriptional activity (Moreno et al., 2013). Corroborating the importance of the glutamatergic system in mediating the effects of LSD, it was recently reported that LSD elicits prosocial effects while potentiating the excitatory neurotransmission in the mPFC via 5-HT$_{2A}$ and AMPA receptors, as measured by an increased responsiveness of neurons in vivo to the microiontophoretic ejection of quisqualate and DOI following repeated LSD administration (De Gregorio et al., 2021b). Another important biological mechanism recently uncovered which might mediate the therapeutic effects of LSD is its epigenetic effects on neurotrophic-related DNA methylation and protein expression in the PFC (Inserra et al., 2022). Thus, via modulating the responsivity of the mPFC to social stimuli as also observed in humans (Duerler et al., 2020), LSD could potentially prove useful in the treatment psychiatric and neurodevelopmental disorders including of Autism Spectrum Disorder (Markopoulos et al., 2022). This possibility is supported by the fact that LSD also affects the hippocampal endocannabinoid system, as well gut microbiome composition (Inserra, Giorgini et al., 2023), both of which are atypical in autism.

While at low doses, LSD does not directly impact the dopaminergic system, at higher doses, it decreases DA firing in the VTA via a multi-receptorial mechanism involving D2, 5-HT$_{1A}$, and TAAR1 receptors. Confirming a dopaminergic involvement in the mechanism of action of LSD, in vitro studies reported that LSD binds the human and murine D1, D2, and D4 receptors.

Aside from interactions with the serotonergic, dopaminergic, and glutamatergic system, LSD and its analogue DOI also modulates prefrontal GABAergic neurotransmission via a 5-HT$_{2A}$-mediated excitation of interneurons and increased GABA release, an effect thought to be involved in the inhibition of specific cortical pyramidal networks (Marek & Aghajanian, 1994), which might take part in the modulation of default mode network activity by LSD and other psychedelics, inducing the phenomenon of "ego death" and "boundlessness" (Tagliazucchi et al., 2016). Further supporting an involvement of the GABAergic system in the effects of LSD, another crucial interaction was recently reported between LSD and inhibitory networks which takes place in the reticular thalamus, a thin sheet of GABAergic neurons projecting exclusively to the thalamus, which by means of finely spatiotemporally tuned inhibition of thalamocortical relay neurons, gates information flow within the cortico-striato-thalamo-cortical circuit (Pinault, 2004). In fact, LSD modulates RT activity in a dual dose-response fashion, inhibiting one subpopulation of RT neurons while exciting another (Inserra et al., 2021b). These effects might disrupt thalamic gating and "open the doors" of consciousness via allowing unconstrained information flow within the cortico-thalamocortical circuit (Müller et al., 2017; Preller et al., 2019).

Concerning noradrenergic neurotransmission, LSD is a 2-adrenoceptor agonist, although in humans it does not alter the urinary NE excretion. Preclinical studies support the possibility that the NE system might be important for the LSD-induced effects. In fact, NE-depleted rats have an attenuated behavioral phenotype in response to LSD. LSD was previously reported to acutely modulate the stress-related hormones cortisol, cortisone, corticosterone, prolactin, oxytocin, and epinephrine in humans. Interestingly, LSD also elicits antiinflammatory effects in vitro via downregulating IL2, IL4, and IL6 (House et al., 1994), suggesting that its fast onset anxiolytic and antidepressant outcomes might be at least partially mediated by its antiinflammatory effects, given the depressogenic effects of pro-inflammatory signaling in the brain (Dantzer et al., 2008; Wong et al., 2016).

Circulating brain-derived neurotrophic factor (BDNF) levels have been shown to increase following the administration of microdoses (5–20 µg) in healthy volunteers. Seemingly corroborating preclinical evidence suggests that LSD and its analog DOI modulates neurotrophic-related gene expression, inducing neurotrophic outcomes. LSD also modulates the sigma-1 receptor (S1R) agonist dehydroepiandrosterone (DHEA), the most abundant neurosteroid in the central nervous system which affects synaptic activity and neurogenesis while ameliorating drug-induced cognitive impairments.

Prosocial effects of LSD were recently reported in mice when administered repeatedly, which are accompanied by a potentiation of the 5-HT$_{2A}$- and AMPA-mediated excitatory neurotransmission in the prefrontal cortex (De Gregorio et al., 2021b).

Clinical Trials

A systematic review of RCTs investigating the use of LSD in psychiatry considered 11 randomized placebo-controlled clinical trials for a total of 567 patients that received LSD ranging from 20 to 800 µg (respectively, a microdose and a very high dose) and found that the strongest potential for LSD in psychiatry is in the treatment of alcoholism, followed by anxiety, depression and psychosomatic diseases, substance abuse, and anxiety and distress associated with a terminal illness (Fuentes et al., 2020). Out of these studies, two serious adverse events were reported: the first was the onset of tonic-clonic seizure (LEO E. Hollister et al., 1969), and the other the onset of a long-lasting psychotic episode in a 21 years old with a previous history of psychosis (Savage & McCabe, 1973). No other serious adverse events were reported for the remainder 565 individuals taken into consideration in this systematic review (Fuentes et al., 2020). Importantly, in individuals treated for anxiety, depression, and distress associated with a terminal illness that survive the illness, improvements are still appreciable for 12 months after a single administration of the drug (Gasser et al., 2014).

A meta-analysis of six studies investigating the potential of LSD as a therapy for alcohol addiction (total sample size $n = 536$) reported low heterogeneity across studies, and reduction in alcohol misuse in the short- (2–3 months) and medium- (6 months) term, but not long- (12 months) term (OR 1.96, 95% CI =1.36, 2.84, I^2 = 0%, $p = 0.0003$), after a single administration of LSD

(Krebs & Johansen, 2012). Other potential applications of LSD in psychiatry revolve around the treatment of neurodevelopmental disorders such as autism spectrum disorder (Sigafoos et al., 2007). Indeed, early clinical trials from the 1950–1970s reported clinical improvements in autistic children treated with LSD, although side effects of various magnitude, such as aggression and self-harm were also reported (Sigafoos et al., 2007; Markopoulos et al., 2020).Clinical trials are ongoing or planned to study 1) the effects of LSD on neuroplasticity 2) the effects of SERT inhibition on the subjective effects of LSD, 3) the effects of LSD on cluster headache, 4) the effects of LSD for alcohol use disorder and 5) the prosocial effects of LSD (see clinicaltrials.gov).

3,4-Methylenedioxy methamphetamine (MDMA)

Mechanism of Action

3,4-Methylenedioxymethamphetamine (MDMA) is absorbed in the intestinal tract, after which it is metabolized by several pathways and reaches plasmatic concentrations peak after about 2 h. MDMA is a 5-HT$_{2A}$ receptor agonist, an interaction thought to be responsible for the MDMA-induced mesolimbic DA release. Acute MDMA administration induces a transient, MDA-mediated, dose-related increase in extracellular 5-HT in the mPFC, striatum, NAc, and hippocampus via inhibiting SERT and thus 5-HT reuptake, and inhibiting tryptophan hydroxylase, the rate-limiting enzyme of 5-HT biosynthesis. MDMA also inhibits monoamine oxidase A and monoamine oxidase B, further increasing monoamine availability at the synapse. MDMA is a weak 5-HT$_{1A}$ receptor agonist, and postsynaptic upregulation of this receptor is observable in the cortex and hypothalamus 1 week after acute administration. Remarkably, the effects of MDMA on SERT within the NAc are sufficient to elicit the prosocial effects of MDMA (Heifets et al., 2019).

MDMA is also a dopaminergic agent via weak D1 receptor affinity, responsible for the MDMA-induced hyperlocomotion in rodents, while D2 receptors are involved in the dopaminergic toxicity induced by high doses of MDMA. Accordingly, MDMA enhances DA release in the striatum, reverses the direction of the DA transporter, and inhibits nigrostriatal and mesolimbocortical DA reuptake, thus increasing synaptic DA. MDMA-induced DA release is mediated at least partially by GABAergic interactions. For example, MDMA enhances GABA efflux in the VTA, which dampens the MDMA-mediated DA release in the NAc shell and decreases GABA efflux in the substantia nigra. Repeated MDMA administration decreases the number of PV+ GABA neurons in the dentate gyrus, an effect attenuated by pretreatment with the anti-inflammatory ketoprofen, suggesting that cyclooxygenase-mediated pathways are responsible for this GABAergic decline following repeated, high doses of MDMA (Anneken et al., 2013). MDMA also increases glutamate levels in the anteromedial striatum and dorsal hippocampus. Importantly, MDMA is an α2-adrenergic receptor agonist and β3-adrenergic receptor agonist, and these effects might mediate hyperthermia-induced rhabdomyolysis. MDMA also increases NE release and inhibits its reuptake.

MDMA influences both the innate and adaptive arms of the immune system. Acute MDMA in humans increases cortisol, prolactin, and the inflammatory mediators hydroxyeicosatetraenoic acid, dihydroxyeicosatetraenoic acid, and octadecadienoic acid, while decreasing circulating CD4$^+$ helper T cells and increasing natural killer cells, potentially as a result of increased HPA axis activation. Increases in the production of the immunosuppressive Th2 cytokines IL4 and IL10, and decreases in the production of proinflammatory Th1 cytokines, such as IL2 and IFN-γ, have been reported. Therefore, MDMA might decrease the immune reactivity and consequent damaging potential of immune cells in PTSD patients, an effect which could be involved in the reduction of clinical symptoms.

Recently, it has been reported that MDMA leads to an OT-dependent reopening of long-term depression in the NAc, a critical mechanism for social reward learning (Nardou et al., 2019). It is thus possible that the creation of a temporary neuroplastic window in which trauma can be reprogrammed, and neural circuits reorganized accordingly, underlies the robust improvements in PTSD symptoms observed after MDMA-augmented psychotherapy.

MDMA appears to have a higher risk of abuse and long-term adverse neurobiological outcomes compared to serotonergic psychedelics. Binge abuse can lead to hyperthermia, cardiac arrhythmias, seizures, intracranial hemorrhage, hepatotoxicity, renal failure, and rhabdomyolysis. MDMA abuse induces neurotoxic effects on 5-HT and DA neurons increasing the likelihood of developing psychiatric disorders. Indeed, SERT homeostasis is disrupted by heavy MDMA use (Baumann et al., 2007; Müller et al., 2019), while decrease striatal DAT is observed in recreational MDMA users who also use amphetamines. Similarly, repeated MDA administration causes long-lasting depletion of cortical, hippocampal, and striatal 5-HT. Importantly, MDMA during adolescence leads to glutamatergic neuroadaptive changes in corticolimbic structures (Kindlundh-Högberg et al., 2008), suggesting that the developing brain is particularly vulnerable to the neurotoxic effects of MDMA.

Clinical Trials

The main clinical interest concerning the application of MDMA in psychotherapy augmentation revolves around its empathogenic effects and the induction of a psychological state in which trauma can be re-accessed and processed, becoming an ideal candidate for treatment-resistant PTSD. A recent meta-analysis and systematic review reported that MDMA is a safe, efficacious, and durable treatment for individuals with chronic, treatment-refractory PTSD which had lasted 7–22 years (Bahji et al., 2020). In controlled clinical settings, MDMA induces a relaxed and euphoric state, feelings of openness, empathy, and disinhibition, creating a cathartic moment in which trauma can be integrated. In the five trials included in the meta-analysis and systematic review, totaling 106 participants with high incidence of anxiety and depression comorbidity, a high rate of clinical response (OR 3.47, 95% CI = 1.70, 7.06, I^2 = 0%), remission (OR 2.63, 95% CI = 1.37, 5.02, I^2 = 0%), with a large effect size at reducing the symptoms of PTSD (SMD 1.30, 95% CI = 0.66, 1.94, I^2 = 34%) was observed in individuals receiving MDMA (2–3 administrations, 50–125 mg per session) (Bahji et al., 2020). Available evidence indicates that MDMA was well-tolerated, with few serious adverse events reported across studies (Bahji et al., 2020) and long-lasting therapeutic improvements for up to 1 year after administration (Mithoefer et al., 2013). In one study, four serious adverse events were reported, which included increased depressive symptoms and suicidal ideation; however, three out of four instances were deemed unrelated to the study drug (Bahji et al., 2020).

These findings have recently been confirmed by a multicenter Phase II clinical study which found significant reductions in scales for PTSD. Specifically, MDMA was found to induce significant and robust attenuation in Clinician-Administered PTSD Scale score compared with placebo ($P < 0.0001$, $d = 0.91$) and to significantly decrease the Sheehan Disability Scale total score ($P = 0.0116$, $d = 0.43$). The mean change in CAPS-5 scores in participants completing treatment was -24.4 (s.d. 11.6) in the MDMA group and -13.9 (s.d. 11.5) in the placebo group (Mitchell et al., 2021).

Aside from the employment of MDMA-augmented psychotherapy for PTSD survivors, other trials are investigating safety and efficacy in the treatment of 1) autism spectrum disorder (Danforth et al., 2018), 2) alcohol abuse disorder and brain circuits mediating improvement 3) OCD, 4) social anxiety and social motivation, 5) stress disorders in healthcare workers, 6) PTSD and opioid use disorder after childbirth, 7) eating disorders, 8) adjustment disorder in patients with cancer and their significant other and 9) comorbid PTSD and alcohol use disorder. Importantly, a crucial aspect of MDMA-assisted psychotherapy is the prosocial effects and increased therapeutic alliance engendered between the therapists and the patients, the latter of which report experiencing increased vulnerability, empathy, connection, and sociability, emotional states that can facilitate therapeutic improvement (Bahji et al., 2020).

The FDA granted Breakthrough Therapy status for MDMA treatment of PTSD after reviewing pooled analysis from six Phase 2 clinical trials (Mithoefer et al., 2019).

Ayahuasca

Mechanism of Action

Ayahuasca is a psychoactive brew used for religious, spiritual, and healing purposes by cultures indigenous of the Amazon basin. The decoction is obtained by boiling *Psychotria viridis* which contain the endogenous-occurring *N,N*-Dimethyl-tryptamine (DMT) and *Banisteriopsis caapi* (Ayahuasca) which contains the β-carboline alkaloids harmine, harmaline, and tetrahydroharmine which function as monoamine oxidase inhibitors (MAOIs) to block DMT degradation, rendering it orally active. A close relative of DMT, 5-MeO-DMT, is also used for therapeutic and ceremonial purposes, and found in several plants (such as *Anadenanthera peregrine*) and certain toads species (such as *Incilius alvarius*).

DMT is a $5\text{-}HT_{2A/2C}$ receptor agonist, an effect which stimulates 5-HT release and inhibits its reuptake via interacting with SERT, suggesting that it might be actively taken up and stored by cells (Blough et al., 2014; Cozzi et al., 2009; Rickli et al., 2016). DMT is also an inhibitor of the vesicular monoamine transporter 2 (VMAT2, which is also involved in the reuptake of monoamines) where it acts as a substrate, further supporting the possibility that DMT might be uptaken by neurons. DMT also interacts with the $5\text{-}HT_{1A}$, $5\text{-}HT_{1D}$, $5\text{-}HT_{1E}$, $5\text{-}HT_{2B}$, $5\text{-}HT_{5A}$, $5\text{-}HT_6$, and $5\text{-}HT_7$ receptors. Pharmacological antagonism of the $5\text{-}HT_{1A}$ receptor potentiates the hallucinogenic effects of DMT, suggesting that $5\text{-}HT_{1A}$ receptor blockade can enhance $5\text{-}HT_{2A}$ receptor–mediated hallucinogenic effects. Like the other serotonergic psychedelics, DMT inhibits 5-HT firing in the DRN. Harmine and harmaline act as reversible monoamine oxidase A inhibitors, whereas tetrahydroharmine binds 5-HT$_{2A}$ and 5-HT$_{2C}$ receptors but not the presynaptic 5-HT$_{1A}$ receptor as previously hypothesized. Preclinical studies suggest potent and region-specific modulation of neurotransmitter release and turnover, such as an increase in 5-HT release and turnover in the hippocampus and amygdala. A similar study found increased whole brain 5-HT levels but no significant changes in 5-HT turnover (Colaço et al., 2020).

Aside from modulating the serotonergic system, Ayahuasca acutely increases amygdalar DA levels and decreases its turnover (de Castro-Neto et al., 2013). Earlier studies reported that DMT modulates striatal, forebrain, and whole-brain DA synthesis, although lacking direct dopaminergic activity. A recent study reported that repeated Ayahuasca administration increases the concentration of the DA metabolites 3,4-dihydroxyphenylacetic acid in whole-brain rodent homogenates corroborating earlier reports of increased DA metabolites after 1 month of treatment. These results suggest that the repeated use might accelerate DA turnover potentially as a mechanism to compensate the increased DA synthesis.

Concerning the noradrenergic system, DMT potentiates the effects of 5-HT and NE in the facial nucleus in rats, while Ayahuasca acutely increases NE levels in the amygdala but not the hippocampus of rats. The level of 4-hydroxy-3-methoxy mandelic acid, a product of NE metabolism, was decreased in the rat amygdala and hippocampus. On the contrary, repeated Ayahuasca administration did not alter the NE metabolite 3-methoxy-4-hydrohyphenylglycol. In humans, Ayahuasca increases the urinary excretion of normetanephrine. Depletion of NE blocks the analgesic effects of 5-MeO-DMT, thus suggesting that the analgesic effects elicited by 5-MeO-DMT might take effect through the NEergic system.

Preclinical findings suggest that Ayahuasca increases hippocampal BDNF levels and decreases anxiety in the offspring, while 5-MeO-DMT increases neurogenesis in the dentate gyrus. These findings have particular translational relevance given that non-medicated depressed patients have decreased BDNF levels and decreased hippocampal neurogenesis, and conventional antidepressants for depression elicit symptoms improvement at least partially through increasing neurogenesis via upregulating BDNF signaling. Sigma-1 receptor (S1R) modulation might represent a synergistic therapeutic mechanism to 5-HT$_{2A}$ receptor modulation by Ayahuasca and DMT (Inserra, 2018), given that DMT is a S1R agonist which mediates the upregulation of neurotrophic factors. In fact, S1Rs stimulation induces adaptive neuroplasticity. Accordingly, the β-carbolines harmol, harmine, harmaline, and tetrahydroharmine (found in *B. caapi*), which inhibit the metabolism of DMT to render it orally active induce differentiation, proliferation, and migration of neural precursors (Morales-García et al., 2017). This strengthens the notion that Ayahuasca alkaloids might be useful for neurodegenerative disorders.

In vitro evidence suggests that 5-MeO-DMT might be beneficial for drug and alcohol addiction. For example, 5-MeO-DMT downregulates metabotropic glutamate receptor (mGluR) 5 in human cerebral organoids, and this gene is involved in alcohol- and drug-induced rewards. Clinical studies corroborate this notion, with self-reported decreases in drug and alcohol consumption following the administration of 5-MeO-DMT in naturalistic settings, which are accompanied by antinflammatory effects (Uthaug et al., 2020). Accordingly, S1R activation by DMT and 5-MeO-DMT induces strong anti-inflammatory effects in immune-challenged, human monocyte–derived dendritic cells, decreasing IL1b, IL6, IL8, and TNF-a, and increasing the antinflammatory cytokine IL10 (Szabo et al., 2014, 2016).

Clinical Trials

A recent international cross-sectional study using data from the "Global Ayahuasca Project" considering almost 12,000 individuals, reported that individuals that self-reported depression and anxiety prior to drinking Ayahuasca, reported that their symptoms were either "very much improved" (respectively 46% and 54%) or "completely resolved" (respectively 32% and 16%), while 2.7% and 4.5% reported that their symptoms had worsened after drinking Ayahuasca (Sarris et al., 2021). In controlled clinical settings, Ayahuasca rapidly relieves depression symptoms in treatment-refractory depression in controlled clinical settings, potentially via the Ayahuasca-induced increasing circulating BDNF levels. Another study reported a statistical trend toward decreased suicidality in patients from these trials. Four weekly sessions of Ayahuasca have been described to increase "acceptance capacities" scores in the non-judging subscale of the Five Facet Mindfulness Questionnaire. A 12-day Ayahuasca retreat in traditional settings was shown to significantly reduce neuroticism scores for up to 6 months in an inversely proportional fashion to the intensity and quality of the mystical experiences associated.

Given that certain religious groups in South America (such as Santo Daime, Barquinha, and União do Vegetal) and more recently worldwide use Ayahuasca as a sacrament weekly or bi-weekly (Labate & Feeney, 2012), these groups represent an ideal, if unique, cohort to study the long-term effects of repeated administration of a serotonergic psychedelic. Generally, regular Ayahuasca users score lower in psychometric tests for depression and confusion and higher for agreeableness, openness, and life quality, changes accompanied by decreased alcohol use and abuse (Barbosa et al., 2018; Garcia-Romeu et al., 2019). Physiological changes arising from the repeated use of Ayahuasca in humans include platelet SERT upregulation, potentially as a result of repeated MAOI ingestion and/or repeated exposure to high levels of DMT. Another study observed pronounced neuromorphologic changes in long-term Ayahuasca users such as increased anterior cingulate cortex (ACC) and decreased posterior cingulate cortex (PCC) thickness which were not associate with any negative outcome in psycho-attitudinary tests and which were in fact associated with lower drug abuse rates and increased spirituality compared to individuals who do not ingest Ayahuasca. Another study found that long-term Ayahuasca users performed better in neuropsychological tasks, suggesting a state of cognitive enhancement (Bouso et al., 2012). Interestingly, in the Amazon basin, adolescents are allowed and encouraged to partake in these ceremonies, generating another naturalistic cohort which could inform potential negative sequalae of Ayahuasca ingestion during teenage. These individuals scored lower for psychiatric symptoms such as anxiety, body dysmorphism, and attentional problems compared with controls, suggesting no psychiatric sequelae from regular consumption of Ayahuasca during adolescence (Da Silveira et al., 2005). Ongoing or planned clinical trials are investigating the safety, tolerability, and pharmacokinetics of single or repeated use of DMT or 5-MeO-DMT in healthy individuals as well as the combination of DMT and mindfulness-based therapies.

Ongoing or planned clinical trials are investigating the safety, tolerability, and pharmacokinetics of single or repeated use of DMT or 5-MeO-DMT in healthy individuals as well as the combination of DMT and mindfulness-based therapies.

The use of DMT and 5-MeO-DMT in naturalistic settings is receiving considerable attention given the rapid improvements elicited in several mental health domains accompanied by sustained decreases in drug and alcohol use and abuse and an increase in quality of life. Inhalation of 5-MeO-DMT has been reported to occasion mystical experiences associated with almost immediate improvements in depression and anxiety scores with enduring positive effects and decreased drug use. A potentially severe side effect that has been reported following the consumption of ayahuasca and related compounds is the switch to mania in patients with bipolar disorder.

Case Reports- BD Switch to Mania

One such case is of a patient with BD with a current depressive episode who switched to a manic episode after a 4-day Ayahuasca ritual. The patient had already experienced hypomanic episodes, one of which 10 days prior to the ritual. The man was admitted to hospital and treated with the medication for psychosis (antipsychotic) risperidone and the benzodiazepine clonazepam for 1 month, after which he became asymptomatic and was discharged (Szmulewicz et al., 2015).

Another case report is of a male BD type 1 treatment-refractory psychiatrist who had attempted to self-medicate with smoked DMT (up to 1 g daily), phenelzine, and clonazepam and who was admitted to hospital after developing a hypomanic psychotic episode with highly disturbed and agitated behavior 2 to 3 days after interrupting his self-medication schedule (Brown et al., 2017). These reports highlight the pressing need to assess the likelihood that patients with BD might switch to mania if receiving psychedelic therapy and to identify the most suited antimanic pharmacological approaches to treat such patients should the switch take place.

Conclusion

In conclusion, both psychedelics and cannabinoids represent promising novel avenues for the treatment of mental diseases, which are the largest causes of morbidity worldwide. Although they are not approved by governmental agencies and present potential for abuse and side effects, including psychosis or mania, these drugs can represent a model for developing novel therapies in psychopharmacology by exploiting their mechanism of action, receptorial affinity, and neuronal circuitries. Larger clinical studies are needed to confirm their efficacy and safety, and more drug discovery efforts are needed to develop new, safer drugs mimicking their pharmacology. However, these drugs remain of paramount interest in the world of mental health therapeutics since unlike other fields of medicine still few targets are available.

Conflict of Interest D.D.G. is a consultant at Diamond Therapeutics Inc., Toronto, ON, Canada. G.G. and D.D.G. are inventors of a provisional patent regarding the use of LSD.

References

Abel, E. L. (1975). Cannabis: Effects on hunger and thirst. *Behavioral Biology, 15*, 255–281.

Abrams, D. I., Jay, C., Shade, S., Vizoso, H., Reda, H., Press, S., Kelly, M., Rowbotham, M., & Petersen, K. (2007). Cannabis in painful HIV-associated sensory neuropathy: A randomized placebo-controlled trial. *Neurology, 68*, 515–521.

Allsop, D. J., Copeland, J., Lintzeris, N., Dunlop, A. J., Montebello, M., Sadler, C., Rivas, G. R., Holland, R. M., Muhleisen, P., Norberg, M. M., Booth, J., & McGregor, I. S. (2014). Nabiximols as an agonist replacement therapy during cannabis withdrawal: A randomized clinical trial. *JAMA Psychiatry, 71*, 281–291.

Andreae, M. H., Carter, G. M., Sharapin, N., Suslov, K., Ellis, R. J., Ware, M. A., Abrams, D. I., Prasad, H., Wilsey, B., Indyk, D., Johnson, M., & Sacks, H. S. (2015). Inhaled cannabis for chronic neuropathic pain: A meta-analysis of individual patient data. *The Journal of Pain, 16*, 1221–1232.

Andries, A., Frystyk, J., Flyvbjerg, A., & Støving, R. K. (2014). Dronabinol in severe, enduring anorexia nervosa: A randomized controlled trial. *The International Journal of Eating Disorders, 47*, 18–23.

Anneken, J. H., Cunningham, J. I., Collins, S. A., Yamamoto, B. K., & Gudelsky, G. A. (2013). MDMA increases glutamate release and reduces parvalbumin-positive GABAergic cells in the dorsal hippocampus of the rat: Role of cyclooxygenase. *Journal of Neuroimmune Pharmacology, 8*, 58–65.

Artigas, F., Romero, L., de Montigny, C., & Blier, P. (1996). Acceleration of the effect of selected antidepressant drugs in major depression by 5-HT1A antagonists. *Trends in Neurosciences, 19*, 378–383.

Bahji, A., Forsyth, A., Groll, D., & Hawken, E. R. (2020). Efficacy of 3,4-methylenedioxymethamphetamine (MDMA)-assisted psychotherapy for posttraumatic stress disorder: A systematic review and meta-analysis. *Progress in Neuro-Psychopharmacology and Biological Psychiatry, 96*, 109735.

Baker, D., Pryce, G., Davies, W. L., & Hiley, C. R. (2006). In silico patent searching reveals a new cannabinoid receptor. *Trends in Pharmacological Sciences, 27*, 1–4.

Bambico, F. R., Katz, N., Debonnel, G., & Gobbi, G. (2007). Cannabinoids elicit antidepressant-like behavior and activate serotonergic neurons through the medial prefrontal cortex. *The Journal of Neuroscience, 27*, 11700–11711.

Bambico, F. R., Nguyen, N. T., Katz, N., & Gobbi, G. (2010). Chronic exposure to cannabinoids during adolescence but not during adulthood impairs emotional behaviour and monoaminergic neurotransmission. *Neurobiology of Disease, 37*, 641–655.

Bambico, F. R., Hattan, P. R., Garant, J.-P., & Gobbi, G. (2012). Effect of delta-9-tetrahydrocannabinol on behavioral despair and on pre-and postsynaptic serotonergic transmission. *Progress in Neuro-Psychopharmacology and Biological Psychiatry, 38*, 88–96.

Bandelow, B., Zohar, J., Hollander, E., Kasper, S., Möller, H.-J., & Wfsbp Task Force On Treatment Guidelines For Anxiety Obsessive-Compulsive Post-Traumatic Stress D. (2008). World Federation of Societies of Biological Psychiatry (WFSBP) guidelines for the pharmacological treatment of anxiety, obsessive-compulsive and post-traumatic stress disorders – First revision. *The World Journal of Biological Psychiatry, 9*, 248–312.

Barbosa, P. C. R., Tófoli, L. F., Bogenschutz, M. P., Hoy, R., Berro, L. F., Marinho, E. A. V., Areco, K. N., & Winkelman, M. J. (2018). Assessment of alcohol and tobacco use disorders among religious users of Ayahuasca. *Frontiers in Psychiatry, 9*, 136.

Barrett, F. S., Krimmel, S. R., Griffiths, R. R., Seminowicz, D. A., & Mathur, B. N. (2020). Psilocybin acutely alters the functional connectivity of the claustrum with brain networks that support perception, memory, and attention. *NeuroImage, 218*, 116980.

Baumann, M. H., Wang, X., & Rothman, R. B. (2007). 3,4-Methylenedioxymethamphetamine (MDMA) neurotoxicity in rats: A reappraisal of past and present findings. *Psychopharmacology, 189*, 407–424.

Bergamaschi, M. M., Queiroz, R. H., Chagas, M. H., de Oliveira, D. C., De Martinis, B. S., Kapczinski, F., Quevedo, J., Roesler, R., Schröder, N., Nardi, A. E., Martín-Santos, R., Hallak, J. E., Zuardi, A. W., & Crippa, J. A. (2011). Cannabidiol reduces the anxiety induced by simulated public speaking in treatment-naïve social phobia patients. *Neuropsychopharmacology, 36*, 1219–1226.

Berger, M., Li, E., Rice, S., Davey, C.G., Ratheesh, A., Adams, S., Jackson, H., Hetrick, S., Parker, A., Spelman, T. and Kevin, R. (2022). Cannabidiol for treatment-resistant anxiety disorders in young people: an open-label trial. *The Journal of Clinical Psychiatry, 83*(5), p.42111.

Bisogno, T., Hanuš, L., De Petrocellis, L., Tchilibon, S., Ponde, D. E., Brandi, I., Moriello, A. S., Davis, J. B., Mechoulam, R., & Di Marzo, V. (2001). Molecular targets for cannabidiol and its synthetic analogues: Effect on vanilloid VR1 receptors and on the cellular uptake and enzymatic hydrolysis of anandamide. *British Journal of Pharmacology, 134*, 845–852.

Black, N., Stockings, E., Campbell, G., Tran, L. T., Zagic, D., Hall, W. D., Farrell, M., & Degenhardt, L. (2019). Cannabinoids for the treatment of mental disorders and symptoms of mental disorders: A systematic review and meta-analysis. *Lancet Psychiatry, 6*, 995–1010.

Blier, P., Lista, A., & De Montigny, C. (1993). Differential properties of pre- and postsynaptic 5-hydroxytryptamine1A receptors in the dorsal raphe and hippocampus: I. Effect of spiperone. *The Journal of Pharmacology and Experimental Therapeutics, 265*, 7–15.

Blough, B. E., Landavazo, A., Decker, A. M., Partilla, J. S., Baumann, M. H., & Rothman, R. B. (2014). Interaction of psychoactive tryptamines with biogenic amine transporters and serotonin receptor subtypes. *Psychopharmacology, 231*, 4135–4144.

Bogenschutz, M. P., Ross, S., Bhatt, S., et al. (2022). Percentage of Heavy Drinking Days Following Psilocybin-Assisted Psychotherapy vs Placebo in the Treatment of Adult Patients With Alcohol Use Disorder: A Randomized Clinical Trial. *JAMA Psychiatry 79*(10): 953–962. https://doi.org/10.1001/jamapsychiatry.2022.2096

Boggs, D. L., Surti, T., Gupta, A., Gupta, S., Niciu, M., Pittman, B., Schnakenberg Martin, A. M., Thurnauer, H., Davies, A., D'Souza, D. C., & Ranganathan, M. (2018). The effects of cannabidiol (CBD) on cognition and symptoms in outpatients with chronic schizophrenia a randomized placebo controlled trial. *Psychopharmacology, 235*, 1923–1932.

Bonaccorso, S., Ricciardi, A., Zangani, C., Chiappini, S., & Schifano, F. (2019). Cannabidiol (CBD) use in psychiatric disorders: A systematic review. *Neurotoxicology, 74*, 282–298.

Bouso, J. C., González, D., Fondevila, S., Cutchet, M., Fernández, X., Ribeiro Barbosa, P. C., Alcázar-Córcoles, M. Á., Araújo, W. S., Barbanoj, M. J., Fábregas, J. M., & Riba, J. (2012). Personality, psychopathology, life attitudes and neuropsychological performance among ritual users of Ayahuasca: A longitudinal study. *PLoS One, 7*, e42421.

Boyaji, S., Merkow, J., Elman, R. N. M., Kaye, A. D., Yong, R. J., & Urman, R. D. (2020). The role of Cannabidiol (CBD) in chronic pain management: An assessment of current evidence. *Current Pain and Headache Reports, 24*, 4.

Brown, T., Shao, W., Ayub, S., Chong, D., & Cornelius, C. (2017). A physician's attempt to self-medicate bipolar depression with *N,N*-dimethyltryptamine (DMT). *Journal of Psychoactive Drugs, 49*, 294–296.

Burch, R. C., Buse, D. C., Lipton, R. B. (2019) Migraine: Epidemiology, burden, and comorbidity. Neurol. Clin. 37, 631–649.

Burch, R., Rizzoli, P., Loder, E. (2021) The prevalence and impact of migraine and severe headache in the United States: Updated age, sex, and socioeconomic-specific estimates from government health surveys. *Headache* 61, 60–68.

Calignano, A., La Rana, G., Giuffrida, A., & Piomelli, D. (1998). Control of pain initiation by endogenous cannabinoids. *Nature, 394*, 277–281.

Carhart-Harris, R. L., Roseman, L., Bolstridge, M., Demetriou, L., Pannekoek, J. N., Wall, M. B., Tanner, M., Kaelen, M., McGonigle, J., Murphy, K., Leech, R., Curran, H. V., & Nutt, D. J. (2017). Psilocybin for treatment-resistant depression: fMRI-measured brain mechanisms. *Scientific Reports, 7*, 13187.

Carhart Harris et al., 2021: N Engl J Med 384:1402–1411. https://doi.org/10.1056/NEJMoa2032994

Carrier, E. J., Auchampach, J. A., & Hillard, C. J. (2006). Inhibition of an equilibrative nucleoside transporter by cannabidiol: A mechanism of cannabinoid immunosuppression. *Proceedings of the National Academy of Sciences, 103*, 7895–7900.

Catlow, B. J., Song, S., Paredes, D. A., Kirstein, C. L., & Sanchez-Ramos, J. (2013). Effects of psilocybin on hippocampal neurogenesis and extinction of trace fear conditioning. *Experimental Brain Research, 228*, 481–491.

Colaço, C. S., Alves, S. S., Nolli, L. M., Pinheiro, W. O., de Oliveira, D. G. R., Santos, B. W. L., Pic-Taylor, A., Mortari, M. R., & Caldas, E. D. (2020). Toxicity of ayahuasca after 28 days daily exposure and effects on monoamines and brain-derived neurotrophic factor (BDNF) in brain of Wistar rats. *Metabolic Brain Disease, 35,* 739–751.

Console-Bram, L., Marcu, J., & Abood, M. E. (2012). Cannabinoid receptors: Nomenclature and pharmacological principles. *Progress in Neuro-Psychopharmacology and Biological Psychiatry, 38,* 4–15.

Cozzi, N. V., Gopalakrishnan, A., Anderson, L. L., Feih, J. T., Shulgin, A. T., Daley, P. F., & Ruoho, A. E. (2009). Dimethyltryptamine and other hallucinogenic tryptamines exhibit substrate behavior at the serotonin uptake transporter and the vesicle monoamine transporter. *Journal of Neural Transmission (Vienna), 116,* 1591–1599.

Crippa, J. A., Derenusson, G. N., Ferrari, T. B., Wichert-Ana, L., Duran, F. L., Martin-Santos, R., Simões, M. V., Bhattacharyya, S., Fusar-Poli, P., Atakan, Z., Santos Filho, A., Freitas-Ferrari, M. C., McGuire, P. K., Zuardi, A. W., Busatto, G. F., & Hallak, J. E. (2011). Neural basis of anxiolytic effects of cannabidiol (CBD) in generalized social anxiety disorder: A preliminary report. *Journal of Psychopharmacology, 25,* 121–130.

D'Souza, D. C., Abi-Saab, W. M., Madonick, S., Forselius-Bielen, K., Doersch, A., Braley, G., Gueorguieva, R., Cooper, T. B., & Krystal, J. H. (2005). Delta-9-tetrahydrocannabinol effects in schizophrenia: Implications for cognition, psychosis, and addiction. *Biological Psychiatry, 57,* 594–608.

Da Silveira, D. X., Grob, C. S., de Rios, M. D., Lopez, E., Alonso, L. K., Tacla, C., & Doering-Silveira, E. (2005). Ayahuasca in adolescence: A preliminary psychiatric assessment. *Journal of Psychoactive Drugs, 37,* 129–133.

Danforth, A. L., Grob, C. S., Struble, C., Feduccia, A. A., Walker, N., Jerome, L., Yazar-Klosinski, B., & Emerson, A. (2018). Reduction in social anxiety after MDMA-assisted psychotherapy with autistic adults: A randomized, double-blind, placebo-controlled pilot study. *Psychopharmacology, 235,* 3137–3148.

Dantzer, R., O'Connor, J. C., Freund, G. G., Johnson, R. W., & Kelley, K. W. (2008). From inflammation to sickness and depression: When the immune system subjugates the brain. *Nature Reviews Neuroscience, 9,* 46–56.

Davis, A. K., Barrett, F. S., May, D. G., Cosimano, M. P., Sepeda, N. D., Johnson, M. W., Finan, P. H., & Griffiths, R. R. (2020). Effects of psilocybin-assisted therapy on major depressive disorder: A randomized clinical trial. *JAMA Psychiatry, 78,* 481–489.

Daynes, R. A., & Jones, D. C. (2002). Emerging roles of PPARs in inflammation and immunity. *Nature Reviews Immunology, 2,* 748–759.

de Castro-Neto, E. F., da Cunha, R. H., da Silveira, D. X., Yonamine, M., Gouveia, T. L., Cavalheiro, E. A., Amado, D., & Naffah-Mazzacoratti Mda, G. (2013). Changes in aminoacidergic and monoaminergic neurotransmission in the hippocampus and amygdala of rats after ayahuasca ingestion. *World Journal of Biological Chemistry, 4,* 141–147.

De Gregorio, D., Posa, L., Ochoa-Sanchez, R., McLaughlin, R., Maione, S., Comai, S., & Gobbi, G. (2016). The hallucinogen d-lysergic diethylamide (LSD) decreases dopamine firing activity through 5-HT1A, D2 and TAAR1 receptors. *Pharmacological Research, 113,* 81–91.

De Gregorio, D., Manchia, M., Carpiniello, B., Valtorta, F., Nobile, M., Gobbi, G., & Comai, S. (2019a). Role of palmitoylethanolamide (PEA) in depression: Translational evidence: Special section on "Translational and Neuroscience Studies in Affective Disorders". Section Editor, Maria Nobile MD, PhD. This section of JAD focuses on the relevance of translational and neuroscience studies in providing a better understanding of the neural basis of affective disorders. The main aim is to briefly summaries relevant research findings in clinical neuroscience with particular regards to specific innovative topics in mood and anxiety disorders. *Journal of Affective Disorders, 255,* 195–200.

De Gregorio, D., McLaughlin, R. J., Posa, L., Ochoa-Sanchez, R., Enns, J., Lopez-Canul, M., Aboud, M., Maione, S., Comai, S., & Gobbi, G. (2019b). Cannabidiol modulates serotonergic transmission and reverses both allodynia and anxiety-like behavior in a model of neuropathic pain. *Pain, 160,* 136.

De Gregorio, D., Dean Conway, J., Canul, M. L., Posa, L., Bambico, F. R., & Gobbi, G. (2020). Effects of chronic exposure to low doses of Δ9- tetrahydrocannabinol in adolescence and adulthood on serotonin/norepinephrine neurotransmission and emotional behaviors. *The International Journal of Neuropsychopharmacology, 23,* 751–761.

De Gregorio, D., Aguilar-Valles, A., Preller, K. H., Heifets, B. D., Hibicke, M., Mitchell, J., & Gobbi, G. (2021a). Hallucinogens in mental health: Preclinical and clinical studies on LSD, psilocybin, MDMA, and ketamine. *The Journal of Neuroscience, 41,* 891–900.

De Gregorio, D., Popic, J., Enns, J. P., Inserra, A., Skalecka, A., Markopoulos, A., Posa, L., Lopez-Canul, M., Qianzi, H., Lafferty, C. K., Britt, J. P., Comai, S., Aguilar-Valles, A., Sonenberg, N., & Gobbi, G. (2021b). Lysergic acid diethylamide (LSD) promotes social behavior through mTORC1 in the excitatory neurotransmission. *Proceedings of the National Academy of Sciences of the United States of America, 118,* e2020705118.

De Gregorio, D., Inserra, A., Enns, J. P., Markopoulos, A., Pileggi, M., El Rahimy, Y., Lopez-Canul, M., Comai, S., & Gobbi, G. (2022). Repeated lysergic acid diethylamide (LSD) reverses stress-induced anxiety-like behavior, cortical synaptogenesis deficits and serotonergic neurotransmission decline. *Neuropsychopharmacology, 47,* 1188–1198.

Deiana, S., Watanabe, A., Yamasaki, Y., Amada, N., Arthur, M., Fleming, S., Woodcock, H., Dorward, P.,

Pigliacampo, B., & Close, S. (2012). Plasma and brain pharmacokinetic profile of cannabidiol (CBD), cannabidivarine (CBDV), Δ 9-tetrahydrocannabivarin (THCV) and cannabigerol (CBG) in rats and mice following oral and intraperitoneal administration and CBD action on obsessive–compulsive behaviour. *Psychopharmacology, 219*, 859–873.

Devinsky, O., Cross, J. H., Laux, L., Marsh, E., Miller, I., Nabbout, R., Scheffer, I. E., Thiele, E. A., & Wright, S. (2017). Trial of Cannabidiol for drug-resistant seizures in the Dravet syndrome. *New England Journal of Medicine, 376*, 2011–2020.

Devinsky, O., Patel, A. D., Cross, J. H., Villanueva, V., Wirrell, E. C., Privitera, M., Greenwood, S. M., Roberts, C., Checketts, D., VanLandingham, K. E., & Zuberi, S. M. (2018a). Effect of Cannabidiol on drop seizures in the Lennox–Gastaut syndrome. *New England Journal of Medicine, 378*, 1888–1897.

Devinsky, O., Patel, A. D., Thiele, E. A., Wong, M. H., Appleton, R., Harden, C. L., Greenwood, S., Morrison, G., & Sommerville, K. (2018b). Randomized, dose-ranging safety trial of cannabidiol in Dravet syndrome. *Neurology, 90*, e1204–e1211.

Di Marzo, V., Melck, D., Orlando, P., Bisogno, T., Zagoory, O., Bifulco, M., Vogel, Z., & de Petrocellis, L. (2001). Palmitoylethanolamide inhibits the expression of fatty acid amide hydrolase and enhances the anti-proliferative effect of anandamide in human breast cancer cells. *Biochemical Journal, 358*, 249–255.

Diana, M., Melis, M., & Gessa, G. L. (1998). Increase in meso-prefrontal dopaminergic activity after stimulation of CB1 receptors by cannabinoids. *European Journal of Neuroscience, 10*, 2825–2830.

Duerler, P., Schilbach, L., Stämpfli, P., Vollenweider, F. X., & Preller, K. H. (2020). LSD-induced increases in social adaptation to opinions similar to one's own are associated with stimulation of serotonin receptors. *Scientific Reports, 10*, 12181.

Evans, F. J. (1991). Cannabinoids: The separation of central from peripheral effects on a structural basis. *Planta Medica, 57*, S60–S67.

Facci, L., Dal Toso, R., Romanello, S., Buriani, A., Skaper, S., & Leon, A. (1995). Mast cells express a peripheral cannabinoid receptor with differential sensitivity to anandamide and palmitoylethanolamide. *Proceedings of the National Academy of Sciences, 92*, 3376–3380.

Flachenecker, P., Henze, T., & Zettl, U. K. (2014). Nabiximols (THC/CBD oromucosal spray, Sativex®) in clinical practice-results of a multicenter, non-interventional study (MOVE 2) in patients with multiple sclerosis spasticity. *European Neurology, 71*, 271–279.

Foltin, R. W., Fischman, M. W., & Byrne, M. F. (1988). Effects of smoked marijuana on food intake and body weight of humans living in a residential laboratory. *Appetite, 11*, 1–14.

Frank, B., Serpell, M. G., Hughes, J., Matthews, J. N. S., & Kapur, D. (2008). Comparison of analgesic effects and patient tolerability of nabilone and dihydrocodeine for chronic neuropathic pain: Randomised, crossover, double blind study. *BMJ (Clinical Research ed.), 336*, 199–201.

Fuentes, J. J., Fonseca, F., Elices, M., Farré, M., & Torrens, M. (2020). Therapeutic use of LSD in psychiatry: A systematic review of randomized-controlled clinical trials. *Frontiers in Psychiatry, 10*, 943.

Ganley, O. H., Graessle, O. E., & Robinson, H. J. (1958). Anti-inflammatory activity of compounds obtained from egg yolk, peanut oil, and soybean lecithin. *The Journal of Laboratory and Clinical Medicine, 51*, 709–714.

Garcia-Romeu, A., Davis, A. K., Erowid, F., Erowid, E., Griffiths, R. R., & Johnson, M. W. (2019). Cessation and reduction in alcohol consumption and misuse after psychedelic use. *Journal of Psychopharmacology, 33*, 1088–1101.

Gasser, P., Kirchner, K., & Passie, T. (2014). LSD-assisted psychotherapy for anxiety associated with a life-threatening disease: A qualitative study of acute and sustained subjective effects. *Journal of Psychopharmacology, 29*, 57–68.

Ghazizadeh-Hashemi, M., Ghajar, A., Shalbafan, M.-R., Ghazizadeh-Hashemi, F., Afarideh, M., Malekpour, F., Ghaleiha, A., Ardebili, M. E., & Akhondzadeh, S. (2018). Palmitoylethanolamide as adjunctive therapy in major depressive disorder: A double-blind, randomized and placebo-controlled trial. *Journal of Affective Disorders, 232*, 127–133.

Gobbi, G. (2019). A role for cannabidiol in psychiatry? Keep calm and follow the drug development rules. *The World Journal of Biological Psychiatry, 20*, 98–100.

Gobbi, G., Bambico, F., Mangieri, R., Bortolato, M., Campolongo, P., Solinas, M., Cassano, T., Morgese, M. G., Debonnel, G., & Duranti, A. (2005). Antidepressant-like activity and modulation of brain monoaminergic transmission by blockade of anandamide hydrolysis. *Proceedings of the National Academy of Sciences, 102*, 18620–18625.

Goodwin et al, 2022: N Engl J Med 387:1637–1648. https://doi.org/10.1056/NEJMoa2206443

Goldberg, S. B., Pace, B. T., Nicholas, C. R., Raison, C. L., & Hutson, P. R. (2020). The experimental effects of psilocybin on symptoms of anxiety and depression: A meta-analysis. *Psychiatry Research, 284*, 112749.

Golombek, P., Müller, M., Barthlott, I., Sproll, C., & Lachenmeier, D. W. (2020). Conversion of cannabidiol (CBD) into psychotropic cannabinoids including tetrahydrocannabinol (THC): A controversy in the scientific literature. *Toxics, 8*, 41.

Grassin-Delyle, S., Naline, E., Buenestado, A., Faisy, C., Alvarez, J. C., Salvator, H., Abrial, C., Advenier, C., Zemoura, L., & Devillier, P. (2014). Cannabinoids inhibit cholinergic contraction in human airways through prejunctional CB1 receptors. *British Journal of Pharmacology, 171*, 2767–2777.

Griffiths, R. R., Johnson, M. W., Carducci, M. A., Umbricht, A., Richards, W. A., Richards, B. D., Cosimano, M. P., & Klinedinst, M. A. (2016).

Psilocybin produces substantial and sustained decreases in depression and anxiety in patients with life-threatening cancer: A randomized double-blind trial. *Journal of Psychopharmacology (Oxford, England), 30*, 1181–1197.

Grotenhermen, F., Russo, E., & Zuardi, A. W. (2017). Even high doses of oral cannabidiol do not cause THC-like effects in humans: Comment on Merrick et al. Cannabis and Cannabinoid Research 2016; 1 (1): 102–112; https://doi.org/10.1089/can. 2015.0004. *Cannabis and Cannabinoid Research, 2*, 1–4.

Guida, F., Luongo, L., Boccella, S., Giordano, M., Romano, R., Bellini, G., Manzo, I., Furiano, A., Rizzo, A., & Imperatore, R. (2017). Palmitoylethanolamide induces microglia changes associated with increased migration and phagocytic activity: Involvement of the CB2 receptor. *Scientific Reports, 7*, 1–11.

Hampson, A., Grimaldi, M., Axelrod, J., & Wink, D. (1998). Cannabidiol and(−) Δ9-tetrahydrocannabinol are neuroprotective antioxidants. *Proceedings of the National Academy of Sciences, 95*, 8268–8273.

Hanuš, L. O., Meyer, S. M., Muñoz, E., Taglialatela-Scafati, O., & Appendino, G. (2016). Phytocannabinoids: A unified critical inventory. *Natural Product Reports, 33*, 1357–1392.

Hasler, F., Grimberg, U., Benz, M. A., Huber, T., & Vollenweider, F. X. (2004). Acute psychological and physiological effects of psilocybin in healthy humans: A double-blind, placebo-controlled dose-effect study. *Psychopharmacology, 172*, 145–156.

Heifets, B. D., Salgado, J. S., Taylor, M. D., Hoerbelt, P., Cardozo Pinto, D. F., Steinberg, E. E., Walsh, J. J., Sze, J. Y., & Malenka, R. C. (2019). Distinct neural mechanisms for the prosocial and rewarding properties of MDMA. *Science Translational Medicine, 11*, eaaw6435.

Herkenham, M., Lynn, A. B., Little, M. D., Johnson, M. R., Melvin, L. S., De Costa, B. R., & Rice, K. C. (1990). Cannabinoid receptor localization in brain. *Proceedings of the National Academy of Sciences, 87*, 1932–1936.

Hill, M. N., Hillard, C. J., Bambico, F. R., Patel, S., Gorzalka, B. B., & Gobbi, G. (2009a). The therapeutic potential of the endocannabinoid system for the development of a novel class of antidepressants. *Trends in Pharmacological Sciences, 30*, 484–493.

Hill, M. N., Miller, G. E., Carrier, E. J., Gorzalka, B. B., & Hillard, C. J. (2009b). Circulating endocannabinoids and *N*-acyl ethanolamines are differentially regulated in major depression and following exposure to social stress. *Psychoneuroendocrinology, 34*, 1257–1262.

Ho, W. S., Barrett, D., & Randall, M. (2008). 'Entourage' effects of *N*-palmitoylethanolamide and *N*-oleoylethanolamide on vasorelaxation to anandamide occur through TRPV1 receptors. *British Journal of Pharmacology, 155*, 837–846.

Hofmann, M. E., & Frazier, C. J. (2013). Marijuana, endocannabinoids, and epilepsy: Potential and challenges for improved therapeutic intervention. *Experimental Neurology, 244*, 43–50.

Hofmann, A., Heim, R., Brack, A., & Kobel, H. (1958). Psilocybin, a psychotropic substance from the Mexican mushroom Psilicybe mexicana Heim. *Experientia, 14*, 107–109.

Hollister, L. E., Shelton, J., & Krieger, G. (1969). A controlled comparison of lysergic acid diethylamide (LSD) and Dextroamphetamine in alcoholics. *American Journal of Psychiatry, 125*, 1352–1357.

Hong, P., & Liu, Y. (2017). Calcitonin gene-related peptide antagonism for acute treatment of migraine: A meta-analysis. *International Journal of Neuroscience, 127*, 20–27.

House, R. V., Thomas, P. T., & Bhargava, H. N. (1994). Immunological consequences of in vitro exposure to lysergic acid diethylamide (LSD). *Immunopharmacology and Immunotoxicology, 16*, 23–40.

Howlett, A. C., Barth, F., Bonner, T. I., Cabral, G., Casellas, P., Devane, W. A., Felder, C. C., Herkenham, M., Mackie, K., Martin, B. R., Mechoulam, R., & Pertwee, R. G. (2002). International Union of Pharmacology. XXVII. Classification of cannabinoid receptors. *Pharmacological Reviews, 54*, 161–202.

Iannotti, F. A., Hill, C. L., Leo, A., Alhusaini, A., Soubrane, C., Mazzarella, E., Russo, E., Whalley, B. J., Di Marzo, V., & Stephens, G. J. (2014). Nonpsychotropic plant cannabinoids, cannabidivarin (CBDV) and cannabidiol (CBD), activate and desensitize transient receptor potential vanilloid 1 (TRPV1) channels in vitro: Potential for the treatment of neuronal hyperexcitability. *ACS Chemical Neuroscience, 5*, 1131–1141.

Inserra, A. (2018). Hypothesis: The psychedelic ayahuasca heals traumatic memories via a sigma 1 receptor-mediated epigenetic-mnemonic process. *Frontiers in Pharmacology, 9*, 330.

Inserra, A., De Gregorio, D., & Gobbi, G. (2021a). Psychedelics in psychiatry: Neuroplastic, immunomodulatory, and neurotransmitter mechanisms. *Pharmacological Reviews, 73*, 202–277.

Inserra, A., De Gregorio, D., Rezai, T., Lopez-Canul, M. G., Comai, S., & Gobbi, G. (2021b). Lysergic acid diethylamide differentially modulates the reticular thalamus, mediodorsal thalamus, and infralimbic prefrontal cortex: An in vivo electrophysiology study in male mice. *Journal of Psychopharmacology (Oxford, England), 35*, 469–482. https://doi.org/10.1177/0269881121991569

Inserra, A., Campanale, A., Cheishvili, D., Dymov, S., Wong, A., Marcal, N., Syme, R. A., Taylor, L., De Gregorio, D., Kennedy, T. E., Szyf, M., Gobbi, G. (2022). Modulation of DNA methylation and protein expression in the prefrontal cortex by repeated administration of D-lysergic acid diethylamide (LSD): Impact on neurotropic, neurotrophic, and neuroplasticity signaling. *Prog Neuropsychopharmacol Biol Psychiatry 119*:110594. https://doi.org/10.1016/j.pnpbp.2022. 110594. Epub 2022 Jun 28. PMID: 35777526.

Inserra A., Giorgini, G., Lacroix, S., Bertazzo, A., Choo, J., Markopolous, A., Grant, E., Abolghasemi, A., De Gregorio, D., Flamand, N., Rogers, G., Comai, S., Silvestri, C., Gobbi, G., & Di Marzo, V. (2023). Effects of repeated lysergic acid diethylamide (LSD) on the mouse brain endocannabinoidome and gut microbiome. *Br J Pharmacol 180*(6), 721–739. https://doi.org/10.1111/bph.15977. Epub 2022 Dec 12. PMID: 36316276.

Jefsen, O. H., Elfving, B., Wegener, G., & Müller, H. K. (2020). Transcriptional regulation in the rat prefrontal cortex and hippocampus after a single administration of psilocybin. *Journal of Psychopharmacology (Oxford, England), 35*, 483–493. https://doi.org/10.1177/0269881120959614

Jetly, R., Heber, A., Fraser, G., & Boisvert, D. (2015). The efficacy of nabilone, a synthetic cannabinoid, in the treatment of PTSD-associated nightmares: A preliminary randomized, double-blind, placebo-controlled cross-over design study. *Psychoneuroendocrinology, 51*, 585–588.

Kathmann, M., Flau, K., Redmer, A., Tränkle, C., & Schlicker, E. (2006). Cannabidiol is an allosteric modulator at mu- and delta-opioid receptors. *Naunyn-Schmiedeberg's Archives of Pharmacology, 372*, 354–361.

Khoury, J. M., MdCLd, N., Roque, M. A. V., Queiroz, D. A. B., Corrêa de Freitas, A. A., de Fátima, Â., Moreira, F. A., & Garcia, F. D. (2019). Is there a role for cannabidiol in psychiatry? *The World Journal of Biological Psychiatry, 20*, 101–116.

Kim, Y.-K., & Won, E. (2017). The influence of stress on neuroinflammation and alterations in brain structure and function in major depressive disorder. *Behavioural Brain Research, 329*, 6–11.

Kim, J., Farchione, T., Potter, A., Chen, Q., & Temple, R. (2019). Esketamine for treatment-resistant depression – first FDA-approved antidepressant in a new class. *The New England Journal of Medicine, 381*, 1–4.

Kindlundh-Högberg, A. M., Blomqvist, A., Malki, R., & Schiöth, H. B. (2008). Extensive neuroadaptive changes in cortical gene-transcript expressions of the glutamate system in response to repeated intermittent MDMA administration in adolescent rats. *BMC Neuroscience, 9*, 39.

Krall, C. M., Richards, J. B., Rabin, R. A., & Winter, J. C. (2008). Marked decrease of LSD-induced stimulus control in serotonin transporter knockout mice. *Pharmacology, Biochemistry, and Behavior, 88*, 349–357.

Krebs, T. S., & Johansen, P.-Ø. (2012). Lysergic acid diethylamide (LSD) for alcoholism: Meta-analysis of randomized controlled trials. *Journal of Psychopharmacology, 26*, 994–1002.

Kuypers, K. P., Ng, L., Erritzoe, D., Knudsen, G. M., Nichols, C. D., Nichols, D. E., Pani, L., Soula, A., & Nutt, D. (2019). Microdosing psychedelics: More questions than answers? An overview and suggestions for future research. *Journal of Psychopharmacology, 33*, 1039–1057.

Kyzar, E. J., Stewart, A. M., & Kalueff, A. V. (2016). Effects of LSD on grooming behavior in serotonin transporter heterozygous (Sert+/−) mice. *Behavioural Brain Research, 296*, 47–52.

Labate, B. C., & Feeney, K. (2012). Ayahuasca and the process of regulation in Brazil and internationally: Implications and challenges. *The International Journal on Drug Policy, 23*, 154–161.

Lago, E. D., & Fernandez-Ruiz, J. (2007). Cannabinoids and neuroprotection in motor-related disorders. *CNS & Neurological Disorders-Drug Targets (Formerly Current Drug Targets-CNS & Neurological Disorders), 6*, 377–387.

Laprairie, R., Bagher, A., Kelly, M., & Denovan-Wright, E. (2015). Cannabidiol is a negative allosteric modulator of the cannabinoid CB1 receptor. *British Journal of Pharmacology, 172*, 4790–4805.

Leonard, H. L., Rapoport, J. L. (1987). Relief of obsessive-compulsive symptoms by LSD and psilocin. The American *Journal of Psychiatry 144*(9):1239–40.

Leweke, F. M., Piomelli, D., Pahlisch, F., Muhl, D., Gerth, C. W., Hoyer, C., Klosterkötter, J., Hellmich, M., & Koethe, D. (2012). Cannabidiol enhances anandamide signaling and alleviates psychotic symptoms of schizophrenia. *Translational Psychiatry, 2*, e94.

Lichtman, A., Sheikh, S., Loh, H., & Martin, B. (2001). Opioid and cannabinoid modulation of precipitated withdrawal in Δ9-tetrahydrocannabinol and morphine-dependent mice. *Journal of Pharmacology and Experimental Therapeutics, 298*, 1007–1014.

Linares, I. M. P., Guimaraes, F. S., Eckeli, A., Crippa, A. C. S., Zuardi, A. W., Souza, J. D. S., Hallak, J. E., & Crippa, J. A. S. (2018). No acute effects of Cannabidiol on the sleep-wake cycle of healthy subjects: A randomized, double-blind, placebo-controlled, crossover study. *Frontiers in Pharmacology, 9*, 315.

Lintzeris, N., Bhardwaj, A., Mills, L., Dunlop, A., Copeland, J., McGregor, I., Bruno, R., Gugusheff, J., Phung, N., Montebello, M., Chan, T., Kirby, A., Hall, M., Jefferies, M., Luksza, J., Shanahan, M., Kevin, R., Allsop, D., & group ftARfCDs. (2019). Nabiximols for the treatment of cannabis dependence: A randomized clinical trial. *JAMA Internal Medicine, 179*, 1242–1253.

Lyons, T., & Carhart-Harris, R. L. (2018). Increased nature relatedness and decreased authoritarian political views after psilocybin for treatment-resistant depression. *Journal of Psychopharmacology, 32*, 811–819.

Marek, G. J., & Aghajanian, G. K. (1994). Excitation of interneurons in piriform cortex by 5-hydroxytryptamine: Blockade by MDL 100,907, a highly selective 5-HT2A receptor antagonist. *European Journal of Pharmacology, 259*, 137–141.

Markopoulos, A., Inserra, A., De Gregorio, D., & Gobbi, G. (2022). Evaluating the potential use of serotonergic psychedelics in autism spectrum disorder. *Frontiers in Pharmacology, 12*, 749068.

Martin, B. R., Compton, D. R., Thomas, B. F., Prescott, W. R., Little, P. J., Razdan, R. K., Johnson, M. R., Melvin, L. S., & Mechoulam, R. (1991). Behavioral, biochemical, and molecular modeling evaluations of cannabinoid analogs. *Pharmacology Biochemistry and Behavior, 40*, 471–478.

Martín-Moreno, A. M., Reigada, D., Ramírez, B. G., Mechoulam, R., Innamorato, N., Cuadrado, A., & de Ceballos, M. L. (2011). Cannabidiol and other cannabinoids reduce microglial activation in vitro and in vivo: Relevance to Alzheimer's disease. *Molecular Pharmacology, 79*, 964–973.

Matsuda, L. A., Lolait, S. J., Brownstein, M. J., Young, A. C., & Bonner, T. I. (1990). Structure of a cannabinoid receptor and functional expression of the cloned cDNA. *Nature, 346*, 561–564.

McGuire, P., Robson, P., Cubala, W. J., Vasile, D., Morrison, P. D., Barron, R., Taylor, A., & Wright, S. (2018). Cannabidiol (CBD) as an adjunctive therapy in schizophrenia: A multicenter randomized controlled trial. *The American Journal of Psychiatry, 175*, 225–231.

Mitchell, J. M., Bogenschutz, M., Lilienstein, A. et al. (2021). MDMA-assisted therapy for severe PTSD: a randomized, double-blind, placebo-controlled phase 3 study. *Nat Med 27*, 1025–1033. https://doi.org/10.1038/s41591-021-01336-3

Mithoefer, M. C., Wagner, M. T., Mithoefer, A. T., Jerome, L., Martin, S. F., Yazar-Klosinski, B., Michel, Y., Brewerton, T. D., & Doblin, R. (2013). Durability of improvement in post-traumatic stress disorder symptoms and absence of harmful effects or drug dependency after 3,4-methylenedioxymethamphetamine-assisted psychotherapy: A prospective long-term follow-up study. *Journal of Psychopharmacology (Oxford, England), 27*, 28–39.

Mithoefer, M. C., Feduccia, A. A., Jerome, L., Mithoefer, A., Wagner, M., Walsh, Z., Hamilton, S., Yazar-Klosinski, B., Emerson, A., & Doblin, R. (2019). MDMA-assisted psychotherapy for treatment of PTSD: Study design and rationale for phase 3 trials based on pooled analysis of six phase 2 randomized controlled trials. *Psychopharmacology, 236*, 2735–2745.

Morales-García, J. A., de la Fuente, R. M., Alonso-Gil, S., Rodríguez-Franco, M. I., Feilding, A., Perez-Castillo, A., & Riba, J. (2017). The alkaloids of Banisteriopsis caapi, the plant source of the Amazonian hallucinogen Ayahuasca, stimulate adult neurogenesis in vitro. *Scientific Reports, 7*, 5309.

Moreno, J. L., Holloway, T., Rayannavar, V., Sealfon, S. C., & Gonzalez-Maeso, J. (2013). Chronic treatment with LY341495 decreases 5-HT(2A) receptor binding and hallucinogenic effects of LSD in mice. *Neuroscience Letters, 536*, 69–73.

Moreno, F. A., Wiegand, C. B., Taitano, E. K., Delgado, P. L. (2006). Safety, tolerability, and efficacy of psilocybin in 9 patients with obsessive-compulsive disorder. *The Journal of Clinical Psychiatry 67*(11):1735–40.

Müller, F., Lenz, C., Dolder, P., Lang, U., Schmidt, A., Liechti, M., & Borgwardt, S. (2017). Increased thalamic resting-state connectivity as a core driver of LSD-induced hallucinations. *Acta Psychiatrica Scandinavica, 136*, 648–657.

Müller, F., Brändle, R., Liechti, M. E., & Borgwardt, S. (2019). Neuroimaging of chronic MDMA ("ecstasy") effects: A meta-analysis. *Neuroscience & Biobehavioral Reviews, 96*, 10–20.

Naef, M., Curatolo, M., Petersen-Felix, S., Arendt-Nielsen, L., Zbinden, A., & Brenneisen, R. (2003). The analgesic effect of oral delta-9-tetrahydrocannabinol (THC), morphine, and a THC-morphine combination in healthy subjects under experimental pain conditions. *Pain, 105*, 79–88.

Nahler, G., Grotenhermen, F., Zuardi, A. W., & Crippa, J. A. (2017). A conversion of oral cannabidiol to delta9-tetrahydrocannabinol seems not to occur in humans. *Cannabis and Cannabinoid Research, 2*, 81–86.

Narang, S., Gibson, D., Wasan, A. D., Ross, E. L., Michna, E., Nedeljkovic, S. S., & Jamison, R. N. (2008). Efficacy of dronabinol as an adjuvant treatment for chronic pain patients on opioid therapy. *The Journal of Pain, 9*, 254–264.

Nardou, R., Lewis, E. M., Rothhaas, R., Xu, R., Yang, A., Boyden, E., & Dölen, G. (2019). Oxytocin-dependent reopening of a social reward learning critical period with MDMA. *Nature, 569*, 116–120.

Nicholson, A. N., Turner, C., Stone, B. M., & Robson, P. J. (2004). Effect of Δ-9-tetrahydrocannabinol and cannabidiol on nocturnal sleep and early-morning behavior in young adults. *Journal of Clinical Psychopharmacology, 24*, 305–313.

Okusanya, B. O., Lott, B. E., Ehiri, J., McClelland, J., & Rosales, C. (2022). Medical cannabis for the treatment of migraine in adults: a review of the evidence. *Frontiers in Neurology, 13*.

O'Sullivan, S. E., Kendall, D. A., & Randall, M. D. (2005). The effects of Delta9-tetrahydrocannabinol in rat mesenteric vasculature, and its interactions with the endocannabinoid anandamide. *British Journal of Pharmacology, 145*, 514–526.

Pacher, P., & Mechoulam, R. (2011). Is lipid signaling through cannabinoid 2 receptors part of a protective system? *Progress in Lipid Research, 50*, 193–211.

Pacher, P., Bátkai, S., & Kunos, G. (2006). The endocannabinoid system as an emerging target of pharmacotherapy. *Pharmacological Reviews, 58*, 389–462.

Pertwee, R. G. (1988). The central neuropharmcology of psychotropic cannabinoids. *Pharmacology & Therapeutics, 36*, 189–261.

Pertwee, R. G., & Cascio, M. G. (2014). Known pharmacological actions of delta-9-tetrahydrocannabinol and of four other chemical constituents of cannabis that activate cannabinoid receptors. In *Handbook of cannabis* (p. 115). Oxford University Press.

Pertwee, R. G., Howlett, A. C., Abood, M. E., Alexander, S. P., Di Marzo, V., Elphick, M. R., Greasley, P. J., Hansen, H. S., Kunos, G., & Mackie, K. (2010a). International Union of basic and clinical pharmacology. LXXIX. Cannabinoid receptors and their ligands: Beyond CB1 and CB2. *Pharmacological Reviews, 62*, 588–631.

Pertwee, R. G., Howlett, A. C., Abood, M. E., Alexander, S. P., Di Marzo, V., Elphick, M. R., Greasley, P. J., Hansen, H. S., Kunos, G., Mackie, K., Mechoulam, R., & Ross, R. A. (2010b). International Union of Basic and Clinical Pharmacology. LXXIX. Cannabinoid receptors and their ligands: Beyond CB$_1$ and CB$_2$. *Pharmacological Reviews, 62*, 588–631.

Pinault, D. (2004). The thalamic reticular nucleus: Structure, function and concept. *Brain Research. Brain Research Reviews, 46*, 1–31.

Portenoy, R. K., Ganae-Motan, E. D., Allende, S., Yanagihara, R., Shaiova, L., Weinstein, S., McQuade, R., Wright, S., & Fallon, M. T. (2012). Nabiximols for opioid-treated cancer patients with poorly-controlled chronic pain: A randomized, placebo-controlled, graded-dose trial. *The Journal of Pain, 13*, 438–449.

Prasad, B., Radulovacki, M., & Carley, D. (2013). Proof of concept trial of Dronabinol in obstructive sleep apnea. *Frontiers in Psychiatry, 4*, 1.

Preller, K. H. (2019). The effects of low doses of lysergic acid diethylamide in healthy humans: Demystifying the microdosing of psychedelics. *Biological Psychiatry, 86*, 736–737.

Preller, K. H., Razi, A., Zeidman, P., Stampfli, P., Friston, K. J., & Vollenweider, F. X. (2019). Effective connectivity changes in LSD-induced altered states of consciousness in humans. *Proceedings of the National Academy of Sciences of the United States of America, 116*, 2743–2748.

Raso, G. M., Russo, R., Calignano, A., & Meli, R. (2014). Palmitoylethanolamide in CNS health and disease. *Pharmacological Research, 86*, 32–41.

Rickli, A., Moning, O. D., Hoener, M. C., & Liechti, M. E. (2016). Receptor interaction profiles of novel psychoactive tryptamines compared with classic hallucinogens. *European Neuropsychopharmacology, 26*, 1327–1337.

Rog, D. J., Nurmikko, T. J., Friede, T., & Young, C. A. (2005). Randomized, controlled trial of cannabis-based medicine in central pain in multiple sclerosis. *Neurology, 65*, 812–819.

Roseman, L., Demetriou, L., Wall, M. B., Nutt, D. J., & Carhart-Harris, R. L. (2018). Increased amygdala responses to emotional faces after psilocybin for treatment-resistant depression. *Neuropharmacology, 142*, 263–269.

Rosenbaum, D., Weissman, C., Anderson, T., Petranker, R., Dinh-Williams, L.-A., Hui, K., & Hapke, E. (2020). Microdosing psychedelics: Demographics, practices, and psychiatric comorbidities. *Journal of Psychopharmacology, 34*, 612–622.

Russo, E. B. (2011). Taming THC: Potential cannabis synergy and phytocannabinoid-terpenoid entourage effects. *British Journal of Pharmacology, 163*, 1344–1364.

Russo, E. B. (2016). Beyond cannabis: Plants and the endocannabinoid system. *Trends in Pharmacological Sciences, 37*, 594–605.

Russo, E. B. (2017). Cannabis and epilepsy: An ancient treatment returns to the fore. *Epilepsy & Behavior, 70*, 292–297.

Russo, E., & Guy, G. W. (2006). A tale of two cannabinoids: The therapeutic rationale for combining tetrahydrocannabinol and cannabidiol. *Medical Hypotheses, 66*, 234–246.

Russo, E. B., Burnett, A., Hall, B., & Parker, K. K. (2005). Agonistic properties of cannabidiol at 5-HT1a receptors. *Neurochemical Research, 30*, 1037–1043.

Sarris, J., Perkins, D., Cribb, L., Schubert, V., Opaleye, E., Bouso, J. C., Scheidegger, M., Aicher, H., Simonova, H., Horák, M., Galvão-Coelho, N. L., Castle, D., & Tófoli, L. F. (2021). Ayahuasca use and reported effects on depression and anxiety symptoms: An international cross-sectional study of 11,912 consumers. *Journal of Affective Disorders Reports, 4*, 100098.

Sastre-Garriga, J., Vila, C., Clissold, S., & Montalban, X. (2011). THC and CBD oromucosal spray (Sativex®) in the management of spasticity associated with multiple sclerosis. *Expert Review of Neurotherapeutics, 11*, 627–637.

Savage, C., & McCabe, O. L. (1973). Residential psychedelic (LSD) therapy for the narcotic addict: A controlled study. *Archives of General Psychiatry, 28*, 808–814.

Schmid, Y., Enzler, F., Gasser, P., Grouzmann, E., Preller, K. H., Vollenweider, F. X., Brenneisen, R., Müller, F., Borgwardt, S., & Liechti, M. E. (2015). Acute effects of lysergic acid diethylamide in healthy subjects. *Biological Psychiatry, 78*, 544–553.

Schneider T, Zurbriggen, L., Dieterle, M., Mauermann, E., Frei, P., Mercer-Chalmers-Bender, K., Ruppen, W. (2022). Pain response to cannabidiol in induced acute nociceptive pain, allodynia, and hyperalgesia by using a model mimicking acute pain in healthy adults in a randomized trial (CANAB I). *Pain 163*(1), e62–e71.

Scuderi, C., Filippis, D. D., Iuvone, T., Blasio, A., Steardo, A., & Esposito, G. (2009). Cannabidiol in medicine: A review of its therapeutic potential in CNS disorders. *Phytotherapy Research: An International Journal Devoted to Pharmacological and Toxicological Evaluation of Natural Product Derivatives, 23*, 597–602.

Sigafoos, J., Green, V. A., Edrisinha, C., & Lancioni, G. E. (2007). Flashback to the 1960s: LSD in the treatment of autism. *Developmental Neurorehabilitation, 10*, 75–81.

Skrabek, R. Q., Galimova, L., Ethans, K., & Perry, D. (2008). Nabilone for the treatment of pain in fibromyalgia. *The Journal of Pain, 9*, 164–173.

Solorzano, C., Zhu, C., Battista, N., Astarita, G., Lodola, A., Rivara, S., Mor, M., Russo, R., Maccarrone, M., & Antonietti, F. (2009). Selective *N*-acylethanolamine-hydrolyzing acid amidase inhibition reveals a key role for endogenous palmitoylethanolamide in inflammation. *Proceedings of the National Academy of Sciences, 106*, 20966–20971.

Strassman, R. (2001). *DMT: The spirit molecule: A doctor's revolutionary research into the biology of near-death and mystical experiences*. Park Street Press.

Strassman, R. J., & Qualls, C. R. (1994). Dose-response study of *N,N*-dimethyltryptamine in humans: I. Neuroendocrine, autonomic, and cardiovascular effects. *Archives of General Psychiatry, 51*, 85–97.

Strassman, R. J., Qualls, C. R., Uhlenhuth, E. H., & Kellner, R. (1994). Dose-response study of *N,N*-dimethyltryptamine in humans: II. Subjective effects and preliminary results of a new rating scale. *Archives of General Psychiatry, 51*, 98–108.

Suraev, A., Grunstein, R. R., Marshall, N. S., D'Rozario, A. L., Gordon, C. J., Bartlett, D. J., Wong, K., Yee, B. J.,

Vandrey, R., Irwin, C., Arnold, J. C., McGregor, I. S., & Hoyos, C. M. (2020). Cannabidiol (CBD) and Δ9-tetrahydrocannabinol (THC) for chronic insomnia disorder ('CANSLEEP' trial): Protocol for a randomised, placebo-controlled, double-blinded, proof-of-concept trial. *BMJ Open, 10*, e034421.

Szabo, A., Kovacs, A., Frecska, E., & Rajnavolgyi, E. (2014). Psychedelic *N,N*-dimethyltryptamine and 5-methoxy-*N,N*-dimethyltryptamine modulate innate and adaptive inflammatory responses through the sigma-1 receptor of human monocyte-derived dendritic cells. *PLoS One, 9*, e106533.

Szabo, A., Kovacs, A., Riba, J., Djurovic, S., Rajnavolgyi, E., & Frecska, E. (2016). The endogenous hallucinogen and trace amine *N,N*-dimethyltryptamine (DMT) displays potent protective effects against hypoxia via sigma-1 receptor activation in human primary iPSC-derived cortical neurons and microglia-like immune cells. *Frontiers in Neuroscience, 10*, 423–423.

Szmulewicz, A. G., Valerio, M. P., & Smith, J. M. (2015). Switch to mania after ayahuasca consumption in a man with bipolar disorder: A case report. *International Journal of Bipolar Disorders, 3*, 4.

Tagliazucchi, E., Roseman, L., Kaelen, M., Orban, C., Muthukumaraswamy Suresh, D., Murphy, K., Laufs, H., Leech, R., McGonigle, J., Crossley, N., Bullmore, E., Williams, T., Bolstridge, M., Feilding, A., Nutt David, J., & Carhart-Harris, R. (2016). Increased global functional connectivity correlates with LSD-induced ego dissolution. *Current Biology, 26*, 1043–1050.

Thiele, E. A., Marsh, E. D., French, J. A., Mazurkiewicz-Beldzinska, M., Benbadis, S. R., Joshi, C., Lyons, P. D., Taylor, A., Roberts, C., Sommerville, K., Gunning, B., Gawlowicz, J., Lisewski, P., Mazurkiewicz Beldzinska, M., Mitosek Szewczyk, K., Steinborn, B., Zolnowska, M., Hughes, E., McLellan, A., Benbadis, S., Ciliberto, M., Clark, G., Dlugos, D., Filloux, F., Flamini, R., French, J., Frost, M., Haut, S., Joshi, C., Kapoor, S., Kessler, S., Laux, L., Lyons, P., Marsh, E., Moore, D., Morse, R., Nagaraddi, V., Rosenfeld, W., Seltzer, L., Shellhaas, R., Sullivan, J., Thiele, E., Thio, L. L., Wang, D., & Wilfong, A. (2018). Cannabidiol in patients with seizures associated with Lennox-Gastaut syndrome (GWPCARE4): A randomised, double-blind, placebo-controlled phase 3 trial. *The Lancet, 391*, 1085–1096.

Thomas, A., Baillie, G., Phillips, A., Razdan, R., Ross, R. A., & Pertwee, R. G. (2007). Cannabidiol displays unexpectedly high potency as an antagonist of CB1 and CB2 receptor agonists in vitro. *British Journal of Pharmacology, 150*, 613–623.

Trigo, J. M., Lagzdins, D., Rehm, J., Selby, P., Gamaleddin, I., Fischer, B., Barnes, A. J., Huestis, M. A., & Le Foll, B. (2016). Effects of fixed or self-titrated dosages of Sativex on cannabis withdrawal and cravings. *Drug and Alcohol Dependence, 161*, 298–306.

Trigo, J. M., Soliman, A., Quilty, L. C., Fischer, B., Rehm, J., Selby, P., Barnes, A. J., Huestis, M. A., George, T. P., Streiner, D. L., Staios, G., & Le Foll, B. (2018). Nabiximols combined with motivational enhancement/cognitive behavioral therapy for the treatment of cannabis dependence: A pilot randomized clinical trial. *PLoS One, 13*, e0190768.

Uthaug, M. V., Lancelotta, R., Szabo, A., Davis, A. K., Riba, J., & Ramaekers, J. G. (2020). Prospective examination of synthetic 5-methoxy-*N,N*-dimethyltryptamine inhalation: effects on salivary IL-6, cortisol levels, affect, and non-judgment. *Psychopharmacology, 237*, 773–785.

Vargas, A. S., Luís, Â., Barroso, M., Gallardo, E., & Pereira, L. (2020). Psilocybin as a new approach to treat depression and anxiety in the context of life-threatening diseases – A systematic review and meta-analysis of clinical trials. *Biomedicine, 8*, 331.

Vaysse, P. J., Gardner, E. L., & Zukin, R. S. (1987). Modulation of rat brain opioid receptors by cannabinoids. *The Journal of Pharmacology and Experimental Therapeutics, 241*, 534–539.

Wacker, D., Wang, S., McCorvy, J. D., Betz, R. M., Venkatakrishnan, A. J., Levit, A., Lansu, K., Schools, Z. L., Che, T., Nichols, D. E., Shoichet, B. K., Dror, R. O., & Roth, B. L. (2017). Crystal structure of an LSD-bound human serotonin receptor. *Cell, 168*, 377–389.e312.

Wade, D. T., Collin, C., Stott, C., & Duncombe, P. (2010). Meta-analysis of the efficacy and safety of Sativex (nabiximols), on spasticity in people with multiple sclerosis. *Multiple Sclerosis Journal, 16*, 707–714.

Wallace, M. S., Marcotte, T. D., Umlauf, A., Gouaux, B., & Atkinson, J. H. (2015). Efficacy of inhaled cannabis on painful diabetic neuropathy. *The Journal of Pain, 16*, 616–627.

Whiting, P. F., Wolff, R. F., Deshpande, S., Di Nisio, M., Duffy, S., Hernandez, A. V., Keurentjes, J. C., Lang, S., Misso, K., Ryder, S., Schmidlkofer, S., Westwood, M., & Kleijnen, J. (2015). Cannabinoids for medical use: A systematic review and meta-analysis. *JAMA, 313*, 2456–2473.

Williams, S., Hartley, J., & Graham, J. (1976). Bronchodilator effect of delta1-tetrahydrocannabinol administered by aerosol of asthmatic patients. *Thorax, 31*, 720–723.

Wilsey, B. L., Deutsch, R., Samara, E., Marcotte, T. D., Barnes, A. J., Huestis, M. A., & Le, D. (2016). A preliminary evaluation of the relationship of cannabinoid blood concentrations with the analgesic response to vaporized cannabis. *Journal of Pain Research, 9*, 587–598.

Wong, M. L., Inserra, A., Lewis, M. D., Mastronardi, C. A., Leong, L., Choo, J., Kentish, S., Xie, P., Morrison, M., Wesselingh, S. L., Rogers, G. B., & Licinio, J. (2016). Inflammasome signaling affects anxiety- and depressive-like behavior and gut microbiome composition. *Molecular Psychiatry, 21*, 797–805.

Yamaori, S., Koeda, K., Kushihara, M., Hada, Y., Yamamoto, I., & Watanabe, K. (2011). Comparison in the in vitro inhibitory effects of major phytocannabinoids and polycyclic aromatic hydrocarbons contained in marijuana smoke on cytochrome P450 2C9 activity. *Drug Metabolism and Pharmacokinetics, 27*, 294–300. 1112080289-1112080289.

Treatments for Medication-Induced Movement Disorders

Shih-Ku Lin

Contents

Abstract

Movement disorders induced by medications for psychosis (antipsychotics), including parkinsonism, akathisia, acute dystonia, and tardive dyskinesia, mainly occur due to the blockade of dopamine in the extrapyramidal system. These side effects more commonly occur in patients receiving dopamine blockers (first-generation antipsychotics) than in those receiving dopamine serotonin blockers or partial agonists (second-generation antipsychotics) and can lead to poor adherence or even refusal of medication in patients if not appropriately managed. Neuroleptic malignant syndrome, also caused by medications for psychosis, is a medical emergency with a high mortality rate if not appropriately treated. This chapter describes the symptomatology, rating scales, and treatment regimens including specific medications for each extrapyramidal symptom and neuroleptic malignant syndrome.

Keywords

Extrapyramidal symptom · Parkinsonism · Akathisia · Acute dystonia · Tardive dyskinesia · Neuroleptic malignant syndrome

Introduction

It is well documented that patients with schizophrenia exhibited some movement disorders such as abnormal mannerisms, tics, stereotypy, catalepsy,

S.-K. Lin (✉)
Department of Psychiatry, Taipei Chang Gung Memorial Hospital, Taipei, Taiwan
e-mail: sklintcpc@gmail.com

© Springer Nature Switzerland AG 2024
A. Tasman et al. (eds.), *Tasman's Psychiatry*,
https://doi.org/10.1007/978-3-030-51366-5_142

and echopraxia as well as orofacial movements that resemble tardive dyskinesia (TD) before the emergence of neuroleptics (Perju-Dumbrava & Kempster, 2020).

In the initial phase of treatment with medications for psychosis, patients were observed to demonstrate some associated movement abnormalities. These movement abnormalities were proposed to be essential for the efficacy of medications for psychosis; however, some authors have reported that these drug-induced extrapyramidal reactions were not necessary for the efficacy of medications for psychosis (Faurbye et al., 1964). Modern work clearly demonstrates that exceeding 80% dopamine D2 receptor occupancy does not increase antipsychotic efficacy but does increase likelihood of extrapyramidal reaction (Pani et al., 2007). Neuroleptics were hypothesized to exert a therapeutic effect on schizophrenia through dopamine blockade in the mesolimbic system, whereas the striatum was proposed to result in extrapyramidal side effects (Crow et al., 1976). Many terms have been used to describe drug-induced movement disorders such as parkinsonism, extrapyramidal parkinsonian syndrome, pseudoparkinsonism, parkinson-like syndrome, extrapyramidal disorder, extrapyramidal symptom (EPS), and extrapyramidal syndrome (Chien & Dimascio, 1967).

Psychiatric clinicians should be vigilant regarding abnormal movements in patients induced by medications for psychosis, especially those stemming from the use of dopamine blockers. In this chapter, I focus on movement disorders and their treatment from the psychiatric point of view.

Parkinsonism

The clinical manifestations of parkinsonism induced by medications for psychosis are similar to the symptoms of Parkinson's disease. These symptoms include slow movement, stiffness, cog-wheeling, resting tremor, mask face and sluggish expression, buckling posture, shuffling gait, lack of postural reflexes, and drooling. These symptoms usually occur within a few weeks of treatment and are more likely to occur in older patients or those using a high-potency FGA such as haloperidol. When these side effects are severe, patients may develop complete dyskinesia, which is similar to catatonia. Patients often develop side effects after receiving high doses of high affinity dopamine antagonists; however, they may infrequently experience side effects at therapeutic doses of lower potency agents.

The Simpson–Angus Scale (SAS) was the first rating scale developed to assess the severity of extrapyramidal side effects (Simpson & Angus, 1970). The SAS is a 5-point scoring scale including the following 10 items, which are assigned a score from 0 to 4: gait, arm dropping, shoulder shaking, elbow rigidity, wrist rigidity, leg pendulousness, head dropping, glabella tap, tremor, and salivation; these symptoms represent frequently occurring motor muscle abnormalities after medications for psychosis. The Extrapyramidal Symptom Rating Scale (ESRS) is a more comprehensive tool developed to examine four types of drug-induced movement disorders: parkinsonism, akathisia, dystonia, and TD (Chouinard & Margolese, 2005). The SAS and ESRS are widely used in the clinical trials of new medications for psychosis to monitor movement-related side effects. The Drug-Induced Extrapyramidal Symptom Scale (DIEPSS) is a more concise rating scale that includes eight individual items, namely gait, bradykinesia, sialorrhea, rigidity, tremor, akathisia, dystonia, and dyskinesia, and one global item (Inada et al., 2003). The DIEPSS was developed in the era of SGAs and can be used for evaluating a low incidence of EPSs. The aforementioned scales can be used to monitor the degree of symptom improvement when a patient is undergoing treatment.

The treatment of patients with parkinsonism may involve reducing dose or switching from FGAs to SGAs. However, if the originally administered drugs must be maintained, then a fixed dose of antiparkinsonism medications can be added during treatment, and the dose of the medication for psychosis can be minimized. When the symptoms of parkinsonism are severe, administration of medications for psychosis should be temporarily discontinued until the symptoms improve, after which a lower dose of medications for psychosis may be initiated.

Antihistamines and anticholinergics have been used to treat parkinsonism for a long time. EPSs were postulated to occur due to interference with the normal functioning of central catecholamines, particularly dopamine, and central histamine (McGeer et al., 1961). Both histamine and acetylcholine are present in high concentrations in the caudate nucleus. Antihistamines such as diphenhydramine (25 mg TID) and anticholinergics such as benztropine (1–2 mg TID), biperidine (2 mg TID), and trihexyphenidyl (2 mg TID) can relieve these side effects. Amantadine, an antiviral drug and a dopamine agonist, acts by releasing dopamine from dopaminergic nerve terminals can ameliorate parkinsonism at a dose of 100 mg TID. Interestingly, there are no high-quality studies examining the efficacy of anticholinergics for drug-induced parkinsonism (Dickenson et al., 2017; Bergman & Soares-Weiser, 2018), and anticholinergics may worsen psychosis (Das et al., 2020), contribute to dementia (Coupland et al., 2019), and are already being deprescribed in Parkinson's disease patients (Barrett et al., 2021). Temporary use of benzodiazepine (1 mg TID of lorazepam or 0.5 mg TID of clonazepam) can relieve severe rigidity.

The use of prophylactic antiparkinsonism medications during medications for psychosis treatment remains controversial (Chien et al., 1979), and a consensus statement by the World Health Organization (WHO, 1990) recommended against the use of prophylactic anticholinergics because the overprescription of anticholinergics can result in additional side effects. A review (Lavin & Rifkin, 1991) suggested that initial prophylaxis can be beneficial for most patients who are starting neuroleptic treatment. However, the occurrence of side effects differs by race.

One study (Potkin et al., 1984) determined that Asian patients had a higher plasma concentration of haloperidol than did non-Asian patients at the same dose; this finding might explain the increased sensitivity of Asian patients to neuroleptic-induced side effects. Another study (Lin et al., 1988) compared differences in prolactin response after haloperidol treatment between Asian and Caucasian patients and suggested that pharmacodynamic factors (e.g., dopamine receptor-mediated responses) contribute to differences in responses and side effects between the two races. A study indicated that Indian patients are likely to experience EPSs when receiving conventional medications for psychosis; thus, initial prophylaxis with antiparkinsonism drugs should be more carefully considered on a routine basis for this population (Dhavale et al., 2004).

Because anticholinergics can cause peripheral side effects, such as dry mouth and constipation, and central adverse effects, such as cognitive impairment (Ruxton et al., 2015) and contributing to the development or worsening of TD, the lowest dose and the shortest duration should be applied in clinical practice, especially for SGAs (Desmarais et al., 2012; Ogino et al., 2014).

L-dopa, a standard drug for Parkinson's disease, can exacerbate psychotic symptoms; however, it was not suitable for treating parkinsonism induced by medications for psychosis. A clinical trial (Chouinard et al., 1987) suggested that L-dopa was effective in a subgroup of patients. Recently, a case report indicated the use of L-dopa was successful in treating long-lasting EPSs after multiple injections of paliperidone palmitate (Takada et al., 2018), indicating that L-dopa can be used for treating refractory parkinsonism.

Akathisia

Patients with akathisia experience an uncomfortable sensation that requires movement to relieve, are unable to sit still, and develop anxiety, agitation, or both. Sometimes, differentiating akathisia from psychosis-related anxiety in clinical practice can be challenging; thus, increased restlessness after the use of medications for psychosis should be considered akathisia. Although akathisia is usually grouped with parkinsonism, it can present as a purely subjective clinical complaint without overt movement abnormalities. Recent evidence shows that the prevalence of akathisia may be higher than previously thought (possibly due to underdiagnosis or misdiagnosis), and akathisia may be present in most patients, especially those receiving new SGAs such as aripiprazole, lurasidone, cariprazine, and brexpiprazole, because their parkinsonism is not evident (Demyttenaere et al., 2019). Akathisia can cause considerable suffering, reducing patients' willingness to receive medication. In addition, akathisia may even cause patients to feel hopeless and may lead to suicide in severe cases. Because patients with akathisia often present with vague and non-specific symptoms such as nervousness and discomfort, the differentiated diagnoses are anxiety, agitation, or restless leg syndrome. In general, anxiety refers to an intrapsychic worrisome or fearful state, and in severe cases may show sweating, feeling restless, tense, and palpitation. When people are agitated, they may also show restless, fidgeting, and pacing, but the major symptoms are feelings of annoyance, inner tension, or even anger, and usually provoked or when under stressful situation. Restless leg syndrome is a condition that causes an overwhelming and uncontrollable urge to move legs, usually because of an uncomfortable sensation, and typically happens in the evening or nighttime hours when sitting or lying down. It can also cause an unpleasant crawling or creeping sensation in the feet, calves, and thighs, and can be temporally relieved with movement, such as stretching, jiggling legs, pacing or walking.

Akathisia is not included in the rating item of SAS. The Barnes Akathisia Rating Scale (BARS) was developed to assess the severity of akathisia (Barnes, 1989). The BARS is used to examine the following symptoms: objectively fidgety or restless movements such as shuffling or tramping movements of the legs and feet, and the subjective awareness of restlessness; it also includes a global assessment. A study indicated that patients presenting with the motor features typical of akathisia but not experiencing subjective discomfort or that lack subjective awareness should be considered to have pseudoakathisia (Havaki-Kontaxaki et al., 2000). Pseudoakathisia has been suggested as a form of delayed dyskinesia or a clinical progression from akathisia, and the treatment of pseudoakathisia is similar to that of akathisia.

Akathisia can be managed by administering the lowest dose of medications for psychosis.

Benzodiazepines (1 mg TID of lorazepam), beta-adrenal blockers (10–20 mg TID of propranolol), or alpha-2 adrenergic receptor agonist clonidine (0.1–0.3 mg TID) may all ameliorate akathisia. Furthermore, trazodone (50 mg daily), an antagonist for 5-HT2A and 5-HT2C postsynaptic receptors, may be effective in treating akathisia (Shams-Alizadeh et al., 2020; Stryjer et al., 2010). Mirtazapine (7.5 mg or 15 mg once daily), a 5-HT 2A antagonist, can be considered a treatment option for akathisia, particularly for patients with contraindications or intolerability to beta-blockers and those with comorbid depression or negative symptoms (Poyurovsky & Weizman, 2020; Hieber et al., 2008). In addition, amantadine can be used to treat akathisia associated with parkinsonism. Anticholinergics are usually not effective for akathisia (Rathbone & Soares-Weiser, 2006). A recently developed second-generation medication for psychosis, iloperidone, which has very high affinity to the noradrenergic alpha1 receptor where it is inhibitory, has rates of akathisia that are similar to placebo in acute and relapse prevention trials (Weiden et al., 2008).

Acute Dystonia

Acute dystonia most frequently occurs within a week beginning antipsychotic treatment. Moreover, acute dystonia is more likely to occur in young patients and those receiving a high dose of medications for psychosis. Acute dystonia is characterized by involuntary contractions of the muscles of the extremities, larynx, face, neck, abdomen, and pelvis in either sustained or intermittent patterns that lead to abnormal movements or postures. Occasionally, patients can present with a thick sensation in the tongue and difficulty in swallowing. These symptoms gradually develop in 3–6 h. In addition, opisthotonos and oculogyric crisis may occur. Patients with acute dystonia may feel uncomfortable or fearful and often experience sequelae. Severe muscle spasms leading to joint dislocation can also occur. In addition, acute dystonia can result in the inability to move the larynx, thus hindering respiratory function. Acute dystonic reaction is believed to be caused by an imbalance of the dopaminergic, cholinergic, and GABAergic systems in the basal ganglia (Termsarasab et al., 2016).

The differential diagnoses of acute dystonia include conversion disorder, tetanus, focal seizure, hypocalcemia, anticholinergic toxicity, meningitis, and trauma (e.g., temporomandibular joint dislocation, mandibular fracture, and orbital fracture).

Acute dystonia can be treated with antihistamines such as diphenhydramine (50 mg IM or IV) or anticholinergics such as benztropine or biperidine (2 mg IM or IV). This regimen can usually ameliorate dystonia rapidly. If the effect of this regimen is not observed after 20 min, then the injection can be repeated. Second-line therapy with benzodiazepines, such as lorazepam (2 mg IM or IV), may be considered for patients who fail to completely respond to anticholinergic therapy. Patients who experience respiratory symptoms or require supportive oxygen should be observed for 12–24 h following the resolution of symptoms to monitor for recurrence. After the improvement of acute dystonia, medications for psychosis can be subsequently administered; however, anticholinergics should be coadministered for at least 2 weeks.

The incidence of acute dystonia is considered to be lower after the administration of SGAs (Satterthwaite et al., 2008; Tural Hesapcioglu et al., 2020); however, clinicians should be aware that any dopamine-blocking agent can lead to acute dystonia.

Tardive Dyskinesia (TD)

After the initial introduction of medications for psychosis, an increasing number of case reports have described the occurrence of a new type of neurological disorder in patients with mental illness, and the large-scale use of phenothiazines or similar drugs was reported to be the main reason for this neurological disorder (Crane, 1968). The term "tardive dyskinesia" was adopted because the condition occurred only after treatment with medications for psychosis for a long period. TD is seldom observed during the first 6 months of treatment and may even occur after several years of treatment in patients after symptom resolution. The characteristics of TD are (1) involuntary and repeated movements of a body part, such as in the face, trunk, extremities, and sometimes the respiratory system; (2) late occurrence in the course of treatment and often after the discontinuation of treatment with medications for psychosis; (3) persistence of disabling manifestations for a long duration of up to months or years; and (4) poor response to any type of therapy. Although distinguishing these movements from the stereotypic movements of patients with some chronic mental diseases is sometimes difficult, patients with TD appear to be less active or exhibit chorea-like involuntary movements. TD usually occurs when patients are taking medications for psychosis, and increased doses of medications for psychosis may mask the symptoms of TD. Abnormal involuntary movements may occur only when the medication is stopped or reduced in some patients. If TD-like movements occur after a dose reduction or cessation of medications for psychosis and disappear within a few days or weeks, it should be defined as "withdrawal or emergent dyskinesia." If the duration is considerably long, it should be defined as TD (persistent for a month after the discontinuation of medication according to DSM-5 criteria).

The involuntary movements of TD can reduce patients' quality of life and lead them to feel embarrassed, thus resulting in social withdrawal. Clinicians should be vigilant regarding the occurrence of TD and screen all patients receiving medications for psychosis because early intervention can result in recovery. Table 1 lists the differential diagnoses of TD. Although no study has indicated that TD can occur again in these patients if medications for psychosis are used again, the use of medications for psychosis should be avoided in these patients to the extent possible.

Patients treated with SGAs had a lower risk of TD than did those treated with FGAs. The prevalence of TD is estimated to be around 25–30% in patients receiving FGAs, around 20% in patients receiving SGAs (Jeste & Caligiuri, 1993; Carbon et al., 2017), and only 7.2% in FGA-naïve patients treated with only SGAs (Carbon et al., 2017).

Table 1 Differential diagnosis of tardive dyskinesia

Neurologic disorders
Wilson's disease
Huntington disease
Brain neoplasms
Fahr's syndrome
Idiopathic dystonia (including blepharospasm, mandibular dystonia, and facial tics)
Meige's syndrome (spontaneous oral dyskinesias)
Torsion dystonia (familial disorder without psychiatric symptoms)
Postanoxic or postencephalitic extrapyramidal symptoms
Drugs and other toxicities
Antidepressants
Lithium
Anticholinergics
Phenytoin
L-dopa and dopamine agonists
Amphetamines and related stimulants
Magnesium and other heavy metals

However, the observed lower incidence may be a reflection of the relatively recent introduction to second-generation antipsychotics; additional time will help determine this. Risk factors for TD are older age, female sex, history of EPSs and akathisia, use of FGAs, exposure to anticholinergics, and a high dose of medications for psychosis (Correll et al., 2017).

TD is considered to be a type of basal ganglia disease, and dopaminergic hypersensitivity resulting from long-term dopamine receptor blockade was proposed to be its pathophysiology (Gerlach et al., 1974). An advanced hypothesis is that the sensitization of the dopamine receptor and the altered function of the N-methyl-D-aspartate receptor result in maladaptive synaptic plasticity, which allows the encoding of abnormal motor programs (Teo et al., 2012). It appears clear that increased dopamine release, increased postsynaptic receptor expression, and increase in the number of dopaminergic synapses all conspire to cause TD (Ali et al., 2020). In addition to dopamine receptors, animal studies have shown that serotonin receptors (particularly 5-HT2 receptors) interact with dopaminergic neurotransmission and their blockade reduces D2 receptor upregulation (Kusumi et al., 2000). The elucidation of TD

pathophysiology can facilitate the development of treatment modalities.

After the diagnosis of TD, clinicians should manage it appropriately. The Abnormal Involuntary Movement Scale was developed by Guy (1976) to evaluate the severity of TD. This scale consists of 10 items including symptoms and attitudes, and these items are assigned a score from 0 to 4 to assess the severity of dyskinesia (particularly orofacial, extremity, and truncal movements). Additional items assess the overall severity, incapacitation, and patient's level of awareness of movements and the distress associated with them. This scale should be administered periodically to monitor the improvement of TD.

The first step in the treatment of TD is to discontinue medications for psychosis that have high affinity to D2 dopamine receptors by tapering their dosage gradually. The sooner the TD-causing medication is discontinued, the more likely it is that TD will gradually resolve (Kiriakakis et al., 1998). If discontinuing medications for psychosis is not possible due to patients' symptomatology, switching to another medication for psychosis with a lower TD risk (i.e., lower dopamine D2 receptor blockade), including clozapine, or quetiapine, or partial agonists is the next preferred option (Lee et al., 2019). Since lumateperone receptor occupancy does not exceed 50%, and it may act as a presynaptic partial agonist, it is theoretically also a good option, but it still does not have sufficient data to confirm this (Sleem & El-Mallakh, 2022). Other medications that can be used to alleviate symptoms include clonazepam, gingko biloba, amantadine, and tetrabenazine (Bhidayasiri et al., 2013). The use of anticholinergics may exacerbate TD symptoms (Ward & Citrome, 2018). However, the link between the use of anticholinergics and the emergence of TD maybe only an epiphenomenon because people who develop EPSs are believed to be more likely to receive anticholinergics and develop TD (Bergman & Soares-Weiser, 2018).

In 2017, valbenazine and deutetrabenazine were approved by the FDA to treat TD in adults. Both drugs are reversible vesicular monoamine transporter type 2 (VMAT2) inhibitors and are related to tetrabenazine, a medication used to treat

Huntington disease. The normal function of VMAT2 is to transport monoamines, such as dopamine, into synaptic vesicles for release into the synaptic cleft when the neuron fires. VMAT2 inhibitors block the transport of dopamine into synaptic vesicles, thereby reducing the amount of dopamine released into the synapse, particularly in the dorsal striatum, a key area of the brain that controls motor movements (Citrome & Saklad, 2020). Deutetrabenazine is a hexahydro-dimethoxybenzoquinolizine derivative and a deuterated tetrabenazine. The presence of deuterium in deutetrabenazine increases the half-lives of the active metabolite and prolongs their pharmacological activity and allows less frequent dosing and a lower daily dose with improvement in tolerability. Valbenazine is a valine amino acid conjugated with tetrabenazine. The active compound is generated with the amino acid is removed with a peptidase. This allows for the slow continuous release of tetrabenazine. Stahl (2018) compared the pharmacologic mechanism of action for the VMAT2 inhibitors valbenazine and deutetrabenazine and suggested that deutetrabenazine has more dosing options to titrate for patients requiring careful dosing titration but a shorter half-life (9–10 h), thus necessitating twice-daily administration with food. The starting dose is 6 mg once daily. Titrate up at weekly intervals by 6 mg per day to a tolerated dose that reduces chorea, up to a maximum recommended daily dosage of 48 mg (24 mg twice daily). Valbenazine has fewer dosing options and a lower need for titration but a longer half-life (15–22 h), allowing for once-a-day administration. The suggested dosing regimen is 40 mg once daily for 1 week. After the first week, 80 mg is recommended once daily; however, continuing 40 mg, or using 60 mg, may be appropriate. The common side effects of valbenazine are somnolence, QT prolongation, and anticholinergic effects, and those of deutetrabenazine are nasopharyngitis and sleep disturbance (insomnia or somnolence), akathisia, and hyperprolactinemia. Long-term use of these medications has demonstrated continued efficacy and favorable tolerability, even when used in combination with baseline dopamine receptor-blocking agents. Because valbenazine and deutetrabenazine reduce the amount of dopamine in the synapse, there are ongoing studies that they may also be effective for treatment-resistant psychosis.

Neuroleptic Malignant Syndrome

Neuroleptic malignant syndrome (NMS) is a severe and life-threatening specific reaction associated with the use of dopamine receptor antagonists or the rapid withdrawal of dopaminergic medication. NMS is more commonly reported in patients receiving high-potency FGAs such as haloperidol and fluphenazine; however, its occurrence is not uncommon in patients receiving SGAs. A review article suggested that SGA-induced NMS is characterized by lower incidence, lower clinical severity, and less frequent lethal outcomes than FGA-induced NMS (Belvederi Murri et al., 2015). A systematic review of risk factors such as sex and age revealed that NMS is 50% more likely to occur in male patients and in young adulthood (Gurrera, 2017). Other risk factors include a high dose of medications for psychosis, polypharmacy, physical restraint, dehydration, and high temperature. The differential diagnoses of NMS include central anticholinergic syndrome, serotonin syndrome, lithium toxicity, heat stroke, central nervous system (CNS) infection, and psychotic catatonia. The most common signs of catatonia are immobility, mutism, withdrawal and refusal to eat, staring, negativism, posturing, rigidity, waxy flexibility/catalepsy, stereotypy, echolalia or echopraxia, and verbigeration.

The main symptoms of NMS are muscle rigidity, autonomic nervous system instability, fever, and delirium. The typical development of symptoms usually occurs within a few hours to a few days. At first, muscle stiffness primarily develops, followed by fever and autonomic nervous system instability occurring within a few hours. Fever can be considerably high, usually up to 41 °C or °106 F. Lead pipe rigidity is quite typical. In some cases, the increase in muscle tension can lead to muscle necrosis. Patients may experience dehydration, leading to severe myoglobinuria and renal failure. The instability of the autonomic nervous system causes unstable blood pressure (too high or too low), palpitations, diaphoresis or cold sweating, and pale skin. Arrhythmia may also occur. In addition to stiffness, motor abnormalities may present as complete motor inability, shaking, and involuntary movements. Patients are usually confused and unable to speak. The degree of consciousness disturbance may vary from agitated to stuporous, and seizure and coma may also occur.

No specific laboratory abnormalities or typical pathological changes were observed after mortality in autopsy study. Because of muscle necrosis, the creatine phosphokinase (CPK) level is usually high, and liver function tests will show abnormal findings, including increased transaminase and lactic dehydrogenase levels. In addition, the white blood cell count may increase (12,000–30,000/mm^3 with or without a left shift). NMS is a clinical diagnosis, and its severity varies considerably. Because no clear diagnostic criteria are available, it is difficult to determine the morbidity and mortality of NMS, particularly for milder cases. The incidence rate of NMS is estimated to range from 0.02% to 3%, and its mortality rate is approximately 10–20%. The causes of death are respiratory failure (due to reduced capacity of chest muscles responsible for breathing or aspiration pneumonia), cardiovascular failure, heart arrhythmia, renal failure, or pulmonary embolism.

The first step in treating NMS is to cease administering the NMS-causing medication for psychosis and other potential contributing psychotropic agents (lithium, anticholinergics, and serotonergic agents). Administration of comprehensive and appropriate supportive therapy for NMS is crucial; this may entail efforts to replenish body fluids, using ice blankets during high fever, helping a patient to turn over to prevent bedsores, and monitoring cardiac function, urine output, and renal function. When renal failure occurs, dialysis may be required. Dantrolene, a postsynaptic direct-acting muscle relaxant, can reduce stiffness, body temperature, heart rate, and the degree of muscle necrosis, and its effect often persists for several hours. However, its appropriate dosage has not yet been established. The general recommended dosage 1–3 mg/kg/day in four

doses orally or intravenously appears to be the most appropriate. Doses exceeding 10 mg/kg/day may cause liver toxicity. Stiffness may improve within a few hours, although complete recovery may require several days of treatment. In addition, bromocriptine, an ergoline derivative and a dopamine agonist used in the treatment of pituitary tumors, Parkinson's disease, and hyperprolactinemia, may reduce the central pathology of NMS in some patients (Mueller et al., 1983; Dhib-Jalbut et al., 1987). The suggested dose of bromocriptine is 7.5–45 mg/day in three divided doses. The combination use of dantrolene and bromocriptine can yield satisfactory results. The treatment duration has not yet been established; however, this regimen should be continued for a week after the alleviation of NMS symptoms. In some severe cases, such as those experiencing fatal stiffness or high fever, general anesthesia with propofol can be administered to paralyze muscles to save lives (Rupreht et al., 1991).

In addition to pharmacotherapy, electroconvulsive therapy (ECT) is an alternative treatment option for NMS (Addonizio & Susman, 1987). ECT is theorized to be effective because of the clinical similarity of NMS to malignant catatonia (for which ECT is the treatment of choice), and ECT is also effective in treating the underlying psychiatric illness for which medications are not acutely tolerated. A review article examining 15 case series (Morcos et al., 2019) indicated that the bitemporal ECT mode was well tolerated and effective in treating NMS refractory to pharmacological interventions, especially if the underlying condition was also responsive to ECT.

Another clinical concern is the reuse of medications for psychosis after the remission of NMS symptoms. Clozapine is usually the first choice because the incidence of NMS is the lowest after experiencing NMS. In patients who develop NMS after receiving clozapine, rechallenge with either clozapine or another medication for psychosis is acceptable (Lally et al., 2019). Not all patients who develop NMS will relapse, even if they are still using the same medication for psychosis that caused NMS. The key principle is to titrate the medication for psychosis and use the lowest effective dosage to reduce the risk of NMS recurrence.

During the COVID-19 pandemic, patients receiving medications for psychosis have had a higher risk of NMS after infection or vaccination (Kajani et al., 2020; Alfishawy et al., 2021; Espiridion et al., 2021). Acute medical illnesses, including organic brain disease, are considered to be risk factors for NMS (Pelonero et al., 1998). Hypothetically, SARS-CoV-2 binds to the angiotensin-converting enzyme 2 receptor and invades human cells to cause COVID-19 related pneumonia. COVID-19 may affect the CNS either through direct mechanisms such as neuronal retrograde dissemination and hematogenous dissemination or through indirect mechanisms, resulting in various CNS manifestations such as encephalitis, acute necrotizing encephalopathy, diffuse leukoencephalopathy, stroke (both ischemic and hemorrhagic), venous sinus thrombosis, meningitis, and NMS (Soltani et al., 2021). In patients with COVID-19 receiving medications for psychosis, the higher risk of NMS may not only be due to the infection or vaccination but also due to the drug-drug interaction of treatment medications with medications for psychosis (Borah et al., 2021).

The mortality rate of NMS has declined over the past 30 years, most likely due to the use of newer medications for psychosis, early recognition of the syndrome, and appropriate intervention. Nonetheless, clinicians, especially primary care practitioners who are using this class of drugs more often for adjunctive treatments, must be cognizant of this syndrome and the implications of their use (Ware et al., 2018).

Conclusion

Though the movement disorders induced by medications for psychosis are milder in SGAs than FGAs, the management of these side effects is clinically crucial, not only to ease patients' suffering but also to increase compliance or adherence to medications for psychosis. Table 2 summarizes the suggested medications for each movement disorder. The prophylactic treatment is controversial, and a minimal dosage and the shortest possible duration are suggested since these medications

Table 2 Suggested medications for each movement disorder

Disorder	Suggested medications
Parkinsonism	Antihistamines, anticholinergics, dopamine agonist (amantadine); benzodiazepines for severe and L-dopa for refractory cases
Akathisia	Benzodiazepines, beta-adrenal blockers, or α-1A agonist clonidine
Acute dystonia	Antihistamines diphenhydramine (50 mg IM or IV) or anticholinergics benztropine or biperidine (2 mg IM or IV)
Tardive dyskinesia	VMAT2 inhibitors valbenazine, deutetrabenazine
Neuroleptic malignant syndrome	Dantrolene, bromocriptine; propofol for severe cases

VMAT vesicular monoamine transporter

carry their own side effects. NMS is rare under SGAs treatment, but still clinicians should be cautiously aware and treat it appropriately.

References

Addonizio, G., & Susman, V. L. (1987). ECT as a treatment alternative for patients with symptoms of neuroleptic malignant syndrome. *The Journal of Clinical Psychiatry, 48*(3), 102–105.

Alfishawy, M., Bitar, Z., Elgazzar, A., et al. (2021). Neuroleptic malignant syndrome following COVID-19 vaccination. *The American Journal of Emergency Medicine, 49*, 408–409.

Ali, Z., Roque, A., & El-Mallakh, R. S. (2020). A unifying theory for the pathoetiologic mechanism of tardive dyskinesia. *Medical Hypotheses, 140*, 109682.

Barnes, T. R. (1989). A rating scale for drug-induced akathisia. *The British Journal of Psychiatry, 154*, 672–676.

Barrett, J. J., Sargent, L., Nawaz, H., et al. (2021). Antimuscarinic anticholinergic medications in Parkinson disease: To prescribe or deprescribe? *Movement Disorders, 8*(8), 1181–1188.

Belvederi Murri, M., Guaglianone, A., Bugliani, M., et al. (2015). Second-generation antipsychotics and neuroleptic malignant syndrome: Systematic review and case report analysis. *Drugs in R&D, 15*(1), 45–62.

Bergman, H., & Soares-Weiser, K. (2018). Anticholinergic medication for antipsychotic-induced tardive dyskinesia. *Cochrane Database of Systematic Reviews, 1*(1), Cd000204.

Bhidayasiri, R., Fahn, S., Weiner, W. J., et al. (2013). Evidence-based guideline: Treatment of tardive syndromes: Report of the Guideline Development Sub-committee of the American Academy of Neurology. *Neurology, 81*(5), 463–469.

Borah, P., Deb, P. K., Chandrasekaran, B., et al. (2021). Neurological consequences of SARS-CoV-2 infection and concurrence of treatment-induced neuropsychiatric adverse events in COVID-19 patients: Navigating the uncharted. *Frontiers in Molecular Biosciences, 8,* 627723.

Carbon, M., Hsieh, C. H., Kane, J. M., et al. (2017). Tardive dyskinesia prevalence in the period of second-generation antipsychotic use: A meta-analysis. *The Journal of Clinical Psychiatry, 78*(3), e264–e278.

Chien, C. P., & Dimascio, A. (1967). Drug-induced extra-pyramidal symptoms and their relations to clinical efficacy. *The American Journal of Psychiatry, 123*(12), 1490–1498.

Chien, C. P., Castaldo, V., Thornton, A., et al. (1979). Prophylactic usage of antiparkinsonian drugs for akinesia [proceedings]. *Psychopharmacology Bulletin, 15*(2), 75–78.

Chouinard, G., & Margolese, H. C. (2005). Manual for the Extrapyramidal Symptom Rating Scale (ESRS). *Schizophrenia Research, 76*(2–3), 247–265.

Chouinard, G., Annable, L., Mercier, P., et al. (1987). Long-term effects of L-dopa and procyclidine on neuroleptic-induced extrapyramidal and schizophrenic symptoms. *Psychopharmacology Bulletin, 23*(1), 221–226.

Citrome, L., & Saklad, S. R. (2020). Revisiting tardive dyskinesia: Focusing on the basics of identification and treatment. *The Journal of Clinical Psychiatry, 81*(2), TV18059AH3C. https://doi.org/10.4088/JCP. TV18059AH3C.

Correll, C. U., Kane, J. M., & Citrome, L. L. (2017). Epidemiology, prevention, and assessment of tardive dyskinesia and advances in treatment. *The Journal of Clinical Psychiatry, 78*(8), 1136–1147.

Coupland, C. A. C., Hill, T., Dening, T., et al. (2019). Anticholinergic drug exposure and the risk of dementia: A nested case-control study. *JAMA Internal Medicine, 179*(8), 1084–1093. https://doi.org/10.1001/jamainternmed.2019.0677

Crane, G. E. (1968). Tardive dyskinesia in patients treated with major neuroleptics: A review of the literature. *The American Journal of Psychiatry, 124*((8), Suppl), 40–48.

Crow, T. J., Johnstone, E. C., Deakin, J. F., et al. (1976). Dopamine and schizophrenia. *Lancet, 2*(7985), 563–566.

Das, S., Chatterjee, S. S., & Malathesh, B. C. (2020). Anticholinergic medications even in therapeutic range can cause recurrence of psychosis. *General Psychiatry, 33*(4), e100235. https://doi.org/10.1136/gpsych-2020-100235

Demyttenaere, K., Detraux, J., Racagni, G., et al. (2019). Medication-induced akathisia with newly approved antipsychotics in patients with a severe mental illness: A systematic review and meta-analysis. *CNS Drugs, 33*(6), 549–566.

Desmarais, J. E., Beauclair, L., & Margolese, H. C. (2012). Anticholinergics in the era of atypical antipsychotics: Short-term or long-term treatment? *Journal of Psychopharmacology, 26*(9), 1167–1174.

Dhavale, H. S., Pinto, C., Dass, J., et al. (2004). Prophylaxis of antipsychotic-induced extrapyramidal side effects in east Indians: Cultural practice or biological necessity? *Journal of Psychiatric Practice, 10*(3), 200–202.

Dhib-Jalbut, S., Hesselbrock, R., Mouradian, M. M., et al. (1987). Bromocriptine treatment of neuroleptic malignant syndrome. *The Journal of Clinical Psychiatry, 48*(2), 69–73.

Dickenson, R., Momcilovic, S., & Donnelly, L. (2017). Anticholinergics vs placebo for neuroleptic-induced Parkinsonism. *Schizophrenia Bulletin, 43*(1), 17. https://doi.org/10.1093/schbul/sbw087

Espiridion, E. D., Mani, V., & Oladunjoye, A. O. (2021). Neuroleptic malignant syndrome after re-introduction of atypical antipsychotics in a COVID-19 patient. *Cureus, 13*(2), e13428.

Faurbye, A., Rasch, P. J., Petersen, P. B., et al. (1964). Neurological symptoms in pharmacotherapy of psychoses. *Acta Psychiatrica Scandinavica, 40*(1), 10–27.

Gerlach, J., Reisby, N., & Randrup, A. (1974). Dopaminergic hypersensitivity and cholinergic hypofunction in the pathophysiology of tardive dyskinesia. *Psychopharmacologia, 34*(1), 21–35.

Gurrera, R. J. (2017). A systematic review of sex and age factors in neuroleptic malignant syndrome diagnosis frequency. *Acta Psychiatrica Scandinavica, 135*(5), 398–408.

Guy, W. (1976). *ECDEU assessment manual for psychopharmacology.* US Department of Health, Education and Welfare.

Havaki-Kontaxaki, B. J., Kontaxakis, V. P., & Christodoulou, G. N. (2000). Prevalence and characteristics of patients with pseudoakathisia. *European Neuropsychopharmacology, 10*(5), 333–336.

Hieber, R., Dellenbaugh, T., & Nelson, L. A. (2008). Role of mirtazapine in the treatment of antipsychotic-induced akathisia. *The Annals of Pharmacotherapy, 42*(6), 841–846.

Inada, T., Beasley, C. M. Jr, Tanaka, Y., & Walker, D. J. (2003). Extrapyramidal symptom profiles assessed with the Drug-Induced Extrapyramidal Symptom Scale: comparison with Western scales in the clinical double-blind studies of schizophrenic patients treated with either olanzapine or haloperidol. *International Clinical Psychopharmacology, 18*(1), 39–48.

Jeste, D. V., & Caligiuri, M. P. (1993). Tardive dyskinesia. *Schizophrenia Bulletin, 19*(2), 303–315.

Kajani, R., Apramian, A., Vega, A., et al. (2020). Neuroleptic malignant syndrome in a COVID-19 patient. *Brain, Behavior, and Immunity, 88,* 28–29.

Kiriakakis, V., Bhatia, K. P., Quinn, N. P., et al. (1998). The natural history of tardive dystonia. A long-term follow-up study of 107 cases. *Brain, 121*(Pt 11), 2053–2066.

Kusumi, I., Takahashi, Y., Suzuki, K., et al. (2000). Differential effects of subchronic treatments with atypical antipsychotic drugs on dopamine D2 and serotonin 5-HT2A receptors in the rat brain. *Journal of Neural Transmission (Vienna), 107*(3), 295–302.

Lally, J., McCaffrey, C., O'Murchu, C., et al. (2019). Clozapine rechallenge following neuroleptic malignant syndrome: A systematic review. *Journal of Clinical Psychopharmacology, 39*(4), 372–379.

Lavin, M. R., & Rifkin, A. (1991). Prophylactic antiparkinson drug use: I. Initial prophylaxis and prevention of extrapyramidal side effects. *Journal of Clinical Pharmacology, 31*(8), 763–768.

Lee, D., Baek, J. H., Bae, M., et al. (2019). Long-term response to clozapine and its clinical correlates in the treatment of tardive movement syndromes: A naturalistic observational study in patients with psychotic disorders. *Journal of Clinical Psychopharmacology, 39*(6), 591–596.

Lin, K. M., Poland, R. E., Lau, J. K., et al. (1988). Haloperidol and prolactin concentrations in Asians and Caucasians. *Journal of Clinical Psychopharmacology, 8*(3), 195–201.

McGeer, P. L., Boulding, J. E., Gibson, W. C., et al. (1961). Drug-induced extrapyramidal reactions. Treatment with diphenhydramine hydrochloride and dihydroxyphenylalanine. *JAMA, 177*, 665–670.

Morcos, N., Rosinski, A., & Maixner, D. F. (2019). Electroconvulsive therapy for neuroleptic malignant syndrome: A case series. *The Journal of ECT, 35*(4), 225–230.

Mueller, P. S., Vester, J. W., & Fermaglich, J. (1983). Neuroleptic malignant syndrome. Successful treatment with bromocriptine. *JAMA, 249*(3), 386–388.

Ogino, S., Miyamoto, S., Miyake, N., et al. (2014). Benefits and limits of anticholinergic use in schizophrenia: Focusing on its effect on cognitive function. *Psychiatry and Clinical Neurosciences, 68*(1), 37–49.

Pani, L., Pira, L., & Marchese, G. (2007). Antipsychotic efficacy: Relationship to optimal D2-receptor occupancy. *European Psychiatry, 22*(5), 267–275. https://doi.org/10.1016/j.eurpsy.2007.02.005

Pelonero, A. L., Levenson, J. L., & Pandurangi, A. K. (1998). Neuroleptic malignant syndrome: A review. *Psychiatric Services, 49*(9), 1163–1172.

Perju-Dumbrava, L., & Kempster, P. (2020). Movement disorders in psychiatric patients. *BMJ Neurology Open, 2*(2), e000057.

Potkin, S. G., Shen, Y., Pardes, H., et al. (1984). Haloperidol concentrations elevated in Chinese patients. *Psychiatry Research, 12*(2), 167–172.

Poyurovsky, M., & Weizman, A. (2020). Treatment of antipsychotic-induced akathisia: Role of serotonin 5-HT(2a) receptor antagonists. *Drugs, 80*(9), 871–882.

Rathbone, J., & Soares-Weiser, K. (2006). Anticholinergics for neuroleptic-induced acute akathisia. *Cochrane Database of Systematic Reviews, 2006*(4), Cd003727.

Rupreht, J., Verkaaik, A. P., & Erdmann, W. (1991). Propofol safely used in a neuroleptic malignant syndrome patient. *Anaesthesiologie und Reanimation, 16*(5), 329–332.

Ruxton, K., Woodman, R. J., & Mangoni, A. A. (2015). Drugs with anticholinergic effects and cognitive impairment, falls and all-cause mortality in older adults: A systematic review and meta-analysis. *British Journal of Clinical Pharmacology, 80*(2), 209–220.

Satterthwaite, T. D., Wolf, D. H., Rosenheck, R. A., et al. (2008). A meta-analysis of the risk of acute extrapyramidal symptoms with intramuscular antipsychotics for the treatment of agitation. *The Journal of Clinical Psychiatry, 69*(12), 1869–1879.

Shams-Alizadeh, N., Maroufi, A., Asadi, Z., et al. (2020). Trazodone as an alternative treatment for neuroleptic-associated akathisia: A placebo-controlled, double-blind, clinical trial. *Journal of Clinical Psychopharmacology, 40*(6), 611–614.

Simpson, G. M., & Angus, J. W. (1970). A rating scale for extrapyramidal side effects. *Acta Psychiatrica Scandinavica. Supplementum, 212*, 11–19.

Sleem, A., & El-Mallakh, R. S. (2022). Adaptive changes to antipsychotics: Their consequences and how to avoid them. *Current Psychiatry, 21*(7), 46–50, 52. https://doi.org/10.12788/cp.0262

Soltani, S., Tabibzadeh, A., Zakeri, A., et al. (2021). COVID-19 associated central nervous system manifestations, mental and neurological symptoms: A systematic review and meta-analysis. *Reviews in the Neurosciences, 32*(3), 351–361.

Stahl, S. M. (2018). Comparing pharmacologic mechanism of action for the vesicular monoamine transporter 2 (VMAT2) inhibitors valbenazine and deutetrabenazine in treating tardive dyskinesia: Does one have advantages over the other? *CNS Spectrums, 23*(4), 239–247.

Stryjer, R., Rosenzcwaig, S., Bar, F., et al. (2010). Trazodone for the treatment of neuroleptic-induced acute akathisia: A placebo-controlled, double-blind, crossover study. *Clinical Neuropharmacology, 33*(5), 219–222.

Takada, R., Yamamuro, K., & Kishimoto, T. (2018). Long-lasting extrapyramidal symptoms after multiple injections of paliperidone palmitate to treat schizophrenia. *Neuropsychiatric Disease and Treatment, 14*, 2541–2544.

Teo, J. T., Edwards, M. J., & Bhatia, K. (2012). Tardive dyskinesia is caused by maladaptive synaptic plasticity: A hypothesis. *Movement Disorders, 27*(10), 1205–1215.

Termsarasab, P., Thammongkolchai, T., & Frucht, S. J. (2016). Medical treatment of dystonia. *Journal of Clinical Movement Disorders, 3*, 19.

Tural Hesapcioglu, S., Ceylan, M. F., Kandemir, G., et al. (2020). Frequency and correlates of acute dystonic reactions after antipsychotic initiation in 441 children and adolescents. *Journal of Child and Adolescent Psychopharmacology, 30*(6), 366–375.

Ward, K. M., & Citrome, L. (2018). Antipsychotic-related movement disorders: Drug-induced parkinsonism vs. tardive dyskinesia-key differences in pathophysiology and clinical management. *Neurology and Therapy, 7*(2), 233–248.

Ware, M. R., Feller, D. B., & Hall, K. L. (2018). Neuroleptic malignant syndrome: Diagnosis and management. *The Primary Care Companion for CNS Disorders*, *20*(1), 17r02185. https://doi.org/10.4088/PCC.17r02185.

Weiden, P. J., Cutler, A. J., Polymeropoulos, M. H., & Wolfgang, C. D. (2008). Safety profile of iloperidone: A pooled analysis of 6-week acute-phase pivotal trials. *Journal of Clinical Psychopharmacology, 28*(2), S12–S19.

WHO. (1990). Prophylactic use of anticholinergics in patients on long-term neuroleptic treatment. A consensus statement. World Health Organization heads of centres collaborating in WHO co-ordinated studies on biological aspects of mental illness. *The British Journal of Psychiatry, 156*, 412.

Maria Muzik, Samantha Shaw, Sophie Grigoriadis,
Kristina M. Deligiannidis, Angelika Wieck, Prabha S. Chandra,
Manisha Murugesan, Cara Anne Poland, and
Nancy Renn-Bugai

Contents

M. Muzik (✉)
Departments of Psychiatry, Obstetrics & Gynecology,
University of Michigan- Michigan Medicine, Ann Arbor,
MI, USA

Zero To Thrive & Women and Infant Mental Health
Program, Ann Arbor, MI, USA

Perinatal Psychiatry Service, Ann Arbor, MI, USA

MC3 Perinatal Psychiatry Assess Program Michigan, Ann
Arbor, MI, USA
e-mail: muzik@med.umich.edu

S. Shaw
Department of Psychiatry, Rachel Upjohn Building, Ann
Arbor, MI, USA

Department of Psychiatry, University of Michigan-
Michigan Medicine, Ann Arbor, MI, USA
e-mail: samantsh@med.umich.edu

S. Grigoriadis
University of Toronto, Toronto, ON, Canada

Women's Mood and Anxiety Clinic: Reproductive
Transitions, Toronto, ON, Canada

Department of Psychiatry, Sunnybrook Health Sciences
Centre, Toronto, ON, Canada

Sunnybrook Research Institute, Toronto, ON, Canada

Women's College Research Institute, Toronto, ON, Canada
e-mail: Sophie.Grigoriadis@sunnybrook.ca

K. M. Deligiannidis
Psychiatry and Obstetrics & Gynecology, Zucker School of
Medicine at Hofstra/Northwell, Hempstead, NY, USA

Feinstein Institutes for Medical Research, Manhasset, NY,
USA

Katz Institute for Women's Health, Queens, NY, USA

Psychiatry, University of Massachusetts Medical School,
Worcester, MA, USA

Women's Behavioral Health, Zucker Hillside Hospital,
Northwell Health, Glen Oaks, NY, USA
e-mail: kdeligian1@northwell.edu

A. Wieck
Greater Manchester Mental Health NHS Foundation Trust
& University of Manchester, Manchester, UK

Laureate House, Manchester, UK

Wythenshawe Hospital, Manchester, UK

© Springer Nature Switzerland AG 2024
A. Tasman et al. (eds.), *Tasman's Psychiatry*,
https://doi.org/10.1007/978-3-030-51366-5_143

Abstract

The pharmacological management of psychopathology among women, specifically during their reproductive years, is highly complex and requires specialized knowledge regarding medication efficacy and safety. In this chapter, the reader is introduced to the complex guidelines for treating women in the peripartum period with psychotropic medicines, elaborating on common mental disorders including depression, bipolar illness, anxiety, posttraumatic stress and other trauma-related conditions, ADHD, and substance use disorders. It is highlighted that mental illness during the peripartum, that is, pregnancy and the first year after the infant's birth, is highly prevalent. While the pharmacological management of mental health conditions during the peripartum is often comparable to treating such conditions at any other period in a woman's life, decisions regarding medication selection require not only consideration of side effects for the mother but must also consider potential adverse effects on the pregnancy, the fetus, delivery, and long-term child development. Thus, critical to reproductive psychiatry is a personalized risk-benefit analysis accounting for a broad range of considerations. In this text the aim is to provide the reader with the foundational knowledge needed to support the patient and family in this complex decision-making process. Treatment algorithms for the pharmacologic management of premenstrual and perimenopausal mental health disorders are also presented. The overall goal of this chapter is to offer comprehensive, yet user-friendly, state-of-the-art information on medication management in reproductive psychiatry.

Keywords

Peripartum · Pregnancy · Postpartum · Premenstrual · Perimenopause · Psychotropics · Treatment guidelines · Risk-benefit analysis · Safety profiles · Breastfeeding

P. S. Chandra
Department of Psychiatry, National Institute of Mental Health and Neurosciences, Bangalore, India

International Association for Women's Mental Health, Bangalore, India
e-mail: chandra@nimhans.ac.in

M. Murugesan
Women's Mental Health, Bangalore, India

Department of Psychiatry, NIMHANS, Bangalore, India

C. A. Poland
Department of Obstetrics, Gynecology and Reproductive Biology, College of Human Medicine, Michigan State University, East Lansing, MI, USA
e-mail: polandc2@MSU.edu

N. Renn-Bugai
Corewell Health, Grand Rapids, MI, USA
e-mail: Nancy.Renn-Bugai@spectrumhealth.org

What Is Reproductive Psychiatry?

Reproductive psychiatry concerns the treatment of psychiatric disorders in women – specifically during the reproductive years. This specialty field of psychiatry concerns a complex area of psychiatric care, requiring unique knowledge and research. The major themes of reproductive psychiatry correspond to age- and stage-related fluctuations of reproductive hormones across women's life cycles, i.e., hormonal changes during the premenstrual, peripartum (i.e., pregnancy and postpartum), and the perimenopausal periods of women's life trajectories.

Most women who present with reproductive psychopathology are pregnant or postpartum; in fact, one in seven women in the peripartum period suffer from mood or anxiety disorders (Wisner et al., 2013; Fairbrother et al., 2015). Given the high prevalence of these peripartum psychopathologies and the known benefits to psychotropic treatment, a major focus of reproductive psychiatry concerns the safety of medicines during this time of pregnancy and postpartum when breastfeeding for the mother and her infant.

Given the stated clinical relevance, this chapter will primarily focus on the pharmacologic management of reproductive psychopathology during the peripartum period, elaborating on treatment algorithms for mental health disorders including depression, bipolar illness, anxiety, posttraumatic stress, ADHD, as well as substance use disorders. It is believed that students, residents, and clinicians working in psychiatry and/or primary care serving peripartum women will benefit from this comprehensive, yet user-friendly, summary of state-of-the-art treatment algorithms for the most common mental health disorders in pregnancy and postpartum. In addition, the pharmacologic management of premenstrual and perimenopausal mental health disorders are also briefly reviewed. Primarily focus of the authors is on the pharmacological management of the various psychopathologies, and the reader is referred to other chapters in this volume for a more in-depth exploration of specific psychiatric conditions (e.g., Premenstrual Dysphoric Disorder and Peripartum Depression and Psychiatric Conditions During Peripartum and Perimenopause).

Pharmacologic Management of Reproductive Psychopathology During the Peripartum

Peripartum Management of Depression, Anxiety, and Insomnia

Introduction

The pharmacological management of perinatal depression, anxiety, and insomnia is comparable to treating such conditions at any other time point; however, consideration for choice of drug has to be expanded beyond the consideration of side effect profile only to also include that of potential adverse effects on the pregnancy, delivery outcomes, the fetus, and the potential longer-term effects on child development. Thus, the decision to use pharmacotherapy during pregnancy requires a risk-benefit analysis that is personalized to the patient and includes considerations regarding active symptom severity, past course of illness, family history, type of prior medication response, experiences with medication taper, and patient (and her family) preference as well as comfort level with pharmacotherapy.

It is imperative to understand that it may not be possible to have zero risk to both the mother and the fetus. A universal algorithm is not possible given the complexity of psychiatric illness, patient preference, and the rapidly evolving outcomes data. To date, it remains unethical to randomize pregnant women in the initial clinical trials studying new compounds, and thus limitations in evaluating potential side effects or adverse fetal or child outcomes will remain. Most data on medication safety have been obtained from observational studies, clinical case reports, or registry data, and such reports carry the risk for confounds and bias. A major obstacle prohibiting conclusions is confounding by indication, meaning that depression or anxiety itself may be associated with the adverse outcomes on the child and not the exposure to the medicines (Fischer Fumeaux et al., 2019). Many studies also do not account for factors such as polytherapy, the use of other or illicit drugs or alcohol, or other medical illness. Studies may find statistical significance of adverse maternal-fetal outcomes, but that does not

necessarily translate to clinical significance or relevance. The positive statistical associations that have been found with medication use in pregnancy tend to be low, typically below an odds ratio (OR) of 2. Moreover, a major limitation of observational studies is that a positive association does not establish causality.

Generally, it has been recommended to use a medication that was previously effective and, if this information is not available, to use one with the most safety data in pregnancy and while breastfeeding. If medication is used, it is imperative that the woman is treated to remission; otherwise both she and the fetus will have been exposed to both the illness and the medication. Monotherapy is preferred to polytherapy as well as using the minimal effective dose. Lastly, there should be clear documentation in the patient's chart that a risk-benefit discussion regarding pharmacotherapy and the specific medicine chosen was conducted with the patient and if possible inclusive of other family of the patient's choice. The goal is to derive an individualized best-practice approach for the particular patient, her child, and her family. In the following section, pharmacological management of unipolar depression, anxiety, and the very common symptom of insomnia in pregnant and breastfeeding women are elaborated.

Depression

Pharmacological management for the treatment of depression in pregnancy is typically reserved for moderate to severe illness or for those women who choose to use medication regardless of severity following a risk-benefit consideration, while psychotherapy is recommended for mild to moderate perinatal depression (Yonkers et al., 2009). There are no head-to-head randomized controlled trials comparing one antidepressant to another nor compared to placebo in pregnant women. Generally, the drug that was effective prior to pregnancy is used, or if a woman who is already being treated with a specific medication becomes pregnant, assuming she is euthymic, most guidelines recommend continuing the medicine already used and not switching. Whenever starting any medication in a childbearing woman, the possibility of pregnancy must be considered as close to half of pregnancies are unplanned (Finer & Zolna, 2016). With gestational growth, there are many changes that can affect drug response including a higher volume of distribution (Costantine, 2014) and enhanced renal drug clearance. Thus, higher doses of medication may be required to maintain mood stability. However, perinatal clinicians usually recommend titrating medication dose based on symptoms as there is yet insufficient evidence to change the dosing in the second or third trimester in the absence of symptoms.

In general, selective serotonin reuptake inhibitors (SSRIs) of the serotonin transporter (SERT) are considered first-line medications for depression based on safety data reported in the most recent critical review of the best available evidence on SSRI use in pregnancy and lactation (Fischer Fumeaux et al., 2019). The SSRI of choice by most guidelines is sertraline, followed by citalopram and escitalopram, and with lesser use fluoxetine and paroxetine (both of which are potent inhibitors of several CYP enzymes including CYP3A4 which is necessary for propter metabolism of estrogen). See Table 1 for a summary of the SSRIs with specific suggestions regarding start and target dose, titration regimen, and unique considerations for prescribing in this time period.

SSRI medications for depression belong to the most studied medicines regarding their safety profile. As the medicines can cross the placenta, they have a potential to affect the development of the embryo and fetus across the pregnancy and also be associated with drug concentrations in the newborn leading to withdrawal or toxicity. There is conflicting evidence whether exposure to SSRIs is associated with spontaneous abortion, and some authors suggest it may be the underlying illness and not the medicine carrying the risk (Johansen et al., 2015).

In animal studies, many antidepressants, including SSRIs, show some adverse effects to fetal pup development including increased pup mortality and low birth weight; however, comparisons to human studies are difficult due to multiple factors including excessive in utero exposures in animal studies (see review by Creeley & Denton, 2019).

Table 1 SSRI use in pregnancy and lactation

Generic name	S: start dose T: target dose	Titration schedule	Special considerations (side effects/risks/lactation)
SSRIs			
Sertraline	S: 25–50 mg T: 100–200 mg For very anxious patients S: 12.5 mg	Increase by 25 mg or 50 mg every 1–2 weeks as tolerated	Can be activating or sedating or cause emotional numbing, more GI effects than others Negligible amounts in breast milk (RID 0.5–3%)
Fluoxetine	S: 10 mg T: 20–80 mg	Increase by 10 mg every 1–2 weeks as tolerated	Longer half-life → withdrawal less likely if doses are missed, but also longer to get out of the system if there are adverse effects, can be activating; RID ~10 + %
Citalopram	S: 10 mg T: 20–40 mg	Increase by 10 mg or 20 mg every 1–2 weeks as tolerated	FDA Drug Safety Communication warnings about increased QTc interval at doses >40 mg, may consider getting EKG at doses >40 mg; RID ~10 + %
Escitalopram	S: 5–10 mg T: 10–20 mg	Increase by 5 mg or 10 mg every 1–2 weeks as tolerated	Safety profile like citalopram, quite well tolerated; very compatible with lactation; RID <6%
Paroxetine	S: 10 mg T: 20–40 mg CR: 25 mg	Increase by 10 mg every 1–2 weeks as tolerated CR: increase by 12.5 mg every 2 weeks as tolerated	Can be sedating, tablet formulary can cause withdrawal effects due to short half-life, CR form less likely to cause withdrawal when tapered; RID 0.5–3%
Fluvoxamine	S: 25 mg T: 25–150 mg	Increase by 25 mg every 1–2 weeks as tolerated	More often used for treatment of obsessive-compulsive disorder; RID <2%

Note: *RID* relative infant dose, i.e., the infant dose per kg body weight expressed as a percentage of the maternal dose per kg body weight. When the RID is below 10%, the exposure generally can be considered negligible. The 10% limit has been accepted by organizations such as the American Academy of Pediatrics

The overall risk for major congenital malformations of SSRIs seems low. This includes older reports on unique observations such as increased risk for malformations such as clubfoot or Hirschsprung's disease; such observations have not been replicated and may have been spurious. In contrast, there is evidence for some elevated risk for cardiovascular malformations, especially atrial and ventricular septal defects (Zhang et al., 2017) in infants exposed in the first trimester to SERTs. However, the meta-analysis also included studies that failed to adjust for confounders limiting somewhat its interpretation. Older studies had singled out paroxetine (and less clearly fluoxetine) exposure to confer increased risk for cardiac malformation, but a large registries-based study including over 500,000 pregnancies found slightly increased risk for cardiac malformations (OR 1.5, 95% CI, 1.06–2.11) associated with first trimester exposure to SSRIs as a group even when adjusting for socioeconomic status as confounder (Jordan et al., 2016). Despite this significant odds ratio, the authors pointed out that the absolute increase in risk for cardiac malformation was no more than 0.4% above and beyond the basic cardiac risk of 0.8% and that most malformations were asymptomatic or spontaneously resolving (often by chance detection in ultrasound performed with higher frequency given elevated vigilance when taking medicines in pregnancy), ultimately making this cardiac malformation finding clinically irrelevant. In addition, prior work by Huybrechts et al. (2014), also on a very large data set of over 60,000 women using SSRIs during the first trimester, found that associations between antidepressant use and cardiac defects were attenuated with increasing levels of adjustment for confounding and basically disappeared when models were fully adjusted for potential confounders. The research group also did not find any significant association between the use of specific SSRIs (such as paroxetine and sertraline) and ventricular septal defects or outflow tract obstructions (Huybrechts et al., 2014).

Adverse pregnancy and delivery outcomes have also been investigated. Regarding maternal outcomes, consistent findings have not emerged. However, outcomes such as hypertension or pre-eclampsia, postpartum hemorrhage, as well as higher rates of cesarean section have been reported although the results have not always been replicated (Fischer Fumeaux et al., 2019).

One of the effects with more consistent replication is the association between SSRI exposure and preterm birth (birth before 37 full weeks of gestation); in a recent meta-analysis including 1.3 million pregnancies, authors found a small increased risk for preterm birth with SSRI exposure compared to control women with depression who were not exposed to SSRI and control women without depression or exposure (adjusted OR 1.24; 95% CI, 1.09–1.41; Eke et al., 2016). Low birth weight following SSRI exposure has also been noted although not consistently (Cantarutti et al., 2016) as the illness itself appeared to drive the association. Despite the positive associations, again, the clinical relevance has been questioned, as the absolute risk in terms of gestational age is -2 to -5 days and of -200 to -70 grams for birth weight (Ross et al., 2013). Similarly, positive significant associations with SSRI exposure and low APGAR scores have been described, with clinical irrelevance(-0.4 points difference in exposed; Ross et al., 2013). Among exposed infants, roughly 14% get admitted to the NICU compared to 8% unexposed (adjusted OR 1.5, 95% CI, 1.4–1.5; Nörby et al., 2016), and main reasons are respiratory and neurological concerns, hypoglycemia, feeding difficulties, and pulmonary arterial hypertension (see below for more); however, interpretation of this elevated risk is limited as it lacks control for confounders (e.g., severity of underlying maternal disease). Up to 30% of exposed infants present with "neonatal adaptation syndrome" (NAS), a typically, short-lived adverse syndrome of the central nervous system (irritability, sleep/wake disturbances, and high-pitched cry), the respiratory system (respiratory distress, tachypnea), the motor system (tremors in the extremities, jitteriness, poor muscle tone, hyper-reflexia), and the GI system (hypoglycemia, feeding difficulties, vomiting). More severe symptoms such as seizures have been described in severe cases but are rare. NAS typically resolves within 2 weeks (Grigoriadis et al., 2013) although there is variation. The etiology for NAS is still being discussed, and both a withdrawal from and toxicity to serotonergic substrate have been suggested as mechanism. Some have speculated that a severe form of NAS is persistent pulmonary hypertension of the newborn (PPHN) which occurs at a rate of 2/1000 to about 5/1000 following late pregnancy SSRI exposure (Huybrechts et al., 2015). The mortality of PPHN is 10–20% making this a serious adverse event; however, according to currently available review of data, the absolute risk of PPHN in late pregnancy exposure appears to be small. Paroxetine and fluoxetine have been most commonly implicated.

Longer-term outcomes studied in infants and children with prenatal SSRI exposure have evaluated effects on brain development, growth, cognition, language development, risk for autism spectrum disorder (ASD), temperament, and attention deficit hyperactivity disorder (ADHD), among others. Overall results have not been consistent, and clinical significance of findings remains unknown. The association of SSRI pregnancy exposure to ASD, a neurodevelopmental pervasive disorder of communication and social behaviors, has been of major interest in the past decade. There have been several meta-analyses with conflicting results (e.g., Kaplan et al., 2017), and overall, many studies did not control for potential confounders (e.g., familial genetic risk for ASD, maternal and/or paternal age, or other exposures) limiting interpretation. If an association does exist, it would be small explaining no more than 1% absolute risk increase. Animal studies, however, consistently associate adult anxiety with perinatal SSRI exposure (Olivier et al., 2011).

The class of serotonin-norepinephrine reuptake inhibitors (SNRIs) is less studied. This group of medicines is an important alternative in the treatment of perinatal depressive symptoms when nonresponsive to SSRIs or if the patient has previously done well on these medicines. SNRIs have a mixed mechanism of action on both serotonin and norepinephrine and have demonstrated good efficacy for treatment-refractory depression. In terms of safety data, there are fewer studies on SNRIs in pregnancy and lactation, but available data support a comparable safety profile as SSRIs, and thus use can be recommended. Substrates in this group comprise venlafaxine, duloxetine, and desvenlafaxine, with most safety data available for venlafaxine (Kimmel et al., 2018). See Table 2 for a summary

Table 2 SNRIs and other medications for depression

Generic name	S: start dose T: target dose	Titration schedule	Special considerations (side effects/risks/lactation)
SNRIs			
Venlafaxine	S: 75 mg T: 150–300 mg	IR: incr 75 mg every 1–2 weeks as tolerated XR: 37.5 mg every 1–2 weeks as tolerated	May cause hypertension. XR less likely to cause withdrawal when tapered; less lactation data; RID 10 + %
Duloxetine	S: 30 mg T: 60–120 mg	Incr 30 mg every 1–2 weeks as tolerated	For chronic/neuropathic pain; less lactation data; RID <1%
Desvenlafaxine	S: 25 mg T: 25–400 mg		No studies currently available on use in pregnancy examining neither teratogenic risks nor available data about long-term developmental outcomes. No evidence >50 mg is helpful. RID 10 + %
Other			
Bupropion	S: 75–150 mg T: 150–450 mg SR BID dosing XR QAM dosing	Increase 75–150 mg every 2 weeks as tolerated	Activating AD; can increase anxiety and lowers seizure threshold for seizure in those with a history of seizure or those engaging in purging behaviors Not to exceed 450 mg (seizure risk!) Helpful for smoking cessation in pregnancy. May help ADHD and other addictive disorders, such as overeating in pregnancy; RID 2%
Mirtazapine	S: 7.5 mg T: 15–45 mg	Increase by 7.5 mg every 2 weeks	Causes sedation and increased appetite — helpful for anxious/depressed patients with insomnia who are not eating; watch out weight gain/diabetogenic: RID 0.5–3%
Monoamine enhancers (tricyclic antidepressants, TCAs)			
Desipramine, nortriptyline	Dose varies for each	Blood levels are possible to obtain	Less anticholinergic, so less orthostatic hypotension and constipation, which are common in pregnancy
Imipramine, doxepin, clomipramine, trimipramine, amitriptyline, protriptyline	Dose varies for each	Blood levels are possible to obtain	More cholinergic → orthostatic hypotension as side effect
Monoamine oxidase inhibitors (MAOIs)			
Isocarboxazid, phenelzine, selegiline, tranylcypromine	Dose varies for each MAOI		Requires special diet, interacts with some medications to cause life-threatening hypertensive crisis

Note: *RID* relative infant dose, i.e., the infant dose per kg body weight expressed as a percentage of the maternal dose per kg body weight. When the RID is below 10%, the exposure generally can be considered negligible. The 10% limit has been accepted by organizations such as the American Academy of Pediatrics of SNRIs with specific suggestions regarding titration regimen, target dose, and unique considerations when prescribing in peripartum.

Bupropion and mirtazapine have very unique mechanistic pathways. Bupropion is the only medication for depression to act on norepinephrine and dopamine reuptake inhibition and release simultaneously (reuptake inhibitor (NET, DAT) and releaser (NE, DA)) and is more activating, whereas mirtazapine is a specific noradrenergic and serotonergic medication for depression (NE alpha-2, 5-HT$_2$, 5-HT$_3$ receptor antagonist) with sedating property (via histamine blockade). Overall, there is a small literature for bupropion, stemming both from the literature as antidepressant medicine and for smoking cessation, and studies do not account for confounders, but there is some potential small risk increase for cardiac malformation (Kimmel et al., 2018). For bupropion, exposure in pregnancy does not seem to cause any major positive or negative impacts on the rate of congenital abnormalities, birth weight, or premature birth (Turner et al., 2019). Bupropion exposure has also been associated with ADHD in the offspring, but the sample size of exposed children was very small making interpretations difficult (Figueroa, 2010). A review of the literature on mirtazapine in pregnancy yields less than 400 cases, but the conclusion so far is that mirtazapine is not associated with increased risk of malformations or maternal complications such as hypertension, preeclampsia, or postpartum hemorrhage (Kimmel et al., 2018). Given mirtazapine's anti-nausea property, this medicine is used in pregnancy as intravenous or oral treatment for hyperemesis gravidarum. See Table 2 for a summary.

Finally, the use of monoamine enhancers (tricyclic antidepressants) and monoamine oxidase inhibitors during the peripartum is extremely limited. Tricyclic antidepressants (TCAs), first discovered in the 1950s, revolutionized the treatment of depression and preceded SSRIs, but because of their unfavorable side effect profile and toxicity in overdose, their current use is limited. Monoamine enhancers (TCAs) have a similar safety record as SSRIs in pregnancy and lactation. Monoamine oxidase inhibitors (MAOIs) were also first discovered in the 1950s and are used for treatment-resistant depression treatment. MAOIs are rarely prescribed in pregnancy or lactation as their use is complicated by the need for a special diet, the risk for drug-drug interactions causing life-threatening hypertensive crises, and extremely limited human data (Kimmel et al., 2018). See Table 2 for a summary of medicines in this group.

Anxiety

Pharmacological management of anxiety disorders is also for moderate to severe illness. The drugs of choice in most guidelines once again fall within the SSRI class. Benzodiazepines are reserved for severe agitation and anxiety, and occasional or short-term use is preferred to chronic use. Benzodiazepines function as positive allosteric modulators of the GABA-$_A$ receptor and are categorized based on half-life time and speed of onset of anxiolytic effect. In pregnancy the most commonly used are lorazepam, clonazepam, and alprazolam.

The dose should be kept as low as possible, and intermediate-acting drugs (e.g., lorazepam, clonazepam, alprazolam) are preferred over very short- (e.g., triazolam or temazepam or midazolam) or long-acting (e.g., diazepam) drugs. They are used at times concurrently with medications for depression at the beginning of treatment, but this should be done cautiously, and monotherapy is preferred during pregnancy. Benzodiazepine exposure at any time in pregnancy has not been found associated with overall congenital malformations (odds ratio [OR] = 1.13; 95% CI, 0.99–1.30) in a recent meta-analysis that pooled eight studies (as well as first trimester separately) nor with cardiac malformations when used as monotherapy (odds ratio [OR] = 1.08; 95% CI, 0.93–1.25; Grigoriadis et al., 2019). However, concurrent use of benzodiazepine and medications for depression (antidepressants) during pregnancy was associated with a significantly increased risk of congenital malformations (OR = 1.40; 95% CI, 1.09–1.80, P = 0.008; three studies), suggesting that caution with this combination is warranted. In terms of negative delivery outcomes, benzodiazepine exposure has been associated with negative delivery outcomes including preterm birth, low birth weight, and NICU admission (Grigoriadis et al., 2020). It has also been noted that the combination of an SSRI and a benzodiazepine will more likely lead to the development of NAS lasting for a longer duration than typical (Salisbury et al., 2016). Diazepam crosses the placenta and accumulates in fetal tissue, therefore use is discouraged, whereas lorazepam crosses the placenta less readily, and newborns can clear lorazepam. Buspirone, a serotonin receptor partial agonist (5-HT$_{1A}$ with 50% intrinsic activity), has not been studied in pregnancy and thus should at this time be avoided until more data is gathered.

Insomnia

Sedating antihistamines can be used for insomnia and may be preferred because they have been used safely during pregnancy for decades. For example, doxylamine is a medication used in the management and treatment of nausea and vomiting of pregnancy (NVP and hyperemesis gravidarum) and of insomnia. It is in the first-generation histamine receptor H1 antagonist class of medications. Doxylamine in combination with pyridoxine (vitamin B6) has been prescribed for nausea to millions of women for decades, and it has been deemed effective and safe for that use without documented adverse effects. The FDA approved it for use for NVP in 2013, but in Canada it had been used since the late 1950s. Similarly, other antihistamines such as diphenhydramine or hydroxyzine can also be used although they have been less studied. As with other medications, monotherapy is recommended, and in particular antihistamines should not be combined with benzodiazepines.

The hypnotic benzodiazepine receptor agonists – zolpidem, zopiclone, and zaleplon ("z drugs") – have not been as well studied as the benzodiazepines, but data to date do not associate them with malformations (Chaudhry & Susser, 2018). However, preterm birth and low birth weight have been reported (Huitfeldt et al., 2020). Medications for depression with sedating side effects are also options, and these include trazodone, mirtazapine, and amitriptyline among others (McLafferty et al.,

2018). Trazodone is a medication for depression with a chemical structure unrelated to other classes and at low doses has weak antidepressant yet strong hypnotic property. Trazodone has been available in the USA since 1982 and is commonly prescribed as a medication for insomnia (hypnotic), often in conjunction with other medications for depression. Trazodone has limited safety data but does not seem to confer risk for malformations (McLafferty et al., 2018).

Many women take "natural products" as sleep aids; these have not been studied in pregnancy. Data on melatonin safety are limited and controversial, and most data come from animal studies. Certain studies show that melatonin does not cause adverse outcomes in offspring and in some cases may be protective through its antioxidant properties. Regardless, some animal studies demonstrate possible risks of melatonin exposure (decreased birth weight and prolonged gestation) despite it not affecting the fetal hormonal system (McLafferty et al., 2018).

Lactation Considerations

Breastfeeding provides tremendous physical, developmental, and psychological benefits for mothers and infants. The American Academy of Pediatrics and the Academy of Breastfeeding Medicine recommend exclusive breastfeeding through 6 months followed by 1 year or longer of continued breastfeeding as complimentary foods are introduced. Human milk represents the ideal primary source of nutrients and provides better immunological and antioxidant protection than do milk substitutes. Therefore, women are encouraged to breastfeed when possible (Berle & Spigset, 2011). The dilemma in the treatment of breastfeeding mothers is weighing the potential risk to the infant of exposure to medications for depression through breast milk against the disadvantage of not receiving mother's milk.

Psychotropic drugs are secreted into breast milk but in differing amounts and much less than what is transferred in utero. In general, breastfeeding is encouraged, and most guidelines recommend continuation of the SSRI (or other medications for depression) during lactation. Drug concentrations in milk parallel those in maternal plasma, but with a slight delay. Due to the lipophilicity of the drugs, milk levels are typically somewhat higher than the levels in maternal plasma. The milk drug concentration can be used to estimate the daily drug dose ingested by the infant, assuming an average milk intake of 150 ml per kg body weight per day. The infant dose per kg body weight can then be expressed as a percentage of the maternal dose per kg body weight. When the relative infant dose (RID) is below 10%, the exposure generally can be considered negligible. The 10% limit has subsequently also been accepted by organizations such as the American Academy of Pediatrics. The RIDs are close to or above 10% for citalopram, fluoxetine, and venlafaxine. RIDs are lower for escitalopram ($<6\%$) and are low for fluvoxamine ($<2\%$), paroxetine (0.5–3%), sertraline (0.5–3%), duloxetine ($<1\%$), reboxetine (1–3%), bupropion (2%), and mirtazapine (0.5–3%) (Berle & Spigset, 2011; Hale, 2020).

Sertraline has been the most studied and found in low concentrations (RID 0.5–3%) in breast milk and is thus favored by most guidelines. Fluoxetine is found in higher concentrations compared to other SSRIs because of its active metabolites (Berle & Spigset, 2011; Hale, 2020) and is generally not preferred. Factors that can affect the decision include prematurity and end organ issues such as kidney, liver, neurological, or circulatory disorders. For minimizing amounts even further, the mother can take once daily dosing and breastfeed just before taking the dose. Most guidelines do recommend the infant be monitored for changes in behavior such as sedation, irritability, and crying. The antihistamines described above as medications for insomnia are also compatible with breastfeeding and have been found in low levels in breast milk (So et al., 2010); nevertheless, the infant should be monitored for sedation. Regarding benzodiazepines, intermediate half-life drugs with lower transferred levels into the breast milk (lorazepam) are preferred. Sedation can be one of the adverse effects reported (Kelly et al., 2012), especially in infants who are known to poorly metabolize drugs, for example, infants born prematurely. Zolpidem has been found in low levels in breast milk. Although some have described using the "pump and dump" method where breast milk is pumped out of the breast at the time of peak medication concentration and discarded, this is not a method supported by scientific literature and adds unnecessary burden for the mother and thus should be discouraged.

Peripartum Management of Bipolar Disorder and Primary Psychosis

Introduction

In population-based studies, pregnancy seems associated with a modest reduction in risk of mania or depression in women with bipolar disorder (Munk-Olsen et al., 2006); however, clinic-based studies provide conflicting results reporting high recurrence rates in pregnancy, particularly in women who discontinue prophylactic medication (Viguera et al., 2007). In contrast, the postnatal period is clearly associated with a surge of new onsets, exacerbations, or recurrences, especially in women with preexisting bipolar disorder or other psychotic disorders (Jones et al., 2014). Nevertheless, clinical studies have shown that medications for psychosis (also called "antipsychotics") and drugs for bipolar disorder (also called "mood stabilizers") are often discontinued in the early weeks of pregnancy because it is feared that they may harm the fetus. In the following sections, it will be shown that significant reproductive risks reported in the older literature for some medications for psychosis and for bipolar disorder (mood-stabilizing medications) are now generally found to be much lower or completely accounted for by confounding factors. This is due to considerable improvements in the recording of clinical data, study design, and statistical methods and greater availability of large databases. However, in several areas high-quality research is still lacking, and considerable uncertainty continues. Guidelines and recommendations for pharmacological management in the perinatal period will be discussed.

Medications for Psychosis or "Antipsychotics"

As these medicines can also cross the placenta, they have similarly potential to adversely affect the development of the embryo and fetus across the pregnancy leading to congenital malformations and poor neonatal outcomes. Previous studies examining the risk for congenital malformations after first trimester exposure to medications for psychosis usually reported up to twofold increases of all congenital malformations taken together and sometimes found a specific increase of cardiovascular malformations. Two main problems made these results unreliable: small sample sizes and insufficient accounting for confounding factors that are now known to be associated with congenital malformations, such as the underlying psychiatric disorder, its severity, maternal body mass index, nicotine and other substance misuse across pregnancy, concomitant medication use, and other factors.

Significant progress was recently made with a cohort study of women enrolled in the Medicaid program in the USA by Huybrechts et al. (2016) which used propensity scoring to control for underlying psychiatric disorder and other potential confounders. It included the largest sample to date with over 1.3 million controls, over 700 women who had filled a prescription for dopamine blockers (first-generation antipsychotic (FGA) or "typical" antipsychotic), and over 9000 women who had filled a prescription of a dopamine serotonin blocker (second-generation antipsychotic (SGA) or "atypical" antipsychotic) in the first trimester. This allowed for the first time to assess the safety of individual atypical antipsychotics within one database. Except for a small increase in the overall risk for congenital malformations after exposure to risperidone, no associations were found for any other typical or atypical medication for psychosis neither to elevated risk for overall malformations nor specific cardiac malformations. Risperidone exposure was associated with small relative risk increase for overall malformations (RR, 1.26; 95% CI, 1.02–1.56) and for cardiac malformations (RR, 1.26; 95% CI, 0.88–1.81) even after controlling for measured confounders. The specific pathway for risperidone is difficult to explain with a known biological mechanism. Nonetheless, overall current evidence suggests that antipsychotic drugs are not major teratogens and that further studies

are needed to clarify the finding for risperidone (Betcher et al., 2019).

Since the endocrine and metabolic changes accompanying pregnancy enhance women's vulnerability to dysglycemia, antipsychotic therapy may increase the risk of gestational diabetes. There is a well-recognized association between treatment with some medications for psychosis, especially those in the atypical category, and metabolic side effects including weight gain and diabetes in the general population, and thus, the US Food and Drug Administration (FDA) has required all manufacturers of these medications to add a warning for risk of hyperglycemia and diabetes to their labels since 2003. Given these facts, the metabolic safety of antipsychotic drugs for pregnant women is particularly poignant. A recent large cohort study on over 1.5 million pregnancies provided somewhat reassuring data. Park et al. (2018) assessed pregnancy exposures among users of several medications for psychosis (the number of exposed users was as follows: 1924 aripiprazole; 673 ziprasidone; 4533 quetiapine; 1425 olanzapine; and 1824 risperidone) and found that quetiapine and olanzapine were associated with small and moderate, respectively, risk increase for gestational diabetes (adjusted relative risk, RR, 1.28; 95% CI = 1.01–1.62 and 1.61; 95% CI = 1.13–2.29), whereas other antipsychotics (aripiprazole, ziprasidone, or risperidone) were not. Gestational diabetes increases the risk of poor infant outcomes (i.e., preterm birth, macrosomia). Thus, monitoring weight gain and body mass index as well as reinforcing healthy diet and exercise throughout pregnancy and monitoring for metabolic and glycemic markers is critical to help mitigate risk. In the UK it is recommended (and common practice in the USA) to routinely screen all pregnant women who are taking antipsychotics in pregnancy for gestational diabetes (National Institute for Health and Care Excellence [NICE], 2014).

Some studies have reported high rates of "neonatal adaptation syndrome" (NAS) to antipsychotic medication in late pregnancy, but the study by Vigod et al. (2015) suggests that such results may be attributable to other factors, such as increased rates of concomitant medication use, substance use, or alcohol use. Data on the neurodevelopment of children following intra-uterine exposure to antipsychotics are limited. In their systematic review and meta-analysis, Poels et al. (2018) concluded that the most consistent finding was a delay in motor development that occurred within the first few months of life and appeared to be transient.

Finally, a word regarding dosing adjustments in pregnancy. In the first systematic study of plasma concentrations of three atypical antipsychotics across the perinatal period, Westin et al. (2018) found no changes for olanzapine but progressive decreases for quetiapine and aripiprazole that reached -76% ($p < 0.001$) and -52% ($p < 0.001$) in the third trimester. These results are consistent with the pregnancy-induced changes in activity of those hepatic P450 enzymes that are involved with the breakdown of the three agents (CYP 1A2, CYP 3A4, and CYP 2D6/3A4, respectively). Clinical consequences are yet unknown, but women's mental state should be particularly closely monitored during this time and dosing adjusted if clinically indicated. If available, measuring plasma levels in the perinatal period, preferably beginning before conception, is recommended and repeating in intervals until the first 2 weeks postpartum.

Breastfeeding is compatible with intake of antipsychotics in postpartum. As described in more detail in the section on SSRI exposure and breastfeeding, it is an important consideration that the daily drug dose ingested by the infant is a percentage of the maternal plasma dose and that a relative infant dose (RID) below 10% maternal plasma dose has been regarded as acceptable by organizations such as the American Academy of Pediatrics. Small case numbers to date suggest low or very low transfer for the majority of antipsychotic drugs into breast milk. Higher values have been found for haloperidol, amisulpride, and risperidone where the estimated daily dose for a fully breastfed infant can approach or exceed 10% of the maternal dose (Uguz, 2021; Hale, 2020). In the limited published clinical observations of breastfed infants, relatively few adverse effects have been reported (Uguz, 2016). However, some authors advise against clozapine therapy

during breastfeeding because there is little clinical experience and sedation and adverse hematologic effects have been reported in the infants (Drugs and Lactation Database, 2021).

See Table 3 for a summary of most used antipsychotics in peripartum with specific suggestions regarding starting and target dose, titration regimen, and unique considerations when prescribing in peripartum, as well as for their mechanism of action. No or very sparse data are yet available for antipsychotics that have become available more recently such as paliperidone, asenapine, lurasidone, and cariprazine.

Medication for Bipolar Disorder ("Mood Stabilizers")

Several drugs targeting the GABA/glutamate neurotransmission system in the brain with varied known or proposed modes of action fall into this group, including valproate, carbamazepine and oxcarbazepine, lamotrigine, and lithium. See Table 3 for a summary of drugs for mood relapse prevention with specific suggestions regarding starting and target dose, titration regimen, and unique considerations when prescribing in peripartum, as well as for their mechanism of action.

Valproate's precise mechanism of action is unclear, but the proposed mechanisms include affecting GABA (γ-aminobutyric acid) levels in the CNS, blocking voltage-gated ion channels in an activity-dependent fashion, and also inhibiting histone deacetylase, all of which "calm" the firing of neurons (thus an anti-seizure medicine). Valproate was first made in 1881 and came into medical use in 1962 and is now primarily used to treat epilepsy (absence, partial, and generalized seizures) and bipolar disorder and to prevent migraine headaches. There is now consistent and compelling evidence that valproate prescribed in pregnancy causes widespread harm to the embryo and fetus (Wieck & Jones, 2018). Exposure in early pregnancy increases the rate of major congenital malformations threefold compared to the general population. The defects tend to be significant and affect several organ systems, including neural tube defects. Neurodevelopment is also compromised with cognitive developmental delay, autism, and autism spectrum disorder

Table 3 Antipsychotics and mood stabilizers

Generic name	S: start dose T: target dose	Titration schedule	Special considerations (side effects/risks/lactation)
Medications for psychosis or "antipsychotics"			
Quetiapine	S: 25 mg T: 25–300 mg	Increase by 25–50 mg every 3–5 days	Dose for mood stabilization (100–300 mg), smaller dose for sleep (25–50 mg) or anxiety (12.5–25 mg); sedating; weight gain/metabolic risk; most + data on malformation safety; RID 1–3%
Olanzapine	S: 2.5 mg T: 2.5–10 mg	Increase by 2.5–5 mg every 3–5 days	Sedating; weight gain/metabolic risk; positive malformation safety data; RID 1.5–4%
Aripiprazole	S: 1–2 mg T: 2–15 mg	Increase by 1–5 mg every 3–5 days	Positive malformation safety data; not sedating; low weight/metabolic risk, but akathisia possible. RID 1–8%; may decrease breast milk supply due to hypoprolactinemia (aripiprazole is partial agonist at D2 receptor)
Ziprasidone	S: 20 mg daily T: 20–60 mg twice daily	Increase by 20 mg twice daily every 3–5 days	Less data available, but so far + malformation safety; not sedating; low weight/metabolic risk. RID 1–2%
Risperidone	S: 0.5–1 mg T: 1–6 mg	Increase by 0.5–1 mg every 3–5 days	Risperidone may be associated with small relative risk increase for total major congenital malformations (RR, 1.26); smaller weight/metabolic risk; RID 2–10%
Medication for bipolar disorder ("mood stabilizers")			
Valproic acid (VPA)	Do not prescribe to women with childbearing potential if not on highly reliable contraceptives		Increased risk of total malformations, and several specific defects, including neural tube defects, cognitive impairment, ASD, ADHD; breastfeeding not recommended as woman could become pregnant, RID 1–4%, monitor infant for hepatotoxicity and thrombocytopenia
Carbamazepine Oxcarbazepine	Do not prescribe to pregnant women or women with childbearing potential if not on highly reliable contraceptives		Risk of neural tube and craniofacial defects, fetal vitamin K deficiency/bleeding risk, IUGR, neonatal toxicity; breastfeeding not recommended as woman could become pregnant, RID 1–3%, monitor infant for hepatotoxicity and thrombocytopenia
Lamotrigine	S: 25 mg T: 200 mg (as 100 mg twice daily)	25 mg/day × 2wks, then 50 mg/day × 2 weeks, then 100 mg/day × 2 weeks, then 100 mg twice daily	Large database indicating it is not a teratogen, not sedating; low weight/metabolic risk; but Stevens-Johnson syndrome

(continued)

Table 3 (continued)

Generic name	S: start dose T: target dose	Titration schedule	Special considerations (side effects/risks/lactation)
			(allergic rash); RID up to 50%, infant monitoring required
Lithium			
Lithium carbonate	S: 150–300 mg T: 900–1200 mg blood level 0.6–1.2 mEq/	Increase 150–300 mg every 3–7 days Narrow therapeutic window, monitor blood level and kidney function in pregnancy	Lithium associated with relative risk increase for cardiac anomaly (RR, 1.65); risk is dose-dependent (>900 mg daily); Li toxicity with NSAIDs; thyroid and GI side effects possible. RID up to 50%, infant monitoring required

Note: *IUGR* intrauterine growth retardation, *GI* gastrointestinal; *RID* relative infant dose, i.e., the infant dose per kg body weight expressed as a percentage of the maternal dose per kg body weight. When the RID is below 10%, the exposure generally can be considered negligible. The 10% limit has been accepted by organizations such as the American Academy of Pediatrics

being 3–7 times more common. There are also preliminary indications that in fact most children have at least mild cognitive problems with an average IQ drop of nearly 10 points below expected (Baker et al., 2015).

In view of this mounting evidence and a slow decline in the numbers of exposed pregnancies in European countries, the European Commission passed a law in 2018 that aimed to reduce prescribing in pregnancy. It states that (1) valproate should not be used in pregnant girls or women who have a mental illness and (2) valproate should only be prescribed to women or girls who have childbearing potential if other treatments are ineffective or not tolerated and a pregnancy prevention program is followed. A similar call for action has been made in the USA (Gotlib et al., 2016).

Carbamazepine (and oxcarbazepine) are also activity-dependent voltage-gated sodium channel blockers, thus preventing repetitive and sustained firing of the most active neurons in the brain, an action that may contribute to possible reduction of glutamate excitotoxicity. Carbamazepine was discovered in the 1950s and oxcarbazepine in 1969. They are used as anti-epilepsy medicines, for neuropathic pain, and as mood stabilizers in bipolar illness. Current evidence suggests that children exposed in early pregnancy to carbamazepine are twice as likely to be born with major congenital malformations than children of untreated mothers with epilepsy or healthy controls, but exposure at various times of fetal development is not thought to cause ADHD or autism spectrum disorder (Medicines and Healthcare Products Regulatory Authority [MHRA], 2021). Evidence on intellectual impairment is conflicting. NICE (2014) guidelines recommend that carbamazepine should not be offered to women who are pregnant, breastfeeding, or planning a pregnancy but do not explicitly advise against its use by other women with childbearing potential. Should this drug be considered for a woman with childbearing potential who does not plan a pregnancy, she should be informed about the reproductive risks and advised on contraception.

Lamotrigine's mode of action is also voltage-gated sodium channel blockage, which stabilizes presynaptic neuronal membranes and inhibits glutamate release and glutamate-induced neurotoxicity. On the market in Europe and the USA since the 1990s, it has been used as drug for specific seizure types and for bipolar depression (Yatham et al., 2018). A large amount of data show that treatment with lamotrigine at recommended doses is neither associated with an increased risk of major congenital malformations nor abnormal fetal growth or development (Medicines and Healthcare Products Regulatory Authority [MHRA], 2021). Although smaller in volume currently available clinical studies on neurodevelopmental outcomes also do not suggest that children exposed to lamotrigine in pregnancy are at an increased risk of neurodevelopmental disorders or delays (MHRA, 2021).

The high levels of estradiol during pregnancy cause substantial increases in lamotrigine clearance in most women and returns to pre-pregnancy levels by about 3 weeks postpartum (Polepally et al., 2014). It is currently not known whether the resulting decreases in plasma levels affect mood stability in women with bipolar disorder. Ideally, blood serum concentrations should be measured preconception, in each trimester of pregnancy and in the early postnatal period. The results will support decisions about dosing in combination with close monitoring of a woman's mental state and consideration of other mood-stabilizing interventions.

Breastfed infants whose mothers are taking lamotrigine have relatively high plasma lamotrigine levels which can reach half of the maternal value (i.e., the RID can be up to 50% of the maternal plasma dose), and there have been occasional reports of adverse effects. These include withdrawal symptoms after abrupt weaning and mild and asymptomatic thrombocytosis (Drugs and Lactation Database, 2021). One infant, who had a high plasma level in the context of a high maternal dose of lamotrigine (850 mg daily), was reported to have experienced a severe apneic episode (Drugs and Lactation Database, 2021). During maternal therapy with lamotrigine, breastfed infants should be carefully monitored for side effects such as apnea, skin rash, excessive drowsiness, and poor sucking (Drugs and Lactation Database, 2021). If there is any concern, the infant's plasma level should be measured, and breastfeeding discontinued until the cause of the problem has been identified. Monitoring of the platelet count and liver function may also be advisable (Drugs and Lactation Database, 2021).

Lithium

Lithium had been introduced for medical use in 1949 by John Cade in Australia who published an article about the observed antimanic action of lithium. Intrigued by these early findings, Mogens Schou, a research-clinician at Aarhus University in Denmark, confirmed lithium's property for mood stabilization in a double-blind placebo-controlled study and got lithium established in psychiatry. Since, it has been used extensively as lithium compounds (salts) for the treatment of bipolar illness, major depression, and suicidality. The exact mechanism of action is still unknown. In the late 1970s, a study, which was subsequently criticized for its methodological flaws, reported a large increase in the incidence of cardiovascular congenital anomalies and in particular the exceedingly rare Ebstein anomaly (anomaly of the tricuspid valve – the valve between the right atrium and the right ventricle of the heart) in children exposed to lithium in the first trimester of pregnancy. Later studies suggested that these observations were most likely a large overestimate. Nevertheless, these original findings have influenced clinical practice until the present time.

Results from two recent studies that assessed the teratogenic risk with much larger samples and improved methodology should be highlighted. A European registry-based study covering 5.6 million deliveries found no cases of lithium exposure among 173 babies with otherwise unexplained Ebstein's (Boyle et al., 2017). In the retrospective cohort study by Patorno et al. (2017), after adjustment for several important confounders, the adjusted risk ratio for cardiac malformations among infants exposed to lithium as compared with unexposed infants was 1.65 (95% confidence interval [CI], 1.02–2.68). The risk ratio was 1.11 (95% CI, 0.46–2.64) for a daily dose of 600 mg or less, 1.60 (95% CI, 0.67–3.80) for 601–900 mg, and 3.22 (95% CI, 1.47–7.02) for more than 900 mg. Thus, the effect was dose related and only significant when a daily lithium dose of more than 900 mg was prescribed. At this dose level, the risk was increased threefold. The likelihood of confounding by indication was low because a similar difference was found when a comparison was made with children who had been exposed to lamotrigine in the first trimester. However, since lithium and lamotrigine therapies tend to be used for different bipolar disorder subtypes and severities, residual confounding may still play a role in this finding.

There have been few studies of other pregnancy outcomes and no systematic studies of the physical health in newborns who were exposed to lithium in pregnancy. A variety of problems, mostly consistent with side effects seen in adults, but no neonatal deaths, have been described in case reports. There is 1 5-year prospective study of 60 children who showed no excess of physical or mental anomalies compared to their unexposed siblings.

In view of this evidence and the greater prominence of lithium in recent guidelines for the long-term treatment of bipolar 1 disorder, the need to switch medication from lithium to an antipsychotic before conception or in early pregnancy is now less compelling. Lithium treatment during pregnancy, however, requires some monitoring by the prescriber. Since the lithium serum concentration during pregnancy is subject to the dynamic changes of renal function across the three trimesters, it needs to be checked more frequently. The NICE guidelines (2014) recommend once a month lithium level checks until gestational week 36 and then once weekly until delivery. Because of the risk of preeclampsia and other conditions with renal impairment, creatinine levels should always be included. In order to prevent toxic lithium levels, the antenatal care plan should include urgent action by psychiatric and maternity services in case of deteriorating renal function.

Because lithium levels are stable in the week before and after delivery, lowering the dosage or discontinuing lithium prior to delivery is not recommended (Molenaar et al., 2021). This is particularly important in view of the very high risk of bipolar recurrence in the early postnatal period. However, because of redistribution of fluids in the immediate postnatal time period with potential risk for acute lithium toxicity (El-Mallakh, 1986), monitoring is warranted. During delivery, standard recommendation is to monitor lithium level 12 hours after last lithium dose and to repeat level on the first postnatal day. If the result is not at the aimed target level, dose adjustments should be continued in intervals until this is achieved.

Finally, historically lithium intake and breastfeeding were seen as contraindicated. Lithium easily enters the breast milk and, in several cases, reports breastfed infants had serum levels up to 50% of the maternal value (Drugs and Lactation Database, 2021). Several instances of infant health problems were described, including suspected lithium intoxication, abnormal thyroid function tests, slow weight gain, and delay in motor development (Drugs and Lactation Database, 2021). It has also been highlighted (Drugs and Lactation Database, 2021; Hale, 2020) that numerous reports exist of infants without any signs of lithium toxicity or developmental problems and that maternal lithium therapy should not be an absolute contraindication to breastfeeding of full-term infants. Thus, nowadays a more nuanced approach is called for. Exercising great caution is recommended when making decisions about breastfeeding during lithium therapy. Some authors recommend monitoring infant serum lithium and renal and thyroid function in intervals, but these precautionary measures may not be possible in some mental health or primary care services. Breastfeeding should be discontinued immediately, and the infant evaluated if she/he appears restless or lethargic or has feeding problems.

Peripartum Management of Trauma-Related Disorders

Introduction

As an overarching concept, a traumatic event may include shocking or overwhelming experiences that involve real or perceived threat of bodily harm, injury, or death. Importantly, objective and subjective ratings of events as "traumatizing" do not necessarily align, and memory and cognitive factors have an important role in subjective appraisals of trauma. Not everyone impacted by a traumatic event develops long-lasting psychopathology, but some do. Women in the peripartum period are vulnerable to develop either new onset, reoccurrence, or exacerbation of existing trauma-related psychopathology, especially if they carry vulnerability due to prior trauma. The range of trauma events in a woman's life leading up to greater vulnerability for peripartum trauma-related disorders is plentiful, ranging from prior perinatal loss or traumatic childbirth, the history of childhood adversity, past or ongoing interpersonal violence, or exposure to natural disasters, accidents, discrimination, or other societal threats. Thus, perinatal trauma-related conditions are not an uncommon occurrence, yet often overlooked because of failure to screen for it in the perinatal care context. If these conditions remain unrecognized and untreated in peripartum, the impacts on mother and child are detrimental (Erickson et al., 2019). In this section, three manifestations of trauma-related disorders, posttraumatic stress disorder (PTSD), complex PTSD (cPTSD), and borderline personality disorder (BPD), are highlighted, and pharmacologic management strategies during the peripartum period are discussed.

To meet clinical criteria for posttraumatic stress disorder (PTSD) in the fifth edition of the *Diagnostic and Statistical Manual of Mental Disorders* (DSM-5), a traumatic event must include death, serious injury, or sexual trauma that is either actual or threatened; individuals must also exhibit a combination of symptoms that include reexperiencing, avoidance, negative changes in mood/cognitions, and increased arousal (American Psychiatric Association, 2013). Approximately 3–7% of perinatal women suffer from perinatal PTSD; however, the condition remains often undetected and untreated because of lack in screening and awareness. More recently, complex PTSD (cPTSD or complex trauma) has been established, denoting a disorder that develops after repeated or chronic exposure to traumatic events where escape is not possible, such as childhood sexual and physical abuse, exposure to human trafficking, enslavement, bondage, or torture, and this diagnosis has been formally recognized in the International Classification of Diseases, 11th edition (ICD-11, release date 2018). Core symptom clusters of cPTSD include those associated with PTSD, such as avoidance, hyper-vigilance, and intrusive symptoms, in addition to three symptom clusters which collectively represent disturbances in self-organization such as affective dysregulation, negative self-concept, and disturbances in relationships. Finally, there is some overlap to another disturbance of self-organization and relationship coherence, and that

is the diagnosis of borderline personality disorder (BPD), a condition characterized by a chronic pattern of unstable relationships, distorted sense of self, strong emotional reactions and/or chronic feelings of emptiness, engagement in self-harm and other dangerous behaviors, chronic suicidality, fear of abandonment, and at times detachment from reality. The origins of BPD are multidetermined including genetic and neurological risks, but adverse life events (especially childhood abuse) appear to play a major role, and women are 3 times as likely as men to be diagnosed with this condition.

Pharmacologic Management of PTSD

Current treatment guidelines recommend psychological interventions as the first line of treatment and pharmacological management as second. However, in circumstances where psychological interventions cannot be provided or the person is unwilling or not in a position to engage in therapy, pharmacological agents are recommended as first line.

Among pharmacological agents, current evidence of efficacy is highest for serotonin reuptake inhibitors (SSRIs) such as fluoxetine, paroxetine, and sertraline. Among serotonin-norepinephrine reuptake inhibitors (SNRIs), venlafaxine is conditionally recommended for treatment of PTSD. These drugs are also highly beneficial for associated symptoms of hyperarousal, dissociation, and comorbid depression (National Institute for Health and Care Excellence [NICE], 2018). Medication use should be also considered in reducing the risk of suicides. Although clinical trials showing efficacy or added benefits of combined medication use and psychotherapy in PTSD are lacking, it is expected to be beneficial similar to the combined treatment regimen followed for treatment of depression and anxiety disorders.

There are few randomized clinical trials involving SSRIs as augmentation with prolonged exposure therapy versus placebo, showing superior results with augmentation strategies (Brady et al., 2000). Common clinical practice also involves prescribing a pharmacological agent such as SSRI combined with psychotherapy for PTSD, with or without comorbid depression. In case of non-tolerability to SSRI, other agents can be considered. In women, the specific considerations regarding side effects of SSRIs include risk of menorrhagia, postpartum hemorrhage, and sexual dysfunction (Uguz et al., 2012). Please refer to the section on pharmacological management of depression and anxiety for more information on the use of SSRIs during pregnancy and breastfeeding. In cases of PTSD with featured symptoms of agitation, aggression, and psychotic symptoms, the use of antipsychotics such as quetiapine (dopamine, serotonin receptor antagonist (D2, 5-HT2) and norepinephrine reuptake inhibitor [NRI]) or risperidone (dopamine, serotonin, norepinephrine receptor antagonist [D2, 5-HT2, NE alpha-2]) has been suggested (Villarreal et al., 2016). Please refer to the previous section of this chapter on the safety of these medications in pregnancy and postpartum.

There are novel agents that are being tried successfully for PTSD symptom clusters. For example, the synthetic cannabinoid nabilone in doses of 0.5–2 mg daily has been evaluated in an open-label trial and shown improvements in sleep disturbance and nightmares among patients with PTSD (Orsolini et al., 2019); however, none of these studies have been done or applied to pregnant or postpartum women. Similarly, while the antihypertensive drug prazosin (NE alpha-1 receptor antagonist) has been shown effective to treat PTSD-associated nightmares and sleep disturbances with moderate to large effect size, there is a lack of safety date for the use in pregnancy or when breastfeeding. A recent review discussed current safety date on prazosin in pregnancy (Davidson et al., 2021) and concluded that currently adequate reproductive safety data that would support use are missing and more research is needed. The authors outline, that while prazosin is used commonly in pregnancy to treat maternal hypertension, there are no data on use for peripartum PTSD. Reports on adverse outcomes such as intrauterine growth restriction or intrauterine death, when using prazosin to treat hypertension, are attributed to the underlying illness itself (hypertension or severe preeclampsia) rather than to in utero exposure to prazosin. However, there is concern of prescribing this drug to

normotensive women with PTSD, risking maternal hypotension and subsequent adverse fetal effects. There are no reports examining effects of prazosin during lactation.

Ketamine (in doses up to 0.5 mg/kg body weight) is being actively researched, and many ongoing trials study ketamine's effect on pain relief, depressive symptoms, and treatment of chronic PTSD. Rapid symptom reduction for depression and PTSD symptoms within 24 hours has been reported (Liriano et al., 2019). To date, there are no studies on the safety of ketamine or esketamine (delivered as nasal spray) on pregnancy and lactation. The only data available are from ketamine use for anesthesia and/or analgesia during vaginal and cesarean deliveries, and reports on neonatal outcomes have been so far favorable, although there are dose-dependent, transient PNAS symptoms possible.

Pharmacological Management of cPTSD

Since cPTSD is a relatively new diagnosis introduced into the classificatory system, there are no current recommended guidelines specific to cPTSD. Most of prior treatment trials on PTSD included patients with cPTSD, and history of childhood maltreatment has been found a predictor of poor treatment response (Karatzias & Cloitre, 2019). The multidimensional nature of complex trauma presentations with challenging comorbidities such as physical illnesses, dissociation, substance abuse, ongoing domestic violence, and other maltreatments makes it more complicated to plan research trials and study the various clinical outcomes. Hence treatment plan for this condition has to be largely adopted from guidelines for treatment of PTSD, borderline personality disorder, and dissociative disorders. The first-line drugs for PTSD (fluoxetine, paroxetine, sertraline, and venlafaxine) in conjunction with evidence-based psychotherapies are considered the first-line treatment for cPTSD.

Pharmacological Management of BPD

Women presenting for obstetric services with a clinical diagnosis of BPD experience considerable risk during prenatal care and in birthing. They anticipate birth as traumatic and frequently requested early delivery; often there is high comorbidity with multiple other psychiatric conditions (including depression, PTSD, and substance use disorder) complicating the course of pregnancy and labor, associated with greater likelihood of negative birth outcomes such as lowered Apgar scores, prematurity, and child protection involvement when compared with controls (Blankley et al., 2015). Thus, early detection and treatment are critical. First line of treatment for this disorder is evidence-based therapy, most commonly practiced dialectical behavioral therapy (DBT), with manuals emerging specifically for the peripartum population (unpublished manual "Michigan Model DBT-Pregnancy and Postpartum" Bresky et al., 2021). The pharmacological management of BPD generally focuses on specific symptom dimensions as the target of treatment. There is no drug authorized specific for BPD by the Food and Drug Administration (FDA) in the USA and by UK market authorization. However, the naturalistic studies on prescription patterns reveal that polypharmacy is a common practice (Yadav, 2020). Although no medications have been formally approved for the treatment of borderline personality disorder, people with this condition receive prescriptions for large amounts of medication with as many as 90% receiving psychiatric drugs. Most of the time, medications are prescribed for co-occurring psychiatric presentations such as anxiety, depression, bipolar disorder or mood dysregulation, PTSD/dissociation symptoms, impulsivity, and comorbid substance use disorders (Stoffers & Lieb, 2015). Majority of the patients are prescribed an antidepressant or an antipsychotic, particularly for affective dysregulation. Psychiatric drugs are used to target cognitive-perceptual symptoms (e.g., distorted thinking, paranoia, mini-psychotic breaks), impulsive behaviors and loss of control (e.g., self-harm, rage attacks, para-suicidal gestures), and affective dysregulation (e.g., mood lability, panic attacks, acute depressive episodes).

SSRIs, such as fluoxetine and sertraline, are the first-line drugs recommended targeting impulsive behaviors and affective dysregulation. MAOI [serotonin, norepinephrine, dopamine, histamine, and melatonin monoamine oxidase enzyme

inhibitor (MAO-A and -B), releaser (DA, NE)] agents such as phenelzine and tranylcypromine followed by medications for mood stabilization such as lamotrigine and lithium are recommended as second line. Lithium has strong beneficial evidence for impulse control and reduction of acute suicidality. SSRIs and MAOI agents are also recommended for the treatment of cognitive-perceptual symptoms; however, antipsychotic medicines (such as aripiprazole, risperidone, quetiapine, and others) are also widely used to treat these symptoms despite the lack of strong evidence based on the clinical trials. Safety of these drugs in pregnancy and when breastfeeding has been reviewed in prior sections of this chapter.

Naltrexone (25–50 mg per day) and clonidine (75–150 micrograms) are agents with good trials-based evidence for reduction in self-injurious behaviors and dissociative symptoms (Moghaddas et al., 2017). Both have also good safety data in pregnancy (Towers et al., 2020); especially clonidine has been used as antihypertensive drug during pregnancy for some time (Leavitt et al., 2019). Naltrexone is also compatible with breastfeeding. Although available data is limited, naltrexone shows minimal excretion into breast milk, and hence breastfeeding can be carried on. In contrast, clonidine has negative effects on lactation and high serum levels are found in breast milk; thus breastfeeding is discouraged if this drug is required.

Peripartum Management of Attention Deficit and Hyperactivity Disorder (ADHD)

Introduction
Over the years, awareness and treatment of ADHD have increased substantially. It is now not uncommon to see women of reproductive age who are diagnosed with ADHD and prescribed medications to manage their symptoms. Treatment of ADHD in the perinatal period is important to preserve overall functioning of the patient, as well as, in some cases, to effectively control symptoms of disorders commonly comorbid with ADHD such as depression and anxiety.

Additionally, the risk of tobacco dependence is increased among those with untreated ADHD (Kollins et al., 2005), and treatment of ADHD with a stimulant decreases the risk of development of substance use disorder in individuals with ADHD (Groenman et al., 2013). Therefore, proper management of ADHD in the perinatal patient serves not only to preserve functioning, but to protect patients from comorbid psychiatric conditions that increase the risk of pregnancy and postpartum complications. Finally, the astute clinician is aware that more mild cases of ADHD may not be diagnosed until later in life when an individual's capacity to manage themselves and their time is put to the test by participation in higher-level education or increasing life demands and responsibilities (including having children). Patients reporting significant difficulties with concentration and focus on the postpartum period should be screened for ADHD in order to avoid missing this important diagnosis. If this diagnosis is overlooked, it can be challenging to effectively treat the comorbid diagnosis/diagnoses for which the patient initially sought treatment.

Use of Monoamine Releasers (Stimulants) (Prescription Amphetamine Salts and Methylphenidate)

The medications belonging in this group (amphetamine, dextroamphetamine, lisdexamfetamine, methylphenidate, and dexmethylphenidate) are all dopamine and norepinephrine reuptake inhibitors (DRI, NRI) and releaser (DA, NE), both mechanisms leading to enhanced release of these neurotransmitters in the synaptic cleft. The safety profiles of these prescription monoamine releasers for ADHD in pregnancy are fairly reassuring. The data do have some limitations, however. One is that much of the early data about the safety of these compounds in pregnancy was derived from women who were abusing illicit monoamine releasers such as methamphetamine and therefore likely engaging in other behaviors that had significant detrimental effects to the health of their fetus and overall pregnancy. However, data specifically about the safety profile of monoamine releasers prescribed at appropriate doses for the treatment of ADHD are growing. The other significant

limitation is lack of data regarding long-term neurodevelopmental outcomes of children exposed to prescribed monoamine releasers in utero and/or via breast milk. The latter is an important point of discussion when educating patients about the risks and benefits of stimulants.

Studies so far have not found evidence of teratogenicity for these medications when used in pregnancy (Diav-Citrin et al., 2016; Nörby et al., 2017). Things become less clear when considering other pregnancy complications associated with these agents. Nörby et al. (2017) reported heightened risk for preterm birth, for NICU admission, and increased risk of a CNS disorder (e.g., seizure) following stimulant exposures, and a systematic review found increased risk for adverse placental outcomes such as preeclampsia, growth restriction, or placental abruption (Peterson et al., 2008). Additionally, a multicenter cohort study with 382 women exposed to methylphenidate during pregnancy reported on increased risk for "neonatal adaptation syndrome" (NAS) in the newborns (Diav-Citrin et al., 2016). In summary, based on today's level of evidence, there is some risk for adverse effects on pregnancy outcomes when exposed to prescription monoamine releasers for ADHD during pregnancy, yet overall, the risks are reasonably small and do not confer teratogenicity, and many of the studies reporting these risks had significant confounding variables (Baker & Freeman, 2018). As such, it is recommended that prescribers continue to treat patients with these medications when necessary, but that, when used, routine vital signs be checked throughout pregnancy, as well as intermittent assessment of fetal growth. At this time, there is no data regarding long-term neurodevelopmental outcomes for children exposed to prescription stimulants in utero.

Data collection about the safety of monoamine releasers while breastfeeding is still underway. However, data so far are reassuring. Preliminary data suggests that the relative infant dose (RID) is about 0.2–0.4% for methylphenidates and about 2.5–7.3% for amphetamines (Hale, 2020). At this time, data is not available regarding long-term neurodevelopmental outcomes for children exposed to monoamine releasers via breast milk.

Because of the effects of monoamine releasers on dopamine levels (increase), prolactin levels may decrease in mothers taking these medications and therefore may lead them to have decreased production of breast milk. Mothers should be counseled about this.

Use of Non-stimulants

Bupropion, with dual action on norepinephrine/dopamine reuptake inhibition (NDRI), has a fair amount of data demonstrating its safety in pregnancy and breastfeeding (Turner et al., 2019; Kimmel et al., 2018). It has utility both for the treatment of ADHD and nicotine addiction (an issue to which those with untreated ADHD are predisposed). The RID is about 0.1–2% (Hale, 2020). Because of its effect on dopamine, it may decrease breast milk production.

Atomoxetine is a norepinephrine reuptake inhibitor (NRI) that is sometimes prescribed for ADHD. There is currently insufficient data regarding its safety in human pregnancy or lactation, so it should be avoided in the perinatal period (Ornoy, 2018).

Use of NE Alpha-2 Receptor Agonists

There is a significant amount of data regarding the safety of clonidine in pregnancy as this was previously evaluated as it pertained to the control of blood pressure in pregnant women (Leavitt et al., 2019). It has not been shown to be teratogenic (Rothberger et al., 2010). Regarding lactation, the RID is estimated to be between 1% and 7%, and no symptoms of dry mouth or sedation were seen in nine infants nursed by mothers taking clonidine. However, more data is needed. Mothers should be counseled that clonidine may decrease breast milk supply as it may reduce prolactin (Hale, 2020).

Guanfacine has not been shown teratogenic in rodent research, but safety in human pregnancy and lactation has not been established. Therefore, it should be avoided in the perinatal period.

General Guidelines

Answering the question of whether or not a woman should be continued on monoamine releasers for management of ADHD during pregnancy involves a discussion of risks and benefits. For those who have mild to moderate symptoms of ADHD and can generally function without these medicines if supports are in place/behavioral strategies are optimized, it is reasonable to attempt to discontinue the monoamine releaser. Substitution of a monoamine releaser with bupropion should also be considered. However, for those with moderate to severe symptoms of ADHD who have significant impairment in functioning when not treated with a monoamine releaser (either due to the ADHD symptoms themselves or worsened symptoms of comorbid disorders), continuing the lowest effective dose during pregnancy is reasonable. When a monoamine releaser medication is continued in pregnancy, closer obstetrical monitoring should be in place to assess for the development of pregnancy complications such as hypertension and low fetal weight (Baker & Freeman, 2018). The decision about whether or not to breastfeed while taking monoamine releasers also involves a discussion with the patient about risks and benefits. So far, there are no reports of short-term adverse effects on infants exposed to these medications via breast milk; however, data regarding longer-term neurodevelopmental outcomes is lacking. In all of these decisions, joint decision-making with the patient is recommended.

Perinatal Management of Substance Use Disorders

Introduction

Substance use during pregnancy carries risk for poor health outcomes such as preterm labor and complications during delivery and birth (NIDA, 2020). Family support, social issues, medical, and maternal health comorbidities are instrumental regarding a pregnant women's recovery and birth outcomes. People involved in illicit drug use are also more likely to be exposed to STDs, such as HCV, with rates as high as 50–62%. Other medical issues that may complicate pregnancy such as sepsis, endocarditis, HIV, and hepatitis B are related to IV drug use. It is estimated that about 5% of pregnant people use one or more substances of misuse (Wendell, 2013). Babies born exposed to substance misuse during pregnancy are at risk for neonatal abstinence syndrome (NAS) and poor neonatal outcomes. Although best studied for opioid exposure, substances like alcohol, benzodiazepines, nicotine, and even caffeine can cause NAS (Hudak and Tan, 2012). NAS is believed to be a function of the newborn's withdrawal from substances that the fetus was exposed to in utero. Thus, NAS is not a sign of addiction in the newborn; rather, it indicates that the pregnant person used substances during pregnancy and that these transferred to the baby through the placenta and the newborn is experiencing transient and treatable, physiological withdrawal. The type and severity of withdrawal depends on the drug(s) used, frequency and amount of use, and gestational age at delivery. Polysubstance use, a combination of multiple classes of drug use, during pregnancy is common and can act as confounding factors when assessing the impact of each drug, may escalate severity of NAS, and complicates pharmacological treatment algorithms to support mother and baby.

In the following section, state-of-the-art treatment considerations for the most frequently used drugs in pregnancy and postpartum are outlined.

Pharmacological Management of Alcohol Use Disorder

A study in 2016 investigated monthly estimates of alcohol drinking during pregnancy and found that in the time frame of 2002–2011, approximately 42.4% of pregnant women consumed alcohol during the first month of their pregnancy. The rate of use dropped to 17% by the second month gestation, and by month 3 of pregnancy, this rate dropped to 3%. Many women are not aware they are pregnant in the first few weeks or months of pregnancy so counseling for future pregnancies is imperative (Alshaarawy et al., 2016). Overall, the rate of alcohol consumption in pregnant women relative to their nonpregnant peers is lower, and in fact, alcohol consumption among nonpregnant women is rising in the past decade. Screening for alcohol use disorder (AUD) in pregnancy is critical, and the T-ACE was developed by an obstetrician specifically for use in obstetric-gynecologic

practices, and its three yes/no questions (Have people annoyed you by criticizing your drinking? Have you felt that you ought to cut down on your drinking? Have you had to have a drink first thing in the morning to steady your nerves or get rid of a hangover – eye opener?) and one quantifying question (Tolerance: How many drinks does it take to make you feel high?) take less than 1 minute to administer, making it practical and efficient. The T-ACE and variant have been recommended by both the American College of Obstetrics and Gynecology and the National Institute on Alcohol Abuse and Alcoholism for AUD screening in pregnant women (Jones et al. 2013a, b).

Perinatal alcohol use can cause craniofacial abnormalities, growth retardation, neurological abnormalities, cognitive impairment, and birth defects. Fetal alcohol spectrum disorder (FASD), with a prevalence rate in North America of 2–5%, is caused by alcohol use while pregnant. The clinical presentation in affected children is consistent with difficulties in motor coordination, emotion regulation, schoolwork, socialization, and subsequently adult cognitive heavy task performance. Neuroimaging shows abnormalities in brain structure, cortical development, white matter microstructure, and functional connectivity which result in difficulties with motor coordination, emotional control, executive function, memory, vision, hearing, motor skills, and social adaptation (Wozniak et al., 2019).

There is little evidence to support the use of pharmacological interventions for AUD in pregnancy (DeVido et al., 2015). Similarly, there are few data to guide management of alcohol detoxification in pregnant women, and the use of benzodiazepines (the mainstay of most alcohol detoxification protocols) in pregnant women is controversial. Despite a lack of robust data to guide management of AUDs in pregnancy, clinicians must nonetheless make management decisions when confronted with these challenging situations. When considering pharmacologic management of both AUDs and alcohol withdrawal in pregnant women, the following treatment options can be considered.

Currently, there are three medications approved for the treatment of AUDs by the Food and Drug Administration (FDA) in the USA: naltrexone, disulfiram, and acamprosate.

All three drugs have no/limited safety data in human pregnancies with known concerning effects on fetal development from rodent studies. Given the lack of study of these medications in pregnancy, the risk and benefits of continued alcohol use must be carefully weighed against the risks of the medications themselves (DeVido et al., 2015).

Naltrexone is a mu- and kappa-opioid receptor antagonist. It is available in both oral and long-acting injectable formulations. Naltrexone is also indicated for the treatment of opioid use disorders (OUDs), and there is data available from studies of pregnant women with OUD treated with naltrexone. These studies have not shown negative birth outcomes, but the long-term effects on the exposed child are yet unknown (DeVido et al., 2015). Disulfiram is an aldehyde dehydrogenase inhibitor that causes a severe reaction with autonomic instability, flushing, nausea, and forceful vomiting when alcohol is consumed. There is inconsistent evidence showing increased risk of fetal malformations in the first trimester. Finally, acamprosate which modulates glutamate neurotransmission (exact mechanism unknown) may also bear possible teratogenic effect and use is discouraged, but no relevant human data is available.

There is limited evidence to guide decisions regarding management of acute alcohol withdrawal in pregnancy. Acute alcohol withdrawal poses a threat to both mother and fetus and should be treated as a medical emergency. Pregnant women may be uniquely vulnerable to the deleterious effects of alcohol withdrawal and may require more intensive monitoring and ongoing evaluation by obstetrical experts. The standard medical treatment for alcohol withdrawal is use of benzodiazepines. The data on the safety of benzodiazepines in pregnancy are scant and conflicting, and clinical decision-making is complicated as alcohol detoxification may necessitate short-term, high-dose benzodiazepine taper. A meta-analysis of 14 cohort studies by Enato et al. (2011) showed no difference in major malformation rates between exposed and nonexposed groups to benzodiazepines in pregnancy among

these cohort studies. At the same time, a review of case-control studies and studies investigating specific benzodiazepines (e.g., lorazepam) yielded some small but positive association to malformations, though there was no controlling for confounders making overall interpretation difficult (reviewed in DeVido et al., 2015). Another possible adverse effect of late pregnancy benzodiazepine exposure is the "floppy infant syndrome" or NAS due to benzodiazepine exposure, characterized by sedation, hypotonia/hypertonia, hyperreflexia, restlessness, irritability, abnormal sleep patterns, inconsolable crying, tremors or jerking of the extremities, bradycardia, cyanosis, suckling/feeding difficulties, apnea, risk of aspiration of feeds, and diarrhea and vomiting. The etiology is either physiologic effects of intoxication or withdrawal from benzodiazepines in the newborn. There are no studies suggesting that using a long-acting (i.e., clonazepam) versus a short-acting (i.e., lorazepam) benzodiazepine reduces the risk of developing benzodiazepine NAS (DeVido et al., 2015).

Pharmacological Management of Opioid Use Disorder

Opioid use during pregnancy can result in neonatal opioid withdrawal syndrome (NOWS). Depending on the half-life of the opioid used, withdrawal can take as long as 2 weeks to fully manifest itself but is commonly seen within the first 5 days of birth. Symptoms include excessive crying, high-pitched cry, irritability, seizures, gastrointestinal distress, poor feeding, and hyperactive autonomic system (Coyle et al., 2018). NOWS is a treatable self-limited consequence of prenatal opioid exposure and has not been associated with long-term adverse consequences (SAMHSA, 2018). There is unclear evidence on the long-term effects of opioid exposure in children with a potential increased risk of behavioral disturbances like attention deficit and hyperactivity disorder and difficulty with executive functioning and emotional regulation. However, studies have not been directly linked to medications for an opioid use disorder (OUD).

Medication for OUD (MOUD) is recommended as the standard of care for persons with an OUD, pregnant or not. During pregnancy, there are some special considerations (ACOG, 2017; SAMHSA, 2018). Methadone is a full mu-opioid receptor agonist with a long half-life. It is the most studied substrate and has over a 50-year history of use for treatment of an opioid use disorder (OUD). In the USA, methadone cannot be prescribed and must be dispensed daily to monthly in a federally licensed opioid treatment program (Tran et al., 2017). Access to this medication in some regions, particularly rural communities, can be limited if at all possible. There are many additional barriers to accessing methadone. However, retention in treatment during pregnancy is improved with methadone compared to buprenorphine (see below).

Buprenorphine (BT) is a partial mu-opioid receptor agonist with high affinity to this receptor. This property allows buprenorphine to act similarly to full opioid agonists at lower doses (mainly in intolerant individuals), reaching a ceiling/plateau at higher doses after which no further increase in typical opioid effects (therapeutic or recreational) occurs. This behavior is responsible for buprenorphine's ability to block most mu-opioid agonists and the phenomenon of precipitated withdrawal when used in persons with full agonists currently active.

BT comes in both a mono-product and in a dual product with naloxone, a mu-opioid receptor antagonist. Naloxone is commonly used to counter decreased breathing in opioid overdose and combined with buprenorphine to decrease the risk of misuse. In pregnancy, the mono-product is most well studied. However, studies looking at risk of exposure from cord blood indicate that naloxone does not enter through the cord blood at physiologically active amounts (Wiegand et al., 2016). Increasingly, treatment is done with the buprenorphine/naloxone dual product with a growing body of evidence pointing toward the efficacy of this approach (Nguyen et al., 2018). NOWS rates in neonates prenatally exposed to buprenorphine are lower than those prenatally exposed to methadone. There is moderately strong evidence that buprenorphine treatment also results in a lower risk of preterm birth, greater birth weight, and larger head circumference with no greater harms (Zedler et al., 2016).

Naltrexone, a mu- and kappa-opioid receptor antagonist, in pregnancy as a treatment for OUD is not well studied, as is true for this medication in treatment of AUD. It requires detoxification and an opioid-free period that may carry its own risks to the pregnant person and health of the pregnancy. There is limited research on the efficacy of this medication during pregnancy, and careful consideration between the pregnant person and her care team to determine the appropriateness of continuing this medication should be done (Jones et al. 2013a, b).

Whether a pregnant woman is treated with methadone or buprenorphine, the risk of intrauterine growth restriction is of some concern. Due to this, ACOG recommends an additional ultrasound in the mid-second trimester for growth and an early ultrasound for dating accuracy. During labor, people should be offered epidural anesthesia to treat pain. Opioid agonist-antagonist drugs such as butorphanol, nalbuphine, and pentazocine should be avoided because they will result in precipitated withdrawal for anyone taking an opioid agonist. For women on buprenorphine, post-surgical pain after a cesarean section will often need to be treated with increased levels of opioid analgesia due to tolerance to their treatment and hypersensitivity to pain. Use of short-acting opioids and anti-inflammatory medication, such as intravenous ketorolac, has been shown to be highly effective for postpartum and post-cesarean pain (ACOG, 2017).

Pharmacological Management of Nicotine Use Disorder

During pregnancy, approximately 20% of people use nicotine-containing products, with approximately 10% of all pregnant people smoking cigarettes in the past month (SAMHSA, 2020). Despite a lack of evidence, pregnant people view e-cigarettes as being safer than tobacco cigarettes (Wagner et al., 2017). There continues to be a rise in e-cigarette use during pregnancy with the prevalence between 0.6% and 15%. The amount of nicotine consumed by both e-cigarettes and tobacco cigarettes is similar. Nicotine crosses the placenta barrier, and the effects of nicotine exposure during fetal development are well-known and are expected to be consistent across delivery devices (Whittington et al., 2018).

Smoking tobacco increases the risk of premature birth, miscarriage, low birth weight, and infant death. Newborns can show signs of stress and drug withdrawal shortly after birth. There are also associative risks related to sudden infant death syndrome (SIDS), learning and behavioral concerns, an increased risk of obesity, and increased risk of tobacco use in childhood (Rydell et al., 2014).

Treatment for nicotine use disorders during pregnancy is indicated. It can be accomplished by starting nicotine replacement therapy (NRT) to replace the nicotine currently used and then easing into tapering once the person has reduced motivation to smoke. NRT comes as gum, transdermal patches, nasal spray, inhaler, and sublingual tablets/lozenges. It increased the rate of quitting by 50–60% (Hartmann-Boyce et al., 2018). In terms of using bupropion and varenicline in pregnancy, a meta-analysis showed no evidence that they might be harmful in pregnancy but also no strong evidence of safety. For bupropion, exposure in pregnancy does not seem to cause any major positive or negative impacts on the rate of congenital abnormalities, birth weight, or premature birth (Turner et al., 2019). Overall, there is a small literature for bupropion and studies did not account for confounders, but there is some potential small risk increase for cardiac malformation (Kimmel et al., 2018). Bupropion exposure has also been associated with ADHD in the offspring, but the sample size was small. In general, there is little research into the efficacy and safety of pharmacotherapy use for smoking cessation, and additional randomized trials are needed (Claire et al., 2020).

Pharmacological Management of Cannabis Use Disorder

From 2010 to 2017, cannabis use more than doubled among pregnant people in the USA. With recent legalization of marijuana, there has been a concomitant increase in neonatal exposure (Volkow et al., 2019). The ACOG committee opinion on cannabis use during pregnancy and breastfeeding cautions that increased exposure may lead to restrictions in fetal growth (ACOG,

2017). It recommends discontinuation of cannabis use in favor of alternative treatments for pain, mental health, or nausea that have been studied in pregnancy. Although animal studies indicate an increased risk for miscarriage early in pregnancy, there is no human research connecting cannabis use to miscarriage. Cannabis use during pregnancy may be associated with growth restriction, stillbirth, spontaneous preterm birth, and neonatal intensive care unit admission (Metz & Borgelt, 2018). There are also association studies to long-term adverse child outcomes in exposed offspring showing an increased risk for developmental and hyperactivity disorders (Goldschmidt et al., 2000) in children and increased use of cannabis as a young adult (Sonon et al., 2015).

More and more women use cannabis in pregnancy to counteract nausea and vomiting in pregnancy with the prevalence increasing annually (Young-Wolff et al., 2019). It is important to note that, in fact, marijuana use can worsen nausea and vomiting in pregnancy.

There are no known medications to treat cannabis use disorder in pregnancy or otherwise. Supportive care should be provided including advising patients to cut back or discontinue cannabis use during pregnancy. Psychosocial treatments and support are the mainstay of treatment.

Pharmacological Management of Stimulant Use Disorder

This group encompasses substance misuse of methamphetamines, cocaine, and ecstasy and of prescription stimulants for ADHD. Stimulants are a widely used and misused group of substances during pregnancy with about 750,000 cocaine-exposed pregnancies each year in the USA (Smid et al., 2019). Stimulant use during pregnancy is commonly combined with other risk behavior that may affect fetal health including other drug misuse, poor nutrition, and insufficient prenatal care (Cain et al., 2013). Due to the vasoconstrictive effects of cocaine, pregnant women who use cocaine are more likely to have seizures, premature membrane rupture, placental abruption (Wendell, 2013), hypertension (including preeclampsia), spontaneous miscarriage, preterm labor, and difficult delivery (Cain et al., 2013)

Complications affecting the baby include low birth weight and small head circumference. Withdrawal can include irritability, hyperactivity, tremors, high-pitched cry, and excessive sucking at birth (Bauer et al., 2005). Throughout childhood, methamphetamine-exposed children show risk of emotional, attention, and cognitive difficulties (Diaz et al., 2014).

There are no known medications to treat stimulant use disorder in pregnancy or otherwise. Supportive care should be provided with monitoring and follow-up for the safety of the pregnancy. Psychosocial treatments and support are the mainstay of treatment.

General Considerations for Breastfeeding

Breastfeeding should generally be encouraged in persons for whom it is safe. Mother should be made aware of the risks of breastfeeding while actively misusing substances. For example, nicotine is passed through breast milk so nicotine use can impact the infant's development (Mennella et al., 2007). Persons on MOUD, in the absence of other drug use, should be encouraged to continue to breastfeed. Persons who breastfeed are also more likely to stay in treatment for an opioid use disorder; thus, conveying the message to the mother and healthcare providers that breastfeeding with MOUDs is not only safe but also encouraged is imperative (Ray-Griffith et al., 2021). Although there is a common myth that alcohol can increase milk supply, this is not true. In fact, it may disrupt the child's sleep cycle (Mennella, 2001). The American Academy of Pediatrics recommends alcohol use should be minimized during breastfeeding (AAP, 2012).

Novel Developments in the Treatment of Postpartum Depression (PPD)

Neuroactive Steroids and Brexanolone

Neuroactive steroids (NASs) refer to steroids which, regardless of their origin (i.e., central nervous system (CNS) or periphery), modify neural activity. NASs bind and modulate different types of neural membrane receptors, and one of the most extensively studied is the gamma-aminobutyric acid (GABA) receptor complex, but others are modulated as well (e.g., glutamate receptors). GABA is the predominant inhibitory neurotransmitter in the CNS, and many NASs are positive allosteric modulators (PAMs) of the GABA-A receptor (GABA$_A$R), facilitating negatively charged chloride ion flow into the post-synaptic neuron facilitating neuronal inhibition. Approximately 20 different GABA$_A$R subunits have been isolated in humans, and the subunit composition of GABA$_A$R subunits on a particular neuron contributes to divergent receptor properties. GABAergic inhibition in the CNS is either phasic or tonic, and type of inhibition is determined which GABA$_A$R subunits are present and bound to. Depending upon their exact subunit composition and arrangement, GABA$_A$Rs may be able to bind barbiturates, benzodiazepines, general anesthetics, alcohol, and NAS. When NASs bind to GABA$_A$R subunits, they can modulate both the tonic and phasic GABAergic neurotransmission, resulting in modulation of the balance between neuronal excitation and inhibition of the neural networks (for a review of detailed mechanism, see Zorumski et al. (2019)). Allopregnanolone (AlloP) is a potent and effective positive allosteric modulator (PAM) of GABA$_A$Rs.

NAS, especially AlloP, serves a critical role in the regulation of mood states (depression, anxiety), stress responses to different stress stimuli, memory processes, and other important phenomena related to mental health. However, the mechanisms how AlloP and other NASs exert their effect on depression are to date unclear. It is highly likely that the GABA$_A$R effects contribute to antidepressant actions. However, other GABA-enhancing drugs (e.g., benzodiazepines) do not appear to be effective medications for depression, so NASs seem to be unique among GABA$_A$R PAMs. This suggests that NAS may have additional (or alternative) targets that are important for antidepressant effects. However, NASs, in contrast to benzodiazepines, augment GABA via multiple GABA$_A$Rs, which may be critical. Moreover, some evidence also suggests that changes in endogenous NAS levels could participate in the etiology of some depressive illnesses.

In the past years, the role of AlloP and other NASs has been heavily studied in the development of postpartum depression (PPD) (Deligiannidis et al., 2019). During pregnancy, levels of AlloP increase dramatically and rapidly decline following the birth of the baby. Rodent models provided strong support for the idea that the increased levels of AlloP during pregnancy likely lead to $GABA_AR$-mediated tonic inhibition, which then when the levels of AlloP plummet in the postpartum period lead to a sudden imbalance in the ratio of excitation to inhibition with a sudden switch to a hyper-excitable state. This is believed to manifest as stress-like and depressive-like behaviors in the rodents. These behaviors were corrected in experiments by agents that increase tonic inhibition such as NAS. It is this research base which led to the exploration of synthetic NAS and their analogs as potential therapeutics for PND (Zorumski et al., 2019).

In 2019, the US Food and Drug Administration (FDA) approved the first medication brexanolone, a synthetic AlloP and potent, selective, PAM of $GABA_AR$ with an indication for unipolar postpartum depression in adult women. Brexanolone is administered as a postpartum 60-hour peripheral intravenous (IV) infusion using a programmable infusion pump. Treatment is initiated with a dosage of 30 µg/kg/hour (0–4 hours), then increased to 60 µg/kg/hour (4–24 hours), and then increased to 90 µg/kg/hour (24–52 hours). A reduction in dosage to 60 µg/kg/hour may be considered during this time period for women who do not tolerate 90 µg/kg/hour. During 52–56 hours, the dosage is reduced/continued to/at 60 µg/kg/hour and reduced to 30 µg/kg/hour during 56–60 hours, and then the infusion is terminated at hour 60.

Brexanolone had been approved in the USA for PPD as of March 2019 and is the only drug approved by the FDA for this indication. Its onset of action is quick (within about 60 hours of infusion) and effects have been reported to last for about 30 days. The rapid improvement of women with PPD following IV infusions of brexanolone is consistent with the hypothesis that directly replenishing AlloP is highly effective in treating this disorder. Unfortunately, the women in the initial studies were not followed for longer and thus

long-term data are needed. Breastfeeding is discouraged as data are lacking but the levels found in breast milk were low. There are limitations however including the high cost and the need for the patient to be in hospital for the IV administration as it is permitted within a special program because of the potential for adverse reactions to the drug such as sedation and loss of consciousness.

A series of placebo-controlled randomized clinical trials (RCTs) (Meltzer-Brody et al., 2018) of brexanolone demonstrated rapid reduction of postpartum depression symptoms at the 60-hour mark. Women involved in these studies were all medically healthy, within 6 months postpartum with onset of a major depressive episode no earlier than third trimester, and showed moderate to severe depression. In these studies women had to be either antidepressant-free or on a stable dose of a traditional antidepressant throughout the trial. Both dosing regimens of brexanolone infusion (90 µg/kg/hour and 60 µg/kg/hour, respectively) and placebo reduced significantly the depression score at 60 hours post IV. Significant differences from placebo were already observed at hour 24 (both dose groups) and maintained through day 30 (Meltzer-Brody et al., 2018).

The most common adverse reactions were sedation/somnolence, dry mouth, loss of consciousness, and flushing/hot flush. In 5% of brexanolone-treated patients, the severity of sedation and somnolence required dose interruption, and 4% of the brexanolone-treated patients had loss of consciousness or altered state of consciousness during the brexanolone IV for up to 60 minutes. These adverse findings led to labeling brexanolone a Schedule IV medication with black box warning for excessive sedation and sudden loss of consciousness and the requirement to only deliver through medically supervised settings with a Risk Evaluation and Mitigation Strategy (REMS; e.g., continuous pulse oximetry, signs/symptoms of sedation every 2 hours during planned non-sleep periods, patients enrolled in a registry) to reduce the risk of serious adverse events.

While in the RCTs women had stopped lactating or had temporarily ceased breastfeeding while receiving brexanolone until 4 days after the end of infusion, it is assumed that brexanolone is compatible with breastfeeding as the RID during infusion is 1–2% of the maternal dose. However, more data on safety while breastfeeding are needed.

Investigational Drugs: Zuranolone and Ganaxolone

Zuranolone is an investigational product under development for unipolar postpartum depression (PPD). Zuranolone is also a potent PAM for $GABA_AR$. The advantage of zuranolone is that it can be delivered orally instead of IV. Zuranolone was recently evaluated in an outpatient randomized, double-blind, placebo-controlled phase 3 trial in severe unipolar PPD (Deligiannidis et al., 2021). One hundred fifty-three adult women were randomized to receive 14 days of either zuranolone 30 mg or placebo and then were followed for 4 weeks. Similar to the brexanolone RCTs, the main outcome in this RCT is a significant reduction of depression scores in the zuranolone group compared to the placebo group. Starting day 3 of the trial and sustained through day 45, 4 weeks after zuranolone cessation, the treatment group showed lower depression score. Most common side effects were sedation, headache, and dizziness, but no reports of loss of consciousness. These findings are highly encouraging, and the oral delivery is a huge advantage; more studies to confirm the positive findings and also to evaluate lactation safety are now urgently needed.

Another AlloP analog, ganaxolone, is currently being investigated for the treatment of PPD, which is also a PAM for $GABA_AR$. Ganaxolone is currently undergoing clinical trial testing in both IV and oral forms, but at the time of writing, published results are not available.

Pharmacologic Management of Psychopathology During the Premenstrual Period

Introduction

Premenstrual dysphoric disorder (PMDD) is a severe mood disorder, characterized by cognitive-affective and physical symptoms lasting 5 days before and up to 3 days after the onset of menses

affecting many women, and recently being recognized as a disorder in DSM-5 (APA, 2013). The exact mechanism by which fluctuations in levels of estrogen and progesterone cause symptoms of PMDD in certain women is still unknown. However, there are some promising hypotheses that seem to align with data from studies assessing the efficacy of different psychopharmacologic treatments of PMDD. The main metabolite of progesterone, allopregnanolone (AlloP), is thought to play an important role in PMDD. Progesterone and AlloP increase in the luteal phase and decrease quickly around menses. AlloP is a modulator of the $GABA_AR$ receptors and in such exerts GABA-ergic inhibitory action. It is thought that women with PMDD may be de-sensitized to the GABA-ergic inhibitory effects of AlloP and thus more likely to experience dysphoria in the luteal phase (Hantsoo & Epperson, 2015). This potential mechanism may be an important avenue for novel treatment development for PMDD.

Estrogen may be another important player contributing to PMDD. Estrogen has potent effects on multiple neurotransmitter systems, most importantly to the serotonin system, involved in the regulation of mood, cognition, sleep, eating, and other aspects of behavior. Estrogen "regulates synthesis, metabolism, and receptor density or activity of both serotonin and norepinephrine" (Lokuge et al., 2011) and therefore can significantly impact mood and anxiety.

Treatment Guidelines

Medications for depression are the first-line treatment for PMDD (Ismaili et al., 2016). Selective serotonin reuptake inhibitors (SSRIs) have been studied more thoroughly for this indication than selective serotonin-norepinephrine reuptake inhibitors (SNRIs). The effective dose of these medications for treatment of PMDD tends to be at the low to midranges of doses used for treatment of major depression (Reid & Soares, 2018). There are three main dosing algorithms:

1. *Intermittent* dosing: medication is started at the beginning of the luteal phase and discontinued the first few days following start of menstruation.

2. *Symptom onset* dosing: medication is started at the onset of PMDD symptoms and discontinued the first few days following start of menstruation.
3. *Continuous* dosing: medication is taken daily regardless of place in menstrual cycle.

Interestingly, women do not report discontinuation symptoms when stopping their antidepressant in the first days after the onset of menses in intermittent and symptom onset regimens (Yonkers et al., 2008). It is possible that the rapid onset of action and lack of discontinuation symptoms from medications for depression for PMDD are based on the mechanism that their primary effect is mediated by interactions with AlloP and GABA and less dependent on longer-term downregulation of postsynaptic receptors (Reid & Soares, 2018).

Second-line treatment of PMDD is oral contraceptive pills (OCPs). The most studied progesterone containing oral contraceptive drug for this indication is drospirenone, which has an FDA indication for PMDD; however, RCTs have not been done to compare the efficacy of this formulation to that of others. The FDA has added a warning to OCPs containing drospirenone regarding concern for developing blood clots, so a thorough individual assessment is required to determine eligibility for this method of treatment (Yonkers et al., 2008). OCPs may also be added as augmentation in cases where patients show improvement with medications for depression, but fail to show symptom remission (Joffe et al., 2007).

A short course of an anxiolytic agent can also be helpful for patients with pronounced premenstrual anxiety and irritability. Alprazolam has been found helpful for these symptoms when used in the luteal phase; however, concern for abuse or dependence makes longer-acting benzodiazepines such as lorazepam and clonazepam more attractive options (although formal studies of these specific benzodiazepines have not been conducted). There is some evidence to support the use of buspirone, a serotonin receptor partial agonist ($5-HT_{1A}$ with 50% intrinsic activity), for patients with a preponderance of premenstrual anxiety (Pearlstein, 2012).

For patients with severe symptoms who have failed the above treatments, use of a gonadotrophin-releasing hormone agonist (GnRHa) analog is sometimes considered (Wyatt et al., 2004). These drugs suppress ovarian function, which is believed to be the trigger for premenstrual syndrome/premenstrual dysphoric disorder. Due to the low serum estradiol concentrations induced by GnRHa, side effects that mimic menopause are common (e.g., bone demineralization). To mitigate this, small amounts of estrogen with or without progestin must be supplemented. As one might expect, some patients experience a resurgence of PMDD symptoms with supplementation of progestin (Reid & Soares, 2018). Trial of treatment with a GnRHa analog should be completed before a patient is considered for definitive surgical treatment of PMDD via oophorectomy (Nevatte et al., 2013).

Pharmacologic Management of Psychopathology During Perimenopause

Introduction

Symptoms of depression in perimenopause are quite similar to those of major depressive disorder (MDD); however, particular symptoms may be more prominent (sleep disturbance, weight fluctuation, poor concentration). Additionally, these symptoms can be accompanied by bothersome vasomotor symptoms such as hot flashes and night sweats. Medications such as SERT and SERT and NET can be helpful for the depressive and vasomotor symptoms that can accompany the perimenopausal period. It is important to note that if vasomotor symptoms persist after perimenopausal depression has been treated to remission, the clinician should encourage the patient to address these, as there is evidence that untreated vasomotor symptoms can contribute to insomnia which in turn can exacerbate depression (Joffe et al., 2016).

Treatment Guidelines

Medications for Perimenopausal Depression

Treatment for perimenopausal depression is quite similar to that of MDD. Maki et al. (2019) provided comprehensive guidelines for evaluation and treatment of perimenopausal depression, which are summarized here as follows. In brief, both SSRI and SNRI have been shown effective for treatment of this condition. At this time the only agent that has been tested in a large, randomized placebo-controlled trial on a well-defined sample of perimenopausal women is desvenlafaxine. Other antidepressants such as sertraline, fluoxetine, citalopram, escitalopram, duloxetine, and venlafaxine have been shown effective in open-label trials. Bupropion has not been studied for efficacy in treatment of perimenopausal depression but is sometimes added to SSRI and SNRI to help manage side effects from these agents (i.e., sexual side effects) or other specific perimenopause-related symptoms (i.e., weight gain). Because there is evidence of efficacy for many antidepressants treating perimenopausal depression, when choosing an agent, the first consideration should be whether the patient had benefit from a particular antidepressant in the past. If so, this agent should be trialed first (Maki et al., 2019).

Hormone Therapy for Perimenopausal Depression

There is some evidence for the benefit of hormonal therapies in perimenopausal depression, particularly estrogen therapy. Two randomized controlled trials have found a positive effect of estradiol on perimenopausal depression (Schmidt et al., 2000; Morgan et al., 2005). Interestingly, this effect seems to persist even after hormone therapy has been discontinued and vasomotor symptoms may have returned (Soares et al., 2001). There are risks associated with treatment with unopposed estrogen (thromboembolism, breast cancer), so the patient should carefully weigh the risks and benefits of this therapy with their gynecologist.

Data regarding the effectiveness of combining antidepressants with estrogen therapy are conflicting (Maki et al., 2019). For patients who achieve benefit but not remission of their depression from antidepressants, the addition of estrogen therapy may be considered (Morgan et al., 2005).

Treatment of Vasomotor Symptoms

Hormone therapy is the most effective treatment for vasomotor symptoms in the perimenopausal period (ACOG Practice Bulletin, 2014). Patients may benefit from discussing treatment of vasomotor symptoms with hormone therapy with their gynecologist. As previously mentioned, hormone therapies are associated with other health risk; therefore they are not always ideal. As mentioned above, SERT and SERT and NET have some ability to reduce vasomotor symptoms. Clonidine, a NE alpha-2 receptor agonist, has been shown of some benefit to control vasomotor symptoms at doses of 0.1 mg/day (Nelson et al., 2006), but is not FDA approved for this indication. Gabapentin, a voltage-gated calcium channel blocker of glutamate transmission, is another medication that, while not FDA approved for the indication, has been shown to reduce vasomotor symptoms related to perimenopause at a dose of 900 mg/day (Guttuso Jr et al., 2003).

Conclusion

In summary, reproductive psychiatry is a complex field in psychiatric care, and the pharmacological management during this period requires special attention, knowledge, and relationship building.

Although the extent of literature, especially around safety of medicines in pregnancy and when breastfeeding, has significantly increased over the past decade and notably there is more access to large-scale population-based studies, the gold standard randomized controlled trials comparing medication efficacy and safety profiles to placebo or other active compounds are missing. The absence of randomization constrains interpretation of findings and leaves a gray zone for the clinician and patient. Thus, pre-pregnancy planning and collaborative sessions between provider and patient (and family) are not only welcomed, but necessary, as a careful discussion of potential risks and benefits, acknowledging what is known along with the limits of extant data, is required for informed decision-making. Ideally, such planning should be initiated many months before attempting pregnancy, leaving sufficient time to make medication changes and ensure stability prior to conception, underscoring the value of a stable and trusting provider-patient relationship prior to pregnancy. Similarly, a strong provider-patient relationship, built upon a foundation of trust, can facilitate ongoing support and monitoring over the course of pregnancy and postpartum, potentially over several pregnancies, and possibly into perimenopause. Thus, reproductive psychiatry is a relationship-focused psychiatry, and best practice follows a patient and supports her unfolding psychiatric needs as she moves across the reproductive life cycle – from pre-pregnancy, peripartum, and perimenopause.

With these reflections in mind, it is concluded that reproductive psychiatry, with its.

distinct patient population and ever-increasing specialized body of medical knowledge and psychopharmacological guidelines, is worthy of recognition as an official subspecialty of psychiatry, emphasizing an informed and thoughtful approach to the management of women's psychiatric disorders across the lifespan.

References

AAP. American Academy of Pediatrics. Section on Breastfeeding. (2012). Breastfeeding and the use of human milk. *Pediatrics, 129*(3), e827–e841.

ACOG (American College of Obstetricians and Gynecologists). (2017). Opioid use and opioid use disorder in pregnancy. Committee Opinion No. 711. *Obstetrics and Gynecology, 130*, e81–e94.

ACOG Practice Bulletin. (2014). https://www.acog.org/clinical/clinical-guidance/practice-bulletin/articles/2014/01/management-of-menopausal-symptoms. Accessed on April 27, 2021.

Alshaarawy, O., Breslau, N., & Anthony, J. C. (2016). Monthly estimates of alcohol drinking during pregnancy: United States, 2002–2011. *Journal of Studies on Alcohol and Drugs, 77*(2), 272–276.

American Psychiatric Association. (2013). *Diagnostic and statistical manual of mental disorders* (5th ed.). American Psychiatric Publishing.

Baker, A. S., & Freeman, M. P. (2018). Management of attention deficit hyperactivity disorder during pregnancy. *Obstetrics and Gynecology Clinics of North America, 45*(3), 495–509.

Baker, G. A., Bromley, R. L., Briggs, M., Cheyne, C. P., Cohen, M. J., García-Fiñana, M., Gummery, A., Kneen, R., Loring, D. W., Mawer, G., Meador, K. J., Shallcross, R., & Clayton-Smith, J. (2015). IQ at 6 years after in utero exposure to antiepileptic drugs. *Neurology, 84*(4), 382–390. https://doi.org/10.1212/WNL.0000000000001182

Bauer, C. R., Langer, J. C., Shankaran, S., Bada, H. S., Lester, B., Wright, L. L., Krause-Steinrauf, H., Smeriglio, V. L., Finnegan, L. P., Maza, P. L., & Verter,

J. (2005). Acute neonatal effects of cocaine exposure during pregnancy. *Archives of Pediatrics & Adolescent Medicine, 159*(9), 824–834.

Berle, J. O., & Spigset, O. (2011). Antidepressant use during breastfeeding. *Current women's health reviews, 7*(1), 28–34.

Betcher, H. K., Montiel, C., & Clark, C. T. (2019). Use of antipsychotic drugs during pregnancy. *Current Treatment Options in Psychiatry, 6*(1), 17–31.

Blankley, G., Galbally, M., Snellen, M., Power, J., & Lewis, A. J. (2015). Borderline personality disorder in the perinatal period: Early infant and maternal outcomes. *Australasian Psychiatry: Bulletin of Royal Australian and New Zealand College of Psychiatrists, 23*(6), 688–692.

Boyle, B., Garne, E., Loane, M., Addor, M. C., Arriola, L., Cavero-Carbonell, C., Gatt, M., Lelong, N., Lynch, C., Nelen, V., Neville, A. J., O'Mahony, M., Pierini, A., Rissmann, A., Tucker, D., Zymak-Zakutnia, N., & Dolk, H. (2017). The changing epidemiology of Ebstein's anomaly and its relationship with maternal mental health conditions: A European registry-based study. *Cardiology in the Young, 27*(4), 677–685.

Brady, K., Pearlstein, T., Asnis, G. M., Baker, D., Rothbaum, B., Sikes, C. R., & Farfel, G. M. (2000). Efficacy and safety of sertraline treatment of post-traumatic stress disorder: A randomized controlled trial. *JAMA, 283*(14), 1837–1844.

Cain, M. A., Bornick, P., & Whiteman, V. (2013). The maternal, fetal, and neonatal effects of cocaine exposure in pregnancy. *Clinical Obstetrics and Gynecology, 56*(1), 124–132.

Cantarutti, A., Merlino, L., Monzani, E., Giaquinto, C., & Corrao, G. (2016). Is the risk of preterm birth and low birth weight affected by the use of antidepressant agents during pregnancy? A population-based investigation. *PloS One, 11*(12), e0168115.

Chaudhry, S. K., & Susser, L. C. (2018). Considerations in treating insomnia during pregnancy: A literature review. *Psychosomatics, 59*(4), 341–348.

Claire, R., Chamberlain, C., Davey, M. A., Cooper, S. E., Berlin, I., Leonardi-Bee, J., & Coleman, T. (2020). Pharmacological interventions for promoting smoking cessation during pregnancy. *The Cochrane Database of Systematic Reviews, 3*(3), CD010078.

Coyle, M. G., Brogly, S. B., Ahmed, M. S., Patrick, S. W., & Jones, H. E. (2018). Neonatal abstinence syndrome. *Nature Reviews. Disease Primers, 4*(1), 47.

Costantine, M. M. (2014). Physiologic and pharmacokinetic changes in pregnancy. *Frontiers in Pharmacology, 5*, 65.

Creeley, C. E., & Denton, L. K. (2019). Use of prescribed psychotropics during pregnancy: A systematic review of pregnancy, neonatal, and childhood outcomes. *Brain Science, 9*(9), 235. https://doi.org/10.3390/brainsci9090235

Davidson, A. D., Bhat, A., Chu, F., Rice, J. N., Nduom, N. A., & Cowley, D. S. (2021). A systematic review of the use of prazosin in pregnancy and lactation. *General Hospital Psychiatry, S0163-8343*(21), 00048–00047. Advance online publication.

Deligiannidis, K. M., Fales, C. L., Kroll-Desrosiers, A. R., Shaffer, S. A., Villamarin, V., Tan, Y., Hall, J. E., Frederick, B. B., Sikoglu, E. M., Edden, R. A., Rothschild, A. J., & Moore, C. M. (2019). Resting-state functional connectivity, cortical GABA, and neuroactive steroids in peripartum and peripartum depressed women: A functional magnetic resonance imaging and spectroscopy study. *Neuropsychopharmacology, 44*(3), 546–554.

Deligiannidis, K. M., Meltzer-Brody, S., Gunduz-Bruce, H., Doherty, J., Jonas, J., Li, S., Sankoh, A. J., Silber, C., Campbell, A. D., Werneburg, B., Kanes, S. J., & Lasser, R. (2021). Effect of zuranolone vs placebo in postpartum depression: A randomized clinical trial. *JAMA Psychiatry, 78*, 951.

DeVido, J., Bogunovic, O., & Weiss, R. D. (2015). Alcohol use disorders in pregnancy. *Harvard Review of Psychiatry, 23*(2), 112–121.

Diav-Citrin, O., Shechtman, S., Arnon, J., Wajnberg, R., Borisch, C., Beck, E., Richardson, J. L., Bozzo, P., Nulman, I., & Ornoy, A. (2016). Methylphenidate in pregnancy: A multicenter, prospective, comparative, observational study. *The Journal of Clinical Psychiatry, 77*(9), 1176–1181.

Diaz, S. D., Smith, L. M., LaGasse, L. L., Derauf, C., Newman, E., Shah, R., Arria, A., Huestis, M. A., Della Grotta, S., Dansereau, L. M., Neal, C., & Lester, B. M. (2014). Effects of prenatal methamphetamine exposure on behavioral and cognitive findings at 7.5 years of age. *The Journal of Pediatrics, 164*(6), 1333–1338.

Drugs and Lactation Database (LactMed) [Internet]. (2021). *Bethesda (MD): National Library of Medicine (US)*. https://www.ncbi.nlm.nih.gov/books/NBK501922/. Accessed April 16, 2021.

Eke, A. C., Saccone, G., & Berghella, V. (2016). Selective serotonin reuptake inhibitor (SSRI) use during pregnancy and risk of preterm birth: A systematic review and meta-analysis. *BJOG: An International Journal of Obstetrics and Gynaecology, 123*(12), 1900–1907.

El-Mallakh, R. S. (1986). Acute lithium neurotoxicity. *Psychiatric Developments, 4*, 311–328.

Enato, E., Moretti, M., & Koren, G. (2011). The fetal safety of benzodiazepines: An updated meta-analysis. *Journal of Obstetrics and Gynaecology Canada: JOGC = Journal d'obstetrique et gynecologie du Canada: JOGC, 33*(1), 46–48.

Erickson, N., Julian, M., & Muzik, M. (2019). Perinatal depression, PTSD, and trauma: Impact on mother-infant attachment and interventions to mitigate the transmission of risk. *International Review of Psychiatry (Abingdon, England), 31*(3), 245–263.

Fairbrother, N., Young, A. H., Janssen, P., Antony, M. M., & Tucker, E. (2015). Depression and anxiety during the perinatal period. *BMC Psychiatry, 15*, 206.

Figueroa, R. (2010). Use of antidepressants during pregnancy and risk of attention-deficit/hyperactivity disorder in the offspring. *Journal of Developmental and Behavioral Pediatrics: JDBP, 31*(8), 641–648.

Finer, L. B., & Zolna, M. R. (2016). Declines in unintended pregnancy in the United States, 2008–2011. *The New England Journal of Medicine, 374*(9), 843–852.

Fischer Fumeaux, C. J., Morisod Harari, M., Weisskopf, E., Eap, C. B., Epiney, M., Vial, Y., Csajka, C., Bickle Graz, M., & Panchaud, A. (2019). Risk-benefit balance assessment of SSRI antidepressant use during pregnancy and lactation based on best available evidence – an update. *Expert Opinion on Drug Safety, 18*(10), 949–963.

Gotlib, D., Perelstein, E., Kurlander, J., Zivin, K., Riba, M., & Muzik, M. (2016). Guideline adherence for mentally ill reproductive-aged women on treatment with valproic acid: A retrospective chart review. *The Journal of Clinical Psychiatry, 77*(4), 527–534.

Goldschmidt, L., Day, N. L., & Richardson, G. A. (2000). Effects of prenatal marijuana exposure on child behavior problems at age 10. *Neurotoxicology and Teratology, 22*(3), 325–336.

Grigoriadis, S., Graves, L., Peer, M., Mamisashvili, L., Dennis, C. L., Vigod, S. N., Steiner, M., Brown, C., Cheung, A., Dawson, H., Rector, N., Guenette, M., & Richter, M. (2019). Benzodiazepine use during pregnancy alone or in combination with an antidepressant and congenital malformations: Systematic review and meta-analysis. *The Journal of Clinical Psychiatry, 80*(4), 18r12412.

Grigoriadis, S., Graves, L., Peer, M., Mamisashvili, L., Ruthirakuhan, M., Chan, P., Hennawy, M., Parikh, S., Vigod, S. N., Dennis, C. L., Steiner, M., Brown, C., Cheung, A., Dawson, H., Rector, N., Guenette, M., & Richter, M. (2020). Pregnancy and delivery outcomes following benzodiazepine exposure: A systematic review and meta-analysis. *Canadian Journal of Psychiatry. Revue Canadienne de Psychiatrie, 65*(12), 821–834.

Grigoriadis, S., VonderPorten, E. H., Mamisashvili, L., Eady, A., Tomlinson, G., Dennis, C. L., Koren, G., Steiner, M., Mousmanis, P., Cheung, A., & Ross, L. E. (2013). The effect of prenatal antidepressant exposure on neonatal adaptation: A systematic review and meta-analysis. *The Journal of Clinical Psychiatry, 74*(4), e309–e320.

Groenman, A. P., Oosterlaan, J., Rommelse, N. N., Franke, B., Greven, C. U., Hoekstra, P. J., Hartman, C. A., Luman, M., Roeyers, H., Oades, R. D., Sergeant, J. A., Buitelaar, J. K., & Faraone, S. V. (2013). Stimulant treatment for attention-deficit hyperactivity disorder and risk of developing substance use disorder. *The British Journal of Psychiatry: The Journal of Mental Science, 203*(2), 112–119.

Hartmann-Boyce, J., Chepkin, S. C., Ye, W., Bullen, C., & Lancaster, T. (2018). Nicotine replacement therapy versus control for smoking cessation. *The Cochrane Database of Systematic Reviews, 5*(5), CD000146.

Guttuso, T., Jr., Kurlan, R., McDermott, M. P., & Kieburtz, K. (2003). Gabapentin's effects on hot flashes in postmenopausal women: A randomized controlled trial. *Obstetrics and Gynecology, 101*(2), 337–345.

Hale's Medications & Mothers' Milk. (2020). *A manual of lactational pharmacology* (19th ed. Dr. Thomas W. Hale, PhD.). Springer.

Hantsoo, L., & Epperson, C. N. (2015). Premenstrual dysphoric disorder: Epidemiology and treatment. *Current Psychiatry Reports, 17*(11), 87.

Hudak, M. L., Tan, R. C., Committee on Drugs, Committee on Fetus and Newborn, & American Academy of Pediatrics. (2012). Neonatal drug withdrawal. *Pediatrics, 129*(2), e540–e560.

Huitfeldt, A., Sundbakk, L. M., Skurtveit, S., Handal, M., & Nordeng, H. (2020). Associations of maternal use of benzodiazepines or benzodiazepine-like hypnotics during pregnancy with immediate pregnancy outcomes in Norway. *JAMA Network Open, 3*(6), e205860.

Huybrechts, K. F., Palmsten, K., Avorn, J., Cohen, L. S., Holmes, L. B., Franklin, J. M., Mogun, H., Levin, R., Kowal, M., Setoguchi, S., & Hernández-Díaz, S. (2014). Antidepressant use in pregnancy and the risk of cardiac defects. *The New England Journal of Medicine, 370*(25), 2397–2407. https://doi.org/10.1056/NEJMoa1312828. PMID: 24941178; PMCID: PMC4062924.

Huybrechts, K. F., Bateman, B. T., Palmsten, K., Desai, R. J., Patorno, E., Gopalakrishnan, C., Levin, R., Mogun, H., & Hernandez-Diaz, S. (2015). Antidepressant use late in pregnancy and risk of persistent pulmonary hypertension of the newborn. *JAMA, 313*(21), 2142–2151.

Huybrechts, K. F., Hernández-Díaz, S., Patorno, E., Desai, R. J., Mogun, H., Dejene, S. Z., Cohen, J. M., Panchaud, A., Cohen, L., & Bateman, B. T. (2016). Antipsychotic use in pregnancy and the risk for congenital malformations. *JAMA Psychiatry, 73*(9), 938–946.

Ismaili, E., Walsh, S., O'Brien, P., Bäckström, T., Brown, C., Dennerstein, L., Eriksson, E., Freeman, E. W., Ismail, K., Panay, N., Pearlstein, T., Rapkin, A., Steiner, M., Studd, J., Sundström-Paromma, I., Endicott, J., Epperson, C. N., Halbreich, U., Reid, R., Rubinow, D., et al. (2016). Fourth consensus of the International Society for Premenstrual Disorders (ISPMD): Auditable standards for diagnosis and management of premenstrual disorder. *Archives of Women's Mental Health, 19*(6), 953–958.

Johansen, R. L., Mortensen, L. H., Andersen, A. M., Hansen, A. V., & Strandberg-Larsen, K. (2015). Maternal use of selective serotonin reuptake inhibitors and risk of miscarriage – Assessing potential biases. *Paediatric and Perinatal Epidemiology, 29*(1), 72–81.

Joffe, H., Crawford, S. L., Freeman, M. P., White, D. P., Bianchi, M. T., Kim, S., Economou, N., Camuso, J., Hall, J. E., & Cohen, L. S. (2016). Independent contributions of nocturnal hot flashes and sleep disturbance to depression in estrogen-deprived women. *The Journal of Clinical Endocrinology and Metabolism, 101*(10), 3847–3855.

Joffe, H., Petrillo, L. F., Viguera, A. C., Gottshcall, H., Soares, C. N., Hall, J. E., & Cohen, L. S. (2007). Treatment of premenstrual worsening of depression with adjunctive oral contraceptive pills: A preliminary report. *The Journal of Clinical Psychiatry, 68*(12), 1954–1962.

Jones, I., Chandra, P. S., Dazzan, P., & Howard, L. M. (2014). Bipolar disorder, affective psychosis, and schizophrenia in pregnancy and the post-partum period. *Lancet, 384*(9956), 1789–1799. https://doi.org/10.1016/S0140-6736(14)61278-2

Jones, H. E., Chisolm, M. S., Jansson, L. M., & Terplan, M. (2013a). Naltrexone in the treatment of opioid-dependent pregnant women: The case for a considered and measured approach to research. *Addiction (Abingdon, England), 108*(2), 233–247.

Jones, T. B., Bailey, B. A., & Sokol, R. J. (2013b). Alcohol use in pregnancy: Insights in screening and intervention for the clinician. *Clinical Obstetrics and Gynecology, 56*(1), 114–123.

Jordan, S., Morris, J. K., Davies, G. I., Tucker, D., Thayer, D. S., Luteijn, J. M., Morgan, M., Garne, E., Hansen, A. V., Klungsøyr, K., Engeland, A., Boyle, B., & Dolk, H. (2016). Selective Serotonin Reuptake Inhibitor (SSRI) antidepressants in pregnancy and congenital anomalies: Analysis of linked databases in Wales, Norway and Funen, Denmark. *PloS One, 11*(12), e0165122.

Kaplan, Y. C., Keskin-Arslan, E., Acar, S., & Sozmen, K. (2017). Maternal SSRI discontinuation, use, psychiatric disorder and the risk of autism in children: A meta-analysis of cohort studies. *British Journal of Clinical Pharmacology, 83*(12), 2798–2806.

Karatzias, T., & Cloitre, M. (2019). Treating adults with complex posttraumatic stress disorder using a modular approach to treatment: Rationale, evidence, and directions for future research. *Journal of Traumatic Stress, 32*(6), 870–876.

Kelly, L. E., Poon, S., Madadi, P., & Koren, G. (2012). Neonatal benzodiazepines exposure during breastfeeding. *The Journal of Pediatrics, 161*(3), 448–451.

Kimmel, M. C., Cox, E., Schiller, C., Gettes, E., & Meltzer-Brody, S. (2018). Pharmacologic treatment of perinatal depression. *Obstetrics and Gynecology Clinics of North America, 45*(3), 419–440.

Kollins, S. H., McClernon, F. J., & Fuemmeler, B. F. (2005). Association between smoking and attention-deficit/hyperactivity disorder symptoms in a population-based sample of young adults. *Archives of General Psychiatry, 62*(10), 1142–1147.

Leavitt, K., Občian, S., & Yankowitz, J. (2019). Treatment and prevention of hypertensive disorders during pregnancy. *Clinics in Perinatology, 46*(2), 173–185. https://doi.org/10.1016/j.clp.2019.02.002. Epub 2019 Mar 30.

Liriano, F., Hatten, C., & Schwartz, T. L. (2019). Ketamine as treatment for post-traumatic stress disorder: A review. *Drugs in Context, 8*, 212305.

Lokuge, S., Frey, B. N., Foster, J. A., Soares, C. N., & Steiner, M. (2011). Depression in women: Windows of vulnerability and new insights into the link between estrogen and serotonin. *The Journal of Clinical Psychiatry, 72*(11), e1563–e1569.

Maki, P. M., Kornstein, S. G., Joffe, H., Bromberger, J. T., Freeman, E. W., Athappilly, G., Bobo, W. V., Rubin, L. H., Koleva, H. K., Cohen, L. S., & Soares, C. N. (2019). Guidelines for the evaluation and treatment of perimenopausal depression: Summary and recommendations. *Journal of Women's Health (2002), 28*(2), 117–134.

McLafferty, L. P., Spada, M., & Gopalan, P. (2018). Pharmacologic treatment of sleep disorders in pregnancy. *Sleep Medicine Clinics, 13*(2), 243–250.

Medicines and Healthcare Products Regulatory Authority. (2021). *Public Assessment Report of antiepileptic drugs: Review of safety of use during pregnancy.* https://www.gov.uk/government/publications/public-assessment-report-of-antiepileptic-drugs-review-of-safety-of-use-during-pregnancy. Accessed April 16, 2021.

Meltzer-Brody, S., Colquhoun, H., Riesenberg, R., Epperson, C. N., Deligiannidis, K. M., Rubinow, D. R., Li, H., Sankoh, A. J., Clemson, C., Schacterle, A., Jonas, J., & Kanes, S. (2018). Brexanolone injection in post-partum depression: Two multicentre, double-blind, randomised, placebo-controlled, phase 3 trials. *Lancet (London, England), 392*(10152), 1058–1070.

Mennella, J. (2001). Alcohol's effect on lactation. *Alcohol Research & Health: The Journal of the National Institute on Alcohol Abuse and Alcoholism, 25*(3), 230–234.

Mennella, J. A., Yourshaw, L. M., & Morgan, L. K. (2007). Breastfeeding and smoking: Short-term effects on infant feeding and sleep. *Pediatrics, 120*(3), 497–502.

Metz, T. D., & Borgelt, L. M. (2018). Marijuana use in pregnancy and while breastfeeding. *Obstetrics and Gynecology, 132*(5), 1198–1210.

Michigan Model DBT-Pregnancy and Postpartum. (2021). Unpublished Manual by Katie Bresky, LMSW, Natalie Burns, LMSW, Maria Muzik, MD, MSc, and Kate Rosenblum, PhD, ABPP. Zero To Thrive, Department of Psychiatry, University of Michigan.

Moghaddas, A., Dianatkhah, M., Ghaffari, S., & Ghaeli, P. (2017). The potential role of naltrexone in borderline personality disorder. *Iranian Journal of Psychiatry, 12*(2), 142–146.

Molenaar, N. M., Poels, E., Robakis, T., Wesseloo, R., & Bergink, V. (2021). Management of lithium dosing around delivery: An observational study. *Bipolar Disorders, 23*(1), 49–54.

Morgan, M. L., Cook, I. A., Rapkin, A. J., & Leuchter, A. F. (2005). Estrogen augmentation of antidepressants in perimenopausal depression: A pilot study. *The Journal of Clinical Psychiatry, 66*(6), 774–780.

Munk-Olsen, T., Laursen, T. M., Pedersen, C. B., Mors, O., & Mortensen, P. B. (2006). New parents and mental disorders: A population-based register study. *JAMA, 296*, 2582–2589.

Nelson, H. D., Vesco, K. K., Haney, E., Fu, R., Nedrow, A., Miller, J., Nicolaidis, C., Walker, M., & Humphrey, L. (2006). Nonhormonal therapies for menopausal hot flashes: Systematic review and meta-analysis. *JAMA, 295*(17), 2057–2071.

Nevatte, T., O'Brien, P. M., Bäckström, T., Brown, C., Dennerstein, L., Endicott, J., Epperson, C. N., Eriksson, E., Freeman, E. W., Halbreich, U., Ismail, K., Panay, N., Pearlstein, T., Rapkin, A., Reid, R., Rubinow, D., Schmidt, P., Steiner, M., Studd, J., Sundström-Poromaa, I., et al. (2013). ISPMD consensus on the management of premenstrual disorders. *Archives of Women's Mental Health, 16*(4), 279–291.

Nguyen, L., Lander, L. R., O'Grady, K. E., Marshalek, P. J., Schmidt, A., Kelly, A. K., & Jones, H. E. (2018). Treating women with opioid use disorder during pregnancy in Appalachia: Initial neonatal outcomes following buprenorphine + naloxone exposure. *The American Journal on Addictions, 27*(2), 92–96. https://doi.org/10.1111/ajad.12687

NICE. (2014). *Antenatal and Postnatal Mental Health* [NG192]. https://www.nice.org.uk/guidance/cg192/resources/antenatal-and-postnatal-mental-health-clinical-management-and-service-guidance-pdf-35109869806789. Accessed April 16, 2021.

NICE. (2018). *Post-traumatic stress disorder* [NG116] Published: December 05, 2018. https://www.nice.org.uk/guidance/ng116. Accessed April 16, 2021.

NIDA. (2020, June 6). *Substance use while pregnant and breastfeeding.* https://www.drugabuse.gov/publications/research-reports/substance-use-in-women/substance-use-while-pregnant-breastfeeding. Accessed April 16, 2021.

Nörby, U., Forsberg, L., Wide, K., Sjörs, G., Winbladh, B., & Källén, K. (2016). Neonatal morbidity after maternal use of antidepressant drugs during pregnancy. *Pediatrics, 138*(5), e20160181.

Nörby, U., Winbladh, B., & Källén, K. (2017). Perinatal outcomes after treatment with ADHD medication during pregnancy. *Pediatrics, 140*(6), e20170747.

Olivier, J. D. A., Vallès, A., van Heesch, F., Afrasiab-Middelman, A., Roelofs, J. J. P. M., Jonkers, M., Peeters, E. J., Korte-Bouws, G. A. H., Dederen, J. P., Kiliaan, A. J., Martens, G. J., Schubert, D., & Homberg, J. R. (2011). Fluoxetine administration to pregnant rats increases anxiety-related behavior in the offspring. *Psychopharmacology, 217*(3), 419–432. https://doi.org/10.1007/s00213-011-2299-z

Ornoy, A. (2018). Pharmacological treatment of attention deficit hyperactivity disorder during pregnancy and lactation. *Pharmaceutical Research, 35*(3), 46.

Orsolini, L., Chiappini, S., Volpe, U., Berardis, D., Latini, R., Papanti, G. D., & Corkery, A. (2019). Use of medicinal cannabis and synthetic cannabinoids in Post-Traumatic Stress Disorder (PTSD): A systematic review. *Medicina (Kaunas, Lithuania), 55*(9), 525.

Park, Y., Hernandez-Diaz, S., Bateman, B. T., Cohen, J. M., Desai, R. J., Patorno, E., Glynn, R. J., Cohen, L. S., Mogun, H., & Huybrechts, K. F. (2018). Continuation of atypical antipsychotic medication during early pregnancy and the risk of gestational diabetes. *The American Journal of Psychiatry, 175*(6), 564–574.

Patorno, E., Huybrechts, K. F., Bateman, B. T., Cohen, J. M., Desai, R. J., Mogun, H., Cohen, L. S., & Hernandez-Diaz, S. (2017). Lithium use in pregnancy and the risk of cardiac malformations. *The New England Journal of Medicine, 376*(23), 2245–2254.

Pearlstein, T. (2012). Psychotropic medications and other non-hormonal treatments for premenstrual disorders. *Menopause International, 18*(2), 60–64.

Peterson, K., McDonagh, M. S., & Fu, R. (2008). Comparative benefits and harms of competing medications for adults with attention-deficit hyperactivity disorder: A systematic review and indirect comparison meta-analysis. *Psychopharmacology, 197*(1), 1–11.

Poels, E., Schrijver, L., Kamperman, A. M., Hillegers, M., Hoogendijk, W., Kushner, S. A., & Roza, S. J. (2018). Long-term neurodevelopmental consequences of intra-uterine exposure to lithium and antipsychotics: A systematic review and meta-analysis. *European Child & Adolescent Psychiatry, 27*(9), 1209–1230. https://doi.org/10.1007/s00787-018-1177-1

Polepally, A. R., Pennell, P. B., Brundage, R. C., Stowe, Z. N., Newport, D. J., Viguera, A. C., Ritchie, J. C., & Birnbaum, A. K. (2014). Model-based lamotrigine clearance changes during pregnancy: Clinical implication. *Annals of Clinical and Translational Neurology, 1*(2), 99–106.

Ray-Griffith, S., Tharp, E., Coker, J. L., Catlin, D., Knight, B., & Stowe, Z. N. (2021). Buprenorphine medication for opioid use disorder: A study of factors associated with postpartum treatment retention. *The American Journal on Addictions, 30*(1), 43–48.

Reid, R. L., & Soares, C. N. (2018). Premenstrual dysphoric disorder: Contemporary diagnosis and management. *Journal of Obstetrics and Gynaecology Canada: JOGC = Journal d'obstetrique et gynecologie du Canada: JOGC, 40*(2), 215–223.

Ross, L. E., Grigoriadis, S., Mamisashvili, L., Vonderporten, E. H., Roerecke, M., Rehm, J., Dennis, C. L., Koren, G., Steiner, M., Mousmanis, P., & Cheung, A. (2013). Selected pregnancy and delivery outcomes after exposure to antidepressant medication: A systematic review and meta-analysis. *JAMA Psychiatry, 70*(4), 436–443.

Rothberger, S., Carr, D., Brateng, D., Hebert, M., & Easterling, T. R. (2010). Pharmacodynamics of clonidine therapy in pregnancy: A heterogeneous maternal response impacts fetal growth. *American Journal of Hypertension, 23*(11), 1234–1240.

Rydell, M., Magnusson, C., Cnattingius, S., Granath, F., Svensson, A. C., & Galanti, M. R. (2014). Exposure to maternal smoking during pregnancy as a risk factor for tobacco use in adult offspring. *American Journal of Epidemiology, 179*(12), 1409–1417.

Salisbury, A. L., O'Grady, K. E., Battle, C. L., Wisner, K. L., Anderson, G. M., Stroud, L. R., Miller-Loncar, C. L., Young, M. E., & Lester, B. M. (2016). The roles of maternal depression, serotonin reuptake inhibitor treatment, and concomitant benzodiazepine use on infant neurobehavioral functioning over the first

postnatal month. *The American Journal of Psychiatry, 173*(2), 147–157.

SAMHSA. (2018). Substance Abuse and Mental Health Services Administration. *Clinical guidance for treating pregnant and parenting women with opioid use disorder and their infants.* HHS Publication No. (SMA) 18-5054. Substance Abuse and Mental Health Services Administration. https://store.samhsa.gov/sites/default/files/d7/priv/sma18-5054.pdf. Accessed on March 27, 2021.

SAMHSA. (2020). *Center for Behavioral Health Statistics and Quality. Results from the 2019 National Survey on Drug Use and Health: Detailed Tables.* Substance Abuse and Mental Health Services Administration. https://www.samhsa.gov/data/report/2019-nsduh-detailed-tables. Accessed on March 27, 2021.

Schmidt, P. J., Nieman, L., Danaceau, M. A., Tobin, M. B., Roca, C. A., Murphy, J. H., & Rubinow, D. R. (2000). Estrogen replacement in perimenopause-related depression: A preliminary report. *American Journal of Obstetrics and Gynecology, 183*(2), 414–420.

Smid, M. C., Metz, T. D., & Gordon, A. J. (2019). Stimulant use in pregnancy: An under-recognized epidemic among pregnant women. *Clinical Obstetrics and Gynecology, 62*(1), 168–184.

So, M., Bozzo, P., Inoue, M., & Einarson, A. (2010). Safety of antihistamines during pregnancy and lactation. *Canadian family physician Medecin de famille canadien, 56*(5), 427–429.

Soares, C. N., Almeida, O. P., Joffe, H., & Cohen, L. S. (2001). Efficacy of estradiol for the treatment of depressive disorders in perimenopausal women: A double-blind, randomized, placebo-controlled trial. *Archives of General Psychiatry, 58*(6), 529–534.

Sonon, K. E., Richardson, G. A., Cornelius, J. R., Kim, K. H., & Day, N. L. (2015). Prenatal marijuana exposure predicts marijuana use in young adulthood. *Neurotoxicology and Teratology, 47*, 10–15.

Stoffers, J. M., & Lieb, K. (2015). Pharmacotherapy for borderline personality disorder–current evidence and recent trends. *Current Psychiatry Reports, 17*(1), 534.

Tran, T. H., Griffin, B. L., Stone, R. H., Vest, K. M., & Todd, T. J. (2017). Methadone, buprenorphine, and naltrexone for the treatment of opioid use disorder in pregnant women. *Pharmacotherapy, 37*(7), 824–839.

Turner, E., Jones, M., Vaz, L. R., & Coleman, T. (2019). Systematic review and meta-analysis to assess the safety of bupropion and varenicline in pregnancy. *Nicotine & Tobacco Research: Official Journal of the Society for Research on Nicotine and Tobacco, 21*(8), 1001–1010.

Towers, C. V., Katz, E., Weitz, B., & Visconti, K. (2020). Use of naltrexone in treating opioid use disorder in pregnancy. *American Journal of Obstetrics and Gynecology, 222*(1), 83.e1–83.e8. https://doi.org/10.1016/j.ajog.2019.07.037. Epub 2019 Jul 31. PMID: 31376396.

Uguz, F., Sahingoz, M., Kose, S. A., Ozbebit, O., Sengul, C., Selvi, Y., Sengul, C. B., Ayhan, M. G., Dagistanli,

A., & Askin, R. (2012). Antidepressants and menstruation disorders in women: A cross-sectional study in three centers. *General Hospital Psychiatry, 34*(5), 529–533.

Uguz, F. (2016). Second-generation antipsychotics during the lactation period: A comparative systematic review on infant safety. *Journal of Clinical Psychopharmacology, 36*(3), 244–252.

Uguz, F. (2021). Neonatal and childhood outcomes in offspring of pregnant women using antidepressant medications: A critical review of current meta-analyses. *Journal of Clinical Pharmacology, 61*(2), 146–158.

Vigod, S. N., Gomes, T., Wilton, A. S., Taylor, V. H., & Ray, J. G. (2015). Antipsychotic drug use in pregnancy: High dimensional, propensity matched, population based cohort study. *BMJ (Clinical Research ed.), 350*, h2298.

Viguera, A. C., Whitfield, T., Baldessarini, R. J., et al. (2007). Risk of recurrence in women with bipolar disorder during pregnancy: Prospective study of mood stabilizer discontinuation. *The American Journal of Psychiatry, 164*, 1817–1824.

Villarreal, G., Hamner, M. B., Cañive, J. M., Robert, S., Calais, L. A., Durklaski, V., Zhai, Y., & Qualls, C. (2016). Efficacy of quetiapine monotherapy in posttraumatic stress disorder: A randomized, placebo-controlled trial. *The American Journal of Psychiatry, 173*(12), 1205–1212.

Volkow, N. D., Han, B., Compton, W. M., & McCance-Katz, E. F. (2019). Self-reported medical and nonmedical cannabis use among pregnant women in the United States. *JAMA, 322*(2), 167–169.

Wagner, N. J., Camerota, M., & Propper, C. (2017). Prevalence and perceptions of electronic cigarette use during pregnancy. *Maternal and Child Health Journal, 21*(8), 1655–1661.

Wendell, A. D. (2013). Overview and epidemiology of substance abuse in pregnancy. *Clinical Obstetrics and Gynecology, 56*(1), 91–96.

Westin, A. A., Brekke, M., Molden, E., Skogvoll, E., Castberg, I., & Spigset, O. (2018). Treatment with antipsychotics in pregnancy: Changes in drug disposition. *Clinical Pharmacology and Therapeutics, 103*(3), 477–484.

Whittington, J. R., Simmons, P. M., Phillips, A. M., Gammill, S. K., Cen, R., Magann, E. F., & Cardenas, V. M. (2018). The use of electronic cigarettes in pregnancy: A review of the literature. *Obstetrical & Gynecological Survey, 73*(9), 544–549.

Wieck, A., & Jones, S. (2018). Dangers of valproate in pregnancy. *BMJ (Clinical Research ed.), 361*, k1609.

Wiegand, S. L., Swortwood, M. J., Huestis, M. A., Thorp, J., Jones, H. E., & Vora, N. L. (2016). Naloxone and metabolites quantification in cord blood of prenatally exposed newborns and correlations with maternal concentrations. *AJP Reports, 6*(4), e385–e390.

Wisner, K. L., Sit, D. K., McShea, M. C., Rizzo, D. M., Zoretich, R. A., Hughes, C. L., Eng, H. F., Luther, J. F., Wisniewski, S. R., Costantino, M. L., Confer, A. L., Moses-Kolko, E. L., Famy, C. S., & Hanusa, B. H. (2013). Onset timing, thoughts of self-harm, and diagnoses in postpartum women with screen-positive depression findings. *JAMA Psychiatry, 70*(5), 490–498.

Wozniak, J. R., Riley, E. P., & Charness, M. E. (2019). Clinical presentation, diagnosis, and management of fetal alcohol spectrum disorder. *The Lancet. Neurology, 18*(8), 760–770.

Wyatt, K. M., Dimmock, P. W., Ismail, K. M., Jones, P. W., & O'Brien, P. M. (2004). The effectiveness of GnRHa with and without 'add-back' therapy in treating premenstrual syndrome: A meta analysis. *BJOG: An International Journal of Obstetrics and Gynaecology, 111*(6), 585–593.

Yadav, D. (2020). Prescribing in borderline personality disorder–the clinical guidelines. *Progress in Neurology and Psychiatry, 24*(2), 25–30.

Yatham, L. N., Kennedy, S. H., Parikh, S. V., Schaffer, A., Bond, D. J., Frey, B. N., Sharma, V., Goldstein, B. I., Rej, S., Beaulieu, S., Alda, M., MacQueen, G., Milev, R. V., Ravindran, A., O'Donovan, C., McIntosh, D., Lam, R. W., Vazquez, G., Kapczinski, F., McIntyre, R. S., et al. (2018). Canadian Network for Mood and Anxiety Treatments (CANMAT) and International Society for Bipolar Disorders (ISBD) 2018 guidelines for the management of patients with bipolar disorder. *Bipolar Disorders, 20*(2), 97–170.

Yonkers, K. A., O'Brien, P. M., & Eriksson, E. (2008). Premenstrual syndrome. *Lancet (London, England), 371*(9619), 1200–1210.

Yonkers, K. A., Wisner, K. L., Stewart, D. E., Oberlander, T. F., Dell, D. L., Stotland, N., Ramin, S., Chaudron, L., & Lockwood, C. (2009). The management of depression during pregnancy: A report from the American Psychiatric Association and the American College of Obstetricians and Gynecologists. *General Hospital Psychiatry, 31*(5), 403–413.

Young-Wolff, K. C., Sarovar, V., Tucker, L. Y., Avalos, L. A., Alexeeff, S., Conway, A., Armstrong, M. A., Weisner, C., Campbell, C. I., & Goler, N. (2019). Trends in marijuana use among pregnant women with and without nausea and vomiting in pregnancy, 2009–2016. *Drug and Alcohol Dependence, 196*, 66–70.

Zedler, B. K., Mann, A. L., Kim, M. M., Amick, H. R., Joyce, A. R., Murrelle, E. L., & Jones, H. E. (2016). Buprenorphine compared with methadone to treat pregnant women with opioid use disorder: A systematic review and meta-analysis of safety in the mother, fetus and child. *Addiction (Abingdon, England), 111*(12), 2115–2128.

Zhang, T. N., Gao, S. Y., Shen, Z. Q., Li, D., Liu, C. X., Lv, H. C., Zhang, Y., Gong, T. T., Xu, X., Ji, C., & Wu, Q. J. (2017). Use of selective serotonin-reuptake inhibitors in the first trimester and risk of cardiovascular-related malformations: A meta-analysis of cohort studies. *Scientific Reports, 7*, 43085.

Zorumski, C. F., Paul, S. M., Covey, D. F., & Mennerick, S. (2019). Neurosteroids as novel antidepressants and anxiolytics: GABA-A receptors and beyond. *Neurobiology of Stress, 11*, 100196.

Wendell A. D. (2013). Overview and epidemiology of substance abuse in pregnancy. *Clinical obstetrics and gynecology, 56*(1), 91–96.

Somatic Treatments and Neuromodulation in Psychiatry

136

Paul B. Fitzgerald and Martijn Arns

Contents

This chapter is an update from the 4th edition. Previous edition authors were Mustafa M. Husain, Sarah H. Lisanby and Jerald Kay

P. B. Fitzgerald (✉)
School of Medicine and Psychology, Australian National University, Canberra, ACT, Australia
e-mail: paul.fitzgerald@anu.edu.au

M. Arns
Brainclinics Foundation, Nijmegen, The Netherlands

Department of Cognitive Neuroscience, Faculty of Psychology and Neuroscience, Maastricht University, Maastricht, The Netherlands
e-mail: martijn@brainclinics.com

© Springer Nature Switzerland AG 2024
A. Tasman et al. (eds.), *Tasman's Psychiatry*,
https://doi.org/10.1007/978-3-030-51366-5_144

Abstract

Brain stimulation, neuromodulation, and other physical therapies have played a relatively small but continuous role in the treatment of mental health conditions since the mid-twentieth century. For many years, the mainstay of non-pharmacological physical treatments was electroconvulsive therapy (ECT). The safety and application of ECT has improved substantially over recent decades and continues to have a significant role, especially in the treatment of patients with severe and treatment-resistant major depressive disorder (MDD). The range of physical and neuromodulation treatments for mental health conditions, however, is relatively rapidly expanding at the moment. Repetitive transcranial magnetic stimulation (rTMS) – characterized by its safety and efficacy – is fast becoming a core part of the treatment of MDD and its use is expanding into other disorders such as obsessive-compulsive disorder (OCD). There is also emerging evidence for the potential use of other forms of non-invasive brain stimulation such as transcranial direct current stimulation (tDCS) and cranial electrical stimulation (CES). Additionally, there are an increasing range of approaches under active investigation including novel seizure therapy approaches such as magnetic seizure therapy (MST). In addition, there are several forms of brain stimulation that require a surgical procedure for implantation. Vagal nerve stimulation (VNS) appears to have a slow acting but persistent benefit in some patients with treatment-resistant MDD, and deep brain stimulation has an emerging role in the management of some patients with severe OCD. Beyond neuromodulation, an increasing range of research studies have been conducted investigating the use of forms of biofeedback, especially as it might pertain to the treatment of attention deficit hyperactivity disorder (ADHD) and there is modest but consistent evidence supporting the use of bright light therapy in seasonal affective disorder and also potentially in MDD. Physical therapies, including those utilizing brain stimulation methods, clearly have an important and growing role to play as the third tier of therapeutic interventions alongside pharmacotherapy and psychotherapy in the management of patients with mental health problems.

Keywords

Electroconvulsive therapy (ECT) · Repetitive transcranial magnetic stimulation (rTMS) · Transcranial direct current stimulation (tDCS) · Magnetic seizure therapy (MST) · Vagal nerve stimulation (VNS) · Deep brain stimulation (DBS) · Biofeedback · Bright light therapy (BLT) · Transcranial alternating current stimulation (tACS) · Transcranial random noise stimulation (tRNS) · Cranial electrotherapy (CES)

Introduction

There is a progressively expanding range of brain stimulation and other physical therapy tools that are being used in the treatment of mental health conditions or which are emerging as a potential therapeutic intervention. These range from quite non-invasive tools such as transcranial direct current stimulation (tDCS) and transcranial magnetic stimulation (TMS), to convulsive therapies such as electroconvulsive therapy (ECT), cranial electric stimulation (CES), and magnetic seizure therapy (MST) to more invasive treatment such as vagus nerve stimulation (VNS) and deep brain stimulation (DBS) (Fig. 1 and Table 1). In addition to these classical neuromodulation tools, there is emerging evidence for the potential use of other non-pharmacological somatic therapies such as light therapy and neurofeedback.

Since the introduction of ECT in 1938, brain stimulation has been in continuous use in psychiatry for over 75 years (Fig. 2). As with other medical procedures, the ECT technique has

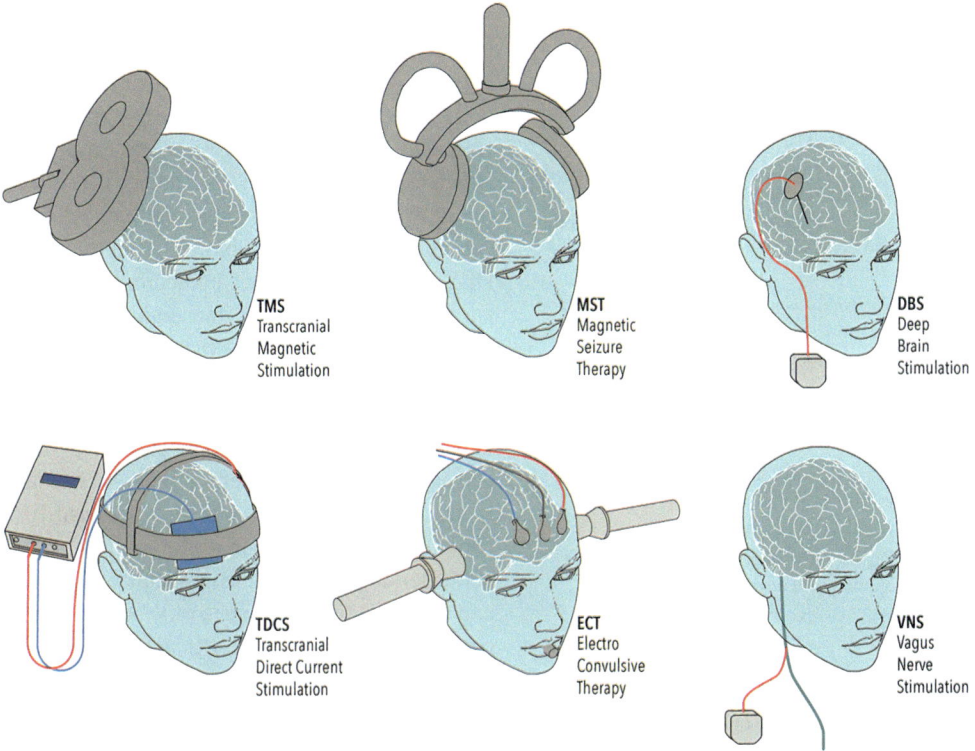

Fig. 1 Graphical overview of the various brain stimulation tools that will be covered in this chapter

evolved over this time, becoming progressively safer and better tolerated with innovations in anesthesia and ECT treatment parameters. Almost 50 years after the inception of ECT, a second form of brain stimulation, this time using pulsed magnetic fields to stimulate the brain (TMS) began to emerge (Barker et al., 1985). Initially used in neurology for neurophysiology studies, TMS rapidly caught the attention of psychiatrists, who were eager to explore whether TMS could represent a less invasive, nonconvulsive alternative to ECT. The therapeutic value of TMS in the treatment of depression has now been firmly established and there is increasing evidence that it may be therapeutically valuable in other conditions including obsessive-compulsive disorder (OCD), post-traumatic stress disorder, addictive disorders, schizophrenia, and chronic pain.

The successful development of TMS as an antidepressant therapy has been followed by an expanding interest in the potential application of other forms of brain stimulation. This has also led to considerable research developing TMS applications in other disorders leading to an expanding profile of clinical indications. There is now emerging evidence for the potential use of tDCS in the treatment of depression and MST, which uses a very strong form of TMS to induce seizures, is being evaluated as a form of convulsive therapy that may have similar efficacy to ECT but with fewer cognitive side effects. Focal electrically administered seizure therapy (FEAST) is another new form of convulsive therapy currently under evaluation.

At the more invasive end of the spectrum, psychiatry has borrowed from neurology in adapting several forms of implantable neuromodulation. There has been increasing interest in the potential use of DBS in the therapy of mental health conditions. DBS has been approved on a humanitarian basis for the treatment of refractory OCD (Alpert, 2011; Dougherty, 2018), and it has been subject to considerable study in the treatment of depression. VNS was also approved by the FDA for clinical use, in the treatment of epilepsy, in 1997 (Fig. 2). VNS, first approved in both Europe and Canada in

Table 1 Characteristics of brain stimulation modalities

Technique[a]	Characteristics
CES	Superficial stimulation of the ear lobe with relatively low levels of electricity. Approved by FDA, but evidence base is sparse
DBS	A wire electrode with multiple stimulating sites is surgically implanted into the brain and chronically administers electrical stimulation
ECT	Electricity is applied transcranially via electrodes placed on the scalp to induce a seizure under anesthesia
MST	High intensities of TMS are applied to induce focal seizures under anesthesia, as a less invasive form of ECT
tDCS	Weak direct electrical currents are applied to the scalp via sponge electrodes to polarize underlying brain tissue
TMS	Rapidly alternating magnetic fields are applied to the scalp to induce small, focused electrical currents in the superficial cortex
VNS	An implanted pacemaker administers electrical pulses to the vagus nerve in the neck, activating the vagal afferents

[a]*CES* cranial electric stimulation, *DBS* deep brain stimulation, *ECT* electroconvulsive therapy, *MST* magnetic seizure therapy, *tDCS* transcranial direct current stimulation, *TMS* transcranial magnetic stimulation, *VNS* vagus nerve stimulation

2001, subsequently received FDA approval for the treatment of chronic, medication-refractory depression in 2005 (Fig. 2) (O'Reardon et al., 2006).

As this timeline suggests, the evolution of brain stimulation techniques in psychiatry appears to be gaining momentum. This chapter reviews the current state of the evidence for each of the brain stimulation modalities.

Brain Stimulation and Physical Therapies in Psychiatry: A Unique Family of Interventions

Mechanisms of Action

Electrical Stimulation of the Brain

The various established brain stimulation modalities share the common characteristic that they stimulate the brain electrically. This electrical stimulation may be applied directly or induced indirectly via magnetic fields. The direct application of electrical stimulation can be performed transcranially as in the case of ECT, transcranial electrical stimulation (TES), CES, and tDCS. However, electricity can also be applied directly (e.g., VNS), epidurally (e.g., epidural electrical stimulation), or intracerebrally (as in the case of DBS). The epidural and intracerbral modalities are more focal than the transcranial application of electricity because they bypass the impedance of the scalp and skull.

The magnetic modalities (TMS and MST) induce electricity in the brain indirectly via the principle of electromagnetic induction. Magnetic fields will induce an electrical current in a conducting medium, such as the brain. The magnetic modalities bypass the impedance of the scalp and skull, but they are able to only stimulate a thin cortical layer of cells. They achieve enhanced focality noninvasively, without the need to implant an electrode within the skull as is required for the direct application of electricity. Therefore, the magnetic modalities offer an unparalleled degree of focality without the need for surgery. In addition, there is emerging evidence that the brain can be successfully stimulated through the use of external focused ultrasound, which induces firing of neurons through a non-electrical stimulation process. Interestingly, ultrasound may eventually be used to clear plaques from brains suffering from Alzheimer's dementia although this capacity will need to be demonstrated in human trials (Leinenga & Götz, 2015).

Comparison with Psychopharmacology

Psychopharmacology and brain stimulation differ as therapeutic modalities in several clinically relevant respects (Table 2). While psychopharmacology relies on molecular action at presynaptic or postsynaptic receptors or intracellular targets, brain stimulation can induce action potentials via electrical depolarization of axons in a manner that is independent of receptor binding. In this sense, brain stimulation has the potential to work when medications have failed, and also the potential to act synergistically with medications, because their sites and mechanisms of action are distinct.

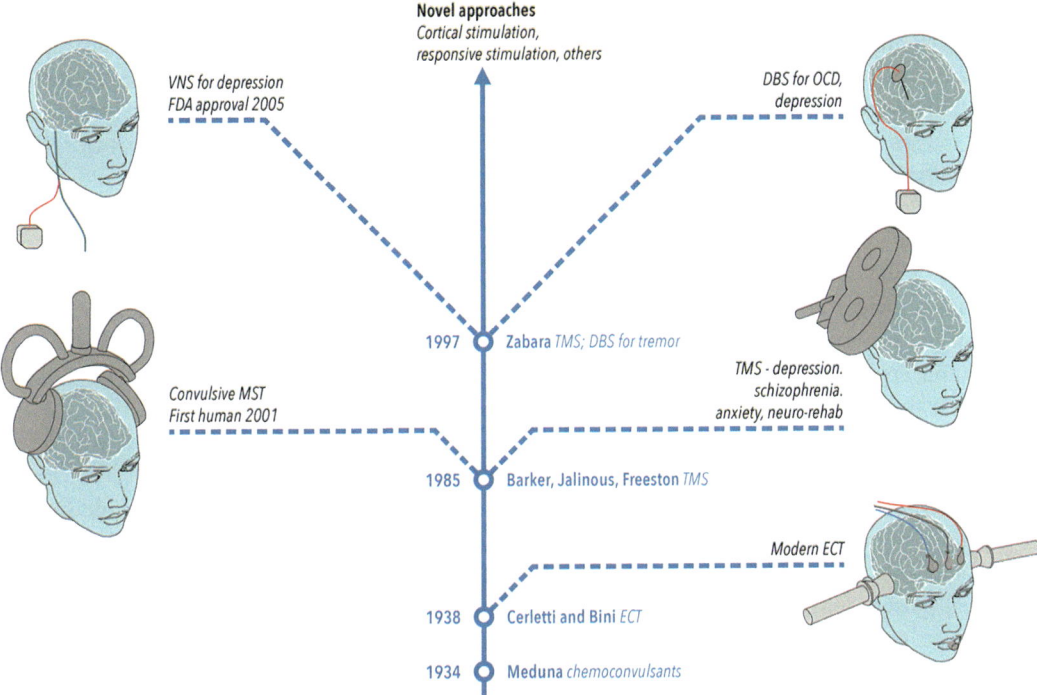

Fig. 2 Evolution of brain stimulation in psychiatry

Table 2 Comparisons between psychopharmacology

Therapeutic agent temporal characteristics	Psychopharmacology	Brain stimulation
Characteristics	Neurochemical	Electromagnetic
Time course	Tonic	Phasic
Delivery	Dependent upon patient compliance	Automated or clinician administered
Site of action	Molecular target	Neuronal depolarization
Distribution	Systemic	Focally applied to the brain
Duration of availability	Typical half-life hours to weeks	Milliseconds to seconds
Dosage parameters	Milligrams, blood level	Intensity, frequency, inter train interval, number of pulses, site of stimulation
Metabolism	Hepatic	None
Clearance	Renal	None
Interactions with metabolism of other agents	Possible	None known

Because stimulation is applied directly to the brain, it avoids the systemic side effects seen with medications that are systemically distributed and may be metabolized to biologically active compounds. Brain stimulation has no metabolites, does not undergo clearance, is not protein bound, and has no known interactions with the metabolism or clearance of psychopharmacological agents.

Whereas medications continue to be biologically available for hours to weeks, depending on their half-life, brain stimulation is applied in a phasic or continuous fashion but regardless, because it is electrical, it is not subject to metabolism or clearance and its direct effects terminate immediately on the cessation of stimulation. Brain stimulation modalities that administer alternating current (AC) (e.g., ECT, DBS, VNS) deliver pulses that typically last 1 ms or less. Direct current stimulation (tDCS) is typically applied for up to 30 min. The transient nature of the immediate effects of brain stimulation poses a challenge to the induction of lasting effects that will persist beyond the period of stimulation. In this context, it is important to consider how much stimulation, applied how frequently, is required to induce lasting neuroplastic changes in brain activity.

Acute Versus Chronic Effects

The actions of brain stimulation can be separated into acute and chronic effects (Fig. 3). Single electrical pulses, delivered at sufficient intensity, can induce neuronal depolarization and trigger trans-synaptic action, resulting in the activation of a functional circuit. For example, a single TMS pulse applied to the hand area of the primary motor cortex can activate the cortical spinal tract and induce a visible twitch in the contralateral hand muscle. These acute effects of brain stimulation can induce positive effects, as in the case of a muscle twitch or phosphenes (sensation of seeing light), or the acute effects can be disruptive, as in the case of visual masking.

Repeating pulses at regular intervals can exert even more powerful acute effects on brain function.

For example, repetitive TMS (rTMS) to the language-dominant hemisphere can induce speech arrest (Epstein et al., 1999), and speech returns to normal after the stimulation is terminated.

Some brain stimulation modalities, such as DBS, deliver repetitive pulses continuously, into specific brain targets, in which case the therapeutic action may rely on acute effects of the stimulation. However, the less invasive and nonsurgical modalities typically involve intermittent stimulation, which presumably requires the induction of some form of neuroplasticity, which persists beyond the period of stimulation.

Neuroplasticity

Repeated stimulation of a circuit can induce lasting changes in the subsequent activity of that circuit. Neuroplasticity is thought to occur through dynamic alterations in synaptic efficacy. For example, repeatedly electrically stimulating the perforant pathway at high frequency induces a lasting increase in the efficacy of that circuit, whereas low frequencies depress it. These phenomena are termed long-term potentiation (LTP) and long-term depression (LTD), respectively. There is evidence that TMS may act in ways that resemble these aspects of neuroplasticity. Specifically, low-frequency rTMS ($\leq$1 Hz) induces a sustained depression in measures of the excitability of the motor cortex, whereas high-frequency rTMS ($\geq$5 Hz) increases motor excitability (Chen et al., 1997; Wang et al., 1996), although such

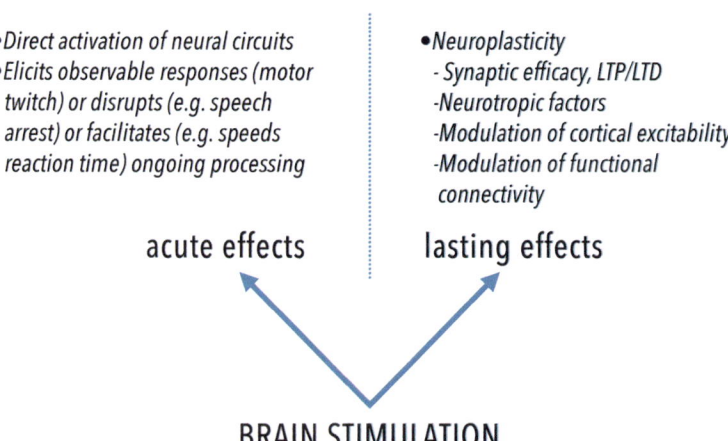

Fig. 3 Mechanisms of action of brain stimulation

- Direct activation of neural circuits
- Elicits observable responses (motor twitch) or disrupts (e.g. speech arrest) or facilitates (e.g. speeds reaction time) ongoing processing

- Neuroplasticity
  - Synaptic efficacy, LTP/LTD
  - Neurotropic factors
  - Modulation of cortical excitability
  - Modulation of functional connectivity

acute effects   lasting effects

BRAIN STIMULATION

frequency specific effects might not generalize to other cortical areas such as the frontal cortex (Fitzgerald & Daskalakis, 2011).

Impact of Development and Illness on the Brain's Response to Stimulation

Because brain stimulation acts via neuronal depolarization or through changing the likelihood of neuronal firing, and possibly via alterations in synaptic efficacy, it is not surprising that factors that influence neural transmission and plasticity may influence the brain's response to exogenous stimulation. For example, before the process of myelination is complete, infants and children have higher thresholds for activation of circuits via single pulses of TMS (Mall et al., 2004). Likewise, it has been reported that patients with schizophrenia fail to show inhibitory effects of low-frequency rTMS over the motor cortex, which may reflect abnormalities in glutamatergic functioning in that illness (Fitzgerald et al., 2004).

Relative Focality and Invasiveness of Brain Stimulation Modalities

The various brain stimulation modalities differ in their relative degrees of focality and invasiveness (Fig. 4). Non-invasive techniques clearly have the advantage that they may be used with fewer concerns regarding adverse effects, making them more acceptable earlier in the course of illness. Focality is an advantage for brain mapping studies that seek to identify brain regions involved in selected functions. Focality is also a therapeutic advantage if the specific therapeutic target is known, and if stimulation of adjacent regions might be counter-therapeutic. However, if the therapeutic target is broad, or not precisely known, then focality could have drawbacks with respect to efficacy.

DBS is the most focal technique, since the electrode is implanted directly into the target brain region, but is also the most invasive. In contrast, tDCS is the least invasive, but also one

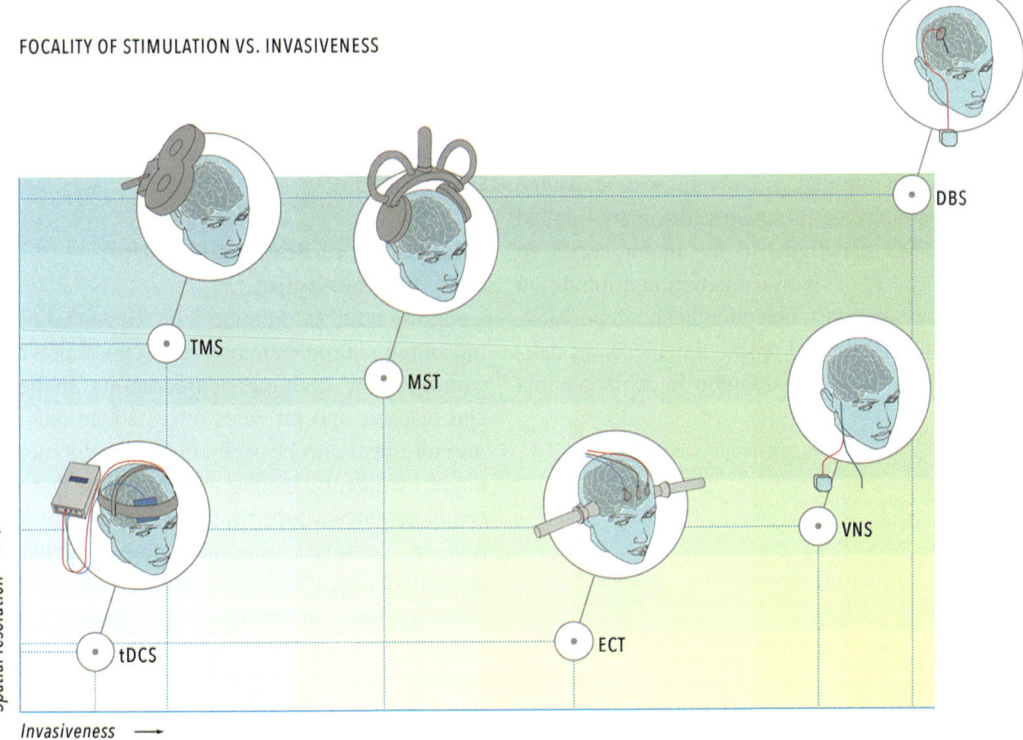

Fig. 4   Brain stimulation landscape: Focality (Spatial Resoution) of stimulation versus invasiveness

of the least focal modalities, as it involves the transcranial application of electricity. Also non-focal is ECT, for the same reasons of scalp and skull impedance faced by tDCS. ECT is more invasive, however, by virtue of the necessity for seizure induction. TMS and MST are more focal than their electrical counterparts because a magnetic field passes through the scalp and skull without impedance, allowing more control over site and extent of stimulation. The degree of focality of TMS can vary considerably with differing TMS coil configurations. VNS is less invasive than DBS, but also less focal because it stimulates the vagal afferents, which are widely distributed.

Non-Therapeutic Applications of Brain Stimulation in Psychiatry

Alongside the therapeutic uses of brain stimulation that will be discussed at length in this chapter, there are uses of these tools in the investigation and study of brain activity. Brain stimulation modalities have in fact contributed to our field in multiple parallel ways: (1) as tools of discovery to examine normal brain function, (2) as tools to examine the pathophysiology of psychiatric disorders, and (3) as novel therapeutic agents (Fig. 5).

The broadest range of initial studies that have utilized brain stimulation techniques to study the pathophysiology of psychiatric disorders applied various TMS paradigms to the motor cortex to explore aspects of brain excitation and inhibition in disorders such as schizophrenia and depression. These studies provided repeated evidence of deficits of GABA-related inhibition in schizophrenia

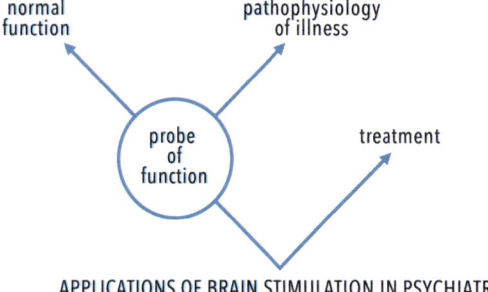

APPLICATIONS OF BRAIN STIMULATION IN PSYCHIATRY

Fig. 5 Applications of brain stimulation in psychiatry

(Fitzgerald et al., 2002) in particular and linked these deficits in inhibition to functional problems in motor cortical plasticity (Fitzgerald et al., 2004).

Combining TMS and Electroencephalography

In more recent years these lines of research have been extended to non-motor regions of the brain as techniques have developed to combine TMS stimulation with EEG to allow for the recording of immediate electrical responses to TMS pulses (Rogasch & Fitzgerald, 2013). The combination of TMS and EEG can be used to study electrically evoked responses, which are not confounded by motivation or other behavioral variables that can affect traditional evoked response EEG studies. TMS – EEG can also be used to study oscillatory responses to TMS pulses as well as network activity by mapping the spread of electrical activity through specific circuits.

These types of TMS – EEG methods have now been applied to explore brain activity in an increasing range of psychiatric disorders. For example, they have provided evidence that patients with schizophrenia have functional GABA-related deficits of inhibition in prefrontal brain areas (Farzan et al., 2012). There is also initial evidence that these tools may be used to predict response to therapeutic brain stimulation (Sun et al., 2018).

Integration of Brain Stimulation with Neuroimaging

Coupling brain stimulation with functional imaging and/or neurophysiology represents a powerful paradigm for studying neurocircuitry in health and disease, and provides intermediate outcome measures that may be useful in the development of novel treatments. Unlike neuroimaging that passively acquires information regarding brain function in various conditions, brain stimulation actively changes brain function and induces measurable effects on brain activity. Because neuroimaging is a passive modality, the information collected in traditional studies represents correlates of behavior, but cannot test causal relationships. Brain stimulation, on the other hand, is an intervention that can test causal relationships between brain and behavior. As such, brain stimulation can test hypotheses generated via functional neuroimaging and extend on these.

Progressively over time TMS has been successfully combined with positron emission tomography (PET) (Paus et al., 1997) and functional magnetic resonance imaging (fMRI) (Bohning et al., 1999). A common strategy is to perform the imaging or physiological measures before and after the TMS intervention. However, it is now also possible to perform simultaneous stimulation and imaging.

Simultaneous imaging is useful for examining acute effects of brain stimulation. For example, Strafella et al. (Strafella et al., 2001) used dopamine receptor PET to demonstrate that prefrontal TMS releases dopamine in the caudate, providing a functional probe of the activity of frontostriatal dopaminergic regulation. On the other hand, the pre/post paradigm may be useful for identifying lasting and chronic effects of stimulation. For example, Speer et al. (Speer et al., 2000) found that a course of daily rTMS given to depressed patients induced lasting increases in regional cerebral blood flow in bilateral frontal, limbic, and paralimbic regions implicated in depression. Such results could inform the mechanisms of action of brain stimulation modalities, and could be used to screen for stimulation paradigms predicted to have therapeutic effects based on their ability to modulate activity in the circuitry implicated in the illness. A concurrent TMS-fMRI study demonstrated that single TMS pulses applied to the left dorsal prefrontal cortex produced significant activity in distal brain regions including the anterior cingulate, caudate and thalamus validating the circuit level effects of TMS stimulation (Dowdle et al., 2018).

Therapeutic Applications of Brain Stimulation Tools in Psychiatry

Electroconvulsive Therapy

Introduction

ECT is a treatment method in which an electric current is passed through the brain via electrodes placed against the head to produce a seizure. A series of treatments constitute an acute course of ECT therapy, which is used to treat several psychiatric conditions when other treatments either have failed, cannot be tolerated, or are not expected to produce satisfactory results quickly enough. The procedure is performed under anesthesia and muscle paralysis to block the motor convulsion that would otherwise occur in the absence of a muscle relaxant, and to insure adequate oxygenation throughout the procedure. Typically, ECT will relieve the most profound symptoms of depression, the most commonly treated illness, within 2–3 weeks, with treatments usually given 2–3 times per week during the acute course of ECT. Continuation ECT refers to the continued application of ECT in the weeks immediately following the end of the acute course, with a tapered schedule, to stabilize remission. Maintenance ECT refers to the continued use of ECT over the months following a successful ECT course to prevent relapse.

History

Initially, convulsive therapy was performed via chemo-convulsant agents, a procedure introduced by Meduna in 1934. The first electroconvulsive treatment was performed on a human patient by the Italian team of Ugo Cerletti and Lucino Bini in 1938. From its inception, the procedure remained almost the exclusive way of treating serious mental illness and steadily gained in popularity until around the early 1960s, when its popularity started to wane, both as a result of the introduction of novel medications and as concerns regarding the side effects of the treatment increased. In the pre-anesthesia era, in addition to memory impairment, ECT side effects occasionally included bone fractures secondary to motor convulsion. Fractures are now mitigated by the use of a muscle relaxant. The introduction of the use of anesthesia in combination with a muscle relaxant in the 1970s in many ways revolutionized the safety profile of ECT therapy. Further innovations that progressively improved its risk-to-benefit ratio in the subsequent decades included the change from sine wave to brief pulse ECT in the 1980s and the introduction of right unilateral (RUL) ECT (Table 3). Today, ECT plays a prominent role in

Table 3 Innovations in ECT practice

Advance	Impact on the field
Anesthesia	Improved medical safety of ECT
Brief pulse	Replaced sine-wave ECT and significantly reduced cognitive side effects
Unilateral electrode placement	Provided an alternative to BL electrode placement, with fewer cognitive side effects
Seizure threshold titration	Provided another means to individualize dosage, in addition to age-based dosing to optimize safety and efficacy
Ultra-brief pulse	Shorter pulse widths appear to carry an even lower risk of side effects

the treatment of medication-resistant disorders. ECT has been in continuous use in psychiatry since its inception owing to its unparalleled efficacy in helping with the most treatment-resistant forms of depression, bipolar mania, catatonia, and affective episodes in schizophrenia.

Mechanisms of Action

Much is known about the neurobiological effects of ECT although there is still a lack of complete understanding of its mechanism of action. For example, ECT exerts a range of neurobiological effects on functional brain activation, neurotransmitter systems, hypothalamic–pituitary–adrenal (HPA) axis (normalizing the dexamethasone suppression test), and neuroplasticity. ECT is also a powerful anticonvulsant.

The ECT procedure involves the administration of anesthesia, the application of electricity, and the induction of a seizure. There is general agreement that it is the seizure itself, rather than the anesthesia or electrical stimulation in the absence of the seizure, which is responsible for the beneficial clinical effects of ECT. ECT-induced seizures result in a collection of neurochemical, neuroendocrine, and neurophysiological effects. Seizures release neurotransmitters, alter neurotransmitter receptor expression, alter gene expression, exert neurotrophic effects, affect cerebral blood flow and metabolism, and trigger anticonvulsant effects that terminate the seizure (Baldinger et al., 2014; Kato, 2009).

Which constellation of these neurobiological effects is necessary and sufficient for the beneficial effects of ECT, and which contribute to side effects, are a topic under active study.

A clue to the mechanisms of action of ECT comes from the clinical observation that different forms of ECT [e.g., RUL versus bilateral temporal (BT), low dosage versus high dosage relative to seizure threshold] result in differential clinical effects. For example, high-dosage RUL provides better antidepressant efficacy than low-dosage RUL, but without proportional increases in memory loss (Lisanby, 2000). This dissociation demonstrates that efficacy and amnesia can be uncoupled, supporting the view that memory loss is not essential to the mechanisms of action of ECT. Differences in electrode placement and dosage result in seizures that have different patterns of expression. These studies suggest that the efficacy and side effects of ECT are heavily dependent upon the topography of the seizure itself. For example, distinct patterns of effects involving anterior frontal regions have been associated with antidepressant response, whereas others involving temporal regions have been more strongly linked to the cognitive side effects of ECT (Luber et al., 2000; Nobler, 1994). These observations have been used in the development of more focal forms of convulsive therapy that might in future retain the superior efficacy of ECT, but with fewer side effects.

Medications for depression have been reported to increase cell proliferation, neuronal viability, and neurogenesis in rodents and primates. Likewise, ECT exerts neuroplastic effects, increases neurotropic factors, upregulates genes related to neuroplasticity (Erraji-Benchekroun et al., 2007), and increases cell proliferation (Perera et al., 2007; Scalia et al., 2007) in the hippocampus of rodents and nonhuman primates. Unlike medications for depression, ECT induces mossy fiber sprouting. The functional significance of these changes, and their potential relationship to antidepressant response, are a topic under investigation.

The anticonvulsant actions of ECT are so powerful that they have been successfully harnessed to treat medication-refractory epilepsy and status epilepticus (Lisanby et al., 2001a).

The massive anticonvulsant action of ECT has been hypothesized to be related to effects on GABA neurotransmission.

Depression is reported to be associated with deficient intracortical inhibition and cortical GABA levels, whereas ECT increases intracortical inhibition (Bajbouj et al., 2006) and cortical GABA (Sanacora et al., 2003) providing another clue to possible mechanisms of action.

Side Effects

ECT involves a number of risks that need to be evaluated prior to starting the treatment (also see Table 4). The risk of death with ECT is relatively low (about 1 per 10,000 patients), comparable to the rate expected from a series of brief anesthetic procedures alone. The most common cause of death with ECT is cardiovascular. Other causes include prolonged apnea, status epilepticus, and cerebral herniation (e.g., in unrecognized cases of brain tumor).

It is especially important to consider the cognitive side effects of ECT in the informed consent process and in the selection of ECT treatment parameters. Acutely, ECT causes anterograde amnesia from which patients typically recover soon after stopping ECT. Retrograde amnesia is the most persistent adverse effect of ECT. Shortly after ECT, many patients have gaps in memory for events that occurred close in time to ECT, but in some cases, retrograde amnesia may extend back months to years. Memory for autobiographical material is relatively less affected by ECT than

memory for events of an impersonal nature (Lisanby, 2000). Although retrograde amnesia often improves during the first few months following ECT, for many patients recovery is incomplete, with persisting amnesia for events that occurred close to the time of treatment (Donahue, 2000) and anecdotally occasionally from much earlier in life. Preexisting cognitive impairment is one of the few predictors of ECT's adverse effects (Mulsant et al., 1991).

Much research has been dedicated to this topic and it has been shown that the placement of the electrodes (RUL versus BL) and also technical characteristics of the electrical current applied (pulse shape and width) will affect the degree of risk of memory impairment (Sackeim et al., 1986, 1993, 2000) with the least risk demonstrated while using unilateral electrode placement in conjunction with ultra-brief pulse delivery (Loo et al., 2010; Sackeim et al., 2008). The Consortium for Research on ECT-sponsored multi-center trial demonstrated the equivalency of RUL placement (Kellner et al., 2010). Utilizing a lower-pulse amplitude (below 0.8 A) may also decrease cognitive side effects (Deng et al., 2011). Sine-wave ECT is no longer considered standard of care, owing to the higher risk of causing long-lasting memory problems.

Other side effects of ECT include headache, muscle aches, nausea, and fatigue. ECT produces a transient increase in pulse and blood pressure; occasionally, it may affect the cardiac rhythm. Therefore, a detailed cardiac history is important to identify individuals at increased risk such as in the immediate post-myocardial infarction setting, unstable angina, or intractable arrhythmias. ECT can be given in the presence of cardiac pacemakers and implanted defibrillators, but cardiology consultation is important in such cases.

The pre-ECT workup should include complete medical and neurological screening to detect factors such as a history of acute myocardial infarction, space-occupying brain lesion or other cause of increased intracranial pressure, unstable aneurysm or vascular malformation, poorly controlled diabetes mellitus, carcinoma, renal failure, or hepatic failure. Although ECT is not contraindicated during pregnancy, careful consideration of risks and

Table 4 Notable adverse consequences or side effects of ECT

Adverse effect	Frequency
Headache	Common/transient
Nausea post anesthesia	Common/transient
Cardiorespiratory complications of anesthesia	Rare
Anterograde amnesia[a]	Common
Retrograde amnesia[a]	Less common
Autobiographical memory impairment[a]	Rare

[a]Frequency of memory impairment is dependent on form of ECT used

benefits is important, as is consultation with obstetrics specialists and the use of fetal monitoring. Careful neuropathological examination of non-human primates receiving electroconvulsive shock has revealed no evidence of neuropathological lesions (Dwork et al., 2004).

Forms of ECT

There are currently several forms of ECT commonly used in clinical practice. These include right unilateral ECT (RUL), bitemporal ECT (BF), and bifrontal ECT (BT). ECT administration can also vary in regards to the stimulation pulse width. RUL ECT is provided either at an ultra-brief pulse width (0.25–0.3 ms) or at a brief pulse width (0.5–2.0 ms), whereas bilateral forms of ECT are usually applied at a brief pulse width. Most studies investigating the use of brief pulse width ECT have used 1.0 ms. 0.5 ms is a length increasingly used in clinical practice but with only limited clinical trial support.

Clinical Studies

There have been numerous studies that describe ECT as a safe and effective treatment for severe or treatment-resistant depression, bipolar disorder (both manic and depressive phases), and schizophrenia. Depression is the most frequently treated diagnosis.

The Consortium for Research in ECT (CORE) reported a 75% remission rate among 217 patients who completed an acute course of ECT (Husain et al., 2004). A systematic review conducted by the UK ECT Review Group (2003) found ECT to be significantly more effective than sham (six trials, $n = 256$ patients, effect size $= 0.91$) and pharmacotherapy (18 trials, $n = 1144$, effect size $= 0.80$). Patients with psychotic subtype of depression respond to ECT at higher rates than those without psychosis (Petrides et al., 2001). The efficacy of ECT is technique dependent, with remission rates ranging from 20% to over 80% depending upon how the treatment is performed. Double-masked, randomized controlled trials demonstrated powerful interactions between electrode placement and dosage relative to seizure threshold in the efficacy and side effects of ECT (Sackeim et al., 2000). The UK ECT

Review Group (2003) found BL to be moderately more effective than RUL ECT (22 trials, $n = 1408$, effect size $= 0.32$), although it should be noted that the efficacy of RUL ECT is dosage sensitive with dosage typically defined as the stimulus intensity relative to seizure threshold for the individual patient. Several studies have found that high-dose RUL (typically 6 times seizure threshold) ECT is as effective as BL ECT, with the advantages of lower cognitive side effects, especially at long-term follow-up (McCall et al., 2000; Sackeim et al., 2000).

Although with optimal technique ECT can be highly effective, success rates in the community hospital setting have been less robust (30–47%, roughly half that seen in clinical trials) (Prudic et al., 2004). Some of these discrepancies may be related to diagnostic comorbidity, but some may also be related to the tendency to halt ECT prior to complete remission (often to mitigate side effects). Only 23.4% of ECT non-remitters immediately following the acute treatment course were found to have sustained remission at 6 months (Prudic et al., 2004), underscoring the need to "treat to wellness" and not terminate prematurely the acute course of ECT.

A considerable body of research has explored the relative efficacy of different forms of ECT. BT ECT has traditionally been considered the most effective form of ECT, but with the greatest side effect burden. Response rates with RUL ECT approach that of BT ECT when a sufficiently high intensity is used. Ultra-brief RUL ECT may be slightly less effective, or take longer to produce clinical response, than standard pulse width RUL when applied at equivalent intensity relative to seizure threshold.

Following an effective course of ECT, the lowest rates of relapse are likely to be experienced by patients receiving maintenance therapy with a medication for depression and medications for bipolar disorder, or in some cases maintenance ECT. Without any form of maintenance treatment, patients relapse at a rate of 84% by 6 months post-ECT (Sackeim et al., 2001a). Monotherapy with nortriptyline reduced the relapse rate to 60%, but combination therapy with nortriptyline and lithium reduced it to 39% (Sackeim et al., 2001a).

Low rates of relapse were seen in elderly patients in the PRIDE study with the combination of lithium and venlafaxine (Rasmussen, 2017).

Current Status in Treatment Algorithms

ECT is usually used under one of two circumstances: either (1) early in the course of treatment for patient who presents with severe depression with psychotic features or severe melancholic depression including where acute suicidal ideation or other risk symptoms indicate a rapid antidepressant response is required or (2) in medication-resistant cases. In the later, ECT is usually considered where several adequate prior drug trials have failed to relieve symptoms sufficiently. Usually a trial of two or more single-action medications for depression [e.g., selective serotonin reuptake inhibitors (SSRIs)] is followed by a dual action medications for depression [e.g., serotonin norepinephrine reuptake inhibitors (SNRIs)] trial and possibly a trial of a monoamine enhancer (tricyclic antidepressant; TCA) or monoamine oxidase inhibitor (MAOI) pharmacotherapy, with or without augmentation strategies and/or TMS. However, in cases of severe depression with psychotic features or severe melancholic depression, ECT should be considered much earlier.

Most practice guidelines for depression recommend ECT for difficult-to-treat cases, often as a fifth, sixth, or seventh step. In routine clinical practice, however, ECT is often used even later in the course of treatment, owing both to its stigma and to its cognitive side effects. The National Institute for Health and Clinical Excellence (NICE) guidance on the use of ECT refers to its use in patients who have failed to in respond "to multiple drug treatments and psychological treatment." The clinical use of ECT remains supported by other professional organisations around the world including the American Psychiatric Association (APA).

Patient Selection

There are several criteria used in selecting ECT as a treatment modality. The first and most obvious one is the diagnosis. ECT has been demonstrated as a very effective treatment for depression, where response rates for depression with psychotic features can approach 90%. It appears the more severe the depression, the better is the treatment response. For example, dysthymic disorder or depressive symptoms in the setting of a personality disorder are likely to be less responsive to ECT. Persistent catatonia, both phases of bipolar disorder, and also positive and particularly affective symptoms of schizophrenia, have all responded well to ECT.

A second criterion in selecting ECT is the severity of the symptoms; ECT may and should be considered as an early treatment modality in psychiatric emergencies, particularly where medications may not be as effective or take a long time to be beneficial in changing the course of the illness. Prior response to treatment is another criterion to be taken into account and ECT may be lifesaving in cases of recurrent severe depression that has previously responded only to ECT in the past. Another important criterion is the medical safety of the treatment. The risks and benefits need to be carefully weighed for each case, especially if the patient has health conditions that may necessitate modifications to the procedure, such as morbid obesity (with respiratory effects complicating anesthesia and airway protection), central nervous system lesions increasing intracranial pressure, or severe cardiovascular disease.

There is a limited but growing literature suggesting that ECT may be used successfully in older adolescent patients with depression, but it should be rarely used in younger children. ECT is effective in older adults and is safe and well tolerated in this age group. The recent PRIDE study found a 62% remission rate in elderly depressed patients with RUL ultra-brief ECT (Rasmussen, 2017). Systematic reviews have suggested that ECT can be used safely in pregnancy but this is based on limited data.

Dosing

ECT dosage planning involves the selection of electrode placement, electrical dosage, and methods for individualizing the dosage. Selection among the electrode placements involves weighing their relative efficacy and risks. For example, RUL or BF may be selected to lower side-effect burden, whereas BT may be selected if RUL or BF is unlikely to be effective (as in the case of a prior failure).

Dosage may be individualized by titrating seizure threshold or by using an age-based method although the former has progressively become the recommended approach in guidelines. Age is a predictor of seizure threshold, but it accounts for only a small percentage of the variance in seizure threshold, which is highly variable across individuals. An advantage of empirical seizure threshold titration is that it provides a sensitive means of adjusting dosage relative to seizure threshold that has been useful in maximizing the efficacy of RUL electrode placement, which is highly dosage sensitive in its efficacy.

Following the determination of electrode placement and the establishment of a seizure threshold, the intensity of stimulation relative to threshold forms part of the dosing procedure. Studies suggest that higher relative dosage with the RUL condition enhances its efficacy, whereas higher dosage within the BL condition enhances its side effects. For RUL ECT, the shorter the pulse width used, the higher the intensity required to achieve adequate therapeutic response. For RUL applied at a 0.5 ms pulse width, stimulation at $5\times$ seizure threshold is recommended and at $6\times$ threshold for RUL ultra-brief ECT (see Table 5) although some studies have used up to $8\times$ seizure threshold.

Future Directions

Future directions in the field of ECT include identifying more effective ways of preventing relapse following an effective course of ECT and

reducing the cognitive side effects of the treatment. The CORE group found that only 46% of patients who remit following ECT remain well at 6 months when receiving maintenance treatment with combination pharmacotherapy or with maintenance ECT (the two groups did not differ) (Kellner et al., 2006). In the community setting, the percentage remaining well drops to 36%, perhaps owing to less aggressive maintenance treatment or premature termination of the ECT course in the presence of residual symptoms (Prudic et al., 2004). More effective strategies for relapse prevention might involve the combined use of maintenance ECT with combination pharmacotherapy, and a more structured format for adapting the maintenance ECT schedule to fluctuations in patient status.

Regarding the cognitive side effect burden of ECT, beyond electrode placement, shorter pulse widths have been reported to reduce amnesia from ECT. Other approaches have included improving the focality of seizure induction using magnetic fields which can be targeted more precisely than electrical stimulation owing to the lack of impedance (see the MST section later) and with differing ECT electrode sizes.

FEAST (Focal Electrically Assisted Seizure Therapy) is a more recently developed form of ECT with improved spatial targeting of the prefrontal regions of the brain that are felt to be the most important therapeutic targets for ECT treatment, while avoiding stimulation of areas of the temporal lobe implicated in ECT-related memory impairment. FEAST methods improve targeting through several methodological modifications. First, stimulation is applied with a novel

Table 5 Typical ECT stimulation parameters

Electrode placement	Pulse width (ms)	Intensity relative to seizure threshold	Notes
Right Unilateral	0.3	6	
	0.5	5–6	Limited evidence
	1.0	3–6	Effective across this range but greater efficacy at higher intensity
Bifrontal	0.5	1.5–2.5	
	1.0	1.5	
Bitemporal	0.5	1.5–2.5	
	1.0	1.5	

combination of a large posterior cathode and a small anterior anode over the anterior regions of the right frontal cortex. This concentrates the electrical current under the smaller electrode over the frontal cortex. Second, FEAST uses a unidirectional current, whereas conventional ECT is bidirectional. This increases the selectivity of neurons stimulated and has the effect of lowering the seizure threshold. FEAST has been evaluated in a series of small open label studies to date (for example, Sahlem et al. (2020)) following initial reports in non-human primates. These studies suggest FEAST has antidepressant activity but have not yet provided any indication as to its relative efficacy compared to other forms of ECT. Research is now commencing to address this question.

Repetitive Transcranial Magnetic Stimulation

Introduction

TMS is a process that involves the induction of small electrical currents in superficial layers of the cerebral cortex by applying rapidly alternating magnetic fields to the head. TMS is administered using an electromagnetic coil that is held on the scalp. The coil emits strong, pulsed magnetic fields. Magnetic fields pass through the scalp and skull freely and enter the brain without the impedance encountered by the direct application of electricity. This makes it easier to focus TMS to smaller regions of the brain than is possible with ECT or tDCS. The rapid change in the strength of the magnetic field induces electrical current in the brain through the principles of electromagnetic induction. TMS has been referred to as "electrodeless" electrical stimulation because it uses magnetic fields to induce electrical pulses indirectly. TMS can be given to awake subjects without the need for anesthesia. As such, it represents a noninvasive means of stimulating focal regions of the brain.

"Single-pulse TMS" refers to TMS when it is given one pulse at a time often at an irregular rhythm and very low rates (≤ 0.3 Hz). rTMS refers to stimulation that is repeated at regular intervals.

The frequency of this repetition generally ranges from 1 to 25 Hz. "Low-frequency rTMS" refers to repetition rates of ≤ 1 Hz, whereas "high-frequency rTMS" refers to stimulation at higher frequencies (usually 5–20 Hz). Theta burst stimulation (TBS) is a patterned form of rTMS involving the application of pulse triplets at very high frequency (usually 50 Hz) repeated at "theta" frequency: most commonly, 5 Hz. Intermittent TBS (iTBS) involves the application of 2 s trains of TBS usually with an 8 s into training interval. Continuous TBS (cTBS) involves the uninterrupted application of TBS usually for a period of 40 s.

History

The first human use of pulsed magnetic fields to stimulate the human brain was reported by Barker, Jalinous, and Freeston at the University of Sheffield in 1985 (Barker et al., 1985). The first-generation TMS stimulators were capable of low repetition rates (typically <0.3 Hz). Single TMS pulses proved useful in the study of motor conduction and other basic brain processes. In early work, these very low frequencies of TMS were applied in exploratory studies on the treatment of depression and schizophrenia (Belmaker & Fleischmann, 1995; Geller et al., 1997) using large non-focal round coils.

In the mid-1990s, repetitive stimulators capable of higher repetition rates (up to 20 Hz) became available. Higher frequency stimulation was able to influence higher brain functions such as language, mood, and memory. Early studies with high-frequency rTMS suggested that it might have antidepressant properties when applied to the dorsolateral prefrontal cortex (DLPFC) (George et al., 1995; Pascual-Leone et al., 1996). Subsequently, the potential therapeutic value of focally applied rTMS, targeted to the neural circuitry underlying depression and other disorders, has attracted considerable and systematic attention.

Mechanisms of Action

Alternating magnetic fields induce electrical currents in the superficial cortex underlying the TMS coil. These electrical currents are called "eddy currents" because they are circular in shape, and are oriented in the plane perpendicular to the plane of the TMS coil. At sufficient intensity, electrical currents will stimulate neuronal depolarization, which can result in an action potential. For example, when the TMS coil is positioned over the hand area of the motor strip, it induces local currents at the site of stimulation that cause the neurons in the primary motor area (M1) to fire. This stimulation activates the polysynaptic corticospinal tract and results in a twitch in the contralateral hand muscle. Thus, TMS uses magnetic fields to induce focal electrical currents indirectly in the brain, which in turn trigger the firing of functional neuronal circuits and can lead to the activation of distal brain regions and observable behavioral effects. By moving the TMS coil to the cortical representation of neighboring muscle groups, single TMS pulses can be used to map the homunculus and study the excitability of the corticospinal system.

When moved to other cortical areas, single TMS pulses can exert other effects, such as transient scotoma, or "blind spots," when positioned over the primary visual cortex (V1). This paradigm is called "visual masking" (Luber et al., 2007b). Thus, TMS can transiently disrupt functions. This mode of action has been termed the "virtual lesion" technique.

These various actions of single-pulse TMS represent the acute effects of the activation of neural circuits triggered by TMS-induced neuronal depolarization (Fig. 3). Generally, the effects of single TMS pulses are immediate and short-lived. However, TMS can exert longer lasting effects when the pulses are repeated at regular intervals (rTMS), or when they are paired with other forms of stimulation, such as in the paired-associative stimulation paradigm (Quartarone et al., 2006) where TMS pulses are paired with electrical stimulation of a peripheral nerve, or when TMS is paired with audiovisual stimuli as in the example of classical conditioning of the brain response to TMS (Luber et al., 2007a). The mechanisms underlying these lasting effects of TMS are thought to be related to neuroplasticity and alterations in synaptic efficacy, as reviewed above.

Because rTMS has the ability to induce lasting changes in cortical excitability at the site of stimulation and in connected regions within neural circuits, its application in the treatment of psychiatric disorders has been driven by attempts to alter focal abnormalities in cortical excitability linked to illness. For example, studies point to reduced activity of the DLPFC (especially on the left) and also other regions that form a distributed network, during clinical depression. Therefore, studies have applied high frequencies of rTMS, which has been reported to increase excitability, to the left DLPFC in an attempt to normalize activity in this region. Furthermore, some theories implicate abnormal interhemispheric balance in activation between the right and left DLPFC. Building upon these theories, some studies have applied low-frequency rTMS, which has been reported to be inhibitory, to the right DLPFC, in an attempt to normalize that interhemispheric balance. Although there is a logical progression from the identification of local cortical changes in activity and the successful application of rTMS, it remains unclear whether the therapeutic effects of stimulation in disorders like depression relate to changes in local activity or in effects around the connected brain circuits.

Further supporting the potential therapeutic role of focally modulating cortical excitability, animal studies suggest that rTMS can exert effects on neurotransmitter release, receptor expression, and gene expression that are similar to some of the effects of medications for depression and ECT (Lisanby & Belmaker, 2000). It should be noted, however, that animal models of the effects of rTMS do not perfectly replicate the effects of rTMS on the human brain, owing to size differences and how they impact the strength and distribution of the induced electric fields.

When applied in clinical treatment, rTMS has most commonly been utilized using a figure of eight shaped coil which produces relatively focal stimulation with a penetration of up to 2 cm into the cortex. Treatment may also be applied with a coil that produces a deeper penetration of the magnetic field. Deeper stimulation can be produced with figure-of-eight coils that are angulated or with specifically developed 'H' coils.

Side Effects

Although TMS is clearly less invasive than ECT, it is not entirely without risk. The most serious known risk of TMS is the indication of a seizure. The relative risk of seizure with TMS depends on the form of TMS (single-pulse versus rTMS), the dosage (intensity, frequency, train duration, inter-train interval), and subject factors that may place the individual at increased risk (such as the presence of a neurological disorder or seizure-lowering medications).

Single-pulse TMS is generally considered to have minimal risk when administered to adults lacking risk factors for seizures. rTMS can induce seizures in individuals without predisposing conditions when given at sufficiently high dosages. Safety guidelines (Fig. 6) were developed to mitigate the risk of seizure (Wassermann, 1998; Rossi et al., 2009, 2020). When given within the safety guidelines, the risk of rTMS induced seizures appears to be very low. However, since seizure is a possibility, subjects should always be advised of this potential. More practically, the likelihood of a seizure occurring as reported in the research literature is 0.007% whereas in clinical practice this is reported to be 0.003%, and both seizure rates are lower as would be expected for medications for depression (George et al., 2013), therefore when adhering the above safety guidelines, rTMS is generally considered a safe treatment. It is also recommended that rTMS be administered under medical supervision, and that procedures be in place to screen and monitor patients medically for changes in clinical status, for example, alcohol or benzodiazepine withdrawal, that could affect seizure risk. The treating clinician should have the proper training to provide an appropriate first response to a seizure should one occur (Belmaker et al., 2003).

The most common side effects of TMS are scalp discomfort and headache. These effects are attributable to scalp muscle stimulation and stimulation of the facial nerve. Earplugs are worn by the patient and administrator to protect hearing. The risks of TMS during pregnancy have not been thoroughly studied but no systematic risks have been seen in studies so far and modelling of the electrical and magnetic fields produced by a TMS coil on the head indicates that a fetus would not experience significant exposure.

Clinical Studies

Major depression is the disorder that has been most thoroughly studied with rTMS to date. The antidepressant efficacy of rTMS has now been established with numerous small and large clinical trials, the results of which have been summarized in multiple positive meta-analysis. Clinical trials and meta-analyses have confirmed the efficacy of high-frequency rTMS applied to the left DLPFC, low frequency rTMS applied to the right DLPFC and sequential bilateral rTMS combining both left and right sided stimulation (Fitzgerald, 2020). Clinical trials and meta-analyses have also

Hz	% MOTOR THRESHOLD				
	100%	110%	120%	130%	140%
1	>1800	>1800	360	>50	>50
5	>10	>10	>10	<10	7.6
10	>5	>5	4.2	4.2	1.3
20	2.05	1.6	1.0	1.0	0.35
25	1.28	0.84	0.4	0.4	0.2

Table shows the duration in seconds that could be administered at each combination of frequency and intensity prior to the emergency of spread of excitation, a warning sign of seizure.
Values preceded by > indicate that this is the longest duration that was tested.

Fig. 6 rTMS safety guidelines as published by Rossi et al. (2009)

indicated that these three forms of stimulation are equally effective. In addition to these three treatment approaches, a substantial clinical trial established that deep TMS, using a specific "H – coil" is also more effective than sham stimulation in the treatment of depression (Levkovitz et al., 2015). There is no substantive evidence as to whether standard forms of rTMS or deep TMS is more effective.

In addition, recent studies have also explored the use of various forms of theta burst stimulation (TBS) in the treatment of depression (Chen et al., 2019). Several small studies have indicated that iTBS applied to the left DLPFC, cTBS applied to the right DLPFC or the combination of both of these approaches are likely to have antidepressant effects. The recent 3D study found antidepressant equivalence between high-frequency left-sided rTMS and left sided iTBS (Blumberger et al., 2018). TBS approaches have a significant potential advantage over traditional forms of rTMS in that they can be applied in a much shorter period of time, 3 min for iTBS compared to roughly 20–30 min for 10 Hz rTMS.

As well as studies comparing rTMS therapy to sham or placebo stimulation, a number of studies have directly compared antidepressant effects of rTMS and ECT. Unfortunately, these studies have generally been underpowered to show meaningful differences between two active treatments or had fairly unbalanced treatment groups, for example, comparing an unlimited number of unilateral and bilateral ECT to a limited number of one type of rTMS. There is a general belief that ECT response rates remain higher than those for rTMS but that the patients accessing these treatments are only partially overlapping groups: the clinical trials comparing ECT and rTMS have mostly been systematically biased towards a likelihood of showing a benefit of ECT, and the ECT advantage largely disappeared in a meta-analysis when limited only to non-psychotic depression (Berlim et al., 2013). rTMS has not been systematically evaluated to any meaningful degree in patients with psychotic depression.

There is considerably less evidence exploring the use of rTMS in longer-term treatment than that supporting its acute clinical use. While most studies report durability of the clinical benefit of rTMS, e.g., 62.5–66% retained response after 6 months (Donse et al., 2018; Dunner et al., 2014), these are based on open-label effectiveness studies and have both used some level of maintenance treatment. Studies have relatively clearly found that the reintroduction of a course of rTMS following relapse of depression is highly likely to result in a second or subsequent clinical response (for example, (Fitzgerald et al., 2006)). Studies have explored the use of regular rTMS sessions (often starting once per week and gradually reducing in frequency) (Benadhira et al., 2017) and the use of clustered rTMS sessions (usually five or six sessions applied over 2 to 3 days every 4 weeks) (Fitzgerald et al., 2013a) in the prevention of depressive relapse. Studies to date have generally supported the value of maintenance rTMS but have not definitively established its effectiveness. The largest study of maintenance rTMS conducted to date demonstrated a significant benefit of clustered maintenance rTMS as a monotherapy or combined with medications for depression, over antidepressant treatment alone, but did so in a population of patients who initially responded to medication rather than rTMS treatment in their index episode (Wang et al., 2017).

Current Status in Treatment Algorithms

The clinical use of rTMS in the treatment of depression is now supported widely in clinical practice guidelines informing the treatment of depression (Perera et al., 2016). Clinical studies have established that rTMS is effective in patients who are receiving, and in patients who are not receiving, medications for depression. Studies have indicated that it is effective in patients who have failed one or more medications for depression. rTMS should be considered after one or more failed medication trials and is likely to have greater efficacy than medication in patients who have failed more than 2 drugs.

Patient Selection

rTMS should be used in patients who have failed one or more medications for depression trials (Fitzgerald, 2020) (see Fig. 7). There is evidence

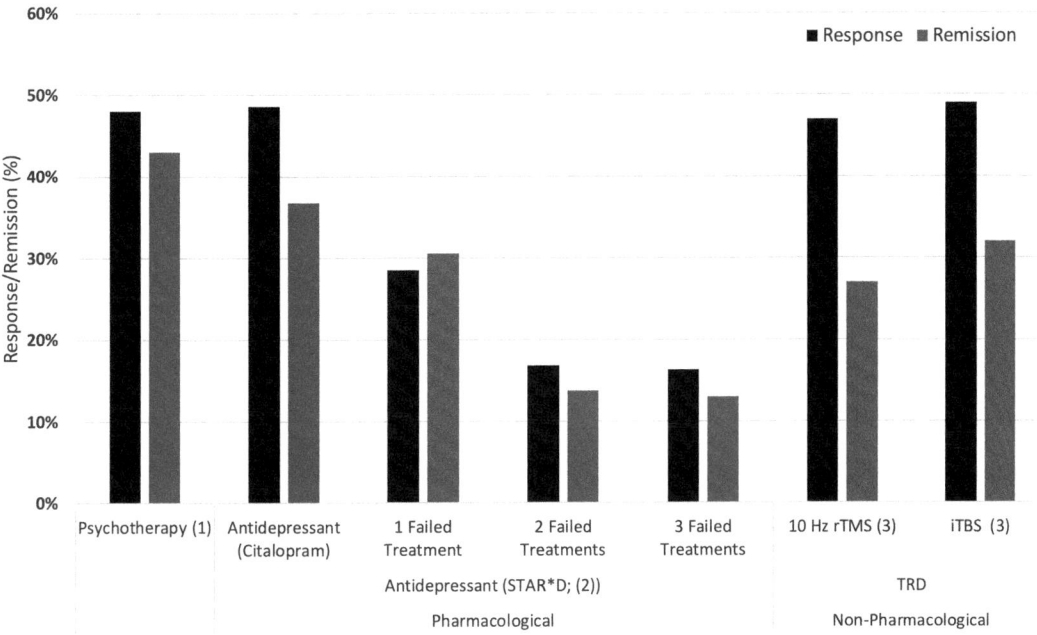

Fig. 7 Response and remission rates of various antidepressant treatments based on the largest effectiveness studies and datasets available. Note the relative higher response and remission rates for both forms of rTMS (10 Hz and iTBS), especially relative to patients that have had two or three prior treatment failures (i.e., treatment-resistant depression), which is the typical population rTMS treatment is currently indicated for (Cuijpers et al., 2014; Rush et al., 2006; Blumberger et al., 2018)

that rTMS may be effective in patients with long histories of persistent depression and many failed medication trials although response rates are likely to be higher in patients earlier in their illness course. It may be used in patients who are continuing with treatment with medications for depression or as a monotherapy and in patients with unipolar or bipolar depression (Perera et al., 2016). rTMS can be used safely and effectively in elderly patients with depression but at this stage there is limited evidence supporting its use in children or adolescents. For the reasons listed in Table 2, TMS may be a reasonable and perhaps safer option than pharmacologic treatments in pregnant women (Kim et al., 2011; Pridmore et al., 2021).

Dosing

There are many aspects to TMS dosing (Table 6) that are usually determined by the prescribing psychiatrist prior to therapy and during an initial evaluation of the patient's resting motor threshold (RMT) (Perera et al., 2016): the minimum simulator intensity required to produce a muscle twitch in the contralateral hand. Standard rTMS treatment is typically applied at 120% of the RMT but may be effective at lower intensity if this level is not tolerated (Fitzgerald, 2020). High-frequency rTMS it is usually applied at a frequency of 10 Hz in 4 s trains. Stimulation in longer trains at 120% of the RMT is associated with significantly increased risk of seizure induction. Seventy-five 4 s trains are most typically applied during left DLPFC treatment with an inter-train interval that can vary from 11 to 25 s. Low-frequency right-sided DLPFC stimulation is usually provided in a single 20-min train of pulses at 1 Hz.

Various methods have been developed to determine the site of stimulation in the prefrontal cortex. Initially, a procedure was developed that involved moving the coil 5 cm anterior to the hotspot for inducing a muscle twitch in the contralateral hand in the sagittal plane. Studies have

Table 6 TMS dosage parameters

Approach	Frequency (Hz)	Coil	Train details
High frequency left DLPFC	10	Figure of 8	75 trains 4 s duration 15–26 s ITI 120% RMT
Low frequency right DLPFC	1	Figure of 8	1 single train 20 min duration 120% RMT
Sequential Bilateral stimulation	10 on left 1 of right	Figure of 8	Left most commonly 50 trains of 4 s duration Right – one train of 20 min duration Both – 120% RMT
Deep TMS	18	H Coil	2 s trains 20 s ITI 55 trains 120% RMT
iTBS – left DLPFC	50/5	Figure of 8	50 Hz triplets repeated at 5 Hz for 2 s in one train 20 trains 8 s inter-train interval 70–120% RMT

DLPFC Dorsolateral prefrontal cortex, *ITI* intertrain interval, *RMT* resting motor threshold, *iTBS* intermittent theta burst stimulation, *ITI* intertrain interval

shown that this typically results in coil placement posterior and potentially somewhat medial to the DLPFC and this method does not take variation of head size into account (Fitzgerald et al., 2009b). Some clinical trials have been conducted using modifications of this involving moving the coil 5.5–6 cm forward from the motor hot spot with or without moving 1 cm in a lateral direction. An alternative approach is to locate stimulation based on the F3 electrode site in the 10–20 EEG system. Software is available to simplify the location of this process, most commonly using the Beam-F3 method (Beam et al., 2009). The most accurate method for site localization involves one of a number of forms of MRI-based neuronavigation where individual patient scans are used to localize the stimulation site, and this may result in improved antidepressant effects of rTMS compared to the 5 cm rule (Fitzgerald et al., 2009a); however, it is unknown if that also holds for the 6–7 cm rule or the Beam-F3 method. Stimulation can be based upon anatomical landmarks – the most common site being at the junction between the anterior and middle third of the middle frontal gyrus – or on standardized coordinates. Recent research has suggested that improved outcomes might be obtained by localization using resting state functional MRI imaging determining the DLPFC site where activity is most highly anti-correlated with activation in the subgenual anterior cingulate cortex (Weigand et al., 2018). Finally, a recently developed approach has been developed that utilizes changes in heart rate produced during stimulation of the DLPFC (Neuro-Cardiac-Guided TMS) to identify the optimal site for producing changes in descending vagal tone (Iseger et al., 2017). Further research is required to prospectively evaluate these methods and if they enhance clinical efficacy relative to the simpler methods.

Other Applications

A large body of research has explored the potential use of rTMS across a number of additional psychiatric disorders (Lefaucheur et al., 2014; Rehn et al., 2018). Obsessive-compulsive disorder (OCD) was the first application other than depression to achieve formal FDA approval in the USA. Approval has been granted for the use of deep TMS using both a H coil as well as an angulated figure-of-eight coil (double-cone coil). The protocol utilized for OCD treatment involves high-frequency stimulation of the bilateral medial frontal cortex (20 Hz trains) applied immediately following a brief process of the induction of distressing OCD symptoms. While the FDA approval was of high-frequency rTMS, low-frequency TMS applied to the supplementary motor area and orbitofrontal have been more consistently investigated and found effective in a number of small single site studies (Rehn et al., 2018).

A number of both high- and low-frequency rTMS protocols applied to the left and right DLPFC have been evaluated in post-traumatic stress disorder but a clearly superior treatment approach has not yet emerged with evidence emerging of the effectiveness of both high and low frequency stimulation applied to the right DLPFC (Cirillo et al., 2019; Yan et al., 2017).

A large number of clinical trials have also been conducted using rTMS in the treatment of various symptoms of schizophrenia (Kennedy et al., 2018). Low-frequency stimulation of the left temporo-parietal cortex appears to have capacity to produce lasting amelioration of persistent auditory hallucinations with several positive meta-analysis but an absence of large multi-site trials. There is also slowly emerging evidence of short-term improvements in negative symptoms with high-frequency frontal stimulation and interest in developing rTMS as a way of improving cognitive function in patients with schizophrenia.

Many studies have also been conducted using rTMS to alleviate chronic pain syndromes with a focus on the stimulation of motor cortex or prefrontal regions but no definitive studies have been published (Cardenas-Rojas et al., 2020). A similar literature has explored the use of rTMS in the treatment of addiction with deep TMS approved by the FDA in the USA for the treatment of smoking cessation in 2020 following the conduct of a pivotal industry sponsored trial.

Future Directions

Applications of rTMS in mental health conditions remain a focus of intensive research with studies underway exploring efficacy in many disorders. In the context of depression, considerable research is underway to establish whether accelerated or intensive rTMS protocols can be utilized to produce a more rapid antidepressant responses potentially making rTMS a useful intervention for patients who are highly suicidal or where the prolonged

period of treatment currently required is inconvenient or clinically problematic. Improved methods for MRI-based neuronavigation are also currently an area of considerable research focus. Further research is required to better understand the role of TMS treatment in adolescent patients with depression, during pregnancy and as a maintenance treatment as well as the role of rTMS in the treatment of other mental health conditions.

Magnetic Seizure Therapy

Definition

MST is a convulsive treatment that uses an alternating magnetic field to cross the scalp and skull and induce a more localized electric current in targeted regions of the cerebral cortex than is possible with ECT. The aim is to produce a seizure whose focus and patterns of spread may be controlled. MST attempts to marry the focality provided by magnetic fields with the powerful antidepressant efficacy of ECT, while possibly limiting the often reported cognitive side effects associated with ECT (Lisanby & Peterchev, 2007).

Like ECT, MST is performed under general anesthesia. MST is given using a modified TMS device that can administer higher output than conventional TMS devices. The MST procedure is also performed with a muscle relaxant and requires very similar staffing and infrastructure resources to those of ECT. Should MST prove to be effective, its uptake is likely to be relatively efficient as the treatment infrastructure for providing convulsive therapy is quite widely available. However, it is important to note that the efficacy of MST is still under evaluation and is not currently approved by the FDA in the USA or any regulatory authority around the world.

History

The first MST-induced seizures, induced in rhesus monkeys, were performed in 1998 using a custom-modified TMS device (Lisanby et al., 2001b). Following safety testing in nonhuman primates (Dwork et al., 2004), the first human case of a 20-year-old inpatient with a medication-resistant major depressive episode treated in Berne, Switzerland, was published in 2001 (Lisanby et al., 2001c). The treatments were well tolerated, resulted in a 50% decrease in the depression scores and had no effect on Mini Mental State Exam scores, which remained at 30 throughout the treatment course. The first trial of MST in the United States, performed at Columbia University and published in 2003, reported fewer acute cognitive side effects with MST than with ECT in a within-subject cross-over blinded trial (Lisanby et al., 2003a).

Mechanisms of Action

Like ECT, MST induces a seizure that is hypothesized to be central to its mechanism of action. However, seizures induced by MST are distinct from those induced by ECT (Lisanby et al., 2003b). MST-induced seizures show less impact on parasympathetic outflow, less generalization to hippocampus and deeper brain structures, and result in less serum prolactin surge.

Side Effects

The risks with MST are similar to those with ECT and are largely connected to the risks associated with anesthesia and generalized seizure. In addition, the MST coil produces a loud clicking noise that may potentially affect hearing and, to prevent any cumulative damage, earplugs should be worn by both the patient and members of the treating team. With respect to memory, studies have suggested that MST results in less retrograde amnesia than ECT (Lisanby et al., 2003a; Fitzgerald et al., 2018), although this result should be replicated in larger trials. The lack of impact of MST related seizures on cognitive function also has resulted in some patients reporting a return of awareness prior to the cessation of muscle paralysis. This can be countered by a reduction in dose of the muscle relaxant or the provision of a small additional dose of anesthetic following cessation of the MST-induced seizure.

Clinical Studies

In the early 2000s a series of small trials were conducted to provide proof of principle of the potential antidepressant effects of MST and provide intial safety data (Lisanby et al., 2003a; White et al., 2006). These early trials demonstrated the feasibility of MST in a clinical setting. However, the maximum stimulation possible with the equipment available at the time (50 Hz, 400 pulses per session) was estimated to be on average only 1.3 times the magnetic seizure threshold, potentially limiting efficacy. Subsequent trials have adopted MST devices capable of stimulation at up to 100 Hz, at 100% of stimulator output, for up to 10 s trains.

A number of clinical trials have now been completed with these high-powered MST devices. An open label trial of 100 Hz MST in 13 depressed patients reported significant antidepressant responses with no evidence of stimulation-related memory impairment (Fitzgerald et al., 2013b). In a small comparative trial, 20 patients with treatment-resistant depression received either MST or right unilateral ECT at three times the seizure threshold. Equivalent antidepressant responses and no cognitive side effects were reported in both groups. A second comparative trial of 37 patients also found no difference in antidepressant effects between MST and right unilateral ECT and MST had a slightly better cognitive profile (Fitzgerald et al., 2018).

Recent research has begun to investigate whether the clinical responses to MST are dependent on stimulation frequency but a clearly optimal frequency is not yet apparent. Recent studies have also reported significant reductions in suicidal ideation with MST (Sun et al., 2018) and demonstrated preliminary evidence of possible benefit in patients with bipolar depression, OCD, and schizophrenia.

Current Status in Treatment Algorithms

MST is not approved as a therapy in any jurisdiction at this time and therefore does not have a specific place in current treatment algorithms. More substantive research is required to establish whether it has sufficient efficacy to allow it to be used as a genuine alternative to ECT. As general anesthesia is part of the MST procedure, it will still likely be reserved for medication treatment-resistant patients.

Patient Selection

MST is being tested in the population of patients referred for ECT, the primary indication for which is depression.

Future Directions

MST is at the early phases of clinical testing. Unanswered questions about dosimetry, optimal coil placement, patient selection, and mechanisms of action are the topics of future studies.

Transcranial Direct Current Stimulation

Definition

tDCS it is a form of non-invasive electrical stimulation that uses very weak (1–2 mA) electric fields applied to the scalp through saline-soaked sponges to change activity in the brain. The fields are applied using direct current (DC), rather than the AC used in ECT, VNS, and DBS. tDCS can be administered with quite small battery-operated and portable devices.

History

tDCS is an old technology that has experienced a modern resurgence of interest. There has been interest in the use of electrical currents to try to change brain activity going back over decades, if not centuries, and there was interest in the use of technologies analogous to tDCS especially in the 1950s and 1960s. These fell out of favor, presumably as pharmacological treatments for mental health conditions became common, but interest in this area has dramatically expanded over the last decade (Rothwell, 2018).

Side Effects

tDCS it is generally well tolerated, with no known serious adverse effects. Common side effects include a tingling or itching sensation at the site of stimulation and some cases of skin irritation. Some patients may also experience headache, fatigue, and/or nausea (Rosa & Lisanby, 2012).

Mechanisms of Action

The mechanisms of action of tDCS is not fully understood. It does not produce direct depolarization of neurons in the same manner of TMS but there appears to be an increase in cortical excitability under the anode (positive electrode) and a corresponding reduction in cortical excitability under the cathode (negative electrode) as a result of changes in the resting membrane potential incurred by the electrical tDCS field. These changes would alter the likelihood of task-related neural firing. The persistent effects of tDCS may be produced through NMDA-related effects on neuroplasticity (Nitsche et al., 2003).

Clinical Studies

Over the last 15 years, tDCS has been the subject of a substantial and progressively increasing body of preclinical and clinical research evaluating its use across multiple mental health conditions. The most consistent body of research has been conducted evaluating its use in the treatment of patients with depression. Note that in contrast to rTMS, ECT and DBS studies that are mostly focused on treatment resistant depression, tDCS studies have more broadly focused on depression, also as a first line of treatment. To date there have been over 20 randomized control trials typically comparing tDCS to a form of sham stimulation where the stimulation device is turned on and off relatively quickly to try to produce some scalp sensation. tDCS is most commonly given for 20–30 min per day for 5 days a week for 15 to 20 sessions. Most trials have used a 2 mA current with the anode over the F3 EEG point targeting left DLPFC and the cathode on the right above the orbit or over the F4 or F8 EEG sites. Many of these studies have produced positive – albeit small – effects although some studies including a multisite international trial have failed to find any difference between active and sham stimulation.

A meta-analysis of 23 controlled trials including 1092 participants found that active tDCS produced a greater rate of clinical response and remission than sham although the overall remission rates were relatively modest (19.12% for active and 9.78% for sham) (Razza et al., 2020). Dropout rates were about 14% in both groups. It is notable that the relative duration of treatment in these tDCS studies has been short and as it has been seen over time with rTMS, overall clinical outcomes may well prove better with longer durations of therapy. To date, minimal research has evaluated whether tDCS can be used as a strategy to prevent depressive relapse despite the potential low-cost and convenience of longer-term home-based tDCS therapy.

Studies have also investigated the use of tDCS in conditions other than depression. In schizophrenia, a number of small clinical trials have investigated the use of anodal stimulation to the temporoparietal cortex in the treatment of auditory hallucinations with inconsistent results. In a recent trial of 100 patients, anodal stimulation of the left DLPFC was reported to produce a greater reduction in negative symptoms than sham despite a short treatment duration (Valiengo et al., 2019). There is a very small emerging literature addressing the potential use of tDCS in OCD and anxiety disorders. For example, in a recent trial of 10 tDCS sessions in 40 participants with PTSD, anodal stimulation of the left DLPFC and cathodal stimulation of the right DLPFC resulted in significant improvement in PTSD symptoms compared to sham (Ahmadizadeh et al., 2019). This included a short-term reduction in specific PTSD symptoms such as re-experiencing and hyperarousal as well as improvement in general anxiety and mood.

Of note, there is a substantial body of research exploring whether tDCS can improve cognition, broadly or in specific domains such as working memory. This research is now gradually translating into studies exploring its use in the treatment of dementia but insufficient research has been conducted so far to allow meaningful conclusions to be made about efficacy.

Current Status in Treatment Algorithms

The role of tDCS in clinical practice is gradually changing. It is not yet FDA approved for treatment of any condition in the USA, but tDCS devices have achieved regulatory approval or listing for use in some countries. Given the excellent safety profile of tDCS, it seems to be a relatively promising option as an antidepressant intervention in patients who are not responding to, struggling to tolerate, or unwilling to try medications for depression. In addition, tDCS could represent a safe and cheap alternative that could reach communities with less access to technological advancements. Given its non-invasiveness and lack of known serious side effects, it would be considered prior to the relatively more invasive technologies.

Future Directions

Considerable research has recently been undertaken to explore whether more complex electrode arrangements can produce more focal and potentially more effective tDCS stimulation. These studies have produced quite promising data but have not yet led to the use of more complex stimulation arrangements in significant clinical trials. In the clinical domain, there is a pressing need for significant research exploring whether longer courses of tDCS produce more robust antidepressant responses and whether tDCS can be used as a viable long-term maintenance treatment for depression or anxiety conditions.

Other Forms of Non-Invasive Electrical Stimulation

In addition to tDCS, there are several other forms of transcranial electrical stimulation which have been developed to modulate brain activity and which are increasingly being evaluated for potential antidepressant or other therapeutic efficacy.

Transcranial Alternating Current Stimulation (tACS)

tACS has similarities with tDCS in that it is applied using a similar cathode–anode arrangement, but it involves a regular, usually sine wave, alternating current rather than the direct unidirectional current that characterizes tDCS (Herrmann et al., 2013). As such, there is no expectation that tACS will produce differential effects under one or the other electrode and as such, its mechanism of effect is likely to be significantly different than that seen with tDCS. tACS has been shown to produce transient and persistent effects on brain activity, but its mechanism of action is not fully clear. tACS induces entrainment where neural firing continues to occur at the frequency of exogenous stimulation after the end of the stimulation train. Studies have suggested that entrainment has some role in the action of tACS but that neuroplastic effects are likely to be induced through other mechanisms.

The selection of the stimulation frequency for a tACS paradigm may be critical as stimulation can be provided at a frequency that is specific to an element of cognition or brain function that one intends to modify or it may be individualized based on the pattern of oscillations in an individual subject's brain as characterized by EEG. In this later example, an individual subjects default frequency of oscillation in a specific frequency band (for example their individual alpha frequency) may be measured and used as the basis of the stimulation frequency. Studies have shown that entrainment is significantly more likely to occur when exogenous stimulation is provided close to one's intrinsic frequency.

A limited number of clinical studies have explored the use of tACS in the treatment of mental health conditions and have provided preliminary evidence that it may be able to produce antidepressant effects or modulate the intensity of auditory hallucinations in patients with schizophrenia or symptoms of OCD (Alexander et al., 2019; Klimke et al., 2016; Mellin et al., 2018).

Transcranial Random Noise Stimulation (tRNS)

tRNS also uses the same type of electrode arrangement as tDCS but involves the delivery of a randomly fluctuating current, usually at relatively high frequency. Since tRNS was first described in 2008 (Terney et al., 2008), a series of studies have demonstrated that it can change cortical excitability and a number of aspects of cognitive function (for example, Abe et al. (2019), Murphy et al. (2020)), although its mechanism of action remains unclear. It has been proposed that tRNS may work through repeatedly opening sodium channels or potentially through stochastic resonance (Antal & Herrmann, 2016). The latter refers to the idea that if noise is added to a subthreshold oscillation, this can be amplified: thus, tRNS could potentially amplify subthreshold neural oscillations.

Few studies have attempted to use tRNS in a therapeutic context to date. In the first study exploring its potential use in the treatment of depression, 69 patients were randomized to receive either 20 active or sham tRNS sessions over a 4 week period in a double blind fashion (Nikolin et al., 2020). No significant differences were seen between active and sham stimulation. Clearly, there are no current clinical indications for the use of tRNS but this may change with emerging research over coming decades.

Cranial Electrotherapy Stimulation (CES)

CES refers to a broad category of non-invasive neurostimulation devices which provide low voltage pulsed or alternating current electrical stimulation usually via electrodes placed on either side of the head on the earlobes, mastoid processes, or temples. CES devices typically use currents below 4 mA, which can be square or sinusoid will and which may be proprietary in nature to a specific device.

Forms of CES were initially used in the treatment of insomnia, which was at the time called electrosleep therapy. CES began to regain popularity from the mid-2000s onwards with a progressive number of devices available in both the USA and international markets. CES devices are classified as class III medical devices in the USA requiring prescription by a licensed healthcare practitioner, and the FDA has proposed future reclassification to class II.

At this time, the mechanism of action of CES devices remains unclear and despite sometimes quite marked claims about their potential efficacy, there is insufficient evidence to support their widespread use in any psychiatric disorder. Recent reviews have failed to find significant evidence for the benefit of CES in depression (Kavirajan et al., 2014) as well as in other conditions such as fibromyalgia, headache, insomnia, and musculoskeletal pain (Shekelle et al., 2018) with reviews indicating the absence of substantial clinical trials in these indications.

Implanted Stimulation Approaches

Vagal Nerve Stimulation

Definition

VNS is a process that involves the stimulation of various branches of the vagus nerve to modulate afferent neural signals to the brain. In its most developed form, it entails the direct, intermittent electrical stimulation of the left cervical vagus nerve via a pulse generator implanted in the left chest wall. The electrode is wrapped around the left vagus nerve in the neck and is connected to the generator subcutaneously. Intermittent left vagal nerve stimulation sends afferent signals to the nucleus tractus solitarius and connected limbic and cortical areas (George et al., 2000). Implantation surgery involves two incisions – one for the generator in the chest and another in the neck for the electrode. Surgery is usually performed under brief general anesthesia, as day surgery or with an overnight stay. Stimulation parameters are adjusted with a programming wand that communicates with the generator. Patients may turn off the stimulation when needed by holding a magnet over the generator.

As well as implantable forms of VNS, various devices are in development to produce non-invasive stimulation of the vagal nerve in the neck or around the external ear.

History

VNS was originally approved by the FDA for the treatment of resistant epilepsy, and in 2005 was approved for the adjunctive treatment of chronic, treatment-resistant depression. However, FDA approval was not followed by widespread funding approval limiting access and uptake.

Mechanisms of Action

The left vagus nerve contains 80% afferent fibers, hence stimulating it activates predominantly vagal afferents. Chronic stimulation of the vagal afferents has been shown to change activity in brainstem nuclei (such as the nucleus of the solitary tract), and from there neighboring nuclei (e.g., Raphe) that alter serotonergic activity in cortical and limbic structures. Chronic stimulation

of the vagal afferents has anticonvulsant effects, and this effect appears to be dependent upon the locus ceruleus.

Side Effects

VNS is generally well tolerated. Voice alteration, cough, dysphagia, dyspnea, and neck pain are the most frequently reported adverse events (Rush et al., 2005; George et al., 2005). The surgical implantation carries the risks of infection, vocal cord paralysis, and bradycardia or asystole [all <0.5% in the US pivotal (D02) study]. VNS does not cause apparent cognitive side effects. In contrast, neurocognitive performance was significantly improved with VNS – apparently due to the improvement in depression (Sackeim et al., 2001b).

Contraindications to VNS include a history of bilateral or left cervical vagotomy and use of short-wave diathermy, microwave diathermy, or therapeutic ultrasound diathermy. Patients with VNS implanted cannot receive routine magnetic resonance imaging (MRI) scans, but can receive MRI with a special "send/receive" coil.

Clinical Studies

Implantable VNS was first tested in a group of 30 treatment-resistant depressed patients with a response rate of 40% after 10 weeks of open-label, adjunctive VNS treatment (Rush et al., 2000). Clinical response was sustained and remission rates gradually improved from 17% to 29% over an additional 9 months of VNS, and significant improvements in function were reported with long-term VNS treatment. A similar pattern of a slow progressive improvement in outcomes, was seen in a subsequent report on an extended sample ($n = 59$) followed for 2 years (Sackeim et al., 2001b).

These promising studies were followed by a 10-week randomized, controlled, and masked multicenter trial comparing adjunctive VNS with sham treatment in 235 outpatients with treatment-resistant unipolar or bipolar depression. This failed to show a significant difference on the primary outcomes between active and sham treatment (Rush et al., 2005).

A naturalistic follow-up study of long-term (1 year) VNS treatment of the same cohort

($n = 202$) revealed progressively increasing improvement with a 12-month response rate of 27.2% and a remission rate of 15.8%. These results, when compared with a comparably matched but nonrandomized control group who received treatment as usual, indicated VNS was associated with greater improvement than pharmacological treatment alone (27% vs. 13%) (George et al., 2005). An open study that followed remitters (37%) and responders (17%) within 90 days found that 44% suffered no relapse at 12 months (Schlaepfer et al., 2008). A more recent randomized trial compared outcomes in three groups of patients (total $N = 331$) receiving a low, medium, or high dose of stimulation (defined by current strength) (Aaronson et al., 2013). All three groups showed antidepressant effects across the acute 22-week treatment phase but the medium and higher dose groups had a more durable response and overall current charge was associated with response to treatment. Finally, a recent registry study examined 5 year outcomes of patients who had received VNS ($n = 494$) compared to a population of patients receiving treatment as usual ($n = 301$) (Aaronson et al., 2017). Patients receiving VNS had a greater rate of cumulative response and remission, and there were substantive response and remission rates in both past ECT responders and non-responders.

Current Status in Treatment Algorithms

The FDA label in the US states that VNS is indicated for the adjunctive long-term treatment of chronic or recurrent depression in patients aged 18 years or older who are experiencing a major depressive episode (unipolar or bipolar) and have not had an adequate response to four or more adequate antidepressant treatments. Consultation with another clinician experienced with treatment-resistant depression and VNS is recommended. Acute success rates with VNS are considerably lower than with ECT and onset of action is comparatively slow. However, the limited data available suggests that VNS may produce a sustained degree of response in a subset of patients who have had very limited response to many other treatments (Aaronson & Conway, 2018). Therefore, VNS may be worth considering when patients have failed to respond to less invasive treatments, ECT was ineffective, or post-ECT relapse cannot be prevented with less invasive means. There is hope that VNS might be helpful with longer-term relapse prevention, but results of controlled trials would be useful to guide practice.

Patient Selection

Steps in the evaluation of a patient for VNS are outlined in Fig. 8. As indicated above, VNS is approved by the FDA for the adjunctive long-

DOES THIS PATIENT MEET FDA LABELING CRITERIA?
Unipolar or bipolar disorder in an MDE
Longterm MDE (≥2 years)
Recurrent depression
Inadequate response to 4 or more adequate treatments

HAVE OTHER CAUSES OF MED RESISTANCE BEEN RULED OUT?
Medical, substance, Axis II
Compliance with adequate dosages and durations

CAN THE PATIENT TOLERATE A SLOW-ACTING THERAPY?

DOES THE PATIENT NEED ACUTE RESPONSE NOW?
ECT may be a better option for the acutely suicidal

GET A SECOND OPINION

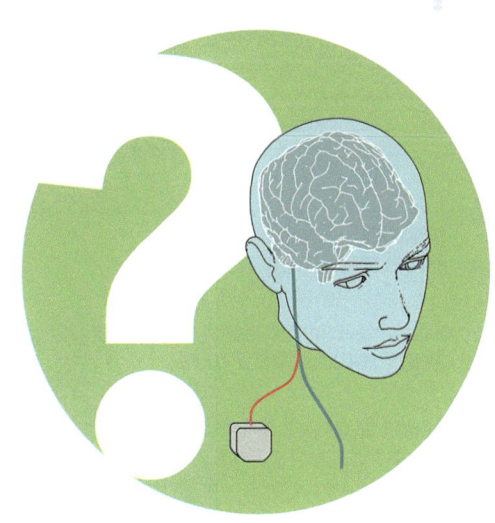

Fig. 8 Process of Evaluation for VNS

term treatment for chronic or recurrent depressive episodes in adults with a major depressive episode who have not had an adequate response to four or more adequate antidepressant trials. The efficacy of VNS in other disorders is unknown. Determination of what constitutes an "adequate" trial requires careful assessment of whether adequate dosages were prescribed (and taken) for an adequate period of time. Optimally, this should be substantiated by careful review of treatment records, pharmacy reports, and blood levels where appropriate. Failed trials should include different medication classes (i.e., not just SSRIs or SNRIs, but also potentially monoamine enhancers or MAOIs), or various augmentation strategies with or without ECT. A second opinion from a clinician specializing in treatment-resistant depression and knowledgeable regarding VNS is advisable.

Sample clinical scenarios when ECT might be favored over VNS, and vice versa, are outlined in Fig. 9. For example, VNS does not exert rapid antidepressant action, thus acutely suicidal patients and others in need of rapid response would be more appropriately treated with other strategies (Rush & Siefert, 2009). VNS is indicated as a long-term treatment option only for those with chronic or recurrent course of illness (Mohr et al., 2011).

ECT can be safely used in patients with an implant if the VNS generator is turned off during delivery of ECT, to avoid anticonvulsant effects of VNS. Whether VNS would be useful in relapse prevention post-ECT deserves study.

Dosing

The optimal dose of VNS is unknown. Based on the one meaningful dosing study (Aaronson et al., 2013), a moderate current strength of at least 0.5–1.0 mA (250 uS pulse width) would seem a sensible starting point with consideration of higher doses (1.25–1.5 mA, 250 uS pulse width) considered in the circumstances of inadequate response. A duration of at least 3–6 months if not longer is likely to be required to achieve optimal effects. The side effects of VNS are known to be dose dependent (e.g., lowering the pulse width reduces neck pain, allowing patients to tolerate higher currents).

Future Directions

More research is needed to define the dose-response relationships for VNS, investigate

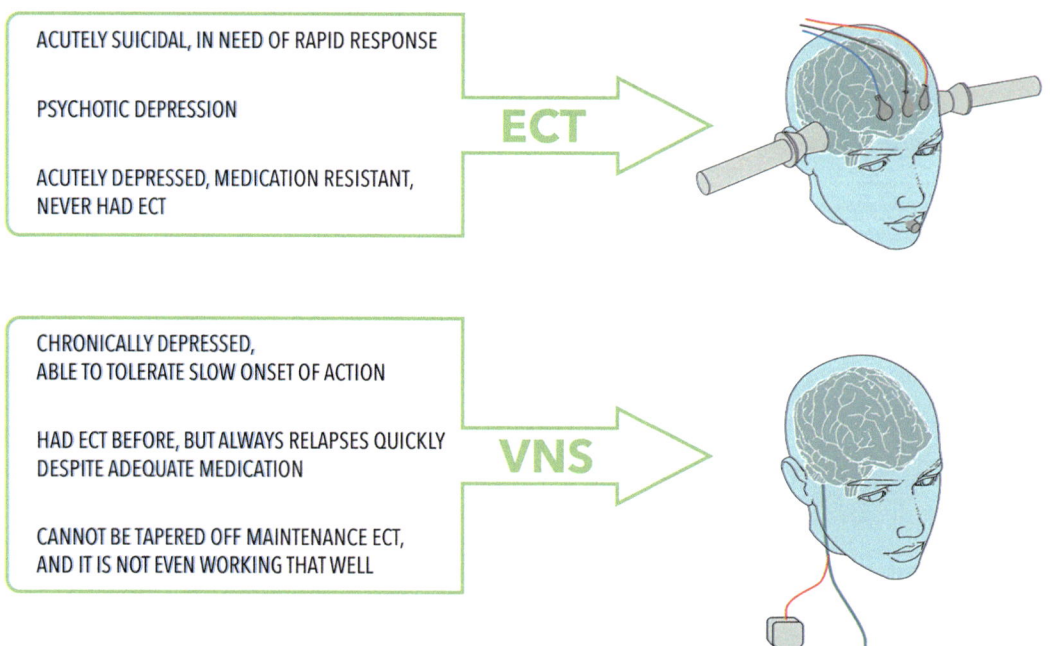

Fig. 9 VNS or ECT?

optimal medication strategies to augment responses, test the potential role of VNS for long-term relapse prevention, and study its mechanisms of action.

In addition, further research is required to investigate the role of non-implanted forms of VNS. The first (non-blinded) study of transcutaneous auricular vagus nerve stimulation reported significant antidepressant effects (Rong et al., 2016), and this form of stimulation has demonstrated effects on inflammatory markers and depression-related neural network activity.

Deep Brain Stimulation

Introduction
DBS involves the implantation of intracranial electrodes during a neurosurgical procedure followed by chronic stimulation of a small-targeted brain region. DBS was initially considered a potential replacement for lesional forms of neurosurgery as the effects of stimulation may be reversible and adjustable. Over time, there has been increasing interest in the potential functional relevance of the stimulation itself and reconsideration of it as a form of functional neurostimulation rather than a method of producing a reversible lesion.

History
DBS was originally developed as a treatment for movement disorders and has been approved by the FDA in the USA, and in many other countries, for the treatment of Parkinson's disease, as well as in some jurisdictions for the management of other conditions such as refractory tremor and persistent dystonia. There is interest in its use in a number of other neuropsychiatric conditions such as epilepsy and Tourette's syndrome.

In regards to mental health, the main interest in DBS has been its potential use in the treatment of severe and treatment-resistant OCD and major depression.

Side Effects
The side effects and complications that can result from DBS can be considered in regards to both the surgery and to the brain stimulation itself. In regards to the operative procedure, the most serious complication is a hemorrhage and infections, and post-implantation seizures have been reported. In regards to stimulation effects, these are generally reversible by the discontinuation of stimulation and are determined by the site of stimulation. A range of effects including anxiety/agitation, headache, mania, and worsening of depression have been reported.

Clinical Studies
To date, DBS for OCD has been assessed in eight randomized controlled trials as well as in a series of over 35 case reports and case series but the total number of patients studied is only slightly more than 300. These studies have included implantation at a variety of sites including the anterior limb of the internal capsule, the ventral caudate/ventral striatum, and the nucleus accumbens. In a recent meta-analysis, the randomized controlled trials showed a difference between active and sham stimulation and the mean reduction in OCD symptoms seen across the active groups was around 39%, considered to be a significant response given the chronic and severe nature of symptoms in these patients (Vicheva et al., 2020). It is notable that five patients (6.3%) in these trials experienced significant surgical-related adverse events including hemorrhage and infection.

The studies evaluating the use of DBS for depression are somewhat more complicated to interpret, especially given an even greater variation in the implantation sites. The most common sites have been the subgenual anterior cingulate cortex and the anterior limb of the internal capsule/ventral caudate/ventral striatal region. Small investigator initiated studies at multiple implantation sites have reported response rates of around 50% in highly treatment-resistant patients with a recent meta analyses finding a significant benefit of active versus sham stimulation (Hitti et al., 2020). However, it is somewhat challenging to interpret this meta-analysis given the pooling of studies involving implantation at substantially divergent sites. In contrast to the results of this meta-analysis, two industry sponsored multisite trials have failed to support antidepressant efficacy of DBS at the ventral capsule/ventral striatum or the subgenual anterior cingulate (Dougherty et al., 2015; Holtzheimer et al., 2017).

Current Status in Treatment Algorithms

The FDA in the USA has granted a Humanitarian Device Exemption (HDE) approval for DBS in the treatment of OCD. HDE approval is intended for situations where clinical trials may not be able to be conducted due to the rarity of a clinical condition, considered to be conditions affecting less than 4000 people per year in the USA. There has been considerable debate whether this is appropriately applied to OCD (Garnaat et al., 2014). It is clearly a much more common condition than this but only a fraction of patients with OCD are likely to be suitable for DBS. These are typically patients who have severe and persistent OCD and who have tried multiple serotonergic medications for depression, including clomipramine, adjuvant treatment with a medication for psychosis, and a benzodiazepine and adequate exposure and response prevention (Garnaat et al., 2014). DBS in the management of depression is still considered experimental and does not have a place in current treatment algorithms.

Outside of the USA, there has been very limited clinical use of DBS for any psychiatric indication, with the vast majority of use occurring within clinical trials. However, patients have been known to travel to countries where there is limited DBS device regulation to access surgery. The ethics of this practice have been questioned as has its clinical appropriateness. Specifically, the conduct of DBS surgery is only the first step of patient management and it should not be divorced from the ongoing patient care and device programming, which is critical in the long term. Ethical issues have also been raised and discussed around the appropriate use of an invasive treatment technique such as DBS, especially given the history in psychiatry of the inappropriate and excessive use of forms of psychosurgery. DBS is being used as a treatment procedure in voluntary patients and some jurisdictions require independent review of each case prior to the provision of permission for the surgery to proceed.

Patient Selection

Given the invasive nature of DBS and the risks associated with surgery, it is likely to remain a treatment of relatively last resort for patients with OCD and to have a similar status should it become more established in the management of depression or other mental health conditions.

Dosing

Stimulation with DBS involves several elements of dosing. Initially, choice needs to be made in regards to whether stimulation is bipolar (between two or more of the stimulation electrodes on each implant wire) or unipolar (between one or more of the electrodes and the implanted pulse generator case in the chest). The choice of stimulation electrode(s) will determine the location and shape of the stimulation field. Stimulation parameters that need to be determined include pulse width, frequency, and the current strength or voltage. Stimulation is frequently applied between 90 and 130 Hz and at a pulse width of between 120 and 210uS. Stimulation intensity/voltage can vary quite dramatically from ~1–7 volts.

Future Directions

Ongoing clinical trials are exploring the use of DBS in depression and other conditions such as anorexia nervosa. In depression, there is an emphasis on exploring new stimulation sites and methods of stimulation localization using techniques such as electrical field modelling and mapping of white matter pathways with diffusion tensor imaging (Roet et al., 2020). For example, promising data has been published of the outcomes of implantation in the medial forebrain bundle. Future research is likely to take advantage of new closed-loop DBS systems that allow for the sampling of electrical activity at the site of implantation to be used to help determine stimulation parameters.

Other Forms of Brain Stimulation Currently Under Development

Low Intensity Forms of Magnetic Stimulation

There are several approaches using low intensity forms of magnetic stimulation that have been developed to try and produce therapeutic benefits in patients with psychiatric conditions. These do not produce the type of high-intensity time variable magnetic fields used in TMS to produce neuronal depolarization. Low-field magnetic stimulation (LFMS) is an approach that uses an echoplanar magnetic resonance spectroscopy imaging waveform to try and produce low intensity stimulation across the brain. LFMS initially showed transient immediate antidepressant effects after a single session in a group of patients with bipolar depression and then in a small group of patients with unipolar depression. However, a multisite trial of four LFMS treatment sessions compared to sham failed to find meaningful clinical benefits (Fava et al., 2018). A smaller more recent study did find some antidepressant effects in patients with treatment resistant depression of three 20 min LFMS sessions applied 48 h apart with antidepressant benefits only seen following the third treatment session (Dubin et al., 2019).

Transcranial pulsed electromagnetic field (T-PEMF) is a second form of magnetic stimulation using a set of coils to produce a low intensity magnetic field of insufficient strength to cause neuronal depolarization. An initial study showed some antidepressant effects in a study of 50 patients over a 5 week period of treatment (Martiny et al., 2010) and a second 8 week study found antidepressant effects in 65 patients with treatment resistant depression (Straaso et al., 2014).

A third low intensity form of magnetic stimulation providing an alternating magnetic field synchronized to an individual patient's alpha frequency has been referred to as synchronized TMS (sTMS). In an initial multisite clinical trial, sTMS showed benefits in a per protocol, but not in an intention-to-treat analysis (Leuchter et al., 2015). The results of a follow-up study to this were not positive but have not yet been published.

Finally, Magnetic e-Resonance Therapy (MeRT) is a form of low-intensity TMS with the individualization of stimulation parameters based on EEG recordings in a proprietary manner. No randomized trials demonstrating efficacy of this approach could be identified.

Overall, these other forms of magnetic brain stimulation using mostly weaker magnetic currents or static magnetic fields cannot be recommended for clinical practice yet.

Trigeminal Nerve Stimulation (TNS)

TNS involves stimulation of the supraorbital and supra-trochlear afferent fibers of the trigeminal nerve through electrodes placed on the forehead. Initial open label studies have suggested that TNS may have benefits in depression and anxiety with antidepressant effect supported in a several small sham-controlled randomized trials (Generoso et al., 2019; Shiozawa et al., 2015).

Transcranial Near-Infra-Red or Photobiomodulation Therapy

Photobiomodulation therapy involves the application of near-infrared light through the scalp which penetrate several centimeters into cortical tissue (Askalsky & Iosifescu, 2019). This form of light has been proposed to affect mitochondrial metabolic pathways and to have potential therapeutic value in depression. Initial small double-blind randomized studies have showed some evidence of antidepressant effects (Cassano et al., 2015, 2018).

Neurofeedback

Neurofeedback is a therapeutic technique that seeks to modulate and retrain brain function to address neurological and/or psychological symptoms of concern. One of the original demonstrations of the potency of neurofeedback involved what is termed the Sensori-Motor Rhythm (SMR), an EEG rhythm in the low beta range (12–15 Hz) derived from the EEG from the region of the scalp located over the sensori-motor strip. Sterman and Friar (Sterman & Friar, 1972) demonstrated the first anticonvulsant effects in epilepsy. Lubar and Shouse (Lubar & Shouse, 1976) then described the successful application of this technique in a child with hyperkinetic syndrome, a condition closely resembling what is now termed ADHD. Several years later, these findings were replicated in a larger study (Shouse & Lubar, 1979) and extended to the use of a modified protocol involving not only training of SMR but also of a slower rhythm called theta (4–7 Hz). This revised protocol was named Theta/Beta Neurofeedback (Lubar & Lubar, 1984). Subsequently, a slightly different form of neurofeedback was described, called Slow Cortical Potential (SCP) neurofeedback that was shown to not only have anticonvulsive properties in epilepsy (Rockstroh et al., 1993) but also clinical effects in ADHD (Heinrich et al., 2004). SCP's are DC shifts related to positive or negative shifts in broad sheets of glial cells, representing increased activation (negativity) or decreased activation (positivity) and these very slow oscillations in the EEG associated with readiness that transfer into daily life during learning. Not all EEG frequencies being trained have been shown to be efficacious in these conditions. For example, training of the posterior alpha rhythm (8–13 Hz) has failed to show clinical benefit in either hyperkinetic syndrome (Nall, 1973) and epilepsy (Rockstroh et al., 1993), suggesting *specificity* in the EEG parameter trained for successful neurofeedback. Therefore, the first three well-investigated protocols (SMR, TBR, and SCP) have also been termed "standard neurofeedback protocols." For a more detailed overview of the history of neurofeedback and these "standard neurofeedback protocols" (see Arns et al. (2014)).

Four multicenter RCTs, employing standard neurofeedback protocols demonstrated significant superiority to semi-active control groups (ranging from EMG biofeedback to attention training), with medium–large pre-post effect sizes end of treatment or follow-up and remission rates of 32–47% in patients with ADHD. Regarding effectiveness, three open-label neurofeedback studies demonstrate similar or better efficacy compared to the multicenter RCTs, demonstrating the effects of neurofeedback translate well into clinical practice (see Arns et al. (2020) for overview).

In two independent systematic reviews and meta-analyses, significant efficacy of standard neurofeedback protocols was confirmed for both parent and teacher-rated symptoms (Cortese et al., 2016), with a small-medium between-group effect size. Between-group analysis resulted in small-medium ES (Cortese et al., 2016; Van Doren et al., 2018), whereas within-group analysis resulted in large ES and effects were sustained at 6–12 months follow-up (Van Doren et al., 2018).

No signs of publication bias have been found and no significant neurofeedback-specific side effects have been reported in any prior study (Arns et al., 2020).

Recently, the results of the first multicenter double-blind placebo-controlled study using a standard protocol (Theta/Beta neurofeedback) in patients with ADHD were published. While this study demonstrated significant pre-post improvement with large effect sizes ($d = 1.5$), no significant difference on the primary outcome measure was found after treatment (Neurofeedback Collaborative, 2020). The active neurofeedback group did use significantly less medication and showed significantly higher remission rates after 13-month follow-up, compared to the sham group. While this study casts some doubt on the specificity of the EEG Theta/Beta control as the sole working mechanism, the large clinical benefit associated with this multifactorial treatment that was maintained after 13 months follow-up, suggests a clinically relevant change comparable to behavioral and pharmacological interventions for ADHD (Arns et al., 2020). Furthermore, these results cannot be generalized to SCP and SMR neurofeedback.

Several other – nonstandard – forms of neurofeedback have also been proposed such as Z-Score neurofeedback, LORETA neurofeedback, and Infraslow neurofeedback; however, a recent systematic review failed to find sufficient evidence to rate the clinical efficacy and effectiveness for these forms, thus no conclusions can be made about these specific forms of neurofeedback (Coben et al., 2019). Some evidence exists for the use of these approaches in other disorders such as the treatment of epilepsy, autism, and insomnia, albeit more limited relative to the treatment for ADHD. Furthermore, much research is ongoing and focused on employing other imaging techniques to provide real-time feedback, most notably fMRI neurofeedback, also see Sitaram et al. (2017) and Watanabe et al. (2017) for further review of the technologies used. However, most fMRI neurofeedback studies to date have been proof-of-concept and relatively small studies, so no firm conclusions can be drawn regarding clinical applications of fMRI neurofeedback.

Bright Light Therapy

Light therapy has been the subject of interest and clinical evaluation since the 1980s for its potential to be used in the treatment of mood disorders, seasonal affective disorder (SAD), and to understand its impacts on the biological clock. These effects are dependent on the wavelength of the light used and the timing of light exposure during the day. Our biological clock or master circadian oscillator is located to the suprachiasmatic nuclei (SCN) of the anterior hypothalamus, and the SCN clock entrains to the light/dark cycle via photic information transferred to the SCN by the retinohypothalamic tract, which is a monosynaptic input to the SCN from a subset of retinal ganglion cells (Pail et al., 2011). Besides rods and cone photoreceptor cells in the retina, there are also more recently discovered retinal photoreceptor cells that are responsible for the non-image forming perception of light intensity. These are the M1-type intrinsically photosensitive Retinal Ganglion Cells (ipRGCs), which modulate, among others, the pupillary reflex and the release of melatonin, and also project to the SCN. The photopigment melanopsin in these ipRGCs is most sensitive to visible light of shorter wavelengths (400–500 nm, the blue light range of the spectrum). As the SCN entrains to the appropriate environmental zeitgebers, such as blue range light, it then synchronizes multiple brain and peripheral clocks that together give rise to whole-organism circadian rhythms in behavior and physiology.

The exact effects of light therapy are determined by the circadian phase during which these cues are presented; for example, the impact of evening light is different to that of early morning light, which is different to the effect of light during the afternoon. Morning bright light advances the phase (phase-advance) of the circadian clock, whereas evening bright light phase-delays the circadian clock leading to later sleep onset time (Pail et al., 2011). Light therapy aimed at resetting the circadian rhythm usually consists of bright white or blue light boxes with an intensity of 10,000 lux, at a distance of 20–40 cm from the eyes. Light therapy aimed at phase-advancing the circadian rhythm should comprise 2–3 weeks of light exposure for 30 min every morning at the same time, preferably between 7 and 8 am (Pail et al., 2011).

Knowing these mechanisms of specific wavelengths and timing, special consideration should be made to more recent developments such as use of mobile phones, tablets, and electric lighting that preferentially triggers the ipRGCs (i.e., "blue" 400–500 nm light, mostly LEDs) in the evening, since such exposure is associated with circadian phase-delays, and subsequent delayed sleep phase, hence careful consideration should be given to the use light-emitting devices in the evening as a potential source of such delays.

Although bright light therapy has been explored as a treatment in a number of other disorders, for example, in eating disorders (Beauchamp & Lundgren, 2016), the major focus of research has been on exploring its use in the treatment of forms of depression and seasonal affective disorder (SAD). A series of clinical trials have established the efficacy of bright light therapy in the treatment of seasonal affective disorder with these effects supported in meta-analyses and systematic reviews (Golden et al., 2005; Terman et al., 1989). However, a recent Cochrane review failed to find significant evidence that bright light therapy can be used to prevent episodes of seasonal mood change (May, 2020). A number of clinical trials have also evaluated the use of bright light therapy in addition to medications for depression demonstrating augmentation effects with a modest effect size (Penders et al., 2016) and a number of studies have also suggested that the effects of light therapy alone may be similar to those seen with medications for depression (Geoffroy et al., 2019). A recent meta-analysis found a similar positive effect in bipolar depression across for clinical trials including 190 patients (Hirakawa et al., 2020).

Although these findings are quite consistent, bright light therapy has not been taken up widely as a core treatment in clinical programs around the world, perhaps because of the lack of commercial incentives or a failure to establish clear pathways for smooth integration into clinical practice.

Conclusions and Future Directions

Brain stimulation and other physical forms of treatment are an important but unrecognized part of the treatment landscape for the management of patients with psychiatric conditions. There has been a relatively rapid expansion of the role of brain stimulation treatments, especially with the development of applications of rTMS, and this remains an area of substantial ongoing interest and research.

The available and emerging technologies differ in their evidence basis, focality, mechanisms, and clinical spectrum. Figure 10 summarizes the state of the evidence for several of these. ECT has unmatched efficacy, but the challenges for the future are to improve its safety and develop methods to minimize relapse an enhanced long-term treatment outcomes. TMS is very safe and is effective in patients with treatment-resistant

depression who have a limited likelihood of responding to medication treatments. Future challenges exist in enhancing its efficacy and improving clinical utility. MST likewise has a good safety profile, and current needs are to optimize and test its efficacy against the gold-standard ECT. VNS has been approved for chronic illness management in the USA but its role as an adjunctive therapy for treatment-resistant depression awaits verification with real-world clinical experience and further research before it will be more widely used in clinical practice. DBS offers the promise of placing the electrode precisely in the target circuitry, but this also poses the challenge to identify just what that circuitry is to ensure efficacy. The invasive nature of DBS is likely to limit its long-term application to a small number of the most treatment-resistant patients. tDCS and other forms of electrical stimulation have considerable potential, especially as home-based applications

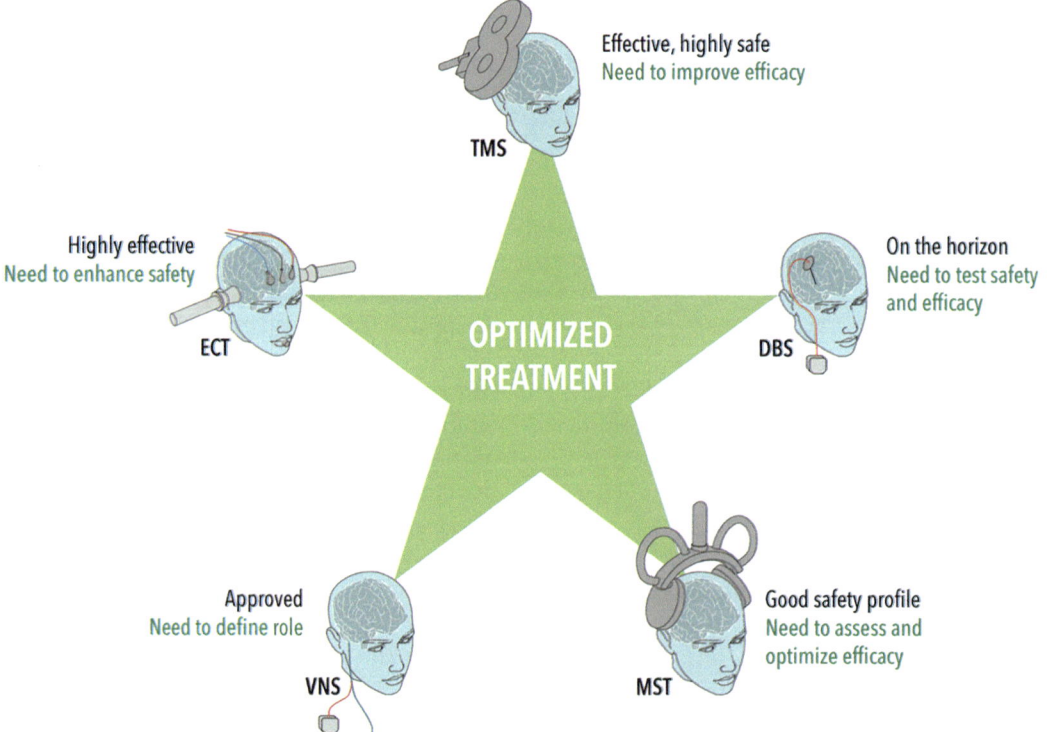

Fig. 10 State of the evidence

that may have considerable long-term relapse pre-vention utility. However, significant research and development is required to establish and improve efficacy and to establish long-term treatment par-adigms. There is an emerging literature supporting the use of forms of neurofeedback in ADHD especially. Finally, bright light therapy shows significant potential for antidepressant effects beyond seasonal affective disorder and is probably underutilized in mental health care as an adjunct treatment in the management of treatment resistant depression.

The potential for brain stimulation in psychia-try has never been more exciting. There is now a collection of technologies that may address differ-ent phases of depression (Fig. 11). Along with this exciting promise, there is also a great need for research in this field to elucidate the science of how intermittent stimulation affects brain func-tion. A better understanding of that science could guide developments in this field. Indeed, some day the application of best practice used in the treatment of adults may become applicable to children and adolescents. Rather than taking off-the-shelf technology designed for neurological applications, a basic science of brain stimulation

optimized for psychiatric disorders offers the promise of creating purpose-built devices tailored to the specific needs of patients. The stages of drug development (target identification, pharma-cokinetics, pharmacodynamics, safety, and toxi-cology) are well known. Less well developed are the device analogs to these stages (Fig. 12). As clinicians approach rational device development in psychiatry, the device analogs of these stages will need to be addressed through basic and pre-clinical research.

The advances in brain stimulation have pro-mpted a very recent call for consideration of a new subspecialty called interventional psychiatry. This proposal, modeled after similar programs such as interventional neurology and interventional cardi-ology, would establish a comprehensive training program with requisite courses in theory, mecha-nisms of action, and the skill-based experiences necessary for administering both noninvasive and invasive brain stimulation techniques. The broader goals of this new area of subspecialization include the establishment and provision of evidence-based safe practices and implementation of a nationally recognized credentialing policy for those clinicians providing brain stimulation.

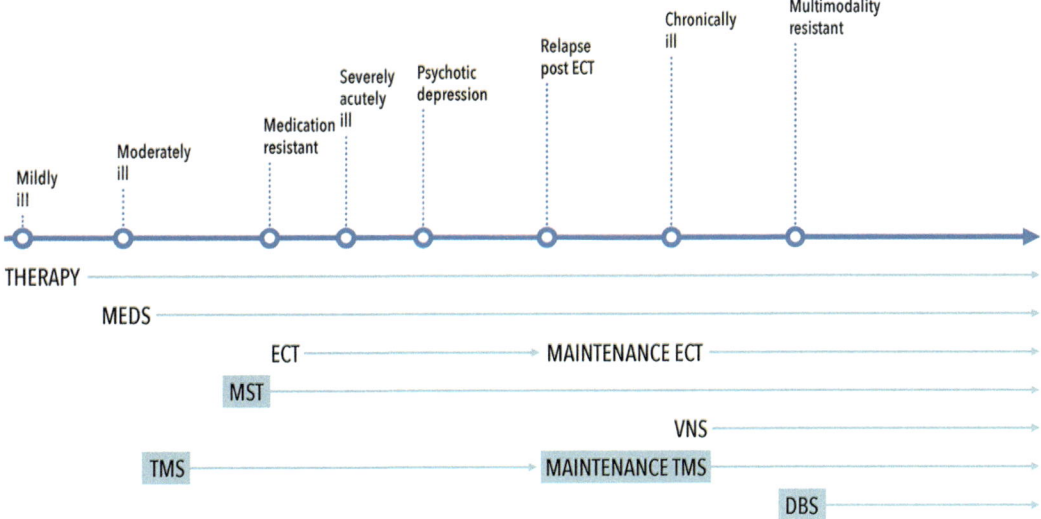

Fig. 11 Putting it all together: what to use and when

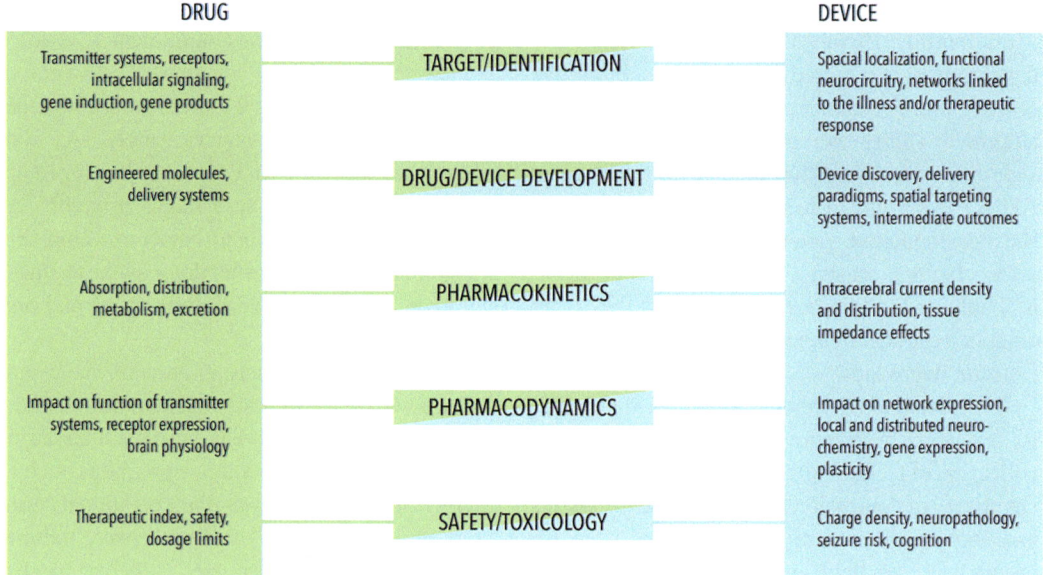

Fig. 12 Intervention development stages

References

Aaronson, S. T., & Conway, C. R. (2018). Vagus nerve stimulation: Changing the paradigm for chronic severe depression? *The Psychiatric Clinics of North America, 41*(3), 409–418. https://doi.org/10.1016/j.psc.2018.05.001

Aaronson, S. T., Carpenter, L. L., Conway, C. R., Reimherr, F. W., Lisanby, S. H., Schwartz, T. L., . . . Bunker, M. (2013). Vagus nerve stimulation therapy randomized to different amounts of electrical charge for treatment-resistant depression: Acute and chronic effects. *Brain Stimulation, 6*(4), 631–640. https://doi.org/10.1016/j.brs.2012.09.013

Aaronson, S. T., Sears, P., Ruvuna, F., Bunker, M., Conway, C. R., Dougherty, D. D., . . . Zajecka, J. M. (2017). A 5-year observational study of patients with treatment-resistant depression treated with vagus nerve stimulation or treatment as usual: Comparison of response, remission, and suicidality. *The American Journal of Psychiatry, 174*(7), 640–648. https://doi.org/10.1176/appi.ajp.2017.16010034

Abe, T., Miyaguchi, S., Otsuru, N., & Onishi, H. (2019). The effect of transcranial random noise stimulation on corticospinal excitability and motor performance. *Neuroscience Letters, 705*, 138–142. https://doi.org/10.1016/j.neulet.2019.04.049

Ahmadizadeh, M. J., Rezaei, M., & Fitzgerald, P. B. (2019). Transcranial direct current stimulation (tDCS) for post-traumatic stress disorder (PTSD): A randomized, double-blinded, controlled trial. *Brain Research Bulletin, 153*, 273–278. https://doi.org/10.1016/j.brainresbull.2019.09.011

Alexander, M. L., Alagapan, S., Lugo, C. E., Mellin, J. M., Lustenberger, C., Rubinow, D. R., & Frohlich, F. (2019). Double-blind, randomized pilot clinical trial targeting alpha oscillations with transcranial alternating current stimulation (tACS) for the treatment of major depressive disorder (MDD). *Translational Psychiatry, 9*(1), 106. https://doi.org/10.1038/s41398-019-0439-0

Alpert, S. (2011). A humanitarian device exemption for deep brain stimulation. *Health Affairs (Millwood), 30*(6), 1212; author reply 1212. https://doi.org/10.1377/hlthaff.2011.0425

Antal, A., & Herrmann, C. S. (2016). Transcranial alternating current and random noise stimulation: Possible mechanisms. *Neural Plasticity, 2016*, 3616807. https://doi.org/10.1155/2016/3616807

Arns, M., Heinrich, H., & Strehl, U. (2014). Evaluation of neurofeedback in ADHD: The long and winding road. *Biological Psychology, 95*, 108–115. https://doi.org/10.1016/j.biopsycho.2013.11.013

Arns, M., Clark, C. R., Trullinger, M., deBeus, R., Mack, M., & Aniftos, M. (2020). Neurofeedback and attention-deficit/hyperactivity-disorder (ADHD) in children: Rating the evidence and proposed guidelines. *Applied Psychophysiology and Biofeedback, 45*(2), 39–48. https://doi.org/10.1007/s10484-020-09455-2

Askalsky, P., & Iosifescu, D. V. (2019). Transcranial photobiomodulation for the management of depression: Current perspectives. *Neuropsychiatric Disease and Treatment, 15*, 3255–3272. https://doi.org/10.2147/NDT.S188906

Bajbouj, M., Lang, U., Niehaus, L., Hellen, F., Heuser, I., & Neu, P. (2006). Effects of right unilateral electroconvulsive therapy on motor cortical excitability in depressive patients. *Journal of Psychiatric Research, 40*(4), 322–327. https://doi.org/10.1016/j.jpsychires.2005.07.002

Baldinger, P., Lotan, A., Frey, R., Kasper, S., Lerer, B., & Lanzenberger, R. (2014). Neurotransmitters and electroconvulsive therapy. *The Journal of ECT, 30*(2), 116–121. https://doi.org/10.1097/YCT.0000000000000138

Barker, A. T., Jalinous, R., & Freeston, I. L. (1985). Non-invasive magnetic stimulation of human motor cortex. *The Lancet, 325*(8437), 1106–1107. https://doi.org/10.1016/S0140-6736(85)92413-4

Beam, W., Borckardt, J. J., Reeves, S. T., & George, M. S. (2009). An efficient and accurate new method for locating the F3 position for prefrontal TMS applications. *Brain Stimulation, 2*(1), 50–54. https://doi.org/10.1016/j.brs.2008.09.006

Beauchamp, M. T., & Lundgren, J. D. (2016). A systematic review of bright light therapy for eating disorders. *The Primary Care Companion for CNS Disorders, 18*(5). https://doi.org/10.4088/PCC.16r02008

Belmaker, R. H., & Fleischmann, A. (1995). Transcranial magnetic stimulation: A potential new frontier in psychiatry. *Biological Psychiatry, 38*(7), 419–421. https://doi.org/10.1016/0006-3223(95)92243-B

Belmaker, B., Fitzgerald, P., George, M. S., Lisanby, S. H., Pascual-Leone, A., Schlaepfer, T. E., & Wassermann, E. (2003). International Society for Transcranial Stimulation Consensus Statement: Managing the risks of repetitive transcranial stimulation. *CNS Spectrums, 8*(7), 489–489. https://doi.org/10.1017/S1092852900018952

Benadhira, R., Thomas, F., Bouaziz, N., Braha, S., Andrianisaina, P. S., Isaac, C., ... Januel, D. (2017). A randomized, sham-controlled study of maintenance rTMS for treatment-resistant depression (TRD). *Psychiatry Research, 258*, 226–233. https://doi.org/10.1016/j.psychres.2017.08.029

Berlim, M. T., Van den Eynde, F., & Daskalakis, Z. J. (2013). Efficacy and acceptability of high frequency repetitive transcranial magnetic stimulation (rTMS) versus electroconvulsive therapy (ECT) for major depression: A systematic review and meta-analysis of randomized trials. *Depression and Anxiety, 30*(7), 614–623. https://doi.org/10.1002/da.22060

Blumberger, D. M., Vila-Rodriguez, F., Thorpe, K. E., Feffer, K., Noda, Y., Giacobbe, P., ... Downar, J. (2018). Effectiveness of theta burst versus high-frequency repetitive transcranial magnetic stimulation in patients with depression (THREE-D): A randomised non-inferiority trial. *Lancet, 391*(10131), 1683–1692. https://doi.org/10.1016/S0140-6736(18)30295-2

Bohning, D., Shastri, A., McConnell, K., Nahas, Z., Lorberbaum, J., Roberts, D., ... George, M. (1999). A combined TMS/fMRI study of intensity-dependent TMS over motor cortex. *Biological Psychiatry, 45*(4), 385–394. https://doi.org/10.1016/S0006-3223(98)00368-0

Cardenas-Rojas, A., Pacheco-Barrios, K., Giannoni-Luza, S., Rivera-Torrejon, O., & Fregni, F. (2020). Noninvasive brain stimulation combined with exercise in chronic pain: A systematic review and meta-analysis.

Expert Review of Neurotherapeutics, 20(4), 401–412. https://doi.org/10.1080/14737175.2020.1738927

Cassano, P., Cusin, C., Mischoulon, D., Hamblin, M. R., De Taboada, L., Pisoni, A., ... Iosifescu, D. V. (2015). Near-infrared transcranial radiation for major depressive disorder: Proof of concept study. *Psychiatry Journal, 2015*, 352979. https://doi.org/10.1155/2015/352979

Cassano, P., Petrie, S. R., Mischoulon, D., Cusin, C., Katnani, H., Yeung, A., ... Iosifescu, D. V. (2018). Transcranial photobiomodulation for the treatment of major depressive disorder. The ELATED-2 Pilot trial. *Photomedicine and Laser Surgery, 36*(12), 634–646. https://doi.org/10.1089/pho.2018.4490

Chen, R., Classen, J., Gerloff, C., Celnik, P., Wassermann, E. M., Hallett, M., & Cohen, L. G. (1997). Depression of motor cortex excitability by low-frequency transcranial magnetic stimulation. *Neurology, 48*(5), 1398–1403.

Chen, L., Chung, S. W., Hoy, K. E., & Fitzgerald, P. B. (2019). Is theta burst stimulation ready as a clinical treatment for depression? *Expert Review of Neurotherapeutics, 19*(11), 1089–1102. https://doi.org/10.1080/14737175.2019.1641084

Cirillo, P., Gold, A. K., Nardi, A. E., Ornelas, A. C., Nierenberg, A. A., Camprodon, J., & Kinrys, G. (2019). Transcranial magnetic stimulation in anxiety and trauma-related disorders: A systematic review and meta-analysis. *Brain and Behavior: A Cognitive Neuroscience Perspective, 9*(6), e01284. https://doi.org/10.1002/brb3.1284

Coben, R., Hammond, D. C., & Arns, M. (2019). 19 channel Z-score and LORETA neurofeedback: Does the evidence support the hype? *Applied Psychophysiology and Biofeedback, 44*(1), 1–8. https://doi.org/10.1007/s10484-018-9420-6

Cortese, S., Ferrin, M., Brandeis, D., Holtmann, M., Aggensteiner, P., Daley, D., ... Stringaris, A. (2016). Neurofeedback for attention-deficit/hyperactivity disorder: Meta-analysis of clinical and neuropsychological outcomes from randomized controlled trials. *Journal of the American Academy of Child & Adolescent Psychiatry, 55*(6), 444–455.

Cuijpers, P., Karyotaki, E., Weitz, E., Andersson, G., Hollon, S. D., & van Straten, A. (2014). The effects of psychotherapies for major depression in adults on remission, recovery and improvement: A meta-analysis. *Journal of Affective Disorders, 159*, 118–126. https://doi.org/10.1016/j.jad.2014.02.026

Deng, Z.-D., Lisanby, S. H., & Peterchev, A. V. (2011). Electric field strength and focality in electroconvulsive therapy and magnetic seizure therapy: A finite element simulation study. *Journal of Neural Engineering, 8*(1), 016007. https://doi.org/10.1088/1741-2560/8/1/016007

Donahue, A. B. (2000). Electroconvulsive therapy and memory loss: A personal journey. *The Journal of ECT, 16*(2), 133–143. https://doi.org/10.1097/00124509-200006000-00005

Donse, L., Padberg, F., Sack, A. T., Rush, A. J., & Arns, M. (2018). Simultaneous rTMS and psychotherapy in major depressive disorder: Clinical outcomes and predictors from a large naturalistic study. *Brain Stimulation, 11*(2), 337–345. https://doi.org/10.1016/j.brs.2017.11.004

Dougherty, D. D. (2018). Deep brain stimulation: Clinical applications. *The Psychiatric Clinics of North America, 41*(3), 385–394. https://doi.org/10.1016/j.psc.2018.04.004

Dougherty, D. D., Rezai, A. R., Carpenter, L. L., Howland, R. H., Bhati, M. T., O'Reardon, J. P., . . . Malone, D. A., Jr. (2015). A randomized sham-controlled trial of deep brain stimulation of the ventral capsule/ventral striatum for chronic treatment-resistant depression. *Biological Psychiatry, 78*(4), 240–248. https://doi.org/10.1016/j.biopsych.2014.11.023

Dowdle, L. T., Brown, T. R., George, M. S., & Hanlon, C. A. (2018). Single pulse TMS to the DLPFC, compared to a matched sham control, induces a direct, causal increase in caudate, cingulate, and thalamic BOLD signal. *Brain Stimulation, 11*(4), 789–796. https://doi.org/10.1016/j.brs.2018.02.014

Dubin, M. J., Ilieva, I. P., Deng, Z. D., Thomas, J., Cochran, A., Kravets, K., . . . Gunning, F. M. (2019). A double-blind pilot dosing study of low field magnetic stimulation (LFMS) for treatment-resistant depression (TRD). *Journal of Affective Disorders, 249*, 286–293. https://doi.org/10.1016/j.jad.2019.02.039

Dunner, D. L., Aaronson, S. T., Sackeim, H. A., Janicak, P. G., Carpenter, L. L., Boyadjis, T., . . . Demitrack, M. A. (2014). A multisite, naturalistic, observational study of transcranial magnetic stimulation for patients with pharmacoresistant major depressive disorder: Durability of benefit over a 1-year follow-up period. *The Journal of Clinical Psychiatry, 75*(12), 1394–1401. https://doi.org/10.4088/JCP.13m08977

Dwork, A. J., Arango, V., Underwood, M., Ilievski, B., Rosoklija, G., Sackeim, H. A., & Lisanby, S. H. (2004). Absence of histological lesions in primate models of ECT and magnetic seizure therapy. *American Journal of Psychiatry, 161*(3), 576–578. https://doi.org/10.1176/appi.ajp.161.3.576

Epstein, C. M., Meador, K. J., Loring, D. W., Wright, R. J., Weissman, J. D., Sheppard, S., . . . Davey, K. R. (1999). Localization and characterization of speech arrest during transcranial magnetic stimulation. *Clinical Neurophysiology, 110*(6), 1073–1079.

Erraji-Benchekroun, L., Lisanby, S. H., Arango, V., Galfalvy, H., Pavlidis, P., & Underwood, M. D. (2007). Effect of electroconvulsive shock and magnetic seizure on gene expression profiles in the prefrontal cortex of the rhesus monkey. *The Journal of ECT, 23*(1), 53. https://doi.org/10.1097/01.yct.0000264356.75257.80

Farzan, F., Barr, M. S., Sun, Y., Fitzgerald, P. B., & Daskalakis, Z. J. (2012). Transcranial magnetic stimulation on the modulation of gamma oscillations in schizophrenia. *Annals of the New York Academy of Sciences, 1265*, 25–35. https://doi.org/10.1111/j.1749-6632.2012.06543.x

Fava, M., Freeman, M. P., Flynn, M., Hoeppner, B. B., Shelton, R., Iosifescu, D. V., . . . Papakostas, G. I. (2018). Double-blind, proof-of-concept (POC) trial of Low-Field Magnetic Stimulation (LFMS) augmentation of antidepressant therapy in treatment-resistant depression (TRD). *Brain Stimulation, 11*(1), 75–84. https://doi.org/10.1016/j.brs.2017.09.010

Fitzgerald, P. B. (2020). An update on the clinical use of repetitive transcranial magnetic stimulation in the treatment of depression. *Journal of Affective Disorders, 276*, 90–103. https://doi.org/10.1016/j.jad.2020.06.067

Fitzgerald, P. B., & Daskalakis, Z. J. (2011). The effects of repetitive transcranial magnetic stimulation in the treatment of depression. *Expert Review of Medical Devices, 8*(1), 85–95. https://doi.org/10.1586/erd.10.57

Fitzgerald, P. B., Brown, T. L., Daskalakis, Z. J., & Kulkarni, J. (2002). A transcranial magnetic stimulation study of inhibitory deficits in the motor cortex in patients with schizophrenia. *Psychiatry Research, 114*(1), 11–22. https://doi.org/10.1016/s0925-4927(02)00002-1

Fitzgerald, P. B., Brown, T. L., Marston, N. A., Oxley, T., De Castella, A., Daskalakis, Z. J., & Kulkarni, J. (2004). Reduced plastic brain responses in schizophrenia: A transcranial magnetic stimulation study. *Schizophrenia Research, 71*(1), 17–26. https://doi.org/10.1016/j.schres.2004.01.018

Fitzgerald, P. B., Benitez, J., de Castella, A. R., Brown, T. L., Daskalakis, Z. J., & Kulkarni, J. (2006). Naturalistic study of the use of transcranial magnetic stimulation in the treatment of depressive relapse. *The Australian and New Zealand Journal of Psychiatry, 40*(9), 764–768. https://doi.org/10.1080/j.1440-1614.2006.01881.x

Fitzgerald, P. B., Hoy, K., McQueen, S., Maller, J. J., Herring, S., Segrave, R., . . . Daskalakis, Z. J. (2009a). A randomized trial of rTMS targeted with MRI based neuro-navigation in treatment-resistant depression. *Neuropsychopharmacology, 34*(5), 1255–1262. https://doi.org/10.1038/npp.2008.233. npp2008233 [pii].

Fitzgerald, P. B., Maller, J. J., Hoy, K. E., Thomson, R., & Daskalakis, Z. J. (2009b). Exploring the optimal site for the localization of dorsolateral prefrontal cortex in brain stimulation experiments. *Brain Stimulation, 2*(4), 234–237. https://doi.org/10.1016/j.brs.2009.03.002

Fitzgerald, P. B., Grace, N., Hoy, K. E., Bailey, M., & Daskalakis, Z. J. (2013a). An open label trial of clustered maintenance rTMS for patients with refractory depression. *Brain Stimulation, 6*(3), 292–297. https://doi.org/10.1016/j.brs.2012.05.003

Fitzgerald, P. B., Hoy, K. E., Herring, S. E., Clinton, A. M., Downey, G., & Daskalakis, Z. J. (2013b). Pilot study of the clinical and cognitive effects of high-frequency magnetic seizure therapy in major depressive disorder. *Depression and Anxiety, 30*(2), 129–136. https://doi.org/10.1002/da.22005

Fitzgerald, P. B., Hoy, K. E., Elliot, D., McQueen, S., Wambeek, L. E., Chen, L., . . . Daskalakis, Z. J.

(2018). A pilot study of the comparative efficacy of 100 Hz magnetic seizure therapy and electroconvulsive therapy in persistent depression. *Depression and Anxiety, 35*(5), 393–401. https://doi.org/10.1002/da.22715

Garnaat, S. L., Greenberg, B. D., Sibrava, N. J., Goodman, W. K., Mancebo, M. C., Eisen, J. L., & Rasmussen, S. A. (2014). Who qualifies for deep brain stimulation for OCD? Data from a naturalistic clinical sample. *The Journal of Neuropsychiatry and Clinical Neurosciences, 26*(1), 81–86. https://doi.org/10.1176/appi.neuropsych.12090226

Geller, V., Grisaru, N., Abarbanel, J. M., Lemberg, T., & Belmaker, R. H. (1997). Slow magnetic stimulation of prefrontal cortex in depression and schizophrenia. *Progress in Neuro-Psychopharmacology and Biological Psychiatry, 21*(1), 105–110. https://doi.org/10.1016/S0278-5846(96)00161-3

Generoso, M. B., Taiar, I. T., Garrocini, L. P., Bernardon, R., Cordeiro, Q., Uchida, R. R., & Shiozawa, P. (2019). Effect of a 10-day transcutaneous trigeminal nerve stimulation (TNS) protocol for depression amelioration: A randomized, double blind, and sham-controlled phase II clinical trial. *Epilepsy & Behavior, 95*, 39–42. https://doi.org/10.1016/j.yebeh.2019.03.025

Geoffroy, P. A., Schroder, C. M., Reynaud, E., & Bourgin, P. (2019). Efficacy of light therapy versus antidepressant drugs, and of the combination versus monotherapy, in major depressive episodes: A systematic review and meta-analysis. *Sleep Medicine Reviews, 48*, 101213. https://doi.org/10.1016/j.smrv.2019.101213

George, M. S., Wassermann, E. M., Williams, W. A., Callahan, A., Ketter, T. A., Basser, P., ... Post, R. M. (1995). Daily repetitive transcranial magnetic stimulation (rTMS) improves mood in depression. *Neuroreport, 6*(14), 1853–1856. https://doi.org/10.1097/00001756-199510020-00008

George, M. S., Sackeim, H. A., Rush, A. J., Marangell, L. B., Nahas, Z., Husain, M. M., ... Ballenger, J. C. (2000). Vagus nerve stimulation: A new tool for brain research and therapy*. *Biological Psychiatry, 47*(4), 287–295. https://doi.org/10.1016/S0006-3223(99)00308-X

George, M. S., Rush, A. J., Marangell, L. B., Sackeim, H. A., Brannan, S. K., Davis, S. M., ... Goodnick, P. (2005). A one-year comparison of vagus nerve stimulation with treatment as usual for treatment-resistant depression. *Biological Psychiatry, 58*(5), 364–373. https://doi.org/10.1016/j.biopsych.2005.07.028

George, M. S., Taylor, J. J., & Short, E. B. (2013). The expanding evidence base for rTMS treatment of depression. *Current Opinion in Psychiatry, 26*(1), 13–18. https://doi.org/10.1097/YCO.0b013e32835ab46d

Golden, R. N., Gaynes, B. N., Ekstrom, R. D., Hamer, R. M., Jacobsen, F. M., Suppes, T., ... Nemeroff, C. B. (2005). The efficacy of light therapy in the treatment of mood disorders: A review and meta-analysis of the evidence. *The American Journal of Psychiatry,*

162(4), 656–662. https://doi.org/10.1176/appi.ajp.162.4.656

Heinrich, H., Gevensleben, H., Freisleder, F. J., Moll, G. H., & Rothenberger, A. (2004). Training of slow cortical potentials in attention-deficit/hyperactivity disorder: Evidence for positive behavioral and neurophysiological effects. *Biological Psychiatry, 55*(7), 772–775. https://doi.org/10.1016/j.biopsych.2003.11.013

Herrmann, C. S., Rach, S., Neuling, T., & Struber, D. (2013). Transcranial alternating current stimulation: A review of the underlying mechanisms and modulation of cognitive processes. *Frontiers in Human Neuroscience, 7*, 279. https://doi.org/10.3389/fnhum.2013.00279

Hirakawa, H., Terao, T., Muronaga, M., & Ishii, N. (2020). Adjunctive bright light therapy for treating bipolar depression: A systematic review and meta-analysis of randomized controlled trials. *Brain and Behavior: A Cognitive Neuroscience Perspective*, e01876. https://doi.org/10.1002/brb3.1876

Hitti, F. L., Yang, A. I., Cristancho, M. A., & Baltuch, G. H. (2020). Deep brain stimulation is effective for treatment-resistant depression: A meta-analysis and meta-regression. *Journal of Clinical Medicine, 9*(9). https://doi.org/10.3390/jcm9092796

Holtzheimer, P. E., Husain, M. M., Lisanby, S. H., Taylor, S. F., Whitworth, L. A., McClintock, S., ... Mayberg, H. S. (2017). Subcallosal cingulate deep brain stimulation for treatment-resistant depression: A multisite, randomised, sham-controlled trial. *Lancet Psychiatry, 4*(11), 839–849. https://doi.org/10.1016/S2215-0366(17)30371-1

Husain, M. M., Rush, A. J., Fink, M., Knapp, R., Petrides, G., Rummans, T., ... Kellner, C. H. (2004). Speed of response and remission in major depressive disorder with acute electroconvulsive therapy (ECT): A Consortium for Research in ECT (CORE) report. *The Journal of Clinical Psychiatry, 65*(4), 485–491. https://doi.org/10.4088/JCP.v65n0406

Iseger, T. A., Padberg, F., Kenemans, J. L., Gevirtz, R., & Arns, M. (2017). Neuro-Cardiac-Guided TMS (NCG-TMS): Probing DLPFC-sgACC-vagus nerve connectivity using heart rate – First results. *Brain Stimulation, 10*(5), 1006–1008. https://doi.org/10.1016/j.brs.2017.05.002

Kato, N. (2009). Neurophysiological mechanisms of electroconvulsive therapy for depression. *Neuroscience Research, 64*(1), 3–11. https://doi.org/10.1016/j.neures.2009.01.014

Kavirajan, H. C., Lueck, K., & Chuang, K. (2014). Alternating current cranial electrotherapy stimulation (CES) for depression. *Cochrane Database of Systematic Reviews*, (7), CD010521. https://doi.org/10.1002/14651858.CD010521.pub2

Kellner, C. H., Knapp, R. G., Petrides, G., Rummans, T. A., Husain, M. M., Rasmussen, K., ... Fink, M. (2006). Continuation electroconvulsive therapy vs pharmacotherapy for relapse prevention in major depression: A multisite study from the Consortium for Research in

Electroconvulsive Therapy (CORE). *Archives of General Psychiatry, 63*(12), 1337. https://doi.org/10.1001/archpsyc.63.12.1337

Kellner, C. H., Knapp, R., Husain, M. M., Rasmussen, K., Sampson, S., Cullum, M., … Petrides, G. (2010). Bifrontal, bitemporal and right unilateral electrode placement in ECT: Randomised trial. *British Journal of Psychiatry, 196*(3), 226–234. https://doi.org/10.1192/bjp.bp.109.066183

Kennedy, N. I., Lee, W. H., & Frangou, S. (2018). Efficacy of non-invasive brain stimulation on the symptom dimensions of schizophrenia: A meta-analysis of randomized controlled trials. *European Psychiatry, 49*, 69–77. https://doi.org/10.1016/j.eurpsy.2017.12.025

Kim, D. R., Epperson, N., Paré, E., Gonzalez, J. M., Parry, S., Thase, M. E., Cristancho, P., Sammel, M. D., & O'Reardon, J. P. (2011). An open label pilot study of transcranial magnetic stimulation for pregnant women with major depressive disorder. *Journal of Women's Health (2002), 20*(2), 255–261. https://doi.org/10.1089/jwh.2010.2353

Klimke, A., Nitsche, M. A., Maurer, K., & Voss, U. (2016). Case report: Successful treatment of therapy-resistant OCD with application of transcranial alternating current stimulation (tACS). *Brain Stimulation, 9*(3), 463–465. https://doi.org/10.1016/j.brs.2016.03.005

Lefaucheur, J. P., Andre-Obadia, N., Antal, A., Ayache, S. S., Baeken, C., Benninger, D. H., … Garcia-Larrea, L. (2014). Evidence-based guidelines on the therapeutic use of repetitive transcranial magnetic stimulation (rTMS). *Clinical Neurophysiology, 125*(11), 2150–2206. https://doi.org/10.1016/j.clinph.2014.05.021

Leinenga, G., & Götz, J. (2015). Scanning ultrasound removes amyloid-β and restores memory in an Alzheimer's disease mouse model. *Science Translational Medicine, 7*(278), 278ra33.

Leuchter, A. F., Cook, I. A., Feifel, D., Goethe, J. W., Husain, M., Carpenter, L. L., … George, M. S. (2015). Efficacy and safety of low-field synchronized transcranial magnetic stimulation (sTMS) for treatment of major depression. *Brain Stimulation, 8*(4), 787–794. https://doi.org/10.1016/j.brs.2015.05.005

Levkovitz, Y., Isserles, M., Padberg, F., Lisanby, S. H., Bystritsky, A., Xia, G., … Zangen, A. (2015). Efficacy and safety of deep transcranial magnetic stimulation for major depression: A prospective multicenter randomized controlled trial. *World Psychiatry, 14*(1), 64–73. https://doi.org/10.1002/wps.20199

Lisanby, S. H. (2000). The effects of electroconvulsive therapy on memory of autobiographical and public events. *Archives of General Psychiatry, 57*(6), 581–590. https://doi.org/10.1001/archpsyc.57.6.581

Lisanby, S. H., & Belmaker, R. H. (2000). Animal models of the mechanisms of action of repetitive transcranial magnetic stimulation (RTMS): Comparisons with electroconvulsive shock (ECS). *Depression and Anxiety, 12*(3), 178–187. https://doi.org/10.1002/1520-6394(2000)12:3<178::AID-DA10>3.0.CO;2-N

Lisanby, S. H., & Peterchev, A. V. (2007). Magnetic seizure therapy for the treatment of depression. In M. A. Marcolin & F. Padberg (Eds.), *Advances in biological psychiatry* (pp. 155–171). KARGER.

Lisanby, S. H., Bazil, C. W., Resor, S. R., Nobler, M. S., Finck, D. A., & Sackeim, H. A. (2001a). ECT in the treatment of status epilepticus. *The Journal of ECT, 17*(3), 210–215. https://doi.org/10.1097/00124509-200109000-00013

Lisanby, S. H., Luber, B., Sackeim, H. A., Finck, A. D., & Schroeder, C. (2001b). Deliberate seizure induction with repetitive transcranial magnetic stimulation in nonhuman primates. *Archives of General Psychiatry, 58*(2), 199. https://doi.org/10.1001/archpsyc.58.2.199

Lisanby, S. H., Schlaepfer, T. E., Fisch, H.-U., & Sackeim, H. A. (2001c). Magnetic seizure therapy of major depression. *Archives of General Psychiatry, 58*(3), 303. https://doi.org/10.1001/archpsyc.58.3.303

Lisanby, S. H., Luber, B., Schlaepfer, T. E., & Sackeim, H. A. (2003a). Safety and feasibility of Magnetic Seizure Therapy (MST) in major depression: Randomized within-subject comparison with electroconvulsive therapy. *Neuropsychopharmacology, 28*(10), 1852–1865. https://doi.org/10.1038/sj.npp.1300229

Lisanby, S. H., Moscrip, T., Morales, O., Luber, B., Schroeder, C., & Sackeim, H. A. (2003b). Chapter 9 Neurophysiological characterization of magnetic seizure therapy (MST) in non-human primates. In *Supplements to clinical neurophysiology* (Vol. 56, pp. 81–99). Elsevier.

Loo, C. K., Kaill, A., Paton, P., & Simpson, B. (2010). The difficult-to-treat electroconvulsive therapy patient – Strategies for augmenting outcomes. *Journal of Affective Disorders, 124*(3), 219–227. https://doi.org/10.1016/j.jad.2009.07.011

Lubar, J. O., & Lubar, J. F. (1984). Electroencephalographic biofeedback of SMR and beta for treatment of attention deficit disorders in a clinical setting. *Applied Psychophysiology and Biofeedback, 9*(1), 1–23.

Lubar, J. F., & Shouse, M. N. (1976). EEG and behavioral changes in a hyperkinetic child concurrent with training of the sensorimotor rhythm (SMR): A preliminary report. *Biofeedback and Self-Regulation, 1*(3), 293–306.

Luber, B., Nobler, M. S., Moeller, J. R., Katzman, G. P., Prudic, J., Devanand, D. P., … Sackeim, H. A. (2000). Quantitative EEG during seizures induced by electroconvulsive therapy: Relations to treatment modality and clinical features. II. Topographic analyses. *The Journal of ECT, 16*(3), 229–243. https://doi.org/10.1097/00124509-200009000-00003

Luber, B., Kinnunen, L. H., Rakitin, B. C., Ellsasser, R., Stern, Y., & Lisanby, S. H. (2007a). Facilitation of performance in a working memory task with rTMS stimulation of the precuneus: Frequency- and time-dependent effects. *Brain Research, 1128*, 120–129. https://doi.org/10.1016/j.brainres.2006.10.011

Luber, B., Stanford, A. D., Malaspina, D., & Lisanby, S. H. (2007b). Revisiting the backward masking deficit in schizophrenia: Individual differences in performance and modeling with transcranial magnetic stimulation. *Biological Psychiatry, 62*(7), 793–799. https://doi.org/10.1016/j.biopsych.2006.10.007

Mall, V., Berweck, S., Fietzek, U. M., Glocker, F. X., Oberhuber, U., Walther, M., ... Heinen, F. (2004). Low level of intracortical inhibition in children shown by transcranial magnetic stimulation. *Neuropediatrics, 35*(2), 120–125. https://doi.org/10.1055/s-2004-815834

Martiny, K., Lunde, M., & Bech, P. (2010). Transcranial low voltage pulsed electromagnetic fields in patients with treatment-resistant depression. *Biological Psychiatry, 68*(2), 163–169. https://doi.org/10.1016/j.biopsych.2010.02.017

May, I. C. (2020). Light therapy for preventing seasonal affective disorder: Summary of a Cochrane review. *Explore (New York, N.Y.), 16*(2), 133–134. https://doi.org/10.1016/j.explore.2019.12.004

McCall, W. V., Reboussin, D. M., Weiner, R. D., & Sackeim, H. A. (2000). Titrated moderately suprathreshold vs fixed high-dose right unilateral electroconvulsive therapy: Acute antidepressant and cognitive effects. *Archives of General Psychiatry, 57*(5), 438. https://doi.org/10.1001/archpsyc.57.5.438

Mellin, J. M., Alagapan, S., Lustenberger, C., Lugo, C. E., Alexander, M. L., Gilmore, J. H., ... Frohlich, F. (2018). Randomized trial of transcranial alternating current stimulation for treatment of auditory hallucinations in schizophrenia. *European Psychiatry, 51*, 25–33. https://doi.org/10.1016/j.eurpsy.2018.01.004

Mohr, P., Rodriguez, M., Slavíčková, A., & Hanka, J. (2011). The application of vagus nerve stimulation and deep brain stimulation in depression. *Neuropsychobiology, 64*(3), 170–181. https://doi.org/10.1159/000325225

Mulsant, B. H., Rosen, J., Thornton, J. E., & Zubenko, G. S. (1991). A prospective naturalistic study of electroconvulsive therapy in late-life depression. *Topics in Geriatrics, 4*(1), 3–12. https://doi.org/10.1177/089198879100400102

Murphy, O. W., Hoy, K. E., Wong, D., Bailey, N. W., Fitzgerald, P. B., & Segrave, R. A. (2020). Transcranial random noise stimulation is more effective than transcranial direct current stimulation for enhancing working memory in healthy individuals: Behavioural and electrophysiological evidence. *Brain Stimulation, 13*(5), 1370–1380. https://doi.org/10.1016/j.brs.2020.07.001

Nall, A. (1973). Alpha training and the hyperkinetic child -is it effective? *Intervention in School and Clinic, 9*(1), 5–19. https://doi.org/10.1177/105345127300900101

Neurofeedback Collaborative, G. (2020). Double-blind placebo-controlled randomized clinical trial of neurofeedback for attention-deficit/hyperactivity disorder with 13-month follow-up. *Journal of the American Academy of Child and Adolescent Psychiatry, 60*, 841. https://doi.org/10.1016/j.jaac.2020.07.906

Nikolin, S., Alonzo, A., Martin, D., Galvez, V., Buten, S., Taylor, R., ... Loo, C. K. (2020). Transcranial random noise stimulation for the acute treatment of depression: A randomized controlled trial. *The International Journal of Neuropsychopharmacology, 23*(3), 146–156. https://doi.org/10.1093/ijnp/pyz072

Nitsche, M. A., Fricke, K., Henschke, U., Schlitterlau, A., Liebetanz, D., Lang, N., ... Paulus, W. (2003). Pharmacological modulation of cortical excitability shifts induced by transcranial direct current stimulation in humans. *The Journal of Physiology, 553*(Pt 1), 293–301. https://doi.org/10.1113/jphysiol.2003.049916

Nobler, M. S. (1994). Regional cerebral blood flow in mood disorders, III: Treatment and clinical response. *Archives of General Psychiatry, 51*(11), 884. https://doi.org/10.1001/archpsyc.1994.03950110044007

O'Reardon, J. P., Cristancho, P., & Peshek, A. D. (2006). Vagus Nerve Stimulation (VNS) and treatment of depression: To the brainstem and beyond. *Psychiatry (Edgmont), 3*(5), 54–63.

Pail, G., Huf, W., Pjrek, E., Winkler, D., Willeit, M., Praschak-Rieder, N., & Kasper, S. (2011). Bright-light therapy in the treatment of mood disorders. *Neuropsychobiology, 64*(3), 152–162. https://doi.org/10.1159/000328950

Pascual-Leone, A., Rubio, B., Pallardó, F., & Catalá, M. D. (1996). Rapid-rate transcranial magnetic stimulation of left dorsolateral prefrontal cortex in drug-resistant depression. *The Lancet, 348*(9022), 233–237. https://doi.org/10.1016/S0140-6736(96)01219-6

Paus, T., Jech, R., Thompson, C. J., Comeau, R., Peters, T., & Evans, A. C. (1997). Transcranial magnetic stimulation during positron emission tomography: A new method for studying connectivity of the human cerebral cortex. *The Journal of Neuroscience, 17*(9), 3178–3184. https://doi.org/10.1523/JNEUROSCI.17-09-03178.1997

Penders, T. M., Stanciu, C. N., Schoemann, A. M., Ninan, P. T., Bloch, R., & Saeed, S. A. (2016). Bright light therapy as augmentation of pharmacotherapy for treatment of depression: A systematic review and meta-analysis. *The Primary Care Companion for CNS Disorders, 18*(5). https://doi.org/10.4088/PCC.15r01906

Perera, T. D., Coplan, J. D., Lisanby, S. H., Lipira, C. M., Arif, M., Carpio, C., ... Dwork, A. J. (2007). Antidepressant-induced neurogenesis in the hippocampus of adult nonhuman primates. *Journal of Neuroscience, 27*(18), 4894–4901. https://doi.org/10.1523/JNEUROSCI.0237-07.2007

Perera, T., George, M. S., Grammer, G., Janicak, P. G., Pascual-Leone, A., & Wirecki, T. S. (2016). The clinical TMS society consensus review and treatment recommendations for TMS therapy for major depressive disorder. *Brain Stimulation, 9*(3), 336–346. https://doi.org/10.1016/j.brs.2016.03.010

Petrides, G., Fink, M., Husain, M. M., Knapp, R. G., Rush, A. J., Mueller, M., ... Kellner, C. H. (2001). ECT remission rates in psychotic versus nonpsychotic depressed patients: A report from CORE. *The Journal of ECT, 17*(4), 244–253. https://doi.org/10.1097/00124509-200112000-00003

Pridmore, S., Turnier-Shea, Y., Rybak, M., & Pridmore, W. (2021). Transcranial Magnetic Stimulation (TMS) during pregnancy: A fetal risk factor. *Australasian Psychiatry, 29*(2), 226–229. https://doi.org/10.1177/1039856221992636. Epub 2021 Mar 2.

Prudic, J., Olfson, M., Marcus, S. C., Fuller, R. B., & Sackeim, H. A. (2004). Effectiveness of electroconvulsive therapy in community settings. *Biological Psychiatry, 55*(3), 301–312. https://doi.org/10.1016/j.biopsych.2003.09.015

Quartarone, A., Rizzo, V., Bagnato, S., Morgante, F., Sant'Angelo, A., Girlanda, P., & Roman Siebner, H. (2006). Rapid-rate paired associative stimulation of the median nerve and motor cortex can produce long-lasting changes in motor cortical excitability in humans: Repetitive paired associative stimulation of human motor cortex. *The Journal of Physiology, 575*(2), 657–670. https://doi.org/10.1113/jphysiol.2006.114025

Rasmussen, K. G. (2017). The PRIDE study and the conduct of electroconvulsive therapy: Questions answered and unanswered. *The Journal of ECT, 33*(4), 225–228. https://doi.org/10.1097/YCT.0000000000000430

Razza, L. B., Palumbo, P., Moffa, A. H., Carvalho, A. F., Solmi, M., Loo, C. K., & Brunoni, A. R. (2020). A systematic review and meta-analysis on the effects of transcranial direct current stimulation in depressive episodes. *Depression and Anxiety, 37*(7), 594–608. https://doi.org/10.1002/da.23004

Rehn, S., Eslick, G. D., & Brakoulias, V. (2018). A meta-analysis of the effectiveness of different cortical targets used in repetitive transcranial magnetic stimulation (rTMS) for the treatment of obsessive-compulsive disorder (OCD). *The Psychiatric Quarterly, 89*(3), 645–665. https://doi.org/10.1007/s11126-018-9566-7

Rockstroh, B., Elbert, T., Birbaumer, N., Wolf, P., Düchting-Röth, A., Reker, M., ... Dichgans, J. (1993). Cortical self-regulation in patients with epilepsies. *Epilepsy Research, 14*(1), 63–72.

Roet, M., Boonstra, J., Sahin, E., Mulders, A. E. P., Leentjens, A. F. G., & Jahanshahi, A. (2020). Deep brain stimulation for treatment-resistant depression: Towards a more personalized treatment approach. *Journal of Clinical Medicine, 9*(9). https://doi.org/10.3390/jcm9092729

Rogasch, N. C., & Fitzgerald, P. B. (2013). Assessing cortical network properties using TMS-EEG. *Human Brain Mapping, 34*(7), 1652–1669. https://doi.org/10.1002/hbm.22016

Rong, P., Liu, J., Wang, L., Liu, R., Fang, J., Zhao, J., ... Kong, J. (2016). Effect of transcutaneous auricular vagus nerve stimulation on major depressive disorder: A nonrandomized controlled pilot study. *Journal of Affective Disorders, 195*, 172–179. https://doi.org/10.1016/j.jad.2016.02.031

Rosa, M. A., & Lisanby, S. H. (2012). Somatic treatments for mood disorders. *Neuropsychopharmacology, 37*(1), 102–116. https://doi.org/10.1038/npp.2011.225

Rossi, S., Hallett, M., Rossini, P. M., Pascual-Leone, A., & Safety of, T. M. S. C. G. (2009). Safety, ethical considerations, and application guidelines for the use of transcranial magnetic stimulation in clinical practice and research. *Clinical Neurophysiology, 120*(12), 2008–2039. https://doi.org/10.1016/j.clinph.2009.08.016

Rossi, S., Antal, A., Bestmann, S., Bikson, M., Brewer, C., Brockmoller, J., ... basis of this article began with a Consensus Statement from the Ifcn Workshop on "Present, F. o. T. M. S. S. E. G. S. O. u. t. A". (2020). Safety and recommendations for TMS use in healthy subjects and patient populations, with updates on training, ethical and regulatory issues: Expert guidelines. *Clinical Neurophysiology, 132*, 269. https://doi.org/10.1016/j.clinph.2020.10.003

Rothwell, J. (2018). Transcranial brain stimulation: Past and future. *Brain and Neuroscience Advances, 2*, 2398212818818070. https://doi.org/10.1177/2398212818818070

Rush, A. J., & Siefert, S. E. (2009). Clinical issues in considering vagus nerve stimulation for treatment-resistant depression. *Experimental Neurology, 219*(1), 36–43. https://doi.org/10.1016/j.expneurol.2009.04.015

Rush, A. J., George, M. S., Sackeim, H. A., Marangell, L. B., Husain, M. M., Giller, C., ... Goodman, R. (2000). Vagus nerve stimulation (VNS) for treatment-resistant depressions: A multicenter study. *Biological Psychiatry, 47*(4), 276–286

Rush, A. J., Marangell, L. B., Sackeim, H. A., George, M. S., Brannan, S. K., Davis, S. M., ... Cooke, R. G. (2005). Vagus nerve stimulation for treatment-resistant depression: A randomized, controlled acute phase trial. *Biological Psychiatry, 58*(5), 347–354. https://doi.org/10.1016/j.biopsych.2005.05.025

Rush, A. J., Trivedi, M. H., Wisniewski, S. R., Nierenberg, A. A., Stewart, J. W., Warden, D., ... Fava, M. (2006). Acute and longer-term outcomes in depressed outpatients requiring one or several treatment steps: A STAR*D report. *The American Journal of Psychiatry, 163*(11), 1905–1917. https://doi.org/10.1176/ajp.2006.163.11.1905

Sackeim, H. A., Portnoy, S., Neeley, P., Steif, B. L., Decina, P., & Malitz, S. (1986). Cognitive consequences of low-dosage electroconvulsive therapy. *Annals of the New York Academy of Sciences, 462*-(1 Electroconvul), 326–340. https://doi.org/10.1111/j.1749-6632.1986.tb51267.x

Sackeim, H. A., Prudic, J., Devanand, D. P., Kiersky, J. E., Fitzsimons, L., Moody, B. J., ... Settembrino, J. M. (1993). Effects of stimulus intensity and electrode placement on the efficacy and cognitive effects of electroconvulsive therapy. *New England Journal of Medicine, 328*(12), 839–846. https://doi.org/10.1056/NEJM199303253281204

Sackeim, H. A., Prudic, J., Devanand, D. P., Nobler, M. S., Lisanby, S. H., Peyser, S., ... Clark, J. (2000). A prospective, randomized, double-blind comparison of bilateral and right unilateral electroconvulsive therapy at different stimulus intensities. *Archives of General Psychiatry, 57*(5), 425. https://doi.org/10.1001/archpsyc.57.5.425

Sackeim, H. A., Haskett, R. F., Mulsant, B. H., Thase, M. E., Mann, J. J., Pettinati, H. M., … Prudic, J. (2001a). Continuation pharmacotherapy in the prevention of relapse following electroconvulsive therapy: A randomized controlled trial. *JAMA, 285*(10), 1299. https://doi.org/10.1001/jama.285.10.1299

Sackeim, H. A., Rush, A. J., George, M. S., Marangell, L. B., Husain, M. M., Nahas, Z., … Goodman, R. R. (2001b). Vagus nerve stimulation (VNS) for treatment-resistant depression: Efficacy, side effects, and predictors of outcome. *Neuropsychopharmacology, 25*(5), 713–728. https://doi.org/10.1016/S0893-133X(01)00271-8

Sackeim, H. A., Prudic, J., Nobler, M. S., Fitzsimons, L., Lisanby, S. H., Payne, N., … Devanand, D. P. (2008). Effects of pulse width and electrode placement on the efficacy and cognitive effects of electroconvulsive therapy. *Brain Stimulation, 1*(2), 71–83. https://doi.org/10.1016/j.brs.2008.03.001

Sahlem, G. L., McCall, W. V., Short, E. B., Rosenquist, P. B., Fox, J. B., Youssef, N. A., … Sackeim, H. A. (2020). A two-site, open-label, non-randomized trial comparing Focal Electrically-Administered Seizure Therapy (FEAST) and right unilateral ultrabrief pulse electroconvulsive therapy (RUL-UBP ECT). *Brain Stimulation, 13*(5), 1416–1425. https://doi.org/10.1016/j.brs.2020.07.015

Sanacora, G., Mason, G. F., Rothman, D. L., Hyder, F., Ciarcia, J. J., Ostroff, R. B., … Krystal, J. H. (2003). Increased cortical GABA concentrations in depressed patients receiving ECT. *American Journal of Psychiatry, 160*(3), 577–579. https://doi.org/10.1176/appi.ajp.160.3.577

Scalia, J., Lisanby, S. H., Dwork, A. J., Johnson, J. E., Bernhardt, E. R., Arango, V., & McCall, W. V. (2007). Neuropathologic examination after 91 ECT treatments in a 92-year-old woman with late-onset depression. *The Journal of ECT, 23*(2), 96–98. https://doi.org/10.1097/YCT.0b013e31804bb99d

Schlaepfer, T. E., Frick, C., Zobel, A., Maier, W., Heuser, I., Bajbouj, M., … Hasdemir, M. (2008). Vagus nerve stimulation for depression: Efficacy and safety in a European study. *Psychological Medicine, 38*(5), 651–661. https://doi.org/10.1017/S0033291707001924

Shekelle, P. G., Cook, I. A., Miake-Lye, I. M., Booth, M. S., Beroes, J. M., & Mak, S. (2018). Benefits and harms of cranial electrical stimulation for chronic painful conditions, depression, anxiety, and insomnia: A systematic review. *Annals of Internal Medicine, 168*(6), 414–421. https://doi.org/10.7326/M17-1970

Shiozawa, P., da Silva, M. E., Netto, G. T., Taiar, I., & Cordeiro, Q. (2015). Effect of a 10-day trigeminal nerve stimulation (TNS) protocol for treating major depressive disorder: A phase II, sham-controlled, randomized clinical trial. *Epilepsy & Behavior, 44*, 23–26. https://doi.org/10.1016/j.yebeh.2014.12.024

Shouse, M. N., & Lubar, J. F. (1979). Operant conditioning of EEG rhythms and ritalin in the treatment of hyperkinesis. *Biofeedback and Self-Regulation, 4*(4), 299–312.

Sitaram, R., et al. (2017). Closed-loop brain training: The science of neurofeedback. *Nature Reviews. Neuroscience, 18*, 86–100.

Speer, A. M., Kimbrell, T. A., Wassermann, E. M., Repella, J. D., Willis, M. W., Herscovitch, P., & Post, R. M. (2000). Opposite effects of high and low frequency rTMS on regional brain activity in depressed patients. *Biological Psychiatry, 48*(12), 1133–1141. https://doi.org/10.1016/S0006-3223(00)01065-9

Sterman, M. B., & Friar, L. (1972). Suppression of seizures in an epileptic following sensorimotor EEG feedback training. *Electroencephalography and Clinical Neurophysiology, 33*(1), 89–95.

Straaso, B., Lauritzen, L., Lunde, M., Vinberg, M., Lindberg, L., Larsen, E. R., … Bech, P. (2014). Dose-remission of pulsating electromagnetic fields as augmentation in therapy-resistant depression: A randomized, double-blind controlled study. *Acta Neuropsychiatr, 26*(5), 272–279. https://doi.org/10.1017/neu.2014.5

Strafella, A. P., Paus, T., Barrett, J., & Dagher, A. (2001). Repetitive transcranial magnetic stimulation of the human prefrontal cortex induces dopamine release in the caudate nucleus. *The Journal of Neuroscience, 21*(15), RC157.

Sun, Y., Blumberger, D. M., Mulsant, B. H., Rajji, T. K., Fitzgerald, P. B., Barr, M. S., … Daskalakis, Z. J. (2018). Magnetic seizure therapy reduces suicidal ideation and produces neuroplasticity in treatment-resistant depression. *Translational Psychiatry, 8*(1), 253. https://doi.org/10.1038/s41398-018-0302-8

Terman, M., Terman, J. S., Quitkin, F. M., McGrath, P. J., Stewart, J. W., & Rafferty, B. (1989). Light therapy for seasonal affective disorder. A review of efficacy. *Neuropsychopharmacology, 2*(1), 1–22. https://doi.org/10.1016/0893-133x(89)90002-x

Terney, D., Chaieb, L., Moliadze, V., Antal, A., & Paulus, W. (2008). Increasing human brain excitability by transcranial high-frequency random noise stimulation. *The Journal of Neuroscience, 28*(52), 14147–14155. https://doi.org/10.1523/JNEUROSCI.4248-08.2008

Valiengo, L., Goerigk, S., Gordon, P. C., Padberg, F., Serpa, M. H., Koebe, S., … Brunoni, A. R. (2019). Efficacy and safety of transcranial direct current stimulation for treating negative symptoms in schizophrenia: A randomized clinical trial. *JAMA Psychiatry, 77*, 121. https://doi.org/10.1001/jamapsychiatry.2019.3199

Van Doren, J., Arns, M., Heinrich, H., Vollebregt, M. A., Strehl, U., & Loo, S. K. (2018). Sustained effects of neurofeedback in ADHD: A systematic review and meta-analysis. *European Child & Adolescent*

Psychiatry, 28, 293. https://doi.org/10.1007/s00787-018-1121-4

Vicheva, P., Butler, M., & Shotbolt, P. (2020). Deep brain stimulation for obsessive-compulsive disorder: A systematic review of randomised controlled trials. *Neuroscience and Biobehavioral Reviews, 109*, 129–138. https://doi.org/10.1016/j.neubiorev.2020.01.007

Wang, H., Wang, X., & Scheich, H. (1996). LTD and LTP induced by transcranial magnetic stimulation in auditory cortex. *Neuroreport, 7*(2), 521–525.

Wang, H. N., Wang, X. X., Zhang, R. G., Wang, Y., Cai, M., Zhang, Y. H., ... Zhang, Z. J. (2017). Clustered repetitive transcranial magnetic stimulation for the prevention of depressive relapse/recurrence: A randomized controlled trial. *Translational Psychiatry, 7*(12), 1292. https://doi.org/10.1038/s41398-017-0001-x

Wassermann, E. M. (1998). Risk and safety of repetitive transcranial magnetic stimulation: Report and suggested guidelines from the International Workshop on the Safety of Repetitive Transcranial Magnetic Stimulation, June 5–7, 1996. *Electroencephalography and Clinical Neurophysiology/Evoked Potentials*

Section, 108(1), 1–16. https://doi.org/10.1016/S0168-5597(97)00096-8

Watanabe, T., Sasaki, Y., Shibata, K., & Kawato, M. (2017). Advances in fMRI real-time neurofeedback. *Trends in Cognitive Sciences, 21*, 997–1010.

Weigand, A., Horn, A., Caballero, R., Cooke, D., Stern, A. P., Taylor, S. F., ... Fox, M. D. (2018). Prospective validation that subgenual connectivity predicts antidepressant efficacy of transcranial magnetic stimulation sites. *Biological Psychiatry, 84*(1), 28–37. https://doi.org/10.1016/j.biopsych.2017.10.028

White, P. F., Amos, Q., Zhang, Y., Stool, L., Husain, M. M., Thornton, L., ... Lisanby, S. H. (2006). Anesthetic considerations for magnetic seizure therapy: A novel therapy for severe depression. *Anesthesia & Analgesia, 103*(1), 76–80. https://doi.org/10.1213/01.ane.0000221182.71648.a3

Yan, T., Xie, Q., Zheng, Z., Zou, K., & Wang, L. (2017). Different frequency repetitive transcranial magnetic stimulation (rTMS) for posttraumatic stress disorder (PTSD): A systematic review and meta-analysis. *Journal of Psychiatric Research, 89*, 125–135. https://doi.org/10.1016/j.jpsychires.2017.02.021

Jerome Sarris, Patricia L. Gerbarg, Richard P. Brown, and Philip R. Muskin

Contents

J. Sarris (✉)
NICM Health Research Institute, Western Sydney
University, Westmead, NSW, Australia
e-mail: J.Sarris@westernsydney.edu.au

P. L. Gerbarg
Department of Psychiatry, New York Medical College,
Valhalla, NY, USA

R. P. Brown
Department of Psychiatry, Columbia University Vagelos
College of Physicians and Surgeons, New York, NY, USA

P. R. Muskin
Department of Psychiatry, Columbia University Irving
Medical Center, New York, NY, USA
e-mail: prm1@cumc.columbia.edu; PRM1@columbia.edu

© Springer Nature Switzerland AG 2024
A. Tasman et al. (eds.), *Tasman's Psychiatry*,
https://doi.org/10.1007/978-3-030-51366-5_145

Abstract

This chapter focuses on Integrative and Complementary Medicine (ICM) treatments with plausible mechanisms of action, supported by evidence of safety and efficacy in clinical trials. The key mental illness disorders covered included: mood, anxiety, stress-related, trauma, sleep, psychotic disorders, attention and hyperactivity disorder, and neurocognitive disorders. The major interventions covered include herbal and nutritional medicines, mind-body approaches (e.g., yoga, tai chi, qigong, breathwork), and acupuncture. All of the interventions discussed have relatively low side-effect profiles and high margins of safety, particularly when compared to prescription medications. There is reason to believe that certain herbal and nutraceutical agents exert significant neurobiological effects, in addition to supporting healthier functioning of the nervous system. Although some valuable ICM treatments have not been generally used by physicians in the United States or approved for specific indications by the FDA, based on the available evidence of safety and efficacy, they are well worth consideration by clinicians who want to explore more options for improving their patients' health and wellbeing. Clinicians need to be mindful of differences in quality and standardization among supplements produced by different companies. Further, in clinical practice, it may require several trials to design effective integrative treatments. A well-informed clinician is best qualified to help patients and families develop a safe, evidence-based integrative treatment plan.

Keywords

Integrative medicine · Nutrition · Nutraceutical · Herbal medicine · Breathwork · Mind-body medicine

Introduction

Integrative and Complementary Medicine (ICM), also known as Complementary, Alternative, and Integrative Medicine (CAIM), contributes to patient care and modern research. Clinicians and healthcare consumers are concerned about the high costs and adverse side effects of pharmaceuticals, as well as nonresponse or suboptimal response to prescription medications and other conventional treatments. Estimates of ICM use among psychiatric patient populations range from 44% to 63% (Massoumi, 2017). Within psychiatry, ICM supports treatment approaches that integrate the evidence-based use of medicinal herbs, nutrients, mind-body medicine, and other non-mainstream modalities with psychological therapies and pharmacotherapy. In addition, ICM expresses a preventative philosophy, incorporating optimal nutrition, exercise, mind-body medicine, and lifestyle modification (Muskin, 2000).

The Evidence: Quality and Quantity

Advances in neuroscience, immunology, molecular biology, and genomics enrich our understanding of how ICM can be used to modulate the neurophysiological processes that underly psychiatric disorders. The quantity of research on ICM is increasing exponentially. Assessing the quality of the research, particularly for mind-body interventions, such as yoga, qigong, tai chi, and meditation, is challenging because, unlike pharmacological studies, it is impossible to use an identical placebo control or to attain subject blinding. Consequently, research reviews and meta-analyses tend to downgrade the quality of ICM research, undermining confidence in the findings. The Consolidated Standards of Reporting Trials (CONSORT) Statement revised its recommendations with an extension intended to provide guidelines to improve the quality of reporting for nonpharmacological studies (Boutron et al., 2017). Although ICM research continues to vary in quality, more and more studies are adhering to higher quality standards, for example, better descriptions of subject selection, randomization, and intervention details. Nevertheless, to date, reviewers show little indication of taking into account the differences between mind-body research and drug trials when they rate the quality of studies.

Clinical Issues

ICM is particularly useful for patients who have incomplete responses or undesirable side effects with prescription medications. In general, the side effects of ICM treatments are far less frequent and less severe than those associated with prescription pharmaceuticals. Another concern has been the potential for harmful contaminants in herbal preparations. Improvements in regulation and product testing have reduced but not completely removed this risk. Surveillance by the US Food and Drug Administration (FDA) and notifications of adverse

events are increasing the overall safety of supplements. For more details, see Box 1 and Blumenthal's discussion of the regulation, quality, and safety of herbal medicines (Blumenthal, 2017).

The majority of patients using ICM also take prescription medications. Knowledge about potential herb-drug interactions is important. Although there is considerable data from *in vitro* testing herb-drug interactions, more studies in humans are needed to assess the actual risks. Fortunately, most herbs and nutrients used in psychiatric practice do not have clinically significant interactions with pharmaceuticals, even if in vitro studies find interactions, for example, with P450 enzymes. When most herbal extracts undergo digestion and metabolism, the resulting secondary metabolites do not show adverse interactions with prescription medications. It is important to understand the potential adverse effects of each of the component herbs when using supplements containing mixtures of herbs, particularly Oriental formulas. A review of the adverse-event data of Japanese Kampo formulations, which utilize many of the same herbs as Chinese medicines, identified the following herbs as responsible for most of the problems: Scutellariae radix, Glycyrrhizae Radix, Gardeniae Fructus, and Ephedrae Herba (Shimada et al., 2019). More information is provided on the risks of herb-drug interactions in the section on Mood Disorders below, Tables 1 and 2, and within the discussion of each treatment covered in this chapter.

Liability Issues

Liability concerns have prevented many doctors from discussing ICM with their patients. Taking the time to educate the patient and obtain informed consent helps minimize liability risks (Gerbarg et al., 2017b). For informed consent, the physician should clearly explain why an alternative treatment is being recommended, what is known about how it works, and its potential risks and benefits. This information should be weighed against the risks and benefits of standard treatment options. The discussion and the rationale for the treatment approach should be documented in the chart, including symptoms that have not

Table 1 Treatment guidelines: Herbs and nutrients for mood disorders

ICM	Clinical uses	Doses	Side effects and drug interactions
St. John's wort (*Hypericum perforatum*)	Depression	300–600 mg TID (depending on standardization of hyperforin or hypericin)	Nausea, heartburn, loose bowels, jitteriness, insomnia, fatigue, bruxism, phototoxic rash, mania in bipolar. Affects CYP 450 and Pgp: ↓ digoxin, warfarin, indivir, cyclosporine, theophylline, birth control pills. D/C: surgery, pregnancy
Saffron (*Crocus sativus*)	Depression	20–30 mg/day	Mild stimulation or anxiety, dyspepsia if taken on empty stomach. At high doses slight anticoagulant effect
S-Adenosyl-L-methionine (SAMe)	Depression Cognitive dysfunction	400–3200 mg/day	Mild nausea, loose bowels, activation, anxiety, mania in bipolar, headache, occasional palpitations
Arctic root (*Rhodiola rosea*)	Depression with fatigue	150–900 mg/day	Agitation, insomnia, anxiety, headache, palpitations, chest pain
Crocus sativa	Depression	30 mg/day	Anxiety, tachycardia, nausea, dyspepsia, changes in appetite
B-vitamins	Depression	B_{12} 1000 µg/day B-complex	Rare: activation
Inositol	Panic disorder	12–20 g/day	Gas, loose bowels, mania
Omega-3 fatty acids (EPA)	Depression (unipolar or bipolar)	1–2 g/day (EPA)	Reflux, loose stools
Choline	Mania	2000–7200 mg/day	Excess doses: muscle twitches, GI upset

GI, gastrointestinal side effects; Pgp, P-glycoprotein

responded to standard treatment attempts, all of the treatment options covered in the discussion, and the specific indications and considerations involved in the decision to use particular herbs, nutrients, or mind-body treatments. This may include the patient's philosophy, cultural background, or the fact that they have had difficulty tolerating conventional agents.

> **Box 1: Quality of Natural Products**
> Clinicians and consumers are often concerned about the quality of herbal extracts and other supplements. Advances in biochemistry have enhanced the stability of many compounds and improved the assessment of the purity and concentration of active constituents. Nevertheless, researchers, clinicians, and the public have ongoing concerns about both quality of products (standardization, bioactivity, raw material, and manufacturing standards), the risk of adulterants, and the rigor of clinical studies. The use

of high-quality raw materials, pharmaceutical-grade manufacturing, further rigorous research, and reliance on substantiated claims, increases confidence in using ICM (Sarris, 2012). Data on adulteration and fraud, collected by the Botanical Adulterants Program is used to advise industry laboratories on best detection methods (Blumenthal, 2017). To navigate the abundance of available products and to identify those of high quality, physicians need to stay current using unbiased sources of evidence-based product information. For example, the presence of a Good Manufacturing Practices (GMP) seal on supplements indicates adherence to manufacturing standards. Independent evaluation of many brands with updates can be found at www.consumerlab.com or www.supplementwatch.com. Guides to finding quality supplements are provided in *How to*

(continued)

Table 2 Treatment guidelines: Herbs and nutrients for anxiety, stress- and trauma-related disorders, and insomnia

ICM	Clinical uses	Dose	Side effects and drug interactions
Kava (*Piper methysticum*)	Anxiety, insomnia	60–120 mg kavalactones BID	GI complaints, skin dryness/scaliness (from high chronic consumption), headache, photosensitivity occasional: ↓ energy, drowsiness, tremor, restlessness, ↓ effects of levodopa, hepatitis, potential very rare liver issues, depression. D/C: pregnancy
Passionflower (*Passiflora incarnata*)	Anxiety	1–2 g TID	Generally safe
Galphimia (*Galphimia glauca*)	Anxiety	3.5–7 g BID	No serious adverse effects noted. Mild side effects: tiredness, nausea, GI complaints, headache.
Chamomile (*Matricaria recutita*)	Anxiety	1–3 g TID	Ragweed family – allergic reactions. D/C: pregnancy
Lemon balm (*Melissa officinalis*)	Anxiety	1–2 g TID	No serious side effects. Mild GI complaints.
Valerian (*Valeriana officinalis*)	Chronic insomnia	450–900 mg HS	Occasional GI complaints, headaches, minimal hangover on high doses >600 mg. D/C: pregnancy, hepatic disease
Melatonin	Sleep	1–12 mg HS	Occasional agitation, abdominal cramps, fatigue, dizziness, headache, vivid dreams. D/C: pregnancy
N-Acetylcysteine, NAC	Obsessive-compulsive Disorder, Hair-pulling	1200–2400 mg/day	Occasional GI complaints, Heartburn

Box 1: Quality of Natural Products (continued) *Use Herbs, Nutrients, and Yoga in Mental Health Care* (Brown et al., 2009) and *Herbal Treatment of Major Depression* (Mendelson, 2020). Information about product labelling, manufacturing guidelines, safety, potential adverse effects, adulteration, toxicity, recalls and warnings is available at https://www.fda.gov/food/dietary-supplements, http://www.fda.gov/medwatch, and www.fda.gov/Food/RecallsOutbreaksEmergencies/SafetyAlertsAdvisories/default.htm.

Diagnostic and Statistical Manual (DSM-5) Categories

This chapter is divided into the following sections in accord with Diagnostic and Statistical Manual-5 (DSM-5): Mood Disorders (including major depressive disorder (MDD) and bipolar disorder) and Herb-Drug Interactions; Anxiety Disorders (e.g., generalized anxiety disorder (GAD), panic disorder, and obsessive-compulsive disorder (OCD)) and Trauma- and Stressor-related Disorders (e.g., posttraumatic stress disorder (PTSD)); Sleep-Wake Disorders; Schizophrenia Spectrum and other Psychotic Disorders; Neurocognitive Disorders (NCDs) (e.g., Alzheimer's disease (AD), vascular dementia, Parkinson's disease (PD), mild NCDs); and Attention-deficit/hyperactivity Disorder (ADHD) (American Psychiatric Association, 2015).

Each section focuses on treatments that are useful in psychiatric practice and that are supported by clinical studies demonstrating safety and efficacy: herbal medicines, nutrients, mind-body medicine (e.g., paced breathing, yoga, tai chi, qigong, and meditation), acupuncture, neurotherapy, and cranial electrotherapy stimulation. Evidence regarding plausible mechanism of action is also discussed.

Tables 1, 2, and 3 show treatment guidelines for the herbs and nutrients described in each section. The term "significant" denotes a statistical significance of $p < 0.05$. For more detailed information on the ICMs discussed in this chapter, as well as additional treatments not covered due to space constraints, see Complementary and Integrative Treatments in Psychiatric Practice published by the American Psychiatric Association Publishing (Gerberg et al., 2017b).

Rigorous controlled studies have been completed for many ICM treatments, but the evidence is modest or of low quality for others. Nevertheless, this should not deter physicians interested in exploring the widest range of therapeutic possibilities from considering ICMs for appropriate patients, especially for those who have not achieved full remission using conventional treatments. Unfortunately, many potentially worthwhile

Table 3 Treatment guidelines for neurocognitive disorders – herbs and nutrients

Nutraceutical	Clinical uses	Dose	Side effects and drug interactions
Acetyl-L-carnitine	AD: slowed progression. TBI and CVA	1500 mg BID	Mild gastric upset. Take with food
B-vitamins	Cognitive enhancement, TBI	B-complex	None
S-Adenosyl-L-methionine (SAMe)	AD, dementia, TBI PD	800–1600 mg/day 400–4000 mg/day	Mild occasional GI, agitation, anxiety, insomnia; rare palpitations. Mania in bipolar disorder. Take 30 min before breakfast and lunch
Arctic root (*Rhodiola rosea*)	Cognitive enhancement, memory, TBI	150–600 mg/day	Activation, agitation, insomnia, jitteriness, mania. Rare: ↑ BP, angina, bruising. Avoid in bipolar I. Take 20 min before breakfast and lunch.
Ginkgo (*Ginkgo biloba*)	AAMI, MCI, AD, CVD	120–240 mg/day	Minimal, headache, ↓ platelet aggregation. D/C: prior to surgery
Korean ginseng (*Panax ginseng*)	Dementia, neurasthenia	400–800 mg/day	Activation GI, anxiety, insomnia, headache, tachycardia, ↓ platelet aggregation
Ba Wei Di Huang Wan/Hachimi-Jio-Gan Rhemannia glutinosa Libosh, Cornus officinalis Sieb et Zucc, Dioscorea batatas Dence, Alisma orientale Juzep, Poria cocos Wolf, Paeonia suffruticosa Andr., Cinnamomum cassia Blume, Aconitum carmichaeli Debx.	Dementia ADL, cognitive	2 g (20 pills) TID	GI symptoms, Rash
Yi-Gan San/ Yokukansan *Atractylodes lancea, Poria, Cnidium rhizome, Uncaria uncus, Angelica acutiloba radix, Bupleuri radix, and Glycyrrhiza radix*	Dementia ADL, cognitive	7.5 g BID	*Glycyrrhiza* – Pseudoaldosteronism, HT, hypokalaemia, heart failure, arrhythmia, liver dysfunction, pneumonia, rhambdomyolysis (muscle injury)
Yokukansankachimpihange YKS, Citrus unshiu peel, Pinellia tuber	Dementia Aggression, sleep	7.5 g BID	*Glycyrrhiza* – Pseudoaldosteronism, HT, hypokalaemia, heart failure, arrhythmia, liver dysfunction, pneumonia, rhambdomyolysis (muscle injury)

AD, Alzheimer's disease; ADL, Activities of Daily Living; BID, twice a day; CVD, cerebrovascular disease; AAMI, Age Associated Memory Impairment; MCI, Mild Cognitive Impairment; BP, blood pressure; GI, gastrointestinal side effects; HT, hypertension; PD, Parkinson's disease; TBI, traumatic brain injury; CVA, cerebrovascular accident; YKS, Yokukansan natural treatments have not attracted the large research investment necessary to obtain US FDA approval for use in specific disorders, even though they are regulated and approved for sale as general supplements. As more clinicians discover the potential benefits of ICM, interest in supporting larger controlled studies should continue to develop.

The Importance of the Doctor-Patient Relationship

Aspects of the importance of the doctor-patient relationship are considered in Box 2.

> **Box 2: The Importance of the Doctor-Patient Relationship**
> Communication between a patient and their physician is the cornerstone of all therapeutic interactions. It is particularly important in light of potential interactions between ICM treatments, which can be either advantageous or adverse. Surveys indicate that only 25–70% of patients inform their physicians about ICM use. Furthermore, patients often use ICM without consulting a qualified ICM practitioner (Wang et al., 2020). Asking a cashier at the health-food store may be the extent of their "consultation." Some patients are, however, well informed about ICM from their own research, or from having consulted an experienced herbalist or other ICM expert. We are practicing in a time of rapidly growing ICM use, yet a minority of physicians are trained to prescribe phytomedicines, nutrients, mind-body medicine, or other ICM therapies. Although many physicians are scrambling to meet this challenge, others are ignoring the need to gain competence in this emerging field. Ignorance and/or insecurity about ICM treatments may contribute to a potential countertransference enactment in which the physician discounts *all* ICM approaches as unfounded or misinterprets the patient's interest in ICM as a form of resistance (Kenny et al., 2001). This can

create a situation in which the patient does not tell the doctor about their use of ICM. Patients are more likely to share information with a physician who is open-minded and nonjudgmental. Open dialogue between patient and physician helps prevent adverse outcomes (Varteresian and Lavretsky, 2018).

A Place for Integrative and Complementary Medicine in Global Mental Health

The emerging field of ICM or *Integrative Mental Health* (IMH) seeks to unite conventional biomedical and psychological treatments with evidence based ICM. IMH adopts the bio-psycho-socio-spiritual model, including mind-body medicine and health-promoting lifestyle modifications within different cultures (Lake et al., 2012; Sarris et al., 2013a). Many of the medicines used today are derived from botanicals known to herbalists for centuries. Similarly, mind-body therapies derive from ancient spiritual and tribal practices. The contributions of diverse cultures and indigenous peoples should be honored.

Evidence suggests that mind-body interventions can positively impact public health because, compared to conventional treatments, they are inexpensive, accessible, low in adverse effects, non-stigmatizing, cross-culturally accepted, and can be provided as group interventions in-person, online, or by tele-medicine to remote and underserved locations (Bhargav, 2020; Gerberg & Brown, 2021; Gerberg et al., 2021; Lavretsky & Feldman, 2021).

Mood Disorders

The Sequential Treatment Alternatives to Relieve Depression (STAR*D) trial highlighted the benefits and limitations of combining or switching antidepressants (medications for depression) (Rush et al., 2006). Pharmacological approaches alone are often insufficient for remission of depression, leaving patients struggling with residual symptoms and at higher risk for relapse. Although prescription antidepressants are highly effective in many cases, they are still associated with high levels of treatment resistance, incomplete response, loss of effectiveness, withdrawal syndromes, acute and long-term side effects, and noncompliance due to side effects. ICM provides therapeutic options that may work through different mechanisms as either adjunctive or stand-alone treatments. Among ICM treatments, stronger evidence of antidepressant activity supports the use of St. John's wort (SJW), *S*-adenosylmethionine (SAMe), saffron, Arctic root, and omega-3 fatty acids (omega-3 FAs). The evidence for mind-body treatments as adjuncts to standard care for depression is growing. Because most of the studies reporting benefits using mindfulness meditation, yoga, and acupuncture are of varying methodological quality, additional randomized controlled trials (RCTs) are needed.

Herbal Medicines

Herb-Drug Interactions

Among the herbs discussed in this chapter, very few have significant potential for adverse herb–drug interactions (HDIs); however, herbs (such as SJW) that affect cytochrome P450 (CYP) drug metabolizing enzymes or the Permeability-glycoprotein (Pgp) transmembrane pump (increased expression of Pgp, which impacts Pgp substrates) which can alter serum levels of numerous medications (Cott, 2002). Case reports of significant changes in heart rate and blood pressure in patients taking SJW, ginkgo, and Korean ginseng have led to recommendations that patients stop phytomedicines 2–3 weeks before surgery (Voelker, 1999). Herbs that can affect bleeding time (e.g., ginkgo) should not be used with warfarin and should be stopped 10 days to 3 weeks prior to surgery. The use of SJW with medications that have significant action on the serotonin system (SSRIs or MAOIs) should be avoided because of limited testing for safety. On the other hand, many herbal extracts can be used to enhance the benefits of psychotropics or to reduce medication side effects such as fatigue or SSRI-induced sexual dysfunction (Schweitzer et al., 2009) (see Table 2). Valerian and kava should be used cautiously with other sedating medications, such as benzodiazepines, because of the potential for additive sedative effects.

Although there have been case reports of increased bleeding with ginkgo, in most cases, the patients were taking numerous medications concurrently, including anticoagulants. A review by Diamond and Bailey noted that an analysis of RCTs of adults with dementia, peripheral artery disease, or diabetes mellitus show positive effects of standardized ginkgo extract on perfusion via reduced blood viscosity, but no significant effects on adenosine $5'$-diphosphate-induced platelet aggregation, fibrinogen aggregation, activated partial thromboplastin time, or prothrombin time, indicating no increase in risks for bleeding (Diamond & Bailey, 2013). A more conservative approach would suggest that ginkgo might increase the risk of bleeding and therefore should not be taken before surgery or be used by anyone who is taking anticoagulants, including heparin, coumadin, or other medications that interfere with normal blood clotting (Sarris, 2013a).

Ginseng may affect blood glucose levels and should be stopped prior to surgery. Herbal laxatives and licorice, by depleting potassium, may affect the therapeutic action of digoxin, beta-blockers, and diuretics (for more extensive information on interactions between natural products and prescription medications, see Gerberg & Brown, 2013). When patients taking standard antidepressants are given activating compounds such as Arctic root, Korean ginseng, ginkgo, or SAMe, they may experience overstimulation and/or induction of hypomania in rare cases. Nevertheless, these agents are extremely well tolerated overall (for an approach to risk management of herb-drug interactions, see Brown et al., 2009; Gerberg et al., 2017b).

Hypericum perforatum (St. John's wort, SJW)

SJW became popular in the United State after publication of a 1996 meta-analysis of 23 RCTs for depression (Linde et al., 1996), 13 of which showed that 55% of patients on SJW improved compared with 22% on placebo, with a side-effect rate similar to that of placebo. However, the studies were limited by vague definitions of depression, less rigorous methodology, unclear outcome measures, low doses of the comparison antidepressant, and short length of trials. A more recent meta-analysis of 27 randomized placebo-controlled trials (RPCTs) (total $n = 3808$) comparing SJW versus placebo revealed a significant reduction of Hamilton Depression Inventory (HAM-D) scores in favor of SJW (pooled SMD -0.068, 95% CI -0.127 to 0.021, ns) (Ng et al., 2017). The analysis also showed efficacy equivalent to antidepressants in reducing depressive symptoms (pooled RR 0.983, 95% CI 0.924–1.042, ns) and in remission rate (pooled RR 1.013, 95% CI 0.892–1.134, ns), with a significantly lower discontinuation/dropout rate (pooled OR 0.587, 95% CI 0.478–0.697) compared to standard SSRIs. However, not all studies have been positive, including the much-publicized 2002 National Institute of Mental Health (NIMH)-funded trial by the Hypericum Depression Trial Study Group (2002). This three-arm, double-blind, randomized controlled trial (DBRCT), which involved 340 adults with MDD, revealed that at week 8 endpoint, on the primary outcome measure, neither SJW nor sertraline were significantly different from placebo in reducing HAM-D scores. A more recent analysis of the 2002 study found that at the conclusion of the follow-up part of the study (week 26), mean HAM-D scores for completers were sertraline 7.1, SJW 6.6, and placebo 5.7, with no significant difference among treatment arms (Sarris et al., 2012a). As in so many recent depression studies, the negative findings may not be due to inefficacy of SJW or antidepressant medications, but more to the marked "placebo response" that can occur when participants entering studies have milder non-biological depression of "psychosocial causation" (Sarris et al., 2013c).

It is a common false belief among clinicians that SJW should not be used to treat severe depression. Contrary to that belief, a 6-week DBRCT ($n = 251$) in moderate-severe depression

(HAM-D >22) showed SJW extract WS 5570 (900 mg/day) as not inferior to placebo and as statistically superior to paroxetine (20 mg/day) (Szegedi et al., 2005). Interestingly, the SJW group had a 14.4-point decrease on HAM-D compared with an 11.4-point decrease for paroxetine. In a 6-week DBRCT study comparing 1800 mg/day of SJW LI 160 extract to 150 mg/day of imipramine in severe depression, there was no statistical difference in HAM-D scores in the severely depressed patients (Vorbach et al., 1997). Although these clinical trials support using SJW in non-suicidal patients with more severe depression, it is currently not recommended as a first-line treatment in cases of serious mental illness or where significant suicidal ideation is present. Some authors have suggested that SJW is effective in 50–70% of patients with mild depression, particularly with wintertime seasonal affective disorder (SAD).

The side-effect profile of SJW at the higher doses used for severe depression may be similar to that of selective serotonin reuptake inhibitors (SSRIs). Common SJW side effects are nausea, heartburn, loose bowels, jitteriness, insomnia, and fatigue. High doses may cause sexual dysfunction and bruxism. Phototoxic rash occurs in less than 1% of people taking the usual dose (900 mg/day) but may be more frequent at higher doses. Patients should be advised to protect against excess sun exposure. Four possible cases of SJW-induced mania in patients with bipolar disorder were reported (Nierenberg et al., 1999). SJW may provide reuptake inhibition of serotonin (5-HT), norepinephrine (NE), and dopamine (DA) (Muller et al., 1998), and cause a change in monocyte cytokine production of interleukin-6 (leading to decreased corticotrophin-releasing hormone) (Thiele et al., 1994). There have been six cases of serotonin syndrome reported with SJW in combination with both SSRIs and nefazodone (Gelenberg, 2000; Medical Letter, 2000).

A review of adverse effects of SJW from 35 DBRCTs found that the reported rates of SJW side effects were similar to those of placebo and lower than those of comparison antidepressants (Knüppel & Linde, 2004). Evidence suggests that SJW extracts are well tolerated and safe if taken under the guidance of a physician who is aware of the potential risks, clearly informs the patient, and monitors the patient for changes in medications. SJW has the potential for serious interactions with a range of medications. Although concerns exist over interactions between SJW and pharmaceuticals, this involves extracts containing higher amounts of hyperforin, which is responsible for inducing CYP P450 pathways and the P-glycoprotein drug efflux pump; thereby reducing drug serum levels (Izzo, 2004). To clarify this further, a systematic review of 19 studies revealed that high-dose hyperforin extracts (>10 mg/day) had outcomes consistent with CYP3A induction, while studies using low-dose hyperforin extracts (<4 mg/day) showed no significant effect on CYP3A (Whitten et al., 2006). Breakthrough bleeding has been reported in women on birth control pills who take SJW, hence caution is advised. (CYP3A4 is required for metabolism of estrogen in birth control formulations. Induction of this enzyme may reduce estrogen levels and cause bleeding or undesired pregnancy.) Anesthesiologists anecdotally report changes in heart rate and blood pressure, particularly in patients taking SJW (in addition to ginkgo and Korean ginseng). These herbs should be discontinued 2–3 weeks before surgery (Voelker, 1999).

Two DBRPC studies suggest that clinical response rates and changes in brain waves are related to the concentration of "hyperforin," not "hypericin" (Muller et al., 1998; Schellenberg et al., 1998). Most independent analyses of SJW brands report the amount of hypericin and/or hyperforin (which often are variable), but do not test for quality or activity. Although a concentration of hypericin of 0.3% of the extract is the usual standard, the measurement of hyperforin 0.3–0.5% may be more relevant. Independent testing of SJW brands by ConsumerLab.com found that 14 out of 21 met their requirements for having the amount of active compound claimed on product labels and a minimal amount of cadmium (a carcinogenic contaminant) (ConsumerLab.com, 2013). Another independent study by www.vitacost.com found that only two out of eight brands contained adequate hyperforin

(Schardt, 2000). DBRCTs and studies showing activity in vitro for particular products provide more compelling evidence of efficacy. Research trials have shown the following SJW preparations to be effective: Kira (LI 160), Remotiv (ZE 117), STW3, and WS 5570.

In summary, current evidence supports the use of high-grade, standardized SJW extracts for the treatment of mild to moderate depression. Although not reviewed in this chapter, there is currently a lack of evidence for in the use of SJW in anxiety disorders or other psychiatric illnesses (Sarris, 2013b). Clinicians should be mindful of potential drug interactions if using products with higher (>4 mg) hyperforin, and in co-administration with other psychotropic medications. Caution is advised regarding use in bipolar disorder for those whose mood swings are not well controlled by mood stabilizers (medications used for bipolar disorder).

Rhodiola rosea (Arctic Root)

The earliest recorded use of Rhodiola rosea was found in the Bronze Age Greek myth of Jason and the Argonauts (Brown & Garbarg, 2004). Arctic root has been used for centuries in Russian and Scandinavian folk medicine as a tonic, anti-fatigue, and anti-bacterial properties. It was used to increase strength, performance, endurance, and fertility by people living at high altitudes and under harsh, cold conditions. Starting from the 1960s, the former Soviet Union conducted extensive animal and human studies. Herbal formulas containing Arctic root, Eleuthero (also called Siberian ginseng, although it is not in the ginseng Family), and Schizandra were used to enhance physical and intellectual performance in military personnel, Olympic athletes, college students, cosmonauts on space missions, and scientists (Brown & Garbarg, 2004; Brown et al., 2002, 2009).

Extracts from Rhodiola rosea rhizomes (roots) contain bioactive alkaloids, polyphenols, and phenylpropanoids, including tyrosol and salidroside (Brown et al., 2002; Panossian et al., 2010). High-performance liquid chromatography (HPLC) found cinnamyl alcohol betavicianidines, rosavin, rosin, and rosarin to be unique to R. rosea and therefore usable as identifying marker compounds (Dubichev et al., 1991). Arctic root extracts should be standardized to a minimum of 1% salidroside and 3% rosavin and should be free of drying agents such as maltodextrin or other carriers. Unfortunately, using these two biomarkers is not sufficient to assure purity because they can now be synthesized and added to the extract concealing the use of less than full potency extracts. Clinicians can find higher quality products by using those proven to be effective in published clinical studies and those recommended by more experience ICM practitioners (Brown et al., 2009).

Arctic root is classed as an herbal adaptogen. Adaptogens are defined as substances that increase resistance against multiple stressors including biological, chemical, or physical insults. They also normalize physiology (whether a body parameter was too high or too low, the adaptogen would bring it towards normal) and should not perturb normal body functions more than necessary to improve resistance (Panossian & Wagner, 2005). DBRPC studies have shown that Rhodiola rosea enhances intellectual work capacity, abstract thinking, and reaction time in healthy subjects while reducing mental and physical fatigue (Shevtsov et al., 2003; Gerbarg & Brown, 2017c). There may be some risk of inducing mania in bipolar patients. Arctic root increases 5-hydroxytryptamine (5-HT), probably by increasing transport of tryptophan and 5-HT into the brain, and may reduce the breakdown of 5-HT by carboxy-O-methyltransferase (COMT) inhibition (Saratikov & Krasnov, 1987b).

Rhodiola rosea has been used to treat depression. An open study of 128 depressed patients given 150 mg TID of Arctic root or placebo showed that two-thirds of those on Arctic root improved significantly (Saratikov & Krasnov, 1987a, b). A DBRPC study of a standardized extract of Arctic root (SHR-5) as a monotherapy for 89 adults with mild to moderate depression (baseline HAM-D ranging from 12 to 31) found that those given Arctic root SHR-5 (340 or 680 mg/day) for 6 weeks showed significantly greater improvements in mean HAM-D compared with the placebo group (Darbinyan et al., 2007). A 12-week DBRPCT in mild to moderate MDD

($n = 57$) assessed the antidepressant effect of a standardized Arctic root extract (SHR-5) versus sertraline and placebo (Mao et al., 2016). The results revealed a non-significant difference among the three groups for depression and anxiety outcomes; a non-significant trend favoring sertraline was evident (although the side effects were far less with Artic root compared to the sertraline). As with many recent depression studies that fail to find significant differences between placebo and treatment groups, the subjects with mild depression probably included individuals with heterogeneous diagnoses who did not have a biologically based depression and who were less likely to show a robust response to antidepressant treatments. Arctic root may be a useful activating agent to augment the treatment of depression, particularly for target symptoms of fatigue, apathy, psychomotor retardation, anhedonia, memory impairment, poor concentration, and loss of libido (Gerbarg & Brown, 2017c). In clinical practice, authors Gerbarg and Brown find Rhodiola rosea particularly beneficial for depression in perimenopausal women with fatigue, "brain fog," and mild memory retrieval difficulties. The improvements in energy, mood, mental sharpness and memory may be related to activating effects, increased mitochondrial energy (Panossian & Amsterdam, 2017), and its potential as a selective estrogen receptor modulator (SERM) (Gerbarg & Brown, 2016a). The use of Rhodiola rosea as a monotherapy or augmentation for moderate to severe depression warrants further study.

Crocus sativus (Saffron)

Saffron is native to Western Asia. The stigma of the flower has been used in traditional medicine to treat mood disorders, muscular spasms, pain, menstrual disorders, and other conditions (Schmidt et al., 2007). Saffron stigmata contain an estimated 40–50 active therapeutic constituents, with high-quality saffron containing approximately 30% crocins, 5–15% picrocrocin, and over 5% volatile compounds including safranal (Schmidt et al., 2007). A meta-analysis revealed a significant reduction in depressive symptoms ($n = 14$, SMD = 0.99, 95% CI 0.23 to 0.74; $I^2 = 82\%$) in favor of saffron (Marx et al. 2019). The preparations studied were commonly standardized to crocin or other bioactive compounds, including safranal and crocin isomers. Trials were predominantly conducted in Iran with study lengths between 4 and 8 weeks, sample sizes ranging between 30 and 68 participants. Some studies of note include two RCTs using 30 mg/day of saffron demonstrated significant improvement of depression over placebo on the HAM-D (Akhondzadeh et al., 2005; Moshiri et al., 2006). The Akhondzadeh RCT using a stamen extract had a large effect size. The Moshiri study, which used the less expensive petal extract, revealed a similarly large effect size. Equivalent effects on HAM-D occurred in three RCTs comparing saffron with imipramine and fluoxetine (Akhondzadeh et al., 2004a, 2007; Noorbala et al., 2005). Clinical trials noted anxiety, tachycardia, nausea, dyspepsia, and changes in appetite as possible side effects (nonsignificant statistical trend against placebo) (Akhondzadeh et al., 2005; Moshiri et al., 2006). This conforms with traditional knowledge of adverse reactions (Schmidt et al., 2007). Although the results are encouraging, conclusions are limited by the short duration of trials (4–6 weeks) and small sample sizes ($n =$ 30–40). Replication of these findings by other research groups would further validate these promising findings. With respect to potential mechanisms of action, preclinical animal models using ethanolic extracts of saffron and its constituents, safranal and crocin, have shown antidepressant, anxiolytic, and hypnotic effects (Hosseinzadeh & Noraei, 2009). Crocin's antidepressant activity purportedly acts via reuptake inhibition of dopamine and norepinephrine, and safranal via serotonin reuptake inhibition (Schmidt et al., 2007).

Nutritional Medicines

S-Adenosyl-L-methionine (SAMe)

SAMe has been widely used in Europe for over 30 years, primarily for treatment of depression, arthritis, and liver disease. Tested in more than 80 clinical trials involving over 24,000 people, SAMe is a prescription medication in some countries and is sold over the counter in others. The US Department of Health and Human Services Agency for Healthcare Research and Quality (2002) concluded that treatment with SAMe was equivalent to standard pharmacotherapy for depression and osteoarthritis, but noted poor study design. SAMe is a physiologically essential molecule in all living cells.

A review by a Workgroup of the American Psychiatric Association Council on research, which included 17 open-label trials, 19 DBRPCs, and 21 trials comparing SAMe to prescription antidepressants, concluded the following (Sharma et al., 2017).

This review of the role of S-adenosylmethionine in the treatment of Major Depressive Disorder found encouraging evidence of efficacy and safety of SAMe as a monotherapy and as an augmentation for other antidepressants. Since the US FDA Agency for Healthcare Research and Quality (AHRQ) review, *S-Adenosyl-L-Methionine for Treatment of Depression, Osteoarthritis, and Liver Disease* (2002), additional studies have generally supported SAMe as efficacious for treatment of MDD and comparable to several prescription antidepressants, though comparisons against newer generation antidepressants are lacking and growing placebo response rates in antidepressant trials present an obstacle to our full understanding of the efficacy of SAMe. Evidence suggests that the greatest benefits may be in more severe MDD, and more systematic inquiry into this question may be of value. In addition to depression, this review found supportive early evidence for SAMe in certain neurocognitive, substance use, and psychotic disorders (see supplemental). Additional clinical studies are needed to further delineate the role of SAMe in neuropsychiatric conditions.

The human diet supplies only part of the body's needs for SAMe, generated by de novo synthesis from methionine and adenosine triphosphate (ATP); the liver is the largest producer (3 g/day). Oral SAMe supplementation is the easiest way to boost SAMe levels. Although most concentrated in the brain and liver, SAMe is an active metabolite in all tissues of the body and is crucial for three central metabolic pathways involving more than 200 different biochemical reactions. Transmethylation increases 5-HT, DA, and NE levels (Otero-Losada & Rubio, 1989a, b), probably contributing to antidepressant activity (Czyrak et al., 1992). Donation of methyl groups protect catecholamine neurons. SAMe improves nerve cell membrane uptake of phospholipids for a more fluid lipid bilayer, allowing the coupling of protein receptors to second messengers and enhancing transmission of impulses by neurons (Bottiglieri et. al., 2017).

SAMe was found to be a safe and effective treatment for depression in 16 uncontrolled trials (660 patients), 13 DBRPC trials (535 patients), and 19 positive-controlled trials, comparing it with other antidepressant medications including imipramine, amitriptyline, clomipramine, nomifensine, minaprine, and desipramine (1134 patients) (Brown et al., 2000), and three trials using SAMe to augment response to imipramine. All but one study showed that SAMe was more effective than placebo and equivalent to the comparison antidepressants for major depression. The study that did not find SAMe superior to placebo used an unstable form of SAMe and had a high placebo response rate (Fava et al., 1992).

SAMe was successfully combined with psychoactive medications in a 6-month study that included 350 patients on tricyclic antidepressants (TCAs)), 500 on benzodiazepines, 60 on monoamine oxidase inhibitors (MAOIs), 445 on anticonvulsants, and 18 alcoholic patients on antidepressants or anticonvulsants (Torta et al., 1988). No adverse effects were reported. Patients on SAMe/TCA combination responded in 7–10 days compared with the 10–15-day period for the TCA-only group. Forty-eight of the SAMe/imipramine patients improved on 30% lower doses of imipramine. In all cases, SAMe reversed or prevented the elevation of γ-glutamyltranspeptidase (GGT), an indicator of liver toxicity, that occurred in the patients taking MAOIs or anticonvulsants and in the alcoholic group (Torta et al., 1988).

After 2005, a change occurred in the methodology, outcomes, and reporting on SAMe for mild-to-moderate depression studies. For example, the conclusions of several DBRPC and 3-arm antidepressant comparison studies stated that SAMe failed to outperform placebo without mentioning that the comparator prescription antidepressant also did no better than placebo. The recurring problem of including individuals with milder forms of depression is discussed under Rhodiola rosea above. A major problem was that in study after study, the placebo response rates were 50% or greater, potentially invalidating the findings. Another concurrent change was that in the USA, companies that had been selling the 1,4 butanedisulfonate form of SAMe began selling the less expensive forms, tosylate and toluenate. Also, researchers in the USA and other countries began using these less expensive formulations in their depression studies. For example, a 12-week three-arm, two-site DBRCT ($n = 144$) using SAMe monotherapy (1600–3200 mg/day) versus SSRI escitalopram (10–20 mg/day) and placebo in adults with diagnosed MDD reported no significant difference in outcomes among the three groups (Mischoulon et al., 2014). Analysis of one site from this study revealed a significant difference between SAMe and placebo from baseline to week 12 (Sarris et al., 2014). In this one site, the remission rates (≤ 7 HAM-D) were 34% for SAMe, 23% for escitalopram, and 6% for placebo, significantly in favor of SAMe. The nutraceutical was found to be superior to placebo from week 1, and superior to escitalopram during weeks 2, 4, and 6. Although the possibility that differences in the proportion of male and female subjects between the sites affected outcomes, no conclusions could be drawn. However, the importance of patient selection in depression studies became evident. The patients for the parent study had been recruited between 2005 and 2009 (Sarris et al., 2014).

A 6-week double-blind RCT ($n = 73$) (Papakostas et al., 2010) involving MDD patients nonresponsive to SSRIs found response (HAM-D >50% reduction) and remission rates were significantly higher for those treated with adjunctive SAMe (46.1 and 35.8%, respectively) compared with adjunctive placebo (17.6 and 11.7%, respectively).

Since the review of SAMe by the APA Work Group (Sharma et al., 2017), the following DBRPC studies of SAMe for MDD found no significant difference between SAMe, placebo, or comparator antidepressant. However, these studies had such high placebo response rates that caution should be used in interpreting the results.

An 8-week, double-blind, randomized controlled trial tested 800 mg/day of SAMe monotherapy versus placebo in 49 patients with MDD (Montgomery-Åsberg Depression Rating Scale (MADRS)) score 14–25. Analyses showed that SAMe was effective in reducing depression amongst participants with milder depression (MADRS ≤ 22). Response was not moderated by brain-derived neurotrophic factor (BDNF), single-nucleotide polymorphisms, or one-carbon cycle biomarkers, although increased folate concentrations were correlated with improved symptoms in the SAMe group ($r = -0.57$). The treatment was safe and well tolerated. Despite showing effectiveness in reducing depression scores, no conclusions could be drawn because a high placebo response rate of 53% occurred (Sarris et al., 2020). Also, the dose of SAMe was subtherapeutic except for mild MDD patients. It is noteworthy that 800 mg twice daily is recommended for moderate to severe MDD, and a higher dose (1600 mg–3200 mg) is required for some patients (Sakurai et al., 2020).

An Australian multi-site, 8-week, DBRCT of 158 outpatients with MDD compared a nutraceutical combination: SAMe; Folinic acid; Omega-3 FAs; Zinc picolinate, 5-hydroxytryptophan (5-HTP), and relevant co-factors versus placebo. No significant differences emerged between groups. The finding of a placebo response rate was 51%, meaning that no conclusions can be drawn from this study (Sarris et al., 2019a).

A 6-week DBRPC antidepressant augmentation study found no statistically significant differences between MSI-195, a proprietary formulation of SAMe, purported to have greater bioavailability, versus placebo in 234 acutely depressed subjects with MDD who had inadequate response to ongoing antidepressant treatment (ADT). In the first half of the study, MSI-195 augmentation was significantly better than placebo on the HAM-D and MADRS, but not in the second half. Demographic and clinical characteristics were significantly different between subjects enrolled in the first and second halves of the study, including body mass index, pre-randomization symptom severity fluctuation, number of lifetime depressive episodes, and

anxious depression sub-type. These post-hoc findings highlight challenges in subject selection. Attention to enrollment criteria and moderating factors affecting treatment outcomes is needed in future MDD studies. The favorable safety profile and clinical benefit observed with MSI-195 in the first half of this study is notable (Targum et al., 2018).

Although the size and methodology of recent trials of SAMe for MDD have improved, questions about the clinical usefulness of these studies persist. First, the high placebo response rate has not been addressed. When a placebo response rate is 50% or higher, the validity of the results is questionable. Second, the use of SAMe formulations that may be less potent, that have less bioactivity, or that are administered in subtherapeutic doses could account for weak results. Until these issues are addressed, studies that find no difference in effects of SAMe versus placebo or antidepressant medications should be taken with a grain of salt (Bottiglieri et al., 2017). Many clinicians, including authors Brown, Gerbarg, and Muskin continue to find SAMe (1,4-butanedisulfonate, brand Azendus™) to be one of the safest and most effective treatments for depression and cognitive dysfunction in clinical practice (see section on NCDs below). DBRPC and antidepressant comparison studies of best quality SAMe 1,4-butanedisulfonate, recruitment of subjects from different populations (e.g., moderate-to-severe MDD), the use of biomarkers of biological depression, and studies with a placebo run-in that reduces the number of placebo responders, are needed to reflect the antidepressant potential of SAMe. This is particularly important because compared to prescription antidepressants, SAMe is safer with fewer adverse effects or drug interactions. Evidence suggests that it may be safe for use during pregnancy and in breastfeeding mothers.

In general, SAMe side-effects are mild and usually transient, including headaches, anxiety, agitation, insomnia, loose bowels, upset stomach, and, occasionally, palpitations. SAMe does not cause weight gain or sexual dysfunction. Contraindicated in patients with a personal or family history of bipolar disorder, SAMe can induce mania (although smaller doses combined with a mood stabilizer may be considered with close monitoring). There is no evidence that SAMe has adverse interactions with any other drugs. It does not interfere with cytochrome P450 metabolism and does not displace other drugs from protein binding. There are no clear cases of serotonin syndrome even when SAMe is combined with prescription antidepressants, including SSRIs and MAOIs. Nevertheless, until more information is available, SAMe and MAOIs should only be combined under close medical supervision. The usual dosage of SAMe is 400–800 mg/day for mild depression and 800–1600 mg/day for moderate to severe depression. Higher doses (up to 3200 mg/day) may be needed for severe or treatment-resistant depression, for example, in patients with PD. SAMe is best absorbed on an empty stomach (at least 30 min before meals) and best prescribed in the morning to early afternoon in one or two doses. The dose can be raised by 200–400 mg every 3–7 days if over stimulation or other side effects do not occur.

SAMe oxidizes rapidly when exposed to air. The manufacture and packaging of SAMe tablets are critical to assure full potency. Few companies test their products for shelf-life. Bargain brands may lose 50–100% potency before use; clinicians and consumers should look for brands containing over 70% active SAMe isomers with enteric-coated tablets in individual blister packs. SAMe methylation pathways require vitamin B_{12} and folate as cofactors (Bottiglieri, 2013), hence the addition of vitamin B_{12} (200 µg/day) and folic acid (400–800 µg/day) may potentially enhance the antidepressant effect of SAMe.

Omega-3 Fatty Acids (Omega-3 FAs)

Omega-3 FAs have a critical role in neurological activity and in treating depression, especially if an inflammatory causation is present (a link between inflammation and depression has been demonstrated) (Raison et al., 2006). Epidemiological studies have shown that increased risk of depressive symptoms may be correlated with lower dietary omega-3 fish oil (eicosapentaenoic acid (EPA) and docosahexaenoic acid (DHA))

(Hibbeln, 1998). Studies have also demonstrated that people with depressed mood have a tendency towards a higher ratio of serum arachidonic acid to essential fatty acids, and an overall lower serum level of omega-3 FAs compared with healthy controls (Sanchez-Villegas et al., 2007). Evidence suggests that omega-3 FAs exert antidepressant activity via modulation of neurotransmitter (norepinephrine, dopamine, and serotonin) reuptake, degradation, synthesis, and receptor binding; anti-inflammatory effects; and enhancement of cell membrane fluidity (Chalon, 2006; Williams et al., 2006).

Dozens of human clinical trials have assessed the efficacy of EPA, DHA, or a combination of both in depression. A meta-review of nutraceutical meta-analyses in the psychiatric field reviewed 13 independent RCTs involving 1233 people with MDD (Firth et al., 2019). Omega-3 supplements (mean: 1422 mg/day of EPA) significantly reduce depressive symptoms (SMD $=$ 0.398, 95% CI: 0.114–0.682), with no evidence of publication bias. When used as an adjunctive to antidepressants in MDD, omega-3 supplements (930–4400 mg/day of EPA) also provided some reduction of depressive symptoms ($n = 448$; $n = 11$, SMD $= 0.608$, 95% CI: 0.154–1.062, $I^2 = 82\%$, ns). In bipolar depression, a meta-analysis of five pooled datasets ($n = 291$), found a significant effect of omega-3 FAs in reducing depression in bipolar disorder, with a moderate effect size (Sarris et al., 2012b). Depletion of omega-3 FAs is common during pregnancy and may contribute to depression. Omega-3 FAs are one of the few interventions known to be safe and potentially beneficial to neural development during pregnancy and breast-feeding (Bourre, 2006).

Recent practice guidelines have been developed by a subcommittee of the International Society for Nutritional Psychiatry Research, to provide clinical direction for omega-3 FA use in depression (Guu et al., 2020). The key practice guidelines contend that: "(1) clinicians and other practitioners are advised to conduct a clinical interview to validate diagnoses, physical conditions, and measurement based psychopathological assessments in the therapeutic settings when recommending omega-3 FAs in depression treatment; (2) with respect to formulation and dosage, both pure EPA or an EPA/DHA combination of a ratio higher than 2 (EPA/DHA >2) are considered effective, and the recommended dosages should be 1–2 g of net EPA daily, from either pure EPA or an EPA/DHA (> 2: 1) formula; (3) the quality of omega-3 FAs may affect therapeutic activity; and (4) potential adverse effects, such as gastrointestinal and dermatological conditions, should be monitored, as well as obtaining comprehensive metabolic panels. The expert consensus panel has agreed on using omega-3 FAs in MDD treatment for pregnant women, children, and the elderly, and prevention in high-risk populations. Personalizing the clinical application of omega-3 FAs in subgroups of MDD with a low omega-3 Index or high levels of inflammatory markers might be regarded as areas that deserve future research."

Zinc

The element zinc, one of the most prevalent trace elements in the amygdala, hippocampus, and neocortex, is involved with hippocampal neurogenesis via upregulation of BDNF and modulates N-methyl-D-aspartate (NMDA) and glutamate activity (Szewczyk et al., 2011). Zinc modulates the hypothalamic-pituitary adrenal (HPA) axis and is neuroprotective in animal models (Takeda, 2011). A low zinc serum level is associated with depression risk and correlated with an increased activation of immune system biomarkers, suggesting that this effect may result in part from a depression-related alteration in the immune-inflammatory system (Szewczyk et al., 2011). Supplementation with zinc has been found to attenuate inflammation via inhibition of TNFα and IL-1β (Prasad, 2009). A cross-sectional study of zinc carried out in 402 postgraduate students (Yary & Aazami, 2012) revealed an inverse relationship between dietary intake of zinc and depression.

Emerging evidence supports zinc for improving depressed mood. A meta-analysis of 4 RCTs revealed a pooled effect estimate of adjunctive supplementation of 25 mg/d of zinc in antidepressant treatment being statistically significant to placebo (SMD = −0.66; [−1.06; 0.26]) (Schefft et al., 2017). This evidence could be interpreted as suggesting potential benefits of zinc as a stand-alone intervention or as an adjunct to conventional therapy with antidepressants for depression. However, in geographic areas with low soil zinc levels, there can be a higher incidence of zinc deficiency. Therefore, it would be necessary for RCTs in the same population to include pre-and post-measures of serum zinc levels before extrapolating the findings to other geographic locations.

B Vitamins

Low levels of vitamin B_{12} (cyanocobalamin) and folate have been associated with disorders of mood, memory, and cognition, particularly in depressed patients and older adults (Bottiglieri, 2013). Studies have demonstrated that B vitamins can improve mood and cognitive functions (Riggs et al., 1996). In a DBRPC study of 14 depressed geriatric inpatients on TCAs, 10 mg each of vitamins B_1, B_2, and B_6 improved depression and cognitive function (Bell et al., 1992). A study of 129 healthy adults given seven B vitamins, vitamin C, and vitamin E at 10 times the FDA-recommended daily allowance for 1 year resulted in a significantly greater improvement in mood than in a placebo group. Higher serum levels of vitamins B_2 and B_6 correlated with improved mood in men, whereas higher levels of vitamin B_1 correlated with better mood in women (Benton et al., 1995). Vitamin B_{12} deficiency (associated with high methylmalonic acid) more than doubled the risk of severe depression in 700 disabled women over the age of 65 years (Penninx et al., 2000). In a study of dietary folate and B vitamins, 121 out of 865 Japanese women developed postpartum depression. Consumption of vitamin B_2 in the third quartile decreased the risk of postpartum depression (Miyake et al., 2006).

Low levels of folate have been associated with nonresponse to antidepressant treatment (Fava et al., 1997; Lerner et al., 2006). In a DBRPC study, 127 women on fluoxetine 20 mg/day were given either folic acid 500 IU/day or placebo. In the fluoxetine/folic acid group, 94% of the women had a good response versus 61% in the fluoxetine/placebo group. In women treated with fluoxetine/folic acid, plasma folate increased, and homocysteine decreased (Coppen & Bailey, 2000). Low folate levels were associated with delayed response to fluoxetine in a study of 110 patients with MDD (Papakostas, 2006). The recommended active forms are either folinic acid or L-methylfolate (7.5 or 15 mg). This is important as folinic acid and L-methylfolate do not require the methylhydrofolate reductase (MTHFR) enzyme, which can be deficient in the presence of MTHFR C677T polymorphisms (Farah, 2009).

5-Hydroxytryptophan (5-HTP)

5-HTP, an essential monoamine precursor required for the synthesis of serotonin, was studied extensively in the latter half of the twentieth century as medication for depression (Shaw et al., 2002). Eight controlled augmentation studies using L-tryptophan or 5-HTP with antidepressants provided encouraging evidence of increased response to the antidepressants (phenelzine sulfate, clomipramine, tranylcypromine, and fluoxetine) (Sarris et al., 2009b). A systematic review and meta-analysis (Shaw et al., 2002) of two studies meeting criteria (pooled $n = 64$) suggested that 5-HTP and L-tryptophan are better than placebo at alleviating depression. Efficacy of 5-HTP has been shown in several neurological conditions, fibromyalgia, and insomnia. 5-HTP is well absorbed from an oral dose and easily crosses the blood-brain barrier to increase effectively the central nervous system (CNS) synthesis of serotonin (Birdsall, 1998). In addition to increasing the synthesis of serotonin, 5-HTP may also affect noradrenergic and dopaminergic transmission. Side effects include increased risk of serotonin syndrome when combined with serotonergic antidepressants.

Inositol

A vitamin in the B_2 complex, inositol boosts cyclic adenosine monophosphate (cAMP), a second messenger in neurons. Inositol 12–20 g/day was superior to placebo in randomized trials for depression (Benjamin et al., 1995a), panic disorder (Benjamin et al., 1995b), and OCD (Fux et al., 1996) (for a review, see: Bender, 2000). However, later studies have revealed non-superiority to placebo in depression, thereby not supporting such application (Firth et al., 2019). A study from the NIMH Systematic Treatment Enhancement Program for Bipolar Disorder (STEP-BP) compared augmentation with antidepressants: lamotrigine, inositol, and risperidone for refractory depression in 66 patients with bipolar disorder. The differences in recovery rates with lamotrigine (24%), inositol (17%), and risperidone (5%) were statistically similar (Nierenberg et al., 2006). There is a possible role for inositol augmentation in bipolar depressed patients who do not respond to lamotrigine, who develop side effects (allergic rash or photosensitivity), or who cannot afford the high cost of lamotrigine. Gastrointestinal side effects include gas and loose bowels. Because a therapeutic dose requires at least six large inositol 650 mg capsules TID, compliance may be a problem. In one case, mania was induced by 3 g/day. In a review of seven RCTs in depression ($n = 242$), inositol had marginally more responders than placebo and a trend towards greater efficacy for depressive symptoms in patients with pre-menstrual dysphoric disorder; in four RCTs there were no statistically significant effects of inositol on anxiety or OCD. Inositol caused more gastrointestinal upset compared with placebo (Mukai et al., 2014). The meta-analysis was limited due to the small number of studies.

Acetyl-L-carnitine (ALCAR)

Acetyl-L-carnitine (ALCAR) contains carnitine and acetyl moieties, both of which have neurochemical properties that may affect brain energy, phospholipid metabolism, neurotrophic factors, and synaptic transmission (Pettegrew et al., 2000). These properties may be beneficial in MDD. ALCAR 500 mg BID was compared with amisulpride 50 mg TID in a 12-week DBRPC study of 204 patients with pure dysthymia. Both groups showed improvement on HAM-D with no significant difference between groups. ALCAR caused fewer side effects (Zanardi & Smeraldi, 2006).

Treatment guidelines for disorders of mood, anxiety, and sleep are summarized in Table 1.

Mind-Body Medicine

Yoga, Qigong, Tai Chi, and Breathing Techniques

Mind-body medicine include ancient systems such as yoga, Qigong, Tai Chi, and meditation, and also their modern derivatives which may use science and technology to adapt selected practices to diverse treatment and self-development programs. Traditional yoga is a comprehensive program of eight limbs that integrates religious philosophy with physical and mental practices. Extensive discussions of the history, philosophy, and practice of yoga can be found in *The Yoga Tradition* (Feuerstein, 1998) and a review by Telles and Singh (2013). Modern secular yoga often emphasizes postures with breathing techniques and meditation. Discussion of the neurophysiological effects of mind-body medicine on psychological states can be found in the subsection on Anxiety Disorders below. Decades of yoga research consisted mostly of case studies and small open trials. More recently, yoga benefits are being documented in RCTs using standardized measures, and biological markers (e.g., heart rate variability, cortisol, neurohormones, inflammatory markers, and BDNF). Yoga researchers are applying advanced technologies, such as brain imaging and genomics to explore the underlying neurophysiological mechanisms involved in mind-body medicine (Bhargav et al., 2021a, b).

A RCT of the effects of a 12-week yoga-and-meditation-based lifestyle intervention (YMLI) on depression severity and biomarkers of neuroplasticity in 58 MDD patients on routine drug treatment was given 2 h per day 5 days per week (Tolahunase et al., 2018a). The program included physical postures (*asanas*), breathing exercises (*pranayama*), and meditation (*dhyana*). The 25 min of breathing practices included 6 techniques of slow breathing between 4 and 6 cpms, with or without creating resistance on either the inhale (*Shitali*) or the exhale (*Ujjayi, Brahmamdra*). There was a significant decrease [difference between means, (95% CI)] in Beck Depression Inventory (Second Edition; BDI-II) score $[-5.83(-7.27, -4.39), p < 0.001]$ and significant increase in BDNF (ng/ml) [5.48 (3.50, 7.46), $p < 0.001$] after YMLI compared to the control. YMLI significantly increased dehydroepiandrosterone sulfate, sirtuin 1, and telomerase activity levels, decreased cortisol, and IL-6 levels, reduced DNA damage, and balanced oxidative stress. This evidence suggests that the decrease

in depression scores after YMLI is associated with improved systemic biomarkers of neuroplasticity in MDD.

The same group published a 12-week RCT in which MDD patients ($n = 178$) were randomized to receive YBLI or drug therapy (Tolahunase et al., 2018b). The short form of the serotonin transporter (5-HTTLPR) variant of the serotonin transporter gene (*SLC6A4*) and *MTHFR* 677C>T polymorphisms have been linked to MDD and antidepressant treatment response. Multivariate logistic regression models for remission including either 5-HTTLPR or *MTHFR* 677C>T genotypes showed statistically significant odds of remission in yoga arm vs. drug arm. Neither 5-HTTLPR nor *MTHFR* 677C>T genotype showed any influence on remission in the group given the YBLI. Childhood adversity interacted with 5-HTTLPR and *MTHFR* 677C>T polymorphisms to decrease treatment response in the drug treatment arm, but not in Yoga intervention arm. YBLI provided MDD remission in those who have susceptible 5-HTTLPR and *MTHFR* 677C>T polymorphisms and are resistant to SSRIs treatment. The others concluded that YBLI may be therapeutic for MDD independent of heterogeneity in its etiopathogenesis (Tolahunase et al., 2018b).

Studies of yoga and Qigong programs that include substantial amounts of breathing practices show significant improvements in mood in individuals diagnosed with depression and increased mental and physical energy, alertness, enthusiasm, and positive mood in healthy individuals Tsang et al., 2006). See additional reviews of clinical yoga research (Brown et al., 2013; Rao et al., 2013; Seshadri et al., 2020; Telles & Singh, 2013; Vancampfort et al., 2021).

In a RCT dosing study of a 12-week intervention using yoga and coherent breathing (paced at 5 breaths per minute) in patients with MDD, Streeter et al. (2020a) found a significant reduction in depression and suicidal ideation (Nyer et al., 2018). Magnetic Resonance Spectroscopy (MRS) detected changes in thalamic gamma-aminobutyric (GABA) acid levels. Low baseline thalamic GABA, characteristic of MDD, increased over the 12 weeks (Streeter et. al., 2020). Evidence suggested that at least once weekly practice would be required to maintain the elevated GABA levels. This preliminary evidence suggests that one mechanism by which the practice of yoga and coherent breathing may improve mood is through increased GABA transmission.

In an RCT of 70 patients with MDD stabilized on antidepressants, those who participated in a 12-week yoga therapy improved significantly more on depression scores (HAM-D) than the control. The group with add-on yoga treatment showed significantly greater improvement in cortical inhibition as indicated by a greater increase in the cortical silent period (CSP). MDD is associated with shortening of the CSP involving deficits in GABA receptor-mediated neurotransmission (Bhargav et al., 2021b).

Breath Practices: Adverse Effects and Contraindications

Slow-paced breathing practices tend to be calming and usually safe regardless of psychiatric diagnosis. However, rapid or forceful breathing, for example, Bellows Breath (Bhastrika), Breath of Fire (Kapalabhati), or rapid cyclical breathing (Kriya, holotropic breathing, or chaotic breathing) can be activating and therefore, as with activating antidepressant medications, they can trigger hypomanic or manic episodes in people on the bipolar spectrum (Gerbarg et al., 2011; Brown et al., 2013). Prolonged, intense breath practices have also been associated with psychotic episodes. In addition, when rapid, forceful breathing reaches the level of hyperventilation, proximal tubular reabsorption of lithium is reduced, leading to possible increased lithium excretion and decreased serum lithium levels (Gerbarg et al., 2011). A detailed discussion of potential adverse effects can be found in the subsection on Anxiety Disorders later. Hence, patients with bipolar disorder being treated with lithium alone should not engage in rapid yoga breathing. Patients on lithium in combination with other medications for bipolar disorder (mood stabilizers) who begin yoga practices should have lithium levels monitored. Bipolar II patients whose mood swings are under control on medication may do yoga breathing under supervision if they avoid rapid or forceful breathing forms. Nevertheless, if they become agitated or anxious, the yoga breathing should be discontinued and the patient should be monitored until stable.

Acupuncture

The use of acupuncture to treat depressive disorders is documented in traditional Chinese medicine (TCM) texts (Maciocia, 1994). A 2018 Cochrane Review meta-analysis (Smith et al., 2018) included 64 acupuncture depression studies (7104 participants). The results revealed that acupuncture (manual and electro-) versus waitlist/no treatment/treatment as usual may moderately reduce the severity of depression by end of treatment (SMD -0.66, 95% CI -1.06 to -0.25, five trials, 488 participants). A weaker effect was found for acupuncture versus sham-acupuncture being associated with a small reduction in the severity of depression. Acupuncture interacts with opioid pathways and substances that modulate these pathways have antidepressant activity (Cabyoglu et al., 2006; Wang et al., 2008). Other possible antidepressant mechanisms include increased release of serotonin and norepinephrine, as well as cortisol modulation.

Anxiety Disorders, Trauma- and Stress-Related Disorders

Substantial evidence indicates anxiolytic effects of plant-based medicines and mind-body techniques, with a smaller number of studies providing evidence for nutrients. Studies using *Piper methysticum* (kava), *Galphimia glauca* (galphimia), *Passiflora incanata* (passionflower), *Matricaria recutita* (chamomile), *Melissa officinalis* (lemon balm), and *Valeriana officinalis* (valerian) have been reviewed (Modabbernia & Akhondzadeh, 2013; Sarris et al., 2021). DBRCTs indicate that SJW is not effective for any anxiety disorder. Individuals with anxiety disorders are vulnerable to substance abuse and dependence on prescription anxiolytics (medications for anxiety). Use of select nutraceuticals and mind-body medicine may provide significant relief without the risk of drug dependence.

Autonomic dysregulation and dysfunction of the HPA axis are characteristic of anxiety disorders and trauma- and stressor-related disorders. Breath-centered practices have been used effectively to restore sympathovagal balance for both acute and long-term reduction in symptoms. Considerable evidence supports the benefits of mind-body techniques in the treatment of anxiety, PTSD, insomnia, stress-related medical conditions, and caregiver stress. Additional mechanisms contributing to the effects of voluntarily regulated breathing practices (VRBPs) are described below.

Herbal Medicine

Piper methysticum (Kava)

The South Pacific medical plant kava (*Piper methysticum*), a social and ceremonial drink in the Pacific Islands, has intoxicant, sedative, anxiolytic, anticonvulsant, and analgesic properties (Sarris et al., 2011b). The active constituents involved in the therapeutic activity of kava involve the six major lipophilic kavalactones, of which kawain and dihydrokawain have the strongest anxiolytic activity (Sarris et al., 2011b). Kava exerts anxiolysis primarily by modulation of GABA pathways via blockade of voltage-gated sodium ion channels, reduced excitatory neurotransmitter release due to blockade of calcium ion channels, and enhanced ligand binding to GABA type A receptors (LaPorte et al., 2011). Other neurochemical effects include reversible inhibition of monoamine oxidase B, inhibition of cyclooxygenase, and reduced neuronal reuptake of dopamine and noradrenalin (prefrontal cortex) (LaPorte et al., 2011).

A meta-analysis of six studies ($n = 473$) revealed no significant difference between kava and placebo for anxiety disorders (Barić, Đorđević et al., 2018). Notably, a 3-week DBRCT (crossover) recruited 60 adult participants (aged 18–65 years) with 1 month or more of elevated generalized anxiety (defined as a Beck Anxiety Inventory (BAI) Score of >10) (Sarris et al., 2009a). Participants were given five kava tablets totaling 250 mg of kavalactones per day or matching placebo. Intention-to-treat analysis revealed that the aqueous extract of kava reduced participants' Hamilton Anxiety Rating Scale (HAM-A) score significantly over placebo, with a substantial effect size. Pooled analyses also revealed highly significant relative reductions in MADRS scores. The aqueous extract was found to be safe in this short-term trial (3 weeks) with no serious adverse effects reported. Another DBRCT involved the chronic administration of kava or placebo over 6 weeks (in addition to a 1-week placebo run-in phase and a 1-week single-blind, post-study observation phase) (Sarris et al., 2013d). Adult participants (aged 18–65 years) with diagnosed GAD were given kava tablets were standardized to contain 60 mg of kavalactones per tablet for a total dose of 120 mg of kavalactones for the first 3-week controlled phase, being titrated to 240 mg of kavalactones in nonresponders at the 3-week mark for the second 3-week controlled phase. Intention-to-treat analysis was conducted on 58 participants with post-baseline data available for analysis. The results revealed a significant reduction in HAM-A scores in favor of kava over placebo, with a moderate effect size. For participants with a moderate to severe level anxiety, the anxiolytic effect of kava was more pronounced. Although neither of these studies found any negative effects on liver function tests and no signs of hepatotoxicity, these adverse effects are unlikely to appear in such small samples. It is important to note, however, that a large multicenter 16-week RCT revealed non-superiority of Kava to placebo in 171 participants with diagnosed GAD (Sarris et al., 2019b). Similarly, a DBRPC Internet-based study of kava and valerian reported no significant differences between kava and placebo for anxiety (Jacobs et al., 2005). While not effective for GAD, Kava has potential benefits for acute or short-term management of anxiety symptoms (Sarris, 2016).

In respect to safety considerations, it was revealed that after kava was introduced to indigenous people in Northern Australia, long-term users developed facial swelling, scaly rash, increased patellar reflexes, and dyspnea (Mathews et al., 1988). Low levels of albumin, increased gamma-glutamyl transpeptidase, abnormal complete blood counts with decreased white cell and platelet counts, and hematuria were also found. A report on 62 indigenous kava users showed increased liver enzymes, which reversed after several weeks' abstinence (Clough et al., 2003). Two post-marketing studies of over 6000 patients found a 1.5% and 2.3% incidence of side effects, mostly gastrointestinal, allergic reactions, headache, and light sensitivity (Ernst, 2002). Less commonly, restlessness, drowsiness, lack of energy, and tremor were reported. In four cases, kava induced dystonic reactions, oral/lingual dyskinesias, and worsening Parkinsonian symptoms in a woman on levodopa. These effects are consistent with DA blockade. In one case, a 45-year-old woman developed severe Parkinsonism after kava treatment (Meseguer et al., 2002). No studies of long-term safety, teratogenicity, or mutagenicity beyond 6 months have been reported. Combining kava with alcohol or other sedatives such as alprazolam or muscle relaxants has resulted in coma, though combining in lower therapeutic doses is highly unlikely to have this effect (Almeida & Grimsley, 1996). Kavalactones have been found to inhibit P450 enzymes involved in the metabolism of most pharmaceuticals (CYP1A2, 2C9, 2C19, 2D6, 3A4, and 4A9/11) in vitro (Mathews et al., 2002). However, in a 28-day study of 12 subjects who orally ingested 1 g BID kava root extract, metabolism of the CYP2E1 substrate chlorzoxazone was inhibited by 40%, but no effect was found for substrates of CYP1A2, 2C9, 2C19, 2D6, or 3A4 (Gurley et al., 2005).

Decades ago, serious liver toxicities including liver failure requiring transplantation have been found to be associated with kava extracted with acetone or alcohol. The FDA issued a Letter to Health Care Professionals and a Consumer Advisory about the potential risk of liver injury with kava (Kraft et al., 2001; FDA, 2002). However, based on case study data (i.e., very low quality data), kava hepatotoxicity appears to be extremely rare, with probable causation established in only a few cases (Teschke et al., 2011). Further, such causative links between kava and liver injury have since not been obviously raised by pharmacovigilance in the past two decades. Previous liver issues found with German kava products may be due in part to the use of non-water-soluble chemical extraction techniques (the traditional solvent is water), use of aerial parts and root and stem peelings, and poorly prepared raw material (Teschke et al., 2011). The World Health Organization commissioned a report assessing the risk of kava products (Coulter, 2007). Recommendation 2.1.3 from this report suggested that products from water-based suspensions be developed and tested in clinical trials and that these preparations be used preferentially, rather than acetonic and ethanolic extracts. Aqueous kava extracts have been found to not affect liver function in a rat study (Singh & Devkota, 2003). The use of a standardized formulation of kava from peeled rootstock of a noble cultivar (such cultivars are higher in kawain and dihydrokawain, and lower in dihydromethysticin) was also recommended. With respect to prescriptive advice, kava may be used intermittently, within the therapeutic dose range, and not in combination with alcohol or benzodiazepines. Kava is contraindicated in patients known to have problems with substance abuse. Occasional liver function tests can be performed if there is any concern about liver dysfunction.

Melissa officinalis (Lemon Balm) and *Valeriana officinalis* (Valerian)

In a DBRPC crossover study using a 20-min laboratory-induced psychological stressor (Defined Intensity Stressor Simulation (DISS) battery), 20 healthy adults were given single doses of lemon balm (600, 1000, and 1600 mg) or placebo (Kennedy et al., 2003). The 600 mg dose significantly improved self-ratings of "calmness" on Bond-Lader scales. A combination of lemon balm and valerian in three doses (600, 1200, and 1800 mg) was given to 24 healthy volunteers in a DBRPC crossover trial using the 20-min DISS battery (Kennedy et al., 2004). Mood and anxiety were evaluated at baseline and after 1, 3, and 6 h. The 600 mg dose significantly reduced anxiety ratings, whereas the 1800 mg dose somewhat increased anxiety ratings. A DBRPC Internet-based study of kava and valerian reported no significant differences between kava and placebo for anxiety, or between valerian and placebo for sleep disturbance (Jacobs et al., 2005). Subject selection was based on State-Trait Anxiety Inventory state and on self-report of "a problem going to sleep or staying asleep over the past 2 weeks" rather than on standard diagnostic criteria. The negative results may have been due to the selection of a heterogeneous population, comorbid conditions, inadequate doses of herbs, or other limitations of Internet surveys. A large uncontrolled study of a lemon balm and valerian combination product was tested in 918 children under the age of 12 years with restlessness and dyssomnia (Muller & Klement, 2006). Substantial improvement occurred in 80.9% of the children with dyssomnia and 70.4% of those with restlessness. Additional controlled trials would be useful in assessing the safety and efficacy of herbs for children.

Other more recent DBRPC studies involving assessing anxiety in people with cardiac pain, palpitations, burn victims, post-surgery, and in adolescent girls with pre-menstrual syndrome, have all revealed positive reduction of anxiety on validated scales (Heydari et al., 2018; Soltanpour et al., 2019).

Passiflora incarnata (Passionflower)
Passionflower contains a dihydroflavone, chrysin, which binds to benzodiazepine receptors (Appel et al., 2011). A pilot RCT using passionflower extract revealed equivalent efficacy to oxazepam (30 mg/day) in reducing anxiety, with fewer side effects (Akhondzadeh et al., 2001). A non-statistical trend towards decreased sedation and less impaired job performance occurred in the herbal medicine group. Although the results were positive, no definitive conclusion can be drawn, as anxiety conditions are notorious for high placebo response, and no placebo arm was used. An acute study using passionflower versus placebo control for presurgical anxiety revealed a significant reduction in anxiety in favor of the herb on a numerical rating scale (Movafegh et al., 2008). Importantly, no difference in sedation levels (with anesthetic) occurred between groups. Most recently, this effect was observed in a double-blind RCT of 40 dental surgery patients (Dantas et al., 2017), showing Passionflower (260 mg) orally administered 30 min before dental surgery reduced subjective anxiety

along with physiological indicators (i.e. blood pressure, heart rate). This occurred to the same extent as midazolam, however with fewer cognitively impairing side effects. Analysis of side effects of passionflower studies in a Cochrane Review revealed no current evidence of any safety concerns (Miyasaka et al., 2007). Although the results are encouraging, an overall limitation to these studies is the use of different doses and preparations of extracts and the lack of longer-term RCTs.

Galphimia glauca (Galphimia)
In traditional Mexican and Central American cultures, the leaves and stem from galphimia are used for a variety of ailments, including nervous disorders. Galphimine B, regarded as an active constituent, has serotonergic effects (Jiménez-Ferrer et al., 2011). In a 4-week RCT, an aqueous extract of galphimia was administered to 153 patients with GAD (Herrera-Arellano et al., 2007). The positive control group received 1 mg of the benzodiazepine lorazepam in capsule form twice daily versus the galphimia group receiving 310 mg of the extract in capsule form twice daily. The results revealed a significant effect for both interventions with no significant difference between groups. Galphimia reduced HAM-A-rated anxiety by 61.2% compared with 60.29% for lorazepam. No significant side effects were noted in the galphimia group, whereas 21.33% of subjects in the lorazepam group experienced excessive sedation. In a follow-up study (Herrera-Arellano et al., 2012), 191 patients with GAD were administered two to four capsules (at the clinician's discretion) containing 3.48 g of dried galphimia per capsule (0.35 or 0.7 mg/day of galphimine B) versus 1–2 mg of lorazepam over a 15-week period. A significantly greater reduction in HAM-A score was observed for galphimia treatment in comparison with lorazepam. In both studies, galphimia was well tolerated. A standardized extract of galphimia was administered daily versus alprazolam for 10 weeks in patients ($n = 167$) with moderate or severe GAD (Romero-Cerecero et al., 2019). A significant reduction on the HAM-A occurred with no significant difference between treatments. After 10 weeks, the average score in the alprazolam

group was 4.6 ± 6.5 points while the HAM-A score was even lower in the galphimia group, 3.5 ± 5.5 points (HAM-A ≤ 7 considered remission). Twenty-two percent treated with alprazolam reported daytime sleepiness, compared to 5% in the galphimia group. While these three studies show efficacy equivalent to benzodiazepines, a placebo-controlled study is needed to confirm efficacy.

Matricaria recutita (Chamomile)

The flowering tops of chamomile are widely consumed throughout Europe, with a long history of traditional use for its perceived calming effect. Preclinical research has found effects on the GABA system (Awad et al., 2007). One of the active constituents is the flavone apigenin, which is a benzodiazepine receptor ligand with anxiolytic activity (Salgueiro et al., 1997). An 8-week DBRCT studied a chamomile extract standardized to 1.2% apigenin for anxiety symptoms in GAD ($n = 57$) (Amsterdam et al., 2009). The study used a dose escalation design, doses ranging from 220 to 1100 mg, depending on response. Chamomile reduced anxiety symptoms significantly more on HAM-A compared with placebo. Two further trials reported both short and long term (i.e., 8 and 38-weeks) effects of 1500 mg of Chamomile (500 mg capsule 3 times daily) in 179 patients with GAD. At week 8, 58% of patients met criteria for clinical response in the chamomile group, with significant reductions in mean anxiety (GAD-7) across the entire sample (Keefe et al., 2016). The clinical responders from this study were randomly assigned to a long-term study, continuing their active or placebo treatment for an additional 26 weeks. GAD relapse rates were 26% in the placebo group, and only 15% in the continued treatment group (Mao et al., 2016); however, neither the difference in relapse rates nor the HAM-A or BAI symptom reductions reached statistical significance. Chamomile was very well tolerated and showed no increase in adverse events at higher doses compared with placebo.

Xiao-Tan-Jie-Yu-Fang

Xiao-Tan-Jie-Yu-Fang (XTJYF), a Chinese herbal medicine, contains *Bupleuri*, *Angelicae sinensis*, *Poria cocus*, *Atractylodis macrocephalae*, *Paeonia*

alba, *Herba menthaeGlycyrrhizae*, *Coptidis*, *Pinelliae preparatae*, *Pericarpium citri reticulatae*, *Os draconis*, *Concha ostreae*, *Rhizoma rhei preparatae*, *Acori graminei*. An RDBPCT compared the XTJYF, to placebo in 245 Sichuan earthquake survivors with PTSD. Compared to the placebo group, the XTJYF group was significantly improved on Symptom Checklist-90-Revised (SCL-90-R) global indices ($P = 0.001 \sim 0.028$). The XTJYF group performed significantly better in SCL factors: somatization, obsessive-compulsive behavior, depression, anxiety, hostility, and sleep quality. No serious adverse events were reported (Meng et al., 2012). *Glycyrrhizae* is associated with increased risk of hypokalemia, hypertension, heart failure, and arrhythmia, particularly in elderly patients.

Saikokeishikankyoto

The Japanese herbal formula, Saikokeishikankyoto, contains a mixture of dried herbs: *Bupleuri Radix* (6 g), *Trichosanthis kirilowii* (3 g), *Cinnamomi bassiae* (3 g), *Scutellariae baicalensis* (3 g), *Concha ostreae* (3 g), *Glycyrrhizae radix* (2 g), and dried *Zingiberis rhizoma* (2 g). Two years after of the Great East Japan Earthquake and Tsunami, the effects of Saikokeishikankyoto (extract TJ-11 by Tsumura, Tokyo) on symptoms of PTSD were assessed in 48 traumatized survivors who participated in a randomized observer-blinded study. Those given Saikokeishikankyoto, showed significantly lower scores on the Impact of Event Scale-Revised (IES-R) than those who were not (Numata et al., 2014). Limitations of this study include the lack of placebo or comparison treatment. Two components of this formula have been associated with increased risk of adverse events: *Glycyrrhizae radix* – hypokalemia, hypertension, heart failure, arrhythmia; *Scutellariae baicalensis* – liver and lung injury.

Nutrients for Obsessive/Compulsive Symptoms

N-Acetylcysteine (NAC)

N-Acetylcysteine (NAC), an amino acid-based nutraceutical, known to attenuate synaptic release of glutamate in subcortical brain regions (Dean et al., 2012). Furthermore, it restores the extracellular glutamate concentration in the nucleus accumbens, offering a critical pharmacological action in reducing compulsive behavior. NAC also reduces inflammation, enhances mitochondrial energy generation and neurogenesis, and is neuroprotective in a variety of stress models (Dean et al., 2012). A 12-week DBRCT involving 50 individuals with trichotillomania (obsessive hair pulling) found that NAC (dose range 1200–2400 mg/day) significantly reduced symptoms on the Massachusetts General Hospital Hair Pulling Scale and the Psychiatric Institute Trichotillomania Scale (Grant et al., 2009). It was found that 56% of patients were "much or very much improved" with NAC use compared with 16% taking placebo.

A 12-week DBRCT involving 48 patients with OCD who failed to respond to SSRIs revealed that 2400 mg/day of NAC was more effective than placebo in reducing symptoms on the Yale-Brown Obsessive Compulsive Scale (Y-BOCS) (Afshar et al., 2012). A more recent 16-week DBPCT using NAC (1.5 grams twice a day) in 44 adults with DSM-5-diagnosed OCD, revealed a non-significant time x treatment interaction for the YBOCS scale total score (Sarris et al., 2015). NAC was well tolerated, aside from more cases of heartburn compared to placebo. Based on currently available evidence, it is not possible to make recommendations on the use of NAC for OCD.

Mind-Body Medicine

Neurophysiological Mechanisms Associated with Mind-Body Medicine

The evolving science of *interoception* is fundamental for understanding mind-body medicine. Interoceptors, molecular sensors or receptors in neurons detect signals (e.g., stretch, pressure, chemical, vibrational, visceral) and transduce these into electrical impulses or non-neural messengers (e.g., hormones, peptides). Interoception has been viewed as the process by which an organism senses, interprets, and integrates information about the internal state of the body by the

ascending transmission of signals from physiological systems outside the central nervous system to the brain (Craig, 2008). Chen and colleagues propose the inclusion of regulation through descending signals (Chen et al., 2021). This controversial expansion of the concept of interoception would include the nervous system, as well as the vascular, endocrine, and immune systems (Chen et al., 2021).

Mind-body medicine is a means to voluntarily generate and shape interoceptive messages. For example, postures and movement sequences (e.g., Yoga, Tai Chi and Qigong) activate stretch receptors and proprioceptors. Breathing practices, found in most mind-body, religious, and tribal traditions, activate respiratory mechanoreceptors, chemoreceptors and baroreceptors.

The vagus nerves (10th cranial nerves) contain critical pathways for mind-body communication. About 20% of vagal fibers transmit efferent messages from the brain to the body, automatically regulating the organs and glands. About 80% of vagal fibers bring afferent messages from peripheral receptors to the brain where they exert widespread effects on how we perceive, think, feel, and respond to the environment and other people (Brown & Gerbarg, 2012b; Porges & Carter, 2017).

According to the Polyvagal Theory, *neuroception* refers to the conscious and unconscious capacities of neural circuits to distinguish whether situations or people are safe, dangerous, or life-threatening. Interoceptions, contribute to assessments of safety or danger through higher level processing in the insular and cingulate cortical networks. Interoceptions, for example, respiratory signals, arising from slow breathing techniques that increase parasympathetic tone, art postulated to recruit social engagement networks, and reduce defensive behaviors (Porges & Carter, 2017; Gerbarg et al., 2019). Social engagement networks enable communication through facial expression, vocalization, perceptions of vocal tone and prosody, and physical postures (Porges & Carter, 2017). When the myelinated vagal pathways are optimally functioning in social interactions and inhibiting sympathetic excitation (that promotes fight-or-flight behaviors), emotions are well regulated and the autonomic state supports calm, positive social behaviors such as flexibility, bonding, empathy, compassion, and cooperation (Brown & Gerbarg, 2017; Gerbarg et al., 2019). Conversely, positive social interactions influence vagal function, induce feelings of safety, and reduce the effects of stress on physiological states.

Voluntarily Regulated Breathing Practices: Neurophysiological Mechanisms

Breathing is the only autonomic function that can be easily controlled voluntarily, thus providing a portal of entry into the brain-body communication system. The respiratory tract, from the nasal passages to the alveoli, the lung tissues, the diaphragm, and the thorax are laden with mechanoreceptors (stretch receptors), chemoreceptors, and pressure receptors. For example, the wall of each alveolus contains three types of mechanoreceptors activated by every inhalation and exhalation. By changing the pattern of breathing, we can change the messages sent from the respiratory system to the brain, rapid messages, with powerful effects on brain functions, including emotion regulation, attention, and cognitive function (Gerbarg & Brown, 2016b; Gerbarg et al., 2019; Mason et. al., 2021; Philippot et. al., 2002).

Numerous studies show that voluntarily regulated breathing practices (VRBPs) modulate autonomic nervous system functions, stress responses, cardiac vagal tone, Heart Rate Variability (HRV), vigilance, attention, central nervous system excitation, and neuroendocrine functions (Brown & Gerbarg, 2009; Brown et al., 2013; Zaccaro et al., 2018). Psychological and behavioral changes associated with these modulations are increased comfort, relaxation, positive emotions, energy, alertness, and connectedness, as well as decreased arousal, anxiety, depression, anger, and negative thinking. Coherent breathing or resonance frequency breathing at 4.0–7.0 breaths per min with equal inhalation and exhalation has been shown to optimally balance sympathovagal response for most adults (Vaschillo et al., 2004; Lehrer et al., 2010). Depression, anxiety, PTSD, ADHD, autism, schizophrenia, and other psychiatric disorders have been associated with autonomic imbalances, decreased parasympathetic nervous system (PNS) activity, and elevated or erratic sympathetic nervous system (SNS) activity.

Mind-body medicine, including VRBPs, may exert calming effects via interoceptive activation of vagal pathways, correction of autonomic nervous system (ANS) imbalances, and improved functioning of the HPA axis, as evidenced by reduced cortisol levels (Granath et al., 2006), and possibly by increasing GABAergic transmission (Streeter et al., 2020a). Central afferent autonomic pathways also project to the anterior cingulate, an area involved in evaluation, decision-making, emotion regulation, and fear extinction (Hopper et al., 2007).

Chanting is a form of resistance breathing in which vocal cord contractions increase resistance to airflow, creating sound, vibration, and vagal afferent stimulation. For example, the "Om" chant entails slow breathing against airway resistance, which induces psychological and physiologic relaxation (Telles et al., 1993). In a functional magnetic resonance imaging (fMRI) study, Om chanting significantly reduced activity in the limbic system (Kalyani et al., 2011).

Studies using direct intracranial electroencephalogram (iEEG) recordings in humans, correlated respiratory cycles with electrical activity across a widespread network of cortical and limbic structures (Herrero et al., 2017). Compared to automatic breathing, voluntarily controlled breathing increased iEEG-breath coherence (locking or entrainment) in the anterior cingulate, insula, amygdala, premotor, olfactory, and frontotemporal cortices. In addition, the conscious focusing of attention on breathing further increased iEEG-breath locking in the anterior cingulate, premotor, insular, and hippocampal cortices. A magnetoencephalography study in humans found that the conscious voluntary regulation of slow-paced breathing modulated cortical alpha activity in a well-structured pattern over wide areas of the brain (Hsu et al., 2020). In effect, voluntarily regulated breathing practices, especially slow-paced breathing with attention focused on the breath, can set the electrical activity of brain areas and neural networks involved in perception, assessment, executive function, cognition, and emotion regulation.

Qigong and Tai Chi

Qigong (qi gong, chi kung, or chi gung) and Tai Chi (T'ai chi ch'üan or Tàijí quán) are ancient Taoist and Chinese mind-body systems, akin to yoga, used for health and longevity. Both are low-impact, moderate-intensity exercises suitable for diverse patient populations with a wide range of ages and health conditions. Abbot and Lavretsky described Tai Chi and Qigong as involving sequences of flowing movements coupled with changes in mental focus (mindfulness), breathing, coordination, and relaxation (Abbot & Lavretsky, 2013). Characterized as "meditative movements," Tai Chi and Qigong are relatively safe and promote relaxation (Payne & Crane-Godreau, 2013). Physiological correlates include reduced sympathetic output, anti-inflammatory pathway activation, reduced expression of inflammation-related genes, reduce inflammatory markers, improved immune function, decreased levels of adrenocorticotropin hormone and cortisol, and increase serum endorphins. Electroencephalography (EEG) studies revealed increased frontal EEG theta activity, consistent with increased relaxation and attentiveness (Abbot & Lavretsky, 2013; Abbott et al., 2017).

A systematic review of Tai Chi interventions found 31 Tai Chi RCTs published from 2002 to 2007 and 11 RCTs from 1992 to 2001 (Li et al., 2011a). Only 23% of RCTs provided adequate details of the Tai Chi intervention. Problematic issues include small sample size, different styles of Tai Chi and Qigong, significant variance in practice duration and frequency, and differences in study durations. Most of the research has been on patients with medical problems and secondary psychological issues. Results from RCTs evaluating Tai Chi and Qigong for specific mental disorder suggest that they may be effective for reducing stress, anxiety, and depression, as well as for improving self-esteem and general psychosocial well-being. Methodologically robust studies of Tai Chi and Qigong for mental disorders are needed.

Internal Qigong primarily uses concentrated attention with inhalation, exhalation, and breath holds to increase and stimulate the circulation of blood and the "vital life energy" or "Qi." External Qigong focuses on physical movements with internal quiescence. Most older studies of Qigong are of limited value because of methodological limitations and small sample sizes. However, more recently, the effects of Qigong on HRV (used as a measure of sympathovagal activity) were assessed in several studies comparing healthy sedentary adults with healthy Qigong trainees. During controlled respiration, Qi training increased high-frequency (HF) power and decreased the low-frequency/high-frequency (LF/HF) power ratio of HRV, indicating increased cardiac parasympathetic tone. Moreover, experienced Qigong practitioners had higher HRV than age-matched sedentary controls (Lee et al., 2005a, b). A review of Qigong-induced mental disorders (Ng, 1999) concluded that incorrect or excessive practice of Qigong could induce abnormal psychosomatic responses and mental disorders. Most cases occurred in individuals with underlying bipolar or psychotic disorders in whom Qigong may have been too activating.

As low-intensity exercises with minimal risk of adverse reactions, Tai Chi and Qigong have shown benefits as adjunctive treatments to reduce psychological and physical symptoms of stress, sleep disorders, and anxiety, in diverse populations including geriatric patients, military personnel, and veterans with mild traumatic brain injury (TBI) (Abbot et al., 2017).

Yoga and Qigong for Anxiety and Posttraumatic Stress Disorder (PTSD)

The practice of yoga is multifaceted. The Eight Limbs of Yoga encompass moral and ethical disciplines (yama), positive spiritual observances (niyama), physical postures (asanas), breath control (voluntarily regulated breathing) (pranayama), conscious sensory withdrawal (pratyahara), focused (single point), concentration (dharana), meditation (awareness without focus) (dhyana), and a state of bliss, enlightenment, oneness with the universe (samadhi). Most Western research focuses on physical postures, breathing techniques, and meditative or relaxation practices.

Brown and Gerbarg reviewed open and controlled trials of breath practices as well as multicomponent interventions that emphasize breath practices (Brown et al., 2013; Brown & Gerbarg, 2017). Four RCTs and one open study of slow-paced breathing showed significant improvements in symptoms of stress, anxiety, anger, exhaustion, and depression associated with increased HRV, indicating activation of vagal PNS pathways. In addition, seven studies of anxiety disorders, GAD, and insomnia, consisting of two RCTs, two controlled trials, and three open pilots, found significant reductions in anxiety with breath-focused yoga programs. Two of the pilot studies evaluated yoga programs for severe treatment-resistant GAD with comorbidities (PTSD, OCD, and phobias). The patients were highly symptomatic despite trials of anxiolytic medications, group therapy, individual therapy, and Mindfulness Based Stress Reduction (MBSR). In the first study ($n = 31$) using the 22-h Sudarshan kriya yoga (SKY) program, the response rate was 73% and the remission rate was 41% (Katzman et al., 2012). The second study of severe treatment-resistant GAD ($n = 20$) using the Breath-Body-Mind 12-h program of breath practices (Coherent Breathing, Resistance Breathing, Breath Moving, "Ha" Breath, Qigong, and meditation with Open Focus Attention Training (Fehmi et al., 2017) resulted in significant improvements in anxiety, sleep, and depression (Gerbarg et al., 2019)). Significant reductions occurred in the mean BAI and BDI-II scores. Changes were also noted on the Anxiety Sensitivity Index and Penn State Worry Questionnaire (PSWQ). The results of this pilot study suggest that the Breath-Body-Mind Workshop represents a potentially valuable adjunctive to standard therapy and medication in patients with GAD and warrants further investigation.

Studies of yoga and Qigong programs emphasizing breath practices for PTSD (three RCTs, one controlled study, and two open pilots) found significant improvements in PTSD, anxiety, and depression among survivors of partner abuse, mass disasters, and military service. Multicomponent programs that emphasize breath practices have been used to relieve psychological

suffering among large groups of civilians and military personnel in the wake of natural and "man-made" disasters such as floods, earthquakes, industrial accidents, wars, and terrorist attacks (Gerbarg et al., 2011). For example, in a comparison-controlled study, 183 coastal village survivors of the 2004 Asian Tsunami who had been living for 9 months in refugee camps in Nagapattinam and who scored above 50 on the Posttraumatic Stress Disorder Checklist-17 (PCL-17), were assigned to one of three groups. The first was given Breath Water Sound (BWS), an 8-h group program of Sudarshan Kriya Yoga (SKY) breath techniques. The second group received the same 8-h BWS program followed 1 week later by Trauma Incident Reduction (TIR) in which a trained facilitator encouraged the subject to describe repeatedly the trauma over one to four individual sessions until they achieve emotional relief. The third group served as a wait-list control for 6 weeks and then received BWS. On post-test at week 1, the BWS group very significantly reduced their scores on the PCL-17 by more than 60% and reduced their scores on the Beck Depression Inventory (BDI) by more than 90%. It is noteworthy that after the intervention, the benefits were sustained with further improvement on test measures at 6-week, 3-month, and 6-month follow-ups. No significant change occurred in the wait-list group over 6 weeks. The addition of TIR showed no further impact on PCL-17 scores, possibly due to a floor effect, the scores having dropped so low after BWS (Descilo et al., 2010).

Singh, Telles, and Balkrishna reviewed eleven RCTs and the underlying neurophysiological mechanism of yoga interventions for PTSD (Singh et al., 2017). For example, among 100 subjects with combat-related PTSD, scores on the Posttraumatic Stress Disorder Checklist Civilian (PCLC) for those assigned to 16 weeks of Satyana yoga improved significantly more ($p < 0.05$) compared to those given mandatory assistance protocol (Colombian Agency for Reintegration). The greatest improvement occurred in the re-experiencing of symptom cluster (Quiñones et al., 2015).

In a rater-blinded RCT, 31 Australian Vietnam veterans on medication, who were classified as

100% disabled (for over 35 years) due to treatment-resistant PTSD, unresponsive to multiple medication trials and individual and group therapies, were randomized to either a 6-day 22-h SKY intervention (adapted for veterans) followed by 2-h monthly group sessions or a 6-week wait-list control. The intervention group showed a significant decrease in Clinician Administered PTSD Scale (CAPS) scores at 6 weeks compared with no change in the control group. After week 6, the control group received the same 6-day SKY intervention and subsequently showed comparably significant improvements on CAPS and other measures. In both groups, the benefits were sustained with further improvements at 6-month follow-up. Although the study was small, it yielded a large effect size, suggesting that multicomponent interventions with yoga breath techniques may offer a valuable adjunctive treatment for veterans with PTSD (Carter et al., 2013).

Meditation

In reviewing yoga-based meditation and mind-body regulation, Telles and Singh (2013) noted that early studies of transcendental meditation and later studies employing open monitor meditation (OM) reported shifts toward parasympathetic dominance with a wakeful but relaxed mental state. In contrast, some studies of meditation techniques using focused attention (FA) showed increases in sympathetic activity.

Meditative and yogic practices are being integrated into mindfulness-based stress reduction (MBSR) and mindfulness-based cognitive behavioral therapy (MBCBT) (Stratyner, 2006). These programs are becoming more sophisticated in their approaches to affect regulation and awareness of internal somatic experience. Numerous meditation practices include breath-centered work, such as slow breathing, focused attention on the breath, or deep abdominal breathing.

Enhancement of alpha and theta activity and phase synchrony (Satyanarayana et al., 1992; Lehmann et al., 2001; Hebert et al., 2005), interhemispheric asymmetry in alpha and beta activity, and high-voltage gamma synchrony (Benson et al., 1990; Davidson et al., 2003; Lutz et al., 2004) occur with meditation and yoga breathing. Data from EEG, positron emission tomography (PET), and fMRI studies are being used to study the effects of mind-body medicine on emotional and cognitive processing, connectivity, neuroplasticity, and brain aging (Lazar et al., 2005). The long-term practice of meditation and yoga breathing has been associated with sustained changes in resting EEG, which may reflect changes in brain structures in areas being exercised during those practices. Repeated experience may lead to neuroplastic development in specific cortical regions. The insula or interoceptive cortex has been associated with interoception, the perception of sensations from inside the body (see above). Lazar et al. (2005) found that this area showed an increased thickness in long-term meditators compared with controls. Awareness of breathing is fundamental to the practice of meditation and yoga. Most practitioners learn to focus on the breath as the first interoceptive exercise. Attending to the breath leads to slowing of the respiratory rate. In the Lazar study, the respiratory rate correlated with cortical thickness in the inferior occipitotemporal visual cortex and the right anterior insula.

A review of meditation for anxiety disorders (Chen et al., 2012) identified 36 RCTs. Of these, 25 reported statistically superior improvements in anxiety in the meditation group compared with control, with moderate effect sizes. However, in most of these studies, anxiety was a secondary outcome measure, and although the subjects had anxiety, they were not diagnosed as having an anxiety disorder. No adverse effects from meditation were reported. Because most of the studies measured only improvements in anxiety symptoms, further studies are needed before these findings can be generalized to patients who have been diagnosed with anxiety disorders.

MBSR techniques have been incorporated into a program developed by Kabat-Zinn (Kabat-Zinn et al., 1992), who performed mindfulness meditation for GAD and panic disorder in 24 participants. In this open study, the intervention included an 8-week course in meditation, followed by a weekly 2-h class of daily meditation of up to 45 min, and a 7.5-h silent retreat 2 weeks before the program ended. Meditation included body scan, sitting meditation, and mindful hatha yoga (postures with breath control). Significant declines were found in scores on HAM-D, HAM-A, and BAI, and frequency of panic attacks over the course of the intervention and 3-month follow-up. A follow-up study of 18 of the original subjects showed that most continued mindfulness practices and maintained improvements on HAM-A and BAI (Miller et al., 1995). This study supports the short- and long-term benefits of the intervention but requires validation by larger RCTs.

In a review of mindfulness meditation, MBSR and mindfulness-based cognitive therapy (MBCT), Marchand (2013) offered guidelines for the use of these approaches in psychiatric disorders. He noted that although a substantial number of studies suggest benefits in psychiatric disorders, most are limited by poor methodology, nonrandomization, lack of controls, and small sample sizes. The strongest evidence supports MBCT for reducing residual symptoms and the rate of relapse in patients with unipolar depression who have had three or more prior episodes (Marchand, 2013). Evidence also supports the use of MBSR and MBCT as adjunctive treatments for anxiety, pain management, and stress management.

Meditative Movement

The term "meditative movement" (MM) has been proposed to denote forms of exercise that use movement with meditative attention to body sensations, such as interoception, proprioception, and kinesthesis. (Payne & Crane-Godreau, 2013) Thus, it includes Qigong, Tai Chi, some forms of yoga, and Western somatic practices. A review of studies found encouraging evidence for benefits in anxiety and depression, but conclusions were limited by generally poor methodological quality of studies (Payne & Crane-Godreau, 2013). The authors extended the discussion of modes of action to include smooth rhythmic motions, grounding and posture, motor imagery, and the possible role of the basal ganglia. The evidence reviewed for possible mechanisms of action was consistent with the evidence and theoretical models previously proposed by various groups (Brown & Gerbarg, 2005; Streeter et al., 2012; Rao et al., 2013).

Technology Assisted Mind-Body Interventions

Technology-assisted mind-body interventions provide cues, for example, sounds or visual displays to pace the breath rate. In addition, some include real-time feedback based on physiological parameters such as HRV. Muench and Kumar reviewed the growing use of digital health technologies for guided meditation, mindfulness, cognitive restructuring, breath training, physiological assessment and feedback, and entrainment (Muench & Kumar, 2017). Studies using HRV feedback in military and first responder populations found that resiliency training using breath pacing at 5 or 6 cpms was associated with reduced HRV during cognitive stress. Covariants found to affect the HRV response to stress included: timing of HRV application; mental and physical health scores; coping and posttraumatic growth indicators; being open to new possibilities; and emotional support. Further studies of effects of relaxation breathing on stress, respiratory sinus arrhythmia, and low frequency HRV should take into account potential covariates (Hourani et al., 2019). Another challenge was that 90% of trainees stopped using their mobile apps within 30 days. In a small sham-controlled study of veterans with PTSD, device-guided slow breathing reduced muscle sympathetic reactivity to cognitive stress (Fonkoue et al., 2020).

Technology-assisted breathing and guided meditation using small home units or downloadable apps for self-regulation are increasingly popular and can be beneficial in mental health treatment. Initial in-person training with ongoing support for continued practice can improve results. Research on the efficacy of specific protocols in psychiatric disorders and on how to maintain user engagement for long-term benefits is needed.

Mind-Body Medicine, Clinical Issues, Contraindications

In recommending mind-body medicine to patients, clinicians are advised to prescribe programs that can achieve significant results with a relatively small amount of practice time. It may be difficult to motivate most patients to do more than 20 min of any daily practice. Some techniques are best learned in courses with certified instructors. However, there are beginner techniques that therapists (or their staff) who obtain training can easily integrate into their treatment approaches with patients. Most people are able to participate in yoga breathing and meditation courses without difficulty.

Clinicians should be aware of the following precautions: pregnant women and patients with uncontrolled hypertension, migraine headaches, severe chronic obstructive pulmonary disease, acute asthma symptoms, recent injuries of the neck, shoulders, or chest, or recent myocardial infarction should not do breath holds, Bhastrika, Kapalabhati, or any rapid or forceful yoga breathing. Coherent Breathing, gentle resistance breathing, such as "Ocean Breath" or "Ujjayi" (without breath-holds), and meditation are safe and soothing. Rapid yoga breathing is generally contraindicated in seizure disorders, while slow calming practices, such as Coherent Breathing, may reduce seizure frequency (Streeter et al., 2020b). Bipolar patients may be triggered to become manic, particularly from rapid cycle breathing, Bhastrika, "Ha" Breath, Kapalabhati, or any other activating practice. Psychosis is a contraindication for intense yoga breathing. However, patients with schizophrenia have been shown to benefit from slow, gentle Coherent Breathing, alternate nostril breathing, and gentle yoga postures (see the section on Mind-Body Medicine in Schizophrenia).

Clinicians who learn about mind-body medicine by taking classes themselves are better equipped to make appropriate referrals, prepare patients, and help overcome obstacles to daily practice. Although there are immediate gains, long-term benefits may not appear for 6–12 months, a time course similar to vagal nerve stimulation and biofeedback. Practitioners who become proficient in yoga breathing can learn how to teach the simpler practices to their patients, providing a safe, non-addicting, inexpensive method to reduce anxiety, insomnia, and other symptoms of hyperarousal. Some psychiatric patients are unable or unwilling to undertake yoga classes because of social anxiety, disbelief in the benefits, discomfort with classes that include Eastern religious elements, time constraints, or financial limitations. In such cases, it can be very helpful to teach the patient one or two basic yoga breath techniques in the office. For example, when skillfully taught, Coherent Breathing will often provide an immediate experience of calmness, relaxation, quieting of the mind, and relief from distress.

Acupuncture

A widely used traditional Asian Acupuncture technique uses fine needles inserted in specific "acupoints" for treating "energetic imbalances." The acupuncturist may manually manipulate the needles or connect the needles to an electrical current. A review of 10 randomized and two non-randomized trials of acupuncture for anxiety included four RCTs of patients with GAD or anxiety neurosis, and six RCTs of patients with acute perioperative anxiety. These 4- to 6- week trials found similar efficacies comparing acupuncture with pharmacotherapy (Pilkington et al., 2007). However, small sample sizes restricted generalizability of the findings. A systematic review and meta-analysis of 20 RCTs of acupuncture for GAD, including 18 studies published in Mandarin, found that acupuncture was more effective than the control condition, with a standard mean effect size of $- 0.41$ (95% CI $- 0.50$ to 0.31; $p < 0.001$), and that acupuncture was well tolerated and safe in treating GAD (Yang et al., 2021).

In summary, although acupuncture appears effective in short-term trials for anxiety symptoms, clear recommendations for specific anxiety disorders cannot be formulated at present owing to methodological weaknesses of most studies and the variability of techniques. In reviewing acupuncture research, Errington-Evans (2012) agreed with the above limitations but pointed out that despite the generally poor methodological quality and a wide range of treatment protocols, the volume of literature consistently reports statistically significant results in a wide range of conditions, suggesting real, positive outcomes using a treatment method preferred by many individuals who resist conventional medicines. From this perspective, acupuncture has a place among integrative treatments while the evidence base is being fortified by higher quality studies.

Schizophrenia and Other Psychotic Disorders

Well-designed studies of ICM therapies used in combination with conventional antipsychotics (medications for psychosis) have yielded positive findings for omega-3 essential FAs, especially EPA, Withania somnifera, *Ginkgo biloba*, yokukansan, the amino acid glycine, and some mind-body medicine. Less robust evidence supports ginseng and L-theanine in the treatment of psychosis, agitation, and anxiety. Medicinal herbs and nutrients have also been found to reduce side effects of antipsychotic medications, including sedation, lethargy, akithesia, and extrapyramidal symptoms (Brown et al., 2009; Sarris et al., 2009b). In patients with schizophrenia, mind-body medicine, such as yoga, are valuable add-on therapies that can improve quality of life, cognitive symptoms, negative symptoms, and physical health (Govindaraj et al., 2020).

Herbal Medicines

Withania somnifera (Ashwaganda)

A 12-week DBRPCT in 66 patients with schizophrenia or schizoaffective disorder found that a standardized extract of Withania somnifera (1000 mg/day) added to their antipsychotic medication significantly reduced Positive and Negative Syndrome Scale (PANSS) negative, general, and total symptoms, as well as perceived stress (PSS) compared to placebo (Chengappa et al., 2018). Secondary analysis of this study revealed significant improvements in indicators of depression and anxiety-depression cluster subscale scores on the PANSS (Gannon et al., 2019). Since the PANSS measures a wide range of symptoms, this sub-analysis suggests that the improvement is in in psychosis.

Panax quinquefolium (American Ginseng)

Individuals diagnosed with schizophrenia frequently show severe deficits in working memory, which interfere with functional improvements. In a 4-week DBRCT involving 64 stable individuals with schizophrenia, maintained on treatment as usual, those who were given a proprietary American ginseng extract showed significant improvements in visual working memory versus those given placebo (Chen & Hui, 2012). Of note, individuals taking American ginseng reported significantly fewer adverse neurological effects after 4 weeks compared with the placebo group. Further studies are needed to confirm these findings and determine optimum dosing of American ginseng as an adjuvant to antipsychotic medications.

Ginkgo biloba (Ginkgo)

A recent meta-analysis of eight DBRPC trials found supportive evidence for a potential role for ginkgo to improve negative symptoms, tardive dyskinesia (TD), and cognitive functions in patients with schizophrenia (Brondino et al., 2013; Zhang et . al., 2011). A DBRPC comparison study of 109 patients with schizophrenia found that those given haloperidol plus a ginkgo extract (EGb-761 360 mg/day) had significant reductions in Scales for Assessment of Negative Symptoms (SANS) and Scales for Assessment of Positive Symptoms (SAPS) compared with patients treated with haloperidol alone (Zhang et al., 2001). In addition, the group treated with EGb-761 had fewer "extrapyramidal side effects." These benefits were hypothetically attributed to EGb scavenging of free radicals. For a review of ginkgo adverse effects see the section on Herb-Drug Interactions under Herbal Medicines above.

Yokukansen

Yokukansan (YKS) is an herbal formula in the Japanese Kampo medicine derived from Traditional Chinese medicine. YKS (TJ-54 extract) contains *Atractylodes lancea, Poria, cnidium rhizome, Uncaria uncus, Angelica acutiloba radix, Bupleuri radix, and Glycyrrhiza radix*. In a DBRPCT, 120 patients with treatment-resistant schizophrenia on standard medications were randomized to receive either TJ-54 (Yokukansan) 7.5 g/day or placebo for 4 weeks. Compared to placebo, YKS showed a greater, but statistically nonsignificant reduction in all PANSS scores. The decreased score reached statistical significance with YKS only on PANSS excitement/hostility factor scores (Miyaoka et al., 2015a, b). No substantial adverse effects were reported. When YKS was traditionally used to treat children, few adverse events were reported. However, since it is increasingly prescribed for elderly patients with dementia, the incidence of adverse reactions has increased, mainly due to the effects of licorice, *Glycyrrhiza radix*: pseudoaldosteronism, hypokalemia, hypertension, rhabdomyolysis, heart failure, arrhythmia, liver dysfunction, and interstitial pneumonia (Shimada et al., 2017).

Camellia sinensis (L-Theanine Derived from Tea)

Tea contains the amino acid L-theanine, which may reduce state anxiety by increasing alpha activity and increasing the synthesis of GABA (Vuong et al., 2011). Noticeable anxiety reduction is generally achieved within 30–40 min with effective doses between 200 and 800 mg/day. Preliminary findings suggest that L-theanine may be an effective adjuvant in the management of psychotic disorders. In an 8-week double blind, placebo-controlled study, 60 patients diagnosed with schizophrenia or schizoaffective disorder were randomized to L-theanine versus placebo while continuing their antipsychotic medications (Ritsner et al., 2011). Forty patients who completed the study experienced moderate reductions in symptoms of psychosis and anxiety; however, there were no differences in quality-of-life measures between the L-theanine and placebo groups. Larger DBRPCTs are needed to replicate these findings, assess safety issues, and determine optimum dosing before L-theanine can be recommended as an adjunctive treatment for psychosis in acute mania or schizophrenia.

Nutrients in Treatment of Schizophrenia and Other Psychotic Disorders

Omega-3 Fatty Acids

Preliminary findings suggest that omega-3 FAs reduce the risk of progression to psychotic disorders in young people experiencing subthreshold symptoms of psychosis. In a 12-week placebo-controlled study, 81 individuals considered to be at high risk for developing schizophrenia were randomized to omega-3 FAs (1.2 g/day) versus placebo (Amminger et al., 2010). For the 76 individuals who completed the intervention, 11 (27.5%) in the placebo group transitioned to a psychotic disorder versus only 2 (5%) subjects in the omega-3 group. Those who received omega-3 FAs had significantly fewer positive and negative psychotic symptoms and improved global functioning compared with the placebo group. A larger follow-up 6-month study ($n = 304$) of omega-3 FAs (1.4 g/day) with Cognitive Behavioral Therapy (CBT) did not reduced the risk of transition to psychosis in an ultra-high-risk youth cohort (Amminger et al., 2020) in the initial analysis. However, the 6-month follow-up analysis did show that relative increased serum levels of omega-3 FAs were associated with better mental functioning. In a small DBRPCT, EPA had significant antipsychotic efficacy when used as a monotherapy in individuals naïve to antipsychotic medications experiencing a first psychotic episode (Peet et al., 2001). Of note, individuals taking adjunctive EPA had fewer positive symptoms such as hallucinations, paranoia, and delusions by the end of the study compared with the placebo group. A larger 3-month study (Berger, 2007) on EPA augmentation in first-episode schizophrenia found no significant differences between EPA and placebo. However, patients treated with EPA required lower doses of antipsychotics for symptom control and reported fewer sexual and neurologic adverse effects compared with individuals taking a placebo with antipsychotics (Berger, 2007). These findings suggest that EPA augmentation may be a reasonable option for improving the tolerability of conventional antipsychotics, especially when treating first episodes.

Findings on omega-3 FAs in established cases of schizophrenia are less promising. A meta-analysis of placebo-controlled studies on EPA augmentation in patients with chronic schizophrenia or other psychotic disorders found no evidence for greater benefit of EPA augmentation over antipsychotics alone (Fusar-Poli & Berger, 2012). A more recent meta-analysis also yielded no supportive evidence of efficacy in treating schizophrenia (with mixed data showing potential to prevent transition of at-risk youth to schizophrenia) (Cakici et al., 2019; Hsu et al., 2020). Possible explanations of negative findings include small study size; absence of clinical efficacy of EPA supplementation in schizophrenia; a "ceiling effect" on the response to EPA augmentation associated with concurrent treatment with antipsychotics; combinations of different fatty acids may be more effective than EPA alone; dietary differences or other sociodemographic factors; and omega-3 augmentation may be effective only as a preventive agent or following a first psychotic episode. EPA supplementation has beneficial effects on health and may accelerate treatment response and reduce the size of the antipsychotic dose required for efficacy in some cases, thereby resulting in fewer neurologic, metabolic (Kris-Etherton et al., 2002), and sexual adverse effects (Berger et al., 2007). Pending the results of future studies, it is reasonable to recommend omega-3 FA supplementation, particularly EPA 1–3 g/day, to individuals diagnosed with schizophrenia.

N-Acetylcysteine (NAC)

A recent statistically significant meta-analysis of seven DBRPC trials ($n = 440$) showed supportive evidence for reducing negative symptoms in schizophrenia (Yolland et al., 2020). For example, in a 6-month DBRPC study of 140 patients with treatment-refractory schizophrenia, the group receiving NAC (1000 mg BID) had significant improvements in negative symptoms, global function, abnormal movements, and akathisia compared with those receiving placebo (Berk et al., 2008). Blinded thematic analysis of the qualitative data from the study also revealed beneficial effects from NAC compared with placebo in improving several affective domains (Berk et al., 2011). The neuroprotective effects of NAC may also be useful in treating schizophrenia and reducing medication-induced side effects (Akhondzadeh et al., 2013).

Glycine

High doses of glycine, the smallest amino acid, may enhance global brain functioning in psychosis by helping to correct abnormal N-methyl-D-aspartate receptor (NMDAR) functioning implicated in the pathogenesis of schizophrenia. In a 6-week double-blind crossover study, 22 patients with treatment-resistant schizophrenia on conventional antipsychotics, given augmentation with glycine 0.8 g/kg (up to 60 g/day) showed significant reductions in negative symptoms and improved global functioning compared with those given placebo (Heresco-Levy et al., 1999). The large doses of glycine required for response may not be practical for many chronically psychotic patients. A meta-analysis of 29 DBRCTs (1253 cases) concluded that glycine (and additionally D-serine and sarcosine) has therapeutic benefit as adjuncts to non-clozapine antipsychotics in the treatment of negative and total symptoms of chronic schizophrenia. Although glycine improves positive and total symptoms as an adjuvant to non-clozapine antipsychotics, it worsens symptoms when added to clozapine (Singh & Singh, 2011).

Vitamin E

Vitamin E was thought to improve symptoms of tardive dyskinesia (TD) via its antioxidant capability. A systematic review of RPCTs of vitamin E in the treatment of antipsychotic medication-induced TD, an often-irreversible neurologic disorder, found no evidence that vitamin E improves symptoms of TD. Evidence from poorly reported RCTs suggests that it may slow the rate of deterioration (Soares-Weiser et al., 2018).

Melatonin

There is evidence that melatonin, a pituitary neurohormone that regulates circadian rhythms, may enhance the efficacy of antipsychotics via anti-inflammatory and antioxidative effects (Dodd et al., 2013). Research suggests that melatonin may improve sleep disturbances commonly seen in schizophrenia, while also lessening the severity of TD and metabolic syndrome associated with antipsychotics (Dodd et al., 2013). Incidentally, an analysis of patient data from Cleveland Clinic's COVID-19 registry suggested that melatonin usage was associated with a nearly 30% reduced likelihood of testing positive for SARS-CoV-2 after adjusting for age, race, smoking history and various disease comorbidities (Zhou et al., 2020).

While an interesting finding, it is too early to reach any conclusions about its true therapeutic impact on reducing COVID-19.

Mind-Body Medicine

A review of studies on mind-body approaches in schizophrenia and other psychotic disorders by Helgason and Sarris (2013) found substantial evidence for music therapy, meditation and mindfulness techniques, multimodal approaches combining different mind-body techniques, and yoga, with less evidence for relaxation training. A more recent review of the effects of yoga add-on treatment for schizophrenia from the National Institute of Mental Health & Neurosciences (NIMHANS) of India found improvement in quality of life, cognitive symptoms and negative symptoms (pooled mean effect size 0.8, 0.6, and 0.4, respectively (Govindaraj et al., 2020). Evidence also indicates yoga benefits for quality of life and physical health. Small effects on positive symptoms were noted. Proposed mechanisms for yoga effects include anti-inflammatory action as well as HPA and autonomic modulation. However, research is needed on neurobiological mechanisms. Findings for hypnosis, thermal or electromyography biofeedback, dance therapy, drama therapy, and art therapy were inconclusive. Most studies were limited by small size, absence of randomization, nonequivalent controls, and lack of standardized measures. Unstructured meditation is not recommended in patients with schizophrenia or psychosis because it may lead to increased psychotic thinking and undermine the sense of contact with reality.

Rao et al. (2013) of NIMHANS reviewed clinical studies in schizophrenia. In a 4-month RCT of 61 moderately ill patients with schizophrenia, stabilized on antipsychotic medications, those assigned to yoga had significantly greater improvements in scores for negative symptoms and social function than those assigned to exercise (Duraiswamy et al., 2007). An 8-week randomized wait-list control study of 18 clinically stable patients with schizophrenia demonstrated significant improvements in PANSS and quality of life

scores in those given a supervised yoga program (Visceglia & Lewis, 2011). The participants in a three-arm rater-blinded study of 119 outpatients with schizophrenia attended approximately 75% of the 25 treatment sessions consisting of 45 min of either yoga or exercise for one month. The yoga group showed significantly greater improvements after 1 month in negative symptoms compared with the exercise group and an inactive wait-list group. The yoga intervention group also significantly improved their scores on a measure of facial emotion recognition compared with the other groups (Varambally et al., 2012).

An open study of yoga for patients with schizophrenia showed significant improvement in cognitive domains, specifically speed, accuracy, and efficiency (Bhatia et al., 2012). A subsequent 3-month (1-h sessions three times weekly) rater-blinded open study the effects of a yoga or Qigong plus breathing practices on 36 chronic schizophrenic patients stabilized on psychotropics and living in residences at the Nathan Kline Institute in New York demonstrated significant improvements in Cognitive Scores on Repeatable Battery for Assessment of Neuropsychological Status (RBANS) Total Scores and sub-scores for Attention, Delayed Memory, Figure Copy, Visual-Spatial Construction, Semantic Fluency, and Language Index (Smith et al., 2013). There were no significant changes in PANSS scores, although there were trends for decreases in Depression Factor, General Factor, and others.

An RCT involving 43 stabilized medicated patients with schizophrenia found that those who participated in a 1-month yoga program had significant elevations in endogenous plasma oxytocin levels compared with a wait-list control group. Approximately half of the yoga time was used for slow breathing practices. In addition, their facial emotion recognition improved, suggesting possible support for the Social Engagement System (Jayaram et al., 2013). Although the current evidence for mind-body techniques in the treatment of schizophrenia has limitations, it is encouraging. These approaches may be safely used in conjunction with antipsychotics without increasing the burden of side effects (Visceglia, 2007). The design of programs should take into account the physical condition of the patients, many of whom are sedentary, obese, and have difficulty balancing or maintaining static postures. Walking and gentle fluid movements may be more suitable and entail less risk of injury, particularly in heavily medicated or sedentary patients (Smith et al., 2013). Larger controlled studies are needed to confirm preliminary findings, to identify effective, feasible mind-body approaches, and to determine the optimum frequency and duration of treatment for schizophrenia and other psychotic disorders.

Sleep Disorders

An analysis of the United States National Health Interview Survey data from 2002 by Pearson et al. (2006) found that of the 17.4% of adults ($n = 93,386$) reporting insomnia or regular sleep disturbance in the preceding month, 4.5% used ICM to enhance sleep. This result extrapolated to 1.6 million noninstitutionalized civilian US citizens. Biologically based products such as herbal or nutritional medicine, and mind-body therapies such as Tai Chi or yoga, were among the most commonly used interventions. A total of 72% stated that they believed that ICM helped their insomnia "a great deal" or "some" (49% and 33%, respectively). The most studied ICM treatments for sleep disorders are acupuncture, acupressure, melatonin, and valerian (Sarris & Byrne, 2011). Systematic reviews and meta-analyses support the use of acupuncture or acupressure (Gooneratne, 2008; Yeung et al., 2009). However, these reviews all point to concerns over inconsistent results and heterogeneous methodology.

Valeriana officinalis (Valerian)

The active constituents from the pungent valerian roots are believed to interact with GABA-A receptors (Benke et al., 2009). One meta-analysis found that in 9 out of 16 RCTs valerian extract did not improve sleep quality (Bent et al., 2006). Another meta-analysis of 18 RCTs found that valerian reduced sleep latency more than placebo by only 0.70 min (not significant) (Fernández-San-Martín

et al., 2010). A more recent meta-analysis concerning valerian alone, evaluated its effectiveness to improve subjective sleep quality (10 studies, $n = 1$) (Shinjyo et al., 2020). Results revealed mixed findings, with only weak support for use in insomnia, though there were no severe adverse events associated with valerian use. One DBRPC crossover study of 16 mild insomniacs using polysomnography reported a significant decrease in slow-wave sleep (SWS) onset latency and an increase in the percentage of SWS time (Donath et al., 2000). Valerian is generally safe, but some patients dislike the unpleasant taste and odor of valerian tea and tablets. The effect of valerian often improves over time, and maximum benefit may take 2 weeks may take 2 weeks to achieve. One advantage of valerian over some sedative/hypnotics (medications for insomnia) is that there have been no cases of habituation or abuse and only one case of possible withdrawal symptoms. Valerian should be avoided in pregnancy.

Melatonin

Melatonin (N-acetyl-5-methoxytryptamine), a neurohormone secreted by the pineal gland, regulates circadian rhythms by interacting with melatonin receptors. In patients with insomnia, some DBRPC trials show that melatonin reduces sleep onset latency (Kayumov et al., 2001) and increases sleep efficiency (Zhdanova et al., 2001), although others have failed to find beneficial effects on sleep. Melatonin has a direct action on sleep and on circadian rhythm; this dual activity makes it particularly useful in treating delayed sleep phase syndrome, insomnia in dementia, and jet lag. Numerous short-and long-term studies of melatonin for insomnia show improvements in sleep, mood, and memory; reduced sundowning; and delay in cognitive deterioration (for a review, see Modabbernia, 2017). A DBRPC crossover trial of 40 patients with PD revealed significant improvements in sleep using melatonin (50 and 5 mg) versus placebo (Dowling et al., 2005). Melatonin is a powerful antioxidant for the dopamine system, and therefore is a promising treatment for insomnia in patients with PD. Melatonin significantly improved sleep in a DBRPC cross-over trial of people with schizophrenia (Shamir et al., 2000) and in a pilot study of 11 patients with bipolar disorder (Bersani & Garavini, 2000).

A review of melatonin describes benefits for counteracting metabolic effects of atypical antipsychotics. Mechanisms of action include restoring circadian rhythms; enhancing oxidative phosphorylation; and interaction with the appetite-regulating hormone, leptin. Small studies suggest that melatonin can reduce insulin resistance and weight gain associated with antipsychotics (Modabbernia, 2017).

In a DBRCT involving 320 people followed for 4 days after a long airplane trip, the group given 5 mg of fast-release melatonin slept better, fell asleep faster, and were more awake and energetic during the day than groups given 5 mg of slow-release melatonin, 0.5 mg of fast-release melatonin, or placebo (Suhner et al., 1998). Another small DBRCT found that airline crews were more rested using 10 mg of fast-release melatonin compared with placebo. The benefits of melatonin were equivalent to zopiclone, without next-day impairment of cognition (Paul et al., 2001). People who travel across five or more time zones benefit most from using melatonin to reduce jetlag (Arendt, 1997).

Melatonin is very low in side effects when taken in therapeutic doses of 1 to 9 mg per day. Further information on melatonin and clinical guidelines are given in Box 3.

> **Box 3: Melatonin: Safety/Efficacy Issues and Clinical Guidelines**
>
> The heterogeneity of melatonin studies has been attributed to variability in product quality and onset of action (slow versus fast release), diagnostic differences, comorbid conditions, and timing of administration. Consumers should choose supplements from mainstream companies with pharmaceutical-grade melatonin. Side effects, such as cramps, fatigue, dizziness, headache, and irritability, are infrequent and usually mild. Melatonin is a safe short-term alternative to prescription hypnotics for insomnia. In therapeutic doses, 1–9 mg, it has few side effects. Elderly patients who are susceptible to cognitive/memory impairment and daytime drowsiness from benzodiazepines may benefit from melatonin without risk of these adverse effects. Melatonin is best given 1–2 h before bedtime. Patients should be instructed to turn lights out for at least 30 min before bedtime to achieve the best effect. Taking melatonin during the bright light of daytime may sensitize the eyes to light damage. The safety of high-dose melatonin has not been established. The possibility exists that very high doses (above 50 mg/day) might have long-term effects on testosterone or prolactin levels. Some patients may experience nightmares. Clinicians who prescribe chronic melatonin should monitor patients for adverse effects (Bellon, 2006).

Mind-Body Techniques

Techniques such as yoga, particularly slow-paced breathing practices, may be safer than medication for treating insomnia, particularly in older adults. In an RCT, 69 residents in a home for the elderly (over the age of 60 years) were randomly assigned to three groups: a 6-week yoga program (postures, relaxation, breath practice, and lectures on yoga philosophy); an herbal Ayurveda preparation, Rasayana Kalpa (composed of *Withania somnifera* root 2 g, *Emblica officinalis* 1 g, *Sida cordifolia* 0.25 g, and *Piper longum* 0.5 g); or a wait-list control (no intervention) (Manjunath & Telles, 2005). On a self-rating sleep questionnaire, the yoga group showed a significant decrease in the time to fall asleep (average decrease of 10 min), an increase in the number of hours slept (average increase 60 min), and improved feelings of being rested compared with no change in the Ayurveda or wait-list control groups. Findings from subjective self-assessments suggest that yoga programs may be safe and effective for sleep disorders in the elderly.

Neurocognitive Disorders (NCDs)

Interventions to enhance neurocognitive function should begin in utero and continue throughout the lifecycle by providing (1) nutrients for the developing brain, (2) neuroprotective supplements for the aging brain, and (3) treatments for NCDs (Businaro et al., 2021). Phytomedicinals and nutraceuticals have the potential to improve a range of cognitive domains (e.g., working memory, attention, vigilance), particularly when suboptimal functioning is due to organic deficits or nutrient-deficiencies. The DSM-5 classification, NCDs, now include age-associated and mild to severe disorders of memory and cognitive function (American Psychiatric Association, 2015).

Herbal Cognitive Enhancers

Traditional herbal medicines are being studied for cognitive enhancement (Akhondzadeh & Abassi, 2006; Kwon and Lee, 2021). Extensive preclinical herbal studies document effects that could contribute to improvements in brain function, thus providing new avenues for development of the next generation of nootropics (brain-enhancing medicines) (Kennedy and Scholey, 2006). However, high-quality clinical studies with well-defined patient diagnoses are needed to strengthen the evidence base. Howes and Perry (2011) extensively reviewed neuropathological processes that could be targeted by herbal extracts and the role of phytomedicines in the treatment and prevention of dementia. Brown and Gerbarg (2011) reviewed the evidence for the use of herbs, nutrients, neurohormones, and nootropics in brain injury and provided guidelines for strategic combinations of herbs and nutrients that can be most effective in improving cognitive function.

Curcuma longa (Curcumin)
Curcumin, a polyphenolic compound derived from the rhizome of Cucuma longa, is the Indian herb used in curry powder. Although preclinical studies demonstrated anti-inflammatory, antioxidants, and anti-amyloid effects, human studies for cognitive enhancement have not shown significant results, possibly due to limited bio-availability. Building on previous trials of more bioavailable curcumin in non-demented adults, an 18-month DBRCT using a more bioavailable form (Theracurmin®) in non-demented adults showed significant improvements in long-term retrieval, visual memory, and attention compared to placebo. (Small et al., 2018) Brain imaging FDDNP-PET showed decreased binding in the amygdala compared to placebo, indicating reduced accumulation of amyloid and tau (associated with dementia progression). Neuroprotective effects of curcumin have been documented in preclinical studies, but human studies in patients with cognitive decline or dementia are needed.

Ginkgo biloba (Ginkgo)
Standardized ginkgo extracts (such as EGb 761 and G115) have been studied for cognitive-enhancing properties. Ginkgo may exert beneficial cognitive-enhancing effects via modulation of cholinergic and monoamine pathways, antioxidant effects, anti-inflammatory effects, GABAergic effects, and nitric oxide activity (Sarris, 2013a), having a potential additive effect with donepezil (Lautenschlager et al., 2012). In a recent review, Diamond and Bailey (2013) concluded that ginkgo has shown potential as a safe, adjunctive treatment for cognitive decline, memory decline, and neuro-psychiatric symptoms associated with AD and cerebrovascular disease. They suggested that additional controlled trials are needed with more sensitive, standardized measure and more detailed patient descriptions and diagnoses.

In the 5-year GuidAge clinical trial involving 2854 adults aged 70 years or older who spontaneously reported memory complaints to their primary-care physician were prescribed either 120 mg standardized EGb 761 twice per day or placebo. After 5 years, the standardized ginkgo extract did not reduce the risk of dementia compared with placebo (Vellas et al., 2012). Long-term memory on associational learning improved in a large DBRCT of 55–57-year-old subjects (Burns et al., 2006). In a 52-week DBRCT of 236 patients with AD, ginkgo extract EGb 761 120 mg/day improved cognitive performance in patients with mild-to-moderate cognitive impairment and slowed deterioration in patients with severe impairment (Le Bars et al., 2002). A multicenter DBRPC study of 410 patients aged 50 years or older with mild-to-moderate dementia (AD or vascular dementia) plus neuropsychiatric symptoms found that those who were treated with ginkgo improved significantly on Neuropsychiatric Inventory and measures of cognition function and quality of life compared with those given placebo (Herrschaft et al., 2012). Side effects from ginkgo are rare and can be minimized by starting at 60 mg/day and increasing gradually to 120 mg BID. Occasionally nausea, headaches, and skin rashes occur (see section on "Herb-Drug Interactions" above).

Panax ginseng (Korean Ginseng, Asian Ginseng)

Numerous preclinical studies of ginsenosides from Korean ginseng have demonstrated anti-inflammatory effects, neuroprotection against glutamate, amyloid-β, and ischemia, increased production of nitric oxide by endothelial cells (essential for blood flow and oxygen delivery), and neurogenesis (Kim et al., 2013). However, clinical studies of cognitive effects in healthy adults and in those with AD or vascular dementia have been small, variable in methodology, and inconclusive. In an 8-week DBRPC study of 112 healthy volunteers over 40 years of age, those given Korean ginseng 400 mg/day showed significantly better abstract thinking and a tendency to faster reaction time, but no significant differences in memory or concentration (Sorensen & Sonne, 1996). Another DBRPC crossover study of 32 healthy adults (aged 18–40 years) found that Korean ginseng (300–800 mg/day of G115) significantly improved reaction time, accuracy, calmness, and working memory. Side effects include overstimulation, anxiety, insomnia, tachycardia, gastrointestinal disturbances, headache, and reduced platelet aggregation (Reay et al., 2010).

Panax quinquefolium (American ginseng)

Another *Panax* species containing ginsenosides, with a differing constituent profile to *P. ginseng* is American ginseng. The ginsenoside Rb$_1$ is a cholinergic modulator which is found to be more highly expressed in American ginseng compare to Korean ginseng. Two DBRCTs have reported enhanced working memory following acute administration of *American ginseng* in healthy samples. A 2010 study, involving 20 middle-aged participants, investigated the acute cognitive effects of American ginseng administration on steady state visually evoked potentials (SSVEPs) during completion of working memory and continuous performance tasks (Scholey et al., 2010). Participants were administered different doses in a crossover design (100 mg, 200 mg, 400 mg) versus placebo on two separate testing sessions. Results showed that compared to placebo (at 6-h post administration), American ginseng significantly reduced prefrontal SSVEP latency. A later follow up DBRCT (White et al., 2020) involving 32 healthy young adults, assessed the acute effects of three doses (100, 200, 400 mg) of American ginseng on mood and neurocognitive outcomes (measured 1, 3, and 6 h following administration). Results revealed that there was a significant improvement of working memory performance associated with American ginseng. Corsi block performance was improved by all doses at all testing times, while choice reaction time accuracy and self-rated "calmness" were significantly improved at the 100 mg dose.

Crocus sativa (Saffron)

A 16-week DBRPC study of 46 patients with mild to moderate AD showed that those give 15 mg twice daily capsule of saffron (IMPIRAN) attained significantly better improvements on the AD Assessment Scale-Cognitive Subscale (ADAS-cog) and on the Clinical Dementia Rating Scale (CDR-SG) compared to placebo (Akhondzadeh et al., 2010). Adverse events were not significantly different between groups. Of note, two patients in the treatment group experienced mild hypomania. This may have been related to the antidepressant effects of saffron. Preclinical studies have indicated anti-inflammatory, antioxidant, and anti-amyloidogenic activity, relevant to cognitive enhancement.

Melissa officinalis (Lemon Balm), *Salvia officinalis* (Sage), *Salvia lavandulaefolia* (Spanish Sage), and *Lavendula officianalis* (Lavender)

In a 4-month DBRCT in 42 patients with mild-to-moderate AD, those given lemon balm (*Melissa officinalis*) oil aromatherapy showed significant cognitive benefit and reduced agitation compared with placebo (Akhondzadeh et al., 2003). A 4-week DBRPC trial of lemon balm aromatherapy in 71 patients with severe dementia documented decreased agitation and withdrawal with improvements in activities of daily living (Ballard et al., 2002). Effects of lemon balm lotion for dementia have been inconsistent (Howes & Perry, 2011).

Another 4-month DBRCT in 42 patients with mild-to-moderate AD found that sage improved scores for cognitive function significantly more than placebo (Modabbernia & Akhondzadeh, 2013). A DBRPC crossover trial of sage in 30 healthy adults found that a single dose of sage improved mood, anxiety, and cognitive performance significantly more than placebo (Kennedy et al., 2006).

In four DBRPC crossover studies, healthy adults (total $n = 90$) were given single doses of Spanish sage (*S. lavandulaefolia*) (Tildesley et al., 2005; Scholey et al., 2008; Kennedy et al., 2011). Doses ranged from 50 to 150 µL and from 167 to 1332 mg. Compared with placebo, Spanish sage was associated with significantly more improvements in cognitive performance, attention, accuracy, processing speed, and memory.

A review of aromatherapy noted three studies (1 RPCT, 1 RCT cross-over, 1 open trial of three essential oils) in which lavender oil aromatherapy reduced agitation in elderly patients with dementia (Scuteri et al., 2017).

Yi-Gan San Chinese Medicine and *Yokukansan* Japanese Kampo Medicine

The Chinese formula, Yi-Gan San (YGS), contains *Atractylodes lancea rhizoma, Poria sclerotium, Cnidium rhizoma, Uncaria uncis, Angelica radix, Bupleuri radix*, and *Glycyrrhiza radix*. It has been used to improve behavioral and psychological symptoms and activities of daily living in dementia patients. In a 4-week observer-blind RCT 52 patients with mild-to-severe dementia were randomly assigned to receive YGS or no drug. No significant changes were noted in the control group. In comparison, those who took YGS had significant improvements in mean scores on the Neuropsychiatric Inventory and Barthel Index for activities of daily living. No adverse effects were reported with YGS (Iwasaki et al., 2005). DBRCTs are warranted to support this preliminary finding.

A similar formula in Japanese Kampo medicine, Yokukansan is a mixture of, 4 g Atractylodes lanceae rhizoma, 4 g Poria, 3 g Cnidii rhizome, 3 g Angelicae radix, 2 g Bupleuri radix, 1.5 g Glycyrrhizae radix, and 3 g Uncariae uncis cum ramulus. In an open study, a Japanese formula, 7.5 g twice a day of Yokukansankachimpihange (Yokukansan plus *Citrus unshiu* peel and *Pinellia* tuber) was administered to 32 Alzheimer's disease patients for 8 weeks. During the study, patients continued their regular medications. Mean scores on the Behavioral Pathology in Alzheimer's Disease Rating Scale (Behave-AD) decreased from 20.8 at baseline to 13.1 at week 8. The greatest improvement occurred in symptoms of aggressiveness and sleep disorder, the symptoms that significantly burden caregivers of dementia patients, No side effects were reported. These encouraging findings need verification with larger DBRCTs (Katsumoto et al., 2021).

Nutrients

Folate and B Vitamins

Adequate vitamin and nutrient concentrations are necessary for neurological functioning. Low levels of folate and B vitamins are associated with poorer memory and cognitive function, particularly in subjects with elevated baseline plasma homocysteine (>11.3 µM) (Bottiglieri, 2013). Folate and vitamin B_{12} are essential for methylation pathways. Bottiglieri (2013) reviewed long-term (18–36 month) DBRPC studies of the effects of treatment with folic acid, methylhydrofolate (MTHF), vitamin B_{12}, and vitamin B_6 on cognitive function including three studies of community-dwelling adults aged 50–74 years (total $n = 1994$), two studies of adults with mild cognitive impairment (MCI) ($n = 434$), and two studies of subjects with AD (total $n = 480$). He concluded that in subjects with MCI, folate and B vitamin supplementation significantly improved global cognition, episodic memory, and semantic memory. Furthermore, reduction of homocysteine by folate and B vitamins was associated with significant slowing of brain atrophy (Smith et al., 2010).

The Cache County study of 4740 elderly subjects followed over 3 years showed that combined use of vitamin C plus E supplements reduced the prevalence and incidence of AD (Zandi et al., 2004). A meta-analysis of 10 RCTs ($n = 3200$) (duration ≥ 1 month) found that multivitamins were effective in improving immediate free recall memory, but not verbal fluency or delayed free recall memory (Grima et al., 2012). Further studies are needed to clarify specific vitamin and antioxidant combinations that would be most beneficial for addressing cognitive dysfunctions in different patient populations.

Given the relationship between B vitamins and cognitive function, folate, vitamin B_{12} and vitamin B_6 supplements may be beneficial in delaying brain atrophy, mild NCDs, and progression of AD. In addition, these vitamins can reduce the levels of serum homocysteine, a known risk factor for cardiovascular disease and cognitive dysfunction. Routine testing for MTHFR polymorphisms can identify those nonresponders who need to be treated with L-methylfolate rather than folic acid.

Omega-3 Fatty Acids

Omega-3 FAs, especially DHA, are important for neuronal development (Bourre, 2005, 2006). DBRCTs show that infants given an omega-3 FA-enriched formula have better brain and eye development (Jensen et al., 1996), better problem-solving ability at 10 months (Willatts et al., 1998), and an increase in scores on the mental development index (Birch et al., 2000). Increased free radical damage with loss of membrane fatty acids occurs with age and is associated with neurodegenerative disorders such as PD and AD. Abnormally low levels of n-3FAs have been found in patients with dementia and AD (Conquer et al., 2000) and a subset of patients with ADHD (Stevens et al., 1995). A prospective study of 1200 elderly subjects found that after 8 years, those with low DHA serum levels had a 67% greater chance of developing AD than those with high DHA levels (Kyle et al., 1998). Neurological symptoms improved in AD patients given DHA

(Nidecker, 1997). In a DBRCT of 174 AD patients, 600 mg EPA slowed cognitive decline in very mildly impaired patients over a 6-month period (Freund-Levi et al., 2006). In a 3.5-year observational study, a cohort ($n = 8085$) of non-demented patients aged ≥ 65 years, frequent weekly fish consumption was associated with reduced AD risk and, in ApoE $\varepsilon 4$ non-carriers, a reduced risk of developing dementia (Barberger-Gateau et al., 2007).

Although more research needs to be done in this area, supplementation with antioxidants and omega-3 FAs may protect against neurodegenerative processes (Youdim et al., 2000). A review of micronutrient studies concluded that definitive dietary recommendations are not yet plausible. Nevertheless, increased consumption of fats derived from fish, vegetable oils, non-starchy vegetables, low glycemic fruits, and reduced consumption of sugar and alcoholic beverages are recommended to reduce risks for dementia and other diet-sensitive diseases (Cardoso et al., 2013). Daily fish oil supplements without pentachlorophenol, mercury, or other contaminants are preferable for those who wish to limit fish consumption to once weekly. High intake of unsaturated, nonhydrogenated fats may lower the risk for AD, whereas high consumption of saturated or *trans-fats* may increase risk (Morris et al., 2003).

Phosphatidylserine (PS)

Phosphatidylserine (PS) is a small component of the inner phospholipid layer that contributes to nerve cell membrane fluidity. Derived from bovine neural tissue, PS has shown modest benefits in studies of AAMI (Caffarra & Santamaria, 1987; Crook et al., 1992) and AD (Amaducci, 1988). In an open 12-week trial in 30 adults with memory complaints (aged 50–90 years), soybean-derived PS (SB-PS) (300 mg/day) significantly improved cognitive performance, memory recognition, memory recall, executive functions, mental flexibility, total learning, and immediate recall. Also, SB-PS significantly reduced systolic and diastolic blood pressure and was well tolerated with no serious adverse events (Richter et al., 2013).

Products derived from cow, pig, or deer neuronal and glandular tissues can transmit prions (abnormally folded protein organisms), which are not destroyed by heating or digestion. Although no cases of Creutzfeldt-Jakob or "mad cow" disease, caused by prions, have been reported with bovine PS, further research is needed to determine if this is a possibility. Considering the risk of acquiring prions from bovine tissue, at present it is advisable to use only products derived from soybeans or from animals kept in New Zealand or Australia where the surveillance for prion-related diseases is highly regulated.

Treatment guidelines for NCDs are summarized in Table 3.

Clinical issues in using cognitive-enhancing agents are summarized in Box 4.

Box 4: Clinical Considerations in Using Cognitive-Enhancing Agents

The main considerations in choosing nutraceuticals are the target symptoms, the mechanism of action of the intervention, the side-effect profile, cost, and availability. For example, memory deficits respond well to cholinergic agents. Patients who have comorbid fatigue or depression in addition to cognitive decline may do better with stimulating agents. Patients with a prominent vascular component, as in multi-infarct or vascular dementia (by history, SPECT scan, or other blood flow studies), may benefit most from herbs that increase cerebral blood flow, such as ginkgo, and cerebrovascular vasodilators, such as picamilon or vinpocetine (Brown & Gerberg, 2011). Patients who, under psychological testing or on EEG, show overactivation of one cerebral hemisphere compared with the other, and who seem to have impaired transfer of information across the corpus callosum (on neurological testing), may respond better to one of the racetam agents in combination with a cholinergic agent. The cholinergic compounds for which there is most evidence of efficacy in cognitively impaired populations or in animal models are galantamine, centrophenoxine, acetyl-L-carnitine, and CDP-choline. Centrophenoxine may work best in combination with the racetam agents. Although piracetam has been the most widely used and studied, aniracetam may be more effective (Brown & Gerberg, 2011).

Attention-Deficit/Hyperactivity Disorder (ADHD)

ADHD is a constellation of symptoms that appears by age 7 years and often persists into adult life. Symptoms derive from inability to inhibit impulsiveness and/or to focus attention and have been linked to dopaminergic, noradrenergic, and cholinergic neurotransmitter systems (Biederman & Faraone, 2005). Stimulant medications, such as methylphenidate are the mainstay of conventional ADHD treatments; however, limitations in tolerance and concerns about side effects have spurred the investigation of alternative approaches. More than 50% of parents of children diagnosed with ADHD treat their children's symptoms with one or more ICMs, but few disclose this to their pediatrician (Chan et al., 2003). One study of 822 children with diagnosed or suspected ADHD found that 12% had used ICMs (Bussing et al., 2002). For reviews of ICM for ADHD, see Sarris et al. (2011a) and Brown and Gerberg (2012a).

Nutritional Medicines

Omega-3 Fatty Acids

Deficiencies of omega-3 FAs have been found in a subgroup of ADHD boys (Antalis et al., 2006). Omega-3 FAs improved behavior and cognitive function in children with developmental coordination disorder (ADHD is often part of this syndrome) (Richardson, 2006). A meta-analysis of 10 trials involving 699 children found that omega-3 FA supplementation demonstrated a small but significant improvement in ADHD symptoms (Bloch & Qawasmi, 2011). The EPA dose within supplements was significantly corre-lated with efficacy. This finding was further con-firmed in a newer meta-analysis involving seven studies ($n = 534$) (Chang et al., 2018); however, four more recent RCTs showed negative or null results, thereby the balance of evidence is not in favor of this approach for ADHD.

Difficulty in interpreting the data derives from the mixed methodological quality, patient selec-tion, and the use of different quantities and ratios of fatty acids, EPA, and DHA. EPA may be more beneficial than DHA in ameliorating ADHD symptoms (Hirayama et al., 2004). One 15-week DBRCT found improvement on hyperactivity and inattention subscales using 558 mg/day EPA and 174 mg/day DHA compared with placebo in 132 children (Sinn & Bryan, 2007).

Phosphatidylserine (PS)

PS is a phospholipid component of cell mem-branes (see the previous discussion of PS under NCDs). A study of 15 ADHD children, aged 6–12 years, indicated that 200 mg/day of soy-derived PS for 2 months significantly improved ADHD symptoms (Hirayama et al., 2013). Enrichment of PS with 300 mg of EPA plus DHA for children with ADHD improved visual sustained attention performance and hyper-activity, especially in the more pronounced hyper-active/impulsive behavior (Vaisman et al., 2008). Another 2-month DBRPC study of 36 children (aged 4–14 years) showed that PS 200 mg/day resulted in significant improvements in ADHD symptoms, short-term auditory memory, working memory, mental performance with visual stimuli, inattention, and impulsivity compared with pla-cebo. PS was well tolerated and showed no adverse effects (Hirayama et al., 2013).

Other Nutrients: Zinc, Iron, Acetyl-L-Carnitine

In a DBRCT, 400 children and adolescents ran-domized to a high dose of zinc (150 mg/day) expe-rienced significant improvements over placebo in hyperactivity and impulsivity (but not inattention) (Bilici et al., 2004). The high dropout rate found in the study may, however, place limits on the signif-icance of the findings. Another study in which zinc

(55 mg/day) was added to psychostimulant medi-cations (1 mg/kg/day) in 44 children with ADHD resulted in greater improvement in symptoms than with psychostimulants alone (Akhondzadeh et al., 2004b). In contrast, an RCT of 52 children with ADHD for 13 weeks (8 weeks of monotherapy then 5 weeks with added D-amphetamine) showed no significant improvements with zinc supplemen-tation compared with placebo (Arnold et al., 2011). In a DBRCT involving 23 non-anemic ADHD children with abnormally low serum ferritin levels [with a small placebo group ($n = 5$)], the 18 who were given ferrous sulfate 80 mg/day showed pro-gressive improvements in ADHD symptoms ver-sus placebo (Konofal et al., 2008).

In a multisite DBRCT of 112 ADHD children, ALCAR treatment (from 1000 to 3000 mg/day depending on the weight of the child) was associated with significantly greater symptom improvement in children with predominantly inat-tentive type ADHD (but not combined type ADHD) compared with placebo (Arnold et al., 2007). ALCAR affects cell membranes, energy reserves, cholinergic function, and omega-3 FA utilization (Pettegrew et al., 2000). A novel study using ALCAR in 51 children with ADHD and a genetic disorder (Fragile X syndrome) found that after one-year ALCAR produced greater improve-ments in ADHD symptoms compared with pla-cebo (Torrioli et al., 2008).

Dietary Elimination Strategies

Dietary elimination strategies have been supported by some scientific evidence since the 1970s. However, in cases of marked neurological dysfunction, such strategies tend to have minor benefit. Anecdotally, a typical responder (a sizable minority of ADHD children) may be a pre-schooler with insomnia, irritability, atopy, physi-cal symptoms, behavioral problems, and sometimes high copper levels. Most families have difficulty maintaining the "few foods" diet; however, eliminating the more suspicious food items may help certain children. Methodological problems and investigator bias in sugar elimina-tion studies have left this issue unresolved (Kidd,

2000). The role of toxicity from heavy metals (lead, aluminum) and organic chemical pollutants (pesticides, dioxins, polychlorinated biphenols, hydrocarbons, etc.) in the developing brain needs more research.

Food additives, for example, azo dyes (red and yellow dyes), can cause histamine release, even in the absence of a skin rash. Certain polymorphisms of the histamine degradation gene (*HNMT*) result in a reduced capacity to degrade and clear histamine from cells, leading to rapid accumulation of histamine. In mouse models, histamine H_3 receptor activity increased hyperactivity and interfered with inhibition learning (Stevenson et al., 2010). A DBRPC within-subject crossover study of a community sample of children including 153 3-year-olds and 144 8–9-year-olds found that the adverse effects (increased hyperactive and impulsive behaviors) associated with azo dyes and other food additives were significantly moderated by the present of the *HNMT* polymorphism in both age groups. Approximately 30% of the children had this polymorphism (Stevenson et al., 2010). A review of published RCTs of nonpharmacological interventions (Sonuga-Barke et al., 2013) concluded that artificial food color exclusion produced larger benefits in individuals selected for food sensitivities. Stronger evidence of efficacy from DBRPC trials is needed.

Herbal Medicines

Herbal medicines for ADHD may be used for cognitive activation or for calming effects. An open trial in 36 children with ADHD given 400 mg/day of American ginseng plus 100 mg/day of ginkgo for 4 weeks found that 74% improved significantly on Conner's ADHD scale and 44% improved on a social problems measure (Lyon et al., 2001). Pycnogenol (an extract from French maritime pine bark) was superior to placebo in a 1-month DBRCT of 61 children with ADHD. On standardized measures and teacher and parent ratings, students on pycnogenol had significantly greater improvements in hyperactivity, attention, concentration, and visual-motor coordination (Trebatická et al.,

2006). In a 12-week DBRCT, 36 children with ADHD given *Bacopa monnieri* (bacopa) showed better sentence repetition, logical memory, and paired associate learning than those given placebo. Improvements persisted 4 weeks after bacopa was discontinued (Negi et al., 2000). An RCT found that ginkgo (80–120 mg/day) had no comparable benefit to methylphenidate in 50 children with ADHD (Salehi et al., 2010), while a more recent 6-week study ($n = 66$) involving children and adolescents with ADHD given methylphenidate (20–30 mg/day) plus either *Ginkgo biloba* (80–120 mg/day) or placebo revealed significant reductions beyond the psychostimulants alone on ADHD scales (Shakibaei et al., 2015). A rigorous 8-week DBRCT of SJW (900 mg/day) found no benefits in the treatment of ADHD (Weber et al., 2008). In an 8-week, double-blind, randomized, methylphenidate-controlled study of 72 children with ADHD, a traditional Chinese herbal medicine (Ningdong: NDG) (5 mg/kg/day) significantly reduced ADHD symptoms with fewer side effects compared with methylphenidate (1 mg/kg/day) (Li et al., 2011b).

Mind-Body Medicine

Meditation

Meditation, which affects EEG rhythms similarly to theta/beta biofeedback training, improved attention, particularly in the classroom, in two controlled studies in ADHD children. Engaging inattentive people with ADHD in meditation can be challenging. Zylowska has developed methods for teaching mindfulness to adolescents and adults with ADHD (Zylowska, 2012).

Yoga

In a pilot study of 19 boys with ADHD aged 8–13 years, a program of yoga postures, yoga breathing, and relaxation led to significant improvements on standardized ADHD tests such as the Conners scale (Jensen & Kenny, 2004). Nineteen children with ADHD were randomized to a yoga program or conventional motor exercises. For all outcome measures, including scores of attention and parent ratings, yoga training was superior with a medium-to-large effect size (Haffner et al., 2006). More studies are warranted.

Neurotherapy

Traditional neurotherapy (neurofeedback or EEG biofeedback) trains patients to notice and influence their state of alertness based on EEG measures. This form of operant conditioning requires many treatments and patient cooperation (for a review of neurotherapy in ADHD, see Holtmann & Stadler, 2006). A modern innovation of neurotherapy uses the International 10–20 system of brain mapping and therapeutic procedures based on changing the frequency or amplitude of "brain waves" at specific sites on that system (Robbins, 2000). It requires little or no effort on the part of the patient. Quantitative EEGs (QEEGs), measuring frequencies and amplitudes over the entire brain, have improved diagnostic and treatment protocols. Inhibition of theta and reinforcement of sensorimotor rhythm (SMR) (immobile attention with alertness) have been postulated to explain improvements in attention, behavior, and impulsivity.

EEG-driven stimulation uses the patient's EEG to provide feedback to the brain via light-emitting diodes. One technique, the low frequency neurotherapy system, has been beneficial in numerous disorders, including ADHD (Larsen, 2006). Other systems are used to reduce theta and enhance alpha and/or beta frequencies for children with ADHD (Beauregard & Levesque, 2006). In a review of neurotherapy for ADHD, most non-RCTs found positive results with medium-to-large effect sizes, but this effect is reduced when only RCTs are included (Moriyama et al., 2012). Treatment protocols have been developed but individual variables preclude precise response prediction. Nevertheless, neurotherapists can identify individual variations by brain mapping, keep detailed records of response to each treatment, and adjust the protocols as needed, based on the individual's unique patterns of response. Experienced neurotherapists gain skill in interpreting the varied patient responses. As in psychotherapy, in which the therapist uses concepts and techniques, the actual form and course of treatment depend on the patient's capacities and responses. Some patients are sensitive and experience side effects such as headaches, nausea, or dizziness even with brief treatments. In such cases, the intensity, duration, and frequency of treatments may need to be adjusted to tolerable levels.

Summary

This chapter focused on treatments supported by evidence of safety and efficacy in clinical trials and evidence of known and plausible mechanisms of action. All of the interventions discussed have relatively low side-effect profiles and high margins of safety, particularly when compared to prescription medications. There is reason to believe that some phytomedicinal and nutraceutical agents have neuroprotective effects in addition to supporting healthier functioning of the nervous system. Although some valuable ICM treatments have not been generally used by physicians in the United States or approved for specific indications by the FDA, based on the available evidence of safety and efficacy, they are well worth consideration by clinicians who want to explore more options for improving their patients' health and wellbeing. In clinical practice, it may require several treatment trials to design effective integrative treatments. A well-informed clinician is best qualified to help patients and their families develop a safe, evidence-based integrative treatment plan.

Acknowledgments Prof Jerome Sarris is funded by an Australian National Health & Medical Research Council fellowship (NHMRC funding ID 1125000).

References

Abbot, R., Chang, D. D., Eyre, H., & Lavretsky, H. (2017). Mind-body practices Tai Chi and Qigong in the treatment and prevention of psychiatric disorders. In P. L. Gerbarg, R. P. Brown, & P. R. Muskin (Eds.), *Complementary and integrative treatments in psychiatric practice* (pp. 261–279). American Psychiatric Association Publishing.

Abbott, R., & Lavretsky, H. (2013). Tai Chi and Qigong for the treatment and prevention of mental disorders. *Psychiatric Clinics of North America, 36*(1), 109–120.

Afshar, H., Roohafza, H., et al. (2012). *N*-acetylcysteine add-on treatment in refractory obsessive-compulsive disorder: A randomized, double-blind, placebo-controlled trial. *Journal of Clinical Psychopharmacology, 32*(6), 797–803.

Akhondzadeh, S., & Abbasi, S. H. (2006). Herbal medicine in the treatment of Alzheimer's disease. *American Journal of Alzheimer's Disease and Other Dementias, 21*, 113–118.

Akhondzadeh, S., Naghavi, H. R., Vazirian, M., et al. (2001). Passionflower in the treatment of generalized anxiety: A pilot double-blind randomized controlled trial with oxazepam. *Journal of Clinical Pharmacy and Therapeutics, 26*, 363–367.

Akhondzadeh, S., Noroozian, M., Mohammadi, M., et al. (2003). *Melissa officinalis* extract in the treatment of patients with mild to moderate Alzheimer's disease: A double blind, randomised, placebo-controlled trial. *Journal of Neurology, Neurosurgery, and Psychiatry, 74*(7), 863–866.

Akhondzadeh, S., Fallah-Pour, H., Afkham, K., et al. (2004a). Comparison of *Crocus sativus* L. and imipramine in the treatment of mild to moderate depression: A pilot double-blind randomized trial. *BMC Complementary and Alternative Medicine, 4*, 12.

Akhondzadeh, S., Mohammadi, M. R., & Khademi, M. (2004b). Zinc sulfate as an adjunct to methylphenidate for the treatment of attention deficit hyperactivity disorder in children: A double blind and randomized trial. *BMC Psychiatry, 4*, 9.

Akhondzadeh, S., Tahmacebi-Pour, N., Noorbala, A. A., et al. (2005). *Crocus sativus* L. in the treatment of mild to moderate depression: A double-blind, randomized and placebo-controlled trial. *Phytotherapy Research, 19*, 148–151.

Akhondzadeh, B., Moshiri, E., et al. (2007). Comparison of petal of *Crocus sativus* L. and fluoxetine in the treatment of depressed outpatients: A pilot double-blind randomized trial. *Progress in Neuro-Psychopharmacology and Biological Psychiatry, 30*(2), 439–442.

Akhondzadeh, S., Sabet, M. S., Harirchian, M. H., et al. (2010). Saffron in the treatment of patients with mild to moderate Alzheimer's disease: A 16-week, randomized and placebo-controlled trial. *Journal of Clinical Pharmacology and Therapeutics, 35*, 581–588.

Akhondzadeh, S., Brown, R. P., & Gerbarg, P. L. (2013). Nutrients for prevention and treatment of mental health disorders. *Psychiatric Clinics of North America, 36*(1), 25–36.

Almeida, J. C., & Grimsley, E. W. (1996). Coma from the health food store: Interaction between kava and alprazolam. *Annals of Internal Medicine, 125*, 940–941.

Amaducci, L. (1988). Phosphatidylserine in the treatment of Alzheimer's disease: Results of a multicenter study. *Psychopharmacology Bulletin, 24*, 130–134.

American Psychiatric Association. (2015). *Diagnostic and statistical manual of mental disorders* (5th ed.). American Psychiatric Press.

Amminger, G. P., Schafer, M. R., Papageorgiou, K., et al. (2010). Long-chain omega-3 fatty acids for indicated prevention of psychotic disorders: A randomized, placebo-controlled trial. *Archives of General Psychiatry, 67*(2), 146–154.

Amminger, G. P., Nelson, B., Markulev, C., et al. (2020). The NEURAPRO biomarker analysis: Long-chain Omega-3 fatty acids improve 6-month and 12-month outcomes in youths at ultra-high risk for psychosis. *Biological Psychiatry, 87*(3), 243–252.

Amsterdam, J. D., Li, Y., et al. (2009). A randomized, double-blind, placebo-controlled trial of oral *Matricaria recutita* (chamomile) extract therapy for generalized anxiety disorder. *Journal of Clinical Psychopharmacology, 29*(4), 378–382.

Antalis, C. J., Stevens, L. J., Campbell, M., et al. (2006). Omega-3 fatty acid status in attention-deficit/hyperactivity disorder. *Prostaglandins, Leukotrienes, and Essential Fatty Acids, 75*, 299–308.

Appel, K., Rose, T., Fiebich, B., et al. (2011). Modulation of the γ-aminobutyric acid (GABA) system by *Passiflora incarnata* L. *Phytotherapy Research, 25*(6), 838–843.

Arendt, J. (1997). Jet lag/night shift, blindness and melatonin. *Transactions of the Medical Society of London, 114*, 7–9.

Arnold, L. E., Amato, A., et al. (2007). Acetyl-L-carnitine (ALC) in attention-deficit/hyperactivity disorder: A multi-site, placebo-controlled pilot trial. *Journal of Child and Adolescent Psychopharmacology, 17*(6), 791–801.

Arnold, L. E., Disilvestro, R. A., & Bozzolo, D. (2011). Zinc for attention-deficit/hyperactivity disorder: Placebo-controlled double-blind pilot trial alone and combined with amphetamine. *Journal of Child and Adolescent Psychopharmacology, 1*(1), 1–19.

Awad, R., Levac, D., et al. (2007). Effects of traditionally used anxiolytic botanicals on enzymes of the γ-aminobutyric acid (GABA) system. *Canadian Journal of Physiology & Pharmacology, 85*(9), 933–942.

Ballard, C. G., O'Brien, J. T., Reichelt, K., et al. (2002). Aromatherapy as a safe and effective treatment for the management of agitation in severe dementia: The results of a double-blind, placebo-controlled trial with Melissa. *Journal of Clinical Psychiatry, 63*(7), 553–558.

Barberger-Gateau, P., Raffaitin, C., Letenneur, L., et al. (2007). Dietary patterns and risk of dementia: The Three-City cohort study. *Neurology, 69*(20), 1921–1930.

Beauregard, M., & Levesque, J. (2006). Functional magnetic resonance imaging investigation of the effects of neurofeedback training on the neural bases of selective attention and response inhibition in children with attention-deficit/hyperactivity disorder. *Applied Psychophysiology and Biofeedback, 31*(1), 3–20.

Bell, I. R., Morrow, F. D., Read, M., et al. (1992). Low thyroxine levels in female psychiatric inpatients with riboflavin deficiency: Implications for folate-dependent methylation. *Acta Psychiatrica Scandinavica, 85*, 360–363.

Bellon, A. (2006). Searching for new options for treating insomnia: Are melatonin and ramelteon beneficial? *Journal of Psychiatric Practice, 12*, 229–243.

Bender, K. J. (2000). Investigating inositol in psychiatry. *Psychiatric Times*, 41–46.

Benjamin, J., Agam, G., Levine, J., et al. (1995a). Inositol treatment in psychiatry. *Psychopharmacology Bulletin, 31*, 167–175.

Benjamin, J., Levine, J., Fux, M., et al. (1995b). Double-blind, placebo-controlled, crossover trial of inositol treatment for panic disorder. *American Journal of Psychiatry, 152*, 1084–1086.

Benke, D., Barberis, A., Kopp, S., et al. (2009). GABA A receptors as in vivo substrate for the anxiolytic action of valerenic acid, a major constituent of valerian root extracts. *Neuropharmacology, 56*(1), 174–181.

Benson, H., Malhotra, M. S., Goldman, R. F., et al. (1990). Three case reports of the metabolic and electroencephalographic changes during advanced Buddhist meditation techniques. *Behavioral Medicine, 16*, 90–95.

Bent, S., Padula, A., Moore, D., et al. (2006). Valerian for sleep: A systematic review and meta-analysis. *American Journal of Medicine, 119*(12), 1005–1012.

Benton, D., Haller, J., & Fordy, J. (1995). Vitamin supplementation for 1 year improves mood. *Neuropsychobiology, 32*, 98–105.

Berger, G. E., Proffitt, T. M., McConchie, M., et al. (2007). Ethyl-eicosapentaenoic acid in first-episode psychosis: A randomized, placebo-controlled trial. *Journal of Clinical Psychiatry, 68*(12), 1867–1875.

Berk, M., Copolov, D., Dean, O., et al. (2008). *N*-Acetylcysteine as a glutathione precursor for schizophrenia – A double-blind, randomized, placebo-controlled trial. *Biological Psychiatry, 64*(5), 361–368.

Berk, M., Munib, A., Dean, O., et al. (2011). Qualitative methods in early-phase drug trials: Broadening the scope of data and methods from an RCT of *N*-acetylcysteine in schizophrenia. *Journal of Clinical Psychiatry, 72*(7), 909–913.

Bersani, G., & Garavini, A. (2000). Melatonin add-on in manic patients with treatment resistant insomnia. *Progress in Neuro-Psychopharmacology and Biological Psychiatry, 24*, 185–191.

Bhargav, H. (2020). Tele-yoga for stress management: Need of the hour during the COVID-19 pandemic and beyond? *Asian Journal of Psychiatry, 54*, 1–3.

Bhargav, H., George, S., Varambally, S., & Gangadhar, B. N. (2021a). Yoga and psychiatric disorders: A review of biomarker evidence. *International Review of Psychiatry, 33*(1–2), 162–169.

Bhargav, P. H., Reddy, P. V., Govindaraj, R., et al. (2021b). Impact of a course of add-on supervised yoga on cortical inhibition in major depressive disorder: A randomized controlled trial. *The Canadian Journal of Psychiatry, 66*(2), 179–181.

Bhatia, T., Agarwal, A., Shah, G., et al. (2012). Adjunctive cognitive remediation for schizophrenia using yoga: An open, non-randomized trial. *Acta Neuropsychiatrica, 24*, 91–100.

Biederman, J., & Faraone, S. V. (2005). Attention-deficit hyperactivity disorder. *The Lancet, 366*(9481), 237–248.

Bilici, M., Yildirim, F., Kandil, S., et al. (2004). Double-blind, placebo-controlled study of zinc sulfate in the treatment of attention deficit hyperactivity disorder. *Progress in Neuro-Psychopharmacology and Biological Psychiatry, 28*, 181–190.

Birch, E. E., Garfield, S., Hoffman, D. R., et al. (2000). A randomized controlled trial of early dietary supply of long-chain polyunsaturated fatty acids and mental development in term infants. *Developmental Medicine and Child Neurology, 42*, 174–181.

Birdsall, T. C. (1998). 5-Hydroxytryptophan: A clinically effective serotonin precursor. *Alternative Medicine Review, 3*(4), 271–280.

Bloch, M. H., & Qawasmi, A. (2011). Omega-3 fatty acid supplementation for the treatment of children with attention-deficit/hyperactivity disorder symptomatology: Systematic review and meta-analysis. *Journal of the American Academy of Child and Adolescent Psychiatry, 50*(10), 991–1000.

Blumenthal, M. (2017). Issues in phytomedicine related to psychiatric practice. In P. L. Gerbarg, R. P. Brown, & P. R. Muskin (Eds.), *Complementary and integrative treatments in psychiatric practice* (pp. 105–112). American Psychiatric Association Publishing.

Bottiglieri, T. (2013). Folate, vitamin B_{12}, and *S*-adenosylmethionine. *Psychiatric Clinics of North America, 36*(1), 1–14.

Bottiglieri, T., Gerbarg, P. L., & Brown, R. P. (2017). S-Adenosylmethionine. In P. L. Gerbarg, R. P. Brown, & P. R. Muskin (Eds.), *Complementary and integrative treatments in psychiatric practice* (pp. 41–52). American Psychiatric Association Publishing.

Bourre, J. M. (2005). Dietary omega-3 fatty acids and psychiatry: Mood, behaviour, stress, depression, dementia and aging. *Journal of Nutrition, Health and Aging, 9*, 31–38.

Bourre, J. M. (2006). Effects of nutrients (in food) on the structure and function of the nervous system: Update on dietary requirements for brain. Part 2. Macronutrients. *Journal of Nutrition Health and Aging, 10*, 386–399.

Boutron, I., Altman, D. G., Moher, D., for the CONSORT NPT Group, et al. (2017). CONSORT statement for randomized trials of nonpharmacologic treatments: A 2017 update and a CONSORT extension for nonpharmacologic trial. *Annals of Internal Medicine, 167*, 40–47.

Brown, R. P., & Garbarg, P. L. (2004). *The Rhodiola revolution*. Rodale Press.

Brown, R. P., & Gerbarg, P. L. (2005). Sudarshan Kriya yoga breathing in the treatment of stress, anxiety, and depression: Part I–Neurophysiological model. *Journal of Complementary and Alternative Medicine, 11*, 189–201.

Brown, R. P., & Gerbarg, P. L. (2009). Yoga breathing, meditation, and longevity. *Annals of the New York Academy of Sciences, 1172*, 54–62.

Brown, R. P., & Gerbarg, P. L. (2011). Complementary and integrative treatments in brain injury. In J. M. Silver, T. W. McAllister, & S. C. Yudofsky (Eds.), *Textbook of traumatic brain injury* (2nd ed., pp. 599–622). American Psychiatric Publishing.

Brown, R. P., & Gerbarg, P. L. (2012a). *Non-drug treatments for ADHD: New options for kids, adults, and clinicians*. W.W. Norton.

Brown, R. P., & Gerbarg, P. L. (2012b). *The healing power of the breath: Simple techniques to reduce stress and anxiety, enhance concentration, and balance your emotions*. Shambhala Publications.

Brown, R. P., & Gerbarg, P. L. (2017). Breathing techniques in psychiatric treatment. In P. L. Gerbarg, R. P. Brown, & P. R. Muskin (Eds.), *Complementary and integrative treatments in psychiatric practice* (pp. 241–250). American Psychiatric Association Publishing.

Brown, R. P., Gerbarg, P. L., & Bottiglieri, T. (2000). S-adenosylmethionine (SAMe) in the clinical practice of psychiatry, neurology, and internal medicine. *Clinical Practice of Alternative Medicine, 1*(4), 230–241.

Brown, R. P., Gerbarg, P. L., & Ramazanov, Z. (2002). A phythomedicinal review of *Rhodiola rosea*. *HerbalGram, 56*, 40–62.

Brown, R. P., Gerbarg, P. L., & Muskin, P. R. (2009). *How to use herbs, nutrients, and yoga in mental health care*. W.W. Norton.

Brown, R. P., Gerbarg, P. L., & Muench, F. (2013). Breathing practices for treatment of psychiatric and stress-related medical conditions. *Psychiatric Clinics of North America, 36*(1), 121–140.

Burns, N. R., Bryan, J., & Nettelbeck, T. (2006). *Ginkgo biloba*: No robust effect on cognitive abilities or mood in healthy young or older adults. *Human Psychopharmacology, 21*, 27–37.

Businaro, R., Vauzour, D., Sarris, J., et al. (2021). Therapeutic opportunities for food supplements in neurodegenerative diseases and depression. *Frontiers in Nutrition: Section Nutrition and Brain Health*. (accepted).

Bussing, R., Zima, B. T., et al. (2002). Use of complementary and alternative medicine for symptoms of attention-deficit hyperactivity disorder. *Psychiatric Services, 53*(9), 1096–1102.

Cabyoglu, M. T., Ergene, N., et al. (2006). The mechanism of acupuncture and clinical applications. *International Journal of Neuroscience, 116*(2), 115–125.

Caffarra, C., & Santamaria, V. (1987). The effects of phosphatidylserine in patients with mild cognitive decline: An open trial. *Clinical Trials Journal, 24*, 109–111.

Cakici, N., Van Beveren, N., Judge-Hundal, G., et al. (2019). An update on the efficacy of anti-inflammatory agents for patients with schizophrenia: A meta-analysis. *Psychological Medicine, 49*(14), 2307–2319.

Cardoso, B. R., Cominetti, C., & Cozzolino, S. M. F. (2013). Importance and management of micronutrient deficiencies in patients with Alzheimer disease. *Clinical Interventions in Aging, 8*, 531–542.

Carter, J., Gerbarg, P. L., Brown, R. P., et al. (2013). Multicomponent yoga breath program for Vietnam veteran post-traumatic stress disorder: Randomized controlled trial. *Journal of Traumatic Stress Disorders & Treatment, 2*(3), 1–10.

Chalon, S. (2006). Omega-3 fatty acids and monoamine neurotransmission. *Prostaglandins, Leukottrienes, and Essential Fatty Acids, 75*(4–5), 259–269.

Chan, E., Rappaport, L. A., et al. (2003). Complementary and alternative therapies in childhood attention and hyperactivity problems. *Journal of Developmental and Behavioral Pediatrics, 24*(1), 4–8.

Chang, J. P. C., Su, K.-P., Mondelli, V., et al. (2018). Omega-3 polyunsaturated fatty acids in youths with attention deficit hyperactivity disorder: A systematic review and meta-analysis of clinical trials and biological studies. *Neuropsychopharmacology, 43*(3), 534–545.

Chen, E. Y., & Hui, C. L. (2012). HT1001, a proprietary North American ginseng extract, improves working memory in schizophrenia: A double-blind, placebo-controlled study. *Phytotherapy Research, 26*(8), 1166–1172.

Chen, K. W., Berger, C. C., Manheimer, E., et al. (2012). Meditative therapies for reducing anxiety: A systematic review and meta-analysis of randomized controlled trials. *Depression and Anxiety, 29*(7), 545–562.

Chen, W. G., Schloesser, D., Arensdorf, A. M., et al. (2021). The emerging science of interoception: Sensing, integrating, interpreting, and regulating signals within the self. *Trends in Neuroscience, 44*(1), 3–16.

Chengappa, K. N. R., Brar, J. S., Gannon, J. M., et al. (2018). Adjunctive use of a standardized extract of *Withania somnifera* (Ashwagandha) to treat symptom exacerbation in schizophrenia: A randomized, double-blind, placebo-controlled study. *Journal of Clinical Psychiatry, 79*(5), 17m11826.

Clough, A. R., Bailie, R. S., & Currie, B. (2003). Liver function test abnormalities in users of aqueous kava extracts. *Journal of Toxicology. Clinical Toxicology, 41*, 821–829.

Conquer, J. A., Tierney, M. C., Zecevic, J., et al. (2000). Fatty acid analysis of blood plasma of patients with Alzheimer's disease, other types of dementia, and cognitive impairment. *Lipids, 35*, 1305–1312.

ConsumerLab.com. (2013). Product review: St. John's Wort supplements review. https://www.consumerlab.com/reviews/St_Johns_Wort/stjohnswort/. Accessed 13 Jan 2014.

Coppen, A., & Bailey, J. (2000). Enhancement of the antidepressant action of fluoxetine by folic acid: A randomised, placebo-controlled trial. *Journal of Affective Disorders, 60*, 121–130.

Cott, J. M. (2002). Herb-drug interactions: Focus on pharmacokinetics. *CNS Spectrums, 6*, 827–832.

Coulter, D. (2007). *Assessment of the risk of hepatotoxicity with kava products*. WHO Appointed Committee. World Health Organization.

Craig, A. D. (2008). Interoception and emotion. In M. Lewis, J. M. Haviland-Jones, & L. F. Barrett (Eds.), *Handbook of emotions* (3rd ed., pp. 272–288). Guilford Press.

Crook, T., Petrie, W., Wells, C., et al. (1992). Effects of phosphatidylserine in Alzheimer's disease. *Psychopharmacology Bulletin, 28*, 61–66.

Czyrak, A., Rogoz, Z., Skuza, G., et al. (1992). Antidepressant activity of *S*-adenosyl-L-methionine in mice and rats. *Journal of Basic and Clinical Physiology and Pharmacology, 3*, 1–17.

Darbinyan, V. A., Aslanyan, G., Embroyan, E., et al. (2007). SHR-5 and the treatment of depression. *Nordic Journal of Psychiatry, 61*, 5.

Davidson, R. J., Kabat-Zinn, J., Schumacher, J., et al. (2003). Alterations in brain and immune function produced by mindfulness meditation. *Psychosomatic Medicine, 65*(4), 564–570.

Dean, O. M., Bush, A., et al. (2012). Translating the Rosetta Stone of *N*-acetylcysteine. *Biological Psychiatry, 71*(11), 935–936.

Descilo, T., Vedamurthachar, A., Gerbarg, P. L., et al. (2010). Effects of a yoga-breath intervention alone and in combination with an exposure therapy for PTSD and depression in survivors of the 2004 Southeast Asia tsunami. *Acta Psychiatrica Scandinavica, 121*(4), 289–300.

Di Rocco, A., Rogers, J. D., Brown, R., et al. (2000). *S*-Adenosyl-methionine improves depression in patients with Parkinson's disease in an open-label clinical trial. *Movement Disorders, 15*(6), 1225–1229.

Diamond, B. J., & Bailey, M. R. (2013). *Ginkgo biloba* indications, mechanisms, and safety. *Psychiatric Clinics of North America, 36*(1), 73–83.

Dodd, S., Maes, M., Anderson, G., et al. (2013). Putative neuroprotective agents in neuropsychiatric disorders. *Progress in Neuro-Psychopharmacology and Biological Psychiatry, 42*, 135–145.

Donath, F., Quispe, S., Diefenbach, K., et al. (2000). Critical evaluation of the effect of valerian extract on sleep structure and sleep quality. *Pharmacopsychiatry, 33*, 47–53.

Dowling, G. A., Mastick, J., Colling, E., et al. (2005). Melatonin for sleep disturbances in Parkinson's disease. *Sleep Medicine, 6*, 459–466.

Dubichev, A. G., Kurkin, B. A., Zapesochnaya, G. G., et al. (1991). Study of *Rhodiola rosea* root chemical composition using HPLC. *Chemico-Pharmaceutical Journal, 2*, 188–193.

Duraiswamy, G., Thirthalli, J., Nagendra, H. R., et al. (2007). Yoga therapy as an add-on treatment in the management of patients with schizophrenia: A randomized controlled trial. *Acta Psychiatrica Scandinavica, 116*, 226–232.

Ernst, E. (2002). The risk-benefit profile of commonly used herbal therapies: Ginkgo, St. John's wort, ginseng, Echinacea, saw palmetto, and kava. *Annals of Internal Medicine, 136*, 42–53.

Errington-Evans, N. (2012). Acupuncture for anxiety. *CNS Neuroscience and Therapeutics, 18*(4), 277–284.

Farah, A. (2009). The role of L-methylfolate in depressive disorders. *CNS Spectrums, 14*(1 Suppl 2), 2–7.

Fava, M., Rosenbaum, J. F., Birnbaum, R., et al. (1992). The thyrotropin response to thyrotropin-releasing hormone as a predictor of response to treatment in depressed outpatients. *Acta Psychiatrica Scandinavica, 86*, 42–45.

Fava, M., Borus, J. S., Alpert, J. E., et al. (1997). Folate, vitamin B12, and homocysteine in major depressive disorder. *American Journal of Psychiatry, 154*(3), 426–428.

FDA. (2002). Kava-containing dietary supplements may be associated with severe liver injury. US Food and Drug Administration Center for Food Safety and Applied Nutrition. http://www.fda.gov/food/resourcesforyou/consumers/ucm085482.htm. Accessed 13 Jan 2014.

Fehmi, L. G., Edward, E. T., & Shor, S. B. (2017). Open focus training for stress, pain, and psychosomatic illness. In P. L. Gerbarg, R. P. Brown, & P. R. Muskin (Eds.), *Complementary and integrative treatments in psychiatric practice* (pp. 293–304). American Psychiatric Association Publishing.

Fernández-San-Martín, M. I., Masa-Font, R., Palacios-Soler, L., et al. (2010). Effectiveness of valerian on insomnia: A meta-analysis of randomized placebo-controlled trials. *Sleep Medicine, 11*(6), 505–511.

Feuerstein, G. (1998). *The yoga tradition: Its history, literature, philosophy and practice*. Hohm Press.

Firth, J., Teasdale, S. B., Allott, K., et al. (2019). The efficacy and safety of nutrient supplements in the treatment of mental illness: A meta-synthesis and appraisal of 33 meta-analyses of randomized controlled trials. *World Psychiatry, 18*(3), 308–324.

Fonkoue, I. T., Hu, Y., Jones, T., et al. (2020). Eight weeks of device-guided slow breathing decreases sympathetic nervous reactivity to stress in posttraumatic stress disorder. *American Journal of Physiology-Regulatory, Integrative and Comparative Physiology, 319*(4), R466–R475.

Freund-Levi, Y., Eriksdotter-Jonhagen, M., Cederholm, T., et al. (2006). Omega-3 fatty acid treatment in 174 patients with mild to moderate Alzheimer disease: OmegAD study: A randomized double-blind trial. *Archives of Neurology, 63*, 1402–1408.

Fusar-Poli, P., & Berger, G. (2012). Eicosapentaenoic acid interventions in schizophrenia: Meta-analysis of randomized, placebo-controlled studies. *Journal of Clinical Psychopharmacology, 32*(2), 179–185.

Fux, M., Levine, J., Aviv, A., et al. (1996). Inositol treatment of obsessive-compulsive disorder. *American Journal of Psychiatry, 153*(9), 1219–1221.

Gannon, J. M., Brar, J., Rai, A., et al. (2019). Effects of a standardized extract of *Withania somnifera* (Ashwagandha) on depression and anxiety symptoms in persons with schizophrenia participating in a randomized, placebo-controlled clinical trial. *Annals of Clinical Psychiatry, 31*(2), 123–129.

Gelenberg, A. J. (2000). St. John's wort update. *Biological Therapies in Psychiatry, 23*, 22–24.

Gerbarg, P. L., & Brown, R. P. (2013). Phytomedicines for prevention and treatment of mental health disorders. *Psychiatric Clinics of North America, 36*(1), 37–48.

Gerbarg, P. L., & Brown, R. P. (2016a). Pause menopause with *Rhodiola rosea*, a natural selective estrogen receptor modulator. *Phytomedicine, 23*(9), 763–769.

Gerbarg, P. L., & Brown, R. P. (2016b). Neurobiology and neurophysiology of breath practices in psychiatric care. *Psychiatric Times, 33*(11), 22–25.

Gerbarg, P. L., & Brown, R. P. (2021). Non-western interventions for stress reduction and resilience. *British Journal of Psychiatric Advances, 27*(3), 198–200.

Gerbarg, P. L., Wallace, G., & Brown, R. P. (2011). Mass disasters and mind-body solutions: Evidence and field insights. *International Journal of Yoga Therapy, 21*, 97–107.

Gerbarg, P. L., Brown, R. P., & Muskin, P. R. (2017a). Complementary and integrative medicine, DSM-5, and clinical decision making. In P. L. Gerbarg, R. P. Brown, & P. R. Muskin (Eds.), *Complementary and integrative treatments in psychiatric practice* (pp. 9–20). American Psychiatric Association Publishing.

Gerbarg, P. L., Muskin, P. R., & Brown, R. P. (Eds.). (2017b). *Complementary and integrative treatments in psychiatric practice.* American Psychiatric Association Publishing.

Gerbarg, P. L. & Brown, R. P. (2017c). Integrating Rhodiola rosea into Clinical Practice. In P. L. Gerbarg, R. P. Brown, & P. R. Muskin (Eds.), *Complementary and integrative treatments in psychiatric practice* (pp. 135–141). American Psychiatric Association Publishing.

Gerbarg, P. L., Brown, R. P., Streeter, C. C., et al. (2019). Breath practices for survivor and caregiver stress, depression, and post-traumatic stress disorder: Connection, co-regulation, compassion. *Integrative and Complementary Medicine OBM, 4*(3), 1–24.

Gerbarg, P. L., Brown, R. P., Mansur, S., & Steidle, K. (2021). Survivors of mass disasters: Breath-based mind-body interventions and global platforms. In S. Okpaku (Ed.), *Innovations in global mental health* (pp. 1557–1579). Switzerland AG, Springer Nature.

Gooneratne, N. S. (2008). Complementary and alternative medicine for sleep disturbances in older adults. *Clinics in Geriatric Medicine, 24*(1), 121–138.

Govindaraj, R., Varambally, S., Rao, N. P., et al. (2020). Does Yoga have a role in schizophrenia management? *Current Psychiatry Reports, 22*(12), 78.

Granath, J., Ingvarsson, S., von Thiele, U., et al. (2006). Stress management: A randomized study of cognitive behavioural therapy and yoga. *Cognitive Behaviour Therapy, 35*, 3–10.

Grant, J. E., Odlaug, B. L., et al. (2009). *N*-Acetylcysteine, a glutamate modulator, in the treatment of trichotillomania: A double-blind, placebo-controlled study. *Archives of General Psychiatry, 66*(7), 756–763.

Grima, N. A., Pase, M. P., et al. (2012). The effects of multivitamins on cognitive performance: A systematic review and meta-analysis. *Journal of Alzheimer's Disease, 29*(3), 561–569.

Guu, T. W., Mischoulon, D., Sarris, J., et al. (2020). A multi-national, multi-disciplinary Delphi consensus study on using omega-3 polyunsaturated fatty acids (n-3 PUFAs) for the treatment of major depressive disorder. *Journal of Affective Disorders, 265*, 233–238.

Haffner, J., Roos, J., Goldstein, N., et al. (2006). The effectiveness of body oriented methods of therapy in the treatment of attention-deficit hyperactivity disorder (ADHD): Results of a controlled pilot study. *Zeitschrift für Kinder-und Jugendpsychiatrie und Psychotherapie, 34*, 37–47. (in German).

Hebert, R., Lehmann, D., Tan, G., et al. (2005). Enhanced EEG alpha time domain phase synchrony during transcendental meditation: Implications for cortical integration theory. *Signal Processing, 85*(11), 2213–2232.

Helgason, C., & Sarris, J. (2013). Mind-body medicine for schizophrenia and psychotic disorders. *Clinical Schizophrenia and Related Psychoses, 7*(3), 138–148.

Heresco-Levy, U., Javitt, D., Ermilov, M., et al. (1999). Efficacy of high-dose glycine in the treatment of enduring negative symptoms of schizophrenia. *Archives of General Psychiatry, 56*, 29–36.

Herrera-Arellano, A., Jiménez-Ferrer, E., et al. (2007). Efficacy and tolerability of a standardized herbal product from *Galphimia glauca* on generalized anxiety disorder. A randomized, double-blind clinical trial controlled with lorazepam. *Planta Medica, 73*(8), 713–717.

Herrera-Arellano, A., Jiménez-Ferrer, J. E., et al. (2012). Therapeutic effectiveness of *Galphimia glauca* vs. lorazepam in generalized anxiety disorder. A controlled 15-week clinical trial. *Planta Medica, 78*(14), 1529–1535.

Herrero, J. L., Khuvis, S., & Mehta, A. D. (2017). Breathing above the brainstem: Volitional control and attentional modulation in humans. *Journal of Neurophysiology, 119*(1), 145–159.

Herrschaft, H., Nacu, A., & Likhachev, S. (2012). *Ginkgo biloba* extract EGb 761® in dementia with neuropsychiatric features: A randomised, placebo-controlled trial to confirm the efficacy and safety of a daily dose of 240 mg. *Journal of Psychiatric Research, 46*(6), 716–723.

Hibbeln, J. R. (1998). Fish consumption and major depression. *The Lancet, 351*(9110), 1213.

Hirayama, S., Hamazaki, T., et al. (2004). Effect of docosahexaenoic acid-containing food administration on symptoms of attention-deficit/hyperactivity disorder – A placebo-controlled double-blind study. *European Journal of Clinical Nutrition, 58*(3), 467–473.

Hirayama, S., Terasawa, K., Rabeler, R., et al. (2013). The effect of phosphatidylserine administration on memory and symptoms of attention-deficit hyperactivity disorder: A randomised, double-blind, placebo-controlled clinical trial. *Journal of Human Nutrition and Dietetics* Suppl. 2:284–291.

Holtmann, M., & Stadler, C. (2006). Electroencephalographic biofeedback for the treatment of attention-deficit hyperactivity disorder in childhood and adolescence. *Expert Review of Neurotherapeutics, 6*, 533–540.

Hopper, J. W., Frewen, P. A., van der Kolk, B. A., et al. (2007). Neural correlates of reexperiencing, avoidance, and dissociation in PTSD: Symptom dimensions and emotion dysregulation in responses to script-driven trauma imagery. *Journal of Traumatic Stress, 20*(5), 713–725.

Hosseinzadeh, H., & Noraei, N. B. (2009). Anxiolytic and hypnotic effect of *Crocus sativus* aqueous extract and its constituents, crocin and safranal, in mice. *Phytotherapy Research, 23*(6), 768–774.

Hourani, L. L., Davila, M. I., Morgan, J., et al. (2019). Mental health, stress, and resilience correlates of heart rate variability among military reservists, guardsmen, and first responders. *Physiology and Behavior, 1*(214), 112734.

Howes, M. J., & Perry, E. (2011). The role of phytochemicals in the treatment and prevention of dementia. *Drugs and Aging, 28*(6), 439–468.

Hsu, S.-M., Tseng, C.-H., Hsieh, C.-H., et al. (2020). Slow-paced inspiration regularizes alpha phase dynamics in the human brain. *Journal of Neurophysiology, 123*(1), 289–299.

Hypericum Depression Trial Study Group. (2002). Effect of *Hypericum perforatum* (St. John's wort) in major depressive disorder: A randomized controlled trial. *JAMA, 287*(14), 1807–1814.

Iwasaki, K., Satoh-Nakagawa, T., & Maruyama, M. (2005). A randomized, observer-blind, controlled trial of the traditional Chinese medicine Yi-Gan San for improvement of behavioral and psychological symptoms and activities of daily living in dementia patients. *The Journal of Clinical Psychiatry, 66*(2), 248–252.

Izzo, A. A. (2004). Drug interactions with St. John's wort (*Hypericum perforatum*): A review of the clinical evidence. *International Journal of Clinical Pharmacology and Therapeutics, 42*(3), 139–148.

Jacobs, B. P., Bent, S., Tice, J. A., et al. (2005). An Internet-based randomized, placebo-controlled trial of kava and valerian for anxiety and insomnia. *Medicine (Baltimore), 84*, 197–207.

Jayaram, N., Varambally, S., Behere, R. V., et al. (2013). Effect of yoga therapy on plasma oxytocin and facial emotion recognition deficits in patients of schizophrenia. *Indian Journal of Psychiatry, 55*(Suppl 3), S409–S413.

Jensen, P. S., & Kenny, D. T. (2004). The effects of yoga on the attention and behavior of boys with attention-deficit/hyperactivity disorder (ADHD). *Journal of Attention Disorders, 7*, 205–216.

Jensen, C. L., Chen, H., Fraley, J. K., et al. (1996). Biochemical effects of dietary linoleic/alpha-linolenic acid ratio in term infants. *Lipids, 31*, 107–113.

Jiménez-Ferrer, E., Herrera-Ruiz, M., et al. (2011). Interaction of the natural anxiolytic galphimine-B with serotonergic drugs on dorsal hippocampus in rats. *Journal of Ethnopharmacology, 137*(1), 724–729.

Kabat-Zinn, J., Massion, A. O., Kristeller, J., et al. (1992). Effectiveness of a meditation-based stress reduction program in the treatment of anxiety disorders. *American Journal of Psychiatry, 149*(7), 936–943.

Kalyani, B. G., Venkatasubramanian, G., Arasappa, R., et al. (2011). Neurohemodynamic correlates of 'OM' chanting: A pilot functional magnetic resonance imaging study. *International Journal of Yoga, 4*(1), 3–6.

Katsumoto, E., Ishida, T., Kinoshita, K., et al. (2021). Yokukansankachimpihange is useful to treat behavioral/psychological symptoms of dementia. *Frontiers in Nutrition, 7*, 529390.

Katzman, M. A., Vermani, M., Gerbarg, P. L., et al. (2012). A multicomponent yoga-based, breath intervention program as an adjunctive treatment in patients suffering from Generalized Anxiety Disorder (GAD) with or without comorbidities. *International Journal of Yoga, 5*(1), 57–65.

Kayumov, L., Brown, G., Jindal, R., et al. (2001). A randomized, double-blind, placebo-controlled crossover study of the effect of exogenous melatonin on delayed sleep phase syndrome. *Psychosomatic Medicine, 63*, 40–48.

Keefe, J. R., Mao, J. J., Soeller, I., et al. (2016). Short-term open-label chamomile (*Matricaria chamomilla L*) therapy of moderate to severe generalized anxiety disorder. *Phytomedicine, 23*, 1699–1705.

Kennedy, D. O., & Scholey, A. B. (2006). The psychopharmacology of European herbs with cognition-enhancing properties. *Current Pharmacuetical Design, 12*(35), 4613–4623.

Kennedy, D. O., Wake, G., Savelev, S., et al. (2003). Modulation of mood and cognitive performance following acute administration of single doses of *Melissa officinalis* (lemon balm) with human CNS nicotinic and muscarinic receptor-binding properties. *Neuropsychopharmacology, 28*, 1871–1881.

Kennedy, D. O., Little, W., & Scholey, A. B. (2004). Attenuation of laboratory-induced stress in humans after acute administration of *Melissa officinalis* (lemon balm). *Psychosomatic Medicine, 66*, 607–613.

Kennedy, D. O., Pace, S., Haskell, C., et al. (2006). Effects of cholinesterase inhibiting sage (*Salvia officinalis*) in mood, anxiety and performance on a psychological stressor battery. *Neuropsychopharmacology, 31*(4), 845–852.

Kennedy, D. O., Dodd, F. L., Robertson, B. C., et al. (2011). Monoterpenoid extract of sage (*Salvia lavandulaefolia*) with cholinesterase inhibiting properties improves cognitive performance and mood in healthy adults. *Journal of Psychopharmacology, 25*(8), 1088–1100.

Kenny, E., Muskin, P. R., Brown, R., et al. (2001). What the general psychiatrist should know about herbal medicine. *Current Psychiatry Reports, 3*, 226–234.

Kidd, P. M. (2000). Attention deficit/hyperactivity disorder (ADHD) in children: Rationale for its integrative management. *Alternative Medicine Review, 5*, 402–428.

Kim, H. J., Kim, P., & Shin, C. Y. (2013). A comprehensive review of the therapeutic and pharmacological effects of ginseng and ginsenosides in central nervous system. *Journal of Ginseng Research, 37*(1), 8–29.

Knüppel, L., & Linde, K. (2004). Adverse effects of St. John's wort: A systematic review. *Journal of Clinical Psychiatry, 65*, 1470–1479.

Ko, R. J. (1998). Adulterants in Asian patent medicines. *New England Journal of Medicine, 339*, 847.

Konofal, E., Lecendreux, M., et al. (2008). Effects of iron supplementation on attention deficit hyperactivity disorder in children. *Pediatric Neurology, 38*(1), 20–26.

Kraft, M., Spahn, T. W., Menzel, J., et al. (2001). Fulminant liver failure after administration of the herbal antidepressant kava-kava. *Deutsche Medizinische Wochenschrift, 126*, 970–972. (in German).

Kris-Etherton, P. M., Harris, W. S., Appel, L. J., & for the Nutrition Committee. (2002). Fish consumption, fish oil, omega-3 fatty acids, and cardiovascular disease. *Circulation, 106*, 2747–2757.

Kwon, C.-Y., & Lee, B. (2021). Herbal medicine for behavioral and psychological symptoms of dementia: A systematic review and meta-analysis. *Frontiers in Pharmacology, 12*, 713287.

Kyle, D. J., Schaefer, E., & Patton, G. (1998). Low serum docosahexanoic acid is a significant risk factor for Alzheimer's dementia. Presented at the 3rd ISSFAL Congress, 1–5 June 1998, Lyon.

Lake, J., Helgason, C., & Sarris, J. (2012). Integrative mental health (IMH): Paradigm, research, and clinical practice. *Explore (New York, N.Y.), 8*(1), 50–57.

LaPorte, E., Sarris, J., et al. (2011). Neurocognitive effects of kava (*Piper methysticum*): A systematic review. *Human Psychopharmacology, 26*(2), 102–111.

Larsen, S. (2006). *The healing power of neurofeedback.* The Healing Press.

Lautenschlager, N. T., Ihl, R., et al. (2012). *Ginkgo biloba* extract EGb 761® in the context of current developments in the diagnosis and treatment of age-related cognitive decline and Alzheimer's disease: A research perspective. *International Psychogeriatrics, 24*(Suppl 1), S46–S50.

Lavretsky, H., & Feldman, J. L. (2021). Precision medicine for breath-focused mind-body therapies for stress and anxiety: Are we ready yet? *Global Advances in Health and Medicine, 10*, 1–4.

Lazar, S. W., Kerr, C. E., Wasserman, R. H., et al. (2005). Meditation experience is associated with increased cortical thickness. *Neuroreport, 16*, 1893–1897.

Le Bars, P. L., Velasco, F. M., Ferguson, J. M., et al. (2002). Influence of the severity of cognitive impairment on the effect of the *Gnkgo biloba* extract EGb 761 in Alzheimer's disease. *Neuropsychobiology, 45*, 19–26.

Lee, M. S., Kim, M. K., & Lee, Y. H. (2005a). Effects of Qi-therapy (external Qigong) on cardiac autonomic tone: A randomized placebo-controlled study. *International Journal of Neuroscience, 115*(9), 1345–1350.

Lee, M. S., Rim, Y. H., Jeong, D. M., et al. (2005b). Nonlinear analysis of heart rate variability during Qi therapy (external Qigong). *American Journal of Chinese Medicine, 33*(4), 579–588.

Lehmann, D., Faber, P. L., Achermann, P., et al. (2001). Brain sources of EEG gamma frequency during volitionally meditation-induced, altered states of consciousness, and experience of the self. *Psychiatry Research, 108*, 111–121.

Lehrer, P., Karavidas, M. K., Lu, S. E., et al. (2010). Voluntarily produced increases in heart rate variability modulate autonomic effects of endotoxin induced systemic inflammation: An exploratory study. *Applied Psychophysiology Biofeedback, 35*(4), 303–315.

Lerner, V., Kanevsky, M., Dwolatzky, T., et al. (2006). Vitamin B_{12} and folate serum levels in newly admitted psychiatric patients. *Clinical Nutrition, 25*, 60–67.

Li, J. Y., Zhang, Y. F., Smith, G. S., et al. (2011a). Quality of reporting of randomized clinical trials in tai chi interventions – A systematic review. *Evidence-based Complementary and Alternative Medicine, 2011*, 383245.

Li, J. J., Li, Z. W., Wang, S. Z., et al. (2011b). Ningdong granule: A complementary and alternative therapy in the treatment of attentiondeficit/hyperactivity disorder. *Psychopharmacology, 216*(4), 501–509.

Linde, K., Ramirez, G., Mulrow, D. F., et al. (1996). St. John's wort for depression – An overview and meta-analysis of randomized clinical trials. *BMJ, 313*(7052), 253–258.

Linde, K., Berner, M., et al. (2008). St John's wort for major depression. *Cochrane Database of Systematic Reviews,* (4), CD000448.

Lutz, A., Greischar, L. L., Rawlings, N. B., et al. (2004). Long-term meditators self-induce high-amplitude gamma synchrony during mental practice. *Proceedings of the National Academy of Sciences of The United States of America, 101*, 16369–16373.

Lyon, M. R., Cline, J. C., Totosy de Zepetnek, J., et al. (2001). Effect of the herbal extract combination *Panax quinquefolium* and *Ginkgo biloba* on attention-deficit hyperactivity disorder: A pilot study. *Journal of Psychiatry and Neuroscience, 26*, 221–228.

Maciocia, G. (1994). *The practice of Chinese medicine: The treatment of diseases with acupuncture and Chinese herbs.* Churchill Livingstone.

Manjunath, N. K., & Telles, S. (2005). Influence of Yoga and Ayurveda on self rated sleep in a geriatric population. *Indian Journal of Medical Research, 121*, 683–690.

Mao, J. J., Xie, S. X., Keefe, J. R., et al. (2016). Long-term chamomile (*Matricaria chamomilla* L.) treatment for generalized anxiety disorder: A randomized clinical trial. *Phytomedicine, 23*(14), 1735–1742.

Marchand, W. R. (2013). Mindfulness meditation practices as adjunctive treatments for psychiatric disorders. *Psychiatric Clinics of North America, 36*(1), 141–152.

Marx, W., Lane, M., Rocks, T., et al. (2019). Effect of saffron supplementation on symptoms of depression and anxiety: A systematic review and meta-analysis. *Nutrition Review, 28,* nuz023.

Mason, H., Gerbarg, P. L., & Brown, R. P. (2021). Psychophysiology: Healing effects of voluntarily regulated breathing practices. In S. Telles & R. K. Gupt (Eds.), *Handbook of research on evidence-based perspectives on the psychophysiology of yoga and its applications* (pp. 24–48). IGI Global.

Massoumi, L. (2017). The growth of complementary and integrative medicine. In P. L. In Gerbarg, R. P. Brown, & P. R. Muskin (Eds.), *Complementary and integrative treatments in psychiatric practice* (pp. 1–3). American Psychiatric Association Publishing.

Mathews, J. D., Riley, M. D., Fejo, L., et al. (1988). Effects of the heavy usage of kava on physical health: Summary of a pilot survey in an aboriginal community. *Medical Journal of Australia, 148,* 548–555.

Mathews, J. M., Etheridge, A. S., Black, S. R., et al. (2002). Inhibition of human cytochrome P450 activities by kava extract and kavalactones. *Drug Metabolism and Disposition, 30,* 1153–1157.

Mendelson, S. D. (2020). *Herbal treatment of major depression.* CRC Press Taylor & Francis Group.

Meng, X.-Z., Wu, F., Wei, P.-K., et al. (2012). A Chinese herbal formula to improve general psychological status in posttraumatic stress disorder: A randomized placebo-controlled trial on Sichuan earthquake survivors. *Evidence-based Complementary and Alternative Medicine, 2012,* 691258.

Meseguer, E., Taboada, R., Sanchez, V., et al. (2002). Life-threatening parkinsonism induced by kava-kava. *Movement Disorders, 17*(1), 195–196.

Miller, L. G. (1998). Herbal medicinals: Selected clinical considerations focusing on known or potential drug-herb interactions. *Archives of Internal Medicine, 158,* 2200–2211.

Miller, J. J., Fletcher, K., & Kabat-Zinn, J. (1995). Three-year follow-up and clinical implications of a mindfulness-based stress reduction intervention in the treatment of anxiety disorders. *General Hospital Psychiatry, 17,* 192–200.

Mischoulon, D., Price, L. H., Carpenter, L. L., et al. (2014). A double-blind, randomized, placebo-controlled clinical trial of *S*-adenosyl-L-methionine (SAMe) versus escitalopram in major depressive disorder. *Journal of Clinical Psychiatry, 75*(4), 370–376.

Miyake, Y., Sasaki, S., Tanaka, K., et al. (2006). Dietary folate and vitamins B_{12}, B_6, and B_2 intake and the risk of postpartum depression in Japan: The Osaka Maternal and Child Health Study. *Journal of Affective Disorders, 96,* 133–138.

Miyaoka, T., Furuya, M., Horiguchi, J., et al. (2015a). Efficacy and safety of yokukansan in treatment-resistant schizophrenia: A randomized, double-blind, placebo-controlled trial (a Positive and Negative Syndrome Scale, five-factor analysis). *Psychopharmacology, 232*(1), 155–164.

Miyaoka, T., Furuya, M., Horiguchi, J., et al. (2015b). Efficacy and safety of yokukansan in treatment-resistant schizophrenia: A randomized, multicenter, double-blind, placebo-controlled trial. *Evidence-based Complementary and Alternative Medicine, 2015,* 1–11.

Miyasaka, L. S., Atallah, A. N., et al. (2007). Passiflora for anxiety disorder. *Cochrane Database of Systematic Reviews,* (1), CD004518.

Modabbernia, A. (2017). Melatonin and melatonin analogues for psychiatric disorders. In P. L. Gerbarg, R. P. Brown, & P. R. Muskin (Eds.), *Complementary and integrative treatments in psychiatric practice* (pp. 211–220). American Psychiatric Association Publishing.

Modabbernia, A., & Akhondzadeh, S. (2013). Saffron, passionflower, valerian, and sage for mental health. *Psychiatric Clinics of North America, 36*(1), 85–92.

Moriyama, T. S., Polanczyk, G., et al. (2012). Evidence-based information on the clinical use of neurofeedback for ADHD. *Neurotherapeutics, 9*(3), 588–598.

Morris, M. C., Evans, D. A., Bienias, J. L., et al. (2003). Dietary fats and the risk of incident Alzheimer disease. *Archives of Neurology, 60,* 194–200.

Moshiri, E., Basti, A. A., et al. (2006). *Crocus sativus* L. (petal) in the treatment of mild-to-moderate depression: A double-blind, randomized and placebo-controlled trial. *Phytomedicine, 13*(9–10), 607–611.

Movafegh, A., Alizadeh, R., et al. (2008). Preoperative oral *Passiflora incarnata* reduces anxiety in ambulatory surgery patients: A double-blind, placebo-controlled study. *Anesthesia and Analgesia, 106*(6), 1728–1732.

Muench, F., & Kumar, D. S. (2017). Using technology-based mind-body tools in clinical practice. In P. L. Gerbarg, R. P. Brown, & P. R. Muskin (Eds.), *Complementary and integrative treatments in psychiatric practice* (pp. 353–366). American Psychiatric Association Publishing.

Mukai, T., Kishi, T., Mats, Y., et al. (2014). A meta-analysis of inositol for depression and anxiety disorders. *Human Psychopharmacology Clinical and Experimental, 29,* 55–63.

Muller, S. F., & Klement, S. (2006). A combination of valerian and lemon balm is effective in the treatment of restlessness and dyssomnia in children. *Phytomedicine, 13,* 383–387.

Muller, W. E., Singer, A., Wonnemann, M., et al. (1998). Hyperforin represents the neurotransmitter reuptake inhibiting constituent of hypericum extract. *Pharmacopsychiatry, 31*(Suppl 1), 16–21.

Murdoch, I., Perry, E. K., Court, J. A., et al. (1998). Cortical cholinergic dysfunction after human head injury. *Journal of Neurotrauma, 15,* 295–305.

Muskin, P. R. (Ed.). (2000). *Complementary and alternative medicine and psychiatry* (Review of psychiatry) (Vol. 19). American Psychiatric Press.

Negi, K. S., Singh, Y. D., Kushwaha, K. P., et al. (2000). Clinical evaluation of memory enhancing properties of Memory Plus in children with attention deficit hyperactivity disorder (Abstract). *Indian Journal of Psychiatry, 42*(Suppl 2), 4.

Ng, B. Y. (1999). Qigong-induced mental disorders: A review. *Australian and New Zealand Journal of Psychiatry, 33*, 197–206.

Ng, Q. X., Venkatanarayanan, N., & Ho, C. Y. X. (2017). Clinical use of *Hypericum perforatum* (St John's wort) in depression: A meta-analysis. *Journal of Affective Disorders, 210*, 211–221.

Nidecker, A. (1997). Probing genes, drugs, and fatty acids in dementia. *Clinical Psychiatry News*, December, 4.

Nierenberg, A. A., Burt, T., Matthews, J., et al. (1999). Mania associated with St. John's wort. *Biological Psychiatry, 46*, 1707–1708.

Nierenberg, A. A., Ostacher, M. J., Calabrese, J. R., et al. (2006). Treatment-resistant bipolar depression: A STEP-BD equipoise randomized effectiveness trial of antidepressant augmentation with lamotrigine, inositol, or risperidone. *American Journal of Psychiatry, 163*, 210–216.

Noorbala, A. A., Akhondzadeh, S., Tahmacebi-Pour, N., et al. (2005). Hydro-alcoholic extract of *Crocus sativus* L. versus fluoxetine in the treatment of mild to moderate depression: A double-blind, randomized pilot trial. *Ethnopharmacology, 97*(2), 281–284.

Numata, T., Gunfan, S., Takayama, S., et al. (2014). Treatment of posttraumatic stress disorder using the traditional Japanese herbal medicine saikokeishikankyoto: A randomized observer-blinded controlled trial in survivors of the great East Japan earthquake and tsunami. *Evidence-Based Complementary and Alternative Medicine, 2014*, 683293.

Nyer, M., Gerbarg, P. L., Liveri, M. M., Johnston, J., Scott, T. M., Nauphal, M., Owen, L., Nielsen, G. H., Mischoulon, D., Brown, R. P., Fava, M., & Streeter, C. C. (2018). A randomized controlled dosing study of Iyengar yoga and coherent breathing for the treatment of major depressive disorder: Impact on suicidal ideation and safety findings. *Complementary Therapies in Medicine, 37*, 136–142.

Otero-Losada, M. E., & Rubio, M. C. (1989a). Acute changes in 5-HT metabolism after *S*-adenosyl-L-methionine administration. *General Pharmacology, 20*, 403–406.

Otero-Losada, M. E., & Rubio, M. C. (1989b). Acute effects of *S*-adenosyl-L-methionine on catecholaminergic central function. *European Journal of Pharmacology, 163*, 353–356.

Panossian, A., & Amsterdam, D. (2017). Adaptogens in psychiatric practice. In P. L. Gerbarg, R. P. Brown, & P. R. Muskin (Eds.), *Complementary and integrative treatments in psychiatric practice* (pp. 113–134). American Psychiatric Association Publishing.

Panossian, A., & Wagner, H. (2005). Stimulating effect of adaptogens: An overview with particular reference to their efficacy following single dose administration. *Phytotherapy Research, 19*, 819–838.

Panossian, A., Wikman, G., et al. (2010). Rosenroot (*Rhodiola rosea*): Traditional use, chemical composition, pharmacology and clinical efficacy. *Phytomedicine, 17*(7), 481–493.

Papakostas, G. I. (2006). Treatment-resistant major depressive disorder: New developments in 2006. *Psychiatric Times, 23*, 17–20.

Papakostas, G. I., Mischoulon, D., et al. (2010). *S*-Adenosyl-methionine (SAMe) augmentation of serotonin reuptake inhibitors for antidepressant nonresponders with major depressive disorder: A double-blind, randomized clinical trial. *American Journal of Psychiatry, 167*(8), 942–948.

Paul, M. A., Brown, G., Buguet, A., et al. (2001). Melatonin and zopiclone as pharmacologic aids to facilitate crew rest. *Aviation, Space, and Environmental Medicine, 72*(11), 974–984.

Payne, P., & Crane-Godreau, M. A. (2013). Meditative movement for depression and anxiety. *Frontiers in Psychiatry, 4*(71), 1–15.

Pearson, N. J., Johnson, L. L., et al. (2006). Insomnia, trouble sleeping, and complementary and alternative medicine: Analysis of the 2002 national health interview survey data. *Archives of Internal Medicine, 166*(16), 1775–1782.

Peet, M., Brind, J., Ramchand, C. N., et al. (2001). Two double-blind placebo-controlled pilot studies of eicosapentaenoic acid in the treatment of schizophrenia. *Schizophrenia Research, 49*(3), 243–251.

Penninx, B. W., Guralnik, J. M., Ferrucci, L., et al. (2000). Vitamin B_{12} deficiency and depression in physically disabled older women: Epidemiologic evidence from the Women's Health and Aging Study. *American Journal of Psychiatry, 157*, 715–721.

Pettegrew, J. W., Levine, J., & McClure, R. J. (2000). Acetyl-L-carnitine physical-chemical, metabolic, and therapeutic properties: Relevance for its mode of action in Alzheimer's disease and geriatric depression. *Molecular Psychiatry, 5*, 616–632.

Philippot, P., Gaetane, C., & Blairy, S. (2002). Respiratory feedback in the generation of emotion. *Cognition and Emotion, 16*, 605–607.

Pilkington, K., Kirkwood, G., et al. (2007). Acupuncture for anxiety and anxiety disorders –a systematic literature review. *Acupuncture in Medicine, 25*(1–2), 1–10.

Porges, S. W., & Carter, S. S. (2017). Polyvagal theory and the social engagement system. Neurophysiological bridge between connectedness and health. In P. L. Gerbarg, O. R. Muskin, & R. P. Brown (Eds.), *Complementary and integrative treatments in psychiatric practice* (pp. 221–240). American Psychiatric Association Publishing.

Prasad, A. S. (2009). Zinc: Role in immunity, oxidative stress and chronic inflammation. *Current Opinion in Clinical Nutrition and Metabolic Care, 12*(6), 646–652.

Quiñones, N., Maquet, Y. G., Vélez, D. M., et al. (2015). Efficacy of a Satyananda Yoga intervention for reintegrating adults diagnosed with posttraumatic stress disorder. *International Journal of Yoga Therapy, 25*(1), 89–99.

Rahimi, R., Nikfar, S., et al. (2009). Efficacy and tolerability of *Hypericum perforatum* in major depressive disorder in comparison with selective serotonin reuptake inhibitors: A meta-analysis. *Progress in Neuro Psychopharmacology and Biological Psychiatry, 33*(1), 118–127.

Raison, C. L., Capuron, L., & Miller, A. H. (2006). Cytokines sing the blues: Inflammation and the pathogenesis of depression. *Trends in Immunology, 27*, 24–31.

Rao, N. P., Varambally, S., & Gangadhar, B. N. (2013). Yoga school of thought and psychiatry: Therapeutic potential. *Indian Journal of Psychiatry, 55*(Suppl 2), S145–S149.

Reay, J. L., Scholey, A. D., & Kennedy, D. O. (2010). *Panax ginseng* (G115) improves aspects of working memory performance and subjective ratings of calmness in healthy young adults. *Human Psychopharmacology, 25*(6), 462–471.

Richardson, A. J. (2006). Omega-3 fatty acids in ADHD and related neurodevelopmental disorders. *International Review of Psychiatry, 18*, 155–172.

Richter, Y., Herzog, Y., Lifshitz, Y., et al. (2013). The effect of soybean-derived phosphatidylserine on cognitive performance in elderly with subjective memory complaints: A pilot study. *Clinical Interventions in Aging, 8*, 557–563.

Riggs, K., Spiro, A., III, Tucker, K., et al. (1996). Relations of vitamin B-12, vitamin B-6, folate, and homocysteine to cognitive performance in the Normative Aging Study. *American Journal of Clinical Nutrition, 63*, 306–314.

Ritsner, M. S., Miodownik, C., Ratner, Y., et al. (2011). L-Theanine relieves positive, activation, and anxiety symptoms in patients with schizophrenia and schizoaffective disorder: An 8-week, randomized, double-blind, placebo-controlled, 2-center study. *Journal of Clinical Psychiatry, 72*(1), 34–42.

Robbins, J. (2000). *A symphony in the brain: The evolution of the new brain wave biofeedback*. Grove Press.

Romero-Cerecero, O., Islas-Garduño, A. L., Zamilpa, A., et al. (2019). Galphimine-B standardized extract versus alprazolam in patients with generalized anxiety disorder: A ten-week, double-blind, randomized clinical trial. *BioMed Research International, 2019*, 1037036.

Rush, A. J., Trivedi, M. H., Wisniewski, S. R., et al. (2006). Acute and longer-term outcomes in depressed outpatients requiring one or several treatment steps: A STAR*D report. *American Journal of Psychiatry, 163*, 1905–1917.

Sakurai, H., Carpenter, L. L., & Tyrka, A. R. (2020). Dose increase of *S*-Adenosyl-methionine and escitalopram in a randomized clinical trial for major depressive disorder. *Journal of Affective Disorders, 262*, 118–125.

Salehi, B., Imani, R., et al. (2010). *Ginkgo biloba* for attention-deficit/hyperactivity disorder in children and adolescents: A double blind, randomized controlled trial. *Progress in Neuro-Psychopharmacology and Biological Psychiatry, 34*(1), 76–80.

Salgueiro, J. B., Ardenghi, P., et al. (1997). Anxiolytic natural and synthetic flavonoid ligands of the central benzodiazepine receptor have no effect on memory tasks in rats. *Pharmacology, Biochemistry, and Behavior, 58*(4), 887–891.

Sanchez-Villegas, A., Henriquez, P., et al. (2007). Long chain omega-3 fatty acids intake, fish consumption and mental disorders in the SUN cohort study. *European Journal of Nutrition, 46*(6), 337–346.

Saratikov, A. S., & Krasnov, E. A. (1987a). Effect of *Rhodiola rosea* on the central nervous system. In A. S. Saratikov & E. A. Krasnov (Eds.), *Rhodiola Rosea is a valuable medicinal plant (Golden root)* (pp. 150–179). Tomsk State University Press.

Saratikov, A. S., & Krasnov, E. A. (1987b). Clinical studies of *Rhodiola*. In A. S. Saratikov & E. A. Krasnov (Eds.), *Rhodiola Rosea is a valuable medicinal plant (Golden root)* (pp. 216–227). Tomsk State University Press.

Sarris, J. (2012). Current challenges in appraising complementary medicine evidence. *Medical Journal of Australia, 196*(5), 310–311.

Sarris, J. (2013a). Ginkgo. In K. Bone & S. Mills (Eds.), *Principles and practice of phytotherapy* (2nd ed.). Churchill Livingstone Elsevier.

Sarris, J. (2013b). St John's wort for the treatment of psychiatric disorders. *Psychiatric Clinics of North America, 36*(1), 65–72.

Sarris, J., & Byrne, G. J. (2011). A systematic review of insomnia and complementary medicine. *Sleep Medicine Reviews, 15*(2), 99–106.

Sarris, J., Kavanagh, D. J., Byrne, G., et al. (2009a). The Kava Anxiety Depression Spectrum Study (KADSS): A randomized, placebo-controlled crossover trial using an aqueous extract of *Piper methysticum*. *Psychopharmacology, 205*(3), 399–407.

Sarris, J., Kavanagh, D. J., & Byrne, G. (2009b). Adjuvant use of nutritional and herbal medicines with antidepressants, mood stabilizers and benzodiazepines. *Journal of Psychiatric Research, 44*(1), 32–41.

Sarris, J., Schoendorfer, N., et al. (2009c). Major depressive disorder and nutritional medicine: A review of monotherapies and adjuvant treatments. *Nutrition Reviews, 67*(3), 125–131.

Sarris, J., Kean, J., Schweitzer, I., et al. (2011a). Complementary medicines (herbal and nutritional products) in the treatment of Attention Deficit Hyperactivity Disorder (ADHD): A systematic review of the evidence. *Complementary Therapies in Medicine, 19*(4), 216–227.

Sarris, J., LaPorte, E., & Schweitzer, I. (2011b). Kava: A comprehensive review of efficacy, safety, and psychopharmacology. *Australian and New Zealand Journal of Psychiatry, 45*(1), 27–35.

Sarris, J., Fava, M., et al. (2012a). St. John's wort (*Hypericum perforatum*) versus sertraline and placebo in major depressive disorder: Continuation data from a 26 week RCT. *Pharmacopsychiatry, 45*(7), 275–278.

Sarris, J., Mischoulon, D., et al. (2012b). Omega-3 for bipolar disorder: Meta-analyses of use in mania and bipolar depression. *Journal of Clinical Psychiatry, 73*(1), 81–86.

Sarris, J., Glick, R., Hoenders, R. J., et al. (2013a). Integrative mental healthcare White Paper: Establishing a new paradigm through research, education, and clinical guidelines. *Advances in Integrative Medicine, 1*(1), 9–16.

Sarris, J., McIntyre, E., & Carnfield, D. A. (2013b). Plant-based medicines for anxiety disorders. Part 2: A systematic review of clinical studies with supporting preclinical evidence. *CNS Drugs, 27*(4), 301–319.

Sarris, J., Nierenberg, A. A., Schweitzer, I., et al. (2013c). Conditional probability of response or nonresponse of placebo compared with antidepressants or St John's wort in major depressive disorder. *Journal of Clinical Psychopharmacology, 33*(6), 827–830.

Sarris, J., Stough, C., et al. (2013d). Kava in the treatment of generalized anxiety disorder: A double-blind, randomized, placebo-controlled study. *Journal of Clinical Psychopharmacology, 33*(5), 643–648.

Sarris, J., Papakostas, G., & Vitelos, et al. (2014). S-Adenosylmethionine (SAMe) versus escitalopram and placebo in major depression: Efficacy and effects of histamine and carnitine as moderators of response. *Journal of Affective Disorders, 164*, 76–81.

Sarris, J., Oliver, G., Camfield, D. A., et al. (2015). *N*-acetyl cysteine (NAC) in the treatment of obsessive-compulsive disorder: A 16-week, double-blind, randomised, placebo-controlled study. *CNS Drugs, 29*(9), 801–809.

Sarris, J., Byrne, G. J., Stough, C., et al. (2019a). Nutraceuticals for major depressive disorder- more is not merrier: An 8-week double-blind, randomised, controlled trial. *Journal of Affective Disorders, 245*, 1007–1015.

Sarris, J., Byrne, G. J., Bousman, C. A., et al. (2019b). Kava for generalised anxiety disorder: A 16-week double-blind, randomized, placebo-controlled study. *The Australian and New Zealand Journal of Psychiatry, 54*(3), 288–229.

Sarris, J., Murphey, S. C., et al. (2020). S-Adenosylmethionine (SAMe) monotherapy for depression: An 8-week double-blind, randomised, controlled trial. *Psychopharmacology, 237*(1), 209–218.

Sarris, J., Wolfgang, M., Ashton, M. M., et al. (2021). Plant-based medicines (Phytoceuticals) in the treatment of psychiatric disorders: A meta-review of meta-analyses of randomized controlled trials. *The Canadian Journal of Psychiatry* (in press).

Satyanarayana, M., Rajeswari, K. R., Rani, N. J., et al. (1992). Effect of Santhi Kriya on certain psychophysiological parameters: A preliminary study. *Indian Journal of Physiology and Pharmacology, 36*(2), 88–92.

Schardt, D. (2000). St. John's worts and all. *Nutrition Action Healthletter*, September, 6–8.

Schaumburg, H. H., & Berger, A. (1992). Alopecia and sensory polyneuropathy from thallium in a Chinese herbal medication. *JAMA, 268*, 3430–3431.

Schefft, C., Kilarski, L. L., Bschor, T., et al. (2017). Efficacy of adding nutritional supplements in unipolar depression: A systematic review and meta-analysis. *European Neuropsychopharmacology, 27*(11), 1090–1109.

Schellenberg, R., Sauer, S., & Dimpfel, W. (1998). Pharmacodynamic effects of two different hypericum extracts in healthy volunteers measured by quantitative EEG. *Pharmacopsychiatry, 31*(Suppl 1), 44–53.

Schmidt, M., Betti, G., & Hensel, A. (2007). Saffron in phytotherapy: Pharmacology and clinical uses. *Wiener Medizinische Wochenschrift, 157*(13–14), 315–319.

Scholey, A. B., Tildesley, N. T., Ballard, C. G., et al. (2008). An extract of *Salvia* (sage) with anticholinesterase properties improves memory and attention in healthy older volunteers. *Psychopharmacology, 198*(1), 127–139.

Scholey, A., Ossoukhova, A., Owen, L., Ibarra, A., Pipingas, A., He, K., Roller, M., & Stough, C. (2010). Effects of American ginseng (*Panax quinquefolius*) on neurocognitive function: An acute, randomised, double-blind, placebo-controlled, crossover study. *Psychopharmacology, 212*(3), 345–356. https://doi.org/10.1007/s00213-010-1964-y. Epub 2010 Jul 31. PMID: 20676609; PMCID: PMC2952762.

Schweitzer, I., Maguire, K., et al. (2009). Sexual side-effects of contemporary antidepressants: Review. *Australian and New Zealand Journal of Psychiatry, 43*(9), 795–808.

Scuteri, D., Morrone, L. A., Rombola, L., et al. (2017). Aromatherapy and aromatic plants for the treatment of behavioural and psychological symptoms of dementia in patients with Alzheimer's disease: Clinical evidence and possible mechanisms. *Evidence-based Complementary and Alternative Medicine, 2017*, 1–30.

Seshadri, A., Adaji, A., Orth, S. S., et al. (2020). Exercise, Yoga, and Tai Chi for treatment of major depressive disorder in outpatient settings: A systematic review and meta-analysis. *Primary Care Companion Central Nervous System Disorders, 23*(1), 20r02722.

Shakibaei, F., Radmanesh, M., Salari, E., et al. (2015). Ginkgo biloba in the treatment attention-deficit/hyperactivity disorder in children and adolescents. A randomized, placebo-controlled trial. *Complementary Therapies in Clinical Practice, 21*(2), 61–67.

Shamir, E., Laudon, M., Barak, Y., et al. (2000). Melatonin improves sleep quality of patients with chronic schizophrenia. *Journal of Clinical Psychiatry, 61*, 373–377.

Sharma, A., Gerbarg, P., Bottiglieri, T., et al. (2017). S-Adenosylmethionine (SAMe) for neuropsychiatric disorders: A clinician-oriented review of research. *Journal of Clinical Psychiatry, 78*(6), e656–e667.

Shaw, K., Turner, J., & Del Mar, C. (2002). Are tryptophan and 5-hydroxytryptophan effective treatments for depression? A meta-analysis. *Australian and New Zealand Journal of Psychiatry, 36*(4), 488–491.

Shevtsov, V. A., Zholus, B. I., Shervarly, V. I., et al. (2003). A randomized trial of two different doses of a SHR-5 *Rhodiola rosea* extract versus placebo and control of capacity for mental work. *Phytomedicine, 10*(2–3), 95–105.

Shimada, Y., Arai, T., Tamaoka, A., et al. (2017). Liquorice-induced hypokalaemia in patients treated with Yokukansan preparations: Identification of the risk factors in a retrospective cohort study. *BMJ Open, 7*, e014218.

Shimada, Y., Fujimoto, M., Nogami, T., et al. (2019). Adverse events associated with ethical Kampo formulations: Analysis of domestic adverse-event data reports of the Ministry of Health, Labor, and Welfare in Japan. *Evidence-based Complementary and Alternative Medicine, 2019*, 1–14, 1643804.

Shinjyo, N., Waddell, G., & Green, J. (2020). Valerian root in treating sleep problems and associated disorders-a systematic review and meta-analysis. *Journal of Evidence Based Integrative Medicine, 25*, 2515690X20967323.

Singh, Y. N., & Devkota, A. K. (2003). Aqueous kava extracts do not affect liver function tests in rats. *Planta Medica, 69*, 496–499.

Singh, S. P., & Singh, V. (2011). Meta-analysis of the efficacy of adjunctive NMDA receptor modulators in chronic schizophrenia. *CNS Drugs, 25*(10), 859–885.

Singh, S. P., Telles, S., & Balkrishna. (2017). Use of yoga in managing posttraumatic stress disorder. In P. L. Gerbarg, R. P. Brown, & P. R. Muskin (Eds.), *Complementary and integrative treatments in psychiatric practice* (pp. 251–260). American Psychiatric Association Publishing.

Sinn, N., & Bryan, J. (2007). Effect of supplementation with polyunsaturated fatty acids and micronutrients on learning and behavior problems associated with child ADHD. *Journal of Developmental and Behavioral Pediatrics, 28*(2), 82–91.

Small, G. W., Siddarth, P., Zhaoping, L., et al. (2018). Memory and brain amyloid and tau effects of a bioavailable form of curcumin in non-demented adults: A double-blind placebo-controlled 18-month trail. *American Journal of Geriatric Psychiatry, 23*(3), 266–277.

Smith, A. D., Smith, A. M., & de Jager, C. A. (2010). Homocysteine-lowering by B vitamins slows the rate of accelerated brain atrophy in mild cognitive impairment: A randomized controlled trial. *PLoS One, 5*(9), e12244.

Smith, R. C., Boules, S., Maayan, L., et al. (2013). Effects of yoga on cognition, psychiatric symptoms, and epigenetic changes in chronic schizophrenic patients. Presented at the 14th international schizophrenia congress, Orlando, FL, 22 April 2013.

Smith, C. A., Armour, M., Lee, M. S., et al. (2018). Acupuncture for depression (review). *Cochrane Database of Systematic Reviews, 3*(3), CD004046.

Soares-Weiser, K., Maayan, N., & Bergman, H. (2018). Vitamin E for neuroleptic-induced tardive dyskinesia. *Cochrane Database of Systematic Reviews*, (1), CD000209.

Sonuga-Barke, E. J., Brandeis, D., Cortese, S., et al. (2013). Nonpharmacological interventions for ADHD: Systematic review and meta-analyses of randomized controlled trials of dietary and psychological treatments. *American Journal of Psychiatry, 170*(3), 275–289.

Sorensen, H., & Sonne, J. (1996). A double-masked study of the effects of ginseng on cognitive functions. *Current Therapeutic Research, 57*, 959–968.

Stevens, L. J., Zentall, S. S., Deck, J. L., et al. (1995). Essential fatty acid metabolism in boys with attention-deficit hyperactivity disorder. *American Journal of Clinical Nutrition, 62*, 761–768.

Stevenson, J., Sonuga-Barke, E., McCann, D., et al. (2010). The role of histamine degradation gene polymorphisms in moderating the effects of food additives on children's ADHD symptoms. *American Journal of Psychiatry, 167*(9), 1108–1115.

Stratyner, H. B. (2006). Multi-factorial approaches to substance use disorders and addiction. *CNS Spectrums, 11*, 828.

Streeter, C. C., Gerbarg, P. L., Saper, R. B., et al. (2012). Effects of yoga on the autonomic nervous system, gamma-aminobutyric acid, and allostasis in epilepsy, depression, and posttraumatic stress disorder. *Medical Hypotheses, 78*(5), 571–579.

Streeter, C. C., Gerbarg, P. L., Brown, R. P., et al. (2020a). Thalamic gamma aminobutyric acid level changes in major depressive disorder after a 12-week Iyengar yoga and coherent breathing intervention. *Journal of Alternative and Complementary Medicine, 26*(3), 190–197.

Streeter, C. C., Gerbarg, P. L., Nyer, M. B., et al. (2020b). Yoga: Part of a treatment plan for uncontrolled epilepsy. *Journal of Neurological Disorders and Stroke, 7*(1), 1156–1162.

Suhner, A., Schlagenhauf, P., Johnson, R., et al. (1998). Comparative study to determine the optimal melatonin dosage form for the alleviation of jet lag. *Chronobiology International, 15*(6), 655–666.

Szegedi, A., Kohnen, R., Dienel, A., et al. (2005). Acute treatment of moderate to severe depression with hypericum extract WS 5570 (St John's wort): Randomised controlled double blind non-inferiority trial versus paroxetine. *BMJ, 330*, 503.

Szewczyk, B., Kubera, M., et al. (2011). The role of zinc in neurodegenerative inflammatory pathways in depression. *Progress in Neuro Psychopharmacology and Biological Psychiatry, 35*(3), 693–701.

Takeda, A. (2011). Zinc signaling in the hippocampus and its relation to pathogenesis of depression. *Molecular Neurobiology, 44*(2), 166–174.

Targum, S. D., Cameron, B. R., & Ferreira, et al. (2018). An augmentation study of MSI-195 (*S*-adenosylmethionine) in major depressive disorder. *Journal of Psychiatric Research, 107*, 86–96.

Telles, S., & Singh, N. (2013). Science of the mind: Ancient yoga texts and modern studies. *Psychiatric Clinics of North America, 36*(1), 93–108.

Telles, S., Nagarathna, R., Nagendra, H. R., et al. (1993). Physiological changes in sports teachers following 3 months of training in yoga. *Indian Journal of Medical Science, 47*(10), 235–238.

Teschke, R., Sarris, J., et al. (2011). Kava, the anxiolytic herb: Back to basics to prevent liver injury? *British Journal of Clinical Pharmacology, 71*(3), 445–448.

Thiele, B., Brink, I., & Ploch, M. (1994). Modulation of cytokine expression by hypericum extract. *Journal of Geriatric Psychiatry and Neurology, 7*(Suppl 1), S60–S62.

Tildesley, N. T., Kennedy, D. O., Perry, E. K., et al. (2005). Positive modulation of mood and cognitive performance following administration of acute doses of *Salvia lavandulaefolia* essential oil to healthy young volunteers. *Physiology and Behavior, 83*(5), 699–709.

Tolahunase, M., Sager, R., Faiq, M., et al. (2018a). Yoga- and meditation-based lifestyle intervention increases neuroplasticity and reduces severity of major depressive disorder: A randomized controlled trial. *Restorative Neurology and Neuroscience, 36*(3), 423–442.

Tolahunase, M., Sager, R., Dada, R., et al. (2018b). 5-HTTLPR and *MTHFR 677C>T* polymorphisms and response to a yoga-based lifestyle intervention in major depressive disorder: A randomized active-controlled trial. *Indian Journal of Psychiatry, 60*(4), 410–426.

Torrioli, M. G., Vernacotola, S., et al. (2008). A double-blind, parallel, multicenter comparison of L-acetylcarnitine with placebo on the attention deficit hyperactivity disorder in fragile X syndrome boys. *American Journal of Medical Genetics, Part A, 146*(7), 803–812.

Torta, R., Zanalda, F., Rocca, P., et al. (1988). Inhibitory activity of *S*-adenosyl-L-methionine on serum gamma-glutamyl-transpeptidase increase induced by psychodrugs and anticonvulsants. *Current Therapeutic Research, 44*, 144–159.

Trebatická, J., Kopasova, S., Hradecna, Z., et al. (2006). Treatment of ADHD with French maritime pine bark extract, pycnogenol. *European Child and Adolescent Psychiatry, 15*, 329–335.

Tsang, H. W., Fung, K. M., Chan, A. S., et al. (2006). Effect of a qigong exercise programme on elderly with depression. *International Journal of Geriatric Psychiatry, 21*(9), 890–897.

US Department of Health and Human Services Agency for Healthcare Research and Quality. (2002). *S*-Adenosyl-L-methionine for treatment of depression, osteoarthritis, and liver disease. Evidence report/technology assessment no. 64. https://www.ncbi.nlm.nih.gov/books/NBK11886/. Accessed 17 Feb 2022.

Vaisman, N., Kaysar, N., & Zaruk-Adasha, Y. (2008). Correlation between changes in blood fatty acid composition and visual sustained attention performance in children with inattention: Effect of dietary n-3 fatty acids containing phospholipids. *American Journal of Clinical Nutrition, 87*(5), 1170–1180.

Vancampfort, D., Stubbs, B., Van Damme, T., et al. (2021). The efficacy of meditation-based mind-body interventions for mental disorders: A meta-review of 17 meta-analyses of randomized controlled trials. *Journal of Psychiatric Research, 134*, 181–191.

Varambally, S., Gangadhar, B. N., Thirthalli, J., et al. (2012). Therapeutic efficacy of add-on yogasana intervention in stabilized outpatient schizophrenia: Randomized controlled comparison with exercise and waitlist. *Indian Journal of Psychiatry, 54*, 227–232.

Varteresian, T., & Lavretsky, H. (2018). Complementary and integrative therapies in psychiatry. *Focus, 16*(1), 54–56.

Vaschillo, E., Vaschillo, B., & Lehrer, P. (2004). Heartbeat synchronizes with respiratory rhythm only under specific circumstances. *Chest, 126*, 1385–1386.

Vellas, B., Coley, N., Ousset, P. J., et al. (2012). Long-term use of standardised *Ginkgo biloba* extract for the prevention of Alzheimer's disease (GuidAge): A randomised placebo-controlled trial. *Lancet Neurology, 11*(10), 851–859.

Visceglia, E. (2007). Healing mind and body: Using therapeutic yoga in the treatment of schizophrenia. *International Journal of Yoga Therapy, 17*, 95–102.

Visceglia, E., & Lewis, S. (2011). Yoga therapy as an adjunctive treatment for schizophrenia: A randomized, controlled pilot study. *Journal of Alternative and Complementary Medicine, 17*, 601–607.

Voelker, R. (1999). Herbs and anesthesia. *JAMA, 281*, 1882.

Vorbach, E. U., Arnoldt, K. H., & Hubner, W. D. (1997). Efficacy and tolerability of St. John's wort extract LI 160 versus imipramine in patients with severe depressive episodes according to ICD-10. *Pharmacopsychiatry, 30*(Suppl 2), 81–85.

Vuong, Q. V., Bowyer, M. C., et al. (2011). L-Theanine: Properties, synthesis and isolation from tea. *Journal of the Science of Food and Agriculture, 91*(11), 1931–1939.

Wang, S. M., Kain, Z. N., et al. (2008). Acupuncture analgesia: I. The scientific basis. *Anesthesia and Analgesia, 106*(2), 602–610.

Wang, C., Li, K., Seo, D.-C., et al. (2020). Use of complementary and alternative medicine in children with ADHD: Results from the 2012 and 2017 National Health Interview Survey. *Complementary Therapies in Medicine, 49*, 102352.

Weber, W., Vander Stoep, A., et al. (2008). *Hypericum perforatum* (St John's wort) for attention-deficit/hyperactivity disorder in children and adolescents: A randomized controlled trial. *JAMA, 299*(22), 2633–2641.

White, D. J., Camfield, D. A., Ossoukhova, A., Savage, K., Le Cozannet, R., Fança-Berthon, P., & Scholey, A. (2020). Effects of *Panax quinquefolius* (American ginseng) on the steady state visually evoked potential during cognitive performance. *Human Psychopharmacology, 35*(6), 1–6. https://doi.org/10.1002/hup.2756.

Epub 2020 Sep 8. PMID: 32896022; PMCID: PMC7685123.

Whitten, D., Myers, D., et al. (2006). The effect of St John's wort extracts on CYP3A: A systematic review of prospective clinical trials. *British Journal of Clinical Pharmacology, 62*(5), 512–526.

Willatts, P., Forsyth, J. S., DiModugno, M. K., et al. (1998). Effect of long-chain polyunsaturated fatty acids in infant formula on problem solving at 10 months of age. *The Lancet, 352*, 688–691.

Williams, A. L., Katz, D., et al. (2006). Do essential fatty acids have a role in the treatment of depression? *Journal of Affective Disorders, 93*(1–3), 117–123.

Yang, X.-Y., Yang, N.-B., & Huang, F. F. (2021). Effectiveness of acupuncture on anxiety disorder: A systematic review and meta-analysis of randomised controlled trials. *Annals of General Psychiatry, 20*(9), 1–14.

Yary, T., & Aazami, S. (2012). Dietary intake of zinc was inversely associated with depression. *Biological Trace Element Research, 145*(3), 286–290.

Yeung, W. F., Chung, K. F., & Leung, Y. K. (2009). Traditional needle acupuncture treatment for insomnia: A systematic review of randomized controlled trials. *Sleep Medicine, 10*(7), 694–704.

Yolland, C. O., Phillipou, A., & Castle, D. (2020). Improvement of cognitive function in schizophrenia with *N*-acetylcysteine: A theoretical review. *Nutrition Neuroscience, 23*(2), 139–148.

Youdim, K. A., Martin, A., & Joseph, J. A. (2000). Essential fatty acids and the brain: Possible health implications. *International Journal of Developmental Neuroscience, 18*, 383–399.

Zaccaro, A., Piarulli, A., Laurino, M., et al. (2018). How breath-control can change your life: A systematic review on psycho-physiological correlates of slow breathing. *Frontiers in Human Neuroscience, 12*, 353–369.

Zanardi, R., & Smeraldi, E. (2006). A double-blind, randomised, controlled clinical trial of acetyl-L-carnitine vs. amisulpride in the treatment of dysthymia. *European Neuropsychopharmacology, 16*, 281–287.

Zandi, P. P., Anthony, J. C., Khachaturian, A. S., et al. (2004). Reduced risk of Alzheimer disease in users of antioxidant vitamin supplements: The Cache County Study. *Archives of Neurology, 61*, 82–88.

Zhang, X. Y., Zhou, D. F., Zhang, P. Y., et al. (2001). A double-blind, placebo-controlled trial of extract of *Ginkgo biloba* added to haloperidol in treatment-resistant patients with schizophrenia. *Journal of Clinical Psychiatry, 62*, 878–883.

Zhdanova, I. V., Wurtman, R. J., Regan, M. M., et al. (2001). Melatonin treatment for age-related insomnia. *Journal of Clinical Endocrinology and Metabolism, 86*, 4727–4730.

Zhou, Y., Hou, Y., Shen, J., et al. (2020). A network medicine approach to investigation and population-based validation of disease manifestations and drug repurposing for COVID-19. *PLoS Biology, 18*(11), e3000970.

Zylowska, L. (2012). *The mindfulness prescription for adult ADHD*. Trumpeter Books.

Placebo and Nocebo Effects: Biological and Cultural Aspects

138

Seetal Dodd, Malcolm Forbes, and Michael Berk

Contents

S. Dodd (✉)
IMPACT The Institute for Mental and Physical Health and Clinical Translation, Deakin University, Geelong, VIC, Australia

Centre for Youth Mental Health, University of Melbourne, Parkville, VIC, Australia

Department of Psychiatry, University of Melbourne, Parkville, VIC, Australia
e-mail: seetal.dodd@barwonhealth.org.au

M. Forbes
IMPACT The Institute for Mental and Physical Health and Clinical Translation, Deakin University, Geelong, VIC, Australia

Department of Psychiatry, University of Melbourne, Parkville, VIC, Australia

M. Berk
IMPACT The Institute for Mental and Physical Health and Clinical Translation, Deakin University, Geelong, VIC, Australia

Centre for Youth Mental Health, University of Melbourne, Parkville, VIC, Australia

Department of Psychiatry, University of Melbourne, Parkville, VIC, Australia

Orygen, the National Centre for Excellence in Youth Mental Health, Parkville, VIC, Australia

Florey Institute for Neuroscience and Mental Health, Parkville, VIC, Australia
e-mail: michael.berk@deakin.edu.au

© Springer Nature Switzerland AG 2024
A. Tasman et al. (eds.), *Tasman's Psychiatry*,
https://doi.org/10.1007/978-3-030-51366-5_146

Abstract

Improvement or benefit in a patient treated with a placebo is a placebo response and worsening or adverse effects in a patient treated with a placebo is a nocebo response. Both placebos and nocebos are inert treatments that have no known mechanism of action for the condition being treated. Curiously, the administration of a placebo may activate the same biological pathways targeted by some verum treatments. Placebo and nocebo phenomena are not rare occurrences that result from specific conditions or characteristics of treatment. Rather, they permeate the treatment environment and moderate treatment outcomes. Outcomes can be affected by the treatment setting, the discourse between the clinician and the patient, and the actions, manner, and professionalism of the clinician, described as "the treatment ritual." All forms of treatment involve a treatment ritual, and this ritual is believed to be the trigger of the placebo or nocebo response. The treating clinician can deploy or mitigate placebo and nocebo effects by taking into consideration the expectations and perception of the patient to enhance the benefits of treatment and minimize adverse effects.

Keywords

Placebo · Nocebo · Therapy · Active treatment · Verum · Response

Introduction

Placebos are routinely administered in clinical trials as a comparator to a pharmaceutical that is being tested. They are used as the control arm of the trial, the inert do-nothing, no-effect, pill compared to which the efficacy and safety of the test pill, the *verum*, can be measured. But as the name implies *Placebo – I will please*, placebos are anything but inert. How can something that does nothing please you?

So what exactly is a placebo? Surprisingly, even the definition is controversial and has evolved as our understanding of the placebo phenomenon has developed to encompass all the elements of treatment outside the pharmaceutical mechanism of action of the verum pill. These include expectancy and conditioning, which will be discussed later in this chapter, the treatment environment, and the clinician-patient relationship. In fact, the elements of treatment that are outside of the pharmaceutical mechanism of action are so broad that they are problematic to use as a practical definition of placebo. A definition needs to be both accurate and consistent with the way that the word is used in everyday communication. An inert pill is not an accurate definition and a definition with too broad a scope is also problematic and overlaps with other concepts. The most accepted definition of the placebo is the "vehicle" – that is, the product, chemical, or form in which the active compound of the experiment is carried. It is expected that the placebo is an agent that has no known mechanism of action for the condition being treated. A placebo can be a pill, injectable, transdermal patch, inhaled agent, or a sham procedure. Placebo researchers may also investigate waiting list or other control arms of non-pharmacological studies and therapies without specific biological mechanisms of action such as alternative therapies, but these are an extension of the conventional use of the term.

Definitions are also required for the placebo effect and the placebo response, which are not the same. The placebo effect is a benefit or improvement that is directly attributable to the administration of a placebo. The placebo effect is commonly but not necessarily investigated using deception, such as in pain experiments with a crossover design where, on one side of the cross, test subjects receive no treatment, and on the other side of the cross, test subjects are told that they are receiving an analgesic but are actually receiving a placebo. In this case, the placebo effect is the measured analgesic efficacy of the placebo. The placebo response on the other hand is the improvement from baseline or pretreatment in test subjects who have been treated with a placebo. For example, a placebo response is measured in placebo-controlled clinical trials, where it is not possible to determine whether improvement is due to natural fluctuations in symptom severity, the natural course of illness, the placebo effect, or other aspects of the treatment environment, or combinations of all these factors (see Table 1).

By corollary, equivalent definitions exist for the nocebo effect and nocebo response, where instead of benefit or improvement, the administration of a nocebo is associated with harm, worsening of symptoms or adverse effects. Nocebo will be discussed in greater detail later in this chapter.

In this chapter, the aim is to (i) provide an update on current knowledge of the placebo and nocebo phenomena, including the underlying biological mechanisms and (ii) discuss key issues around the use of placebos in psychiatry as a research tool and as a clinical treatment in its own right.

Table 1 Current definitions

Term	Definition
Placebo	An agent given as a treatment that has no known mechanism of action for the condition being treated
Placebo effect	The benefit or improvement that is directly attributable to the administration of a placebo
Placebo response	Improvement from baseline in subjects who have been treated with a placebo and may be due to the placebo effect or due to other factors
Nocebo effect	The worsening of a condition or development of adverse effects that occurs due to experiencing medical treatment but cannot be considered due to the specific treatment used
Nocebo response	The worsening of a condition or development of adverse effects that occurs at the same time as experiencing medical treatment but cannot be considered due to the specific treatment used and may be due to the nocebo effect or due to other factors

Biological Underpinnings

A placebo response can be observed in clinical trials across a broad range of illnesses, from psychiatric illness to hypotension, suggesting that the biological mechanism of action also varies depending on the illness being treated. This also suggests that different biological pathways can be triggered by a placebo and may include pathways that overlap with the known mechanisms of action of established pharmaceutical agents. It may be difficult to disentangle if a neurobiological response is applicable to the placebo or nocebo phenomenon in general or applicable only to a specific context such as a specific illness and treatment (Dodd et al., 2017).

Neuroanatomical Regions

The thalamus, primary and secondary somatosensory cortex, anterior cingulate cortex (ACC), amygdala, basal ganglia, and right lateral prefrontal cortex have been identified in functional magnetic resonance imaging (fMRI) studies as brain regions where response to pain stimulus was attenuated by placebo analgesia. Similarly, the rostral ACC, amygdala, insula, thalamus, prefrontal cortex, nucleus accumbens, and periaqueductal gray were identified by positron emission tomography (PET) as regions with perturbed dopamine (D2 and D3) receptor occupancy during placebo analgesia (Dodd et al., 2017). The involvement of these brain regions was demonstrated in placebo models among healthy participants using MRI and PET scans. Different brain regions may be involved when patients are treated with placebos in clinical or other settings. There is evidence to suggest that brain changes can result in loss of placebo mechanisms, with individuals with reduced functional connectivity secondary to Alzheimer's disease (measured by frontal assessment battery scores) showing a reduced placebo response from analgesia (Benedetti et al., 2006).

Neurochemical Processes

A study in a cohort of placebo responders confirmed the hypothesis that endorphin release mediates placebo analgesia by using naloxone to block the analgesic effect of placebo for dental postoperative pain (Levine et al., 1978). Further research, primarily in pain studies but also in a study of major depressive disorder, has implicated endorphin and dopamine release and opioid and dopamine receptors, and demonstrated increased μ-opioid neurotransmission (Pecina et al., 2015). Elsewhere, a pain study using PET found increased dopamine and opioid neurotransmission in multiple brain regions with placebo analgesia and decreased dopamine and opioid neurotransmission in the same brain regions for nocebo responders (Scott et al., 2008). Other studies have shown analgesic placebo effects to be associated with the release of endocannabinoids, oxytocin, and vasopressin (Colloca & Barsky, 2020). Similar to clinical studies and behavioral studies, studies of the biological underpinnings of placebo typically demonstrate considerable interindividual variation in response. Furthermore, there is a paucity of research in this area and further investigations are required, especially in clinical areas other than placebo analgesia.

Do Placebo Studies Make an Important Contribution to Modern Medicine?

We can all be intrigued when a test subject responds to a placebo that they believe is a powerful pharmaceutical. However, there are important lessons that can be learnt from placebo studies. Placebo studies show us that it is not only the pharmacological mechanism of action of a medication that contributes towards treatment outcomes but also the treatment environment, the context, the beliefs, perceptions and experience of both the patient and the person or team offering treatment, the therapeutic alliance, and many other factors that influence treatment outcomes. While factors other than pharmacological mechanisms of action are widely accepted as important to treatment outcomes, their contribution towards treatment outcomes is often underestimated and sometimes ignored altogether. Modern placebo studies demonstrate that the contribution of these factors is considerable, especially when treating

illnesses that are known from clinical trials to have large placebo response rates. Large placebo response rates of 35–40% have been reported in depression (Furukawa et al., 2016); however, it can be difficult to isolate the role of the placebo effect from spontaneous remission and regression to the mean (Hengartner, 2020).

The key to eliciting a placebo response is not deception. Patients and research participants do not respond to placebo because they believe they have received the verum, and this somehow triggers a biological response. Research started over a decade ago has demonstrated that placebos can work without deception (Kaptchuk et al., 2010). Instead, the placebo effect works by triggering something which is already in the patient or research participant. Neurobiological pathways are activated by triggering beliefs and expectations already held by the person being treated with the placebo.

The modern understanding of why placebos work encompasses factors such as the importance of bedside manner and professionalism, which have been understood since antiquity and were discussed by Hippocrates. This extends to the doctor-patient relationship or therapeutic alliance, which considers both patient and clinician and the engagement between them (Ardito & Rabellino, 2011). Nevertheless, these concepts suggest that a key part of a clinicians' training or preexisting manners contributes to beneficial outcomes.

To placebo researchers, the therapeutic act is a ritual that is culturally specific and requires mutual understanding between all parties involved in the ritual (Kaptchuk, 2011). A model of ritual healing has been proposed where the core requirements for healing are: a predisposition to be healed, an experience of empowerment where the healing ritual provides the experience of being empowered with culturally appropriate treatment or transformation, and finally a concrete perception of transformation, relief, or recovery (Csordas, 1983). These rituals influence clinically significant neurobiological processes. There are many and varied healing rituals, including traditional medicine, faith healers, complementary and alternative medicine, and mainstream modern medicine.

Current biomedical treatments are grounded in science and an evidence-based approach, but modern medical practice also has extensive rituals that significantly affect treatment outcomes: the setting of care; the treatment frame; and the doctor, with their unique communication style and own beliefs and attitudes around treatment. Different doctors and different treatment environments can have different therapeutic outcomes, even when they prescribe identical medications. Placebo research teaches us that the efficacy of health practitioners can be improved by improving our understanding of all factors that contribute towards treatment outcomes.

As well as improving our understanding of how to maximize benefit of treatment and minimize adverse effects, there may also be a direct role for using placebos as an ethical and effective treatment in some circumstances, which is addressed below.

The Placebo Phenomenon

Several theories have been postulated to explain why placebos work. All theories presented here have been investigated and are supported by evidence and they all *contribute* towards understanding why placebos work. However, their importance and the strength of their effect may vary considerably depending on the situation and type of condition treated. As with biological pathways, the psychological mechanisms associated with the placebo effect are many and varied, leading to the observation that "there is not one single placebo effect, but many" (Benedetti, 2014).

The Expectancy Hypothesis

Expectancy is a belief about a likelihood of something that may occur in the future and has adaptive advantages at a species level. Being able to predict factors about a predator, or something you are hunting, has obvious benefits even if those predictions may not always be accurate. *Expectancy theory* was originally developed within the context of motivational theory, where the expectation

of an outcome positively influences behavior to achieve a desired outcome (Vroom, 1964). Furthermore, expectation of a specific outcome may also elicit cognitive and emotional responses. Expectancy can influence outcomes of placebo treatment and of all medical treatments. There may be an expectation that treatment will result in changes in symptoms and/or there may be an expectation that treatment will result in adverse effects.

Expectancy can influence treatment outcomes both positively and negatively and can be caused by many factors including the treatment environment and the actions of the treating clinician. There is stimulant expectancy, which is anticipation of external events, and response expectancy, where a response to an external event is predicted. Expectancy can be manipulated in both clinical and research environments. Kirsch (2018) noted that, "the greater the ambiguity of a stimulus, the larger the role of expectancy in the perception of it," reinforcing the benefit of clearly explaining procedures to patients in a clinical environment. A patient with limited understanding of the process being told that he might benefit from ECT may have very inaccurate perceptions of what to expect. Poor prior treatment experiences or adverse events may create expectations that prime patients for nocebo responses to future treatments. Extinction, the loss of expectancy, may occur when an expected outcome does not result and may also influence treatment outcomes. Expectancy and extinction can be investigated in research environments and placebos are a useful tool for this purpose.

Experiments investigating "enhanced" placebo responses can be devised by manipulating expectancy. For example, a stronger effect is reported for participants receiving multiple or higher doses or bigger pills (of placebo with deception) (Kirsch, 2018). Similarly, a placebo injection has a stronger effect than a placebo pill (Kaptchuk et al., 2006).

Expectancy can also play a role in the researchers performing the study. Clinical trials which contain several active arms are associated with failure to separate from placebo (Dumitrescu et al., 2019). This is believed to be related, in part, to researchers' expectations that study participants are more likely to be on active medications.

The Conditioning Hypothesis

Conditioned behaviors are reactions that occur through learning or experience. Classical conditioning pairs a stimulus or cue to an involuntary response: Pavlov's dog who salivates when a bell is rung. Operant conditioning involves changing of behavior through reinforcement or punishment. Conditioning is related to expectancy in that a predictable outcome is generated by the conditioning process.

Every time a person interacts with the health system or medical treatment, they have an experience from which they learn. They learn that an intervention may bring relief or discomfort or both. Conditioning may be an important driver of nocebo effects. A patient who has previously experienced adverse effects may be more likely to complain of adverse effects from a new treatment. Equally, people who responded well or poorly to a prior treatment are more likely to expect a subsequent treatment to behave similarly. Disclosure of adverse effect information, for example, on a package insert or participant information and consent form, may also be associated with a likelihood of experiencing those listed adverse effects, although evidence is limited, and any effect is weak at a population level.

Counterconditioning is where an undesirable response to a stimulus or cue is replaced by a favorable response. Extinction is where a stimulus or cue does not result in the expected outcome. With reinforcement or punishment, and with repetition, the involuntary response eventually diminishes or ceases to occur.

Other Factors Associated with the Placebo Response

Rabkin et al. (1987) investigated the characteristics of depressed participants in a clinical trial who had responded to placebo treatment during a 10-day placebo washout phase and found that they were more mildly ill, more chronic, contained fewer cases of primary depression, and had fewer illness precipitants. Personality types have been investigated with mixed results. There is some evidence associating an optimistic personality type with a greater placebo response (Kern et al., 2020). Weimer et al. (2015a) investigated predictors of placebo response and concluded that younger age may contribute to the placebo response in some conditions but sex does not. Lower symptom severity at baseline was the most robust predictor of placebo response.

In addition to focusing on the characteristics of the person responding to the placebo, contextual factors should be considered. Investigations of the role of social and interpersonal factors suggest that the person administering the treatment can be as important and the person receiving the treatment. Factors such as social isolation (loneliness) and the patient and treating clinician's perceptions of each other are predictors of the placebo response (Necka & Atlas, 2018).

There is good evidence that the placebo response is stronger for some illnesses compared to others, with the best evidenced being pain, depression, Parkinson's disease, fatigue, allergies, and immune deficiencies (Evers et al., 2018).

In addition to predictors of a placebo response, the magnitude of the placebo response is also an area of investigation. Placebo has long been known to be an effective treatment while seemingly paradoxically used as an inert control when evaluating new treatments in clinical trials. The magnitude of the placebo effect can be assessed in experiments conducted on healthy test subjects, such as placebo analgesia experiments where participants are tested in crossover design pain experiments with or without placebo; however, measuring the placebo in clinical treatment is fraught as it would require disentangling the placebo effect from other factors, including genuine symptom fluctuation.

The Nocebo Phenomenon

The nocebo phenomenon is the flipside of the placebo phenomenon. In addition to worsening of illness, this also includes adverse events. Adverse event reporting in placebo arms of clinical trials is very common, although it is difficult to separate nocebo events from commonly occurring incidental adverse events. For example, if a participant in a placebo arm of a clinical trial reports a headache, it may not be possible to determine if the headache is a response to placebo treatment or coincidental. Meta-analyses of placebo-arm data from clinical trials have identified a spike in adverse events within the first 2 weeks of placebo treatment, suggesting that adverse events that occur shortly after commencing placebo treatment are more likely to be due to the nocebo effect, whereas adverse events occurring during placebo maintenance treatment are more likely to be coincidental (Dodd et al., 2019). Furthermore, more adverse events were reported in the placebo arms of early and phase II trials of olanzapine than in later and phase IV trials, possibly because earlier phase studies are perceived as carrying greater risk (Dodd et al., 2015).

As with the placebo effect, patient characteristics associated with the nocebo effect have been reported but they are weak associations, sometimes observed in one analysis but not replicated. Trait anxiety has been associated with a greater nocebo response (Kern et al., 2020). Meta-analyses have reported the following participant characteristics: not being treatment naïve, younger age, being obese, being a participant in a previous study, and having previously been treated with *Hypericum perforatum*, and regarding study characteristics: studies conducted in the USA (Dodd et al., 2015, 2019).

Although the association between the nocebo effect and predictive factors may be weak, the nocebo effect can be very powerful as exemplified by phenomena including "voodoo death" and "pointing the bone," although these examples are not well documented. An example clearly recorded in scientific literature was that of a 26-year-old male participant in a placebo-controlled trial with medications for depression who consumed 29 capsules and experienced hypotension requiring intravenous fluids but recovered quickly when informed that the capsules were placebo (Reeves et al., 2007). More generally, adverse events are common causes of treatment nonadherence and the nocebo effect may have a significant role in this. Nocebo type

drivers also underlie group psychosomatic phenomena occurring in mass hysteria events.

Techniques to reduce the nocebo response include providing patients with enhanced information, optimizing patient-clinician communication and relationships, and psychoeducation on coping skills to manage patient expectations (Manai et al., 2019). Informing people about the nocebo effect using a short information sheet resulted in study participants reporting fewer nocebo symptoms compared to a control group (Pan et al., 2019).

Should Placebos Be Used to Treat Patients?

There is expert consensus that educating clinicians about placebo and nocebo effects and training clinicians to maximize placebo and minimize nocebo effects is likely to result in improved treatment outcomes (Evers et al., 2018). More controversial is whether actual placebos should be administered as treatments to patients with certain conditions known to be amenable to placebo treatments. The key concerns are whether placebo treatments may be unethical and that placebo treatment may be less effective compared to other treatments.

In counterargument to these concerns, it is noted that placebos can have similar effect sizes compared to treatments (Howick et al., 2013); furthermore, some illnesses do not have any proven pharmacotherapies. Most verum treatments carry real risks. Also, concerns about the ethical implications of deceptively administering a placebo rather than verum treatment are countered by recent research demonstrating the efficacy of placebo administration without deception, where open-labeled placebo with an accompanying narrative was superior to no treatment for irritable bowel syndrome (Kaptchuk et al., 2010). Arguably, the ethics of deception may be more of a concern to ethicists than to patients, with evidence to suggest that some patients may have a more pragmatic view of the use of placebos (Koteles & Ferentzi, 2012).

Placebo researchers have proposed that there is a role for open-labeled placebos and dose-extending placebos as routine clinical treatments (Colloca & Howick, 2018). Dose-extending placebos have been proposed as a way of weaning patients off medications on which they have become dependent. Dose-extending placebos – the term used to refer to the pairing of placebo pills with a physiologically active treatment – can also be open labeled and are believed to reinforce conditioned responses learnt during verum treatment. The placebo pill can be used to "extend" the efficacy of the medication it was paired with and enhancing the effect of subsequent placebo administration. Other advantages of placebo treatment are that placebos are inexpensive and well tolerated.

Should Placebo-Treated Controls Be Used in Clinical Trials?

Aside from the ethical debate around the use of placebos in clinical trials, there is also debate regarding whether placebo-controlled trials are always the best study design for determining the safety and efficacy of new treatments. Randomized, placebo-controlled trials (RCTs) have been the accepted methodology for decades, but even the pioneers of this method recognized the problems of disregarding the placebo effect, stating that "placebos can mimic in many ways the characteristics of active drugs" (Lasagna, 1962). Indeed, they recognize other shortfalls of the RCT method, such as the difficulty with recruiting representative samples. However, placebo-controlled RCTs rapidly became standard because they were seen as superior to the uncontrolled studies that were common prior to the 1960s, and current expert consensus remains supportive of placebo-controlled studies (Similon et al., 2022).

More recently, other problems with the RCT method have begun to emerge. It has been observed that the effect size of the placebo response appears to be growing. A study of the placebo response in depression showed a growth rate of 7% per year (Walsh et al., 2002). This has been attributed to changes in study design such as an increase in multiple arm study designs, where expectation is increased due to the increased probability of allocation to a verum arm (Dumitrescu et al., 2019), and the greater inclusion of participants with lower symptom severity scores and shorter index episodes, although other factors may also contribute (Weimer et al., 2015b). While growth of the placebo response effect size is limited, a strong response in placebo arms of clinical trials highlights a major problem for placebo-controlled, randomized clinical trials, with many trials resulting in the verum arm failing to separate from the placebo arm. Furthermore, a large nocebo response in the placebo arm may provide inaccurate measures of risk while equally masking some genuine adverse effects in the verum arm. The problem of clinical trials failing to identify uncommon adverse events is well known.

Several strategies have been considered to attempt to address problems inherent with the use of the placebo control, with limited success. Attempts to characterize and exclude placebo and nocebo responders have been largely unsuccessful. The use of a placebo lead-in phase for an RCT may reduce placebo and nocebo responses during the treatment phase, but may have other disadvantages with regards to practicality and sample size. A 2:1 randomization ratio of placebo to verum will decrease the probability of being randomized to verum and may decrease the placebo response through decreased expectancy, but will also result in an increase in the required sample size and perhaps difficulties with participant recruitment. An unexpectedly low placebo response may suggest inadequacy of the blind.

In spite of the inherent difficulties with the use of placebo as a control, it remains the method of choice for drug evaluation because the alternatives have greater limitations (Similon et al., 2022). Trialing new treatments using established treatments as a control is an option for illnesses that have established treatments of demonstrated efficacy, although some argue that a placebo control is required to calculate effect sizes and detect adverse effects of the verum, notwithstanding that placebos also have their own effect sizes and adverse effects. Such trials typically require non-inferiority designs that statistically need to be much larger and hence are both more expensive and expose more participants to the risks of the experimental treatment.

Specific Considerations for Psychiatric Practice

There appears to be greater recognition of the placebo effect in psychiatry than in other medical specialties. A 2011 Canadian study found that, compared to other medical specialists, psychiatrists were more likely to recognize that placebos could exert a powerful effect and 38% of psychiatrists reported giving subtherapeutic doses of medication to their patients compared to 6% of nonpsychiatrists (Raz et al., 2011). A small German study found that physicians (mainly psychiatrists, neurologists, and general practitioners) recognized the physician-patient relationship and patient expectations and experiences as determinants of the effectiveness of treatment with medications for depression. They were less likely to believe that informing the patient about known side effects could induce side effects (Kampermann et al., 2017).

There are several clinical risk factors for the nocebo response, including a history of adverse effects and a history of unsatisfactory experiences with medical care can increase risk, due to reasons outlined above. Patients with high levels of neuroticism, somatization, and hypochondriasis are also at elevated risk of a nocebo responses (Data-Franco & Berk, 2013).

Psychiatrists can incorporate an understanding of placebo and nocebo effects to benefit patient care. Adopting an unconditional positive regard for the patient and maintaining an empathic stance can predispose patients toward recovery. Attention to cognitive biases can be utilized to help nudge patients towards favorable expectations of treatment. Patients must always provide informed consent to treatment, but the information can be framed in a way to bring attention to benefits rather than losses to avoid the natural consequences of availability bias (Data-Franco & Berk, 2013).

Obtaining a detailed medication history is an important first step with treatment. During this medication history, it is essential to collect information about past experiences of treatment and expectations of future treatment. These expectations can be addressed and considered when choosing medications. Pharmacological treatment, in the absence of clinical urgency, should be prescribed in a "start low and go slow" fashion – starting a low dose and titrating up gradually. There should be consideration of administering medication in a positive context, considering multisensory cues (e.g., sight, smell, and taste stimulations associated with the medication) to promote conditioning (Colloca & Barsky, 2020).

Greater understanding of the nocebo effect may be useful in counselling patients about treatment adverse effects. A systematic review of 56 placebo-controlled RCTs of monotherapy with medications for depression found that adverse effects more common with medications for depression versus placebos include neurological, sexual, and anticholinergic effects. Negative thoughts, emotions, and behavior, as well as pain, including headaches, appeared to be driven largely by nocebo effects (Sinyor et al., 2020). It is important for clinicians not to discount adverse effects from patients but to be aware that nonspecific nocebo responses are common with psychotropic medications.

Shared decision-making is key to aligning treatment selection with patient expectations. In an open-label RCT of chamomile extract for generalized anxiety disorder, there was a positive relationship found between patient expectations of a response and a reduction in anxiety symptoms, consistent with other studies in the area (Keefe et al., 2017). Spending the time with a patient to assess their beliefs about a particular psychotherapeutic or pharmacological treatment, and correcting these where necessary, may influence the subsequent effectiveness of the treatment.

The use of pharmacogenetic support tools, both by facilitating the prescription of a medication concordant with a patient's pharmacogenetic profile and also via response expectancy (the patient believes the medication is "right" for them and is thus more likely to expect a positive response with less side effects), is one promising strategy for harnessing the placebo effect.

Pharmacogenetic support tools can also support shared decision-making in clinical practice (Arandjelovic et al., 2017).

The placebo effect is not limited to pharmacological treatment in psychiatry. The use of placebos may also enhance the effect of psychotherapy. Jurinec and Schienle (2020) randomized 126 patients with depression to one of three groups: cognitive-behavioral therapy (CBT) course group, a CBT course plus placebo group, and a waiting list group. The placebo was sunflower oil and was introduced as a natural medicine to help patients focus on their inner strengths. The placebo group demonstrated improved CBT homework compliance and reduced depressive symptoms.

There is a need for more research into the placebo and nocebo response in psychiatry to maximize the efficacy of existing treatments and identify those are heightened risk of adverse medication effects. The identification of biomarker signatures related to placebo effects may allow identification of neurotransmitter systems involved in symptom improvement and open the door to novel treatment targets (Huneke et al., 2020). It is imperative that new treatment options in psychiatry, for example, the renewed interests in psychedelics, be evaluated in light of their vulnerability to powerful placebo responses (Olson et al., 2020).

Conclusions

The placebo has powerful psychological and biological effects and is an inseparable component of the treatment ritual. The placebo effect unites modern, evidence-based medicine with all other versions of therapy including ancient, traditional, and alternative therapies. By understanding the causes of the placebo and nocebo effects, the modern clinician should aim to manipulate their therapeutic environment and treatment rituals to maximize placebo effects and minimize nocebo effects, and this should be customized for individual patients taking into consideration their expectations, perceptions, and previous experience.

There is strong evidence to suggest that administering actual placebos as treatment may benefit some patients and this should be done taking ethical considerations into account. Placebo experts suggest that placebos could be used in clinical practice, as they are well tolerated and effective therapies with effect sizes equal to some verum treatments, and their use in clinical trials needs to be seen in context that they are active treatments rather than an inert comparator.

References

Arandjelovic, K., Eyre, H. A., Lenze, E., Singh, A. B., Berk, M., & Bousman, C. (2019). The role of depression pharmacogenetic decision support tools in shared decision making. *J Neural Transm (Vienna)*, *126*(1), 87–94. https://doi.org/10.1007/s00702-017-1806-8. Epub 2017 Oct 29. PMID: 29082439.

Ardito, R. B., & Rabellino, D. (2011). Therapeutic alliance and outcome of psychotherapy: Historical excursus, measurements, and prospects for research. *Frontiers in Psychology*, *2*, 270. https://doi.org/10.3389/fpsyg.2011.00270

Benedetti, F. (2014). Placebo effects: From the neurobiological paradigm to translational implications. *Neuron*, *84*(3), 623–637. https://doi.org/10.1016/j.neuron.2014.10.023

Benedetti, F., Arduino, C., Costa, S., Vighetti, S., Tarenzi, L., Rainero, I., & Asteggiano, G. (2006). Loss of expectation-related mechanisms in Alzheimer's disease makes analgesic therapies less effective. *Pain*, *121*(1–2), 133–144. https://doi.org/10.1016/j.pain.2005.12.016

Colloca, L., & Barsky, A. J. (2020). Placebo and nocebo effects. *The New England Journal of Medicine*, *382*(6), 554–561. https://doi.org/10.1056/NEJMra1907805

Colloca, L., & Howick, J. (2018). Placebos without deception: Outcomes, mechanisms, and ethics. *International Review of Neurobiology*, *138*, 219–240. https://doi.org/10.1016/bs.irn.2018.01.005

Csordas, T. J. (1983). The rhetoric of transformation in ritual healing. *Culture, Medicine and Psychiatry*, *7*(4), 333–375. https://doi.org/10.1007/BF00052238

Data-Franco, J., & Berk, M. (2013). The nocebo effect: A clinicians guide. *The Australian and New Zealand Journal of Psychiatry*, *47*(7), 617–623. https://doi.org/10.1177/0004867412464717

Dodd, S., Schacht, A., Kelin, K., Duenas, H., Reed, V. A., Williams, L. J., . . . Berk, M. (2015). Nocebo effects in the treatment of major depression: Results from an individual study participant-level meta-analysis of the placebo arm of duloxetine clinical trials. *The Journal of Clinical Psychiatry*, *76*(6), 702–711. https://doi.org/10.4088/JCP.13r08858

Dodd, S., Dean, O. M., Vian, J., & Berk, M. (2017). A review of the theoretical and biological understanding of the nocebo and placebo phenomena. *Clinical Therapeutics*, *39*(3), 469–476. https://doi.org/10.1016/j.clinthera.2017.01.010

Dodd, S., Walker, A. J., Brnabic, A. J. M., Hong, N., Burns, A., & Berk, M. (2019). Incidence and characteristics of the nocebo response from meta-analyses of the placebo arms of clinical trials of olanzapine for bipolar disorder. *Bipolar Disorders, 21*(2), 142–150. https://doi.org/10.1111/bdi.12662

Dumitrescu, T. P., McCune, J., & Schmith, V. (2019). Is placebo response responsible for many phase III failures? *Clinical Pharmacology and Therapeutics, 106*(6), 1151–1154. https://doi.org/10.1002/cpt.1632. Erratum in: Clin Pharmacol Ther, 2020; *107*(6), 1457.

Evers, A. W. M., Colloca, L., Blease, C., Annoni, M., Atlas, L. Y., Benedetti, F., . . . Kelley, J. M. (2018). Implications of placebo and nocebo effects for clinical practice: Expert consensus. *Psychotherapy and Psychosomatics, 87*(4), 204–210. https://doi.org/10.1159/000490354

Furukawa, T. A., Cipriani, A., Atkinson, L. Z., Leucht, S., Ogawa, Y., Takeshima, N., . . . Salanti, G. (2016). Placebo response rates in antidepressant trials: A systematic review of published and unpublished double-blind randomised controlled studies. *The Lancet Psychiatry, 3*(11), 1059–1066. https://doi.org/10.1016/S2215-0366(16)30307-8

Hengartner, M. P. (2020). Is there a genuine placebo effect in acute depression treatments? A reassessment of regression to the mean and spontaneous remission. *BMJ Evidence Based Medicine, 25*(2), 46–48. https://doi.org/10.1136/bmjebm-2019-111161

Howick, J., Friedemann, C., Tsakok, M., Watson, R., Tsakok, T., Thomas, J., . . . Heneghan, C. (2013). Are treatments more effective than placebos? A systematic review and meta-analysis. *PLoS One, 8*(5), e62599. https://doi.org/10.1371/journal.pone.0062599

Huneke, N. T. M., van der Wee, N., Garner, M., & Baldwin, D. S. (2020). Why we need more research into the placebo response in psychiatry. *Psychological Medicine, 50*(14), 2317–2323. https://doi.org/10.1017/S0033291720003633

Jurinec, N., & Schienle, A. (2020). Utilizing placebos to leverage effects of cognitive-behavioral therapy in patients with depression. *Journal of Affective Disorders, 277*, 779–784. https://doi.org/10.1016/j.jad.2020.08.087

Kampermann, L., Nestoriuc, Y., & Shedden-Mora, M. C. (2017). Physicians' beliefs about placebo and nocebo effects in antidepressants – An online survey among German practitioners. *PLoS One, 12*(5), e0178719. https://doi.org/10.1371/journal.pone.0178719

Kaptchuk, T. J. (2011). Placebo studies and ritual theory: A comparative analysis of Navajo, acupuncture and biomedical healing. *Philosophical Transactions of the Royal Society of London. Series B, Biological Sciences, 366*(1572), 1849–1858. https://doi.org/10.1098/rstb.2010.0385

Kaptchuk, T. J., Stason, W. B., Davis, R. B., Legedza, A. R., Schnyer, R. N., Kerr, C. E., . . . Goldman, R. H. (2006). Sham device v inert pill: Randomised controlled trial of two placebo treatments. *BMJ,* 332(7538), 391–397. https://doi.org/10.1136/bmj.38726.603310.55

Kaptchuk, T. J., Friedlander, E., Kelley, J. M., Sanchez, M. N., Kokkotou, E., Singer, J. P., . . . Lembo, A. J. (2010). Placebos without deception: A randomized controlled trial in irritable bowel syndrome. *PLoS One, 5*(12), e15591. https://doi.org/10.1371/journal.pone.0015591

Keefe, J. R., Amsterdam, J., Li, Q. S., Soeller, I., DeRubeis, R., & Mao, J. J. (2017). Specific expectancies are associated with symptomatic outcomes and side effect burden in a trial of chamomile extract for generalized anxiety disorder. *Journal of Psychiatric Research, 84*, 90–97. https://doi.org/10.1016/j.jpsychires.2016.09.029

Kern, A., Kramm, C., Witt, C. M., & Barth, J. (2020). The influence of personality traits on the placebo/nocebo response: A systematic review. *Journal of Psychosomatic Research, 128*, 109866. https://doi.org/10.1016/j.jpsychores.2019.109866

Kirsch, I. (2018). Response expectancy and the placebo effect. *International Review of Neurobiology, 138*, 81–93. https://doi.org/10.1016/bs.irn.2018.01.003

Koteles, F., & Ferentzi, E. (2012). Ethical aspects of clinical placebo use: What do laypeople think? *Evaluation & the Health Professions, 35*(4), 462–476. https://doi.org/10.1177/0163278712453993

Lasagna, L. (1962). Controlled trials: Nuisance or necessity? *Methods of Information in Medicine, 1*(3), 79–82. Retrieved from https://www.ncbi.nlm.nih.gov/pubmed/21735876

Levine, J. D., Gordon, N. C., & Fields, H. L. (1978). The mechanism of placebo analgesia. *Lancet, 2*(8091), 654–657. https://doi.org/10.1016/s0140-6736(78)92762-9

Manai, M., van Middendorp, H., Veldhuijzen, D. S., Huizinga, T. W. J., & Evers, A. W. M. (2019). How to prevent, minimize, or extinguish nocebo effects in pain: A narrative review on mechanisms, predictors, and interventions. *Pain Reports, 4*(3), e699. https://doi.org/10.1097/PR9.0000000000000699

Necka, E. A., & Atlas, L. Y. (2018). The role of social and interpersonal factors in placebo analgesia. *International Review of Neurobiology, 138*, 161–179. https://doi.org/10.1016/bs.irn.2018.01.006

Olson, J. A., Suissa-Rocheleau, L., Lifshitz, M., Raz, A., & Veissiere, S. P. L. (2020). Tripping on nothing: Placebo psychedelics and contextual factors. *Psychopharmacology, 237*(5), 1371–1382. https://doi.org/10.1007/s00213-020-05464-5

Pan, Y., Kinitz, T., Stapic, M., & Nestoriuc, Y. (2019). Minimizing drug adverse events by informing about the nocebo effect-an experimental study. *Frontiers in Psychiatry, 10*, 504. https://doi.org/10.3389/fpsyt.2019.00504

Pecina, M., Bohnert, A. S., Sikora, M., Avery, E. T., Langenecker, S. A., Mickey, B. J., & Zubieta, J. K. (2015). Association between placebo-activated neural systems and antidepressant responses: Neurochemistry of placebo effects in major depression. *JAMA Psychiatry, 72*(11), 1087–1094. https://doi.org/10.1001/jamapsychiatry.2015.1335

Rabkin, J. G., Stewart, J. W., McGrath, P. J., Markowitz, J. S., Harrison, W., & Quitkin, F. M. (1987). Baseline characteristics of 10-day placebo washout responders in antidepressant trials. *Psychiatry Research, 21*(1), 9–22. https://doi.org/10.1016/0165-1781(87)90057-6

Raz, A., Campbell, N., Guindi, D., Holcroft, C., Dery, C., & Cukier, O. (2011). Placebos in clinical practice: Comparing attitudes, beliefs, and patterns of use between academic psychiatrists and nonpsychiatrists. *Canadian Journal of Psychiatry, 56*(4), 198–208. https://doi.org/10.1177/070674371105600403

Reeves, R. R., Ladner, M. E., Hart, R. H., & Burke, R. S. (2007). Nocebo effects with antidepressant clinical drug trial placebos. *General Hospital Psychiatry, 29*(3), 275–277. https://doi.org/10.1016/j.genhosppsych.2007.01.010

Scott, D. J., Stohler, C. S., Egnatuk, C. M., Wang, H., Koeppe, R. A., & Zubieta, J. K. (2008). Placebo and nocebo effects are defined by opposite opioid and dopaminergic responses. *Archives of General Psychiatry, 65*(2), 220–231. https://doi.org/10.1001/archgenpsychiatry.2007.34

Similon, M. V. M., Paasche, C., Krol, F., et al. (2022). Expert consensus recommendations on the use of randomized clinical trials for drug approval in psychiatry- comparing trial designs. *European Neuropsychopharmacology, 60,* 91–99. https://doi.org/10.1016/j.euroneuro.2022.05.002

Sinyor, M., Cheung, C. P., Abraha, H. Y., Lanctot, K. L., Saleem, M., Liu, C. S., … Schaffer, A. (2020). Antidepressant-placebo differences for specific adverse events in major depressive disorder: A systematic review. *Journal of Affective Disorders, 267,* 185–190. https://doi.org/10.1016/j.jad.2020.02.013

Vroom, V. H. (1964). *Work and motivation.* Wiley.

Walsh, B. T., Seidman, S. N., Sysko, R., & Gould, M. (2002). Placebo response in studies of major depression: Variable, substantial, and growing. *JAMA, 287*(14), 1840–1847. https://doi.org/10.1001/jama.287.14.1840

Weimer, K., Colloca, L., & Enck, P. (2015a). Age and sex as moderators of the placebo response - an evaluation of systematic reviews and meta-analyses across medicine. *Gerontology, 61*(2), 97–108. https://doi.org/10.1159/000365248

Weimer, K., Colloca, L., & Enck, P. (2015b). Placebo effects in psychiatry: Mediators and moderators. *Lancet Psychiatry, 2*(3), 246–257. https://doi.org/10.1016/S2215-0366(14)00092-3

Section X

Consultation Liaison and Collaborative Care in Medical Illness and Comorbid Psychiatric and Medical Illnesses

Jessie Whitfield, Daniela Heddaeus, Martin Härter, and Jürgen Unützer

Contents

This chapter is an update from the 4th edition. Previous edition authors were Jürgen Unützer and Wayne Bentham

J. Whitfield · J. Unützer (✉)
Department of Psychiatry and Behavioral Sciences, University of Washington, Seattle, WA, USA
e-mail: jwhitfi@uw.edu; unutzer@uw.edu

D. Heddaeus · M. Härter
Department of Medical Psychology, University Medical Center Hamburg-Eppendorf, Hamburg, Germany
e-mail: d.heddaeus@uke.de;
m.haerter@uke.uni-hamburg.de

Abstract

Mental health and substance use disorders are leading causes of morbidity and disability around the world. Despite the existence of effective and affordable treatments for these disorders, the majority of those who need mental health care do not receive it, especially in low- and middle-income countries (LMICs) and rural areas of high-income countries, where a major barrier to effective mental health

© Springer Nature Switzerland AG 2024
A. Tasman et al. (eds.), *Tasman's Psychiatry*,
https://doi.org/10.1007/978-3-030-51366-5_65

treatment is a critical shortage of mental health specialists. Embedding mental health specialists into general medical settings helps address limited access to scarce mental health resources by supporting primary care clinicians to provide mental health treatments. This type of health care delivery model, called integrated care, has been shown to have several advantages, including increasing access to care, enhancing quality of mental health care, creating more cost-effective care, and improving outcomes of comorbid physical conditions. In this chapter, the rationale and evidence for integrated care models from around the world is reviewed, with a focus on a highly coordinated integrated care model called collaborative care (CoCM). While the evidence base for CoCM from controlled trials is strong, research regarding facilitators and barriers to the implementation of CoCM in real-world settings is still growing as worldwide implementation efforts arise across diverse health care systems. This shift toward integrated care creates increasing demand for psychiatrists who are experienced in this model, and offers an outstanding opportunity for psychiatrists to leverage their skill set to reach large patient populations who currently do not have access to effective mental health care.

Keywords

Collaborative care · CoCM · Integrated care · Mental health care · Mental health · Depression · Anxiety

Introduction

Mental health and substance use disorders are leading causes of morbidity and disability around the world (World Health Organization, 2017). These conditions are also associated with substantial mortality; individuals living with severe mental illness lose on average 20 years of life in part due to poorly treated comorbid medical conditions and deaths from suicide and drug overdose (WHO, 2017; McGinty & Daumit, 2020).

Effective and affordable treatments exist for mental health and substance use disorders, but the majority of those in need do not receive effective care (Thornicroft et al., 2017). Both high- and low-income countries struggle to meet the demand for mental health treatment, though this gap is most severe in low- and middle-income countries (LMICs) and rural areas of high-income countries (HIC). Estimates of those receiving necessary mental health services range from 1.6–40% (WHO, 2017; Mendenhall et al., 2014; Wang et al., 2005, 2007).

One of the major barriers to effective mental health treatment is a critical shortage of mental health specialists such as psychiatrists, psychiatric nurses, and psychotherapists. This shortage continues despite urgent calls for action from national and international organizations over the past 15 years (Collins et al., 2011; WHO, 2017; Patel et al., 2013, 2018). In some countries (such as many in sub-Saharan Africa), there are fewer than one psychiatrist per million inhabitants (WHO, 2017) and current estimates suggest that more than 1 million additional mental health professionals are needed to meet the needs in LMICs (Araya et al., 2018; Kakuma et al., 2011).

Traditional referrals to specialty mental health care practices with psychotherapists and psychiatrists often result in long wait times and failure to connect to treatment. In the USA, only two in ten adults with common mental disorders receive care from a mental health specialist in any given year (Wang et al., 2005). Almost half of those referred to mental health specialists in the USA do not follow through with a referral, and the mean number of visits among those who do pursue a specialty mental health referral is two (Grembowski et al., 2002; Simon et al., 2012). Rates of care are often even lower in LMICs; in India, only 5% of people with mental health conditions sought treatment from specialty mental health providers (Araya et al., 2018; Kessler et al., 2005; Wells et al., 2013). Due in part to these shortages of mental health professionals, individuals with mental health needs are more likely to be seen by a primary care provider than by a psychiatrist or psychotherapist (Wang et al., 2007b).

General medical care practices present widely available and uniquely positioned opportunities to facilitate identification and treatment of common mental health and substance use conditions (Table 1). Embedding mental health specialists into general medical settings helps address limited access to scarce mental health resources by supporting primary care clinicians to provide mental health treatments, or providing direct mental health care alongside primary care services. Effective collaboration of psychiatrists and other mental health professionals with general medical providers offers several advantages. First, it can substantially improve access to and quality of mental health care, especially for those who have been unable or unwilling to engage in specialty mental health care previously (Davies & Lund, 2017). Specialist consultation to primary care can lead to substantial improvements in the quality of treatment and improved clinical outcomes (Unützer et al., 2002; Rush et al., 2004; Bower et al., 2006; Sowa et al., 2018). Second, integrated care creates opportunities for a more patient-centered health care experience for patients who often have both mental health and acute or chronic medical problems (Katon, 2003; Moussavi et al., 2007). Individuals living with chronic physical conditions have significantly higher rates of depression and other common mental disorders, creating important opportunities for general medical practitioners to identify and engage people with comorbid medical and mental health disorders in care (Wang et al., 2007; Härter et al., 2007; Baumeister & Härter, 2007; Petersen et al., 2018).

Table 1 Rationale for integrating mental health care into primary care, as proposed by WHO and Wonca (2008)

Rationale for integrating mental health into primary care
The burden of mental health disorders is great
Mental and physical health problems are interwoven
The treatment gap for mental disorders is enormous
Primary care for mental health enhances access
Primary care for mental health promotes respect of human rights
Primary care for mental health is affordable and cost-effective
Primary care for mental health improves clinical outcomes

When comorbid mental health problems are effectively treated with specialist consultation, patients experience improved quality of life, self-care, adherence to medical and mental health treatments, and overall health outcomes (Coates et al., 2020). Third, improving access to mental health care via primary care can also reach vulnerable populations who might avoid specialty mental health care due to stigma associated with mental illness and substance use issues. Such stigma varies by country and culture, but continues to be widespread and is an oft-cited factor that contributes patients' discomfort with seeking specialty mental health care (Wang et al. 2007). Lastly, recent studies suggest integrating behavioral health care into primary care may contribute to reduced health care disparities and particularly improve access to care for racial and ethnic minorities (Lee-Tauler et al., 2018).

Integrating mental health care into general medical practices can take many forms, ranging from traditional psychiatric services colocated in primary care settings to fully integrated collaborative care teams. Integrated care models involve an interdisciplinary team of medical and behavioral health clinicians who provide coordinated and patient-centered care to a population of patients, with the goal of improving outcomes for behavioral health, substance use, as well as acute and chronic medical conditions (Huang et al., 2014; Bree Collaborative Behavioral Health Integration Report, 2017). Such programs are not designed to replace other forms of available specialty mental health care, especially for patients with severe and persistent mental disorders; instead, they are intended to increase access within primary care and provide proper triaging to higher levels of specialized care when needed (Fig. 1).

The past two decades have seen an increasing recognition of integrated mental health care as a solution to the global mental health burden, and several large-scale efforts have attempted to improve access to effective mental health services in primary care. In 2008, the WHO launched the Mental Health Gap Action Program (mhGAP), a set of guidelines for health-care providers to facilitate scaling up of mental health care, particularly

Fig. 1 Integrated care within a broader health care system context. (Copyright: UW AIMS Center)

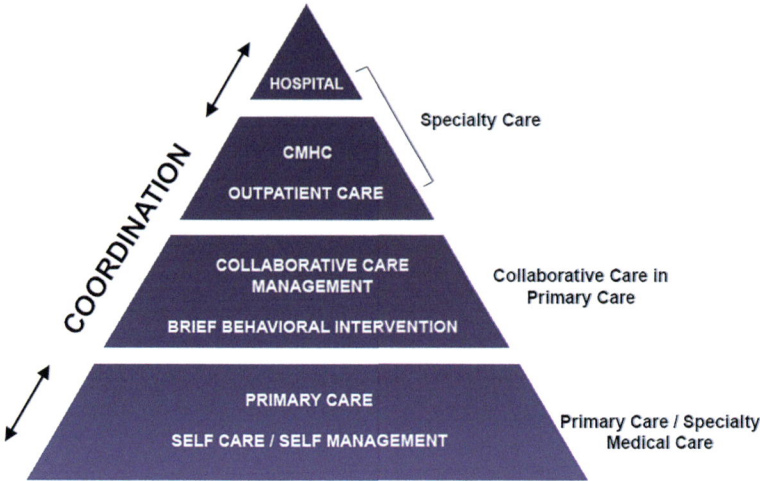

through incorporating mental health services into primary care (WHO, 2014; Davies & Lund, 2017). Organizations such as WHO and the American Medical Association (AMA) have called for integrating mental health into primary care as a feasible, scalable, and effective solution for addressing the global mental health burden. A recent Lancet Commission listed integrating mental health services into primary care delivery as one of the five grand challenges in global mental health (Patel et al., 2018; Collins et al., 2011), and in 2020, there were renewed calls for action to implement proven models of integrated and collaborative care as ready solutions to the worsening mental health crisis causes by the COVID-19 pandemic (Carlo et al., 2021).

In this chapter, the history and evidence for integrated behavioral health care is reviewed, with a focus on collaborative care, an approach that has a substantial evidence base for its effectiveness. Core elements of collaborative care programs, real-world challenges, and practical applications for psychiatrists, psychotherapists, and other mental health professionals are also reviewed.

History of Integrated Care

Early approaches to integrating mental health and primary care were described in the UK in the 1970s and 1980s. Michael Shepherd championed the general practitioner's role in treating mental disorders in general medical settings (Shepherd & Wilkinson, 1988; Shepherd, 1991). David Goldberg also pioneered integration of mental health services in primary care setting, in part by improving training for general medical practitioners to screen and diagnose mental disorders in primary care (Blackwell & Goldberg, 1968; Goldberg et al., 1980).

In the late 1990s, a team of researchers in the University of Washington developed a patient-centered approach called the collaborative care model (CoCM). This approach was informed by earlier work on the high prevalence of depression and other common mental disorders in primary care and the chronic care model developed by Wagner and colleagues (Wagner et al., 1996). CoCM is an evidence-based, systematic approach in which primary care and mental health providers work closely together to deliver effective treatment for depression and other common mental disorders to defined populations of patients in primary care and other medical settings. In the largest trial of CoCM to date, the IMPACT study, depressed patients who were randomly assigned to CoCM had more than twice the rate of improvement in depression than patients in primary care as usual (Unützer et al., 2002). The IMPACT CoCM also reduced physical pain (Lin et al., 2003), and improved social and physical functioning and overall quality of life when compared to care as usual. Patients and primary care providers reported high rates of satisfaction with CoCM (Unützer et al., 2002; Levine et al., 2005), and subsequent studies demonstrated the effectiveness of the IMPACT program in depressed adolescents (Richardson et al., 2009), depressed cancer patients (Ell et al., 2008), and diabetic patients (Katon et al., 2008), including low-income Spanish-speaking patients (Gilmer et al., 2008).

Also in the early 2000s, researchers in Chile demonstrated the effectiveness of a collaborative care model that would become a model for implementing and improving mental health service delivery in primary care settings in LMICs. Introduced in 2001, the National Depression Detection and Treatment Program (Programa Nacional de Diagnóstico y Tratamiento de la Depresión, or PNDTD) was a stepped-care program that featured a mental health team led by a trained nonmedical health worker (a social worker or nurse), and focused on delivery of brief behavioral interventions in group sessions, systematic monitoring of clinical progress, and medication management directed by a primary care provider who followed evidence-based treatment guidelines for depression (Araya et al., 2003). Participants in the program were found to have significantly better depression outcomes than those in usual care and the program was highly cost-effective (Siskind et al., 2010; Araya et al., 2006). By 2003, this model spread to a national program, comprised of more than 500 primary care centers nationwide, and by 2007 it was treating over 170,000 patients annually (Araya et al., 2012).

Examples of Integrated Care

A number of approaches to integrating behavioral health care have been developed to fit diverse patient demographics, population health needs, and primary care practice contexts. A 2020 systematic review found 37 unique models of integrated care in published literature, and a review by Cubillos et al. in 2021 describes a typology of six different models of integrated care in LMICs

(Coates et al., 2020; Cubillos et al., 2021). Of these models, CoCM is one of the few with significant research evidence supporting its effectiveness across multiple conditions as well as clear published guidelines for its implementation (Archer et al., 2012; Gilbody et al., 2006a; AIMS Center, 2020a). Here, two other widely used, evidence-based integrated care models are briefly reviewed before focusing further on the CoCM framework and evidence base.

Primary Care Behavioral Health Model

One widely utilized integrated care model is primary care behavioral health (PCBH). This model is a team-based approach to managing behavioral health problems that arise in primary care populations, with a goal of addressing both behavioral health conditions as well as health-related behaviors that could impact the management of comorbid medical conditions (Reiter et al., 2018). Although there has been some debate about the structure and components of the PCBH model over the last 20 years, in the USA it generally consists of incorporating a mental health consultant or clinician (oftentimes a clinical psychologist) in primary care. The PCBH model provides important opportunities to reach populations often not served by mental health services otherwise, especially patients whose mental health symptoms negatively impact health behaviors and chronic medical conditions. Mental health clinicians in the PCBH model address a wide range of mental health symptoms and health behaviors in patients across the age spectrum. They aim to remain available to provide same day services in primary care and impact a large number of patients by maintaining short visits (15–30 mins) that focus on management symptoms and improving function. These mental health clinicians work as consultants to the PCP until symptoms improve, at which point the PCP continues managing the patient on his/her own and may refer patients with treatment-resistant symptoms to more traditional specialty mental health services (Reiter et al., 2018).

Task Shifting

Task shifting, or task sharing, is another approach that can help improve access to mental health services, particularly in low-resource settings. This approach aims to address unmet mental health needs by shifting tasks from highly trained individuals, such as primary care providers and psychologists, to other members of the primary care team (such as nurses, peer support workers, community health workers, and other non-specialist mental health workers) who are trained to provide basic mental health services or care coordination (Davies & Lund, 2017). Task shifting approaches that were pioneered for treating HIV/AIDS in Africa were later found effective in mental health service scale up models in India, Uganda, and Pakistan (Grant et al., 2018; Hoeft et al., 2018). This approach can help reduce the demands on the scarce resource of highly trained mental health professionals and increase access to mental health care by shifting services to more widely available non specialists, with appropriate training and supervision (Grant et al., 2018; Hoeft et al., 2018; Padmanathan & De Silva, 2013). A 2012 qualitative study in five countries (Ethiopia, India, Nepal, South Africa, and Uganda) participating in the Program for Improving Mental Health Care (PRIME) found that task sharing was feasible and acceptable to primary care providers, particularly if it included key elements such as increased personnel, improved access to medications, continual structured supervision, and adequate training and compensation for task-sharing workers (Mendenhall et al., 2014).

The nature and staffing used in task sharing models vary widely. A few examples of this approach include the following:

- In the USA, home visits from a nurse and community health worker team for pregnant women to target stress and manage behavioral health conditions significantly improved contacts for prenatal care (Roman et al., 2007).
- In Germany, the SMADS trial evaluated a primary care–based, nurse-led intervention to promote self-management in patients with

anxiety, depressive, or somatic symptoms and found that this approach increased perceived self-efficacy and self-management of patients with psychosomatic symptoms, although it did not prove to be cost-effective relative to routine care (Zimmermann et al., 2016; Grochtdreis et al., 2018).

- In Ethiopia, an ongoing randomized clinical trial is testing the cost-effectiveness of task sharing for patients with severe mental disorders, with care delivered by nonspecialist mental health workers who are integrated in primary care. This study is part of PRIME, a research consortium evaluating task sharing in LMICs (Davies & Lund, 2017; Fekadu et al., 2016).

Collaborative Care

CoCM is a model of integrated care that provides team-based mental health care in primary care and other general medical settings by adding a psychiatric consultant and care manager to the primary care team. The program leverages limited mental health specialty resources to reach as many patients as possible through coordinated team activities, such as regular clinical reviews and indirect psychiatric consultations on a defined caseload of patients treated in primary care. This approach helps reach more patients in need than a traditional approach of colocating a psychiatrist in primary care which limits the number of patients seen and leads to long wait times for patients. Research suggests that colocating mental health specialists in primary care settings without also enhancing collaboration between primary and mental health-care providers does not go far enough toward improving outcomes (Uebelacker et al., 2009). CoCM is designed to address this issue by having psychiatrists who are working with a designated care manager review and consult on a larger caseload of patients on a regular basis.

In CoCM as practiced in the USA, the principal care providers include the primary care provider, a designated mental health care manager (often a social worker, nurse, or psychologist) and a

Fig. 2 Collaborative care team, including primary medical provider, care manager, psychiatric consultant, and patient. (Developed by AIMS Center. Copyright 2020 AIMS Center and University of Washington)

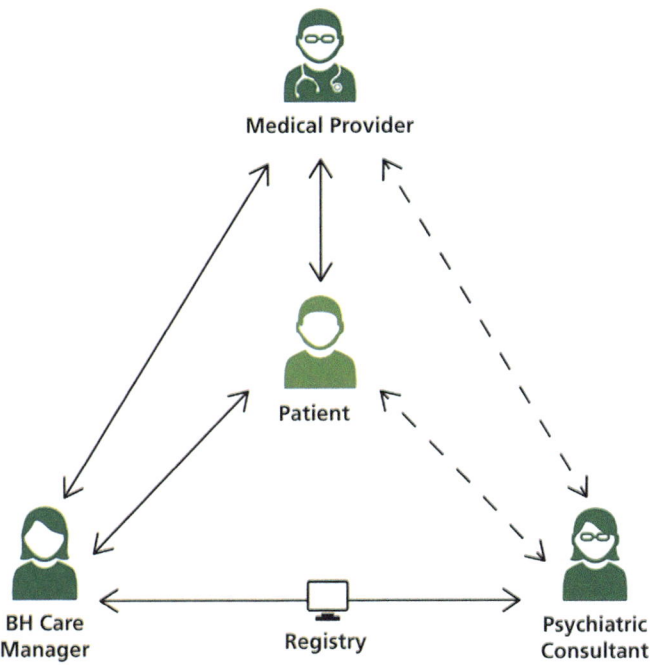

psychiatric consultant, who all work in a coordinated patient-centered team approach to provide effective pharmacologic and behavioral interventions that support the primary care team. Figure 2 represents how care providers and services collaborate in this model:

The composition of team members on such collaborative care teams varies somewhat depending on the needs of the patient population and the types of professionals available in diverse settings.

CoCM Team Roles

The *primary care provider* (PCP), usually a generalist physician, nurse practitioner, or other medical provider, makes the initial diagnosis of a mental health problem (often using screening measures such as the Patient Health Questionnaire [PHQ9]) and engages the rest of the collaborative care team (Kroenke et al., 2001). With recommendations and support from the psychiatric consultant and care manager, the PCP is the leader of the care team, maintaining active ownership of the patient's overall care, mental health care, and psychotropic medication management. The PCPs maintain regular patient contact and appointments

to track the progress of care and make treatment adjustments as necessary.

The *care manager (CM) is most often* a nurse, clinical social worker, counselor, psychologist, or medical assistant who is based in primary care and trained to provide evidence-based care coordination, brief behavioral interventions, and to support treatments such as medications initiated by the PCP (see Table 2). These care management services may be provided in the patient's primary care clinic, by telephone, or other telehealth modalities. A typical care manager carries an active caseload of 50–100 patients. The CM works to keep patients engaged in care, to track clinical outcomes, facilitates access to psychiatric consultation and other specialty services as needed, and provides direct care such as brief evidence-based counseling (e.g., problem-solving treatment, behavioral activation, and cognitive behavioral therapy) under the supervision of the psychiatric consultant.

A *psychiatric consultant* works closely with the care manager to review the clinical progress of all patients on the caseload, and offers suggestions for diagnostic clarification or adjustments in treatment based on patients' clinical progress.

Table 2 Example of care manager responsibilities

- Screening for depression and other common mental disorders, or for medical conditions in patients with serious mental illnesses
- Patient engagement and education
- Close and proactive follow-up focusing on treatment adherence, treatment effectiveness, and treatment side effects
- Brief counseling using established and evidence-based techniques such as motivational interviewing, behavioral activation, and problem-solving treatment in primary care
- Regular (usually weekly) review of caseload with the psychiatric consultant
- Facilitation of communication between the PCP and the psychiatric consultant
- Facilitation of referrals to and coordination with outside mental health specialty care or medical specialty care, substance abuse services, and social services
- Development and communication of a relapse prevention plan once patient is improved

Developed by Jürgen Unützer and colleagues. Copyright 2013 University of Washington

Because clinical recommendations often involve management of psychotropic medications, psychiatrists, and other practitioners who are experienced and licensed to prescribe psychiatric medications typically provide these services. Psychiatric consultation responsibilities include regular (usually weekly) reviews of the care manager's caseload of patients, focusing primarily on patients who present diagnostic challenges or who are not showing clinical improvements. Regular contact and an effective working relationship and partnership between the psychiatric consultant and the care manager are crucial to ensuring that such caseload consultation is productive. The level of effort for psychiatric consultants is typically 3 h per week for each care manager's primary care caseload. A psychiatric consultant may support a number of care coordinators, depending upon the size of the program. Psychiatric consultants can also have the option of focused direct psychiatric evaluations of patients who present particular diagnostic or therapeutic challenges (Table 3).

Another essential component of the CoCM model is a registry, or a list of patients that helps team members track engagement in care and

Table 3 Psychiatric consultant responsibilities

- Regular (usually weekly) caseload review and consultation with care manager
- Ad hoc consultation with care manager or primary care provider as needed for urgent cases
- Support and training for primary care–based providers
- If possible, direct consultations for patients who require further diagnostic clarification or with complex presentations; in person or by telehealth technology
- Recommendations for diagnostic evaluations or treatment changes for patients who are not improving as expected

Developed by Jürgen Unützer and colleagues. Copyright 2020 University of Washington

clinical outcomes for all patients in the program. This registry may be a digital spreadsheet or panel that exists within the medical record that is shared between the care manager and the psychiatric consultant as a critical tool to facilitate their caseload reviews. Information tracked in this registry typically includes patient scores on clinical outcome measures such as the PHQ9, last point of contact, and the last time the patient was reviewed with the psychiatrist. Regular review of this data by the care manager and psychiatric consultant helps the team to closely monitor patient progress, identify patients who need treatment adjustments, and ensure that patients do not "fall through the cracks."

Core Principles of CoCM

Evidence-based CoCM follows five core principles (Table 4):

- *Patient-centered team care*: Primary care and behavioral health providers collaborate effectively using shared care plans that incorporate patient goals and deliver this treatment plan with a focus on increasing patient engagement and agency in their own behavioral and physical care plan.
- *Population-based care*: Through use of a registry and routine caseload reviews, the care manager and psychiatric consultant track a defined group of patients to ensure timely engagement, minimize patients lost to follow-up, and target recommendations and treatment titration to patients not improving. Clinic-wide

Table 4 CoCM core principles and corresponding roles for CoCM team members

Processes	Roles	
	Behavioral health care manager	Psychiatric consultant
1. Population-based care	Patient engagement, education, and self-management support Close follow-up to make sure patients do not "fall through the cracks" and to identify patients who are not improving	Regular systematic caseload review Consultation on cases presenting diagnostic challenges
2. Measurement-based treatment to target	Support medication management by primary care provider (PCP)	Consultation focused on patients not improving as expected
3. Evidence-based treatment	Brief counseling (e.g., behavioral activation or problem-solving treatment in primary care)	Recommendations to PCP for additional treatment or referral to specialty care
4. Patient-centered team care	Facilitate patient referral to mental health specialty care as needed Relapse prevention planning Eliciting patient goals for treatment	
5. Accountable care	Regularly reviewing caseload outcomes and care delivery metrics	Regularly reviewing caseload outcomes and care delivery metrics

Developed by Jürgen Unützer and colleagues. Copyright 2020 University of Washington

protocols for depression and anxiety screening as part of the CoCM referral workflow can improve the reach of the CoCM program and include patients who might be otherwise missed.

- *Measurement-based treatment to target*: For each patient, clinical outcomes are tracked by evidence-based rating scales like the PHQ-9 and used to adjust treatments (including therapy or medication) if patients are not improving as expected, until the patient's treatment goals are achieved. Such systematic treatment to target can overcome the "clinical inertia" that is often responsible for ineffective treatments of mental disorders in primary care (Henke et al., 2009).
- *Evidence-based care*: As part of their treatment plan, patients receive treatments with evidence of efficacy in treating their mental health condition, including evidence-based psychotherapies, such as problem-solving treatment, behavioral activation, or other short-term evidence-based psychotherapies. Psychiatric medication management is based on clinical guidelines and standards of practice.
- *Accountable care*: All team members involved in patient care are accountable to the outcomes of the population of patients they are working with and are responsible for regularly reviewing

metrics that may indicate areas of improvement in care delivery. (AIMS Center, 2020)

Effective CoCM programs also follow the principles of stepped care, a systematic approach to intensifying treatment if patients are not improving (Wagner et al., 1996; Von Korff & Tiemens, 2000). If patients are not improving as expected, the PCP adjusts treatment with input from the psychiatric consultant and care manager. Patients who do not respond to treatment, have an acute crisis, or require a higher level of care are referred to mental health specialty care, as are patients who seek specialty mental health referral. Below are examples of stepped care activities within CoCM (Table 5).

Systematic Caseload Review

The systematic caseload review is a key feature of CoCM that connects many of the core principles of CoCM, such as measurement-based treatment to target, population-based care, and accountable care. In these caseload reviews, the care manager and the psychiatric consultant discuss the progress of patients in the program and develop or adjust the treatment plans based on that progress (Bauer et al., 2019). At the beginning of the consultation session, the care manager and the psychiatric consultant use the registry to prioritize which patients

Table 5 Stepped care activities for depression in evidence-based collaborative care programs

Stepped care activity	Depression	Local considerations
Patient identification/screening	Provider referral or screening with / PHQ-9	• Who provides key treatment components: primary care provider, mental health staff in primary care, substance abuse counselor, care manager, or other?
Patient education and engagement	Patient education	
	Shared goal setting	
	Screening for bipolar disorder, substance use, psychosis, and suicide risk	
Close follow-up and behavioral activation	Close telephone or in-person follow-up	• Where are treatments provided? (e.g., what can happen in primary care clinic vs. in referral center?)
	Support of medical management	
	Behavioral activation strategies used at each contact	• How do team members communicate and collaborate effectively?
Medication management prescribed by primary care provider with guidance from psychiatric consultant	Antidepressant medication regimens (including both monotherapy and augmentation strategies)	
	Anxiolytics as clinically indicated	• Who is responsible for tracking patient and program outcomes?
		• How is psychiatric caseload review and consultation provided?
Evidence-based psychotherapy	Problem-solving treatment in primary care (PST-PC) or other evidence-based psychotherapy (e.g., cognitive behavioral therapy or interpersonal therapy)	• What kinds of training and ongoing clinical support are needed for the program to succeed?
Community-based resources	Local community–based resources and support services	
Specialty mental health/substance abuse referral and treatment	Mental health specialty care for severe or persistent depression or comorbidities (e.g., suicidal, dual diagnosis, and psychotic)	

Developed by Jürgen Unützer and colleagues. Copyright 2013 University of Washington

will be discussed. A typical ranking of priority would be: new patients, patients who are not improving, and then patients with other diagnostic or therapeutic challenges. Research suggests that following best practices in caseload review, such as focusing on initial patient engagement and on patients who are not improving within 8 weeks of starting the program, can improve patient care and outcomes (Sowa et al., 2018; Bao et al., 2011, 2016). To maximize the psychiatric consultant's reach and capacity, it is also suggested that 6–10 patients be discussed in a one-hour caseload review session. Access to accurate and reliable clinical information greatly facilitates efficient and effective consultations and recommendations.

Effective psychiatric consultants recognize that primary care providers typically have extremely busy and variable practices and provide brief, focused, and specific recommendations that include a brief clinical rationale for recommendations and attention to high-risk situations such as the risk of suicide or potential adverse effects associated with particular treatment choices. Recommendations may include interventions to be done by the primary care provider (e.g., medications and additional laboratory testing), by the care manager (tracking symptoms and providing brief psychosocial interventions), or other services (referral for nutrition services). Both care managers and psychiatric consultants are responsible to make sure that psychiatric recommendations are effectively communicated to the patient's PCP or other relevant team members. Typically, it is also expected that the psychiatric consultant is available to the PCP during the week by pager to answer questions about recommendations made, to address urgent questions that arise in between caseload review sessions, and to support and educate the care managers and the PCPs they support.

Psychiatrists who are new to this model of care may express concerns regarding liability for making recommendations without having personally done an evaluation of the patient. Likewise, primary care providers may express concern about liability for care. If a CoCM program is systematically and effectively implemented, however, such a program should reduce the risk of an adverse outcome compared with typical primary care practice in which primary care providers are largely on their own managing patients with mental health conditions, or where they refer to specialty services without a guarantee that the patient will follow-up reliably. In a well-run CoCM program, patients are closely followed by a care manager trained to systematically inquire about thoughts of death or suicide. This increases the chances of identifying patients who are at risk for bad outcomes and to obtain additional consultation and/or referrals as needed. Both the care manager and the PCP have access to a designated psychiatric consultant if there are urgent or emergent concerns or if patients are not improving as expected.

The Psychiatric Consultant as Educator

Psychiatric consultants also have important opportunities as educators. They may provide information and education about the CoCM model of care to PCPs, care managers, or other clinic staff or scheduled "in-services" or lunch talks on the management of common mental disorders in primary care or the management of high-risk clinical situations. They may also provide important support for care managers who may be new to a busy primary care setting and going through their own adjustment about working in primary care. Each case reviewed and each set of treatment recommendations conveyed represent an opportunity for education that is directly relevant to the PCPs' patients. By focusing this education on a specific clinical challenge experienced by the PCPs' own patient rather than a theoretical discussion of different mental disorders and their

treatments, psychiatric consultants often find their PCP colleagues building new mental health knowledge and skill sets over time. This opportunity to serve as an educator to the primary care team can be one of the most gratifying aspects of being a psychiatric consultant in this model.

Care managers and other mental health specialists on the team, such as psychotherapists, have similar opportunities to educate PCPs and other team members. Psychiatric consultants have a lot to learn from their colleagues on the team as well, including about the real-life impact of psychiatric symptoms on patients' lives, the day-to-day culture and management of the primary care clinic(s), the availability of community services (e.g., substance use treatment, transportation, and vocational rehabilitation services), the particulars of mental health coverage, and other important contextual factors for the effective management of patients' mental health. Over time, all members of the team will learn from each other, significantly improving the capacity and skills of individual team members and the entire CoCM team.

Research Evidence

Although many forms of psychiatric consultation and integrated care are typically welcomed by primary care providers and other medical colleagues, the strongest evidence base exists thus far for CoCM (Archer et al., 2012; Coates et al., 2020). In the last two decades, published evidence for CoCM has originated from a number of countries; more than 80 randomized controlled trials have established robust evidence base for CoCM across the world (Archer et al., 2012). While the majority of CoCM has occurred in the USA, CoCM programs have been successfully launched and studied in Europe, Africa, Asia, and South America as well (Farooq, 2013; Coates et al., 2020). Here some of the research evidence for CoCM around the world is briefly reviewed. This is not a systematic review, but some examples to illustrate the feasibility and effectiveness of CoCM in a wide range countries, cultures, and health care systems are used.

The UK

In the UK, where only one in four depressed patients receive adequate pharmacological treatment and less than one in ten receive effective psychotherapy, early research on CoCM found moderate to large effects of collaborative care in treating depression versus usual care (Richards et al., 2008, 2013). Other trials, including the Collaborative Interventions for Circulation and Depression (COINCIDE) trial, found CoCM to be effective for improving depression in patients with comorbid diabetes and/or cardiovascular disease and associated with higher patient satisfaction with care (Coventry et al., 2015). Studies of longer-term effectiveness and in special populations have also demonstrated good results for CoCM (Camacho et al., 2016, 2018). Cost analyses in the UK demonstrate CoCM as a potentially cost-effective approach compared to usual care (Green et al., 2014). In both the CASPER and COINCIDE trials, cost analyses suggested collaborative care was associated with better value compared to usual care (Bosanquet et al., 2017; Lewis et al., 2017; Camacho et al., 2016).

Since establishing an evidence base for CoCM in the UK, the National Institute for Health and Care Excellence (NICE) has recommended that people with moderate to severe depression with comorbid chronic illness and functional impairment should be managed with CoCM (NICE, 2009).

The USA

Randomized controlled trials and other studies of CoCM in the USA have been conducted in diverse health care settings with varying financing mechanisms, practice sizes, and patient populations (Gilbody et al., 2006a; Archer et al., 2012; Community Preventive Services Task Force, 2012; Thota et al., 2012), including patients with depression (Gilbody et al., 2006a; Simon, 2009), anxiety disorders (Muntingh et al., 2016), substance use disorders (Watkins et al., 2017), and more serious conditions such as bipolar disorder and PTSD (Reilly et al., 2012; Woltmann et al., 2012;

Engel et al., 2008). Across this extensive literature, CoCM has consistently demonstrated higher effectiveness than usual care (Roy-Byrne et al., 2005; Archer et al., 2012; Community Preventive Services Task Force, 2012; Thota et al., 2012; Watkins et al., 2017). A study at the Mayo Clinic demonstrated that patients in CoCM improved up to 500 days faster in depression treatment than those in usual primary care (Garrison et al., 2016).

CoCM has also been demonstrated to improve patient outcomes for comorbid physical conditions, such as diabetes and cardiovascular disease, in large trials such as Care of Mental, Physical, and Substance Use Syndromes (COMPASS), TEAMCare, and the Hopeful Heart trial (Coleman et al., 2017a, b; Rossom et al., 2017; Lin et al., 2012; Katon et al., 2010; McGregor et al., 2011; Herbeck Belnap et al., 2019). In one meta-analysis of CoCM for comorbid depression and congestive heart disease, CoCM interventions were more likely to achieve depression remission, though did not lead to significant reductions in major adverse cardiac events in the long term (Tully & Baumeister, 2015). Another meta-analysis of 31 RCTs comparing CoCM to usual care in adults with depression suggested that CoCM is effective for depression alone as well as depression comorbid with other medical conditions (Panagioti et al., 2016).

Two large trials of CoCM for perinatal depression, the Depression Attention for Women Now (DAWN) trial and MOMCare, found CoCM was effective in reducing perinatal depression and improving quality of care in both rural and urban disadvantaged populations (Bhat et al., 2018; LaRocco-Cockburn et al., 2013; Grote et al., 2015). Evidence is also growing for CoCM effectiveness and financial feasibility in adolescent populations for depression, as evidenced by the Reaching Out to Adolescents in Distress (ROAD) and Early Management and Evidence-Based Recognition of Adolescents Living with Depression (EMERALD) studies (Richardson et al., 2009, 2014; Shippee et al., 2018; Wright et al., 2016), as well as post-concussive syndrome (McCarty et al., 2016). There are now growing calls for trials of CoCM in school-based settings to further increase access to behavioral health for at-risk youth and adolescents, though the implementation of these programs is effective (Lyon et al., 2019). Several studies have demonstrated that CoCM programs are highly effective in safety-net patients and patients from ethnic minority groups (Miranda et al., 2003, 2004; Arean et al., 2005; Ell et al., 2009, 2010a, b; Lee-Tauler et al., 2018; Davis et al., 2011).

Several studies have demonstrated that collaborative care for depression is not only more effective but also more cost-effective than usual care (Gilbody et al., 2006b; Glied et al., 2010). Long-term (four-year) cost analyses found that patients in the CoCM intervention arm in the IMPACT study had substantially lower overall health care costs than those in usual care (Unützer et al., 2008). These findings from research studies are consistent with published data from large integrated health care systems, including Kaiser Permanente and Intermountain Healthcare that have implemented CoCM programs (Grypma et al., 2006; Reiss-Brennan et al., 2010).

Additional economic benefits from CoCM include improved productivity and workforce participation (Wang et al., 2008; Katon, 2009). Adults with depression have substantially lower personal income than those without depression (Dismuke & Egede, 2010) and individuals who retire early due to depression face long-term financial disadvantages compared with people who are treated and able to remain employed (Schofield et al., 2011). Research has shown that the systematic implementation of CoCM for depression can reduce many of these negative economic effects of depression, resulting in improved personal income, employment (Wells et al., 2001; Schoenbaum et al., 2002), and other workplace outcomes (Wang et al., 2007).

Spain

In Spain, the INterventions for Depression Improvement (INDI) project has overseen the widescale study and implementation of CoCM adapted to the public health care system in Catalan, Spain (Aragonès et al., 2012). In addition to developing the role of care managers in clinics and supporting psychiatric consultation directly with PCPs, this approach also provided enhanced depression training for general medical providers and a depression treatment algorithm for optimizing antidepressant medication prescription and titration. In the initial study comparing the INDI model to usual primary care depression management randomized at the clinic level, depression response and remission at 12 months were significantly higher in the INDI model versus usual care (Aragonès et al., 2014). Subsequently, efforts to scale up implementation of CoCM nationwide began in 2017 (Aragonès et al., 2017).

Germany

In Germany, where most of primary and specialty care providers are organized in outpatient privately owned practices, one of the first randomized controlled trials on feasibility and effectiveness of CoCM was the Primary Care Monitoring for Depressive Patients Trial (PRoMPT) (Gensichen et al., 2009). In this study, a case manager was introduced and trained in chronic disease management for primary care (without psychiatric specialist supervision). Compared with usual care, the case management intervention recipients had lower mean PHQ-9 scores, higher patient satisfaction with care, and improved treatment adherence at 12 months (Petersen et al., 2014). A later randomized controlled trial compared a CoCM program with usual care for adult primary care patients with depression. This model included a multi-professional network of PCP, psychotherapists, psychiatrists, and clinics. The CoCM program demonstrated significantly higher odds of depression remission and response based on symptom reduction in the PHQ-9 at 12 months, although no evidence for cost-effectiveness to treatment as usual was found (Härter et al., 2018; Brettschneider et al., 2020). Further studies currently underway examine the effectiveness of a stepped CoCM for primary care patients suffering from depressive, anxiety, somatoform, and/or alcohol abuse disorders in one model

(Heddaeus et al., 2019), and a cross-sectoral network with managed care, comprehensive psychological, somatic and social diagnostics, crisis resolution, and a general structure of four severity levels, each with assigned evidence-based therapy models (e.g., assertive community treatment) and therapies as part of the RECOVER trial (Lambert et al., 2020).

The effectiveness of CoCMs adapted for geriatric populations has also been studied in Germany. One randomized controlled trial evaluated a CoCM intervention adapted to dementia care management that included nurse care managers visiting patients' homes for late-life depression and dementia care. This approach was found to significantly decrease symptoms of dementia, increase chance of neurocognitive-targeting pharmacotherapy, improve quality of life for patients living with others, and perceived caregiver burden when compared with usual care (Thyrian et al., 2017). A cost-effectiveness analysis found a lower hospitalization rate associated with the CoCM intervention (Michalowsky et al., 2019). The German IMPACT trial found CoCM likely to be cost-effective compared to usual care within primary care (Grochtdreis et al., 2019; Wernher et al., 2014). A third study of a model of CoCM adapted for elderly populations, LoChro-Care, is currently underway and will evaluate impacts of CoCM on depression symptoms, functionality, and satisfaction with care as well as cost-effectiveness of CoCM in this population and health care system (Frank et al., 2019).

The Netherlands

A Dutch randomized controlled trial of CoCM in primary care centers and general medical provider practices found CoCM to be effective for depression treatment response (Huijbregts et al., 2013). Other studies found CoCM to be associated with shorter wait times for referrals, higher satisfaction among primary care providers, lower treatment costs, shorter time in treatment, and more cost-effective (van Orden et al., 2009; Goorden et al., 2015). A trial specifically in workers out of work due to depression found that CoCM had a shorter

time to treatment response by an average of almost 3 months, though no difference was found in time to remission, leading study authors to conclude CoCM may not be effective in occupational health care settings (Vlasveld et al., 2013).

India

In India, where extremely scarce mental health resources include only 0.3 psychiatrists per 100,000 population, the Integrating Depression and Diabetes Treatment (INDEPENDENT) trial demonstrated CoCM was effective in treating both depression and chronic medical illness. This randomized controlled trial found that for patients with poorly controlled diabetes and moderate to severe depression, the CoCM intervention lead to improvement in a composite depression and cardiometabolic outcome compared to usual care (Ali et al., 2020). The collaborative care intervention was based on both WHO and IMPACT models of integrated care. As patients with depression and diabetes have worse glycemic control and higher mortality, these findings suggest a sustainable approach to managing the detrimental impacts of comorbid depression on chronic illness management in a low-resource health care system (Lin et al., 2004).

South Africa

South Africa has a relatively high prevalence of mental health disorders and substantial gaps in access to care (Wang et al., 2007; Petersen et al., 2018). As part of the Program for Improving Mental Health Care (PRIME), South Africa developed task-sharing integrated care models to address comorbid depression and alcohol use disorders that could be feasibly integrated into existing primary care clinics (Petersen et al., 2018; Davies & Lund, 2017). A 2019 cohort study demonstrated a significant increase in the identification of depression and alcohol use disorders, as well as higher depression response and remission at 12 months follow-up (Petersen et al.,

2019). Further trials are ongoing to test a similar collaborative care approach in screening and treating depression in adults receiving antiretroviral therapy (Fairall et al., 2018).

Chile

Following on earlier work establishing evidence for CoCM in Chile, two recent studies have added to the evidence for remote CoCM and CoCM in adolescent populations. A randomized control trial of primary care centers in the remote Araucanía Region of Chile found that though there was no difference in depressive symptoms between CoCM and usual care, adolescents and primary care providers were more satisfied with care in the CoCM arm (Martínez et al., 2018). Another study of CoCM provided remotely (with an on-site primary care team supported by psychiatrists at academic center in Santiago, Chile) found that patients in CoCM program had higher satisfaction and treatment adherence rates compared to usual care, though again depressive symptoms did not differ between groups (Rojas et al., 2018).

Implementing Effective Integrated Care Programs

Implementing effective integrated programs requires substantial practice change, and such efforts can encounter a number of barriers such as workforce, financial, and leadership challenges. Effective implementation of an integrated care program is best achieved by employing a structured implementation approach, with a continuous quality improvement strategy that iterates on identified inefficiencies to facilitate program sustainment.

One implementation framework that is widely used in launching of evidence-based clinical programs is the Exploration, Preparation, Implementation, and Sustainment (EPIS) framework. This framework is comprised of four well-defined phases, as well as the identification of contextual factors that may impact the implementation

process (such as extraorganizational influences of changes in national health care policy, or intraorganizational shifts in clinic workforce resources, or the interplay between insurance infrastructure change and program financial sustainability) (Moullin et al., 2019). In the USA, the AIMS Center at the University of Washington has developed a similar and widely used implementation framework that is specific to CoCM. This implementation guide, depicted in Fig. 3, is available at the AIMS website along with a set of team-building tools that help guide organizations that wish to implement evidence-based CoCM programs (www.aims.uw.edu). Psychiatric consultants are well positioned to take a leadership role not only in the clinical work of the CoCM team but also in the process of developing, implementing, and supporting such a team in primary care.

Using strategic implementation frameworks, a number of health care organizations and health plans throughout the world have successfully implemented CoCM. In the USA, these include state-wide implementations with both commercially insured adults and also with urban and rural low-income/safety-net populations. Several health care organizations have undertaken large-scale implementations including national and regional health plans, such as Kaiser Permanente, Intermountain Health, and the Veterans Affairs Health Care System (Grypma et al., 2006; Reiss-Brennan et al., 2010). RESPECT-MIL is a CoCM program implemented in over 100 Army primary care clinics in the USA and abroad (Engel et al., 2008); in New York state, the Department of Health and Office of Mental Health supported the integration of collaborative care in 19 academic medical centers as part of the NYS Collaborative Care Initiative (NYS-CCI) (Sederer et al., 2016); in Minnesota, CoCM was implemented at 75 primary care clinics as part of the Depression Improvement Across Minnesota Offering a New Direction (DIAMOND) initiative (Solberg et al., 2013); and the COMPASS trial, which was implemented in more than 150 clinics across 8 states and 18 medical groups that differed in size, institutional structure, patient populations, and payment systems (including commercial,

Step by Step Collaborative Care (CoCM) Implementation Guide

Lay the Foundation	Plan for Clinical Practice Change	Build Your Clinical Skills	Launch Your Care	Nurture Your Care
CoCM is a new way of doing medicine and requires an openness to creating a new vision that everyone supports.	*Clearly define team roles, create a patient-centered workflow, and decide how to track patient treatment & outcomes.*	*Effective CoCM creates a team in which all providers work together using evidence-based treatments.*	*Is your team in place? Are they ready to use evidence- based interventions for primary care? All systems go? Time to launch!*	*Now is the time to see the results of your efforts as well as to think about ways to improve.*
✓ Develop an understanding of CoCM, its history and guiding principles.	✓ Identify all CoCM team members and organize them for training.	✓ Describe CoCM's key tasks, including patient engagement and identification, treatment initiation, outcome tracking, treatment adjustment and relapse prevention.	✓ Implement a patient engagement plan.	✓ Implement the care team monitoring plan to ensure effective team collaborations.
✓ Develop strong advocacy for CoCM within leadership and among the clinical team.	✓ Develop a clinical flowchart and detailed action plan for the care team.		✓ Manage the enrollment and tracking of patients in a registry.	✓ Update your program vision and workflows as needed.
✓ Create a CoCM vision with your overall mission and QI efforts in mind.	✓ Identify a population-based tracking system for your organization.	✓ Equip care team with the knowledge necessary to successfully implement CoCM.	✓ Develop a care team monitoring plan to ensure effective collaborations.	✓ Implement advanced training and support where necessary.
✓ Assess the difference between your current model and CoCM.	✓ Plan for funding, space, human resource, and other administrative needs.	✓ Develop care team skills in evidence-based interventions appropriate for primary care (e.g. PST, BA).	✓ Develop clinical skills to help patients throughout treatment, including a relapse prevention plan.	
	✓ Plan to merge CoCM monitoring and reporting outcomes into an existing quality improvement plan.			

Fig. 3 Step-by-step collaborative care implementation guide. (Copyright AIMS Center)

Medicare, and Medicaid insurances) and achieved target goals for population improvement in depression, hypertension, and diabetes (Coleman et al., 2017a; Chauhan & Niazi, 2017).

In Washington State, the Mental Health Integration Program (MHIP), sponsored by the Community Health Plan of Washington and Seattle King County Public Health, implemented CoCM in a partnership with more than 140 community health centers and over 30 community mental health centers, serving managed Medicaid patients with medical and behavioral health needs. This program incorporated an innovative pay-for-performance component, in which a proportion of the payment for the program is tied to effective treatment. Performance is assessed on a number of quality indicators, including timely follow-up with patients, demonstration of improved patient outcomes, or systematic consultation and treatment adjustment for patients who are not improving. After the pay-for-performance component was introduced in 2008, the effectiveness of the program has substantially improved; for instance, the median time to improvement in depression was cut by more than half after implementation of the pay-for-performance incentive payment (Unützer et al., 2012).

What makes some implementations of CoCM more successful than others is an ongoing research question. Factors at the level of policy, organization, provider, and patient may facilitate or hinder collaborative care effectiveness and implementation efforts. Implementation science has identified some facilitators, as summarized in Table 6. In addition to these facilitators, several studies have associated certain implementation factors directly with improved patient outcomes. At least one study has demonstrated that higher fidelity to core principles and components of CoCM (discussed previously) are tied to improved depression scores (Belsher et al., 2018), while one systemic review with meta-analysis found that CoCM programs that included brief behavioral interventions (regardless of antidepressant use) had better depression outcomes than those without these psychotherapy-based treatments (Coventry et al., 2014). Protecting care manager time was also viewed as an important factor tied directly to improved patient outcomes; in the COMPASS trial, patients of care managers who had more time protected for COMPASS-related tasks had higher rates of depression response and remission, as well as improved rates of blood pressure control

Table 6 Facilitators of collaborative care implementation, stratified by organization, team, and provider level. (Adapted from Beck et al., 2018; Moise et al., 2018; Overbeck et al., 2017; Whitebird et al., 2014; Coleman et al., 2017a,b; Bjerregaard et al., 2018; Wood et al., 2017; Unutzer et al., 2020)

Organization/Administration	Clinical team	Provider
Prior implementation experience	Strong team building and cohesion	Strong PCP engagement and buy-in
Use of quality improvement principles and continuous outcome monitoring	Well-defined care manager role	Strong PCP champion
Strong leadership support for CoCM	Use of patient tracking registries	Strong social and professional skills in care manager role
Financial flexibility to take on operating costs	Warm handoffs between PCP and care manager	Strong care manager training
Reimbursement for PCPs' additional work	Thoughtful selection of patients based on scope/goal of program	Accessible and on-site care manager
Clearly defined workflows and care pathways	Well-developed workflows and opportunities for clear team communication	Engaged psychiatrist
External/eEnhanced implementation support	Administrative support for care manager time devoted to CoCM role	

(Coleman et al., 2017b). A recent analysis of CoCM implementation in 130 primary care practices in nine US states showed substantial variation in the effectiveness of the programs implemented and found that the level of implementation support was strongly associated with successful clinical outcomes (Unützer et al., 2020b).

Pragmatic Challenges

While the evidence base for CoCM from randomized controlled trials is strong, less is known about the effectiveness of CoCM in real-world settings. Early studies of large-scale CoCM implementations outside of research trials suggest that clinical outcomes for CoCM may fall short of results from more rigorously controlled trials, a phenomenon that has been observed in different types of service settings (Rossom et al., 2017; Unutzer et al., 2020; Solberg et al., 2015). A number of authors have reported on factors that could contribute to this variation in effectiveness across clinics and health systems, as well as barriers to implementation.

In the USA, variation in depression outcomes of CoCM programs has been found to be associated with factors at the patient level (initial symptom burden), clinic factors (the duration of CoCM

program existence), and the amount of external implementation support, which was thought to be a particularly important factor in impacting the effectiveness of a CoCM program (Carlo et al., 2020; Unützer et al., 2020).

Research about barriers to implementation of CoCM comes from a number of countries, including HICs and LMICs. These barriers include policy, health care system, as well as organizational and provider factors. For provider factors, securing buy-in among primary care providers was found to be a crucial but challenging step of implementation; at least one study indicated that buy-in was negatively impacted by programs that limited the target population to one mental health condition or insurance provider (Overbeck et al., 2016; Kathol et al., 2010). Staff attitudes toward integrated care, absence of a strong provider champion, poor understanding of CoCM, and poor interprofessional communication were also common pitfalls at the provider level (Coupe et al., 2014; Wood et al., 2017; Kathol et al., 2010; Watzke et al., 2020; Heddaeus et al., 2018; Maehder et al., 2020). At the organizational level, programs tended to struggle most if they did not build shared access to electronic medical records across providers, lacked administrator support for time needed for providing training, education, and team building, did not have time dedicated to building new workflows, had poor outcome monitoring and data collection, had weak or absent administrator champions or leadership, and managed growth poorly or too rapidly or had minimal prior experience with implementing clinic-level change (Kathol et al., 2010). At the patient level, high levels of stigma around mental illness also remains an impactful barrier (Breuer et al., 2016). Larger health care system structure also impacted implementation of CoCM; in the Netherlands, one cited barrier to implementation and scale up of CoCM was the large proportion of privately owned primary care practices employing a single provider (de Jong et al., 2009).

One of the main inhibitors of sustainable CoCM implementation, as well as integrated care more broadly, has long been the challenge in building financially maintainable programs (Kathol et al., 2010; Brettschneider et al., 2020). In the USA, for CoCM specifically, clinics and organizations have often struggled to achieve financial sustainability using traditional fee-for-service reimbursement methods, or have relied on grants to financially support programs; even clinics in large scale, well-supported implementations, such as the DIAMOND trial, opted to discontinue their programs due to difficulties with ongoing financial sustainability (Moise et al., 2018; Solberg et al., 2015). Since these trials, however, additional reimbursement strategies have emerged in the USA, when in 2017 the Centers for Medicare and Medicaid Services (CMS) introduced new CoCM billing codes, similar to other chronic care management codes, that are designed to capture nonphysician or licensed health professional services, based on time the CoCM team spends in coordinating or delivering care for a patient per month (Carlo et al., 2019a). While some organizations have been successful using these codes to build financially sustainable programs, their implementation may require workflow adjustments that are challenging or prohibitive for some organizations (Carlo et al., 2019).

Research on barriers to implementing CoCM as well as other models of integrated care has also come from LMICs, and shares similarities with barriers already discussed, including financial constraints and lack of clear leadership and provider champions (Tapia-Muñoz et al., 2015).

Additional barriers from these implementations included reliable supplies of medication, shortage of specialist mental health professionals to perform CoCM roles and supervise lay community workers, low mental health literacy and awareness in the community, and high stigma around mental illness and treatment (Hanlon et al., 2014; Lund et al., 2012; Davies & Lund, 2017; Li et al., 2020; Farooq, 2013). There remains a need for further implementation studies to elucidate facilitators and barriers to implementation of integrated care to better construct effective implementation guidance. In Germany, the I-COMET study currently underway aims at evaluating facilitators and barriers to implementation in the German health care system. First, qualitative data from interviews with psychotherapists of the COMET trial found deficient resources and remuneration as well as a perceived lack of esteem by other medical specialties and unfavorable regional distributions between the involved care providers as main barriers for the implementation of collaborative care (Heddaeus et al., 2019; Maehder et al., 2020, 2021).

Opportunities for Psychiatrists and Other Mental Health Professionals

The movement toward integrated care (and particularly collaborative care) has gained momentum across multiple settings and diverse health care systems around the world as the evidence base for integrating behavioral health care into primary care settings has grown. This shift is creating increasing opportunities and demand for psychiatrists who are experienced in integrated care. For practicing psychiatrists or psychotherapists, this is an outstanding opportunity to develop rewarding relationships with patients, primary care providers, other medical specialists, and other mental health providers and to leverage their skill set, effectively reaching large patient populations who currently do not have access to good mental health care and improving the capacity of their primary care colleagues who provide a substantial portion of mental health services around the world.

References

Advancing Integrated Mental Health (AIMS) Center. (2020a). *Welcome to the collaborative care implementation guide.* Implementation Guide, University of Washington AIMS Center, https://aims.uw.edu/collaborative-care/implementation-guide

Advancing Integrated Mental Health (AIMS) Center. (2020b). *Principles of collaborative care.* Collaborative Care Principles, University of Washington AIMS Center, https://aims.uw.edu/collaborative-care/principles-collaborative-care

Ali, M. K., Chwastiak, L., Poongothai, S., et al. (2020). Effect of a collaborative care model on depressive symptoms and glycated hemoglobin, blood pressure, and serum cholesterol among patients with depression and diabetes in India: The INDEPENDENT Randomized Clinical Trial. *JAMA, 324*(7), 651–662.

Aragonès, E., Lluís Piñol, J., Caballero, A., et al. (2012). Effectiveness of a multi-component programme for managing depression in primary care: A cluster randomized trial. The INDI project. *Journal of Affective Disorders, 142*, 297–305.

Aragonès, E., Caballero, A., Piñol, J.-L., et al. (2014). Persistence in the long term of the effects of a collaborative care programme for depression in primary care. *Journal of Affective Disorders, 166*, 36–40.

Aragonès, E., Palao, D., López-Cortacans, G., Caballero, A., Cardoner, N., Casaus, P., Cavero, M., Monreal, J. A., Pérez-Sola, V., Cirera, M., Loren, M., Bellerino, E., Tomé-Pires, C., & Palacios, L. (2017). Development and assessment of an active strategy for the implementation of a collaborative care approach for depression in primary care (the INDI·i project). *BMC Health Services Research, 17*(1), 821.

Araya, R., Rojas, G., Fritsch, R., Gaete, J., Rojas, M., Simon, G., et al. (2003). Treating depression in primary care in low-income women in Santiago, Chile: A randomised controlled trial. *Lancet, 361*(9362), 995–1000.

Araya, R., Flynn, T., Rojas, G., Fritsch, R., & Simon, G. (2006). Cost-effectiveness of a primary care treatment program for depression in low-income women in Santiago, Chile. *The American Journal of Psychiatry, 163*(8), 1379–1387.

Araya, R., Alvarado, R., Sepúlveda, R., & Rojas, G. (2012). Lessons from scaling up a depression treatment program in primary care in Chile. *Revista Panamericana de Salud Pública, 32*(3), 234–240.

Araya, R., Zitko, P., Markkula, N., Rai, D., & Jones, K. (2018). Determinants of access to health care for depression in 49 countries: A multilevel analysis. *Journal of Affective Disorders, 234*, 80–88.

Archer, J., Bower, P., Gilbody, S., et al. (2012). Collaborative care for depression and anxiety problems. *Cochrane Database of Systematic Reviews, 10*, CD006525.

Arean, P. A., Ayalon, L., Hunkeler, E., et al. (2005). Improving depression care for older, minority patients in primary care. *Medical Care, 43*(4), 381–390.

Bao, Y., Alexopoulos, G. S., Casalino, L. P., Ten Have, T. R., Donohue, J. M., Post, E. P., Schackman, B. R., & Bruce, M. L. (2011). Collaborative depression care management and disparities in depression treatment and outcomes. *Archives of General Psychiatry, 68*(6), 627–636.

Bao, Y., Druss, B. G., Jung, H. Y., Chan, Y. F., & Unützer, J. (2016). Unpacking collaborative care for depression: Examining two essential tasks for implementation. *Psychiatric Services, 67*(4), 418–424.

Bauer, A. M., Williams, M. D., Ratzliff, A., & Unützer, J. (2019). Best practices for systematic case review in collaborative care. *Psychiatric Services, 70*(11), 1064–1067.

Baumeister, H., & Härter, M. (2007). Prevalence of mental disorders based on general population surveys. *Social Psychiatry and Psychiatric Epidemiology, 42*(7), 537–546.

Beck, A., Boggs, J. M., Alem, A., Coleman, K. J., Rossom, R. C., Neely, C., Williams, M. D., Ferguson, R., & Solberg, L. I. (2018). Large-scale implementation of collaborative care management for depression and diabetes and/or cardiovascular disease. *Journal of American Board of Family Medicine, 31*(5), 702–711.

Belsher, B. E., Evatt, D. P., Liu, X., Freed, M. C., Engel, C. C., Beech, E. H., & Jaycox, L. H. (2018). Collaborative care for depression and posttraumatic stress disorder: Evaluation of collaborative care fidelity on symptom trajectories and outcomes. *Journal of General Internal Medicine, 33*(7), 1124–1130.

Bhat, A., Reed, S., Mao, J., Vredevoogd, M., Russo, J., Unger, J., Rowles, R., & Unützer, J. (2018). Delivering perinatal depression care in a rural obstetric setting: A mixed methods study of feasibility, acceptability and effectiveness. *Journal of Psychosomatic Obstetrics and Gynaecology, 39*(4), 273–280.

Bjerregaard, F., Zech, J., Frank, F., Hüll, M., Stieglitz, R. D., & Hölzel, L. (2018). Implementation of the GermanIMPACT collaborative care program: A qualitative study on the perspective of care managers and supervisors. *Zeitschrift für Evidenz, Fortbildung und Qualität im Gesundheitswesen, 134*, 42–48.

Blackwell, B., & Goldberg, D. P. (1968). Psychiatric interviews in general practice. *British Medical Journal, 4*(5623), 99–101.

Bosanquet, K., Adamson, J., Atherton, K., Bailey, D., Baxter, C., Beresford-Dent, J., Birtwistle, J., Chew-Graham, C., Clare, E., Delgadillo, J., Ekers, D., Foster, D., Gabe, R., Gascoyne, S., Haley, L., Hamilton, J., Hargate, R., Hewitt, C., Holmes, J., Keding, A., Lewis, H., McMillan, D., Meer, S., Mitchell, N., Nutbrown, S., Overend, K., Parrott, S., Pervin, J., Richards, D. A., Spilsbury, K., Torgerson, D., Traviss-Turner, G., Trépel, D., Woodhouse, R., & Gilbody, S. (2017). CollAborative care for Screen-Positive EldeRs with major depression (CASPER plus): A multicentred randomised controlled trial of clinical effectiveness and cost-effectiveness. *Health Technology Assessment, 21*(67), 1–252.

Bower, P., Gilbody, S., Richards, D., Fletcher, J., & Sutton, A. (2006). Collaborative care for depression in primary care: Making sense of a complex intervention: Systematic review and meta-regression. *The British Journal of Psychiatry, 189*, 484–493.

Bree Collaborative Behavioral Health Integration Report. (2017). http://www.breecollaborative.org/topic-areas/previous-topics/behavioral-health/. Accessed 15 Nov 20.

Brettschneider, C., Heddaeus, D., Steinmann, M., et al. (2020). Cost-effectiveness of guideline-based stepped and collaborative care versus treatment as usual for patients with depression – A cluster-randomized trial. *BMC Psychiatry, 20*, 427.

Camacho, E. M., Ntais, D., Coventry, P., Bower, P., Lovell, K., Chew-Graham, C., Baguley, C., Gask, L., Dickens, C., & Davies, L. M. (2016). Long-term cost-effectiveness of collaborative care (vs usual care) for people with depression and comorbid diabetes or cardiovascular disease: A Markov model informed by the COINCIDE randomised controlled trial. *BMJ Open, 6*(10), e012514.

Camacho, E. M., Davies, L. M., Hann, M., Small, N., Bower, P., Chew-Graham, C., Baguely, C., Gask, L., Dickens, C. M., Lovell, K., Waheed, W., Gibbons, C. J., & Coventry, P. (2018). Long-term clinical and cost-effectiveness of collaborative care (versus usual care) for people with mental-physical multimorbidity: Cluster-randomised trial. *The British Journal of Psychiatry, 213*(2), 456–463.

Carlo, A. D., Corage Baden, A., McCarty, R. L., & Ratzliff, A. D. H. (2019a). Early health system experiences with collaborative care (CoCM) billing codes: A qualitative study of leadership and support staff. *Journal of General Internal Medicine, 34*(10), 2150–2158.

Carlo, A. D., Jeng, P. J., Bao, Y., & Unützer, J. (2019b). The learning curve after implementation of collaborative care in a state mental health integration program. *Psychiatric Services, 70*(2), 139–142.

Carlo, A. D., Barnett, B. S., & Unützer, J. (2021). Harnessing collaborative care to meet mental health demands in the era of COVID-19. *JAMA Psychiatry, 78*(4):355–356. https://doi.org/10.1001/jamapsychiatry.2020.3216. PMID: 33084852.

Chauhan, M., & Niazi, S. K. (2017). Caring for patients with chronic physical and mental health conditions: Lessons from TEAMcare and COMPASS. *Focus (Am Psychiatr Publ), 15*(3), 279–283.

Coates, D., Coppleson, D., & Schmied, V. (2020). Integrated physical and mental healthcare: An overview of models and their evaluation findings. *International Journal of Evidence-Based Healthcare, 18*(1), 38–57.

Coleman, K. J., Magnan, S., Neely, C., Solberg, L., Beck, A., Trevis, J., Heim, C., Williams, M., Katzelnick, D., Unützer, J., Pollock, B., Hafer, E., Ferguson, R., & Williams, S. (2017a). The COMPASS initiative: Description of a nationwide collaborative approach to the care of patients with depression and diabetes and/or cardiovascular disease. *General Hospital Psychiatry, 44*, 69–76.

Coleman, K. J., Hemmila, T., Valenti, M. D., Smith, N., Quarrell, R., Ruona, L. K., Brandenfels, E., Hann, B., Hinnenkamp, T., Parra, M. D., Monkman, J., Vos, S., & Rossom, R. C. (2017b). Understanding the experience of care managers and relationship with patient outcomes: The COMPASS initiative. *General Hospital Psychiatry, 44*, 86–90.

Collins, P. Y., Patel, V., Joestl, S. S., et al. (2011). Grand challenges in global mental health. *Nature, 475*(7354), 27–30.

Community Preventive Services Task Force. (2012). Recommendation from the Community Preventive Services Task Force for use of collaborative care for the management of depressive disorders. *American Journal of Preventive Medicine, 42*(5), 521–524.

Coupe, N., Anderson, E., Gask, L., Sykes, P., Richards, D. A., & Chew-Graham, C. (2014). Facilitating professional liaison in collaborative care for depression in UK primary care; a qualitative study utilising normalisation process theory. *BMC Family Practice, 15*, 78.

Coventry, P. A., Hudson, J. L., Kontopantelis, E., Archer, J., Richards, D. A., Gilbody, S., Lovell, K., Dickens, C., Gask, L., Waheed, W., & Bower, P. (2014). Characteristics of effective collaborative care for treatment of depression: A systematic review and meta-regression of 74 randomised controlled trials. *PLoS One, 9*(9), e108114.

Coventry, P., Lovell, K., Dickens, C., Bower, P., Chew-Graham, C., McElvenny, D., Hann, M., Cherrington, A., Garrett, C., Gibbons, C. J., Baguley, C., Roughley, K., Adeyemi, I., Reeves, D., Waheed, W., & Gask, L. (2015). Integrated primary care for patients with mental and physical multimorbidity: Cluster randomised controlled trial of collaborative care for patients with depression comorbid with diabetes or cardiovascular disease. *BMJ, 350*, h638.

Cubillos, L., Bartels, S. M., Torrey, W. C., Naslund, J., Uribe-Restrepo, J. M., Gaviola, C., Díaz, S. C., John, D. T., Williams, M. J., Cepeda, M., Gómez-Restrepo, C., & Marsch, L. A. (2021). The effectiveness and cost-effectiveness of integrating mental health services in primary care in low- and middle-income countries: Systematic review. *BJPsych Bulletin, 45*(1), 40–52. https://doi.org/10.1192/bjb.2020.35. PMID: 32321610; PMCID: PMC8058938.

Davies, T., & Lund, C. (2017). Integrating mental health care into primary care systems in low- and middle-income countries: Lessons from PRIME and AFFIRM. *Global Mental Health, 4*, e7.

Davis, T. D., Deen, T., Bryant-Bedell, K., Tate, V., & Fortney, J. (2011). Does minority racial-ethnic status moderate outcomes of collaborative care for depression? *Psychiatric Services, 62*(11), 1282–1288.

de Jong, F. J., van Steenbergen-Weijenburg, K. M., Huijbregts, K. M., Vlasveld, M. C., Van Marwijk, H. W., Beekman, A. T., van der Feltz-Cornelis, C. M., & The Depression Initiative. (2009). Description of a collaborative care model for depression and of the factors influencing its implementation in the primary care setting in the Netherlands. *International Journal of Integrated Care, 9*, e81.

Dismuke, C. E., & Egede, L. E. (2010). Association between major depression, depressive symptoms and personal income in US adults with diabetes. *General Hospital Psychiatry, 32*(5), 484–491.

Ell, K., Xie, B., Quon, B., et al. (2008). Randomized controlled trial of collaborative care management of depression among low-income patients with cancer. *Journal of Clinical Oncology, 26*(27), 4488–4496.

Ell, K., Katon, W., Cabassa, L. J., et al. (2009). Depression and diabetes among low-income Hispanics: Design elements of a socioculturally adapted collaborative care model randomized controlled trial. *International Journal of Psychiatry in Medicine, 39*(2), 113–132.

Ell, K., Aranda, M. P., Xie, B., et al. (2010a). Collaborative depression treatment in older and younger adults with physical illness: Pooled comparative analysis of three randomized clinical trials. *American Journal of Geriatric Psychiatry, 18*(6), 520–530.

Ell, K., Katon, W., Xie, B., et al. (2010b). Collaborative care management of major depression among low-income, predominantly Hispanic subjects with diabetes: A randomized controlled trial. *Diabetes Care, 33*(4), 706–713.

Engel, C. C., Oxman, T., Yamamoto, C., Gould, D., Barry, S., Stewart, P., Kroenke, K., Williams, J. W., Jr., & Dietrich, A. J. (2008). RESPECT-Mil: Feasibility of a systems-level collaborative care approach to depression and post-traumatic stress disorder in military primary care. *Military Medicine, 173*(10), 935–940.

Fairall, L., Petersen, I., Zani, B., Folb, N., Georgeu-Pepper, D., Selohilwe, O., Petrus, R., Mntambo, N., Bhana, A., Lombard, C., Bachmann, M., Lund, C., Hanass-Hancock, J., Chisholm, D., MCcrone, P., Carmona, S., Gaziano, T., Levitt, N., Kathree, T., & Thornicroft, G. (2018). CobALT research team. Collaborative care for the detection and management of depression among adults receiving antiretroviral therapy in South Africa: Study protocol for the CobALT randomised controlled trial. *Trials, 19*(1), 193.

Farooq, S. (2013). Collaborative care for depression: A literature review and a model for implementation in developing countries. *International Health, 5*(1), 24–28.

Fekadu, A., Hanlon, C., Medhin, G., Alem, A., Selamu, M., Giorgis, T. W., Shibre, T., Teferra, S., Tegegn, T., Breuer, E., Patel, V., Tomlinson, M., Thornicroft, G., Prince, M., & Lund, C. (2016). Development of a scalable mental healthcare plan for a rural district in Ethiopia. *The British Journal of Psychiatry, 208 Suppl 56*(Suppl 56), s4–s12.

Frank, F., Bjerregaard, F., Bengel, J., Bitzer, E. M., Heimbach, B., Kaier, K., Kiekert, J., Krämer, L., Kricheldorff, C., Laubner, K., Maun, A., Metzner, G., Niebling, W., Salm, C., Schütter, S., Seufert, J., Farin, E., & Voigt-Radloff, S. (2019). Local, collaborative, stepped and personalised care management for older people with chronic diseases (LoChro): Study protocol of a randomised comparative effectiveness trial. *BMC Geriatrics, 19*(1), 64.

Garrison, G. M., Angstman, K. B., O'Connor, S. S., Williams, M. D., & Lineberry, T. W. (2016). Time to remission for depression with Collaborative Care Management (CCM) in primary care. *Journal of American Board of Family Medicine, 29*(1), 10–17.

Gensichen, J., von Korff, M., Peitz, M., et al. (2009). Case management for depression by health care assistants in small primary care practices: A cluster randomized trial. *Annals of Internal Medicine, 151*, 369–378.

Gilbody, S., Bower, P., Fletcher, J., et al. (2006a). Collaborative care for depression: A cumulative meta-analysis and review of longer-term outcomes. *Archives of Internal Medicine, 166*(21), 2314–2321.

Gilbody, S., Bower, P., & Whitty, P. (2006b). Costs and consequences of enhanced primary care for depression: Systematic review of randomised economic evaluations. *British Journal of Psychiatry, 189*, 297–308.

Gilmer, T. P., Walker, C., Johnson, E. D., et al. (2008). Improving treatment of depression among Latinos with diabetes using project Dulce and IMPACT. *Diabetes Care, 31*(7), 1324–1326.

Glied, S., Herzog, K., & Frank, R. (2010). Review: The net benefits of depression management in primary care. *Medical Care Research and Review: MCRR, 67*(3), 251–274.

Goldberg, D. P., Steele, J. J., Smith, C., & Spivey, L. (1980). Training family doctors to recognise psychiatric illness with increased accuracy. *Lancet, 2*(8193), 521–523.

Goorden, M., Huijbregts, K. M., van Marwijk, H. W., Beekman, A. T., van der Feltz-Cornelis, C. M., & Hakkaart-van, R. L. (2015). Cost-utility of collaborative care for major depressive disorder in primary care in the Netherlands. *Journal of Psychosomatic Research, 79*(4), 316–323.

Grant, K. L., Simmons, M. B., & Davey, C. G. (2018). Three nontraditional approaches to improving the capacity, accessibility, and quality of mental health services: An overview. *Psychiatric Services, 69*(5), 508–516.

Grembowski, D.E., Martin, D., Patrick, DL., et al. (2002). Managed care, access to mental health specialists, and outcomes among primary care patients with depressive symptoms. *Journal of General Internal Medicine, 17*(4), 258–269.

Green, C., Richards, D. A., Hill, J. J., Gask, L., Lovell, K., Chew-Graham, C., Bower, P., Cape, J., Pilling, S., Araya, R., Kessler, D., Bland, J. M., Gilbody, S., Lewis, G., Manning, C., Hughes-Morley, A., & Barkham, M. (2014). Cost-effectiveness of collaborative care for depression in UK primary care: Economic evaluation of a randomised controlled trial (CADET). *PLoS One, 9*(8), e104225.

Grochtdreis, T., Zimmermann, T., Puschmann, E., Porzelt, S., Dams, J., Scherer, M., & König, H. (2018). Cost-utility of collaborative nurse-led self-management support for primary care patients with anxiety, depressive or somatic symptoms: A cluster-randomized controlled trial (the SMADS trial). *International Journal of Nursing Studies, 80*, 67–75.

Grochtdreis, T., Brettschneider, C., Bjerregaard, F., Bleich, C., Boczor, S., Härter, M., Hölzel, L. P., Hüll, M., Kloppe, T., Niebling, W., Scherer, M., Tinsel, I., & König, H. H. (2019). Cost-effectiveness analysis of collaborative treatment of late-life depression in primary care (German IMPACT). *European Psychiatry, 57*, 10–18.

Grote, N. K., Katon, W. J., Russo, J. E., Lohr, M. J., Curran, M., Galvin, E., & Carson, K. (2015). Collaborative care for perinatal depression in socioeconomically disadvantaged women: A randomized trial. *Depression and Anxiety, 32*(11), 821–834.

Grypma, L., Haverkamp, R., Little, S., et al. (2006). Taking an evidence-based model of depression care from research to practice: Making lemonade out of depression. *General Hospital Psychiatry, 28*(2), 101–107.

Hanlon, C., Luitel, N. P., Kathree, T., Murhar, V., Shrivasta, S., Medhin, G., et al. (2014). Challenges and opportunities for implementing integrated mental healthcare: A district level situation analysis from five low- and middle-income countries. *PLoS One, 9*, e88437.

Härter, M., Baumeister, H., Reuter, K., Jacobi, F., Höfler, M., Bengel, J., & Wittchen, H.-U. (2007). Increased 12-month prevalence rates of mental disorders in patients with chronic somatic diseases. *Psychotherapy and Psychosomatics, 76*(6), 354–360.

Härter, M., Watzke, B., Daubmann, A., Wegscheider, K., König, H. H., Brettschneider, C., Liebherz, S., Heddaeus, D., & Steinmann, M. (2018). Guideline-based stepped and collaborative care for patients with depression in a cluster-randomised trial. *Scientific Reports, 8*(1), 9389.

Heddaeus, D., Steinmann, M., Daubmann, A., Härter, M., & Watzke, B. (2018). Treatment selection and treatment initialization in guideline-based stepped and collaborative care for depression. *PLoS One, 13*(12), e0208882.

Heddaeus, D., Dirmaier, J., Brettschneider, C., Daubmann, A., Grochtdreis, T., von dem Knesebeck, O., König, H. H., Löwe, B., Maehder, K., Porzelt, S., Rosenkranz, M., Schäfer, I., Scherer, M., Schulte, B., Wegscheider, K., Weigel, A., Werner, S., Zimmermann, T., & Härter, M. (2019). Study protocol for the COMET study: A cluster-randomised, prospective, parallel-group, superiority trial to compare the effectiveness of a collaborative and stepped care model versus treatment as usual in patients with mental disorders in primary care. *BMJ Open, 9*(11), e032408.

Henke, R. M., Zaslavsky, A. M., McGuire, T. G., et al. (2009). Clinical inertia in depression treatment. *Medical Care, 47*(9), 959–967.

Herbeck Belnap, B., Anderson, A., Abebe, K. Z., Ramani, R., Muldoon, M. F., Karp, J. F., & Rollman, B. L. (2019). Blended collaborative care to treat heart failure and comorbid depression: Rationale and Study Design of the Hopeful Heart Trial. *Psychosomatic Medicine, 81*(6), 495–505.

Hoeft, T. J., Fortney, J. C., Patel, V., & Unützer, J. (2018). Task-sharing approaches to improve mental health care in rural and other low-resource settings: A systematic review. *The Journal of Rural Health, 34*(1), 48–62.

Huang, H., Meller, W., Kishi, Y., & Kathol, R. G. (2014). What is integrated care? *International Review of Psychiatry, 26*(6), 620–628. https://doi.org/10.3109/09540261.2014.964189

Huijbregts, K. M., de Jong, F. J., van Marwijk, H. W., Beekman, A. T., Adèr, H. J., Hakkaart-van Roijen, L., Unützer, J., & van der Feltz-Cornelis, C. M. (2013). A target-driven collaborative care model for Major Depressive Disorder is effective in primary care in the Netherlands. A randomized clinical trial from the depression initiative. *Journal of Affective Disorders, 146*(3), 328–337.

Kakuma, R., Minas, H., van Ginneken, N., Dal Poz, M. R., Desiraju, K., Morris, J. E., & Scheffler, R. M. (2011). Human resources for mental health care: Current situation and strategies for action. *Lancet, 378*(9803), 1654–1663.

Kathol, R. G., Butler, M., McAlpine, D. D., & Kane, R. L. (2010). Barriers to physical and mental condition integrated service delivery. *Psychosomatic Medicine, 72*(6), 511–518. https://doi.org/10.1097/PSY.0b013e3181e2c4a0

Katon, W. J. (2003). Clinical and health services relationships between major depression, depressive symptoms, and general medical illness. *Biological Psychiatry, 54*(3), 216–226.

Katon, W. J. (2009). The impact of depression on workplace functioning and disability costs. *American Journal of Managed Care, 15*(11 Suppl), S322–S327.

Katon, W. J., Russo, J. E., Von Korff, M., et al. (2008). Long-term effects on medical costs of improving depression outcomes in patients with depression and diabetes. *Diabetes Care, 31*(6), 1155–1159.

Katon, W. J., Lin, E. H., Von Korff, M., et al. (2010). Collaborative care for patients with depression and chronic illnesses. *The New England Journal of Medicine, 363*(27), 2611–2620.

Kessler, R. C., Demler, O., Frank, R. G., Olfson, M., Pincus, H. A., Walters, E. E., . . . Zaslavsky, A. M. (2005). Prevalence and treatment of mental disorders, 1990 to 2003. *New England Journal of Medicine, 352*, 2515–2523.

Kroenke, K., Spitzer, R. L., & Williams, J. B. (2001). The PHQ-9: Validity of a brief depression severity measure. *Journal of General Internal Medicine, 16*(9), 606–613.

Lambert, M., Karow, A., Gallinat, J., Lüdecke, D., Kraft, V., Rohenkohl, A., Schröter, R., Finter, C., Siem, A. K., Tlach, L., Werkle, N., Bargel, S., Ohm, G., Hoff, M., Peter, H., Scherer, M., Mews, C., Pruskil, S., Lüke, J., Härter, M., Dirmaier, J., Schulte-Markwort, M., Löwe, B., Briken, P., Peper, H., Schweiger, M., Mösko, M., Bock, T., Wittzack, M., Meyer, H. J., Deister, A., Michels, R., Herr, S., Konnopka, A., König, H., Wegscheider, K., Daubmann, A., Zapf, A., Peth, J., König, H. H., & Schulz, H. (2020). Study protocol for a randomised controlled trial evaluating an evidence-based, stepped and coordinated care service model for mental disorders (RECOVER). *BMJ Open, 10*(5), e036021.

LaRocco-Cockburn, A., Reed, S. D., Melville, J., et al. (2013). Improving depression treatment for women: Integrating a collaborative care depression intervention into OB-GYN care. *Contemporary Clinical Trials, 36*(2), 362–370.

Lee-Tauler, S. Y., Eun, J., Corbett, D., & Collins, P. Y. (2018). A systematic review of interventions to improve initiation of mental health care among racial-ethnic minority groups. *Psychiatric Services, 69*(6), 628–647.

Levine, S., Unützer, J., Yip, J. Y., et al. (2005). Physicians' satisfaction with a collaborative disease management program for late-life depression in primary care. *General Hospital Psychiatry, 27*(6), 383–391.

Lewis, H., Adamson, J., Atherton, K., Bailey, D., Birtwistle, J., Bosanquet, K., Clare, E., Delgadillo, J., Ekers, D., Foster, D., Gabe, R., Gascoyne, S., Haley, L., Hargate, R., Hewitt, C., Holmes, J., Keding, A., Lilley-Kelly, A., Maya, J., McMillan, D., Meer, S., Meredith, J., Mitchell, N., Nutbrown, S., Overend, K., Pasterfield, M., Richards, D., Spilsbury, K., Torgerson, D., Traviss-Turner, G., Trépel, D., Woodhouse, R., Ziegler, F., & Gilbody, S. (2017). Collaborative care and active surveillance for Screen-Positive EldeRs with subthreshold depression (CASPER): A multicentred randomised controlled trial of clinical effectiveness and cost-effectiveness. *Health Technology Assessment, 21*(8), 1–196.

Li, L. W., Xue, J., Conwell, Y., Yang, Q., & Chen, S. (2020). Implementing collaborative care for older people with comorbid hypertension and depression in rural China. *International Psychogeriatrics, 32*(12), 1457–1465. https://doi.org/10.1017/S104161021 9001509. Epub 2019 Oct 21. PMID: 31630703; PMCID: PMC7170762.

Lin, E. H., Katon, W., Von Korff, M., et al. (2003). Effect of improving depression care on pain and functional outcomes among older adults with arthritis: A randomized controlled trial. *JAMA, 290*(18), 2428–2429.

Lin, E. H., Katon, W., Von Korff, M., et al. (2004). Relationship of depression and diabetes self-care, medication adherence, and preventive care. *Diabetes Care, 27*(9), 2154–2216.

Lin, E. H., Von Korff, M., Ciechanowski, P., Peterson, D., Ludman, E. J., Rutter, C. M., Oliver, M., Young, B. A., Gensichen, J., McGregor, M., McCulloch, D. K., Wagner, E. H., & Katon, W. J. (2012). Treatment adjustment and medication adherence for complex patients with diabetes, heart disease, and depression: A randomized controlled trial. *Annals of Family Medicine, 10*(1), 6–14.

Lund, C., Tomlinson, M., De Silva, M., Fekadu, A., Shidhaye, R., Jordans, M., Petersen, I., Bhana, A., Kigozi, F., Prince, M., et al. (2012). PRIME: A programme to reduce the treatment gap for mental disorders in five low- and middle-income countries. *PLoS Medicine, 9*(12), e1001359.

Lyon, A. R., Whitaker, K., Richardson, L. P., French, W. P., & McCauley, E. (2019). Collaborative care to improve access and quality in school-based behavioral health. *The Journal of School Health, 89*(12), 1013–1023.

Maehder, K., Löwe, B., Härter, M., Heddaeus, D., von dem Knesebeck, O., & Weigel, A. (2020). Psychotherapists' perspectives on collaboration and stepped care in outpatient psychotherapy – A qualitative study. *PLoS One, 15*, 2.

Maehder, K., Werner, S., Weigel, A., Löwe, B., Heddaeus, D., Härter, M., & von dem Knesebeck, O. (2021). How do care providers evaluate collaboration? – Qualitative process evaluation of a cluster-randomized controlled trial of collaborative and stepped care for patients with mental disorders. *BMC Psychiatry, 21*(1), 296.

Martínez, V., Rojas, G., Martínez, P., Zitko, P., Irarrázaval, M., Luttges, C., & Araya, R. (2018). Remote collaborative depression care program for adolescents in Araucanía Region, Chile: Randomized controlled trial. *Journal of Medical Internet Research, 20*(1), e38.

McCarty, C. A., Zatzick, D., Stein, E., Wang, J., Hilt, R., Rivara, F. P., & Seattle Sports Concussion Research Collaborative. (2016). Collaborative care for adolescents with persistent postconcussive symptoms: A randomized trial. *Pediatrics, 138*(4), e20160459.

McGinty, E. E., & Daumit, G. L. (2020). Integrating mental health and addiction treatment into general medical care: The role of policy. *Psychiatric Services, 71*(11), 1163–1169.

McGregor, M., Lin, E. H., & Katon, W. J. (2011). TEAMcare: An integrated multicondition collaborative care program for chronic illnesses and depression. *The Journal of Ambulatory Care Management, 34*(2), 152–162.

Mendenhall, E., De Silva, M. J., Hanlon, C., Petersen, I., Shidhaye, R., Jordans, M., Luitel, N., Ssebunnya, J., Fekadu, A., Patel, V., Tomlinson, M., & Lund, C. (2014). Acceptability and feasibility of using non-specialist health workers to deliver mental health care: Stakeholder perceptions from the PRIME district sites in Ethiopia, India, Nepal, South Africa, and Uganda. *Social Science & Medicine, 118*, 33–42.

Michalowsky, B., Xie, F., Eichler, T., Hertel, J., Kaczynski, A., Kilimann, I., Teipel, S., Wucherer, D., Zwingmann, I., Thyrian, J. R., & Hoffmann, W. (2019). Cost-effectiveness of a collaborative dementia care management-Results of a cluster-randomized controlled trial. *Alzheimers Dement, 15*(10), 1296–1308.

Miranda, J., Duan, N., Sherbourne, C., et al. (2003). Improving care for minorities: Can quality improvement interventions improve care and outcomes for depressed minorities? Results of a randomized controlled trial. *Health Services Research, 38*(2), 613–630.

Miranda, J., Schoenbaum, M., Sherbourne, C., et al. (2004). Effects of primary care depression treatment on minority patients' clinical status and employment. *Archives of General Psychiatry, 61*(8), 827–834.

Moise, N., Shah, R. N., Essock, S., Jones, A., Carruthers, J., Handley, M. A., Peccoralo, L., & Sederer, L. (2018). Sustainability of collaborative care management for depression in primary care settings with academic affiliations across New York State. *Implementation Science, 13*(1), 128.

Moullin, J. C., Dickson, K. S., Stadnick, N. A., Rabin, B., & Aarons, G. A. (2019). Systematic review of the Exploration, Preparation, Implementation, Sustainment (EPIS) framework. *Implementation Science, 14*(1), 1.

Moussavi, S., Chatterji, S., Verdes, E., et al. (2007). Depression, chronic diseases, and decrements in health: Results from the World Health Surveys. *The Lancet, 370*(9590), 851–858.

Muntingh, A. D., van der Feltz-Cornelis, C. M., van Marwijk, H. W., Spinhoven, P., & van Balkom, A. J. (2016). Collaborative care for anxiety disorders in primary care: A systematic review and meta-analysis. *BMC Family Practice, 17*, 62.

National Institute for Health and Care Excellence. (2009). Depression in adults with a chronic physical health problem: Recognition and management. Published: 28 October 2009. www.nice.org.uk/guidance/cg91

Overbeck, G., Davidsen, A. S., & Kousgaard, M. B. (2016). Enablers and barriers to implementing collaborative care for anxiety and depression: A systematic qualitative review. *Implementation Science, 11*, 165.

Padmanathan, P., & De Silva, M. (2013). The acceptability and feasibility of task-sharing for mental healthcare in low- and middle-income countries: A systematic review. *Social Science & Medicine, 97*, 82–86.

Panagioti, M., Bower, P., Kontopantelis, E., Lovell, K., Gilbody, S., Waheed, W., Dickens, C., Archer, J., Simon, G., Ell, K., Huffman, J. C., Richards, D. A., van der Feltz-Cornelis, C., Adler, D. A., Bruce, M., Buszewicz, M., Cole, M. G., Davidson, K. W., de Jonge, P., Gensichen, J., Huijbregts, K., Menchetti, M., Patel, V., Rollman, B., Shaffer, J., Zijlstra-Vlasveld, M. C., & Coventry, P. A. (2016). Association between chronic physical conditions and the effectiveness of collaborative care for depression: An individual participant data meta-analysis. *JAMA Psychiatry, 73*(9), 978–989.

Patel, V., Belkin, G. S., Chockalingam, A., Cooper, J., Saxena, S., & Unützer, J. (2013). Grand challenges: Integrating mental health services into priority health care platforms. *PLoS Medicine, 10*(5), e1001448.

Patel, V., Saxena, S., Lund, C., Thornicroft, G., Baingana, F., Bolton, P., Chisholm, D., Collins, P. Y., Cooper, J. L., Eaton, J., Herrman, H., Herzallah, M. M., Huang, Y., Jordans, M. J. D., Kleinman, A., Medina-Mora, M. E., Morgan, E., Niaz, U., Omigbodun, O., Prince, M., Rahman, A., Saraceno, B., Sarkar, B. K., De Silva, M., Singh, I., Stein, D. J., Sunkel, C., & UnÜtzer, J. (2018). The Lancet Commission on global mental health and sustainable development. *Lancet, 392*(10157), 1553–1598.

Petersen, J. J., König, J., Paulitsch, M. A., Mergenthal, K., Rauck, S., Pagitz, M., Schmidt, K., Haase, L., Gerlach, F. M., & Gensichen, J. (2014). Long-term effects of a collaborative care intervention on process of care in family practices in Germany: A 24-month follow-up study of a cluster randomized controlled trial. *General Hospital Psychiatry, 36*(6), 570–574.

Petersen, I., Bhana, A., Folb, N., Thornicroft, G., Zani, B., Selohilwe, O., Petrus, R., Mntambo, N., Georgeu-Pepper, D., Kathree, T., Lund, C., Lombard, C., Bachmann, M., Gaziano, T., Levitt, N., & Fairall, L. (2018). PRIME-SA research team. Collaborative care for the detection and management of depression among adults with hypertension in South Africa: Study protocol for the PRIME-SA randomised controlled trial. *Trials, 19*(1), 192.

Petersen, I., Bhana, A., Fairall, L. R., Selohilwe, O., Kathree, T., Baron, E. C., Rathod, S. D., & Lund, C. (2019). Evaluation of a collaborative care model for integrated primary care of common mental disorders comorbid with chronic conditions in South Africa. *BMC Psychiatry, 19*(1), 107.

Reilly, S., Planner, C., Gask, L., et al. (2012). Collaborative care approaches for people with severe mental illness (Protocol). *Cochrane Database of Systematic Reviews, 1*, CD009531.

Reiss-Brennan, B., Briot, P. C., Savitz, L. A., et al. (2010). Cost and quality impact of Intermountain's mental health integration program. *Journal of Healthcare Management, 55*(2), 97–113; discussion 113–114.

Reiter, J. T., Dobmeyer, A. C., & Hunter, C. L. (2018). The primary care behavioral health (PCBH) model: An overview and operational definition. *Journal of Clinical Psychology in Medical Settings, 25*(2),109–126.

Richards, D. A., Lovell, K., Gilbody, S., Gask, L., Torgerson, D., Barkham, M., Bland, M., Bower, P., Lankshear, A. J., Simpson, A., Fletcher, J., Escott, D., Hennessy, S., & Richardson, R. (2008). Collaborative care for depression in UK primary care: A randomized controlled trial. *Psychological Medicine, 38*(2), 279–287.

Richards, D. A., Hill, J. J., Gask, L., et al. (2013). Clinical effectiveness of collaborative care for depression in UK primary care (CADET): Cluster randomised controlled trial. *BMJ, 347*, f4913.

Richardson, L., McCauley, E., & Katon, W. (2009). Collaborative care for adolescent depression: A pilot study. *General Hospital Psychiatry, 31*(1), 36–45.

Richardson, L. P., Ludman, E., McCauley, E., Lindenbaum, J., Larison, C., Zhou, C., Clarke, G., Brent, D., & Katon, W. (2014). Collaborative care for adolescents with depression in primary care: A randomized clinical trial. *JAMA, 312*(8), 809–816.

Rojas, G., Guajardo, V., Martínez, P., Castro, A., Fritsch, R., Moessner, M., & Bauer, S. (2018). A remote collaborative care program for patients with depression living in rural areas: Open-label trial. *Journal of Medical Internet Research, 20*(4), e158.

Roman, L. A., Lindsay, J. K., Moore, J. S., et al. (2007). Addressing mental health and stress in Medicaid-insured pregnant women using a nurse-community health worker home visiting team. *Public Health Nursing, 24*, 239–248.

Rossom, R. C., Solberg, L. I., Magnan, S., Crain, A. L., Beck, A., Coleman, K. J., Katzelnick, D., Williams, M. D., Neely, C., Ohnsorg, K., Whitebird, R., Brandenfels, E., Pollock, B., Ferguson, R., Williams, S., & Unützer, J. (2017). Impact of a national collaborative care initiative for patients with depression and diabetes or cardiovascular disease. *General Hospital Psychiatry, 44*, 77–85.

Roy-Byrne, P. P., Craske, M. G., Stein, M. B., et al. (2005). A randomized effectiveness trial of cognitive-behavioral therapy and medication for primary care panic disorder. *Archives of General Psychiatry, 62*(3), 290–298.

Rush, A. J., Trivedi, M., Carmody, T. J., et al. (2004). One-year clinical outcomes of depressed public sector outpatients: A benchmark for subsequent studies. *Biological Psychiatry, 56*(1), 46–53.

Schoenbaum, M., Unützer, J., MCcaffrey, D., & etal. (2002). The effects of primary care depression treatment on patients' clinical status and employment. *Health Services Research, 37*(5), 1145–1158.

Schofield, D. J., Kelly, S. J., Shrestha, R. N., et al. (2011). How depression and other mental health problems can affect future living standards of those out of the labour force. *Aging and Mental Health, 15*(5), 654–662.

Sederer, L. I., Derman, M., Carruthers, J., & Wall, M. (2016). The New York State Collaborative Care Initiative: 2012–2014. *The Psychiatric Quarterly, 87*(1), 1–23.

Shepherd, M. (1991). Primary care psychiatry: The case for action. *The British Journal of General Practice, 41*(347), 252–255.

Shepherd, M., & Wilkinson, G. (1988). Primary care as the middle ground for psychiatric epidemiology. *Psychological Medicine, 18*(2), 263–267.

Shippee, N. D., Mattson, A., Brennan, R., Huxsahl, J., Billings, M. L., & Williams, M. D. (2018). Effectiveness in regular practice of collaborative care for depression among adolescents: A Retrospective Cohort Study. *Psychiatric Services, 69*(5), 536–541.

Simon G (2009) Collaborative care for mood disorders. *Current Opinion in Psychiatry, 22*(1), 37–41.

Simon, G. E., Ding, V., Hubbard, R., et al. (2012). Early dropout from psychotherapy for depression with group- and network-model therapists. *Administration and Policy in Mental Health, 39*(6), 440–447.

Siskind, D., Araya, R., & Kim, J. (2010). Cost-effectiveness of improved primary care treatment of depression in women in Chile. *The British Journal of Psychiatry: The Journal of Mental Science, 197*(4), 291–296.

Solberg, L. I., Crain, A. L., Jaeckels, N., Ohnsorg, K. A., Margolis, K. L., Beck, A., et al. (2013). The DIAMOND initiative: Implementing collaborative care for depression in 75 primary care clinics. *Implementation Science, 8*, 135.

Solberg, L. I., Crain, A. L., Maciosek, M. V., Unützer, J., Ohnsorg, K. A., Beck, A., Rubenstein, L., Whitebird, R. R., Rossom, R. C., Pietruszewski, P. B., Crabtree, B. F., Joslyn, K., Van de Ven, A., & Glasgow, R. E. (2015). A stepped-wedge evaluation of an initiative to spread the collaborative care model for depression in primary care. *Annals of Family Medicine, 13*(5), 412–420.

Sowa, N. A., Jeng, P., Bauer, A. M., Cerimele, J. M., Unützer, J., Bao, Y., & Chwastiak, L. (2018). Psychiatric case review and treatment intensification in collaborative care management for depression in primary care. *Psychiatric Services, 69*(5), 549–554.

Tapia-Muñoz, T., Mascayano, F., & Toso-Salman, J. (2015). Collaborative care models to address late-life depression: Lessons for low-and-middle-income countries. *Frontiers in Psychiatry, 6*, 64.

Thornicroft, G., Chatterji, S., Evans-Lacko, S., et al. (2017). Undertreatment of people with major depressive disorder in 21 countries. *British Journal of Psychiatry, 210*, 119–124.

Thota, A. B., Sipe, T. A., Byard, G. J., et al. (2012). Collaborative care to improve the management of depressive disorders: A community guide systematic review and meta-analysis. *American Journal of Preventive Medicine, 42*(5), 525–538.

Thyrian, J. R., Hertel, J., Wucherer, D., Eichler, T., Michalowsky, B., Dreier-Wolfgramm, A., Zwingmann, I., Kilimann, I., Teipel, S., & Hoffmann, W. (2017). Effectiveness and safety of dementia care management in primary care: A randomized clinical trial. *JAMA Psychiatry, 74*(10), 996–1004.

Tully, P. J., & Baumeister, H. (2015). Collaborative care for comorbid depression and coronary heart disease: A systematic review and meta-analysis of randomised controlled trials. *BMJ Open, 5*(12), e009128.

Uebelacker, L. A., Smith, M., Lewis, A. W., et al. (2009). Treatment of depression in a low-income primary care setting with colocated mental health care. *Families, Systems and Health: The Journal of Collaborative Family Healthcare, 27*(2), 161–171.

Unützer, J., Katon, W., Callahan, C. M., et al. (2002). Collaborative-care management of late-life depression in the primary care setting. *JAMA, 288*(22), 2836–2845.

Unützer, J., Katon, W. J., Fan, M. Y., et al. (2008). Long-term cost effects of collaborative care for late-life depression. *American Journal of Managed Care, 14*(2), 95–100.

Unützer, J., Chan, Y. F., Hafer, E., et al. (2012). Quality improvement with pay-for-performance incentives in integrated behavioral health care. *American Journal of Public Health, 102*(6), e41–e45.

Unützer, J., Carlo, A. D., & Collins, P. Y. (2020a). Leveraging collaborative care to improve access to mental health care on a global scale. *World Psychiatry, 19*(1), 36–37.

Unützer, J., Carlo, A. C., Arao, R., Vredevoogd, M., Fortney, J., Powers, D., & Russo, J. (2020b). Variation in the effectiveness of collaborative care for depression: Does it matter where you get your care? *Health Affairs (Millwood), 39*(11), 1943–1950.

van Orden, M., Hoffman, T., Haffmans, J., Spinhoven, P., & Hoencamp, E. (2009). Collaborative mental health care versus care as usual in a primary care setting: A randomized controlled trial. *Psychiatric Services, 60*(1), 74–79.

Vlasveld, M. C., van der Feltz-Cornelis, C. M., Adèr, H. J., Anema, J. R., Hoedeman, R., van Mechelen, W., & Beekman, A. T. (2013). Collaborative care for sick-listed workers with major depressive disorder: A randomised controlled trial from the Netherlands Depression Initiative aimed at return to work and depressive symptoms. *Occupational and Environmental Medicine, 70*(4), 223–230.

Von Korff, M., & Tiemens, B. (2000). Individualized stepped care of chronic illness. *Western Journal of Medicine, 172*(2), 133–137.

Wagner, E. H., Austin, B. T., & Von Korff, M. (1996). Organizing care for patients with chronic illness. *The Milbank Quarterly, 74*(4), 511–544.

Wang, P. S., Lane, M., Olfson, M., et al. (2005). Twelve-month use of mental health services in the United States: Results from the National Comorbidity Survey Replication. *Archives of General Psychiatry, 62*(6), 629–640.

Wang, P. S., Demler, O., Olfson, M., et al. (2006). Changing profiles of service sectors used for mental health care in the United States. *American Journal of Psychiatry, 163*(7), 1187–1198.

Wang, P. S., Simon, G. E., Avorn, J., et al. (2007a). Telephone screening, outreach, and care management for depressed workers and impact on clinical and work productivity outcomes: A randomized controlled trial. *JAMA, 298*(12), 1401–1411.

Wang, P. S., Aguilar-Gaxiola, S., Alonso, J., et al. (2007b). Use of mental health services for anxiety, mood, and substance disorders in 17 countries in the WHO world mental health surveys. *Lancet, 370*(9590), 841–850.

Wang, P. S., Simon, G. E., & Kessler, R. C. (2008). Making the business case for enhanced depression care: The National Institute of Mental Health-Harvard Work Outcomes Research and Cost-Effectiveness Study. *Journal of Occupational and Environmental Medicine, 50*(4), 468–475.

Watkins, K. E., Ober, A. J., Lamp, K., Lind, M., Setodji, C., Osilla, K. C., Hunter, S. B., MCcullough, C. M., Becker, K., Iyiewuare, P. O., Diamant, A., Heinzerling, K., & Pincus, H. A. (2017). Collaborative care for opioid and alcohol use disorders in primary care: The SUMMIT randomized clinical trial. *JAMA Internal Medicine, 177*(10), 1480–1488.

Watzke, B., Heddaeus, D., Steinmann, M., Daubmann, A., Wegscheider, K., & Härter, M. (2020). Does symptom severity matter in stepped and collaborative care for depression? *Journal of Affective Disorders, 277*, 287–295.

Wells, K., Klap, R., Koike, A., et al. (2001). Ethnic disparities in unmet need for alcoholism, drug abuse, and mental health care. *American Journal of Psychiatry, 158*(12), 2027–2032.

Wells, J. E., Browne, M. O., Aguilar-Gaxiola, S., et al. (2013). Drop out from out-patient mental healthcare in the World Health Organization's World Mental Health Survey initiative. *The British Journal of Psychiatry, 202*, 42–49.

Wernher, I., Bjerregaard, F., Tinsel, I., et al. (2014). Collaborative treatment of late-life depression in primary care (GermanIMPACT): Study protocol of a cluster-randomized controlled trial. *Trials, 15*, 351.

Whitebird, R. R., Solberg, L. I., Jaeckels, N. A., Pietruszewski, P. B., Hadzic, S., Unützer, J., Ohnsorg, K. A., Rossom, R. C., Beck, A., Joslyn, K. E., & Rubenstein, L. V. (2014). Effective implementation of collaborative care for depression: What is needed? *The American Journal of Managed Care, 20*(9), 699–707.

Woltmann, E., Grogan-Kaylor, A., Perron, B., et al. (2012). Comparative effectiveness of collaborative chronic care models for mental health conditions across primary, specialty, and behavioral health care settings: Systematic review and meta-analysis. *American Journal of Psychiatry, 169*(8), 790–804.

Wood, E., Ohlsen, S., & Ricketts, T. (2017). What are the barriers and facilitators to implementing collaborative care for depression? A systematic review. *Journal of Affective Disorders, 214*, 26–43.

World Health Organization. (2014). mhGAP intervention guide for mental, neurological and substance use disorders in non-specialised health settings: Guideline update, 31 December 2014.

World Health Organization. (2017). *Depression and other common mental disorders: Global Health estimates*. World Health Organization.

World Health Organization, & WONCA. (2008). *Integrating mental health into primary care: A global perspective*. World Health Organization.

Wright, D. R., Haaland, W. L., Ludman, E., McCauley, E., Lindenbaum, J., & Richardson, L. P. (2016). The costs and cost-effectiveness of collaborative care for adolescents with depression in primary care settings: A randomized clinical trial. *JAMA Pediatrics, 170*(11), 1048–1054.

Zimmermann, T., Puschmann, E., van den Bussche, H., Wiese, B., Ernst, A., Porzelt, S., Daubmann, A., & Scherer, M. (2016). Collaborative nurse-led self-management support for primary care patients with anxiety, depressive or somatic symptoms: Cluster-randomised controlled trial (findings of the SMADS study). *International Journal of Nursing Studies, 63*, 101–111.

Pediatric Consultation and Liaison Psychiatry

140

Jessica E. Becker, Joshua R. Smith, Claire De Souza, and Eric P. Hazen

Contents

J. E. Becker · E. P. Hazen (✉)
Division of Child and Adolescent Psychiatry, Department of Psychiatry, Massachusetts General Hospital, Boston, MA, USA

Harvard Medical School, Boston, MA, USA
e-mail: Eric.Hazen@MGH.HARVARD.EDU

J. R. Smith
Department of Psychiatry and Behavioral Sciences, Division of Child and Adolescent Psychiatry, Vanderbilt University Medical Center, Nashville, TN, USA
e-mail: joshua.r.smith@vumc.org

C. De Souza
Department of Psychiatry, The Hospital for Sick Children, Toronto, ON, Canada

Department of Psychiatry, University of Toronto, Toronto, ON, Canada
e-mail: claire.desouza@sickkids.ca

© Springer Nature Switzerland AG 2024
A. Tasman et al. (eds.), *Tasman's Psychiatry*,
https://doi.org/10.1007/978-3-030-51366-5_8

Abstract

Pediatric consultation-liaison (C-L) psychiatry is a dynamic and growing subspecialty. Pediatric C-L psychiatrists serve a variety of roles at the intersection of physical and psychiatric illness in children, including supporting the psychiatric well-being of children with chronic physical illness, assessing for psychiatric manifestations of physical illness, and treating children with primary psychiatric illness who require medical intervention or are awaiting inpatient psychiatric placement. After a brief introduction to the field, this chapter discusses the role of the pediatric C-L psychiatrist; highlights the unique aspects of pediatric C-L psychiatry, including a discussion of the developmental approach to chronic illness; explores common reasons for pediatric C-L psychiatry requests; discusses evolving systems of C-L care; and reviews legal and ethical concerns encountered in pediatric C-L psychiatry. While the role of pediatric C-L psychiatry is expanding, this chapter focuses primarily on pediatric C-L in the inpatient pediatric setting, a primary location for the practice of pediatric C-L psychiatry.

Keywords

Pediatric consultation-liaison psychiatry ·
Child psychiatry · Physical illness · Physically
ill children · Pediatrics · Consultation ·
Liaison · Development · Systems of care

Introduction

Pediatric consultation-liaison (C-L) psychiatry is a diverse and growing subspecialty through which psychiatrists care for children with physical illness. In recent years, tertiary care hospitalizations for children with psychiatric diagnoses have been growing rapidly (Zima et al., 2016), as have pediatric psychiatry consultation requests (Shaw et al.,

2016). Pediatric C-L psychiatrists can play an integral role in the care for children with physical and psychiatric illnesses alike, liaising with a multidisciplinary healthcare team and increasingly treating children with primary psychiatric presentations who are awaiting placement in a primary psychiatric unit (Becker et al., 2020). While pediatric C-L psychiatry has been a longstanding fixture of many inpatient pediatric settings, there have been increasing efforts to integrate psychiatric care into primary and specialty pediatric outpatient clinics in order to increase access to mental health treatment and to improve physical health outcomes, in part due to shifting trends in healthcare payment models to population health and risk-sharing approaches.

In this chapter, a brief survey of this dynamic sub-specialty is provided, focusing primarily on the inpatient setting, including a discussion of the unique aspects of pediatric C-L psychiatry, the importance of a developmental approach in pediatric C-L psychiatry care, common reasons for pediatric psychiatry consultation, and shifting trends in systems of care. Given the subspecialty nature of pediatric C-L psychiatry, this chapter focuses on approaches in settings with access to specialty-trained child and adolescent or pediatric C-L psychiatrists. While modifications to the approach would be necessary in settings with fewer resources, it is anticipated that the general principles discussed in this chapter will be useful in caring for pediatric patients with medical and psychiatric needs, regardless of the provider's field of expertise.

Role of the Pediatric C-L Psychiatrist

Inpatient

The role of the pediatric C-L psychiatrist can be both broad and deep. A primary setting for pediatric consultation-liaison psychiatry is the inpatient pediatric floor. This chapter will focus primarily on the role of the inpatient pediatric C-L psychiatrist. As described in the subsequent sections, the inpatient pediatric C-L psychiatrist may be consulted for a variety of reasons at the interface of medical and psychiatric care, including emotional adjustment to medical illness, psychiatric manifestations of medical illness or treatment, and primary psychiatric conditions. The systems-focused training of pediatric psychiatry lends well to the pediatric C-L approach, which often requires a holistic, multifaceted approach to understanding a child's illness and devising an appropriate treatment plan. Moreover, liaising – between medical teams, between team members, and between families and medical providers – is an integral role of the C-L psychiatrist. Indeed, given the complexity and severity of illness in many children with comorbid medical and psychiatric conditions, these children are at times perceived as "difficult" patients and can generate substantial countertransference from the multidisciplinary team providers. C-L psychiatrists can be crucial in providing psychoeducation to team members, assisting team members in understanding their countertransference, and advocating for patients' needs, leading ultimately to better outcomes for patients and providers alike. C-L psychiatrists can help to facilitate multidisciplinary team or family meetings to streamline care for more complicated patients as well. Given the complexity of the pediatric C-L psychiatry patient population, the pediatric C-L psychiatrist will often need a substantial handle on engaging external services for the child. C-L psychiatrists often must work closely with Social Workers, Child Protection services, and sometimes Bioethics, in order to provide the most robust, thoughtful, and supportive care for a sick child.

Outpatient and School Consultation Models

While this chapter focuses primarily on inpatient psychiatric consultation for children, there has been an increasing impetus to integrate mental healthcare into outpatient primary care and pediatric specialty clinics (Richardson et al., 2014; Samsel et al., 2017), particularly in light of emphasis on population health payment models. Integrated

mental health services encompass a wide range of practice models, ranging from care coordination, to collaborative care models, to embedded psychiatric care. While data evaluating these programs are limited, early evidence indicates their benefit for managing mental health conditions (Asarnow et al., 2015; Richardson et al., 2014). More work is needed to assess outpatient integrated mental health services for children in the primary care and specialty setting. Schools are another common setting for children to receive mental health services (Merikangas et al., 2011), and there is often a role for pediatric psychiatrists to consult with schools about general resource issues as well as to provide individually tailored recommendations.

Unique Aspects of Pediatric Consultation

Unique Aspects of Pediatric Consultation
1. Systems approach to care
   (a) Team based approach of the consulting service
   (b) Interdisciplinary nature of pediatric care team
   (c) Family and caregiver system
2. Developmental considerations
   (a) Age- and development-based approach

*See Table 1

Systems-Based Care

There are several unique features of pediatric consultation that distinguish it from working with adults. First is the consideration that children are not autonomous; they are dependent on multiple caregivers and systems of care. The pediatric C-L team itself represents a system and includes a wide range of providers. Based on a 2016 survey by Shaw et al., nearly all teams have an attending psychiatrist and child psychiatry fellow (Shaw et al., 2016). Approximately half have a child

Table 1 Developmental stages (Campbell et al., 2017; Crain, 2011; Gaillard et al., 2018; Pao & Raza, 2006)

Age and developmental stage	Social/emotional development	Cognitive development	Clinical application
0–2 years Sensorimotor	Development of attachment, awareness and recognition of others	Differentiation between self and other occurs, resulting in object permanence	Medical trauma or caregiver separation during this time may increase risk for attachment difficulties and failure to thrive
2–7 years Pre-operational	Improved ability to separate from caregivers Ego-centric view of the world Developing a sense of guilt and responsibility	Child can use internal symbols for representation in their mind, but thought process is illogical	"Magical thinking" can occur, leading a child to have false beliefs about medical interventions which may worsen anxiety May inappropriately attribute blame to themselves for medical condition and/or view illness or treatment as punishment There may be parental overprotection and initiative may be discouraged There may be regression in previously acquired skills for feeding and elimination
7–11 years Concrete operations	Better to separate from caregivers Ego-centrism resolves Developing sense of competency	Child develops systemic thinking, but is unable to think abstractly	Increased ability to understand illness and treatment in rudimentary presentation Ability to accomplish goals may be impaired due to medical illness and stagnate development Highest risk for regression compared to other stages
12–18 years Adolescence	Individuation begins Sense of identity is developing Focus on growth, appearance, and relationships	Formal operations, increased focus on morals, ethics, and the good of others	Begin to consider treatment adherence as a component to care Careful evaluation of cognitive capacities can determine how to present information around illness and implications Medical illness and their consequences may be integrated into part of the child's identity Gentle, structured transitioning toward adult care

psychologist, and a quarter have social work support on service. Other team members are often present, including advanced practice nursing, general psychiatry residents, and psychology interns (Shaw et al., 2016).

In the hospital setting, other other providers include the primary medical team, and clinical supports such as child life specialists, complementary therapies, social work, child protection teams, nursing, and sub-specialty services. As children usually do not request their own psychiatric consultation, obtaining collateral and coordinating care from all involved parties enhances the quality of

history-taking, assessment, and treatment planning. Children and families may also develop rapport with various providers, making them an integral part of the evaluation (Guerrero et al., 2018).

Beyond the acute nature of the child's hospitalization, consideration of the child's place within the family system and developmental history are key aspects of the assessment. Illness of a child may often result in strain on the members of their family. In cases of pediatric cancer, children and parents experience increased distress, lower quality of life, and reduced psychosocial function immediately and in the months after receiving a diagnosis. Involving the family together in consultation and interviewing individual family members as indicated is a pivotal aspect of rapport building, assessment, and treatment (Kazak et al., 2015).

Developmental Considerations

Of similar importance is taking a developmentally informed approach. Children are constantly growing and developing in a process that relies on the success of earlier developmental stages. Acknowledgment and recognition of developmental stagnation may assist in predicting the onset of regressive behaviors under stress associated with hospitalization (Crain, 2011). Thus, knowledge of a child's developmental stage and history at the time of evaluation is critical, as it provides a blueprint for providers and families on engaging with the child and crafting appropriate expectations of the child's reaction to medical care. Consultation for assistance in navigating complex family dynamics, improving communication between the family and medical providers, and conducting a developmental assessment has the opportunity to improve a child's medical admission and outcomes (Guerrero et al., 2018).

Children adapt to illness based on their cognitive, social, and emotional developmental stages (Pao & Raza, 2006). These changes are summarized in Table 1. Children ages 0–2 years progress through development as they interact with the world around them, attach to caregivers, and differentiate between themselves and others (Crain, 2011). Thus, medical trauma or caregiver separation during hospitalization may result in attachment difficulties, stranger anxiety, and failure to thrive (Gaillard et al., 2018). Children 2–7 years of age typically view the world in an egocentric manner. They develop the ability to cognitively process the world around them through the use of internal symbols and are beginning to develop a sense of responsibility and guilt (Crain, 2011). Thus, illness during this stage may result in egocentric and catastrophic fantasies about benign medical procedures, which may increase anxiety.

Initiative may be discouraged with parental overprotection, and there may be a regression in previously acquired skills. Family involvement, developmentally guided patient education, and allowing pre-procedural interaction with medical equipment with the help of a child life specialist can be effective in managing challenges (Gaillard et al., 2018).

Compared to younger children, those 7–11 years of age have an increased ability to understand medical interventions. Cognitively, egocentrism resolves at this stage; the child's thought process becomes systemic, yet remains concrete as they are unable to engage in cognitive abstraction (Crain, 2011). Therefore, while they are usually able to understand cause and effect and no longer wrongly attribute responsibility for their medical condition to themselves, the concrete nature of their understanding increases the risk of regression. Children of this age may also be unable to accomplish developmentally appropriate goals due to their medical condition, resulting in poor self-esteem and stagnant development. When considering intervention, education through storytelling can be a helpful therapeutic approach. When children are able to form a narrative about their medical interventions and discuss with adults, their anxiety is often reduced. Additionally, if a child draws pictures or creates a book for themselves, this can be accessed in the future if anxiety returns (Gaillard et al., 2018).

In adolescence, the overarching goal of development is separation and individuation. This is accompanied by an increased focus on the child's peer group rather than their family (Crain, 2011). Medical illness may disrupt this process as it often results in increased reliance on parents and concerns that one may be different from their peer group. Adolescents are also undergoing identity and body image development. Thus, bodily injury can be exceptionally distressing during this time compared to other developmental stages, and adolescents may grapple with a sick role identity. Given rapid advancement of intellectual abilities in this stage, careful examination of cognitive capacities is key in delivery of patient education, addressing treatment adherence, and setting expectations for the patient, family, and treatment team (Gaillard et al., 2018). Overall, the most frequently used intervention for helping children and adolescents to cope with medical illness is cognitive behavioral therapy (CBT) and extensions thereof (Eccleston et al., 2015). Specific coping strategies that have been shown to be helpful in coping with illness also include journaling, diaphragmatic breathing, and art, music, or play therapies (Coakley & Wihak, 2017).

The transition to adulthood can be particularly challenging for youth with chronic illness. In addition to the developmental challenges of separation and individuation, the young adult must transition from the traditionally gentle and supportive pediatric care setting to the adult care setting, which can be perceived as less personalized or warm and which may have fewer wraparound supports. This shift can be associated with patients lost to follow-up, treatment non-adherence, and poor medical outcomes (White et al., 2018). To aid with this potentially jarring transition and assist in the development of autonomy in medical decision-making, the American Academy of Pediatrics (AAP) recommends thoughtful, structured transitions from pediatric to adult care, starting with initial discussions with children and families at age 12 and continuing with ultimate transition of providers through young adulthood; the AAP recommends that a similar process be undertaken even if the youth is staying with the same provider, such as a family medicine or medicine/pediatrics-trained provider (White et al., 2018). A consulting child and adolescent psychiatrist can assist patients and families with planning for this transition, help to support patients in developing independence as they take on incremental responsibility over their medical care, and provide specific additional support for youth with comorbid psychiatric illness, who may be at particular risk of decompensation during the transition period (White et al., 2018).

Reasons for Consultation

In recent years, requests for inpatient pediatric C-L consultations have been rapidly growing (Shaw et al., 2016). The 2016 Shaw and colleagues survey

highlighted common reasons for inpatient pediatric C-L consultation, including safety assessments, medically unexplained symptoms, adjustment to illness, medication assessments, delirium, treatment non-adherence, and medical boarding (Shaw et al., 2016). In this section, these and other common reasons for pediatric psychiatry consultation are discussed.

Common Reasons for Inpatient Pediatric Psychiatry Consultations
- Suicide risk assessment
- Agitation
- Autism spectrum disorder
- Mental status changes in medically ill children
  - Delirium
  - Catatonia
  - New episode psychosis
  - Autoimmune and other emerging presentations
- Adjustment to illness
- Treatment non-adherence
- Trauma
- Medically unexplained symptoms
- Eating disorders
- Substance use
- Boarding

Suicide Risk Assessment

Per recent literature in the field, the most frequent reason for psychiatric consultation to pediatrics is suicide risk assessment (Shaw et al., 2016). This report aligns with other research, as risk of suicide is elevated in persons with severe medical illness (Substance Abuse and Mental Health Services Administration, 2017) and psychopathology is present in 90% of suicide decedents ("Sentinel Event Alert Issue 56: Detecting and treating suicide ideation in all settings," 2016). Additionally, suicide risk is increased in the 72 h after an inpatient or emergency department evaluation. In a 2016 report, The Joint Commission (TJC) identified sub-par suicide risk assessment as the most common root cause; TJC now recommends screening all patients for suicidal ideation with a brief, standardized, and evidence-based screening tool ("Sentinel Event Alert Issue 56: Detecting and treating suicide ideation in all settings," 2016). Therefore, for the pediatrician and pediatric C-L psychiatrist, knowledge and utilization of such tools is an essential aspect of clinical practice. Examples include the Ask Suicide-Screening Questions (ASQ) toolkit (Horowitz et al., 2012), Patient Health Questionnaire-9 (PHQ-9) (Rossom et al., 2017), Columbia-Suicide Severity Rating Scale (C-SSRS) (Posner et al., 2011), and Suicide Behavior Questionnaire-Revised (SBQ-R) (Chitavi et al., 2018), all of which have evidence supporting their use in the pediatric patients (Thom et al., 2019a). Using these tools along with current data in the field, Brahmbhatt and colleagues developed streamlined clinical pathways for pediatric hospitals to implement efficient and evidence-based youth suicide assessment strategies in hospitalized patients (Brahmbhatt et al., 2019). To provide comprehensive consultation to institutions and treatment teams, pediatric C-L psychiatrists should be familiar with both the clinical utility of these recommendations and possible increases in staff burden, time spent in screening, and documentation requirements. If a patient screens positive and is determined to be at an elevated risk for self-harm and suicide, interventions include multidisciplinary safety planning on the medical unit, constant observation, admitting the patient to a room designed to maintain safety, and restricting internet and telephone access to minimize the impact of external psychosocial stressors.

Evidence-Based Screening Tools for Suicidal Ideation in Pediatric Patients
- Ask Suicide-Screening Questions (ASQ) toolkit (Horowitz et al., 2012)
- Patient Health Questionnaire-9 (PHQ-9) (Rossom et al., 2017)
- Columbia-Suicide Severity Rating Scale (C-SSRS) (Posner et al., 2011)
- Suicide Behavior Questionnaire-Revised (SBQ-R) (Chitavi et al., 2018)

Agitation

Pediatric C-L psychiatrists are commonly called upon to assist with managing agitation in the medical setting. Agitation can occur for a variety of reasons, whether related to a primary psychiatric condition, pain or discomfort, altered mental status, or a medical or neurological condition (Dalmau & Graus, 2018; Gerson et al., 2019; Schieveld & Janssen, 2014; Sorg et al., 2018). An estimated 6–10% of children presenting to the emergency department require restraint, which should be used only when less restrictive options have failed, given their potential for physical and emotional harm to a child (Gerson et al., 2019). To address this issue, the American Association for Emergency Psychiatry and the American Academy of Child and Adolescent Psychiatry (AACAP) Emergency Child Psychiatry Committee recently released consensus guidelines for management of agitation in pediatric patients in the emergency department setting, which can be extrapolated to the general pediatric setting (Gerson et al., 2019). These guidelines highlight the importance of understanding the etiology of agitation, as etiology guides treatment; utilizing non-pharmacologic approaches to de-escalation, including addressing physical and environmental needs and using clear communication; and judiciously using medication intervention in concert with non-pharmacologic interventions to minimize risk of harm (Gerson et al., 2019). While there are few studies evaluating pharmacologic management of agitation in youth, these guidelines provide consensus medication recommendations (Gerson et al., 2019). Special considerations should also be made for children with certain disorders, such as autism spectrum disorder, who may be particularly prone to developing agitation in the medical setting (R. P. Thom et al., 2019b), as well as for agitation in the context of altered mental status, where the etiology may prohibit the use of certain medications, as discussed more below.

Autism Spectrum Disorder

The prevalence of autism spectrum disorder (ASD) has been rising for decades (Kogan et al.,

2018) and is associated with higher rates of medical illness when compared to the general population (Gurney et al., 2006). Thus, children with ASD are more frequently evaluated in the ED (Liu et al., 2017) or medical inpatient setting (Croen et al., 2006) and have longer lengths of stay compared to typically developing patients. Due to core features of the disorder, children with ASD and their families often face obstacles when interacting with the medical system and are frequently dissatisfied with care (Liptak et al., 2006; Zablotsky et al., 2014). Specific symptoms and possible difficulties include: impairments in social and emotional communication resulting in misunderstandings between the patient, family, and treatment team; sensory sensitivities, which are often exacerbated by the hospital environment and can make preparation for procedures more challenging; and comfort found in strict adherence to routines, which cannot be followed during a medical admission (Kopecky et al., 2013; R. P. Thom et al., 2019b). Pediatric C-L psychiatrists can assist the treatment team by providing a developmentally informed assessment, identifying historic or active psychiatric symptoms, and evaluating for risk of challenging behaviors including agitation, self-injury, or aggression (R. P. Thom et al., 2019b). Once this information is obtained, plans for communication, sensory-based trigger reduction strategies, self-soothing techniques, and psychopharmacologic approaches can be developed to prevent distress and plan for difficult behaviors before they occur (Broder-Fingert et al., 2016; Lokhandwala et al., 2012; Sakai et al., 2014).

> **Special Considerations for Children with Autism Spectrum Disorder in the Medical Inpatient Setting**
> - Social and emotional communication style
> - Sensory sensitivities and their exacerbation in the stimulating hospital setting
> - Importance of routine
> - History of psychiatric symptoms
> - Risk for agitation
> - Risk of self-injury

> - Techniques for self-soothing
> - Prior response to psychopharmacologic intervention

Mental Status Changes in Medically Ill Children

Myriad systemic medical conditions can cause mental status alteration in children, including the following: epilepsy, endocrinopathies, autoimmune diseases, genetic and metabolic disorders, infection, and nutritional deficiencies (Freudenreich et al., 2007; Giannitelli et al., 2018). Changes in the mental status of children can impact long-term outcomes and increase the risk of post-traumatic stress disorder (PTSD) in affected children and their families (Nelson & Gold, 2012). Careful consideration of etiology, high-risk patient screening, preventative strategies, as well as behavioral and pharmacologic interventions, is of paramount clinical importance for the consulting pediatric psychiatrist. In this section, several common considerations related to altered mental status in children are reviewed.

> **Common Medical Causes of Mental Status Changes in Children**
> - Epilepsy
> - Endocrinopathies
> - Autoimmune diseases
> - Genetic and metabolic disorders
> - Infection
> - Nutritional deficiencies

Delirium

Delirium, defined as an acute neuropsychiatric condition secondary to a general medical illness with cardinal features of fluctuating consciousness and impaired attention (Turkel, 2017), is common but under-recognized in children (Traube et al., 2014). Infants and children with developmental delays are at an increased risk for delirium (Alvarez et al., 2018; Nelson & Gold, 2012; Traube et al., 2017). Multiple screening tools are available for pediatric patients. Commonly used tools include the Pediatric Confusion Assessment Method for the Intensive Care Unit (pCAM-ICU) and the Cornell Assessment for Pediatric Delirium (CAPD/CAPD-R). The CAPD is derived from the Pediatric Anesthesia Emergence Delirium (PAED) scale (Turkel, 2017), contains specific screening questions for hypoactive delirium, and can be used in children of all ages with or without developmental delay (Traube et al., 2014). The pCAM and its preschool version for children under 5 years old (psCAM-ICU) are derived from the adult CAM rating scale and require a Richmond Agitation Sedation Scale (RASS) score above −4 for assessment (Smith et al., 2011; Turkel, 2017).

At present, there are no FDA-approved medications for the treatment of delirium in pediatric or adult patients; available evidence for management is primarily focused on adults. However, evidence-based treatment recommendations can be made from the growing body of literature on pediatric delirium and extrapolation from adult studies. Antipsychotics including risperidone, olanzapine, and quetiapine have been shown to improve agitation, sleep-wake disturbance, and symptom severity with equivalent safety profiles (Carlton et al., 2017; Joyce et al., 2015; Turkel & Hanft, 2014). Very low dosages can be used to treat infants and toddlers, while higher dosages are often needed for adolescents (Turkel, 2017; Turkel et al., 2013). Alpha-2 agonists including

clonidine and dexmedetomidine have been safely administered in the intensive care setting and are associated with a decreased risk of delirium by reducing use of deliriogenic medications, as well as duration and incidence of emergence delirium (Afshari, 2019; Shi et al., 2019; Turkel, 2017; Wang et al., 2017). Due to suppressed melatonin secretion from the pineal gland associated with critical illness, melatonin supplementation is commonly used to assist with sleep cycle regulation in delirious patients (Calandriello et al., 2018) (Table 2).

Catatonia

Catatonia is a dangerous and potentially life-threatening condition that demonstrates significant phenotypic variability. Widely accepted and specific symptoms of catatonia include but are not limited to rigidity, posturing, waxy flexibility, stupor, unresponsiveness, negativism, echopraxia, echolalia, verbigeration, mutism, physical agitation, and repetitive or stereotypic movements (Dhossche & Wachtel, 2010; Sorg et al., 2018). Malignant catatonia may also develop, which is accompanied by life-threatening symptoms of delirium as well as autonomic and thermoregulatory instability (Sorg et al., 2018). In pediatric patients, catatonia is often under-diagnosed and under-treated due to symptom overlap with other pediatric presentations. For example, children with developmental disorders such as autism spectrum disorder are at greater risk for catatonia; however, stereotypic movements and mutism are seen in both catatonia and developmental

Table 2 Scales for assessment of pediatric delirium (Bajwa et al., 2010; Silver et al., 2012; Smith et al., 2011; Traube et al., 2014; Turkel, 2017)

Scale	Ages	Clinically significant scores	Clinical factors of consideration
CAPD	0–21	Greater than or equal to 9	
CAPD-R	0–16		No patient interaction <2 min to administer Specific assessment of hypoactive delirium Can be used in patients with developmental delay
PAED	1–17	>12	Requires no patient interaction Low sensitivity for hypoactive delirium
PCAM-ICU	>5	0–2: no delirium 3–5: mild to moderate delirium 6–7: severe delirium	Patient interaction required <2 min to administer
PsCAM-ICU	<5		

disorders (Dhossche & Wachtel, 2010). Similarly, urinary incontinence and behavioral agitation may present in catatonia as well as a myriad of diagnoses in children. Thus, in the assessment of pediatric catatonia, the etiology of specific symptoms must be carefully delineated (Dhossche & Wachtel, 2010).

Once diagnosed, first-line treatment of catatonia includes intravascular or intramuscular lorazepam, along with the removal and subsequent avoidance of exacerbating medications, including D2-receptor blocking agents. Should symptoms remain refractory to first-line treatment, N-methyl-D-aspartate antagonist medications, such as amantadine or memantine, can be used as augmentation (Sorg et al., 2018). If symptoms persist or the patient develops life-threatening symptoms of malignant catatonia, electroconvulsive therapy (ECT) should be considered; however, the practitioner should be mindful that some states have age restrictions and regulations in place that may limit access to ECT for pediatric patients (Sorg et al., 2018).

New Episode Psychosis

When encountering new onset psychotic symptoms in pediatric patients, a robust medical work up along with careful clinical consideration and history taking is warranted. Numerous medical conditions and drug exposures may result in pediatric psychosis; thus, while there is no standard protocol for medical work up, investigation into possible causes is critical prior to establishing the diagnosis of a primary psychotic disorder (Freudenreich et al., 2007; Stevens et al., 2014; Tyson et al., 2020). Examples of potential studies are broad and include urinalysis, electroencephalogram, brain imaging, including magnetic resonance imaging or computed tomography of the brain, complete metabolic panel, complete blood count, calcium, phosphorus, liver function tests, ceruloplasmin, toxicology studies, vitamin B12, folate, thyroid stimulating hormone, antinuclear antibodies, erythrocyte sedimentation rate, FTA-Abs, human immunodeficiency virus screening, and autoimmune encephalitis studies (Freudenreich et al., 2007; Ross et al., 2020; Stevens et al., 2014). When assessing a patient by

history and clinical interview, one must also consider that young children and some individuals with developmental disorders may report beliefs or perceptions that are not firmly grounded in reality and may resemble psychotic symptoms (Tyson et al., 2020). Additionally, in cases of primary psychotic disorder, a prodromal phase characterized by irregular social and cognitive development is often present before the onset of psychotic symptoms (Ruhrmann et al., 2003). Thus, obtaining a history of a child's behavior prior to the onset of psychosis is critical in the clinical assessment as rapid changes in mental status in the absence of prodromal symptoms may provide additional evidence toward a medical etiology. In cases of secondary psychosis, treatment involves addressing the underlying medical condition and use of antipsychotics for acute symptom alleviation during stabilization (Stevens et al., 2014).

Emerging Areas and Topics of Interest

Autoimmune Encephalitis (AE) is a recently recognized disease entity, treatable by immunotherapy, which can present with delirium, mental status changes, and/or catatonia Symptoms can include irritability, agitation, psychosis, changes in memory and cognition, altered levels of consciousness, abnormal movements, seizures, and dysautonomia. Anti-NMDA receptor (anti-NMDAR) encephalitis is the most commonly recognized of these conditions and primarily presents in children and young adults. While most identified patients with anti-NMDAR encephalitis progress to having autonomic instability and seizures, some patients present with purely psychiatric symptoms (Kayser et al., 2013). Thus, anti-NMDAR encephalitis should be considered in young patients with acute onset of psychiatric symptoms, including new onset of psychosis or sudden and severe behavioral change without a clear stressor (Ross et al., 2020). These patients are also susceptible to neuroleptic malignant syndrome; the consulting pediatric C-L psychiatrist must consider this risk in the treatment of associated psychiatric symptoms (Dalmau & Graus, 2018). Finally, there can be diagnostic and treatment challenges which require collaborative discussions with the treating teams.

Other conditions of emerging consideration are pediatric autoimmune neuropsychiatric disorders associated with *Streptococcus* (PANDAS) and the more general pediatric acute-onset neuropsychiatric syndrome (PANS). These syndromes are thought to present rapidly (24–48 h) in children with a *Streptococcus* infection in cases of PANDAS or any infection in cases of PANS. Specific symptoms include those similar to obsessive compulsive disorder (OCD) with tics and possible cognitive changes, food refusal, frequent urination, sleep disturbance, psychosis, sensory changes, or deterioration of fine motor skills (Murphy et al., 2014). Treatment includes antibiotics, immunotherapy, plasma exchange, anti-inflammatories, and occupational therapy in cases of fine motor sequelae, along with standard treatment for OCD and tics (Chiarello et al., 2017; Murphy et al., 2014).

Adjustment to Illness

Adjustment to illness is one of the most common reasons for pediatric psychiatry consultation (Shaw et al., 2016). Given the potential life-changing nature of some medical diagnoses, the pediatric C-L psychiatrist may be called upon to provide support and assist a child with coping. Indeed, as medicine has advanced and individuals with chronic illness are living longer than before, assistance with coping and adjustment to a potentially lifelong illness is imperative. As discussed above, children benefit most from a family system and developmentally based approach, and coping looks different for children at different ages.

In evaluating a child with chronic illness, it is important to differentiate between normative grief and psychopathology for both the child and family. The time from diagnosis, stage of illness, and context during which an evaluation occurs must be considered. When informed of a medical illness, parental and patient distress is expected. However, monitoring of persistent symptoms in children and parents is important, as prolonged symptoms may increase risk of poor coping by all members of the family. Additionally, the psychological assessment tool (PAT) is a parent-completed screening tool which can be used for screening and triage of patients and families at risk for ongoing symptomology (Gaillard et al., 2018). A child's ability to cope with illness is impacted, both positively and negatively, by family involvement. Conflict, financial strain, and parental distress can interfere with a child's ability to cope (Campbell et al., 2017), though early involvement of parents in treatment planning and modeling of coping behaviors for children can ease a child's adjustment (Coakley & Wihak, 2017).

Despite interventions to assist in the coping process, children with chronic medical conditions are at greater risk for anxiety and depression due to biological mechanisms related to a specific medical illness, response to treatment in a medical hospital, and exposure to pain or medical procedures early in life (Bennet, 1994; Pao & Bosk, 2011; Pinquart & Shen, 2011). Psychiatric disorders that co-occur with medical conditions may result in serious consequences (Pao & Bosk, 2011). Therefore, appropriate long-term treatment, such as outpatient psychotherapy and psychopharmacologic management, should be arranged.

Treatment Non-adherence

In addition to assisting with adjustment to medical diagnosis, pediatric C-L psychiatrists may be asked to assist in the management of patients who have difficulty adhering to their medical treatment. In such cases, the pediatric C-L psychiatrist can assist with a comprehensive psychiatric assessment to determine whether there may be a psychiatric component to the adherence challenges, such as depression or demoralization. Given the pediatric C-L psychiatrist's developmental- and family systems-based approach, the psychiatrist may be able to help illuminate developmental or systems-based challenges interfering with the child's adherence. Family, school, and outpatient medical systems may need to be leveraged for a complete understanding of the child's non-adherence and for devising a plan to improve adherence. Moreover, treatment adherence concerns may lead to questions about the patient's legal capacity to refuse treatment; depending on the patient's age and local regulations regarding the age of consent to medical treatment, a capacity assessment may need to be undertaken. While any physician can complete a capacity assessment, the C-L psychiatrist may be able to offer particular support in this regard to a primary medical team, given experience in this area. Capacity assessment and its interaction with local legal regulations are discussed further in the Legal and Ethical Considerations section.

Trauma

Adverse childhood events (ACEs) or exposure to traumatic events are common in children and families presenting for medical care; presenting an opportunity for screening (Kia-Keating et al., 2019; Marsac et al., 2016). For children with medical illness, medical interventions and procedures are often an inciting trauma. Approximately 25–30% of medically ill children develop sub-syndromal symptoms of PTSD, 10–20% will meet full criteria for the disorder, and their parents are at an increased risk for developing post-traumatic symptoms. PTSD symptoms and ACEs are associated with medical non-adherence, psychiatric co-morbidity, and negative long-term physical and mental health (Forgey & Bursch, 2013; Kia-Keating et al., 2019; Supelana et al., 2016). The majority of patients and families will have resolution of traumatic symptoms without formal intervention (Forgey & Bursch, 2013). However, in cases of persistent symptomology, trauma-informed psychotherapeutic interventions and approaches (Marsac et al., 2016; Supelana et al., 2016) with a focus on psychosocial stressors and social determinants of health can be of clinical benefit (Berry et al., 2020; Hoysted et al., 2018).

Medically Unexplained Symptoms

Another common reason for psychiatric consultation to the inpatient pediatric setting is to assist with symptoms that are considered medically unexplained – inconsistent with understood pathology or etiology (Malas et al., 2017). Two important considerations in this context are somatic symptoms and related disorders and medical child abuse.

Somatization, defined as physical expression of emotional discomfort, is common in children (Ibeziako et al., 2019). It is estimated that up to 50% of pediatric primary care visits have a somatic component (Malas et al., 2017). However, when somatic symptoms become severe enough to impact a child's daily functioning, such as limiting the child's ability to attend school or impairing a child's ability to walk or eat, such symptoms become a disorder. In the *Diagnostic and Statistical Manual of Mental Disorders Fifth Edition* (*DSM-5*), these disorders are categorized into a new heading, Somatic Symptom and Related Disorders (SSRDs), which includes somatic symptom disorder, illness anxiety disorder, functional neurologic symptom disorder (or conversion disorder), and factitious disorder (*American Psychiatric Association: Diagnostic and Statistical Manual of Mental Disorders, Fifth Edition*, 2013). While somatic symptom and functional neurological symptom disorders, which are unconscious in nature, are common in children, factitious disorder and factitious disorder imposed on another (formerly "by proxy") are important considerations as well.

SSRDs require an integrated, multidisciplinary team approach for diagnosis and treatment. As discussed in a helpful pathway of care devised by Ibeziako et al., SSRDs should be an early consideration in the differential diagnosis for medically unexplained symptoms, rather than a diagnosis of exclusion (Ibeziako et al., 2019); however, SSRDs commonly co-occur in children with known medical disorders, which makes a thorough medical work-up crucial. In evaluation, the pediatric C-L psychiatrist has a multifaceted role, including assessing for ongoing psychiatric symptoms, which could contribute to an SSRD presentation; gleaning a history of trauma or trauma-related symptoms, which may be – but frequently are not – present; and attempting to understand the child and family's means of expressing emotion. A child who frequently internalizes or intellectualizes or a family who is more comfortable discussing physical rather than emotional discomfort may point toward consideration of an SSRD diagnosis (Malas et al., 2017).

Once the diagnosis is made, the C-L psychiatrist can assist the primary team with arranging a multidisciplinary family meeting to provide psychoeducation around the diagnosis (Ibeziako et al., 2019). Subsequently, the C-L psychiatrist can play an important role in ensuring that the child has appropriately integrated, multidisciplinary treatment support, including any necessary medical treatment, physical therapy, occupational therapy, behavioral therapy, and psychiatric medication, if indicated. Children with SSRDs and their families can often cause substantial countertransference in medical providers or suffer stigma by medical professionals; the C-L psychiatrist can also serve an important role in liaising with the medical team to provide psychoeducation and support in these areas to improve the experience of patients and providers in managing this complex population. The American Academy of Child and Adolescent Psychiatry, the AboutKidsHealth website developed by the Hospital for Sick Children in Toronto, and the Kelty Mental Health website developed by BC Children's Hospital in Vancouver provide useful resources for families to understand the diagnosis and treatment of SSRDs (Ibeziako et al., 2019; "Kelty Mental Health Resource Centre," n.d.; "Physical Symptoms of Emotional Distress: Somatic Symptoms and Related Disorders," 2020; SickKids staff, 2019).

> **Multidisciplinary Treatment Considerations for Children Diagnosed with Somatic Symptom and Related Disorders**
> - Necessary medical treatment
> - Physical therapy
> - Occupational therapy
> - Behavioral therapy
> - Psychiatric medication, as indicated

Another consideration for children presenting with medically unexplained symptoms is medical child abuse. Medical child abuse, or caregiver-fabricated illness in a child, is an encompassing term for a child receiving unnecessary and possibly harmful medical care due to a caregiver falsifying or inducing physical or psychological symptoms in the child (Flaherty & MacMillan, 2013); this includes but is not limited to factitious disorder imposed on another. As compared to prior terminology, the terms medical child abuse and caregiver-fabricated illness in a child attempt to acknowledge the broad range of possible behaviors and motivations by the perpetrator and to focus more on the direct impact on the child (Flaherty & MacMillan, 2013). When medical child abuse is suspected, the hospital child protection service should be involved, as well as appropriate local authorities. The pediatric C-L psychiatrist can assist with assessing for report of trauma, symptoms of trauma-related disorders, and arranging for ongoing psychiatric or psychological support for the child.

Eating Disorders

Another common reason for pediatric psychiatry consultation is assessment and management of eating disorders, given the complex mix of medical and psychiatric sequelae of these conditions. The incidence of eating disorders in adolescents has been rising in recent years (Bhattacharya et al., 2020). Eating disorders, including anorexia nervosa, bulimia nervosa, binge-eating disorder, and avoidant/restrictive food intake disorder (ARFID), are common among children with serious medical illnesses as well as other psychiatric illnesses, and they are associated with high rates of morbidity and mortality due to medical complications and suicide (Bhattacharya et al., 2020). ARFID was newly defined as a diagnosis in the *DSM-5;* Individuals with ARFID have inadequate energy or nutritional intake, but do not have a distorted perception of their body image or a strong compulsion to lose weight (*American Psychiatric Association: Diagnostic and Statistical Manual of Mental Disorders, Fifth Edition*, 2013).

In children and adolescents, eating disorders often arise at particular developmental transitions, namely, the transition from childhood into adolescence and the transition from adolescence to young adulthood. Eating disorders may stall both the physical and psychological development that normally takes place during these often tumultuous times of transition and can lead to an adolescent's ongoing dependence on family rather than growth toward independence (Dalle Grave et al., 2020). Indeed, common psychological theories about the development of eating disorders emphasize family systems with substantial enmeshment and the need for an adolescent to feel in control when faced with a sense of ineffectiveness (Marks, 2019). Eating disorders may occur in youth who have a family history of eating disorder, which can be an important consideration for family-based treatments, as discussed below.

There are many medical complications of eating disorders, including electrolyte abnormalities, bradycardia, hypotension, liver enzyme elevation, decreased bone density, and delayed gastric emptying; severe complications include arrhythmias and sudden cardiac death (Bhattacharya et al., 2020). Medical admission may be required for youth with eating disorders in the setting of acute electrolyte disturbance or dehydration, substantial weight loss or severely low body mass index, or cardiac complications. While hospitalized, eating disorder patients may require a nutritional deficiency protocol to assist with weight restoration, and phosphorus levels should be closely monitored and repleted to avoid refeeding syndrome. Pediatric C-L psychiatrists can assist the medical team in assessing the underlying cause of the patient's eating changes; collaborating closely with specialists, such as Adolescent Medicine and Nutrition, in ensuring the patient is working toward weight restoration; and working to connect the child with eating disorder and other psychiatric treatments once medically stabilized. Structured, evidence-based therapies for eating disorders, and particularly family-based therapy, are important considerations when linking to outpatient care (Marks, 2019).

> **Potential Medical Complications of Eating Disorders**
> - Electrolyte abnormalities
> - Bradycardia
> - Hypotension
> - Liver enzyme elevation
> - Decreased bone density
> - Delayed gastric emptying
> - Refeeding syndrome
> - Cardiac arrhythmias
> - Sudden cardiac death

Substance Use

Substance use in pediatric patients is increasing in prevalence (Moss et al., 2018) and is associated with an increased risk of chronic disease, earlier age of death (Schulte & Hser, 2013), risk of life long dependence (Hingson et al., 2006), and an acute exacerbation of multiple medical conditions (Barritt et al., 2018; Maung et al., 2018; Myrvik et al., 2012; Zhabenko et al., 2016). In the medical inpatient setting, approximately half of pediatric and young adult patients admitted for a substance-use-related condition will present with co-morbid psychopathology, nearly double the rate seen in adults ("Hospital Stays for Harm Caused by Substance Use Among Youth Age 10 to 24, September 2019," 2019). In addition, approximately half of adolescents will not see their primary care physician in a 12-month period and therefore cannot be screened for substance use or substance use disorders (Rand & Goldstein, 2018). Furthermore, early pharmacological treatment of pediatric substance use is associated with higher retention rates (Hadland, 2019), which may improve long-term outcomes in this patient population. Thus, the C-L psychiatrist is uniquely positioned to intervene in the management and care of this patient population. As a part of the treatment team, the CL psychiatrist may assist in the following ways: screen for substance use and consult on evidence-based screening practices, evaluate patients for co-morbid psychopathology, begin pharmacologic or psychotherapeutic treatment, and refer to outpatient services.

Boarding

Increasingly, pediatric C-L psychiatrists are being asked to assist with caring for children who are awaiting inpatient psychiatric placement.

Boarding is defined by TJC as the temporary holding of a patient in the hospital while awaiting an alternative placement (Gallagher et al., 2017). In the United States, pediatric emergency room visits for psychiatric reasons have dramatically increased in recent years; at the same time, the number of inpatient psychiatric beds for children has declined (Carubia et al., 2016). As a result, increasing numbers of children must wait days or even weeks in the emergency department or on a pediatric medical floor before a psychiatric placement becomes available.

While hospitals have different practices in terms of determining the primary caregiver for boarding patients, the pediatric C-L psychiatrist often plays an integral role in management. The C-L psychiatrist can assist with thorough history-gathering, collateral-gathering, and devising a robust treatment plan. Moreover, while multi-disciplinary resources may be limited in the emergency department or inpatient pediatric settings, the C-L psychiatric team can provide pharmacologic recommendations, agitation recommendations, and brief psychotherapeutic intervention. Indeed, brief psychotherapy may lead to improvements in the patient's status even prior to transferring to the inpatient psychiatric setting (Gallagher et al., 2017).

The emergency department and pediatric inpatient setting also pose particular challenges for children facing psychiatric crises; unlike the inpatient psychiatric unit, these settings may not be locked and may not secure potentially harmful objects. As such, the C-L psychiatrist may need to assist the primary team in ensuring the patient is cared for as safely as possible, such as enlisting an observer to monitor the patient and helping to limit access to potentially dangerous items.

Evolving Systems of Care

Inpatient Utilization and Costs

From a systems perspective, recent studies have identified the complexity of comorbid medical and psychiatric diseases in children. Over a recent 10-year period, medical hospitalizations at tertiary care centers for children with comorbid psychiatric conditions increased five times more than among children without psychiatric conditions (Zima et al., 2016). Medically ill children with comorbid mental health conditions tend to have higher overall healthcare costs, longer hospital lengths of stay, and increased hospital readmission rates when compared to children without mental health conditions (Doupnik et al., 2016, 2018; Dunbar et al., 2019; Perrin et al., 2019). Psychiatric consultation, which can assist in management of these children to improve their trajectory, has promise for reducing overall healthcare utilization and costs for children with comorbid conditions. Indeed, a recent study by Bujoreanu and colleagues demonstrated that earlier pediatric psychiatry consultation for pediatric inpatients was associated with shorter hospital lengths of stay and lower hospital charges (Bujoreanu et al., 2015), while other studies focused on particular medical conditions have demonstrated that psychiatric intervention can improve overall healthcare utilization (Ellis et al., 2005; Keerthy et al., 2016). More work is needed to explore these findings on a broader scale.

Telepsychiatry

In recent years, telepsychiatry has gained popularity, particularly given the flexibility that it allows patients and providers and its promise to extend mental health care to underserved areas, given shortages of child psychiatrists (Sossong et al., 2019). In the setting of the COVID-19 pandemic, telemedicine has become particularly important in pediatric C-L psychiatry, as hospitals have rapidly scaled up remote consultations to inpatient settings in order to minimize traffic in hospital settings (J. E. Becker et al., 2020). Novel remote methods may allow for increased access to pediatric C-L psychiatry in hospitals that do not have on-site pediatric C-L psychiatry consultation. Moreover, technology is playing an increasing role in assisting the longer-term management of psychiatric disease with Internet-based cognitive-behavioral and other skills-based therapy modalities (J. Becker, 2017). One upcoming study plans to examine the use of Internet- and mobile-based cognitive behavioral therapy for youth with chronic medical illness and comorbid anxiety and depression (Lunkenheimer et al., 2020), which has the potential to impact medical and psychiatric outcomes alike. Future work should help to evaluate novel technological interventions to improve the care of children with comorbid medical and psychiatric diseases.

Legal and Ethical Concerns

There are a number of legal and ethical issues frequently encountered by pediatric C-L psychiatrists. Perhaps chief among them are issues related to the pediatric patient's right to make decisions regarding their medical care, particularly when those decisions are not aligned with the recommendations of the medical team. In ethical terms, these situations represent a perceived conflict between the principles of autonomy, or right to make decisions for oneself, and beneficence, or doing what is believed to be in the patient's best interests. Given that pediatric patients can span a broad range of developmental levels, with corresponding variability in their capacity to understand the important aspects of their treatment, including risks and benefits of a proposed intervention, as well as their capacity to reason abstractly in their decision-making, these issues can become quite complicated. An infant clearly is not capable of making medical decisions, but an intelligent 19-year-old who has been living and managing a chronic illness independently generally can and should be able to make his/her own medical decisions. The more challenging cases arise between these two extremes.

The role of the psychiatric consultant in navigating these ethical issues is often to first listen carefully to the patient's perspective, assessing their understanding of the situation. Sometimes, perceived refusals or resistance to care can be addressed by understanding and addressing the patient's concerns, by clarifying misunderstandings, or by performing a liaison role between the patient and the medical team to reach a reasonable compromise. For example, an important but non-urgent medical procedure could be delayed until after an important event in an adolescent's life. When, after thoughtful discussion, a patient continues to refuse treatment, there are important legal and ethical considerations in determining whether a medical intervention should occur against the patient's wishes.

For patients who are considered to be adults, providers must seek the patient's informed consent before proceeding with a medical intervention. For minors, the medical provider seeks the informed consent of the patient's legal guardian, often a parent. While a minor patient, by definition, cannot give informed consent, they should be asked to give their assent for treatment, indicating that they are in agreement (Committee on Bioethics and American Academy of Pediatrics, 1995). If the legal guardian gives informed consent but the pediatric patient does not assent, the medical team and the patient's legal guardian (s) must weigh several important factors in determining whether to proceed, including consideration of the patient's developmental level and understanding of the situation, as well as the risks of withholding the proposed treatment. A life-saving treatment for which the guardian has given consent should move forward even if a developmentally immature minor patient is refusing to assent based on a limited understanding of the illness and its treatment. However, many cases can be more complicated and nuanced than this example, and in these cases, consultation by an institutional ethics committee can be helpful. Even when proceeding with treatment without assent, providers should attempt to keep the patient informed and involved in the treatment as much as is developmentally appropriate (Grootens-Wiegers et al., 2017).

It should be noted that the legal definition of age to consent varies by location. In the United States, this relates to the age of majority, which is 18 years of age. A number of countries note age to consent to treatment as 16 years, including England, Wales, Scotland, Ireland, Australia, and New Zealand. In Canada, the age to consent to treatment varies by province and includes no age in some provinces. The Canadian Pediatric

Society Bioethics Committee 2016 has recommended that children and adolescents be appropriately involved in decisions affecting them according to their stage of development, and once they have sufficient decision-making capacity (the ability to understand and appreciate the risks and benefits of treatment versus no treatment), they should become the principal decision maker (Harrison & Canadian Paediatric Society Bioethics Committee, 2004). International standards including the 1989 UN Convention on the Rights of the Child and the 2006 UN Convention on the Rights of Persons with Disabilities specify ethical principles to include the evolving capacities of the child, the mature minor principle, best interests, concurrent consent, and autonomy of children (United Nations Department of Economic and Social Affairs Disability, 2006; United Nations Human Rights Office of the High Commissioner, 1989; World Medical Association, 1948).

When medical care or interventions are refused by patients who meet the legal definition of majority, e.g., patients over the age of 18 years in the United States who are their own legal guardians, the psychiatric consultant is often asked to perform a capacity assessment. Evaluation of capacity is discussed in more detail elsewhere in this textbook. Briefly, determination of capacity is specific to a particular time point and a particular decision and can be made by any physician. The elements of capacity include an understanding of the relevant information related to a choice, an appreciation of the significance of the choice being made and its implications, the ability to use reason in weighing treatment options, and the ability to express a clear choice (Appelbaum, 2007). Standardized assessment tools can assist the clinician in capacity assessment (Barstow et al., 2018). A patient's capacity to make a medical decision can change over time; for example, when a patient's delirium clears, he may regain capacity. When a patient is deemed not to have the capacity to make a particular decision, a substitute decision-maker must be sought. This is usually a family member, though in emergency situations a physician can make a decision based upon what the physician believes a reasonable person would choose in the situation.

Conclusion

Pediatric consultation-liaison psychiatry is a rapidly growing field encompassing a diverse array of conditions. Pediatric C-L psychiatrists must be well-equipped to assist in the diagnosis and management of complex disease presentations, and they play an important role in supporting children, families, and medical teams. The pediatric C-L psychiatric approach includes particular attention to developmental and systems-related forces that may impact a child's medical and psychiatric presentation; these principles help guide the pediatric C-L psychiatrist in managing a variety of illnesses. As children with medical complexity live longer lives and pediatric psychiatry consultation requests continue to grow, increased resources will need to be made available to support this important work. Novel approaches to C-L psychiatric care, including care pathways, integrated care, and telemedicine, may help to address this burden and support this field aiming to improve pediatric medical and psychiatric wellbeing alike.

References

Afshari, A. (2019). Clonidine in pediatric anesthesia: The new panacea or a drug still looking for an indication? *Current Opinion in Anaesthesiology, 32*(3), 327–333. https://doi.org/10.1097/ACO.0000000000000724

Alvarez, R. V., Palmer, C., Czaja, A. S., Peyton, C., Silver, G., Traube, C., ... Kaufman, J. (2018). Delirium is a common and early finding in patients in the pediatric cardiac intensive care unit. *The Journal of Pediatrics, 195*, 206–212. https://doi.org/10.1016/j.jpeds.2017.11.064

American Psychiatric Association: Diagnostic and Statistical Manual of Mental Disorders, Fifth Edition. (2013). American Psychiatric Association.

Appelbaum, P. (2007). Assessment of patients' competence to consent to treatment. *The New England Journal of Medicine, 357*(18), 1834–1840. https://doi.org/10.1056/NEJMcp074045. PMID: 17978292.

Asarnow, J. R., Rozenman, M., Wiblin, J., & Zeltzer, L. (2015). Integrated medical-behavioral care compared with usual primary care for child and adolescent behavioral health: A meta-analysis. *JAMA Pediatrics, 169*(10), 929–937.

Bajwa, S., Costi, D., & Cyna, A. (2010). A comparison of emergence delirium scales following general anesthesia in children. *Pediatric Anesthesia, 20*(8), 704–711.

Barritt, A. S., Lee, B., Runge, T., Schmidt, M., & Jhaveri, R. (2018). Increasing prevalence of hepatitis C among hospitalized children is associated with an increase in substance abuse. *The Journal of Pediatrics, 192*, 159–164. https://doi.org/10.1016/j.jpeds.2017.09.016

Barstow, C., Shahn, B., & Roberts, M. (2018). Evaluating medical decision-making capacity in practice. *American Family Physician, 98*(1), 40–46.

Becker, J. (2017). Telemental health modalities: Videoconferencing, store-and-forward, web-based, and mHealth. In H. Jefee-Bahloul, A. Barkil-Oteo, & E. F. Augusterfer (Eds.), *Telemental health in resource-limited global settings*. Oxford University Press.

Becker, J. E., Smith, J. R., & Hazen, E. P. (2020). Pediatric consultation-liaison psychiatry: An update and review. *Psychosomatics, 61*(5), 467–480.

Bennet, D. S. (1994). Depression among children with chronic medical problems: A meta-analysis. *Journal of Pediatric Psychology, 19*(2), 149–169.

Berry, J. G., Harris, D., Coller, R. J., Chung, P. J., Rodean, J., Macy, M., & Linares, D. E. (2020). The interwoven nature of medical and social complexity in US children. *JAMA Pediatrics*, e200280. https://doi.org/10.1001/jamapediatrics.2020.0280

Bhattacharya, A., Defilipp, L., & Timko, C. A. (2020). Feeding and eating disorders. In R. Lanzenberger, G. Kranz, & I. Savic (Eds.), *Handbook of clinical neurology, Vol. 175 (3rd series) Sex differences in neurology and psychiatry* (pp. 387–403). Elsevier B.V). https://doi.org/10.1016/B978-0-444-64123-6.00026-6

Brahmbhatt, K., Kurtz, B. P., Afzal, K. I., Giles, L. L., Kowal, E. D., Johnson, K. P., . . . Horowitz, L. M. (2019). Suicide risk screening in pediatric hospitals: Clinical pathways to address a global health crisis. *Psychosomatics, 60*(1), 1–9. https://doi.org/10.1016/j.psym.2018.09.003

Broder-Fingert, S., Shui, A., Ferrone, C., Iannuzzi, D., Cheng, E. R., Giauque, A., . . . Kuhlthau, K. (2016). A pilot study of autism-specific care plans during hospital admission. *Pediatrics*. https://doi.org/10.1542/peds.2015-2851R

Bujoreanu, S., White, M. T., Gerber, B., & Ibeziako, P. (2015). Effect of timing of psychiatry consultation on length of pediatric hospitalization and hospital charges. *Hospital Pediatrics, 5*(5), 269–275. https://doi.org/10.1542/hpeds.2014-0079

Calandriello, A., Tylka, J., & Patwari, P. (2018). Sleep and delirium in pediatric critical illness: What is the relationship? *Medical Science, 6*(4), 90. https://doi.org/10.3390/medsci6040090

Campbell, L., DiLorenzo, M., Atkinson, N., & Riddell, R. (2017). Systematic review: A systematic review of the interrelationships among children's coping responses, children's coping outcomes, and parent

cognitive-affective, behavioral, and contextual variables in the needle-related procedures context. *Journal of Pediatric Psychology, 42*(6), 611–621.

Carlton, E. F., Mahowald, M. K., & Malas, N. (2017). Management of multifactorial infant delirium with intravenous haloperidol in the setting of over sedation and poor enteral absorption. *Journal of Child and Adolescent Psychopharmacology, 27*(3), 289–290. https://doi.org/10.1089/cap.2016.0103

Carubia, B., Becker, A., & Levine, B. H. (2016). Child psychiatric emergencies: Updates on trends, clinical care, and practice challenges. *Current Psychiatry Reports, 18*(41). https://doi.org/10.1007/s11920-016-0670-9

Chiarello, F., Spitoni, S., Hollander, E., Matucci Cerinic, M., & Pallanti, S. (2017). An expert opinion on PANDAS/PANS: Highlights and controversies. *International Journal of Psychiatry in Clinical Practice, 21*(2), 91–98.

Chitavi, S., Paul, S., Wells, E., Bronk, K., Castro, G., Glossa, D., . . . Baker, D. (2018). Suicide Prevention Resources to support Joint Commission Accredited organizations implementation of NPSG 15.01.01. Retrieved from https://www.jointcommission.org/-/media/tjc/documents/resources/patient-safety-topics/suicide-prevention/suicide_prevention_resources_to_support_npsg150101_nov201821.pdf

Coakley, R., & Wihak, T. (2017). Evidence-based psychological interventions for the management of pediatric chronic pain: New directions in research and clinical practice. *Children, 4*(2), 9. https://doi.org/10.3390/children4020009

Committee on Bioethics, & American Academy of Pediatrics. (1995). Informed consent, parental permission, and assent in pediatric practice. *Pediatrics, 95*(2), 314–317.

Crain, W. (2011). Cognitive-developmental theory. In *Theories of development: Concepts and applications* (6th ed., pp. 117–156). Psychology Press.

Croen, L. A., Najjar, D. V., Ray, G. T., Lotspeich, L., & Bernal, P. (2006). A comparison of health care utilization and costs of children with and without autism spectrum disorders in a large group-model health plan. *Pediatrics*. https://doi.org/10.1542/peds.2006-0127

Dalle Grave, R., Sartirana, M., Sermattei, S., & Calugi, S. (2020). Treatment of eating disorders in adults versus adolescents: Similarities and differences. *Clinical Therapeutics, 43*(1), 70–84.

Dalmau, J., & Graus, F. (2018). Antibody-mediated encephalitis. *New England Journal of Medicine, 378*, 840–851. https://doi.org/10.1056/NEJMra1708712

Dhossche, D. M., & Wachtel, L. E. (2010). Catatonia is hidden in plain sight among different pediatric disorders: A review article. *Pediatric Neurology, 43*(5), 307–315. https://doi.org/10.1016/j.pediatrneurol.2010.07.001

Doupnik, S. K., Lawlor, J., Zima, B. T., Coker, T. R., Bardach, N. S., Hall, M., & Berry, J. G. (2016). Mental health conditions and medical and surgical hospital utilization. *Pediatrics, 138*(6), e20162416–e20162416. https://doi.org/10.1542/peds.2016-2416

Doupnik, S. K., Lawlor, J., Zima, B. T., Coker, T. R., Bardach, N. S., Rehm, K. P., … Berry, J. G. (2018). Mental health conditions and unplanned hospital readmissions in children. *Journal of Hospital Medicine, 13*(7), 445–452. https://doi.org/10.12788/jhm.2910.

Dunbar, P., Hall, M., Gay, J. C., Hoover, C., Markham, J. L., Bettenhausen, J. L., … Berry, J. G. (2019). Hospital readmission of adolescents and young adults with complex chronic disease. *JAMA Network Open, 2*(7), e197613. https://doi.org/10.1001/jamanetworkopen.2019.7613

Eccleston, C., Fisher, E., Law, E., Bartlett, J., & Palermo, T. (2015). Psychological interventions for parents of children and adolescents with chronic illness. *Cochrane Database of Systematic Reviews.* https://doi.org/10.1002/14651858.CD009660.pub3

Ellis, D. A., Frey, M. A., Naar-King, S., Templin, T., Cunningham, P., & Cakan, N. (2005). Use of multisystemic therapy to improve regimen adherence among adolescents with type 1 diabetes in chronic poor metabolic control: A randomized controlled trial. *Diabetes Care, 28*(7), 1604–1610.

Flaherty, E. G., & MacMillan, H. L. (2013). Caregiver-fabricated illness in a child: A manifestation of child maltreatment. *Pediatrics, 132*(3), 590–597. https://doi.org/10.1542/peds.2013-2045

Forgey, M., & Bursch, B. (2013). Assessment and management of pediatric iatrogenic medical trauma. *Current Psychiatry Reports, 15*, 340. https://doi.org/10.1007/s11920-012-0340-5

Freudenreich, O., Holt, D. J., Cather, C., & Goff, D. C. (2007). The evaluation and management of patients with first-episode schizophrenia: A selective, clinical review of diagnosis, treatment, and prognosis. *Harvard Review of Psychiatry, 15*(5), 189–211. https://doi.org/10.1080/10673220701679804

Gaillard, S., Tamaye, H., Wong, J., & Hirsch, W. (2018). Adjustment to medical illness. In R. Koli, A. Guerrero, P. Lee, & N. Skokauskas (Eds.), *Pediatric consultation-liaison psychiatry: A global, healthcare systems-focused, and problem-based approach.* Springer International Publishing.

Gallagher, K. A. S., Bujoreanu, I. S., Cheung, P., Choi, C., Golden, S., Brodziak, K., … Ibeziako, P. (2017). Psychiatric boarding in the pediatric inpatient medical setting: A retrospective analysis. *Hospital Pediatrics, 7*(8), 444–450.

Gerson, R., Malas, N., Feuer, V., Silver, G. H., Prasad, R., & Mroczkowski, M. M. (2019). Best practices for evaluation and treatment of agitated children and adolescents (BETA) in the emergency department: Consensus statement of the American Association for Emergency Psychiatry. *Western Journal of Emergency Medicine, 20*(2), 409–418. https://doi.org/10.5811/westjem.2019.1.41344

Giannitelli, M., Consoli, A., Raffin, M., Jardri, R., Levinson, D. F., Cohen, D., & Laurent-Levinson, C. (2018). An overview of medical risk factors for childhood psychosis: Implications for research and treatment. *Schizophrenia Research, 192*, 39–49. https://doi.org/10.1016/j.schres.2017.05.011

Grootens-Wiegers, P., Hein, I., van den Broek, J., & de Vries, M. (2017). Medical decision-making in children and adolescents: Developmental and neuroscientific aspects. *BMC Pediatrics, 17*(1), 120. https://doi.org/10.1186/s12887-017-0869-x

Guerrero, A., Skokauskas, N., Lee, P., Fishman, H., Bell, C., & Keifer, J. (2018). General principles of pediatric consultation-liaison psychiatry. In A. P. S. Guerrero, P. C. Lee, & N. Skokauskas (Eds.), *Pediatric consultation-liaison psychiatry: A global, healthcare systems-focused, and problem-based approach* (pp. 5–20). Springer International Publishing. https://doi.org/10.1007/978-3-319-89488-1

Gurney, J. G., McPheeters, M. L., & Davis, M. M. (2006). Parental report of health conditions and health care use among children with and without autism: National survey of children's health. *Archives of Pediatrics and Adolescent Medicine.* https://doi.org/10.1001/archpedi.160.8.825

Hadland, S. (2019). How clinicians caring for youth can address the opioid-related overdose crisis. *The Journal of Adolescent Health, 65*(2), 177–180. https://doi.org/10.1016/j.jadohealth.2019.05.008

Harrison, C., & Canadian Paediatric Society Bioethics Committee. (2004). Treatment decisions regarding infants, children and adolescents. *Paediatrics & Child Health, 9*(2), 99–103.

Hingson, R. W., Heeren, T., & Winter, M. R. (2006). Age at drinking onset and alcohol dependence: Age at onset, duration, and severity. *Archives of Pediatrics & Adolescent Medicine, 160*(7), 739–746. https://doi.org/10.1001/archpedi.160.7.739

Horowitz, L. M., Bridge, J. A., Teach, S. J., Ballard, E., Klima, J., Rosenstein, D. L., … Pao, M. (2012). Ask suicide-screening questions (ASQ): A brief instrument for the pediatric emergency department. *Archives of Pediatrics & Adolescent Medicine, 166*(12), 1170. https://doi.org/10.1001/archpediatrics.2012.1276

Hospital Stays for Harm Caused by Substance Use Among Youth Age 10 to 24, September 2019. (2019). Canadian Institute for Health Information.

Hoysted, C., Babl, F., Kassam-Adams, N., Landolt, M., Jobson, L., Van Der Westhuizen, C., … Alisic, E. (2018). Knowledge and training in paediatric medical traumatic stress and trauma-informed care among emergency medical professionals in low- and middle-income countries. *European Journal of Psychotraumatology, 9*(1), 1468703. https://doi.org/10.1080/20008198.2018.1468703

Ibeziako, P., Brahmbhatt, K., Chapman, A., De Souza, C., Giles, L., Gooden, S., … Plioplys, S. (2019). Developing a clinical pathway for somatic symptom and related disorders in pediatric hospital settings. *Hospital Pediatrics, 9*(3), 147–155. https://doi.org/10.1542/hpeds.2018-0205

Joyce, C., Witcher, R., Herrup, E., Kaur, S., Mendez-Rico, E., Silver, G., et al. (2015). Evaluation of the safety of quetiapine in treating delirium in critically ill children: A retrospective review. *Journal of Child and Adolescent Psychopharmacology, 25*(9), 666–670. https://doi.org/10.1089/cap.2015.0093

Kayser, M. S., Titulaer, M. J., Gresa-Arribas, N., & Dalmau, J. (2013). Frequency and characteristics of isolated psychiatric episodes in anti-N-mehtyl-D-aspartate receptor encephalitis. *JAMA Neurology, 70*(9), 1133–1139.

Kazak, A. E., Abrams, A. N., Banks, J., Christofferson, J., DiDonato, S., Grootenhuis, M. A., . . . Kupst, M. J. (2015). Psychosocial assessment as a standard of care in pediatric cancer. *Pediatric Blood & Cancer, 62*(S5), S426–S459. https://doi.org/10.1002/pbc.25730

Keerthy, D., Youk, A., Srinath, A. I., Malas, N., Bujoreanu, S., Bousvaros, A., . . . Szigethy, E. M. (2016). Effect of psychotherapy on healthcare utilization in children with inflammatory bowel disease and depression. *Journal of Pediatric Gastroenterology and Nutrition, 63*(6), 658–664. https://doi.org/10.1097/MPG.0000000000001207. Corresponding

Kelty Mental Health Resource Centre. (n.d.). Retrieved December 15, 2020, from https://keltymentalhealth.ca/

Kia-Keating, M., Barnett, M., Liu, S., & Ruth, A. (2019). Trauma-responsive care in a pediatric setting: Feasibility and acceptability of screening for adverse childhood experiences. *American Journal of Community Psychology, 64*(3–4), 286–297. https://doi.org/10.1002/ajcp.12366

Kogan, M. D., Vladutiu, C. J., Schieve, L. A., Ghandour, R. M., Blumberg, S. J., Zablotsky, B., . . . Lu, M. C. (2018). The prevalence of parent-reported autism spectrum disorder among US children. *Pediatrics*. https://doi.org/10.1542/peds.2017-4161

Kopecky, K., Broder-Fingert, S., Iannuzzi, D., & Connors, S. (2013). The needs of hospitalized patients with autism spectrum disorders: A parent survey. *Clinical Pediatrics*. https://doi.org/10.1177/0009922813485974

Liptak, G. S., Orlando, M., Yingling, J. T., Theurer-Kaufman, K. L., Malay, D. P., Tompkins, L. A., & Flynn, J. R. (2006). Satisfaction with primary health care received by families of children with developmental disabilities. *Journal of Pediatric Health Care*. https://doi.org/10.1016/j.pedhc.2005.12.008

Liu, G., Pearl, A. M., Kong, L., Leslie, D. L., & Murray, M. J. (2017). A profile on emergency department utilization in adolescents and young adults with autism spectrum disorders. *Journal of Autism and Developmental Disorders*. https://doi.org/10.1007/s10803-016-2953-8

Lokhandwala, T., Khanna, R., & West-Strum, D. (2012). Hospitalization burden among individuals with autism. *Journal of Autism and Developmental Disorders*. https://doi.org/10.1007/s10803-011-1217-x

Lunkenheimer, F., Domhardt, M., Geirhos, A., Kilian, R., Mueller-Stierlin, A. S., Holl, R. W., . . . COACH consortium. (2020). Effectiveness and cost-effectiveness of guided internet- and mobile-based CBT for adolescents and young adults with chronic somatic conditions and comorbid depression and anxiety symptoms (youthCOACH CD): Study protocol for a multicentre randomized contro. *Trials, 21*(253). https://doi.org/10.1186/s13063-019-4041-9

Malas, N., Ortiz-Aguayo, R., Giles, L., & Ibeziako, P. (2017). Pediatric somatic symptom disorders. *Current Psychiatry Reports, 19*(11). https://doi.org/10.1007/s11920-017-0760-3

Marks, A. (2019). The evolution of our understanding and treatment of eating disorders over the past 50 years. *Journal of Clinical Psychology, 75*(8), 1380–1391. https://doi.org/10.1002/jclp.22782

Marsac, M., Kassam-Adams, N., Hildenbrand, A., Nicholls, E., Winston, F., Leff, S., & Fein, J. (2016). Implementing a trauma-informed approach in pediatric health care networks. *JAMA Pediatrics, 170*(1), 70–77. https://doi.org/10.1001/jamapediatrics.2015.2206

Maung, A. A., Becher, R. D., Schuster, K. M., & Davis, K. A. (2018). When should screening of pediatric trauma patients for adult behaviors start? *Trauma Surgery & Acute Care Open, 3*(1), e000181. https://doi.org/10.1136/tsaco-2018-000181

Merikangas, K. R., He, J., Burstein, M., Swendsen, J., Avenevoli, S., Case, B., . . . Olfson, M. (2011). Service utilization for lifetime mental disorders in U.S. adolescents: Results of the National Comorbidity Survey-Adolescent supplement (NCS-A). *Journal of the American Academy of Child and Adolescent Psychiatry, 50*(1), 32–45. https://doi.org/10.1016/j.jaac.2010.10.006

Moss, J., Liu, B., & Ziu, L. (2018). State prevalence and ranks of adolescent substance use: Implications for cancer prevention. *Preventing Chronic Disease, 15*(170345). https://doi.org/10.5888/pcd15.170345

Murphy, T. K., Gerardi, D. M., & Leckman, J. F. (2014). Pediatric acute-onset neuropsychiatric syndrome. *Psychiatric Clinics of North America, 37*, 353–374. https://doi.org/10.1016/j.psc.2014.06.001

Myrvik, M. P., Campbell, A. D., Davis, M. M., & Butcher, J. L. (2012). Impact of psychiatric diagnoses on hospital length of stay in children with sickle cell anemia. *Pediatric Blood & Cancer, 58*(2), 239–243. https://doi.org/10.1002/pbc.23117

Nelson, L. P., & Gold, J. I. (2012). Posttraumatic stress disorder in children and their parents following admission to the pediatric intensive care unit: A review. *Pediatric Critical Care Medicine, 13*(3), 338–347. https://doi.org/10.1097/PCC.0b013e3182196a8f

Pao, M., & Bosk, A. (2011). Anxiety in medically ill children/adolescents. *Depression and Anxiety, 28*(1), 40–49.

Pao, M., & Raza, H. (2006). Essential issues in Pediatric psychosomatic medicine. *Psychiatric Times, 23*(6). Retrieved from https://www.psychiatrictimes.com/

view/essential-issues-pediatric-psychosomatic-medicine

Perrin, J. M., Asarnow, J. R., Stancin, T., Melek, S. P., & Fritz, G. K. (2019). Mental health conditions and health care payments for children with chronic medical conditions. *Academic Pediatrics, 19*(1), 44–50. https://doi.org/10.1016/j.acap.2018.10.001

Physical Symptoms of Emotional Distress: Somatic Symptoms and Related Disorders. (2020). Retrieved from https://www.aacap.org/AACAP/Families_and_Youth/Facts_for_Families/FFF-Guide/Physical_Symptoms_of_Emotional_Distress-Somatic_Symptoms_and_Related_Disorders-124.aspx

Pinquart, M., & Shen, Y. (2011). Depressive symptoms in children and adolescents with chronic physical illness: An updated meta-analysis. *Journal of Pediatric Psychology, 36*(4), 375–384.

Posner, K., Brown, G. K., Stanley, B., Brent, D. A., Yershova, K. V., Oquendo, M. A., … Mann, J. J. (2011). The Columbia–suicide severity rating scale: Initial validity and internal consistency findings from three multisite studies with adolescents and adults. *American Journal of Psychiatry, 168*(12), 1266–1277. https://doi.org/10.1176/appi.ajp.2011.10111704

Rand, C., & Goldstein, N. (2018). Patterns of primary care physician visits for US adolescents in 2014: Implications for vaccination. *Academic Pediatrics, 18*(2), S72–S78. https://doi.org/10.1016/j.acap.2018.01.002

Richardson, L. P., Ludman, E., McCauley, E., Lindenbaum, J., Larison, C., Zhou, C., … Katon, W. (2014). Collaborative care for adolescents with depression in primary care: A randomized clinical trial. *JAMA, 312*(8), 809–816. https://doi.org/10.1001/jama.2014.9259

Ross, E., Becker, J., Linnoila, J., & Soeteman, D. (2020). Cost-effectiveness of routine screening for autoimmune encephalitis in patients with first-episode psychosis in the United States. *The Journal of Clinical Psychiatry, 82*(1), 19m13168.

Rossom, R. C., Coleman, K. J., Ahmedani, B. K., Beck, A., Johnson, E., Oliver, M., & Simon, G. E. (2017). Suicidal ideation reported on the PHQ9 and risk of suicidal behavior across age groups. *Journal of Affective Disorders, 215*, 77–84. https://doi.org/10.1016/j.jad.2017.03.037

Ruhrmann, S., Schultze-Lutter, F., & Klosterkotter, J. (2003). Early detection and intervention in the initial prodromal phase of schizophrenia. *Pharmacopsychiatry, 36*(Suppl 3), S162–S167.

Sakai, C., Miller, K., Brussa, A. K., Macpherson, C., & Augustyn, M. (2014). Challenges of autism in the inpatient setting. *Journal of Developmental and Behavioral Pediatrics.* https://doi.org/10.1097/DBP.0000000000000024

Samsel, C., Ribeiro, M., Ibeziako, P., & DeMaso, D. R. (2017). Integrated behavioral health care in pediatric subspecialty clinics. *Child and Adolescent Psychiatric Clinics of North America, 26*(4), 785–794. https://doi.org/10.1016/j.chc.2017.06.004

Schieveld, J. N. M., & Janssen, N. J. J. F. (2014). Delirium in the pediatric patient: On the growing awareness of its clinical interdisciplinary importance. *JAMA Pediatrics, 168*(7), 595. https://doi.org/10.1001/jamapediatrics.2014.125

Schulte, M., & Hser, Y. (2013). Substance use and associated health conditions throughout the lifespan. *Public Health Reviews, 35*(2), 3. https://doi.org/10.1007/BF03391702

Sentinel Event Alert Issue 56: Detecting and treating suicide ideation in all settings. (2016, February). The Joint Commission.

Shaw, R. J., Pao, M., Holland, J. E., & DeMaso, D. R. (2016). Practice patterns revisited in pediatric psychosomatic medicine. *Psychosomatics, 57*(6), 576–585. https://doi.org/10.1016/j.psym.2016.05.006

Shi, M., Miao, S., Gu, T., Wang, D., Zhang, H., & Liu, J. (2019). Dexmedetomidine for the prevention of emergence delirium and postoperative behavioral changes in pediatric patients with sevoflurane anesthesia: A double-blind, randomized trial. *Drug Design, Development and Therapy, 13*, 897–905. https://doi.org/10.2147/DDDT.S196075

SickKids staff. (2019). Somatization: How to help your child or teen at home. Retrieved from https://www.aboutkidshealth.ca/article?contentid=3770&language=english

Silver, G., Traube, C., Kearney, J., Kelly, D., Yoon, M. J., Nash Moyal, W., … Ward, M. J. (2012). Detecting pediatric delirium: Development of a rapid observational assessment tool. *Intensive Care Medicine, 38*(6), 1025–1031. https://doi.org/10.1007/s00134-012-2518-z

Smith, H. A. B., Boyd, J., Fuchs, D. C., Melvin, K., Berry, P., Shintani, A., … Ely, E. W. (2011). Diagnosing delirium in critically ill children: Validity and reliability of the pediatric confusion assessment method for the intensive care unit*. *Critical Care Medicine, 39*(1), 150–157. https://doi.org/10.1097/CCM.0b013e3181feb489

Sorg, E. M., Chaney-Catchpole, M., & Hazen, E. P. (2018). Pediatric catatonia: A case series-based review of presentation, evaluation, and management. *Psychosomatics, 59*, 531–538. https://doi.org/10.1016/j.psym.2018.05.012

Sossong, A. D., Zhrebker, L., Becker, J. E., Chaudhary, N. P., & Rubin, D. H. (2019). The use of telepsychiatry in caring for youth and families: Overcoming the shortages in child and adolescent psychiatrists. In E. V. Beresin & C. K. Olson (Eds.), *Child and adolescent psychiatry and the media.* Elsevier.

Stevens, J., Prince, J., Prager, L., & Stern, T. (2014). Psychotic disorders in children and adolescents. *The Primary Care Companion for CNS Disorders, 16*(2).

Substance Abuse and Mental Health Services Administration. (2017). *National strategy for suicide prevention implementation assessment report.* HHS Publication No. SMA17– 5051. Rockville, MD: Center for Mental Health Services, Substance Abuse and Mental Health Services Administration.

Supelana, C., Ra, A., Kaplan, D., Helcer, J., & Ml, S. (2016). PTSD in solid organ transplant recipients: Current understanding and future implications. *Pediatric Transplantation, 20*, 23–33. https://doi.org/10.1111/petr.12628

Thom, R., Hogan, C., & Hazen, E. (2019a). Suicide risk screening in the hospital setting: A review of brief validated tools. *Psychosomatics*. https://doi.org/10.1016/j.psym.2019.08.009

Thom, R. P., McDougle, C. J., & Hazen, E. P. (2019b). Challenges in the medical care of patients with autism spectrum disorder: The role of the consultation-liaison psychiatrist. *Psychosomatics*. https://doi.org/10.1016/j.psym.2019.04.003

Traube, C., Silver, G., Kearney, J., Patel, A., Atkinson, T. M., Yoon, M. J., … Greenwald, B. (2014). Cornell assessment of pediatric delirium: A valid, rapid, observational tool for screening delirium in the PICU*. *Critical Care Medicine, 42*(3), 656–663. https://doi.org/10.1097/CCM.0b013e3182a66b76

Traube, C., Silver, G., Gerber, L. M., Kaur, S., Mauer, E. A., Kerson, A., … Greenwald, B. M. (2017). Delirium and mortality in critically ill children: Epidemiology and outcomes of pediatric delirium*. *Critical Care Medicine, 45*(5), 891–898. https://doi.org/10.1097/CCM.0000000000002324.

Turkel, S. B. (2017). Pediatric delirium: Recognition, management, and outcome. *Current Psychiatry Reports, 19*(12), 101. https://doi.org/10.1007/s11920-017-0851-1

Turkel, S. B., & Hanft, A. (2014). The pharmacologic management of delirium in children and adolescents. *Pediatric Drugs, 16*(4), 267–274. https://doi.org/10.1007/s40272-014-0078-0

Turkel, S. B., Jacobson, J. R., & Tavaré, C. J. (2013). The diagnosis and management of delirium in infancy. *Journal of Child and Adolescent Psychopharmacology, 23*(5), 352–356. https://doi.org/10.1089/cap.2013.0001

Tyson, J., House, E., & Donovan, A. (2020). Assessing youth with psychotic experiences: A phenomenological approach. *Child and Adolescent Psychiatric Clinics of North America, 29*(1), 1–13.

United Nations Department of Economic and Social Affairs Disability. (2006). *Convention on the Rights of Persons with Disabilities*. Retrieved from https://www.un.org/development/desa/disabilities/convention-on-the-rights-of-persons-with-disabilities/convention-on-the-rights-of-persons-with-disabilities-2.html

United Nations Human Rights Office of the High Commissioner. (1989). *Convention on the Rights of the Child*. Retrieved from https://www.ohchr.org/en/professionalinterest/pages/crc.aspx

Wang, X., Deng, Q., Liu, B., & Yu, X. (2017). Preventing emergence agitation using ancillary drugs with sevoflurane for pediatric anesthesia: A network meta-analysis. *Molecular Neurobiology, 54*(9), 7312–7326. https://doi.org/10.1007/s12035-016-0229-0

White, P. H., Cooley, W. C., Transitions Clinical Report Authoring Group, American Academy of Pediatrics, American Academy of Family Physicians, & American College of Physicians. (2018). Supporting the health care transition from adolescence to adulthood in the medical home. *Pediatrics, 142*(5), e20182587. https://doi.org/10.1542/peds.2018-2587

World Medical Association. (1948). *WMA Declaration of Geneva*. Retrieved from https://www.wma.net/policies-post/wma-declaration-of-geneva/

Zablotsky, B., Kalb, L. G., Freedman, B., Vasa, R., & Stuart, E. A. (2014). Health care experiences and perceived financial impact among families of children with an autism spectrum disorder. *Psychiatric Services*. https://doi.org/10.1176/appi.ps.201200552

Zhabenko, O., Austic, E., Conroy, D. A., Ehrlich, P., Singh, V., Epstein-Ngo, Q., … Walton, M. A. (2016). Substance use as a risk factor for sleep problems among adolescents presenting to the emergency department: *Journal of Addiction Medicine, 10*(5), 331–338. https://doi.org/10.1097/ADM.0000000000000243

Zima, B. T., Rodean, J., Hall, M., Bardach, N. S., Coker, T. R., & Berry, J. G. (2016). Psychiatric disorders and trends in resource use in pediatric hospitals. *Pediatrics, 138*(5), e20160909. https://doi.org/10.1542/peds.2016-0909

Psychiatric Conditions During Peripartum and Perimenopause

141

Jennifer L. Payne and Susan G. Kornstein

Contents

This chapter is an update from the 4th edition. Previous
edition authors were Jennifer L. Payne and Susan G.
Kornstein

J. L. Payne (✉)
Department of Psychiatry and Neurobehavioral Sciences,
University of Virginia, Charlottesville, VA, USA
e-mail: jlp4n@uvahealth.org

S. G. Kornstein
Department of Psychiatry and Institute for Women's
Health, Virginia Commonwealth University School of
Medicine, Richmond, VA, USA
e-mail: susan.kornstein@vcuhealth.org

© Springer Nature Switzerland AG 2024
A. Tasman et al. (eds.), *Tasman's Psychiatry*,
https://doi.org/10.1007/978-3-030-51366-5_4

Abstract

Growing interest in women's mental health has
paralleled the expanding clinical and research
focus on women's health issues over the past
two decades. Research has shown that prior to
puberty, both genders are equally at risk for
developing mood disorders; however, with
the onset of menstruation, women begin to
show an increased risk for unipolar depression
compared with men. Although studies have not
yet revealed a consistent association between
female reproductive hormones and psychiatric
illness, there appears to be a subgroup of
women at increased risk for psychiatric

symptoms during times of hormonal fluctuation, such as premenstrually, postpartum, and perimenopause. Further, the management of mood disorders during pregnancy must take into account not only the risks of medication exposure to the unborn child but also the risks associated with untreated psychiatric illness, given the high relapse rate in women who stop their psychiatric medications for pregnancy. Psychosocial risk factors and intervening life events are also key factors in increasing women's risk for psychiatric illness. This chapter focuses on psychiatric conditions during important reproductive transitions in women, including pregnancy, postpartum, and perimenopause; premenstrual syndrome and premenstrual dysphoric disorder are thoroughly discussed in a separate chapter.

Keywords

Pregnancy · Postpartum · Postpartum depression · Postpartum psychosis · Antidepressants · Perimenopause · Perimenopausal depression

Introduction

With more than 50% of patients with psychiatric illness being women, psychiatrists will routinely evaluate women who are in reproductive transitions such as pregnancy, postpartum, or perimenopause. Treatment during these critical life transitions is not straightforward as during pregnancy and the postpartum time-period, the provider must think not only about the woman but also about potential exposures to the child and during perimenopause psychiatric illness is frequently exacerbated or complicated by hormonal fluctuations, hot flashes, and sleep disruption. Although in many countries there are few resources available to treat psychiatric illness in general, much less during pregnancy, postpartum, and during perimenopause, this topic is important to address. This chapter will provide practical clinical guidance on the treatment of psychiatric disorders in women during and after pregnancy and during the perimenopause transition.

Pregnancy

Pregnancy can be a stressful and high-risk time for mood instability, particularly in women with a past history of mood disorder (Viguera et al., 2000; Cohen et al., 2006a). While there has been an increasing research and public health focus on postpartum depression (PPD), a significant number of women with PPD have symptoms that begin in pregnancy (Evan et al., 2001). Perinatal depression (occurring during and/or after pregnancy) has an estimated prevalence rate of at least 10% (O'Hara & Swain, 1996; Yonkers et al., 2001; Gaynes et al., 2005; Dietz et al., 2007) and this rate approaches 25% or higher (Payne et al., 2007; Viguera et al., 2011) in women with a history of major depressive disorder (MDD) or other mood disorder. Far fewer studies have looked at the prevalence and course of bipolar disorder, anxiety, and psychotic disorders during and after pregnancy. Clearly, many pregnant women, particularly those with a history of psychiatric illness, have worsening symptoms during pregnancy, as a result of either psychosocial risk factors, discontinuing or decreasing medications, changes in medication blood levels, or other physiologic changes that occur during the course of pregnancy (Wisner et al., 1993; Altshuler & Hendrick, 1996). In 2016, the US Preventative Services Task Force (USPSTF) recommended routine screening for depression including both the general and perinatal populations (Siu et al., 2016), and current American College of Obstetrics and Gynecology guidelines (ACOG Committee Opinion, 2018) recommend screening for psychiatric illness, including depression and anxiety, at least once during the perinatal period, particularly at the comprehensive postpartum visit. Screening and treatment during pregnancy may not only help symptomatic expectant mothers but may also minimize their risk of worsening mood and anxiety after delivery. This section outlines a framework for approaching patients with significant psychiatric symptoms during and after pregnancy.

Evaluation of Pregnant Women with Psychiatric Symptoms

Depression and anxiety during pregnancy frequently go undetected by providers (Kelly et al., 2001; Ko et al., 2012). Timely diagnosis and treatment are often hampered by confusion over whether somatic symptoms, such as fatigue and changes in sleep and appetite, are normal consequences of pregnancy or evolving mood or anxiety disorders. A study demonstrated that 65.9% of pregnant women who met DSM-IV criteria for a major depressive episode (MDE) went undiagnosed despite regular contact with the medical system (Ko et al., 2012). Routine formalized screening for major axis I diagnoses would assist clinicians in identifying and tracking patients during pregnancy (Spitzer et al., 2000). In addition, screening for the following risk factors may help identify women at risk for depression during and after pregnancy: prior history of depression or bipolar disorder, young age, limited social support, marital strain, and ambivalence about pregnancy (Altshuler et al., 2000). Examples of screening tools for psychiatric disorders that can be used during the perinatal period include: the Edinburgh Postnatal Depression Scale (EPDS) (Cox et al., 1987), the Patient Health Questionnaire (PHQ-9) (Spitzer et al., 2000), the Hospital Anxiety and Depression Scale (HADS) (Zigmond et al., 1983), and the Mood Disorder Questionnaire (MDQ) (Hirschfeld, 2002).

Treatment of Pregnant Women with Psychiatric Symptoms

Treatment of psychiatric illness during pregnancy requires a thorough discussion about the risks of untreated illness and the risks and benefits of medication. Many pregnant women and their physicians overestimate the risk of medication and underestimate the risks of untreated depression, anxiety, and other psychiatric symptoms. It is important for clinicians to put the risks of medication into context. For instance, pregnancy itself carries many risks, including spontaneous abortion and congenital defects. Untreated psychiatric

illness can compound these risks by contributing to poor self-care, decreased prenatal compliance, increased nicotine and substance misuse (Hanna et al., 1994; Zhu & Valbo, 2002), poor obstetric outcomes (Kurki et al., 2000; Chung et al., 2001), and increased risk of PPD (Beck, 2003). A first step in approaching pregnant or soon-to-be pregnant patients with mood disorders involves reviewing the course of their illness, severity of past mood episodes, and response to different treatments, including both medications and psychotherapy. Part of this risk-benefit discussion should also include a review of the available data regarding the risks associated with psychotropic medication use during pregnancy together with a review of the risks associated with discontinuing medication.

Major Risks Associated with the Discontinuation of Psychiatric Medications for Pregnancy

The available data suggest that there is a high relapse rate in women who discontinue their psychiatric medications for pregnancy. Cohen et al. (2006a) followed 201 women with a history of recurrent MDD and found an increased risk of relapse in those who discontinued antidepressant medication compared with those who continued (68% vs. 26%). For women with bipolar disorders, Viguera et al. (2000) found that approximately 55% of women who stopped lithium for pregnancy relapsed over the course of 40 weeks of pregnancy. In a more recent study, they found that 85.5% of women who discontinued mood stabilizers for pregnancy relapsed (Viguera et al., 2007). These high relapse rates during pregnancy are actually similar to the rate and time course of relapse in nonpregnant women who discontinue their psychiatric medications. In contrast, the risk increased by 2.9-fold during the postpartum time period (Viguera et al., 2000).

Exposure to depression in utero is also associated with poor outcomes for the infant (Yonkers et al., 2009). Antenatal depression has been associated with increased rates of preterm birth (Li et al., 2009; Yonkers et al., 2009), low birth weight (Yonkers et al., 2009), cigarette, alcohol, other substance use (Zuckerman et al., 1989), ambivalence about the pregnancy, and overall worse health status (Orr et al., 2007). In addition, children exposed to perinatal (either during pregnancy or postpartum) maternal depression have higher cortisol levels than infants of nondepressed mothers (Ashman et al., 2002; Essex et al., 2002; Diego et al., 2004; Halligan et al., 2004). These elevated cortisol levels have been found to continue through adolescence (Halligan et al., 2004). Notably, maternal treatment of MDD during pregnancy normalizes infant cortisol levels (Brennan et al., 2008). While the long-term effects of elevated cortisol levels remain unclear, these findings provide a mechanism for an increased vulnerability to psychopathology in children of mothers with antenatal MDD (O'Connor et al., 2005). Finally, depression during pregnancy is also one of the strongest predictors of PPD, which has repeatedly been shown to have effects on the language and intellectual development of the exposed child (Cuijpers et al., 2008; O'Hara, 2009; Hirst & Moutier, 2010; Soufia et al., 2010; Breese McCoy, 2011; Field, 2011). Hence exposure to maternal mental illness must also be considered a risk for the developing infant.

Major Reproductive Risks Associated with Psychotropic Medications

Understanding the Limitations of the Literature and the FDA Categories

When reading the literature on the safety of psychotropic medication in pregnancy, it is important to understand that the literature is complicated by the fact that many studies do not control for underlying psychiatric illness and related factors. Women with psychiatric disorders have other associated health-related behaviors, comorbid illnesses, and risk factors that complicate the interpretation of results in studies that do not control for these factors. These confounding factors may influence the outcomes of studies attempting to examine the risks of in utero exposure of psychotropic medications to a fetus or infant. For

example, diabetes, obesity, smoking, and substance use are more common in patients with psychiatric illness than in the general population. Studies that have not controlled for the underlying psychiatric illness and its confounding behaviors and characteristics may find associations between psychotropic medications and outcomes that are not directly caused by exposure to the medication itself, but by characteristics and behaviors that are more highly prevalent in the population of patients who take psychotropic medications during pregnancy. The following sections will note which studies were well controlled and which were not.

It is also important to note that the FDA pregnancy categories are being phased out in favor of including a label summary of all available information on the safety of the medication during pregnancy and lactation in order to better help the clinician weigh the risks and benefits of prescribing a drug during pregnancy or breastfeeding. Pregnancy Category in the text below is occasionally referred to from a historical perspective.

Congenital Malformations

Teratogenesis is the malformation of fetal organs, leading to structural or functional anomalies. Drugs ingested in pregnancy are considered teratogenic if they are associated with an increased frequency of congenital malformations above the baseline risk of 3–4%. Most studies of tricyclic antidepressants (TCAs) and selective serotonin reuptake inhibitors (SSRIs) have been reassuring and shown no increased risk of major congenital malformations (Hendrick et al., 2003; Malm et al., 2005; Yonkers et al., 2009). However, there is great disparity in the amount of published information on different antidepressants within these drug classes, and it remains unclear whether all drugs within a class have the same reproductive risk. For example, GlaxoSmithKline, the manufacturer of paroxetine, issued a warning that two studies found a possible association between first trimester paroxetine exposure and increased risk for cardiac defects, particularly atrial and ventral septal defects (www.gsk.com/media/paroxetine_

pregnancyhtm). In 2005, the US Food and Drug Administration (FDA) issued a similar warning that first trimester paroxetine use was associated with an increased risk of major malformations (4% vs. 3%), particularly cardiac malformations (2% vs. 1%). Although the purported association between paroxetine use and cardiac defects contradicts other published studies (Kulin et al., 1998; Ericson et al., 1999; Einarson & Einarson, 2005), the manufacturer and the FDA changed the pregnancy labeling from category C to D, indicating that published studies in pregnant women have demonstrated a risk to the fetus. However, the joint report on the management of depression during pregnancy from the American Psychiatric Association and the American College of Obstetricians and Gynecologists (Yonkers et al., 2009) encourages an individualized approach to treatment, since in some cases the benefits of paroxetine may outweigh the risks of discontinuing the medication.

Other studies of SSRIs as a class are reassuring for lack of an association between major organ malformations and in utero SSRI exposure. For example, a study with a sample size of more than 900,000 women did not find an association between first trimester antidepressant exposure and cardiac malformations when the statistical analyses controlled for MDD by comparing the outcomes of women with MDD who took antidepressants to outcomes of women with MDD who did not take antidepressants (Huybrechts, et al., 2014). A meta-analysis of prospective cohort studies found no association between SSRI use in the first trimester and heart defects when comparing women with MDD who took SSRIs in the first trimester with women with MDD who did not take antidepressants in pregnancy (Wang et al., 2015).

Although there have been fewer reported cases of prenatal exposures to non-SSRI antidepressants, the limited data available have not shown an increased risk of congenital malformations with venlafaxine (Einarson et al., 2001; Lassen et al., 2016), duloxetine (Lassen et al., 2016), bupropion (Chun-Fai-Chan et al., 2005; Hendrick et al., 2017), mirtazapine (Djulus et al., 2006;

Smit et al., 2016), or trazodone (Einarson et al., 2003). Data on the newest antidepressants, desvenlafaxine, vortioxetine, and vilazodone, are limited to case reports or not available. A study using the Swedish Medical Birth Registry identified 732 women who used serotonin norepinephrine reuptake inhibitors (SNRIs) or norepinephrine reuptake inhibitors (NRIs) in early pregnancy and found no increased risk for congenital malformation (Lennestål & Källén, 2007).

Perinatal Effects Following Late Pregnancy Antidepressant Exposure

Poor neonatal adaptation syndrome (PNAS) refers to transient symptoms in the neonate following in utero exposure to medication, such as antidepressants, antipsychotics, and benzodiazepines. These symptoms can include irritability, tremulousness, insomnia, poor feeding, temperature dysregulation, increased or decreased muscle tone, and/or respiratory distress. These perinatal syndromes are likely related to medication withdrawal or toxicity (or a combination) and have been described with many SSRIs and SNRIs (Källén, 2004; Zeskind & Stephens, 2004; Moses-Kolko et al., 2005; Lennestål & Källén, 2007). Using a formal screening tool, one study found that 30% of the newborns exposed to SSRIs in late pregnancy developed PNAS symptoms (Levinson-Castiel et al., 2006). In general, PNAS has been found to be transient and not life threatening. In an attempt to minimize the risk of PNAS in 2004, the FDA initially recommended that antidepressants be tapered in the last weeks of pregnancy, although there was no evidence to support this recommendation as safe or effective. In 2010, Warburton and colleagues published a study comparing the outcomes of infants exposed to SSRIs during the last 14 days of gestation to infants who were exposed to SSRIs earlier during gestation (Warburton, et al., 2010). While respiratory distress was more common in infants exposed to SSRIs during the last 14 days of gestation, when the analyses were controlled for maternal illness severity, there were no differences between the two groups. This finding again emphasizes the need to control for confounding factors, including severity of psychiatric illness, in studies examining the risks of psychiatric medication exposure during pregnancy. This finding also indicates that other factors outside of in utero antidepressant exposure may influence the severity of PNAS in antidepressant-exposed newborns. The FDA no longer recommends tapering antidepressants in the third trimester; given that there is currently no evidence that tapering decreases the risk of PNAS, this course of action would not be prudent in patients with significant depression or highly recurrent illness, who are at risk for postpartum depression.

Persistent pulmonary hypertension of the newborn (PPHN) is a serious and rare condition, affecting 1–2 out of 1000 live births and leading to death in 10–20%. Chambers et al. (2006) published a study in which they found a possible association between PPHN and SSRI exposure after 20 weeks' gestation. Although this study was retrospective in design and involved only a small number of affected infants, its findings were concerning and again highlight the uncertainty that patients and providers must assume in choosing to use antidepressants or any psychotropic medications during pregnancy (see Clinical Vignette 1). To date, studies have shown mixed results, with some studies finding an association (for example: Chambers et al., 2006; Källén & Olausson, 2008; Kieler et al., 2012) and others not (for example: Andrade et al., 2009; Wichman et al., 2009; Wilson et al., 2011). To complicate interpretation of these studies, some of the known risk factors for PPHN, such as maternal smoking, maternal obesity, maternal diabetes, and delivery by caesarean section have higher rates in the psychiatric population (Occhiogrosso et al., 2012). Two meta-analyses (Grigoriadis et al., 2014; Ng et al., 2019) have concluded that there appears to be a significant, though very small, increase in the odds of developing PPHN in antidepressant-exposed newborns. It is important to note that even with the small increased risk, more than 99% of exposed infants will *not* develop this serious complication because it is such a rare phenomenon (Yonkers et al., 2009).

Clinical Vignette 1

Kathy is a 35-year-old attorney with a history of recurrent MDD. She is happily married and has been stable for 2 years on an antidepressant. She recently learned that she is 7 weeks pregnant and wants to know whether to discontinue her medication. Her psychiatrist reviews with her the available studies regarding her elevated risk of recurrence of depression if she stops the medication, given her past history of two major depressive episodes. In addition, her psychiatrist reviews with her the available data regarding the reproductive safety profile of different antidepressant treatment options. Kathy recalls how much her past depressive episodes cost her emotionally and professionally and is not willing to place her stability at risk by discontinuing her antidepressant. After a thorough risk-benefit discussion, she consents to continuing her SSRI and also begins a course of interpersonal psychotherapy to address conflicts with her husband. As she enters her third trimester, she and her psychiatrist again review the risk of perinatal toxicity and pulmonary hypertension of the newborn associated with late trimester SSRI exposure. Kathy feels that the benefit she has gained from her antidepressant far outweighs the relatively slight risk of these adverse outcomes. Kathy also decides that it would be unwise to withdraw medication prior to entering the postpartum period, a time of heightened risk for many women with recurrent depression. She and her psychiatrist therefore decide to maintain her current dose as she heads into the last part of pregnancy.

Long-Term Neurobehavioral Effects

Neurobehavioral teratogenesis refers to the long-term neurologic or behavioral effects of in utero drug exposure. In a study by Nulman et al. (2002), preschool children exposed in utero to TCAs or fluoxetine were found to have no difference in temperament, language, and cognitive development compared with nonexposed children. Another study by Nulman et al. (2012) compared the children (aged 3–6 years) of mothers who took venlafaxine with children of mothers who took SSRIs, with children of mothers who stopped taking antidepressants prior to conception, and with children of healthy controls. Full-scale IQs were significantly lower in the groups of children exposed to antidepressants than in the children of healthy controls and were not different from the group of children whose mothers were depressed but taking no medication. Children from all three groups exposed to maternal depression had more problematic behaviors. More work needs to be done in this area.

Risks Associated with Other Psychotropic Medications

Lithium use during the first trimester has been associated with an increased risk of a serious congenital heart defect known as Ebstein's anomaly, which occurs in approximately 1 out of 1000 live births. Lithium has also been associated with perinatal toxicity, including case reports of hypotonia, cyanosis, neonatal goiter, and diabetes insipidus. For women with severe bipolar illness (i.e., recurrent mania or depression, history of psychosis, etc.) and who have selective response to lithium historically, the risk of recurrence during pregnancy may overshadow the relatively small risk of Ebstein's anomaly (Yonkers et al., 2004). For such women, maintenance lithium therapy during pregnancy may be the most appropriate course. On the other hand, for women with significant periods of euthymia and few past mood episodes, slowly tapering off lithium and reintroducing lithium after the first trimester may help reduce the risk of relapse during the postpartum, though this technique has not been formally evaluated.

Lamotrigine is another highly effective mood stabilizer, particularly for bipolar depression. According to the manufacturer-sponsored Lamotrigine Pregnancy Registry and other published studies (Cunnington et al., 2005),

there appeared to be no increased risk of congenital defects above the baseline risk with lamotrigine monotherapy; however, when combined with valproic acid in pregnancy, the risk estimate was found to be elevated to above 10%. Although these initial findings seemed to offer women a relatively safe alternative to other anticonvulsants in pregnancy, the North American Antiepileptic Drug Pregnancy Registry (www2. massgeneral.org/aed/) found that infants who are exposed to lamotrigine monotherapy during pregnancy have a much higher risk of oral cleft defects. In this study, 564 children exposed to lamotrigine monotherapy had a prevalence rate of major malformations of 2.7%; however, five infants had oral cleft lip and palate, yielding a prevalence rate of 9 out of 1000 births compared with a baseline prevalence of 0.5–2.0 per 1000 in unexposed infants (Holmes et al., 2008). A larger study failed to find this association (Dolk et al., 2008): in this study, the association between oral clefts and exposure to lamotrigine was assessed using a population-based case-control design utilizing data from the EUROCAT congenital malformation registries. The study population included 3.9 million births from 19 registries from 1995 to 2005. Registrations included congenital anomalies among live births, stillbirths, and terminations of pregnancy following prenatal diagnosis. The authors identified 5511 cases of nonsyndromic oral cleft. The control group consisted of 80,052 cases of nonchromosomal, nonoral cleft malformations. There were 72 lamotrigine-exposed (40 mono-and 32 poly-therapy) cases. In this study, there was no evidence of an increased risk of isolated oral clefts relative to other malformations. There have been a number of studies since then, and a recent systemic review and meta-analysis of 21 studies concluded that there was no association between prenatal lamotrigine monotherapy and birth defects (Pariente et al., 2017). Even if future prospective studies confirm an association between first trimester lamotrigine use and oral cleft lip and palate, the overall risk appears to be low and may be overshadowed by the high risk of recurrent illness in some women with bipolar disorder. In the setting of rising estrogen levels, lamotrigine levels may decrease over the course of pregnancy and thus should be followed and adjusted if needed (Fotopoulou et al., 2009). Clinicians should encourage all pregnant patients taking any anticonvulsant in pregnancy to participate in the North American Antiepileptic Drug Pregnancy Registry by calling 1-888-233-2334, as this growing database will continue to offer valuable new information.

Valproic acid and *carbamazepine* have well-established risks of neural tube defects of 5% and 1%, respectively. Carbamazepine is also a competitive inhibitor of prothrombin precursors and may increase the risk of neonatal hemorrhage. Newer anticonvulsants (e.g., *gabapentin*, *oxcarbazepine*, and *topiramate*) have limited reproductive safety data to guide their use in pregnancy. The data so far are reassuring for gabapentin, although a number of the available studies involved small sample sizes. For example, a prospective study of 223 pregnant women exposed to gabapentin, found no increased risk for major malformations although there were higher rates of preterm birth, low birth weight, and admission to a neonatal intensive care unit (NICU) (Fujii et al., 2013). A recent study of gabapentin-exposed pregnancies using the US Medicaid Analytic eXtract (MAX) dataset examined outcomes in 4642 pregnancies exposed in the first trimester, 3745 exposed in early pregnancy only, 556 exposed in late pregnancy only, and 1275 exposed in both early and late pregnancy (Patorno et al., 2020). They found no increased risk of major organ malformations with early pregnancy gabapentin exposure, but late pregnancy use was associated with a higher risk of preterm birth, small for gestational age, and NICU admission. Topiramate has been associated with a higher risk of cleft palate (Margulis et al., 2012), and higher doses may increase this risk (Hernandez-Diaz et al., 2018). Although anticonvulsants would not be the first choice in pregnancy, they may be indicated for particular pregnant women with refractory bipolar illness and a history of good response to these medications. Providers should encourage pregnant women who elect to continue any anticonvulsant to take high-dose folate (4 mg per day) for the theoretical benefit of reducing the risk of neural tube defects. In addition, pregnant women should undergo a second trimester Level II ultrasound to screen for major congenital anomalies.

Antipsychotics, like other psychotropic medications, are often essential during pregnancy to help minimize risks associated with psychosis, such as noncompliance with prenatal care, increased drug and alcohol use, and other high-risk behaviors. High-and medium-potency neuroleptics, such as haloperidol and perphenazine, are generally considered to be safe during pregnancy with no demonstrated teratogenic risk, whereas low-potency phenothiazines have shown increased risk of congenital malformations and should be avoided. There have been a number of case reports and case series reports on the use of the atypical antipsychotic medications during pregnancy and most have been reassuring. For example, one study (Habermann et al., 2013) prospectively followed the pregnancies of 561 women exposed to second-generation antipsychotic agents (SGAs) (study cohort) and compared these with 284 pregnant women exposed to first-generation antipsychotic agents (FGAs) (comparison cohort I) and with 1122 pregnant women using drugs known as not harmful to the unborn (comparison cohort II). Major malformation rates in the SGA-exposed cohort were higher compared with comparison cohort II (adjusted odds ratio 2.17, 95% confidence interval 1.20–3.91), possibly reflecting a detection bias concerning atrial and ventricular septal defects. Postnatal disorders, defined as any "peculiarity of the newborn, which was fatal, life-threatening, resulting in persistent or significant disability/incapacity, or in necessity of therapy or hospitalization" occurred significantly more often in infants prenatally exposed to SGAs (15.6%) and FGAs (21.6%) compared with comparison cohort II (4.2%). The numbers of stillbirths and neonatal deaths were within the reference range. Preterm birth and low birth weight were more common in infants exposed to FGAs. The authors concluded that the study did not reveal a major teratogenic risk for SGAs, making the older and better studied drugs of this group a treatment option during pregnancy. A more recent, Medicaid-based study was also reassuring and found that antipsychotics (as a group) used early in pregnancy did not increase the risk for congenital malformations or cardiac malformations when analyses controlled for confounding factors, though there was a small increased risk for congenital malformations with risperidone (Huybrechts et al., 2016). Note that many antipsychotics are associated with excessive maternal weight gain, increased infant birth weight, and increased risk of gestational diabetes (Newport et al., 2007; Newham et al., 2008). Routine ultrasound monitoring of fetal size in late pregnancy for women taking these medications is warranted (Newham et al., 2008).

Studies of *benzodiazepine* use during pregnancy have been contradictory and controversial. Benzodiazepine use during pregnancy has been associated with case reports of perinatal toxicity, including temperature dysregulation, apnea, depressed APGAR scores, hypotonia, and poor feeding. In addition, early studies revealed an elevated risk of oral cleft palate defects compared with the baseline risk in the general population. However, more recent studies have shown that the overall risk of cleft lip and palate with benzodiazepine use in pregnancy is probably quite low (Iqbal et al., 2002; Lin et al., 2004). Infants exposed to an SSRI in combination with a benzodiazepine may have a higher incidence of congenital heart defects even when controlling for maternal illness characteristics (Oberlander et al., 2008). In considering the risks and benefits of benzodiazepines, clinicians should also consider the risks of untreated insomnia and anxiety in pregnancy, which may lead to physiologic effects and also diminished self-care, worsening mood, and impaired functioning. Given the consequences of untreated psychiatric symptoms and the limited and controversial risks associated with benzodiazepine use, some women with overwhelming anxiety symptoms or sleep disturbance may find that the benefits outweigh any theoretical risks.

Psychotherapy

Although concerns about the reproductive safety of medications often prompt women to seek psychiatric consultation during pregnancy, it is important to note that psychotherapy is often a useful alternative (or adjunct) to medication. Establishing a therapeutic alliance, exploring feelings about the pregnancy, and taking an inventory of strengths, supports, and stressors can help target interventions, which may include assistance with basic needs for food, shelter, or financial support. Studies have demonstrated that depressed pregnant women can also benefit greatly from interpersonal psychotherapy (IPT), a brief structured therapy focused on certain areas of particular relevance to pregnant women, such as role transitions, grief, and interpersonal disputes (Spinelli & Endicott, 2003). One version of IPT, Partner Assisted IPT, was specifically developed for perinatal depression and demonstrated promising results in a preliminary proof of concept trial in women who were depressed either during or after pregnancy (Brandon et al., 2012). Supportive and cognitive behavioral psychotherapy have also been shown anecdotally to help with depression and anxiety during pregnancy (Dimidjian & Goodman, 2009).

Light Therapy

Light therapy (LT) has been established as a treatment for seasonal affective disorder (Lam & Levitt, 1999; American Psychiatric Association, 2000) but has also been demonstrated to be safe and effective in the treatment of nonseasonal MDD (Kripke et al., 1992; Yamada et al., 1995; Kripke, 1998). LT has also been demonstrated to be effective in depressed pregnant women. Oren et al. (2002) conducted an open-label trial of bright light therapy in an A–B–A design in 16 pregnant patients with MDD who were depressed. They found that mean depression ratings improved by 49%. There have been two randomized trials of LT in antepartum depression. Epperson et al. (2004) found that women randomized to LT (10,000 lux) had a 60% improvement in depression ratings compared with 41% in the placebo group (500 lux). The difference was not statistically different in this small sample ($n = 10$). A larger randomized trial of LT (10,000 lux) in comparison with sham LT (70 lux red light) in 27 depressed pregnant women found that LT was superior, with 81% of the LT sample responding compared with 45% of the placebo group (Wirz-Justice et al., 2011).

Repetitive Transcranial Magnetic Stimulation

Repetitive transcranial magnetic stimulation (rTMS) delivers a focused magnetic pulse to specific areas of the cerebral cortex (Gershon et al., 2003). It is currently approved by the FDA for treatment in adults with MDD who have failed at least one antidepressant trial in the current depressive episode. A few case reports and open-label studies have provided preliminary support for the utility of rTMS in treating depression in pregnant women (Nahas et al., 1999; Klirova et al., 2008; Zhang & Hu, 2009; Zhang et al., 2010; Kim et al., 2011). The benefits of using rTMS to treat major depression during pregnancy include a minimal side-effect profile, good tolerability, and decreased fetal exposure to medications. In one open-label study, Kim et al. (2011) administered 20 daily sessions of rTMS, each lasting for 10 min and targeted to the right dorsolateral prefrontal cortex, in 10 pregnant women with MDD who were clinically depressed. In this study, 70% of the patients responded to treatment and 30% met criteria for remission. No adverse pregnancy or fetal outcomes were reported.

Electroconvulsive Therapy

Electroconvulsive therapy (ECT) is an underutilized treatment during pregnancy despite studies supporting its safety. For high-risk situations, such as psychotic depression or mania, which require prompt relief of psychiatric symptoms to protect both mother and fetus, ECT is the treatment of choice. The safe and effective use of ECT during pregnancy requires coordination of care among the patient's psychiatrist, obstetrician, and anesthesiologist (O'Reardon et al., 2011).

Postpartum Psychiatric Illness

Postpartum psychiatric symptoms are extremely common and range in quality and degree of functional impairment. Although the risk for suicide deaths and attempts is lower during and after pregnancy than in the general population of women, suicides account for up to 20% of all postpartum deaths and represent one of the leading causes of peripartum mortality (Lindahl et al., 2005). Untreated postpartum psychiatric symptoms can not only lead to maternal suffering, but also affect how mothers care for their infants. In a prospective study of 570 postpartum women and their 3-month-old infants, mother-infant dyads with depressed mothers had less vocal and visual communications (Righetti-Veltema et al., 2002). Other studies have shown that maternal depression was associated with impaired mother-infant attachment (Brockington et al., 2001; Moehler et al., 2006), higher risk of child cognitive and behavioral problems (Murray et al., 1996; Brennan et al., 2000), and lower IQ in the child (Grace et al., 2003).

Studies examining the pathophysiologic basis of postpartum psychiatric disorders have focused on the dramatic shifts in reproductive hormones during and after pregnancy, in addition to the hypothalamic-pituitary–adrenal axis and changes in cortisol, thyroid hormone, prolactin, melatonin, oxytocin, and vasopressin (Hendrick et al., 1998; Parry et al., 2003; Kammerer et al., 2006). Although there is no compelling evidence to support most hormonal theories, a sensitivity to the normal postpartum hormonal fluctuations appear to contribute to PPD in a subgroup of women at risk (Bloch et al., 2000). This observation is further supported by evidence of the neuromodulatory effects of estrogen and progesterone, which in turn affect mood regulation (Rubinow et al., 1998).

Heightened emotional sensitivity and sleep disturbance are normal experiences during the postpartum transition. In fact, eight out of ten new mothers experience *postpartum blues* (PPB), a self-limited constellation of symptoms that usually begin 2–3 days postpartum and can consist of mood lability, anxiety, insomnia, and appetite disturbance. This is a nonpathologic condition that resolves by 2 weeks postpartum. For some women, however, these symptoms can persist and evolve into *postpartum depression* (PPD), a condition that affects 10–20% of all postpartum women in the general population, with higher rates in low-income or disadvantaged populations and also in women with a history of a mood disorder (Payne et al., 2007). Although previous versions of the DSM used the term postpartum to describe major depression that occurs within 4 weeks postpartum, the most recent version, DSM-5, used the term "peripartum," defined as a major depressive episode that begins during pregnancy or within the first 4 weeks postpartum. However, many new mothers develop PPD more than 2–3 months after delivery. Clinical features include all the signs and symptoms of major depression. In addition, *postpartum anxiety* symptoms are extremely common and can include panic attacks, intense anxiety about not getting enough sleep, obsessive worry about the baby's health or safety, and intrusive thoughts or mental images of hurting the baby (Brandes et al., 2004). New mothers with intrusive thoughts are often ashamed of these ego-dystonic images and may develop behaviors to diffuse some of their anxiety and fear, such as avoiding sharp objects or compulsively checking their infants' breathing.

Postpartum psychosis (PPP) is a rare condition that occurs in 1–2 of every 1000 postpartum women. PPP can begin acutely within the first 48–72 h postpartum and may include delirium, memory impairment, irritability, and mood lability. Psychosis can also occur in new mothers with a history of a chronic psychotic disorder or as part of a major depressive or manic episode. Psychotic symptoms in a new mother require immediate intervention in order to protect both mother and infant from harm, and hospitalization is generally the rule.

Evaluation of Postpartum Women with Psychiatric Risk Factors or Symptoms

In assessing any woman with significant mood or anxiety symptoms after delivery, it is important to rule out possible organic causes of low mood, insomnia, and appetite disturbance, such as anemia and thyroid dysfunction. Clinicians should also inquire about the status of patients' physical recovery from delivery, since prolonged pain following a complicated vaginal delivery or cesarean section can certainly contribute to sleep and mood symptoms. In screening for common Axis I

diagnoses, such as major depression or anxiety disorders, clinicians should also screen for bipolar disorder, given the heightened risk of postpartum mood episodes in women with bipolar illness (Viguera et al., 2000, 2007; Payne et al., 2007). In addition, clinicians should ask new mothers about intrusive thoughts regarding their infants' safety or violent images or thoughts about their infants (see Clinical Vignette 2). These intrusive thoughts are common in women with postpartum depression or anxiety, while acting them out is thankfully very rare (Brandes et al., 2004). Nonetheless, specifically asking distressed postpartum mothers about these thoughts and images is extremely important, since psychotic thinking can at times be subtle. Similarly, direct questioning about suicidal or homicidal thoughts is important to establish safety and need for hospitalization.

Clinical Vignette 2

Mary is a 32-year-old married nurse with a history of panic attacks that have been well controlled for years. She presents 3 months postpartum, following a difficult pregnancy complicated by severe hyperemesis gravidarum and dysphoria in addition to traumatic delivery with a third-degree perineal tear. Mary now complains of crying spells, decreased appetite, insomnia, and obsessive worry over the baby's health. She feels isolated from her husband, who is overwhelmed by her emotional needs and tends to retreat to work. Her psychiatrist completes a thorough diagnostic assessment and determines that Mary is suffering from major depression and generalized anxiety disorder. When she is feeling particularly overwhelmed, she also has violent and intrusive thoughts of throwing her infant against a wall. These thoughts are ego dystonic and very distressing for Mary, who has no evidence of psychosis and no history of violence. She begins a course of cognitive behavioral therapy to address her anxiety symptoms and negative thinking. She also

(continued)

enlists the support of a babysitter several mornings per week to allow her additional rest, which improves her perineal pain and gives her some much needed time for herself. In addition, she begins an SSRI, and within a few weeks she starts to feel much more composed and connected to her baby.

Systematic screening for risk factors in addition to current symptoms facilitates the prevention, detection, and treatment of postpartum psychiatric syndromes. The Edinburgh Postnatal Depression Scale (EPDS) is a widely recommended, cost-effective means of screening for PPD (Wisner et al., 2002). The EPDS is a 10-item self-rating scale that has been validated in Spanish and English and asks about depressive symptoms in the preceding week, including crying spells, decreased interest and pleasure, increased guilt, anxiety, sleeping problems, and thoughts of self-harm (Cox et al., 1987). To improve prevention interventions, clinicians could also screen for risk factors both during and after pregnancy. These risk factors include a personal or family history of mood or anxiety disorder, a family history of postpartum psychiatric illness, psychiatric symptoms during pregnancy, limited social support, interpersonal conflicts, and negative life events during and after the pregnancy (O'Hara et al., 1991; Stowe & Nemeroff, 1995; Robertson et al., 2004; Murphy-Eberenz et al., 2006; Payne et al., 2008). In addition, a history of extreme premenstrual irritability may be a possible marker for increased vulnerability during times of hormonal fluctuation, such as the dramatic hormonal shifts after delivery (Stewart & Boydell, 1993; Payne et al., 2007).

Treatment Considerations in Postpartum Women with Psychiatric Symptoms

Support and Psychotherapy

All women would benefit from education about the normalcy of certain psychiatric symptoms after delivery and review of the differences between the self-limited PPB and more serious conditions such as PPD. Referrals to mothers' groups, childcare resources, or financial assistance agencies can also be extremely therapeutic. In the United Kingdom, postpartum support includes nurses with advanced training, who provide home visits to screen and assist postpartum mothers. In the United States, clinicians can refer women to Postpartum Support International (www.postpartum.net), an organization with many local chapters that offers support and resources for women with postpartum psychiatric illness. Other options for surrogate family assistance include the services of a doula, a professional caregiver who can assist not only with childbirth but also during the first postpartum weeks by helping with childcare or household tasks. Finally, as there is evidence that sleep deprivation may trigger mood episodes in women with bipolar disorder (Sharma, 2003), educating family members who can provide assistance to minimize sleep disturbance and who can also identify mood changes early is an important preventative measure.

Studies of IPT suggest that this time-limited therapy is helpful in treating mild-to-moderate postpartum major depression (O'Hara et al., 2000; Reay et al., 2006). This form of therapy focuses on specific problem areas, such as role transitions or interpersonal disputes. Zlotnick et al. (2001) found that pregnant women receiving group IPT compared with regular care had decreased rates of PPD at 3 months postpartum. As noted earlier, Partner Assisted IPT has also demonstrated efficacy for PPD (Brandon et al., 2012). A small study comparing cognitive behavioral therapy (CBT) and fluoxetine showed that CBT is also an effective treatment for PPD (Appleby et al., 1997). The USPSTF concluded that counseling interventions can be effective in preventing PPD in high-risk groups, and that two interventions, IPT and CBT, had the greatest effect (Curry et al., 2019).

Antidepressants

Despite the high prevalence of postpartum psychiatric illness, there have been only a limited number of antidepressant treatment studies. In small studies, fluoxetine (Appleby et al., 1997), venlafaxine (Cohen et al., 2001), and bupropion (Nonacs et al., 2005) have shown efficacy in treating PPD, and sertraline has shown benefit in both preventing and treating PPD (Wisner et al., 2004, 2006). In contrast, nortriptyline did not separate from placebo in a blinded, randomized trial (Wisner et al., 2001) on the prevention of PPD, and it demonstrated equivalent though slightly slower onset of efficacy to sertraline in the treatment of PPD (Wisner et al., 2006). Women at increased risk for PPD should consider initiating prophylactic antidepressants in either late pregnancy or early postpartum. Alternatively, women may elect for a wait-and-see approach; however, patients, their loved ones, and psychiatrists should be vigilant for early signs of recurrence in order to institute prompt treatment. Close follow-up psychiatric care in those at risk should be the rule during the postpartum period.

The choice of an antidepressant is largely guided by the patient's depressive symptoms, past history of medication response, and medication side effects. If a patient plans to breastfeed, clinicians must facilitate a careful risk-benefit discussion about taking psychotropic medication during lactation (see later). However, if the infant was exposed to an antidepressant in utero, the dosage of exposure will be lower in breast milk than the dosage in utero, hence there is usually no reason to limit breastfeeding. To temper the tremendous changes in estrogen and progesterone following birth, some studies have looked at hormone therapy (HT). Several studies of estrogen alone or as an adjunct to antidepressants concluded that estrogen improved depressive symptoms; however, these studies were small and have limited generalizability. Randomized trials examining the efficacy of HT in the postpartum population are currently ongoing.

Brexanolone

Brexanolone is a synthetic version of the natural neurosteroid and progesterone metabolite, allopregnanolone. Allopregnanolone levels, like estrogen and progesterone, drop precipitously in the immediate postpartum time period, and some research suggests that women with lower levels of allopregnanolone during pregnancy have an increased risk for postpartum depression. Brexanolone has been studied in two multicenter, randomized, placebo-controlled trials and has been shown to be effective in moderate to severe postpartum depression (Meltzer-Brody et al., 2018), Brexanolone is given as an intravenous infusion over the course of 60 h and can produce rapid (12–24 h) relief of depressive symptoms. Common adverse effects include somnolence, dizziness, and headache. A small number of patients required cessation of the infusion due to excessive sedation or loss of consciousness, therefore a Risk Evaluation and Mitigation Strategies (REMS) program is in place to minimize adverse events. Currently, brexanolone infusion requires inpatient admission with 24-h pulse-oximetry monitoring.

Breastfeeding

All psychotropic medications are secreted into breast milk. Mothers on psychotropic medications therefore require a thorough risk-benefit discussion of breastfeeding and taking medication to treat psychiatric symptoms.

1. *Antidepressants*. One analysis of the available studies of antidepressants during lactation revealed that sertraline, paroxetine, and nortriptyline are the least likely to lead to accumulation in the infant (Weissman et al., 2004). Note that a patient who is doing well on a particular antidepressant during pregnancy should generally not be switched to one that has lower levels during breastfeeding, as this would increase the number of medication exposures for the baby. Studies of TCAs and SSRIs have been reassuring and shown no consistent association between any particular antidepressant and problems in nursing newborns. There have been isolated case reports of elevated infant levels and toxicity with breastfeeding mothers taking doxepin or fluoxetine. Little is known about the safety of other antidepressants during lactation, such as venlafaxine, buproprion, and mirtazapine.

2. *Mood stabilizers*. For women with bipolar illness, the choice to breastfeed while taking a

mood stabilizer is even less clear-cut than with antidepressants. Lithium can quickly accumulate in the nursing infant and lead to levels exceeding 50% of the maternal level in the setting of dehydration. Given this risk of lithium toxicity in the nursing infant, breastfeeding while on lithium is generally not recommended, although a recent small case series demonstrated that infant blood levels were 10–17% of the maternal level (Bogen et al., 2012). More recent studies support an individualized approach depending upon maternal and infant factors that may influence outcomes such as prematurity of the infant and/or disorganization in the mother (Newmark et al., 2019). Although not absolutely contraindicated in nursing mothers, valproate has been associated with infant anemia and thrombocytopenia, and carbamazepine has been associated with transient hepatic dysfunction. In contrast, lamotrigine has been used successfully during breastfeeding in both epileptic and psychiatric populations (Newport et al., 2008).

In general, clinicians should advise nursing women on psychotropic medications to monitor infants for behavioral changes, such as excessive sedation, jitteriness, or inconsolable crying. Infants who develop any of these symptoms should be evaluated by their pediatrician for possible drug toxicity. In the meantime, mothers can consider temporarily pumping and storing/ discarding their breast milk and using formula to see if their infants' symptoms resolve. In infants who are preterm or have any medical problems, mothers on psychotropic medication could also consider pumping and storing/discarding breast milk and introducing nursing later when the infant is healthy and can presumably metabolize medication more efficiently.

Perimenopause

Perimenopause refers to the 5–10-year transition from regular menstrual functioning to menopause, which is defined by the absence of menses for 1 year or longer. Hormonal changes during this period include significant fluctuations in estrogen and progesterone levels until menstrual cycles cease and reproductive hormones remain stably low. Most women naturally transition to menopause, but other women experience a more abrupt "chemical" or "surgical" menopause as a result of exogenous hormone treatment or following bilateral oophorectomy.

Some women experience significant mood symptoms during the menopausal transition. As with premenstrual syndrome and postpartum psychiatric symptoms, perimenopausal hormonal fluctuations may increase mood symptoms in a subset of at-risk women. Several studies have found that perimenopausal depressive symptoms are likely associated with reproductive hormonal changes rather than abnormally low levels of estrogen or androgens (Schmidt et al., 2000). Furthermore, several trials have shown that estradiol reduced depressive symptoms or a major depressive episode in perimenopausal women, and that baseline or post-treatment estradiol levels did not predict therapeutic response (Rubinow et al., 2015; Schmidt et al., 2000; Soares et al., 2001). Other theories of perimenopausal depression have highlighted the "domino" effect of somatic symptoms associated with estrogen withdrawal, such as hot flashes, which in turn lead to sleep and mood disturbance. In addition, psychosocial factors have also been shown to increase risk for midlife depression in women, including negative life events, health complaints, and relationship difficulties (Schmidt et al. 2004). The confluence of other developmental midlife transitions, such as having adult children leave home or aging and dying parents, may also play a role in increasing risk for psychological distress; however, studies of this so-called *empty nest* phenomenon have been contradictory (Greene & Cooke, 1980; Dennerstein et al., 2002; Schmidt et al. 2004). Although interpreting the available data regarding the link between perimenopause and depression is hampered by the lack of a standard definition of menopausal status in different studies, both biological and psychosocial factors appear to play a role. This multifactorial etiology of perimenopausal depression was supported by two prospective studies, in which premenopausal women with no history of major depression were found to have an increased risk of depression during the menopausal transition (Cohen et al., 2006b; Freeman et al., 2006a). A previous depression history, vasomotor symptoms (e.g., hot flashes), negative life events, and variability in estradiol levels were also associated with increased risk for depression. These findings are consistent with the notion that both biological and psychosocial factors increase the risk for depression during the menopausal transition.

Approach to the Patient with Perimenopausal Mood Symptoms

Perimenopausal depression requires a biopsychosocial assessment and treatment approach (see Clinical Vignette 3). Psychiatric evaluation should include screening for major Axis I diagnoses and a history of psychiatric symptoms during other reproductive events. In addition, clinicians should take inventory of recent stressors, personal coping strengths and liabilities, and the extent of social supports. Screening for alcohol and substance abuse can also help identify possible reasons for worsening psychiatric symptoms. Medical evaluation should include assessment of reproductive endocrine status by taking a menstrual history and screening for perimenopausal vasomotor symptoms, such as hot flashes and vaginal dryness. Hormonal changes in menopause include a decrease in estrogen with subsequent elevations of luteinizing hormone and follicle-stimulating hormone. Clinicians should also rule out possible medical causes of mood symptoms that are common in midlife women, including thyroid and cardiac disease.

Clinical Vignette 3

Diane is a 50-year-old physician with no psychiatric history until this past year, when she experienced many life changes, including separating from her husband and caring for her ailing parents. In addition, over the last 6 months, her periods have become irregular and she has had recurrent hot flashes that have interrupted her sleep. She presents with symptoms consistent with major depression, including low mood, irritability, decreased interests, poor concentration, sleep and appetite disturbance, and feelings of hopelessness. From Diane's report of vasomotor symptoms and irregular menses, it is clear that she is in the menopausal transition. A trial of an antidepressant improves her depressive symptoms and also alleviates her hot flashes. In addition, a course of psychodynamic psychotherapy helps her begin to understand the connection between her early life experiences and her current anxiety, anger, and sadness as she grieves the end of her marriage and the possible loss of her debilitated parents.

Treatment Considerations in Women with Perimenopausal Mood Symptoms

Guidelines on the evaluation and treatment of perimenopausal depression were recently published (Maki et al., 2019). Studies have clearly shown that HT alleviates vasomotor symptoms, such as hot flashes (Warren, 2004). Some studies have also found that estrogen may have antidepressant properties, possibly as a direct neuromodulatory effect and also a result of the psychological relief that depressed women experience as their vasomotor symptoms diminish (Schmidt et al., 2000; Soares et al., 2001). For perimenopausal women with significant vasomotor symptoms, short-term HT may alleviate mild mood symptoms along with insomnia, vaginal dryness, and hot flashes. Prior to recommending HT, however, clinicians should consider the Women's Health Initiative study results, in which HT was associated with an increased risk of breast cancer, coronary heart disease, and thromboembolic events in postmenopausal women (Rossouw et al., 2002; Manson et al., 2003). It remains to be seen whether these risks are generalizable to younger perimenopausal women on HT and whether or not the benefits of short-term HT for depressed perimenopausal women outweigh the risks.

For perimenopausal women who present with major depression, clinicians should consider standard treatment, including antidepressant medication and psychotherapy. There is some evidence that menopausal status may affect the efficacy of different classes of antidepressants (Kornstein et al., 2000; Thase et al., 2005; Pinto-Meza et al., 2006). For instance, in one study, premenopausal women with major depression were found to respond better to sertraline than imipramine, whereas postmenopausal women responded similarly to both agents (Kornstein et al., 2000). Few studies have prospectively evaluated the efficacy of antidepressants in peri- or postmenopausal women and until recently most were small, open-label trials (Joffe et al., 2001, 2007; Soares et al., 2003, 2006; Ladd et al., 2005; Freeman et al., 2006b). Two large, randomized double-blind, placebo controlled studies have demonstrated the efficacy of desvenlafaxine for the treatment of MDD in peri-and postmenopausal women (Kornstein et al., 2010; Clayton et al., 2013). Some psychotropic medications such as the SSRIs (Stearns et al., 2003; Gordon et al., 2006; Soares et al., 2006), venlafaxine (Evans et al., 2005; Ladd et al., 2005), and gabapentin (Reddy et al., 2006) have also been shown to alleviate hot flashes, although only paroxetine has an FDA indication for this use. Among the botanical and dietary supplements, St. John's wort and black cohosh appear to be the most beneficial for mood and anxiety changes during menopause (Geller & Studee, 2005) though evidence is limited (Maki et al., 2019). In addition, psychotherapy, both alone and in combination with medication, is highly beneficial for women coping with the challenges associated with midlife. Two clinical trials have demonstrated that CBT is efficacious in improving depression related to the menopause transition (Khoshbooii et al., 2011; Brandon et al., 2013), while a pilot study has provided evidence that CBT can be helpful in reducing both the emotional and physical symptoms of perimenopause (Green et al., 2013).

Conclusion

Women have higher rates of major depression than men in every age group, with peak rates during the reproductive years. This observation has inspired a field of research examining psychiatric conditions associated with reproductive events, such as menses, pregnancy, and menopause. The transition to motherhood and, later, to menopause can be fraught with challenges as women adapt to shifting roles and other life changes. In addition, the hormonal changes after delivery and during the perimenopausal transition contribute to mood and anxiety problems in a subset of women who are prone to psychiatric symptoms during hormonal shifts. Recent studies questioning the reproductive safety of certain psychotropic medications and also the thromboembolic and cancer risks associated with postmenopausal HT underscore the fact that reproductive psychiatry is a dynamic and evolving area of research. Whether treating perinatal or perimenopausal patients, clinicians therefore must adopt an individualized treatment approach that addresses biopsychosocial factors and also incorporates an up-to-date discussion of the risks and benefits of medication and psychotherapy options.

References

ACOG Committee Opinion no. 757. (2018). Screening for perinatal depression. *Obstetrics and Gynecology, 132*(5), e208–e212.

Altshuler, L. L., & Hendrick, V. (1996). Pregnancy and psychotropic medication: Changes in blood levels [Letter]. *Journal of Clinical Psychopharmacology, 16*, 78–80.

Altshuler, L. L., Hendrick, V., & Cohen, L. S. (2000). An update on mood and anxiety disorders during pregnancy and the postpartum period. *Primary Care Companion to the Journal of Clinical Psychiatry, 2*, 217–222.

American Psychiatric Association. (2000). *Practice guidelines for the treatment of psychiatric disorders. Compendium 2000.* American Psychiatric Association Press.

Andrade, S. E., McPhillips, H., Loren, D., et al. (2009). Antidepressant medication use and risk of persistent pulmonary hypertension of the newborn. *Pharmacoepidemiology and Drug Safety, 18*(3), 246–252.

Appleby, L., Warner, R., Whitton, A., et al. (1997). A controlled study of fluoxetine and cognitive-behavioural counseling in the treatment of postnatal depression. *BMJ (Clinical Research Edition), 314*, 932–936.

Ashman, S. B., Dawson, G., Panagiotides, H., et al. (2002). Stress hormone levels of children of depressed mothers. *Development and Psychopathology, 14*(2), 333–349.

Beck, C. T. (2003). Postpartum depression predictors inventory (revised). *Advances in Neonatal Care, 3*, 47–48.

Bloch, M., Schmidt, P. J., Danaceau, M., et al. (2000). Effects of gonadal steroids in women with a history of postpartum depression. *American Journal of Psychiatry, 157*, 924–930.

Bogen, D. L., Sit, D., Genovese, A., et al. (2012). Three cases of lithium exposure and exclusive breastfeeding. *Archives of Women's Mental Health, 15*(1), 69–72.

Brandes, M., Soares, C. N., & Cohen, L. S. (2004). Postpartum onset obsessive-compulsive disorder: Diagnosis and management. *Archives of Women's Mental Health, 7*(2), 99–110.

Brandon, A. R., Ceccotti, N., Hynan, L. S., et al. (2012). Proof of concept: Partner-assisted interpersonal psychotherapy for perinatal depression. *Archives of Women's Mental Health, 15*(6), 469–480.

Brandon, A. R., Minhajuddin, A., Thase, M. E., & Jarrett, R. B. (2013). Impact of reproductive status and age on response of depressed women to cognitive therapy. *Journal of Women's Health (Larchmt), 22*, 58–66.

Breese McCoy, S. J. (2011). Postpartum depression: An essential overview for the practitioner. *Southern Medical Journal, 104*(2), 128–132.

Brennan, P. A., Hammen, C., Andersen, M. J., et al. (2000). Chronicity, severity, and timing of maternal depressive symptoms: Relationships with child outcomes at age 5. *Developmental Psychology, 36*, 759–766.

Brennan, P. A., Pargas, R., Walker, E. F., et al. (2008). Maternal depression and infant cortisol: Influences of timing, comorbidity and treatment. *Journal of Child Psychology and Psychiatry, 49*(10), 1099–1107.

Brockington, I. F., Oates, J., George, S., et al. (2001). A screening questionnaire for mother-infant bonding disorders. *Archives of Women's Mental Health, 3*, 133–140.

Chambers, C. D., Hernandez-Diaz, S., Van Marter, L. J., et al. (2006). Selective serotonin-reuptake inhibitors and risk of persistent pulmonary hypertension of the newborn. *New England Journal of Medicine, 354*, 579–587.

Chun-Fai-Chan, B., Koren, G., Fayez, I., et al. (2005). Pregnancy outcome of women exposed to bupropion during pregnancy: A prospective comparative study. *American Journal of Obstetrics and Gynecology, 192*, 932–936.

Chung, T. K., Lau, T. K., Yip, A. S., et al. (2001). Antepartum depressive symptomatology is associated with adverse obstetric and neonatal outcomes. *Psychosomatic Medicine, 63*, 830–834.

Clayton, A. H., Kornstein, S. G., Dunlop, B. W., et al. (2013). Efficacy and safety of desvenlafaxine 50 mg/d in a randomized, placebo-controlled study of perimenopausal and postmenopausal women with major depressive disorder. *The Journal of Clinical Psychiatry, 74*, 1010–1017.

Cohen, L. S., Viguera, A. C., Bouffard, S. M., et al. (2001). Venlafaxine in the treatment of postpartum depression. *Journal of Clinical Psychiatry, 62*, 529–596.

Cohen, L. S., Altshuler, L. L., Harlow, B. L., et al. (2006a). Relapse of major depression during pregnancy in women who maintain or discontinue antidepressant treatment. *JAMA, 295*, 499–507.

Cohen, L. S., Soares, C. N., Vitonix, A. F., et al. (2006b). Risk for new onset of depression during the menopausal transition: The Harvard Study of Moods and Cycles. *Archives of General Psychiatry, 63*, 385–390.

Cox, J. L., Holden, J. M., & Sagovsky, R. (1987). Detection of postnatal depression: Development of the 10-item Edinburgh Postnatal Depression Scale. *British Journal of Psychiatry, 150*, 782–786.

Cuijpers, P., Brannmark, J. G., & Van Straten, A. (2008). Psychological treatment of postpartum depression: A meta-analysis. *Journal of Clinical Psychology, 64*(1), 103–118.

Cunnington, M., Tennis, P., & International Lamotrigine Pregnancy Registry Scientific Advisory Committee. (2005). Lamotrigine and the risk of malformations in pregnancy. *Neurology, 64*, 955–960.

Curry, S. I., Preventive Services Task Force, U. S., Krist, A. H., et al. (2019). Interventions to prevent perinatal depression. US Preventative Services Task Force recommendation statement. *JAMA, 321*(6), 580–587.

Dennerstein, L., Dudley, E., & Guthrie, J. (2002). Empty nest or revolving door? A prospective study of women's quality of life in midlife during the phase of children leaving and re-entering the home. *Psychological Medicine, 32*, 545–550.

Diego, M. A., Field, T., Hernandez-Reif, M., et al. (2004). Prepartum, postpartum, and chronic depression effects on newborns. *Psychiatry, 67*(1), 63–80.

Dietz, P. M., Williams, S. B., Callaghan, W. M., et al. (2007). Clinically identified maternal depression before, during, and after pregnancies ending in live births. *American Journal of Psychiatry, 164*(10), 1515–1520.

Dimidjian, S., & Goodman, S. (2009). Nonpharmacologic intervention and prevention strategies for depression during pregnancy and the postpartum. *Clinical Obstetrics and Gynecology, 52*, 498–515.

Djulus, J., Koren, G., Einarson, T. R., et al. (2006). Exposure to mirtazapine during pregnancy: A prospective, comparative study of birth outcomes. *Journal of Clinical Psychiatry, 67*, 1280–1284.

Dolk, H., Jentink, J., Loane, M., et al. (2008). Does lamotrigine use in pregnancy increase orofacial cleft risk relative to other malformations? *Neurology, 71*(10), 714–722.

Einarson, T. R., & Einarson, A. (2005). Newer antidepressants in pregnancy and rates of major malformations: A meta-analysis of prospective comparative studies. *Pharmacoepidemiology and Drug Safety, 14*, 823–827.

Einarson, A., Fatoye, B., Sarkar, M., et al. (2001). Pregnancy outcome following gestational exposure to venlafaxine: A multicenter prospective controlled study. *American Journal of Psychiatry, 158*, 1728–1730.

Einarson, A., Bonari, L., Voyer-Lavigne, S., et al. (2003). A multicentre prospective controlled study to determine the safety of trazadone and nefazodone use during pregnancy. *Canadian Journal of Psychiatry. Revue Canadienne de Psychiatrie, 48*, 106–110.

Epperson, C. N., Terman, M., Terman, J. S., et al. (2004). Randomized clinical trial of bright light therapy for antepartum depression: Preliminary findings. *Journal of Clinical Psychiatry, 65*, 421–425.

Ericson, A., Kallen, B., & Wilholm, B. E. (1999). Delivery outcome after the use of antidepressants in early pregnancy. *European Journal of Clinical Pharmacology, 55*, 503–508.

Essex, M. J., Klein, M. H., Cho, E., et al. (2002). Maternal stress beginning in infancy may sensitize children to later stress exposure: Effects on cortisol and behavior. *Biological Psychiatry, 52*(8), 776–784.

Evan, J., Heron, J., Francomb, H., et al. (2001). Cohort study of depressed mood during pregnancy and after childbirth. *BMJ (Clinical Research Edition), 323*, 257–260.

Evans, M. L., Pritts, E., Vittinghoff, E., et al. (2005). Management of postmenopausal hot flushes with venlafaxine hydrochloride: A randomized, controlled trial. *Obstetrics and Gynecology, 105*, 161–166.

Field, T. (2011). Prenatal depression effects on early development: A review. *Infant Behavior and Development, 34*(1), 1–14.

Fotopoulou, C., Kretz, R., Bauer, S., et al. (2009). Prospectively assessed changes in lamotrigine-concentration in women with epilepsy during pregnancy, lactation and the neonatal period. *Epilepsy Research, 85*(1), 60–64.

Freeman, E. W., Sammel, M. D., Lin, H., et al. (2006a). Associations of hormones and menopausal status with depressed mood in women with no history of depression. *Archives of General Psychiatry, 63*, 375–382.

Freeman, M. P., Hill, R., & Brumbach, B. H. (2006b). Escitalopram for perimenopausal depression: An open-label pilot study. *Journal of Women's Health (Larchmont), 15*, 857–861.

Fujii, H., Goel, A., Bernard, N., et al. (2013). Pregnancy outcomes following gabapentin use: Results of a prospective comparative cohort study. *Neurology, 80*(17), 1565–1570.

Gaynes, B. N., Gavin, N., Meltzer-Brody, S., et al. (2005). Perinatal depression: Prevalence, screening accuracy, and screening outcomes. *Evidence Report/Technology Assessment (Summary), 119*, 1–8.

Geller, S. E., & Studee, L. (2005). Botanical supplements for menopausal symptoms: What works, what does not work. *Journal of Women's Mental Health, 14*, 634–649.

Gershon, A. A., Dannon, P. N., & Grunhaus, N. (2003). Transcranial magnetic stimulation in the treatment of depression. *American Journal of Psychiatry, 160*(5), 835–845.

Gordon, P. R., Kerwin, J. P., Boessen, K. G., et al. (2006). Sertraline to treat hot flashes: A randomized controlled, double-blind, crossover trial in a general population. *Menopause, 13*, 546–548.

Grace, S. L., Evindar, A., & Stewart, D. E. (2003). The effect of postpartum depression on child cognitive development and behavior: A review and critical analysis of the literature. *Archives of Women's Mental Health, 6*(4), 263–274.

Green, S. M., Haber, E., McCabe, R. E., et al. (2013). Cognitive-behavioral group treatment for menopausal symptoms: A pilot study. *Archives of Women's Mental Health, 16*, 325–332.

Greene, J. G., & Cooke, D. J. (1980). Life stress and symptoms at the climacterium. *British Journal of Psychiatry, 136*, 486–491.

Grigoriadis, S., Vonderporten, E. H., Mamisashvili, L., et al. (2014). Prenatal exposure to antidepressants and persistent pulmonary hypertension of the newborn: Systematic review and meta-analysis. *BMJ, 348*, f6932.

Habermann, F., Fritzsche, J., Fuhlbruck, F., et al. (2013). Atypical antipsychotic drugs and pregnancy outcome: A propective, cohort study. *Journal of Clinical Psychopharmacology, 33*, 453–462.

Halligan, S. L., Herbert, J., Goodyer, I. M., et al. (2004). Exposure to postnatal depression predicts elevated cortisol in adolescent offspring. *Biological Psychiatry, 55*(4), 376–381.

Hanna, E. Z., Faden, V. B., & Dufour, M. C. (1994). The motivational correlates of drinking, smoking, and illicit drug use during pregnancy. *Journal of Substance Abuse, 6*, 155–167.

Hendrick, V., Altschuler, L. L., & Suri, R. (1998). Hormonal changes in the postpartum and implications for postpartum depression. *Psychosomatics, 39*, 93–101.

Hendrick, V., Smith, L. M., Suri, R., et al. (2003). Birth outcomes after prenatal exposure to antidepressant medication. *American Journal of Obstetrics and Gynecology, 188*, 812–815.

Hendrick, V., Suri, R., Gitlin, M. J., & Ortiz-Portillo, E. (2017). Bupropion use during pregnancy: A systematic review. *Primary Care Companion for CNS Disorders, 19*(5), 17r02160.

Hernandez-Diaz, S., Huybrechts, K. F., Desai, R. J., et al. (2018). Topiramate use early in pregnancy and the risk of oral clefts: A pregnancy cohort study. *Neurology, 90*(4), e342–e351.

Hirschfeld, R. M. (2002). The mood disorder questionnaire: A simple, patient-rated screening instrument for bipolar disorder. *Primary Care Companion to the Journal of Clinical Psychiatry, 4*(1), 9–11.

Hirst, K. P., & Moutier, C. Y. (2010). Postpartum major depression. *American Family Physician, 82*(8), 926–933.

Holmes, L. B., Baldwin, E. J., Smith, C. R., et al. (2008). Increased frequency of isolated cleft palate in infants exposed to lamotrigine during pregnancy. *Neurology, 70*(22 Pt 2), 2152–2158.

Huybrechts, K. F., Palmsten, K., Avorn, J., et al. (2014). Antidepressant use in pregnancy and the risk of cardiac defects. *The New England Journal of Medicine, 370*(25), 2397–2407.

Huybrechts, K. F., Hernández-Díaz, S., Patorno, E., et al. (2016). Antipsychotic use in pregnancy and the risk for congenital malformations. *JAMA Psychiatry, 73*(9), 938–946.

Iqbal, M. M., Sobhan, T., & Ryais, T. (2002). Effects of commonly used benzodiazepines on the fetus, the neonate, and the nursing infant. *Psychiatric Services, 53*, 39–49.

Joffe, H., Groninger, H., Soares, C. N., et al. (2001). An open trial of mirtazapine in menopausal women with depression unresponsive to estrogen replacement therapy. *Journal of Women's Health and Gender-Based Medicine, 10*, 999–1004.

Joffe, H., Soares, C. N., Petrillo, L. F., et al. (2007). Treatment of depression and menopause-related symptoms with the serotonin-norepinephrine reuptake inhibitor duloxetine. *Journal of Clinical Psychiatry, 68*, 943–950.

Källén, B. (2004). Neonate characteristics after maternal use of antidepressants late in pregnancy. *Archives of Pediatrics and Adolescent Medicine, 158*, 312–316.

Källén, B., & Olausson, P. O. (2008). Maternal use of selective serotonin reuptake inhibitors and persistent pulmonary hypertension of the newborn. *Pharmacoepidemiology and Drug Safety, 17*(8), 801–806.

Kammerer, M., Taylor, A., & Glover, V. (2006). The HPA axis and perinatal depression: A hypothesis. *Archives of Women's Mental Health, 9*, 187–196.

Kelly, R. H., Zatzick, D. F., & Anders, T. F. (2001). The detection and treatment of psychiatric disorders and substance use among pregnant women cared for in obstetrics. *American Journal of Psychiatry, 158*, 213–219.

Khoshbooii, R., Hassan, S. A. B., Hamzah, M. S. G. B., & Bini Baba, M. (2011). Effectiveness of group cognitive therapy on depression among Iranian women around menopause. *Australian Journal of Basic and Applied Sciences, 5*, 991–995.

Kieler, H., Artama, M., Engeland, A., et al. (2012). Selective serotonin reuptake inhibitors during pregnancy and risk of persistent pulmonary hypertension in the newborn: Population based cohort study from the five Nordic countries. *BMJ, 344*, d8012.

Kim, D. R., Epperson, N., Paré, E., et al. (2011). An open label pilot study of transcranial magnetic stimulation for pregnant women with major depressive disorder. *Journal of Women's Health, 20*(2), 255–261.

Klirova, M., Novak, T., Kopecek, M., et al. (2008). Repetitive transcranial magnetic stimulation (rTMS) in major depressive episode during pregnancy. *Neuroendocrinology Letters, 29*(1), 69–70.

Ko, J. Y., Farr, S. L., Dietz, P. M., et al. (2012). Depression and treatment among U.S. pregnant and nonpregnant women of reproductive age, 2005–2009. *Journal of Women's Health, 21*, 830–836.

Kornstein, S. G., Schatzberg, A. F., Thase, M. E., et al. (2000). Gender differences in treatment response to sertraline versus imipramine in chronic depression. *American Journal of Psychiatry, 157*, 1445–1452.

Kornstein, S. G., Jiang, Q., Reddy, S., et al. (2010). Short-term efficacy and safety of desvenlafaxine in a randomized, placebo-controlled study of peri-and postmenopausal women with major depressive disorder. *Journal of Clinical Psychiatry, 71*, 1088–1096.

Kripke, D. F. (1998). Light treatment for nonseasonal depression: Speed, efficacy, and combined treatment. *Journal of Affective Disorders, 49*, 109–117.

Kripke, D. F., Mullaney, D. J., Klauber, M. R., et al. (1992). Controlled trial of bright light for nonseasonal major depressive disorders. *Biological Psychiatry, 31*, 119–134.

Kulin, N. A., Pastuszak, A., Sage, S., et al. (1998). Pregnancy outcome following maternal use of the new selective serotonin reuptake inhibitors: A prospective controlled multicenter study. *JAMA, 279*, 609–610.

Kurki, T., Hiilesmaa, V., Raitasalo, R., et al. (2000). Depression and anxiety in early pregnancy and risk for preeclampsia. *Obstetrics and Gynecology, 95*, 487–490.

Ladd, C. O., Newport, D. J., Ragan, K. A., et al. (2005). Venlafaxine in the treatment of depressive and vasomotor symptoms in women with perimenopausal depression. *Depression and Anxiety, 22*(2), 94–97.

Lam, R. W., & Levitt, A. J. (1999). *Canadian consensus guidelines for the treatment of seasonal affective disorder.* Clinical and Academic Publishing.

Lassen, D., Ennis, Z. N., & Damkier, P. (2016). First-trimester exposure to venlafaxine or duloxetine and risk of major congenital malformations: A systematic review. *Basic & Clinical Pharmacology & Toxicology, 118*(1), 32–36.

Lennestål, R., & Källén, B. (2007). Delivery outcome in relation to maternal use of some recently introduced antidepressants. *Journal of Clinical Psychopharmacology, 27*(6), 607–613.

Levinson-Castiel, R., Merlob, P., Linder, N., et al. (2006). Neonatal abstinence syndrome after in utero exposure to selective serotonin reuptake inhibitors in term infants. *Archives of Pediatrics and Adolescent Medicine, 160*, 173–176.

Li, D., Liu, L., & Odouli, R. (2009). Presence of depressive symptoms during early pregnancy and the risk of preterm delivery: A prospective cohort study. *Human Reproduction, 24*(1), 146–153.

Lin, A. E., Peller, A. J., Westgate, M. N., et al. (2004). Clonazepam use in pregnancy and the risk of malformations. *Birth Defects Research Part A: Clinical and Molecular Teratology, 70*, 534–536.

Lindahl, V., Pearson, J. L., & Colpe, L. (2005). Prevalence of suicidality during pregnancy and the postpartum. *Archives of Women's Mental Health, 8*(2), 77–87.

Maki, P. M.*, Kornstein, S. G.*, Joffe, H., et al. (2019). Guidelines for the evaluation and treatment of perimenopausal depression: Summary and recommendations. *Journal of Women's Health, 28*, 117–134. (*co-first authors).

Malm, H., Klaukka, T., & Neuvonen, P. J. (2005). Risks associated with selective serotonin reuptake inhibitors in pregnancy. *Obstetrics and Gynecology, 106*, 1289–1296.

Manson, J. E., Hsia, J., Johnson, K. C., et al. (2003). Estrogen plus progestin and the risk of coronary heart disease. *New England Journal of Medicine, 349*, 523–534.

Margulis, A. V., Mitchell, A. A., Gilboar, S. M., et al. (2012). Use of topiramate in pregnancy and risk of oral clefts. *American Journal of Obstetrics and Gynecology, 207*(5), 405.e1–405.e7.

Meltzer-Brody, S., Colquhoun, H., Riesenberg, R., et al. (2018). Brexanolone injection in post-partum depression: Two multicentre, double-blind, randomised, placebo-controlled, phase 3 trials. *Lancet, 392*(10152), 1058–1070.

Moehler, E., Brunner, R., Wiebel, A., et al. (2006). Maternal depressive symptoms in the postnatal period are associated with long-term impairment of mother-child bonding. *Archives of Women's Mental Health, 9*, 273–278.

Moses-Kolko, E. L., Bogen, D., Perel, J., et al. (2005). Neonatal signs after late in utero exposure to serotonin reuptake inhibitors: Literature review and implications for clinical applications. *JAMA, 293*, 2372–2383.

Murphy-Eberenz, K., Zandi, P. P., March, D., et al. (2006). Is perinatal depression familial? *Journal of Affective Disorders, 90*(1), 49–55.

Murray, L., Hipwell, A., Hooper, R., et al. (1996). The cognitive development of 5-year-old children of postnatally depressed mothers. *Journal of Child Psychology and Psychiatry, and Allied Disciplines, 37*, 927–935.

Nahas, Z., Bohning, D. E., Molloy, M. A., et al. (1999). Safety and feasibility of repetitive transcranial magnetic stimulation in the treatment of anxious depression in pregnancy: A case report. *Journal of Clinical Psychiatry, 60*(1), 50–52.

Newham, J. J., Thomas, S. H., MacRitchie, K., McElhatton, P. R., et al. (2008). Birth weight of infants after maternal exposure to typical and atypical antipsychotics: Prospective comparison study. *The British Journal of Psychiatry, 192*(5), 333–337.

Newmark, R. L., Bogen, D. L., Wisner, K. L., et al. (2019). Risk-benefit assessment of infant exposure to lithium through breast milk: A systematic review of the literature. *International Review of Psychiatry, 31*(3), 295–304.

Newport, D. J., Calamaras, M. R., DeVane, C. L., et al. (2007). Atypical antipsychotic administration during late pregnancy: Placental passage and obstetrical outcomes. *The American Journal of Psychiatry, 164*(8), 1214–1220.

Newport, D. J., Pennell, P. B., Calamaras, M. R., et al. (2008). Lamotrigine in breast milk and nursing infants: Determination of exposure. *Pediatrics, 122*(1), e223–e231.

Ng, Q. X., Venkatanarayanan, N., Ho, C. Y. X., et al. (2019). Selective serotonin reuptake inhibitors and persistent pulmonary hypertension of the newborn: An update meta-analysis. *Journal of Women's Health (2002), 28*(3), 331–338.

Nonacs, R. M., Soares, C. N., Viguera, A. C., et al. (2005). Bupropion SR for the treatment of postpartum depression: A pilot study. *International Journal of Neuropsychopharmacology, 8*(3), 445–449.

Nulman, I., Rovet, J., Stewart, D. E., et al. (2002). Child development following exposure to tricyclic antidepressants or fluoxetine throughout fetal life: A prospective, controlled study. *American Journal of Psychiatry, 159*, 1889–1895.

Nulman, I., Koren, G., Rovet, J., et al. (2012). Neurodevelopment of children following prenatal exposure to venlafaxine, selective serotonin reuptake inhibitors or untreated maternal depression. *American Journal of Psychiatry, 169*, 1165–1174.

O'Connor, T. G., Ben-Shlomo, Y., Heron, J., et al. (2005). Prenatal anxiety predicts individual differences in cortisol in pre-adolescent child. *Biological Psychiatry, 58*(3), 211–217.

O'Hara, M. W. (2009). Postpartum depression: What we know. *Journal of Clinical Psychology, 65*(12), 1258–1269.

O'Hara, M. W., & Swain, A. M. (1996). Rates and risk of postpartum depression – A meta-analysis. *International Review of Psychiatry, 8*(1), 37–54.

O'Hara, M. W., Schlecte, J. A., Lewis, D. A., et al. (1991). Controlled prospective study of postpartum mood disorders: Psychological, environmental and hormonal variables. *Journal of Abnormal and Social Psychology, 100*, 63–73.

O'Hara, M. W., Stuart, S., Gorman, L. L., et al. (2000). Efficacy of interpersonal psychotherapy for postpartum depression. *Archives of General Psychiatry, 57*, 1039–1045.

O'Reardon, J. P., Cristancho, M. A., von Andreae, C. V., et al. (2011). Acute and maintenance electroconvulsive therapy for treatment of severe major depression during the second and third trimesters of pregnancy with infant follow-up to 18 months: Case report and review of the literature. *Journal of ECT, 27*, e23–e26.

Oberlander, T. F., Warburton, W., Misri, S., et al. (2008). Major congenital malformations following prenatal exposure to serotonin reuptake inhibitors and benzodiazepines using population-based health data. *Birth*

Defects Resesrch Part B: Developmental and Reproductive Toxicology, 83(1), 68–76.

Occhiogrosso, M., Omran, S. S., & Altemus, M. (2012). Persistent pulmonary hypertension of the newborn and selective serotonin reuptake inhibitors: Lessons from clinical and translational studies. *American Journal of Psychiatry, 169*(2), 134–140.

Oren, D. A., Wisner, K. L., Spinelli, M., et al. (2002). An open trial of morning light therapy for treatment of antepartum depression. *American Journal of Psychiatry, 159,* 666–669.

Orr, S. T., Blazer, D. G., James, S. A., et al. (2007). Depressive symptoms and indicators of maternal health status during pregnancy. *Journal of Women's Health, 16*(4), 535–542.

Pariente, G., Leibson, T., Shulman, T., et al. (2017). Pregnancy outcomes following in utero exposure to lamotrigine: A systematic review and meta-analysis. *CNS Drugs, 31,* 439–450.

Parry, B. L., Sorenson, D. L., Meliska, C. J., et al. (2003). Hormonal basis of mood and postpartum disorders. *Current Women's Health Reports, 3,* 230–235.

Patorno, E., Hernandez-Diaz, S., Huybrechts, K. F., et al. (2020). Gabapentin in pregnancy and the risk of adverse neonatal and maternal outcomes: A population-based cohort study nested in the US Medicaid Analytic eXtract dataset. *PLoS Medicine, 17*(9), e1003322.

Payne, J. L., Roy, P. S., Murphy-Eberenz, K., et al. (2007). Reproductive cycle-associated mood symptoms in women with major depression and bipolar I disorder. *Journal of Affective Disorders, 99,* 221–229.

Payne, J. L., MacKinnon, D. F., Mondimore, F. M., et al. (2008). Familial aggregation of postpartum mood symptoms in families with bipolar disorder. *Journal of Bipolar Disorders, 10*(1), 38–44.

Pinto-Meza, A., Usall, J., Serrano-Blanco, A., et al. (2006). Gender differences in response to antidepressant treatment prescribed in primary care. Does menopause make a difference? *Journal of Affective Disorders, 93,* 53–60.

Reay, R., Fisher, Y., Robertson, M., et al. (2006). Group interpersonal psychotherapy for postnatal depression: A pilot study. *Archives of Women's Mental Health, 9,* 31–39.

Reddy, S. Y., Warner, H., Guttuso, T., Jr., et al. (2006). Gabapentin, estrogen, and placebo for treating hot flushes: A randomized controlled trial. *Obstetrics and Gynecology, 108,* 41–48.

Righetti-Veltema, M., Conne-Perreard, E., Bousquet, A., et al. (2002). Postpartum depression and mother-infant relationship at 3 months old. *Journal of Affective Disorders, 70,* 291–306.

Robertson, E., Grace, S., Wallington, T., et al. (2004). Antenatal risk factors for postpartum depression: A synthesis of recent literature. *General Hospital Psychiatry, 26,* 289–295.

Rossouw, J. E., Anderson, G. L., Prentice, R. L., et al. (2002). Risks and benefits of estrogen plus progestin in health postmenopausal women: Principal results from the Women's Health Initiative randomized controlled trial. *JAMA, 288,* 321–333.

Rubinow, D. R., Schmidt, P. J., & Roca, C. A. (1998). Estrogen-serotonin interactions: Implications for affective regulation. *Biological Psychiatry, 44,* 839–850.

Rubinow, D. R., Johnson, S. L., Schmidt, P. J., Girdler, S., & Gaynes, B. (2015). Efficacy of estradiol in perimenopausal depression: So much promise and so few answers. *Depression and Anxiety, 32,* 539–549.

Schmidt, P. J., Nieman, L., Danaceau, M. A., et al. (2000). Estrogen replacement in perimenopause-related depression: A preliminary report. *American Journal of Obstetrics and Gynecology, 183,* 414–420.

Schmidt, P. J., Murphy, J. H., Haq, N., et al. (2004). Stressful life events, personal losses, and perimenopause-related depression. *Archives of Women's Mental Health, 7,* 19–26.

Sharma, V. (2003). Role of sleep loss in the causation of puerperal psychosis. *Medical Hypotheses, 61*(4), 477–481.

Siu, A. L., US Preventive Services Task Force (USPSTF), Bibbins-Domingo, K., et al. (2016). Screening for depression in adults: US preventive services task force recommendation statement. *JAMA, 315*(4), 380–387.

Smit, M., Dolman, K. M., & Honig, A. (2016). Mirtazapine in pregnancy and lactation- A systematic review. *European Neuropsychopharmacology, 26*(1), 126–135.

Soares, C. N., Almeida, O. P., Joffe, H., et al. (2001). Efficacy of estradiol for the treatment of depressive disorders in perimenopausal women: A doubleblind, randomized, placebo-controlled trial. *Archives of General Psychiatry, 58,* 529–534.

Soares, C. N., Poitras, J. R., Prouty, J., et al. (2003). Efficacy of citalopram as a monotherapy or as an adjunctive treatment to estrogen therapy for perimenopausal and postmenopausal women with depression and vasomotor symptoms. *Journal of Clinical Psychiatry, 64,* 473–479.

Soares, C. N., Arsenio, H., Joffe, H., et al. (2006). Escitalopram versus ethinyl estradiol and norethindrone acetate for symptomatic peri-and postmenopausal women: Impact on depression, vasomotor symptoms, sleep, and quality of life. *Menopause, 13,* 780–786.

Soufia, M., Aoun, J., Gorsane, M. A., et al. (2010). SSRIs and pregnancy: A review of the literature. *Encephale, 36*(6), 513–516.

Spinelli, M. G., & Endicott, J. (2003). Controlled clinical trial of interpersonal psychotherapy versus parenting education program for depressed pregnant women. *American Journal of Psychiatry, 160,* 555–562.

Spitzer, R. L., Williams, J. B., Kroenke, K., et al. (2000). Validity and utility of the PRIME-MD patient health questionnaire in assessment of 3000 obstetric/gynecologic patients: The PRIME-MD Patient Health Questionnaire Obstetric-Gynecology Study. *American Journal of Obstetrics and Gynecology, 183,* 759–769.

Stearns, V., Beebe, K. L., Iyengar, M., et al. (2003). Paroxetine controlled release in the treatment of menopausal hot flashes: A randomized controlled trial. *JAMA, 289*, 2827–2834.

Stewart, D. E., & Boydell, K. M. (1993). Psychologic distress during menopause: Associations across the reproductive life cycle. *International Journal of Psychiatry in Medicine, 23*, 157–162.

Stowe, Z. N., & Nemeroff, C. B. (1995). Women at risk for postpartum-onset major depression. *American Journal of Obstetrics and Gynecology, 173*, 639–645.

Thase, M. E., Entsuah, R., Cantillon, M., et al. (2005). Relative antidepressant efficacy of venlafaxine and SSRIs: Sex-age interactions. *Journal of Women's Health, 14*, 609–616.

Viguera, A. C., Nonacs, R., Cohen, L. S., et al. (2000). Risk of recurrence of bipolar disorder in pregnant and nonpregnant women after discontinuing lithium maintenance. *American Journal of Psychiatry, 157*, 179–184.

Viguera, A. C., Whitfield, T., Baldessarini, R. J., et al. (2007). Risk of recurrence in women with bipolar disorder during pregnancy: Prospective study of mood stabilizer discontinuation. *American Journal of Psychiatry, 164*(12), 1817–1824.

Viguera, A. C., Tondo, L., Koukopoulos, A. E., et al. (2011). Episodes of mood disorders in 2,252 pregnancies and postpartum periods. *American Journal of Psychiatry, 168*(11), 1179–1185.

Wang, S., Yang, L., Wang, L., et al. (2015). Selective serotonin reuptake inhibitors (SSRIs) and the risk of congenital heart defects: A meta-analysis of prospective cohort studies. *Journal of the American Heart Association, 4*(5), e001681.

Warburton, W., Hertzman, C., & Oberlander, T. F. (2010). A register study of the impact of stopping third trimester selective serotonin reuptake inhibitor exposure on neonatal health. *Acta Psychiatrica Scandinavica, 121*(6), 471–479.

Warren, M. P. (2004). A comparative review of the risks and benefits of hormone replacement therapy regimen. *American Journal of Obstetrics and Gynecology, 190*, 1141–1167.

Weissman, A. M., Levy, B. T., Hartz, A. J., et al. (2004). Analysis of antidepressant levels in lactating mothers, breast milk, and nursing infants. *American Journal of Psychiatry, 161*, 1066–1078.

Wichman, C. L., Moore, K. M., Lang, T. R., et al. (2009). Congenital heart disease associated with selective serotonin reuptake inhibitor use during pregnancy. *Mayo Clinic Proceedings, 84*(1), 23–27.

Wilson, K. L., Zelig, C. M., Harvey, J. P., et al. (2011). Persistent pulmonary hypertension of the newborn is associated with mode of delivery and not with maternal use of selective serotonin reuptake inhibitors. *American Journal of Perinatology, 28*(1), 19–24.

Wirz-Justice, A., Bader, A., Frisch, U., et al. (2011). A randomized, double-blind, placebo-controlled study of light therapy for antepartum depression. *Journal of Clinical Psychiatry, 72*(7), 986–993.

Wisner, K. L., Perel, J. M., & Wheller, S. B. (1993). Tricyclic dose requirements across pregnancy. *American Journal of Psychiatry, 150*, 1541–1542.

Wisner, K. L., Perel, J. M., Peindl, K. S., et al. (2001). Prevention of recurrent postpartum depression: A randomized clinical trial. *Journal of Clinical Psychiatry, 62*, 82–86.

Wisner, K. L., Parry, B. L., & Piontek, C. M. (2002). Postpartum depression. *New England Journal of Medicine, 347*, 194–199.

Wisner, K. L., Perel, J. M., Peindl, K. S., et al. (2004). Prevention of postpartum depression: A pilot randomized clinical trial. *American Journal of Psychiatry, 161*, 1290–1292.

Wisner, K. L., Hanusa, B. H., Perel, J. M., et al. (2006). Postpartum depression: A randomized trial of sertraline versus nortriptyline. *Journal of Clinical Psychopharmacology, 26*, 353–360.

Yamada, N., Martin-Iverson, M. T., Daimon, K., et al. (1995). Clinical and chronobiological effects of light therapy on nonseasonal affective disorders. *Biological Psychiatry, 37*, 866–873.

Yonkers, K. A., Ramin, S. M., Rush, A. J., et al. (2001). Onset and persistence of postpartum depression in an inner-city maternal health clinic system. *American Journal of Psychiatry, 158*(11), 1856–1863.

Yonkers, K. A., Wisner, K. L., Stowe, Z., et al. (2004). Management of bipolar disorder during pregnancy and the postpartum period. *American Journal of Psychiatry, 161*(4), 608–620.

Yonkers, K. A., Wisner, K. L., Steward, D. E., et al. (2009). The management of depression during pregnancy: A report from the American Psychiatric Association and the American College of Obstetricians and Gynecologists. *Obstetrics and Gynecology, 114*, 703–713.

Zeskind, P. S., & Stephens, L. E. (2004). Maternal selective serotonin reuptake inhibitor use during pregnancy and newborn neurobehavior. *Pediatrics, 113*, 368–375.

Zhang, D., & Hu, Z. (2009). RTMS may be a good choice for pregnant women with depression. *Archives of Women's Mental Health, 12*(3), 189–190.

Zhang, X., Liu, K., Sun, J., et al. (2010). Safety and feasibility of repetitive transcranial magnetic stimulation (rTMS) as a treatment for major depression during pregnancy. *Archives of Women's Mental Health, 13*(4), 369–370.

Zhu, S. H., & Valbo, A. (2002). Depression and smoking during pregnancy. *Addictive Behaviors, 27*, 649–658.

Zlotnick, C., Johnson, S. L., Miller, I. W., et al. (2001). Postpartum depression in women receiving public assistance: Pilot study of an interpersonal-therapy-oriented group intervention. *American Journal of Psychiatry, 158*, 638–640.

Zuckerman, B., Amaro, H., Bauchner, H., et al. (1989). Depressive symptoms during pregnancy: Relationship to poor health behaviors. *American Journal of Obstetrics and Gynecology, 160*(5 Pt 1), 1107–1111.

Consultation-Liaison Psychiatry and Psychological Factors Affecting Other Medical Conditions

142

Daniel C. McFarland and Yesne Alici

Contents

D. C. McFarland (✉)
Department of Medicine, Northwell Health Cancer
Institute, Lenox Hill Hospital, Manhattan Eye Ear Throat
Hospital, New York, NY, USA

Department of Psychiatry, Division of Collaborative Care
and Wellness, Department of Medicine, Division of
Hematology and Oncology, University of Rochester
Medical Center, Rochester, NY, USA
e-mail: Daniel_McFarland@URMC.Rochester.edu

Y. Alici
Department of Psychiatry and Behavioral Sciences,
Memorial Sloan Kettering Cancer Center, New York, NY,
USA
e-mail: aliciy@mskcc.org

© Springer Nature Switzerland AG 2024
A. Tasman et al. (eds.), *Tasman's Psychiatry*,
https://doi.org/10.1007/978-3-030-51366-5_90

Abstract

Consult-liaison (CL) psychiatry addresses complex medically ill patients who experience psychiatric symptoms. Additionally, CL psychiatry may address psychological factors affecting other medical conditions (PSAOMC) since these psychological factors not only impair quality of life but also worsen medical outcomes. This chapter explores the underlying concept of PSAOMC in relation to worsened medical outcomes and the role of the CL psychiatrist in mitigating these relationships. A history of CL psychiatry is reviewed for background context as a distinct discipline and its role with integrating psychiatric care into medical settings. As one particular application, PSAOMC is explored in the context of various medical settings such as oncology, cardiology, nephrology, endocrinology, dermatology, pulmonology, and reproductive medicine. These various settings reveal the cross-sectional applicability of CL psychiatry and how addressing psychological factors can concomitantly improve medical outcomes.

Keywords

Psychological factors · Consult-liaison psychiatry · Medical comorbidity · Depression and anxiety

Introduction

Consultation-liaison (C-L) psychiatry is the newest psychiatric subspecialty approved by the American Board of Medical Specialties. There have been many other names for this specialized field, most recently referred to as psychosomatic medicine, medical-surgical psychiatry, psychological medicine, or psychiatric care of the complex medically ill (Rundell, 2018). C-L psychiatrists have special expertise in the diagnosis and treatment of psychiatric conditions in complex medically ill patients. Working closely with physicians in primary care and other specialties, its practitioners treat and study four general groups of patients, sometimes referred to as the "complex medically ill": those with comorbid psychiatric and general medical illnesses complicating each other's management; those with psychiatric disorders that are the direct consequence of a primary medical condition or its treatment, such as delirium, major neurocognitive disorders, or other secondary mental disorders (e.g., anxiety disorder due to hyperthyroidism); those with complex illness behavior such as somatoform and functional disorders; and patients with acute psychopathology admitted to medical-surgical units, such as after attempted suicide.

C-L psychiatrists are trained to deliver services in general medical settings working with the complex medically ill. C-L psychiatrists are primarily hospital-based, or in medical-psychiatric inpatient units, or in settings in which mental health services are integrated with primary care or medical specialties to provide collaborative care. Thus, the field's name reflects the fact that it exists at the interface of psychiatry and other medical or surgical disciplines dedicated to the interaction at the interface of psychiatry and the other disciplines (Boland et al., 2018). C-L psychiatrists are keenly interested in how medical or surgical issues influence psychiatric or psychological phenomena and the application of mental health expertise to various medical or surgical settings.

In addition to providing an overview of C-L psychiatry as a sub-discipline, this chapter provides an in-depth look at psychological factors that affect other medical conditions (PSAOMC), which is found in the DSM-5 under somatization disorders. The reader should obtain an understanding of the concept along with its applicability to various medical and surgical modalities. Other textbooks provide comprehensive perspectives on C-L psychiatry (Levenson, 2019).

History

The term "psychosomatic" was introduced by Johann Heinroth in 1818, and "psychosomatic medicine" by Felix Deutsch around 1922 (Lipsitt, 2001). Psychoanalysts and psychophysiologists pioneered the study of mind-body interactions from very different perspectives, each contributing to the growth of the field as a clinical and scholarly one. The modern history of the field in the United States (see Table 1) perhaps starts with the Rockefeller Foundation's funding of psychosomatic medicine units in several US teaching hospitals in 1935. The National Institute of Mental Health (NIMH) made it a priority to foster the growth of C-L psychiatry, the name of the field at the time, through training grants and a research development program.

The integration of C-L psychiatry into the core of psychiatric residency training began in the 1960s at individual institutions. By the 1980s, all psychiatry residency programs were required to provide substantial clinical experience in C-L psychiatry. In the United States, subspecialty fellowship training in C-L psychiatry has been available since the 1990s but only became an accredited fellowship program at the beginning of the twenty-first century.

In 2001, the Academy of Consult-Liaison Psychiatry (ACLP), then called Academy of Psychosomatic Medicine (APM), applied to the American Board of Psychiatry and Neurology (ABPN) for the recognition of "psychosomatic medicine" as the newest subspecialty of psychiatry. Formal approval was granted by the American

Table 1 Key Dates in the Modern History of Consultation-Liaison Psychiatry

1935	Rockefeller Foundation opens first C-L Psychosomatic Units at Massachusetts General Hospital, Duke University, and the University of Colorado
1936	American Psychosomatic Society founded
1939	First issue of *Psychosomatic Medicine*
1953	First issue of *Psychosomatics*
1954	Academy of Psychosomatic Medicine (APM) founded
1975	NIMH Training Grants for C-L Psychiatry
1985	NIMH Research Development Program for C-L Psychiatry
1991	APM-recognized fellowships number 55
2001	Subspecialty application for Psychosomatic Medicine
2003	Approval as subspecialty by American Board of Medical Specialties
2005	First subspecialty examination in Psychosomatic Medicine, American Board of Psychiatry and Neurology
2012	European Association of Psychosomatic Medicine founded
2017	APM changed its name to Academy of Consultation Liaison Psychiatry
2018	Subspecialty name changed to Consultation-Liaison Psychiatry
2022	The number of National Residency Matching Program participating C-L programs have risen to 65.

Source: Adapted from The American Psychiatric Publishing Textbook of Psychosomatic Medicine Psychiatric Care of the Medically Ill, Second Edition, Edited by James L. Levenson, 2011, American Psychiatric Association

Psychiatric Association, the ABPN, the Residency Review Commission (RRC) of the Accreditation Committee for Graduate Medical Education (ACGME), and the American Board of Medical Specialties (ABMS). The first certifying examination was administered in June 2005 to almost 500 psychiatrists. As of May 2021, the ACGME had accredited 62 fellowship-training programs, which increased from 52 reported in May of 2013 and 29 programs back in 2006.

An impediment to the field's growth has been the split between general medical healthcare and mental healthcare, with major adverse effects on the quality of medical service delivery and patient- oriented care. This split is reflected in disparities and disintegration in the reimbursement of patient care (carve outs, lack of parity in coverage), and also in research funding mechanisms in silos that do not promote cross-disease studies. C-L psychiatry can help mend the fragmentation in healthcare, but its continued growth may depend on addressing healthcare financing, proactive consultations, and research sponsorship (Oldham et al., 2021).

International Developments in Consultation-Liaison Psychiatry

A large collaborative study of C-L psychiatry services in Europe in the 1990s by the European Consultation Liaison Workgroup demonstrated significant variation in how C-L services were delivered among countries and played a vital role in the evolution of the field in Europe (Sollner et al., 2007). In 2012, the European Association of Psychosomatic Medicine was formed by the merger of the European Association of Consultation Liaison Psychiatry and the European Conference on Psychosomatic Research (Sollner & Schussler, 2012). In the United Kingdom's National Health Service (NHS), there is integration between primary, specialty, and inpatient care. Many general hospitals have "liaison psychiatry services," but only a minority actually have a dedicated psychiatric consultant. In Germany, psychiatry and neurology did not become

separate specialties until 1992 (Diefenbacher, 2005). C-L psychiatry services in general hospitals are provided by psychiatrists, while most psychosomatic practitioners are internists who work in outpatient private practice. In Australia, competition for limited resources with psychiatric services for psychosis jeopardized the future of C-L psychiatry, but the most recent national mental health plan does recognize its importance. The Japanese Society of Psychosomatic Medicine was founded in 1959, and may be the world's largest with over 3600 members (Kubo, 2013). While insufficient funding has inhibited the development of C-L services in many countries, there nevertheless are enthusiastic growing C-L movements in The Netherlands, Spain, Mexico, Italy, Brazil, China, and elsewhere in the international community (Gala et al., 1999; Herrmann-Lingen, 2017).

Consultation-Liaison Psychiatry as a Scholarly Discipline

The foundation of C-L psychiatry is a specialized body of scientific knowledge regarding psychiatric aspects of medical illness. This has been articulated in classic (Table 2) and contemporary textbooks (Levenson, 2019); journals (Table 3); and the regular scientific meetings of national (Table 4) and international (Table 5) societies. A cadre of scholars and researchers has emerged involved in a wide spectrum of investigations looking at the medical illness-psychiatry interface. Important contributions have occurred in the interface between psychiatry and HIV-AIDS, cancer, transplantation, cardiology, neurology, endocrinology, pulmonary, renal, and gastrointestinal diseases, obstetrics-gynecology, and geriatric medicine. Most recently, C-L psychiatrists have played an integral role in the assessment and management of COVID-19 associated delirium, and post-COVID psychiatric syndromes.

Medical and psychiatric comorbidity has been documented extensively, especially in the complex medically ill (Katon, 2003). Psychiatric comorbidity leads to increased mortality and

Table 2 Selected Classic Texts in Consultation-Liaison Psychiatry

1935	*Emotions and Body Change* (Dunbar)
1943	*Psychosomatic Medicine* (Weiss and English)
1950	*Psychosomatic Medicine* (Alexander)
1968	*Handbook of Psychiatric Consultation* (Schwab)
1978	*Organic Psychiatry* (Lishman)
1978	*MGH Handbook of General Hospital Psychiatry* (Hackett and Cassem)
1993	*Psychiatric Care of the Medical Patient* (Stoudemire and Fogel)
2005	*American Psychiatric Publishing Textbook of Psychosomatic Medicine*
2011	*The American Psychiatric Publishing Textbook of Psychosomatic Medicine Psychiatric Care of the Medically Ill, 2nd edition, Edited by James L. Levenson.*
2019	*The American Psychiatric Publishing Textbook of Psychosomatic Medicine and Consultation-Liaison Psychiatry, 3rd edition, Edited by James L. Levenson.*

Source: Adapted from The American Psychiatric Publishing Textbook of Psychosomatic Medicine Psychiatric Care of the Medically Ill, Third Edition, Edited by James L. Levenson, 2019, American Psychiatric Association

Table 3 Selected Journals in Consultation-Liaison

Journal	Date of Initial Publication
Psychosomatic Medicine	1939
Psychosomatics	1953
Psychotherapy and Psychosomatics	1953
Psychophysiology	1954
Journal of Psychosomatic Research	1956
Advances in Psychosomatic Medicine	1960
International Journal of Psychiatry in Medicine	1970
General Hospital Psychiatry	1979
Journal of Psychosomatic Obstetrics and Gynecology	1982
Journal of Psychosocial Oncology	1983
Stress Medicine	1985
Psycho-Oncology	1986

Source: Adapted from The American Psychiatric Publishing Textbook of Psychosomatic Medicine Psychiatric Care of the Medically Ill, Third Edition, Edited by James L. Levenson, 2019, American Psychiatric Association

morbidity, loss of quality of life, and excess healthcare utilization through several psychosocial and biological mechanisms, including

Table 4 US National Organizations

Academy of Consultation Liaison Psychiatry
American Psychosomatic Society
American Association for General Hospital Psychiatry
Society for Liaison Psychiatry Association for Academic Psychiatry – C-L Section
American Neuropsychiatric Association
American Psychosocial Oncology Society
North American Society for Psychosomatic Obstetrics and Gynecology
Association for Medicine and Psychiatry

Source: Adapted from The American Psychiatric Publishing Textbook of Psychosomatic Medicine Psychiatric Care of the Medically Ill, Third Edition, Edited by James L. Levenson, 2019, American Psychiatric Association

Table 5 International Organizations

European Association of Psychosomatic Medicine (EAPM)
World Psychiatric Association: section on (1) psychiatry, medicine and primary care and (2) psycho-oncology and palliative care
International College of Psychosomatic Medicine
International Neuropsychiatric Association International Psycho-Oncology Society
International Society of Consultation Liaison Nurses

Source: Adapted from The American Psychiatric Publishing Textbook of Psychosomatic Medicine Psychiatric Care of the Medically Ill, Third Edition, Edited by James L. Levenson, 2019, American Psychiatric Association

promotion of risk factors, amplification of somatic symptoms, noncompliance, adverse effects on the doctor–patient relationship and systems of care, and pathophysiologic effects (Katon, 2003; Santos & Pyter, 2018). High prevalence rates of a broad range of psychiatric disorders have been documented in a variety of medical illnesses, and also in patients with abnormal illness behavior such as those with unexplained physical complaints. Inpatients in general hospitals have the highest rates of psychiatric disorders but rates are also increased in medical outpatients compared with the general population. Depression, major neurocognitive disorder, delirium, and substance abuse are the most common psychiatric disorders in the general hospital, while depression, anxiety disorders, substance abuse, and somatoform disorders are very common in medical outpatient settings.

The impact of psychiatric illness on medical morbidity is substantial in terms of quality of life deprecation, healthcare cost and utilization, morbidity, and mortality. That is, depression portends increased risk of morbidity and death from myocardial infarction, risk of stroke in hypertensive patients, and worse glycemic control in diabetic patients. Yet, treatable psychiatric disorders in the complex medically ill are frequently not addressed. They remain underdiagnosed and undertreated as a "silent" comorbidity. Patients are frequently afraid to raise their concerns; perhaps they do not want to distract the attention from medical concerns, feel stigmatized, or do not realize that these valid concerns ought to be addressed. Clinicians are often afraid to ask.

The Clinical Practice of C-L Psychiatry

The major goal of C-L psychiatry is to improve the psychiatric, psychological care of medically ill patients with and without chronic mental illness (i.e., serious mental illness). Most patents evaluated by C-L psychiatrists do not have chronic mental illness but are psychiatrically or psychologically compromised by medical illness; its treatment and sequelae. Non-mental health clinicians may recognize the large percentage of patients with psychiatric symptoms (e.g., insomnia, anxiety, distress, depressive symptoms) and provide de facto mental health care (Snell-Rood et al., 2019). But de facto non-psychiatric care does not provide consistent attention to psychiatric symptoms because non-mental health clinicians are not trained to diagnose and treat the psychiatric symptomatology. At the same time, general psychiatrists may not be adept at addressing psychiatric symptom issues in the complex medically ill without specific CL training. Therefore, despite the higher concentration of patients with psychiatric disorders in medical settings, many medical patients with psychiatric comorbidity do not receive recognition, appropriate diagnosis, initiation of adequate treatment, and/or referral for the follow-up psychiatric care they need (Druss & Walker, 2011). Failure to identify, evaluate, diagnose, treat, or achieve

symptom resolution results in preventable adverse outcomes and increased costs as identified in specialty specific studies (Walker et al., 2014).

Access to psychiatric and CL-specific psychiatric expertise is not readily accessible in many locales, especially rural areas (Morales et al., 2020). To disseminate psychiatric knowledge into primary care and other specialty clinics, efforts are made to co-locate C-L trained psychiatrists in their clinic, which takes considerable coordination and flexibility (Ramanuj et al., 2019). This is an ideal model that is not implemented often enough. Collaborative care models offer a similar integration and extension of services usually using non-psychiatrically trained clinicians (Gillies et al., 2015). Other efforts to bridge this access gap have been pursued through tele-psychiatry. Consensus statements clearly indicate that CL-psychiatry led collaborative models are optimal care for patients with complex medical illness such as cancer and HIV or for those patients being considered for a transplant or gastric bypass (Huijbregts et al., 2013). These efforts require specialty-specific collaboration with available psychiatrically trained professionals.

Psychiatric liaison, in which psychiatrists are integrated members of a specialized care team, is an even more advanced model, with greater ability to provide early detection and prevention of psychiatric symptoms or decompensation (Abouljoud et al., 2018). These types of specialty specific liaison services are limited to large teaching and specialty hospitals where they are a value-added service. Financial reimbursement remains problematic under managed care plans despite studies showing cost benefits (Green et al., 2014; Walker et al., 2014). Hence, C-L psychiatrists are mostly found in tertiary care sites, and are not evenly distributed geographically.

There is a shortage of C-L psychiatrists in the United States and internationally (Bishop et al., 2016). Additionally, C-L psychiatry service models of care delivery are quite variable. They vary in the following: (1) location (e.g., medical inpatients vs. outpatients); (2) involved disciplines (e.g., psychiatry, psychology, psychiatric nursing, social work); (3) integration with nonpsychiatric services and systems; (4) their relative devotion to consultation, liaison, education, and continued psychiatric care; and (5) the extent to which they provide emergency psychiatric services.

C-L psychiatry is essential and should be part of the standard of care for the medically ill in all care settings. Professional organizations such as ACLP continue to advocate for patients and families of the medically ill, educate clinicians of all disciplines, and support the research efforts in the field to move toward wider availability of evidence-based C-L psychiatry across the nation and the globe.

Psychiatric Diagnosis in the Medically Ill

Psychiatric disorders are frequently underdiagnosed in the medically ill for a number of reasons. Yet, certain physical symptoms are often overly attributed to psychiatric illness by non-psychiatric professionals. The role of the CL-psychiatrist should be diagnostically inclined, but also aware of the normative role that is often requested by medical and surgical services. Frequently, they need assistance to ascertain the implications of symptom severity and to rule out psychiatric complications. Consulting teams may not be able to accurately attribute symptoms, which may be underly or overly emphasized.

Psychiatric symptoms can mimic medical illness and vice versa. As a result, somatic symptoms need to be considered in context, but can still be problematic diagnostically. For example, a medical service may attribute fatigue, anorexia, and weight loss to depression when a medical illness such as cancer is not found to explain these symptoms but the patient does not endorse depression or anhedonia (i.e., false positive). Or the medical team may see a patient who is apathetic and slow to respond as depressed when they have hypoactive delirium. At the same time, medical teams may mistakenly attribute symptoms of depression to an underlying medical situation that is perhaps comorbid (e.g., Parkinson disease) or one that is not yet diagnosed (i.e., false negative).

A variety of approaches have been proposed to limit confounding medical symptoms in the diagnosis of depression. In an "exclusive" and "etiologic" approach, symptoms that are judged by the clinician to be etiologically related to a general medical condition are excluded from the diagnostic criteria for major depressive disorder (MDD) (Spitzer et al., 1978). However, symptoms that are clearly related to medical illness versus psychiatrically related are not always easily delineated. In a "substitutive" approach, symptoms most likely confused with medical illness, such as fatigue and weight loss, are substituted with symptoms that are more likely to be affective in origin, such as irritability and social withdrawal. Such substitution eliminates the need to distinguish symptoms of medical illness from those of depression, but it also excludes some somatic symptoms that are core manifestations of depression. Furthermore, valid criteria to determine which symptoms should be substituted have not been established. An "inclusive" approach applies the unmodified diagnostic criteria, which makes it unlikely that depression will be missed, but at the price of more false positives (Saracino et al., 2018).

Aside from acting as diagnostician, CL psychiatrists are often called upon to determine normal versus abnormal psychological reactions to illness or disease-related situations. It can be difficult to differentiate a normal psychological reaction from a psychiatric disorder. Normal emotional reactions occur in response to uncertainty about diagnosis or prognosis, loss of body parts or body functions, fear of death, impact of illness on identity and livelihood, dependency, reactions to strangers, and being alone in the hospital, and may include anxiety, sadness, guilt, denial, avoidance, anger, hopelessness, and obsessive control. Normal adaptive emotional and behavioral reactions to illness, maladaptive responses, and psychiatric disorder are on a spectrum, and their differentiation is one of degree, informed by clinical experience. Finally, it must be recognized that a given symptom, for example, tachycardia or poor concentration, may simultaneously result from a normal reaction, a psychiatric disorder, and the medical disorder itself all in the same patient.

Given this complexity, this chapter evaluates and reviews this key dimension of mind-body interactions, encapsulated in the DSM-5 diagnosis of psychological factors affecting other medical conditions (PFAOMC).

Psychological Factors Affecting Other Medical Conditions

Overview

CL psychiatry and related fields such as health psychology focus on the effects of the mind on health and the effects of physical health on the mind. A patient's illness experience is rooted in various inter-related concepts applicable to the clinical encounter. Acute and chronic stress, personality types, coping strategies, and defense mechanism are interwoven in the complexity of an individual's emotional response to illness. The biopsychosocial model introduced in the 1970s by George Engel encapsulates the holistic approach to diagnostic and treatment considerations attuned to medical patients with psychiatric symptoms (Engel, 1997). In addition, the thorough understanding of the biological stress response originally proposed by Hans Selyles (Szabo et al., 2012) and the subsequently overlaid theory of stress and coping developed by Lazarus and Folkman are highly relevant (Lazarus, 1974). Understanding these emotional and psychological responses is highly applicable to CL-psychiatry where the clinician can help patients of varying coping ability interact with their illness, especially within the confines of complicated health systems. Both adaptive and maladaptive patient responses become heightened in times of stress, or under duress to make medical decisions, or secondary to mortality salience. As a CL-psychiatrist, understanding the PFAOMC is integral to being able to operationalize support in making medical decisions and management medical issues.

At the same time, research places increasing emphasis on social determinates to health (e.g., demographic, and socioeconomic variables) which have clearly demonstrated long-term consequences in health outcomes (Braveman & Gottlieb, 2014). Socioeconomic status, education, racial inequities, and poverty wreak havoc on health and consequences of disease in parallel with poor mental health (Alegria et al., 2018). For example, it is increasingly clear that factors such as parental education predict health outcome, along with one's childhood neighborhood (Woolf & Braveman, 2011). These social determinates of health are operationally complex and not necessarily in the purview of CL psychiatry, but their effects are mediated to some extent by stress and epigenetic phenomena as well as health-related decisions (Perez-Tejada et al., 2019). They factor into the overall psychological makeup of a person who may be under the stress of a physical illness or coping with this loss for the first time or as a successes of losses. In other words, people who are facing a health crisis have degrees of psychological stamina, coping reserve, or ability. They also have various levels of social and personal support (e.g., financial, educational, personal), all of which is factored into the consultant's appraisal.

Clinical medicine tends to focus on disease with its physical or physiological (biomedical) attributes without paying enough attention to the psychological and emotional aspects of patients' well-being and its root causes of disturbance. Physical and mental health are both composite structures of general well-being and affect each other tremendously but to varying individualistic degrees. This interrelationship is increasingly recognized in all its complexity and has an ever-increasing role in optimizing short- and long-term outcomes for patients with various medical maladies and calamities.

The origins of psychosomatic medicine and the mind body connection stem from antiquity. In the modern era, Franz Alexander described the effects of depression on cardiac outcomes and peptic ulcer disease. Research has expanded upon these original concepts and can now describe many of the basic mechanisms by which stress and emotional phenomena affect health. In general, psychological phenomena can affect health directly or indirectly through physiology or through behavior and healthcare choices (e.g., medication adherence). Personality and defensive structures mediate patients' responses to illness as well.

PFAOMC integrates psychodynamic concepts of character style and psychic defenses with other psychological concepts such as stress and coping. The ability or inability to cope is individual and subjectively experienced although objective observation also plays a role in assessment. In this way, patient's ability to cope and deal with adversity should be appreciated to the same extent as the presence of maladaptive coping and behaviors. The CL psychiatrist should appreciate that maladaptive coping does not usual require psychiatric consultation and that maladaptive coping can be episodic. A spectrum of coping spanning time and severity should be appreciated as well as the inherent reluctance in discussing emotional matters with physicians due to stigma or for other reasons.

There is no one correct way to characterize a psychological response to illness – the clinician should focus on its positive and negative aspects as well as longitudinal, historical, and predictive factors. For example, a 55-year-old business executive with substernal chest pain suspicious for cardiogenic illness may make excuses for recurrence of this intermittent pain and forego medical assessment because "he's too busy" even though he has also kept up to date with all of his age-appropriate cancer screening. Perhaps he is not able to confront that that his work stress is exacerbates his symptoms. Multiple avenues of supportive therapy or motivational interviewing may help motivate him to seek help-supportive therapy to understand his reluctance, stress modification (e.g., mindfulness-based stress reduction), cognitive behavioral therapy, and couples or family therapy. Many of these types of interventions may be short term, done over the phone or at the bedside. In this way, CL psychiatry plays an integral role in enhancing medical care and extending it to vulnerable populations and people.

The effects of PFAOMC are mediated directly at a pathophysiologic level (e.g., psychological stress inducing myocardial ischemia) or through the patient's behavior (e.g., noncompliance) (Perez-Tejada et al., 2019). The effects of psychological factors on medical conditions have become the focus of extensive research. There is growing interest in the basic disease mechanisms (e.g., psychoneuroimmunology) and a growing interest in improving both the psychosocial outcomes and the efficiency of healthcare delivery (Danese & Lewis, 2017). In addition to quality-of-life benefits, data reveal that survival related to the primary medical disease (e.g., cancer, coronary artery disease, rheumatologic) is shortened with concomitant depression and other serious mental illness (e.g., schizophrenia, bipolar disorder, substance abuse) (Pratt et al., 2016). Not only do these serious mental illnesses negatively affect outcome in medical illness (e.g., survival), they increase adverse events such as hospital length of stay and healthcare costs (Konig et al., 2019; Prina et al., 2015).

The clinical approach to PFAOMC is individualistic and based on the patient presentation, risk, and mediating factors (adaptive and maladaptive coping). The primary interventional approach is to recognize its presence and provide a variable level of support. This can be accomplished through myriad modalities and approaches, which depend on the patient's presentation. For primary care practices, the focus is on making a diagnosis through screening or some other mechanism. For mental health specialty and CL psychiatry practices, addressing PFAOMC requires a thorough diagnostic assessment to rule out other diagnoses and a subsequently patient-centric approach (Katon et al., 2010b). Collaborative care models have demonstrated effectiveness in various primary care settings, but their clinical application remains under-utilized. For that reason, upcoming studies are focusing on dissemination and implementation so that clinician understand how to operationalize these multidisciplinary approaches (Katon et al., 2010a).

Assessment and Diagnosis

A diagnosis of PFAOMC can only be established if psychological factors influence the course of the medical condition, interfered with its treatment, contributed to health risks, or physiologically aggravated the medical condition. The DSM-5 requires that the presence of a general medical symptom or disorder; psychological or behavioral factors negatively affect the medical condition in one or more of the following ways: (A) factors pose additional health risks for the patient; (B) factors aggravate the underlying pathophysiology of a medical condition and precipitate or exacerbate symptoms; (C) factors affect the course of the medical condition as manifested by a close temporal relationship between the factors and the onset or exacerbation or the medical condition; (D) factors disrupt treatment of the general medical condition, including not seeking medical care, nonadherence with follow-up visits or prescribed treatment, or maladaptive modifications to treatment by the patient or family; and (E) not better explained by the presence of other psychological or behavioral factors.

It should be noted that a close temporal relationship between abnormal psychological and behavioral factors with a general medical illness is not sufficient to diagnose PFAOMC. There must be evidence that PFAOMC factors exert a clinically significant adverse effect upon the illness. The diagnosis of PFAOMC should be recorded with the general medical condition and may be graded by severity as follows: mild (e.g., increases medical risk): moderate (i.e., exacerbates general medical condition); severe (i.e., causes hospitalization); or extreme, if life threatening (e.g., ignoring myocardial infarction symptoms).

PFAOMC is classified among somatic symptom and related disorders along with conversion disorder, factitious disorders, illness anxiety disorder, and somatic symptom disorder. This section was added to the DSM5 edition and offers a perspective of overlap between medical and psychiatric illness. Psychological and behavioral factors affecting disorders or diseases classified elsewhere are those that may adversely affect the manifestation, treatment, or course of a condition classified in another chapter of the ICD. These factors may adversely affect the manifestation, treatment, or course of the disorder or disease classified in another chapter by the following: interfering with the treatment of the disorder or disease; affecting treatment adherence or care

seeking; constituting an additional health risk; or influencing the underlying pathophysiology to precipitate or exacerbate symptoms or otherwise necessitate medical attention. This diagnosis should be assigned only when the factors increase the risk of suffering, disability, or death and represent a focus of clinical attention, and should be assigned together with the diagnosis for the relevant other condition. PFAOMC helps supplant the controversial somatoform disorders that appeared in the DSM-IV as issues detrimental to medical outcomes such as disease phobia, persistent somatization, conversion symptoms, illness denial, demoralization, and irritable mood need to be described for the advancement of CL psychiatry (Fava et al., 2007, 2017).

Psychological Factors That Influence Medical Conditions

Maladaptive behaviors, coping or attachment styles, and common "immature" defense mechanisms negatively affect general medical illness. When viewed as a life stressor, illness calls upon the patient to gather resources and institute coping mechanisms. Psychological factors can negatively influence this process when coping has historically been maladaptive or coping skills were never available for the patient in the first place.

Attachment styles are formed in childhood. Its vestiges persistent into adulthood and may become more prominent under the stress of illness. Attachment styles can be categorized as positive or negative views of oneself, as well as positive or negative expectations of others. Coping styles for managing stress of medical illness may include hostile or aggressive actions to change the situation, distancing oneself psychological from the situation, avoidance, self-control of one's actions or feelings, seeking social support, accepting responsibility, viewing the situation positively, or problem solving. Most patients use multiple strategies simultaneously to approach complex situation. Clearly, certain coping mechanisms work better than others for given stressors.

Defense mechanisms reflect intrapsychic unconscious processes that patients use to defend themselves against life stressors and trauma. These mechanisms work to allay anxiety and conflict, persevere self-esteem, and maintain a sense of control. For example, regression is a partial or complete return to immature patterns of behavior, including dependency upon others. It is common for medically ill patients to regress and seek comfort, care, and release from their responsibilities. There is a spectrum of regression which may be adaptive and functional. For example, acutely and chronically ill patients receive subtle and overt cues to give up control and/or accept dependence on others. This may be functional up to a certain point where too much regression undermines the patients' participation in care or life events. This may manifest as a minimal tolerance for frustration, symptoms, or being alone, and patients may seem excessively needy or needlessly upset. Some patients who become regressed may make excessive demands or avoid responsibility for their care, which can frustrate the doctor-patient relationship. Regression can manifest in patients who are overly perfectionistic as well when the refuse to relinquish control and insist on specific treatment types and are intolerant of any delay in care for example. Another form of regression may be witnessed in patients who appear entitled and are oblivious to the needs of other patients. Regression can damage patients' relationships with clinicians and caregivers and inadvertently cause their care to suffer. Other defense mechanisms can be seen on a spectrum from adaptive to maladaptive.

Anxiety is ubiquitous in medical settings. It exists along a continuum that includes normal psychological reactions that tend to fluctuate in intensity and do not render patients unable to function to pathological, unremitting anxiety. Clinicians can become desensitized to the patient's experience of serious illness and lose sight of what constitutes normal anxiety reactions given the wide variability of reaction to medical illness in terms of stress and anxiety. It should be noted that while some anxiety may be expected, even a normalized level of anxiety may have adverse effects. For example, even episodic moderate anxiety may

exacerbate an acute myocardial infarction, asthma exacerbation, or cause prolonged recovery. There are many different forms of anxiety that are particular to the illness experience. For example, phobias (e.g., needle phobia), death anxiety, or fear of cancer recurrence are particular to certain medical diagnoses. PTSD is now included under Trauma- and Stressor-Related Disorders in DSM5 of which physical illness may be causal. At the same time, too little anxiety makes one vulnerable to minimizing the risks of the disease and need for treatment. In some cases, the apparent absence of anxiety is misleading, because the patient may be exceptionally anxious without being aware of it. They may use defenses like denial to avoid feeling overwhelmed.

Depressive symptoms, dysphoria, or minor depression can take the form of sadness, hopelessness, emptiness, demoralization, grief, irritability, or mood instability. These symptoms readily interfere for executive functioning and cognitive abilities and undermine participation in medical care. Clear associations with poor outcomes are noted in patients with coronary heart disease, stroke, diabetes, end-stage renal disease, and rheumatoid arthritis. A circular pattern may develop where patients feel dejected, discouraged, and helpless in the face of disease and disability that their disease worsens furthering morbidity and even mortality. At the same time, patients are often aware of the relentlessness of positive thinking from well-meaning friends, family, and colleagues and they feel guilty when their internal thinking does not live up to those encouragements. Acceptance of realistic feelings about the condition are often most helpful.

PFAOMC should be considered when (1) psychological factors influence the course of a general medical condition (e.g., cardiac symptoms, diabetes); (2) psychological factors interfere with treatment (e.g., seeking medical attention, follow-up appointments, medication adherence); (3) behaviors contribute to health risk (e.g., tobacco use); and (4) medical treatments are not responding as expected (e.g., unexpected outcomes, treatment resistance). The latter category is particularly challenging since the clinician needs to avoid inciting blame or stigma for inferior medical outcomes. The patient may otherwise feel labeled as having a psychogenic illness or issue, which ultimately undermines the therapeutic relationship. Yet, psychological factors that undermine effective medical treatment should be addressed. Collateral information and the use of screening instruments may be helpful to begin conversations about the effects of illness on psychosocial well-being.

For example, an adolescent with brittle diabetes who is admitted to the hospital for diabetic ketoacidosis may encounter incredulous clinicians who do not trust the patient's adherence with her diabetes treatment regimen. Psychological issues that are typical of this population include the desire to be "normal" combined with an age-specific tendency to reject authority and life-style restrictions, a feeling of invulnerability, and ambivalence about the need for nurturance. Yet, adolescent diabetes tends to be challenging medically based on disease characteristics as well and a given adolescent patient may not be influenced by the previously mentioned psychological factors to the same degree as others.

Differential Diagnosis

The key distinction in determining the presence of PFAOMC is the direct relationship with worsening of the medical condition as opposed to an affective response to the medical condition. The latter is consistent with an adjustment disorder. This directional relationship should be clear when making the diagnosis of PFAOMC and sets it apart from other psychiatric disorders in response to stress or medical illness. In other words, with co-occurring psychiatric symptoms and medical illness, the mental symptoms may be the result of the medical condition (i.e., the causality is in a direction opposite from that of PFAOMC) Therefore the timing of psychological factors followed by worsening medical condition is key. When a medical condition is judged to be pathophysiologically causing the mental disorder (e.g., hypothyroidism causing depressive symptoms), the correct diagnosis is the appropriate

mental disorder due to another medical condition (e.g., depressive disorder due to hypothyroidism, with depressive features). In PFAOMC, the psychological or behavioral factors are judged to precipitate or aggravate the medical condition. Of note, close temporal association between the psychiatric symptoms and medical condition does not always reflect PFAOMC because the psychological factors and worsening medical condition may be coincidental as well.

Substance abuse may complicate or exacerbate psychiatric symptoms and complicate medical treatments. Diagnostically, the substance abuse disorder diagnosis can be used to account for psychiatric symptoms and medical complications given its primacy. Patients with somatic symptom disorder (SSD) present most commonly with multiple physical symptoms, accompanied by excessive and persistent abnormal thoughts, feelings, and/or behaviors related to the physical symptoms. In SSD, the somatic symptoms may or may not be medically explained, and the abnormal thoughts, feelings, and behavior are a reaction to the somatic symptoms. In conversion disorder (i.e., a functional neurological symptom disorder), patients present with neurological symptoms that may mimic a neurological disorder, but the neurological symptoms are accounted for by the psychiatric disorder. That is the medical condition is entirely the result of the psychiatric presentation. In principle, it might seem that conversion disorder is easily distinguished from PFAOMC, because PFAOMC requires the presence of a diagnosable medical condition. The distinction in practice is sometimes difficult because the patient may have both conversion disorder and a neurological disorder. For example, a patient with seizures regularly precipitated by emotional stress might have true epilepsy aggravated by stress (PFAOMC), pseudoseizures (conversion disorder), or both.

PFAOMC is common but even rough estimates of prevalence are difficult to gauge due to the variety of possible interactions between a wide range of psychological traits, states, and behaviors on the one hand and the full range of medical diseases on the other.

Etiology and Pathophysiology

Psychological factors may promote and couple with behavioral risk factors that are associated with or worsen medical conditions. For example, tobacco use, substance abuse, overuse of analgesics, sedentary lifestyle, poor diet, obesity, poor sleep hygiene, and unsafe sex have negative health outcomes associated with the behavior. The choice to engage in the behavior may be initiated, promoted, and sustained by PFAOMC undermining prevention and increasing health risk.

Psychological factors impact the course of medical illness by influencing how patients respond to their symptoms, including whether they seek medical care and in what capacity. For example, the defense mechanism of denial may lead an individual to ignore anginal chest pain, attribute it to indigestion, delay seeking medical attention, or minimize the pain when describing it to a physician. This results in treatment delay after the acute onset of coronary symptoms, with consequently greater morbidity and mortality. Alternatively, patients may opt for alternative forms of treatment rather than undergoing a diagnostic work up. Anxiety is a common cause of avoidance or delay of healthcare; phobic fears of needles, sight of blood, or surgery and other healthcare phobias are common (Noyes et al., 2000). Patients may also fail to seek medical care promptly because of personality traits such as procrastination, catastrophizing, or extreme frugality.

The physician-patient relationship may be unduly influenced by complex psychopathology, which undermines the patient's illness trajectory. Psychological factors influence both patients' health behaviors and physicians' diagnostic and treatment decisions. A substantial proportion of the excess mortality experienced by individuals with mental disorders is explained by their receiving poorer quality medical care (Druss et al., 2001).

Psychological factors reduce a patient's adherence to diagnostic and therapeutic recommendations. PFAOMC can undermine needed lifestyle changes and interfere with recovery and rehabilitation by compromising motivation, understanding, optimism, or tolerance. A meta-analysis found that patients with significant depressive symptoms are three times more likely to be non-compliant with medical treatment than patients without such symptoms (DiMatteo et al., 2000).

Psychological factors undermine stress homeostasis and affect medical outcomes via pathophysiological mechanisms. The understanding of peripheral inflammation, psychological symptoms (e.g., depression, anxiety, distress), and centrally mediated inflammation-based changes (e.g., upregulation of indolamine 2,3-dioxygenase leading to increased neurotoxic metabolites and decreased production γ-aminobutyric acid [GABA] and monoamines like serotonin and dopamine through depletion of tryptophan and L-tyrosine) are increasingly described (Dantzer et al., 2008). These physiological changes influence depression, anxiety, and distress and in turn impact health outcomes (McFarland et al., 2021b; Miller et al., 2008). Repeated and chronic stressors imprint epigenetically – another pathway that can undermine biological functioning (Gescher et al., 2018). Disruptions are seen in not only mood regulation but sleep architecture, other circadian rhythms, and neuroendocrine secretion and feedback (e.g., upregulation of cortisol) (Perez-Tejada et al., 2019). While these changes are eloquently described, their clinical import has yet to be directly actionable. Nonetheless, these biological changes hold unique therapeutic promise for addressing neuropsychiatric symptoms (Miller & Raison, 2016).

Treatment

Management of PFAOMC should be tailored both to the particular psychological factor of relevance and to the medical outcome of concern. While treatment options vary, its recognition may be therapeutic in and of itself. The physician, whether in primary care or a specialty, should not ignore apparent psychological distress or maladaptive behavior. Unfortunately, this occurs all too often because of discomfort, stigma, lack of training, or disinterest. At the same time, mental health referral alone by itself does not fix the problem because the physician will still need to attend to its potential impact on the patient's medical illness. Similarly, psychiatrists and other mental health practitioners should not ignore coincident medical disease or assume that referral to a nonpsychiatric physician absolves them of all responsibility for the patient's medical problems.

A collaborative approach is ideal and takes shape in many different forms based on the practice setting. In general, it is an exchange of ideas in the service of patient care. Inroads to a collaborative partnership matter for patients' outcomes. The use of diagnostic screening tools and approaches (e.g., distress or depression screening) may create the need for collaboration and creativity. These measures are often sensitive but not specific for capturing psychological factors affecting medical care. They are an entry point of discussion in the primary care setting and are helped greatly by colocation of mental health clinicians.

In general, evidence-based treatments are lacking for psychological factors that do not meet DSM cut-off threshold and are therefore considered "subsyndromal." This is unfortunate given the large prevalence of comorbid symptomatology. But subsyndromal symptoms and PFAOMC may respond to approaches used for similar DSM-5 psychiatric disorders, with appropriate modifications.

Behavior modification and motivational interviewing may be attempted as well as other non-pharmacologic treatment modalities as adjunctive or supportive interventions. They may include the provision of counseling, peer or caregiver support, psychoeducation, relaxation training and stress management techniques, and professional psychotherapy. In patients with comorbid medical illness, psychotherapy often requires modification and flexibility regarding the length and frequency of appointments, and deviations from standard therapeutic abstinence and neutrality. For example, it may be appropriate to provide advice, or act on a patient's behalf, which are not generally in the patient's best interest in other psychotherapeutic contexts (Perry & Viederman, 1981). Also, psychotherapists treating patients with PFAOMC will be much more active in communicating with other healthcare professionals caring for the patient (with the patient's consent) than is usually the case in psychotherapy. If the psychiatrist decides to institute pharmacotherapy for symptom management (e.g., anxiety, insomnia), she should assess the potential for drug-drug interactions, changes in pharmacokinetics (absorption, protein binding, metabolism, and excretion), and medication side effects in combination with medical illness (e.g., anticholinergic medication will exacerbate postoperative ileus).

If the patient has an entrenched maladaptive personality or coping style, the psychiatrist should modify the patient's treatment accordingly rather than addressing personality factors directly – this would be challenging and all too time consuming in the setting of pertinent medical illness. For example, patients who tend to be suspicious or mistrustful should receive careful explanations and reassurance, particularly before invasive or anxiety-provoking procedures. With narcissistic patients, the psychiatrist should avoid relating in ways that may seem excessively paternalistic or authoritarian to the patient. With some dependent patients, it may be advisable to be more directive, without overdoing it and fostering excessive dependency.

Any intervention directed by the psychiatrist at a particular patient's psychological symptoms or behavior should be grounded in exploratory discussion with the patient. Interventions without such grounding at best tend to seem superficial and artificial, and at worst are entirely off the mark. For example, if the psychiatrist wrongly presumes to know why a particular patient seems anxious without asking, the patient is likely to feel misunderstood. Also, nonspecific reassurance can undermine the physician–patient relationship because the patient is likely to feel that the psychiatrist is out of touch with and not really interested in the patient's experience. It is especially important with discouraged or demoralized patients that psychiatrists avoid premature or unrealistic reassurance or an overly cheerful attitude; this tends to alienate them, making the psychiatrist seem insensitive and lacking in understanding and empathy. Physicians *should* provide specific and realistic reassurance, emphasizing a constructive treatment plan, and mobilize the patient's support system.

Medical clinicians should be aware of the effect of the patient's personality style on the physician-patient relationship and modify management to fit the patient better. The liaison part of CL psychiatry is instrumental in helping non-psychiatrist clinicians understand these patient interactions and how modifications to the relationship may be possible. For example, a type A personality patient with time urgency issues may instill a feeling of being rushed and annoyance. Attention to scheduling issues, avoiding long wait times, and planning ahead are appropriate adjustments to be considered. Active coping may be enhanced with group therapy, especially for patients with serious medical illnesses such as cancer, heart disease, and renal failure. These support groups are designed to be broadly generalizable rather than targeted to one particular trait or style.

The modification of maladaptive health behaviors is an area of research with many promising approaches. For example, cognitive behavioral therapy for insomnia (CBT-i) is efficacious for chronic insomnia and could be incorporated into therapy with medically ill patients as needed (Raglan et al., 2019). Other behavioral therapies are useful for smoking cessation along with psychopharmacologic therapies like varenicline or bupropion, and nicotine replacement (M. B. Steinberg et al., 2009). Behavioral strategies such as motivational interviewing are also useful in promoting better dietary practices, sleep hygiene, safe sex, and exercise (Griffith, 2008; Pollak et al., 2011). Biofeedback, relaxation techniques, hypnosis, and other stress management interventions (e.g., mindfulness based stress reduction) have been helpful in reducing stress-induced exacerbations of medical illness, including cardiac, gastrointestinal, headache, and other symptoms (Cramer et al., 2012). For some patients, change can be achieved efficiently through support groups, whereas others change more effectively through a one-to-one relationship with a healthcare professional.

Psychological Factors in Specific Medical Disorders

An understanding of discipline specific nuances is important for interpreting how PSAOMC will affect medical or surgical outcomes. The stressors placed on patients vary based on illness specificity, physical morbidity, requirements for treatment, and impact on role (e.g., work), implications for family and support, and other variables. Not only are the implications of some diagnoses more pressing and cause more psychological strife, but clearly requirements for medical care and treatment differ. In other words, psychological factors may affect one medical diagnosis more than another. For this reason, the CL psychiatrist should understand the basic principles of disease and care among commonly encountered co-morbidities and collaborate when the diagnosis or treatment course is not understood.

Considering the enormity of PSAOMC, much more data are needed generally to inform specific, context dependent treatments. Unfortunately, many earlier research efforts were limited by various methodological flaws (e.g., biased samples, lack of adequate control groups, and retrospective study designs). Research efforts have improved generally and are now evident in most areas of CL psychiatry.

Psychological Factors in Oncology

Despite recent advances in oncology, cancer continues to instill dread and fear. Psychological factors are intricately intertwined with cancer; its diagnosis, treatment, recurrence, and even its cure (i.e., survivorship) are fraught with psychological issues (Holland, 2001). At the same time, it should be appreciated that cancer as a diagnosis means different things to the individual patient or family. For some patients, there may be an opportunity to be "living with cancer" and for all patients, it is a lifestyle change, and its presence never disappears entirely. This adversity may also draw out positive changes, renewed coping styles, and meaningful social supports, for example, which may be encompassed under Post Traumatic Growth.

The most common psychological factors in oncology are psychological distress, anxiety, and depression. The prevalence of psychological symptoms varies based on cancer type, its treatment, and its trajectory. Most psychological symptoms are time-limited but all symptoms may interfere with medical care, psychosocial, and cancer outcomes. Since psychological symptoms are expected, the assessment of psychological symptoms is inherently normative; but how much it attributes to issues with cancer or other medical care may be quantified to some extent. This makes a diagnosis of PFAOMC intrinsically useful for the CL psychiatrist seeing a patient with cancer.

Some psychological factors are unique to the cancer setting and may become persistent. For example, fear of cancer recurrence, existential anxiety, reactions to end of life, specific phobia (e.g., needle phobia), role changes, self and body image, and their effects on medical adherence (e.g., office visits, treatment completion). Other relevant psychological factors include demoralization, anger, embitterment, grief, and bereavement. Current research has also focused on the effects of cancer on caregivers and clinicians who work with patients with cancer (e.g., burnout, compassion fatigue). The mental health of clinicians who care for patients with cancer can affect their quality of life and sense of well-being.

Mental Disorders and Psychological Symptoms Affecting Cancer

Historically, many myths were perpetuated regarding the influence of psychological factors on cancer. At one time it was believed that certain characterological factors would lead to cancer or that cancer can be cured with "mind over body" techniques. For this reason, literature in this area coined the phrase "tyranny of positive thinking" because so many patients have been subjected to these pressures to adjust their psychology accordingly. At present, behavior, and chemical exposures due to lifestyle, but not psychology, are associated with cancer initiation. That is, psychological factors such as personality or depression

are not linked to the development of cancer (Garssen, 2004).

Psycho-oncology as a discipline developed out of many of these concerns while cancer treatments (e.g., surgery, radiotherapy, and systemic) began to be effective. The field continues to evolve alongside cancer management with innovations in providing comprehensive cancer care (Holland, 2018). Studies have focused on prevalence of distress, anxiety, and depression and the impact of affective states on disease outcome in cancer patients (Holland et al., 2015). Many studies are in the breast cancer setting specifically but all major cancer types have been studied. Affect symptoms may depend on timing. For example, following a cancer diagnosis is a pivotal time when patients experience extreme distress that tends to ameliorate with time. Patients may experience psychological decompensation when the cancer progresses or at recurrence, which may incur even higher levels of distress and anxiety (Andersen et al., 2005). Of these three psychological symptoms, depression appears to be most closely associated with inferior quality of life and even shortened cancer-related survival times (Satin et al., 2009). Interestingly, depression remission is associated with a return to baseline survival equivalent to non-depressed patients, but survival is decreased in patients with persistent depression (Sullivan et al., 2016). Of note, a study of 199 hematologic oncology patients following stem cell transplantation showed that a diagnosis of major depressive disorder predicted significantly higher 1- and 3-year mortality whereas subsyndromal or minor depression had no statistically significant association with mortality (Prieto et al., 2005). Besides survival, depression in cancer patients may result in poorer pain control, poorer compliance, and an increased desire for hastened death (Die Trill, 2012; DiMatteo et al., 2000; O'Mahony et al., 2005).

Personality Traits or Coping Style Affecting Cancer

Coping, attachment style, and personality traits play a role in maintaining quality of life during the cancer trajectory and into survivorship. In general, a more expressive communicative style may be associated with an improved quality of life (Temoshok, 1987). Some studies, which vary based on definition and study population, have not supported an influence of this personality style (Cassileth et al., 1985). Neuroticism has been studied with regard to its influence on cancer outcomes with no conclusive evidence to show an effect on survival outcomes (Dahl, 2010). Although, neuroticism may be associated with long-term morbidity even into survivorship (Grov et al., 2009) and complications at the end of life (Chochinov et al., 2006).

At one time, there was great interest in the idea of the "fighting spirit" and its effect on cancer outcomes. Although some studies revealed benefit to this attitude, the so-called "positive psychology" interventions have not been scientifically proven to impact disease progression in cancer patients (Watson et al., 2005).

Other Psychological Factors in Cancer

An enormous literature documents the adverse effects of maladaptive health behaviors as risk factors for the development of various cancers, especially smoking but also excessive alcohol use, unsafe sex, and dietary practices (Glenn et al., 2018). Relatively less research has examined the effects of interpersonal variables on cancer, but there is some evidence that the quality of relationships may affect cancer onset and its course (Weihs et al., 2008). Social relations and social support and their effects on cancer patients (as with other diseases) are complex phenomena and may vary with cancer site and extent of disease (Decker, 2007). Another equivocal association with cancer initiation is that of a stressful life event but many studies have failed to find an association (Saito-Nakaya et al., 2012).

Psychosocial Intervention and Cancer Outcome

At one time, a controversy existed regarding a potential survival benefit associated with supportive-expressive group therapy, a meaningful intervention that improves quality of life (Spiegel et al., 1981). Replication studies did not demonstrate an association with improved survival even though benefits were sustained in terms of quality of life and other psychosocial outcomes (Goodwin et al., 2001). A Cochrane Review pooled data from studies examining psychological interventions in metastatic breast cancer and concluded that psychological interventions may demonstrate an improvement in short-term (12-month) but not long-term survival (Mustafa et al., 2013).

Despite questionable cancer-related outcomes from psychosocial interventions, data are clear that psychiatric symptoms, namely depression, are associated with inferior overall survival (Giese-Davis et al., 2011; Satin et al., 2009) It should be noted that these associations are with psychological symptoms and not necessarily meeting diagnostic criteria or cut-off points. Longitudinal studies have provided an increasingly amount of support for this relationship. A large epidemiological study of lung cancer demonstrated that survival outcomes returned to non-depressed baseline for depressed lung cancer patients whose depression remitted (Sullivan et al., 2016). A smaller study demonstrated the same association that only persistent depression was associated with worse survival (McFarland et al., 2021a). Psychosocial interventions that consistently improve depression may also benefit multiple cancer-related outcomes.

Psychological Factors in Cardiology

Coronary Disease

Depression is a strong risk factor that affects both the onset and course of coronary artery disease (CAD) (Kent & Shapiro, 2009). Type A personality, which is described as driven and hardworking, has received a great deal of attention for its association with CAD. But depression appears to have a stronger association with CAD than Type A personality. The weighting of depression as an independent risk factor in CAD must be adjusted for its interrelationships with other risk factors, especially lifestyle factors such as smoking and activity level, and also comorbid medical conditions such as hypertension and diabetes. The relationship between depression and CAD is not accounted for by other factors and demonstrates an approximately fourfold increase in mortality 6 month after a myocardial infarction in patients with major depression compared to cardiac patients without depression (Carney et al., 2003). In a large epidemiologic study, major depression tripled the relative risk of cardiac mortality in those without heart disease, and quadrupled it in those who did have heart disease (Penninx et al., 2001). Other epidemiologic studies have found that even subthreshold depressive symptoms (i.e., those not meeting criteria for major depressive disorder) can impart a significant risk for cardiac mortality after myocardial infarction (Bush et al., 2001). Severity of depressive symptoms has been found to predict disability to a greater extent than even the number of stenosed coronary arteries (Sullivan et al., 1997). In the Heart and Soul Study of over 1000 outpatients with stable CAD, depressive symptoms were associated with poor health status (e.g., symptom burden, physical limitations, and reduced quality of life) to a greater extent than measures of ischemia or ejection fraction (Ruo et al., 2003).

Patients with severe mental disorders have about twice the prevalence of the classic risk factors for CAD (Birkenaes et al., 2006). The exact mechanism linking depression to CAD are complex, multifactorial, and still incompletely understood (Penninx et al., 2001). But it appears that depression is associated with changes in coronary pathophysiology and microcirculation (Vaccarino et al., 2020). Thus, increased morbidity and mortality may be related to depression's adverse effects on heart rate variability, autonomic imbalance and arrhythmia, and platelet activation (Jangpangi et al., 2016; Sanner & Frazier, 2011). Proinflammatory mediators associated with depression may also be a contributor to CAD (Mason & Libby, 2015; Miller & Raison, 2016). In CAD, depression also reduces functional capacity, amplifies somatic symptoms (especially pain), and reduces motivation and compliance with medication, lifestyle change, and cardiac rehabilitation (Gehi et al., 2005). Multicenter intervention studies have demonstrated the efficacy of treating depression in CAD but have not as yet shown that such treatment can improve medical outcomes [e.g., the Sertraline Antidepressant Heart Attack Randomized Trial (SADHART) (Glassman et al., 2002), and the Myocardial Infarction and Depression Intervention Trial (MIND-IT) (Honig et al., 2007)].

One specific mechanism by which psychological factors can affect CAD has been demonstrated experimentally. Silent myocardial ischemia (ischemic changes on the electrocardiogram without symptoms of angina) can be precipitated by acute mental stress. Those who experience it are twice as likely to have major cardiac events than those who do not (Conti et al., 2012; Jiang et al., 1996). Silent ischemia may be partly a consequence of cognitive or defensive traits such as denial, hyposensitivity to somatic sensation, or systematic misperception of angina (Barsky et al., 1990). Psychological stress also changes the balance between procoagulation and fibrinolysis (von Kanel et al., 2001). Psychological factors may also affect outcome in CAD via differences in healthcare received. After myocardial infarction, patients with mental disorders are less likely to undergo cardiac catheterization and coronary revascularization than those without mental disorder (Druss & Walker, 2011).

Although the diagnosis and treatment of anxiety in patients with cardiac symptoms have attracted much attention, there has been less examination of anxiety as a risk factor affecting CAD. Increases in myocardial infarction and/or sudden death have been documented in epidemiologic studies of populations undergoing missile attacks, earthquakes, and other disasters (Krantz et al., 2000). Generalized anxiety disorder is associated with CAD risk factors even after controlling for depression (Barger & Sydeman, 2005). Furthermore, patients with generalized anxiety disorder and CAD appear to have an increased incidence of adverse cardiovascular events. Phobic anxiety predicts risk of ventricular arrhythmias and deaths from coronary heart disease. Posttraumatic stress disorder has also been shown to be associated with CAD and a resultant increase in mortality, independent of other risk factors including lifestyle habits and common medical comorbidities (Ahmadi et al., 2011). The INTERHEART study of 11,000 patients with first myocardial infarction compared with 13,000 controls from 52 countries found higher rates of stress factors (work, home, financial, and major life events) associated with increased risk of myocardial infarction (Rosengren et al., 2004). Anxiety following myocardial infarction may lead to more frequent readmission for unstable angina and more myocardial infarction recurrences, and also higher mortality (Moser & Dracup, 1996). Anxiety's adverse effects on CAD outcome may occur via effects on heart rate variability, QT prolongation, or other autonomically mediated phenomena, such as the stress-induced silent ischemia described earlier.

Denial is another common and significant psychological factor in patients with coronary disease. Denial may prevent individuals from acknowledging acute cardiac symptoms and promptly seeking medical care. The length of delay between the onset of symptoms of a myocardial infarction and hospitalization is a powerful predictor of morbidity and mortality, so denial at the onset of symptoms has an adverse impact on acute coronary disease. Type A behavior pattern is a PFAOMC that may be associated with denial and has been linked to CAD. It is a pattern of complex set of traits most often associated with impatience, hostility, intense achievement drive, and time urgency. Evidence for its relationship with CAD emerged in the 1970s (Rosenman et al., 1975). However, its association remains definitively inconclusive at this point and stronger evidence exists for affective symptoms and CAD. In addition, maladaptive health behaviors are also risk factors in coronary disease while the effects of smoking are better established than those of sedentary lifestyle, obesity, or specific diet. The effects of psychopathology and of smoking on heart disease are easily confounded, as persons with psychiatric disorders are overall twice as likely to smoke than others, with the increased risk found with all the major anxiety, mood, and psychotic disorders (Lasser et al., 2000). There are other additional psychological factors that deserve study. Of note, women with severe marital stress had triple the risk of recurrent coronary events

than those without marital stress (Orth-Gomer et al., 2000).

Arrhythmias

Psychological stressors can play a role in the precipitation of arrhythmia, the most serious of which are the ventricular arrhythmias. Depression is a risk factor for autonomic dysfunction, thereby increasing one's risk for arrhythmias (Wang et al., 2013). Anger and hostility have been associated with a modest increase in the incidence of atrial fibrillation in men. Life-threatening ventricular arrhythmias may occur after stressful events or calamities such as following the attack on the World Trade Center on 11 September 2001 (Steinberg et al., 2004). A review of 96 published studies investigating psychosocial risk factors for arrhythmia found that 92% were positive (Hemingway et al., 2001).

Congestive Heart Failure

Depression is especially common in patients with congestive heart failure (CHF). In patients hospitalized for CHF, major depression is independently associated with increased mortality and readmission 3 and 12 months later (Jiang et al., 2001). The SADHART-CHF trial further supported these findings, showing a significant (7%) increase in 12-week mortality in a cohort of 469 CHF patients with co-morbid depression (O'Connor et al., 2010). Furthermore, this study showed that sertraline had no statistically significant benefit over placebo in reducing depression or improving cardiovascular outcomes (O'Connor et al., 2010). As with CAD, further large-scale, randomized studies are needed to elucidate what, if any, effect depression interventions have on cardiovascular outcomes.

Hypertension

The stress-related physiologic response subcategory of PFAOMC is particularly relevant to hypertension. Blood pressure reactivity to stress is a risk factor for the development of hypertension and may also influence progression of disease. In particular, a 12-year follow-up to the West Scotland Study and the subsequent Dutch Famine Birth Cohort Study further supported the stress reactivity hypothesis, specifically showing that initial systolic blood pressure reactivity to stress is a significant predictor of future hypertension (Carroll et al., 2012). Many studies have examined relationships between personality, coping style, blood pressure reactivity, and hypertension, but conflicting results and methodological limitations have precluded any consensus conclusions.

Psychological Factors in Endocrinology

Diabetes Mellitus

Psychiatric symptoms and psychological factors appear to play a role in glycemic control of patients with diabetes. Although the "diabetic personality" of early psychosomatic literature has been debunked, there does appear to be a relationship between psychological factors and glucose control. Glycemic control is poorer in those diabetic patients who have more perceived stress (Lloyd et al., 1999). Metabolic control is also worse in depressed versus non-depressed children and in adult depressed type 1 and type 2 diabetics based on hemoglobin A1c, which provides a valid interpretable indication of glucose control over 3 months (Dirmaier et al., 2010; Melin et al., 2013). A meta-analysis of 24 studies found that depression predicted increased hyperglycemia in both type 1 and type 2 diabetes leading to greater diabetic morbidity (Lustman et al., 2000). Depression is associated with diabetic complications and a greater severity of diabetic symptoms (Ludman et al., 2004) and greater disability, mortality, and higher healthcare costs compared with other diabetics (Simon et al., 2005). A prospective cohort study of 4623 patients with type 2 diabetes showed that those patients who were diagnosed with major depressive disorder had a significant increase in the risk for micro- and macrovascular complications, independent of diabetes severity (Lin et al., 2010). Although psychiatric illness and its adverse effects on glucose control may be mediated by non-adherence, it is also clear that psychological stress impairs glucose control in both insulin-dependent and non-insulin dependent diabetes (Faulenbach et al., 2012; Hilliard et al., 2016).

Psychological interventions have shown improvements in glucose control in diabetic patients based on several randomized controlled trials. Small RCTs have demonstrated improvements in glucose control in diabetic patients receiving psychological interventions (Zamani-Alavijeh et al., 2018; Zareban et al., 2014). Both psychological interventions and antidepressant medications are effective in treating depressive symptoms in patients with diabetes but have mixed effects on glycemic control (Holt et al., 2014). The Pathways Study, a comprehensive clinical trial of enhanced treatment for depression in diabetic patients, demonstrated improved the quality of care and outcomes for depression but did not result in improved glycemic control (Katon et al., 2004). Antidepressants are effective in the treatment of depression in diabetic patients, but can cause increases or decreases in blood glucose themselves. Glucose control in patients with schizophrenia or other psychotic disorders may be complicated by atypical antipsychotic drugs, which increase blood glucose levels in addition obesity, dietary habits, and poorer access and integration with healthcare overall (Annamalai & Tek, 2015). Early collaboration and office-based interventions are key to diabetes management in patients with schizophrenia (Annamalai & Tek, 2015).

Thyroid Disease

It is well established that too little or too much thyroid hormone can result in disturbances in mood and activity (Fukao et al., 2020). As the terms "Basedow psychosis" and "myxedema psychosis" indicate, thyroid diseases are highly concordant with mood disturbance and other mental disorders. Most commonly, too much or too little thyroid function effects mood, cognition, and activity. In the other direction, the effects of emotion and stress on thyroid function, although long a focus of interest, are not well established. A number of classic studies reported a relationship between antecedent traumatic stress and the onset of thyrotoxicosis, particularly as part of Graves' disease. However, these studies were retrospective, uncontrolled, and too methodologically flawed by current standards to support the validity of such a link. Later studies have supported stressful life events as a risk factor for Graves' disease (Santos et al., 2002). More recently, several studies have found and concluded that psychosocial factors including emotional stress are related to the onset of grave's disease (Mizokami et al., 2004; Winsa et al., 1991). But it is also clear that ongoing depression and psychological stress effects thyroid function in patients with Grave's disease (Fukao et al., 2020).

Psychological Factors in Pulmonary Disease

Although asthma was once regarded as a classic psychosomatic disorder, it is currently viewed as a primary respiratory disease with varying immunological and autonomic pathophysiologic changes. Nonetheless, many psychological symptoms, particularly anxiety, are associated with asthma and psychosocial stress clearly exacerbates asthma morbidity (Yonas et al., 2012). Anxiety is primarily associated with the underlying etiology of reactive airway disease (Goodwin et al., 2003). Respiratory distress itself causes a wide array of anxiety symptoms (panic attacks, generalized and anticipatory anxiety, phobic avoidance), and most of the drugs used to treat asthma have anxiety as a potential side effect (Yonas et al., 2012). In a large population-based sample of adults, asthma was associated with a significantly increased likelihood of anxiety disorders (especially panic, generalized anxiety disorder, and phobias) and affective disorders (Goodwin et al., 2003). A population-based cohort study of 2868 Australian children followed from birth further suggests that the likelihood of developing a psychiatric disorder increases with asthma severity (Goodwin et al., 2013). Clinically, it does appear that anxiety may at times trigger asthma but whether patients with primary anxiety disorders are more likely to develop asthma is yet to be determined despite longitudinal studies. There is no typical personality type susceptible to development of asthma. Anxiety and depression are associated in asthmatic patients with more respiratory symptom complaints but no differences in objective measures of respiratory function. At the same time, asthma symptom severity increased in New York City following the terrorist attacks on 11 September 2001, and posttraumatic stress disorder was a significant predictor of the increase in asthmatic symptoms (Fagan et al., 2003). Psychological factors and psychosocial problems in hospitalized asthmatic patients were a more powerful predictor of which ones required intubation than any other examined variable (e.g., smoking, infection, prior hospitalization) (LeSon & Gershwin, 1996). Psychological morbidity is associated with high levels of denial and delays in seeking medical care, which may be life-threatening in severe asthma (Miles et al., 1997). Furthermore, psychological morbidity in and of itself imparts an independent risk factor for poor asthma control. Not surprisingly, then, psychopathology in severe asthmatic patients is associated with increased healthcare utilization, including hospitalizations, and outpatient and emergency room visits, together with worse quality of life, independent of asthma severity (Yonas et al., 2012).

Similar relationships between anxiety or depression and other chronic obstructive pulmonary diseases (COPD) (e.g., chronic bronchitis, emphysema). In general, depression and anxiety are common in COPD (Yohannes & Alexopoulos, 2014) and overlaps with its association with smoking. Psychological distress in COPD amplifies dyspnea without usually causing changes in objective pulmonary function and leads to lower exercise tolerance, noncompliance with treatment, poorer health status, and increased disability in COPD (Yohannes & Alexopoulos, 2014). In addition, depression and anxiety add to the economic burden of COPD (Dalal et al., 2011). Anxious COPD patients are more likely to be hospitalized and re-hospitalized with all of the attendant morbidity associated with inpatient hospital stay. Importantly, depression in patients with COPD is an independent predictor of mortality even among younger otherwise medically healthy patients (Vikjord et al., 2020). Although depression and anxiety are associated with smoking, these psychological symptoms predict mortality in patients with COPD independent of tobacco use (Lou et al., 2014). Nonetheless, interventions should target psychological factors (e.g., anxiety and depression) and smoking.

Psychological Factors in Rheumatoid Arthritis

Rheumatoid arthritis (RA) is an autoimmune disease that was also thought of as a psychosomatic disorder at one time. There is no particular personality type susceptible to the development of RA. Research suggesting that stressful life events play a role in the development or onset of RA and other autoimmune diseases has yielded mixed results. A large-scale cohort study of 15,357 patients revealed an association between stressful childhood events and the subsequent development of autoimmune diseases in adulthood (Dube et al., 2009). Interestingly, psychological factors and disease manifestations account for comparable proportions of disability in RA (Escalante & del Rincon, 1999). Psychological symptom burden in RA is associated with more pain, poorer quality of life, more joint surgery and interventions, lower adherence, and increased use of healthcare resources (Dickens et al., 2002).

Depression has been the most frequently studied psychological disturbance in RA and is very common resulting in pain aggravation, healthcare utilization, and social isolation (Dickens et al., 2002). A large longitudinal study found depression to be an independent risk factor for mortality in patients with RA (Ang et al., 2005). A number of RCTs of antidepressants in depressed RA patients have demonstrated improvements in pain, morning stiffness, and disability in addition to depression (Ash et al., 1999). However, a Cochrane analysis concluded that there was insufficient evidence to support the widespread use of antidepressants for pain management in this RA (Richards et al., 2011).

Several maladaptive coping strategies have been associated with poorer adjustment to RT. These include passive, avoidant, and emotion-laden coping strategies that include self-blame and wish-fulfilling fantasy, for example. Pro-active, problem-focused coping (e.g., information seeking, cognitive restructuring) is associated with improved RA outcomes (Covic et al., 2000). Patients with RA who demonstrate high levels of helplessness are more likely to receive psychotropic, analgesic, and anti-inflammatory drugs and to be less adherent to treatment than those with low helplessness (Stein et al., 1988). The bidirectional relationship between RA and depression, along with associated worse medical outcomes, may be driven by pro-inflammatory mediators that are heightened in the setting of RA (Lwin et al., 2020). A recent review of various studies looking at different psychotherapeutic interventions in RA has shown consistent benefits from disclosure therapy, self-regulation therapy, and long-term cognitive behavioral therapy (CBT) with maintenance therapy (Knittle et al., 2010).

Psychological Factors in Neurology

Not only is depression commonly encountered after a stroke (e.g., first few weeks to 6 months), studies have demonstrated that post-stroke depression may remain stable for up to 10 years after the stroke (Ayerbe et al., 2013). The presence of post-stroke depression is associated with poorer outcomes, including higher later mortality, poor quality of life, and increased disability (Ellis et al., 2010). Addressing depression improves post-stroke functional status but antidepressants may impart a benefit to post-stoke functioning apart from its direct effect on depression. A double-blind, randomized, placebo-controlled trial of antidepressants in post-stroke depression showed that antidepressant use was associated with a statistically significant improvement in disability independent of depressive symptoms (Mikami et al., 2011). The opposite is also true. A negative attitude after stroke (i.e., feeling there is nothing one can do to help oneself) is associated with decreased survival (Dennis et al., 2000).

Attention has been focused on depression as a complication of stroke but there is also evidence that depression and other psychological factors constitute risks for stroke. A meta-analysis and systematic review found that depression was associated with significantly increased risk of stroke morbidity and mortality (Pan et al., 2011). Perhaps not surprisingly, depression appears to be an independent predictor of increased healthcare use after stroke and is also associated with poorer functional recovery (Jia et al., 2006). As with many other major medical illnesses, stroke patients with extensive social support have better functional outcomes than those who do not have such support.

Depression is common in Parkinson's disease, may antedate the development of motor symptoms, and is associated with cognitive dysfunction (Marsh, 2013). Depression and other psychological factors affect the course and outcome of Parkinson's disease; depression, in particular, results in impairment of functional capacity, but is not contribute to worsening motor function (Holroyd et al., 2005). Depression is also common in patients with multiple sclerosis and is associated with impaired quality of life (Fruehwald et al., 2001).

Studying depression in the context of neurological disease is challenging since depression may be a direct, physiologically mediated consequence of the disease, a psychological reaction to the illness, or a complication of pharmacotherapy (Zorzon et al., 2001). Depression is especially difficult to study in the setting of MS because of its uncertain relationship to the MS-fatigue syndrome.

Patients with chronic migraine headaches have often been described as having a "typical" personality characterized as conscientious, perfectionistic, ambitious, rigid, tense, and resentful, but controlled studies have not supported any consistent conclusion. Specific personality traits in migraine appear more likely to be a consequence rather than a cause of suffering from recurrent headaches (Stronks et al., 1999). Migraine and psychiatric disorders are highly comorbid; specifically, migraine is associated with a higher prevalence of major depression, bipolar disorder, and anxiety disorders (Jette et al., 2008). The relationship between migraine and personality has been of long-standing interest. In 1948, Harold Wolff coined the term "migraine personality," which has not stood the test of time but demonstrates an intuitive and inherent association (Davis et al., 2013). Of note, a community-based survey found more personality disturbance and 2.5 times more psychological distress in migraine sufferers than in matched control subjects, but there was no relationship between headache frequency and the severity of psychological distress or personality abnormality (Davis et al., 2013).

Psychological Factors in End-Stage Renal Disease

Studies of the influence of psychological factors on the course of end-stage renal disease (ESRD) have nearly all focused on depression or non-adherence with therapy. There is a paucity of data describing any association between psychological factors and non-dialysis dependent chronic kidney disease (CKD). The mechanism of depression may be different in ESRD compared with CKD (Shirazian et al., 2017). But for patients on dialysis, depression is associated with shorter survival than non-depressed dialysis dependent ESRD patients and evidence that the relationship is dose dependent with more severe depressive symptoms associated with greater mortality (Farrokhi et al., 2014). A prospective study in patients receiving peritoneal dialysis (vs. hemodialysis) found higher depressive symptoms associated with a greater risk of peritonitis, after controlling for other risk factors (Troidle et al., 2003). Another study also found that depression was independently associated with an increased risk for infection-related mortality (Riezebos et al., 2010). In hemodialysis patients, the level of depression symptoms was a unique and significant predictive risk factor for the subsequent decision to withdraw from dialysis (McDade-Montez et al., 2006).

Psychological Factors in Gastroenterology

Inflammatory Bowel Disease (IBD)

Psychological stress appears to aggravate both symptom complaints and mucosal disease activity in ulcerative colitis (Mawdsley & Rampton, 2005) and depression predicts relapse in IBD (Kurina et al., 2001). In fact, depression is a better predictor of subjective impairment in IBD than inflammatory activity (Cuntz et al., 1999). Disability and distress in patients with IBD are increased by the presence of a concurrent psychiatric disorder. Depression, in particular, has been shown to be a significant predictor of poor quality of life, independent of disease severity in IBD. Although psychotherapy has been shown to improve psychological outcomes in IBD, the evidence as to its efficacy in improving physical symptoms and reducing disease severity remains limited and is mixed (Knowles et al., 2013).

Irritable Bowel Syndrome (IBS)

IBS is a heterogeneous condition with a high frequency of comorbid anxiety (especially panic attacks), depression, and somatization. Although patients with IBS experience psychological distress, there is not a common profile of psychological symptoms or personality traits; however, neuroticism may be more common in patients with IBS (Tayama et al., 2012). Patients with IBS are more likely to have a history of childhood sexual abuse than those with other gastrointestinal disorders in studies of patients seeking care at tertiary referral centers (E. A. Walker et al., 1993). In fact, almost all psychological characteristics and psychopathology thought to be more common in IBS are differentially increased in patients whose symptoms drive them to seek medical care. Although gastrointestinal symptoms seem to be aggravated by stress in patients with IBS, there is no clear evidence that stress causes a different gastrointestinal smooth muscle response than in control subjects (Blanchard et al., 2008). But, the effects of psychological factors on IBS may be mediated by pain perception and other somatic symptoms that lead to healthcare seeking behavior. After an acute episode of infectious gastroenteritis, individuals with more life stress were more likely to develop IBS without any differences in intestinal physiology (Gwee et al., 1999).

Psychological Factors Affecting Infectious Diseases

Clinical evidence supports the notion that psychological factors influence the immune system and its ability to protect the body from infectious disease. For example, viral and bacterial respiratory infections (Cohen et al., 2001; Mehr et al., 2001), genital herpes (Levenson et al., 1987), and recurrent urinary tract infections (Hunt & Waller, 1992) are worsened by stress and psychological factors. A number of studies have convincingly shown that psychological stress suppresses the secondary (but not primary) antibody response to immunization (Cohen et al., 2001).

The relationship between psychological factors and HIV is complex. For example, while there is a distinct relationship, the directionality between severe mental illness and HIV infection is less well understood. That is, patients with severe mental illness acquire and transmit HIV at a higher rate than the general population. Patients with psychiatric illness engage in at-risk behaviors. These include but are not limited to substance misuse (e.g., intravenous drug use) and sexual behaviors (e.g., unprotected sex, prostitution), and are less informed about HIV-related issues (De Hert et al., 2011). At the same time, patients with psychiatric conditions are more likely to be tested for HIV than those without mental illness thus leading to earlier detection than the general population (Yehia et al., 2014). Psychiatric conditions and psychological factors develop due to HIV infection. These are commonly issues with sleep, anxiety, depression, and cognition. Many studies have shown that stress, depression, and trauma may cause disease progression, decreased immune function, and early mortality in patients with HIV infection (Ironson et al., 2005; Leserman et al., 2007; Sewell et al., 2000).

HIV is a neurotropic virus. Late stages of HIV precipitate HIV encephalopathy or HIV-associated dementia. There is a paucity of psychiatric treatment guidelines for HIV-positive individuals (Himelhoch & Medoff, 2005). But systematic reviews of various intervention studies for depression and anxiety in HIV have consistently shown that CBT is an effective treatment modality, often superior to pharmacotherapy (Clucas et al., 2011; Sherr et al., 2011).

Other infectious disease commonly affected by psychological factors include those diseases that cause meningitides such as syphilis, Lyme disease (arthropod borne illness), and tuberculosis. Multi-drug resistant tuberculosis (MDRTB) is especially problematic where scarce resources exist to provide concurrent multidrug therapy. Symptoms of neurosyphilis can be divided into meningovascular (e.g., hemiplegia, seizures, aphasia), parenchymatous (e.g., optic atrophy tremors, papillary disturbances, changes in personality, speech disturbances), and tabes dorsalis (e.g., shooting radicular pains, ataxia, bladder disturbances, cranial neuropathy, positive Romberg sign).

Psychological Factors in Dermatology

Dermatologists observe important relationships between psychological factors and urticaria, angioedema, atopic dermatitis, hyperhidrosis, acne, and psoriasis, although controlled studies are lacking (Kimyai-Asadi & Usman, 2001). Dermatologists routinely observe the effect of psychological factors, especially anxiety, in the aggravation of a wide variety of dermatological conditions. Both anxiety and depression appear to worsen pruritus (itching). Excoriation complicates many dermatological disorders and is aggravated by anxiety, depression, and other behavioral factors. That so many skin diseases appear to be precipitated or exacerbated by psychological stress also suggests a nonspecific impairment of cutaneous function. There is now evidence in both animals and humans that stress negatively affects the skin's function as a permeability barrier (Garg et al., 2001). Various nonpharmacologic interventions, specifically hypnosis and CBT, may be helpful in reducing the severity and symptom burden in various pathologic skin conditions (Shenefelt, 2003).

Psychological Factors in Obstetrics

Although much more attention has been paid to postpartum depression and psychosis, antepartum mood, anxiety, and psychotic disorders also adversely affect pregnancy outcome. In a number of studies, antepartum anxiety, posttraumatic stress disorder, bipolar disorder, schizophrenia, and depression have been associated with growth retardation and premature birth (Boden et al., 2012; Neggers et al., 2006). Whether depression and other psychological dysfunction cause poorer obstetric outcomes through poor nutrition, substance abuse (including tobacco), poor adherence with prenatal care, and/or physiologic (hormonal, vascular) effects require further investigation. Nevertheless, addressing psychosocial variables in pregnancy is imperative. Treatment interventions, particularly pharmacotherapy, in pregnancy require a thorough individualized evaluation of risk versus benefit.

Psychological Factors in Infertility

Psychological factors are likely to affect fertility because frequency and timing of sexual intercourse are important determinants of fertility. Nonconsummation, avoidance of intercourse, vaginismus, and psychogenic amenorrhea are attributable to psychological origins. Stress, but not of psychopathology specifically, has been associated with infertility (Greil, 1997). This topic is complicated by the potential interaction of psychological factors involving the couple and their effects on sexual behavior and fertility

(Cwikel et al., 2004). Psychological factors have also been identified as predictors of dropout from infertility treatment (Smeenk et al., 2004). Most psychological distress seen in infertile couples is a result of, rather than a cause of, infertility. Although in vitro fertilization (IVF) was a predictor of subsequent depression and anxiety, neither condition was a predictor of IVF failure in a prospective cohort study of 202 women undergoing IVF (Pasch et al., 2012).

Clinical Vignette 1

Ms. B a 56-year-old married attorney, was referred for psychiatric evaluation by her gastroenterologist, who follows her for long-standing IBS. She has had IBS since the age of 20 years, with complaints of intermittent constipation, diarrhea, crampy abdominal pain, and bloating. She feels that these symptoms have gradually worsened, particularly in the last month. She describes a highly pressured job and a stressful marriage. She has specifically noticed a precipitous increase in intestinal symptoms immediately after arguments with her husband and when facing deadlines at work. Three months ago, she developed depressed mood, early-morning awakening, anorexia, fatigue, crying spells, impaired concentration, irritability, and preoccupation with thoughts of ill health. Her family physician diagnosed major depression and prescribed amitriptyline, which was discontinued after it caused worsening of her constipation. Her psychiatrist then tried fluoxetine (discontinued because of diarrhea) and trazodone (too sedating). She then responded well to nortriptyline, with disappearance of the symptoms of depression and improvement in her IBS. However, severe IBS symptoms continued to follow the frequent marital arguments. The psychiatrist asked the patient to invite her husband to join one of their sessions so that marital issues could be further explored. He did so, resulting in the discovery that her husband was himself significantly depressed. He was

referred to another psychiatrist for treatment, the marital discord abated, and her IBS symptoms returned to a manageable level.

In this clinical vignette, the patient had features of both depressive symptoms and stress-related physiologic response affecting her medical condition, IBS. Treatment included individual psychotherapy and antidepressant medication plus marital assessment and intervention. Pharmacotherapy required modification because of gastrointestinal sensitivity to side effects.

Clinical Vignette 2

Mr. X is a 60-year-old married pharmaceutical executive with unstable angina and coronary artery disease (CAD). He was referred for psychiatric evaluation by his cardiologist because he declined coronary artery bypass surgery despite strong and repeated recommendations for surgery by the cardiologist. The cardiologist noted that the patient has been well informed about the need for surgical interventions but is confused as to why the patient is not following through with the recommendation.

Mr. X has no reported psychiatric history and denied any acute or chronic psychiatric symptoms. He reports that his CAD is serious and that he has had two myocardial infarctions and an arrhythmia. He continued to have recurrent angina despite maximal medical management; his pain occurred mainly at night as "a predictable consequence of pushing too hard at work" (he typically worked 12-h days). He had also had a stroke 3 months ago, from which he had made a complete recovery with no sequelae. Twenty years earlier, he had surgery for a renal stone that was complicated postoperatively with five pulmonary emboli. He reports several "near-

(continued)

death experiences" and feels lucky to be alive.

Mr. X wanted to discuss his reluctance to have coronary bypass surgery. He had not decided against the surgery but had been unable to reach a decision. He brought a two-page list of arguments for and against surgery and other variables that could influence the decision and outcome. He was aware that he was approaching the question of surgery as if he were weighing a business decision. He worked longer days than his colleagues because he believed it took more time to get his deals just right. His analysis of the pros and cons of surgery, and also intervening factors affecting and affected by the decision, appeared to the psychiatrist to be well informed, accurate, flexible, and appropriate. There was no evidence of rigidity in his thinking, premature closure, or distorted perceptions. Whereas the thought processes were logical, they had not enabled him to reach a decision, despite extensive discussions with the cardiologist over a period of months. He was aware that this was another decision in his life that was taking much longer than average, but he thought it could not be resolved any other way.

This case represents an example of personality trait or coping style affecting another medical condition. His obsessional style was largely adaptive for his business, but it reduced his professional and personal efficiency. Now confronted with a major healthcare decision, and mindful of major complications he had suffered after surgery in the past, the need to weigh all sides of an issue had paralyzed his decision-making. The presence of phobias in his history raised the possibility of these too affecting his decision-making, but he denied feeling fearful of the surgery, anesthesia, intubation, and the like. The anxiety he was experiencing was entirely focused around making the right decision.

References

Abouljoud, M., Ryan, M., Eshelman, A., Bryce, K., & Jesse, M. T. (2018). Leadership perspectives on integrating psychologists into specialty care clinics: An evolving paradigm. *Journal of Clinical Psychology in Medical Settings, 25*(3), 267–277. https://doi.org/10.1007/s10880-017-9532-9

Ahmadi, N., Hajsadeghi, F., Mirshkarlo, H. B., Budoff, M., Yehuda, R., & Ebrahimi, R. (2011). Post-traumatic stress disorder, coronary atherosclerosis, and mortality. *The American Journal of Cardiology, 108*(1), 29–33. https://doi.org/10.1016/j.amjcard.2011.02.340

Alegria, M., NeMoyer, A., Falgas Bague, I., Wang, Y., & Alvarez, K. (2018). Social determinants of mental health: Where we are and where we need to go. *Current Psychiatry Reports, 20*(11), 95. https://doi.org/10.1007/s11920-018-0969-9

Andersen, B. L., Shapiro, C. L., Farrar, W. B., Crespin, T., & Wells-Digregorio, S. (2005). Psychological responses to cancer recurrence. *Cancer, 104*(7), 1540–1547. https://doi.org/10.1002/cncr.21309

Ang, D. C., Choi, H., Kroenke, K., & Wolfe, F. (2005). Comorbid depression is an independent risk factor for mortality in patients with rheumatoid arthritis. *The Journal of Rheumatology, 32*(6), 1013–1019. Retrieved from https://www.ncbi.nlm.nih.gov/pubmed/15940760

Annamalai, A., & Tek, C. (2015). An overview of diabetes management in schizophrenia patients: Office based strategies for primary care practitioners and endocrinologists. *International Journal of Endocrinology, 2015*, 969182. https://doi.org/10.1155/2015/969182

Ash, G., Dickens, C. M., Creed, F. H., Jayson, M. I., & Tomenson, B. (1999). The effects of dothiepin on subjects with rheumatoid arthritis and depression. *Rheumatology (Oxford), 38*(10), 959–967. https://doi.org/10.1093/rheumatology/38.10.959

Ayerbe, L., Ayis, S., Wolfe, C. D., & Rudd, A. G. (2013). Natural history, predictors and outcomes of depression after stroke: Systematic review and meta-analysis. *The British Journal of Psychiatry, 202*(1), 14–21. https://doi.org/10.1192/bjp.bp.111.107664

Barger, S. D., & Sydeman, S. J. (2005). Does generalized anxiety disorder predict coronary heart disease risk factors independently of major depressive disorder? *Journal of Affective Disorders, 88*(1), 87–91. https://doi.org/10.1016/j.jad.2005.05.012

Barsky, A. J., Hochstrasser, B., Coles, N. A., Zisfein, J., O'Donnell, C., & Eagle, K. A. (1990). Silent myocardial ischemia. Is the person or the event silent? *JAMA, 264*(9), 1132–1135. https://doi.org/10.1001/jama.264.9.1132

Birkenaes, A. B., Sogaard, A. J., Engh, J. A., Jonsdottir, H., Ringen, P. A., Vaskinn, A., ...& Andreassen, O. A. (2006). Sociodemographic characteristics and cardiovascular risk factors in patients with severe mental disorders compared with the general population. *The Journal of Clinical Psychiatry, 67*(3), 425–433. https://doi.org/10.4088/jcp.v67n0314.

Bishop, T. F., Seirup, J. K., Pincus, H. A., & Ross, J. S. (2016). Population of US practicing psychiatrists declined, 2003–13, which may help explain poor access to mental health care. *Health Aff (Millwood)*, *35*(7), 1271–1277. https://doi.org/10.1377/hlthaff.2015.1643

Blanchard, E. B., Lackner, J. M., Jaccard, J., Rowell, D., Carosella, A. M., Powell, C., ... & Kuhn, E. (2008). The role of stress in symptom exacerbation among IBS patients. *Journal of Psychosomatic Research, 64*(2), 119–128. https://doi.org/10.1016/j.jpsychores.2007.10.010.

Boden, R., Lundgren, M., Brandt, L., Reutfors, J., Andersen, M., & Kieler, H. (2012). Risks of adverse pregnancy and birth outcomes in women treated or not treated with mood stabilisers for bipolar disorder: Population based cohort study. *BMJ, 345*, e7085. https://doi.org/10.1136/bmj.e7085

Boland, R. J., Rundell, J., Epstein, S., & Gitlin, D. (2018). Consultation-liaison psychiatry vs psychosomatic medicine: What's in a name? *Psychosomatics, 59*(3), 207–210. https://doi.org/10.1016/j.psym.2017.11.006

Braveman, P., & Gottlieb, L. (2014). The social determinants of health: it's time to consider the causes of the causes. *Public Health Reports, 129*(Suppl 2), 19–31. https://doi.org/10.1177/00333549141291S206

Bush, D. E., Ziegelstein, R. C., Tayback, M., Richter, D., Stevens, S., Zahalsky, H., & Fauerbach, J. A. (2001). Even minimal symptoms of depression increase mortality risk after acute myocardial infarction. *The American Journal of Cardiology, 88*(4), 337–341. https://doi.org/10.1016/s0002-9149(01)01675-7

Carney, R. M., Blumenthal, J. A., Catellier, D., Freedland, K. E., Berkman, L. F., Watkins, L. L., ... & Jaffe, A. S. (2003). Depression as a risk factor for mortality after acute myocardial infarction. *The American Journal of Cardiology, 92*(11), 1277–1281. https://doi.org/10.1016/j.amjcard.2003.08.007.

Carroll, D., Ginty, A. T., Painter, R. C., Roseboom, T. J., Phillips, A. C., & de Rooij, S. R. (2012). Systolic blood pressure reactions to acute stress are associated with future hypertension status in the Dutch Famine Birth Cohort Study. *International Journal of Psychophysiology, 85*(2), 270–273. https://doi.org/10.1016/j.ijpsycho.2012.04.001

Cassileth, B. R., Lusk, E. J., Miller, D. S., Brown, L. L., & Miller, C. (1985). Psychosocial correlates of survival in advanced malignant disease? *The New England Journal of Medicine, 312*(24), 1551–1555. https://doi.org/10.1056/NEJM198506133122406

Chochinov, H. M., Kristjanson, L. J., Hack, T. F., Hassard, T., McClement, S., & Harlos, M. (2006). Personality, neuroticism, and coping towards the end of life. *Journal of Pain and Symptom Management, 32*(4), 332–341. https://doi.org/10.1016/j.jpainsymman.2006.05.011

Clucas, C., Sibley, E., Harding, R., Liu, L., Catalan, J., & Sherr, L. (2011). A systematic review of interventions for anxiety in people with HIV. *Psychology, Health & Medicine, 16*(5), 528–547. https://doi.org/10.1080/13548506.2011.579989

Cohen, S., Miller, G. E., & Rabin, B. S. (2001). Psychological stress and antibody response to immunization: A critical review of the human literature. *Psychosomatic Medicine, 63*(1), 7–18. https://doi.org/10.1097/00006842-200101000-00002

Conti, C. R., Bavry, A. A., & Petersen, J. W. (2012). Silent ischemia: Clinical relevance. *Journal of the American College of Cardiology, 59*(5), 435–441. https://doi.org/10.1016/j.jacc.2011.07.050

Covic, T., Adamson, B., & Hough, M. (2000). The impact of passive coping on rheumatoid arthritis pain. *Rheumatology (Oxford), 39*(9), 1027–1030. https://doi.org/10.1093/rheumatology/39.9.1027

Cramer, H., Lauche, R., Paul, A., & Dobos, G. (2012). Mindfulness-based stress reduction for breast cancer-a systematic review and meta-analysis. *Current Oncology, 19*(5), e343–e352. https://doi.org/10.3747/co.19.1016

Cuntz, U., Welt, J., Ruppert, E., & Zillessen, E. (1999). Determination of subjective burden from chronic inflammatory bowel disease and its psychosocial consequences. Results from a study of 200 patients. *Psychotherapie, Psychosomatik, Medizinische Psychologie, 49*(12), 494–500. Retrieved from https://www.ncbi.nlm.nih.gov/pubmed/10634068

Cwikel, J., Gidron, Y., & Sheiner, E. (2004). Psychological interactions with infertility among women. *European Journal of Obstetrics, Gynecology, and Reproductive Biology, 117*(2), 126–131. https://doi.org/10.1016/j.ejogrb.2004.05.004

Dahl, A. A. (2010). Link between personality and cancer. *Future Oncology, 6*(5), 691–707. https://doi.org/10.2217/fon.10.31

Dalal, A. A., Shah, M., Lunacsek, O., & Hanania, N. A. (2011). Clinical and economic burden of depression/anxiety in chronic obstructive pulmonary disease patients within a managed care population. *COPD, 8*(4), 293–299. https://doi.org/10.3109/15412555.2011.586659

Danese, A., & Lewis, S (2017). Psychoneuroimmunology of early-life stress: The hidden wounds of childhood trauma? *Neuropsychopharmacology, 42*(1), 99–114. https://doi.org/10.1038/npp.2016.198.

Dantzer, R., O'Connor, J. C., Freund, G. G., Johnson, R. W., & Kelley, K. W. (2008). From inflammation to sickness and depression: When the immune system subjugates the brain. *Nature Reviews. Neuroscience, 9*(1), 46–56. https://doi.org/10.1038/nrn2297

Davis, R. E., Smitherman, T. A., & Baskin, S. M. (2013). Personality traits, personality disorders, and migraine: A review. *Neurological Sciences, 34*(Suppl 1), S7–S10. https://doi.org/10.1007/s10072-013-1379-8

De Hert, M., Cohen, D., Bobes, J., Cetkovich-Bakmas, M., Leucht, S., Ndetei, D. M., ... & Correll, C. U. (2011). Physical illness in patients with severe mental disorders. II. Barriers to care, monitoring and treatment guidelines, plus recommendations at the system and individual level. *World Psychiatry, 10*(2), 138–151. https://doi.org/10.1002/j.2051-5545.2011. tb00036.x.

Decker, C. L. (2007). Social support and adolescent cancer survivors: A review of the literature. *Psychooncology, 16*(1), 1–11. https://doi.org/10.1002/pon.1073

Dennis, M., O'Rourke, S., Lewis, S., Sharpe, M., & Warlow, C. (2000). Emotional outcomes after stroke: Factors associated with poor outcome. *Journal of Neurology, Neurosurgery, and Psychiatry, 68*(1), 47–52. https://doi.org/10.1136/jnnp.68.1.47

Dickens, C., McGowan, L., Clark-Carter, D., & Creed, F. (2002). Depression in rheumatoid arthritis: A systematic review of the literature with meta-analysis. *Psychosomatic Medicine, 64*(1), 52–60. https://doi.org/10.1097/00006842-200201000-00008

Die Trill, M. (2012). Psychological aspects of depression in cancer patients: An update. *Annals of Oncology, 23*(Suppl 10), x302–x305. https://doi.org/10.1093/annonc/mds350

Diefenbacher, A. (2005). Psychiatry and psychosomatic medicine in Germany: Lessons to be learned? *The Australian and New Zealand Journal of Psychiatry, 39*(9), 782–794. https://doi.org/10.1080/j.1440-1614.2005.01683.X

DiMatteo, M. R., Lepper, H. S., & Croghan, T. W. (2000). Depression is a risk factor for noncompliance with medical treatment: Meta-analysis of the effects of anxiety and depression on patient adherence. *Archives of Internal Medicine, 160*(14), 2101–2107. https://doi.org/10.1001/archinte.160.14.2101

Dirmaier, J., Watzke, B., Koch, U., Schulz, H., Lehnert, H., Pieper, L., & Wittchen, H. U. (2010). Diabetes in primary care: Prospective associations between depression, nonadherence and glycemic control. *Psychotherapy and Psychosomatics, 79*(3), 172–178. https://doi.org/10.1159/000296135

Druss, B. G., Rohrbaugh, R. M., Levinson, C. M., & Rosenheck, R. A. (2001). Integrated medical care for patients with serious psychiatric illness: A randomized trial. *Archives of General Psychiatry, 58*(9), 861–868. https://doi.org/10.1001/archpsyc.58.9.861

Druss, B. G., & Walker, E. R. (2011). Mental disorders and medical comorbidity. *Synth Proj Res Synth Rep, 21*, 1–26. Retrieved from https://www.ncbi.nlm.nih.gov/pubmed/21675009

Dube, S. R., Fairweather, D., Pearson, W. S., Felitti, V. J., Anda, R. F., & Croft, J. B. (2009). Cumulative childhood stress and autoimmune diseases in adults. *Psychosomatic Medicine, 71*(2), 243–250. https://doi.org/10.1097/PSY.0b013e3181907888

Ellis, C., Zhao, Y., & Egede, L. E. (2010). Depression and increased risk of death in adults with stroke. *Journal of Psychosomatic Research, 68*(6), 545–551. https://doi.org/10.1016/j.jpsychores.2009.11.006

Engel, G. L. (1997). From biomedical to biopsychosocial. Being scientific in the human domain. *Psychosomatics,* *38*(6), 521–528. https://doi.org/10.1016/S0033-3182(97)71396-3

Escalante, A., & del Rincon, I. (1999). How much disability in rheumatoid arthritis is explained by rheumatoid arthritis? *Arthritis and Rheumatism, 42*(8), 1712–1721. https://doi.org/10.1002/1529-0131(199908)42:8<1712::AID-ANR21>3.0.CO;2-X

Fagan, J., Galea, S., Ahern, J., Bonner, S., & Vlahov, D. (2003). Relationship of self-reported asthma severity and urgent health care utilization to psychological sequelae of the September 11, 2001 terrorist attacks on the World Trade Center among New York City area residents. *Psychosomatic Medicine, 65*(6), 993–996. https://doi.org/10.1097/01.psy.0000097334.48556.5f

Farrokhi, F., Abedi, N., Beyene, J., Kurdyak, P., & Jassal, S. V. (2014). Association between depression and mortality in patients receiving long-term dialysis: A systematic review and meta-analysis. *American Journal of Kidney Diseases, 63*(4), 623–635. https://doi.org/10.1053/j.ajkd.2013.08.024

Faulenbach, M., Uthoff, H., Schwegler, K., Spinas, G. A., Schmid, C., & Wiesli, P. (2012). Effect of psychological stress on glucose control in patients with Type 2 diabetes. *Diabetic Medicine, 29*(1), 128–131. https://doi.org/10.1111/j.1464-5491.2011.03431.x

Fava, G. A., Cosci, F., & Sonino, N. (2017). Current psychosomatic practice. *Psychotherapy and Psychosomatics, 86*(1), 13–30. https://doi.org/10.1159/000448856

Fava, G. A., Fabbri, S., Sirri, L., & Wise, T. N. (2007). Psychological factors affecting medical condition: A new proposal for DSM-V. *Psychosomatics, 48*(2), 103–111. https://doi.org/10.1176/appi.psy.48.2.103

Fruehwald, S., Loeffler-Stastka, H., Eher, R., Saletu, B., & Baumhackl, U. (2001). Depression and quality of life in multiple sclerosis. *Acta Neurologica Scandinavica, 104*(5), 257–261. https://doi.org/10.1034/j.1600-0404.2001.00022.x

Fukao, A., Takamatsu, J., Arishima, T., Tanaka, M., Kawai, T., Okamoto, Y., ... & Imagawa, A. (2020). Graves' disease and mental disorders. *Journal of Clinical & Translational Endocrinology, 19*, 100207. https://doi.org/10.1016/j.jcte.2019.100207.

Gala, C., Rigatelli, M., De Bertolini, C., Rupolo, G., Gabrielli, F., & Grassi, L. (1999). A multicenter investigation of consultation-liaison psychiatry in Italy. Italian C-L Group. *General Hospital Psychiatry, 21*(4), 310–317. https://doi.org/10.1016/s0163-8343(99)00015-8

Garg, A., Chren, M. M., Sands, L. P., Matsui, M. S., Marenus, K. D., Feingold, K. R., & Elias, P. M. (2001). Psychological stress perturbs epidermal permeability barrier homeostasis: Implications for the pathogenesis of stress-associated skin disorders. *Archives of Dermatology, 137*(1), 53–59. https://doi.org/10.1001/archderm.137.1.53

Garssen, B. (2004). Psychological factors and cancer development: Evidence after 30 years of research.

Clinical Psychology Review, 24(3), 315–338. https://doi.org/10.1016/j.cpr.2004.01.002

Gehi, A., Haas, D., Pipkin, S., & Whooley, M. A. (2005). Depression and medication adherence in outpatients with coronary heart disease: Findings from the heart and soul study. *Archives of Internal Medicine, 165*(21), 2508–2513. https://doi.org/10.1001/archinte.165.21.2508

Gescher, D. M., Kahl, K. G., Hillemacher, T., Frieling, H., Kuhn, J., & Frodl, T. (2018). Epigenetics in personality disorders: Today's insights. *Frontiers in Psychiatry, 9*, 579. https://doi.org/10.3389/fpsyt.2018.00579

Giese-Davis, J., Collie, K., Rancourt, K. M., Neri, E., Kraemer, H. C., & Spiegel, D. (2011). Decrease in depression symptoms is associated with longer survival in patients with metastatic breast cancer: A secondary analysis. *Journal of Clinical Oncology, 29*(4), 413–420. https://doi.org/10.1200/JCO.2010.28.4455

Gillies, D., Buykx, P., Parker, A. G., & Hetrick, S. E. (2015). Consultation liaison in primary care for people with mental disorders. *Cochrane Database of Systematic Reviews, 9*, CD007193. https://doi.org/10.1002/14651858.CD007193.pub2

Glassman, A. H., O'Connor, C. M., Califf, R. M., Swedberg, K., Schwartz, P., Bigger, J. T., Jr., ... & Sertraline Antidepressant Heart Attack Randomized Trial Group (2002). Sertraline treatment of major depression in patients with acute MI or unstable angina. *JAMA, 288*(6), 701–709. https://doi.org/10.1001/jama.288.6.701.

Glenn, B. A., Crespi, C. M., Rodriguez, H. P., Nonzee, N. J., Phillips, S. M., Sheinfeld Gorin, S. N., ... & Group, M. S. (2018). Behavioral and mental health risk factor profiles among diverse primary care patients. *Preventive Medicine, 111*, 21–27. https://doi.org/10.1016/j.ypmed.2017.12.009.

Goodwin, P. J., Leszcz, M., Ennis, M., Koopmans, J., Vincent, L., Guther, H., ... & Hunter, J. (2001). The effect of group psychosocial support on survival in metastatic breast cancer. *The New England Journal of Medicine, 345*(24), 1719–1726. https://doi.org/10.1056/NEJMoa011871.

Goodwin, R. D., Jacobi, F., & Thefeld, W. (2003). Mental disorders and asthma in the community. *Archives of General Psychiatry, 60*(11), 1125–1130. https://doi.org/10.1001/archpsyc.60.11.1125

Goodwin, R. D., Robinson, M., Sly, P. D., McKeague, I. W., Susser, E. S., Zubrick, S. R., ... & Mattes, E. (2013). Severity and persistence of asthma and mental health: A birth cohort study. *Psychological Medicine, 43*(6), 1313–1322. https://doi.org/10.1017/S0033291712001754.

Green, C., Richards, D. A., Hill, J. J., Gask, L., Lovell, K., Chew-Graham, C. ..., & Barkham, M. (2014). Cost-effectiveness of collaborative care for depression in UK primary care: Economic evaluation of a randomised controlled trial (CADET). *PLoS One, 9*(8), e104225. https://doi.org/10.1371/journal.pone.0104225.

Greil, A. L. (1997). Infertility and psychological distress: A critical review of the literature. *Social Science & Medicine, 45*(11), 1679–1704. https://doi.org/10.1016/s0277-9536(97)00102-0

Griffith, L. J. (2008). The Psychiatrist's guide to motivational interviewing. *Psychiatry (Edgmont), 5*(4), 42–47. Retrieved from https://www.ncbi.nlm.nih.gov/pubmed/19727309

Grov, E. K., Fossa, S. D., Bremnes, R. M., Dahl, O., Klepp, O., Wist, E., & Dahl, A. A. (2009). The personality trait of neuroticism is strongly associated with long-term morbidity in testicular cancer survivors. *Acta Oncologica, 48*(6), 842–849. https://doi.org/10.1080/02841860902795232

Gwee, K. A., Leong, Y. L., Graham, C., McKendrick, M. W., Collins, S. M., Walters, S. J., ... & Read, N. W. (1999). The role of psychological and biological factors in postinfective gut dysfunction. *Gut, 44*(3), 400–406. https://doi.org/10.1136/gut.44.3.400.

Hemingway, H., Malik, M., & Marmot, M. (2001). Social and psychosocial influences on sudden cardiac death, ventricular arrhythmia and cardiac autonomic function. *European Heart Journal, 22*(13), 1082–1101. https://doi.org/10.1053/euhj.2000.2534

Herrmann-Lingen, C. (2017). Past, present, and future of psychosomatic movements in an ever-changing world: Presidential address. *Psychosomatic Medicine, 79*(9), 960–970. https://doi.org/10.1097/PSY.0000000000000521

Hilliard, M. E., Yi-Frazier, J. P., Hessler, D., Butler, A. M., Anderson, B. J., & Jaser, S. (2016). Stress and A1c among people with diabetes across the lifespan. *Current Diabetes Reports, 16*(8), 67. https://doi.org/10.1007/s11892-016-0761-3

Himelhoch, S., & Medoff, D. R. (2005). Efficacy of antidepressant medication among HIV-positive individuals with depression: A systematic review and meta-analysis. *AIDS Patient Care and STDs, 19*(12), 813–822. https://doi.org/10.1089/apc.2005.19.813

Holland, J. C. (2001). Improving the human side of cancer care: Psycho-oncology's contribution. *Cancer Journal, 7*(6), 458–471. Retrieved from http://www.ncbi.nlm.nih.gov/pubmed/11769856

Holland, J. C. (2018). Psycho-oncology: Overview, obstacles and opportunities. *Psychooncology, 27*(5), 1364–1376. https://doi.org/10.1002/pon.4692

Holland, J. C., Breitbart, W. S., Butow, P. N., Jacobson, P. B., Loscalzo, M. J., & McCorkle, R. (2015). *Psychooncology.* University of Oxford: Oxford Press.

Holroyd, S., Currie, L. J., & Wooten, G. F. (2005). Depression is associated with impairment of ADL, not motor function in Parkinson disease. *Neurology, 64*(12), 2134–2135. https://doi.org/10.1212/01.WNL.0000165958.12724.0D

Holt, R. I., de Groot, M., & Golden, S. H. (2014). Diabetes and depression. *Current Diabetes Reports, 14*(6), 491. https://doi.org/10.1007/s11892-014-0491-3

Honig, A., Kuyper, A. M., Schene, A. H., van Melle, J. P., de Jonge, P., Tulner, D. M., et al. (2007). Treatment of post-myocardial infarction depressive disorder: A randomized, placebo-controlled trial with mirtazapine. *Psychosomatic Medicine, 69*(7), 606–613. https://doi. org/10.1097/PSY.0b013e31814b260d

Huijbregts, K. M., de Jong, F. J., van Marwijk, H. W., Beekman, A. T., Ader, H. J., Hakkaart-van Roijen, L., . . . & van der Feltz-Cornelis, C. M. (2013). A target-driven collaborative care model for Major Depressive Disorder is effective in primary care in the Netherlands. A randomized clinical trial from the depression initiative. *Journal of Affective Disorders, 146*(3), 328–337. https://doi.org/10.1016/j.jad.2012.09.015.

Hunt, J. C., & Waller, G. (1992). Psychological factors in recurrent uncomplicated urinary tract infection. *British Journal of Urology, 69*(5), 460–464. https://doi.org/10.1111/j.1464-410x.1992.tb15588.x

Ironson, G., O'Cleirigh, C., Fletcher, M. A., Laurenceau, J. P., Balbin, E., Klimas, N., . . . & Solomon, G. (2005). Psychosocial factors predict CD4 and viral load change in men and women with human immunodeficiency virus in the era of highly active antiretroviral treatment. *Psychosomatic Medicine, 67*(6), 1013–1021. https://doi.org/10.1097/01.psy.0000188569.58998.c8.

Jangpangi, D., Mondal, S., Bandhu, R., Kataria, D., & Gandhi, A. (2016). Alteration of heart rate variability in patients of depression. *Journal of Clinical and Diagnostic Research, 10*(12), CM04–CM06. https://doi.org/10.7860/JCDR/2016/22882.9063

Jette, N., Patten, S., Williams, J., Becker, W., & Wiebe, S. (2008). Comorbidity of migraine and psychiatric disorders – a national population-based study. *Headache, 48*(4), 501–516. https://doi.org/10.1111/j.1526-4610.2007.00993.x.

Jia, H., Damush, T. M., Qin, H., Ried, L. D., Wang, X., Young, L. J., & Williams, L. S. (2006). The impact of poststroke depression on healthcare use by veterans with acute stroke. *Stroke, 37*(11), 2796–2801. https://doi.org/10.1161/01.STR.0000244783.53274.a4

Jiang, W., Alexander, J., Christopher, E., Kuchibhatla, M., Gaulden, L. H., Cuffe, M. S., . . . & O'Connor, C. M. (2001). Relationship of depression to increased risk of mortality and rehospitalization in patients with congestive heart failure. *Archives of Internal Medicine, 161*(15), 1849–1856. https://doi.org/10.1001/archinte.161.15.1849.

Jiang, W., Babyak, M., Krantz, D. S., Waugh, R. A., Coleman, R. E., Hanson, M. M., . . . & Blumenthal, J. A. (1996). Mental stress–induced myocardial ischemia and cardiac events. *JAMA, 275*(21), 1651–1656. https://doi.org/10.1001/jama.275.21.1651.

Katon, W., Unutzer, J., Wells, K., & Jones, L. (2010a). Collaborative depression care: History, evolution and ways to enhance dissemination and sustainability. *General Hospital Psychiatry, 32*(5), 456–464. https://doi.org/10.1016/j.genhosppsych.2010.04.001

Katon, W. J. (2003). Clinical and health services relationships between major depression, depressive symptoms, and general medical illness. *Biological Psychiatry,*

54(3), 216–226. Retrieved from http://www.ncbi.nlm.nih.gov/pubmed/12893098

Katon, W. J., Lin, E. H., Von Korff, M., Ciechanowski, P., Ludman, E. J., Young, B., . . . & McCulloch, D. (2010b). Collaborative care for patients with depression and chronic illnesses. *The New England Journal of Medicine, 363*(27), 2611–2620. https://doi.org/10.1056/NEJMoa1003955.

Katon, W. J., Von Korff, M., Lin, E. H., Simon, G., Ludman, E., Russo, J., . . . & Bush, T. (2004). The pathways study: A randomized trial of collaborative care in patients with diabetes and depression. *Archives of General Psychiatry, 61*(10), 1042–1049. https://doi.org/10.1001/archpsyc.61.10.1042.

Kent, L. K., & Shapiro, P. A. (2009). Depression and related psychological factors in heart disease. *Harvard Review of Psychiatry, 17*(6), 377–388. https://doi.org/10.3109/10673220903463333

Kimyai-Asadi, A., & Usman, A. (2001). The role of psychological stress in skin disease. *Journal of Cutaneous Medicine and Surgery, 5*(2), 140–145. https://doi.org/10.1007/BF02737869

Knittle, K., Maes, S., & de Gucht, V. (2010). Psychological interventions for rheumatoid arthritis: Examining the role of self-regulation with a systematic review and meta-analysis of randomized controlled trials. *Arthritis Care & Research (Hoboken), 62*(10), 1460–1472. https://doi.org/10.1002/acr.20251

Knowles, S. R., Monshat, K., & Castle, D. J. (2013). The efficacy and methodological challenges of psychotherapy for adults with inflammatory bowel disease: A review. *Inflammatory Bowel Diseases, 19*(12), 2704–2715. https://doi.org/10.1097/MIB.0b013e318296ae5a

Konig, H., Konig, H. H., & Konnopka, A. (2019). The excess costs of depression: A systematic review and meta-analysis. *Epidemiology and Psychiatric Sciences, 29*, e30. https://doi.org/10.1017/S2045796019000180

Krantz, D. S., Sheps, D. S., Carney, R. M., & Natelson, B. H. (2000). Effects of mental stress in patients with coronary artery disease: Evidence and clinical implications. *JAMA, 283*(14), 1800–1802. https://doi.org/10.1001/jama.283.14.1800

Kubo, C. (2013). New year address on the state of psychosomatic medicine in Japan. *Biopsychosoc Med, 7*(1), 2. https://doi.org/10.1186/1751-0759-7-2

Kurina, L. M., Goldacre, M. J., Yeates, D., & Gill, L. E. (2001). Depression and anxiety in people with inflammatory bowel disease. *Journal of Epidemiology and Community Health, 55*(10), 716–720. https://doi.org/10.1136/jech.55.10.716

Lasser, K., Boyd, J. W., Woolhandler, S., Himmelstein, D. U., McCormick, D., & Bor, D. H. (2000). Smoking and mental illness: A population-based prevalence study. *JAMA, 284*(20), 2606–2610. https://doi.org/10.1001/jama.284.20.2606

Lazarus, R. S. (1974). Psychological stress and coping in adaptation and illness. *International Journal of Psychiatry in Medicine, 5*(4), 321–333. https://doi.org/10.2190/T43T-84P3-QDUR-7RTP

Leserman, J., Pence, B. W., Whetten, K., Mugavero, M. J., Thielman, N. M., Swartz, M. S., & Stangl, D. (2007). Relation of lifetime trauma and depressive symptoms to mortality in HIV. *The American Journal of Psychiatry, 164*(11), 1707–1713. https://doi.org/10.1176/appi.ajp.2007.06111775

LeSon, S., & Gershwin, M. E. (1996). Risk factors for asthmatic patients requiring intubation. III. Observations in young adults. *The Journal of Asthma, 33*(1), 27–35. https://doi.org/10.3109/02770909609077760

Levenson, J. (2019). *Textbook of psychosomatic medicine and consult-liaison psychiatry.* The American Psychiatric Association.

Levenson, J. L., Hamer, R. M., Myers, T., Hart, R. P., & Kaplowitz, L. G. (1987). Psychological factors predict symptoms of severe recurrent genital herpes infection. *Journal of Psychosomatic Research, 31*(2), 153–159. https://doi.org/10.1016/0022-3999(87)90071-7

Lin, E. H., Rutter, C. M., Katon, W., Heckbert, S. R., Ciechanowski, P., Oliver, M. M., ... & Von Korff, M. (2010). Depression and advanced complications of diabetes: A prospective cohort study. *Diabetes Care, 33*(2), 264–269. https://doi.org/10.2337/dc09-1068.

Lipsitt, D. R. (2001). Consultation-liaison psychiatry and psychosomatic medicine: The company they keep. *Psychosomatic Medicine, 63*(6), 896–909. https://doi.org/10.1097/00006842-200111000-00008

Lloyd, C. E., Dyer, P. H., Lancashire, R. J., Harris, T., Daniels, J. E., & Barnett, A. H. (1999). Association between stress and glycemic control in adults with type 1 (insulin-dependent) diabetes. *Diabetes Care, 22*(8), 1278–1283. https://doi.org/10.2337/diacare.22.8.1278

Lou, P., Chen, P., Zhang, P., Yu, J., Wang, Y., Chen, N., ... & Zhao, J. (2014). Effects of smoking, depression, and anxiety on mortality in COPD patients: A prospective study. *Respiratory Care, 59*(1), 54–61. https://doi.org/10.4187/respcare.02487.

Ludman, E. J., Katon, W., Russo, J., Von Korff, M., Simon, G., Ciechanowski, P., ... & Young, B. (2004). Depression and diabetes symptom burden. *General Hospital Psychiatry, 26*(6), 430–436. https://doi.org/10.1016/j.genhosppsych.2004.08.010.

Lustman, P. J., Anderson, R. J., Freedland, K. E., de Groot, M., Carney, R. M., & Clouse, R. E. (2000). Depression and poor glycemic control: A meta-analytic review of the literature. *Diabetes Care, 23*(7), 934–942. https://doi.org/10.2337/diacare.23.7.934

Lwin, M. N., Serhal, L., Holroyd, C., & Edwards, C. J. (2020). Rheumatoid arthritis: The impact of mental health on disease: A narrative review. *Rheumatol Ther, 7*(3), 457–471. https://doi.org/10.1007/s40744-020-00217-4

Marsh, L. (2013). Depression and Parkinson's disease: Current knowledge. *Current Neurology and Neuroscience Reports, 13*(12), 409. https://doi.org/10.1007/s11910-013-0409-5

Mason, J. C., & Libby, P. (2015). Cardiovascular disease in patients with chronic inflammation: Mechanisms underlying premature cardiovascular events in rheumatologic conditions. *European Heart Journal, 36*(8), 482–489c. https://doi.org/10.1093/eurheartj/ehu403

Mawdsley, J. E., & Rampton, D. S. (2005). Psychological stress in IBD: New insights into pathogenic and therapeutic implications. *Gut, 54*(10), 1481–1491. https://doi.org/10.1136/gut.2005.064241

McDade-Montez, E. A., Christensen, A. J., Cvengros, J. A., & Lawton, W. J. (2006). The role of depression symptoms in dialysis withdrawal. *Health Psychology, 25*(2), 198–204. https://doi.org/10.1037/0278-6133.25.2.198

McFarland, D. C., Miller, A. H., & Nelson, C. (2021a). A longitudinal analysis of inflammation and depression in patients with metastatic lung cancer: Associations with survival. *Biological Research for Nursing, 23*(3), 301–310. https://doi.org/10.1177/1099800420959721

McFarland, D. C., Saracino, R. M., Miller, A. H., Breitbart, W., Rosenfeld, B., & Nelson, C. (2021b). Prognostic implications of depression and inflammation in patients with metastatic lung cancer. *Future Oncology, 17*(2), 183–196. https://doi.org/10.2217/fon-2020-0632

Mehr, D. R., Binder, E. F., Kruse, R. L., Zweig, S. C., Madsen, R., Popejoy, L., & D'Agostino, R. B. (2001). Predicting mortality in nursing home residents with lower respiratory tract infection: The Missouri LRI study. *JAMA, 286*(19), 2427–2436. https://doi.org/10.1001/jama.286.19.2427

Melin, E. O., Thunander, M., Svensson, R., Landin-Olsson, M., & Thulesius, H. O. (2013). Depression, obesity, and smoking were independently associated with inadequate glycemic control in patients with type 1 diabetes. *European Journal of Endocrinology, 168*(6), 861–869. https://doi.org/10.1530/EJE-13-0137

Mikami, K., Jorge, R. E., Adams, H. P., Jr., Davis, P. H., Leira, E. C., Jang, M., & Robinson, R. G. (2011). Effect of antidepressants on the course of disability following stroke. *The American Journal of Geriatric Psychiatry, 19*(12), 1007–1015. https://doi.org/10.1097/JGP.0b013e31821181b0

Miles, J. F., Garden, G. M., Tunnicliffe, W. S., Cayton, R. M., & Ayres, J. G. (1997). Psychological morbidity and coping skills in patients with brittle and non-brittle asthma: A case-control study. *Clinical and Experimental Allergy, 27*(10), 1151–1159. https://doi.org/10.1046/j.1365-2222.1997.1080961.x

Miller, A. H., Ancoli-Israel, S., Bower, J. E., Capuron, L., & Irwin, M. R. (2008). Neuroendocrine-immune mechanisms of behavioral comorbidities in patients with cancer. *Journal of Clinical Oncology, 26*(6), 971–982. https://doi.org/10.1200/JCO.2007.10.7805

Miller, A. H., & Raison, C. L. (2016). The role of inflammation in depression: From evolutionary imperative to modern treatment target. *Nature Reviews. Immunology, 16*(1), 22–34. https://doi.org/10.1038/nri.2015.5

Mizokami, T., Wu Li, A., El-Kaissi, S., & Wall, J. R. (2004). Stress and thyroid autoimmunity. *Thyroid, 14*(12), 1047–1055. https://doi.org/10.1089/thy.2004.14.1047

Morales, D. A., Barksdale, C. L., & Beckel-Mitchener, A. C. (2020). A call to action to address rural mental health disparities. *J Clin Transl Sci, 4*(5), 463–467. https://doi.org/10.1017/cts.2020.42

Moser, D. K., & Dracup, K. (1996). Is anxiety early after myocardial infarction associated with subsequent ischemic and arrhythmic events? *Psychosomatic Medicine, 58*(5), 395–401. https://doi.org/10.1097/00006842-199609000-00001

Mustafa, M., Carson-Stevens, A., Gillespie, D., & Edwards, A. G. (2013). Psychological interventions for women with metastatic breast cancer. *Cochrane Database of Systematic Reviews, 6*, CD004253. https://doi.org/10.1002/14651858.CD004253.pub4

Neggers, Y., Goldenberg, R., Cliver, S., & Hauth, J. (2006). The relationship between psychosocial profile, health practices, and pregnancy outcomes. *Acta Obstetricia et Gynecologica Scandinavica, 85*(3), 277–285. https://doi.org/10.1080/00016340600566121

Noyes, R., Jr., Hartz, A. J., Doebbeling, C. C., Malis, R. W., Happel, R. L., Werner, L. A., & Yagla, S. J. (2000). Illness fears in the general population. *Psychosomatic Medicine, 62*(3), 318–325. https://doi.org/10.1097/00006842-200005000-00005

O'Connor, C. M., Jiang, W., Kuchibhatla, M., Silva, S. G., Cuffe, M. S., Callwood, D. D., ... & Investigators, S.-C. (2010). Safety and efficacy of sertraline for depression in patients with heart failure: Results of the SADHART-CHF (sertraline against depression and heart disease in chronic heart failure) trial. *Journal of the American College of Cardiology, 56*(9), 692–699. https://doi.org/10.1016/j.jacc.2010.03.068.

O'Mahony, S., Goulet, J., Kornblith, A., Abbatiello, G., Clarke, B., Kless-Siegel, S., ... & Payne, R. (2005). Desire for hastened death, cancer pain and depression: Report of a longitudinal observational study. *Journal of Pain and Symptom Management, 29*(5), 446–457. https://doi.org/10.1016/j.jpainsymman.2004.08.010.

Oldham, M. A., Desan, P. H., Lee, H. B., Bourgeois, J. A., Shah, S. B., Hurley, P. J., ... & Council on Consultation-Liaison, P. (2021). Proactive consultation-liaison psychiatry: American Psychiatric Association resource document. *J Acad Consult Liaison Psychiatry, 62*(2), 169–185. https://doi.org/10.1016/j.jaclp.2021.01.005.

Orth-Gomer, K., Wamala, S. P., Horsten, M., Schenck-Gustafsson, K., Schneiderman, N., & Mittleman, M. A. (2000). Marital stress worsens prognosis in women with coronary heart disease: The Stockholm Female Coronary Risk Study. *JAMA, 284*(23), 3008–3014. https://doi.org/10.1001/jama.284.23.3008

Pan, A., Sun, Q., Okereke, O. I., Rexrode, K. M., & Hu, F. B. (2011). Depression and risk of stroke morbidity and mortality: A meta-analysis and systematic review. *JAMA, 306*(11), 1241–1249. https://doi.org/10.1001/jama.2011.1282

Pasch, L. A., Gregorich, S. E., Katz, P. K., Millstein, S. G., Nachtigall, R. D., Bleil, M. E., & Adler, N. E. (2012). Psychological distress and in vitro fertilization outcome. *Fertility and Sterility, 98*(2), 459–464. https://doi.org/10.1016/j.fertnstert.2012.05.023

Penninx, B. W., Beekman, A. T., Honig, A., Deeg, D. J., Schoevers, R. A., van Eijk, J. T., & van Tilburg, W. (2001). Depression and cardiac mortality: Results from a community-based longitudinal study. *Archives of General Psychiatry, 58*(3), 221–227. https://doi.org/10.1001/archpsyc.58.3.221

Perez-Tejada, J., Garmendia, L., Labaka, A., Vegas, O., Gomez-Lazaro, E., & Arregi, A. (2019). Active and passive coping strategies: Comparing psychological distress, cortisol, and proinflammatory cytokine levels in breast cancer survivors. *Clinical Journal of Oncology Nursing, 23*(6), 583–590. https://doi.org/10.1188/19.CJON.583-590

Perry, S., & Viederman, M. (1981). Adaptation of residents to consultation-liaison psychiatry. I. Working with the physically ill. *General Hospital Psychiatry, 3*(2), 141–147. https://doi.org/10.1016/0163-8343(81)90056-6

Pollak, K. I., Childers, J. W., & Arnold, R. M. (2011). Applying motivational interviewing techniques to palliative care communication. *Journal of Palliative Medicine, 14*(5), 587–592. https://doi.org/10.1089/jpm.2010.0495

Pratt, L. A., Druss, B. G., Manderscheid, R. W., & Walker, E. R. (2016). Excess mortality due to depression and anxiety in the United States: Results from a nationally representative survey. *General Hospital Psychiatry, 39*, 39–45. https://doi.org/10.1016/j.genhosppsych.2015.12.003

Prieto, J. M., Atala, J., Blanch, J., Carreras, E., Rovira, M., Cirera, E., ... & Gasto, C. (2005). Role of depression as a predictor of mortality among cancer patients after stem-cell transplantation. *Journal of Clinical Oncology, 23*(25), 6063–6071. https://doi.org/10.1200/JCO.2005.05.751.

Prina, A. M., Cosco, T. D., Dening, T., Beekman, A., Brayne, C., & Huisman, M. (2015). The association between depressive symptoms in the community, non-psychiatric hospital admission and hospital outcomes: A systematic review. *Journal of Psychosomatic Research, 78*(1), 25–33. https://doi.org/10.1016/j.jpsychores.2014.11.002

Raglan, G. B., Swanson, L. M., & Arnedt, J. T. (2019). Cognitive behavioral therapy for insomnia in patients with medical and psychiatric comorbidities. *Sleep Medicine Clinics, 14*(2), 167–175. https://doi.org/10.1016/j.jsmc.2019.01.001

Ramanuj, P., Ferenchik, E., Docherty, M., Spaeth-Rublee, B., & Pincus, H. A. (2019). Evolving models of integrated behavioral health and primary care. *Current Psychiatry Reports, 21*(1), 4. https://doi.org/10.1007/s11920-019-0985-4

Richards, B. L., Whittle, S. L., & Buchbinder, R. (2011). Antidepressants for pain management in rheumatoid arthritis. *Cochrane Database of Systematic Reviews, 11*, CD008920. https://doi.org/10.1002/14651858.CD008920.pub2

Riezebos, R. K., Nauta, K. J., Honig, A., Dekker, F. W., & Siegert, C. E. (2010). The association of depressive symptoms with survival in a Dutch cohort of patients with end-stage renal disease. *Nephrology, Dialysis, Transplantation, 25*(1), 231–236. https://doi.org/10.1093/ndt/gfp383

Rosengren, A., Hawken, S., Ounpuu, S., Sliwa, K., Zubaid, M., Almahmeed, W. A., ... Investigators, I. (2004). Association of psychosocial risk factors with risk of acute myocardial infarction in 11119 cases and 13648 controls from 52 countries (the INTERHEART study): Case-control study. *Lancet, 364*(9438), 953–962. https://doi.org/10.1016/S0140-6736(04)17019-0.

Rosenman, R. H., Brand, R. J., Jenkins, D., Friedman, M., Straus, R., & Wurm, M. (1975). Coronary heart disease in Western Collaborative Group Study. Final follow-up experience of 8 1/2 years. *JAMA, 233*(8), 872–877. Retrieved from https://www.ncbi.nlm.nih.gov/pubmed/1173896

Rundell, J. (2018). What's old is new again: Psychosomatic medicine renamed C-L psychiattry. *Psychiatric New*s. https://doi.org/10.1176/appi.pn.2018.5a5

Ruo, B., Rumsfeld, J. S., Hlatky, M. A., Liu, H., Browner, W. S., & Whooley, M. A. (2003). Depressive symptoms and health-related quality of life: The heart and soul study. *JAMA, 290*(2), 215–221. https://doi.org/10.1001/jama.290.2.215

Saito-Nakaya, K., Bidstrup, P. E., Nakaya, N., Frederiksen, K., Dalton, S. O., Uchitomi, Y., ... & Johansen, C. (2012). Stress and survival after cancer: A prospective study of a Finnish population-based cohort. *Cancer Epidemiology, 36*(2), 230–235. https://doi.org/10.1016/j.canep.2011.04.008.

Sanner, J. E., & Frazier, L. (2011). The role of serotonin in depression and clotting in the coronary artery disease population. *The Journal of Cardiovascular Nursing, 26*(5), 423–429. https://doi.org/10.1097/JCN.0b013e3182076a81

Santos, A. M., Nobre, E. L., Garcia e Costa, J., Nogueira, P. J., Macedo, A., De Castro, J. J., & Teles, A. G. (2002). Grave's disease and stress. *Acta Médica Portuguesa, 15*(6), 423–427. Retrieved from https://www.ncbi.nlm.nih.gov/pubmed/12680288

Santos, J. C., & Pyter, L. M. (2018). Neuroimmunology of behavioral comorbidities associated with cancer and cancer treatments. *Frontiers in Immunology, 9*, 1195. https://doi.org/10.3389/fimmu.2018.01195

Saracino, R. M., Rosenfeld, B., & Nelson, C. J. (2018). Performance of four diagnostic approaches to depression in adults with cancer. *General Hospital Psychiatry, 51*, 90–95. https://doi.org/10.1016/j.genhosppsych.2018.01.006

Satin, J. R., Linden, W., & Phillips, M. J. (2009). Depression as a predictor of disease progression and mortality in cancer patients: A meta-analysis. *Cancer, 115*(22), 5349–5361. https://doi.org/10.1002/cncr.24561

Sewell, M. C., Goggin, K. J., Rabkin, J. G., Ferrando, S. J., McElhiney, M. C., & Evans, S. (2000). Anxiety syndromes and symptoms among men with AIDS: A longitudinal controlled study. *Psychosomatics, 41*(4), 294–300. https://doi.org/10.1176/appi.psy.41.4.294

Shenefelt, P. D. (2003). Biofeedback, cognitive-behavioral methods, and hypnosis in dermatology: Is it all in your mind? *Dermatologic Therapy, 16*(2), 114–122. https://doi.org/10.1046/j.1529-8019.2003.01620.x

Sherr, L., Clucas, C., Harding, R., Sibley, E., & Catalan, J. (2011). HIV and depression – systematic review of interventions. *Psychology, Health & Medicine, 16*(5), 493–527. https://doi.org/10.1080/13548506.2011.579990

Shirazian, S., Grant, C. D., Aina, O., Mattana, J., Khorassani, F., & Ricardo, A. C. (2017). Depression in chronic kidney disease and end-stage renal disease: Similarities and differences in diagnosis, epidemiology, and management. *Kidney Int Rep, 2*(1), 94–107. https://doi.org/10.1016/j.ekir.2016.09.005

Simon, G. E., Katon, W. J., Lin, E. H., Ludman, E., VonKorff, M., Ciechanowski, P., & Young, B. A. (2005). Diabetes complications and depression as predictors of health service costs. *General Hospital Psychiatry, 27*(5), 344–351. https://doi.org/10.1016/j.genhosppsych.2005.04.008

Smeenk, J. M., Verhaak, C. M., Stolwijk, A. M., Kremer, J. A., & Braat, D. D. (2004). Reasons for dropout in an in vitro fertilization/intracytoplasmic sperm injection program. *Fertility and Sterility, 81*(2), 262–268. https://doi.org/10.1016/j.fertnstert.2003.09.027

Snell-Rood, C., Feltner, F., & Schoenberg, N. (2019). What role can community health workers play in connecting rural women with depression to the "De Facto" mental health care system? *Community Mental Health Journal, 55*(1), 63–73. https://doi.org/10.1007/s10597-017-0221-9

Sollner, W., Creed, F., & European Association of Consultation-Liaison, P., & Psychosomatics Workgroup on Training in, C.-L. (2007). European guidelines for training in consultation-liaison psychiatry and psychosomatics: Report of the EACLPP Workgroup on Training in Consultation-Liaison Psychiatry and Psychosomatics. *Journal of Psychosomatic Research, 62*(4), 501–509. https://doi.org/10.1016/j.jpsychores.2006.11.003

Sollner, W., & Schussler, G. (2012). New 'European Association of Psychosomatic Medicine' founded. *Journal of Psychosomatic Research, 73*(5), 343–344. https://doi.org/10.1016/j.jpsychores.2012.09.003

Spiegel, D., Bloom, J. R., & Yalom, I. (1981). Group support for patients with metastatic cancer. A randomized outcome study. *Archives of General Psychiatry, 38*(5), 527–533. https://doi.org/10.1001/archpsyc.1980.01780300039004

Spitzer, R. L., Endicott, J., & Robins, E. (1978). Research diagnostic criteria: Rationale and reliability. *Archives of General Psychiatry, 35*(6), 773–782. https://doi.org/10.1001/archpsyc.1978.01770300115013

Stein, M. J., Wallston, K. A., Nicassio, P. M., & Castner, N. M. (1988). Correlates of a clinical classification schema for the arthritis helplessness subscale. *Arthritis and Rheumatism, 31*(7), 876–881. https://doi.org/10.1002/art.1780310708

Steinberg, J. S., Arshad, A., Kowalski, M., Kukar, A., Suma, V., Vloka, M., et al. (2004). Increased incidence of life-threatening ventricular arrhythmias in implantable defibrillator patients after the World Trade Center attack. *Journal of the American College of Cardiology, 44*(6), 1261–1264. https://doi.org/10.1016/j.jacc.2004.06.032

Steinberg, M. B., Greenhaus, S., Schmelzer, A. C., Bover, M. T., Foulds, J., Hoover, D. R., & Carson, J. L. (2009). Triple-combination pharmacotherapy for medically ill smokers: A randomized trial. *Annals of Internal Medicine, 150*(7), 447–454. https://doi.org/10.7326/0003-4819-150-7-200904070-00004

Stronks, D. L., Tulen, J. H., Pepplinkhuizen, L., Verheij, R., Mantel, G. W., Spinhoven, P., & Passchier, J. (1999). Personality traits and psychological reactions to mental stress of female migraine patients. *Cephalalgia, 19*(6), 566–574. https://doi.org/10.1046/j.1468-2982.1999.019006566.x

Sullivan, D. R., Forsberg, C. W., Ganzini, L., Au, D. H., Gould, M. K., Provenzale, D., & Slatore, C. G. (2016). Longitudinal changes in depression symptoms and survival among patients with lung cancer: A National Cohort Assessment. *Journal of Clinical Oncology, 34*(33), 3984–3991. https://doi.org/10.1200/JCO.2016.66.8459

Sullivan, M. D., LaCroix, A. Z., Baum, C., Grothaus, L. C., & Katon, W. J. (1997). Functional status in coronary artery disease: A one-year prospective study of the role of anxiety and depression. *The American Journal of Medicine, 103*(5), 348–356. https://doi.org/10.1016/s0002-9343(97)00167-8

Szabo, S., Tache, Y., & Somogyi, A. (2012). The legacy of Hans Selye and the origins of stress research: A retrospective 75 years after his landmark brief "letter" to the editor# of nature. *Stress, 15*(5), 472–478. https://doi.org/10.3109/10253890.2012.710919

Tayama, J., Nakaya, N., Hamaguchi, T., Tomiie, T., Shinozaki, M., Saigo, T., ... & Fukudo, S. (2012). Effects of personality traits on the manifestations of irritable bowel syndrome. *Biopsychosoc Med, 6*(1), 20. https://doi.org/10.1186/1751-0759-6-20.

Temoshok, L. (1987). Personality, coping style, emotion and cancer: Towards an integrative model. *Cancer Surveys, 6*(3), 545–567. Retrieved from http://www.ncbi.nlm.nih.gov/pubmed/3326661

Troidle, L., Watnick, S., Wuerth, D. B., Gorban-Brennan, N., Kliger, A. S., & Finkelstein, F. O. (2003). Depression and its association with peritonitis in long-term peritoneal dialysis patients. *American Journal of Kidney Diseases, 42*(2), 350–354. https://doi.org/10.1016/s0272-6386(03)00661-9

Vaccarino, V., Badimon, L., Bremner, J. D., Cenko, E., Cubedo, J., Dorobantu, M., ... & Reviewers, E. S. C. S. D. G. (2020). Depression and coronary heart disease: 2018 position paper of the ESC working group on coronary pathophysiology and microcirculation. *European Heart Journal, 41*(17), 1687–1696. https://doi.org/10.1093/eurheartj/ehy913.

Vikjord, S. A. A., Brumpton, B. M., Mai, X. M., Vanfleteren, L., & Langhammer, A. (2020). The association of anxiety and depression with mortality in a COPD cohort. The HUNT study, Norway. *Respiratory Medicine, 171*, 106089. https://doi.org/10.1016/j.rmed.2020.106089

von Kanel, R., Mills, P. J., Fainman, C., & Dimsdale, J. E. (2001). Effects of psychological stress and psychiatric disorders on blood coagulation and fibrinolysis: A biobehavioral pathway to coronary artery disease? *Psychosomatic Medicine, 63*(4), 531–544. https://doi.org/10.1097/00006842-200107000-00003

Walker, E. A., Katon, W. J., Roy-Byrne, P. P., Jemelka, R. P., & Russo, J. (1993). Histories of sexual victimization in patients with irritable bowel syndrome or inflammatory bowel disease. *The American Journal of Psychiatry, 150*(10), 1502–1506. https://doi.org/10.1176/ajp.150.10.1502

Walker, S., Walker, J., Richardson, G., Palmer, S., Wu, Q., Gilbody, S., ... & Sharpe, M. (2014). Cost-effectiveness of combining systematic identification and treatment of co-morbid major depression for people with chronic diseases: The example of cancer. *Psychological Medicine, 44*(7), 1451–1460. https://doi.org/10.1017/S0033291713002079.

Wang, Y., Zhao, X., O'Neil, A., Turner, A., Liu, X., & Berk, M. (2013). Altered cardiac autonomic nervous function in depression. *BMC Psychiatry, 13*, 187. https://doi.org/10.1186/1471-244X-13-187

Watson, M., Homewood, J., Haviland, J., & Bliss, J. M. (2005). Influence of psychological response on breast cancer survival: 10-year follow-up of a population-based cohort. *European Journal of Cancer, 41*(12), 1710–1714. https://doi.org/10.1016/j.ejca.2005.01.012

Weihs, K. L., Enright, T. M., & Simmens, S. J. (2008). Close relationships and emotional processing predict decreased mortality in women with breast cancer: Preliminary evidence. *Psychosomatic Medicine, 70*(1), 117–124. https://doi.org/10.1097/PSY.0b013e31815c25cf

Winsa, B., Adami, H. O., Bergstrom, R., Gamstedt, A., Dahlberg, P. A., Adamson, U., & ... Karlsson, A. (1991). Stressful life events and Graves' disease. *Lancet, 338*(8781), 1475–1479. https://doi.org/10.1016/0140-6736(91)92298-g.

Woolf, S. H., & Braveman, P. (2011). Where health disparities begin: the role of social and economic determinant – and why current policies may make matters worse. *Health Aff (Millwood), 30*(10), 1852–1859. https://doi.org/10.1377/hlthaff.2011.0685

Yehia, B. R., Herati, R. S., Fleishman, J. A., Gallant, J. E., Agwu, A. L., Berry, S. A., ... & Network, H. I. V. R. (2014). Hepatitis C virus testing in adults living with HIV: A need for improved screening efforts. *PLoS One, 9*(7), e102766. https://doi.org/10.1371/journal.pone.0102766.

Yohannes, A. M., & Alexopoulos, G. S. (2014). Depression and anxiety in patients with COPD. *European Respiratory Review, 23*(133), 345–349. https://doi.org/10.1183/09059180.00007813

Yonas, M. A., Lange, N. E., & Celedon, J. C. (2012). Psychosocial stress and asthma morbidity. *Current Opinion in Allergy and Clinical Immunology, 12*(2), 202–210. https://doi.org/10.1097/ACI.0b013e32835090c9

Zamani-Alavijeh, F., Araban, M., Koohestani, H. R., & Karimy, M. (2018). The effectiveness of stress management training on blood glucose control in patients with type 2 diabetes. *Diabetology and Metabolic Syndrome, 10*, 39. https://doi.org/10.1186/s13098-018-0342-5

Zareban, I., Karimy, M., Niknami, S., Haidarnia, A., & Rakhshani, F. (2014). The effect of self-care education program on reducing HbA1c levels in patients with type 2 diabetes. *Journal of Education Health Promotion, 3*, 123. https://doi.org/10.4103/2277-9531.145935

Zorzon, M., de Masi, R., Nasuelli, D., Ukmar, M., Mucelli, R. P., Cazzato, G., . . . Zivadinov, R. (2001). Depression and anxiety in multiple sclerosis. A clinical and MRI study in 95 subjects. *Journal of Neurology, 248*(5), 416–421. https://doi.org/10.1007/s004150170184.

Collaborative Care and Geriatric Psychiatry

Jimmy N. Avari, Alessandra Costanza, Kerstin Weber, and Alessandra Canuto

Contents

J. N. Avari (✉)
Weill Cornell Institute of Geriatric Psychiatry, Weill
Cornell Medical College, White Plains, NY, USA
e-mail: jia9010@med.cornell.edu

A. Costanza · K. Weber
Faculty of Medicine, University of Geneva, Geneva,
Switzerland
e-mail: alessandra.costanza@hcuge.ch;
kerstin.weber@hcuge.ch

A. Canuto
University Hospitals of Geneva, Geneva, Switzerland
e-mail: alessandra.canuto@hcuge.ch

© Springer Nature Switzerland AG 2024
A. Tasman et al. (eds.), *Tasman's Psychiatry*,
https://doi.org/10.1007/978-3-030-51366-5_57

Abstract

Clinicians treating older adults should do multi-morbidity and individualized patient-centered assessment by a multidisciplinary collaborative team, including the patient's individual health history, standardized rating scales, and laboratory workup. This assessment allows for a multi-morbidity patient-centered individualized treatment plan which takes into account both the old age patient's physical and mental health and aims for the same values as the assessment, namely, a patient-centered, population-focused, and measurement-guided medical treatment. This chapter describes the two critical elements of this approach to geriatric psychiatry: (1) Establishing a correct diagnosis by way of targeted foundational aspects of the diagnostic interview, rating scales, and the laboratory workup of older adults and (2) Utilization of evidence-based collaborative care models for the elderly, with a description of the need for collaborative care models in the elderly, the necessary elements, the team structure, the evidence-based models in the elderly, and the limitations of the current state of collaborative care.

Keywords

Diagnostic interview · Rating scales · Liaison psychiatry · Collaborative care

Introduction

Demographics of the global population are rapidly changing. In particular, life expectancy is increasing, and the proportion of the world population older than 60 years is predicted to reach 2 billion people, or 22% of the population, by 2050 (Kanasi et al., 2016). While increased longevity is a worthy goal, older citizens are not necessarily experiencing a high or even adequate quality of life due to the appearance of complex chronic health issues and multi-morbidities. At a systemic level, multi-morbidity is problematic for healthcare providers since there is a positive link between multi-morbidity and higher healthcare spending: the top 10% of healthcare spenders, who account for about two-thirds of all spending, are predominantly patients with long-term multi-morbidity and complex needs (Zulman et al., 2015; Wang et al., 2018). Multi-morbidity patients are high users of all types of health services, and their needs are influenced by non-clinical factors such as functional ability, informal caregiver support, family support, financial circumstances, and community service availability. This increased demand for medical services and the complex care that is required are within a delivery system that is often fragmented, inefficient, and ineffective (Cerra & Brandt, 2015).

A complicating factor for multi-morbidity in the elderly is the inclusion of psychiatric illnesses. About 15% of older citizens live with a mental disorder, with the most common being dementia, unipolar depression, and anxiety (WHO, 2015a). A European study of older adults between 65 and 84 years old found that one in two subjects experienced a lifetime mental disorder, while one-third of them had a mental disorder in the past year. The most prevalent disorders in this large sample were anxiety disorders, affective disorders, and substance-related disorders (Andreas et al., 2017). However, more than half of older adults with mental illness are not treated because of stigma or lack of identification and treatment, and an additional third of medically ill patients remain in hospitals with undiagnosed mental health problems (Karlin & Fuller, 2007; Costanza et al., 2020a). Frequent co-occurrence of somatic and psychiatric disorders in the elderly indicates a need for better detection and treatment in both general and specialist medical settings (Barnett et al., 2012; Fabbri et al., 2015; Costanza et al., 2015, 2020b). Thus, effective collaboration between mental health professionals and other medical services is necessary to overcome current shortcomings in medical delivery to geriatric populations.

Clinicians treating older adults should do a multi-morbidity and individualized patient-centered assessment by a multidisciplinary collaborative team, including the patient's individual health history, standardized rating scales, and laboratory workup. This assessment allows for a multi-morbidity patient-centered individualized treatment plan which takes into account both the old age patient's physical and mental health and aims for the same values as the assessment, namely, a patient-centered, population-focused, and measurement-guided medical treatment. This chapter describes the two critical elements of this approach to geriatric psychiatry: (1) Establishing a correct diagnosis by way of a targeted foundational aspect of the diagnostic interview, rating scales, and the laboratory workup of older adults and (2) Utilization of evidence-based collaborative care models for the elderly, with a description of the need for collaborative care models in the elderly, the necessary elements, the team structure, the evidence-based models in the elderly, and the limitations of the current state of collaborative care.

The Diagnostic Interview of Older Adults in Inpatient Settings

The core feature in the accurate assessment and diagnosis of the elderly remains the diagnostic interview. The interview is often overlooked because of time pressure, overvaluing laboratory workup, and overemphasizing standardized scales. Other barriers that are specific to older adults, like cognitive impairment or sensory deficits, may further limit a clinician's desire to perform a comprehensive diagnostic interview. Effective communication strategies are paramount to perform a complete diagnostic interview with an older adult.

Older patients may present with physical problems that interfere with typical communication strategies. Hearing loss, for example, is commonly seen in those who are aging. Approximately one in three people between the ages of 65 and 74 has hearing loss, and of those 75 years and above, half have difficulty hearing (NIH, 2018). Impairments in sensory systems, such as visual or hearing difficulties, will impact the ease of interview and often lead to clinicians abbreviating the interview or relying on other sources of information. Although that may ultimately be necessary, all efforts should be made to complete a comprehensive interview. The first step should be to find a quiet and private location. If there are external noises, place yourself between the patient and the sources of noise. Physicians should take a stance that allows them to be at eye level and within arm's distance from the patients they are interviewing. They should be looking directly at the patient when speaking. This may help patients read lips and may allow better transmission of the interviewer's voice if hearing is impaired. Also, proximity to patients will help those with visual impairments avoid strain. If hearing is impaired, the interviewer should try to ensure the patient is using assistive listening devices and that the device is being used correctly. If no assistive devices are available and the person's hearing is reduced, simply asking how you can help or if hearing is better on one side may improve communication while simultaneously displaying care and understanding of the patient's experience. The interviewer should speak slowly and clearly, use short sentences, and adjust volume and tone to maximize what is being heard. Sufficient time should be given for patients to answer, and clinicians should be willing to repeat questions as often as necessary.

Visual impairment is also a major concern in the elderly; approximately one in three elderly persons has some form of vision-reducing eye disease by the age of 65 (Quillen, 1999). If patients have visual impairment, clinicians should ensure adequate lighting, glasses, or lenses are being used and that there is a close distance during the interview. Visual acuity will aid patients in responding to nonverbal communication and in recognizing treatment team members that are involved in care.

Obtaining a History in Older Adults

Taking a complete history in older adults is not drastically different from other adults. It should include a history of the present illness, a history of medical and psychiatric illness, a family history, a complete list of medications and supplements, and a social history. Areas that are important and unique to older adults are highlighted. These elements may be necessary not only in diagnosis but also in developing treatment and follow-up plans and in the evaluation of long-term needs. Obtaining collateral history is essential in the elderly and is a fundamental clinical obligation of clinicians. Collateral should be obtained from family, members of the treatment team, hospital records, or any other available sources. Although necessary, collateral information should not replace or be given more weight than the interview. Ensuring patients that collateral is adjunct data is necessary to avoid disempowering patients.

Begin with a thorough review of medical records. This may help guide the inquiry and prove useful for patients with cognitive impairment. Begin with a review of symptoms, exploring each symptom's onset, duration, severity, precipitating factors, diurnal variation, and symptom cluster. With older patients, clinicians must pay attention to functional status and deficits in activities of daily living (ADLs). ADLs can be divided into basic and instrumental categories. Basic ADLs are skills that are typically learned as young children. They include ambulating, feeding, dressing, personal hygiene, continence, and toileting. Instrumental ADLs require more complex thinking and organization skills. These include managing transportation, shopping, meal preparation, finances, housecleaning, home maintenance, communication with others (telephone and mail), and medications. In 2011, the United States National Health Interview Survey determined that 20.7% of adults aged 85 or older, 7% of those aged 75 to 84, and 3.4% of those aged

65 to 74 needed help with ADLs (Wolff et al., 2016; Adams et al., 2012). Assessment should be completed in all clinical domains, including but not limited to mood symptoms, psychotic symptoms, and anxiety symptoms.

A history of medical illnesses should be explored. It is critical to determine whether they led to hospitalization if medications were prescribed and how the illness has progressed. Elderly patients will often omit medical illnesses that they see as unrelated or may forget earlier life illnesses, so an exhaustive medical record review and contact with collaterals may be helpful. In addition to conventional medical illnesses, topics specific to older adults should be explored. These may include nutrition, hearing, vision, continence, falls, and osteoporosis.

A full psychiatric history provides a narrative of previous episodes, possible hospitalization and treatments, and response to all treatments. Often, patients will not recall the exact names of previous medications; it may be useful to contact the pharmacy or to provide common medications prescribed for the illness being reported. In the elderly, medication trials may be remote, so clinicians should familiarize themselves with common medications from previous generations.

Polypharmacy and excessive dosages are common causes of iatrogenic illness in the elderly. Asking patients for a list of medications is the first step, but it may be insufficient. A simple and comprehensive approach is to ask a caregiver to gather all medications and bring them to the clinician. They should also bring a written schedule and pill container if available. They should be told to include over-the-counter (OTC) preparations. By requesting the pills, the schedule, and a pill box, a clinician can cross-check medications are being taken as recommended. Elderly patients will often keep medications that are no longer being prescribed that may be taken in error and have interactions with other medications.

Taking a family history has expanding importance when thinking about older adults. It is necessary to understand familial medical and psychiatric illnesses. In addition, a clinician must understand the family's socioeconomics, family dynamics, and the type and amount of support that is available to the older patient. This should include obtaining information and understanding the family's willingness in areas of transportation, supervision or checking on the patient, meal preparation, financial assistance, and coordination of services. This allows the clinician to engage in family planning interventions with the cooperation and buy-in of the identified family members. Obtaining this degree of information will also help the team identify which members are at risk for caregiver burnout.

Obtaining a developmental and occupational history will provide elements necessary to understand an older patient's experience. This includes performance in school and job history. Other areas of exploration should include the retirement age, reason for retirement, and activities that have followed retirement. A driving history should be obtained and may require intervention by the team or family if concerns arise. Also, exploration of substance use, cultural, religious, and spiritual belief systems will help in understanding the patient and future planning.

Rating Scales

Rating scales and structured interviews have been increasingly used in clinical practice. Standardized scales provide a tool to improve the validity and reliability of psychiatric assessments. The primary roles should be as screening devices, a complement to clinical judgment, to objectively track an illness over time. Depression, cognitive dysfunction, and general assessments emphasizing clinical utility and practicality will be reviewed.

Depression Rating Scales

A variety of depression rating scales have been developed; four that have a specific value for older adults will be reviewed here. The most widely used is the Hamilton Rating Scale for Depression (HDRS). The original version is a semi-structured interview that contains 17 items that assess depressive symptoms over a 1-week interval,

pertaining to symptoms of depression experienced over the past week (Hamilton, 1960). The scale was originally developed with hospital inpatients in mind, so its emphasis is on melancholic and physical symptoms of depression. A later 21-item version (HDRS21) includes four items intended to subtype depression. A limitation of the HDRS is that atypical symptoms of depression are not assessed.

The Geriatric Depression Scale (GDS) is a 30-item questionnaire that is entirely yes or no questions pertaining to how they felt over the past week (Yesavage et al., 1982). It is entirely self-reported and does not have a trained interviewer. A short form of the GDS (Sheikh & Yesavage, 1986) was developed and consists of 15 questions; this version can be administered more easily to physically ill and mildly to moderately demented patients or those with limited attention spans. A limitation of this scale is that it does not assess for suicidality.

The Beck Depression Inventory (BDI) is a 21-item, self-report rating inventory that measures characteristic attitudes and symptoms of depression (Beck et al., 1961). It has been modified into the BDI-1A and BDI-II. A limitation of the BDI is the self-reported nature of this scale, which allows for easy exaggeration or minimization of symptoms.

The Montgomery–Asberg Depression Rating Scale (MADRS) is a clinician-administered scale that concentrates on 10 symptoms of depression (Montgomery & Åsberg, 1979). The included items are apparent sadness, reported sadness, inner tension, reduced sleep, reduced appetite, concentration difficulties, lassitude, inability to feel, pessimistic thoughts, and suicidal thoughts. Unlike the HDRS, it does not focus on somatic symptoms of depression. It is used to assess the severity of depression in patients with a diagnosis of depression and therefore is sensitive to change resulting from antidepressant therapy.

The Cornell Scale for Depression in Dementia (CSDD) is a scale designed to specifically assess depression in patients with dementia. This 19-item clinician-administered scale uses information from interviews with both the patient and the patient's caregiver (Alexopoulos et al., 1988).

It differs in administration methods from other depression scales and has great value when assessing mood in the presence of cognitive impairment.

The Hospital Anxiety and Depression Scale (HADS) is a self-assessment tool that was developed to measure anxiety and depression in a general medical population of patients. The scale is comprised of seven questions for anxiety and seven questions for depression (Zigmond & Snaith, 1983). The HADS only takes 2–5 min to complete. A limitation of the HADS is that it does not include assess areas of appetite, sleep, and self-harm of suicide.

The Composite International Diagnostic Interview 65+ (CIDI65+) is an age-adapted version of the CIDI to assess the needs of the elderly. The CIDI65+ was designed to address the social, cognitive, and psychological abilities and needs of the elderly. The instrument's diagnostics evaluate somatoform, mood, substance use disorders, cognitive impairment, anxiety disorders, and obsessive-compulsive disorders (Wittchen et al., 2015).

Dementia and Cognitive Dysfunction Scales

The Mini-Mental Status Examination (MMSE) is a brief screening tool of 11 questions or tasks that provides a quantitative assessment of cognitive impairment and change over time (Folstein et al., 1975). It assesses the domains of orientation to time, orientation to place, registration of three words, attention and calculation, recall of three words, language, and visual construction. The MMSE relies heavily on verbal response, reading, and writing. This may result in patients with hearing or visual impairment, intubation, or communication disorders being incorrectly scored as more impaired.

The Montreal Cognitive Assessment (MoCA) is a screen to detect cognitive impairment, and more specifically, mild cognitive impairment. It is a 16-item scale that assessed domains of memory, visuospatial abilities, executive function, naming, memory, attention, language, abstract

reason, and orientation. As compared to the MMSE, this scale involves more words, fewer learning trials, and longer delay before recall, which allows better recognition of mild cognitive impairment (Nasreddine et al., 2005).

The clock drawing test (CDT) is a rapidly administered test that is appropriate for primary care practices. Although there are multiple versions of this test, in general, they all ask the patient to draw the face of a clock and then draw the hands to indicate a particular time. This single test may be sensitive to dementia because it involves many cognitive areas that can be affected by dementia, including executive function, visuospatial abilities, motor programming, and attention and concentration (Shulman et al., 1986).

The Blessed Dementia Scale, developed in 1968, was designed to quantify the degree of intellectual and personality deterioration. It is a 22-item scale that measures changes in the performance of everyday activities, self-care, and changes in personality, interests, and drives (Blessed et al., 1968). Ratings are based on information from friends or relatives over a 6-month period. If available, medical records can also be used.

The Neuropsychiatric Inventory (NPI) was developed to assess common behaviors associated with dementia. A structured interview of the caregiver familiar with the patient's behavior assesses 10 behavioral domains: delusions, hallucinations, dysphoria, euphoria, anxiety, agitation/aggression, apathy, irritability/lability, disinhibition, and aberrant motor behavior. The NPI only scores domains that have positive screening responses, to minimize the time of administration (Cummings et al., 1994).

The Cohen-Mansfield Agitation Inventory (CMAI) is a 29-item scale to assess the types and frequency of agitation. The agitated behaviors include physical/aggressive, physical/non-aggressive, verbal/aggressive, and verbal/non-aggressive that are rated on a seven-point scale of frequency. The CMAI is useful for the assessment of agitation in residents of nursing homes (Cohen-Mansfield et al., 1989).

In detecting delirium, the Confusion Assessment Method (CAM) is the most widely used tool available. It was designed to allow non-psychiatric clinicians to diagnose delirium. The CAM assesses the presence, severity, and fluctuation of nine delirium features: acute onset, inattention, disorganized thinking, altered level of consciousness, disorientation, memory impairment, perceptual disturbances, psychomotor agitation or retardation, and an altered sleep-wake cycle (Inouye et al., 1990). Another version, the CAM-ICU, was developed for non-verbal, mechanically ventilated patients (Ely et al., 2001). This scale follows a detailed protocol and with graining can be administered by any member of the ICU staff (Pisani et al., 2007).

The Health of the Nation Outcome Scales 65+ (HoNOS 65+) is a clinician-rated instrument that was developed (Gowers et al., 1999) after the original Health of the Nation Outcome Scale for working-age adults. The HoNOS 65+ is intended to have adequate coverage of clinical and social functions and to be sensitive to change over time. It is comprised of 12 simple scales measuring behavioral disturbance, non-accidental self-injury, problem drinking or drug use, cognitive problems, physical illness or disability problems, problems associated with hallucinations and delusions, problems with depressive symptoms, other mental and behavioral problems, problems with relationships, problems with activities of daily living, problems with living conditions, and problems with work and leisure activities. Each item is rated on a 5-point scale and accounts for the worst manifestation of each behavior over the past 2 weeks, regardless of cause.

Laboratory Workup of Older Adults

Laboratory testing is an essential part of the psychiatric evaluation of older adults. Combining laboratory results, psychiatric interview, rating scales, and physical exam will aid in diagnosis and treatment planning. The goal of a geriatric psychiatrist is to use clinical judgment to select a laboratory workup that factors in the cost and relevance of each requested laboratory. Basic laboratory tests that are necessary in the initial workup of an older patient are outlined.

Laboratory testing in the early stages of diagnosis may aid in assessing a person's general health, determining causes or contributors to a current condition, and getting baseline values for future difficulties.

Basic chemistry and hematology will be done on most patients presenting to facilities. These screening tests may help identify electrolyte abnormalities or sign of infection that may contribute to changes in mental status. Furthermore, they provide baseline results that may require monitoring if medications are initiated. Similarly, urine analysis is inexpensive and adequately screens for urinary tract infections (UTIs). UTIs account for many admissions and varied clinical presentations in the elderly, specifically delirium. Early identification and prompt antibiotic initiation often result in the resolution of psychiatric symptoms without requiring other psychotropic interventions. Toxicology Screen may prove useful in detecting the presence of illegal drugs or prescription medications and should be done in older adults.

Thyroid function screening should be done in older patients. Thyroid disease may present with mood symptoms, psychotic symptoms, anxiety, or impaired cognition. Thyroid Stimulating Hormone (TSH) is a frequently used and excellent screening test. If TSH is abnormal, further examination of specific thyroid hormones and antibodies is indicated.

Vitamin B12 and folate levels should also be collected when completing an assessment in an older individual. Deficiencies may result in mood symptoms, psychotic symptoms, or cognitive impairment.

Testing for syphilis and human immunodeficiency virus is less absolute. Clinicians should use clinical judgment to determine if there are risk factors, atypical symptoms, and an atypical narrative of illness that suggest tertiary syphilis or HIV-associated illnesses.

An electrocardiogram (ECG or EKG) is usually done on admission to a facility. For a geriatric psychiatrist, an ECG provides information necessary to initiate specific medications. Many psychotropic medications have varied cardiac effects that geriatric psychiatrists should familiarize themselves with prior to initiating medications. The corrected QT interval (QTc) has become a significant marker in the EKG and can be prolonged by antipsychotics and antidepressants. Prolongation of the QT interval is considered to be a surrogate marker for the risk of developing a particular type of ventricular tachyarrhythmia called "torsades de pointes."

The Diagnostic Workup of Older Adults

Diagnostic Interview	History of Present Illness – current symptoms, functional status, ADLs, Medical History- medical illnesses, medications, supplements Psychiatric History- previous episodes of psychiatric illness, psychiatric hospitalizations, treatment response, suicide/homicide history Family History- familial psychiatric and medical conditions Social History- developmental history, occupational history, driving history, substance use history, cultural and/or religious belief system
Collateral Information	Family, treatment team, medical records
Laboratory Workup	Chemistry, hematology, urine analysis, urine toxicology, thyroid function tests, B12 level, folate level

Liaison Psychiatry

Such a detailed patient-centered assessment allows for defining a multi-morbidity, individualized treatment plan which takes into account both the old age patient's physical and mental health. Indeed, multi-morbidity in the elderly that includes mental illness demands more effective care coordination.

One approach is providing a type of psychiatric care to hospital wards called liaison psychiatry. Liaison psychiatry has developed over the last 30 years in response to the overlap between physical and mental health problems and the institutional separation between the ways these medical issues are addressed (Mukaetova-Ladinska, 2016). In theory, liaison psychiatry has the potential to better attend to the needs of elderly patients in hospitals for somatic diseases. However, a poor

degree of agreement between primary care physicians' and liaison psychiatrics' evaluation has been repeatedly described. Non-agreement is higher in old age compared to younger patients, and it significantly depends on patients' symptom severity and on their personality traits (Canuto et al., 2015).

Limitations that prevent these mental healthcare practitioners from maximizing their contribution to patient outcomes include: (i) they often operate using a referral model and therefore only attend to a fraction of the patients identified as having a mental illness; (ii) they are not always guided by measurement-driven outcomes such as length of hospital stay; (iii) their contribution to patient care may be limited to consultations and advice; and (iv) they may not be integrated into the patient's clinical team (Walker et al., 2019). To date, clinical services provided by liaison psychiatric professionals have been variable, making them difficult to assess, and current evidence for the efficacy of care and cost-effectiveness is consequently very limited. Systematic reviews indicate that the liaison model is more effective at improving clinical outcomes in the elderly compared to the consultant-led model, but they fail to provide strong evidence to alter current treatment modalities (Draper, 2000; Callaghan et al., 2003).

Collaborative Care

To effectively address complex geriatric health issues, there is a recognized need to go beyond the liaison model in order to improve person-centered care and patient engagement. For example, care fragmentation should be addressed through the integration of diverse medical services alongside seamless digital support (de Bruin et al., 2012). More emphasis should be placed on proactive and preventive practices to improve health outcomes and reduce costs (WHO, 2015b). This collaborative care model, also called integrated care, is a team-based approach for screening, education, treatment, and inclusion of information technology. Ideally, the integration of healthcare delivery would include, depending on need, mental and physical medical practitioners, nurses, pharmacists, social workers, physiotherapists, dieticians, and others such as those involved in the promotion of mental, physical, educational, and spiritual well-being. Thus, this would require team members to participate in multidisciplinary and interdisciplinary training, and possibly development of management positions to oversee these diverse teams. This would lead, according to WHO (2016), to "services that are managed and delivered so that people receive a continuum of health promotion, disease prevention, diagnosis, treatment, disease-management, rehabilitation and palliative care services, coordinated across the different levels and sites of care within and beyond the health sector, and according to their needs throughout the life course."

Essential Elements of Collaborative Care in the Elderly

Patient-Centered Team Care

The team is composed of a multidisciplinary group of professionals that is charged with developing, implementing, and updating a treatment plan. The team is often led by a primary care provider and supported by office staff, managers, nurses, and appropriate specialists, including a geriatrics practitioner. A crucial aspect of the team-driven approach is for internal accountability and follow-up, which requires checks and balances to ensure and maintain the appropriate standard of care (APA, 2016).

Population-Focused

Healthcare can be implemented more effectively for groups of patients with similar morbidities. This will enable clinicians who are typically accustomed to treating one patient at a time to aggregate data on larger groups of patients and allow for identification of trends and delivery system gaps, which can be addressed more expeditiously. Furthermore, population-based care

enables team members to identify patients who are not responding well to their treatments and may require more proactive adjustments.

Measurement-Guided

Each patient's plan should clearly articulate treatment goals and clinical outcomes which are measured by evidence-based tools. Clinical decision-making should be driven by patient-reported outcome measures. For example, symptom rating scales can be used to track the frequency and severity of symptoms that patients are experiencing. This patient- and physician-reported system will help the team monitor and maintain treatment goals.

Evidence-Based

An essential element of collaborative care is treatment guidelines, which ensure that the team is delivering the most accurate and current therapies. The treatment plan must have measurable outcomes that, when achieved, directly lead to an improvement in the quality of life of the patient. Evidence-based care incorporates measurement-guided data into the clinical decision-making process to personalize disease management and allows clinical teams to be confident in their decisions while maintaining cost and efficacy.

Collaborative Care Team Structure

Collaborative care requires a team of professionals with complementary skills who work together to serve a population of patients with common health conditions. It requires a shift in how medicine is practiced, the creation of entirely new workflows, and a dynamic team, which is constantly gaining and losing members. As mentioned earlier, the collaborative team includes the primary-care provider (typically a family physician or nurse practitioner); care management professionals such as a nurse or psychologist who provide coordination and support the delivery of treatment; a consulting psychiatrist who advises the team on any challenges related to mental health; auxiliary health specialists such as pharmacists and dieticians; office support staff and information technology specialists; and social/community support services (McMaster, 2017). With modern information technology, it is not necessary that diverse team members are co-located, with some professionals providing remote support. Although this is the ideal, collaborative care teams from recent studies tend to be much smaller in composition.

For practitioners in solo medical delivery systems, a stepwise approach can guide the integration of a team and its ability to fulfill expectations. This can begin with training sessions and working as a team on simple projects. As familiarity, trust, and experience develop, the team can advance to more challenging endpoints to eventually achieve transformative outcomes. For success, it is essential to have strong leadership, an intimate understanding of personal and team member roles within the group, mechanisms for clear communication, and a shared plan that can be updated with monitoring of patient outcomes.

This narrative review aims to provide an examination of the current state of collaborative care models for elderly populations with mental and somatic multi-morbidities. Additionally, the need for further implementation and testing of collaborative care models is described, and some of the challenges encountered thus far are highlighted.

Unmet Needs in Older Populations with Multi-morbidities

Many earlier studies highlighted the need for a collaborative care model to address the unmet needs of specific elderly patient populations. For example, a US study examined medical files of 132,405 patients with Alzheimer's disease (AD) using the Nationwide Inpatient Sample database (Yen et al., 2020). Logistic regression analysis was used to determine which comorbidities increase the risk of in-hospital mortality. Results showed that congestive cardiac failure and renal failure were significantly associated

with higher risk, and tumors with or without metastasis also increase in-hospital mortality. The authors indicate that a collaborative care model is required to effectively manage comorbidities in AD patients to improve their quality of life and reduce morbidity and mortality.

Borges et al. (2021) investigated whether frailty could explain the association between depression and geriatric disorders in the Netherlands. They hypothesized that since frailty is an accelerated decrease of reserve in several interrelated physiological systems, it can be used to predict the onset of chronic diseases. According to their results, depressive disorders as well as depressive symptom severity were significantly associated with a number of geriatric diseases. Frailty was also significantly associated with the other geriatric syndromes, and this association remained significant even after adjustment for relevant confounding factors. Frailty was consistently observed to be a possible explanatory variable in these associations since frailty models interfered with the strength of the association between depression and other geriatric syndromes. Based on these findings, the authors suggested that collaborative care models are necessary to address somatic and mental comorbidities; frailty screening is important in mental healthcare, especially since it tends to be neglected in this field; and frailty should be prevented and treated to minimize its impact on other chronic diseases in older age groups.

A study by Kim et al. (2018) described how functional deterioration related to aging is not being adequately addressed by the Korean healthcare system. Using data from the Korea Health Panel Study, results showed that those with memory, visual, or hearing impairments were about 1.5 times more likely to have unmet healthcare needs compared to those without the impairments. Depression was similarly associated with this deficit. The authors emphasize that, in addition to socioeconomic and health-related factors, aging should have an important role in risk assessment. Since all factors are linked organically, collaborative care is needed to improve health outcomes among the older generation by addressing chronic diseases holistically and for the promotion of lifelong health advancement.

Collaborative care may be beneficial for geriatric chronic diseases, such as osteoporosis, which can lead to bone fractures. Working in the UK, Clancy et al. (2018) looked at readmission rates and mortality in patients with hip fractures that were either discharged to a bedded immediate care or to home. The authors included inpatient diagnoses of delirium or pre-existing diagnoses of some form of dementia. Results showed that delirium and dementia were associated with higher readmission rates and lower rates of discharge to home. It was concluded that a collaborative care type approach should be established to address the needs of elderly patients with hip fractures.

In addition to specific chronic disease cohorts, collaborative care models may also be useful for specific populations limited by geography and/or socioeconomic status. For example, one study showed a high prevalence of multi-morbidity in a rural South African population (Chang et al., 2019). The most prevalent profiles were combinations of cardiometabolic conditions, depression, HIV, anemia, and combinations of mental disorders. Chang et al. (2019) argued that there should be more awareness of chronic multi-morbidity in this region and more coordinated, long-term collaborative care management that spans across multiple chronic somatic and psychiatric conditions. Other studies in the search results investigated the association between dementia and co-morbid chronic conditions such as frailty, hypertension, and diabetes in a cohort of patients from Singapore (Chua et al., 2019); found loneliness to be a risk factor for the development of depression, dementia, hypertension, cardiovascular disease, and stroke in older citizens in the study from five European countries (Tan et al., 2020); and found that collaborative care models are necessary to address deficits in social support that underlie connectivity between depression and hypertension in rural Chinese villages (Zhu et al., 2019).

Evidence-Based Models in the Elderly

Evidence-based models for collaborative care of older adults with mental illness and physical comorbidities have recently been reviewed (Bartels et al., 2020), and here other models that were identified from the database search will be discussed. Firstly, the 48/6 Model of Care is a collaborative care initiative that was developed for hospitalized seniors in Canada (Uhm et al., 2020). The model begins with screening of six care areas (bowel and bladder management, cognitive functioning, functional mobility, medication management, nutrition and hydration, and pain management). Based on these assessments, healthcare teams can implement a personalized, documented care plan within 48 h of a decision to admit a patient. This model aims to prevent functional decline and in-hospital comorbidity by helping elderly patients to return home sooner and at the level of independence that they had prior to admission. The study by Uhm et al. (2020) investigated the usefulness of the 48/6 Model of Care, including the prevalence of dysfunction in the six care areas and its relationship with self-reported life-space mobility in older adults living independently. Results indicated that screening of functional impairment was feasible, dysfunction in the six areas was highly prevalent among community-dwelling older adults, and dysfunction correlated with the extent of mobility.

In England, the Rapid Assessment Interface and Discharge (RAID) model of collaborative care has been tested in a few different locales (Thacker et al., 2017; Singh et al., 2016). One study was composed of a team lead, a consultant psychiatrist, a consultant geriatrician, liaison nurses, an occupational therapist, a social worker, an old age psychiatry registrar, two secretaries (team and project), and a project manager for a 4-month trial period (Singh et al., 2016). The objective of the collaborative care team was to improve the mental health care of older people admitted to all departments (i.e., acute, medical, surgical, and trauma wards) of a general hospital. Inclusion criteria were patients above 65 years of age who had mental health issues or cognitive problems at the time of their admission and those under 65 years with cognitive problems not caused by delirium or acute intoxication. The RAID trial treated 339 patients with a mean age of 82.18 ± 8.04 years and patients were 60% female.

Results showed that direct discharges were increased by 7% and the mean hospital stay was reduced from 35 to 20 days in the acute setting and from 108 to 47 days in long-stay wards. Improved cost benefits were based on a mean reduction in hospital stay and admission reduction, leading to a total annualized bed savings of 44 days. The authors concluded that prompt mental health assessments for acutely unwell, frail older people were not only cost-effective but also improved clinical outcomes (Singh et al., 2016).

The goal of the Perioperative Optimization of Senior Health (POSH) program at Duke University Hospital was to improve post-operative outcomes for older adults undergoing elective surgeries by incorporating expertise from the fields of geriatrics, general surgery, and anesthesia (McDonald et al., 2018). Geriatrics experts are involved throughout the perioperative period and tasked with specific targeted interventions such as management of comorbidities, reduction of polypharmacy, enhancement of mobility and nutrition, and risk mitigation of delirium. Study results showed that high-risk older adults undergoing elective abdominal surgery who participated in the POSH collaborative care program had fewer complications, shorter hospitalizations, more frequent discharges to home, and fewer readmissions than the control group (McDonald et al., 2018).

The Chinese Older Adult Collaborations in Health (COACH) study is a random controlled trial to test the effectiveness of a collaborative care-based approach to treat hypertension and depression comorbidities in rural China (Li et al., 2020). In this model, a team consisting of a village doctor, elderly worker, and psychiatrist consultant collaborate to enroll and treat subjects for a 12-month study duration. Preliminary results based on qualitative analyses of this ongoing trial suggested that the teams worked well together and that team members regarded the COACH model to be effective in reducing depressive symptoms and improving patient health. Other collaborative care models identified in the literature search also showed varying levels of improvement in patient outcomes (Looman et al., 2016; Franse et al., 2018; Mas et al., 2018; Spoorenberg et al., 2019).

Limitations of the Current State of Collaborative Care

Building and implementing collaborative care systems requires the expenditure of valuable and scarce resources. Long-term investment in these teams by healthcare networks requires proof-of-principle evidence that they provide positive cost benefits in terms of improved patient outcomes and reduced financial output. The database search identified two systemic reviews that focused on collaborative care models; however, these analyses did not yield conclusive evidence that interventions based on inter-professional or multi-professional teamwork significantly improved health determinants for geriatric patients. Platzer et al. (2020) highlighted the main problems associated with cross-study analysis, including the scarcity of data and inconsistencies in the nature of the trials, such as wide differences in assessment tools, disparate worldwide healthcare settings, and heterogenous target populations. A meta-analysis by Frost et al. (2019) that examined non-pharmacological interventions for depression and/or anxiety in older adults with physical comorbidities affecting functioning identified 14 eligible trials. Results showed that collaborative care did not appear to affect depressive symptoms, functioning, or quality of life, and there was inclusive evidence for effects upon remission rates. However, trials were limited by small sample sizes and short follow-up periods.

Conclusions

The diagnostic interview of older adults is the cornerstone of formulating an appropriate diagnosis and treatment plan. This, in combination with the utilization of appropriate standardized instruments, and a proper medical workup are essential components of treating older patients. Most existing health systems manage the complex chronic health issues of older patients in a disconnected and fragmented way, and there is a lack of coordination across care providers and in the timing of care provided. Collaborative care is a common-sense approach to effectively preventing, identifying, and treating multimorbidities across the physical and mental spectra of chronic diseases that emerge with age. For collaborative care to reach its potential for producing positive patient-centered outcomes, a greater number of larger-scale, and well-planned trials need to be conducted that are supported by sufficient expertise in various fields, have buy-in from team members and other stakeholders, proper training for team building, and advanced digital support for team communication, patient monitoring, and bioinformatics.

References

Adams, P. F., Kirzinger, W. K., & Martinez, M. E. (2012). Summary health statistics for the U.S. population: National health interview survey, 2011. *Vital and Health Statistics. Series 10*, (255), 1–110.

Alexopoulos, G. S., Abrams, R. C., Young, R. C., & Shamoian, C. A. (1988). Cornell scale for depression in dementia. *Biological Psychiatry, 23*(3), 271–284.

American Psychiatric Association/Academy of Psychosomatic Medicine. (2016). *Dissemination of integrated care within adult primary care settings: The collaborative care model.* https://www.psychiatry.org/File%20Library/Psychiatrists/Practice/Professional-Topics/Integrated-Care/APA-APM-Dissemination-Integrated-Care-Report.pdf

Andreas, S., Schulz, H., Volkert, J., Dehoust, M., Sehner, S., Suling, A., . . . & Härter, M. (2017). Prevalence of mental disorders in elderly people: The European MentDis_ICF65+ study. *The British Journal of Psychiatry, 210*(2), 125–131.

Barnett, K., Mercer, S. W., Norbury, M., Watt, G., Wyke, S., & Guthrie, B. (2012). Epidemiology of multimorbidity and implications for health care, research, and medical education: A cross-sectional study. *Lancet, 380*, 37–43.

Bartels, S. J., DiMilia, P. R., Fortuna, K. L., & Naslund, J. A. (2020). Integrated care for older adults with serious mental illness and medical comorbidity: Evidence-based models and future research directions. *Clinics in Geriatric Medicine, 36*, 341–352.

Beck, A. T., Ward, C., Mendelson, M., Mock, J., & Erbaugh, J. (1961). Beck depression inventory (BDI). *Archives of General Psychiatry, 4*(6), 561–571.

Blessed, G., Tomlinson, B. E., & Roth, M. (1968). The association between quantitative measures of dementia and of senile change in the cerebral grey matter of elderly subjects. *The British Journal of Psychiatry, 114*(512), 797–811.

Borges, M. K., Voshaar, R. C. O., Romanini, C. F. V., Oliveira, F. M., Lima, N. A., Petrella, M., Costa, D. L., Martinelli, J. E., Mingardi, S. V. B., Siqueira, A., Biela, M., Collard, R., & Aprahamian, I. (2021). Could frailty be an explanatory factor of the association between depression and other geriatric syndromes in later life? *Clinical Gerontologist, 44*, 143–153.

Callaghan, P., Eales, S., Coates, T., & Bowers, L. (2003). A review of research on the structure, process and outcome of liaison mental health services. *Journal of Psychiatric and Mental Health Nursing, 10*, 155–165.

Canuto, A., Gkinis, G., DiGiorgio, S., Arpone, F., Herrmann, F. R., & Weber, K. (2015). Agreement between physicians and liaison psychiatrists on depression in old age patients of a general hospital: Influence of symptom severity, age and personality. *Aging & Mental Health, 20*(10), 1092.

Cerra, F. B., & Brandt, B. F. (2015). Chapter 9 – The growing integration of health professions education. In S. A. Wartman (Ed.), *The transformation of academic health centers* (pp. 81–90). Academic.

Chang, A. Y., Gómez-Olivé, F. X., Payne, C., Rohr, J. K., Manne-Goehler, J., Wade, A. N., Wagner, R. G., Montana, L., Tollman, S., & Salomon, J. A. (2019). Chronic multimorbidity among older adults in rural South Africa. *BMJ Global Health, 4*, e001386.

Chua, X. Y., Ha, N. H. L., Cheong, C. Y., Wee, S. L., & Yap, P. L. K. (2019). The changing profile of patients in a geriatric medicine led memory clinic over 12 years. *The Journal of Nutrition, Health & Aging, 23*, 310–315.

Clancy, U., Brown, M., Alio, Z., Wardle, K., & Pendleton, N. (2018). Older people with hip fracture transferred to intermediate care: Outcomes in an integrated health and social care model. *Future Healthcare Journal, 5*, 58–63.

Cohen-Mansfield, J., Marx, M. S., & Rosenthal, A. S. (1989). A description of agitation in a nursing home. *Journal of Gerontology, 44*(3), M77–M84.

Costanza, A., Baertschi, M., Weber, K., & Canuto, A. (2015). Maladies neurologiques et suicide: de la neurobiologie au manque d'espoir [Neurological diseases and suicide: From neurobiology to hopelessness]. *Revue Médicale Suisse, 11*, 402–405.

Costanza, A., Amerio, A., Radomska, M., Ambrosetti, J., Di Marco, S., Prelati, M., Aguglia, A., Serafini, G., Amore, M., Bondolfi, G., Michaud, L., & Pompili, M. (2020a). Suicidality assessment of the elderly with physical illness in the emergency department. *Frontiers in Psychiatry, 11*, 558974.

Costanza, A., Amerio, A., Aguglia, A., Escelsior, A., Serafini, G., Berardelli, I., Pompili, M., & Amore, M. (2020b). When sick brain and hopelessness meet: Some aspects of suicidality in the neurological patient. *CNS & Neurological Disorders Drug Targets, 19*, 257–263.

Cummings, J. L., Mega, M., Gray, K., Rosenberg-Thompson, S., Carusi, D. A., & Gornbein, J. (1994). The neuropsychiatric inventory: Comprehensive assessment of psychopathology in dementia. *Neurology, 44*(12), 2308–2308.

de Bruin, S. R., Versnel, N., Lemmens, L. C., Molema, C. C. M., Schellevis, F. G., Nijpels, G., & Baan, C. A. (2012). Comprehensive care programs for patients with multiple chronic conditions: A systematic literature review. *Health Policy, 107*, 108–145.

Draper, B. (2000). The effectiveness of old age psychiatry services. *International Journal of Geriatric Psychiatry, 15*, 687–703.

Ely, E. W., Margolin, R., Francis, J., May, L., Truman, B., Dittus, R., ... & Inouye, S. K. (2001). Evaluation of delirium in critically ill patients: Validation of the Confusion Assessment Method for the Intensive Care Unit (CAM-ICU). *Critical Care Medicine, 29*(7), 1370–1379.

Fabbri, E., An, Y., Zoli, M., Simonsick, E. M., Guralnik, J. M., Bandinelli, S., Boyd, C. M., & Ferrucci, L. (2015). Aging and the burden of multimorbidity: Associations with inflammatory and anabolic hormonal biomarkers. *The Journals of Gerontology. Series A, Biological Sciences and Medical Sciences, 70*, 63–70.

Folstein, M. F., Folstein, S. E., & McHugh, P. R. (1975). "Mini-mental state": A practical method for grading the cognitive state of patients for the clinician. *Journal of Psychiatric Research, 12*(3), 189–198.

Franse, C. B., van Grieken, A., Alhambra-Borrás, T., Valía-Cotanda, E., van Staveren, R., Rentoumis, T., Markaki, A., Bilajac, L., Marchesi, V. V., Rukavina, T., Verma, A., Williams, G., Koppelaar, E., Martijn, R., Voorham, A. J. J., Mattace Raso, F., Garcés-Ferrer, J., & Raat, H. (2018). The effectiveness of a coordinated preventive care approach for healthy ageing (UHCE) among older persons in five European cities: A pre-post controlled trial. *International Journal of Nursing Studies, 88*, 153–162.

Frost, R., Bauernfreund, Y., & Walters, K. (2019). Non-pharmacological interventions for depression/anxiety in older adults with physical comorbidities affecting functioning: Systematic review and meta-analysis. *International Psychogeriatrics, 31*, 1121–1136.

Gowers, S. G., Harrington, R. C., Whitton, A., Lelliott, P., Beevor, A., Wing, J., & Jezzard, R. (1999). Brief scale for measuring the outcomes of emotional and behavioural disorders in children. *The British Journal of Psychiatry, 174*(5), 413–416.

Hamilton, M. (1960). A rating scale for depression. *Journal of Neurology, Neurosurgery, and Psychiatry, 23*(1), 56.

Inouye, S. K., van Dyck, C. H., Alessi, C. A., Balkin, S., Siegal, A. P., & Horwitz, R. I. (1990). Clarifying confusion: The confusion assessment method: A new method for detection of delirium. *Annals of Internal Medicine, 113*(12), 941–948.

Kanasi, E., Ayilavarapu, S., & Jones, J. (2016). The aging population: Demographics and the biology of aging. *Periodontology 2000, 72*, 13–18.

Karlin, B. E., & Fuller, J. D. (2007). Meeting the mental health needs of older adults. *Geriatrics, 62*, 26–35.

Kim, Y. S., Lee, J., Moon, Y., Kim, K. J., Lee, K., Choi, J., & Han, S. H. (2018). Unmet healthcare needs of elderly people in Korea. *BMC Geriatrics, 18*, 98.

Li, L. W., Xue, J., Conwell, Y., Yang, Q., & Chen, S. (2020). Implementing collaborative care for older people with comorbid hypertension and depression in rural China. *International Psychogeriatrics, 32*, 1457–1465.

Looman, W. M., Fabbricotti, I. N., de Kuyper, R., & Huijsman, R. (2016). The effects of a pro-active integrated care intervention for frail community-dwelling older people: A quasi-experimental study with the GP-practice as single entry point. *BMC Geriatrics, 16*, 43.

Mas, M., Santaeugènia, S. J., Tarazona-Santabalbina, F. J., Gámez, S., & Inzitari, M. (2018). Effectiveness of a hospital-at-home integrated care program as alternative resource for medical crises care in older adults with complex chronic conditions. *Journal of the American Medical Directors Association, 19*, 860–863.

McDonald, S. R., Heflin, M. T., Whitson, H. E., Dalton, T. O., Lidsky, M. E., Liu, P., Poer, C. M., Sloane, R., Thacker, J. K., White, H. K., Yanamadala, M., & Lagoo-Deenadayalan, S. A. (2018). Association of integrated care coordination with postsurgical outcomes in high-risk older adults: The Perioperative Optimization of Senior Health (POSH) Initiative. *JAMA Surgery, 153*, 454–462.

McMaster Health Forum. (2017). Identifying and assessing core components of collaborative-care models for mental and physical health conditions. https://www.mcmasterforum.org/docs/default-source/product-documents/rapid-responses/identifying-and-assessing-core-components-of-collaborative-care-models.pdf?sfvrsn=2

Montgomery, S. A., & Åsberg, M. A. R. I. E. (1979). A new depression scale designed to be sensitive to change. *The British Journal of Psychiatry, 134*(4), 382–389.

Mukaetova-Ladinska, E. B. (2016). Current and future perspectives of liaison psychiatry services: Relevance for older people's care. *Geriatrics (Basel), 1*, 7.

Nasreddine, Z. S., Phillips, N. A., Bédirian, V., Charbonneau, S., Whitehead, V., Collin, I., ... & Chertkow, H. (2005). The Montreal Cognitive Assessment, MoCA: A brief screening tool for mild cognitive impairment. *Journal of the American Geriatrics Society, 53*(4), 695–699.

National Institute of Health, National Institute of Deafness and Other Communication Disorders. (2018) Hearing Loss and Older Adults. https://www.nidcd.nih.gov/health/hearing-loss-older-adults

Pisani, M. A., Murphy, T. E., Van Ness, P. H., Araujo, K. L., & Inouye, S. K. (2007). Characteristics associated with delirium in older patients in a medical intensive care unit. *Archives of Internal Medicine, 167*(15), 1629–1634.

Platzer, E., Singler, K., Dovjak, P., Wirnsberger, G., Perl, A., Lindner, S., Liew, A., & Roller-Wirnsberger, R. E. (2020). Evidence of inter-professional and multi-professional interventions for geriatric patients: A systematic review. *International Journal of Integrated Care, 20*, 6.

Quillen, D. A. (1999). Common causes of vision loss in elderly patients. *American Family Physician, 60*(1), 99–108.

Sheikh, J. I., & Yesavage, J. A. (1986). Geriatric Depression Scale (GDS): Recent evidence and development of a shorter version. *Clinical Gerontologist: The Journal of Aging and Mental Health, 5*, 165–173.

Shulman, K. I., Shedletsky, R., & Silver, I. L. (1986). The challenge of time: Clock-drawing and cognitive function in the elderly. *International Journal of Geriatric Psychiatry, 1*(2), 135–140.

Singh, I., Fernando, P., Griffin, J., Edwards, C., Williamson, K., & Chance, P. (2016). Clinical outcome and predictors of adverse events of an enhanced older adult psychiatric liaison service: Rapid Assessment Interface and Discharge (Newport). *Clinical Interventions in Aging, 12*, 29–36.

Spoorenberg, S. L., Reijneveld, S. A., Uittenbroek, R. J., Kremer, H. P., & Wynia, K. (2019). Health-related problems and changes after 1 year as assessed with the geriatric ICF core set (GeriatrICS) in community-living older adults who are frail receiving person-centered and integrated care from embrace. *Archives of Physical Medicine and Rehabilitation, 100*, 2334–2345.

Tan, S. S., Fierloos, I. N., Zhang, X., Koppelaar, E., Alhambra-Borras, T., Rentoumis, T., Williams, G., Rukavina, T., van Staveren, R., Garces-Ferrer, J., Franse, C. B., & Raat, H. (2020). The association between loneliness and health related quality of life (HR-QoL) among community-dwelling older citizens. *International Journal of Environmental Research and Public Health, 17*, 600.

Thacker, S., Skelton, M., & Harwood, R. (2017). Psychiatry and the geriatric syndromes – creating constructive interfaces. *BJPsych Bulletin, 41*, 71–75.

Uhm, K. E., Oh-Park, M., Kim, Y. S., Park, J. M., Choi, J., Moon, Y., Han, S. H., Hwang, J. H., Lee, K. S., & Lee, J. (2020). Applicability of the 48/6 model of care as a health screening tool, and its association with mobility in community-dwelling older adults. *Journal of Korean Medical Science, 35*, e43.

Walker, J., Burke, K., Toynbee, M., van Niekerk, M., Frost, C., Magill, N., Walker, S., Sculpher, M., White, I. R., & Sharpe, M. (2019). The HOME Study: Study protocol for a randomised controlled trial comparing the addition of Proactive Psychological Medicine to usual care, with usual care alone, on the time spent in hospital by older acute hospital inpatients. *Trials, 20*, 483.

Wang, L., Si, L., Cocker, F., Palmer, A. J., & Sanderson, K. (2018). A systematic review of cost-of-illness studies of multimorbidity. *Applied Health Economics and Health Policy, 16*, 15–29.

Wittchen, H. U., Strehle, J., Gerschler, A., Volkert, J., Dehoust, M. C., Sehner, S., . . . & Andreas, S. (2015). Measuring symptoms and diagnosing mental disorders in the elderly community: The test–retest reliability of the CIDI65+. *International Journal of Methods in Psychiatric Research, 24*(2), 116–129.

Wolff, J. L., Feder, J., & Schulz, R. (2016). Supporting family caregivers of older Americans. *New England Journal of Medicine, 375*(26), 2513–2515.

World Health Organization. (2015a). *World report on ageing and health.* https://www.who.int/ageing/events/world-report-2015-launch/en/

World Health Organization. (2015b). *WHO global strategy on integrated people-centred health services 2016–2026.* https://www.who.int/servicedeliverysafety/areas/people-centred-care/global-strategy/en/

World Health Organization. (2016). *Framework on integrated people-centred health services.* https://www.who.int/servicedeliverysafety/areas/people-centred-care/framework/en/

Yen, T. Y., Beriwal, N., Kaur, P., Ravat, V., & Patel, R. S. (2020). Medical comorbidities and association with mortality risk in Alzheimer's disease: Population-based study of 132,405 geriatric inpatients. *Cureus, 12*, e8203.

Yesavage, J. A., Brink, T. L., Rose, T. L., Lum, O., Huang, V., Adey, M., & Leirer, V. O. (1982). Development and validation of a geriatric depression screening scale: A preliminary report. *Journal of Psychiatric Research, 17*(1), 37–49.

Zhu, T., Xue, J., & Chen, S. (2019). Social support and depression related to older adults' hypertension control in rural China. *The American Journal of Geriatric Psychiatry, 27*, 1268–1276.

Zigmond, A. S., & Snaith, R. P. (1983). The hospital anxiety and depression scale. *Acta Psychiatrica Scandinavica, 67*(6), 361–370.

Zulman, D. M., Pal Chee, C., Wagner, T. H., Yoon, J., Cohen, D. M., Holmes, T. H., Ritchie, C., & Asch, S. M. (2015). Multimorbidity and healthcare utilisation among high-cost patients in the US Veterans Affairs Health Care System. *BMJ Open, 5*, e007771.

Determination of Decisional Capacity 144

Matthew W. Grover, Amina Z. Ali, and Debra A. Pinals

Contents

Abstract

Determination of decisional capacity is a task commonly requested of psychiatrists by medical colleagues in consultation settings and may be required when pursuing guardianship or in other legal contexts, though all physicians regularly complete such assessments as part of their day-to-day clinical work. Decisional capacity is a requirement for the ability to give informed consent for treatment or to refuse it. One commonly used framework for the assessment of decisional capacity encompasses examination of an individual's abilities across four domains: ability to communicate a choice, factual understanding of the relevant information, appreciation of one's situation

M. W. Grover · D. A. Pinals (✉)
Department of Psychiatry, University of Michigan, Ann Arbor, MI, USA
e-mail: groverma@med.umich.edu;
dpinals@med.umich.edu

A. Z. Ali
University of Toronto, Centre for Addiction and Mental Health, Toronto, ON, Canada
e-mail: Amina.Ali@camh.ca

© Springer Nature Switzerland AG 2024
A. Tasman et al. (eds.), *Tasman's Psychiatry*,
https://doi.org/10.1007/978-3-030-51366-5_66

and the possible consequences of the decision, and the ability to rationally manipulate information and reason out treatment options. Several exceptions to informed consent exist, and both the medical community and the legal system have developed models for addressing the care and treatment of individuals unable to give informed consent. For patients who refuse treatment when involuntarily civilly committed, rights-driven and treatment-driven approaches exist that highlight balancing treatment-goals and respect for patient autonomy. Strategies to optimize and support patient decision-making in advance can provide opportunities for patients to have input into their treatment goals during episodes of subsequent incapacity. Psychiatrists can also be asked to assess capacity regarding non-medical decisions, which may require special consideration of the pertinent legal issues associated with the task at hand.

Keywords

Capacity · Informed consent · Advance directives · Alternate decision maker · Guardianship · Supported decision-making · Right to refuse treatment · Testamentary capacity

Introduction

Psychiatrists are regularly asked to assess a patient's "competency" to understand a proposed medical or surgical intervention, particularly when the patient is refusing the recommended intervention. The term competent is used in different scenarios, but generally, it refers to some minimal mental, cognitive, or behavioral ability, trait, or capability required to perform a particular jural act, or to assume some legal role (Rajput & Bekes, 2002). Although the word is often used by health care professionals, competency is a legal decision requiring a judicial determination regarding an individual's capacities. Common law systems, such as the United States, presume that individuals are competent unless there has been

a legal judgment to the contrary. Although there may be jurisdictional differences, a psychiatrist's assessment of a patient's ability to perform the clinical task of making a medical decision about their treatment generally is referred to as an assessment of decisional capacity.

A psychiatrist's capacity determination may have an ethical impact on the patient's autonomy if there is a finding that the patient lacks capacity, as it means that a patient is unable to provide informed and valid consent. An attempt to honor patient decisions is the default position. Yet, when a patient is unable to provide informed consent, the psychiatrist may have ethical obligations to pursue an alternative course for decision-making to occur. In some instances, the psychiatrist may need to take actions that override the patient's wishes. Depending on the clinical setting or jurisdiction, the need for a judicial determination of an alternate decision maker may be required. For psychiatric patients, efforts have been made to optimize patient autonomy, including the use of psychiatric advance directives and supported decision-making.

This chapter will review the elements of a decisional capacity assessment and describe how it forms the foundation for informed consent. It will also provide an overview of commonly encountered ways in which the lack of decisional capacity can be addressed, including the use of advance directives, guardianship, and models for treating psychiatric inpatients refusing treatment. Lastly, examples in which a psychiatrist may be asked to provide input into an individual's capacity to make non-medical decisions will be provided.

Decisional Capacity Assessments

In most instances, decisional capacity assessments are a task-specific determination. Examples encompass a broad spectrum of proposed interventions, including laboratory tests, medications, and surgical procedures. Some capacity assessments include an evaluation of decision-making related to a procedure and the extensive amount of aftercare required, such as an evaluation that occurs prior to organ transplantation. Psychiatrists are also often asked to determine a patient's ability to leave the hospital against medical advice. Because different interventions require different abilities and understanding, capacity determinations may vary in the level of scrutiny required as part of the assessment. Therefore, clarification of the capacity question from the treatment team is imperative. Additionally, once the clarification of the capacity question is made, the psychiatrist may require additional information to better understand the clinical condition and the proposed intervention, as well as its risks and the risks of foregoing the intervention. In some instances, the treatment team's participation in the assessment may be necessary.

Decisional capacity is based upon a patient's understanding of their underlying medical condition, its prognosis, the intervention proposed by the treatment team, the consequences of treatment, and the consequences of no treatment. It should account for a patient's autonomy, their needs, and their values. Appelbaum and Grisso developed the best-established method of making a determination of an individual's medical decision-making (Appelbaum & Grisso, 1988; Appelbaum, 2007). The four prongs for consideration in a capacity determination include assessing the patient's ability to communicate their choice, to factually understand the relevant information, to appreciate their situation and the possible consequences, and to reason about treatment options and arriving at their decisions. In order for a patient to demonstrate decisional capacity, they must be able to demonstrate all four elements. Some have raised concerns that this method of assessing decision-making capacity is over-reliant upon cognitive abilities and does not fully take into account patient values and cultural issues, but they still recognize this model as the prevailing one espoused and utilized in practice (Palmer & Harmell, 2016).

Practically speaking, decisional capacity is typically questioned when a patient refuses a recommended intervention, as it is not often questioned when a patient agrees with the proposed interventions of the treatment team. However, consideration of a patient's capacity should be a part of a clinician's assessment when any treatment intervention is considered and is a required element of informed consent.

In communicating a choice, the patient must demonstrate a consistent preference over time that enables the treatment intervention to ensue. Patients who cannot communicate a consistent preference regarding a decision are said to lack decisional capacity. To be clear, a patient should not be precluded from changing their mind, though if their decision is vacillating widely to the point of being unable to move forward, or if they are unable or unwilling to communicate a choice at all, this may indicate a lack of capacity in and of itself. Also, communication need not be verbal.

The assessment of a patient's factual understanding of the relevant information is the second prong of a capacity determination. The patient's understanding of the information can be explored by asking them about their condition, the treatment options that have been provided, the risks and benefits of each treatment option, the treatment option recommended by the primary team, and the overall prognosis with treatment. The patient should also understand the prognosis of declining all forms of treatment. If a factual understanding of the relevant information is not demonstrated, the clinician is tasked with determining if the person is able to learn the information. In providing the patient with the relevant information, the primary treatment team may need to be brought into the assessment in order to educate the patient, and if necessary, assist in the identification of education materials consistent with the patient's capabilities. In such situations, it is essential that the consulting psychiatrist be present to observe the interaction between the patient and their treatment team, looking for opportunities to optimize the patient's capabilities. Optimization strategies can include simplifying written information, audiovisual or multimedia programming, test/feedback techniques, or extended discussions. Additionally, special attention should be taken to address limitations in literacy or limited English proficiency, including the use of an interpreter by phone or video teleconference (Schenker et al., 2011). Cultural considerations should also be part of this assessment. A lack of decision-making capacity may be determined if the team has made attempts to inform the patient of the relevant information, and the patient remains unable to learn and retain it.

The third prong of a capacity assessment is whether or not the patient is able to appreciate the significance of the information provided to them by their treatment providers. Such an assessment of information requires the patient to evaluate the implications of their treatment decision, including the ramifications of accepting the recommended treatment, requesting an alternative treatment, or refusing treatment. An individual's appraisal of the information often requires a comparison of statistical findings, including the likelihood of efficacy and rates of side effects with each treatment option. Patients with psychiatric symptoms such as delusions, for example, may be found to be unable to appreciate their situation because they cannot appreciate that they have a mental illness for which the treatment is recommended.

The final prong in determining decisional capacity involves the assessment of the patient's ability with regard to the rational manipulation of information provided to the patient. The clinician should focus on how the decision fits into the patient's prior preferences, decisions, and values and not on the actual decision itself. There are instances in which some decisions appear irrational, though with further inquiry, are the result of long-standing beliefs. The patient's decision to refuse certain interventions related to longstanding religious beliefs is different from the decision to refuse a medical intervention secondary to paranoid delusions. Collateral information from family, close associates, or a patient's primary care provider may be essential in determining whether certain beliefs are longstanding. A clinician should be aware of their individual values, particularly when working to respect a patient with differing values and wishes.

A standardized psychiatric interview, which should include a workup for an underlying neurocognitive disorder or other medical conditions as indicated, should form the basis of any decisional capacity assessment. Examples of commonly asked questions that are part of a decisional capacity assessment are summarized in Table 1 (Appelbaum, 2007). A thorough history is imperative to developing an understanding of the patient, including their psychiatric history, medical history, recent changes in treatment, and the timeline of symptoms associated with their current presentation. Reviewing the hospital

Table 1 Commonly asked questions to assess decisional capacity (from Appelbaum, 2007)

Capacity prong	Examples of questions to assess the respective prong
Communicate a choice	• Can you tell me what decision you have decided to make? • Can you tell me why you are having difficulty making a decision?
Understand the relevant facts	• Can you explain to me what the treatment team has told you about your health? • What treatment have they recommended? • What are the possible benefits of the treatment? • What are the possible side effects of the treatment? • What are the possible risks of treatment? • What other alternative treatments are available? What are the benefits and risks of the alternative treatments? • What are the risks and benefits if you decide not to treat the condition?
Appreciate the situation and its consequences	• What do you think is wrong with your health right now? • Do you think that you need treatment for your health? • What do you think the treatment will do for you? • Why do you think that the treatment will have that effect? • What do you think will happen if you are not treated? • Why do believe that your doctor has recommended this treatment?
Rational manipulation of information	• How did you come to the decision to either accept or reject the recommendation of your treatment team? • What makes the option that you have chosen better than the option the treatment team has recommended?

chart, speaking with the referral source, and talking with staff working with the patient can all be effective ways of obtaining information to identify contributing factors to their current presentation. Collateral information can be a valuable resource in a decisional capacity assessment. Although a psychiatrist must not disclose patient-specific information to a family member without consent from a person with the capacity to do so, obtaining information from family can help to determine the patient's long-standing beliefs and preferences. Additionally, collateral sources assist in ascertaining any changes in the patient's mental state. If it is determined that a patient lacks decisional capacity, the psychiatrist should identify the alternate decision maker to inform them of the current contributing factors to the patient's impaired capacity.

Physical exam findings and laboratory testing can be informative in determining the etiology of the patient's presentation, including the identification of a primary neurological condition or underlying medical condition. Physical exam findings such as focal or lateralized neurological signs are more consistent with primary neurological disorders, which can help guide consideration for neuroimaging or the need for input from the neurology consultation service. Other physical exam findings, such as scleral icterus or asterixis, may help guide laboratory testing. A review of an individual's vital signs over the course of their hospitalization may also help to identify emerging signs and symptoms of substance withdrawal or other medical etiologies such as an infection (Kruse et al., 2017).

Laboratory testing can help to identify causes of delirium; identify medical conditions that may contribute to psychosis, mood changes, or changes in cognition; and rule out the contribution of substances or medication toxicities to the patient's presentation. Commonly ordered tests include serum electrolytes, complete blood count, thyroid function tests, liver function tests, calcium, creatinine, urinalysis, urine culture, therapeutic drug levels, a blood alcohol level, and a urine drug screen. Typically, lumbar punctures and neuroimaging are limited to those patients with ongoing cognitive impairment or delirium

where the etiology remains unclear. Electroencephalograms (EEG) are used for patients who may appear delirious to rule out nonconvulsive status epilepticus or subclinical seizures. EEGs may also be used to confirm certain forms of delirium due to metabolic or infectious causes that present with characteristic EEG patterns (Kruse et al., 2017).

Common Psychiatric Conditions Impacting Decisional Capacity

There are several common psychiatric diagnoses that contribute to incapacity, including cognitive disorders, substance use disorders, and psychotic disorders. Mood disorders, including bipolar and depressive disorders, can contribute to incapacity, especially when an individual is experiencing a mood state rather than during interepisode recovery (Boettger et al., 2015). In a seminal paper examining differences in decision-making related to medical decisions comparing patient populations, persons with schizophrenia, followed by those with depression, fared worse on an assessment of capacity in comparison with patients with cardiac conditions (Grisso & Appelbaum, 1995). Although on the flip side, many patients in these cohorts did not lack decision-making capacity, which highlights the critical importance of the recognition of individualized assessments even within vulnerable populations. Examples of specific conditions where decisional capacity may be compromised are reviewed below.

Although persons with any number of diagnoses may have their capacity called into question, a few conditions will be highlighted below given they represent some of the more commonly encountered conditions in consultation-liaison psychiatry. For example, delirium, which occurs at high levels in both medical and surgical inpatients, is commonly encountered on a consultation-liaison psychiatry service. It is often under-diagnosed but is a main source of decisional capacity consults (Young & Inouye, 2007). Delirium is defined as a disturbance in attention, cognition, and awareness that is a change from the patient's baseline, develops over a short period of time, and can fluctuate in severity over the course of the day (American Psychiatric Assocation, 2022). Delirium is also defined by the fact that the disturbance is a direct physiological consequence of another medical condition, substance intoxication, withdrawal, exposure to a toxin, or multiple etiologies (American Psychiatric Association, 2022). Delirium can be superimposed on an underlying neurocognitive disorder or other psychiatric condition, complicating the overall clinical picture (Boettger et al., 2015, 2016). Identification and treatment of the underlying medical condition results in the improvement of the patient's mental status. Lucid intervals during the hospital course with a return of intact decisional capacity can provide an opportunity to identify treatment goals, patient preferences, advance directives, and health care proxies. Such information can be valuable if the patient's delirium worsens, they lose decisional capacity, and the medical team and alternate decision maker are again facing circumstances that require significant decisions (Bourgeois et al., 2019). Serial assessments of individuals with delirium may be required given their fluctuating cognitive status because a decisional capacity assessment is time and task specific. A patient can lack capacity at one point in time but may be found to regain capacity at a different moment in time. Again, however, the inability to sustain an expressed choice may be in and of itself sufficient to determine incapacity.

Major and Mild Neurocognitive disorders are a second group of conditions that can be associated with compromising individual decisional capacity. Several studies have demonstrated that individuals with neurocognitive disorders score lower on standardized assessment tools of decisional capacity (Appelbaum, 2010; Karlawish et al., 2005). As part of a decisional capacity assessment, psychiatrists should consider including either a Mini-Mental State Examination (MMSE), Montreal Cognitive Assessment (MoCA), or similar tool (Bourgeois et al., 2019). Patients with a superimposed delirium on an underlying neurocognitive disorder may improve their ability to participate in the decision-making

process after successful treatment of the underlying conditions leading to delirium.

Substance use disorders may also lead to impacted decisional capacity. Substance intoxication, withdrawal, and substance-induced delirium may all contribute to the patient's incapacity. If possible, delaying the assessment of decisional capacity until the effects of the substances have resolved can assist in differentiating limited insight, poor judgment, and cognitive deficits in the capacity decision-making process (Hazelton et al., 2003). Co-occurring mood, psychotic, or neurocognitive disorders should be considered when the patient's incapacity persists beyond the period of intoxication, withdrawal, or delirium, as the rates of decisional incapacity in individuals with substance use disorders not actively withdrawing or intoxicated are low (Jeste & Saks, 2006).

Psychotic disorders can contribute to decisional incapacity, though psychiatrists should take the time to remind the patient's treatment team that an individual's psychiatric diagnosis is not the equivalent of incapacity, as was also highlighted in the Grisso and Appelbaum study noted above (1995). In other words, although patients with these conditions may be more likely to have decisional impairments, an individual assessment is required as many will be able to make particular decisions without difficulty. Studies looking at the severity of psychopathology and decisional capacity demonstrated that the psychopathology correlation with impaired capacity was much lower than the correlations looking at cognitive performance and incapacity (Dunn et al., 2007; Palmer & Savla, 2007).

Informed Consent

A patient's decisional capacity is one of three components essential to informed consent. Informed consent is the process by which a health care provider discloses appropriate information about a treatment to a competent patient, allowing that patient to make a voluntary choice for or against treatment. Informed consent is required

Table 2 Components of informed consent

Component	Description
Decisional Capacity	The patient must be able to communicate a choice, understand the relevant information, appreciate the situation and its consequences as it pertains to them, and rationally manipulate the relevant information.
Knowing/ Disclosure	The patient must be engaged in a conversation with the provider in which they obtain the required information regarding the proposed treatment.
Voluntary	The patient must make an autonomous decision without evidence of coercion.

before any intervention is made. A medical intervention made without informed consent is considered battery because it is the unauthorized touching of another person. The three components of informed consent are capacity, disclosure, and voluntariness (Table 2). When a patient lacks one of the three elements, they are unable to provide informed consent. For example, when a patient does not have decisional capacity, the autonomy of that patient is limited (Appelbaum et al., 1987). Similarly, when a patient lacks the required information about a treatment intervention or is unable to make a voluntary decision, informed consent has not been completed. Informed consent standards vary depending on the legal standards specific to the region (Shah et al., 2020).

Knowing/Disclosure

In order to make a decision about a proposed treatment, the patient must possess the required information regarding that treatment or procedure. Obtaining informed consent requires engagement and communication between the provider and the patient. Conversations to obtain informed consent should be accurately documented in the patient chart, and if necessary, depending on the jurisdiction and the type of intervention (e.g., laboratory studies v. surgery), a consent form may need to be signed. Separate from the technical administrative elements, it has been observed that the best method of obtaining informed consent includes

ongoing conversations with the patient to discuss the treatment in question.

There are three commonly used legal standards regarding the necessary disclosure to provide informed consent: the subjective patient standard, the reasonable person standard, and the reasonable medical practitioner standard. Some jurisdictions use a combination of these methods to obtain informed consent (King & Moulton, 2006).

The subjective patient standard includes information that the individual patient would need to know and understand to make an informed decision about the specific treatment. This standard takes into consideration the unique circumstances of that patient. It has been a difficult standard to operationalize, but it is used in some jurisdictions today (Shah et al., 2020). The subjective patient standard addresses particular moral issues that are raised by the two standards described below in that it takes into consideration the different values and experiences of the unique individual making the decision (Dransieka et al., 2017). This standard has been called the preferred moral standard and has been said to protect the individual's rights (Dransieka et al., 2017).

The reasonable medical practitioner standard is best defined by the Kansas Supreme Court case *Natanson v. Kline* (1960). Irma Natanson, the patient, filed for damages after suffering radiation burns from cobalt radiation therapy following a mastectomy. Ms. Natanson stated that she was not informed by Dr. Kline, the radiologist, of the risks of treatment before providing consent. Ms. Natanson had argued that a physician had a duty to make a full disclosure to the patient of all matters within their knowledge effecting the interests of the patient. The Kansas Supreme Court determined that the requested instruction was too broad and that a physician has the duty to disclose only what a reasonable medical practitioner would disclose under similar circumstances. This standard is also known as the clinician-focused standard (Schouten et al., 2016).

The reasonable person standard was defined by the United States Court of Appeals for the District of Columbia Circuit case, *Canterbury v. Spence* (1972), and is used by a substantial minority of states. The patient, Jerry Canterbury, underwent a back surgery by Dr. Spence; however, he was not told about potential risks of the surgery, including paralysis. After the surgery, Mr. Canterbury suffered paralysis in the lower half of his body and incontinence. Mr. Canterbury sought damages for negligence, stating that there was a failure to disclose a risk of serious disability. The Court held that it was a physician's duty to disclose the choices with respect to proposed treatment and the risks involved in that treatment. This standard is also known as the patient-focused standard (Pinals, 2019b).

Regardless of the standard relevant to the particular jurisdiction, there are several categories of generally acceptable information that should be disclosed to the patient when obtaining informed consent. This information includes the diagnosis and nature of the condition, the treatment and benefits of the proposed treatment, the risks and likelihood of possible risks, the inability to know the outcome of treatment, the potential irreversibility of treatment, and the expected risks and benefits of alternatives to treatment or no treatment (American Medical Association, 2017).

Voluntary

Informed consent requires that a treatment decision is given voluntarily by a patient. It is necessary that the patient be able to comprehend the information discussed regarding the treatment and be able to use this information to make an autonomous decision without evidence of coercion (Del Carmen & Joffe, 2005). Limitations can arise in those with mental illnesses, cognitive deficits, and prisoners, to name a few (Tori et al., 2020; Moser et al., 2004). Patients in long-term care facilities, mental health facilities, and prisons are often deemed unable to consent to treatment based on the unequal power between the individual and the institution (Silva et al., 2017). The issues associated with the ability to give a voluntary choice in these populations were raised in *Kaimowitz v. Department of Mental Health for the State of Michigan* (1973), a Michigan Circuit Court case in Wayne County. "John Doe" was a patient who was involuntary committed to a state hospital

under the criminal sexual psychopath laws that existed in Michigan at the time. Drs. Rodin and Gottlieb had proposed a study for uncontrollable aggression that included psychosurgery. Mr. Doe provided informed consent to participate in the study and have the procedure. The plaintiff's attorney, Gabe Kaimowitz, became aware of the study, and a suit was filed. The courts found that it was impossible for an involuntarily committed patient to give consent voluntarily. Since that case, there has been ongoing considerations of this issue. In environments where patients are held involuntarily, coercion by staff or quid quo pro type statements such as "If you agree to this treatment, you will be discharged," should be shifted to "This treatment can help you with symptoms. Your symptoms are contributing to behaviors that make it unsafe to discharge you." In this way, if there is concern for the voluntariness of a patient's choice, adequate time should be taken to explain the treatment/procedures, risks and benefits of the treatment, alternatives to the treatment, risks and benefits of alternatives and accurately assess the patient's understanding of these elements (Tori et al., 2020). If it is determined that the patient is unable to make a voluntary decision, the relevant procedures should be undertaken for an alternate decision maker to be put into place (Shah et al., 2020).

Exceptions to the Informed Consent Requirement

There are several exceptions to the requirements of informed consent. These exceptions include an incapacitated patient, certain emergency medical situations, voluntary-waived consent, and therapeutic privilege.

In emergency situations, not obtaining informed consent is acceptable when it is deemed that failure to treat would result in serious and detrimental deterioration of the patient's condition. In this situation, it is acceptable to treat the patient until they stabilize, at which time informed consent should be obtained. If the physician is aware that the patient's prior wishes would include not receiving treatment, then there may be other considerations that would shift the balance of proceeding with treatment when legally permissible. If at any time the patient is deemed to have capacity to make a decision even in an emergency, their choice, including their choice to refuse treatment, must be upheld (Evans, 2006).

Exceptions to informed consent also arise when it is unclear whether a patient is able to make a decision on their own behalf. In these situations, a psychiatrist may be asked to determine capacity. When a patient is deemed unable to make a decision, depending on the region, an alternate decision maker will be dedicated to making these decisions. If an alternate decision maker cannot be found, a court may appoint a legal guardian (Shah et al., 2020). In those instances in which a patient already has a guardian or an alternate decision maker (e.g., a health care proxy has already been invoked), then the alternate decision maker would give the informed consent on behalf of the incapacitated person.

A voluntary-waived consent is a third exception to informed consent. Patients may voluntarily waive the informed consent process if they are found to be competent to make that waiver decision to forego the informed consent. The assessment of the individual's ability to waive informed consent should be documented (Applebaum et al., 1987) especially as a patient could subsequently argue that they lacked capacity to waive the informed consent and that they would have wanted further discussion before treatment was initiated.

A waiver based upon therapeutic privilege is the fourth exception to informed consent that is frequently described in the literature but in practice should be invoked only rarely. In this situation, the physician would forego informed consent based on an assessment that the consent process itself would worsen the patient's condition. In some instances where this exception is invoked, the physician may obtain consent from an alternate decision maker (Murray, 2012), but this should be legally allowable in the particular jurisdiction. Also, the rationale for foregoing the informed consent for the particular patient should be well documented in the medical record. The therapeutic exception to informed consent should

not be used solely for the reason that the patient may refuse the proposed treatment (Menon et al., 2021). Therapeutic privilege has been viewed as creating conflict between the provider's obligations and allowing for patient autonomy, and its use by physicians has decreased over time.

Informed consent involving children is the final exception. The age at which children can provide informed consent can vary depending on the jurisdiction and the treatment being offered (e.g., psychiatric or substance use treatment), and some jurisdictions do not provide a specific age and instead use maturity as the standard. When a child or adolescent is unable by law to consent to treatment, the parents or guardians provide consent on the child or adolescent's behalf (Evans, 2006).

Advance Directives and Supported Decision-Making

When patients are unable to provide informed consent, there arises the need for alternate decision makers. Unsurprisingly, decisional incapacity in medical settings is common. It is estimated that rates of decisional incapacity reach nearly 40% in adult medical inpatients and residential hospice patients, and it exceeds 90% in intensive care units (Raymont et al., 2004; Sorger et al., 2007; Cohen et al., 2005). Patients who lack decisional capacity require the identification of an alternate decision maker to provide authorization or refusal of medical interventions.

The process by which an alternate decision maker is identified can occur with or without a judicial determination. An advance directive, a document that is created at a time when an individual is competent, can identify an alternate decision maker and/or provide guidance in making decisions when an individual loses decisional capacity at a future time. Examples of advance directives include a health care proxy and a durable power of attorney for healthcare.

The use of advance directives evolved in the context of several prominent court cases of individuals who survived anoxic brain injury but were left in a persistent vegetative state that required artificial nutrition, hydration, and constant care (DeMartino et al., 2017). In the case of *Cruzan v. Director, Missouri Department of Health* (1990), the family of Nancy Cruzan attempted to remove life support after she had been in a persistent vegetative state for several years. The hospital refused in the absence of a court order, and the case reached the United States Supreme Court. In its decision, the Supreme Court ruled that it was constitutional for the State of Missouri to require "clear and convincing evidence" that the decision to terminate life support was what Ms. Cruzan would have wanted were she competent. The Court noted that such a standard was appropriate given the competing interests of the patient's liberty and the state's interest in persevering life.

Following the *Cruzan* case, the Patient Self-Determination Act of 1990 (1995) was passed and signed into law. Since then, hospitals have been legally required to provide patients with information regarding advance directives and to inquire as to whether a patient has an advance directive at the time of hospital admission. Despite these efforts, rates of completed advance directives in the United States population ranges between 20% and 37% (Larson & Eaton, 1997; Rao et al., 2014; Yadav et al., 2017). If a patient does not have an advance directive and lacks decisional capacity, states vary in their procedures for the appointment and contesting of alternate decision makers. For example, the New York Family Health Care Decisions Act creates an alternate decision maker hierarchy, which starts with a legal guardian and then progresses to the spouse or domestic partner, adult child, parent, adult sibling, and lastly a close friend (New York State Public Health Law Article 29-CC, 2010). At least 41 states address the appointment of a default decision maker for at least some health decisions; however, the variability from state to state creates challenges for both patients and healthcare providers. The lack of a uniform national standard can lead to confusion for patients who may live in one state but receive care in a neighboring one. Additionally, healthcare systems that operate in several states may require state-specific knowledge when incapacitated patients require alternate decision makers. Clinicians on the frontlines

would do well working with their local legal staff to understand the rules in their jurisdiction.

Similar high-risk decisions, such as evaluations associated with physician assisted dying and euthanasia, are a rapidly evolving legal landscape. Several states plus the District of Columbia have state statutes or case law around the issues associated with physician assisted death. Some states direct the attending or consulting physician to refer patients for a mental health assessment as part of the evaluation process. Psychiatrists completing such evaluations should follow the legal parameters and requirements of their local jurisdiction should they choose to complete them (Anfang et al., 2017). In Canada, an assessment is carried out consistent with the medical assistance in dying (MAID) law. Specific criteria must be met for a person to qualify for MAID and the assessment is carried out by two independent practitioners (Health Canada, 2021).

In an effort to provide increased autonomy for psychiatric patients, the development of Psychiatric Advance Directives (PADs) has been encouraged. PADs provide competent patients with mental illnesses an opportunity to document their care preferences in times of crisis. Preferences can include the consent or refusal for treatment and authorization for a decision-making proxy. Since the 1990s, over half of the states have enacted legislation that allows for advance mental health care planning (Easter et al., 2017). Although there is interest among patients in PADs, prior studies have indicated that rates of PAD completion ranged between 4% and 13% (Swanson et al., 2006). Barriers from both providers and patients on their understanding of PADs and how to implement them continue to need to be tackled to further their use (Swartz et al., 2021). Optimally, PADs provide patients increased autonomy during periods of decompensation and permit them to communicate preferences obtained during a period in which the individual possessed decisional capacity, thereby optimizing care consistent with their wishes. PADs may also improve communication between patients, psychiatrists, and supports (Murray & Wortzel, 2019).

Several challenges related to PADs exist. Patients may struggle to identify an alternate decision maker due to a lack of family or supports (Joshi, 2003). Additionally, providers may worry that a patient could create a treatment plan that could preclude psychiatric care, including medications, or create a plan incompatible with appropriate standards of care for their condition. Most states allow a healthcare provider to go to court to override the PAD.

Hargrave v. Vermont (2003) outlines other challenges associated with PADs, albeit relevant in only limited jurisdictions. Nancy Hargrave had executed a durable power of attorney, the Vermont equivalent of a PAD, which forbade her alternate decision maker from consenting to any psychiatric medication. When she was hospitalized at Vermont State Hospital, she was medicated in a non-emergent situation and subsequently sued, claiming that her advance directive was ignored due to her mental illness in violation of the Americans with Disability Act (ADA). The State of Vermont argued that mental illness in an involuntarily committed individual was not protected by the ADA because the individual posed "a direct threat," but both the trial court and the United States Court of Appeals for the Second Circuit found in favor of Ms. Hargrave. Although this case is not applicable in a majority of court jurisdictions in the United States, the possibility exists that a patient could create an advance directive that would preclude psychotropic medication and successfully argue in court that it should not be overturned. Still initial fears that this case would have a chilling effect on PADs has not borne out, and the use of PADs is actually being fostered as a way to help patients determine the treatments that they will more likely agree to rather than refuse.

Supported decision-making has been proposed as another alternative to provide psychiatric patients with increased autonomy and offer self-redirected, respectful, person-centered care. Unlike shared decision-making, it assumes that individuals make their own decisions but with the supports needed to make and communicate them (Pinals, 2019a). Originally developed in Canada and Australia for individuals with intellectual and developmental disabilities, literature supporting supported decision-making for persons with mental illness has showed promise,

indicating a need for further study (Browning et al., 2014).

Guardianship

Full guardianship is a responsibility appointed by the Court that gives power to another person to make decisions for an incapacitated person regarding all aspects of their lives including medical care, finances, and property (Chamberlain et al., 2018). A judicial determination is made that the person lacks the capacity to make decisions on their behalf. Most commonly, guardianship is sought in individuals who have a mental, physical, or developmental disability. Laws regarding guardianship for persons with intellectual disability often differ from guardianship laws related to other conditions. Guardianship can also be sought in the elderly who can no longer make decisions regarding property or finances. Guardianship can protect incapacitated individuals in making decisions; however, it can also take away fundamental rights related to the autonomy of that individual, and thus it is pertinent that thorough assessments are conducted prior to appointing guardianship (Moye et al., 2007).

Guardianship laws vary across jurisdictions, and it is pertinent to understand the laws in the region where a psychiatrist resides when seeking guardianship of another individual (Moye et al., 2007). In the United States, each state has specific duties outlining what decisions a guardian can and cannot make. Often guardianship provisions aim to limit the areas of the authority of the guardian, yielding so-called limited guardianships. For example, an individual may be deemed incompetent to make a decision about finances but not about medical decisions. Also, other decision-making provisions may be looked to that would delimit the guardianship issues. In some regions, if a Power of Attorney is in place, a guardian cannot revoke it (American Bar Association Commission on Law and Aging, 2018). In several jurisdictions, guardianship laws explicitly state that a guardian cannot consent to an individual's admission to a mental health facility without following the jurisdiction's formal civil commitment processes. However, in other states, a guardian can consent to having an individual "voluntarily" hospitalized in a psychiatric hospital. Lastly, in some states, a guardian can consent to hospitalization as long as the incapacitated person assents or does not object (Boldt, 2015).

There are different types of guardianship depending on the needs of the individual. Plenary guardianship is sought when an individual is generally unable to make decisions on their own behalf due to the inability to understand or communicate their decisions regarding care, finances, or property (New York Surrogate Court Procedure Act § 1750, 2016). The specific definition of incompetence or the standards for assigning a guardian will be delineated in state statute. In these situations, the guardian is appointed plenary guardianship and is responsible for making all decisions for the incapacitated individual. This type of guardianship is known to be the most restrictive form of guardianship (Moye & Naik, 2011). On the other hand, a limited guardianship grants power to make specific decisions for an individual, and these powers are specified by the court order. The individual will lack some capacity to make decisions in specific areas related to safe self-care, medical decisions, finances, or property, however, still will retain capacity to make other decisions. Limited guardianships are less restrictive but can be more complex to implement (Moye et al., 2007). The use of limited guardianships allows for persons to retain some autonomy. When a limited guardianship is put into place, it is imperative that all parties know and understand what decisions can be made by the individual themselves and those that require decision by the guardian.

Other types of guardianship orders exist, which depend on the jurisdiction. A temporary guardian may be appointed in an emergency to ensure the incapacitated individual receives necessary care or protection (American Bar Association Commission on Law and Aging, 2018). A testamentary guardian is a person named by the parent of a minor or an adult with special needs in the event of the death of the parent(s). This type of guardian would be appointed by the court at the time of a parent's death, and the court will ensure that the parent's choice was appropriate. Successor guardians are individuals appointed to guardianship if a previously appointed guardian passes away, resigns, or becomes incapacitated (Geller, 2014).

The Uniform Probate Code (UPC) was developed by the National Conference of Commissioners on Uniform State Laws (NCCUSL) to create consistency in decision-making regarding inheritances, guardianship, durable powers of attorney, and trust administration across states. It was first drafted in 1969 and revised in 1990. Currently, eighteen states have adopted all or part of the UPC. As a result, inconsistencies in statutes across states remain (Lowder et al., 2008). In Canada, each province has laws specific to adult guardianship. The Enforcement of Canadian Judgments and Decrees Act (ECJDA), passed in 2003, allows one province to register an order from a different province, which could include guardianship, allowing enforcement of laws between provinces. Although this process appears to be available, it has not been enforced and may still allow for gaps between jurisdictions (British Columbia Law Institute, 2005). The Uniform Adult Guardianship and Protective Proceedings Jurisdiction Act (UAGPPJA), an act also drafted by the NCCUSL and approved in 2007, addresses inconsistencies and dilemmas between states relating to guardianship. Guardianship can involve several jurisdictions, which leads to complexities and difficulties across state-lines. The UAGPPJA was implemented to allow for improved communication between jurisdictions. The Act helps to outline processes and procedures for the initial jurisdiction of guardianship, transfer of guardianship, and recognition of guardianship across states, communication issues, and emergencies. As of 2018, 47 states had adopted this act (American Bar Association Commission on Aging, 2018).

Right to Refuse Treatment

Rules governing the right to refuse treatment and the procedural protections around treatment refusal have evolved over the last 50 years as people were committed to psychiatric institutions but then declined medications while there (Pinals & Mossman, 2012). The rules that govern treatment refusal vary across jurisdictions, requiring psychiatrists to become familiar with the regulations, statutes, and case law in the area in which they practice. Across all jurisdictions, voluntary patients cannot be treated against their will unless they are found to be imminently dangerous to themselves or others in an emergency or in a non-emergency context when there are other legal provisions in place to authorize non-emergency involuntary administration of medications. Additionally, involuntary hospitalization due to dangerousness does not establish incompetence to refuse medication. In jurisdictions requiring a judicial finding of incompetence to refuse medication, a bifurcation in the processes of involuntary civil commitment and involuntary treatment exists. Practically speaking, a patient may refuse psychiatric treatment following involuntary psychiatric admission and require a separate judicial or administrative determination to treat the patient over their objection absent an emergency.

Two different legal approaches have emerged in the adjudication of treatment refusal: the so-called treatment-driven model and the rights-driven model. The treatment-driven model emphasizes the need for timely, quality patient care, while the rights-driven model emphasizes a patient's right to object to the proposed treatment (Appelbaum, 1994). As the case law has evolved in these areas, several different arguments have emerged, including the right to free speech by way of the First Amendment; the right to be free of cruel and unusual punishment by way of the Eighth Amendment; the right to due process by way of the Fourteenth Amendment; and the right to privacy derived from the First, Fourth, Fifth, and Ninth Amendments (Ciccone et al., 1990; Hermann, 1990).

Rights-driven models, which emphasize patient autonomy and the right to determine treatment, are at their core driven by informed consent and the right for a competent patient to refuse treatment. It is not an "anti-treatment" model, but rather places the emphasis of decision-making overrides onto objective legal authorities to maximize the rights of the individual whose decision-making is being questioned. In jurisdictions that follow this framework, a formal, legal determination of incompetency is held, and after a judicial determination of the incompetency, the treatment decision is placed in the hands of a guardian or the courts (Appelbaum, 1994; Appelbaum & Gutheil, 2007). Commonly in rights-driven models, the decision is placed in the hands of an objective decision maker, who attempts to decide in the best interests of the patient, though in some jurisdictions the judge will make a substituted judgment decision.

Rogers v. Commissioner of Mental Health (1983), considered the landmark case for rights-driven models of treatment refusal, established an alternative, more rigorous standard for treatment refusal referred to as substituted judgment as determined by a judge. Rubie Rogers brought a class action lawsuit to the Massachusetts Federal District Court in the name of all involuntarily committed patients at Boston State Hospital refusing psychotropic medication. She argued for a constitutional right to refuse treatment. The decision in *Rogers* affirmed a constitutional right to refuse treatment for an involuntarily committed patients unless the patient has been adjudicated incompetent by the court. If the judge determines that the patient is incompetent, the judge must then decide what the incompetent patient would have chosen if they had been competent. *Rogers* outlined the six main factors that a judge has to account for in making a substituted judgment determination: the patient's expressed preferences, any religious convictions, the impact on family of the decision, the probability of side effects of the proposed medications, the prognosis with treatment, and the prognosis without treatment, along with any other relevant factors the judge may wish to consider (Rogers v. Commissioner of Mental Health, 1983). A determination of the patient's wishes can be challenging if there are no family or prior records to determine the patient's preferences when previously competent. Additionally, challenges arise when an individual has never been competent secondary to an underlying condition such as an intellectual disability.

Although the rights-driven approach has been criticized for resulting in delayed treatment, longer hospitalizations, and increased use of court resources, patient rights advocates contend that such protections maintain a patient's rights and respect their autonomy (Biswas et al., 2018).

In contrast, the treatment-driven model relies on professional judgement and approaches the issue from the angle that civil commitment is intended to bring about treatment and is not simply for the purposes of confinement. In a treatment-driven model, the standard for an override of a patient's refusal is placed in the hands of other professionals, including treating psychiatrists, independent psychiatric evaluators, medical boards, or multidisciplinary boards. If the treatment is determined to be appropriate, then the treatment is allowed and can proceed over the patient's objection.

Rennie v. Klein (1983) is the landmark case that exemplifies deference to medical decision makers and the treatment-driven approaches. John Rennie had been involuntarily civilly committed to a New Jersey state psychiatric hospital and objected to being prescribed antipsychotic medication. Mr. Rennie sued, claiming that involuntarily committed patients had a right to refuse antipsychotic treatment. The United States Court of Appeals for the Third Circuit found a constitutional right to refuse treatment existed. However, they stated that the administrative procedures established by the state that required hospital administrative review prior to forcible administration of medications absent an emergency provided adequate patient protection and satisfied the due process requirements. It was also noted that the treatment offered should be the least restrictive treatment alternative available. Psychiatrists generally favor treatment-driven models for adjudicating treatment refusal, as the models emphasizes the role of clinical decision-making in treatment and reduces the opportunities for patients to go without treatment (Pinals et al., 2017). Additionally, such models minimize clinician time in settings (e.g., court) that divert from their patient-care duties.

As noted, in both the rights-driven and treatment-driven models, there are exceptions that allow for interventions of an emergency basis. Most often, the exception to treatment occurs when a patient demonstrates a risk of serious, imminent harm for which no less restrictive intervention to emergency administration of medication is appropriate to the situation.

Other Types of Capacity

Testamentary Capacity

Testamentary capacity is the ability of a person to a make a will outlining how they want their assets to be distributed after death. There are four elements of testamentary capacity that must be present for a will to be valid. These elements require the testator to (1) know the nature and extent of the property, (2) have knowledge of those who would be in line to inherit the property should the testator die, (3) communicate the person who they would like the property to go to should they die, and (4) be able to connect these elements to formulate a plan. The testator must be found to be free of any "insane delusion" from a disorder of the mind at the time of making the will that would directly negate the will (Kennedy, 2012). Laws regarding testamentary capacity also vary depending on jurisdiction; however, most laws internationally are based on the 1870 case, *Banks v. Goodfellow* (1870). In this case, the testator, who had a diagnosis of schizophrenia, was found to have testamentary capacity as he was able to understand and execute a plan and his illness did not impact this ability (Kennedy, 2012). The important factor of this case was that a diagnosis alone does not preclude a person's capacity to make a valid will.

If a person is suffering from a mental disorder or cognitive dysfunction, an expert, such as a forensic psychiatrist, may be asked to assess testamentary capacity. An important factor in making a will is to ensure that the testator was not unduly pressured by any outside sources, or in other words, was not under undue influence, at the time of making the will (Skidmore, 2018). Undue influence occurs when the beneficiary of the undue influence has the ability to impact the mind of the testator, ultimately for the benefit of the person exerting the undue influence (Regan & Gordon, 1997). Consideration of undue influence is particularly important in those suffering from cognitive decline or delusions of the mind. If undue influence were found to be present at the time the will was made, it would void the will, even if the testator had testamentary capacity (Skidmore, 2018). Other factors that should be addressed in assessing testamentary capacity include any dramatic changes from prior wills, the appreciation of distribution especially when natural beneficiaries are excluded, and any delusions that may have impacted the decisions (Shulman et al., 2005).

Testamentary capacity assessments may also be requested after a person has died, and the will is challenged by natural heirs. A testamentary capacity assessment in this instance is retrospective in nature. The psychiatrist completing the assessment will use collateral information including medical records, prior cognitive tests, psychiatric history, financial documents, videos, and forms of electronic communication as well as any other information that may help address the question. In rendering retrospective opinions, it can be pertinent for the assessor to express that they have not evaluated the testator in person (Shulman et al., 2005). Common issues that arise from a retrospective evaluation include failure to presume competence, failure to obtain independent information, focusing on diagnosis versus the task of making a will, misapplying cognitive testing, and misapplying known delusions (Gutheil, 2007). A judge will have the final opinion on the individual's testamentary capacity, and the psychiatrist is often called on as an expert witness to provide evidence on the individual's ability to make a will (Skidmore, 2018).

Capacity to Drive

In some regions, the capacity to drive is assessed in relation to mental illness and cognitive disorders. Although not specifically related to decisional capacity, laws regarding assessments and mandatory or permissive reporting to driving

licensing authorities vary depending on the jurisdiction. With an aging population facing cognitive and medical issues, there is a need to attend to when patients may be precluded from driving (Chihuri et al., 2016). Clinicians must consider their assessment of the patient as well as the particular jurisdiction's requirements related to what might be needed if a patient is potentially an impaired driver (Morgan, 2018). Also, if an individual's decision-making has been removed and assigned to a guardian, a guardian can decide to not allow the individual the keys to the car. Otherwise, a person may need a capacity assessment if there are concerns that their decision-making with regard to driving is impaired. The family physician in this population may have the ability to form an opinion with regard to capacity to decide to drive and even the separate capacity to drive; however, studies have shown that most physicians are not confident in making these decisions (Wolfe & Lehockey, 2016). In some jurisdictions, such as those in Canada, a person with mental illness displaying acute symptoms of their illness is reportable to the ministry of transportation or department of motor vehicles. The entity is then required to investigate this report and make a decision about the person's license and ability to drive. This investigation may include ongoing feedback from the person's physician (Canadian Counsel of Motor Transport Administration, 2021). In the United States, laws regarding capacity to drive vary between states. However, the Uniform Vehicle Code states that a person's driver's license may be denied if the state finds that a person "by reason of physical or mental disability would not be able to operate a motor vehicle with safety upon the highways" (National Committee on Uniform Traffic Laws and Ordinances, 2000; Wood, 2016).

Conclusion

The determination of capacity involves the interaction of clinical, ethical, and legal considerations and is pertinent throughout medicine and psychiatry. It is important for providers to consider an individual's specific capacity related to the task at hand in addition to global capacity issues when relevant. The process of assessing decisional capacity includes ensuring the patient has knowledge of the treatment or task, has appreciation of the decision and situation, is making a voluntary decision, and is able to express their choices. In the event that the patient lacks decision-making capacity, the provider must engage in the process to seek an alternate decision maker, which can include the appointment of a guardian or invocation of previously executed advanced directives. In cases of involuntary treatment of psychiatric inpatients, court orders to override non-emergency rejection of psychiatric medications may or may not be required depending on the jurisdiction. Psychiatrists are also often tasked with evaluating capacity in other areas, including the making of wills. Psychiatrists must be aware of the laws in the jurisdiction in which they practice to understand local issues around capacity determinations, the role of the courts, and their impacts upon patient care.

References

American Bar Association Commission on Law and Aging. (2018). *State adult guardianship legislation summary: Directions of reform.* https://www.americanbar.org/content/dam/aba/administrative/law_aging/2018-adult-guardianship-legislative-summary.pdf

American Medical Association, Counsel on Ethical and Judicial Affairs. (2017). 2.1.1 Informed consent. In *Code of medical ethics of the American Medical Association.* American Medical Association. https://www.ama-assn.org/delivering-care/ethics/informed-consent

American Psychiatric Association. (2022). *Diagnostic and statistical manual of mental disorders* (5th ed., text rev.). American Psychiatric Association. https://doi.org/10.1176/appi.books.9780890425787

Anfang, S., Bonnie, R., Brendel, R., Chen, D., Datta, V., Gandhi, T., Hoge, S. K., & Weinstock, R. (2017). *APA resource document on physician assisted death.* American Psychiatric Association.

Appelbaum, P. S. (1994). *Almost a revolution: Mental health law and the limits of change.* Oxford University Press.

Appelbaum, P. S. (2007). Assessment of patients' competence to consent to treatment. *New England Journal of Medicine, 357*(18), 1834–1840. https://doi.org/10.1056/NEJMcp074045

Appelbaum, P. S. (2010). Consent in impaired populations. *Current Neurology Neuroscience Reports, 10*(5), 367–373. https://doi.org/10.1007/s11910-010-0123-5

Appelbaum, P. S., & Grisso, T. (1988). Assessing patients' capacities to consent to treatment. *The New England Journal of Medicine, 319*(25), 1635–1638. https://doi.org/10.1056/NEJM198812223192504

Appelbaum, P. S., & Gutheil, T. G. (2007). *Clinical handbook of psychiatry and the law* (4th ed.). Lippincott Williams and Wilkins.

Appelbaum, P. S., Lidz, C. W., & Meisel, A. (1987). *Informed consent: Legal theory and clinical practice.* Oxford University Press.

Banks v. Goodfellow, 1870, LR5 QB 549.

Biswas, J., Drogin, E. Y., & Gutheil, T. G. (2018). Treatment delayed is treatment denied. *The Journal of the American Academy of Psychiatry and the Law, 46*(4), 447–453. https://doi.org/10.29158/JAAPL.003786-18

Boettger, S., Bergman, M., Jenewein, J., & Boettger, S. (2015). Assessment of decisional capacity: Prevalence of medical illness and psychiatric comorbidities. *Palliative & Supportive Care, 13*(5), 1275–1281. https://doi.org/10.1017/S1478951514001266

Boettger, S., Bergman, M., Jenewein, J., & Boettger, S. (2016). Advanced age and decisional capacity: The effect of age on the ability to make health care decisions. *Archives of Gerontology and Geriatrics, 66,* 211–217. https://doi.org/10.1016/j.archger.2016.06.011

Boldt, R. C. (2015). The "voluntary" inpatient treatment of adults under guardianship. *Villanova Law Review, 60*(1), 1–52.

Bourgeois, J. A., Tiamson-Kassab, M., Sheehan, K. A., Robinson, D., & Zein, M. (2019). *Resource document on decisional capacity determinations in consultation-liaison psychiatry: A guide for the general psychiatrist.* American Psychiatric Association.

British Columbia Law Institute. (2005, March). *Report on the recognition of adult guardianship orders from outside the province.* http://www.bcli.org/sites/default/files/AGO_Outside_Province_Rep.pdf

Browning, M., Bigby, C., & Douglas, J. (2014). Supported decision making: Understanding how its conceptual link to legal capacity is influencing the development of practice. *Research and Practice in Intellectual and Developmental Disabilities, 1*(1), 34–45. https://doi.org/10.1080/23297018.2014.902726

Canadian Council of Motor Transport Administrators. (2021, February). *National safety code standard 6: Determining driver fitness in Canada.* https://ccmta.ca/web/default/files/PDF/National%20Safety%20Code%20Standard%206%20-%20Determining%20Fitness%20to%20Drive%20in%20Canada%20-%20February%202021%20-%20Final.pdf

Canterbury v. Spence 464 F.2d. 772 (D.C. Cir. 1972).

Chamberlain, S., Baik, S., & Estabrooks, C. (2018). Going it alone: A scoping review of unbefriended older adults. *Canadian Journal on Aging, 37*(1), 1–11. https://doi.org/10.1017/S0714980817000563

Chihuri, S., Mielenz, T. J., Dimaggio, C. J., Betz, M. E., DiGuiseppi, C., Jones, V. C., & Li, G. (2016). Driving cessation and health outcomes in older adults. *Journal of the American Geriatrics Society, 64*(2), 332–341. https://doi.org/10.1111/jgs.13931

Ciccone, J. R., Tokoli, J. F., Clements, C. D., & Gift, T. E. (1990). Right to refuse treatment: Impact of Rivers v

Katz. *Bulletin of the American Academy of Psychiatry and the Law, 18*(2), 203–215.

Cohen, S., Sprung, C., Sjokvist, P., Lippert, A., Ricou, B., Baras, M., Hovilehto, S., Maia, P., Phelan, D., Reinhart, K., Werdan, K., Bulow, H., & Woodcock, T. (2005). Communications of end-of-life decisions in European intensive care units. *Intensive Care Medicine, 31*(9), 1215–1221. https://doi.org/10.1007/s00134-005-2742-x

Cruzan v. Director, Missouri Department of Health, 497 U.S. 261, 110 S. Ct. 2841 (1990).

Del Carmen, M. G., & Joffe, S. (2005). Informed consent for medical treatment and research: A review. *The Oncologist, 10*(8), 636–641. https://doi.org/10.1634/theoncologist.10-8-636

DeMartino, E. S., Dudzinski, D. M., Doyle, C. K., Sperry, B. P., Gregory, S. E., Siegler, M., Sulmasy, D. P., Mueller, P. S., & Kramer, D. B. (2017). Who decides when a patient can't? Statutes on alternate decision makers. *The New England Journal of Medicine, 376*(15), 1478–1482. https://doi.org/10.1056/NEJMms1611497

Dransieka, V., Piasecki, J., & Waligora, M. (2017). Relevant information and informed consent in research: In defense of the subjective standard. *Science and Engineering Ethics, 23*(1), 215–225. https://doi.org/10.1007/s11948-016-9755-4

Dunn, L. B., Palmer, B. W., Appelbaum, P. S., Saks, E. R., Aarons, G. A., & Jeste, D. V. (2007). Prevalence and correlates of adequate performance on a measure of abilities related to decisional capacity: Differences among three standards for the MacCAT-CR in patients with schizophrenia. *Schizophrenia Research, 89*(1–3), 110–118. https://doi.org/10.1016/j.schres.2006.08.005

Easter, M. M., Swanson, J. W., Robertson, A. G., Moser, L. L., & Swartz, M. S. (2017). Facilitation of psychiatric advance directives by peers and clinicians on assertive community treatment teams. *Psychiatric Services, 68*(7), 717–723. https://doi.org/10.1176/appi.ps.201600423

Evans, K. G. (2006). *Consent: A guide for Canadian physicians.* https://www.cmpa-acpm.ca/en/advice-publications/handbooks/consent-a-guide-for-canadian-physicians

Geller, B. (2014). *Handbook for guardians of adults* (11th ed.). https://www.berriencounty.org/DocumentCenter/View/13001/Guardian-Handbook-PDF

Grisso, T., & Appelbaum, P. S. (1995). The MacArthur treatment competence study. III: Abilities of patients to consent to psychiatric and medical treatments. *Law and Human Behavior, 19*(2), 149–174. https://doi.org/10.1007/BF01499323

Gutheil, T. G. (2007). Common pitfalls in the evaluation of testamentary capacity. *The Journal of the American Academy of Psychiatry and the Law, 35*(4), 514–517.

Hargrave v. Vermont, 340 F.3D 27 (2d Cir. 2003).

Hazelton, L. D., Sterns, G. L., & Chisholm, T. (2003). Decision-making capacity and alcohol abuse: Clinical and ethical considerations in personal care choices.

General Hospital Psychiatry, 25(2), 130–135. https://doi.org/10.1016/s0163-8343(03)00005-7

Health Canada. (2021, June 18). *Medical assistance in dying.* Government of Canada. https://www.canada.ca/en/health-canada/services/medical-assistance-dying.html

Hermann, D. H. (1990). Autonomy, self-determination, the right of involuntarily committed persons to refuse treatment, and the use of substituted judgment in medication decisions involving incompetent persons. *International Journal of Law and Psychiatry, 13*, 361–385.

Jeste, D. V., & Saks, E. (2006). Decisional capacity in mental illness and substance use disorders: Empirical database and policy implications. *Behavioral Sciences & the Law, 24*(4), 607–628. https://doi.org/10.1002/bsl.707

Joshi, K. G. (2003). Psychiatric advance directives. *Journal of Psychiatric Practice, 9*(4), 303–306.

Kaimowitz v. Department of Mental Health for the State of Michigan, No.; 73–19434-AW, Michigan Circuit Court, Wayne County (1973).

Karlawish, J. H., Casarett, D. J., James, B. D., Xie, S. X., & Kim, S. Y. (2005). The ability of persons with Alzheimer disease (AD) to make a decision about taking an AD treatment. *Neurology, 64*(9), 1514–1519. https://doi.org/10.1212/01.WNL.0000160000.01742.9D

Kennedy, K. M. (2012). Testamentary capacity: A practical guide to assessment of ability to make a valid will. *Journal of Forensic and Legal Medicine, 19*(4), 191–195. https://doi.org/10.1016/j.jflm.2011.12.029

King, J. S., & Moulton, B. W. (2006). Rethinking informed consent: The case for shared medical decision-making. *American Journal of Law and Medicine, 32*(4), 429–501.

Kruse, J. L., James, M. L., & Guze, B. H. (2017). Medical assessment and laboratory testing in psychiatry. In B. J. Sadock, V. A. Sadock, & P. Ruiz (Eds.), *Kaplan and Sadock's comprehensive textbook of psychiatry* (Vol. 1, 10th ed., pp. 1923–1956). Wolters Kluwer Health.

Larson, E. J., & Eaton, T. A. (1997). The limits of advanced directives: A history and assessment of the patient self-determination act. *Wake Forest Law Review, 32*(2), 249–293.

Lowder, J. L., Buzney, S. J., & Montoni, L. M. (2008). Uniform probate code. In S. Loue & M. Sajatovic (Eds.), *Encyclopedia of aging and public health.* Springer. https://doi.org/10.1007/978-0-387-33754-8_435

Menon, S., Entwistle, V., Campbell, A. V., & Van Delden, J. M. J. (2021). How should the 'privilege' in therapeutic privilege be conceived when considering the decision-making process for patients with borderline capacity? *Journal of Medical Ethics, 47*, 47–50. https://doi.org/10.1136/medethics-2019-105792

Morgan, E. (2018). Driving dilemmas: A guide to driving assessment in primary care. *Clinics in Geriatric Medicine, 34*(1), 107–115. https://doi.org/10.1016/j.jcger.2017.09.006

Moser, D. J., Arndt, S., Kanz, J. E., Benjamin, M. L., Bayless, J. D., Reese, R. L., Paulsen, J. S., & Flaum, M. A. (2004). Coercion and informed consent in research involving prisoners. *Comprehensive Psychiatry, 45*(1), 1–9. https://doi.org/10.1016/j.comppsych.2003.09.009

Moye, J., & Naik, A. D. (2011). Preserving rights for individuals facing guardianship. *JAMA, 305*(9), 936–937. https://doi.org/10.1001/jama.2011.247

Moye, J., Wood, S., Edelstein, B., Armesto, J. C., Bower, E. H., Harrison, J. A., & Wood, E. (2007). Clinical evidence in guardianship of older adults is inadequate: Findings from a tri-state study. *Gerontologist, 47*(5), 604–612. https://doi.org/10.1093/geront/47.5.604

Murray, B. (2012). Informed consent: What must a physician disclose to a patient? *AMA Journal of Ethics, 14*(7), 563–566. https://doi.org/10.1001/virtualmentor.2012.14.7.hlaw1-1207

Murray, H., & Wortzel, H. S. (2019). Psychiatric advance directives: Origins, benefits, challenges, and future directions. *Journal of Psychiatric Practice, 25*(4), 303–307.

Natanson v. Kline, 350 P.2d 1093, 186 (Kan. 1960).

National Committee on Uniform Traffic Laws and Ordinances. (2000). *Uniform vehicle code.* https://iamtraffic.org/wp-content/uploads/2013/01/UVC2000.pdf

New York State Public Health Law, Article 29-CC, (2010). https://www.nysenate.gov/legislation/laws/PBH/A29-CC

New York Surrogate's Court Procedure Act § 1750, (2016). https://www.nysenate.gov/legislation/laws/SCP/1750

Palmer, B. W., & Harmell, A. L. (2016). Assessment of healthcare decision-making capacity. *Archives of Clinical Neuropsychology, 31*(6), 530–540. https://doi.org/10.1093/arclin/acw051

Palmer, B. W., & Savla, G. N. (2007). The association of specific neuropsychological deficits with capacity to consent to research or treatment. *Journal of the International Neuropsychological Society, 13*(6), 1047–1059. https://doi.org/10.1017/S1355617707071299

Patient Self-Determination Act of 1990, 42 USC 1395 cc (a); final rule: 60 CFR 123 at 33294 (1995).

Pinals, D. A. (2019a). *Beyond the borders: Lessons from the international community to improve mental health outcomes.* National Association of State Mental Health Program Directors. https://www.nasmhpd.org/sites/default/files/Paper1International_508C.pdf

Pinals, D. A. (2019b). Informed consent: Canterbury v. Spence (1972). In P. Ash (Ed.), *From Courtroom to clinic: Legal cases that changed mental health treatment* (pp. 69–83). Cambridge University Press.

Pinals, D. A., & Mossman, D. (2012). *Evaluation for civil commitment.* Oxford University Press.

Pinals, D. A., Nesbit, A., & Hoge, S. K. (2017). Treatment refusal in psychiatric practice. In R. Rosner & C. L. Scott (Eds.), *Principles and practice of forensic psychiatry* (3rd ed., pp. 155–164). CRC Press.

Rajput, V., & Bekes, C. E. (2002). Ethical issues in hospital medicine. *The Medical Clinics of North America, 86*(4), 869–886. https://doi.org/10.1016/s0025-7125(02)00013-5

Rao, J. K., Anderson, L. A., Lin, F. C., & Laux, J. P. (2014). Completion of advance directives among U.S. consumers. *American Journal of Preventative Medicine, 46*(1), 65–70. https://doi.org/10.1016/j.amepre.2013.09.008

Raymont, V., Bingley, W., Buchanan, A., David, A. S., Hayward, P., Wessely, S., & Hotopf, M. (2004). Prevalence of mental incapacity in medical inpatients and associated risk factors: Cross-sectional study. *Lancet, 364*(9443), 1421–1427. https://doi.org/10.1016/S0140-6736(04)17224-3

Regan, W. M., & Gordon, S. M. (1997). Assessing testamentary capacity in elderly people. *Southern Medical Journal, 90*(1), 13–15. https://doi.org/10.1097/00007611-199701000-00003

Rennie v. Klein, 720 F.2D 266 (3rd Cir. 1983).

Rogers v. Commissioner of Department of Mental Health, 458 N.E.2D 308 (Mass. 1983).

Schenker, Y., Fernandez, A., Sudore, R., & Schillinger, D. (2011). Interventions to improve patient comprehension in informed consent for medical and surgical procedures: A systemic review. *Palliative & Supportive Care, 31*(1), 151–173. https://doi.org/10.1177/0272989X10364247

Schouten, R., Edersheim, J. G., & Hidalgo, J. A. (2016). Legal and ethical issues in psychiatry I: Informed consent, competency, treatment refusal, and civil commitment. In T. Stern, M. Fava, T. Wilens, & J. Rosenbaum (Eds.), *Massachusetts General Hospital comprehensive clinical psychiatry* (2nd ed., pp. 912–920). Elsevier.

Shah, P., Thornton, I., Turrin, D., & Hipkind, J. E. (2020). *StatPearls: Informed Consent.* https://www.ncbi.nlm.nih.gov/books/NBK430827/

Shulman, K. I., Cohen, C. A., & Hull, I. (2005). Psychiatric issues in retrospective challenges of testamentary capacity. *International Journal of Geriatric Psychiatry, 20*(1), 63–69. https://doi.org/10.1002/gps.1257

Silva, D. S., Matheson, F. I., & Lavery, J. V. (2017). Ethics of health research with prisoners in Canada. *BMC Medical Ethics, 18*(1), 31. https://doi.org/10.1186/s12910-017-0189-6

Skidmore, B. (2018). Litigating the presumption of undue influence based on confidential or fiduciary relations. *Michigan Bar Journal, 937*(11), 34–37.

Sorger, B. M., Rosenfeld, B., Pessin, H., Timm, A. K., & Cimino, J. (2007). Decision-making capacity in elderly, terminally ill patients with cancer. *Behavioral Sciences & the Law, 25*(3), 393–404. https://doi.org/10.1002/bsl.764

Swanson, J., Swartz, M., Ferron, J., Elbogen, E., & Van Dorn, R. (2006). Psychiatric advance directives among public mental health consumers in five U.S. cities: Prevalence, demand, and correlates. *Journal of the American Academy of Psychiatry and the Law, 34*(1), 43–57.

Swartz, M. S., Swanson, J. W., Easter, M. M., & Robertson, A. G. (2021). Implementing psychiatric advance directives: The transmitter and receiver problem and the neglected right to be deemed incapable. *Psychiatric Services, 72*(2), 219–221. https://doi.org/10.1176/appi.ps.202000659

Tori, K., Kalligeros, M., Shehadeh, F., Khader, R., Nanda, A., Van Aalst, R., Chit, A., & Mylonakis, E. (2020). The process of obtaining informed consent to research in long term care facilities (LTCFs). *Medicine, 99*(21). https://doi.org/10.1097/MD.0000000000020225

Wolfe, P. L., & Lehockey, K. A. (2016). Neuropsychological assessment of driving capacity. *Archives of Clinical Neuropsychologist, 31*(6), 517–529. https://doi.org/10.1093/arclin/acw050

Wood, E. F. (2016). Evaluating the capacity to drive. *Bifocal, 37*(4), 79–81.

Yadav, K. N., Gabler, N. B., Conney, E., Kent, S., Kim, J., Herbst, N., Mante, A., Halpern, S. D., & Courtright, K. R. (2017). Approximately one in three US adults completes any type of advance directive for end-of-life care. *Health Affairs, 36*(7), 1244–1251. https://doi.org/10.1377/hlthaff.2017.0175

Young, J., & Inouye, S. K. (2007). Delirium in older people. *British Medical Journal, 334*(7598), 842–846. https://doi.org/10.1136/bmj.39169.706574.AD

Mari Lloyd-Williams and Oscar Rodriguez Mayoral

Contents

Abstract

Palliative care aims to support patients with life-limiting and terminal illnesses to have the best possible quality of life by expertly assessing and managing every symptom, which includes physical and psychosocial symptoms. In this chapter, the integral link between palliative care and pain and psychiatry is explored, in relation to Anxiety, Depression, Delerium, Bereavement, and Pain, including chemical coping and the wish to hasten death. Using clinical vignettes and the literature, the reader is guided as to how to assess patients with palliative care needs and how to manage the psychological and psychiatric symptoms which palliative care patients may present. The focus is largely on the cancer patient as cancer patients access palliative care services more frequently than other patients living with life-limiting illnesses; however, everything in this chapter is equally applicable to patients with other life-limiting conditions, e.g., end-stage neurological, cardiovascular, respiratory, and hepato-renal disease. Palliative care

M. Lloyd-Williams (✉)
Department of Primary Care and Mental Health/Liverpool Marie Curie Hospice and Liverpool Health Partners, University of Liverpool, Liverpool, UK
e-mail: mlw@liverpool.ac.uk

O. R. Mayoral
Palliative Care Unit, Cancer Institute of Mexico, Mexico City, Mexico

© Springer Nature Switzerland AG 2024
A. Tasman et al. (eds.), *Tasman's Psychiatry*,
https://doi.org/10.1007/978-3-030-51366-5_7

is constantly developing and supporting more and more patients and families, and the support of the psychological as well as the physical is as important now as it always has been.

Keywords

Palliative care · Assessment · Management · Pain · Anxiety · Depression · Delerium · Bereavement · Chemical coping

Introduction

There are only two certainties in life for every human, and that is that all are born and at some time all will die. In Western society, death usually happens in old age, and until the COVID-19 pandemic, when increasing numbers of people died at home, death usually occurs in hospitals. Death is seen as something so separate to life and not as part of it, and any connection, including palliative and end-of-life care, is seen as something that must be avoided at all costs. In the United Kingdom, for example, over 60% of adults have not made a will. Thinking about end-of-life wishes includes not only who will receive prized possessions but also where and how you wish to be cared for at the end of life. Palliative care affirms life and neither hastens nor prolongs death.

Palliative care is often thought of as the new specialty founded in the 1960s by Dame Cicely Saunders (Carter, 2003), the founder of the internationally known St Christopher's hospice in London; however, physicians throughout the ages have always aimed to relieve the suffering of their patients. Some of the earliest hospices were located in religious institutions where patients were cared for by religious orders of Nuns. Our Lady's hospice in Harold's Cross Dublin Ireland is one of the earliest hospices in Ireland, and St Joseph's Hospice in London is the earliest in England. To palliate means to cloak or to hide, and palliative care aims to hide or cloak any distressing symptom to alleviate the suffering of patients, including those living with life-limiting illnesses and those at the end of life.

Although there is clearly an emphasis on the palliation of physical pain and symptoms such as nausea and breathlessness, there is also an emphasis on the palliation and relief of emotional pain within palliative care. Psychiatry has and does play a crucial part in supporting patients, staff, and volunteers working in palliative care by assessing patients with severe distress and also providing emotional support to staff and volunteers, along with bereavement support to relatives. Modern-day palliative care is delivered in any setting where the patient is located, which can include the family home, care home, hospital, and hospice. Although there has been considerable debate as to the relatively scant attention paid to palliative care teaching within both the medical and nursing curricular, palliative care skills are essential within every clinical setting. It is also important to note that palliative care is not exclusive to those patients with cancer, and indeed, palliative care supports patients with life-limiting neurological diseases, e.g., Motor Neuron Disease and Multiple Sclerosis, and patients with life-limiting conditions such as end-stage heart failure, end-stage renal failure, and end-stage chronic obstructive pulmonary disease.

In this chapter, the common problems patients present in palliative care and their relation to psychiatry are reviewed, highlighting clinical issues using clinical case vignettes.

Palliative Care Referral and Assessment

Cancers are increasingly curable, and for cancers such as breast and non-melanoma skin cancers, survival can be measured in decades. Successful treatment and prolonged survival depend on early detection and diagnosis of cancer. There is strong evidence that patients from lower SE backgrounds have later diagnoses and at a more advanced stage due to a lack of knowledge regarding symptoms and a lack of access to services. Even when cancer is diagnosed with curable intent, palliative care can still support patients. For example, a patient with severe pain due to a bone tumor could benefit from prompt referral to palliative care services to optimize pain relief while they are receiving chemotherapy/radiotherapy. However, for some cancers, e.g., lung/gastrointestinal cancers, by the time the patient has presented to medical services and received a diagnosis, the cancer is often metastatic, and treatment is aimed at prolonging the quality and quantity of life, but treatment cannot cure cancer. Such patients benefit from prompt referral to palliative care services where a multidisciplinary team can support the patient in terms of physical symptoms, for example, pain, breathlessness, and nausea. The palliative care team can provide aid for the home via the occupational therapist and assistance with mobility and maximizing function with support from the physiotherapy team. For many patients with cancer, there are many associated costs for example in attending appointments, paying for extra heating, and supporting with childcare for those with young children, and the social work team can support patients in applying for benefits in some countries or for charitable support.

A palliative care consultation focuses on the symptoms that the patient reports and on setting goals for improving the patient's quality of life. Pain is a frequent symptom, and many patients and some professionals believe that opiates should not be used until the very end of life. Opiates when prescribed and titrated for pain can be safely used for any patient and need not preclude patients from working, driving, and instead of preventing patients from leading a normal life, opiates allow a patient to lead a normal life. Opiates can often cause nausea and constipation; therefore, prescribing antiemetics and laxatives can be necessary. It is also important to remember that for patients who are receiving treatment, e.g., radiotherapy or chemotherapy to shrink the tumor, the pain will also reduce, and therefore opiate dose needs monitoring with a view of reducing the opiate requirement.

> Opiates when prescribed and titrated for pain can be:
>
> - safely used for any patients
> - need not preclude patients from working, driving
> - allow a patient to lead a normal life

Many patients present to palliative care with psychological distress. The diagnosis of a severe and life-limiting illness is a major psychosocial stressor, but for many patients with support from family, friends, information, and attending support groups, and meeting others in a similar situation, distress will reduce as they adjust to a new sense of normality. Palliative care not only offers support to the patient but also to their family and provides on-going support to families. For some patients, however, anxiety and depression can be enduring and problematic and require specialized support (see the sections on "Anxiety" and "Depression").

For some patients, their spiritual belief gives their life a sense of meaning and having a religious belief can give comfort and hope, as can the support received from being with others at religious meetings. However, the possession of a religious or spiritual belief can be challenged by the diagnosis of a serious life-limiting illness, leading to feelings of loss, betrayal, and abandonment. All palliative care teams have access to a spiritual leader, who may belong to a religious group or may be of no affiliation. Spiritual leaders are available to all patients, their families, and staff and offer support and can liaise with relevant faith/religious leaders if specific support or religious support is needed.

Anxiety

It is normal and understandable for any patient living with a life-limiting illness to be anxious; indeed, it would be abnormal for anxiety not to be present in such situations. However, for some palliative care patients, anxiety becomes all-pervasive and always present. Reasons for uncontrolled anxiety can include the fear of uncontrolled symptoms and of being left alone to die. Anxiety becomes a problem when its duration and severity exceed normal expectations. Patients who are very anxious may either be over-talkative and appear agitated or fidgety, tearful, or may be withdrawn, with little to say – it is important to remember that anxiety and depression frequently co-exist in palliative care patients, with about 10% of patients experiencing both symptoms (Wilson et al., 2007). When assessing a patient, it is important to know the patients' pre-morbid trait state – have they always been unduly anxious, and is this longstanding or related to the current diagnosis? Does the anxiety arise in certain situations, such as attending clinics or in the lead-up to chemotherapy or radiotherapy treatment? Anxiety can be contagious, and it is important also to meet with the family as their anxiety may be making the patient unduly anxious, or vice versa, the patients' anxiety may be making the family overly anxious. Alcohol withdrawal can also lead to symptoms of excessive anxiety, and it is important to take an alcohol history in palliative care as in any other clinical assessment. Stimulant drugs and other non-prescribed drug taking can also manifest as excessive anxiety in a palliative care patient. Patients with anxiety may complain of feeling their heart beating excessively quickly, having a constant dry mouth, intruding thoughts, not being able to sleep, not being able to relax, or even to sit down without having to get up and constantly pace or walk about. A thorough assessment will allow the clinician to know more about the patient and their anxiety, and a joint management plan with the patient and their family should be co-developed. Patients and their families should be given the time to talk about their worries and concerns, and an open and honest conversation alone can do much to alleviate anxiety. Uncontrolled pain, breathlessness, nausea, or any uncontrolled symptom serve as a reminder of the patients' illness and increase anxiety, with some patients believing that things will never improve symptomatically. Excellent attention to detail and the pharmacological management of physical symptoms are very important, as is involving the multidisciplinary team. Relaxation techniques and complementary therapies can help reduce anxiety, as can suggesting the patient engages or re-engages with a hobby or interest as diversional activities focus the mind, giving less space and time for intrusive thoughts and behaviors.

Anxious patients may be:

- over talkative
- appear agitated
- fidgety
- tearful
- withdrawn with little to say

It is important to remember that anxiety and depression frequently co-exist in palliative care patients, with about 10% of patients experiencing both symptoms (Wilson et al., 2007)

If pharmacological interventions are required, restoring sleep and a good sleep pattern is often the first step to improvement, and suggesting a patient try anti-histamines in the first instance, e.g., Chlorpherinamine 4–8 mg to aid restoration of sleep or a sleeping tablet, such as Temazepam 10–20 mg, may provide rapid benefits in improvement. If the anxiety is situational, e.g., attending for chemotherapy, Lorazepam 0.5–1 mg is effective. Citalopram at a dose of 10–20 mg is effective for the treatment of anxiety. Supporting the patient and the family and offering non-pharmacological interventions alongside is also essential. Zweers et al. (2019) found that patients who were unduly anxious within a hospice setting wanted their management to include information, open communication, regaining a sense of control, adequate symptom management, and respect for the patients' own coping strategy. Individualized care is essential. In some circumstances, anxiety is not amenable to treatment and the patients' distress escalates, causing issues such as not attending for treatment or not sleeping, referral to the liaison psychiatry team to assess the patients further and to give a management plan for the patient is required.

Clinical Vignette

Elsie was 71 and a retired English teacher. She had married in her 20s and had one

(continued)

daughter, but the marriage had ended in divorce, and Elsie had been fiercely independent throughout her life. Elsie did not retire until she was 68 and was looking forward to traveling the world and catching up with old friends. Elsie had been a lifelong smoker but gave up in her 50s. She was used to having chest infections, so when one winter she had four infections with hardly a break, she was not surprised, but her doctor sent her for a chest x-ray, and she was found to have a large mass in the right lung. An urgent referral to the respiratory team and further scans revealed local and metastatic spread to the bones, and Elsie was told the cancer was inoperable and she had a prognosis of months. She was offered palliative chemotherapy and radiotherapy. Elsie stopped sleeping; she could not think of anything but her illness; she dreaded being dependent on others and could not cope with the fact that she, who had done absolutely everything herself, may need some help. Elsie was awake all night, mostly pacing around the house, with neighbors commenting that the lights were always on. The day Elsie was due to have radiotherapy for the painful metastases in her spine, she missed the appointment – the doctor asked the nurse to visit, who could not get into the house but could see Elsie sleeping on sofa.

When Elsie eventually woke, she told the nurse it was the first time she had slept for more than a few minutes in a week. She could not keep still, said she felt her heart was beating so fast it would jump out of her chest, and that she was not eating as she felt ill every time she tried. The nurse accessed immediate referral to the palliative care team. They asked Elsie what was worrying her most - she replied that being a burden on anyone and the fact that already with pain in her spine she could hardly sit on the toilet. Elsie was prescribed Lorazepam 0.5–1 mg morning and night, was given another

appointment for radiotherapy to her spine, the occupational therapist visited and fitted a raised toilet seat which Elsie could use, and the hospice linked Elsie with a compassionate neighbor scheme so she had someone she could contact if she needed any help with shopping or to get to appointments. The following day, Elsie felt different – she had a good night's sleep, and she could sit on the toilet without pain, and the day after the compassionate neighbor took Elsie to her new radiotherapy appointment.

Elsie lived for another year, during which time, with her pain controlled and her feeling in control of her life, she visited friends, went on holiday, and felt more content, although sad at her diagnosis. Elsie died peacefully at home with her daughter and the support of the hospice and palliative care team.

Depression

As a clinician working in palliative care and with an interest in depression and coping, if I had received a £1 for every clinician who when referring a patient with depression had also written "this patient is understandably depressed" or "and I would also be depressed if I had their diagnosis," I would have been able to retire comfortably some years ago! Depression is one of the "orphan" symptoms of palliative care. Depression is rarely assessed and, therefore, rarely treated. It is known that depression in advanced cancer patients is common, difficult to assess (Li et al., 2012), and therefore treat. Depression magnifies physical symptoms (Lloyd-Williams et al., 2004; Grotmol et al., 2018) and therefore makes symptom management more difficult and the physical symptom burden for patients who are depressed is greater. Depression adversely affects the quality of life of patients and of their families and is an independent predictor of earlier death in palliative care patients (Lloyd-Williams et al., 2009, 2014). Additionally, research has also reported that depression in patients with advanced cancer is associated with longer hospital stay (Nipp et al., 2017) and that patients with advanced cancer who are also depressed have greater healthcare utilization (Mausbach & Irwin, 2017). Depression not only directly impacts an adverse quality of life and greater symptom burden for patients, impacts on quality of life of their relatives but also has significance for health and social care budgets.

> - Depression magnifies physical symptoms and therefore makes symptom management more difficult.
> - Physical symptom burden for patients who are depressed is greater.

The prevalence of depression varies according to different studies and the criteria used within studies to assess depression but the prevailing works suggest that between 1 in 3 and 1 in 4 patients with advanced cancer will have significant symptoms of depression to warrant further assessment and treatment. Within the palliative care consultation, the predominant reason for referring a patient for a palliative care assessment is a physical symptom, and patients are rarely referred for assessment due to depression alone. During the assessment, if depression is assessed, the predominant trend appears to be for physical symptoms to take priority. Additionally, many patients are reluctant to mention depression or low mood, feeling that they are in some way weak or would be perceived as not coping; therefore, even for a skilled clinician, assessing depression within a palliative care consultation is challenging.

Palliative care uses several assessment tools for physical symptoms, and some tools, e.g., the ESAS (Bruera et al., 1991), enquire about depression and anxiety. Using a rating tool to assess symptoms within palliative care can have many benefits apart from highlighting the presence of symptoms to the clinician. Enquiring about a symptom alerts the patient that the clinician is not totally focused on the physical symptoms and thereby gives patients permission to talk further about their low mood or depression and how it is affecting them. Using an assessment tool can also be helpful in assessing how a patient is responding to any treatment, be it a pharmacological or non-pharmacological intervention. The PHQ-9 is a nine-item self-report instrument which can be scored with validated cut-offs for mild (scores 5–9), moderate (scores 10–14), moderately severe (scores 15–19), and severe depression (scores 20–27) (Kroenke et al., 2001). The PHQ-9 has been validated in cancer populations (Thekkumpurath et al., 2011), with a threshold score of 10 used to identify a depression "case." The Brief Edinburgh Depression Scale (BEDS) (Lloyd-Williams et al., 2007) was developed from the Edinburgh Depression Scale (Cox et al., 1987) and excludes somatic symptoms of depression. The 6-item scale was developed and validated within the palliative care population and a cut-off of 6 suggests a patient may be depressed and requires further evaluation. Further work is being carried out to determine cut-off thresholds for mild, moderate, and severe depression on the BEDs. Clearly, alongside using an assessment tool, a clinical assessment of depression should include past history of depression, including postnatal depression, and patients should be asked about thoughts of self-harm or suicidal intent.

The management of depression in palliative care patients is the same as for other patients who present with depression. Antidepressant medication takes 2–3 weeks to show improvement, and patients and their families often need support and encouragement during this time to continue with the medication, especially for those patients who feel there is little point in doing anything or that nothing can or will improve how they are feeling. Some clinicians routinely prescribe a short dose of steroids for 1 to 2 weeks with the aim of improving well-being whilst the antidepressants are taking effect. Commonly, a regimen of Dexamethasone 4 mg for 1 week followed by Dexamethasone 2 mg for 1 week is prescribed, with regular review of the patient.

Common antidepressants to use within palliative care are Mirtazepine (15–45 mg od), which can be prescribed at night to aid sleep and also has the additional beneficial side effect for patients with advanced cancer in that it can promote weight gain. Sertraline (50–200 mg od) is frequently commenced for older patients, and Citalopram (10–40 mg od) is beneficial when there are symptoms of both anxiety and depression. Alongside antidepressant medication, excellent attention to physical symptom control and referral for complementary therapies or to diversional activities can also be helpful for patients and their families.

- The management of depression in palliative care patients is the same as for other patients present with depression.
- Antidepressant medication takes 2–3 weeks to show improvement.
- Patients and their families often need support and encouragement during this time.
- Some clinicians routinely prescribe a short dose of steroids for 1 to 2 weeks with the aim of improving well-being whilst the antidepressants are taking effect.

There have been several trials of non-pharmacological psychological interventions to support palliative care patients with depression. A Canadian Phase 3 trial of the CALM recruited 305 patients with stage 3 or 4 cancer, a mean PHQ-9 score of 7.45 (mild depression), and a prognosis of 12–18 months. CALM is a psychological intervention delivered by senior clinicians or Social work staff, and the trial reported that patients receiving CALM intervention had significant improvement in the severity of depressive symptoms at 3 months and a greater effect at 6 months compared with usual care (Rodin et al., 2018). A recent UK randomized controlled trial recruited 230 patients with advanced cancer and moderate depression from Oncology clinics, Primary Care and Hospice Services. Patients were randomized for up to 12 sessions of individual CBT over 3 months, or treatment as usual. The trial found no treatment effect for CBT over treatment as usual (Serfaty et al., 2020).

Our research group has developed a program of work to improve the assessment and management of depression in advanced cancer patients with limited prognosis. Interviewing advanced cancer patients with depression and exploring their views on acceptable interventions, a focused narrative intervention for depression was co-developed and refined. A pilot trial of this focused narrative intervention was carried out to determine if the intervention plus usual care, when compared to usual care alone, had an effect on depression (as measured by the PHQ-9) in advanced cancer patients with a prognosis of <12 months (Lloyd-Williams et al., 2018). Palliative care patients with advanced cancer with a prognosis of 12 months or less, attending hospice services, and scoring 10 or above on PHQ-9 (indicating moderate to severe depression) were randomized to a single session of face-to-face focused narrative intervention plus usual care or usual care alone and followed up at 2, 4, and 6 weeks. Of the 284 patients screened, 115 scored 10 or more on PHQ-9 and 57 patients (71% female) with a mean age of 65.1 years were randomized. The mean PHQ-9 score at baseline was 16.4, indicative of moderately severe depression. Sixteen patients died, and death accounted for 55% of the study attrition with being too unwell to complete follow-up as the predominant reason for remainder of attrition. Patients receiving the focused narrative intervention were more likely to have a five-point or more reduction (regarded as clinically significant) in PHQ-9 score, which was sustained at 6-week follow-up. The median survival of all patients in the pilot trial was 142 days, and patients receiving the intervention plus usual care survived a median of 157 days compared to a median of 102 days for usual care alone. These results suggest a focused narrative intervention in advanced cancer patients with limited prognosis and moderate to severe depression that can reduce depression and may extend life more than usual care alone. Currently, this intervention is being developed and trialed.

Depression in palliative care is common and can be difficult to assess and treat. Apart from a thorough assessment, it is important to involve psychiatric services if a patient has very severe depression, is displaying symptoms of psychotic depression, or is expressing suicidal thoughts. Patients with depression can also be challenging for hospice staff, as the control of physical symptoms can be difficult and they may decline to take medication. Hospice staff also benefit from support of the multidisciplinary team, both in managing the patient and for mutual staff support.

Clinical Vignette

Felix was 39 and the father of three small boys, the youngest not yet at school. He worked in a local office, played squash, and spent most of his time at home with his family. Always having had a "funny tummy," he did not take much notice of diarrhea he was having until one morning he woke up and bled profusely into the toilet. He was rushed to hospital, who at first felt he had a diverticular disease, but a colonoscopy revealed a large necrotizing tumor in the bowel. He was referred to the surgeon, and scans showed that there was no metastatic spread, and Felix was offered surgery. He recovered well and had 6 months of chemotherapy and was told he was clear of cancer but would be followed up for 5 years. Three years later, Felix felt more tired and lethargic and was persuaded by his wife to visit the doctor. Blood tests were arranged, and the results showed abnormal liver function and a low hemoglobin. Felix was referred to oncology, who suggested chemotherapy. Felix came home from the appointment and told his wife he felt "totally empty." He went to bed that night and stayed there. He slept intermittently but did not wash, ate very little despite what his wife tried to offer him, told his adored three boys not to come to the bedroom, and became totally withdrawn. His wife feared that he was dying,

and he was referred to palliative care. At the visit to the home, the palliative care doctor sat quietly by Felix's bed, and slowly he began to speak. He told of his total despair that after being told he was cured, the cancer had returned. He told of his terror of what would happen to him when his cancer "took hold," and he told of his utter devotion to his three small boys. He also told of his unbearable pain of leaving them and said as there was no hope he had already given up… to "get it over and done with as soon as possible for my wife and boys – the sooner I am gone the better." He denied suicidal intent but said he did not want to linger. The doctor realized that Felix was very depressed and liaised with the on-call psychiatry team. Both persuaded him to be admitted to the hospice for further assessment and also to give his wife and small children some respite, as the children were visibly upset and disturbed by what was happening.

At the hospice, Felix was gently encouraged to get up into the chair and to accept some help with personal care. He was further assessed by the psychiatrist at the hospice and commenced on Mirtazepine 30 mg nocte along with Dexamethasone 4 mg for 1 week then 2 mg for 1 week. He slowly started eating and walking and felt that, after getting out of bed, he started to feel better. Although he told the nurse that he was not "arty farty," he accepted complementary therapies, which he found relaxing, and also art therapy. During his first art therapy session, he cried for the whole hour, but slowly he began to paint and found he was quite good at art. His children visited, and after 16 days, Felix returned home, still low in mood but feeling he wanted to make the most of what time he had left. Felix received chemotherapy, which reduced the liver metastases sufficiently for him to have a liver resection. Two years later, Felix is

(continued)

well and back at work. The outcome for Felix would have been very different had the depression not been assessed and treated.

Delerium

Delirium can be defined as a state of disturbed consciousness and lack of attention with cognitive impairment. Delerium can have an acute onset and can be fluctuating and is due to physiological cause of disease or treatment. Delerium is common and can affect between 30 and 80% of hospice patients and can have many different presentations. The three main types of delerium are hyperactive when the patient is agitated, hypoactive when the patient is withdrawn and a mixed pattern. It is important to assess the patient and to take a careful history from family or friends who know the patient well.

- Delerium is common.
- It can affect between 30 and 80% of hospice patients.
- Delerium can have many different presentations.
  - Hyperactive when the patient is agitated.
  - Hypoactive when the patient is withdrawn.
  - Mixed pattern.
- Important to assess the patient and to take a careful history from family or friends who know the patient well.

There are several potential causes for delirium. Although delerium can affect any age, it commonly affects older people, and factors such as dehydration, constipation, and urinary retention are common. Drugs, e.g., opiates, steroids, anticholinergic drugs, and bezodiazepines can cause delerium, as can withdrawal of substances, for example, alcohol, nicotine, non-prescribed drugs, and sedatives. Physiological causes can include renal and liver impairment, hypercalcaemia, hyperglycaemia, infection, and hypoxia. Factors such as hearing and visual impairment and dementia can increase the risk of delerium. The assessment of delerium includes a full biochemistry and haematology tests, assessment for infection, checking for factors, e.g., constipation, urinary retention, and review of all drugs. The management of delerium includes managing patients in a quiet environment with a low lighting and familiar objects, e.g., family photos. Ensuring that if patients wear glasses, they are worn and the use of hearing aids can all support recovery as can reassuring the patient and their family and maintaining adequate hydration of the patient is essential. Haloperidol, 500 mcg to 3 mg once a day, is the maintstay of pharmacological treatment with benzodiazepines, e.g., Lorazepam, 0.5–1 mg, if drug or alcohol withdrawal is suspected.

Bereavement

At birth, there is a huge celebration for the welcoming of a new life, and families and friends join together to welcome a new arrival. When a death is expected, there is deep sadness at a life that is coming to an end, but as with birth, this is also a time where families and friends join together to support the one who is dying and to support each other. In most cultures, families and friends wish to visit, stay, and hold the person whose life is ending. Palliative care services and hospices not only provide support to the person who is dying but also to their family members. A bereaved person's needs vary widely, and the UK National Institute for Clinical Excellence (NICE, 2004) has set out a three component approach. Most bereaved people experience a normal reaction to grief and will recover with time (Waller et al., 2016), but may benefit from further information about the normal features of grief and encouragement to have conversations with their pre-existing support network, such as friends and family, self-help groups, or volunteer bereavement support groups.

- A bereaved persons' needs vary widely.
- UK National Institute for Clinical Excellence has set out a three-component approach.
- Most bereaved people experience a normal reaction to grief and will recover with time.
- May benefit from further information about the normal features of grief.
- Encouraged to have conversations with their pre-existing support network such as friends and family, self-help groups, or volunteer bereavement support groups.

Research has suggested that 10% of bereaved adults may develop prolonged or complicated grief, which can include intense grief for 6 months or more following bereavement with separation distress, intrusive thoughts, and meaninglessness (Lundorff et al., 2017). Factors such as not being present at the time of death, older people who are socially isolated, and those living in more socio-economic deprived circumstances can all contribute to prolonged grief disorder. Prolonged grief disorder is typified by an intense longing or persistent preoccupation with the deceased person and includes difficulties in accepting the death, difficulties in continuing with life, an emotional numbness, and avoidance of anything that can serve as a reminder of the deceased. Prolonged grief is also typical in those who believe death could have been avoided and blame themselves or others regarding the death. The grief response will have persisted for an atypically long period of time following the loss (more than 6 months at a minimum) and clearly exceeds expected social, cultural, or religious norms for an individual's culture and context. Grief reactions that have persisted for longer periods that are considered to be within a normative period of grieving

given the person's cultural and religious context are not viewed as an abnormal bereavement responses and are not assigned a diagnosis. The DSM 5 Revised has added Prolonged Grief Disorder to ▶ Chap. 61, "Depressive Disorders," and the American Psychiatric Association approved Prolonged Grief Disorder as a new diagnosis in 2020 (Prigerson et al., 2021a, b).

Many hospices and palliative care services have an annual service, usually at Christmas, where anyone, whether connected with the service or not, can attend and remember a death, even if it happened many years previously. Although there is no published research or evaluations of the impact such services may have on the bereavement process, there is anecdotal evidence that people joining others who are also bereaved and being able to share memories and remember a person who has died can be a therapeutic process and help the grieving process.

Palliative care services have close links with psychiatry and psychology and can refer patients and families to services as required, especially those who have prolonged grief disorder. Some hospices, e.g., St Christopher's Hospice in London, have a psychiatrist leading their bereavement support. Palliative care services in the past were criticized for providing unified bereavement support to all those who had been bereaved. Bereavement support is now more tailored with commonly all families receiving a letter with information regarding support available within the community following a bereavement along with support groups and groups the hospice may facilitate, and information as to how bereavement may affect individuals and in different ways but the reassurance that grief is normal.

Clinical Vignette

Doris was 75 and had been married to her husband John for 50 years. They had no children and had done everything together– both saying they did not need anyone else if they had each other. John had always been a

(continued)

smoker and had developed a hoarse voice. Doris was not worried until one morning when John told her he was coughing blood. A chest x-ray and CT scan revealed inoperable cancer of the larynx and widespread metastases, and John was referred to an oncologist and palliative radiotherapy was offered. John initially responded well, but 3 months later he experienced pain and rapidly deteriorated. John was referred to the local hospice and died 10 days after he was admitted. Doris was bereft and alone as apart from a cousin, she had no family. The hospice chaplain helped her arrange the funeral and visited Doris at home weekly for 6 weeks, offering her the time to speak about John and their life together – Doris could cry and laugh with the chaplain and found she not only valued but enjoyed the visits. After 6 weeks, Doris was invited to the "crazy crafters bereavement group." She was unsure whether to attend, but she had enjoyed crafts and felt she wanted to meet others in a similar situation. On the first night, Doris was welcomed warmly and joined 6 other women, from aged 40 to 80 who had all recently been widowed. After coffee, they were invited to make some cards, but the gentle time and space allowed Doris to speak about John as it allowed everyone else to speak about their loss – many tears were shed but also some laughter, and Doris felt again supported. The crazy crafters were Doris' lifeline, especially approaching Christmas was a time she and John always enjoyed, despite not having a family and being on their own. Doris was roped into helping to decorate the hospice for Christmas with the group and found herself accepting an invitation to spend Christmas day with her cousin after talking it through at the bereavement group and realizing she needed to be with someone and not on her own. Doris remained part of crazy crafters for a year and is now a regular volunteer at the hospice.

Chemical Coping

- In patients with advanced cancer, 35–96% experience pain, one of the most frequent and worrisome symptoms.
- Despite specific treatment guidelines by the WHO, more than 30% of cancer patients are inadequately treated for pain.
- Patient survival is increasing due to innovative therapies, and with this a higher number of patients treated in the palliative care setting.

Most patients with serious diseases, such as cancer, experience multiple physical symptoms. Their prevalence and severity depend on the location and nature of the disease as well as the treatments implemented. In patients with advanced cancer, 35–96% experience pain. This symptom is undoubtedly one of the most frequent and worrisome for patients, their families, and the healthcare team caring for them (Henson et al., 2020).

Cancer pain is commonly treated pharmacologically based on the WHO Analgesic Ladder (Anekar & Cascella, 2021). Where strong opioids are implemented, with morphine being the first choice drug. Despite its importance, pain is inadequately treated in more than 30% of cancer patients (Greco et al., 2014) for many reasons. In some countries, there is poor access to opioid treatments, and it's difficult to access specialized care for pain management, such as palliative care units, which can be scarce (Pastrana et al., 2017).

Early palliative care in patients with incurable cancer is more common (Mah et al., 2021). Receiving innovative treatments, such as immunotherapy, has also led to an increased survival rate in many patients being cared for in palliative care units (Farkona et al., 2016). Recommendations have been made regarding the treatment of pain in palliative care patients, such as identifying patients at risk of aberrant use of opioids (e.g., patients with a history of substance abuse) (Gaertner et al., 2019).

Aberrant opioid behavior is defined as the use of opioids in a way other than prescribed (Passik et al., 2014). When this happens, it may be due to opioid dependence, poor pain control (also known as pseudoaddiction), as a antisocial behavior, or as chemical coping (Kwon et al., 2014).

- Patients at risk for aberrant use of opioids must be identified in palliative care, including those who are undergoing "chemical coping," the use of opioids to cope with emotional distress.
- Chemical coping can be present in up to 18% of adult patients with advanced cancer treated with opioids.
- Risks for chemical coping include history of alcohol and tobacco use, younger age, depression, and anxiety disorders, among others.

Bruera et al. (1995) conducted a study where they observed that, after being sent to palliative care, patients with advanced cancer and alcohol dependency relapsed or increased their alcohol intake as a way of dealing with their situation, giving birth to the term "chemical coping," which has a negative impact on pain control. Kwon et al. (2015) through a consensus of experts, defined: "Chemical coping with opioids is the use of opioids to cope with emotional distress and is characterized by inappropriate and/or excessive opioid use." It's important to differentiate chemical coping with opioids and opioid dependence, since in chemical coping there is no craving for the substance, serious behavioral problems, or significant deterioration associated with opioid consumption – all of which happen with opioid dependence. Chemical coping in its milder form can be observed as the use of opioids to cope with suffering, and in its more severe forms, it can be considered as a component of opioid dependence because the lack of opiates triggers significant anxiety and dysphoria, leading to more serious aberrant drug-taking behaviors.

The incidence of chemical coping was 18% of adult patients with advanced cancer who were receiving opioid treatment and who attended a palliative care unit (Kwon et al., 2015).

Risk factors such as a history of alcohol and tobacco consumption or the use of any other substance by the patient or a relative. Younger patients have a higher risk of developing chemical coping as well as those patients with depression and anxiety disorders, less functionality and greater expression of physical symptoms; a history of sexual abuse, cancer diagnosis, and sexual and personality disorders are also possible factors (Del Fabbro, 2014; Kim et al., 2016). Due to their characteristics, opioid drugs have an effect on the central nervous system, producing a sensation of reward, relieving emotional distress, or generating euphoria, as well as reducing physical pain – which explains why patients use opioids to control anxiety, depression, or insomnia symptoms (Tang & Tanco, 2021). Chemical coping with opioids has negative consequences such as neurotoxicity (myoclonus, excessive sedation, delirium, or even the presence of seizures), overdose, and the use of other licit or illicit substances. But mainly, it complicates symptom control, such as pain, and strains relationships with family/caregivers (Del Fabbro, 2014).

- A diagnosis for chemical coping can be achieved by use of proper instruments, including the SOAPP-R, the Opioid Risk Tool, the Current Opioid Misuse Measure, the Pain Medication Questionnaire, and the CAGE tool.
- It is imperative to provide patients and caregivers with information regarding the effects of opioids, including adverse effects, risks of overdose, and the importance of adherence to the prescription.

A chemical coping diagnosis is challenging for clinicians; however, instruments are also used to search for aberrant use of opioids with the intention of identifying opioid use disorder in high-risk patients: Screener and Opioid Assessment for Patients with Pain-revised (SOAPP-R) and the Opioid Risk Tool (ORT) (Butler et al., 2008; Webster & Webster, 2005). The SOAPP-R instrument consists of 24 questions, with a score of 18 or more indicating that a patient is at high risk for opioid abuse, with a sensitivity of 81% and 68%. On the other hand, the ORT is made up of 5 questions, where a score of 3 or less is considered low risk, 8 or more is a high risk, and intermediate scores are moderate risk. A characteristic of the ORT is that it is easy to use in a clinical setting, but its sensitivity and specificity have not been consistent. Instruments that monitor patients during their opioid treatment have also been developed with the intention of evaluating the progress of a patient and preventing aberrant drug-taking behavior; these must be carried out regularly (Chou et al., 2009). One of these instruments is the Current Opioid Misuse Measure (COMM), which is self-administered and consists of 17 questions with a score ranging from 0 to 68; with scores of 9 or more indicating a high risk and the need for close monitoring. This has a sensitivity and specificity of 71% (Butler et al., 2010). Another instrument is the Pain Medication Questionnaire (PMQ), it is a self-administered questionnaire that consists of 26 items with a score from 0 to 104, where a score of 30 or more indicates a high risk of aberrant use of opioids. Its sensitivity and specificity are 74% and 93%, respectively, although this has not been consistent in different reiterations (Adams et al., 2004; Buelow et al., 2009; Jones et al., 2012). Finally, as the description of chemical coping was initially in patients with alcohol use disorder, and alcohol being one of the main factors, one of the most used tools is the CAGE (Cut down, Annoyed, Guilty, Eye-opener), which was originally used for screening for alcohol use disorder; the CAGE-AID was developed to include other drugs in addition to alcohol. These tools have four questions, and with two affirmative answers, the test is considered positive, with a sensitivity of 90% and a specificity of 95% (Dhalla & Kopec,

2007). The presence of positive CAGE implies one of the most reported risk factors for the incidence of chemical coping.

The recommendations presented in the pain management guidelines include providing information to all patients and their families/caregivers about the effects of opioids, their adverse effects, risks of overdose, as well as the importance of adherence to the medical prescription, and the risk of misuse of opioids. The routine use of screening scales such as those described is also recommended (Del Fabbro, 2014).

Addressing any aberrant use of opioids is a great challenge for all experts in palliative care. In the case of chemical coping, the approach by a mental health expert is important and generates the need for a multidisciplinary treatment focused on a patient's needs by identifying causes of pain and factors that influence the perception of pain, such as major depressive disorder (Rodríguez-Mayoral et al., 2020). It is important to treat the latter, as well as a history of use of licit or illicit substances and personality disorders, with the available means, such as pharmacology, family interventions, or psychotherapeutic processes, as well as social and spiritual interventions (Del Fabbro, 2014).

- Patients identified with chemical coping should be approached by a mental health expert to assess causes of pain and influencing factors, including major depressive disorder, which should be treated hastily.
- Chemical coping should be actively sought and timely treated in patients undergoing opioid treatment in palliative care units.

In palliative care, more patients with an increased survival rate are being treated, many of whom receive opioid treatment and are therefore at risk of developing chemical coping. This phenomenon is a novel construct that gains importance every day, since its presence negatively affects patients, their families, and complicates their control of physical symptoms, mainly pain.

Awareness of Chemical coping, like all aberrant opioid use behaviors, should be present in all health personnel in palliative care units. Until now, the literature is limited on this topic, and further research on prevalence, factors, and treatments is required.

Wish to Hasten Death (WTHD)

Severe and advanced diseases generate multiple effects on patients and their families and/or caregivers. Among those physical symptoms, there are changes in a patients' functionality and mental health, which can lead to wish to hasten death (WTHD).

- Patients with severe and advanced diseases may experience changes in their functionality and mental health, which can lead to a wish to hasten death (WTHD).
- WTHD can be encompassed as a wish to die, hasten death, or end suffering through suicidal behavior, abandoning treatments, and through euthanasia or medically assisted suicide.
- Factors associated with a WTHD include physical symptoms, psychological affectations, social factors, and existential and spiritual suffering.

Wish To Hasten Death, can be encompassed as a wish to die (through suicidal behavior), a wish to hasten death to end suffering (abandoning treatments), or as the request for assisted death (through euthanasia or medically assisted suicide). In the vast majority of cases, the presence of WTHD is proposed as an expression of

suffering and not necessarily a genuine desire to die (Wilson et al., 2007). Balaguer et al. (2016) generated the concept of "wish to hasten death" through a study using the Delphi method, defining it as a "...phenomenon that occurs as a consequence of suffering, generated by a life-threatening disease, and for which the patient can't find a solution other than hasten death; this arises in response to factors such as the presence of uncontrolled physical symptoms, psychological and social stress, existential suffering, and/or the presence of mental disorders. This phenomenon must be distinguished from the acceptance of imminent death and a desire to die naturally in the near future."

Among the main factors are physical symptoms (such as pain, fatigue, dyspnea, etc.), psychological affectations (hopelessness, depression, and anxiety disorders), social factors (the feeling of being a burden for others, economic stressors, and the perception of poor social support); and existential and spiritual suffering, where the loss of autonomy and dignity take on great relevance (Hudson et al., 2006).

The understanding of the wish to hasten death has led to different qualitative investigations as described by Nissim et al. (2009), where he describes WTHD as three different experiences: as a hypothetical exit plan which is persistent; as an expression of despair that is presented in a transitory and brief way; and finally as a manifestation of letting go, which was presented in the last weeks of life. These experiences, primarily the hypothetical exit plan, frequently coexisted with a strong desire to prolong life. This study is very relevant as it makes us reflect on the need to explore this phenomenon, taking these experiences into account, since they have relevant implications for health policies and clinical practice.

- WTHD can occur in any neurodegenerative or end-stage chronic disease, though it has been mainly described to present in approximately 30% of patients with advanced HIV/aids and 3–20% of

> patients with cancer undergoing palliative care.
> - Mental health care is essential in patients with serious/advanced diseases, as patients with a WTHD frequently present with depressive disorders, anxiety, and substance abuse.

Until now, WTHD has been described mainly in patients with HIV-AIDS and advanced cancer (Breitbart et al., 2010; Rosenfeld et al., 1999), and recently in patients with amyotrophic lateral sclerosis (Erdmann et al., 2021). However, it can occur in patients with any neurodegenerative disease and end-stage chronic diseases.

Patients with HIV/AIDS have decreased dramatically in recent decades, thanks to the treatments made available to them; however, it occurs in 30% of patients with this advanced infection (Breitbart et al., 2010; Kolva et al., 2017). It's been found that 11% to 55% of patients treated in cancer palliative care units experience WTHD transiently, while 3% to 20% experience it permanently (Wilson et al., 2016); meanwhile, 34.6% to 44% of patients who were referred for a psychiatric evaluation experience it (Rodríguez-Mayoral et al., 2019). Their recurrence varies due to the population studied and the evaluation methods applied.

Different factors associated with the presence of WTHD have been described, including: female sex; the location of the primary tumor in cancer patients (pancreas, liver, lung, head, and neck); presence of difficult-to-control physical symptoms (such as pain); decreased functionality; impact on physical appearance (Kumar et al., 2017); a higher educational attainment; unstable interpersonal relationships; a poor family support network; fear of physical and psychological suffering (Quill et al., 2016); loss of autonomy and dignity; feeling like a burden for others (Chochinov, 2016); and largely, the presence of psychiatric disorders (Wilson et al., 2016).

Depressive disorders were most frequent in patients with WTHD, followed by anxiety

disorders and in patients with substance abuse and personality disorders (Rodríguez-Mayoral et al., 2019; Wilson et al., 2016). This shows that mental health care is essential for all patients with a serious and advanced disease (Wilson et al., 2016). Since mental disorders are strongly associated with WTHD, this is explained by the nature of depressive and anxiety disorders, as well as by the multiple factors associated with them, such as decreased functionality, intensified physical symptoms, poor therapeutic adherence, longer hospital stays, and a lower quality of life and survival (Rodríguez-Mayoral et al., 2020).

> - Instruments to evaluate WTHD include the Desire for Death Rating Scale and the Schedule of Attitudes Toward Hastened Death, which have been widely used and validated in diverse populations.
> - Approaching patients with a WTHD requires specific training and skills by healthcare professionals, including acknowledging and conversing about the phenomenon, active listening, and generating a specific plan to follow.
> - Further research is required in order to better understand WTHD and to develop specific and effective interventions.

One of the main needs was to create instruments to evaluate WTHD in order to identify, quantify, and compare populations. Different instruments have been developed that evaluate WTHD; the most used are those developed by Chochinov et al. (1995) in a palliative oncological population called Desire for Death Rating Scale (DDRS) and by Rosenfeld et al. (1999, 2000) called The Schedule of Attitudes Toward Hastened Death (SAHD). The latter was initially developed for HIV/AIDS and was later validated for patients with advanced cancer. These instruments have been translated, adapted, and validated in different languages and populations. When these instruments were compared, it was found that the DDRS is more clinically practice-oriented, while the SAHD is mostly research-oriented due to its characteristics. The DDRS consists of an initial question ("Do you ever wish that your illness would progress more rapidly so that you're suffering could be over sooner?"), and if this is positive, another three questions are asked to clarify the severity and persistence of the desire. On the other hand, the SAHD was developed as a self-applied research tool based on various aspects of the "death wish" construct, such as: (1) an active death wish; (2) optimism/pessimism towards the future quality of life; (3) social and personal factors that may influence the willingness to consider assisted suicide or euthanasia; (4) passive hopes to have a hasty death; and (5) behaviors that could reflect a wish to die. The instrument was designed with 20 items and dichotomous choice responses (true/false), and the authors concluded that many specific thoughts or actions would be present or absent (Bellido-Pérez et al., 2018; Rosenfeld et al., 1999). A Spanish version with only five items was recently reported (Monforte-Royo et al., 2017), which facilitates its administration in a clinical context, provided that information on the content of the instrument is provided to the patient. So far, the most reported instrument in the literature is the SAHD.

Patients often spontaneously convey the wish to hasten death to their relatives or doctors. In other instances, it is necessary to explore the topic using an instrument. It is necessary to have skills to react to WTHD since health professionals can often choose to avoid delving into the subject, which negatively influences the patient's experience (Galushko et al., 2016). The appropriate response of a healthcare professional to a patient's expression of a WTHD requires specific skills, considering it is a complex phenomenon where multiple emotions are involved.

There are only a few guidelines or recommendations to approach WTHD. Generally, it's recommended to acknowledge the verbalization of a desire to hasten death and never reject the conversation. To always listen actively, recognizing factors and physical needs, such as uncontrolled symptoms like pain, dyspnea, and fatigue and identifying the presence of mental disorders such as depression and anxiety. It is vital to provide a plan to deal with the stressors that are presented and how they can be modified through all available resources; pharmacological, social, and family intervention, psychological treatments at the time of the identification of WTHD, and generating a specific plan to follow (Kremeike et al., 2018).

In conclusion, the wish to hasten death is a complex and multifactorial phenomenon which is understood as an expression of suffering and a cry for help, where mental health, especially psychiatry, has a relevant role in its identification and treatment. Therefore, it needs to be explored routinely with patients in palliative contexts. Approaching it requires active and holistic medical attention from a multidisciplinary team. It is also important to highlight the limitations experienced by resource-poor settings, where multidisciplinary teams are rarely available. In these cases, resourcefulness by attending physicians is imperative and could include the use of telemedicine in order to discuss cases with specialists located in central areas or other regions. Also, it is important to help raise awareness around existing healthcare personnel so that they can recognize this phenomenon and channel the patients appropriately. It is necessary to develop more research to know all the characteristics related to this phenomenon, as well as effective specific interventions.

- In resource-poor settings, where multidisciplinary team is rarely available, resourcefulness by attending physicians is imperative and could include the use of telemedicine in order to discuss cases with specialists located in central areas or other regions.
- It is also important to help raise awareness around existing healthcare personnel so that they can recognize this phenomenon and channel the patients appropriately.

Conclusions

This chapter aimed to summarize the integral link between palliative care, pain, and psychiatry in relation to Anxiety, Depression, Delerium, Bereavement, Pain including chemical coping, and the wish to hasten death. With the aid of clinical vignettes, some of the issues facing patients at the end of life and how palliative care services along with expertise from Psychiatry can work together to optimize care for patients have been described. Chemical coping and wish to hasten death, and the support and care that are essential for such patients have been explored. The profile of palliative care has increased with the COVID pandemic and the importance of excellent symptom control for patients dying of all conditions has been highlighted. Historically, palliative care has supported patients dying of malignancy, but increasingly, the importance of palliative care for all other life-limiting conditions is being acknowledged. It needs to be ensured that symptom control for all patients encompasses the psychological and psychiatric sequelae of life-limiting illness in order to give all patients the optimum care which they need and for support to be extended to their families during bereavement.

References

Adams, L. L., Gatchel, R. J., Robinson, R. C., Polatin, P., Gajraj, N., Deschner, M., & Noe, C. (2004). Development of a self-report screening instrument for assessing potential opioid medication misuse in chronic pain patients. *Journal of Pain and Symptom Management, 27*(5), 440–459. https://doi.org/10.1016/j.jpainsymman.2003.10.009

Anekar, A. A., & Cascella, M. (2021). WHO Analgesic Ladder. In *StatPearls*. StatPearls Publishing. http://www.ncbi.nlm.nih.gov/books/NBK554435/

Balaguer, A., Monforte-Royo, C., Porta-Sales, J., Alonso-Babarro, A., Altisent, R., Aradilla-Herrero, A., Bellido-Pérez, M., Breitbart, W., Centeno, C., Cuervo, M. A., Deliens, L., Frerich, G., Gastmans, C., Lichtenfeld, S., Limonero, J. T., Maier, M. A., Materstvedt, L. J., Nabal, M., Rodin, G., ... Voltz, R. (2016). An international consensus definition of the wish to hasten death and its related factors. *PLoS One, 11*(1), e0146184. https://doi.org/10.1371/journal.pone.0146184

Bellido-Pérez, M., Crespo, I., Wilson, K. G., Porta-Sales, J., Balaguer, A., & Monforte-Royo, C. (2018). Assessment of the wish to hasten death in patients with advanced cancer: A comparison of 2 different approaches. *Psycho-Oncology, 27*(6), 1538–1544. https://doi.org/10.1002/pon.4689

Breitbart, W., Rosenfeld, B., Gibson, C., Kramer, M., Li, Y., Tomarken, A., Nelson, C., Pessin, H., Esch, J., Galietta, M., Garcia, N., Brechtl, J., & Schuster, M. (2010). Impact of treatment for depression on desire for hastened death in patients with advanced AIDS. *Psychosomatics, 51*(2), 98–105. https://doi.org/10.1016/S0033-3182(10)70669-1

Bruera, E., Kuehn, N., Miller, M. J., Selmser, P., & Macmillan, K. (1991). The Edmonton Symptom Assessment System (ESAS): A simple method for the assessment of palliative care patients. *Journal of Palliative Care, 7*, 6–9.

Bruera, E., Moyano, J., Seifert, L., Fainsinger, R. L., Hanson, J., & Suarez-Almazor, M. (1995). The frequency of alcoholism among patients with pain due to terminal cancer. *Journal of Pain and Symptom Management, 10*(8), 599–603. https://doi.org/10.1016/0885-3924(95)00084-4

Buelow, A. K., Haggard, R., & Gatchel, R. J. (2009). Additional validation of the pain medication questionnaire in a heterogeneous sample of chronic pain patients. *Pain Practice, 9*(6), 428–434. https://doi.org/10.1111/j.1533-2500.2009.00316

Butler, S. F., Fernandez, K., Benoit, C., Budman, S. H., & Jamison, R. N. (2008). Validation of the revised screener and opioid assessment for patients with pain (SOAPP-R). *The Journal of Pain, 9*(4), 360–372. https://doi.org/10.1016/j.jpain.2007.11.014

Butler, S. F., Budman, S. H., Fanciullo, G. J., & Jamison, R. N. (2010). Cross validation of the current opioid misuse measure to monitor chronic pain patients on opioid therapy. *The Clinical Journal of Pain, 26*(9), 770–776. https://doi.org/10.1097/AJP.0b013e3181f195ba

Carter, R. (2003). Cicely Saunders – Founder of the hospice movement: Selected letters 1959–1999. *Journal of the Royal Society of Medicine, 96*, 149–151.

Chochinov, H. M. (2016). Physician-assisted death in Canada. *JAMA, 315*(3), 253–254. https://doi.org/10.1001/jama.2015.17435

Chochinov, H. M., Wilson, K. G., Enns, M., Mowchun, N., Lander, S., Levitt, M., & Clinch, J. J. (1995). Desire for death in the terminally ill. *The American Journal of Psychiatry, 152*(8), 1185–1191. https://doi.org/10.1176/ajp.152.8.1185

Chou, R., Fanciullo, G. J., Fine, P. G., Adler, J. A., Ballantyne, J. C., Davies, P., Donovan, M. I., Fishbain, D. A., Foley, K. M., Fudin, J., Gilson, A. M., Kelter, A., Mauskop, A., O'Connor, P. G., Passik, S. D., Pasternak, G. W., Portenoy, R. K., Rich, B. A., Roberts, R. G., ... American Pain Society-American Academy of Pain Medicine Opioids Guidelines Panel. (2009). Clinical guidelines for the use of chronic opioid therapy in chronic noncancer pain. *The Journal of Pain, 10*(2), 113–130. https://doi.org/10.1016/j.jpain.2008.10.008

Cox, J. L., Holden, J. M., & Sagovsky, R. (1987 Jun). Detection of postnatal depression. Development of the 10-item Edinburgh postnatal depression scale. *The British Journal of Psychiatry, 150*, 782–786.

Del Fabbro, E. (2014). Assessment and management of chemical coping in patients with cancer. *Journal of Clinical Oncology, 32*(16), 1734–1738.

Dhalla, S., & Kopec, J. A. (2007). The CAGE questionnaire for alcohol misuse: A review of reliability and validity studies. *Clinical & Investigative Medicine, 30*(1), 33. https://doi.org/10.25011/cim.v30i1.447

Erdmann, A., Spoden, C., Hirschberg, I., & Neitzke, G. (2021). The wish to die and hastening death in amyotrophic lateral sclerosis: A scoping review. *BMJ Supportive & Palliative Care*, bmjspcare-2020-002640. https://doi.org/10.1136/bmjspcare-2020-002640

Farkona, S., Diamandis, E. P., & Blasutig, I. M. (2016). Cancer immunotherapy: The beginning of the end of cancer? *BMC Medicine, 14*(1), 73. https://doi.org/10.1186/s12916-016-0623-5

Gaertner, J., Boehlke, C., Simone, C. B., II, & Hui, D. (2019). Early palliative care and the opioid crisis: Ten pragmatic steps towards a more rational use of opioids. *Annals of Palliative Medicine, 8*(4), 490–497. https://doi.org/10.21037/apm.2019.08.01

Galushko, M., Frerich, G., Perrar, K. M., Golla, H., Radbruch, L., Nauck, F., Ostgathe, C., & Voltz, R. (2016). Desire for hastened death: How do professionals in specialized palliative care react?: Professionals' responses to desire for hastened death. *Psycho-Oncology, 25*(5), 536.

Greco, M. T., Roberto, A., Corli, O., Deandrea, S., Bandieri, E., Cavuto, S., & Apolone, G. (2014). Quality of cancer pain management: An update of a systematic review of undertreatment of patients with cancer. *Journal of Clinical Oncology, 32*(36), 4149–4154. https://doi.org/10.1200/JCO.2014.56.0383

Grotmol, K. S., Lie, H. C., Loge, J. H., Aass, N., Haugen, D. F., Stone, P. C., Kaasa, S., & Hjermstad, M. J. (2018 Jan). Patients with advanced cancer and depression report a significantly higher symptom burden than non-depressed patients. *Palliative & Supportive Care, 10*, 1–7.

Henson, L. A., Maddocks, M., Evans, C., Davidson, M., Hicks, S., & Higginson, I. J. (2020). Palliative care and the management of common distressing symptoms in advanced cancer: Pain, breathlessness, nausea and vomiting, and fatigue. *Journal of Clinical Oncology, 38*(9), 905–914. https://doi.org/10.1200/JCO.19.00470

Hudson, P. L., Kristjanson, L. J., Ashby, M., Kelly, B., Schofield, P., Hudson, R., Aranda, S., O'Connor, M., & Street, A. (2006). Desire for hastened death in patients with advanced disease and the evidence base of clinical guidelines: A systematic review. *Palliative Medicine, 20*(7), 693–701. https://doi.org/10.1177/0269216306071799

Jones, T., Moore, T., Levy, J. L., Daffron, S., Browder, J. H., Allen, L., & Passik, S. D. (2012). A comparison of various risk screening methods in predicting discharge from opioid treatment. *The Clinical Journal of Pain, 28*(2), 93–100. https://doi.org/10.1097/AJP.0b013e318225da9e

Kim, Y. J., Dev, R., Reddy, A., Hui, D., Tanco, K., Park, M., Liu, D., Williams, J., & Bruera, E. (2016). Association between tobacco use, symptom expression, and alcohol and illicit drug use in advanced cancer patients. *Journal of Pain and Symptom Management, 51*(4), 762–768. https://doi.org/10.1016/j.jpainsymman.2015.11.012

Kolva, E., Rosenfeld, B., Liu, Y., Pessin, H., & Breitbart, W. (2017). Using item response theory (IRT) to reduce patient burden when assessing desire for hastened death. *Psychological Assessment, 29*(3), 349–353. https://doi.org/10.1037/pas0000343

Kremeike, K., Galushko, M., Frerich, G., Romotzky, V., Hamacher, S., Rodin, G., Pfaff, H., & Voltz, R. (2018). The DEsire to DIe in palliative care: Optimization of management (DEDIPOM) – A study protocol. *BMC Palliative Care, 17*(1). https://doi.org/10.1186/s12904-018-0279-3

Kroenke, K., Spitzer, R. L., & Williams, J. B. (2001). The Phq-9 – Validity of a brief depression severity measure. *Journal of General Internal Medicine, 16*(9), 606–613.

Kumar, V., Chaudhary, N., Soni, P., & Jha, P. (2017). Suicide rates in cancer patients in the current era in United States. *American Journal of Psychiatry Residents' Journal, 12*(1), 11–14. https://doi.org/10.1176/appi.ajp-rj.2017.120104

Kwon, J. H., Tanco, K., Hui, D., Reddy, A., & Bruera, E. (2014). Chemical coping versus pseudoaddiction in patients with cancer pain. *Palliative and Supportive Care, 12*(05), 413–417. https://doi.org/10.1017/S1478951513001351

Kwon, J. H., Tanco, K., Park, J. C., Wong, A., Seo, L., Liu, D., Chisholm, G., Williams, J., Hui, D., & Bruera, E. (2015). Frequency, predictors, and medical record documentation of chemical coping among advanced cancer patients. *The Oncologist, 20*(6), 692–697. https://doi.org/10.1634/theoncologist.2015-0012

Li, M., Fitzgerald, P., & Rodin, G. (2012). Evidence-based treatment of depression in patients with cancer. *Journal of Clinical Oncology, 30*(11), 1187–1196.

Lloyd-Williams, M., Dennis, M., & Taylor, F. (2004). A prospective study to determine the association between physical symptoms and depression in patients with advanced cancer. *Palliative Medicine, 18*(6), 558–563.

Lloyd-Williams, M., Shiels, C., & Dowrick, C. (2007). The development of the Brief Edinburgh Depression Scale (BEDS) to screen for depression in patients with advanced cancer. *Journal of Affective Disorders, 99*(1–3), 259–264.

Lloyd-Williams, M., Shiels, C., Taylor, F., & Dennis, M. (2009). Depression – An independent predictor of early death in patients with advanced cancer. *Journal of Affective Disorders, 113*(1), 127–132.

Lloyd-Williams, M., Payne, S., Reeve, J., & Dona, R. K. (2014). Thoughts of self-harm and depression as prognostic factors in palliative care patients. *Journal of Affective Disorders, 166*, 324–329.

Lloyd-Williams, M., Shiels, C., Ellis, J., Abba, K., Gaynor, E., Wilson, K., & Dowrick, C. (2018). Pilot randomised controlled trial of focused narrative intervention for moderate to severe depression in palliative care patients: DISCERN trial. *Palliative Medicine, 32*(1), 206–215.

Lundorff, M., Holmgren, H., Zachariae, R., Farver-Vestergaard, I., & O'Connor, M. (2017). Prevalence of prolonged grief disorder in adult bereavement: A systematic review and meta-analysis. *Journal of Affective Disorders, 212*, 138–149.

Mah, K., Swami, N., O'Connor, B., Hannon, B., Rodin, G., & Zimmermann, C. (2021). Early palliative intervention: Effects on patient care satisfaction in advanced cancer. *BMJ Supportive & Palliative Care*, bmjspcare-2020-002710. https://doi.org/10.1136/bmjspcare-2020-002710

Mausbach, B. T., & Irwin, S. A. (2017). Depression and healthcare service utilization in patients with cancer. *Psychooncology, 26*, 1133–1139.

Monforte-Royo, C., González-de Paz, L., Tomás-Sábado, J., Rosenfeld, B., Strupp, J., Voltz, R., & Balaguer, A. (2017). Development of a short form of the Spanish schedule of attitudes toward hastened death in a palliative care population. *Quality of Life Research, 26*(1), 235–239. https://doi.org/10.1007/s11136-016-1409-0

National Institute for Clinical Excellence. (2004). Supportive and palliative care for adults with cancer. Available at: https://www.nice.org.uk/guidance/csg4. Accessed 20 Apr 2021.

Nipp, R. D., El-Jawahri, A., Moran, S. M., D'Arpino, S. M., Johnson, P. C., Lage, D. E., Wong, R. L., Pirl, W. F., Traeger, L., Lennes, I. T., Cashavelly, B. J., Jackson, V. A., Greer, J. A., Ryan, D. P., Hochberg, E. P., & Temel, J. S. (2017). The relationship between physical and psychological symptoms and health care utilization in hospitalized patients with advanced cancer. *Cancer, 123*(23), 4720–4727.

Nissim, R., Gagliese, L., & Rodin, G. (2009). The desire for hastened death in individuals with advanced cancer: A longitudinal qualitative study. *Social Science & Medicine (1982), 69*(2), 165–171. https://doi.org/10.1016/j.socscimed.2009.04.021

Passik, S. D., Narayana, A., & Yang, R. (2014). Aberrant drug-related behavior observed during a 12-week open-label extension period of a study involving patients taking chronic opioid therapy for persistent pain and fentanyl buccal tablet or traditional short-acting opioid for breakthrough pain. *Pain Medicine, 15*(8), 1365–1372. https://doi.org/10.1111/pme.12431

Pastrana, T., Wenk, R., Radbruch, L., Ahmed, E., & De Lima, L. (2017). Pain treatment continues to be inaccessible for many patients around the globe: Second phase of opioid price watch, a cross-sectional study to monitor the prices of opioids. *Journal of Palliative Medicine, 20*(4), 378–387. https://doi.org/10.1089/jpm.2016.0414

Prigerson, H., Boelen, P., Xu, J., Smith, K., & Maciejewski, P. (2021a). Validation of the new DSM-5-TR criteria for prolonged grief disorder and the PG-13-Revised (PG-13-R) scale. *World Psychiatry, 20*, 96–106.

Prigerson, H., Kakarala, S., Gang, J., & Maciejewski, P. (2021b). History and status of prolonged grief disorder as a psychiatric diagnosis. *Annual Review of Clinical Psychology, 17*, 109–126.

Quill, T. E., Back, A. L., & Block, S. D. (2016). Responding to patients requesting physician-assisted death: Physician involvement at the very end of life. *JAMA, 315*(3), 245–246. https://doi.org/10.1001/jama.2015.16210

Rodin, G., Lo, C., Rydall, A., Shnall, J., Malfitano, C., Chiu, A., Panday, T., Watt, S., An, E., Nissim, R., Li, M., Zimmermann, C., & Hales, S. (2018). Managing Cancer and Living Meaningfully (CALM): A randomized controlled trial of a psychological intervention for patients with advanced cancer. *Journal of Clinical Oncology, 36*(23), 2422–2432.

Rodríguez-Mayoral, O., Ascencio-Huertas, L., Verástegui, E., Delgado-Guay, M. O., & Allende-Pérez, S. (2019). The desire to hasten death in advanced cancer patients at a Mexican palliative care service. *Salud Mental, 42*(3), 103–109. https://doi.org/10.17711/SM.0185-3325.2019.014

Rodríguez-Mayoral, O., Cacho-Díaz, B., Peña-Nieves, A., Monreal-Carrillo, E., Allende-Pérez, S., & Lloyd-Williams, M. (2020). Depressive disorder and clinical factors: Impact on survival in palliative care cancer patients. *General Hospital Psychiatry, 64*, 133–135. https://doi.org/10.1016/j.genhosppsych.2020.03.001

Rosenfeld, B., Breitbart, W., Stein, K., Funesti-Esch, J., Kaim, M., Krivo, S., & Galietta, M. (1999). Measuring desire for death among patients with HIV/AIDS: The schedule of attitudes toward hastened death. *The American Journal of Psychiatry, 156*(1), 94–100. https://doi.org/10.1176/ajp.156.1.94

Rosenfeld, B., Breitbart, W., Galietta, M., Kaim, M., Funesti-Esch, J., Pessin, H., Nelson, C. J., & Brescia, R. (2000). The schedule of attitudes toward hastened death: Measuring desire for death in terminally ill cancer patients. *Cancer, 88*(12), 2868–2875.

Serfaty, M., King, M., Nazareth, I., Moorey, S., Aspden, T., Mannix, K., Davis, S., Wood, J., & Jones, L. (2020). Effectiveness of cognitive-behavioural therapy for depression in advanced cancer: CanTalk randomised controlled trial. *The British Journal of Psychiatry, 216*(4), 213–221.

Tang, M., & Tanco, K. (2021). How to measure pain. *Current Oncology Reports, 23*(1), 6. https://doi.org/10.1007/s11912-020-00999-4

Thekkumpurath, P., Walker, J., Butcher, I., Hodges, L., Kleiboer, A., O'Connor, M., et al. (2011). Screening for major depression in cancer outpatients. *Cancer, 117*(1), 218–227.

Waller, A., Turon, H., Mansfield, E., Clark, K., Hobden, B., & Sanson-Fisher, R. (2016). Assisting the bereaved: A systematic review of the evidence for grief counselling. *Palliative Medicine, 30*, 132–148.

Webster, L. R., & Webster, R. M. (2005). Predicting aberrant behaviors in opioid-treated patients: Preliminary validation of the Opioid Risk Tool. *Pain Medicine (Malden, Mass.), 6*(6), 432–442.

Wilson, K. G., Chochinov, H. M., Graham Skirko, M., Allard, P., Chary, S., Gagnon, P. R., Macmillan, K., De Luca, M., O'Shea, F., Kuhl, D., Fainsinger, R. L., & Clinch, J. J. (2007). Depression and anxiety disorders in palliative cancer care. *Journal of Pain and Symptom Management, 33*(2), 118–129. https://doi.org/10.1016/j.jpainsymman.2006.07.016

Wilson, K. G., Dalgleish, T. L., Chochinov, H. M., Chary, S., Gagnon, P. R., Macmillan, K., De Luca, M., O'Shea, F., Kuhl, D., & Fainsinger, R. L. (2016). Mental disorders and the desire for death in patients receiving palliative care for cancer. *BMJ Supportive & Palliative Care, 6*(2), 170–177. https://doi.org/10.1136/bmjspcare-2013-000604

Zweers, D., de Graeff, A., Duijn, J., de Graff, E., Witteveen, P., & Teunnissen, S. (2019). Patients' needs regarding anxiety management in palliative cancer care: A qualitative study in a hospice setting. *The American Journal of Hospice & Palliative Care, 36*, 947–954.

Death and Bereavement

146

Joseph S. Goveas, M. Katherine Shear, and Naomi M. Simon

Contents

This chapter is an update from the 4th edition. Previous edition authors were M. Katherine Shear and Naomi M. Simon.

J. S. Goveas
Department of Psychiatry and Behavioral Medicine, Medical College of Wisconsin, Milwaukee, WI, USA
e-mail: jgoveas@mcw.edu

M. K. Shear (✉)
Columbia University School of Social Work, Columbia University College of Physicians and Surgeons, New York, NY, USA
e-mail: ks2394@columbia.edu

N. M. Simon
Department of Psychiatry, New York University Grossman School of Medicine, New York, NY, USA
e-mail: naomi.simon@nyulangone.org

Abstract

This chapter provides clinicians with a concise summary on the topics of death and dying, working with terminally ill patients, and an overview of bereavement and grief. Special emphasis is given to explaining grief as the response to losing a loved one, including acute and integrated grief, as well as the new DSM-5-TR diagnosis of prolonged grief disorder. Advances in epidemiology, risk factors, pathophysiology, clinical features, diagnostic methods, and treatment of prolonged grief disorder are summarized. A discussion of other psychiatric disorders that can occur after bereavement, such as major depressive disorder and post-traumatic stress disorder, is also included.

© Springer Nature Switzerland AG 2024
A. Tasman et al. (eds.), *Tasman's Psychiatry*,
https://doi.org/10.1007/978-3-030-51366-5_43

Keywords

Death · Dying · Terminally ill · End-of-life ·
Pain · Bereavement · Grief · Acute grief ·
Integrated grief · Attachment theory ·
Prolonged grief disorder · Complicated grief ·
Prolonged grief disorder treatment

Introduction

Dying and bereavement are among life's most difficult experiences. Yet mortality is a natural part of life. Those who understand this are better able to cope with the pain of facing their death as well as the loss of loved ones. C.S. Lewis provides an eloquent description of death as a natural sequence in a relationship:

> For all pairs of lovers without exception, bereavement is a universal and integral part of our experience of love. It follows marriage as normally as marriage follows courtship or as autumn follows summer. It is not a truncation of the process but one of its phases; not the interruption of the dance, but the next figure. We are "taken out of ourselves" by the loved one while she is here. Then comes the tragic figure of the dance in which we must learn to be still taken out of ourselves though the bodily presence is withdrawn, to love the very Her, and not fall back to loving our past, or our memory, or our sorrow, or our relief from sorrow, or our own love. (Lewis, 1961, p. 63.)

Although Lewis is writing about the loss of a spouse, his ideas apply to the loss of other loved ones. Yet not everyone arrives at this kind of philosophical acceptance, and even those who do, including Lewis, are not spared a prior period of great disruption. As a result, dying patients and their bereaved loved ones often develop psychological and/or physical symptoms. Although distress associated with the end of life and acute bereavement is normative, psychiatrists may be called upon to assist in providing support, solace, and, at times, clinical interventions to dying patients and their caregivers, and to bereaved people following the death of a loved one. This chapter provides a discussion of death, including its rates and causes, current thinking about the management of terminally ill patients, and an overview of bereavement and grief that can inform psychiatric practice.

Death and Dying

"World Death Rate Holding Steady At 100 Percent" heads a satirical article in the internet journal *The Onion* that draws attention to the American cultural attitude of defiance of death and the expectation that it is the job of the medical profession to avert death at all costs:

> World Health Organization officials expressed disappointment Monday at the group's finding that, despite the enormous efforts of doctors, rescue workers, and other medical professionals worldwide, the global death rate remains constant at 100 percent. Death, a metabolic affliction causing total shutdown of all life functions, has long been considered humanity's number one health concern. Responsible for 100 percent of all recorded fatalities worldwide, the condition has no cure . . . Many are suggesting that the high mortality rate represents a massive failure on the part of the planet's health care workers. "The inability of doctors and scientists to adequately address this issue of death is nothing less than a scandal," concerned parent Marcia Gretto said. "Do you have any idea what a full-blown case of death looks like? . . . (*The Onion,* 22 January 1997, Issue 13, No. 2.)

In 2020, provisional mortality data indicate that approximately 3,358,814 deaths occurred in the United States (Ahmad & Anderson, 2021). From 2019 to 2020, provisional estimates indicate that age-adjusted rate increased by 15.9%, from 715.2 to 828.7 deaths per 100,000. COVID-19 was the third leading cause of death in 2020, after heart disease and cancer. The increases in mortality trends between 2019 and 2020 appear to have been directly or indirectly associated with the COVID-19 pandemic. The other leading causes of death included unintentional injuries, stroke, chronic lower respiratory disease, Alzheimer's disease, diabetes, influenza and pneumonia, kidney disease, and suicide (Ahmad & Anderson, 2021).

The attitude satirized in *The Onion* article is still extant among a substantial proportion of the population, especially physicians. Thankfully, there have been important efforts to change this.

For much of the twentieth century, medicine was oriented primarily to preventing death whenever possible. This effort, along with improved quality of life, did succeed in lengthening the average life span considerably. Currently, there is a focus on the quality in addition to the quantity of life, and on assistance with dying, provided by hospice and/or palliative care. The hospice and palliative care movement has a strong foothold and is growing in the United States and elsewhere.

Bedell et al. (2001) called attention to the need for physicians to support family members of their deceased patients as their final responsibility to the patient. They suggested writing a personal letter of condolence while acknowledging barriers to doing so. They recognize that physicians are very busy and sometimes have not seen the patient for a while. The physician may hesitate because of not knowing the patient very well, because of feeling a sense of personal failure regarding the death or simply not knowing what to say. Nevertheless, Bedell et al. emphasized the importance of reaching out to the family. They outlined a possible approach to such a letter and emphasized how helpful such an approach can be for bereaved family members. However, a randomized controlled trial of condolence letters sent to bereaved families after loss of a loved one in the ICU, showed no effect on grief symptoms and was unexpectedly associated with more symptoms of depression and PTSD (Kentish-Barnes et al., 2017). It is unclear how widespread this practice has become. Many physicians are uneasy about maintaining a relationship with a dying patient for whom they have no further healing treatments.

Continuing to provide care to a terminally ill patient is important, even when there are no further curative possibilities. Yet for many physicians, interest in the patient dwindles in parallel with dwindling treatment options. At best, such physicians may refer their patients to a palliative care service, a medical subspecialty that is strong and growing. Psychiatrists may be involved in the terminal care of patients, either through their practice or through work in palliative care. Investigators have begun to document the clinical issues involved in treating terminally ill patients, which may serve as a tool to physicians trying to improve patient care at the end of life (Lyness, 2004; Blinderman & Billings, 2015) (see Table 1).

Work with Terminally Ill Patients

Advance Directives

A major goal in effective end-of-life (EOL) care is to bring treatment as closely as possible into alignment with the wishes of patients and their families (Lyness, 2004; Blinderman & Billings, 2015). Effective communication methods that can help identify the core lifelong values (e.g., those related to culture, religion, etc.), goals, and preferences of terminally ill patients should be undertaken (Blinderman & Billings, 2015). Sensitive discussions about EOL care should begin early

Table 1 End-of Life-Intervention targets

Patient	Caregiver/Family	System
Symptoms	Symptoms	Effective recognition and management of symptoms in the patient and family
Quality of life	Caregiver burden/quality of life	Address quality of life issues
Sense of autonomy	Sense of autonomy	Recognize and promote autonomy in the patient and caregiver
Sense of connectedness	Sense of connectedness	Recognize and promote connectedness of patient/caregiver and others
Participate in decision-making	Participate in decision-making	Organize effective decision-making
Optimal medical care	–	Provide optimal medical care
Satisfaction with care	Satisfaction with care	–

during a severe disabling or life-threatening illness. The clinician should explore the patient's understanding of the medical situation, assess the patient's and family's information-sharing preferences, inquire about the patient's concerns, clarify any misconceptions, ask about his/her views on suffering, medical treatment, and death, and understand the patient's relationship with their family. It should only be with this shared knowledge that an EOL care plan is developed. The physician should also communicate a commitment to assist the patient to die with dignity, have an attitude of respect for the patient's wishes, and provide assurance that she/he will measures may serve as a bridge to recovery in some situations, but in others simply prolong the process of dying. For example, if a person is dying, will cardiopulmonary resuscitation (CPR) be administered? What about antibiotics, dialysis, mechanical ventilation, and artificial nutrition and hydration? The two main types of patient-initiated advanced directives are the living will and the power of attorney for healthcare. A living will specify the medical treatments that a person would want to have in the event of terminal illness and the treatments they would not want to have. It often stipulates that comfort and symptom control will take priority over the prolongation of life through artificial measures and describes the way the person hopes to be cared for at the end of life. A power of attorney for healthcare specifies who a person would want to represent them in meetings with their physician about EOL care if unable to speak for her-or himself. Unlike the usual living will, a power of attorney for healthcare does not require that the signer have a terminal condition. The designated agent advocates for the ill person based not only on their previously expressed wishes but also in response to changing clinical situations. The agent is empowered to make decisions about whether healthcare will be provided, withheld, or withdrawn, not abandon the patient if and when disease-focused treatment is no longer effective (Lyness, 2004; Blinderman & Billings, 2015).

Advance directives provide the legal means for honoring a person's wishes about EOL care in anticipation of the possibility that when dying, people may not be able to speak or speak up for themselves. Advanced directives address which life-sustaining measures will be used and under what circumstances. Life-sustaining respecting the delicate line between recoverable and chronic conditions. The third type of advanced directive is a Physician Orders for Life-Sustaining Treatment, in which a patient discusses their wishes with their doctor, who writes orders that the patient then keeps.

The EPEC (Education in Palliative and End-of-Life Care) program for clinicians (http://epec.net/) provides a comprehensive curriculum in palliative care for healthcare professionals.

Further useful information is available at the following websites:

National Institute on Aging: http://www.nia.nih. gov/health/publication/advance-care-planning

National Hospice and Palliative Care Organization (NHPCO): https://www.nhpco.org/ patients-and-caregivers/

National Cancer Institute: https://www.cancer. gov/about-cancer/managing-care/advance-directives

State-by-state documents can be found at https:// www.nhpco.org/advancedirective/

Management of the Terminally Ill Patient

The clinician's tasks change as a patient's terminal disease progresses. At times, there may be bad news to deliver. This should be done with sensitivity to the patient's readiness to hear the news and openness to engage in a discussion about the information. The clinician needs to monitor the patient's symptoms, including pain, anxiety, and depression, as well as changes in functional abilities and the status of important relationships. Many dying patients experience cognitive impairments and these must be accurately assessed. Over time, the patient and family will be confronted with choices regarding care options, and they will need to make decisions about these. Clinicians should provide accurate information about prognosis and treatment options, handle unrealistic hopes and wishes in a sensitive, empathic, and supportive manner, and assist in accessing appropriate resources and guiding the decision-making process. For the management of pain and other non-psychiatric symptoms experienced by the terminal ill, please refer to the chapter titled *Palliative Care.*

Psychiatric symptoms common among dying patients include depression, anxiety, and cognitive impairment. Suicidal ideation is also seen and needs to be carefully evaluated. Major depression can commonly occur among those with a terminal illness, but diagnosis can be tricky as many depressive symptoms overlap with those of the underlying medical condition (Lyness, 2004; Rosenstein, 2011). However, untreated depression worsens medical prognosis in addition to the quality of life. Many researchers thus use an inclusive approach to diagnosis, but this could lead to over-diagnosis. Current clinical practice for depression management in EOL care is to institute antidepressants to reduce specific symptoms. Extrapolating from treatment guidelines for depression in the medically ill, it is generally assumed that depression in the terminally ill will respond to standard antidepressant medications, though evidence specifically in the latter group is limited (Rosenstein, 2011). The selection of an appropriate antidepressant is mostly based on patient symptoms, preferences, prognosis, and risk of side effects and drug-drug interactions. For instance, mirtazapine, with its sedating, weight gain, and anti-emetic properties, has been an attractive antidepressant choice to treat depression in the terminally ill (Rosenstein, 2011). Psychostimulants and modafanil are also commonly used in palliative care settings to reduce fatigue, abulia, and anhedonia, even in the absence of a full-blown major depressive syndrome (Lyness, 2004; Rosenstein, 2011). Moreover, psychostimulants, due to their rapid onset of action, may have a role in the acute management of depression.

A range of anxiety disorders may be seen in dying patients, including generalized anxiety disorder, focused on worries about the future, about the wellbeing of family members, about disability or loss of resources, or a myriad of other daily life difficulties for ill patients. Anxiety may be focused on aspects of the terminal illness including illness progression, pain management, and/or death. Posttraumatic stress disorder (PTSD) may occur as a reaction to a life-threatening diagnosis and/or painful or frightening medical procedures. Panic attacks may also occur and may be more common among patients with respiratory distress but can also be associated with fear of dying. Treatment of anxiety may be achieved with benzodiazepines, although these should be used cautiously in terminal illness as vulnerability to cognitive impairment is increased. Low-dose antipsychotic medication, especially an atypical antipsychotic, can be very helpful in controlling anxiety, including rumination and agitation. Opioid treatment of respiratory distress and pain can also reduce anxiety related to these problems.

Delirium is very common among dying patients, either because of the underlying illness or as a side effect of medications (e.g., opioids). Older age and pre-existing cognitive impairment can also be risk factors for delirium. While reducing medication that is causing delirium may be useful, it must be balanced against the potential loss of pain control. Validated tools, such as the Confusion Assessment Method, can be used as a delirium screening tool. Nonpharmacological delirium prevention methods, such as educating nursing staff, assessing and adjusting medication regimen, increasing patient mobilization, and improving the patient's environment can reduce delirium. Management of delirium can usually be accomplished by optimizing the patient's environment (i.e., promoting normal sleep-wake cycle, improving safety, decreasing stimulation, and re-orienting frequently), treating reversible causes, and using an antipsychotic medication short-term, particularly to address agitation. Education of family members is often necessary and very helpful in the management of delirium, as this can be an especially difficult problem for caregivers (Lyness, 2004; Blinderman & Billings, 2015; Hosker & Bennett, 2016).

Psychotherapy has a role in the management of terminally ill patients. A supportive therapist can be especially helpful if there is time to develop a strong alliance and if the therapist has known the patient before the onset of the later stages of a terminal illness. In such a situation, the therapist can assist the patient in problem solving, goal setting, and interpersonal interactions. Therapists working with terminally ill patients need to include work with the family to optimize the supportive resources of the family, assist the family in coping with the illness and impending death of a loved one, and address interpersonal toxicity if it occurs. Intervention with distressed family members can help reduce caregiver burden and may play a role in decreasing the likelihood of later bereavement-related problems such as depression or prolonged grief disorder. Cognitive behavioral therapy (CBT) appears to be effective to treat depressive symptoms in terminally ill patients (Rosenstein, 2011; Stagg & Lazenby, 2012). Other approaches such as massage therapy, guided imagery/progressive muscle relaxation, and music therapy may also improve mood when used as an adjunct to antidepressants and/or psychotherapy (Stagg & Lazenby, 2012). Supportive and/or CBT can also be useful for anxious patients who are alert and able to participate in such treatments.

Work with parents of terminally ill children may be especially important, as the loss of a child is the most difficult stressor that a parent can endure. Studies by Meert and colleagues have documented the importance of the availability and attentiveness of hospital staff to parental needs for information and honest, complete communication (Meert et al. 2007; Sharman et al. 2005). Physicians may need help in learning to communicate with parents of terminally ill children in a sensitive, respectful, caring, and empathic manner, while not withholding information. Another study documented that parents often want the opportunity to have a conference with the physician after a child dies and that this occurred in only 13% of deaths in a pediatric intensive care unit.

In summary, physician communication is very important for dying patients and their family members. Psychiatric involvement may facilitate this process. Psychiatrists have a role in providing assessment, pharmacotherapy, and psychotherapy. Effective intervention, including acceptance of death as a normal part of life, and honest, empathic, and sensitive communication about terminal illness and its prognosis, is an important part of effective palliative care.

Bereavement and Grief

> Loss of a loved person is one of the most intensely painful experiences any human being can suffer. And not only is it painful to experience but it is also painful to witness, if only because we are so impotent to help. To the bereaved nothing but the return of the lost person can bring true comfort; should what we provide fall short of that it is felt almost as an insult. (Bowlby, 1980, pp. 7–8.)

These comments draw attention to the intensely stressful nature of bereavement, and the difficulty in trying to help bereaved people (Bowlby, 1980). Because it is a confrontation with death, and because loss of an attachment figure dramatically affects one's sense of safety, whether or not it meets current conventions for formal definitions of a trauma used in PTSD criteria sets, it is appropriate to consider bereavement as a traumatic experience (Shear & Shair, 2005; Shear et al., 2007). However, unlike other traumatic life events, death of a loved one is an expected and universal occurrence. Among the population, bereavement is frequent. Given the yearly US death rate of about 2.5 million people and using an estimated average of nine bereaved people for each death from a recent COVID study (Verdery et al., 2020), almost 22.5 million people lose a close friend or relative in the United States each year. As with other traumatic events, the great majority of people weather the storm of such a death. Some even find themselves stronger, having a greater sense of purpose, renewed faith, or a sense of new meaning in life. Some researchers have studied loss as a type of trauma resulting in this phenomenon termed posttraumatic growth (Triplett et al., 2012). While bereavement is considered a risk for a myriad of health problems, in fact only a minority develop such problems. For this clinically important subgroup, bereavement leads to long-lasting negative outcomes. Clinicians may see bereaved individuals who continue to suffer for many years following the death of a loved one. In order to treat such people, it is important to understand the mechanisms by which problematic consequences follow bereavement, and to know as much as possible about what is happening when things go well.

Grief is the natural biobehavioral response to loss, influenced by cultural input. Grief is the experience of the loss at any moment in time. It changes over time, erratically for a time, but usually evolving gradually and becoming integrated into the life of a bereaved person as they adapt to life without the deceased person. Grief is unique to the relationship lost; responses to loss vary significantly both between and within individuals. Grief is influenced by a variety of factors including the circumstances of the death and the relationship with the deceased. The psychiatric literature pertaining to bereavement and grief sometimes uses terms interchangeably. However, to improve communication and precision of thinking, bereavement researchers have suggested that *bereavement* should be defined as "the objective condition of having lost someone significant" and *mourning* should be defined as "the actions and manner of expressing grief, which often reflect the mourning practices of one's culture" (Stroebe et al., 2001, p. 5; Shear, 2012). *Grief* is the reaction to bereavement (Table 2).

Different designations, in particular, complicated grief, have been used over the years to refer to the clinical syndrome that occurs in an important minority, when the progress of adapting to a loss is blocked. Prior designations have now been replaced by Prolonged Grief Disorder (PGD), a name shared by both the American Psychiatric Association's DSM-5-TR and the International Classification of Diseases 11th revision (APA, 2013; World Health Organization, 2020). Since the loss of an attachment figure is the most important type of bereavement, a summary of current thinking about the nature of these relationships is included.

John Bowlby (1980) introduced the idea that attachment has a biological basis and is important throughout the life span. Subsequently, investigators have identified an attachment figure as a person (1) who is a target of proximity seeking, (2) from whom separation is resisted, (3) to whom a partner turns to for "safe haven" when under stress, and (4) serves as a secure base from which the partner freely interacts in the world, seeking novelty, taking risks, and exploring the unknown (Hazan & Zeifman, 1999; Feeney, 2004).

Mental representations of attachment relationships enable us to maintain a sense of attachment in the absence of physical proximity to a loved one. Also called "working models," such mental representations devolve from repeated lived experiences with the attachment figure. Information provided by a working model is continuously matched with information derived from experiences pertaining to the relationship it represents. However, once established, the working model does not easily change its configuration (Bretherton, 1999). Only in the context of a consistent mismatch is the working model revised. Given this view of working models, these mental representations need to undergo an important change when an attachment figure dies, and this change will not occur quickly. This means that bereavement is associated with a temporary impairment in functioning of the attachment system.

Numerous studies have shown that internalized cognitive representations of attachment relationships are activated upon exposure to stress (e.g., Mikulincer et al., 2003) and that the security of such relationships influence the perception of available social support (Collins & Feeney, 2000). Attachment security influences cognition, emotion, and behavior, especially with respect to interpersonal functioning, and may affect other psychological and physiological functions. Individuals with stable secure attachments have been repeatedly shown to be psychologically healthy and resilient. Those with anxious and fearful

Table 2 Bereavement, grief, and mourning

Terminology	Definition
Bereavement	The objective experience of the death of a loved one
Grief	The natural response to bereavement
Mourning	A psychological process of adapting to the loss, which can be influenced by social, cultural, and religious beliefs. It entails effective emotion regulation and new learning.

attachment often have difficulty with affect regulation (Mikulincer et al., 2003) and experience heightened levels of negative emotions, including sadness, anger, shame, and anxiety, in addition to lower positive emotions. These people tend to experience low self-esteem and low confidence in the esteem of others, often enter dependent relationships, and are especially vulnerable to stress.

Attachment security also contributes to motivation for interactions in the world. Thus, the attachment system modulates the activity of the exploratory system (Elliott et al., 2003). The latter, in turn, motivates active interest in the environment, needed for learning and overall effective functioning in the world. A physically accessible and emotionally responsive caregiver who provides a sense of security facilitates exploration and learning and minimizes fear. In other words, the biobehavioral systems for attachment, fear, and exploration are linked. Considering that the death of a loved one disrupts attachment functioning, there is associated inhibition of the exploratory system in addition to impaired regulation of fear and other emotions for a period after the death of a close attachment.

An important aspect of adult attachment relationships is that caregiving is as important as receiving care (Feeney & Collins, 2001). Among those who lose a loved one, and a child in particular, effects on the caregiving system are likely to be dominant. A very typical, though often neglected, aspect of grief is the occurrence of self-blame, related to this caregiving function. In summary, the experience of acute grief is characterized by aspects of a traumatic response to the loss, impairment in attachment functioning, inhibition of exploratory functioning, impaired emotion regulation, and a tendency for caregiver self-blame. It is useful to think of acute grief as having these components. Over time, as the bereaved person adapts to the loss and its permanence, the trauma is resolved, attachment functioning is reestablished, and interactions with the world resume, along with effective emotion regulation and resolution of self-blame. Grief recedes into the background, lessening in intensity and dominance.

Ideas about grief were outlined by Freud and Bowlby. According to Freud and his followers, effective adjustment to bereavement requires a period of "grief work." Although not explicitly defined, and sometimes misconstrued to mean a necessary period of intense emotional pain, the idea of grief work has some merit. The idea is that through a process of "reality testing" in which there is repeated confrontation with unpleasant reminders of the loss, the bereaved person eventually accepts the reality of the loss, and pervasive feelings of intense sadness and yearning for reunion with the deceased subside. Unless this is accomplished, the bereaved person remains shackled through an ongoing, unrevised attachment to the deceased. Continued longing and searching for the deceased restrict the freedom of a bereaved person, especially the ability to engage deeply with the living and at times to find pleasure and meaning in life after the death. These people remain preoccupied with thoughts of the deceased loved one and are persistently laden with pain and suffering caused by their loss. Grief does not evolve to a state that is less prominent or impairing, in contrast to integrated grief where there may be heightened periods of distress on anniversary dates or with other important reminders, but it is less pervasive.

Most of these ideas hold true, although often a strong sense of connection to the deceased person remains, even after the loss has been integrated and acute grief subsides. Additionally, there is a newer model, supported by some empirical data, which posits that coping with bereavement entails coping, in tandem, with restoration as well as loss (Stroebe & Schut, 1999). Another difference is that the traditional model considers pathological grief to be the consequence of an ambivalent relationship. In contrast to this theory, PGD appears to occur instead when someone has lost a very special and very positive relationship. Table 3 lists some common ideas about grief, and the empirical findings related to them.

Bereavement is widely understood to be a severe stressor. In fact, as Stroebe and Schut (1999) pointed out, loss of a loved one is not one stressor, but rather many simultaneous stressors. Our close friends and family members impact

Table 3 Two forms of usual grief (Shear et al., 2017)

Acute Grief (Painful and Preoccupying)	Integrated Grief (Bittersweet, in the background)
Waxing and waning sense of disbelief, difficulty comprehending and/or accepting reality of the death	Acceptance of the reality of the loss, a changed relationship to the deceased and other life changes
Frequent preoccupying memories, thoughts images of the deceased	Memories, thoughts, images accessible, less frequent and preoccupying
Prominent yearning, longing mix of positive and painful emotions; painful ones usually dominant and intense	Mix of emotions, mostly bittersweet or positive, less intense and frequent
Little interest in ongoing life, sense of unease in interacting with others	Restoration of the capacity for wellbeing, comfort, and confidence in activities and relationships

many areas of our lives. When we lose them, our lives are profoundly changed, and we must adapt to these changes. Some aspects of bereavement are related to the loss. Loved ones provide us with a sense of purpose. We do things to make them feel proud, to bring them happiness, to keep them close, and to help them feel safe. A close relationship is an important source of pleasure and solace. We may attain social status, financial security, and/or self-esteem from our relationships with people we love. These are the people who bolster our confidence and help us feel important and proud of ourselves. Our loved ones provide a sense of balance and completeness. They comfort us and humor us when we are feeling vulnerable or hurt. Bereavement entails loss-related stressors such as these. In addition to psychological support, people with whom we live may help regulate our neuroendocrine system, influence our sleep and social rhythms, and provide a myriad of small cues that can trigger conditioned emotional responses. Places and things may have special meaning because of associations with a loved one. Activities may be satisfying because they are shared.

Other aspects of adjustment to an important loss are related to how we re-envision our lives without the person who died and restore our capacity for wellbeing. This may be accompanied by restoration-related stressors such as managing financial matters and taking on household tasks that were previously done by the deceased person. Social life is reorganized; it may be necessary for someone who was staying at home to become employed. There may be children or other dependents who need to be looked after. Different kinds of plans may need to be changed. Thus, bereavement is not just one, but sometimes a virtual encyclopedia of severe stressors.

The stress of bereavement clearly requires active coping. Some information is available about which coping methods are adaptive and which are not. Folkman outlined models of coping that she and her colleagues have developed. First, they define coping as "the changing thoughts and acts that an individual uses to manage the external or internal demands of stressful situations" (Stroebe et al., 2001, p. 565). They remind readers that coping and mastery are not the same things. Additionally, coping occurs in an environmental context that is a continually unfolding process that may change over time following the stressor. Coping with a specific stressor is usually multidimensional with some problem-focused and some emotion-focused strategies, some approach and some avoidance, and some interpersonal and some intrapersonal.

Bonanno (2001, 2004) employed a social–functional model in considering the emotional response to bereavement. This empirically based model posits an association between negative affect, disrupted social relationships, and later physical and mental health problems across a range of situations. Thus, in contradistinction to traditional ideas about "grief work," a social-functional perspective predicts that internal psychological functioning and adaptive social functioning are impeded by painful negative emotions and enhanced by positive emotions. Studies of the bereavement of AIDS caregivers also support the importance of positive mood states in adjustment to bereavement (Moskowitz et al., 2003). Except for a few weeks immediately after the death, positive mood states were as prevalent among AIDS caregivers as negative ones during both caregiving and bereavement. Active problem solving and positive reappraisal were important in maintaining positive mood states, and these, in turn, helped foster more adaptive coping. It is likely that the presence of positive mood is helpful in adapting to loss and might explain why of a pre-existing or co-occurring mood disorder is a risk factors for poor outcomes after an important loss.

Wortman and Silver (1989) wrote a seminal paper challenging traditional concepts of grief, which they described as myths. They concluded that it was not proven that grief work was essential to enable the bereaved to come to terms with their loss. They identified two patterns of normal grieving, one in which a bereaved person moves from high to low emotional distress over time and the other in which people never experience high distress. A third grief trajectory is that of persistent high distress, continuing for years. This is the pattern of Prolonged Grief Disorder (PGD; previously known as complicated grief.), which has since been demonstrated in multiple longitudinal trajectory studies (Djelantik et al., 2018; Sveen et al., 2018).

Field and colleagues reported a series of studies of continuing bonds (Field et al., 2005; Field, 2006). The results suggested that some types of continued connection with the deceased were associated with more prolonged grief. However, grief was not more intense when the connection was related to thoughts about how a deceased spouse made a difference in the life of the bereaved person or gaining enjoyment from reminiscences about the deceased. They highlighted the distinction between behavioral manifestations of continuing bonds, as compared with feelings of connectedness. The internalized representation seems to allow the deceased to play a sustaining role in the emotional life of the bereaved person that is not inconsistent with the reality of their death. Once integrated in this way, thoughts and memories of the deceased continue to grow and change (Neimeyer et al., 2002).

Sociocultural factors often play a role in grief and mourning. Perhaps more than most stressors, bereavement is a public event that occurs in a larger social context. Most cultures provide rules for the disposition of a deceased body, and some supply procedures for disposing of the belongings of the decease. Cultures frequently contain beliefs about the proper public expression of emotions and private thoughts, and about the need to confront or avoid emotionally provocative situations. Religion has an obvious place in the process of coping.

Death rituals and customs are universal components of religious rites. One way in which cultural/religious practices exert their influence is through directives regarding the expression of emotion after a loss. Formal ceremonies often encourage the bereaved to confront and release emotions. However, the value placed on emotional expression varies in different cultures. For example, in some cultures, emotional expression is thought to be a threat, interfering with good judgment, and affecting others, in addition to oneself. In others, open display of emotions is thought to be needed to cleanse the soul. In some cultures, ancestors, who have the power to do harm to the bereaved, are thought to be offended by excessive shows of emotion. In some cultures, there is a focus on working through grief in order to distance oneself from a deceased loved one, whereas in others, the emphasis is on maintaining a relationship with the dead ancestor who is treated as a living person, being offered food and other things, and seen as remaining present in the life of the bereaved.

Religious beliefs help mourners envision a transition of their loved one to a final resting place. This provides an important form of continuity with the person who died. Being able to locate imaginally a deceased person seems to be helpful in the process of coping with the death, especially if this contains an idea that the spirit is at peace. Religious rituals also focus on marking a separation of the dead. In some cultures, a change in residence is prescribed, along with removing all physical reminders of the dead and avoiding mention of the name of the deceased. Some native American tribes have a 4-day accepted length of mourning, after which a bereaved person is expected to return to normal life and neither speak of the deceased nor discuss personal feelings about the loss. These practices are associated with a belief that too much emotion is offensive to the deceased. Interestingly, and in line with findings from naturalistic studies summarized earlier, it has been reported that members of cultures with such final ceremonies appeared less likely to experience chronic grief. Many religious practices dictate a prescribed period of mourning. For Jews, after a year of mourning there is an unveiling ceremony at the gravesite (i.e., placing of the tombstone). Clothes that were rent at the time of the death are mended and worn as a sign that life, although scarred, must go on. Some research with groups such as with Cambodian refugees who were unable to follow their usual cultural rituals after a death suggests this interference may heighten prolonged grief symptoms (Hinton et al., 2013).

In summary, the usual trajectory of grief following loss of someone very close (see Fig. 1) is for the bereaved person to experience acute grief lasting as short as only a few days or as long as 6–12 months and characterized by symptoms related to (1) trauma (e.g., disbelief, intrusive thoughts and memories, a tendency to avoidance), (2) an activated attachment system (e.g., longing, yearning, searching for the deceased, sense of insecurity or self-doubt, intense sadness), (3) emotional and physiological dysregulation, (4) thoughts and feelings of self-blame, and (5) inhibition of interest and engagement with ongoing life. In the context of cultural rituals and social support, the bereaved person is usually able to accept the reality of the loss and restore their capacity for wellbeing. As they do so, acute grief symptoms recede into the background. Grief is not fully over, but it is accepted as a part of ongoing life. Memories and thoughts of the loved one are still accessible but no longer a central preoccupation. Longing, yearning, and searching lessen. Emotional and physiological regulation is reestablished. Feelings of security and self-confidence are restored, and there is renewed interest and engagement in ongoing life. Metaphorically speaking, the deceased loved one has moved from preoccupying the mind of the bereaved person to residing comfortably in their heart.

Sometimes, this progression does not occur smoothly. Heterogeneous clinical trajectories are identified following bereavement (Maccallum et al., 2015; Lenferink et al., 2020). While most bereaved individuals fall into the resilient category (i.e., minimal to no persistent highly distressing symptoms post-loss), others who experience especially intense symptoms either respond adaptively to the loss or develop a mental or physical condition. Bereavement is a major stressor that can trigger the onset of a range of psychiatric disorders, e.g., a mood or anxiety disorder, PTSD, alcohol use disorder (Keyes et al., 2014). Importantly, DSM-5-TR includes a new disorder of Prolonged Grief Disorder, affecting a clinically significant minority of the bereaved, in whom intense yearning, longing, and/or preoccupation with thoughts and memories of the deceased persists along with other symptoms of pervasive grief that interfere with functioning (Fig. 2).

Bereavement-Related Depression

Depression is an important consequence of bereavement in a minority of individuals. Although grief resembles depression it can be clearly distinguished from depression. A seminal study showed that about one-third of widows and widowers met major depression criteria 1 month following bereavement, 25% met these criteria at 7 months, and approximately 15% at 1 year (Zisook et al., 2010). Bereavement-related major

Fig. 1 The trajectory of usual grief

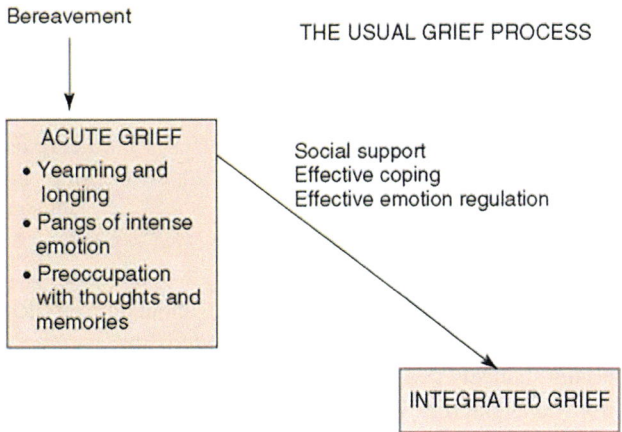

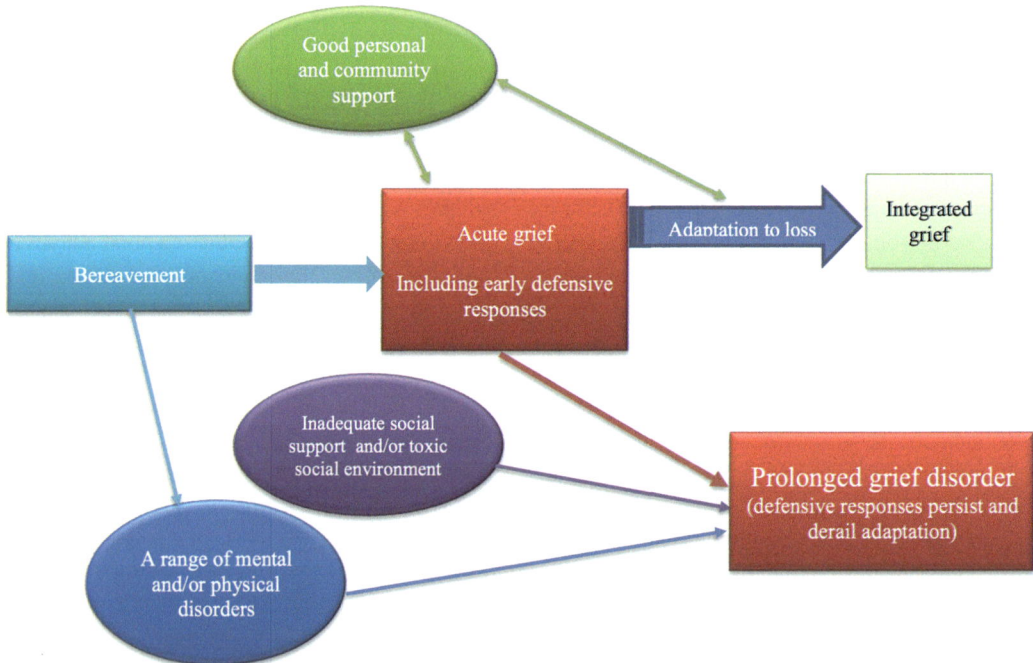

Fig. 2 Bereavement and its consequences

MDD is like non-bereavement related MDD, in terms of clinical, biological, and personality characteristics; comorbidity patterns; disease course; treatment response; and recurrence risk (Zisook et al., 2007). A study showed that depression in the first 2 months of bereavement can be effectively treated with antidepressant medication (Zisook et al., 2001). Although the DSM-IV included an exclusion for major depression in the first 2 months of bereavement, the preponderance of evidence demonstrating that bereavement-related depression is similar to depression following any other life stressor and therefore should be diagnosed and treated (Zisook & Kendler, 2007; Zisook et al., 2007) led to removal of the bereavement exclusion for major depression in DSM-5.

Post-Traumatic Stress Disorder

PTSD criteria in DSM-5 include learning about or witnessing the death of a loved one *only* if this occurs suddenly and unexpectedly and explicitly from unnatural (i.e., violent or accidental) causes. However, again, studies have documented that PTSD does occur in the wake of bereavement

from other events (i.e., chronic illness, unexpected deaths from natural causes, etc.) (Zisook et al., 1998). In fact, in at least one community sample, bereavement was the most common trigger of PTSD (Breslau et al., 1998). A recent study documents the occurrence of threshold-level PTSD symptoms in many individuals seeking treatment for PGD (Na et al., 2021). As discussed above, loss of a close attachment is often experienced as a trauma and typically triggers a response that includes a sense of disbelief, intrusive thoughts, avoidance, numbing, and dysregulated emotions. However, grief is the response to the absence of a close attachment whereas PTSD is a response to the presence of a physical threat. The core feature of grief is intense yearning and longing, not usually a prominent feature of PTSD, where the core symptom is fear and anxiety.

Prolonged Grief Disorder

Clinicians have long recognized prolonged intense grief reactions that occur in a significant minority of bereaved individuals as a clinical condition that is distinct from major depressive disorder and PTSD (Horowitz et al., 1997; Zisook et al., 2010; Simon, 2012). Over the years, this psychiatric condition has been labeled in various ways, including pathological, unresolved, chronic, delayed, traumatic, or complicated grief. In 2013, the DSM-5 included persistent complex bereavement disorder with provisional criteria for this condition (APA, 2013). The International Classification of Diseases included a new diagnosis of Prolonged Grief Disorder (PGD) in the 11th revision (Table 4). Following further research and extensive engagement with the field, DSM-5-TR, followed suit and there is now a new diagnosis of PGD in the DSM-5-TR section on trauma and stressor related disorders (Prigerson et al., 2021a, for review).

Since there was no official criteria set until 2020, there are no epidemiologic studies of prevalence. A reasonable estimate is that PGD affects approximately 2–4% of the general adult population worldwide, and roughly 10% of those grieving a loss from natural causes (Lundorff et al., 2017). Rates of PGD appear to be higher when a death occurs in a sudden, violent way (e.g., suicide, homicide, accident, natural disaster) (Neria et al., 2007; Tal et al., 2017; Kristensen et al., 2020). Rate of PGD is also expected to be higher following the death of a life partner or a child. For those losing a spouse, the estimated rate is 10–20% (Lobb et al., 2010); for those losing a child, it can be as high as 60% (Meert et al. 2010).

Risk factors for PGD are similarly provisional due to the lack of official criteria until this year. However, they can generally be grouped into pre-loss, loss-related, and peri-loss factors (Simon, 2013; Shear, 2015). Pre-loss factors might include female sex, older age, non-Caucasian race, past or coexisting mood, anxiety, trauma-related, alcohol or substance use disorders, higher medical comorbidities, anxious/avoidant attachment style, prior history of loss; loss-related factors include violent, sudden, or unexpected deaths; and peri-loss factors include inadequate social support, unemployment, and interference with mourning rituals due to circumstances, such as the coronavirus pandemic (Goveas & Shear, 2020; Simon et al.,

Table 4 DSM-5-TR and ICD-11 Diagnostic Criteria for Prolonged Grief Disorder

DSM-5-TR PGD criteria	ICD-11 PGD criteria

A. The death of a person close to the bereaved at least 12 months previously
B. Since the death, clinically significant degree of grief response characterized by intense yearning/longing for the deceased person or a preoccupation with thoughts *or* memories of the deceased person, occurring nearly every day for at least the last month
C. At least 3 of the following symptoms that is experienced to a clinically significant degree, nearly every day for at least the last month:
  • Identity disruption
  • Marked sense of disbelief about the death
  • Avoidance of reminders that the person is dead
  • Intense emotional pain (e.g., anger, bitterness, sorrow) related to the death.
  • Difficulty moving on with life
  • Emotional numbness
  • Feeling that life is meaningless.   Intense loneliness
D. The disturbance causes clinically significant distress or impairment in social, occupational, or other important areas of functioning
E. The duration of bereavement reaction clearly exceeds expected social, cultural, or religious norms for the individual's culture and context
F. The symptoms are not better explained by another mental disorder
	A. Death of a partner, parent, child, or other person close to the bereaved
	B. Persistent and pervasive grief response:
	  • Yearning/longing for the deceased *or* persistent preoccupation with the deceased
	  • Accompanied by intense emotional pain:
	    • Sadness, guilt, anger, denial, blame
	    • Difficulty accepting the death
	    • Feeling one has lost a part of one's self
	    • An inability to experience positive mood
	    • Emotional numbness
	    • Difficulty in engaging with social or other activities
	C. Grief response has persisted for an atypically long period following the loss (more than 6 months minimum)
	D. The grief response clearly exceeds expected social, cultural or religious norms for the individual's culture and context
	E. The disturbance causes significant impairment in personal, family, social, educational, occupational or other important areas of functioning.

2020a) that cause interruptions in social interactions, supportive activities and relationships.

A burgeoning body of research is shedding light on possible neurobiological mechanisms underlying PGD. Consistent with a theorized role for attachment and emotion regulation processes, brain structure and function abnormalities in the limbic, prefrontal, cingulate, and reward processing brain regions have been reported in PGD (O'Connor, 2012; Kakarala et al., 2020). Acute grief has also been associated with elevated proinflammatory cytokines, dysregulated hypothalamic-pituitary-adrenal axis activity (e.g., elevated cortisol levels), decreases in natural killer cell activity, and diminished lymphocytic response to pathogens (O'Connor, 2012). Identifying such neurobiological measures is vital as they, in the future, may aid in understanding mechanistic similarities and differences between acute grief and PGD and might help predict those at increased risk for developing PGD. Discovery of key biomarkers may also help identify candidates for early initiation of currently available interventions for PGD, and aid in developing novel interventions to prevent or treat PGD.

The clinical hallmark of PGD is the persistence of pervasive intense grief well beyond the time by which this would be expected to subside within an individual's social, cultural, and religious traditions. It is useful to consider the condition as related to factors that impede the process of adapting to the loss, including biological, psychological, or social factors. Prolonged grief disorder is manifest as persistent pervasive yearning, longing, or preoccupation with the deceased, along with at least 3 of a group of associated symptoms and significant distress and/or impairment present nearly every day at least 12 months after a death. Clinically, patients may describe feeling as though their loved one had died very recently, engage in self-criticism about their grief (e.g., fear, embarrassment, or shame regarding their own emotions) and/or experience impaired attention, concentration, or memory; they may focus on catastrophic misinterpretation of the consequences of separation (e.g., ideas that they cannot survive or function without the person who died; thoughts of betrayal or guilt regarding the death), experience impairing avoidance of reminders of the permanence of the loss and/or compulsive proximity seeking. Patients often experience impaired emotion regulation, and/or social-environmental toxicity or resource depletion (see Fig. 3).

The PGD diagnostic criterion in the DSM-5-TR harmonize well with the ICD-11. However, there are some dissimilarities (Table 4) related to differences in the intent of these different diagnostic systems. ICD-11 is intended as a guide for clinicians and does not include specific algorithms. By contrast, DSM-5-TR diagnosis requires a specific algorithm. Either approach can be used to diagnose PGD. Another difference is that DSM-5-TR requires a minimum of

Fig. 3 Prolonged Grief Disorder: A closed loop

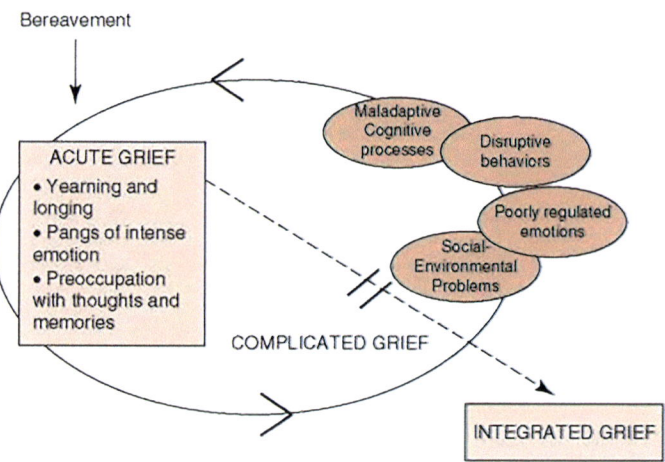

12 months since the death versus 6 months in ICD-11. Despite these minor differences, the inclusion of guidance for establishing a new diagnosis of PGD is a very important step in the ongoing process of understanding and treating PGD, a form of grief that results in immense suffering and chronic disability. PGD has also been associated with serious work and social impairment, long-term health problems (e.g., cardiovascular disease, cancer, etc.), cognitive decline, poor quality of life, functional impairments, substance abuse, and increased risk of suicide and premature mortality (Prigerson et al., 1997; Stroebe et al., 2007; Shear et al., 2011; Perez et al., 2018).

Suicide bereavement deserves special mention in this context. Many suicides occur in a context of ongoing mental health treatment. This means clinicians working with suicidal patients need to be prepared to support loved ones if the patient completes suicide. Suicide bereavement is widely believed to be especially difficult to cope with because of stigma that may surround suicide and the strong likelihood that friends and family members may blame themselves for not preventing the death. A sensitive, available clinician can be helpful in providing effective bereavement support and in mitigating these painful aspects of suicide bereavement.

Research has demonstrated that individuals with PGD have elevated rates of concomitant psychopathology, mainly depression, PTSD, and substance use disorders (Shear et al., 2018). Moreover, bereaved individuals with mood and anxiety disorders are at increased risk for PGD (Simon et al., 2007; Sung et al., 2011; Marques et al., 2013). Research into the psychological characteristics of the condition has also identified a range of significant components such as avoidance behaviors, emotion dysregulation, and deficits in future thinking.

PGD can be identified in clinical practice with self-reported screening tools such as the 5-item Brief Grief Questionnaire (Ito et al., 2012), the 19-item Inventory of Complicated Grief (Prigerson et al., 1995), and the Prolonged Grief Disorder-13 (PG-13) scales, with the PG-13-R having been revised to align with the DSM-5-TR criteria (Prigerson et al., 2021b). The Structured Clinical Interview for Complicated Grief (SCI-CG) is a validated, clinician-administered semi-structured interview that can reliably identify individuals with PGD (Bui et al., 2015; Shear, 2015). In addition, clinicians should also assess for comorbid psychopathology, including depressive, anxiety, trauma- and stressor-related, alcohol and substance use disorders, suicidality, and cognitive impairment (the latter especially in older adults).

While bereaved individuals benefit from support, most do not require clinical intervention, however treatment is indicated for PGD. Psychotherapeutic interventions are the first-line treatment as there is yet no evidence that medication is effective in relief of grief symptoms. The best validated treatment is Prolonged Grief disorder Therapy (PGDT) (also called Complicated Grief Treatment [CGT]). PGDT is a manualized, 16-session protocol that includes an eclectic mix of techniques used in cognitive behavioral therapy, interpersonal psychotherapy (IPT), motivational interviewing, and psychodynamic psychotherapy. PGDT targets adaptation to loss conceptualized as: (1) accepting the reality of the loss (permanence of the loss and grief; changes in the relationship to the deceased and associated changes in the life of the bereaved), and (2) restoring the capacity for wellbeing (e.g., a sense of purpose and meaning in life, sense of competence and feelings of belonging and mattering). These are addressed by supporting seven HEALING milestones and by identifying and addressing common impediments (or DERAILERS) to adaptation (Goveas & Shear, 2020; Center for Prolonged Grief, http://www.complicatedgrief.columbia.edu). Three randomized-controlled trials funded by the National Institute of Mental Health have been conducted to-date: In Study 1 ($n = 95$ middle aged adults), PGDT was found to show superior efficacy over interpersonal psychotherapy: 51% responded to PGDT versus 32% to IPT (Shear et al., 2005). The results of Study 2 ($n = 151$ older adults) were similar (70% PGDT vs. 32% IPT (Shear et al., 2014). Study 3, a 4-site study ($n = 394$) with a range of adults, compared citalopram to placebo, with and without PGDT

(Shear, Reynolds, Simon, Zisook et al., 2016). Results showed no difference between citalopram and placebo alone and no difference in outcome for PGDT when combined with citalopram versus placebo. Response to PGDT was clinically and statistically significantly greater than for those who did not receive this therapy. Of note, though, addition of citalopram to PGDT significantly improved response to depressive symptoms. These findings highlight that PGDT, not antidepressants, is the treatment of choice for PGD; however, antidepressants may aid in effectively treating a co-occurring depressive syndrome. A substudy of Study 3 included showed efficacy for PGDT among suicide-bereaved individuals ($n = 56$) (Zisook et al., 2018). PGDT was also effective when administered in a group format to older adults with PGD (Supiano & Luptak, 2014).

Several different studies of grief-focused CBT also suggest efficacy of this approach. Boelen et al. (2007) found CBT superior to supportive counseling in study completers. Bryant et al. (2014) studied group CBT with and without an exposure component related to the story of the death and found the exposure component resulted in significantly better success in reducing prolonged grief symptoms. CBT that included the exposure component also showed durable benefits over a 2-year follow-up (Bryant et al., 2017). An internet-based PGD-focused CBT was more efficacious than wait-list controls or usual care (Kersting et al., 2013). Boelen et al. (2021) reported a promising study of CBT for PGD in children and adolescents.

Comorbidity of depression or PTSD is common in PGD and does not moderate outcome to PGDT. However, PGD comorbidity does appear to moderate outcome of PTSD (Simon et al., 2020b) and depression treatments (Zisook et al., 2019). In these studies, PGD was associated with poorer treatment response to proven efficacious interventions for PTSD and depression, respectively (Bui et al., 2012). These studies highlight the importance of screening for and adequately managing coexisting PGD in those with PTSD and MDD.

Other interventions have also showed some promise for PGD. Psychological and lifestyle interventions may have a role in preventing the development of bereavement-related complications, such as depression, PTSD, and PGD, in high-risk bereaved individuals (Litz et al., 2014; Stahl et al., 2017; Johannsen et al., 2019), though the evidence for such preventative interventions is currently preliminary. CBT with Eye Movement Desensitization and Reprocessing to reduce PGD and PTSD symptoms in homicidally bereaved individuals are being examined (Lenferink et al., 2017).

In summary, psychiatrists encounter death and bereavement in many aspects of their work. Psychiatrists can be helpful in the assessment and treatment of dying patients and their caregivers. Psychiatrists play a central role in the management of suicidality. There are an estimated 22.5 million bereaved people every year in the United States, and at least 10–20% of these people will suffer from bereavement-related psychiatric illness. It is important that psychiatrists understand how to diagnose and treat people suffering from bereavement-related mood and trauma-related disorders. Importantly, clinicians, researchers, and policymakers need to be aware of the new ICD-11 and DSM5-TR diagnosis of PGD and the evidence-based treatments that are available for this condition. Psychiatrists and other mental health professionals need to learn how to diagnose and treat this condition as well as other mental health problems that occur in the wake of bereavement.

References

Ahmad, F. B., & Anderson, R. N. (2021). The leading causes of death in the US for 2020. *Journal of the American Medical Association, 325*(18), 1829–1830.

American Psychiatric Association. (2013). *Diagnostic and statistical manual of mental disorders* (5th ed., p. DSM-5). American Psychiatric Publishing.

Bedell, S. E., Cadenhead, K., & Graboys, T. B. (2001). The doctor's letter of condolence. *New England Journal of Medicine, 344*(15), 1162–1164.

Blinderman, C. D., & Billings, J. A. (2015). Comfort care for patients dying in the hospital. *New England Journal of Medicine, 373*, 2549–2561.

Boelen, P. A., de Keijser, J., van den Hout, M. A., et al. (2007). Treatment of complicated grief: A comparison between cognitive-behavioral therapy and supportive counseling. *Journal of Consulting and Clinical Psychology, 75*(2), 277–284.

Boelen, P.A., Lenferink, L.I.M., & Spuij, M. (2021). CBT for prolonged grief in children and adolescents: A randomized clinical trial. *American Journal of Psychiatry*, appiajp202020050548.

Bonanno, G. A. (2001). Grief and emotion: A social-functional perspective. In M. S. Stroebe, R. O. Hansson, W. Stroebe, et al. (Eds.), *Handbook of bereavement research: Consequences, coping, and care*. American Psychological Association.

Bonanno, G. A. (2004). Loss, trauma, and human resilience: Have we underestimated the human capacity to thrive after extremely aversive events? *American Psychologist, 59*(1), 20–28.

Bowlby, J. (1980). *Loss*. Basic Books.

Breslau, N., Kessler, R. C., Chilcoat, H. D., Schultz, L. R., Davis, G. C., & Andreski, P. (1998). Trauma and posttraumatic stress disorder in the community: The 1996 Detroit area survey of trauma. *Archives of General Psychiatry, 55*(7), 626–632.

Bretherton, I. (1999). Updating the 'internal working model' construct: Some reflections. *Attachment & Human Development, 1*(3), 343–357.

Bryant, R. A., Kenny, L., Joscelyne, A., et al. (2014). Treating prolonged grief disorder: A randomized clinical trial. *JAMA Psychiatry, 71*(12), 1332–1339.

Bryant, R. A., Kenny, L., Joscelyne, A., et al. (2017). Treating prolonged grief disorder: A 2-year follow-up of a randomized controlled trial. *Journal of Clinical Psychiatry, 78*(9), 1363–1368.

Bui, E., Nadal-Vicens, M., & Simon, N. M. (2012). Pharmacological approaches to the treatment of complicated grief: Rationale and a brief review of the literature. *Dialogues in Clinical Neuroscience, 14*(2), 149–157.

Bui, E., Mauro, C., Robinaugh, D. J., et al. (2015). The structured clinical interview for complicated grief: Reliability, validity, and exploratory factor analysis. *Depression and Anxiety, 32*(7), 485–492.

Collins, N. L., & Feeney, B. C. (2000). A safe haven: An attachment theory perspective on support seeking and caregiving in intimate relationships. *Journal of Personality and Social Psychology, 78*(6), 1053–1073.

Djelantik, A. A. A. M. J., Smid, G. E., Kleber, R. J., & Boelen, P. A. (2018). Early indicators of problematic grief trajectories following bereavement. *European Journal of Psychotraumatology, 8*(sup6), 1423825.

Elliott, R., Watson, J. C., & Goldman, R. N., et al. (2003). Empathy and exploration: The core of process-experiential therapy. In *Learning emotion-focused therapy: The process-experiential approach to change* (pp. 111–140). American Psychological Association.

Feeney, B. C. (2004). A secure base: Responsive support of goal strivings and exploration in adult intimate relationships. *Journal of Personality and Social Psychology, 87*(5), 631–648.

Feeney, B. C., & Collins, N. L. (2001). Predictors of caregiving in adult intimate relationships: An attachment theoretical perspective. *Journal of Personality and Social Psychology, 80*(6), 972–994.

Field, N. P., Gao, B., & Paderna, L. (2005). Continuing bonds in bereavement: An attachment theory based perspective. *Death Studies, 29*(4), 277–299.

Field, N. P. (2006). Continuing bonds in adaptation to bereavement: Introduction. *Death Studies, 30*(8), 709–714.

Goveas, J. S., & Shear, M. K. (2020). Grief and the COVID-19 pandemic in older adults. *American Journal of Geriatric Psychiatry, 28*(10), 1119–1125.

Hazan, C., & Zeifman, D. (1999). Pair bonds as attachments: Evaluating the evidence. In J. Cassidy & P. R. Shaver (Eds.), *Handbook of attachment: Theory, research, and clinical applications* (pp. 336–354). Guilford Press.

Hinton, D. E., Peou, S., Joshi, S., Nickerson, A., & Simon, N. M. (2013). Normal grief and complicated bereavement among traumatized Cambodian refugees: Cultural context and the central role of dreams of the dead. *Culture, Medicine and Psychiatry, 37*(3), 427–464.

Horowitz, M. J., Siegel, B., Holen, A., et al. (1997). Diagnostic criteria for complicated grief disorder. *American Journal of Psychiatry, 154*(7), 904–910.

Hosker, C. M. G., & Bennett, M. I. (2016). Delirium and agitation at the end of life. *British Medical Journal, 353*, i3085.

Ito, M., Nakajima, S., Fujisawa, D., et al. (2012). Brief measure for screening complicated grief: Reliability and discriminant validity. *PLoS One, 7*(2), e31209.

Johannsen, M., Damholdt, M. F., Zachariae, R., Lundorff, M., Farver-Vestergaard, I., & O'Connor, M. (2019). Psychological interventions for grief in adults: A systematic review and meta-analysis of randomized controlled trials. *Journal of Affective Disorders, 253*, 69–86.

Kakarala, S. E., Roberts, K. E., Rogers, M., et al. (2020). The neurobiological reward system in prolonged grief disorder (PGD): A systematic review. *Psychiatry Research: Neuroimaging, 303*, 111135.

Kentish-Barnes, N., Chevret, S., Champigneulle, B., et al. (2017). Effect of a condolence letter on grief symptoms among relatives of patients who died in the ICU: A randomized clinical trial. *Intensive Care Medicine, 43*(4), 473–484.

Kersting, A., Dolemeyer, R., Steinig, J., et al. (2013). Brief internet-based intervention reduces posttraumatic stress and prolonged grief in parents after the loss of a child during pregnancy: A randomized controlled trial. *Psychotherapy and Psychosomatics, 82*(6), 372–381.

Keyes, K. M., Pratt, C., Galea, S., McLaughlin, K. A., Koenen, K. C., & Shear, M. K. (2014). The burden of loss: Unexpected death of a loved one and psychiatric disorders across the life course in a national study. *American Journal of Psychiatry, 171*(8), 864–871.

Kristensen, P., Dyregrov, K., & Gjestad, R. (2020). Different trajectories of Prolonged grief in bereaved family members after terror. *Frontiers in Psychiatry, 11*, 545368.

Lenferink, L. I. M., Piersma, E., de Keijser, J., Smid, G. E., & Boelen, P. A. (2017). Cognitive therapy and eye movement desensitization and reprocessing for reducing psychopathology among disaster-bereaved individuals: Study protocol for a randomized controlled trial. *European Journal of Psychotraumatology, 8*(1), 1388710.

Lenferink, L. I. M., Nickerson, A., de Keijser, J., Smid, G. E., & Boelen, P. A. (2020). Trajectories of grief, depression, and posttraumatic stress in disaster-bereaved people. *Depression and Anxiety, 37*(1), 35–44.

Lewis, C. (1961). *A grief observed*. HarperCollins.

Litz, B. T., Schorr, Y., Delaney, E., et al. (2014). A randomized controlled trial of an internet-based therapist-assisted indicated preventive intervention for prolonged grief disorder. *Behaviour Research and Therapy, 61*, 23–34.

Lobb EA, Kristjanson LJ, Aoun SM, Monterosso L, Halkett GK, Davies A (2010). Predictors of complicated grief: a systematic review of empirical studies. *Death Studies, 34*(8), 673–698.

Lundorff, M., Holmgren, H., Zachariae, R., Farver-Vestergaard, I., & O'Connor, M. (2017). Prevalence of prolonged grief disorder in adult bereavement: A systematic review and meta-analysis. *Journal of Affective Disorders, 212*, 138–149.

Lyness, J. M. (2004). End-of-life care: Issues relevant to the geriatric psychiatrist. *American Journal of Geriatric Psychiatry, 12*(5), 457–472.

Maccallum, F., Galatzer-Levy, I. R., & Bonanno, G. A. (2015). Trajectories of depression following spousal and child bereavement: A comparison of the heterogeneity in outcomes. *Journal of Psychiatric Research, 69*, 72–79.

Marques, L., Bui, E., Leblanc, N., et al. (2013). Complicated grief symptoms in anxiety disorders: Prevalence and associated impairment. *Depression and Anxiety, 30*(12), 1211–1216. https://doi.org/10.1002/da.22093

Meert, K. L., Eggly, S., Pollack, M., et al. (2007). Parents' perspectives regarding a physician-parent conference after their child's death in the pediatric intensive care unit. *Journal of Pediatrics, 151*(1), 50–55.

Meert KL, Donaldson AE, Newth CJ, Harrison R, Berger J, Zimmerman J, Anand KJ, Carcillo J, Dean JM, Willson DF, Nicholson C, Shear K (2010) Eunice Kennedy Shriver National Institute of Child Health and Human Development Collaborative Pediatric Critical Care Research Network. Complicated grief and associated risk factors among parents following a child's death in the pediatric intensive care unit. *Archives of Pediatrics & Adolescent Medicine, 164*(11), 1045–1051

Mikulincer, M., Shaver, P. R., & Pereg, D. (2003). Attachment theory and affect regulation: The dynamics, development, and cognitive consequences of attachment-related strategies. *Motivation and Emotion, 27*(2), 77–102.

Moskowitz, J. T., Folkman, S., & Acree, M. (2003). Do positive psychological states shed light on recovery from bereavement? Findings from a 3-year longitudinal study. *Death Studies, 27*(6), 471–500.

Na, P. J., Adhikari, S., Szuhany, K. L., et al. (2021). Posttraumatic distress symptoms and their response to treatment in adults with prolonged grief disorder. *Journal of Clinical Psychiatry* 2021, 82(3):20m13576.

Neimeyer, R. A., Prigerson, H. G., & Davies, B. (2002). Mourning and meaning. *American Behavioral Scientist, 46*(2), 235–251.

Neria, Y., Gross, R., Litz, B., et al. (2007). Prevalence and psychological correlates of complicated grief among bereaved adults 2.5-3.5 years after September 11th attacks. *Journal of Traumatic Stress, 20*(3), 251–262.

O'Connor, M. F. (2012). Immunological and neuroimaging biomarkers of complicated grief. *Dialogues in Clinical Neuroscience, 14*(2), 141–148.

Perez, H., Ikram, M., Direk, N., & Tiemeier, H. (2018). Prolonged grief and cognitive decline: A prospective population-based study in middle-aged and older persons. *The American Journal of Geriatric Psychiatry, 26*(4), 451–460.

Prigerson, H. G., Maciejewski, P. K., Reynolds, C. F. I. I. I., et al. (1995). Inventory of complicated grief: A scale to measure maladaptive symptoms of loss. *Psychiatry Research, 59*(1–2), 65–79.

Prigerson, H. G., Bierhals, A. J., Kasl, S. V., et al. (1997). Traumatic grief as a risk factor for mental and physical morbidity. *American Journal of Psychiatry, 154*(5), 616–623.

Prigerson, H. G., Kakarala, S., Gang, J., & Maciejewski, P. K. (2021a). History and status of prolonged grief disorder as a psychiatric diagnosis. *Annual Review of Clinical Psychology, 17*, 109–126.

Prigerson, H. G., Boelen, P. A., Xu, J., et al. (2021b). Validation of the new DSM-5-TR criteria for prolonged grief disorder and the PG-13-revised (PG-13-R) scale. *World Psychiatry, 20*(1), 96–106.

Rosenstein, D. L. (2011). Depression and end-of-life care for patients with cancer. *Dialogues in Clinical Neuroscience, 13*, 101–108.

Sharman, M., Meert, K. L., & Sarnaik, A. P. (2005). What influences parents' decisions to limit or withdraw life support? *Pediatric Critical Care Medicine, 6*(5), 513–518.

Shear, K., & Shair, H. (2005). Attachment, loss, and complicated grief. *Developmental Psychobiology, 47*(3), 253–267.

Shear, K., Frank, E., Houck, P. R., et al. (2005). Treatment of complicated grief: A randomized controlled trial. *JAMA, 293*(21), 2601–2608.

Shear, K., Monk, T., Houck, P., et al. (2007). An attachment-based model of complicated grief including the role of avoidance. *European Archives of Psychiatry and Clinical Neuroscience, 257*(8), 453–461.

Shear, M. K., Simon, N., Wall, M., et al. (2011). Complicated grief and related bereavement issues for DSM-5. *Depression and Anxiety, 28*(2), 103–117.

Shear, M. K. (2012). Getting straight about grief. *Depression and Anxiety, 29*(6), 461–464.

Shear, M. K., Wang, Y., Skritskaya, N., Duan, N., Mauro, C., & Ghesquiere, A. (2014). Treatment of complicated grief in elderly persons: A randomized clinical trial. *JAMA Psychiatry, 71*(11), 1287–1295.

Shear, M. K. (2015). Clinical practice. Complicated grief. *New England Journal of Medicine, 372*(2), 153–160.

Shear, M. K., Reynolds, C. F., 3rd, Simon, N. M., et al. (2016). Optimizing treatment of complicated grief: A randomized clinical trial. *JAMA Psychiatry, 73*(7), 685–694.

Shear, M. K., Muldberg, S., & Periyakoil, V. (2017). Supporting patients who are bereaved. *British Medical Journal, 6*(358), j2854.

Shear, K., Reynolds, C., Simon, N., et al. (2018). Grief and bereavement in adults: Clinical features. *Uptodate*. Accessed 17 March 2021.

Simon, N. M., Shear, K. M., Thompson, E. H., et al. (2007). The prevalence and correlates of psychiatric comorbidity in individuals with complicated grief. *Comprehensive Psychiatry, 48*(5), 395–399.

Simon, N. M. (2012). Is complicated grief a post-loss stress disorder? *Depression and Anxiety, 29*(7), 541–544.

Simon, N. M. (2013). Treating complicated grief. *JAMA, 310*(4), 416–423.

Simon, N. M., Saxe, G. N., & Marmar, C. R. (2020a). Mental health disorders related to COVID-19-related deaths. *JAMA, 324*(15), 1493–1494.

Simon, N. M., Hoeppner, S. S., Lubin, R. E., et al. (2020b). Understanding the impact of complicated grief on combat related posttraumatic stress disorder, guilt, suicide, and functional impairment in a clinical trial of post-9/11 service members and veterans. *Depression and. Anxiety, 37*(1), 63–72.

Stagg, E. K., & Lazenby, M. (2012). Best practices for the nonpharmacological treatment of depression at the end of life. *The American Journal of Hospice & Palliative Care, 29*(3), 183–194.

Stahl, S. T., Emanuel, J., Albert, S. M., et al. (2017). Design and rationale for a technology-based healthy lifestyle intervention in older adults grieving the loss of a spouse. *Contemporary Clinical Trials Communications, 8*, 99–105.

Stroebe, M., & Schut, H. (1999). The dual process model of coping with bereavement: Rationale and description. *Death Studies, 23*(3), 197–224.

Stroebe, M. S., Hansson, R. O., Stroebe, W., et al. (Eds.). (2001). *Handbook of bereavement research: Consequences, coping, and care*. American Psychological Association.

Stroebe, M., Schut, H., & Stroebe, W. (2007). Health outcomes of bereavement. *Lancet, 370*(9603), 1960–1973.

Sung, S. C., Dryman, M. T., Marks, E., et al. (2011). Complicated grief among individuals with major depression: Revalence, comorbidity, and associated features. *Journal of Affective Disorders, 134*(1–3), 453–458.

Supiano, K. P., & Luptak, M. (2014). Complicated grief in older adults: A randomized controlled trial of complicated grief group therapy. *Gerontologist, 54*(5), 840–856.

Sveen, J., Bergh Johannesson, K., Cernvall, M., & Arnberg, F. K. (2018). Trajectories of prolonged grief one to six years after a natural disaster. *PLoS One, 13*(12), e0209757.

Tal, I., Mauro, C., Reynolds, C. F., et al. (2017). Complicated grief after suicide bereavement and other causes of death. *Death Studies, 41*(5), 267–275.

Triplett, K. N., Tedeschi, R. G., Cann, A., et al. (2012). Posttraumatic growth, meaning in life, and life satisfaction in response to trauma. *Psychological Trauma Theory Research Practice and Policy, 4*(4), 400–410.

Verdery, A. M., Smith-Greenaway, E., Margolis, R., & Daw, J. (2020). Tracking the reach of COVID-19 kin loss with a bereavement multiplier applied to the United States. *Proceedings of the National Academy of Sciences, 117*(30), 17695–17701.

World Health Organization. (2020). *International statistical classification of diseases and related health problems* (11th ed.). https://icd.who.int/browse11/l-m/en#/ http://id.who.int/icd/entity/1183832314

Wortman, C. B., & Silver, R. C. (1989). The myths of coping with loss. *Journal of Consulting and Clinical Psychology, 57*(3), 349–357.

Zisook, S., Chentsova-Dutton, Y., & Shuchter, S. R. (1998). PTSD following bereavement. *Annals of Clinical Psychiatry, 10*(4), 157–163.

Zisook, S., Shuchter, S. R., Pedrelli, P., et al. (2001). Bupropion sustained release for bereavement: Results of an open trial. *Journal of Clinical Psychiatry, 62*(4), 227–230.

Zisook, S., Shear, K., & Kendler, K. S. (2007). Validity of the bereavement exclusion criterion for the diagnosis of major depressive episode. *World Psychiatry, 6*(2), 38–43.

Zisook, S., & Kendler, K. S. (2007). Is bereavement-related depression different than non-bereavement-related depression? *Psychological Medicine, 37*(6), 779–794.

Zisook, S., Reynolds, C. F., 3rd, Pies, R., et al. (2010). Bereavement, complicated grief, and DSM, part 1: Depression. *Journal of Clinical Psychiatry, 71*(7), 955–956.

Zisook, S., Shear, M. K., Reynolds, C. F., et al. (2018). Treatment of complicated grief in survivors of suicide loss: A HEAL report. *Journal of Clinical Psychiatry, 79*(2), e1–e7.

Zisook, S., Johnson, G. R., Tal, I., et al. (2019). General predictors and moderators of depression remission: A VAST-D report. *American Journal of Psychiatry, 176*(5), 348–357.

9783030513658VOL05